SELECTED SOLUTIONS MANUAL

Kathleen Thrush Shaginaw

Community College of Philadelphia
Particular Solutions, Inc.

Mary Beth Kramer

University of Delaware

CHEMISTRY
A MOLECULAR APPROACH
Third Edition

NIVALDO J. TRO

PEARSON

Boston   Columbus   Indianapolis   New York   San Francisco   Upper Saddle River
Amsterdam   Cape Town   Dubai   London   Madrid   Milan   Munich   Paris   Montréal   Toronto
Delhi   Mexico City   São Paulo   Sydney   Hong Kong   Seoul   Singapore   Taipei   Tokyo

Editor in Chief: Adam Jaworski
Senior Acquisitions Editor: Terry Haugen
Senior Marking Manager: Jonathan Cottrell
Project Editor: Jessica Moro
Assistant Editor: Erin Kneuer
Managing Editor, Chemistry and Geosciences: Gina M. Cheselka
Production Project Manager: Pat Brown
Full-Service Project Management/Composition: PreMediaGlobal

Cover Image Credit: Nanotube sensor: A carbon nanotube treated with a capture agent, in yellow, can bind with and delect the purple-colored target protein—this changes the electrical resistance of the nanotube and creates a sensing device. Artist: Ethan Minot, in Nanotube technology leading to fast, lower-cost medical diagnostics, Oregon State University, @ 2012, 01 pp., http://www.flickr.com/photos/oregonstateuniversity/6816133738/

PEARSON

www.pearsonhighered.com

3 4 5 6 7 8 9 10—**EBM**—16 15 14 13

ISBN-10: 0-321-81364-2; ISBN-13: 978-0-321-81364-0

Contents

Student Guide to Using This Solutions Manual

The vision of this solutions manual is to provide guidance that is useful for both the struggling student and the advanced student.

An important feature of this solutions manual is that answers for the review questions are given. This will help in the review of the major concepts in the chapter.

The format of the solutions very closely follows the format in the textbook. Each mathematical problem includes **Given, Find, Conceptual Plan, Solution**, and **Check** sections.

Given and Find: Many students struggle with taking the written problem, parsing the information into categories, and determining the goal of the problem. It is also important to know which pieces of information in the problem are not necessary to solve the problem and if additional information needs to be gathered from sources such as tables in the textbook.

Conceptual Plan: The conceptual plan shows a step-by-step method to solve the problem. In many cases, the given quantities need to be converted to a different unit. Under each of the arrows is the equation, constant, or conversion factor needed to complete this portion of the problem. In the "Problems by Topic" section of the end-of-chapter exercises, the odd-numbered and even-numbered problems are paired. This allows you to use a conceptual plan from an odd-numbered problem in this manual as a starting point to solve the following even-numbered problem. Students should keep in mind that the examples shown are one way to solve the problems. Other mathematically equivalent solutions may be possible.

5.45   **Given:** $m\,(CO_2) = 28.8$ g, $P = 742$ mmHg, and $T = 22\,°C$ **Find:** V

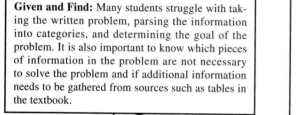

Conceptual Plan: °C → K and mmHg → atm and g → mol then $n, P, T → V$

$$K = °C + 273.15 \qquad \frac{1\ atm}{760\ mmHg} \qquad \frac{1\ mol}{44.01\ g} \qquad PV = nRT$$

Solution: $T_1 = 22\,°C + 273.15 = 295$ K, $P = 742\ \cancel{mmHg} \times \dfrac{1\ atm}{760\ \cancel{mmHg}} = 0.976316$ atm

$n = 28.8\ \cancel{g} \times \dfrac{1\ mol}{44.01\ \cancel{g}} = 0.654397$ mol $\quad PV = nRT$ Rearrange to solve for V.

$$V = \frac{nRT}{P} = \frac{0.654397\ \cancel{mol} \times 0.08206\ \dfrac{L \cdot \cancel{atm}}{\cancel{mol} \cdot \cancel{K}} \times 295\ \cancel{K}}{0.976316\ \cancel{atm}} = 16.2\ L$$

Check: The units (L) are correct. The magnitude of the answer (16 L) makes sense because one mole of an ideal gas under standard conditions (273 K and 1 atm) occupies 22.4 L. Although these are not standard conditions, they are close enough for a ballpark check of the answer. Because this gas sample contains 0.65 mole, a volume of 16 L is reasonable.

Solution: The Solution section walks you through solving the problem after the conceptual plan. Equations are rearranged to solve for the appropriate quantity. Intermediate results are shown with additional digits to minimize round-off error. The units are canceled in each appropriate step.

Check: The Check section confirms that the units in the answer are correct. This section also challenges the student to think about whether the magnitude of the answer makes sense. Thinking about what is a reasonable answer can help uncover errors such as calculation errors.

1 Matter, Measurement, and Problem Solving

Review Questions

1.1 "The properties of the substances around us depend on the atoms, ions, or molecules that compose them" means that the specific types of atoms and molecules that compose something tell us a great deal about which properties to expect from a substance. A material composed of only sodium and chloride ions will have the properties of table salt. A material composed of molecules with one carbon atom and two oxygen atoms will have the properties of the gas carbon dioxide. If the atoms and molecules change, so do the properties that we expect the material to have.

1.3 The scientific approach to knowledge is based on observation and experiment. Scientists observe and perform experiments on the physical world to learn about it. Observations often lead scientists to formulate a hypothesis, a tentative interpretation or explanation of their observations. Hypotheses are tested by experiments, highly controlled procedures designed to generate such observations. The results of an experiment may support a hypothesis or prove it wrong—in which case the hypothesis must be modified or discarded. A series of similar observations can lead to the development of scientific law, a brief statement that summarizes past observations and predicts future ones. One or more well-established hypotheses may form the basis for a scientific theory. A scientific theory is a model for the way nature is and tries to explain not merely what nature does, but why.

The Greek philosopher Plato (427–347 B.C.) took an opposite approach. He thought that the best way to learn about reality was not through the senses, but through reason. He believed that the physical world was an imperfect representation of a perfect and transcendent world (a world beyond space and time). For him, true knowledge came, not through observing the real physical world, but through reasoning and thinking about the ideal one.

1.5 Antoine Lavoisier studied combustion and made careful measurements of the mass of objects before and after burning them in closed containers. He noticed that there was no change in the total mass of material within the container during combustion. Lavoisier summarized his observations on combustion with the Law of Conservation of Mass, which states, "In a chemical reaction, matter is neither created nor destroyed."

1.7 The statement "that is just a theory" is generally taken to mean that there is no scientific proof behind the statement. This statement is the opposite of the meaning in the context of the scientific theory, where theories are tested again and again.

1.9 In solid matter, atoms or molecules pack close to each other in fixed locations. Although the atoms and molecules in a solid vibrate, they do not move around or past each other. Consequently, a solid has a fixed volume and rigid shape.

In liquid matter, atoms or molecules pack about as closely as they do in solid matter, but they are free to move relative to each other, giving liquids a fixed volume but not a fixed shape. Liquids assume the shape of their container.

In gaseous matter, atoms or molecules have a lot of space between them and are free to move relative to one another, making gases compressible. Gases always assume the shape and volume of their container.

1.11 A pure substance is composed of only one type of atom or molecule. In contrast, a mixture is a substance composed of two or more different types of atoms or molecules that can be combined in variable proportions.

1.13   A homogeneous mixture has the same composition throughout, while a heterogeneous mixture has different compositions in different regions.

1.15   Mixtures of miscible liquids (substances that easily mix) can usually be separated by distillation, a process in which the mixture is heated to boil off the more volatile (easily vaporizable) liquid. The volatile liquid is then recondensed in a condenser and collected in a separate flask.

1.17   Changes that alter only state or appearance, but not composition, are called physical changes. The atoms or molecules that compose a substance *do not change* their identity during a physical change. For example, when water boils, it changes its state from a liquid to a gas, but the gas remains composed of water molecules; so this a physical change. When sugar dissolves in water, the sugar molecules are separated from each other, but the molecules of sugar and water remain intact.

In contrast, changes that alter the composition of matter are called chemical changes. During a chemical change, atoms rearrange, transforming the original substances into different substances. For example, the rusting of iron, the combustion of natural gas to form carbon dioxide and water, and the denaturing of proteins when an egg is cooked are examples of chemical changes.

1.19   Chemical energy is potential energy. It is the energy that is contained in the bonds that hold the molecules together. This energy arises primarily from electrostatic forces between the electrically charged particles (protons and electrons) that compose atoms and molecules. Some of these arrangements—such as the one within the molecules that compose gasoline—have a much higher potential energy than others. When gasoline undergoes combustion, the arrangement of these particles changes, creating molecules with much lower potential energy and transferring a great deal of energy (mostly in the form of heat) to the surroundings. A raised weight has a certain amount of potential energy (dependent on the height the weight is raised) that can be converted to kinetic energy when the weight is released.

1.21   The three different temperature scales are Kelvin (K), Celsius (°C), and Fahrenheit (°F). The size of the degree is the same in the Kelvin and the Celsius scales, and they are 1.8 times larger than the degree size for the Fahrenheit scale.

1.23   A derived unit is a combination of other units. Examples of derived units include speed in meters per second (m/s), volume in meters cubed (m^3), and density in grams per cubic centimeter (g/cm^3).

1.25   An intensive property is a property that is independent of the amount of the substance. An extensive property is a property that depends on the amount of the substance.

1.27   In multiplication or division, the result carries the same number of significant figures as the factor with the fewest significant figures.

1.29   When rounding to the correct number of significant figures, round down if the last (or left-most) digit dropped is four or less and round up if the last (or left-most) digit dropped is five or more.

1.31   Random error is error that has equal probability of being too high or too low. Almost all measurements have some degree of random error. Random error can, with enough trials, average itself out. Systematic error is error that tends toward being either too high or too low. Systematic error does not average out with repeated trials.

Problems by Topic

The Scientific Approach to Knowledge

1.33   (a)   This statement is a theory because it attempts to explain why. It is not possible to observe individual atoms.
        (b)   This statement is an observation.
        (c)   This statement is a law because it summarizes many observations and can explain future behavior.
        (d)   This statement is an observation.

1.35   (a)   If we divide the mass of the oxygen by the mass of the carbon, the result is always 4/3.
        (b)   If we divide the mass of the oxygen by the mass of the hydrogen, the result is always 16.
        (c)   These observations suggest that the masses of elements in molecules are ratios of whole numbers [4:3 and 16:1, respectively, for parts (a) and (b)].
        (d)   Atoms combine in small whole number ratios and not as random weight ratios.

The Classification and Properties of Matter

1.37   (a)   Sweat is a homogeneous mixture of water, sodium chloride, and other components.

(b)   Carbon dioxide is a pure substance that is a compound (two or more elements bonded together).

(c)   Aluminum is a pure substance that is an element (element 13 in the periodic table).

(d)   Vegetable soup is a heterogeneous mixture of broth, chunks of vegetables, and extracts from the vegetables.

1.39

Substance	Pure or Mixture	Type (element or compound)
aluminum	pure	element
apple juice	mixture	neither—mixture
hydrogen peroxide	pure	compound
chicken soup	mixture	neither—mixture

1.41   (a)   pure substance that is a compound (one type of molecule that contains two different elements)

(b)   heterogeneous mixture (two different molecules that are segregated into regions)

(c)   homogeneous mixture (two different molecules that are randomly mixed)

(d)   pure substance that is an element (individual atoms of one type)

1.43   (a)   physical property (color can be observed without making or breaking chemical bonds)

(b)   chemical property (must observe by making or breaking chemical bonds)

(c)   physical property (the phase can be observed without making or breaking chemical bonds)

(d)   physical property (density can be observed without making or breaking chemical bonds)

(e)   physical property (mixing does not involve making or breaking chemical bonds, so this can be observed without making or breaking chemical bonds)

1.45   (a)   chemical property (burning involves breaking and making bonds, so bonds must be broken and made to observe this property)

(b)   physical property (shininess is a physical property and so can be observed without making or breaking chemical bonds)

(c)   physical property (odor can be observed without making or breaking chemical bonds)

(d)   chemical property (burning involves breaking and making bonds, so bonds must be broken and made to observe this property)

1.47   (a)   chemical change (new compounds are formed as methane and oxygen react to form carbon dioxide and water)

(b)   physical change (vaporization is a phase change and does not involve the making or breaking of chemical bonds)

(c)   chemical change (new compounds are formed as propane and oxygen react to form carbon dioxide and water)

(d)   chemical change (new compounds are formed as the metal in the frame is converted to oxides)

1.49   (a)   physical change (vaporization is a phase change and does not involve the making or breaking of chemical bonds)

(b)   chemical change (new compounds are formed)

(c)   physical change (vaporization is a phase change and does not involve the making or breaking of chemical bonds)

Units in Measurement

1.51   (a)   To convert from °F to °C, first find the equation that relates these two quantities. $°C = \dfrac{°F - 32}{1.8}$ Now substitute °F into the equation and compute the answer. Note: The number of digits reported in this answer follows significant figure conventions, covered in Section 1.6. $°C = \dfrac{°F - 32}{1.8} = \dfrac{0.}{1.8} = 0.\ °C$

(b)   To convert from K to °F, first find the equations that relate these two quantities.

$K = °C + 273.15$ and $°C = \dfrac{°F - 32}{1.8}$

Because these equations do not directly express K in terms of °F, you must combine the equations and then solve the equation for °F. Substituting for °C:

$K = \dfrac{°F - 32}{1.8} + 273.15$; rearrange $K - 273.15 = \dfrac{°F - 32}{1.8}$;

rearrange $1.8(K - 273.15) = (°F - 32)$; finally, $°F = 1.8(K - 273.15) + 32$. Now substitute K into the equation and compute the answer.

$°F = 1.8(77 - 273.15) + 32 = 1.8(-196) + 32 = -353 + 32 = -321 °F$

Alternatively, you could first convert to K to °C, using the formula $K = 273.15 + °C$, and then convert °C to °F. So, $77 K = 273.15 + °C$. Rearranging to solve for °C we get $°C = 77K - 273.15 = -196.15 °C$. Then substitute the temperature in °C in the formula $°C = \dfrac{°F - 32}{1.8}$ getting $-196.15 °C = \dfrac{°F - 32}{1.8}$. Multiplying both sides by 1.8 we get $-196 °C \times 1.8 = °F - 32$ and so $-353.07 °C = °F - 32$. Adding 32 to both sides of the equation we get $°F = -321.07 °F$ or $-321 °F$, with the proper number of significant figures.

(c)   To convert from °F to °C, first find the equation that relates these two quantities. $°C = \dfrac{°F - 32}{1.8}$
Now substitute °F into the equation and compute the answer.

$°C = \dfrac{-109 °F - 32 °F}{1.8} = \dfrac{-141}{1.8} = -78.3 °C$

(d)   To convert from °F to K, first find the equations that relate these two quantities.

$K = °C + 273.15$ and $°C = \dfrac{°F - 32}{1.8}$

Because these equations do not directly express K in terms of °F, you must combine the equations and then solve the equation for K. Substituting for °C: $K = \dfrac{°F - 32}{1.8} + 273.15$.

Now substitute °F into the equation and compute the answer.

$K = \dfrac{(98.6 - 32)}{1.8} + 273.15 = \dfrac{66.6}{1.8} + 273.15 = 37.0 + 273.15 = 310.2 K$

Alternatively, you could first convert to °F to °C, using the formula $°C = \dfrac{°F - 32}{1.8}$, and then convert °C to K, using $K = 273.15 + °C$. So, $°C = \dfrac{98.6 °F - 32}{1.8} = \dfrac{66.6 °F}{1.8} = 37.0 °C$. Then $K = 273.15 + 37.0 °C = 310.2 K$.

1.53   To convert from °F to °C, first find the equation that relates these two quantities: $°C = \dfrac{°F - 32}{1.8}$. Now substitute °F into the equation and compute the answer. Note: The number of digits reported in this answer follows significant figure conventions, covered in Section 1.6. $°C = \dfrac{-80. °F - 32 °F}{1.8} = \dfrac{-112}{1.8} = -62.2 °C$

Begin by finding the equation that relates the quantity that is given (°C) and the quantity you are trying to find (K). $K = °C + 273.15$. Because this equation gives the temperature in K directly, simply substitute the correct value for the temperature in °C and compute the answer. $K = -62.2 °C + 273.15 = 210.9 K$

1.55   Use Table 2 to determine the appropriate prefix multiplier and substitute the meaning into the expressions.
(a) 10^{-9} is equivalent to "nano," so 1.2×10^{-9} m = 1.2 nanometers = 1.2 nm
(b) 10^{-15} is equivalent to "femto," so 22×10^{-15} s = 22 femtoseconds = 22 fs
(c) 10^{9} is equivalent to "giga," so 1.5×10^{9} g = 1.5 gigagrams = 1.5 Gg
(d) 10^{6} is equivalent to "mega," so 3.5×10^{6} L = 3.5 megaliters = 3.5 ML

1.57   Use Table 2 to determine the appropriate prefix multiplier and substitute the meaning into the expressions.
(a) 10^{-9} is equivalent to "nano," so 4.5 ns = 4.5 nanoseconds = 4.5×10^{-9} s
(b) 10^{-15} is equivalent to "femto," so 18 fs = 18 femtoseconds = 18×10^{-15} s = 1.8×10^{-14} s.
Remember that in scientific notation, the first number should be smaller than 10.
(c) 10^{-12} is equivalent to "pico," so 128 pm = 128×10^{-12} m = 1.28×10^{-10} m.
Remember that in scientific notation, the first number should be smaller than 10.
(d) 10^{-6} is equivalent to "micro," so 35 μm = 35 micrograms = 35×10^{-6} g = 3.5×10^{-5} m.
Remember that in scientific notation, the first number should be smaller than 10.

1.59   (b)   **Given:** 515 km   **Find:** dm

       Conceptual Plan: km → m → dm

$$\frac{1000\ m}{1\ km} \quad \frac{10\ dm}{1\ m}$$

       Solution: $515\ \cancel{km} \times \dfrac{1000\ \cancel{m}}{1\ \cancel{km}} \times \dfrac{10\ dm}{1\ \cancel{m}} = 5.15 \times 10^6\ dm$

       Check: The units (dm) are correct. The magnitude of the answer (10^6) makes physical sense because a decimeter is a much smaller unit than a kilometer.

       Given: 515 km   **Find:** cm

       Conceptual Plan: km → m → cm

$$\frac{1000\ m}{1\ km} \quad \frac{100\ dm}{1\ m}$$

       Solution: $515\ \cancel{km} \times \dfrac{1000\ \cancel{m}}{1\ \cancel{km}} \times \dfrac{100\ cm}{1\ \cancel{m}} = 5.15 \times 10^7\ cm$

       Check: The units (cm) are correct. The magnitude of the answer (10^7) makes physical sense because a centimeter is a much smaller unit than a kilometer or a decimeter.

   (c)   **Given:** 122.355 s   **Find:** ms

       Conceptual Plan: s → ms

$$\frac{1000\ ms}{1\ s}$$

       Solution: $122.355\ \cancel{s} \times \dfrac{1000\ ms}{1\ \cancel{s}} = 1.22355 \times 10^5\ ms$

       Check: The units (ms) are correct. The magnitude of the answer (10^5) makes physical sense because a millisecond is a much smaller unit than a second.

       Given: 122.355 s   **Find:** ks

       Conceptual Plan: s → ks

$$\frac{1\ ks}{1000\ s}$$

       Solution: $122.355\ \cancel{s} \times \dfrac{1\ ks}{1000\ \cancel{s}} = 1.22355 \times 10^{-1}\ ks = 0.122355\ ks$

       Check: The units (ks) are correct. The magnitude of the answer (10^{-1}) makes physical sense because a kilosecond is a much larger unit than a second.

   (d)   **Given:** 3.345 kJ   **Find:** J

       Conceptual Plan: kJ → J

$$\frac{1000\ J}{1\ kJ}$$

       Solution: $3.345\ \cancel{kJ} \times \dfrac{1000\ J}{1\ \cancel{kJ}} = 3.345 \times 10^3\ J$

       Check: The units (J) are correct. The magnitude of the answer (10^3) makes physical sense because a joule is a much smaller unit than a kilojoule.

       Given: 3.345×10^3 J (from above)   **Find:** mJ

       Conceptual Plan: J → mJ

$$\frac{1000\ mJ}{1\ J}$$

       Solution: $3.345 \times 10^3\ \cancel{J} \times \dfrac{1000\ mJ}{1\ \cancel{J}} = 3.345 \times 10^6\ mJ$

       Check: The units (mJ) are correct. The magnitude of the answer (10^6) makes physical sense because a millijoule is a much smaller unit than a joule.

1.61   (a)   **Given:** 254,998 m   **Find:** km

       Conceptual Plan: m → km

$$\frac{1\ km}{1000\ m}$$

       Solution: $254,998\ \cancel{m} \times \dfrac{1\ km}{1000\ \cancel{m}} = 2.54998 \times 10^2\ km = 254.998\ km$

       Check: The units (km) are correct. The magnitude of the answer (10^2) makes physical sense because a kilometer is a much larger unit than a meter.

(b)   **Given:** 254,998 m   **Find:** Mm
Conceptual Plan: m → Mm
$$\frac{1\ \text{Mm}}{10^6\ \text{m}}$$
Solution: $254{,}998\ \cancel{\text{m}} \times \dfrac{1\ \text{Mm}}{10^6\ \cancel{\text{m}}} = 2.54998 \times 10^{-1}\ \text{Mm} = 0.254998\ \text{Mm}$

Check: The units (Mm) are correct. The magnitude of the answer (10^{-1}) makes physical sense because a megameter is a much larger unit than a meter or a kilometer.

(c)   **Given:** 254,998 m   **Find:** mm
Conceptual Plan: m → mm
$$\frac{1000\ \text{mm}}{1\ \text{m}}$$
Solution: $254{,}998\ \cancel{\text{m}} \times \dfrac{1000\ \text{mm}}{1\ \cancel{\text{m}}} = 2.54998 \times 10^{8}\ \text{mm}$

Check: The units (mm) are correct. The magnitude of the answer (10^8) makes physical sense because a millimeter is a much smaller unit than a meter.

(d)   **Given:** 254,998 m   **Find:** cm
Conceptual Plan: m → cm
$$\frac{100\ \text{cm}}{1\ \text{m}}$$
Solution: $254{,}998\ \cancel{\text{m}} \times \dfrac{100\ \text{cm}}{1\ \cancel{\text{m}}} = 2.54998 \times 10^{7}\ \text{cm}$

Check: The units (cm) are correct. The magnitude of the answer (10^7) makes physical sense because a centimeter is a much smaller unit than a meter, but larger than a millimeter.

1.63   **Given:** 1 square meter $(1\ \text{m}^2)$ **Find:** cm^2
Conceptual Plan: $1\ \text{m}^2 \rightarrow \text{cm}^2$
$$\frac{100\ \text{cm}}{1\ \text{m}}$$
Notice that for squared units, the conversion factors must be squared.
Solution: $1\ \cancel{\text{m}^2} \times \dfrac{(100\ \text{cm})^2}{(1\ \cancel{\text{m}})^2} = 1 \times 10^4\ \text{cm}^2$

Check: The units of the answer are correct, and the magnitude makes sense. The unit centimeter is smaller than a meter, so the value in square centimeters should be larger than in square meters.

Density

1.65   **Given:** $m = 2.49\ \text{g};\ V = 0.349\ \text{cm}^3$ **Find:** d in g/cm^3 and compare to pure copper.
Conceptual Plan: $m, V \rightarrow d$
$$d = m/V$$
Compare to the published value: d (pure copper) $= 8.96\ \text{g/cm}^3$. (This value is in Table 4.)
Solution: $d = \dfrac{2.49\ \text{g}}{0.349\ \text{cm}^3} = 7.13\ \dfrac{\text{g}}{\text{cm}^3}$
The density of the penny is much smaller than the density of pure copper $(7.13\ \text{g/cm}^3; 8.96\ \text{g/cm}^3)$, so the penny is not pure copper.

Check: The units (g/cm^3) are correct. The magnitude of the answer seems correct. Many coins are layers of metals, so it is not surprising that the penny is not pure copper.

1.67   **Given:** $m = 4.10 \times 10^3\ \text{g};\ V = 3.25\ \text{L}$ **Find:** d in g/cm^3
Conceptual Plan: $m, V \rightarrow d$ then $L \rightarrow \text{cm}^3$
$$d = m/V \qquad\qquad \frac{1000\ \text{cm}^3}{1\ \text{L}}$$
Solution: $d = \dfrac{4.10 \times 10^3\ \text{g}}{3.25\ \cancel{\text{L}}} \times \dfrac{1\ \cancel{\text{L}}}{1000\ \text{cm}^3} = 1.26\ \dfrac{\text{g}}{\text{cm}^3}$

Check: The units (g/cm^3) are correct. The magnitude of the answer seems correct.

1.69   (a)   **Given:** $d = 1.11 \text{ g/cm}^3$; $V = 417 \text{ mL}$ **Find:** m

             Conceptual Plan: $d, V \rightarrow m$ **then** $\text{cm}^3 \rightarrow \text{mL}$

$$d = m/V \qquad\qquad \frac{1 \text{ mL}}{1 \text{ cm}^3}$$

             Solution: $d = m/V$ Rearrange by multiplying both sides of equation by V. $m = d \times V$

$$m = 1.11 \, \frac{\text{g}}{\text{cm}^3} \times \frac{1 \text{ cm}^3}{1 \text{ mL}} \times 417 \text{ mL} = 4.63 \times 10^2 \text{ g}$$

             Check: The units (g) are correct. The magnitude of the answer seems correct considering that the value of the density is about 1 g/cm³.

     (b)   **Given:** $d = 1.11 \text{ g/cm}^3$; $m = 4.1 \text{ kg}$ **Find:** V in L

             Conceptual Plan: $d, V \rightarrow m$ **then** $\text{kg} \rightarrow \text{g}$ **and** $\text{cm}^3 \rightarrow \text{L}$

$$d = m/V \qquad\qquad \frac{1000 \text{ g}}{1 \text{ kg}} \qquad\qquad \frac{1 \text{ L}}{1000 \text{ cm}^3}$$

             Solution: $d = m/V$ Rearrange by multiplying both sides of equation by V and dividing both sides of the equation by d.

$$V = \frac{m}{d} = \frac{4.1 \text{ kg}}{1.11 \, \frac{\text{g}}{\text{cm}^3}} \times \frac{1000 \text{ g}}{1 \text{ kg}} = 3.7 \times 10^3 \text{ cm}^3 \times \frac{1 \text{ L}}{1000 \text{ cm}^3} = 3.7 \text{ L}$$

             Check: The units (L) are correct. The magnitude of the answer seems correct considering that the value of the density is about 1 g/cm³.

1.71   **Given:** $V = 245 \text{ L}$, $d = 0.821 \text{ g/mL}$ **Find:** m

        Conceptual Plan: $\text{g/mL} \rightarrow \text{g/L}$ **then** $d, V \rightarrow m$

$$\frac{1000 \text{ mL}}{1 \text{ L}} \qquad\qquad d = m/V$$

        Solution: $d = m/V$ Rearrange by multiplying both sides of equation by V. $m = d \times V$

$$m = 245 \text{ L} \times \frac{1000 \text{ mL}}{1 \text{ L}} \times \left(0.821 \, \frac{\text{g}}{\text{mL}} \right) = 2.01 \times 10^5 \text{ g}$$

        Check: The units (g) are correct. The magnitude of the answer seems correct considering the value of the density is less than 1 g/mL and the volume is very large.

The Reliability of a Measurement and Significant Figures

1.73   To obtain the readings, look to see where the bottom of the meniscus lies. Estimate the distance between two markings on the device.

     (a)   73.2 mL—the bottom of the meniscus appears to be sitting just above the 73 mL mark.

     (b)   88.2 °C—the mercury is between the 84 °C mark and the 85 °C mark, but it is closer to the lower number.

     (c)   645 mL—the meniscus appears to be just above the 640 mL mark.

1.75   Remember that

     1.   interior zeros (zeros between two numbers) are significant.

     2.   leading zeros (zeros to the left of the first nonzero number) are not significant. They only serve to locate the decimal point.

     3.   trailing zeros (zeros at the end of a number) are categorized as follows:

         •   Trailing zeros after a decimal point are always significant.

         •   Trailing zeros before an implied decimal point are ambiguous and should be avoided by using scientific notation or by inserting a decimal point at the end of the number.

       (a)   1,050,501 km

       (b)   0.0020 m

       (c)   0.000000000000002 s

       (d)   0.001090 cm

1.77   Remember the following rules from Section 1.7.

    (a)   Three significant figures. The 3, 1, and 2 are significant (rule 1). The leading zeros only mark the decimal place and are therefore not significant (rule 3).

    (b)   Ambiguous. The 3, 1, and 2 are significant (rule 1). The trailing zeros occur before an implied decimal point and are therefore ambiguous (rule 4). Without more information, we would assume three significant figures. It is better to write this as 3.12×10^5 to indicate three significant figures or as 3.12000×10^5 to indicate six (rule 4).

    (c)   Three significant figures. The 3, 1, and 2 are significant (rule 1).

    (d)   Five significant figures. The 1s, 3, 2, and 7 are significant (rule 1).

    (e)   Ambiguous. The 2 is significant (rule 1). The trailing zeros occur before an implied decimal point and are there-fore ambiguous (rule 4). Without more information, we would assume one significant figure. It is better to write this as 2×10^3 to indicate one significant figure or as 2.000×10^3 to indicate four (rule 4).

1.79   (a)   This is not exact because π is an irrational number. The number 3.14 only shows three of the infinite number of significant figures that π has.

    (b)   This is an exact conversion because it comes from a definition of the units and so has an unlimited number of significant figures.

    (c)   This is a measured number, so it is not an exact number. There are two significant figures.

    (d)   This is an exact conversion because it comes from a definition of the units and so has an unlimited number of significant figures.

1.81   (a)   156.9—The 8 is rounded up because the next digit is a 5.

    (b)   156.8—The last two digits are dropped because 4 is less than 5.

    (c)   156.8—The last two digits are dropped because 4 is less than 5.

    (d)   156.9—The 8 is rounded up because the next digit is a 9, which is greater than 5.

Significant Figures in Calculations

1.83   (a) $9.15 \div 4.970 = 1.84$—Three significant figures are allowed to reflect the three significant figures in the least precisely known quantity (9.15).

    (b) $1.54 \times 0.03060 \times 0.69 = 0.033$—Two significant figures are allowed to reflect the two significant figures in the least precisely known quantity (0.69). The intermediate answer (0.03251556) is rounded up because the first nonsignificant digit is a 5.

    (c) $27.5 \times 1.82 \div 100.04 = 0.500$—Three significant figures are allowed to reflect the three significant fig-ures in the least precisely known quantity (27.5 and 1.82). The intermediate answer (0.50029988) is truncated because the first nonsignificant digit is a 2, which is less than 5.

    (d) $(2.290 \times 10^6) \div (6.7 \times 10^4) = 34$—Two significant figures are allowed to reflect the two significant figures in the least precisely known quantity (6.7×10^4). The intermediate answer (34.17910448) is truncated because the first nonsignificant digit is a 1, which is less than 5.

1.85   (a)   43.7

        $\underline{-2.341}$

        $41.359 = 41.4$

    Round the intermediate answer to one decimal place to reflect the quantity with the fewest decimal places (43.7). Round the last digit up because the first nonsignificant digit is 5.

    (b)   17.6

        +2.838

        +2.3

        $\underline{+110.77}$

        $133.508 = 133.5$

    Round the intermediate answer to one decimal place to reflect the quantity with the fewest decimal places (2.3). Truncate nonsignificant digits because the first nonsignificant digit is 0.

(c)   19.6
       +58.33
       −4.974
       ⎯⎯⎯⎯⎯⎯
       72.956 = 73.0

Round the intermediate answer to one decimal place to reflect the quantity with the fewest decimal places (19.6). Round the last digit up because the first nonsignificant digit is 5.

(d)   5.99
       −5.572
       ⎯⎯⎯⎯⎯⎯
       0.418 = 0.42

Round the intermediate answer to two decimal places to reflect the quantity with the fewest decimal places (5.99). Round the last digit up because the first nonsignificant digit is 8.

1.87   Perform operations in parentheses first. Keep track of significant figures in each step by noting the last significant digit in an intermediate result.

(a) $(24.6681 \times 2.38) + 332.58 = 58.\underline{7}10078$
                          +332.58
                        ⎯⎯⎯⎯⎯⎯⎯⎯⎯
                        391.290078 = 391.3

The first intermediate answer has one significant digit to the right of the decimal because it is allowed three significant figures [reflecting the quantity with the fewest significant figures (2.38)]. Underline the most significant digit in this answer. Round the next intermediate answer to one decimal place to reflect the quantity with the fewest decimal places (58.7). Round the last digit up because the first nonsignificant digit is 9.

(b) $\dfrac{(85.3 - 21.489)}{0.0059} = \dfrac{63.\underline{8}11}{0.0059} = 1.\underline{0}81542 \times 10^4 = 1.1 \times 10^4$

The first intermediate answer has one significant digit to the right of the decimal to reflect the quantity with the fewest decimal places (85.3). Underline the most significant digit in this answer. Round the next intermediate answer to two significant figures to reflect the quantity with the fewest significant figures (0.0059). Round the last digit up because the first nonsignificant digit is 8.

(c) $(512 \div 986.7) + 5.44 = 0.\underline{5}189014$
                        +5.44
                       ⎯⎯⎯⎯⎯⎯⎯⎯
                       5.9589014 = 5.96

The first intermediate answer has three significant figures and three significant digits to the right of the decimal, reflecting the quantity with the fewest significant figures (512). Underline the most significant digit in this answer. Round the next intermediate answer to two decimal places to reflect the quantity with the fewest decimal places (5.44). Round the last digit up because the first nonsignificant digit is 8.

(d) $[(28.7 \times 10^5) \div 48.533] + 144.99 = 59\underline{1}35.02$
                                  +144.99
                                 ⎯⎯⎯⎯⎯⎯⎯⎯⎯
                                 59280.01 = 59300 = 5.93 \times 10^4

The first intermediate answer has three significant figures, reflecting the quantity with the fewest significant figures (28.7×10^5). Underline the most significant digit in this answer. Because the number is so large, when the addition is performed, the most significant digit is the hundreds place. Round the next intermediate answer to the hundreds place and put in scientific notation to remove any ambiguity. Note that the last digit is rounded up because the first nonsignificant digit is 8.

Unit Conversions

1.89   (a)   **Given:** 27.8 L   **Find:** cm^3
                Conceptual Plan: L → cm³
                         $\dfrac{1000\ cm^3}{1\ L}$

                Solution: $27.8\ \cancel{L} \times \dfrac{1000\ cm^3}{1\ \cancel{L}} = 2.78 \times 10^4\ cm^3$

                Check: The units (cm^3) are correct. The magnitude of the answer (10^4) makes physical sense because cm^3 is much smaller than a liter; so the answer should go up several orders of magnitude. Three significant figures are allowed because of the limitation of 27.8 L (three significant figures).

(b)   **Given:** 1898 mg   **Find:** kg
Conceptual Plan: mg → g → kg

$$\frac{1\text{ g}}{1000\text{ mg}} \qquad \frac{1\text{ kg}}{1000\text{ g}}$$

Solution: $1898\text{ mg} \times \dfrac{1\text{ g}}{1000\text{ mg}} \times \dfrac{1\text{ kg}}{1000\text{ g}} = 1.898 \times 10^{-3}\text{ kg}$

Check: The units (kg) are correct. The magnitude of the answer (10^{-3}) makes physical sense because a kilogram is a much larger unit than a milligram. Four significant figures are allowed because 1898 mg has four significant figures.

(c)   **Given:** 198 km   **Find:** cm
Conceptual Plan: km → m → cm

$$\frac{1000\text{ m}}{1\text{ m}} \qquad \frac{100\text{ cm}}{1\text{ m}}$$

Solution: $198\text{ km} \times \dfrac{1000\text{ m}}{1\text{ km}} \times \dfrac{100\text{ cm}}{1\text{ m}} = 1.98 \times 10^{7}\text{ cm}$

Check: The units (cm) are correct. The magnitude of the answer (10^{7}) makes physical sense because a kilometer is a much larger unit than a centimeter. Three significant figures are allowed because 198 km has three significant figures.

1.91   (a)   **Given:** 154 cm   **Find:** in
Conceptual Plan: cm → in

$$\frac{1\text{ in}}{2.54\text{ cm}}$$

Solution: $154\text{ cm} \times \dfrac{1\text{ in}}{2.54\text{ cm}} = 60.62992\text{ in} = 60.6\text{ in}$

Check: The units (in) are correct. The magnitude of the answer (60.6) makes physical sense because an inch is a larger unit than a cm. Three significant figures are allowed because 154 cm has three significant figures.

(b)   **Given:** 3.14 kg   **Find:** g
Conceptual Plan: kg → g

$$\frac{1000\text{ g}}{1\text{ kg}}$$

Solution: $3.14\text{ kg} \times \dfrac{1000\text{ g}}{1\text{ kg}} = 3.14 \times 10^{3}\text{ g}$

Check: The units (g) are correct. The magnitude of the answer (10^{3}) makes physical sense because a kilogram is a much larger unit than a gram. Three significant figures are allowed because 3.14 kg has three significant figures.

(c)   **Given:** 3.5 L   **Find:** qt
Conceptual Plan: L → qt

$$\frac{1.057\text{ qt}}{1\text{ L}}$$

Solution: $3.5\text{ L} \times \dfrac{1.057\text{ qt}}{1\text{ L}} = 3.6995\text{ qt} = 3.7\text{ qt}$

Check: The units (qt) are correct. The magnitude of the answer (3.7) makes physical sense because a L is a smaller unit than a qt. Two significant figures are allowed because 3.5 L has two significant figures.

Round the last digit up because the first nonsignificant digit is a 9.

(d)   **Given:** 109 mm   **Find:** in
Conceptual Plan: mm → m → in

$$\frac{1\text{ m}}{1000\text{ mm}} \qquad \frac{39.37\text{ in}}{1\text{ m}}$$

Solution: $109\text{ mm} \times \dfrac{1\text{ m}}{1000\text{ mm}} \times \dfrac{39.37\text{ in}}{1\text{ m}} = 4.29133\text{ in} = 4.29\text{ in}$

Check: The units (in) are correct. The magnitude of the answer (4) makes physical sense because a mm is a much smaller unit than an inch. Three significant figures are allowed because 109 mm has three significant figures.

1.93   **Given:** 10.0 km   **Find:** minutes   **Other:** running pace $= 7.5$ miles per hour
         Conceptual Plan: km → mi → hr → min

$$\frac{0.6214 \text{ mi}}{1 \text{ km}} \quad \frac{1 \text{ hr}}{7.5 \text{ mi}} \quad \frac{60 \text{ min}}{1 \text{ hr}}$$

Solution: $10.0 \text{ km} \times \dfrac{0.6214 \text{ mi}}{1 \text{ km}} \times \dfrac{1 \text{ hr}}{7.5 \text{ mi}} \times \dfrac{60 \text{ min}}{1 \text{ hr}} = 49.712 \text{ min} = 50. \text{ min} = 5.0 \times 10^1 \text{ min}$

Check: The units (min) are correct. The magnitude of the answer (50) makes physical sense because she is running almost 7.5 miles (which would take her 60 min $= 1$ hr). Two significant figures are allowed because of the limitation of 7.5 mi/hr (two significant figures). Round the last digit up because the first nonsignificant digit is a 7.

1.95   **Given:** 17 km/L   **Find:** miles per gallon

Conceptual Plan: $\dfrac{\mathbf{km}}{\mathbf{L}} \rightarrow \dfrac{\mathbf{mi}}{\mathbf{L}} \rightarrow \dfrac{\mathbf{mi}}{\mathbf{gal}}$

$$\frac{0.6214 \text{ mi}}{1 \text{ km}} \quad \frac{3.785 \text{ L}}{1 \text{ gallon}}$$

Solution: $\dfrac{17 \text{ km}}{1 \text{ L}} \times \dfrac{0.6214 \text{ mi}}{1 \text{ km}} \times \dfrac{3.785 \text{ L}}{1 \text{ gal}} = 39.98398 \dfrac{\text{mi}}{\text{gal}} = 40. \dfrac{\text{mi}}{\text{gal}}$

Check: The units (mi/gal) are correct. The magnitude of the answer (40) makes physical sense because the dominating factor is that a liter is much smaller than a gallon; so the answer should go up. Two significant figures are allowed because of the limitation of 17 km/L (two significant figures). Round the last digit up because the first nonsignificant digit is a 9.

1.97   (a)   **Given:** 195 m^2 **Find:** km^2
                 Conceptual Plan: $\mathbf{m^2 \rightarrow km^2}$

$$\frac{(1 \text{ km})^2}{(1000 \text{ m})^2}$$

Notice that for squared units, the conversion factors must be squared.

Solution: $195 \text{ m}^2 \times \dfrac{(1 \text{ km})^2}{(1000 \text{ m})^2} = 1.95 \times 10^{-4} \text{ km}^2$

Check: The units $\left(\text{km}^2\right)$ are correct. The magnitude of the answer $\left(10^{-4}\right)$ makes physical sense because a kilometer is a much larger unit than a meter.

(b)   **Given:** 195 m^2 **Find:** dm^2
        Conceptual Plan: $\mathbf{m^2 \rightarrow dm^2}$

$$\frac{(10 \text{ dm})^2}{(1 \text{ m})^2}$$

Notice that for squared units, the conversion factors must be squared.

Solution: $195 \text{ m}^2 \times \dfrac{(10 \text{ dm})^2}{(1 \text{ m})^2} = 1.95 \times 10^4 \text{ dm}^2$

Check: The units $\left(\text{dm}^2\right)$ are correct. The magnitude of the answer $\left(10^4\right)$ makes physical sense because a decimeter is a much smaller unit than a meter.

(c)   **Given:** 195 m^2 **Find:** cm^2
        Conceptual Plan: $\mathbf{m^2 \rightarrow cm^2}$

$$\frac{(100 \text{ cm})^2}{(1 \text{ m})^2}$$

Notice that for squared units, the conversion factors must be squared.

Solution: $195 \text{ m}^2 \times \dfrac{(100 \text{ cm})^2}{(1 \text{ m})^2} = 1.95 \times 10^6 \text{ cm}^2$

Check: The units $\left(\text{cm}^2\right)$ are correct. The magnitude of the answer $\left(10^6\right)$ makes physical sense because a centimeter is a much smaller unit than a meter.

1.99　　**Given:** 435 acres　**Find:** square miles　**Other:** 1 acre = 43,560 ft²; 1 mile = 5280 ft

Conceptual Plan: acres → ft² → mi²

$$\frac{43560 \text{ ft}^2}{1 \text{ acre}} \quad \frac{(1 \text{ mi})^2}{(5280 \text{ ft})^2}$$

Notice that for squared units, the conversion factors must be squared.

Solution: $435 \text{ acres} \times \dfrac{43560 \text{ ft}^2}{1 \text{ acres}} \times \dfrac{(1 \text{ mi})^2}{(5280 \text{ ft})^2} = 0.6796875 \text{ mi}^2 = 0.680 \text{ mi}^2$

Check: The units (mi²) are correct. The magnitude of the answer (0.7) makes physical sense because an acre is much smaller than a mi²; so the answer should go down several orders of magnitude. Three significant figures are allowed because of the limitation of 435 acres (three significant figures). Round the last digit up because the first nonsignificant digit is a 7.

1.101　　**Given:** 14 lb　**Find:** mL　**Other:** 80 mg/0.80 mL; 15 mg/kg body

Conceptual Plan: lb → kg body → mg → mL

$$\frac{1 \text{ kg body}}{2.205 \text{ lb}} \quad \frac{15 \text{ mg}}{1 \text{ kg body}} \quad \frac{0.80 \text{ mL}}{80 \text{ mg}}$$

Solution: $14 \text{ lb} \times \dfrac{1 \text{ kg body}}{2.205 \text{ lb}} \times \dfrac{15 \text{ mg}}{1 \text{ kg body}} \times \dfrac{0.80 \text{ mL}}{80 \text{ mg}} = 0.9523809524 \text{ mL} = 0.95 \text{ mL}$

Check: The units (mL) are correct. The magnitude of the answer (1 mL) makes physical sense because it is a reasonable amount of liquid to give to a baby. Two significant figures are allowed because of the statement in the problem. Truncate after the last significant digit because the first nonsignificant digit is a 2.

Cumulative Problems

1.103　　**Given:** solar year　**Find:** seconds

Other: 60 seconds/minute; 60 minutes/hour; 24 hours/solar day; 365.24 solar days/solar year

Conceptual Plan: yr → day → hr → min → sec

$$\frac{365.24 \text{ day}}{1 \text{ solar yr}} \quad \frac{24 \text{ hr}}{1 \text{ day}} \quad \frac{60 \text{ min}}{1 \text{ hr}} \quad \frac{60 \text{ sec}}{1 \text{ min}}$$

Solution: $1 \text{ solar yr} \times \dfrac{365.24 \text{ day}}{1 \text{ solar yr}} \times \dfrac{24 \text{ hr}}{1 \text{ day}} \times \dfrac{60 \text{ min}}{1 \text{ hr}} \times \dfrac{60 \text{ sec}}{1 \text{ min}} = 3.1556736 \times 10^7 \text{ sec} = 3.1557 \times 10^7 \text{ sec}$

Check: The units (sec) are correct. The magnitude of the answer (10^7) makes physical sense because each conversion factor increases the value of the answer—a second is many orders of magnitude smaller than a year. Five significant figures are allowed because all conversion factors are assumed to be exact except for the 365.24 days/solar year (five significant figures). Round the last digit up because the first nonsignificant digit is a 7.

1.105　　(a)　Extensive—The volume of a material depends on how much is present.

　　　　(b)　Intensive—The boiling point of a material is independent of how much material you have; so these values can be published in reference tables.

　　　　(c)　Intensive—The temperature of a material is independent of how much is present.

　　　　(d)　Intensive—The electrical conductivity of a material is independent of how much material you have; so these values can be published in reference tables.

　　　　(e)　Extensive—The energy contained in material depends on how much is present. If you double the amount of material, you double the amount of energy.

1.107　　**Given:** 130 °X = 212 °F and 10 °X = 32 °F　**Find:** temperature where °X = °F.

Conceptual Plan: Use data to derive an equation relating °X and °F. Then set °F = °X = z and solve for z.

Solution: Assume a linear relationship between the two temperatures $(y = mx + b)$.

Let $y = $ °F and let $x = $ °X.

The slope of the line (m) is the relative change in the two temperature scales:

$$m = \frac{\Delta \text{ °F}}{\Delta \text{ °X}} = \frac{212 \text{ °F} - 32 \text{ °F}}{130 \text{ °X} - 10 \text{ °X}} = \frac{180 \text{ °F}}{120 \text{ °X}} = 1.5$$

Solve for intercept (b) by plugging one set of temperatures into the equation:

$y = 1.5x + b \rightarrow 32 = (1.5)(10) + b \rightarrow 32 = 15 + b \rightarrow b = 17 \rightarrow {}^\circ F = (1.5)\,{}^\circ X + 17$

Set ${}^\circ F = {}^\circ X = z$ and solve for z.

$z = 1.5z + 17 \rightarrow -17 = 1.5z - z \rightarrow -17 = 0.5z \rightarrow z = -34 \rightarrow -34\,{}^\circ F = -34\,{}^\circ X$

Check: The units (${}^\circ F$ and ${}^\circ X$) are correct. Plugging the result back into the equation confirms that the calculations were done correctly. The magnitude of the answer seems correct because it is known that the result is not between $32\,{}^\circ F$ and $212\,{}^\circ F$. The numbers are getting closer together as the temperature drops.

1.109   1G. $F = ma = kg(m/s^2)$. Let's call it N for Newton. Ten tons $= 20{,}000$ lb $= (1\,kg/2.2\,lb) \times 20{,}000\,lb = 4.4 \times 10 \times 10^4\,kg$,   deceleration $= 55\,mi \times 0.6\,km/mi \times 10^3\,m/km \times 1/3.6 \times 10^3\,s^2$.   Exponents $= 10^4 \times 10^3 \times 10^{-3} = 10^4$. So the kN is convenient.
For one molecule, the mass is $10^{-20}\,kg$ and deceleration is $3 \times 10^8\,m/s^2$. So Exponents $= 10^{-20} \times 10^8 = 10^{-12}$. So the pN is convenient.

1.111   (a) $1.76 \times 10^{-3}/8.0 \times 10^2 = 2.2 \times 10^{-6}$ Two significant figures are allowed to reflect the quantity with the fewest significant figures (8.0×10^2).

(b)   Write all figures so that the decimal points can be aligned:

    0.0187

$+$ 0.0002   All quantities are known to four places to the right of the decimal place;

$-$ 0.0030   so the answer should be reported to four places to the right of the

    0.0159   decimal place, or three significant figures.

(c) $[(136000)(0.000322)/0.082](129.2) = 6.899910244 \times 10^4 = 6.9 \times 10^4$ Round the intermediate answer to two significant figures to reflect the quantity with the fewest significant figures (0.082). Round the last digit up because the first nonsignificant digit is 9.

1.113   (a)   **Given:** cylinder dimensions: length $= 22$ cm; radius $= 3.8$ cm; $d(\text{gold}) = 19.3\,g/cm^3$; $d(\text{sand}) = 3.00\,g/cm^3$ **Find:** $m(\text{gold})$ and $m(\text{sand})$
Conceptual Plan: $l, r \rightarrow V$ then $d, V \rightarrow m$
$$V = l\pi r^2 \qquad d = m/V$$

Solution: $V(\text{gold}) = V(\text{sand}) = (22\,cm)(\pi)(3.8\,cm)^2 = 998.0212\,cm^3 \quad d = m/V$
Rearrange by multiplying both sides of equation by V. $\rightarrow m = d \times V$

$$m(\text{gold}) = \left(19.3\,\frac{g}{cm^3}\right) \times (998.0212\,cm^3) = 1.926181 \times 10^4\,g = 1.9 \times 10^4\,g$$

Check: The units (g) are correct. The magnitude of the answer seems correct considering that the value of the density is ~20 g/cm^3. Two significant figures are allowed to reflect the significant figures in 22 cm and 3.8 cm. Truncate the nonsignificant digits because the first nonsignificant digit is a 2.

$$m(\text{sand}) = \left(3.00\,\frac{g}{cm^3}\right) \times (998.0212\,cm^3) = 2.99406 \times 10^3\,g = 3.0 \times 10^3\,g$$

Check: The units (g) are correct. The magnitude of the answer seems correct considering that the value of the density is 3 g/cm^3. This number is much lower than the gold mass. Two significant figures are allowed to reflect the significant figures in 22 cm and 3.8 cm. Round the last digit up because the first nonsignificant digit is a 9.

(b)   Comparing the two values $1.9 \times 10^4\,g$ versus $3.0 \times 10^3\,g$ shows a difference in weight of almost a factor of 10. This difference should be enough to trip the alarm and alert the authorities to the presence of the thief.

1.115   **Given:** 3.5 lb of titanium   **Find:** volume in in^3 **Other:** density of titanium is $4.51\,g/cm^3$
Conceptual Plan: lb $\rightarrow$ g then $m, d \rightarrow V$ then $cm^3 \rightarrow in^3$
$$\frac{453.6\,g}{1\,lb} \qquad d = m/V \qquad \frac{(1\,in)^3}{(2.54\,cm)^3}$$

Solution: $3.5\,lb \times \dfrac{453.6\,g}{1\,lb} = 1.5876 \times 10^3\,g$

$d = m/V$ Rearrange by multiplying both sides of the equation by V and dividing both sides of the equation by d.

$$V = \frac{m}{d} = \frac{1.5876 \times 10^3 \ \cancel{g}}{4.51 \ \dfrac{\cancel{g}}{cm^3}} = 3.520 \times 10^2 \ cm^3 = 3.5 \times 10^2 \ \cancel{cm^3} \times \frac{(1 \ in)^3}{(2.54 \ \cancel{cm})^3} = 21 \ in^3$$

Check: The units (in^3) are correct. The magnitude of the answer seems correct considering the number of grams. Two significant figures are allowed to reflect the significant figures in 3.5 lb. Truncate the nonsignificant digits because the first nonsignificant digit is a 2.

1.117   **Given:** cylinder dimensions: length $= 2.16$ in; radius $= 0.22$ in; $m = 41$ g   **Find:** density (g/m^3)
          Conceptual Plan: in $\rightarrow$ cm then $l, r \rightarrow V$ then $m, V \rightarrow d$

$$\frac{2.54 \ cm}{1 \ in} \qquad V = l\pi r^2 \qquad d = m/V$$

Solution: $2.16 \ \cancel{in} \times \dfrac{2.54 \ cm}{1 \ \cancel{in}} = 5.4864 \ cm = l \qquad 0.22 \ \cancel{in} \times \dfrac{2.54 \ cm}{1 \ \cancel{in}} = 0.5588 \ cm = r$

$V = l\pi r^2 = (5.4864 \ cm)(\pi)(0.5588 \ cm)^2 = 5.3820798 \ cm^3$

$d = \dfrac{m}{V} = \dfrac{41 \ g}{5.3820798 \ cm^3} = 7.6178729 \ \dfrac{g}{cm^3} = 7.6 \ \dfrac{g}{cm^3}$

Check: The units (g/cm^3) are correct. The magnitude of the answer seems correct considering that the value of the density of iron (a major component in steel) is 7.86 g/cm^3. Two significant figures are allowed to reflect the significant figures in 0.22 in and 41 g. Truncate the nonsignificant digits because the first nonsignificant digit is a 1.

1.119   **Given:** 185 cubic yards (yd^3) of H_2O **Find:** mass of the H_2O (pounds)
          Other: $d(H_2O) = 1.00 \ g/cm^3$ at 4 °C
          Conceptual Plan: $yd^3 \rightarrow m^3 \rightarrow cm^3 \rightarrow g \rightarrow lb$

$$\frac{(1 \ m)^3}{(1.094 \ yd)^3} \quad \frac{(100 \ cm)^3}{(1 \ m)^3} \quad \frac{1.00 \ g}{1.00 \ cm^3} \quad \frac{1 \ lb}{453.59 \ g}$$

Solution: $185 \ \cancel{yd^3} \times \dfrac{(1 \ \cancel{m})^3}{(1.094 \ \cancel{yd})^3} \times \dfrac{(100 \ \cancel{cm})^3}{(1 \ \cancel{m})^3} \times \dfrac{1.00 \ \cancel{g}}{1.00 \ \cancel{cm^3}} \times \dfrac{1 \ lb}{453.59 \ \cancel{g}} = 3.114987377 \times 10^5 \ lb = 3.11 \times 10^5 \ lb$

Check: The units (lb) are correct. The magnitude of the answer (10^5) makes physical sense because a pool is not a small object. Three significant figures are allowed because the conversion factor with the least precision is the density $\left[1.00 \ g/cm^3 - (3 \ \text{significant figures})\right]$ and the initial size has three significant figures. Truncate after the last digit because the first nonsignificant digit is a 4.

1.121   **Given:** 15 liters of gasoline   **Find:** kilometers   **Other:** 52 mi/gal in the city
          Conceptual Plan: L $\rightarrow$ gal $\rightarrow$ mi $\rightarrow$ km

$$\frac{1 \ gal}{3.785 \ L} \quad \frac{52 \ mi}{1 \ gal} \quad \frac{1 \ km}{0.6214 \ mi}$$

Solution: $15 \ \cancel{L} \times \dfrac{1 \ \cancel{gal}}{3.785 \ \cancel{L}} \times \dfrac{52 \ \cancel{mi}}{1 \ \cancel{gal}} \times \dfrac{1 \ km}{0.6214 \ \cancel{mi}} = 3.316327941 \times 10^2 \ km = 3.3 \times 10^2 \ km$

Check: The units (km) are correct. The magnitude of the answer (10^2) makes physical sense because the dominating conversion factor is the mileage, which increases the answer. Two significant figures are allowed because the conversion factor with the least precision is 52 mi/gal (two significant figures) and the initial volume (15 L) has two significant figures. Truncate the last digit because the first nonsignificant digit is a 1. It is best to put the answer in scientific notation so that it is clear how many significant figures are expressed.

1.123   **Given:** radius of nucleus of the hydrogen atom $= 1.0 \times 10^{-13}$ cm; radius of the hydrogen atom $= 52.9$ pm
          Find: percent of volume occupied by nucleus (%)
          Conceptual Plan: cm $\rightarrow$ m then pm $\rightarrow$ m then $r \rightarrow V$ then $V_{atom}, V_{nucleus} \rightarrow$ % $V_{nucleus}$

$$\frac{1 \ m}{100 \ cm} \qquad \frac{1 \ m}{10^{12} \ pm} \qquad V = (4/3)\pi r^3 \qquad \% \ V_{nucleus} = \frac{V_{nucleus}}{V_{atom}} \times 100\%$$

Solution: $1.0 \times 10^{-13} \ \cancel{cm} \times \dfrac{1 \ m}{100 \ \cancel{cm}} = 1.0 \times 10^{-15} \ m$ and $52.9 \ \cancel{pm} \times \dfrac{1 \ m}{10^{12} \ \cancel{pm}} = 5.29 \times 10^{-11} \ m$

$V = (4/3)\pi r^3$ Substitute into % V equation.

$$\% \ V_{nucleus} = \frac{V_{nucleus}}{V_{atom}} \times 100\% \qquad \rightarrow \% \ V_{nucleus} = \frac{(4/3) \ \pi r^3_{nucleus}}{(4/3) \ \pi r^3_{atom}} \times 100\% \qquad \text{Simplify equation.}$$

$$\% \ V_{nucleus} = \frac{r^3_{nucleus}}{r^3_{atom}} \times 100\% \qquad \text{Substitute numbers and calculate result.}$$

$$\% \ V_{nucleus} = \frac{(1.0 \times 10^{-15} \ m)^3}{(5.29 \times 10^{-11} \ m)^3} \times 100\% = (1.\underline{8}90359168 \times 10^{-5})^3 \times 100\% = 6.\underline{7}55118686 \times 10^{-13}$$

$$= 6.8 \times 10^{-13}$$

Check: The units (none) are correct. The magnitude of the answer seems correct (10^{-13}) because a proton is so small. Two significant figures are allowed to reflect the significant figures in 1.0×10^{-13} cm. Round the last digits up because the first nonsignificant digit is a 5.

1.125   **Given:** radius of hydrogen $= 212$ pm; radius of Ping-Pong ball $= 4.0$ cm, 6.02×10^{23} atoms and balls in a row
Find: row length (km)
Conceptual Plan: atoms → pm → m → km and ball → cm → m → km

$$\frac{212 \ pm}{1 \ atom} \quad \frac{1 \ m}{10^{12} \ pm} \quad \frac{1 \ km}{1000 \ m} \qquad\qquad \frac{4.0 \ cm}{1 \ ball} \quad \frac{100 \ cm}{1 \ m} \quad \frac{1 \ km}{1000 \ m}$$

Solution: $6.02 \times 10^{23} \ \text{atoms} \times \dfrac{212 \ pm}{1 \ atom} \times \dfrac{1 \ m}{10^{12} \ pm} \times \dfrac{1 \ km}{1000 \ m} = 1.28 \times 10^{11}$ km

$6.02 \times 10^{23} \ \text{balls} \times \dfrac{4.0 \ cm}{1 \ ball} \times \dfrac{1 \ m}{100 \ cm} \times \dfrac{1 \ km}{1000 \ m} = 2.4 \times 10^{19}$ km

Check: The units (km) are correct. The magnitude of the answers seem correct $(10^{11}$ and $10^{19})$. The answers are driven by the large number of atoms or balls. The Ping-Pong ball row is 108 times longer. Three significant figures are allowed to reflect the significant figures in 212 pm. Two significant figures are allowed to reflect the significant figures in 4.0 cm.

1.127   **Given:** 39.33 g sodium/100 g salt; 1.25 g salt/100 g snack mix; FDA maximum 2.40 g sodium/day
Find: g snack mix
Conceptual Plan: g sodium → g salt → g snack mix

$$\frac{100 \ g \ salt}{39.33 \ g \ sodium} \qquad \frac{100 \ g \ snack \ mix}{1.25 \ g \ salt}$$

Solution: $\dfrac{2.40 \ \text{g sodium}}{1 \ day} \times \dfrac{100 \ \text{g salt}}{39.33 \ \text{g sodium}} \times \dfrac{100 \ \text{g snack mix}}{1.25 \ \text{g salt}} = 488.\underline{1}770$ g snack mix/day

$= 488$ g snack mix/day

Check: The units (g) are correct. The magnitude of the answer seems correct (500) because salt is less than half sodium and there is a little over a gram of salt per 100 grams of snack mix. Three significant figures are allowed to reflect the significant figures in the FDA maximum and in the amount of salt in the snack mix.

1.129   **Given:** 24.0 kg copper wire; wire is a cylinder of radius $= 1.63$ mm   **Find:** resistance (Ω)
Other: $d(\text{copper}) = 8.96 \ g/cm^3$; resistance $= 2.061 \ \Omega/km$
Conceptual Plan: mm → m → cm and kg → g then

$$\frac{10^{-3} \ m}{1 \ mm} \quad \frac{100 \ cm}{1 \ m} \qquad\qquad \frac{1000 \ g}{1 \ kg}$$

d, m → V* then *V, r → l*(cm) → m → km → Ω

$$d = m/V \qquad V = l \ \pi \ r^2 \qquad \frac{1 \ m}{100 \ cm} \quad \frac{1 \ km}{1000 \ m} \quad \frac{2.061 \ \Omega}{1 \ km}$$

Solution: $1.63 \ \text{mm} \times \dfrac{10^{-3} \ m}{1 \ mm} \times \dfrac{100 \ cm}{1 \ m} = 0.163$ cm

$24.0 \ \text{kg} \times \dfrac{1000 \ g}{1 \ kg} = 2.40 \times 10^4$ g; then $d = m/V$. Rearrange by multiplying both sides of equation by V to get

$m = d \times V$; then divide both sides by d. $V = \dfrac{m}{d} = \dfrac{2.40 \times 10^4 \ g}{\dfrac{8.96 \ g}{1 \ cm^3}} = 2.6\underline{7}85714 \times 10^3 \ cm^3$; then $V = l\pi r^2$

Rearrange by dividing both sides of the equation by πr^2 to get

$$l = \frac{V}{\pi r^2} = \frac{2.6785714 \times 10^3 \text{ cm}^3}{\pi(0.163 \text{ cm})^2} = 3.2090623 \times 10^4 \text{ cm}$$

then

$$3.2090623 \times 10^4 \text{ cm} \times \frac{1 \text{ m}}{100 \text{ cm}} \times \frac{1 \text{ km}}{1000 \text{ m}} \times \frac{2.061 \ \Omega}{1 \text{ km}} = 0.6613877 \ \Omega = 0.661 \ \Omega$$

Check: The units (Ω) are correct. The magnitude of the answer seems correct because we expect a small resistance for a material that is commonly used for electrical wiring. Three significant figures are allowed to reflect the significant figures in 1.63 mm and 24.0 kg. Truncate the nonsignificant digits because the first nonsignificant digit is a 3.

1.131 **Given:** $d(\text{liquid nitrogen}) = 0.808 \text{ g/mL}$; $d(\text{gaseous nitrogen}) = 1.15 \text{ g/L}$; 175 L liquid nitrogen; $10.00\text{m} \times 10.00 \text{ m} \times 2.50 \text{ m}$ room **Find:** fraction of room displaced by nitrogen gas

Conceptual Plan: L $\rightarrow$ mL then $V_{\text{liquid}}, d_{\text{liquid}} \rightarrow m_{\text{liquid}}$ then set $m_{\text{liquid}} = m_{\text{gas}}$ then $m_{\text{gas}}, d_{\text{gas}} \rightarrow V_{\text{gas}}$ then

$$\frac{1000 \text{ mL}}{1 \text{ L}} \qquad\qquad d = m/v \qquad\qquad d = m/v$$

calculate the $V_{\text{room}} \rightarrow \text{cm}^3 \rightarrow$ L then calculate the fraction displaced

$$V = l \times w \times h \quad \frac{(100 \text{ cm})^3}{(1 \text{ m})^3} \quad \frac{1 \text{ L}}{1000 \text{ cm}^3} \quad \frac{V_{\text{gas}}}{V_{\text{room}}}$$

Solution: $175 \text{ L} \times \dfrac{1000 \text{ mL}}{1 \text{ L}} = 1.75 \times 10^5 \text{ mL}$. Solve for m by multiplying both sides of the equation by V.

$$m = V \times d = 1.75 \times 10^5 \text{ mL} \times \frac{0.808 \text{ g}}{1 \text{ mL}} = 1.414 \times 10^5 \text{ g nitrogen liquid} = 1.414 \times 10^5 \text{ g nitrogen gas}$$

$d = m/V$ Rearrange by multiplying both sides of the equation by V and dividing both sides of the equation by d.

$$V = \frac{m}{d} = \frac{1.414 \times 10^5 \text{ g}}{1.15 \ \frac{\text{g}}{\text{L}}} = 1.229565 \times 10^5 \text{ L nitrogen gas}$$

$$V_{\text{room}} = l \times w \times h = 10.00 \text{ m} \times 10.00 \text{ m} \times 2.50 \text{ m} \times \frac{(100 \text{ cm})^3}{(1 \text{ m})^3} \times \frac{1 \text{ L}}{1000 \text{ cm}^3} = 2.50 \times 10^5 \text{ L}$$

$$\frac{V_{\text{gas}}}{V_{\text{room}}} = \frac{1.229565 \times 10^5 \text{ L}}{2.50 \times 10^5 \text{ L}} = 0.491826 = 0.492$$

Check: The units (none) are correct. The magnitude of the answer seems correct (0.5) because there is a large volume of liquid and the density of the gas is about a factor of 1000 less than the density of the liquid. Three significant figures are allowed to reflect the significant figures in the densities and the volume of the liquid given.

Challenge Problems

1.133 Force = (mass) $\times$ (acceleration), or $F = ma$, and Pressure = force/area. So if a force of 2.31×10^4 N is applied on an area of 125 cm^2, the

$$\text{Pressure} = \frac{2.31 \times 10^4 \text{ N}}{125 \text{ cm}^2 \times \dfrac{(1 \text{ m})^2}{(100 \text{ cm})^2}} = 1.846 \times 10^6 \frac{\text{N}}{\text{m}^2}.$$

Referring to Chapter 5, $1 \text{ N/m}^2 = 1$ Pa and

$$1 \text{ atm} = 101{,}325 \text{ Pa}; \text{ so } 1.846 \times 10^6 \frac{\text{N}}{\text{m}^2} = 1.846 \times 10^6 \text{ Pa} \times \frac{1 \text{ atm}}{101{,}325 \text{ Pa}} = 18.2 \text{ atm}$$

1.135 Referring to the definition of energy in Chapter 6, 1 Joule $= 1\text{J} = \text{kg} \cdot \text{m}^2/\text{s}^2$. For kinetic energy, if the units of mass are the kilogram (kg) and the units of the velocity are meters/second (m/s), then kinetic energy units $=$

$$m v^2 = \text{kg} \left(\frac{\text{m}}{\text{s}} \right)^2 = \frac{\text{kg} \cdot \text{m}^2}{\text{s}^2} = \text{J}. \text{ Because a Newton (N) is a unit of force and has units of kg} \cdot \text{m/s}^2,$$

Pressure = force / area and has units of N/m^2, and Force = (mass) × (acceleration), or $F = ma$, then

$$3/2 \, PV \text{ units} = \frac{N}{m^2} \cdot m^3 = \frac{kg \cdot m}{s^2} \cdot m = \frac{kg \cdot m^2}{s^2} = J.$$

1.137 **Given:** 15.0 ppm CO; 8-hour period **Find:** milligrams of carbon monoxide
Other: 0.50 L of air per breath; 20 breaths per minute; carbon monoxide has a density of 1.2 g/L; 15.0 ppm CO means 15.0 L CO per 10^6 L air
Conceptual Plan: hr $\rightarrow$ min $\rightarrow$ breaths $\rightarrow$ L_{air} $\rightarrow$ L_{CO} $\rightarrow$ g_{CO} $\rightarrow$ mg_{CO}

$$\frac{60 \text{ min}}{1 \text{ hr}} \quad \frac{20 \text{ breath}}{1 \text{ min}} \quad \frac{0.50 \, L_{air}}{1 \text{ breath}} \quad \frac{15.0 \, L_{CO}}{1 \times 10^6 \, L_{air}} \quad \frac{1.2 \, g_{CO}}{1 \, L_{CO}} \quad \frac{1000 \, mg_{CO}}{1 \, g_{CO}}$$

Solution:

$$8 \text{ hr} \times \frac{60 \text{ min}}{1 \text{ hr}} \times \frac{20 \text{ breath}}{1 \text{ min}} \times \frac{0.50 \, L_{air}}{1 \text{ breath}} \times \frac{15.0 \, L_{CO}}{1 \times 10^6 \, L_{air}} \times \frac{1.2 \, g_{CO}}{1 \, L_{CO}} \times \frac{1000 \, mg_{CO}}{1 \, g_{CO}} = 86.4 \, mg_{CO} = 9 \times 10^1 \, mg_{CO}$$

Check: The units (mg) are correct. The magnitude of the answer (10^2) makes physical sense because there are more than 6 powers of 10 visible in these conversion factors in the numerator and one factor of 10^6 in the denominator. This means that most of the conversions cancel each other out, but there is still some left in the numerator. One significant figure is allowed because the conversion factor with the least precision is 20 breaths/minute (one significant figure); the starting time (8 hours) also has one significant figure. Round the last digit up because the first nonsignificant digit is a 6.

1.139 Because the person weighs 155 lb and has a density of $1.0 \, g/cm^3$, the volume of the person can be calculated as

$$155 \text{ lb} \times \frac{453.59 \, g}{1 \text{ lb}} \times \frac{1 \, cm^3}{1.0 \, g} = 7.0\underline{3}0645 \times 10^4 \, cm^3.$$

Approximating the volume of a person as a cylinder 4.0 feet tall, $V = l\pi r^2$. Rearranging the equation, solve for r.

$$r = \left(\frac{V}{l\pi}\right)^{\frac{1}{2}} = \sqrt{\frac{7.0\underline{3}0645 \times 10^4 \, cm^3}{4.0 \, ft \times \frac{30.48 \, cm}{1 \, ft} \times \pi}} = 1\underline{3}.54831 \, cm$$

The circumference is $2\pi r = 2\pi(1\underline{3}.54831 \, cm) = 8\underline{5}.12655 \, cm$. When the person gains 40.0 lb of fat, the volume increase is $40.0 \text{ lb} \times \frac{453.59 \, g}{1 \text{ lb}} \times \frac{1 \, cm^3}{0.918 \, g} = 1.9\underline{7}643 \times 10^4 \, cm^3.$

Thus, the new volume is $7.0\underline{3}0645 \times 10^4 \, cm^3 + 1.9\underline{7}643 \times 10^4 \, cm^3 = 9.0\underline{0}708 \times 10^4 \, cm^3$. So the new radius is $r = \left(\frac{V}{l\pi}\right)^{\frac{1}{2}} = \sqrt{\frac{9.0\underline{0}708 \times 10^4 \, cm^3}{4.0 \, ft \times \frac{30.48 \, cm}{1 \, ft} \times \pi}} = 1\underline{5}.33485 \, cm$, and the new circumference is

$2\pi r = 2\pi(1\underline{5}.33485 \, cm) = 9\underline{6}.35170 \, cm.$

The percent increase in circumference $\frac{9\underline{6}.35170 \, cm - 8\underline{5}.12655 \, cm}{8\underline{5}.12655 \, cm} \times 100\% = 1\underline{3}.1864\% = 13\%.$

Conceptual Problems

1.141 No. Because the container is sealed, the atoms and molecules can move around but they cannot leave. If no atoms or molecules can leave, the mass must be constant.

1.143 This problem is similar to Problem 1.64 except that the dimension is changed to 7 cm on each edge.
Given: 7 cm on each edge cube **Find:** cm^3
Conceptual Plan: Read the information carefully. The cube is 7 cm on each side.
$l, w, h \rightarrow V$
$V = l \, w \, h$
in a cube $l = w = h$

Solution: $7 \text{ cm} \times 7 \text{ cm} \times 7 \text{ cm} = (7 \text{ cm})^3 = 343 \, cm^3$, or 343 cubes

1.145   Remember that density = mass/volume.

(a)   The darker-colored box has a heavier mass but a smaller volume, so it is denser than the lighter-colored box.

(b)   The lighter-colored box is heavier than the darker-colored box, and both boxes have the same volume; so the lighter-colored box is denser.

(c)   The larger box is the heavier box, so it cannot be determined with this information which box is denser.

1.147   Remember that an observation is the information collected when studying phenomena. A law is a concise statement that summarizes observed behaviors and observations and predicts future observations. A theory attempts to explain why the observed behavior is happening.

(a)   This statement is most like a law because it summarizes many observations and can explain future behavior—many places and many days.

(b)   This statement is a theory because it attempts to explain why (gravitational forces).

(c)   This statement is most like an observation because it is information collected to understand tidal behavior.

(d)   This statement is most like a law because it summarizes many observations and can explain future behavior—many places and many days and months.

2 Atoms and Elements

Review Questions

2.1   Scanning tunneling microscopy is a technique that can image, and even move, individual atoms and molecules. A scanning tunneling microscope works by moving an extremely sharp electrode over a surface and measuring the resulting tunneling current, the electric current that flows between the tip of the electrode and the surface even though the two are not in physical contact.

2.3   The law of conservation of mass states the following: In a chemical reaction, matter is neither created nor destroyed. In other words, when you carry out any chemical reaction, the total mass of the substances involved in the reaction does not change.

2.5   The law of multiple proportions states the following: When two elements (call them A and B) form two different compounds, the masses of element B that combine with 1 g of element A can be expressed as a ratio of small whole numbers. This means that when two atoms (A and B) combine to form more than one compound, the ratio of B in one compound to B in the second compound will be a small whole number. The Law of Definite Proportions refers to the composition of a particular compound, not to a comparison of different compounds. The Law of Definite Proportions states that a particular compound is always made of the same elements in the same ratio.

2.7   In the late 1800s, an English physicist named J. J. Thomson performed experiments to probe the properties of cathode rays. Thomson found that these rays were actually streams of particles with the following properties: They traveled in straight lines, they were independent of the composition of the material from which they originated, and they carried a negative electrical charge. He measured the charge to mass ratio of the particles and found that the cathode ray particle was about 2000 times lighter than hydrogen.

2.9   The plum-pudding model of the atom proposed by J. J. Thomson hypothesized that the negatively charged electrons were small particles electrostatically held within a positively charged sphere.

2.11   Rutherford's nuclear model of the atom has three basic parts: (1) Most of the atom's mass and all of its positive charge are contained in a small core called the nucleus. (2) Most of the volume of the atom is empty space, throughout which tiny negatively charged electrons are dispersed. (3) There are as many negatively charged electrons outside the nucleus as there are positively charged particles within the nucleus, so that the atom is electrically neutral. The revolutionary part of this theory is the idea that matter, at its core, is much less uniform than it appears.

2.13   The three subatomic particles that compose atoms are as follows:
        Protons, which have a mass of 1.67262×10^{-27} kg or 1.00727 amu and a relative charge of $+1$
        Neutrons, which have a mass of 1.67493×10^{-27} kg or 1.00866 amu and a relative charge of 0
        Electrons, which have a mass of 0.00091×10^{-27} kg or 0.00055 amu and a relative charge of -1

2.15   The atomic number, Z, is the number of protons in an atom's nucleus. The atomic mass number A is the sum of the neutrons and protons in an atom.

2.17   Isotopes are atoms with the same number of protons but different numbers of neutrons. The percent natural abundance is the relative amount of each different isotope in a naturally occurring sample of a given element.

2.19   An ion is a charged particle. Positively charged ions are called cations. Negatively charged ions are called anions.

2.21   Metals are found on the left side and in the middle of the periodic table. They are good conductors of heat and electricity; they can be pounded into flat sheets (malleable), they can be drawn into wires (ductile), they are often shiny, and they tend to lose electrons when they undergo chemical changes.

Nonmetals are found on the upper-right side of the periodic table. Their properties are more varied: Some are solids at room temperature, while others are liquids or gases. As a whole, they tend to be poor conductors of heat and electricity and to gain electrons when they undergo chemical changes.

Metalloids lie along the zigzag diagonal line that divides metals and nonmetals. They show mixed properties. Several metalloids are also classified as semiconductors because of their intermediate and temperature-dependent electrical conductivity.

2.23   Main group metals tend to lose electrons, forming cations with the same number of electrons as the nearest noble gas. Main group nonmetals tend to gain electrons, forming anions with the same number of electrons as the nearest following noble gas.

2.25   In a mass spectrometer, the sample is injected into the instrument and vaporized. The vaporized atoms are then ionized by an electron beam. The electrons in the beam collide with the vaporized atoms, removing electrons from the atoms and creating positively charged ions. Charged plates with slits in them accelerate the positively charged ions into a magnetic field, which deflects them. The amount of deflection depends on the mass of the ions—lighter ions are deflected more than heavier ones are. Finally, the ions strike a detector and produce an electrical signal that is recorded.

2.27   A mole is an amount of material. It is defined as the amount of material containing 6.0221421×10^{23} particles (Avogadro's number). The numerical value of the mole is defined as being equal to the number of atoms in exactly 12 grams of pure carbon-12. It is useful for converting number of atoms to moles of atoms and moles of atoms to number of atoms.

Problems by Topic

The Laws of Conservation of Mass, Definite Proportions, and Multiple Proportions

2.29   **Given:** 1.50 g hydrogen; 12.0 g oxygen   **Find:** grams of water vapor
        Conceptual Plan: total mass reactants = total mass products
        Solution: Mass of reactants = 1.50 g hydrogen + 12.0 g oxygen = 13.5 grams
                    Mass of products = mass of reactants = 13.5 grams water vapor

Check: According to the law of conservation of mass, matter is neither created nor destroyed in a chemical reaction. Since water vapor is the only product, the masses of hydrogen and oxygen must combine to form the mass of water vapor.

2.31   **Given:** sample 1: 38.9 g carbon, 448 g chlorine; sample 2: 14.8 g carbon, 134 g chlorine
        Find: consistent with definite proportions
        Conceptual Plan: Determine mass ratio of samples 1 and 2 and compare.
                    $\frac{\text{mass of chlorine}}{\text{mass of carbon}}$

        Solution: sample 1: $\frac{448 \text{ g chorine}}{38.9 \text{ g carbon}} = 11.5$ sample 2: $\frac{134 \text{ g chlorine}}{14.8 \text{ g carbon}} = 9.05$

Results are not consistent with the law of definite proportions because the ratio of chlorine to carbon is not the same.

Check: According to the law of definite proportions, the mass ratio of one element to another is the same for all samples of the compound.

2.33   **Given:** mass ratio sodium to fluorine $= 1.21:1$; sample $= 28.8$ g sodium   **Find:** g fluorine

         Conceptual Plan: g sodium $\rightarrow$ g fluorine

$$\frac{\text{mass of fluorine}}{\text{mass of sodium}}$$

         Solution: $28.8 \; \text{g sodium} \times \dfrac{1 \text{ g fluorine}}{1.21 \text{ g sodium}} = 23.8$ g fluorine

         Check: The units of the answer (g fluorine) are correct. The magnitude of the answer is reasonable because it is less than the grams of sodium.

2.35   **Given:** 1g osmium: sample 1 $= 0.168$ g oxygen; sample 2 $= 0.3369$ g oxygen

         Find: consistent with multiple proportions

         Conceptual Plan: Determine mass ratio of oxygen.

$$\frac{\text{mass of oxygen sample 2}}{\text{mass of oxygen sample 1}}$$

         Solution: $\dfrac{0.3369 \text{ g oxygen}}{0.168 \text{ g oxygen}} = 2.00$ Ratio is a small whole number. Results are consistent with multiple proportions.

         Check: According to the law of multiple proportions, when two elements form two different compounds, the masses of element B that combine with 1 g of element A can be expressed as a ratio of small whole numbers.

2.37   **Given:** sulfur dioxide $= 3.49$ g oxygen and 3.50 g sulfur; sulfur trioxide $= 6.75$ g oxygen and 4.50 g sulfur

         Find: mass oxygen per g S for each compound and then determine the mass ratio of oxygen

$$\frac{\text{mass of oxygen in sulfur dioxide}}{\text{mass of sulfur in sulfur dioxide}} \quad \frac{\text{mass of oxygen in sulfur trioxide}}{\text{mass of sulfur in sulfur trioxide}} \quad \frac{\text{mass of oxyen in sulfur trioxide}}{\text{mass of oxyen in sulfur dioxide}}$$

         Solution: sulfur dioxide $= \dfrac{3.49 \text{ g oxygen}}{3.50 \text{ g sulfur}} = \dfrac{0.997 \text{ g oxygen}}{1 \text{ g sulfur}}$ sulfur trioxide $= \dfrac{6.75 \text{ g oxygen}}{4.50 \text{ g sulfur}} = \dfrac{1.50 \text{ g oxygen}}{1 \text{ g sulfur}}$

         $\dfrac{1.50 \text{ g oxygen in sulfur trioxide}}{0.997 \text{ g oxygen in sulfur dioxide}} = \dfrac{1.50}{1} = \dfrac{3}{2}.$ The ratio is converted from 1.50:1 to 3:2 because the law of multiple proportions states that the ratio is in small whole numbers. Ratio is in small whole numbers and is consistent with multiple proportions.

         Check: According to the law of multiple proportions, when two elements form two different compounds, the masses of element B that combine with 1 g of element A can be expressed as a ratio of small whole numbers.

Atomic Theory, Nuclear Theory, and Subatomic Particles

2.39   (a)   Sulfur and oxygen atoms have the same mass. *Inconsistent* with Dalton's atomic theory because only atoms of the same element have the same mass.

         (b)   All cobalt atoms are identical. *Consistent* with Dalton's atomic theory because all atoms of a given element have the same mass and other properties that distinguish them from atoms of other elements.

         (c)   Potassium and chlorine atoms combine in a 1:1 ratio to form potassium chloride. *Consistent* with Dalton's atomic theory because atoms combine in simple whole number ratios to form compounds.

         (d)   Lead atoms can be converted into gold. *Inconsistent* with Dalton's atomic theory because atoms of one element cannot change into atoms of another element.

2.41   (a)   The volume of an atom is mostly empty space. *Consistent* with Rutherford's nuclear theory because most of the volume of the atom is empty space, throughout which tiny, negatively charged electrons are dispersed.

         (b)   The nucleus of an atom is small compared to the size of the atom. *Consistent* with Rutherford's nuclear theory because most of the atom's mass and all of its positive charge are contained in a small core called the nucleus.

         (c)   Neutral lithium atoms contain more neutrons than protons. *Inconsistent* with Rutherford's nuclear theory because it did not distinguish where the mass of the nucleus came from other than from the protons.

         (d)   Neutral lithium atoms contain more protons than electrons. *Inconsistent* with Rutherford's nuclear theory because there are as many negatively charged particles outside the nucleus as there are positively charged particles in the nucleus.

2.43   **Given:** drop A $= -6.9 \times 10^{-19}$ C;   drop B $= -9.2 \times 10^{-19}$ C;   drop C $= -11.5 \times 10^{-19}$ C;
drop D $= -4.6 \times 10^{-19}$ C
Find: the charge on a single electron
Conceptual Plan: Determine the ratio of charge for each set of drops.
$$\frac{\text{charge on drop 1}}{\text{charge on drop 2}}$$
Solution: $\dfrac{-6.9 \times 10^{-19}\,\text{C drop A}}{-4.6 \times 10^{-19}\,\text{C drop D}} = 1.5$ $\dfrac{-9.2 \times 10^{-19}\,\text{C drop B}}{-4.6 \times 10^{-19}\,\text{C drop D}} = 2$ $\dfrac{-11.5 \times 10^{-19}\,\text{C drop C}}{-4.6 \times 10^{-19}\,\text{C drop D}} = 2.5$

The ratios obtained are not whole numbers, but they can be converted to whole numbers by multiplying by 2. Therefore, the charge on the electron has to be 1/2 the smallest value experimentally obtained. The charge on the electron $= -2.3 \times 10^{-19}$ C.

Check: The units of the answer (Coulombs) are correct. The magnitude of the answer is reasonable because all of the values experimentally obtained are integer multiples of -2.3×10^{-19}.

2.45   **Given:** charge on body $= -15\,\mu$C   **Find:** number of electrons; mass of the electrons
Conceptual Plan: μC $\rightarrow$ C $\rightarrow$ **number of electrons** $\rightarrow$ **mass of electrons**
$$\frac{1\,\text{C}}{10^6\,\mu\text{C}} \quad \frac{1\,\text{electron}}{-1.60 \times 10^{-19}\,\text{C}} \qquad \frac{9.10 \times 10^{-28}\,\text{g}}{1\,\text{electron}}$$
Solution: $-15\,\mu\cancel{C} \times \dfrac{1\,\cancel{C}}{10^6\,\mu\cancel{C}} \times \dfrac{1\,\text{electron}}{-1.60 \times 10^{-19}\,\cancel{C}} = 9.375 \times 10^{13}$ electrons $= 9.4 \times 10^{13}$ electrons

$9.375 \times 10^{13}\,\cancel{\text{electrons}} \times \dfrac{9.10 \times 10^{-28}\,\text{g}}{1\,\cancel{\text{electron}}} = 8.5 \times 10^{-14}$ g

Check: The units of the answers (number of electrons and grams) are correct. The magnitude of the answers is reasonable because the charge on an electron and the mass of an electron are very small.

2.47   (a)   True: Protons and electrons have equal and opposite charges.
(b)   True: Protons and electrons have opposite charges, so they will attract each other.
(c)   True: The mass of the electron is much less than the mass of the neutron.
(d)   False: The mass of the proton and the mass of the neutron are about the same.

2.49   **Given:** mass of proton   **Find:** number of electrons in equal mass
Conceptual Plan: mass of protons $\rightarrow$ **number of electrons**
$$\frac{1.67262 \times 10^{-27}\,\text{kg}}{1\,\text{proton}} \qquad \frac{1\,\text{electron}}{9.10938 \times 10^{-31}\,\text{kg}}$$
Solution: $1.67262 \times 10^{-27}\,\cancel{\text{kg}} \times \dfrac{1\,\text{electron}}{9.10938 \times 10^{-31}\,\cancel{\text{kg}}} = 1.83615 \times 10^{3}$ electrons

Check: The units of the answer (electrons) are correct. The magnitude of the answer is reasonable since the mass of the electron is much less than the mass of the proton.

Isotopes and Ions

2.51   For each of the isotopes determine Z (the number of protons) from the periodic table and determine A (protons + neutrons). Then write the symbol in the form X-A.
(a)   The silver isotope with 60 neutrons: $Z = 47$; $A = 47 + 60 = 107$;   Ag-107
(b)   The silver isotope with 62 neutrons: $Z = 47$; $A = 47 + 62 = 109$;   Ag-109
(c)   The uranium isotope with 146 neutrons: $Z = 92$; $A = 92 + 146 = 238$;   U-238
(d)   The hydrogen isotope with 1 neutron: $Z = 1$; $A = 1 + 1 = 2$;   H-2

2.53   (a) $^{14}_{7}$N: $Z = 7$; $A = 14$; protons $= Z = 7$; neutrons $= A - Z = 14 - 7 = 7$
(b) $^{23}_{11}$Na: $Z = 11$; $A = 23$; protons $= Z = 11$; neutrons $= A - Z = 23 - 11 = 12$
(c) $^{222}_{86}$Rn: $Z = 86$; $A = 222$; protons $= Z = 86$; neutrons $= A - Z = 222 - 86 = 136$
(d) $^{208}_{82}$Pb: $Z = 82$; $A = 208$; protons $= Z = 82$; neutrons $= A - Z = 208 - 82 = 126$

2.55   Carbon-14: $A = 14, Z = 6; {}^{14}_{6}C$ # protons $= Z = 6$ # neutrons $= A - Z = 14 - 6 = 8$

2.57   In a neutral atom, the number of protons = the number of electrons = Z. For an ion, electrons are lost (cations) or gained (anions).

       (a) $Ni^{2+}: Z = 28 =$ protons; $Z - 2 = 26 =$ electrons

       (b) $S^{2-}: Z = 16 =$ protons; $Z + 2 = 18 =$ electrons

       (c) $Br^-: Z = 35 =$ protons; $Z + 1 = 36 =$ electrons

       (d) $Cr^{3+}: Z = 24 =$ protons; $Z - 3 = 21 =$ electrons

2.59   Main group metal atoms will lose electrons to form a cation with the same number of electrons as the nearest previous noble gas. Nonmetal atoms will gain electrons to form an anion with the same number of electrons as the nearest noble gas.

       (a) O^{2-} O is a nonmetal and has eight electrons. It will gain electrons to form an anion. The nearest noble gas is neon with ten electrons, so O will gain two electrons.

       (b) K^+ K is a main group metal and has 19 electrons. It will lose electrons to form a cation. The nearest noble gas is argon with 18 electrons, so K will lose 1 electron.

       (c) Al^{3+} Al is a main group metal and has 13 electrons. It will lose electrons to form a cation. The nearest noble gas is neon with 10 electrons, so Al will lose 3 electrons.

       (d) Rb^+ Rb is a main group metal and has 37 electrons. It will lose electrons to form a cation. The nearest noble gas is krypton with 36 electrons, so Rb will lose 1 electron.

2.61   Main group metal atoms will lose electrons to form a cation with the same number of electrons as the nearest previous noble gas. Atoms in period 4 and higher lose electrons to form the same ion as the element at the top of the group. Nonmetal atoms will gain electrons to form an anion with the same number of electrons as the nearest noble gas.

Symbol	Ion Formed	Number of Electrons in Ion	Number of Protons in Ion
Ca	Ca^{2+}	18	20
Be	Be^{2+}	2	4
Se	Se^{2-}	36	34
In	In^{3+}	46	49

The Periodic Table and Atomic Mass

2.63   (a)   K   Potassium is a metal.

       (b)   Ba   Barium is a metal.

       (c)   I   Iodine is a nonmetal.

       (d)   O   Oxygen is a nonmetal.

       (e)   Sb   Antimony is a metalloid.

2.65   (a)   tellurium   Te is in group 6A and is a main group element.

       (b)   potassium   K is in group 1A and is a main group element.

       (c)   vanadium   V is in group 5B and is a transition element.

       (d)   manganese   Mn is in group 7B and is a transition element.

2.67   (a)   sodium   Na is in group 1A and is an alkali metal.

       (b)   iodine   I is in group 7A and is a halogen.

       (c)   calcium   Ca is in group 2A and is an alkaline earth metal.

       (d)   barium   Ba is in group 2A and is an alkaline earth metal.

       (e)   krypton   Kr is in group 8A and is a noble gas.

2.69   (a)   N and Ni would not be most similar. Nitrogen is a nonmetal; nickel is a metal.

        (b)   Mo and Sn would not be most similar. Although both are metals, molybdenum is a transition metal and tin is a main group metal.

        (c)   Na and Mg would not be most similar. Although both are main group metals, sodium is in group 1A and magnesium is in group 2A.

        (d)   Cl and F would be most similar. Chlorine and fluorine are both in group 7A. Elements in the same group have similar chemical properties.

        (e)   Si and P would not be most similar. Silicon is a metalloid, and phosphorus is a nonmetal.

2.71   **Given:** Ga-69; mass $=$ 68.92558 amu; 60.108%: Ga-71; mass $=$ 70.92470 amu; 39.892%   **Find:** sketch the mass spectrum

        Conceptual Plan: $\rightarrow$ **show relative % abundance versus mass on the graph**

        Solution:

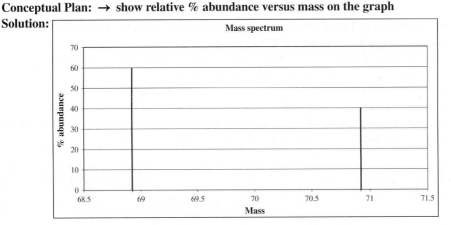

        Check: The mass spectrum is reasonable because it has two mass lines corresponding to the two isotopes and the line at 68.92588 is about 1.5 times larger than the line at 70.92470.

2.73   Fluorine has an isotope F-19 with a very large abundance so that the mass of fluorine is very close to the mass of the isotope and the line in the mass spectrum reflects the abundance of F-19. Chlorine has two isotopes (Cl-35 and Cl-37), and the mass of 35.45 amu is the weighted average of these two isotopes; so there is no line at 35.45 amu.

2.75   **Given:** isotope 1 mass $=$ 120.903 amu, 57.4%; isotope 2 mass $=$ 122.9042 amu

        Find: atomic mass of the element and identify the element

        Conceptual Plan:

        % abundance isotope 2 $\rightarrow$ and then % abundance $\rightarrow$ fraction and then find atomic mass

$$100\% - \% \text{ abundance isotope 1} \qquad \frac{\% \text{ abundance}}{100} \qquad \text{Atomic mass} = \sum_{n}(\text{fraction of isotope } n) \times (\text{mass of isotope } n)$$

        Solution: 100.0% $-$ 57.4% Isotope 1 $=$ 42.6% Isotope 2

$$\text{Fraction isotope 1} = \frac{57.4}{100} = 0.574 \qquad \text{Fraction isotope 2} = \frac{42.6}{100} = 0.426$$

$$\text{Atomic mass} = \sum_{n}(\text{fraction of isotope } n) \times (\text{mass of isotope } n)$$

$$= 0.574(120.9038 \text{ amu}) + 0.426(122.9042 \text{ amu}) = 121.8 \text{ amu}$$

        From the periodic table, Sb has a mass of 121.757 amu; so it is the closest mass, and the element is antimony.

        Check: The units of the answer (amu) are correct. The magnitude of the answer is reasonable because it lies between 120.9038 and 122.9042 and is slightly less than halfway between the two values because the lower value has a slightly greater abundance.

2.77   **Given:** Br-81; mass $=$ 80.9163 amu; 49.31%: atomic mass Br $=$ 79.904 amu

        Find: mass and abundance

        Conceptual Plan: % abundance Br-79 $\rightarrow$ then % abundance $\rightarrow$ fraction $\rightarrow$ mass Br-79

$$100\% - \% \text{ Br-81} \qquad \frac{\% \text{ abundance}}{100} \qquad \text{Atomic mass} = \sum_{n}(\text{fraction of isotope } n) \times (\text{mass of isotope } n)$$

Solution: $100.00\% - 49.31\% = 50.69\%$

Fraction Br-79 $= \dfrac{50.69}{100} = 0.5069$ Fraction Br-81 $= \dfrac{49.31}{100} = 0.4931$

Let x be the mass of Br-79.

Atomic mass $= \sum_{n}(\text{fraction of isotope } n) \times (\text{mass of isotope } n)$

$79.904 \text{ amu} = 0.5069(x \text{ amu}) + 0.4931(80.9163 \text{ amu})$

$x = 78.92 \text{ amu} = \text{mass Br-79}$

Check: The units of the answer (amu) are correct. The magnitude of the answer is reasonable because it is less than the mass of the atom and the second isotope (Br-81) has a mass greater than the mass of the atom.

2.79   **Given:** graph of mass and % abundance of Eu   **Find:** atomic mass of Eu
Conceptual Plan: % abundance $\rightarrow$ relative fraction abundance $\rightarrow$ mass Eu

$$\dfrac{\text{\% abundance of each isotope}}{\text{sum of all \% abundance}} \qquad \text{Atomic mass} = \sum_{n}(\text{fraction of isotope } n) \times (\text{mass of isotope } n)$$

Solution: fraction $^{151}\text{Eu} = \dfrac{91.6}{191.6} = 0.478\underline{1}$ fraction $^{153}\text{Eu} = \dfrac{100.0}{191.6} = 0.521\underline{9}$

Atomic mass $= \sum_{n}(\text{fraction of isotope } n) \times (\text{mass of isotope } n)$

$= 0.478\underline{1}(151 \text{ amu}) + 0.521\underline{9}(153 \text{ amu})$

$= 15\underline{2}.04 \text{ amu} = 152 \text{ amu}$

Check: The units of the answer (amu) are correct. The magnitude of the answer is reasonable because the fraction of each isotope is about 0.5.

The Mole Concept

2.81   **Given:** 5.52 mol sulfur   **Find:** atoms of sulfur
Conceptual Plan: mol S $\rightarrow$ atoms S

$$\dfrac{6.022 \times 10^{23} \text{ atoms}}{\text{mol}}$$

Solution: $5.52 \text{ mol S} \times \dfrac{6.022 \times 10^{23} \text{ atoms S}}{\text{mol S}} = 3.32 \times 10^{24} \text{ atoms S}$

Check: The units of the answer (atoms S) are correct. The magnitude of the answer is reasonable because more than 1 mole of material is present.

2.83   (a)   **Given:** 11.8 g Ar   **Find:** mol Ar
Conceptual Plan: g Ar $\rightarrow$ mol Ar

$$\dfrac{1 \text{ mol Ar}}{39.95 \text{ g Ar}}$$

Solution: $11.8 \text{ g Ar} \times \dfrac{1 \text{ mol Ar}}{39.95 \text{ g Ar}} = 0.295 \text{ mol Ar}$

Check: The units of the answer (mol Ar) are correct. The magnitude of the answer is reasonable because less than the mass of 1 mol is present.

(b)   **Given:** 3.55 g Zn   **Find:** mol Zn
Conceptual Plan: g Zn $\rightarrow$ mol Zn

$$\dfrac{1 \text{ mol Zn}}{65.41 \text{ g Zn}}$$

Solution: $3.55 \text{ g Zn} \times \dfrac{1 \text{ mol Zn}}{65.41 \text{ g Zn}} = 0.0543 \text{ mol Zn}$

Check: The units of the answer (mol Zn) are correct. The magnitude of the answer is reasonable because less than the mass of 1 mol is present.

(c)   **Given:** 26.1 g Ta   **Find:** mol Ta
Conceptual Plan: g Ta → mol Ta
$$\frac{1 \text{ mol Ta}}{180.95 \text{ g Ta}}$$
Solution: $26.1 \text{ g Ta} \times \dfrac{1 \text{ mol Ta}}{180.95 \text{ g Ta}} = 0.144 \text{ mol Ta}$

Check: The units of the answer (mol Ta) are correct. The magnitude of the answer is reasonable because less than the mass of 1 mol is present.

(d)   **Given:** 0.211 g Li   **Find:** mol Li
Conceptual Plan: g Li → mol Li
$$\frac{1 \text{ mol Li}}{6.941 \text{ g Li}}$$
Solution: $0.211 \text{ g Li} \times \dfrac{1 \text{ mol Li}}{6.941 \text{ g Li}} = 0.0304 \text{ mol Li}$

Check: The units of the answer (mol Li) are correct. The magnitude of the answer is reasonable because less than the mass of 1 mol is present.

2.85   **Given:** 3.78 g silver   **Find:** atoms Ag
Conceptual Plan: g Ag → mol Ag → atoms Ag
$$\frac{1 \text{ mol Ag}}{107.87 \text{ g Ag}} \qquad \frac{6.022 \times 10^{23} \text{ atoms}}{\text{mol}}$$
Solution: $3.78 \text{ g Ag} \times \dfrac{1 \text{ mol Ag}}{107.87 \text{ g Ag}} \times \dfrac{6.022 \times 10^{23} \text{ atoms Ag}}{1 \text{ mol Ag}} = 2.11 \times 10^{22} \text{ atoms Ag}$

Check: The units of the answer (atoms Ag) are correct. The magnitude of the answer is reasonable because less than the mass of 1 mol of Ag is present.

2.87   (a)   **Given:** 5.18 g P   **Find:** atoms P
Conceptual Plan: g P → mol P → atoms P
$$\frac{1 \text{ mol P}}{30.97 \text{ g P}} \qquad \frac{6.022 \times 10^{23} \text{ atoms}}{\text{mol}}$$
Solution: $5.18 \text{ g P} \times \dfrac{1 \text{ mol P}}{30.97 \text{ g P}} \times \dfrac{6.022 \times 10^{23} \text{ atoms P}}{1 \text{ mol P}} = 1.01 \times 10^{23} \text{ atoms P}$

Check: The units of the answer (atoms P) are correct. The magnitude of the answer is reasonable because less than the mass of 1 mol of P is present.

(b)   **Given:** 2.26 g Hg   **Find:** atoms Hg
Conceptual Plan: g Hg → mol Hg → atoms Hg
$$\frac{1 \text{ mol Hg}}{200.59 \text{ g Hg}} \qquad \frac{6.022 \times 10^{23} \text{ atoms}}{\text{mol}}$$
Solution: $2.26 \text{ g Hg} \times \dfrac{1 \text{ mol Hg}}{200.59 \text{ g Hg}} \times \dfrac{6.022 \times 10^{23} \text{ atoms Hg}}{1 \text{ mol Hg}} = 6.78 \times 10^{21} \text{ atoms Hg}$

Check: The units of the answer (atoms Hg) are correct. The magnitude of the answer is reasonable because much less than the mass of 1 mol of Hg is present.

(c)   **Given:** 1.87 g Bi   **Find:** atoms Bi
Conceptual Plan: g Bi → mol Bi → atoms Bi
$$\frac{1 \text{ mol Bi}}{208.98 \text{ g Bi}} \qquad \frac{6.022 \times 10^{23} \text{ atoms}}{\text{mol}}$$
Solution: $1.87 \text{ g Bi} \times \dfrac{1 \text{ mol Bi}}{208.98 \text{ g Bi}} \times \dfrac{6.022 \times 10^{23} \text{ atoms Bi}}{1 \text{ mol Bi}} = 5.39 \times 10^{21} \text{ atoms Bi}$

Check: The units of the answer (atoms Bi) are correct. The magnitude of the answer is reasonable because less than the mass of 1 mol of Bi is present.

(d)   **Given:** 0.082 g Sr   **Find:** atoms Sr
Conceptual Plan: g Sr → mol Sr → atoms Sr
$$\frac{1 \text{ mol Sr}}{87.62 \text{ g Sr}} \qquad \frac{6.022 \times 10^{23} \text{ atoms}}{\text{mol}}$$

Solution: $0.082 \text{ g Sr} \times \dfrac{1 \text{ mol Sr}}{87.62 \text{ g Sr}} \times \dfrac{6.022 \times 10^{23} \text{ atoms Sr}}{1 \text{ mol Sr}} = 5.6 \times 10^{20}$ atoms Sr

Check: The units of the answer (atoms Sr) are correct. The magnitude of the answer is reasonable because less than the mass of 1 mol of Sr is present.

2.89   (a)   **Given:** 1.1×10^{23} gold atoms   **Find:** grams Au

       Conceptual Plan: atoms Au → mol Au → g Au

$$\dfrac{1 \text{ mol}}{6.022 \times 10^{23} \text{ atoms}} \qquad \dfrac{196.97 \text{ g Au}}{1 \text{ mol Au}}$$

       Solution: $1.1 \times 10^{23} \text{ atoms Au} \times \dfrac{1 \text{ mol Au}}{6.022 \times 10^{23} \text{ atoms Au}} \times \dfrac{196.97 \text{ g Au}}{1 \text{ mol Au}} = 36$ g Au

       Check: The units of the answer (g Au) are correct. The magnitude of the answer is reasonable because fewer than Avogadro's number of atoms are in the sample.

   (b)   **Given:** 2.82×10^{22} helium atoms   **Find:** grams He

       Conceptual Plan: atoms He → mol He → g He

$$\dfrac{1 \text{ mol}}{6.022 \times 10^{23} \text{ atoms}} \qquad \dfrac{4.002 \text{ g He}}{1 \text{ mol He}}$$

       Solution: $2.82 \times 10^{22} \text{ atoms He} \times \dfrac{1 \text{ mol He}}{6.022 \times 10^{23} \text{ atoms He}} \times \dfrac{4.002 \text{ g He}}{1 \text{ mol He}} = 0.187$ g He

       Check: The units of the answer (g He) are correct. The magnitude of the answer is reasonable because fewer than Avogadro's number of atoms are in the sample.

   (c)   **Given:** 1.8×10^{23} lead atoms   **Find:** grams Pb

       Conceptual Plan: atoms Pb → mol Pb → g Pb

$$\dfrac{1 \text{ mol}}{6.022 \times 10^{23} \text{ atoms}} \qquad \dfrac{207.2 \text{ g Pb}}{1 \text{ mol Pb}}$$

       Solution: $1.8 \times 10^{23} \text{ atoms Pb} \times \dfrac{1 \text{ mol Pb}}{6.022 \times 10^{23} \text{ atoms Pb}} \times \dfrac{207.2 \text{ g Pb}}{1 \text{ mol Pb}} = 62$ g Pb

       Check: The units of the answer (g Pb) are correct. The magnitude of the answer is reasonable because fewer than Avogadro's number of atoms are in the sample.

   (d)   **Given:** 7.9×10^{21} uranium atoms   **Find:** grams U

       Conceptual Plan: atoms U → mol U → g U

$$\dfrac{1 \text{ mol}}{6.022 \times 10^{23} \text{ atoms}} \qquad \dfrac{238.029 \text{ g U}}{1 \text{ mol U}}$$

       Solution: $7.9 \times 10^{21} \text{ atoms U} \times \dfrac{1 \text{ mol U}}{6.022 \times 10^{23} \text{ atoms U}} \times \dfrac{238.029 \text{ g U}}{1 \text{ mol U}} = 3.1$ g U

       Check: The units of the answer (g U) are correct. The magnitude of the answer is reasonable because fewer than Avogadro's number of atoms are in the sample.

2.91   **Given:** 52 mg diamond (carbon)   **Find:** atoms C

     Conceptual Plan: mg C → g C → mol C → atoms C

$$\dfrac{1 \text{ g C}}{1000 \text{ mg C}} \qquad \dfrac{1 \text{ mol C}}{12.011 \text{ g C}} \qquad \dfrac{6.022 \times 10^{23} \text{ atoms}}{\text{mol}}$$

     Solution: $52 \text{ mg C} \times \dfrac{1 \text{ g C}}{1000 \text{ mg C}} \times \dfrac{1 \text{ mol C}}{12.011 \text{ g C}} \times \dfrac{6.022 \times 10^{23} \text{ atoms C}}{1 \text{ mol C}} = 7.2 \times 10^{21}$ atoms C

     Check: The units of the answer (atoms C) are correct. The magnitude of the answer is reasonable because less than the mass of 1 mol of C is present.

2.93   **Given:** 1 atom platinum   **Find:** g Pt

     Conceptual Plan: atoms Pt → mol Pt → g Pt

$$\dfrac{1 \text{ mol}}{6.022 \times 10^{23} \text{ atoms}} \qquad \dfrac{195.08 \text{ g Pt}}{1 \text{ mol Pt}}$$

     Solution: $1 \text{ atom Pt} \times \dfrac{1 \text{ mol Pt}}{6.022 \times 10^{23} \text{ atoms Pt}} \times \dfrac{195.08 \text{ g Pt}}{1 \text{ mol Pt}} = 3.239 \times 10^{-22}$ g Pt

     Check: The units of the answer (g Pt) are correct. The magnitude of the answer is reasonable because only one atom is in the sample.

Cumulative Problems

2.95   **Given:** 7.83 g HCN sample 1: 0.290 g H; 4.06 g N; 3.37 g HCN sample 2   **Find:** g C in sample 2
          Conceptual Plan: g HCN sample 1 $\rightarrow$ g C in HCN sample 1 $\rightarrow$ ratio g C to g HCN $\rightarrow$ g C in HCN sample 2

$$\text{g HCN} - \text{g H} - \text{g N} \qquad \frac{\text{g C}}{\text{g HCN}} \qquad \text{g HCN} \times \frac{\text{g C}}{\text{g HCN}}$$

          Solution: 7.83 g HCN $-$ 0.290 g H $-$ 4.06 g N = 3.48 g C

$$3.37 \text{ g HCN} \times \frac{3.48 \text{ g C}}{7.83 \text{ g HCN}} = 1.50 \text{ g C}$$

          Check: The units of the answer (g C) are correct. The magnitude of the answer is reasonable because the sample size is about half the original sample size and the g C are about half the original g C.

2.97   **Given:** in CO, mass ratio O:C = 1.33:1; in compound X, mass ratio O:C = 2:1   **Find:** formula of X
          Conceptual Plan: Determine the mass ratio of O:O in the two compounds.

          Solution: For 1 gram of C $\dfrac{2 \text{ g O in compound X}}{1.33 \text{ g O in CO}} = 1.5$

             So the ratio of O to C in compound X has to be 1.5:1, and the formula is C_2O_3.

          Check: The answer is reasonable because it fulfills the criteria of multiple proportions and the mass ratio of O:C is 2:1.

2.99   **Given:** $^4\text{He}^{2+}$ = 4.00151 amu   **Find:** charge to mass ratio C/kg
          Conceptual Plan: Determine total charge on $^4\text{He}^{2+}$ and then amu $^4\text{He}^{2+} \rightarrow$ g $^4\text{He}^{2+} \rightarrow$ kg $^4\text{He}^{2+}$.

$$\frac{+1.60218 \times 10^{-19} \text{ C}}{\text{proton}} \qquad \frac{1.66054 \times 10^{-24} \text{ g}}{1 \text{ amu}} \quad \frac{1 \text{ kg}}{1000 \text{ g}}$$

          Solution: $\dfrac{2 \text{ protons}}{1 \text{ atom } ^4\text{He}^{2+}} \times \dfrac{+1.60218 \times 10^{-19} \text{ C}}{\text{proton}} = \dfrac{3.20436 \times 10^{-19} \text{ C}}{\text{atom } ^4\text{He}^{2+}}$

$$\frac{4.00151 \text{ amu}}{1 \text{ atom } ^4\text{He}^{2+}} \times \frac{1.66054 \times 10^{-24} \text{ g}}{1 \text{ amu}} \times \frac{1 \text{ kg}}{1000 \text{ g}} = \frac{6.64466742 \times 10^{27} \text{ kg}}{1 \text{ atom } ^4\text{He}^{2+}}$$

$$\frac{3.20436 \times 10^{-19} \text{ C}}{\text{atom } ^4\text{He}^{2+}} \times \frac{1 \text{ atom } ^4\text{He}^{2+}}{6.64466742 \times 10^{-27} \text{ kg}} = 4.82245 \times 10^7 \text{ C/kg}$$

          Check: The units of the answer (C/kg) are correct. The magnitude of the answer is reasonable when compared to the charge to mass ratio of the electron.

2.101   **Given:** graph of mass and % abundance of Pb   **Find:** atomic mass of Pb
          Conceptual Plan: % abundance $\rightarrow$ relative fraction abundance $\rightarrow$ mass Eu

$$\frac{\text{\% abundance of each isotope}}{\text{sum of all \% abundance}} \qquad \text{Atomic mass} = \sum_n (\text{fraction of isotope } n) \times (\text{mass of isotope } n)$$

          Solution: fraction $^{204}\text{Pb} = \dfrac{2.7}{190.9} = 0.0141$ fraction $^{206}\text{Pb} = \dfrac{46.0}{190.9} = 0.2410$

                fraction $^{207}\text{Pb} = \dfrac{42.2}{190.9} = 0.2211$ fraction $^{208}\text{Pb} = \dfrac{100.0}{190.9} = 0.5238$

        Atomic mass $= \sum_n (\text{fraction of isotope } n) \times (\text{mass of isotope } n)$

                $= 0.0141(204 \text{ amu}) + 0.2410(206 \text{ amu}) + 0.2211(207 \text{ amu}) + 0.5238(208 \text{ amu})$

                $= 207.24 \text{ amu} = 207 \text{ amu}$

          Check: The units of the answer, (amu) are correct. The magnitude of the answer is reasonable because the fraction of the heavier isotopes is greater than that of the lighter isotopes.

2.103 $^{236}_{90}\text{Th}$ A $-$ Z = number of neutrons. $236 - 90 = 146$ neutrons. So any nucleus with 146 neutrons is an isotone of $^{236}_{90}\text{Th}$.
          Some would be $^{238}_{92}\text{U}$, $^{239}_{93}\text{Np}$, $^{241}_{95}\text{Am}$, $^{237}_{91}\text{Pa}$, $^{235}_{89}\text{Ac}$, and $^{244}_{98}\text{Cf}$.

2.105

Symbol	Z	A	Number of Protons	Number of Electrons	Number of Neutrons	Charge
O^{2-}	8	16	8	10	8	2−
Ca^{2+}	20	40	20	18	20	2+
Mg^{2+}	12	25	12	10	13	2+
N^{3-}	7	14	7	10	7	3−

2.107 **Given:** $r_{nucleus} = 2.7$ fm; $r_{atom} = 70$ pm (assume two significant figures)

Find: $V_{nucleus}$; V_{atom}; % $V_{nucleus}$

Conceptual Plan:

$r_{nucleus}(fm) \rightarrow r_{nucleus}(pm) \rightarrow V_{nucleus}$ and then $r_{atom} \rightarrow V_{atom}$ and then % V

$$\frac{10^{-15}\,m}{1\,fm}\quad \frac{1\,pm}{10^{-12}\,m} \qquad V = \frac{4}{3}\pi r^3 \qquad V = \frac{4}{3}\pi r^3 \qquad \frac{V_{nucleus}}{V_{atom}} \times 100\%$$

Solution:

$$2.7\,fm \times \frac{10^{-15}\,m}{fm} \times \frac{1\,pm}{10^{-12}\,m} = 2.7 \times 10^{-3}\,pm \quad V_{nucleus} = \frac{4}{3}\pi (2.7 \times 10^{-3}\,pm)^3 = 8.2 \times 10^{-8}\,pm^3$$

$$V_{atom} = \frac{4}{3}\pi(70\,pm)^3 = 1.4 \times 10^6\,pm^3 \qquad \frac{8.2 \times 10^{-8}\,pm^3}{1.4 \times 10^6\,pm^3} \times 100\% = 5.9 \times 10^{-12}\%\,V$$

Check: The units of the answer (% V) are correct. The magnitude of the answer is reasonable because the nucleus occupies only a very small percentage of the V of the atom.

2.109 **Given:** 6.022×10^{23} pennies **Find:** the amount in dollars; the dollars/person

Conceptual Plan: pennies → dollars → dollars/person

$$\frac{1\,dollar}{100\,pennies} \qquad 6.5\text{ billion people}$$

Solution:

$$6.022 \times 10^{23}\,pennies \times \frac{1\,dollar}{100\,pennies} = 6.022 \times 10^{21}\,dollars \qquad \frac{6.022 \times 10^{21}\,dollars}{6.5 \times 10^9\,people} = 9.3 \times 10^{11}\,dollars/person$$

They are billionaires.

2.111 **Given:** O = 16.00 amu when C = 12.01 amu **Find:** mass O when C = 12.000 amu

Conceptual Plan: Determine ratio O:C for ^{12}C system; then use the same ratio when C = 12.00.

$$\frac{mass\,O}{mass\,C}$$

Solution: Based on ^{12}C = 12.00, O = 15.9994 and C = 12.011; so $\frac{mass\,O}{mass\,C} = \frac{16.00\,amu}{12.01\,amu} = \frac{1.3322\,amu\,O}{1\,amu\,C}$.

Based on C = 12.00, the ratio has to be the same:

$$12.000\,amu\,C \times \frac{1.3322\,amu\,O}{1\,amu\,C} = 15.986\,amu\,O = 15.99\,amu\,O$$

Check: The units of the answer (amu O) are correct. The magnitude of the answer is reasonable because the value for the new mass basis is smaller than the original mass basis; therefore, the mass of O should be less.

2.113 **Given:** Cu sphere: r = 0.935 in; d = 8.96 g/cm³ **Find:** number of Cu atoms

Conceptual Plan: r (inch) → r(cm) → V_{sphere} → g Cu → mol Cu → atoms Cu

$$\frac{2.54\,cm}{1\,in}\quad V = \frac{4}{3}\pi r^3 \quad \frac{8.96\,g}{cm^3}\quad \frac{1\,mol\,Cu}{63.546\,g}\quad \frac{6.022 \times 10^{23}\,atoms}{mol}$$

Solution: $0.935\,in \times \frac{2.54\,cm}{in} = 2.3749\,cm$

$$\frac{4}{3}\pi(2.3749\,cm)^3 \times \frac{8.96\,g}{cm^3} \times \frac{1\,mol\,Cu}{63.546\,g} \times \frac{6.022 \times 10^{23}\,atoms\,Cu}{1\,mol\,Cu} = 4.76 \times 10^{24}\,atoms\,Cu$$

Check: The units of the answer (atoms Cu) are correct. The magnitude of the answer is reasonable because about 8 mol Cu are present.

2.115　　**Given:** Li-6 = 6.01512 amu; Li-7 = 7.01601 amu; B = 6.941 amu
　　　　　Find: % abundance Li-6 and Li-7
　　　　　Conceptual Plan: Let x = **fraction Li-6 then** $1 - x$ = **fraction Li-7 → abundances**

$$\text{Atomic mass} = \sum_n (\text{fraction of isotope } n) \times (\text{mass of isotope } n)$$

　　　　　Solution: Atomic mass $= \sum_n (\text{fraction of isotope } n) \times (\text{mass of isotope } n)$

$$6.941 = (x)(6.01512 \text{ amu}) + (1 - x)(7.01601 \text{ amu})$$
$$0.07501 = 1.00089x$$
$$x = 0.07494 \qquad 1 - x = 0.92506$$
$$\text{Li-6} = 0.07494 \times 100 = 7.494\% \text{ and Li-7} = 0.92506 \times 100 = 92.506\%$$

　　　　　Check: The units of the answer (%, which gives the relative abundance of each isotope) are correct.
The relative abundances are reasonable because Li has an atomic mass closer to the mass of Li-7 than to Li-6.

2.117　　**Given:** Alloy of Au and Pd = 67.2 g; 2.49 × 10²³ atoms　　**Find:** % composition by mass
　　　　　Conceptual Plan: atoms Au and Pd → mol Au and Pd → g Au and Pd → g Au

$$\frac{1 \text{ mol}}{6.022 \times 10^{23} \text{ atoms}} \qquad \frac{196.97 \text{ g Au}}{1 \text{ mol Au}}, \frac{106.42 \text{ g Pd}}{1 \text{ mol Pd}}$$

　　　　　Solution: Let x = atoms Au and y = atoms Pd; develop expressions that will permit atoms to be related to moles and then to grams.

$$(x \text{ atoms Au})\left(\frac{1 \text{ mol Au}}{6.022 \times 10^{23} \text{ atoms Au}}\right) = \frac{x}{6.022 \times 10^{23}} \text{ mol Au}$$

$$(y \text{ atoms Pd})\left(\frac{1 \text{ mol Pd}}{6.022 \times 10^{23} \text{ atoms Pd}}\right) = \frac{y}{6.022 \times 10^{23}} \text{ mol Pd}$$

$$x + y = 2.49 \times 10^{23} \text{ atoms}; \quad y = 2.49 \times 10^{23} - x$$

$$\left(\frac{x}{6.022 \times 10^{23}} \text{ mol Au}\right)\left(\frac{196.97 \text{ g Au}}{\text{mol Au}}\right) = \frac{196.97x}{6.022 \times 10^{23}} \text{ g Au}$$

$$\left(\frac{2.49 \times 10^{23} - x}{6.022 \times 10^{23}} \text{ mol Pd}\right)\left(\frac{106.42 \text{ g Pd}}{\text{mol Pd}}\right) = \frac{106.42(2.49 \times 10^{23} - x)}{6.022 \times 10^{23}} \text{ g Pd}$$

g Au + g Pd = 67.2 g total

$$\frac{196.97x}{6.022 \times 10^{23}} \text{g Au} + \frac{106.42(2.49 \times 10^{23} - x)}{6.022 \times 10^{23}} \text{g Pd} = 67.2 \text{ g}$$

$$X = 1.5426 \times 10^{23} \text{ atoms Au}$$

$$(1.54 \times 10^{23} \text{ atoms Au})\left(\frac{1 \text{ mol Au}}{6.022 \times 10^{23} \text{ atoms Au}}\right)\left(\frac{196.97 \text{ g Au}}{\text{mol Au}}\right) = 50.37 \text{ g Au}$$

$$\left(\frac{50.37 \text{ g Au}}{67.2 \text{ g sample}}\right) \times 100 = 74.96\% \text{ Au} = 75.0\% \text{ Au}$$

% Pd = 100.0% − 75.0% Au = 25.0% Pd

　　　　　Check: Units of the answer (% composition) are correct.

2.119　　**Given:** Ag-107, 51.839%, Ag-109, $\dfrac{\text{mass Ag-109}}{\text{mass Ag-107}} = 1.0187$　　**Find:** mass Ag-107

　　　　　Conceptual Plan: % abundance Ag-107 → % abundance Ag-109 → fraction → mass Ag-107

$$100\% - (\% \text{ Ag-107}) \qquad \frac{\% \text{ abundance}}{100}$$

$$\text{Atomic mass} = \sum_n (\text{fraction of isotope } n) \times (\text{mass of isotope } n)$$

　　　　　Solution: $100.00\% - 51.839\% = 48.161\%$ Ag − 109

$$\text{Fraction Ag-107} = \frac{51.839}{100.00} = 0.51839 \quad \text{Fraction Ag-109} = \frac{48.161}{100.00} = 0.48161$$

Let x be the mass of Ag-107 then mass Ag-109 $= 1.0187x$

$$\text{Atomic mass} = \sum_n (\text{fraction of isotope } n) \times (\text{mass of isotope } n)$$

$107.87 \text{ amu} = 0.51839(x \text{ amu}) + 0.48161(1.0187x \text{ amu})$
$x = 106.9\underline{0}7 \text{ amu} = 106.91 \text{ amu mass Ag-107}$

Check: The units of the answer (amu) are correct. The answer is reasonable because it is close to the atomic mass number of Ag-107.

2.121   **Given:** 0.255 oz 18K Au   **Find:** atoms Au
Conceptual Plan: oz 18K Au → oz pure Au → g Au → mol Au → atoms Au

$$\frac{75 \text{ oz Au}}{100 \text{ oz 18K Au}} \qquad \frac{453.59 \text{ g Au}}{16 \text{ oz Au}} \qquad \frac{1 \text{ mol Au}}{196.97 \text{ g Au}} \qquad \frac{6.022 \times 10^{23} \text{ atoms Au}}{1 \text{ mol Au}}$$

Solution: $0.255 \text{ oz 18K Au} \times \left(\dfrac{75 \text{ oz pure Au}}{100 \text{ oz 18K Au}}\right) \times \left(\dfrac{453.59 \text{ g}}{16 \text{ oz}}\right) \times \left(\dfrac{1 \text{ mol Au}}{196.97 \text{ g Au}}\right) \times \left(\dfrac{6.022 \times 10^{23} \text{ atoms Au}}{1 \text{ mol Au}}\right)$

$= 1.\underline{6}58 \times 10^{22} \text{ atoms Au} = 1.7 \times 10^{22} \text{ atoms Au}$

Check: The units of the answer (atoms Au) are correct. The magnitude of the answer is reasonable because less than 1 mol of Au is in the sample.

Challenge Problems

2.123   **Given:** $d = 1.4 \text{ g/cm}^3$, $r = 7 \times 10^8$ m; 100 billion stars/galaxy; 10 billion galaxies/universe
Find: number of atoms in the universe
Conceptual Plan: $r_{(star)} \text{ (m)} \rightarrow r_{(star)} \text{ (cm)} \rightarrow V_{(star)} \rightarrow \text{g H/star} \rightarrow \text{mol H/star} \rightarrow \text{atoms H/star}$

$$\frac{100 \text{ cm}}{\text{m}} \qquad V = \frac{4}{3}\pi r^3 \qquad \frac{1.4 \text{ g H}}{\text{cm}^3} \qquad \frac{1 \text{ mol H}}{1.008 \text{ g}} \qquad \frac{6.022 \times 10^{23} \text{ atoms}}{\text{mol}}$$

atoms H/star → atoms H/galaxy → atoms H/universe

$$\frac{100 \times 10^9 \text{ stars}}{\text{galaxy}} \qquad \frac{10 \times 10^9 \text{ galaxies}}{\text{universe}}$$

Solution: $7 \times 10^8 \text{ m} \times \dfrac{100 \text{ cm}}{\text{m}} = 7 \times 10^{10} \text{ cm}$

$$\frac{4}{3}\pi \frac{(7 \times 10^{10} \text{ cm})^3}{\text{star}} \times \frac{1.4 \text{ g H}}{\text{cm}^3} \times \frac{1 \text{ mol H}}{1.008 \text{ g H}} \times \frac{6.022 \times 10^{23} \text{ atoms H}}{\text{mol H}} \times \frac{100 \times 10^9 \text{ stars}}{\text{galaxy}} \times \frac{10 \times 10^9 \text{ galaxies}}{\text{universe}}$$

$= 1 \times 10^{78} \text{ atoms/universe}$

Check: The units of the answer (atoms/universe) are correct.

2.125   **Given:** $\dfrac{\text{mass 2 O}}{\text{mass 1 N}} = \dfrac{2.29}{1.00}; \dfrac{\text{mass 3 F}}{\text{mass 1 N}} = \dfrac{4.07}{1.00}$ **Find:** $\dfrac{\text{mass O}}{\text{mass 2 F}}$

Conceptual Plan: Mass O/N and mass F/N → mass O/F → mass O/2F

$$\frac{\text{mass 2 O}}{\text{mass 1 N}} \qquad \frac{\text{mass 3 F}}{\text{mass 1 N}} \qquad \frac{\text{mass 2 O}}{\text{mass 3 F}}$$

Solution: $\dfrac{\text{mass 2 O}}{\text{mass 1 N}} = \dfrac{2.29}{1.00}; \dfrac{\text{mass 3 F}}{\text{mass 1 N}} = \dfrac{4.07}{1.00}$ $\left(\dfrac{2.29 \text{ mass 2 O}}{4.07 \text{ mass 3 F}}\right)\left(\dfrac{1 \text{ O}}{2 \text{ O}}\right)\left(\dfrac{3 \text{ F}}{2 \text{ F}}\right) = \dfrac{0.422 \text{ mass O}}{\text{mass 2 F}}$

Check: Mass ratio of O to F is reasonable because the mass of O is slightly less than the mass of fluorine.

2.127   **Given:** 7.36 g Cu; 0.51 g Zn   **Find:** atomic mass of sample
Conceptual Plan: fraction Cu and Zn → atomic mass

$$\text{Atomic mass} = \sum_n (\text{fraction of atom } n) \times (\text{mass of atom } n)$$

Solution: $7.36 \text{ g Cu} + 0.51 \text{ g Zn} = 7.87 \text{ g sample}$

$$\left(\frac{7.36 \text{ g Cu}}{7.87 \text{ g sample}}\right)\left(\frac{63.55 \text{ g Cu}}{\text{mol Cu}}\right) + \left(\frac{0.51 \text{ g Zn}}{7.87 \text{ g sample}}\right)\left(\frac{65.41 \text{ g Zn}}{\text{mol Zn}}\right) = 63.67 \text{ g/mol}$$

Check: The units of the answer (g/mol) are correct. The magnitude of the answer is reasonable because it is between the mass of Cu (63.55 g/mol) and Zn (65.41 g/mol) and is closer to the mass of Cu.

2.129   **Given:** Mg $= 24.312$ amu, ^{24}Mg $= 23.98504$, 78.99%, ^{26}Mg $= 25.98259$ amu, $\dfrac{\text{abundance } ^{25}\text{Mg}}{\text{abundance } ^{26}\text{Mg}} = \dfrac{0.9083}{1}$

Find: mass ^{25}Mg

Conceptual Plan: abundance of ^{24}Mg and ratio ^{25}Mg/^{26}Mg $\rightarrow$ abundance ^{25}Mg and ^{26}Mg $\rightarrow$ mass ^{25}Mg

$$\text{Atomic mass} = \sum_{n}(\text{fraction of isotope } n) \times (\text{mass of isotope } n)$$

Solution: $100.00\% - \%$ abundance ^{24}Mg $= \%$ abundance ^{25}Mg and ^{26}Mg

$100.00\% - 78.99\% = 21.01\%$ ^{25}Mg and ^{26}Mg

fraction ^{25}Mg and ^{26}Mg $= \dfrac{21.01}{100.0} = 0.2101$

$\dfrac{\text{abundance } ^{25}\text{Mg}}{\text{abundance } ^{26}\text{Mg}} = \dfrac{0.9083}{1}$

Let $x = $ fraction ^{26}Mg, $0.9083x = $ fraction ^{25}Mg

fraction ^{25}Mg and ^{26}Mg $= x + 0.9083x = 0.2101$

$x = {}^{26}$Mg $= 0.1101$, $0.9083x = {}^{25}$Mg $= 0.1000$

$$\text{Atomic mass} = \sum_{n}(\text{fraction of isotope } n) \times (\text{mass of isotope } n)$$

$24.312 = (0.7899)(23.98504 \text{ amu}) + (0.1000)(\text{mass } ^{25}\text{Mg}) + (0.1101)(25.98259 \text{ amu})$

mass ^{25}Mg $= 25.0\underline{5}6$ amu $= 25.06$ amu

Check: The units of the answer (amu) are correct. The magnitude of the answer is reasonable because it is between the masses of ^{24}Mg and ^{26}Mg.

Conceptual Problems

2.131   Li-6 nucleus   Li-7 nucleus

Because Li-6 has an abundance of 7.5%, in a sample of 1000 atoms, there should, on average, be 75 Li-6 atoms.

2.133   If the amu and mole were not based on the same isotope, the numerical values obtained for an atom of material and a mole of material would not be the same. If, for example, the mole was based on the number of particles in C–12 but the amu was changed to a fraction of the mass of an atom of Ne–20 the number of particles and the number of amu that make up one mole of material would no longer be the same. We would no longer have the relationship where the mass of an atom in amu is numerically equal to the mass of a mole of those atoms in grams.

2.135   The different isotopes of the same element have the same number of protons and electrons, so the attractive forces between the nucleus and the electrons is constant and there is no difference in the radii of the isotopes. Ions, on the other hand, have a different number of electrons than the parent atom from which they are derived. Cations have fewer electrons than the parent atom. The attractive forces are greater because there is a larger positive charge in the nucleus than the negative charge in the electron cloud. So cations are smaller than the parent atom from which they are derived. Anions have more electrons than the parent. The electron cloud has a greater negative charge than the nucleus, so the anions have larger radii than the parent.

3 Molecules, Compounds, and Chemical Equations

Problems by Topic

Chemical Formulas and Molecular View of the Elements

3.1   The properties of compounds are generally very different from the properties of the elements that compose them. When two elements combine to form a compound, an entirely new substance results.

3.3   Chemical compounds can be represented by chemical formulas and molecular models. The type of formula or model you use depends on how much information you have about the compound and how much you want to communicate. An empirical formula gives the relative number of atoms of each element in the compound. It contains the smallest whole number ratio of the elements in the compound. A molecular formula gives the actual number of atoms of each element in the compound. A structural formula shows how the atoms are connected. A ball-and-stick model shows the geometry of the compound. A space-filling model shows the relative sizes of the atoms and how they merge together.

3.5   Atomic elements exist in nature with single atoms as their base units. Neon (Ne), gold (Au), and potassium (K) are a few examples of atomic elements.

Molecular elements do not normally exist in nature with single atoms as their base unit; rather, they exist as molecules, two or more atoms of the same element bonded together. Most exist as diatomic molecules [e.g., hydrogen (H_2), nitrogen (N_2), and oxygen (O_2)]. Some exist as polyatomic molecules: phosphorus (P_4) and sulfur (S_8).

Ionic compounds are generally composed of one or more metal cations (usually one type of metal) and one or more nonmetal anions bound together by ionic bonds. Sodium chloride (NaCl) and potassium sulfate (Na_2SO_4) are examples of ionic compounds.

Molecular compounds are composed of two or more covalently bonded nonmetals. Examples are water (H_2O), sulfur dioxide (SO_2), and nitrogen dioxide (NO_2).

3.7   Binary ionic compounds are named using the name of the cation (metal) and the base name of the anion (nonmetal) + the suffix *-ide*. Ionic compounds that contain a polyatomic anion are named using the name of the cation (metal) and the name of the polyatomic anion.

3.9   To name a binary molecular inorganic compound, list the name of the first element with a prefix to indicate the number of atoms in the compound if there is more than one, followed by the base name of the second element with a prefix to indicate the number of atoms in the compound if there is more than one, followed by the suffix *-ide*.

3.11   Binary acids are composed of hydrogen and a nonmetal. The names for binary acids have the form: hydro plus the base name of the nonmetal + *-ic* acid. Oxyacids contain hydrogen and an oxyanion. The names of oxyacids depend on the ending of the oxyanion and have the following forms: oxyanions ending with *-ate*: base name of the oxyanion + *ic* acid; oxyanions ending with *-ite*: base name of the oxyanion + *ous* acid.

3.13   The chemical formula indicates the elements present in the compound and the relative number of atoms of each type. The chemical formula gives the conversion factor between the kind of element and the formula; it also allows the determination of mass percent composition.

3.15   Chemical formulas contain inherent relationships between atoms (or moles of atoms) and molecules (or moles of molecules). For example, the formula CCl_2F_2 tells us that one mole of CCl_2F_2 contains one mole of C atoms, two moles of Cl atoms, and two moles of F atoms.

3.17   The molecular formula is a whole number multiple of the empirical formula. To find the molecular formula, the molar mass of the compound must be known. The molecular molar mass divided by the empirical molar mass gives the whole number multiple used to convert the empirical formula to the molecular formula.

3.19   Organic compounds are composed of carbon; hydrogen; and a few other elements, including nitrogen, oxygen, and sulfur.

3.21   In functionalized hydrocarbons, a functional group—a characteristic atom or group of atoms—has been incorporated into the hydrocarbon. The family or organic compounds known as alcohols have an $-OH$ functional group.

3.23   The chemical formula gives you the kind of atom and the number of each atom in the compound.
    (a) $Mg_3(PO_4)_2$ contains: 3 magnesium atoms, 2 phosphorus atoms, and 8 oxygen atoms
    (b) $BaCl_2$ contains: 1 barium atom and 2 chlorine atoms
    (c) $Fe(NO_2)_2$ contains: 1 iron atom, 2 nitrogen atoms, and 4 oxygen atoms
    (d) $Ca(OH)_2$ contains: 1 calcium atom, 2 oxygen atoms, and 2 hydrogen atoms

3.25   (a)   1 blue $=$ nitrogen, 3 white $=$ hydrogen: NH_3
    (b)   2 black $=$ carbon, 6 white $=$ hydrogen: C_2H_6
    (c)   1 yellow $=$ sulfur, 3 red $=$ oxygen: SO_3

3.27   (a)   Neon is an element, and it is not one of the elements that exists as diatomic molecules; therefore, it is an atomic element.
    (b)   Fluorine is one of the elements that exists as diatomic molecules; therefore, it is a molecular element.
    (c)   Potassium is not one of the elements that exists as diatomic molecules; therefore, it is an atomic element.
    (d)   Nitrogen is one of the elements that exists as diatomic molecules; therefore, it is a molecular element.

3.29   (a) CO_2 is a compound composed of a nonmetal and a nonmetal; therefore, it is a molecular compound.
    (b) $NiCl_2$ is a compound composed of a metal and a nonmetal; therefore, it is an ionic compound.
    (c)   NaI is a compound composed of a metal and a nonmetal; therefore, it is an ionic compound.
    (d) PCl_3 is a compound composed of a nonmetal and a nonmetal; therefore, it is a molecular compound.

3.31   (a)   white $=$ hydrogen: a molecule composed of two of the same elements; therefore, it is a molecular element.
    (b)   blue $=$ nitrogen, white $=$ hydrogen: a molecule composed of a nonmetal and a nonmetal; therefore, it is a molecular compound.
    (c)   purple $=$ sodium: a substance composed of the same atoms; therefore, it is an atomic element.

Formulas and Names for Ionic Compounds

3.33   To write the formula for an ionic compound, do the following: (1) Write the symbol for the metal cation and its charge and the symbol for the nonmetal anion and its charge. (2) Adjust the subscript on each cation and anion to balance the overall charge. (3) Check that the sum of the charges of the cations equals the sum of the charges of the anions.

    (a)   calcium and oxygen: Ca^{2-} O^{2-} CaO   cations 2+, anions 2−
    (b)   zinc and sulfur: Zn^{2+} S^{2-} ZnS   cations 2+, anions 2−
    (c)   rubidium and bromine: Rb^+ Br^- RbBr   cation +, anions −
    (d)   aluminum and oxygen: Al^{3+} O^{2-} Al_2O_3 cation 2(3+) = 6+, anions 3(2−) = 6−

3.35　To write the formula for an ionic compound, do the following: (1) Write the symbol for the metal cation and its charge and the symbol for the polyatomic anion and its charge. (2) Adjust the subscript on each cation and anion to balance the overall charge. (3) Check that the sum of the charges of the cations equals the sum of the charges of the anions. Cation = calcium: Ca^{2+}

 (a) hydroxide: OH^- $Ca(OH)_2$ cation 2+, anion 2(1−) = 2−
 (b) chromate: $CrO_4{}^{2-}$ $CaCrO_4$ cation 2+, anion 2−
 (c) phosphate: $PO_4{}^{3-}$ $Ca_3(PO_4)_2$ cation 3(2+) = 6+, anion 2(3−) = 6−
 (d) cyanide: CN^- $Ca(CN)_2$ cation 2+, anion 2(1−) = 2−

3.37　To name a binary ionic compound, name the metal cation followed by the base name of the anion + -ide.

 (a) Mg_3N_2: The cation is magnesium; the anion is from nitrogen, which becomes nitride: magnesium nitride.
 (b) KF: The cation is potassium; the anion is from fluorine, which becomes fluoride: potassium fluoride.
 (c) Na_2O: The cation is sodium; the anion is from oxygen, which becomes oxide: sodium oxide.
 (d) Li_2S: The cation is lithium; the anion is from sulfur, which becomes sulfide: lithium sulfide.
 (e) CsF: The cation is cesium; the anion is fluorine, which becomes fluoride: cesium fluoride.
 (f) KI: The cation is potassium; the anion is iodine, which becomes iodide: potassium iodide.

3.39　To name these compounds, you must first decide whether the metal cation is invariant or can have more than one charge. Then, name the metal cation followed by the base name of the anion + -ide.

 (a) SnO: Sn can have more than one charge. The charge on Sn must be 2+ for the compound to be charge neutral: the cation is tin(II); the anion is from oxygen, which becomes oxide: tin(II) oxide.
 (b) Cr_2S_3: Cr can have more than one charge. The charge on Cr must be 3+ for the compound to be charge neutral: the cation is chromium(III); the anion is from sulfur, which becomes sulfide: chromium(III) sulfide.
 (c) RbI: Rb is invariant: the cation is rubidium; the anion is from iodine, which becomes iodide: rubidium iodide.
 (d) $BaBr_2$: Ba is invariant: the cation is barium; the anion is from bromine, which becomes bromide: barium bromide.

3.41　To name these compounds, you must first decide whether the metal cation is invariant or can have more than one charge. Then, name the metal cation followed by the name of the polyatomic anion.

 (a) $CuNO_2$: Cu can have more than one charge. The charge on Cu must be 1+ for the compound to be charge neutral: The cation is copper(I); the anion is nitrite: copper(I) nitrite.
 (b) $Mg(C_2H_3O_2)_2$: Mg is invariant: The cation is magnesium; the anion is acetate: magnesium acetate.
 (c) $Ba(NO_3)_2$: Ba is invariant: The cation is barium; the anion is nitrate: barium nitrate.
 (d) $Pb(C_2H_3O_2)_2$: Pb can have more than one charge. The charge on Pb must be 2+ for the compound to be charge neutral: The cation is lead(II); the anion is acetate: lead(II) acetate.

3.43　To write the formula for an ionic compound, do the following: (1) Write the symbol for the metal cation and its charge and the symbol for the nonmetal anion or polyatomic anion and its charge. (2) Adjust the subscript on each cation and anion to balance the overall charge. (3) Check that the sum of the charges of the cations equals the sum of the charges of the anions.

 (a) sodium hydrogen sulfite: Na^+ $HSO_3{}^-$ $NaHSO_3$ cation 1+, anion 1−
 (b) lithium permanganate: Li^+ $MnO_4{}^-$ $LiMnO_4$ cation 1+, anion 1−
 (c) silver nitrate: Ag^+ $NO_3{}^-$ $AgNO_3$ cation 1+, anion 1−
 (d) potassium sulfate: K^+ $SO_4{}^{2-}$ K_2SO_4 cation 2(1+) = 2+, anion 2−
 (e) rubidium hydrogen sulfate: Rb^+ $HSO_4{}^-$ $RbHSO_4$ cation 1+, anion 1−
 (f) potassium hydrogen carbonate: K^+ $HCO_3{}^-$ $KHCO_3$ cation 1+, anion 1−

3.45　Hydrates are named the same way as other ionic compounds with the addition of the term *prefix*hydrate, where the prefix is the number of water molecules associated with each formula unit.

 (a) $CoSO_4 \cdot 7H_2O$ cobalt(II) sulfate heptahydrate
 (b) iridium(III) bromide tetrahydrate $IrBr_3 \cdot 4H_2O$
 (c) $Mg(BrO_3)_2 \cdot 6H_2O$ magnesium bromate hexahydrate
 (d) potassium carbonate dihydrate $K_2CO_3 \cdot 2H_2O$

Formulas and Names for Molecular Compounds and Acids

3.47   (a)   CO   The name of the compound is the name of the first element, *carbon*, followed by the base name of the second element, *ox*, prefixed by *mono-* to indicate 1 and given the suffix *-ide*. Because the prefix ends with "o" and the base name begins with "o," the first "o" is dropped: carbon monoxide.

       (b) NI_3 The name of the compound is the name of the first element, *nitrogen*, followed by the base name of the second element, *iod*, prefixed by *tri-* to indicate 3 and given the suffix *-ide*: nitrogen triiodide.

       (c) $SiCl_4$ The name of the compound is the name of the first element, *silicon*, followed by the base name of the second element, *chlor*, prefixed by *tetra-* to indicate 4 and given the suffix *-ide*: silicon tetrachloride.

       (d) N_4Se_4 The name of the compound is the name of the first element, *nitrogen*, prefixed by *tetra-* to indicate 4 followed by the base name of the second element, *selen*, prefixed by *tetra-* to indicate 4 and given the suffix *-ide*: tetranitrogen tetraselenide.

3.49   (a)   phosphorus trichloride: PCl_3
       (b)   chlorine monoxide: ClO
       (c)   disulfur tetrafluoride: S_2F_4
       (d)   phosphorus pentafluoride: PF_5

3.51   (a)   HI: The base name of I is *iod*, so the name is hydroiodic acid.
       (b) HNO_3:   The oxyanion is *nitrate*, which ends in *-ate*; therefore, the name of the acid is nitric acid.
       (c) H_2CO_3:   The oxyanion is *carbonate*, which ends in *-ate*; therefore, the name of the acid is carbonic acid.

3.53   (a)   hydrofluoric acid: HF
       (b)   hydrobromic acid: HBr
       (c)   sulfurous acid: H_2SO_3

3.55   To use the flowchart, begin by determining the type of compound you are trying to name—ionic, molecular, or acid.

       (a) $SrCl_2$ is composed of a metal and a nonmetal, so it is ionic. Begin at the box labeled "IONIC" at the far left of the flowchart. The metal in the compound can only form one type of ion; therefore, take the left branch in the flowchart. Then name the compound according to the blocks at the end of the path in the flowchart. Write the name of the cation followed by the base name of the anion appended with the ending *-ide*. The full name is strontium chloride.

       (b) SnO_2 is composed of a metal and a nonmetal, so it is ionic. Begin at the box labeled "IONIC" at the far left of the flowchart. The metal in the compound forms more than one type of ion; therefore, take the right branch in the flowchart. Then name the compound according to the blocks at the end of the path in the flowchart. Write the name of the cation followed by the charge of the cation in Roman numerals followed by the base name of the anion appended with the ending *-ide*. The full name is tin(IV) oxide.

       (c) P_2S_5 is composed of two nonmetals and is a molecular compound. Begin at the box labeled "MOLECULAR" in the middle of the flowchart. Write the prefix to indicate the number of atoms of the first element, then the name of the first element followed by the prefix to indicate the number of atoms of the second element, then the base name of the second element followed by the ending *-ide*. The full name is diphorphorus pentasulfide.

       (d) $HC_2H_3O_2(aq)$ is a compound composed of H and one or more nonmetals, so it is an acid. Begin at the box labeled "ACIDS" at the far right of the flowchart. The acid is an oxyacid; therefore, take the right branch in the flowchart. The anion has the ending *-ate*, so take the left branch of the flowchart. Then name the compound according to the blocks at the end of the path in the flowchart. Write the base name of the oxyanion appended with the ending *-ic* followed by *acid*. The name of the oxoanion is acetate so the name of the acid is acetic acid.

3.57   To use the flowchart, begin by determining the type of compound you are trying to name—ionic, molecular, or acid.

       (a) $KClO_3$ is composed of a metal and an oxyanion, so it is ionic. Begin at the box labeled "IONIC" at the far left of the flowchart. The metal in the compound can only form one type of ion; therefore, take the left branch in the flowchart. Then name the compound according to the blocks at the end of the path in the flowchart. Write the name of the cation followed by the base name of the anion appended with the ending *-ate*. ClO_3^- is the polyatomic ion chlorate, so the name of the compound is potassium chlorate.

       (b) I_2O_5 is composed of two nonmetals and is a molecular compound. Begin at the box labeled "MOLECULAR" in the middle of the flowchart. Write the prefix to indicate the number of atoms of the first element, then the name of the first element followed by the prefix to indicate the number of atoms of the second element, then the base name of the second element followed by the ending *-ide*. The full name is diiodine pentoxide.

(c)   PbSO₄ is composed of a metal and an oxyanion, so it is ionic. Begin at the box labeled "IONIC" at the far left of the flowchart. The metal in the compound forms more than one type of ion; therefore, take the right branch in the flowchart. Then name the compound according to the blocks at the end of the path in the flowchart. Write the name of the cation followed by the charge of the cation in Roman numerals followed by the base name of the anion appended with the ending -*ate*. SO₄²⁻ is the polyatomic ion sulfate, so the name of the compound is lead (II) sulfate.

Formula Mass and the Mole Concept for Compounds

3.59   To find the formula mass, sum the atomic masses of each atom in the chemical formula.

(a) NO_2 formula mass $= 1 \times (\text{atomic mass N}) + 2 \times (\text{atomic mass O})$
$= 1 \times (14.01 \text{ amu}) + 2 \times (16.00 \text{ amu})$
$= 46.01 \text{ amu}$

(b) C_4H_{10} formula mass $= 4 \times (\text{atomic mass C}) + 10 \times (\text{atomic mass H})$
$= 4 \times (12.01 \text{ amu}) + 10 \times (1.008 \text{ amu})$
$= 58.12 \text{ amu}$

(c) $C_6H_{12}O_6$ formula mass $= 6 \times (\text{atomic mass C}) + 12 \times (\text{atomic mass H}) + 6 \times (\text{atomic mass O})$
$= 6 \times (12.01 \text{ amu}) + 12 \times (1.008 \text{ amu}) + 6 \times (16.00 \text{ amu})$
$= 180.16 \text{ amu}$

(d) $Cr(NO_3)_3$ formula mass $= 1 \times (\text{atomic mass Cr}) + 3 \times (\text{atomic mass N}) + 9 \times (\text{atomic mass O})$
$= 1 \times (52.00 \text{ amu}) + 3 \times (14.01 \text{ amu}) + 9 \times (16.00 \text{ amu})$
$= 238.03 \text{ amu}$

3.61   (a)   **Given:** 72.5 g CCl₄   **Find:** number of moles
Conceptual Plan: g CCl₄ → mole CCl₄
$$\frac{1 \text{ mol}}{153.81 \text{ g CCl}_4}$$
Solution: $72.5 \text{ g CCl}_4 \times \dfrac{1 \text{ mol CCl}_4}{153.81 \text{ g CCl}_4} = 0.47\underline{1}4 \text{ mol CCl}_4 = 0.471 \text{ mol CCl}_4$

Check: Units of the answer (mole CCl₄) are correct. The magnitude is appropriate because it is less than 1 mole of NO₂.

(b)   **Given:** 12.4 g C₁₂H₂₂O₁₁   **Find:** number of moles
Conceptual Plan: g C₁₂H₂₂O₁₁ → mole KNO₃
$$\frac{1 \text{ mol}}{342.296 \text{ g C}_{12}\text{H}_{22}\text{O}_{11}}$$
Solution: $12.4 \text{ g C}_{12}\text{H}_{22}\text{O}_{11} \times \dfrac{1 \text{ mol C}_{12}\text{H}_{22}\text{O}_{11}}{342.296 \text{ g C}_{12}\text{H}_{22}\text{O}_{11}} = 0.036\underline{2}3 \text{ mol C}_{12}\text{H}_{22}\text{O}_{11} = 0.0362 \text{ mol C}_{12}\text{H}_{22}\text{O}_{11}$

Check: Units of the answer (mole C₁₂H₂₂O₁₁) are correct. The magnitude is appropriate because there is less than 1 mole of C₁₂H₂₂O₁₁.

(c)   **Given:** 25.2 kg C₂H₂   **Find:** number of moles
Conceptual Plan: kg CO₂ → g C₂H₂ → mole C₂H₂
$$\frac{1000 \text{ g C}_2\text{H}_2}{\text{kg C}_2\text{H}_2} \qquad \frac{1 \text{ mol}}{26.036 \text{ g C}_2\text{H}_2}$$
Solution: $25.2 \text{ kg CO}_2 \times \dfrac{1000 \text{ g C}_2\text{H}_2}{\text{kg C}_2\text{H}_2} \times \dfrac{1 \text{ mol C}_2\text{H}_2}{26.036 \text{ g C}_2\text{H}_2} = 96\underline{7}.9 \text{ mol C}_2\text{H}_2 = 968 \text{ mol C}_2\text{H}_2$

Check: Units of the answer (mole C₂H₂) are correct. The magnitude is appropriate because more than a kilogram of C₂H₂ is present.

(d)   **Given:** 12.3 g dinitrogen monoxide   **Find:** number of moles
Conceptual Plan: Dinitrogen monoxide → formula and then g N₂O → mole N₂O
$$\frac{1 \text{ mol}}{44.02 \text{ g N}_2\text{O}}$$
Solution: Dinitrogen monoxide is N₂O.

$12.3 \text{ g N}_2\text{O} \times \dfrac{1 \text{ mol N}_2\text{O}}{44.02 \text{ g N}_2\text{O}} = 0.279\underline{4} \text{ mol N}_2\text{O} = 0.279 \text{ mol N}_2\text{O}$

Check: Units of the answer (mole N₂O) are correct. The magnitude is appropriate because less than a mole of N₂O is present.

3.63   (a)   **Given:** 25.5 g NO_2 **Find:** number of moles

Conceptual Plan: g NO_2 → mole NO_2

$$\frac{1 \text{ mol}}{46.01 \text{ g } NO_2}$$

Solution: $25.5 \text{ g } NO_2 \times \dfrac{1 \text{ mol } NO_2}{46.01 \text{ g } NO_2} = 0.554 \text{ mol } NO_2$

Check: The units of the answer (mole NO_2) are correct. The magnitude is appropriate because it is less than 1 mole of NO_2.

   (b)   **Given:** 1.25 kg CO_2 **Find:** number of moles

Conceptual Plan: kg CO_2 → g CO_2 → mole CO_2

$$\frac{1000 \text{ g } CO_2}{\text{kg } CO_2} \qquad \frac{1 \text{ mol}}{44.01 \text{ g } NO_2}$$

Solution: $1.25 \text{ kg } CO_2 \times \dfrac{1000 \text{ g } CO_2}{\text{kg } CO_2} \times \dfrac{1 \text{ mol } CO_2}{44.01 \text{ g } CO_2} = 28.4 \text{ mol } CO_2$

Check: The units of the answer (mole CO_2) are correct. The magnitude is appropriate because over a kg of CO_2 is present.

   (c)   **Given:** 38.2 g KNO_3 **Find:** number of moles

Conceptual Plan: g KNO_3 → mole KNO_3

$$\frac{1 \text{ mol}}{101.11 \text{ g } KNO_3}$$

Solution: $38.2 \text{ g } KNO_3 \times \dfrac{1 \text{ mol } KNO_3}{101.11 \text{ g } KNO_3} = 0.378 \text{ mol } KNO_3$

Check: The units of the answer (mole KNO_3) are correct. The magnitude is appropriate because there is less than 1 mole of KNO_3.

   (d)   **Given:** 155.2 kg Na_2SO_4 **Find:** number of moles

Conceptual Plan: kg Na_2SO_4 → g Na_2SO_4 → mole Na_2SO_4

$$\frac{1000 \text{ g } Na_2SO_4}{\text{kg } Na_2SO_4} \qquad \frac{1 \text{ mol}}{142.05 \text{ g } Na_2SO_4}$$

Solution: $155.2 \text{ kg } Na_2SO_4 \times \dfrac{1000 \text{ g } Na_2SO_4}{\text{kg } Na_2SO_4} \times \dfrac{1 \text{ mol } Na_2SO_4}{142.05 \text{ g } Na_2SO_4} = 1092 \text{ mol } Na_2SO_4$

Check: The units of the answer (mole Na_2SO_4) are correct. The magnitude is appropriate because over 100 kg of Na_2SO_4 is present.

3.65   (a)   **Given:** 6.5 g H_2O **Find:** number of molecules

Conceptual Plan: g H_2O → mole H_2O → number H_2O molecules

$$\frac{1 \text{ mol}}{18.02 \text{ g } H_2O} \qquad \frac{6.022 \times 10^{23} \text{ } H_2O \text{ molecules}}{\text{mol } H_2O}$$

Solution: $6.5 \text{ g } H_2O \times \dfrac{1 \text{ mol } H_2O}{18.02 \text{ g } H_2O} \times \dfrac{6.022 \times 10^{23} \text{ } H_2O \text{ molecules}}{\text{mol } H_2O} = 2.2 \times 10^{23} \text{ } H_2O \text{ molecules}$

Check: The units of the answer (H_2O molecules) are correct. The magnitude is appropriate; it is smaller than Avogadro's number, as expected, because we have less than 1 mole of H_2O.

   (b)   **Given:** 389 g CBr_4 **Find:** number of molecules

Conceptual Plan: g CBr_4 → mole CBr_4 → number CBr_4 molecules

$$\frac{1 \text{ mol}}{331.6 \text{ g } CBr_4} \qquad \frac{6.022 \times 10^{23} \text{ } CBr_4 \text{ molecules}}{\text{mol } CBr_4}$$

Solution: $389 \text{ g } CBr_4 \times \dfrac{1 \text{ mol } CBr_4}{331.6 \text{ g } CBr_4} \times \dfrac{6.022 \times 10^{23} \text{ } CBr_4 \text{ molecules}}{\text{mol } CBr_4} = 7.06 \times 10^{23} \text{ } CBr_4 \text{ molecules}$

Check: The units of the answer (CBr_4 molecules) are correct. The magnitude is appropriate; it is larger than Avogadro's number, as expected, because we have more than 1 mole of CBr_4.

   (c)   **Given:** 22.1 g O_2 **Find:** number of molecules

Conceptual Plan: g O_2 → mole O_2 → number O_2 molecules

$$\frac{1 \text{ mol}}{32.00 \text{ g } O_2} \qquad \frac{6.022 \times 10^{23} \text{ } O_2 \text{ molecules}}{\text{mol } O_2}$$

Solution: $22.1 \text{ g } \cancel{O_2} \times \dfrac{1 \text{ mol } \cancel{O_2}}{32.00 \text{ g } \cancel{O_2}} \times \dfrac{6.022 \times 10^{23} \text{ O}_2 \text{ molecules}}{\text{mol } \cancel{O_2}} = 4.16 \times 10^{23} \text{ O}_2 \text{ molecules}$

Check: The units of the answer (O_2 molecules) are correct. The magnitude is appropriate; it is smaller than Avogadro's number, as expected, because we have less than 1 mole of O_2.

(d) **Given:** 19.3 g C_8H_{10} **Find:** number of molecules

Conceptual Plan: g $C_8H_{10} \rightarrow$ mole $C_8H_{10} \rightarrow$ number C_8H_{10} molecules

$$\dfrac{1 \text{ mol}}{106.16 \text{ g } C_8H_{10}} \qquad \dfrac{6.022 \times 10^{23} \text{ } C_8H_{10} \text{ molecules}}{\text{mol } C_8H_{10}}$$

Solution:

$$19.3 \text{ g } \cancel{C_8H_{10}} \times \dfrac{1 \text{ mol } \cancel{C_8H_{10}}}{106.16 \text{ g } \cancel{C_8H_{10}}} \times \dfrac{6.022 \times 10^{23} \text{ } \cancel{C_8H_{10} \text{ molecules}}}{\text{mol } C_8H_{10}} = 1.09 \times 10^{23} \text{ } C_8H_{10} \text{ molecules}$$

Check: The units of the answer (C_8H_{10} molecules) are correct. The magnitude is appropriate; it is smaller than Avogadro's number, as expected, because we have less than 1 mole of C_8H_{10}.

3.67   (a) **Given:** 5.94×10^{20} SO_3 molecules   **Find:** mass in g

Conceptual Plan: Number SO_3 molecules $\rightarrow$ mole $SO_3 \rightarrow$ g SO_3

$$\dfrac{1 \text{ mol } SO_3}{6.022 \times 10^{23} \text{ } SO_3 \text{ molecules}} \qquad \dfrac{80.07 \text{ g } SO_3}{1 \text{ mol } SO_3}$$

Solution: $5.94 \times 10^{20} \cancel{SO_3 \text{ molecules}} \times \dfrac{1 \text{ mol } \cancel{SO_3}}{6.022 \times 10^{23} \cancel{SO_3 \text{ molecules}}} \times \dfrac{80.07 \text{ g } SO_3}{1 \text{ mol } \cancel{SO_3}} = 0.0790 \text{ g } SO_3$

Check: The units of the answer (grams SO_3) are correct. The magnitude is appropriate; there is less than Avogadro's number of molecules, so we have less than 1 mole of SO_3.

(b) **Given:** 2.8×10^{22} H_2O molecules   **Find:** mass in g

Conceptual Plan: Number H_2O molecules $\rightarrow$ mole $H_2O \rightarrow$ g H_2O

$$\dfrac{1 \text{ mol } H_2O}{6.022 \times 10^{23} \text{ } H_2O \text{ molecules}} \qquad \dfrac{18.02 \text{ g } H_2O}{1 \text{ mol } H_2O}$$

Solution: $2.8 \times 10^{22} \cancel{H_2O \text{ molecules}} \times \dfrac{1 \text{ mol } \cancel{H_2O}}{6.022 \times 10^{23} \cancel{H_2O \text{ molecules}}} \times \dfrac{18.02 \text{ g } H_2O}{1 \text{ mol } \cancel{H_2O}} = 0.84 \text{ g } H_2O$

Check: The units of the answer (grams H_2O) are correct. The magnitude is appropriate; there is less than Avogadro's number of molecules, so we have less than 1 mole of H_2O.

(c) **Given:** 1 $C_6H_{12}O_6$ molecule   **Find:** mass in g

Conceptual Plan: Number $C_6H_{12}O_6$ molecules $\rightarrow$ mole $C_6H_{12}O_6 \rightarrow$ g $C_6H_{12}O_6$

$$\dfrac{1 \text{ mol } C_6H_{12}O_6}{6.022 \times 10^{23} \text{ } C_6H_{12}O_6 \text{ molecules}} \qquad \dfrac{180.16 \text{ g } C_6H_{12}O_6}{1 \text{ mol } C_6H_{12}O_6}$$

Solution:

$$1 \cancel{C_6H_{12}O_6 \text{ molecule}} \times \dfrac{1 \text{ mol } \cancel{C_6H_{12}O_6}}{6.022 \times 10^{23} \cancel{C_6H_{12}O_6 \text{ molecules}}} \times \dfrac{180.16 \text{ g } C_6H_{12}O_6}{1 \text{ mol } \cancel{C_6H_{12}O_6}} = 2.992 \times 10^{-22} \text{ g } C_6H_{12}O_6$$

Check: The units of the answer (grams $C_6H_{12}O_6$) are correct. The magnitude is appropriate; there is much less than Avogadro's number of molecules, so we have much less than 1 mole of $C_6H_{12}O_6$.

3.69   **Given:** 1.8×10^{17} $C_{12}H_{22}O_{11}$ molecules   **Find:** mass in mg

Conceptual Plan: Number $C_{12}H_{22}O_{11}$ molecules $\rightarrow$ mole $C_{12}H_{22}O_{11} \rightarrow$ g $C_{12}H_{22}O_{11} \rightarrow$ mg $C_{12}H_{22}O_{11}$

$$\dfrac{1 \text{ mol } C_{12}H_{22}O_{11}}{6.022 \times 10^{23} \text{ } C_{12}H_{22}O_{11} \text{ molecules}} \qquad \dfrac{342.3 \text{ g } C_{12}H_{22}O_{11}}{1 \text{ mol } C_{12}H_{22}O_{11}} \qquad \dfrac{1 \times 10^{3} \text{ mg } C_{12}H_{22}O_{11}}{1 \text{ g } C_{12}H_{22}O_{11}}$$

Solution:

$$1.8 \times 10^{17} \cancel{C_{12}H_{22}O_{11} \text{ molecules}} \times \dfrac{1 \text{ mol } \cancel{C_{12}H_{22}O_{11}}}{6.022 \times 10^{23} \cancel{C_{12}H_{22}O_{11} \text{ molecules}}} \times \dfrac{342.3 \text{ g } \cancel{C_{12}H_{22}O_{11}}}{1 \text{ mol } \cancel{C_{12}H_{22}O_{11}}} \times \dfrac{1 \times 10^{3} \text{ mg } \cancel{C_{12}H_{22}O_{11}}}{1 \text{ g } \cancel{C_{12}H_{22}O_{11}}}$$

$$= 0.10 \text{ mg } C_{12}H_{22}O_{11}$$

Check: The units of the answer (mg $C_{12}H_{22}O_{11}$) are correct. The magnitude is appropriate; there is much less than Avogadro's number of molecules, so we have much less than 1 mole of $C_{12}H_{22}O_{11}$.

Composition of Compounds

3.71   (a)   **Given:** CH_4 **Find:** mass percent C

Conceptual Plan: Mass % C $= \dfrac{1 \times \text{molar mass C}}{\text{molar mass } CH_4} \times 100\%$

Solution:

$$1 \times \text{molar mass C} = 1(12.01 \text{ g/mol}) = 12.01 \text{ g C}$$
$$\text{molar mass } CH_4 = 1(12.01 \text{ g/mol}) + 4(1.008 \text{ g/mol}) = 16.04 \text{ g/mol}$$
$$\text{mass \% C} = \dfrac{1 \times \text{molar mass C}}{\text{molar mass } CH_4} \times 100\%$$
$$= \dfrac{12.01 \text{ g/mol}}{16.04 \text{ g/mol}} \times 100\%$$
$$= 74.87\%$$

Check: The units of the answer (%) are correct. The magnitude is reasonable because it is between 0 and 100% and carbon is the heaviest element.

(b)   **Given:** C_2H_6 **Find:** mass percent C

Conceptual Plan: Mass % C $= \dfrac{2 \times \text{molar mass C}}{\text{molar mass } C_2H_6} \times 100\%$

Solution:

$$2 \times \text{molar mass C} = 2(12.01 \text{ g/mol}) = 24.02 \text{ g C}$$
$$\text{molar mass } C_2H_6 = 2(12.01 \text{ g/mol}) + 6(1.008 \text{ g/mol}) = 30.07 \text{ g/mol}$$
$$\text{mass \% C} = \dfrac{2 \times \text{molar mass C}}{\text{molar mass } C_2H_6} \times 100\%$$
$$= \dfrac{24.02 \text{ g/mol}}{30.07 \text{ g/mol}} \times 100\%$$
$$= 79.88\%$$

Check: The units of the answer (%) are correct. The magnitude is reasonable because it is between 0 and 100% and carbon is the heaviest element.

(c)   **Given:** C_2H_2 **Find:** mass percent C

Conceptual Plan: Mass % C $= \dfrac{2 \times \text{molar mass C}}{\text{molar mass } C_2H_2} \times 100\%$

Solution:

$$2 \times \text{molar mass C} = 2(12.01 \text{ g/mol}) = 24.02 \text{ g C}$$
$$\text{molar mass } C_2H_2 = 2(12.01 \text{ g/mol}) + 2(1.008 \text{ g/mol}) = 26.04 \text{ g/mol}$$
$$\text{mass \% C} = \dfrac{2 \times \text{molar mass C}}{\text{molar mass } C_2H_2} \times 100\%$$
$$= \dfrac{24.02 \text{ g/mol}}{26.04 \text{ g/mol}} \times 100\%$$
$$= 92.24\%$$

Check: The units of the answer (%) are correct. The magnitude is reasonable because it is between 0 and 100% and carbon is the heaviest element.

(d)   **Given:** C_2H_5Cl **Find:** mass percent C

Conceptual Plan: Mass % C $= \dfrac{2 \times \text{molar mass C}}{\text{molar mass } C_2H_5Cl} \times 100\%$

Solution:

$$2 \times \text{molar mass C} = 2(12.01 \text{ g/mol}) = 24.02 \text{ g C}$$
$$\text{molar mass } C_2H_5Cl = 2(12.01 \text{ g/mol}) + 5(1.008 \text{ g/mol}) + 1(35.45 \text{ g/mol}) = 64.51 \text{ g/mol}$$
$$\text{mass \% C} = \dfrac{2 \times \text{molar mass C}}{\text{molar mass } C_2H_5Cl} \times 100\%$$

$$= \frac{24.02 \ \cancel{g/mol}}{64.51 \ \cancel{g/mol}} \times 100\%$$

$$= 37.23\%$$

Check: The units of the answer (%) are correct. The magnitude is reasonable because it is between 0 and 100% and chlorine is heavier than carbon.

3.73   **Given:** NH_3 **Find:** mass percent N

Conceptual Plan: Mass % N $= \dfrac{1 \times \text{molar mass N}}{\text{molar mass } NH_3} \times 100\%$

Solution:

$$1 \times \text{molar mass N} = 1(14.01 \ \text{g/mol}) = 14.01 \ \text{g N}$$
$$\text{molar mass } NH_3 = 3(1.008 \ \text{g/mol}) + (14.01 \ \text{g/mol}) = 17.03 \ \text{g/mol}$$
$$\text{mass \% N} = \frac{1 \times \text{molar mass N}}{\text{molar mass } NH_3} \times 100\%$$

$$= \frac{14.01 \ \cancel{g/mol}}{17.03 \ \cancel{g/mol}} \times 100\%$$

$$= 82.27\%$$

Check: The units of the answer (%) are correct. The magnitude is reasonable because it is between 0 and 100% and nitrogen is the heaviest atom present.

Given: $CO(NH_2)_2$ **Find:** mass percent N

Conceptual Plan: Mass % N $= \dfrac{2 \times \text{molar mass N}}{\text{molar mass } CO(NH_2)_2} \times 100\%$

Solution:

$$2 \times \text{molar mass N} = 1(14.01 \ \text{g/mol}) = 28.02 \ \text{g N}$$
$$\text{molar mass } CO(NH_2)_2 = (12.01 \ \text{g/mol}) + (16.00 \ \text{g/mol}) + 2(14.01 \ \text{g/mol}) + 4(1.008 \ \text{g/mol}) = 60.06 \ \text{g/mol}$$
$$\text{mass \% N} = \frac{2 \times \text{molar mass N}}{\text{molar mass } CO(NH_2)_2} \times 100\%$$

$$= \frac{28.02 \ \cancel{g/mol}}{60.06 \ \cancel{g/mol}} \times 100\%$$
$$= 46.65\%$$

Check: The units of the answer (%) are correct. The magnitude is reasonable because it is between 0 and 100% and there are two nitrogens and only one carbon and one oxygen per molecule.

Given: NH_4NO_3 **Find:** mass percent N

Conceptual Plan: Mass % N $= \dfrac{2 \times \text{molar mass N}}{\text{molar mass } NH_4NO_3} \times 100\%$

Solution:

$$2 \times \text{molar mass N} = 2(14.01 \ \text{g/mol}) = 28.02 \ \text{g N}$$
$$\text{molar mass } NH_4NO_3 = 2(14.01 \ \text{g/mol}) + 4(1.008 \ \text{g/mol}) + 3(16.00 \ \text{g/mol}) = 80.05 \ \text{g/mol}$$
$$\text{mass \% N} = \frac{2 \times \text{molar mass N}}{\text{molar mass } NH_4NO_3} \times 100\%$$

$$= \frac{28.02 \ \cancel{g/mol}}{80.05 \ \cancel{g/mol}} \times 100\%$$

$$= 35.00\%$$

Check: The units of the answer (%) are correct. The magnitude is reasonable because it is between 0 and 100%. The mass of nitrogen is less than the mass of oxygen, and there are two nitrogens and three oxygens per molecule.

Given: $(NH_4)_2SO_4$ **Find:** mass percent N

Conceptual Plan: Mass % N $= \dfrac{2 \times \text{molar mass N}}{\text{molar mass } (NH_4)_2SO_4} \times 100\%$

Solution:

$$2 \times \text{molar mass N} = 2(14.01 \text{ g/mol}) = 28.02 \text{ g N}$$

$$\text{molar mass } (NH_4)_2SO_4 = 2(14.01 \text{ g/mol}) + 8(1.008 \text{ g/mol}) + (32.07 \text{ g/mol}) + 4(16.00 \text{ g/mol}) = 132.15 \text{ g/mol}$$

$$\text{mass \% N} = \dfrac{2 \times \text{molar mass N}}{\text{molar mass } (NH_4)_2SO_4} \times 100\%$$

$$= \dfrac{28.02 \text{ g/mol}}{132.15 \text{ g/mol}} \times 100\%$$

$$= 21.20\%$$

Check: The units of the answer (%) are correct. The magnitude is reasonable because it is between 0 and 100% and the mass of nitrogen is less than the mass of oxygen and sulfur.
The fertilizer with the highest nitrogen content is NH_3 with a N content of 82.27%.

3.75   **Given:** 55.5 g CuF_2; 37.42   **Find:** g F in CuF_2
Conceptual Plan: g $CuF_2 \rightarrow$ g F

$$\dfrac{37.42 \text{ g F}}{100.0 \text{ g } CuF_2}$$

Solution: 55.5 g CuF_2 $\times \dfrac{37.42 \text{ g F}}{100.0 \text{ g } CuF_2} = 20.7\underline{7} = 20.8$ g F

Check: The units of the answer (g F) are correct. The magnitude is reasonable because it is less than the original mass.

3.77   **Given:** 150 μg I; 76.45% I in KI   **Find:** μg KI
Conceptual Plan: μg I $\rightarrow$ g I $\rightarrow$ g KI $\rightarrow$ μg KI

$$\dfrac{1 \text{ g I}}{1 \times 10^6 \text{ }\mu\text{g I}} \quad \dfrac{100.0 \text{ g KI}}{76.45 \text{ g I}} \quad \dfrac{1 \times 10^6 \text{ }\mu\text{g KI}}{1 \text{ g KI}}$$

Solution: 150 μg I $\times \dfrac{1 \text{ g I}}{1 \times 10^6 \text{ }\mu\text{g I}} \times \dfrac{100.0 \text{ g KI}}{76.45 \text{ g I}} \times \dfrac{1 \times 10^6 \text{ }\mu\text{g KI}}{1 \text{ g KI}} = 196 \text{ }\mu$g KI

Check: The units of the answer (μg KI) are correct. The magnitude is reasonable because it is greater than the original mass.

3.79   (a)   red = oxygen, white = hydrogen: 2H:O H_2O
        (b)   black = carbon, white = hydrogen: C:4H CH_4
        (c)   black = carbon, white = hydrogen, red = oxygen: 2C:6H:O CH_3CH_2OH or C_2H_6O

3.81   (a)   **Given:** 0.0885 mol C_4H_{10} **Find:** mol H atoms
        Conceptual Plan: Mol $C_4H_{10} \rightarrow$ mole H atom

$$\dfrac{10 \text{ mol H}}{1 \text{ mol } C_4H_{10}}$$

        Solution: 0.0885 mol C_4H_{10} $\times \dfrac{10 \text{ mol H}}{1 \text{ mol } C_4H_{10}} = 0.885$ mol H atoms

        Check: The units of the an.swer (mol H atoms) are correct. The magnitude is reasonable because it is greater than the original mol C_4H_{10}.

        (b)   **Given:** 1.3 mol CH_4 **Find:** mol H atoms
        Conceptual Plan: Mol $CH_4 \rightarrow$ mole H atom

$$\dfrac{4 \text{ mol H}}{1 \text{ mol } CH_4}$$

        Solution: 1.3 mol CH_4 $\times \dfrac{4 \text{ mol H}}{1 \text{ mol } CH_4} = 5.2$ mol H atoms

        Check: The units of the answer (mol H atoms) are correct. The magnitude is reasonable because it is greater than the original mol CH_4.

(c) **Given:** 2.4 mol C_6H_{12} **Find:** mol H atoms
Conceptual Plan: Mol C_6H_{12} → mole H atom

$$\frac{12\ \text{mol H}}{1\ \text{mol } C_6H_{12}}$$

Solution: $2.4\ \cancel{\text{mol } C_6H_{12}} \times \dfrac{12\ \text{mol H}}{1\ \cancel{\text{mol } C_6H_{12}}} = 29\ \text{mol H atoms}$

Check: The units of the answer (mol H atoms) are correct. The magnitude is reasonable because it is greater than the original mol C_6H_{12}.

(d) **Given:** 1.87 mol C_8H_{18} **Find:** mol H atoms
Conceptual Plan: Mol C_8H_{18} → mole H atom

$$\frac{18\ \text{mol H}}{1\ \text{mol } C_8H_{18}}$$

Solution: $1.87\ \cancel{\text{mol } C_8H_{18}} \times \dfrac{18\ \text{mol H}}{1\ \cancel{\text{mol } C_8H_{18}}} = 33.7\ \text{mol H atoms}$

Check: The units of the answer (mol H atoms) are correct. The magnitude is reasonable because it is greater than the original mol C_8H_{18}.

3.83 (a) **Given:** 8.5 g NaCl   **Find:** g Na
Conceptual Plan: g NaCl → mole NaCl → mol Na → g Na

$$\frac{1\ \text{mol NaCl}}{58.44\ \text{g NaCl}} \qquad \frac{1\ \text{mol Na}}{1\ \text{mol NaCl}} \qquad \frac{22.99\ \text{g Na}}{1\ \text{mol Na}}$$

Solution: $8.5\ \cancel{\text{g NaCl}} \times \dfrac{1\ \cancel{\text{mol NaCl}}}{58.44\ \cancel{\text{g NaCl}}} \times \dfrac{1\ \cancel{\text{mol Na}}}{1\ \cancel{\text{mol NaCl}}} \times \dfrac{22.99\ \text{g Na}}{1\ \cancel{\text{mol Na}}} = 3.3\ \text{g Na}$

Check: The units of the answer (g Na) are correct. The magnitude is reasonable because it is less than the original g NaCl.

(b) **Given:** 8.5 g Na_3PO_4 **Find:** g Na
Conceptual Plan: g Na_3PO_4 → mole Na_3PO_4 → mol Na → g Na

$$\frac{1\ \text{mol } Na_3PO_4}{163.94\ \text{g } Na_3PO_4} \qquad \frac{3\ \text{mol Na}}{1\ \text{mol } Na_3PO_4} \qquad \frac{22.99\ \text{g Na}}{1\ \text{mol Na}}$$

Solution: $8.5\ \cancel{\text{g } Na_3PO_4} \times \dfrac{1\ \cancel{\text{mol } Na_3PO_4}}{163.94\ \cancel{\text{g } Na_3PO_4}} \times \dfrac{3\ \cancel{\text{mol Na}}}{1\ \cancel{\text{mol } Na_3PO_4}} \times \dfrac{22.99\ \text{g Na}}{1\ \cancel{\text{mol Na}}} = 3.6\ \text{g Na}$

Check: The units of the answer (g Na) are correct. The magnitude is reasonable because it is less than the original g Na_3PO_4.

(c) **Given:** 8.5 g $NaC_7H_5O_2$ **Find:** g Na
Conceptual Plan: g $NaC_7H_5O_2$ → mole $NaC_7H_5O_2$ → mol Na → g Na

$$\frac{1\ \text{mol } NaC_7H_5O_2}{144.10\ \text{g } NaC_7H_5O_2} \qquad \frac{1\ \text{mol Na}}{1\ \text{mol } NaC_7H_5O_2} \qquad \frac{22.99\ \text{g Na}}{1\ \text{mol Na}}$$

Solution: $8.5\ \cancel{\text{g } NaC_7H_5O_2} \times \dfrac{1\ \cancel{\text{mol } NaC_7H_5O_2}}{144.10\ \cancel{\text{g } NaC_7H_5O_2}} \times \dfrac{1\ \cancel{\text{mol Na}}}{1\ \cancel{\text{mol } NaC_7H_5O_2}} \times \dfrac{22.99\ \text{g Na}}{1\ \cancel{\text{mol Na}}} = 1.4\ \text{g Na}$

Check: The units of the answer (g Na) are correct. The magnitude is reasonable because it is less than the original g $NaC_7H_5O_2$.

(d) **Given:** 8.5 g $Na_2C_6H_6O_7$ **Find:** g Na
Conceptual Plan: g $Na_2C_6H_6O_7$ → mole $Na_2C_6H_6O_7$ → mol Na → g Na

$$\frac{1\ \text{mol } Na_2C_6H_6O_7}{236.1\ \text{g } Na_2C_6H_6O_7} \qquad \frac{2\ \text{mol Na}}{1\ \text{mol } Na_2C_6H_6O_7} \qquad \frac{22.99\ \text{g Na}}{1\ \text{mol Na}}$$

Solution: $8.5\ \cancel{\text{g } Na_2C_6H_6O_7} \times \dfrac{1\ \cancel{\text{mol } Na_2C_6H_6O_7}}{236.1\ \cancel{\text{g } Na_2C_6H_6O_7}} \times \dfrac{2\ \cancel{\text{mol Na}}}{1\ \cancel{\text{mol } Na_2C_6H_6O_7}} \times \dfrac{22.99\ \text{g Na}}{1\ \cancel{\text{mol Na}}} = 1.7\ \text{g Na}$

Check: The units of the answer (g Na) are correct. The magnitude is reasonable because it is less than the original g $Na_2C_6H_6O_7$.

Chemical Formulas from Experimental Data

3.85   (a)   **Given:** 1.651 g Ag; 0.1224 g O   **Find:** empirical formula
Conceptual Plan:
Convert mass to mol of each element → write pseudoformula → write empirical formula

$$\frac{1 \text{ mol Ag}}{107.9 \text{ g Ag}} \quad \frac{1 \text{ mol O}}{16.00 \text{ g O}} \qquad \text{divide by smallest number}$$

Solution: $1.651 \text{ g Ag} \times \dfrac{1 \text{ mol Ag}}{107.9 \text{ g Ag}} = 0.01530 \text{ mol Ag}$

$0.1224 \text{ g O} \times \dfrac{1 \text{ mol O}}{16.00 \text{ g O}} = 0.007650 \text{ mol O}$

$Ag_{0.01530} O_{0.007650}$

$Ag_{\frac{0.01530}{0.007650}} O_{\frac{0.007650}{0.007650}} \rightarrow Ag_2O$

The correct empirical formula is Ag_2O.

(b)   **Given:** 0.672 g Co; 0.569 g As; 0.486 g O   **Find:** empirical formula
Conceptual Plan: Convert mass to mol of each element → write pseudoformula → write empirical formula

$$\frac{1 \text{ mol Co}}{58.93 \text{ g Co}} \quad \frac{1 \text{ mol As}}{74.92 \text{ g As}} \quad \frac{1 \text{ mol O}}{16.00 \text{ g O}} \qquad \text{divide by smallest number}$$

Solution: $0.672 \text{ g Co} \times \dfrac{1 \text{ mol Co}}{58.93 \text{ g Co}} = 0.0114 \text{ mol Co}$

$0.569 \text{ g As} \times \dfrac{1 \text{ mol As}}{74.92 \text{ g As}} = 0.00759 \text{ mol O}$

$0.486 \text{ g O} \times \dfrac{1 \text{ mol O}}{16.00 \text{ g O}} = 0.0304 \text{ mol O}$

$Co_{0.0114} As_{0.00759} O_{0.0304}$

$Co_{\frac{0.0114}{0.00759}} As_{\frac{0.00759}{0.00759}} O_{\frac{0.0304}{0.00759}} \rightarrow Co_{1.5}As_1O_4$

$Co_{1.5}As_1O_4 \times 2 \rightarrow Co_3As_2O_8$

The correct empirical formula is $Co_3As_2O_8$.

(c)   **Given:** 1.443 g Se; 5.841 g Br   **Find:** empirical formula
Conceptual Plan:
Convert mass to mol of each element → write pseudoformula → write empirical formula

$$\frac{1 \text{ mol Se}}{78.96 \text{ g Se}} \quad \frac{1 \text{ mol Br}}{79.90 \text{ g Br}} \qquad \text{divide by smallest number}$$

Solution: $1.443 \text{ g Se} \times \dfrac{1 \text{ mol Se}}{78.96 \text{ g Se}} = 0.01828 \text{ mol Se}$

$5.841 \text{ g Br} \times \dfrac{1 \text{ mol Br}}{79.90 \text{ g Br}} = 0.07310 \text{ mol Br}$

$Se_{0.01828} Br_{0.07310}$

$Se_{\frac{0.01828}{0.01828}} Br_{\frac{0.07310}{0.01828}} \rightarrow SeBr_4$

The correct empirical formula is $SeBr_4$.

3.87   (a)   **Given:** in a 100 g sample of nicotine: 74.03 g C; 8.70 g H; 17.27 g N   **Find:** empirical formula
Conceptual Plan:
Convert mass to mol of each element → write pseudoformula → write empirical formula

$$\frac{1 \text{ mol C}}{12.01 \text{ g C}} \quad \frac{1 \text{ mol H}}{1.008 \text{ g H}} \quad \frac{1 \text{ mol N}}{14.01 \text{ g N}} \qquad \text{divide by smallest number}$$

Solution: $74.03 \text{ g C} \times \dfrac{1 \text{ mol C}}{12.01 \text{ g C}} = 6.164 \text{ mol C}$

$8.70 \text{ g H} \times \dfrac{1 \text{ mol H}}{1.008 \text{ g H}} = 8.63 \text{ mol H}$

$$17.27 \text{ g N} \times \frac{1 \text{ mol N}}{14.01 \text{ g N}} = 1.233 \text{ mol N}$$

$$C_{6.164}H_{8.63}N_{1.233}$$

$$C_{\frac{6.164}{1.233}} H_{\frac{8.63}{1.233}} N_{\frac{1.233}{1.233}} \rightarrow C_5H_7N$$

The correct empirical formula is C_5H_7N.

(b) **Given:** in a 100 g sample of caffeine: 49.48 g C; 5.19 g H; 28.85 g N; 16.48 g O   **Find:** empirical formula
Conceptual Plan:
Convert mass to mol of each element → write pseudoformula → write empirical formula

$$\frac{1 \text{ mol C}}{12.01 \text{ g C}} \quad \frac{1 \text{ mol H}}{1.008 \text{ g H}} \quad \frac{1 \text{ mol N}}{14.01 \text{ g N}} \quad \frac{1 \text{ mol O}}{16.00 \text{ g O}}$$

divide by smallest number

Solution: $49.48 \text{ g C} \times \dfrac{1 \text{ mol C}}{12.01 \text{ g C}} = 4.120 \text{ mol C}$

$5.19 \text{ g H} \times \dfrac{1 \text{ mol H}}{1.008 \text{ g H}} = 5.15 \text{ mol H}$

$28.85 \text{ g N} \times \dfrac{1 \text{ mol N}}{14.01 \text{ g N}} = 2.059 \text{ mol N}$

$16.48 \text{ g O} \times \dfrac{1 \text{ mol O}}{16.00 \text{ g O}} = 1.030 \text{ mol O}$

$$C_{4.120} H_{5.15} N_{2.059} O_{1.030}$$

$$C_{\frac{4.120}{1.030}} H_{\frac{5.15}{1.030}} N_{\frac{2.059}{1.030}} O_{\frac{1.030}{1.030}} \rightarrow C_4H_5N_2O$$

The correct empirical formula is $C_4H_5N_2O$.

3.89   **Given:** in a 100 g sample of ibuprofen 75.69 g C; 8.80 g H; 15.51 g O   **Find:** empirical formula
Conceptual Plan: Convert mass to mol of each element → write pseudoformula → write empirical formula

$$\frac{1 \text{ mol C}}{12.01 \text{ g C}} \quad \frac{1 \text{ mol H}}{1.008 \text{ g H}} \quad \frac{1 \text{ mol O}}{16.00 \text{ g O}}$$

divide by smallest number

Solution: $75.69 \text{ g C} \times \dfrac{1 \text{ mol C}}{12.01 \text{ g C}} = 6.302 \text{ mol C}$

$8.80 \text{ g H} \times \dfrac{1 \text{ mol H}}{1.008 \text{ g H}} = 8.73 \text{ mol H}$

$15.51 \text{ g O} \times \dfrac{1 \text{ mol O}}{16.00 \text{ g O}} = 0.9694 \text{ mol O}$

$$C_{6.302} H_{8.73} O_{0.9694}$$

$$C_{\frac{6.302}{0.9694}} H_{\frac{8.73}{0.9694}} O_{\frac{0.9694}{0.9694}} \rightarrow C_{6.50}H_{9.01}O$$

$$C_{6.50}H_{9.01}O \times 2 = C_{13}H_{18}O_2$$

The correct empirical formula is $C_{13}H_{18}O_2$.

3.91   **Given:** 0.77 mg N; 6.61 mg N_xCl_y **Find:** empirical formula
Conceptual Plan: Find mg Cl → convert mg to g for each element → convert mass to mol of each element →

$$\text{mg } N_xCl_y - \text{mg N} \qquad \frac{1 \text{ g}}{1000 \text{ mg}} \qquad \frac{1 \text{ mol N}}{14.01 \text{ g N}} \qquad \frac{1 \text{ mol Cl}}{35.45 \text{ g Cl}}$$

write pseudoformula → write empirical formula

divide by smallest number

Solution: $6.61 \text{ mg } N_xCl_y - 0.77 \text{ mg N} = 5.84 \text{ mg Cl}$

$0.77 \text{ mg N} \times \dfrac{1 \text{ g N}}{1000 \text{ mg N}} \times \dfrac{1 \text{ mol N}}{14.01 \text{ g N}} = 5.5 \times 10^{-5} \text{ mol N}$

$5.84 \text{ mg Cl} \times \dfrac{1 \text{ g Cl}}{1000 \text{ mg Cl}} \times \dfrac{1 \text{ mol Cl}}{35.45 \text{ g Cl}} = 1.6 \times 10^{-4} \text{ mol Cl}$

$$N_{5.5 \times 10^{-5}} Cl_{1.6 \times 10^{-4}}$$

$$N_{\frac{5.5 \times 10^{-5}}{5.5 \times 10^{-5}}} Cl_{\frac{1.6 \times 10^{-4}}{5.5 \times 10^{-5}}} \rightarrow NCl_3$$

The correct empirical formula is NCl_3.

3.93   (a)   **Given:** empirical formula $= C_6H_7N$; molar mass $= 186.24$ g/mol   **Find:** molecular formula

Conceptual Plan: Molecular formula $=$ empirical formula $\times$ n $n = \dfrac{\text{molar mass}}{\text{empirical formula mass}}$

Solution: empirical formula mass $= 6(12.01$ g/mol$) + 7(1.008$ g/mol$) + 1(14.01$ g/mol$) = 93.13$ g/mol

$$n = \frac{\text{molar mass}}{\text{formula molar mass}} = \frac{186.24 \text{ g/mol}}{93.13 \text{ g/mol}} = 1.9998 = 2$$

molecular formula $= C_6H_7N \times 2$
                        $= C_{12}H_{14}N_2$

(b)   **Given:** empirical formula $= C_2HCl$; molar mass $= 181.44$ g/mol   **Find:** molecular formula

Conceptual Plan: Molecular formula $=$ empirical formula $\times$ n $n = \dfrac{\text{molar mass}}{\text{empirical formula mass}}$

Solution: empirical formula mass $= 2(12.01$ g/mol$) + 1(1.008$ g/mol$) + 1(35.45$ g/mol$) = 60.48$ g/mol

$$n = \frac{\text{molar mass}}{\text{formula molar mass}} = \frac{181.44 \text{ g/mol}}{60.48 \text{ g/mol}} = 3$$

molecular formula $= C_2HCl \times 3$
                        $= C_6H_3Cl_3$

(c)   **Given:** empirical formula $= C_5H_{10}NS_2$; molar mass $= 296.54$ g/mol   **Find:** molecular formula

Conceptual Plan: Molecular formula $=$ empirical formula $\times$ n $n = \dfrac{\text{molar mass}}{\text{empirical formula mass}}$

Solution: empirical formula mass $= 5(12.01$ g/mol$) + 10(1.008$ g/mol$)$
    $+ 1(14.01$ g/mol$) + 2(32.07) = 148.28$ g/mol

$$n = \frac{\text{molar mass}}{\text{formula molar mass}} = \frac{296.54 \text{ g/mol}}{148.28 \text{ g/mol}} = 2$$

molecular formula $= C_5H_{10}NS_2 \times 2$
                        $= C_{10}H_{20}N_2S_4$

3.95   **Given:** 33.01 g CO_2; 13.51 g H_2O **Find:** empirical formula
Conceptual Plan: Mass CO_2, H_2O $\rightarrow$ mol CO_2, H_2O $\rightarrow$ mol C, mol H $\rightarrow$ pseudoformula $\rightarrow$ empirical formula

$$\frac{1 \text{ mol } CO_2}{44.01 \text{ g } CO_2} \quad \frac{1 \text{ mol } H_2O}{18.02 \text{ g } H_2O} \quad \frac{1 \text{ mol } C}{1 \text{ mol } CO_2} \quad \frac{2 \text{ mol } H}{1 \text{ mol } H_2O}$$ divide by smallest number

Solution:

$$33.01 \text{ g } CO_2 \times \frac{1 \text{ mol } CO_2}{44.01 \text{ g } CO_2} = 0.7501 \text{ mol } CO_2$$

$$13.51 \text{ g } H_2O \times \frac{1 \text{ mol } H_2O}{18.02 \text{ g } H_2O} = 0.7497 \text{ mol } H_2O$$

$$0.7501 \text{ mol } CO_2 \times \frac{1 \text{ mol } C}{1 \text{ mol } CO_2} = 0.7501 \text{ mol } C$$

$$0.7497 \text{ mol } H_2O \times \frac{2 \text{ mol } H}{1 \text{ mol } H_2O} = 1.499 \text{ mol } H$$

$$C_{0.7501} H_{1.499}$$

$$C_{\frac{0.7501}{0.7501}} H_{\frac{1.499}{0.7501}} \rightarrow CH_2$$

The correct empirical formula is CH_2.

3.97   **Given:** 4.30 g sample; 8.59 g CO_2; 3.52 g H_2O **Find:** empirical formula

        Conceptual Plan: Mass CO_2, H_2O → mol CO_2, H_2O → mol C, mol H → mass C, mass H, mass O → mol O →

$$\frac{1\ mol\ CO_2}{44.01\ g\ CO_2}\quad \frac{1\ mol\ H_2O}{18.02\ g\ H_2O}\quad \frac{1\ mol\ C}{1\ mol\ CO_2}\quad \frac{2\ mol\ H}{1\ mol\ H_2O}\quad \frac{12.01\ g\ C}{1\ mol\ C}\quad \frac{1.008\ g\ H}{1\ mol\ H}\quad g\ sample - g\ C - g\ H\quad \frac{1\ mol\ O}{16.00\ g\ O}$$

        pseudoformula → empirical formula

                divide by smallest number

        Solution:

$$8.59\ g\ CO_2 \times \frac{1\ mol\ CO_2}{44.01\ g\ CO_2} = 0.195\ mol\ CO_2$$

$$3.52\ g\ H_2O \times \frac{1\ mol\ H_2O}{18.02\ g\ H_2O} = 0.195\ mol\ H_2O$$

$$0.195\ mol\ CO_2 \times \frac{1\ mol\ C}{1\ mol\ CO_2} = 0.195\ mol\ C$$

$$0.195\ mol\ H_2O \times \frac{2\ mol\ H}{1\ mol\ H_2O} = 0.390\ mol\ H$$

$$0.195\ mol\ CO \times \frac{12.01\ mol\ C}{1\ mol\ CO} = 2.34\ g\ C$$

$$0.390\ mol\ H_2O \times \frac{1.008\ g\ H}{1\ mol\ H} = 0.393\ g\ H$$

$$4.30\ g - 2.34\ g - 0.393\ g = 1.57\ g\ O$$

$$1.57\ g\ O \times \frac{1\ mol\ O}{16.00\ g\ O} = 0.0981\ mol\ O$$

$$C_{0.195}\ H_{0.390}\ O_{0.0981}$$

$$C_{\frac{0.195}{0.0981}}\ H_{\frac{0.390}{0.0981}}\ O_{\frac{0.0981}{0.0981}} \rightarrow C_2H_4O$$

    The correct empirical formula is C_2H_4O.

Writing and Balancing Chemical Equations

3.99   **Conceptual Plan: Write a skeletal reaction → balance atoms in more complex compounds → balance elements that occur as free elements → clear fractions**

        Solution: Skeletal reaction: $SO_2(g) + O_2(g) + H_2O(l) \rightarrow H_2SO_4(aq)$

                         Balance O: $SO_2(g) + 1/2\ O_2(g) + H_2O(l) \rightarrow H_2SO_4(aq)$

                         Clear fraction: $2\ SO_2(g) + O_2(g) + 2\ H_2O(l) \rightarrow 2\ H_2SO_4(aq)$

        Check:

left side	right side
2 S atoms	2 S atoms
8 O atoms	8 O atoms
4 H atoms	4 H atoms

3.101   **Conceptual Plan: Write a skeletal reaction → balance atoms in more complex compounds → balance elements that occur as free elements → clear fractions**

        Solution: Skeletal reaction: $Na(s) + H_2O(l) \rightarrow H_2(g) + NaOH(aq)$

                         Balance H: $Na(s) + H_2O(l) \rightarrow 1/2\ H_2(g) + NaOH(aq)$

                         Clear fraction: $2\ Na(s) + 2\ H_2O(l) \rightarrow H_2(g) + 2\ NaOH(aq)$

        Check:

left side	right side
2 Na atoms	2 Na atoms
4 H atoms	4 H atoms
2 O atoms	2 O atoms

3.103   **Conceptual Plan: Write a skeletal reaction → balance atoms in more complex compounds → balance elements that occur as free elements → clear fractions**

   Solution: Skeletal reaction: $C_{12}H_{22}O_{11}(aq) + H_2O(l) \rightarrow C_2H_5OH(aq) + CO_2(g)$

   Balance H: $C_{12}H_{22}O_{11}(aq) + H_2O(l) \rightarrow 4\,C_2H_5OH(aq) + CO_2(g)$

   Balance C: $C_{12}H_{22}O_{11}(aq) + H_2O(l) \rightarrow 4\,C_2H_5OH(aq) + 4\,CO_2(g)$

   Check:

left side	right side
12 C atoms	12 C atoms
24 H atoms	24 H atoms
12 O atoms	12 O atoms

3.105   (a)   **Conceptual Plan: Write a skeletal reaction → balance atoms in more complex compounds → balance elements that occur as free elements → clear fractions**

   Solution: Skeletal reaction: $PbS(s) + HBr(aq) \rightarrow PbBr_2(s) + H_2S(g)$

   Balance Br: $PbS(s) + 2\,HBr(aq) \rightarrow PbBr_2(s) + H_2S(g)$

   Check:

left side	right side
1 Pb atom	1 Pb atom
1 S atom	1 S atom
2 H atoms	2 H atoms
2 Br atoms	2 Br atoms

   (b)   **Conceptual Plan: Write a skeletal reaction → balance atoms in more complex compounds → balance elements that occur as free elements → clear fractions**

   Solution: Skeletal reaction: $CO(g) + H_2(g) \rightarrow CH_4(g) + H_2O(l)$

   Balance H: $CO(g) + 3\,H_2(g) \rightarrow CH_4(g) + H_2O(l)$

   Check:

left side	right side
1 C atom	1 C atom
1 O atom	1 O atom
6 H atoms	6 H atoms

   (c)   **Conceptual Plan: Write a skeletal reaction → balance atoms in more complex compounds → balance elements that occur as free elements → clear fractions**

   Solution: Skeletal reaction: $HCl(aq) + MnO_2(s) \rightarrow MnCl_2(aq) + H_2O(l) + Cl_2(g)$

   Balance Cl: $4\,HCl(aq) + MnO_2(s) \rightarrow MnCl_2(aq) + H_2O(l) + Cl_2(g)$

   Balance O: $4\,HCl(aq) + MnO_2(s) \rightarrow MnCl_2(aq) + 2\,H_2O(l) + Cl_2(g)$

   Check:

left side	right side
4 H atoms	4 H atoms
4 Cl atoms	4 Cl atoms
1 Mn atom	1 Mn atom
2 O atoms	2 O atoms

   (d)   **Conceptual Plan: Write a skeletal reaction → balance atoms in more complex compounds → balance elements that occur as free elements → clear fractions**

   Solution: Skeletal reaction: $C_5H_{12}(l) + O_2(g) \rightarrow CO_2(g) + H_2O(l)$

   Balance C: $C_5H_{12}(l) + O_2(g) \rightarrow 5\,CO_2(g) + H_2O(l)$

   Balance H: $C_5H_{12}(l) + O_2(g) \rightarrow 5\,CO_2(g) + 6\,H_2O(l)$

   Balance O: $C_5H_{12}(l) + 8\,O_2(g) \rightarrow 5\,CO_2(g) + 6\,H_2O(l)$

   Check:

left side	right side
5 C atoms	5 C atoms
12 H atoms	12 H atoms
16 O atoms	16 O atoms

3.107   **Conceptual Plan: Write a skeletal reaction → balance atoms in more complex compounds → balance elements that occur as free elements → clear fractions**

   Solution: Skeletal reaction: $Na_2CO_3(aq) + CuCl_2(aq) \rightarrow CuCO_3(s) + NaCl(aq)$

   Balance Na: $Na_2CO_3(aq) + CuCl_2(aq) \rightarrow CuCO_3(s) + 2\,NaCl(aq)$

Check:

	left side	right side
	2 Na atoms	2 Na atoms
	1 C atom	1 C atom
	3 O atoms	3 O atoms
	1 Cu atom	1 Cu atom
	2 Cl atoms	2 Cl atoms

3.109   (a)   **Conceptual Plan: Balance atoms in more complex compounds → balance elements that occur as free elements → clear fractions**

         Solution: Skeletal reaction: $CO_2(g) + CaSiO_3(s) + H_2O(l) \rightarrow SiO_2(s) + Ca(HCO_3)_2(aq)$

                         Balance C: $2 CO_2(g) + CaSiO_3(s) + H_2O(l) \rightarrow SiO_2(s) + Ca(HCO_3)_2(aq)$

         Check:

	left side	right side
	2 C atoms	2 C atoms
	8 O atoms	8 O atoms
	1 Ca atom	1 Ca atom
	1 Si atom	1 Si atom
	2 H atoms	2 H atoms

   (b)   **Conceptual Plan: Balance atoms in more complex compounds → balance elements that occur as free elements → clear fractions**

         Solution: Skeletal reaction: $Co(NO_3)_3(aq) + (NH_4)_2S(aq) \rightarrow Co_2S_3(s) + NH_4NO_3(aq)$

                         Balance S: $Co(NO_3)_3(aq) + 3 (NH_4)_2S(aq) \rightarrow Co_2S_3(s) + NH_4NO_3(aq)$

                         Balance Co: $2 Co(NO_3)_3(aq) + 3 (NH_4)_2S(aq) \rightarrow Co_2S_3(s) + NH_4NO_3(aq)$

                         Balance N: $2 Co(NO_3)_3(aq) + 3 (NH_4)_2S(aq) \rightarrow Co_2S_3(s) + 6 NH_4NO_3(aq)$

         Check:

	left side	right side
	2 Co atoms	2 Co atoms
	12 N atoms	12 N atoms
	18 O atoms	18 O atoms
	24 H atoms	24 H atoms
	3 S atoms	3 S atoms

   (c)   **Conceptual Plan: Balance atoms in more complex compounds → balance elements that occur as free elements → clear fractions**

         Solution: Skeletal reaction: $Cu_2O(s) + C(s) \rightarrow Cu(s) + CO(g)$

                         Balance Cu: $Cu_2O(s) + C(s) \rightarrow 2 Cu(s) + CO(g)$

         Check:

	left side	right side
	2 Cu atoms	2 Cu atoms
	1 O atom	1 O atom
	1 C atom	1 C atom

   (d)   **Conceptual Plan: Balance atoms in more complex compounds → balance elements that occur as free elements → clear fractions**

         Solution: Skeletal reaction: $H_2(g) + Cl_2(g) \rightarrow HCl(g)$

                         Balance Cl: $H_2(g) + Cl_2(g) \rightarrow 2 HCl(g)$

         Check:

	left side	right side
	2 H atoms	2 H atoms
	2 Cl atoms	2 Cl atoms

Organic Compounds

3.111   (a)   composed of metal cation and polyatomic anion—inorganic compound

      (b)   composed of carbon and hydrogen—organic compound

      (c)   composed of carbon, hydrogen, and oxygen—organic compound

      (d)   composed of metal cation and nonmetal anion—inorganic compound

3.113   (a)   contains a double bond—alkene

      (b)   contains only single bonds—alkane

      (c)   contains triple bond—alkyne

      (d)   contains only single bonds—alkane

3.115   (a)   prop $=$ 3 C, ane $=$ single bonds: $CH_3CH_2CH_3$

        (b)   3 C $=$ prop, single bonds $=$ ane: propane

        (c)   oct $=$ 8 C, ane $=$ single bonds: $CH_3CH_2CH_2CH_2CH_2CH_2CH_2CH_3$

        (d)   5 C $=$ pent, single bonds $=$ ane: pentane

3.117   (a)   contains O: functionalized hydrocarbon: alcohol

        (b)   contains only C and H: hydrocarbon

        (c)   contains O: functionalized hydrocarbon: ketone

        (d)   contains N: functionalized hydrocarbon: amine

Cumulative Problems

3.119   **Given:** 145 mL C_2H_5OH; $d = 0.789$ g/cm^3 **Find:** number of molecules

        Conceptual Plan: cm^3 → mL : mL C_2H_5OH → g C_2H_5OH → mol C_2H_5OH → molecules C_2H_5OH

$$\frac{1\ cm^3}{1\ mL} \qquad \frac{0.789\ g\ C_2H_5OH}{cm^3} \quad \frac{1\ mol\ C_2H_5OH}{46.07\ g\ C_2H_5OH} \quad \frac{6.022 \times 10^{23}\ molecules\ C_2H_5OH}{1\ mol\ C_2H_5OH}$$

        Solution:

$$145\ mL\ C_2H_5OH \times \frac{1\ cm^3}{1\ mL} \times \frac{0.789\ g\ C_2H_5OH}{cm^3} \times \frac{1\ mol\ C_2H_5OH}{46.07\ g\ C_2H_5OH} \times \frac{6.022 \times 10^{23}\ molecules\ C_2H_5OH}{1\ mol\ C_2H_5OH}$$

$$= 1.50 \times 10^{24}\ molecules\ C_2H_5OH$$

        Check: The units of the answer (molecules C_2H_5OH) are correct. The magnitude is reasonable because we had more than two moles of C_2H_5OH and we have more than two times Avogadro's number of molecules.

3.121   (a)   To write the formula for an ionic compound, do the following: (1) Write the symbol for the metal cation and its charge and the symbol for the nonmetal anion or polyatomic anion and its charge. (2) Adjust the subscript on each cation and anion to balance the overall charge. (3) Check that the sum of the charges of the cations equals the sum of the charges of the anions.

           potassium chromate: K^+ CrO_4^{2-}; K_2CrO_4 cation 2(1+) $=$ 2+; anion 2−

           Given: K_2CrO_4 **Find:** mass percent of each element

           Conceptual Plan: % K, then % Cr, then % O

$$mass\ \%\ K = \frac{2 \times molar\ mass\ K}{molar\ mass\ K_2CrO_4} \times 100\% \quad mass\ \%\ Cr = \frac{1 \times molar\ mass\ Cr}{molar\ mass\ K_2CrO_4} \times 100\% \quad mass\ \%\ O = \frac{4 \times molar\ mass\ O}{molar\ mass\ K_2CrO_4} \times 100\%$$

           molar mass of K $=$ 39.10 g/mol; molar mass Cr $=$ 52.00 g/mol; molar mass O $=$ 16.00 g/mol

           Solution: molar mass $K_2CrO_4 =$ 2(39.10 g/mol) $+$ 1(52.00 g/mol) $+$ 4(16.00 g/mol) $=$ 194.20 g/mol

           2 $\times$ molar mass K $=$ 2(39.10 g/mol) $=$ 78.20 g K   1 $\times$ molar mass Cr $=$ 1(52.00 g/mol) $=$ 52.00 g Cr

$$mass\ \%\ K = \frac{2 \times molar\ mass\ K}{molar\ mass\ K_2CrO_4} \times 100\% \qquad mass\ \%\ Cr = \frac{1 \times molar\ mass\ Cr}{molar\ mass\ K_2CrO_4} \times 100\%$$

$$= \frac{78.20\ g/mol}{194.20\ g/mol} \times 100\% \qquad\qquad = \frac{52.00\ g/mol}{194.20\ g/mol} \times 100\%$$

$$= 40.27\% \qquad\qquad\qquad = 26.78\%$$

$$4 \times molar\ mass\ O = 4(16.00\ g/mol) = 64.00\ g\ O$$

$$mass\ \%\ O = \frac{4 \times molar\ mass\ O}{molar\ mass\ K_2CrO_4} \times 100\%$$

$$= \frac{64.00\ g/mol}{194.20\ g/mol} \times 100\%$$

$$= 32.96\%$$

        Check: The units of the answer (%) are correct. The magnitude is reasonable because each is between 0 and 100% and the total is 100%.

        (b)   To write the formula for an ionic compound, do the following: (1) Write the symbol for the metal cation and its charge and the symbol for the nonmetal anion or polyatomic anion and its charge. (2) Adjust the subscript on each cation and anion to balance the overall charge. (3) Check that the sum of the charges of the cations equals the sum of the charges of the anions.

Lead(II)phosphate: Pb^{2+} PO_4^{3-}; $Pb_3(PO_4)_2$ cation $3(2+) = 6+$; anion $2(3-) = 6-$

Given: $Pb_3(PO_4)_2$ **Find:** mass percent of each element

Conceptual Plan: % Pb, then % P, then % O

$$\text{mass \% Pb} = \frac{3 \times \text{molar mass Pb}}{\text{molar mass Pb}_3(\text{PO}_4)_2} \times 100\% \quad \text{mass \% P} = \frac{2 \times \text{molar mass P}}{\text{molar mass Pb}_3(\text{PO}_4)_2} \times 100\% \quad \text{mass \% O} = \frac{8 \times \text{molar mass O}}{\text{molar mass Pb}_3(\text{PO}_4)_2} \times 100\%$$

Solution: molar mass $Pb_3(PO_4)_2 = 3(207.2 \text{ g/mol}) + 2(30.97 \text{ g/mol}) + 8(16.00 \text{ g/mol}) = 811.5 \text{ g/mol}$

$3 \times$ molar mass $Pb = 3(207.2 \text{ g/mol}) = 621.6 \text{ g Pb}$ $2 \times$ molar mass $P = 2(30.97 \text{ g/mol}) = 61.94 \text{ g P}$

$$\text{mass \% Pb} = \frac{3 \times \text{molar mass Pb}}{\text{molar mass Pb}_3(\text{PO}_4)_2} \times 100\% \qquad \text{mass \% P} = \frac{2 \times \text{molar mass P}}{\text{molar mass Pb}_3(\text{PO}_4)_2} \times 100\%$$

$$= \frac{621.6 \text{ g/mol}}{811.5 \text{ g/mol}} \times 100\% \qquad\qquad = \frac{61.94 \text{ g/mol}}{811.5 \text{ g/mol}} \times 100\%$$

$$= 76.60\% \qquad\qquad\qquad\qquad = 7.633\%$$

$4 \times$ molar mass $O = 8(16.00 \text{ g/mol}) = 128.0 \text{ g O}$

$$\text{mass \% O} = \frac{8 \times \text{molar mass O}}{\text{molar mass Pb}_3(\text{PO}_4)_2} \times 100\%$$

$$= \frac{128.0 \text{ g/mol}}{811.5 \text{ g/mol}} \times 100\%$$

$$= 15.77\%$$

Check: The units of the answer (%) are correct. The magnitude is reasonable because each is between 0 and 100% and the total is 100%.

(c)   sulfurous acid: H_2SO_3

Given: H_2SO_3 **Find:** mass percent of each element

Conceptual Plan: % H, then % S, then % O

$$\text{mass \% H} = \frac{2 \times \text{molar mass H}}{\text{molar mass H}_2\text{SO}_3} \times 100\% \quad \text{mass \% S} = \frac{1 \times \text{molar mass S}}{\text{molar mass H}_2\text{SO}_3} \times 100\% \quad \text{mass \% O} = \frac{3 \times \text{molar mass O}}{\text{molar mass H}_2\text{SO}_3} \times 100\%$$

Solution: molar mass $H_2SO_3 = 2(1.008 \text{ g/mol}) + 1(32.07 \text{ g/mol}) + 3(16.00 \text{ g/mol}) = 82.086 \text{ g/mol}$

$2 \times$ molar mass $H = 2(1.008 \text{ g/mol}) = 2.016 \text{ g H}$ $1 \times$ molar mass $S = 1(32.07 \text{ g/mol}) = 32.07 \text{ g S}$

$$\text{mass \% H} = \frac{2 \times \text{molar mass H}}{\text{molar mass H}_2\text{SO}_3} \times 100\% \qquad \text{mass \% S} = \frac{1 \times \text{molar mass S}}{\text{molar mass H}_2\text{SO}_3} \times 100\%$$

$$= \frac{2.016 \text{ g/mol}}{82.086 \text{ g/mol}} \times 100\% \qquad\qquad = \frac{32.07 \text{ g/mol}}{82.086 \text{ g/mol}} \times 100\%$$

$$= 2.456\% \qquad\qquad\qquad\qquad = 39.07\%$$

$3 \times$ molar mass $O = 3(16.00 \text{ g/mol}) = 48.00 \text{ g O}$

$$\text{mass \% O} = \frac{3 \times \text{molar mass O}}{\text{molar mass H}_2\text{SO}_3} \times 100\%$$

$$= \frac{48.00 \text{ g/mol}}{82.086 \text{ g/mol}} \times 100\%$$

$$= 58.48\%$$

Check: The units of the answer (%) are correct. The magnitude is reasonable because each is between 0 and 100% and the total is 100%.

(d)   To write the formula for an ionic compound, do the following: (1) Write the symbol for the metal cation and its charge and the symbol for the nonmetal anion or polyatomic anion and its charge. (2) Adjust the subscript on each cation and anion to balance the overall charge. (3) Check that the sum of the charges of the cations equals the sum of the charges of the anions.

cobalt(II)bromide: Co^{2+} Br^-; $CoBr_2$ cation $2+ = 2+$; anion $2(1-) = 2-$

Given: $CoBr_2$ **Find:** mass percent of each element

Conceptual Plan: % Co, then % Br

$$\text{mass \% Co} = \frac{1 \times \text{molar mass Co}}{\text{molar mass CoBr}_2} \times 100\% \qquad \text{mass \% Br} = \frac{2 \times \text{molar mass Br}}{\text{molar mass CoBr}_2} \times 100\%$$

Solution: molar mass $CoBr_2$ = (58.93 g/mol) + 2(79.90 g/mol) = 218.73 g/mol

2 × molar mass Co = 1(58.93 g/mol) = 58.93 g Co   1 × molar mass Br = 2(79.90 g/mol) = 159.80 g Br

$$\text{mass \% Co} = \frac{1 \times \text{molar mass Co}}{\text{molar mass CoBr}_2} \times 100\% \qquad \text{mass \% Br} = \frac{2 \times \text{molar mass Br}}{\text{molar mass CoBr}_2} \times 100\%$$

$$= \frac{58.93 \text{ g/mol}}{218.73 \text{ g/mol}} \times 100\% \qquad\qquad = \frac{159.80 \text{ g/mol}}{218.73 \text{ g/mol}} \times 100\%$$

$$= 26.94\% \qquad\qquad\qquad\qquad = 73.058\%$$

Check: The units of the answer (%) are correct. The magnitude is reasonable because each is between 0 and 100% and the total is 100%.

3.123   **Given:** 25 g CF_2Cl_2/mo   **Find:** g Cl_2/yr
Conceptual Plan: g CF_2Cl_2/mo → g Cl_2/mo → g Cl_2/yr

$$\frac{70.90 \text{ g Cl}_2}{120.91 \text{ g CF}_2\text{Cl}_2} \qquad \frac{12 \text{ mo}}{1 \text{ yr}}$$

Solution: $\dfrac{25 \text{ g CF}_2\text{Cl}_2}{\text{mo}} \times \dfrac{70.90 \text{ g Cl}_2}{120.91 \text{ g CF}_2\text{Cl}_2} \times \dfrac{12 \text{ mo}}{1 \text{ yr}} = 1.8 \times 10^2 \text{ g Cl}_2/\text{yr}$

Check: The units of the answer (g Cl_2) are correct. Magnitude is reasonable because it is less than the total CF_2Cl_2/yr.

3.125   **Given:** MCl_3; 65.57% Cl   **Find:** identify M
Conceptual Plan: g Cl → mol Cl → mol M → atomic mass M

$$\frac{1 \text{ mol Cl}}{35.45 \text{ g Cl}} \qquad \frac{1 \text{ mol M}}{3 \text{ mol Cl}} \qquad \frac{\text{g M}}{\text{mol M}}$$

Solution: in 100 g sample: 65.57 g Cl; 34.43 g M

$$65.57 \text{ g Cl} \times \frac{1 \text{ mol Cl}}{35.45 \text{ g Cl}} \times \frac{1 \text{ mol M}}{3 \text{ mol Cl}} = 0.6165 \text{ mol M} \qquad \frac{34.43 \text{ g M}}{0.6165 \text{ mol M}} = 55.85 \text{ g/mol M}$$

molar mass of 55.85 = Fe
The identity of M = Fe.

3.127   **Given:** in a 100 g sample of estradiol: 79.37 g C; 8.88 g H; 11.75 g O; molar mass = 272.37 g/mol   **Find:** molecular formula
Conceptual Plan:
Convert mass to mol of each element → pseudoformula → empirical formula → molecular formula

$$\frac{1 \text{ mol C}}{12.01 \text{ g C}} \qquad \frac{1 \text{ mol H}}{1.008 \text{ g H}} \qquad \frac{1 \text{ mol O}}{16.00 \text{ g O}} \qquad \text{divide by smallest number} \qquad \text{empirical formula} \times n$$

Solution: $79.37 \text{ g C} \times \dfrac{1 \text{ mol C}}{12.01 \text{ g C}} = 6.609 \text{ mol C}$

$8.88 \text{ g H} \times \dfrac{1 \text{ mol H}}{1.008 \text{ g H}} = 8.81 \text{ mol H}$

$11.75 \text{ g O} \times \dfrac{1 \text{ mol O}}{16.00 \text{ g O}} = 0.7344 \text{ mol O}$

$C_{6.609} H_{8.81} O_{0.7344}$

$$C_{\frac{6.609}{0.7344}} H_{\frac{8.81}{0.7344}} O_{\frac{0.7344}{0.7344}} \rightarrow C_9H_{12}O$$

The correct empirical formula is $C_9H_{12}O$.

empirical formula mass = 9(12.01 g/mol) + 12(1.008 g/mol) + 1(16.00 g/mol) = 136.19 g/mol

$$n = \frac{\text{molar mass}}{\text{formula molar mass}} = \frac{272.37 \text{ g/mol}}{136.19 \text{ g/mol}} = 2$$

molecular formula   = $C_9H_{12}O \times 2 = C_{18}H_{24}O_2$

3.129   **Given:** 13.42 g sample of equilin; 39.61 g CO_2; 9.01 g H_2O; molar mass $= 268.34$ g/mol
Find: molecular formula
Conceptual Plan:
Mass CO_2, H_2O → mol CO_2, H_2O → mol C, mol H → mass C, mass H, mass O → mol O →

$$\frac{1 \text{ mol } CO_2}{44.01 \text{ g } CO_2} \quad \frac{1 \text{ mol } H_2O}{18.02 \text{ g } H_2O} \quad \frac{1 \text{ mol } C}{1 \text{ mol } CO_2} \quad \frac{2 \text{ mol } H}{1 \text{ mol } H_2O} \quad \frac{12.01 \text{ g } C}{1 \text{ mol } C} \quad \frac{1.008 \text{ g } H}{1 \text{ mol } H} \quad \text{g sample} - \text{g C} - \text{g H} \quad \frac{1 \text{ mol } O}{16.00 \text{ g } O}$$

pseudoformula → empirical formula → molecular formula

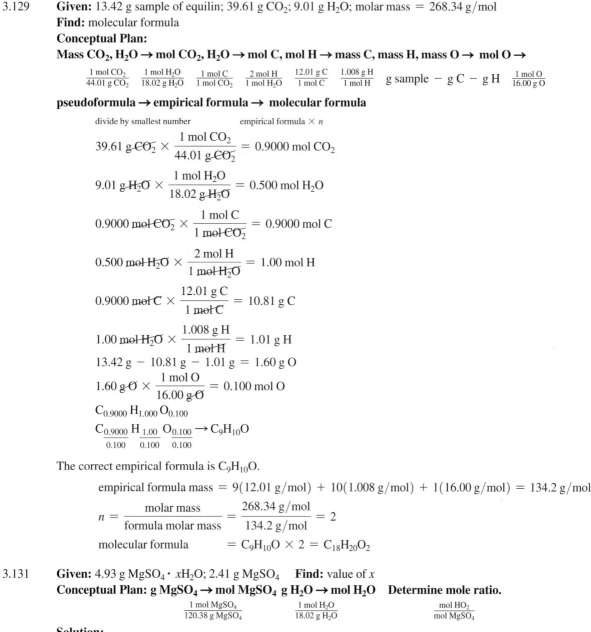

divide by smallest number   empirical formula $\times$ n

$$39.61 \text{ g } CO_2 \times \frac{1 \text{ mol } CO_2}{44.01 \text{ g } CO_2} = 0.9000 \text{ mol } CO_2$$

$$9.01 \text{ g } H_2O \times \frac{1 \text{ mol } H_2O}{18.02 \text{ g } H_2O} = 0.500 \text{ mol } H_2O$$

$$0.9000 \text{ mol } CO_2 \times \frac{1 \text{ mol } C}{1 \text{ mol } CO_2} = 0.9000 \text{ mol } C$$

$$0.500 \text{ mol } H_2O \times \frac{2 \text{ mol } H}{1 \text{ mol } H_2O} = 1.00 \text{ mol } H$$

$$0.9000 \text{ mol } C \times \frac{12.01 \text{ g } C}{1 \text{ mol } C} = 10.81 \text{ g } C$$

$$1.00 \text{ mol } H_2O \times \frac{1.008 \text{ g } H}{1 \text{ mol } H} = 1.01 \text{ g } H$$
$$13.42 \text{ g} - 10.81 \text{ g} - 1.01 \text{ g} = 1.60 \text{ g } O$$

$$1.60 \text{ g } O \times \frac{1 \text{ mol } O}{16.00 \text{ g } O} = 0.100 \text{ mol } O$$

$C_{0.9000} H_{1.000} O_{0.100}$

$$C_{\frac{0.9000}{0.100}} H_{\frac{1.00}{0.100}} O_{\frac{0.100}{0.100}} \rightarrow C_9H_{10}O$$

The correct empirical formula is $C_9H_{10}O$.

$$\text{empirical formula mass} = 9(12.01 \text{ g/mol}) + 10(1.008 \text{ g/mol}) + 1(16.00 \text{ g/mol}) = 134.2 \text{ g/mol}$$

$$n = \frac{\text{molar mass}}{\text{formula molar mass}} = \frac{268.34 \text{ g/mol}}{134.2 \text{ g/mol}} = 2$$

$$\text{molecular formula} = C_9H_{10}O \times 2 = C_{18}H_{20}O_2$$

3.131   **Given:** 4.93 g $MgSO_4 \cdot xH_2O$; 2.41 g $MgSO_4$ **Find:** value of x
Conceptual Plan: g $MgSO_4$ → mol $MgSO_4$ g H_2O → mol H_2O Determine mole ratio.

$$\frac{1 \text{ mol } MgSO_4}{120.38 \text{ g } MgSO_4} \qquad \frac{1 \text{ mol } H_2O}{18.02 \text{ g } H_2O} \qquad \frac{\text{mol } HO_2}{\text{mol } MgSO_4}$$

Solution:

$$2.41 \text{ g } MgSO_4 \times \frac{1 \text{ mol } MgSO_4}{120.38 \text{ g } MgSO_4} = 0.0200 \text{ mol } MgSO_4$$

Determine g H_2O: 4.93 g $MgSO_4 \cdot xH_2O$ − 2.41 g $MgSO_4$ = 2.52 g H_2O

$$2.52 \text{ g } H_2O \times \frac{1 \text{ mol } H_2O}{18.02 \text{ g } H_2O} = 0.140 \text{ mol } H_2O$$

$$\frac{0.140 \text{ mol } H_2O}{0.0200 \text{ mol } MgSO_4} = 7$$

$x = 7$

3.133   **Given:** molar mass $= 177$ g/mol; g C $= 8$(g H)   **Find:** molecular formula
Conceptual Plan: C_xH_yBrO
Solution: in 1 mol compound, let $x =$ mol C and $y =$ mol H, assume mol Br $= 1$, assume mol O $= 1$
        $177 \text{ g/mol} = x(12.01 \text{ g/mol}) + y(1.008 \text{ g/mol}) + 1(79.90 \text{ g/mol}) + 1(16.00 \text{ g/mol})$
        $x(12.01 \text{ g/mol}) = 8[y(1.008 \text{ g/mol})]$

$$177 \text{ g/mol} = 8y(1.008 \text{ g/mol}) + y(1.008 \text{ g/mol}) + 79.90 \text{ g/mol} + 16.00 \text{ g/mol}$$
$$81 = 9y(1.008)$$
$$y = 9 = \text{mol H}$$
$$x(12.01) = 8 \times 9(1.008)$$
$$x = 6 = \text{mol C}$$
$$\text{molecular formula} = \text{C}_6\text{H}_9\text{BrO}$$

Check: molar mass $= 6(12.01 \text{ g/mol}) + 9(1.008 \text{ g/mol}) + 1(79.90 \text{ g/mol}) + 1(16.00 \text{ g/mol}) = 177.0 \text{ g/mol}$

3.135 **Given:** 23.5 mg $\text{C}_{17}\text{H}_{22}\text{ClNO}_4$ **Find:** total number of atoms

Conceptual Plan: mg compound $\rightarrow$ g compound $\rightarrow$ mol compound $\rightarrow$ mol atoms $\rightarrow$ number of atoms

$$\frac{1 \text{ g}}{1000 \text{ mg}} \qquad \frac{1 \text{ mol}}{339.8 \text{ g}} \qquad \frac{45 \text{ mol atoms}}{1 \text{ mol compound}} \qquad \frac{6.022 \times 10^{23} \text{ atoms}}{1 \text{ mol atoms}}$$

Solution: $23.5 \text{ mg} \times \dfrac{1 \text{ g}}{1000 \text{ mg}} \times \dfrac{1 \text{ mol cpd}}{339.8 \text{ g}} \times \dfrac{45 \text{ mol atoms}}{1 \text{ mol cpd}} \times \dfrac{6.022 \times 10^{23} \text{ atoms}}{\text{mol}} = 1.87 \times 10^{21} \text{ atoms}$

Check: The units of the answer (number of atoms) are correct. The magnitude of the answer is reasonable because the molecule is so complex.

3.137 **Given:** MCl_3; 2.395 g sample; 3.606×10^{-2} mol Cl **Find:** atomic mass M

Conceptual Plan: mol Cl $\rightarrow$ g Cl $\rightarrow$ g X

$$\frac{35.45 \text{ g Cl}}{1 \text{ mol Cl}} \quad \text{g sample} - \text{g Cl} = \text{g M}$$

mol Cl $\rightarrow$ mol M $\rightarrow$ atomic mass M

$$\frac{1 \text{ mol M}}{3 \text{ mol Cl}} \qquad \frac{\text{g M}}{\text{mol M}}$$

Solution:

$$3.606 \times 10^{-2} \text{ mol Cl} \times \frac{35.45 \text{ g}}{1 \text{ mol Cl}} = 1.278 \text{ g Cl}$$

$$2.395 \text{ g} - 1.278 \text{ g} = 1.117 \text{ g M}$$

$$3.606 \times 10^{-2} \text{ mol Cl} \times \frac{1 \text{ mol M}}{3 \text{ mol Cl}} = 1.202 \times 10^{-2} \text{ mol M}$$

$$\frac{1.117 \text{ g M}}{0.01202 \text{ mol M}} = 92.93 \text{ g/mol M}$$

$$\text{molar mass of M} = 92.93 \text{ g/mol}$$

3.139 **Given:** $\text{Fe}_x\text{Cr}_y\text{O}_4$; 28.59% O **Find:** x and y

Conceptual Plan: % O $\rightarrow$ molar mass $\text{Fe}_x\text{Cr}_y\text{O}_4$ $\rightarrow$ mass Fe $+$ Cr

$$\frac{\text{mass O}}{\text{molar mass compound}} \times 100\% = \% \text{ O} \quad \text{mass cpd} - \text{mass O} = \text{mass Fe} + \text{Cr}$$

Solution: $\dfrac{28.59 \text{ g O}}{100.0} = \dfrac{64.00 \text{ g O}}{\text{molar mass cpd}}$ molar mass $= 223.9 \text{ g/mol}$

Mass Fe + Cr = molar mass $- (4 \times \text{molar mass O}) = 223.9 - 64.00 = 159.9 \text{ g}$

Molar mass Fe = 55.85; molar mass Cr = 52.00

Because the mass of the two metals is close, the average mass can be used to determine the total moles of Fe and

Cr present in the compound. Average mass of Fe and Cr = 53.9. $\dfrac{159.9 \text{ g}}{53.9 \text{ g/mol}} = 2.97 = 3 \text{ mol metal}$

Let $x = \text{mol Fe}$ and $y = \text{mol Cr}$

$x \text{ mol Fe} + y \text{ mol Cr} = 3 \text{ mol total}$

$x \text{ mol Fe}(55.85 \text{ g Fe/mol}) + y \text{ mol Cr}(52.00 \text{ gCr/mol}) = 159.8$

$y \text{ mol Cr} = 3 - x \text{ mol Fe}$

$x(55.85) + (3 - x)(52.00) = 159.8$

So $x = 1$ and $y = 2$.

Check: Formula $= \text{FeCr}_2\text{O}_4$ would have a molar mass of Fe + 2Cr + 4O $= 55.85 + 2(52.00) + 4(16.00) = 223.85$, and the molar mass of the compound is 223.8.

3.141   **Given:** 0.0552% $NaNO_2$; 8.00 oz bag   **Find:** mass Na in bag
Conceptual Plan: oz bag → g bag → g $NaNO_2$ → g Na

$$\frac{453.6 \text{ g}}{16.00 \text{ oz}} \qquad \frac{0.0552 \text{ g } NaNO_2}{100.0 \text{ g bag}} \qquad \frac{22.99 \text{ g Na}}{69.00 \text{ g } NaNO_2}$$

Solution: $8.00 \text{ oz bag} \times \dfrac{453.6 \text{ g bag}}{16.00 \text{ oz bag}} \times \dfrac{0.0552 \text{ g } NaNO_2}{100.0 \text{ g bag}} \times \dfrac{22.99 \text{ g Na}}{69.00 \text{ g } NaNO_2} \times \dfrac{1000 \text{ mg Na}}{\text{g Na}} = 41.7 \text{ mg Na}$

Check: The units of the answer (mg Na) are correct. The magnitude of the answer is reasonable because only a small percentage of the total mass is Na.

Challenge Problems

3.143   **Given:** g NaCl + g NaBr = 2.00 g; g Na = 0.75 g   **Find:** g NaBr
Conceptual Plan:
   let x = mol NaCl, y = mol NaBr, then x(molar mass NaCl) = g NaCl, y(molar mass NaBr) = g NaBr
Solution: $x(58.4) + y(102.9) = 2.00$
   $x(23.0) + y(23.0) = 0.75 \qquad y = 0.0326 - x$
   $58.4x + 102.9(0.0326 - x) = 2.00$
   $58.4x + 3.354 - 102.9x = 2.00$
   $44.5x = 1.354$
   $\qquad x = 0.03043 \text{ mol NaCl}$
   $\qquad y = 0.0326 - 0.03043 = 0.00217 \text{ mol NaBr}$
   g NaBr = $(0.00217)(102.9 \text{g/mol}) = 0.223$ g NaBr

Check: The units of the answer (g NaBr) are correct. The magnitude is reasonable because it is less than the total mass.

3.145   **Given:** sample of $CaCO_3$ and $(NH_4)_2CO_3$ is 61.9% CO_3^{2-} **Find:** % $CaCO_3$
Conceptual Plan: Let x = $CaCO_3$, y = $(NH_4)_2CO_3$, then x(molar mass $CaCO_3$) = g $CaCO_3$,
   y[molar mass $(NH_4)_2CO_3$] = g $(NH_4)_2CO_3$
   then a 100.0 g sample contains $x(100.0)$ g $CaCO_3$; $y(96.1)$ g $(NH_4)_2CO_3$; 61.9 g CO_3^{2-}
Solution: $x(100.0) + y(96.1) = 100.0$
   $x(60.0) + y(60.0) = 61.9 \qquad y = 1.03167 - x$
   $100.0x + 96.1(1.03 - x) = 100$
   $100.0x + 99.0 - 96.1x = 100$
   $3.9x \quad = 1.0$
   $\qquad x = 0.26 \text{ mol } CaCO_3$
   $\qquad y = 1.032 - 0.26 = 0.77 \text{ mol } (NH_4)_2CO_3$
   g $CaCO_3$ = $(0.26 \text{ mol})(100.0 \text{g/mol}) = 26$ g $CaCO_3$ in a 100 g sample mass % $CaCO_3 = 26\%$

Check: The units of the answer (mass % $CaCO_3$) are correct. The magnitude is reasonable because it is between 0 and 100%.

3.147   **Given:** 1.1 kg CF_2Cl_2/automobile; 25% leaked/year; 100×10^6 automobiles   **Find:** kg Cl/yr
Conceptual Plan: kg CF_2Cl_2/auto → kg CF_2Cl_2 leaked/yr → kg Cl/yr/auto → kg Cl

$$\frac{25 \text{ kg } CF_2Cl_2}{100 \text{ kg } CF_2Cl_2} \qquad \frac{70.9 \text{ kg Cl}}{120.91 \text{ kg } CF_2Cl_2} \qquad 100 \times 10^6 \text{ auto}$$

Solution: $\dfrac{1.1 \text{ kg } CF_2Cl_2}{\text{auto}} \times \dfrac{25 \text{ kg } CF_2Cl_2}{100 \text{ kg } CF_2Cl_2} \times \dfrac{70.9 \text{ kg Cl}}{120.91 \text{ kg } CF_2Cl_2} \times 100 \times 10^6 \text{ auto} = 1.6 \times 10^7 \text{ kg Cl/yr}$

Check: The units of the answer (kg Cl) are correct. The magnitude is reasonable because it is less than the kilogram CF_2Cl_2 leaked per year.

3.149   **Given:** rock contains: 38.0% PbS; 25.0% $PbCO_3$; 17.4% $PbSO_4$ **Find:** kg rock needed for 5.0 metric ton Pb
Conceptual Plan: Determine kg Pb/100 kg rock then ton Pb → kg Pb → kg rock

$$\frac{1000 \text{ kg}}{\text{metric ton}} \qquad \frac{100 \text{ kg rock}}{64.2 \text{ kg rock}}$$

Solution: in 100 kg rock:

$$\left(38.0 \text{ kg PbS} \times \frac{207.2 \text{ kg Pb}}{239.3 \text{ kg PbS}}\right) + \left(25.0 \text{ kg PbCO}_3 \times \frac{207.2 \text{ kg Pb}}{267.2 \text{ kg PbCO}_3}\right) + \left(17.4 \text{ kg PbSO}_4 \times \frac{207.2 \text{ kg Pb}}{303.3 \text{ kg PbSO}_4}\right)$$

$$= 64.2 \text{ kg Pb}$$

$$5.0 \text{ metric ton Pb} \times \frac{1000 \text{ kg Pb}}{\text{metric ton Pb}} \times \frac{100 \text{ kg rock}}{64.2 \text{ kg Pb}} = 7.8 \times 10^3 \text{ kg rock}$$

Check: The units of the answer (kg rock) are correct. Magnitude is reasonable because it is greater than the amount of Pb needed.

3.151 **Given:** molar mass $= 229$ g/mol, 6 times mass C as H   **Find:** molecular formula

Conceptual Plan: Let $x =$ **mass of C, then** $6x =$ **mass of C**

Solution: in 1 mol of the compound: g C + g H + g S + g I = 229 g

Because the molar mass of I $= 127$, there cannot be more than 1 mol of I in the compound; so

$x + 6x + $ g S $+ 127 = 229.$

$x + 6x + $ g S $= 102$

If the compound contains 1 mol S, then $7x = 102 - 32 = 70$ and $x = 10$ g H and $6x = 60$ g C.

$$10 \text{ g H} \times \frac{1 \text{ mol H}}{1.0 \text{ g H}} = 10 \text{ mol H}$$

$$60 \text{ g C} \times \frac{1 \text{ mol C}}{12 \text{ g C}} = 5 \text{ mol C}$$

1 mol I and 1 mol S; so empirical formula is $C_5H_{10}SI$

Check: Molar mass of $C_5H_{10}SI = 5(12) + 10(1.0) + 32 + 127 = 229$ g/mol, which is the mass given.

3.153 **Given:** compound is 1/3 X by mass; atomic mass X is 1/3 atomic mass Y   **Find:** empirical formula

Conceptual Plan: **Mass X and Y → mass ratio X: Y and then g X → mol X and g Y → mol Y and then**

$$\frac{\text{g X}}{\text{atomic mass X}} \qquad \frac{\text{g Y}}{\text{atomic mass Y}}$$

mole ratio

Solution: $\dfrac{\text{mass X}}{\text{mass Y}} = \dfrac{\frac{1}{3}}{\frac{2}{3}} = \dfrac{1}{2}.$ $\text{mol X} = \dfrac{1 \text{ g}}{\text{atomic mass X}}$ and $\text{mol Y} = \dfrac{2 \text{ g}}{\text{atomic mass Y}}$

But atomic mass X $= 1/3$ (atomic mass Y)

$$\text{mol X} = \frac{1 \text{ g}}{1/3 (\text{atomic mass Y})} \text{ and mol Y} = \frac{2 \text{ g}}{\text{atomic mass Y}}$$

$$\frac{\text{mol X}}{\text{mol Y}} = \frac{\dfrac{1 \text{ g}}{1/3 (\text{atomic mass Y})}}{\dfrac{2 \text{ g}}{\text{atomic mass Y}}} = 1/6 \qquad \text{empirical formula} = \text{XY}_6$$

Conceptual Problems

3.155 The sphere in the molecular models represents the electron cloud of the atom. On this scale, the nucleus would be too small to see.

3.157 The statement is incorrect because a chemical formula is based on the ratio of atoms combined, not the ratio of grams combined. The statement should read as follows: The chemical formula for ammonia (NH_3) indicates that ammonia contains three hydrogen atoms to each nitrogen atom.

3.159 H_2SO_4: Atomic mass S is approximately twice atomic mass O; both are much greater than atomic mass H. The order of % mass is % O $>$% S $>$ % H.

4 Chemical Quantities and Aqueous Reactions

Review Questions

4.1 Reaction stoichiometry is the numerical relationships between chemical amounts in a balanced chemical equation. The coefficients in a chemical reaction specify the relative amounts in moles of each of the substances involved in the reaction.

4.3 No, the percent yield would not be different if the actual yield and theoretical yield were calculated in moles. The relationship between grams and moles is the molar mass. This would be the same value for the actual yield and the theoretical yield.

4.5 Molarity is a concentration term. It is the amount of solute (in moles) divided by the volume of solution (in liters). The molarity of a solution can be used as a conversion factor between moles of the solute and liters of the solution.

4.7 Acids are molecular compounds that ionize—form ions—when they dissolve in water. A strong acid completely ionizes in solution. A weak acid does not completely ionize in water. A solution of a weak acid is composed mostly of the nonionized acid.

4.9 The solubility rules are a set of empirical rules that have been inferred from observations on many ionic compounds. The solubility rules allow us to predict whether a compound is soluble or insoluble.

4.11 A precipitation reaction is one in which a solid or precipitate forms upon mixing two solutions. An example is $2 \, KI(aq) + Pb(NO_3)_2(aq) \rightarrow PbI_2(s) + 2 \, KNO_3(aq)$.

4.13 A molecular equation shows the complete neutral formulas for each compound in the reaction as if they existed as molecules. Equations that list individually all of the ions present as either reactants or products in a chemical reaction are complete ionic equations. Equations that show only the species that actually change during the reaction are net ionic equations.

4.15 When an acid and a base are mixed, the $H^+(aq)$ from the acid combines with the $OH^-(aq)$ from the base to form $H_2O(l)$. An example is $HCl(aq) + NaOH(aq) \rightarrow H_2O(l) + NaCl(aq)$.

4.17 Aqueous reactions that form a gas upon mixing two solutions are called gas-evolution reactions. An example is $H_2SO_4(aq) + Li_2S(aq) \rightarrow H_2S(g) + Li_2SO_4(aq)$.

4.19 Oxidation–reduction reactions, or redox reactions, are reactions in which electrons are transferred from one reactant to the other. An example is $4 \, Fe(s) + 3 \, O_2(g) \rightarrow 2 \, Fe_2O_3(s)$.

4.21 To identify redox reactions using oxidation states, begin by assigning oxidation states to each atom in the reaction. A change in oxidation state for the atoms indicates a redox reaction.

4.23 A substance that causes the oxidation of another substance is called an oxidizing agent. A substance that causes the reduction of another substance is called a reducing agent.

Problems by Topic

Reaction Stoichiometry

4.25   **Given:** 7.2 moles C_6H_{14} **Find:** balanced reaction, moles O_2 required
Conceptual Plan: Balance the equation then mol C_6H_{14} → mol O_2

$$2\,C_6H_{14}(g) + 19\,O_2(g) \rightarrow 12\,CO_2(g) + 14\,H_2O(g) \qquad \frac{19\ \text{mol } O_2}{2\ \text{mol } C_6H_{14}}$$

Solution: $7.2\ \cancel{\text{mol } C_6H_{14}} \times \dfrac{19\ \text{mol } O_2}{2\ \cancel{\text{mol } C_6H_{14}}} = 68.\underline{4}\ \text{mol } O_2 = 68\ \text{mol } O_2$

Check: The units of the answer $(\text{mol } O_2)$ are correct. The magnitude is reasonable because much more O_2 is needed than C_6H_{14}.

4.27   (a)   **Given:** 2.5 mol N_2O_5 **Find:** mol NO_2
Conceptual Plan: mol N_2O_5 → mol NO_2

$$\frac{4\ \text{mol } NO_2}{2\ \text{mol } N_2O_5}$$

Solution: $2.5\ \cancel{\text{mol } N_2O_5} \times \dfrac{4\ \text{mol } NO_2}{2\ \cancel{\text{mol } N_2O_5}} = 5.0\ \text{mol } NO_2$

Check: The units of the answer $(\text{mol } NO_2)$ are correct. The magnitude is reasonable because it is greater than mol N_2O_5.

(b)   **Given:** 6.8 mol N_2O_5 **Find:** mol NO_2
Conceptual Plan: mol N_2O_5 → mol NO_2

$$\frac{4\ \text{mol } NO_2}{2\ \text{mol } N_2O_5}$$

Solution: $6.8\ \cancel{\text{mol } N_2O_5} \times \dfrac{4\ \text{mol } NO_2}{2\ \cancel{\text{mol } N_2O_5}} = 13.\underline{6}\ \text{mol } NO_2 = 14\ \text{mol } NO_2$

Check: The units of the answer $(\text{mol } NO_2)$ are correct. The magnitude is reasonable because it is greater than mol N_2O_5.

(c)   **Given:** 15.2 g N_2O_5 **Find:** mol NO_2
Conceptual Plan: g N_2O_5 → mol N_2O_5 → mol NO_2

$$\frac{1\ \text{mol } N_2O_5}{108.02\ \text{g } N_2O_5} \qquad \frac{4\ \text{mol } NO_2}{2\ \text{mol } N_2O_5}$$

Solution: $15.2\ \cancel{\text{g } N_2O_5} \times \dfrac{1\ \cancel{\text{mol } N_2O_5}}{108.02\ \cancel{\text{g } N_2O_5}} \times \dfrac{4\ \text{mol } NO_2}{2\ \cancel{\text{mol } N_2O_5}} = 0.281\underline{4}\ \text{mol } NO_2 = 0.281\ \text{mol } NO_2$

Check: The units of the answer $(\text{mol } NO_2)$ are correct. The magnitude is reasonable because 10 g is about 0.09 mol N_2O_5 and the answer is greater than mol N_2O_5.

(d)   **Given:** 2.87 kg N_2O_5 **Find:** mol NO_2
Conceptual Plan: kg N_2O_5 → g N_2O_5 → mol N_2O_5 → mol NO_2

$$\frac{1000\ \text{g } N_2O_5}{\text{kg } N_2O_5} \qquad \frac{1\ \text{mol } N_2O_5}{108.02\ \text{g } N_2O_5} \qquad \frac{4\ \text{mol } NO_2}{2\ \text{mol } N_2O_5}$$

Solution:

$$2.87\ \cancel{\text{kg } N_2O_5} \times \frac{1000\ \cancel{\text{g } N_2O_5}}{\cancel{\text{kg } N_2O_5}} \times \frac{1\ \cancel{\text{mol } N_2O_5}}{108.02\ \cancel{\text{g } N_2O_5}} \times \frac{4\ \text{mol } NO_2}{2\ \cancel{\text{mol } N_2O_5}} = 53.\underline{1}4\ \text{mol } NO_2 = 53.1\ \text{mol } NO_2$$

Check: The units of the answer $(\text{mol } NO_2)$ are correct. The magnitude is reasonable because 1.55 kg is about 14 mol N_2O_5 and the answer is greater than mol N_2O_5.

4.29   **Given:** 3 mol SiO_2 **Find:** mol C; mol SiC; mol CO
Conceptual Plan: mol SiO_2 → mol C → mol SiC → mol CO

$$\frac{3\ \text{mol C}}{1\ \text{mol } SiO_2} \qquad \frac{1\ \text{mol SiC}}{1\ \text{mol } SiO_2} \qquad \frac{2\ \text{mol CO}}{1\ \text{mol } SiO_2}$$

Solution: $3 \text{ mol SiO}_2 \times \dfrac{3 \text{ mol C}}{1 \text{ mol SiO}_2} = 9 \text{ mol C}$ $\qquad$ $3 \text{ mol SiO}_2 \times \dfrac{1 \text{ mol SiC}}{1 \text{ mol SiO}_2} = 3 \text{ mol SiC}$

$3 \text{ mol SiO}_2 \times \dfrac{2 \text{ mol CO}}{1 \text{ mol SiO}_2} = 6 \text{ mol CO}$

Given: 6 mol C $\quad$ **Find:** mol SiO_2; mol SiC; mol CO

Conceptual Plan: mol C $\rightarrow$ mol SiO$_2$ $\rightarrow$ mol SiC $\rightarrow$ mol CO

$$\dfrac{1 \text{ mol SiO}_2}{3 \text{ mol C}} \qquad \dfrac{1 \text{ mol SiC}}{3 \text{ mol C}} \qquad \dfrac{2 \text{ mol CO}}{3 \text{ mol C}}$$

Solution: $6 \text{ mol C} \times \dfrac{1 \text{ mol SiO}_2}{3 \text{ mol C}} = 2 \text{ mol SiO}_2$ $\qquad$ $6 \text{ mol C} \times \dfrac{1 \text{ mol SiC}}{3 \text{ mol C}} = 2 \text{ mol SiC}$

$6 \text{ mol C} \times \dfrac{2 \text{ mol CO}}{3 \text{ mol C}} = 4 \text{ mol CO}$

Given: 10 mol CO $\quad$ **Find:** mol SiO_2; mol C; mol SiC

Conceptual Plan: mol CO $\rightarrow$ mol SiO$_2$ $\rightarrow$ mol C $\rightarrow$ mol SiC

$$\dfrac{1 \text{ mol SiO}_2}{2 \text{ mol CO}} \qquad \dfrac{3 \text{ mol C}}{2 \text{ mol CO}} \qquad \dfrac{1 \text{ mol SiC}}{2 \text{ mol CO}}$$

Solution: $10 \text{ mol CO} \times \dfrac{1 \text{ mol SiO}_2}{2 \text{ mol CO}} = 5.0 \text{ mol SiO}_2$ $\qquad$ $10 \text{ mol C} \times \dfrac{3 \text{ mol C}}{2 \text{ mol CO}} = 15 \text{ mol C}$

$10 \text{ mol CO} \times \dfrac{1 \text{ mol SiC}}{2 \text{ mol CO}} = 5.0 \text{ mol SiC}$

Given: 2.8 mol SiO_2 $\quad$ **Find:** mol C; mol SiC; mol CO

Conceptual Plan: mol SiO$_2$ $\rightarrow$ mol C $\rightarrow$ mol SiC $\rightarrow$ mol CO

$$\dfrac{3 \text{ mol C}}{1 \text{ mol SiO}_2} \qquad \dfrac{1 \text{ mol SiC}}{1 \text{ mol SiO}_2} \qquad \dfrac{2 \text{ mol CO}}{1 \text{ mol SiO}_2}$$

Solution: $2.8 \text{ mol SiO}_2 \times \dfrac{3 \text{ mol C}}{1 \text{ mol SiO}_2} = 8.4 \text{ mol C}$ $\qquad$ $2.8 \text{ mol SiO}_2 \times \dfrac{1 \text{ mol SiC}}{1 \text{ mol SiO}_2} = 2.8 \text{ mol SiC}$

$2.8 \text{ mol SiO}_2 \times \dfrac{2 \text{ mol CO}}{1 \text{ mol SiO}_2} = 5.6 \text{ mol CO}$

Given: 1.55 mol C $\quad$ **Find:** mol SiO_2; mol SiC; mol CO

Conceptual Plan: mol C $\rightarrow$ mol SiO$_2$ $\rightarrow$ mol SiC $\rightarrow$ mol CO

$$\dfrac{1 \text{ mol SiO}_2}{3 \text{ mol C}} \qquad \dfrac{1 \text{ mol SiC}}{3 \text{ mol C}} \qquad \dfrac{2 \text{ mol CO}}{3 \text{ mol C}}$$

Solution: $1.55 \text{ mol C} \times \dfrac{1 \text{ mol SiO}_2}{3 \text{ mol C}} = 0.517 \text{ mol SiO}_2$ $\qquad$ $1.55 \text{ mol C} \times \dfrac{1 \text{ mol SiC}}{3 \text{ mol C}} = 0.517 \text{ mol SiC}$

$1.55 \text{ mol C} \times \dfrac{2 \text{ mol CO}}{3 \text{ mol C}} = 1.03 \text{ mol CO}$

Mol SiO$_2$	Mol C	Mol SiC	Mol CO
3	9	3	6
2	**6**	2	4
5.0	15	5.0	**10**
2.8	8.4	2.8	5.6
0.517	**1.55**	0.517	1.03

4.31 $\quad$ **Given:** 3.2 g Fe $\quad$ **Find:** g HBr; gH_2

Conceptual Plan: g Fe $\rightarrow$ mol Fe $\rightarrow$ mol HBr $\rightarrow$ g HBr

$$\dfrac{1 \text{ mol Fe}}{55.8 \text{ g Fe}} \qquad \dfrac{2 \text{ mol HBr}}{1 \text{ mol Fe}} \qquad \dfrac{80.9 \text{ g HBr}}{1 \text{ mol HBr}}$$

g Fe $\rightarrow$ mol Fe $\rightarrow$ mol H$_2$ $\rightarrow$ g H$_2$

$$\dfrac{1 \text{ mol Fe}}{55.8 \text{ g Fe}} \qquad \dfrac{1 \text{ mol H}_2}{1 \text{ mol Fe}} \qquad \dfrac{2.02 \text{ g H}_2}{1 \text{ mol H}_2}$$

Solution: $3.2 \text{ g Fe} \times \dfrac{1 \text{ mol Fe}}{55.8 \text{ g Fe}} \times \dfrac{2 \text{ mol HBr}}{1 \text{ mol Fe}} \times \dfrac{80.9 \text{ g HBr}}{1 \text{ mol HBr}} = 9.2\underline{8} \text{ g HBr} = 9.3 \text{ g HBr}$

$3.2 \text{ g Fe} \times \dfrac{1 \text{ mol Fe}}{55.8 \text{ g Fe}} \times \dfrac{1 \text{ mol H}_2}{1 \text{ mol Fe}} \times \dfrac{2.02 \text{ g H}_2}{1 \text{ mol H}_2} = 0.1\underline{16} \text{ g H}_2 = 0.12 \text{ g H}_2$

Check: The units of the answers (g HBr, g H$_2$) are correct. The magnitude of the answers is reasonable because molar mass HBr is greater than Fe and molar mass H$_2$ is much less than Fe.

4.33   (a)   **Given:** 3.67 g Ba   **Find:** g BaCl$_2$
            Conceptual Plan: g Ba → mol Ba → mol BaCl$_2$ → g BaCl$_2$

$\dfrac{1 \text{ mol Ba}}{137.33 \text{ g Ba}} \quad \dfrac{1 \text{ mol BaCl}_2}{1 \text{ mol Ba}} \quad \dfrac{208.23 \text{ g BaCl}_2}{1 \text{ mol BaCl}_2}$

Solution: $3.67 \text{ g Ba} \times \dfrac{1 \text{ mol Ba}}{137.33 \text{ g Ba}} \times \dfrac{1 \text{ mol BaCl}_2}{1 \text{ mol Ba}} \times \dfrac{208.23 \text{ g BaCl}_2}{1 \text{ mol BaCl}_2} = 5.5\underline{6}47 \text{ g BaCl}_2 = 5.56 \text{ g BaCl}_2$

Check: The units of the answer (g BaCl$_2$) are correct. The magnitude of the answer is reasonable because it is larger than grams Ba.

          (b)   **Given:** 3.67 g CaO   **Find:** g CaCO$_3$
            Conceptual Plan: g CaO → mol CaO → mol CaCO$_3$ → g CaCO$_3$

$\dfrac{1 \text{ mol CaO}}{56.08 \text{ g CaO}} \quad \dfrac{1 \text{ mol CaCO}_3}{1 \text{ mol CaO}} \quad \dfrac{100.09 \text{ g CaCO}_3}{1 \text{ mol CaCO}_3}$

Solution:

$3.67 \text{ g CaO} \times \dfrac{1 \text{ mol CaO}}{56.08 \text{ g CaO}} \times \dfrac{1 \text{ mol CaCO}_3}{1 \text{ mol CaO}} \times \dfrac{100.09 \text{ g CaCO}_3}{1 \text{ mol CaCO}_3} = 6.5\underline{5}0 \text{ g CaCO}_3 = 6.55 \text{ g CaCO}_3$

Check: Units of answer (g CaCO$_3$) are correct. The magnitude of the answer is reasonable because it is larger than grams CaO.

          (c)   **Given:** 3.67 g Mg   **Find:** g MgO
            Conceptual Plan: g Mg → mol Mg → mol MgO → g MgO

$\dfrac{1 \text{ mol Mg}}{24.30 \text{ g Mg}} \quad \dfrac{1 \text{ mol MgO}}{1 \text{ mol Mg}} \quad \dfrac{40.30 \text{ g MgO}}{1 \text{ mol MgO}}$

Solution: $3.67 \text{ g Mg} \times \dfrac{1 \text{ mol Mg}}{24.30 \text{ g Mg}} \times \dfrac{1 \text{ mol MgO}}{1 \text{ mol Mg}} \times \dfrac{40.30 \text{ g MgO}}{1 \text{ mol MgO}} = 6.0\underline{8}6 \text{ g MgO} = 6.09 \text{ g MgO}$

Check: The units of the answer (g MgO) are correct. The magnitude of the answer is reasonable because it is larger than grams Mg.

(d)   **Given:** 3.67 g Al   **Find:** g Al$_2$O$_3$
    Conceptual Plan: g Al → mol Al → mol Al$_2$O$_3$ → g Al$_2$O$_3$

$\dfrac{1 \text{ mol Al}}{26.98 \text{ g Al}} \quad \dfrac{2 \text{ mol Al}_2\text{O}_3}{4 \text{ mol Al}} \quad \dfrac{101.96 \text{ g Al}_2\text{O}_3}{1 \text{ mol Al}_2\text{O}_3}$

Solution: $3.67 \text{ g Al} \times \dfrac{1 \text{ mol Al}}{26.98 \text{ g Al}} \times \dfrac{2 \text{ mol Al}_2\text{O}_3}{4 \text{ mol Al}} \times \dfrac{101.96 \text{ g Al}_2\text{O}_3}{1 \text{ mol Al}_2\text{O}_3} = 6.9\underline{3}4 \text{ g Al}_2\text{O}_3 = 6.93 \text{ g Al}_2\text{O}_3$

Check: The units of the answer (g Al$_2$O$_3$) are correct. The magnitude of the answer is reasonable because it is larger than grams Al.

4.35   (a)   **Given:** 4.85 g NaOH   **Find:** g HCl
            Conceptual Plan: g NaOH → mol NaOH → mol HCl → g HCl

$\dfrac{1 \text{ mol NaOH}}{40.01 \text{ g NaOH}} \quad \dfrac{1 \text{ mol HCl}}{1 \text{ mol NaOH}} \quad \dfrac{36.46 \text{ g HCl}}{1 \text{ mol HCl}}$

Solution: $4.85 \text{ g NaOH} \times \dfrac{1 \text{ mol NaOH}}{40.01 \text{ g NaOH}} \times \dfrac{1 \text{ mol HCl}}{1 \text{ mol NaOH}} \times \dfrac{36.46 \text{ g HCl}}{1 \text{ mol HCl}} = 4.42 \text{ g HCl}$

Check: The units of the answer (g HCl) are correct. The magnitude of the answer is reasonable because it is less than grams NaOH.

(b) **Given:** 4.85 g Ca(OH)$_2$ **Find:** g HNO$_3$

Conceptual Plan: g Ca(OH)$_2$ → mol Ca(OH)$_2$ → mol HNO$_3$ → g HNO$_3$

$$\frac{1 \text{ mol Ca(OH)}_2}{74.10 \text{ g Ca(OH)}_2} \quad \frac{2 \text{ mol HNO}_3}{1 \text{ mol Ca(OH)}_2} \quad \frac{63.02 \text{ g HNO}_3}{1 \text{ mol HNO}_3}$$

Solution: $4.85 \text{ g Ca(OH)}_2 \times \dfrac{1 \text{ mol Ca(OH)}_2}{74.10 \text{ g Ca(OH)}_2} \times \dfrac{2 \text{ mol HNO}_3}{1 \text{ mol Ca(OH)}_2} \times \dfrac{63.02 \text{ g HNO}_3}{1 \text{ mol HNO}_3} = 8.25 \text{ g HNO}_3$

Check: The units of the answer (g HNO$_3$) are correct. The magnitude of the answer is reasonable because it is more than g Ca(OH)$_2$.

(c) **Given:** 4.85 g KOH   **Find:** g H$_2$SO$_4$

Conceptual Plan: g KOH → mol KOH → mol H$_2$SO$_4$ → g H$_2$SO$_4$

$$\frac{1 \text{ mol KOH}}{56.11 \text{ g KOH}} \quad \frac{1 \text{ mol H}_2\text{SO}_4}{2 \text{ mol KOH}} \quad \frac{98.09 \text{ g H}_2\text{SO}_4}{1 \text{ mol H}_2\text{SO}_4}$$

Solution: $4.85 \text{ g KOH} \times \dfrac{1 \text{ mol KOH}}{56.11 \text{ g KOH}} \times \dfrac{1 \text{ mol H}_2\text{SO}_4}{2 \text{ mol KOH}} \times \dfrac{98.09 \text{ g H}_2\text{SO}_4}{1 \text{ mol H}_2\text{SO}_4} = 4.24 \text{ g H}_2\text{SO}_4$

Check: The units of the answer (g H$_2$SO$_4$) are correct. The magnitude of the answer is reasonable because it is less than g KOH.

Limiting Reactant, Theoretical Yield, and Percent Yield

4.37 (a) **Given:** 2 mol Na; 2 mol Br$_2$ **Find:** limiting reactant

Conceptual Plan: mol Na → mol NaBr

$$\frac{2 \text{ mol NaBr}}{2 \text{ mol Na}} \quad \to \textbf{ smaller mol amount determines limiting reactant}$$

mol Br$_2$ → mol NaBr

$$\frac{2 \text{ mol NaBr}}{1 \text{ mol Br}_2}$$

Solution: $2 \text{ mol Na} \times \dfrac{2 \text{ mol NaBr}}{2 \text{ mol Na}} = 2 \text{ mol NaBr}$

$2 \text{ mol Br}_2 \times \dfrac{2 \text{ mol NaBr}}{1 \text{ mol Br}_2} = 4 \text{ mol NaBr}$

Na is the limiting reactant.

Check: The answer is reasonable because Na produced the smallest amount of product.

(b) **Given:** 1.8 mol Na; 1.4 mol Br$_2$ **Find:** limiting reactant

Conceptual Plan: mol Na → mol NaBr

$$\frac{2 \text{ mol NaBr}}{2 \text{ mol Na}} \quad \to \textbf{ smaller mol amount determines limiting reactant}$$

mol Br$_2$ → mol NaBr

$$\frac{2 \text{ mol NaBr}}{1 \text{ mol Br}_2}$$

Solution: $1.8 \text{ mol Na} \times \dfrac{2 \text{ mol NaBr}}{2 \text{ mol Na}} = 1.8 \text{ mol NaBr}$

$1.4 \text{ mol Br}_2 \times \dfrac{2 \text{ mol NaBr}}{1 \text{ mol Br}_2} = 2.8 \text{ mol NaBr}$

Na is the limiting reactant.

Check: The answer is reasonable because Na produced the smallest amount of product.

(c) **Given:** 2.5 mol Na; 1 mol Br$_2$ **Find:** limiting reactant

Conceptual Plan: mol Na → mol NaBr

$$\frac{2 \text{ mol NaBr}}{2 \text{ mol Na}} \quad \to \textbf{ smaller mol amount determines limiting reactant}$$

mol Br$_2$ → mol NaBr

$$\frac{2 \text{ mol NaBr}}{1 \text{ mol Br}_2}$$

Solution: $2.5 \; \text{mol Na} \times \dfrac{2 \; \text{mol NaBr}}{2 \; \text{mol Na}} = 2.5 \; \text{mol NaBr}$

$1 \; \text{mol Br}_2 \times \dfrac{2 \; \text{mol NaBr}}{1 \; \text{mol Br}_2} = 2 \; \text{mol NaBr}$

Br_2 is the limiting reactant.

Check: The answer is reasonable because Br_2 produced the smallest amount of product.

(d)   **Given:** 12.6 mol Na; 6.9 mol Br_2 **Find:** limiting reactant

     Conceptual Plan: mol Na → mol NaBr

$$\dfrac{2 \; \text{mol NaBr}}{2 \; \text{mol Na}} \qquad \rightarrow \textbf{ smaller mol amount determines limiting reactant}$$

           mol Br_2 → mol NaBr

$$\dfrac{2 \; \text{mol NaBr}}{1 \; \text{mol Br}_2}$$

Solution: $12.6 \; \text{mol Na} \times \dfrac{2 \; \text{mol NaBr}}{2 \; \text{mol Na}} = 12.6 \; \text{mol NaBr}$

$6.9 \; \text{mol Br}_2 \times \dfrac{2 \; \text{mol NaBr}}{1 \; \text{mol Br}_2} = 13.8 \; \text{mol NaBr} = 14 \; \text{mol NaBr}$

Na is the limiting reactant.

Check: The answer is reasonable because Na produced the smallest amount of product.

4.39   The greatest number of Cl_2 molecules will be formed from reaction mixture (b) and would be 3 molecules Cl_2.

(a)   **Given:** 7 molecules HCl; 1 molecule O_2 **Find:** theoretical yield Cl_2

     Conceptual Plan: molecules HCl → molecules Cl_2

$$\dfrac{2 \; \text{molecules Cl}_2}{4 \; \text{molecules HCl}} \qquad \rightarrow \textbf{ smaller molecule amount determines limiting reactant}$$

           molecules O_2 → molecules Cl_2

$$\dfrac{2 \; \text{molecules Cl}_2}{1 \; \text{molecule O}_2}$$

Solution: $7 \; \text{molecules HCl} \times \dfrac{2 \; \text{molecules Cl}_2}{4 \; \text{molecules HCl}} = 3 \; \text{molecules Cl}_2$

$1 \; \text{molecule O}_2 \times \dfrac{2 \; \text{molecules Cl}_2}{1 \; \text{molecule O}_2} = 2 \; \text{molecules Cl}_2$

theoretical yield = 2 molecules Cl_2

(b)   **Given:** 6 molecules HCl; 3 molecules O_2 **Find:** theoretical yield Cl_2

     Conceptual Plan: molecules HCl → molecules Cl_2

$$\dfrac{2 \; \text{molecules Cl}_2}{4 \; \text{molecules HCl}} \qquad \rightarrow \begin{array}{l} \textbf{smaller molecule amount determines} \\ \textbf{limiting reactant} \end{array}$$

           molecules O_2 → molecules Cl_2

$$\dfrac{2 \; \text{molecules Cl}_2}{1 \; \text{molecule O}_2}$$

Solution: $6 \; \text{molecules HCl} \times \dfrac{2 \; \text{molecules Cl}_2}{4 \; \text{molecules HCl}} = 3 \; \text{molecules Cl}_2$

$3 \; \text{molecules O}_2 \times \dfrac{2 \; \text{molecules Cl}_2}{1 \; \text{molecule O}_2} = 6 \; \text{molecules Cl}_2$

theoretical yield = 3 molecules

(c)   **Given:** 4 molecules HCl; 5 molecules O_2 **Find:** theoretical yield Cl_2

     Conceptual Plan: molecules HCl → molecules Cl_2

$$\dfrac{2 \; \text{molecules Cl}_2}{4 \; \text{molecules HCl}} \qquad \rightarrow \begin{array}{l} \textbf{smaller molecule amount determines} \\ \textbf{limiting reactant} \end{array}$$

           molecules O_2 → molecules Cl_2

$$\dfrac{2 \; \text{molecules Cl}_2}{1 \; \text{molecule O}_2}$$

Solution: $4 \text{ molecules HCl} \times \dfrac{2 \text{ molecules Cl}_2}{4 \text{ molecules HCl}} = 2 \text{ molecules Cl}_2$

$5 \text{ molecules O}_2 \times \dfrac{2 \text{ molecules Cl}_2}{1 \text{ molecule O}_2} = 10 \text{ molecules Cl}_2$

theoretical yield $= 2 \text{ molecules Cl}_2$

Check: The units of the answer (molecules Cl_2) are correct. The answer is reasonable based on the limiting reactant in each mixture.

4.41 (a) **Given:** 4 mol Ti; 4 mol Cl_2 **Find:** theoretical yield $TiCl_4$

Conceptual Plan: mol Ti → mol TiCl$_4$

$\dfrac{1 \text{ mol TiCl}_4}{1 \text{ mol Ti}}$ → **smaller mol amount determines limiting reactant**

mol Cl$_2$ → mol TiCl$_4$

$\dfrac{1 \text{ mol TiCl}_4}{2 \text{ mol Cl}_2}$

Solution: $4 \text{ mol Ti} \times \dfrac{1 \text{ mol TiCl}_4}{1 \text{ mol Ti}} = 4 \text{ mol TiCl}_4$

$4 \text{ mol Cl}_2 \times \dfrac{1 \text{ mol TiCl}_4}{2 \text{ mol Cl}_2} = 2 \text{ mol TiCl}_4$

theoretical yield $= 2 \text{ mol TiCl}_4$

Check: The units of the answer (mol $TiCl_4$) are correct. The answer is reasonable because Cl_2 produced the smallest amount of product and is the limiting reactant.

(b) **Given:** 7 mol Ti; 17 mol Cl_2 **Find:** theoretical yield $TiCl_4$

Conceptual Plan: mol Ti → mol TiCl$_4$

$\dfrac{1 \text{ mol TiCl}_4}{1 \text{ mol Ti}}$ → **smaller mol amount determines limiting reactant**

mol Cl$_2$ → mol TiCl$_4$

$\dfrac{1 \text{ mol TiCl}_4}{2 \text{ mol Cl}_2}$

Solution: $7 \text{ mol Ti} \times \dfrac{1 \text{ mol TiCl}_4}{1 \text{ mol Ti}} = 7 \text{ mol TiCl}_4$

$17 \text{ mol Cl}_2 \times \dfrac{1 \text{ mol TiCl}_4}{2 \text{ mol Cl}_2} = 8.5 \text{ mol TiCl}_4$

theoretical yield $= 7 \text{ mol TiCl}_4$

Check: The units of the answer (mol $TiCl_4$) are correct. The answer is reasonable because Ti produced the smallest amount of product and is the limiting reactant.

(c) **Given:** 12.4 mol Ti; 18.8 mol Cl_2 **Find:** theoretical yield $TiCl_4$

Conceptual Plan: mol Ti → mol TiCl$_4$

$\dfrac{1 \text{ mol TiCl}_4}{1 \text{ mol Ti}}$ → **smaller mol amount determines limiting reactant**

mol Cl$_2$ → mol TiCl$_4$

$\dfrac{1 \text{ mol TiCl}_4}{2 \text{ mol Cl}_2}$

Solution: $12.4 \text{ mol Ti} \times \dfrac{1 \text{ mol TiCl}_4}{1 \text{ mol Ti}} = 12.4 \text{ mol TiCl}_4$

$18.8 \text{ mol Cl}_2 \times \dfrac{1 \text{ mol TiCl}_4}{2 \text{ mol Cl}_2} = 9.40 \text{ mol TiCl}_4$

theoretical yield $= 9.40 \text{ mol TiCl}_4$

Check: The units of the answer (mol $TiCl_4$) are correct. The answer is reasonable because Cl_2 produced the smallest amount of product and is the limiting reactant.

4.43   **Given:** 4.2 mol ZnS; 6.8 mol O_2 **Find:** mole amount of excess reactant left

Conceptual Plan: mol ZnS → mol ZnO

$$\frac{2 \text{ mol ZnO}}{2 \text{ mol ZnS}}$$ → **smaller mol amount determines limiting reactant**

mol O_2 → mol ZnO

$$\frac{2 \text{ mol ZnO}}{3 \text{ mol } O_2}$$

mol limiting reactant → mol excess reactant required → mol excess reactant left

$$\frac{3 \text{ mol } O_2}{2 \text{ mol ZnS}}$$

Solution: $4.2 \text{ mol ZnS} \times \dfrac{2 \text{ mol ZnO}}{2 \text{ mol ZnS}} = 4.2 \text{ mol ZnO}$

$6.8 \text{ mol } O_2 \times \dfrac{2 \text{ mol ZnO}}{3 \text{ mol } O_2} = 4.5 \text{ mol ZnO}$

ZnS is the limiting reactant; therefore, O_2 is the excess reactant.

$$4.2 \text{ mol ZnS} \times \frac{3 \text{ mol } O_2}{2 \text{ mol ZnS}} = 6.3 \text{ mol } O_2 \text{ required}$$

$$6.8 \text{ mol } O_2 - 6.3 \text{ mol } O_2 = 0.5 \text{ mol } O_2 \text{ left}$$

Check: The units of the answer (mol O_2) are correct. The magnitude is reasonable because it is less than the original amount of O_2.

4.45   (a)   **Given:** 2.0 g Al; 2.0 g Cl_2 **Find:** theoretical yield in g $AlCl_3$

Conceptual Plan: g Al → mol Al → mol $AlCl_3$

$$\frac{1 \text{ mol Al}}{26.98 \text{ g Al}} \quad \frac{2 \text{ mol } AlCl_3}{2 \text{ mol Al}}$$ → **smaller mol amount determines limiting reactant**

g Cl_2 → mol Cl_2 → mol $AlCl_3$

$$\frac{1 \text{ mol } Cl_2}{70.90 \text{ g } Cl_2} \quad \frac{2 \text{ mol } AlCl_3}{3 \text{ mol } Cl_2}$$

then mol $AlCl_3$ → g $AlCl_3$

$$\frac{133.3 \text{ g } AlCl_3}{\text{mol } AlCl_3}$$

Solution: $2.0 \text{ g Al} \times \dfrac{1 \text{ mol Al}}{26.98 \text{ g Al}} \times \dfrac{2 \text{ mol } AlCl_3}{2 \text{ mol Al}} = 0.074 \text{ mol } AlCl_3$

$2.0 \text{ g } Cl_2 \times \dfrac{1 \text{ mol } Cl_2}{70.90 \text{ g } Cl_2} \times \dfrac{2 \text{ mol } AlCl_3}{3 \text{ mol } Cl_2} = 0.0188 \text{ mol } AlCl_3$

$0.0188 \text{ mol } AlCl_3 \times \dfrac{133.3 \text{ g } AlCl_3}{\text{mol } AlCl_3} = 2.5 \text{ g } AlCl_3$

Check: The units of the answer (g $AlCl_3$) are correct. The answer is reasonable because Cl_2 produced the smallest amount of product and is the limiting reactant.

(b)   **Given:** 7.5 g Al; 24.8 g Cl_2 **Find:** theoretical yield in g $AlCl_3$

Conceptual Plan: g Al → mol Al → mol $AlCl_3$

$$\frac{1 \text{ mol Al}}{26.98 \text{ g Al}} \quad \frac{2 \text{ mol } AlCl_3}{2 \text{ mol Al}}$$ → **smaller mol amount determines limiting reactant**

g Cl_2 → mol Cl_2 → mol $AlCl_3$

$$\frac{1 \text{ mol } Cl_2}{70.90 \text{ g } Cl_2} \quad \frac{2 \text{ mol } AlCl_3}{3 \text{ mol } Cl_2}$$

then mol $AlCl_3$ → g $AlCl_3$

$$\frac{133.3 \text{ g } AlCl_3}{\text{mol } AlCl_3}$$

Solution: $7.5 \text{ g Al} \times \dfrac{1 \text{ mol Al}}{26.98 \text{ g Al}} \times \dfrac{2 \text{ mol } AlCl_3}{2 \text{ mol Al}} = 0.2780 \text{ mol } AlCl_3$

$24.8 \text{ g } Cl_2 \times \dfrac{1 \text{ mol } Cl_2}{70.90 \text{ g } Cl_2} \times \dfrac{2 \text{ mol } AlCl_3}{3 \text{ mol } Cl_2} = 0.2332 \text{ mol } AlCl_3$

$0.2332 \text{ mol } AlCl_3 \times \dfrac{133.3 \text{ g } AlCl_3}{\text{mol } AlCl_3} = 31.1 \text{ g } AlCl_3$

Check: The units of the answer (g $AlCl_3$) are correct. The answer is reasonable because Cl_2 produced the smallest amount of product and is the limiting reactant.

(c)   **Given:** 0.235 g Al; 1.15 g Cl_2 **Find:** theoretical yield in g $AlCl_3$

Conceptual Plan: g A → mol Al → mol $AlCl_3$

$$\frac{1\ mol\ Al}{26.98\ g\ Al} \quad \frac{2\ mol\ AlCl_3}{2\ mol\ Al} \qquad \rightarrow \textbf{ smaller mol amount determines limiting reactant}$$

g Cl_2 → mol Cl_2 → mol $AlCl_3$

$$\frac{1\ mol\ Cl_2}{70.90\ g\ Cl_2} \quad \frac{2\ mol\ AlCl_3}{3\ mol\ Cl_2}$$

then mol $AlCl_3$ → g $AlCl_3$

$$\frac{133.3\ g\ AlCl_3}{mol\ AlCl_3}$$

Solution: $0.235\ g\ Al \times \dfrac{1\ mol\ Al}{26.98\ g\ Al} \times \dfrac{2\ mol\ AlCl_3}{2\ mol\ Al} = 0.008710\ mol\ AlCl_3$

$1.15\ g\ Cl_2 \times \dfrac{1\ mol\ Cl_2}{70.90\ g\ Cl_2} \times \dfrac{2\ mol\ AlCl_3}{3\ mol\ Cl_2} = 0.01081\ mol\ AlCl_3$

$0.008710\ mol\ AlCl_3 \times \dfrac{133.3\ g\ AlCl_3}{mol\ AlCl_3} = 1.16\ g\ AlCl_3$

Check: The units of the answer (g $AlCl_3$) are correct. The answer is reasonable because Al produced the smallest amount of product and is the limiting reactant.

4.47   **Given:** 22.55 g Fe_2O_3; 14.78 g CO   **Find:** mole amount of excess reactant left

Conceptual Plan: g Fe_2O_3 → mol Fe_2O_3 → mol Fe

$$\frac{1\ mol\ Fe_2O_3}{159.7\ g\ Fe_2O_3} \quad \frac{2\ mol\ Fe}{1\ mol\ Fe_2O_3} \qquad \rightarrow \textbf{ smaller mol amount determines limiting reactant}$$

g CO → mol CO → mol Fe

$$\frac{1\ mol\ CO}{28.01\ g\ CO} \quad \frac{2\ mol\ Fe}{3\ mol\ CO}$$

then mol limiting reactant → mol excess reactant required → mol excess reactant left → g excess

reactant left $\dfrac{3\ mol\ CO}{1\ mol\ Fe_2O_3}$ $\dfrac{1\ mol\ Fe_2O_3}{159.7\ g\ Fe_2O_3}$ or $\dfrac{28.01\ g\ CO}{1\ mol\ CO}$

Solution: $22.55\ g\ Fe_2O_3 \times \dfrac{1\ mol\ Fe_2O_3}{159.7\ g\ Fe_2O_3} \times \dfrac{2\ mol\ Fe}{1\ mol\ Fe_2O_3} = 0.2824\ mol\ Fe$

$14.78\ g\ CO \times \dfrac{1\ mol\ CO}{28.01\ g\ CO} \times \dfrac{2\ mol\ Fe}{3\ mol\ CO} = 0.3518\ mol\ Fe$

Fe_2O_3 is the limiting reactant; therefore, CO is the excess reactant.

$22.55\ g\ Fe_2O_3 \times \dfrac{1\ mol\ Fe_2O_3}{159.7\ g\ Fe_2O_3} \times \dfrac{3\ mol\ CO}{1\ mol\ Fe_2O_3} \times \dfrac{28.01\ g\ CO}{1\ mol\ CO} = 11.865\ g\ CO\ required$

$14.78\ g\ CO - 11.87\ g\ CO = 2.91\ g\ CO\ left$

Check: The units of the answer (g CO) are correct. The magnitude is reasonable because it is less than the original amount of CO.

4.49   **Given:** 28.5 g KCl; 25.7 g Pb^{2+}; 29.4 g $PbCl_2$ **Find:** limiting reactant; theoretical yield $PbCl_2$; % yield

Conceptual Plan: g KCl → mol KCl → mol $PbCl_2$

$$\frac{1\ mol\ KCl}{74.55\ g\ KCl} \quad \frac{1\ mol\ PbCl_2}{2\ mol\ KCl} \qquad \rightarrow \textbf{ smaller mol amount determines limiting reactant}$$

g Pb^{2+} → mol Pb^{2+} → mol $PbCl_2$

$$\frac{1\ mol\ Pb^{2+}}{207.2\ g\ Pb^{2+}} \quad \frac{1\ mol\ PbCl_2}{1\ mol\ Pb^{2+}}$$

then mol $PbCl_2$ → g $PbCl_2$ then determine % yield

$$\frac{278.1\ g\ PbCl_2}{mol\ PbCl_2} \qquad \frac{actual\ yield\ g\ PbCl_2}{theoretical\ yield\ g\ PbCl_2} \times 100\%$$

Solution: $28.5 \text{ g KCl} \times \dfrac{1 \text{ mol KCl}}{74.55 \text{ g KCl}} \times \dfrac{1 \text{ mol PbCl}_2}{2 \text{ mol KCl}} = 0.19\underline{1}1 \text{ mol PbCl}_2$

$25.7 \text{ g Pb}^{2+} \times \dfrac{1 \text{ mol Pb}^{2+}}{207.2 \text{ g Pb}^{2+}} \times \dfrac{1 \text{ mol PbCl}_2}{1 \text{ mol Pb}^{2+}} = 0.1240 \text{ mol PbCl}_2$ Pb^{2+} is the limiting reactant.

$0.12\underline{4}0 \text{ mol PbCl}_2 \times \dfrac{278.1 \text{ g PbCl}_2}{1 \text{ mol PbCl}_2} = 34.\underline{5} \text{ g PbCl}_2$

$\dfrac{29.4 \text{ g PbCl}_2}{34.\underline{5} \text{ g PbCl}_2} \times 100\% = 85.2\%$

Check: The theoretical yield has the correct units (g PbCl_2) and has a reasonable magnitude compared to the mass of Pb^{2+}, the limiting reactant. The % yield is reasonable, under 100%.

4.51   **Given:** 136.4 kg NH_3; 211.4 kg CO_2; 168.4 kg $\text{CH}_4\text{N}_2\text{O}$ **Find:** limiting reactant; theoretical yield $\text{CH}_4\text{N}_2\text{O}$; % yield
Conceptual Plan: kg $\text{NH}_3 \rightarrow$ g $\text{NH}_3 \rightarrow$ mol $\text{NH}_3 \rightarrow$ mol $\text{CH}_4\text{N}_2\text{O}$

$\dfrac{1000 \text{ g}}{1 \text{ kg}}$ $\dfrac{1 \text{ mol NH}_3}{17.03 \text{ g NH}_3}$ $\dfrac{1 \text{ mol CH}_4\text{N}_2\text{O}}{2 \text{ mol NH}_3}$ **$\rightarrow$ smaller amount determines limiting reactant**

kg $\text{CO}_2 \rightarrow$ g $\text{CO}_2 \rightarrow$ mol $\text{CO}_2 \rightarrow$ mol $\text{CH}_4\text{N}_2\text{O}$

$\dfrac{1000 \text{ g}}{1 \text{ kg}}$ $\dfrac{1 \text{ mol CO}_2}{44.01 \text{ g CO}_2}$ $\dfrac{1 \text{ mol CH}_4\text{N}_2\text{O}}{1 \text{ mol CO}_2}$

then mol $\text{CH}_4\text{N}_2\text{O} \rightarrow$ g $\text{CH}_4\text{N}_2\text{O} \rightarrow$ kg $\text{CH}_4\text{N}_2\text{O}$ then determine % yield

$\dfrac{60.06 \text{ g CH}_4\text{N}_2\text{O}}{1 \text{ mol CH}_4\text{N}_2\text{O}}$ $\dfrac{1 \text{ kg}}{1000 \text{ g}}$ $\dfrac{\text{actual yield kg CH}_4\text{N}_2\text{O}}{\text{theoretical yield kg CH}_4\text{N}_2\text{O}} \times 100\%$

Solution: $136.4 \text{ kg NH}_3 \times \dfrac{1000 \text{ g}}{1 \text{ kg}} \times \dfrac{1 \text{ mol NH}_3}{17.03 \text{ g NH}_3} \times \dfrac{1 \text{ mol CH}_4\text{N}_2\text{O}}{2 \text{ mol NH}_3} = 400\underline{4}.7 \text{ mol CH}_4\text{N}_2\text{O}$

$211.4 \text{ kg CO}_2 \times \dfrac{1000 \text{ g}}{1 \text{ kg}} \times \dfrac{1 \text{ mol CO}_2}{44.01 \text{ g CO}_2} \times \dfrac{1 \text{ mol CH}_4\text{N}_2\text{O}}{1 \text{ mol CO}_2} = 480\underline{3}.5 \text{ mol CH}_4\text{N}_2\text{O}$

NH_3 is the limiting reactant.

$400\underline{4}.7 \text{ mol CH}_4\text{N}_2\text{O} \times \dfrac{60.06 \text{ g CH}_4\text{N}_2\text{O}}{1 \text{ mol CH}_4\text{N}_2\text{O}} \times \dfrac{1 \text{ kg}}{1000 \text{ g}} = 240.\underline{5}2 \text{ kg CH}_4\text{N}_2\text{O}$

$\dfrac{168.4 \text{ kg CH}_4\text{N}_2\text{O}}{240.\underline{5}2 \text{ kg CH}_4\text{N}_2\text{O}} \times 100\% = 70.01\%$

Check: The theoretical yield has the correct units $(\text{kg CH}_4\text{N}_2\text{O})$ and has a reasonable magnitude compared to the mass of NH_3, the limiting reactant. The % yield is reasonable, under 100%.

Solution Concentration and Solution Stoichiometry

4.53   (a)   **Given:** 3.25 mol LiCl; 2.78 L solution   **Find:** molarity LiCl
Conceptual Plan: mol LiCl, L solution $\rightarrow$ molarity

$\text{molarity (M)} = \dfrac{\text{amount of solute (in moles)}}{\text{volume of solution (in L)}}$

Solution: $\dfrac{3.25 \text{ mol LiCl}}{2.78 \text{ L solution}} = 1.1\underline{6}9 \text{ M} = 1.17 \text{ M}$

Check: The units of the answer (M) are correct. The magnitude of the answer is reasonable. Concentrations are usually between 0 M and 18 M.

(b)   **Given:** 28.33 g $\text{C}_6\text{H}_{12}\text{O}_6$; 1.28 L solution   **Find:** molarity $\text{C}_6\text{H}_{12}\text{O}_6$
Conceptual Plan: g $\text{C}_6\text{H}_{12}\text{O}_6 \rightarrow$ mol $\text{C}_6\text{H}_{12}\text{O}_6$, L solution $\rightarrow$ molarity

$\dfrac{\text{mol C}_6\text{H}_{12}\text{O}_6}{180.16 \text{ g C}_6\text{H}_{12}\text{O}_6}$ $\text{molarity (M)} = \dfrac{\text{amount of solute (in moles)}}{\text{volume of solution (in L)}}$

Solution: $28.33 \text{ g } \cancel{C_6H_{12}O_6} \times \dfrac{1 \text{ mol } C_6H_{12}O_6}{180.16 \text{ g } \cancel{C_6H_{12}O_6}} = 0.15724 \text{ mol } C_6H_{12}O_6$

$\dfrac{0.15724 \text{ mol } C_6H_{12}O_6}{1.28 \text{ L solution}} = 0.1228 \text{ M} = 0.123 \text{ M}$

Check: The units of the answer (M) are correct. The magnitude of the answer is reasonable. Concentrations are usually between 0 M and 18 M.

(c) **Given:** 32.4 mg NaCl; 122.4 mL solution   **Find:** molarity NaCl
Conceptual Plan: mg NaCl → g NaCl → mol NaCl, and mL solution → L solution then molarity

$$\dfrac{1 \text{ g NaCl}}{1000 \text{ mg NaCl}} \qquad \dfrac{\text{mol NaCl}}{58.45 \text{ g NaCl}} \qquad \dfrac{\text{L solution}}{1000 \text{ mL solution}} \qquad \text{molarity (M)} = \dfrac{\text{amount of solute (in moles)}}{\text{volume of solution (in L)}}$$

Solution: $32.4 \text{ mg } \cancel{NaCl} \times \dfrac{1 \text{ g}}{1000 \text{ mg}} \times \dfrac{1 \text{ mol NaCl}}{58.45 \text{ g } \cancel{NaCl}} = 5.543 \times 10^{-4} \text{ mol NaCl}$

$122.4 \text{ mL solution} \times \dfrac{1 \text{ L}}{1000 \text{ mL}} = 0.1224 \text{ L}$

$\dfrac{5.543 \times 10^{-4} \text{ mol NaCl}}{0.1224 \text{ L}} = 0.004528 \text{ M NaCl} = 0.00453 \text{ M NaCl}$

Check: The units of the answer (M) are correct. The magnitude of the answer is reasonable. Concentrations are usually between 0 M and 18 M.

4.55 (a) **Given:** 0.150 M KNO_3 **Find:** M NO_3^-
Conceptual Plan: M KNO_3 → M NO_3^-

$$\dfrac{1 \text{ M } NO_3^-}{1 \text{ M } KNO_3}$$

Solution: $0.150 \text{ M } \cancel{KNO_3} \times \dfrac{1 \text{ M } NO_3^-}{1 \text{ M } \cancel{KNO_3}} = 0.150 \text{ M } NO_3^-$

Check: The units of the answer, M, are correct. The answer is reasonable because it is greater than or equal to the concentration of KNO_3.

(b) **Given:** 0.150 M $Ca(NO_3)_2$ **Find:** M NO_3^-
Conceptual Plan: M $Ca(NO_3)_2$ → M NO_3^-

$$\dfrac{2 \text{ M } NO_3^-}{1 \text{ M } Ca(NO_3)_2}$$

Solution: $0.150 \text{ M } \cancel{Ca(NO_3)_2} \times \dfrac{2 \text{ M } NO_3^-}{1 \text{ M } \cancel{Ca(NO_3)_2}} = 0.300 \text{ M } NO_3^-$

Check: The units of the answer, M, are correct. The answer is reasonable because it is greater than or equal to the concentration of $Ca(NO_3)_2$.

(c) **Given:** 0.150 M $Al(NO_3)_3$ **Find:** M NO_3^-
Conceptual Plan: M $Al(NO_3)_3$ → M NO_3^-

$$\dfrac{3 \text{ M } NO_3^-}{1 \text{ M } Al(NO_3)_3}$$

Solution: $0.150 \text{ M } \cancel{Al(NO_3)_3} \times \dfrac{3 \text{ M } NO_3^-}{1 \text{ M } \cancel{Al(NO_3)_3}} = 0.450 \text{ M } NO_3^-$

Check: The units of the answer, M, are correct. The answer is reasonable because it is greater than or equal to the concentration of $Al(NO_3)_3$.

4.57 (a) **Given:** 0.556 L; 2.3 M KCl   **Find:** mol KCl
Conceptual Plan: volume solution × M = mol

$$\text{volume solution (L)} \times M = \text{mol}$$

Solution: $0.556 \text{ L solution} \times \dfrac{2.3 \text{ mol KCl}}{\text{L solution}} = 1.3 \text{ mol KCl}$

Check: The units of the answer (mol KCl) are correct. The magnitude is reasonable because it is less than 1 L solution.

(b)   **Given:** 1.8 L; 0.85 M KCl   **Find:** mol KCl
      Conceptual Plan: volume solution × M = mol

$$\text{volume solution (L)} \times \text{M} = \text{mol}$$

Solution: $1.8 \ \text{L solution} \times \dfrac{0.85 \ \text{mol KCl}}{\text{L solution}} = 1.5 \ \text{mol KCl}$

Check: The units of the answer (mol KCl) are correct. The magnitude is reasonable because it is less than 2 L solution.

(c)   **Given:** 114 mL; 1.85 M KCl   **Find:** mol KCl
      Conceptual Plan: mL solution → L solution, then volume solution × M = mol

$$\dfrac{1 \ \text{L}}{1000 \ \text{mL}} \qquad \text{volume solution (L)} \times \text{M} = \text{mol}$$

Solution: $114 \ \text{mL solution} \times \dfrac{1 \ \text{L}}{1000 \ \text{mL}} \times \dfrac{1.85 \ \text{mol KCl}}{\text{L solution}} = 0.211 \ \text{mol KCl}$

Check: The units of the answer (mol KCl) are correct. The magnitude is reasonable because it is less than 1 L solution.

4.59   **Given:** 400.0 mL; 1.1 M $NaNO_3$ **Find:** g $NaNO_3$
       Conceptual Plan: mL solution → L solution, then volume solution × M = mol $NaNO_3$

$$\dfrac{1 \ \text{L solution}}{1000 \ \text{mL solution}} \qquad \text{volume solution (L)} \times \text{M} = \text{mol}$$

then mol $NaNO_3$ → g $NaNO_3$

$$\dfrac{185.00 \ \text{g}}{\text{mol } NaNO_3}$$

Solution: $400.0 \ \text{mL solution} \times \dfrac{1 \ \text{L}}{1000 \ \text{mL}} \times \dfrac{1.1 \ \text{mol } NaNO_3}{\text{L solution}} \times \dfrac{85.00 \ \text{g}}{\text{mol } NaNO_3} = 37.40 \ \text{g } NaNO_3 = 37 \ \text{g } NaNO_3$

Check: The units of the answer (g $NaNO_3$) are correct. The magnitude is reasonable for the concentration and volume of solution.

4.61   **Given:** $V_1 = 123$ mL; $M_1 = 1.1$ M; $V_2 = 500.0$ mL   **Find:** M_2
       Conceptual Plan: mL → L then $V_1, M_1, V_2 → M_2$

$$\dfrac{1 \ \text{L}}{1000 \ \text{mL}} \qquad V_1 M_1 = V_2 M_2$$

Solution: $123 \ \text{mL} \times \dfrac{1 \ \text{L}}{1000 \ \text{mL}} = 0.123 \ \text{L} \qquad 500.0 \ \text{mL} \times \dfrac{1 \ \text{L}}{1000 \ \text{mL}} = 0.5000 \ \text{L}$

$$M_2 = \dfrac{V_1 M_1}{V_2} = \dfrac{(0.123 \ \text{L})(1.1 \ \text{M})}{(0.5000 \ \text{L})} = 0.27 \ \text{M}$$

Check: The units of the answer (M) are correct. The magnitude of the answer is reasonable because it is less than the original concentration.

4.63   **Given:** $V_1 = 50$ mL; $M_1 = 12$ M; $M_2 = 0.100$ M   **Find:** V_2
       Conceptual Plan: mL → L then $V_1, M_1, M_2 → V_2$

$$\dfrac{1 \ \text{L}}{1000 \ \text{mL}} \qquad V_1 M_1 = V_2 M_2$$

Solution: $50 \ \text{mL} \times \dfrac{1 \ \text{L}}{1000 \ \text{mL}} = 0.050 \ \text{L}$

$$V_2 = \dfrac{V_1 M_1}{M_2} = \dfrac{(0.050 \ \text{L})(12 \ \text{M})}{(0.100 \ \text{M})} = 6.0 \ \text{L}$$

Check: The units of the answer (L) are correct. The magnitude of the answer is reasonable because the new concentration is much less than the original; the volume must be larger.

4.65   **Given:** 95.4 mL; 0.102 M $CuCl_2$; 0.175 M Na_3PO_4 **Find:** volume Na_3PO_4
         Conceptual Plan: $mL\ CuCl_2 \rightarrow L\ CuCl_2 \rightarrow mol\ CuCl_2 \rightarrow mol\ Na_3PO_4 \rightarrow L\ Na_3PO_4 \rightarrow mL\ Na_3PO_4$

$$\frac{1\ L}{1000\ mL} \qquad \frac{0.102\ mol\ CuCl_2}{L} \qquad \frac{2\ mol\ Na_3PO_4}{3\ mol\ CuCl_2} \qquad \frac{1\ L}{0.175\ mol\ Na_3PO_4} \qquad \frac{1000\ mL}{L}$$

Solution: $95.4\ mL\ CuCl_2 \times \dfrac{1\ L}{1000\ mL} \times \dfrac{0.102\ mol\ CuCl_2}{1\ L} \times \dfrac{2\ mol\ Na_3PO_4}{3\ mol\ CuCl_2} \times \dfrac{1\ L}{0.175\ mol\ Na_3PO_4} \times \dfrac{1000\ mL}{1\ L}$

$= 37.1\ mL\ Na_3PO_4$

Check: The units of the answer $(mL\ Na_3PO_4)$ are correct. The magnitude of the answer is reasonable because the concentration of Na_3PO_4 is greater.

4.67   **Given:** 25.0 g H_2; 6.0 M H_2SO_4 **Find:** volume H_2SO_4
         Conceptual Plan: $g\ H_2 \rightarrow mol\ H_2 \rightarrow mol\ H_2SO_4 \rightarrow L\ H_2SO_4$

$$\frac{1\ mol\ H_2}{2.016\ g\ H_2} \qquad \frac{3\ mol\ H_2SO_4}{3\ mol\ H_2} \qquad \frac{1\ L}{6.0\ mol\ H_2SO_4}$$

Solution: $25.0\ g\ H_2 \times \dfrac{1\ mol\ H_2}{2.016\ g\ H_2} \times \dfrac{3\ mol\ H_2SO_4}{3\ mol\ H_2} \times \dfrac{1\ L}{6.0\ mol\ H_2SO_4} = 2.1\ L\ H_2SO_4$

Check: The units of the answer $(L\ H_2SO_4)$ are correct. The magnitude is reasonable because there are approximately 12 mol H_2 and the mole ratio is 1:1.

4.69   **Given:** 25.0 mL, 1.20 M KCl; 15.0 mL, 0.900 M $Ba(NO_3)_2$; 2.45 g $BaCl_2$
         Find: limiting reactant, theoretical yield $BaCl_2$, % yield
         Conceptual Plan: volume KCl solution $\times$ M $\rightarrow$ mol KCl $\rightarrow$ mol $BaCl_2$

$$volume\ solution\ (L) \times M = mol \qquad \frac{1\ mol\ BaCl_2}{2\ mol\ KCl} \rightarrow \quad \textbf{smaller mol amount determines}$$
$$\textbf{limiting reactant}$$

volume $Ba(NO_3)_2$ solution $\times$ M $\rightarrow$ mol $Ba(NO_3)_2$ $\rightarrow$ mol $BaCl_2$

$$volume\ solution\ (L) \times M = mol \qquad \frac{1\ mol\ BaCl_2}{1\ mol\ Ba(NO_3)_2}$$

then mol $BaCl_2$ $\rightarrow$ g $BaCl_2$ **then determine % yield**

$$\frac{208.23\ g\ BaCl_2}{mol\ BaCl_2} \qquad \frac{actual\ yield\ g\ BaCl_2}{theoretical\ yield\ g\ BaCl_2} \times 100\%$$

Solution: $25.0\ mL\ solution \times \dfrac{1\ L\ solution}{1000\ mL\ solution} \times \dfrac{1.20\ mol\ KCl}{L\ solution} \times \dfrac{1\ mol\ BaCl_2}{2\ mol\ KCl} = 0.0150\ mol\ BaCl_2$

$15.0\ mL\ solution \times \dfrac{1\ L\ solution}{1000\ mL\ solution} \times \dfrac{0.900\ mol\ Ba(NO_3)_2}{L\ solution} \times \dfrac{1\ mol\ BaCl_2}{1\ mol\ Ba(NO_3)_2} = 0.0135\ mol\ BaCl_2$

$0.0135\ mol\ BaCl_2 \times \dfrac{208.23\ g\ BaCl_2}{1\ mol\ BaCl_2} = 2.811\ g\ BaCl_2$

$\dfrac{2.45\ g\ BaCl_2}{2.811\ g\ BaCl_2} \times 100\% = 87.2\%$

Types of Aqueous Solutions and Solubility

4.71   (a)   CsCl is an ionic compound. An aqueous solution is an electrolyte solution, so it conducts electricity.
         (b) CH_3OH is a molecular compound that does not dissociate. An aqueous solution is a nonelectrolyte solution, so it does not conduct electricity.
         (c) $Ca(NO_2)_2$ is an ionic compound. An aqueous solution is an electrolyte solution, so it conducts electricity.
         (d) $C_6H_{12}O_6$ is a molecular compound that does not dissociate. An aqueous solution is a nonelectrolyte solution, so it does not conduct electricity.

4.73   (a) $AgNO_3$ is soluble. Compounds containing NO_3^- are always soluble with no exceptions. The ions in the solution are $Ag^+(aq)$ and $NO_3^-(aq)$.

    (b) $Pb(C_2H_3O_2)_2$ is soluble. Compounds containing $C_2H_3O_2^-$ are always soluble with no exceptions. The ions in the solution are $Pb^{2+}(aq)$ and $C_2H_3O_2^-(aq)$.

    (c) KNO_3 is soluble. Compounds containing K^+ or NO_3^- are always soluble with no exceptions. The ions in solution are $K^+(aq)$ and $NO_3^-(aq)$.

    (d) $(NH_4)_2S$ is soluble. Compounds containing NH_4^+ are always soluble with no exceptions. The ions in solution are $NH_4^+(aq)$ and $S^{2-}(aq)$.

Precipitation Reactions

4.75   (a) $LiI(aq) + BaS(aq) \rightarrow$ Possible products: Li_2S and BaI_2. Li_2S is soluble. Compounds containing S^{2-} are normally insoluble, but Li^+ is an exception. BaI_2 is soluble. Compounds containing I^- are normally soluble, and Ba^{2+} is not an exception. $LiI(aq) + BaS(aq) \rightarrow$ No Reaction

    (b) $KCl(aq) + CaS(aq) \rightarrow$ Possible products: K_2S and $CaCl_2$. K_2S is soluble. Compounds containing S^{2-} are normally insoluble, but K^+ is an exception. $CaCl_2$ is soluble. Compounds containing Cl^- are normally soluble, and Ca^{2+} is not an exception. $KCl(aq) + CaS(aq) \rightarrow$ No Reaction

    (c) $CrBr_2(aq) + Na_2CO_3(aq) \rightarrow$ Possible products: $CrCO_3$ and $NaBr$. $CrCO_3$ is insoluble. Compounds containing CO_3^{2-} are normally insoluble, and Cr^{2+} is not an exception. $NaBr$ is soluble. Compounds containing Br^- are normally soluble, and Na^+ is not an exception.
$CrBr_2(aq) + Na_2CO_3(aq) \rightarrow CrCO_3(s) + 2\,NaBr(aq)$

    (d) $NaOH(aq) + FeCl_3(aq) \rightarrow$ Possible products: $NaCl$ and $Fe(OH)_3$. $NaCl$ is soluble. Compounds containing Na^+ are normally soluble—no exceptions. $Fe(OH)_3$ is insoluble. Compounds containing OH^- are normally insoluble, and Fe^{3+} is not an exception.
$3\,NaOH(aq) + FeCl_3(aq) \rightarrow 3\,NaCl(aq) + Fe(OH)_3(s)$

4.77   (a) $K_2CO_3(aq) + Pb(NO_3)_2(aq) \rightarrow$ Possible products: KNO_3 and $PbCO_3$. KNO_3 is soluble. Compounds containing K^+ are always soluble—no exceptions. $PbCO_3$ is insoluble. Compounds containing CO_3^{2-} are normally insoluble, and Pb^{2+} is not an exception.
$K_2CO_3(aq) + Pb(NO_3)_2(aq) \rightarrow 2\,KNO_3(aq) + PbCO_3(s)$

    (b) $Li_2SO_4(aq) + Pb(C_2H_3O_2)_2(aq) \rightarrow$ Possible products: $LiC_2H_3O_2$ and $PbSO_4$. $LiC_2H_3O_2$ is soluble. Compounds containing Li^+ are always soluble—no exceptions. $PbSO_4$ is insoluble. Compounds containing SO_4^{2-} are normally soluble, but Pb^{2+} is an exception.
$Li_2SO_4(aq) + Pb(C_2H_3O_2)_2(aq) \rightarrow 2\,LiC_2H_3O_2(aq) + PbSO_4(s)$

    (c) $Cu(NO_3)_2(aq) + MgS(s) \rightarrow$ Possible products: CuS and $Mg(NO_3)_2$. CuS is insoluble. Compounds containing S^{2-} are normally insoluble, and Cu^{2+} is not an exception. $Mg(NO_3)_2$ is soluble. Compounds containing NO_3^- are always soluble—no exceptions.
$Cu(NO_3)_2(aq) + MgS(s) \rightarrow CuS(s) + Mg(NO_3)_2(aq)$

    (d) $Sr(NO_3)_2(aq) + KI(aq) \rightarrow$ Possible products: SrI_2 and KNO_3. SrI_2 is soluble. Compounds containing I^- are normally soluble, and Sr^{2+} is not an exception. KNO_3 is soluble. Compounds containing K^+ are always soluble—no exceptions. $Sr(NO_3)_2(aq) + KI(aq) \rightarrow$ No Reaction

Ionic and Net Ionic Equations

4.79   (a) $H^+(aq) + \cancel{Cl^-}(aq) + \cancel{Li^+}(aq) + OH^-(aq) \rightarrow H_2O(l) + \cancel{Li^+}(aq) + \cancel{Cl^-}(aq)$
$H^+(aq) + OH^-(aq) \rightarrow H_2O(l)$

    (b) $\cancel{Mg^{2+}}(aq) + S^{2-}(aq) + Cu^{2+}(aq) + 2\,\cancel{Cl^-}(aq) \rightarrow CuS(s) + \cancel{Mg^{2+}}(aq) + 2\,\cancel{Cl^-}(aq)$
$Cu^{2+}(aq) + S^{2-}(aq) \rightarrow CuS(s)$

    (c) $\cancel{Na^+}(aq) + OH^-(aq) + H^+(aq) + \cancel{NO_3^-}(aq) \rightarrow H_2O(l) + \cancel{Na^+}(aq) + \cancel{NO_3^-}(aq)$
$H^+(aq) + OH^-(aq) + \quad \rightarrow H_2O(l)$

    (d) $6\,\cancel{Na^+}(aq) + 2\,PO_4^{3-}(aq) + 3\,Ni^{2+}(aq) + 6\,\cancel{Cl^-}(aq) \rightarrow Ni_3(PO_4)_2(s) + 6\,\cancel{Na^+}(aq) + 6\,\cancel{Cl^-}(aq)$
$3\,Ni^{2+}(aq) + 2\,PO_4^{3-}(aq) \rightarrow Ni_3(PO_4)_2(s)$

4.81 $Hg_2^{2+}(aq) + 2\,\cancel{NO_3^-}(aq) + 2\,\cancel{Na^+}(aq) + 2\,Cl^-(aq) \rightarrow Hg_2Cl_2(s) + 2\,\cancel{Na^+}(aq) + 2\,\cancel{NO_3^-}(aq)$
$Hg_2^{2+}(aq) + 2\,Cl^-(aq) \rightarrow Hg_2Cl_2(s)$

Acid–Base and Gas-Evolution Reactions

4.83   Skeletal reaction: $HBr(aq) + KOH(aq) \rightarrow H_2O(l) + KBr(aq)$
                                    acid   base   water   salt
        Net ionic equation: $H^+(aq) + OH^-(aq) \rightarrow H_2O(l)$

4.85   (a)   Skeletal reaction: $H_2SO_4(aq) + Ca(OH)_2(aq) \rightarrow H_2O(l) + CaSO_4(s)$
                                        acid   base   water   salt
              Balanced reaction: $H_2SO_4(aq) + Ca(OH)_2(aq) \rightarrow 2\,H_2O(l) + CaSO_4(s)$
        (b)   Skeletal reaction: $HClO_4(aq) + KOH(aq) \rightarrow H_2O(l) + KClO_4(aq)$
                                        acid   base   water   salt
              Balanced reaction: $HClO_4(aq) + KOH(aq) \rightarrow H_2O(l) + KClO_4(aq)$
        (c)   Skeletal reaction: $H_2SO_4(aq) + NaOH(aq) \rightarrow H_2O(l) + Na_2SO_4(aq)$
                                        acid   base   water   salt
              Balanced reaction: $H_2SO_4(aq) + 2\,NaOH(aq) \rightarrow 2\,H_2O(l) + Na_2SO_4(aq)$

4.87   **Given:** 22.62 mL, 0.2000 M NaOH solution; 25.00 mL $HClO_4$ solution   **Find:** M $HClO_4$ solution
        Conceptual Plan: mL NaOH → L NaOH → mol NaOH → mol $HClO_4$

$$\frac{1\ L}{1000\ mL} \qquad \frac{0.2000\ mol\ NaOH}{1\ L\ NaOH} \qquad \frac{1\ mol\ HClO_4}{1\ mol\ NaOH}$$

mol $HClO_4$, volume $HClO_4$ solution → M

$$M = \frac{1\ mol\ HClO_4}{L\ HClO_4\ solution}$$

Solution: $22.62\ mL\ NaOH \times \dfrac{1\ L}{1000\ mL} \times \dfrac{0.2000\ mol\ NaOH}{1\ L\ NaOH} \times \dfrac{1\ mol\ HClO_4}{1\ mol\ NaOH} = 0.00452\underline{4}\ mol\ HClO_4$

$$\frac{0.00452\underline{4}\ mol\ HClO_4}{25.00\ mL\ HClO_4} \times \frac{1000\ mL}{1\ L} = 0.180\underline{9}6\ M\ HClO_4 = 0.1810\ M\ HClO_4$$

Check: The units of the answer (M $HClO_4$) are correct. The magnitude of the answer is reasonable because it is less than the M of NaOH.

4.89   (a)   Skeletal reaction: $HBr(aq) + NiS(s) \rightarrow NiBr_2(aq) + H_2S(g)$
                                                                        gas
              Balanced reaction: $2\,HBr(aq) + NiS(s) \rightarrow NiBr_2(aq) + H_2S(g)$
        (b)   Skeletal reaction:
              $NH_4I(aq) + NaOH(aq) \rightarrow NH_4OH(aq) + NaI(aq) \rightarrow H_2O(l) + NH_3(g) + NaI(aq)$
                                                        decomposes   gas
              Balanced reaction: $NH_4I(aq) + NaOH(aq) \rightarrow H_2O(l) + NH_3(g) + NaI(aq)$
        (c)   Skeletal reaction: $HBr(aq) + Na_2S(aq) \rightarrow NaBr(aq) + H_2S(g)$
                                                                        gas
              Balanced reaction: $2\,HBr(aq) + Na_2S(aq) \rightarrow 2\,NaBr(aq) + H_2S(g)$
        (d)   Skeletal reaction:
              $HClO_4(aq) + Li_2CO_3(aq) \rightarrow H_2CO_3(aq) + LiClO_4(aq) \rightarrow H_2O(l) + CO_2(g) + LiClO_4(aq)$
                                                        decomposes   gas
              Balanced reaction: $2\,HClO_4(aq) + Li_2CO_3(aq) \rightarrow H_2O(l) + CO_2(g) + 2\,LiClO_4(aq)$

Oxidation–Reduction and Combustion

4.91   (a)   Ag. The oxidation state of Ag = 0. The oxidation state of an atom in a free element is 0.
        (b) Ag^+. The oxidation state of $Ag^+ = +1$. The oxidation state of a monatomic ion is equal to its charge.
        (c) CaF_2. The oxidation state of Ca = +2, and the oxidation state of F = −1. The oxidation state of a group 2A metal always has an oxidation state of +2, and the oxidation of F is −1 because the sum of the oxidation states in a neutral formula unit = 0.

(d) H_2S. The oxidation state of H $= +1$, and the oxidation state of S $= -2$. The oxidation state of H when listed first is $+1$, and the oxidation state of S is -2 because S is in group 6A and the sum of the oxidation states in a neutral molecular unit $= 0$.

(e) CO_3^{2-}. The oxidation state of C $= +4$, and the oxidation state of O $= -2$. The oxidation state of O is normally -2, and the oxidation state of C is deduced from the formula because the sum of the oxidation states must equal the charge on the ion. (C ox state) $+3$(O ox state) $= -2$; (C ox state) $+ 3(-2) = -2$, so C ox state $= +4$.

(f) CrO_4^{2-}. The oxidation state of Cr $= +6$, and the oxidation state of O $= -2$. The oxidation state of O is normally -2, and the oxidation state of Cr is deduced from the formula because the sum of the oxidation states must equal the charge on the ion. (Cr ox state) $+ 4$(O ox state) $= -2$; (Cr ox state) $+ 4(-2) = -2$, so Cr ox state $= +6$.

4.93   (a)   CrO. The oxidation state of Cr $= +2$, and the oxidation state of O $= -2$. The oxidation state of O is normally -2, and the oxidation state of Cr is deduced from the formula because the sum of the oxidation states must $= 0$. (Cr ox state) $+$ (O ox state) $= 0$; (Cr ox state) $+ (-2) = 0$, so Cr $= +2$.

(b) CrO_3. The oxidation state of Cr $= +6$, and the oxidation state of O $= -2$. The oxidation state of O is normally -2, and the oxidation state of Cr is deduced from the formula because the sum of the oxidation states must $= 0$. (Cr ox state) $+ 3$(O ox state) $= 0$; (Cr ox state) $+ 3 (-2) = 0$, so Cr $= +6$.

(c) Cr_2O_3. The oxidation state of Cr $= +3$, and the oxidation state of O $= -2$. The oxidation state of O is normally -2, and the oxidation state of Cr is deduced from the formula because the sum of the oxidation states must $= 0$. 2(Cr ox state) $+ 3$(O ox state) $= 0$; 2(Cr ox state) $+ 3(-2) = 0$, so Cr $= +3$.

4.95   (a)
$$4\,Li(s) + O_2(g) \rightarrow 2\,Li_2O(s)$$
Oxidation states;   0   0   +1 −2
This is a redox reaction because Li increases in oxidation number (oxidation) and O decreases in number (reduction). O_2 is the oxidizing agent, and Li is the reducing agent.

(b)
$$Mg(s) + Fe^{2+}(aq) \rightarrow Mg^{2+}(aq) + Fe(s)$$
Oxidation states;   0   +2   +2   0
This is a redox reaction because Mg increases in oxidation number (oxidation) and Fe decreases in number (reduction). Fe^{2+} is the oxidizing agent, and Mg is the reducing agent.

(c)
$$Pb(NO_3)_2(aq) + Na_2SO_4(aq) \rightarrow PbSO_4(s) + 2\,NaNO_3(aq)$$
Oxidation states;   +2 +5−2   +1+6−2   +2+6−2   +1+5−2
This is a not a redox reaction because none of the atoms undergoes a change in oxidation number.

(d)
$$HBr(aq) + KOH(aq) \rightarrow H_2O(l) + KBr(aq)$$
Oxidation states;   +1−1   +1−2+1   +1−2   +1−1
This is a not a redox reaction because none of the atoms undergoes a change in oxidation number.

4.97   (a)   Skeletal reaction: $S(s) + O_2(g) \rightarrow SO_2(g)$
              Balanced reaction: $S(s) + O_2(g) \rightarrow SO_2(g)$

(b)   Skeletal reaction: $C_3H_6(g) + O_2(g) \rightarrow CO_2(g) + H_2O(g)$
              Balance C: $C_3H_6(g) + O_2(g) \rightarrow 3\,CO_2(g) + H_2O(g)$
              Balance H: $C_3H_6(g) + O_2(g) \rightarrow 3\,CO_2(g) + 3\,H_2O(g)$
              Balance O: $C_3H_6(g) + 9/2\,O_2(g) \rightarrow 3\,CO_2(g) + 3\,H_2O(g)$
              Clear fraction: $2\,C_3H_6(g) + 9\,O_2(g) \rightarrow 6\,CO_2(g) + 6\,H_2O(g)$

(c)   Skeletal reaction: $Ca(s) + O_2(g) \rightarrow CaO(s)$
              Balance O: $Ca(s) + O_2(g) \rightarrow 2\,CaO(s)$
              Balance Ca: $2\,Ca(s) + O_2(g) \rightarrow 2\,CaO(s)$

(d)   Skeletal reaction: $C_5H_{12}S(l) + O_2(g) \rightarrow CO_2(g) + H_2O(g) + SO_2(g)$
              Balance C: $C_5H_{12}S(l) + O_2(g) \rightarrow 5\,CO_2(g) + H_2O(g) + SO_2(g)$
              Balance H: $C_5H_{12}S(l) + O_2(g) \rightarrow 5\,CO_2(g) + 6\,H_2O(g) + SO_2(g)$
              Balance S: $C_5H_{12}S(l) + O_2(g) \rightarrow 5\,CO_2(g) + 6\,H_2O(g) + SO_2(g)$
              Balance O: $C_5H_{12}S(l) + 9\,O_2(g) \rightarrow 5\,CO_2(g) + 6\,H_2O(g) + SO_2(g)$

Cumulative Problems

4.99   **Given:** in 100 g solution, 20.0 g $C_2H_6O_2$; density of solution $= 1.03$ g/mL   **Find:** M of solution
Conceptual Plan: g $C_2H_6O_2$ → mol $C_2H_6O_2$ and g solution → mL solution → L solution

$$\frac{1 \text{ mol } C_2H_6O_2}{62.07 \text{ g } C_2H_6O_2} \qquad\qquad \frac{1.00 \text{ mL}}{1.03 \text{ g}} \qquad \frac{1 \text{ L}}{1000 \text{ mL}}$$

then M $C_2H_6O_2$

$$M = \frac{\text{mol } C_2H_6O_2}{\text{L solution}}$$

Solution:

$$20.0 \text{ g } C_2H_6O_2 \times \frac{1 \text{ mol } C_2H_6O_2}{62.07 \text{ g } C_2H_6O_2} = 0.32\underline{2}2 \text{ mol } C_2H_6O_2$$

$$100.0 \text{ g solution} \times \frac{1.00 \text{ mL solution}}{1.03 \text{ g solution}} \times \frac{1 \text{ L}}{1000 \text{ mL}} = 0.097\underline{0}9 \text{ L}$$

$$M = \frac{0.32\underline{2}2 \text{ mol } C_2H_6O_2}{0.097\underline{0}9 \text{ L}} = 3.32 \text{ M}$$

Check: The units of the answer (M $C_2H_6O_2$) are correct. The magnitude of the answer is reasonable because the concentration of solutions is usually between 0 M and 18 M.

4.101   **Given:** 2.5 g $NaHCO_3$ **Find:** g HCl
Conceptual Plan: g $NaHCO_3$ → mol $NaHCO_3$ → mol HCl → g HCl

$$\frac{1 \text{ mol } NaHCO_3}{84.02 \text{ g } NaHCO_3} \qquad \frac{1 \text{ mol HCl}}{1 \text{ mol } NaHCO_3} \qquad \frac{36.46 \text{ g HCl}}{1 \text{ mol HCl}}$$

Solution: $HCl(aq) + NaHCO_3(aq) \rightarrow H_2O(l) + CO_2(g) + NaCl(aq)$

$$2.5 \text{ g } NaHCO_3 \times \frac{1 \text{ mol } NaHCO_3}{84.02 \text{ g } NaHCO_3} \times \frac{1 \text{ mol HCl}}{1 \text{ mol } NaHCO_3} \times \frac{36.46 \text{ g HCl}}{1 \text{ mol HCl}} = 1.1 \text{ g HCl}$$

Check: The units of the answer (g HCl) are correct. The magnitude of the answer is reasonable because the molar mass of HCl is less than the molar mass of $NaHCO_3$.

4.103   **Given:** 1.0 kg C_8H_{18} **Find:** kg CO_2
Conceptual Plan: kg C_8H_{18} → g C_8H_{18} → mol C_8H_{18} → mol CO_2 → g CO_2 → kg CO_2

$$\frac{1000 \text{ g}}{\text{kg}} \qquad \frac{1 \text{ mol } C_8H_{18}}{114.22 \text{ g } C_8H_{18}} \qquad \frac{16 \text{ mol } CO_2}{2 \text{ mol } C_8H_{18}} \qquad \frac{44.01 \text{ g } CO_2}{1 \text{ mol } CO_2} \qquad \frac{\text{kg}}{1000 \text{ g}}$$

Solution: $2 C_8H_{18}(g) + 25 O_2(g) \rightarrow 16 CO_2(g) + 18 H_2O(g)$

$$1.0 \text{ kg } C_8H_{18} \times \frac{1000 \text{ g}}{\text{kg}} \times \frac{1 \text{ mol } C_8H_{18}}{114.22 \text{ g } C_8H_{18}} \times \frac{16 \text{ mol } CO_2}{2 \text{ mol } C_8H_{18}} \times \frac{44.01 \text{ g } CO_2}{1 \text{ mol } CO_2} \frac{\text{kg}}{1000 \text{ g}} = 3.1 \text{ kg } CO_2$$

Check: The units of the answer (kg CO_2) are correct. The magnitude of the answer is reasonable because the mole ratio of CO_2 to C_8H_{18} is 8:1.

4.105   **Given:** 3.00 mL $C_4H_6O_3$, $d = 1.08$ g/mL; 1.25 g $C_7H_6O_3$; 1.22 g $C_9H_8O_4$ **Find:** limiting reactant; theoretical yield $C_9H_8O_4$; % yield $C_9H_8O_4$
Conceptual Plan: mL $C_4H_6O_3$ → g $C_4H_6O_3$ → mol $C_4H_6O_3$ → mol $C_9H_8O_4$

$$\frac{1.08 \text{ g } C_4H_6O_3}{1.00 \text{ mL } C_4H_6O_3} \quad \frac{1 \text{ mol } C_4H_6O_3}{102.09 \text{ g } C_4H_6O_3} \quad \frac{1 \text{ mol } C_9H_8O_4}{1 \text{ mol } C_4H_6O_3} \quad \rightarrow \quad \begin{array}{l}\textbf{smaller amount determines} \\ \textbf{limiting reactant}\end{array}$$

g $C_7H_6O_3$ → mol $C_7H_6O_3$ → mol $C_9H_8O_4$

$$\frac{1 \text{ mol } C_7H_6O_3}{138.12 \text{ g } C_7H_6O_3} \qquad \frac{1 \text{ mol } C_9H_8O_4}{1 \text{ mol } C_7H_6O_3}$$

then mol $C_9H_8O_4$ → g $C_9H_8O_4$ then determine % yield

$$\frac{180.1 \text{ g } C_9H_8O_4}{\text{mol } C_9H_8O_4} \qquad\qquad \frac{\text{actual yield g } C_9H_8O_4}{\text{theoretical yield g } C_9H_8O_4} \times 100\%$$

Solution:

$$3.00 \text{ mL } C_4H_6O_3 \times \frac{1.08 \text{ g } C_4H_6O_3}{\text{mL } C_4H_6O_3} \times \frac{1 \text{ mol } C_4H_6O_3}{102.09 \text{ g } C_4H_6O_3} \times \frac{1 \text{ mol } C_9H_8O_4}{1 \text{ mol } C_4H_6P_3} = 0.031\underline{7}4 \text{ mol } C_9H_8O_4$$

$$1.25 \text{ g } C_7H_6O_3 \times \frac{1 \text{ mol } C_7H_6O_3}{138.12 \text{ g } C_7H_6O_3} \times \frac{1 \text{ mol } C_9H_8O_4}{1 \text{ mol } C_7H_6O_3} = 0.0090\underline{5}0 \text{ mol } C_9H_8O_4$$

Salicylic acid is the limiting reactant.

$$0.0090\underline{5}0 \text{ mol } C_9H_8O_4 \times \frac{180.1 \text{ g } C_9H_8O_4}{1 \text{ mol } C_9H_8O_4} = 1.6\underline{3}0 \text{ g } C_9H_8O_4$$

$$\frac{1.22 \text{ g } C_9H_8O_4}{1.6\underline{3}0 \text{ g } C_9H_8O_4} \times 100\% = 74.8\%$$

Check: The theoretical yield has the correct units ($g\ C_9H_8O_4$) and has a reasonable magnitude compared to the mass of $C_7H_6O_3$, the limiting reactant. The % yield is reasonable, under 100%.

4.107　**Given:** (a) 11 molecules H_2, 2 molecules O_2; (b) 8 molecules H_2, 4 molecules O_2; (c) 4 molecules H_2, 5 molecules O_2; (d) 3 molecules H_2, 6 molecules O_2　**Find:** loudest explosion based on equation
Conceptual Plan: Loudest explosion will occur in the balloon with the mol ratio closest to the balanced equation that contains the most H_2.
Solution: $2 H_2(g) + O_2(g) \rightarrow H_2O(l)$
Balloon (a) has enough O_2 to react with 4 molecules H_2; balloon (b) has enough O_2 to react with 8 molecules H_2; balloon (c) has enough O_2 to react with 10 molecules H_2; balloon (d) has enough O_2 for 3 molecules of H_2 to react. Balloon (b) also has the proper stoichiometric ratio of $2 H_2:1 O_2$ unlike the other three. Therefore, balloon (b) will have the loudest explosion because it has the most H_2 that will react.

Check: The answer seems correct because it has the most H_2 with enough O_2 in the balloon to react completely.

4.109　(a)　Skeletal reaction:　　$HCl(aq) + Hg_2(NO_3)_2(aq) \rightarrow Hg_2Cl_2(s) + HNO_3(aq)$
　　　　　　Balance Cl:　　　　$2 HCl(aq) + Hg_2(NO_3)_2(aq) \rightarrow Hg_2Cl_2(s) + 2 HNO_3(aq)$
　　　　(b)　Skeletal reaction:　　$KHSO_3(aq) + HNO_3(aq) \rightarrow H_2O(l) + SO_2(g) + KNO_3(aq)$
　　　　　　Balanced reaction:　　$KHSO_3(aq) + HNO_3(aq) \rightarrow H_2O(l) + SO_2(g) + KNO_3(aq)$
　　　　(c)　Skeletal reaction:　　$NH_4Cl(aq) + Pb(NO_3)_2(aq) \rightarrow PbCl_2(s) + NH_4NO_3(aq)$
　　　　　　Balance Cl:　　　　$2 NH_4Cl(aq) + Pb(NO_3)_2(aq) \rightarrow PbCl_2(s) + NH_4NO_3(aq)$
　　　　　　Balance N:　　　　$2 NH_4Cl(aq) + Pb(NO_3)_2(aq) \rightarrow PbCl_2(s) + 2 NH_4NO_3(aq)$
　　　　(d)　Skeletal reaction:　　$NH_4Cl(aq) + Ca(OH)_2(aq) \rightarrow NH_3(g) + H_2O(l) + CaCl_2(aq)$
　　　　　　Balance Cl:　　　　$2 NH_4Cl(aq) + Ca(OH)_2(aq) \rightarrow NH_3(g) + H_2O(l) + CaCl_2(aq)$
　　　　　　Balance N:　　　　$2 NH_4Cl(aq) + Ca(OH)_2(aq) \rightarrow 2 NH_3(g) + H_2O(l) + CaCl_2(aq)$
　　　　　　Balance H:　　　　$2 NH_4Cl(aq) + Ca(OH)_2(aq) \rightarrow 2 NH_3(g) + 2 H_2O(l) + CaCl_2(aq)$

4.111　**Given:** 1.5 L solution; 0.050 M $CaCl_2$; 0.085 M $Mg(NO_3)_2$　**Find:** g Na_3PO_4
Conceptual Plan: $V, M\ CaCl_2 \rightarrow mol\ CaCl_2$ **and** $V, M\ Mg(NO_3)_2 \rightarrow mol\ Mg(NO_3)_2$
　　　　　　　　　　　　　$V \times M = mol$　　　　　　　　　　　　　$V \times M = mol$
then $(mol\ CaCl_2 + mol\ Mg(NO_3)_2) \rightarrow Na_3PO_4 \rightarrow g\ Na_3PO_4$
　　　　　　$\frac{2 \text{ mol } Na_3PO_4}{3 \text{ mol } (CaCl_2 + Mg(NO_3)_2)}$　$\frac{163.97 \text{ g } Na_3PO_4}{1 \text{ mol } Na_3PO_4}$

Solution: $3 CaCl_2(aq) + 2 Na_3PO_4(aq) \rightarrow Ca_3(PO_4)_2(s) + 6 NaCl(aq)$
　　　　　　$3 Mg(NO_3)_2(aq) + 2 Na_3PO_4(aq) \rightarrow Mg_3(PO_4)_2(s) + 6 NaCl(aq)$
　　　　　　$1.5 \text{ L} \times 0.050 \text{ M } CaCl_2 = 0.07\underline{5} \text{ mol } CaCl_2$
　　　　　　$1.5 \text{ L} \times 0.085 \text{ M } Mg(NO_3)_2 = 0.12\underline{7}5 \text{ mol } Mg(NO_3)_2$

$$0.2\underline{0}25 \text{ mol } CaCl_2 \text{ and } Mg(NO_3)_2 \times \frac{2 \text{ mol } Na_3PO_4}{3 \text{ mol } CaCl_2 \text{ and } Mg(NO_3)_2} \times \frac{163.97 \text{ g mol } Na_3PO_4}{\text{mol } Na_3PO_4} = 22 \text{ g mol } Na_3PO_4$$

Check: The units of the answer ($g\ Na_3PO_4$) are correct. The magnitude of the answer is reasonable because it is needed to remove both the Ca and Mg ions.

4.113   **Given:** 1.0 L; 0.10 M OH⁻   **Find:** g Ba

Conceptual Plan: $V, M \rightarrow$ **mol OH⁻ $\rightarrow$ mol Ba(OH)$_2$ $\rightarrow$ mol BaO $\rightarrow$ mol Ba $\rightarrow$ g Ba**

$$V \times M = \text{mol} \quad \frac{1 \text{ mol Ba(OH)}_2}{2 \text{ mol OH}^-} \quad \frac{1 \text{ mol BaO}}{1 \text{ mol Ba(OH)}_2} \quad \frac{1 \text{ mol Ba}}{1 \text{ mol BaO}} \quad \frac{137.3 \text{ g Ba}}{1 \text{ mol Ba}}$$

Solution: $\text{BaO}(s) + \text{H}_2\text{O}(l) \rightarrow \text{Ba(OH)}_2(aq)$

$$1.0 \text{ L} \times \frac{0.10 \text{ mol OH}^-}{\text{L}} \times \frac{1 \text{ mol Ba(OH)}_2}{2 \text{ mol OH}^-} \times \frac{1 \text{ mol BaO}}{1 \text{ mol Ba(OH)}_2} \times \frac{1 \text{ mol Ba}}{1 \text{ mol BaO}} \times \frac{137.3 \text{ g Ba}}{1 \text{ mol Ba}} = 6.9 \text{ g Ba}$$

Check: The units of the answer (g Ba) are correct. The magnitude is reasonable because the molar mass of Ba is large and there are 2 moles hydroxide per mole Ba.

4.115   **Given:** 30.0% NaNO$_3$, $9.00/100 lb; 20.0% (NH$_4$)$_2SO_4$, $8.10/100 lb   **Find:** cost/lb N

Conceptual Plan: mass fertilizer $\rightarrow$ mass NaNO$_3$ $\rightarrow$ mass N $\rightarrow$ cost/lb N

$$\frac{30.0 \text{ lb NaNO}_3}{100 \text{ lb fertilizer}} \quad \frac{16.48 \text{ lb N}}{100 \text{ lb NaNO}_3} \quad \frac{\$9.00}{100 \text{ lb fertilizer}}$$

and mass fertilizer $\rightarrow$ mass (NH$_4$)$_2$SO$_4$ $\rightarrow$ mass N $\rightarrow$ cost/lb N

$$\frac{20.0 \text{ lb (NH}_4)_2\text{SO}_2}{100 \text{ lb fertilizer}} \quad \frac{21.2 \text{ lb N}}{100 \text{ lb (NH}_4)_2\text{SO}_4} \quad \frac{\$8.10}{100 \text{ lb fertilizer}}$$

Solution:

$$100 \text{ lb fertilizer} \times \frac{30.0 \text{ lb NaNO}_3}{100 \text{ lb fertilizer}} \times \frac{16.48 \text{ lb N}}{100 \text{ lb NaNO}_3} = 4.9\underline{4}4 \text{ lb N}$$

$$\frac{\$9.00}{100 \text{ lb fertilizer}} \times \frac{100 \text{ lb fertilizer}}{4.9\underline{4}4 \text{ lb N}} = \$1.82/\text{lb N}$$

$$100 \text{ lb fertilizer} \times \frac{20.0 \text{ lb (NH}_4)_2\text{SO}_4}{100 \text{ lb fertilizer}} \times \frac{21.2 \text{ lb N}}{100 \text{ lb (NH}_4)_2\text{SO}_4} = 4.2\underline{4}0 \text{ lb N}$$

$$\frac{\$8.10}{100 \text{ lb fertilizer}} \times \frac{100 \text{ lb fertilizer}}{4.2\underline{4} \text{ lb N}} = \$1.91/\text{lb N}$$

The more economical fertilizer is the NaNO$_3$ because it costs less/lb N.

Check: The units of the cost ($/lb N) are correct. The answer is reasonable because you compare the cost/lb N directly.

4.117   **Given:** 24.5 g Au; 24.5 g BrF$_3$; 24.5 g KF   **Find:** g KAuF$_4$

Conceptual Plan: g Au $\rightarrow$ mol Au $\rightarrow$ mol KAuF$_4$

$$\frac{1 \text{ mol Au}}{196.97 \text{ g Au}} \quad \frac{2 \text{ mol KAuF}_4}{2 \text{ mol Au}}$$

g BrF$_3$ $\rightarrow$ mol BrF$_3$ $\rightarrow$ mol KAuF$_4$ $\rightarrow$ **smaller mol amount determines limiting reactant**

$$\frac{1 \text{ mol BrF}_3}{136.9 \text{ g BrF}_3} \quad \frac{2 \text{ mol KAuF}_4}{2 \text{ mol BrF}_3}$$

g KF $\rightarrow$ mol KF $\rightarrow$ mol KAuF$_4$

$$\frac{1 \text{ mol KF}}{58.10 \text{ g KF}} \quad \frac{2 \text{ mol KAuF}_4}{2 \text{ mol KF}}$$

then mol KAuF$_4$ $\rightarrow$ g KAuF$_4$

$$\frac{312.07 \text{ g KAuF}_4}{\text{mol KAuF}_4}$$

$$2 \text{ Au}(s) + 2 \text{ BrF}_3(l) + 2 \text{ KF}(s) \rightarrow \text{Br}_2(l) + 2 \text{ KAuF}_4(s)$$

Oxidation states;   0   +3 −1   +1 −1   0   +1 +3 −1

This is a redox reaction because Au increases in oxidation number (oxidation) and Br decreases in number (reduction). BrF$_3$ is the oxidizing agent, and Au is the reducing agent.

Solution:

$$24.5 \text{ g Au} \times \frac{1 \text{ mol Au}}{196.97 \text{ g Au}} \times \frac{2 \text{ mol KAuF}_4}{2 \text{ mol Au}} = 0.124\underline{4} \text{ mol KAuF}_4$$

$$24.5 \text{ g BrF}_3 \times \frac{1 \text{ mol BrF}_3}{136.90 \text{ g BrF}_3} \times \frac{2 \text{ mol KAuF}_4}{2 \text{ mol BrF}_3} = 0.179\underline{0} \text{ mol KAuF}$$

$$24.5 \text{ g KF} \times \frac{1 \text{ mol KF}}{58.10 \text{ g KF}} \times \frac{2 \text{ mol KAuF}_4}{2 \text{ mol KF}} = 0.421\underline{7} \text{ mol KAuF}_4$$

$$0.124\underline{4} \text{ mol KAuF}_4 \times \frac{312.07 \text{ g KAuF}_4}{1 \text{ mol KAuF}_4} = 38.8 \text{ g KAuF}_4$$

Check: The units of the answer (g KAuF$_4$) are correct. The magnitude of the answer is reasonable compared to the mass of the limiting reactant Au.

4.119   **Given:** solution may contain Ag$^+$, Ca^{2+}, and Cu^{2+} **Find:** which ions are present.
Conceptual Plan: Test the solution sequentially with NaCl, Na$_2$SO$_4$, and Na$_2$CO$_3$ and see if precipitates form.
Solution: Original solution + NaCl yields no reaction: Ag$^+$ is not present because chlorides are normally soluble, but Ag$^+$ is an exception.
Original solution with Na$_2$SO$_4$ yields a precipitate and solution 2. The precipitate is CaSO$_4$, so Ca^{2+} is present. Sulfates are normally soluble, but Ca^{2+} is an exception.
Solution 2 with Na$_2$CO$_3$ yields a precipitate. The precipitate is CuCO$_3$, so Cu^{2+} is present. All carbonates are insoluble.
Net Ionic Equations:
$$Ca^{2+}(aq) + SO_4{}^{2-}(aq) \rightarrow CaSO_4(s)$$
$$Cu^{2+}(aq) + CO_3{}^{2-}(aq) \rightarrow CuCO_3(s)$$

Check: The answer is reasonable because two different precipitates formed and all of the Ca^{2+} was removed before the carbonate was added.

4.121   **Given:** 1.00 g NH$_3$ **Find:** g PH$_3$
Conceptual Plan: Determine reaction sequence, then g NH$_3 \rightarrow$ mol NH$_3 \rightarrow$ mol PH$_3 \rightarrow$ g PH$_3$

$$\frac{1 \text{ mol NH}_3}{17.04 \text{ g NH}_3} \qquad \frac{34.00 \text{ g PH}_3}{1 \text{ mol PH}_3}$$

Solution: Balance the reaction sequence:
$$6 \text{ NH}_3 + 7 \tfrac{1}{2} \text{ O}_2 \rightarrow 6 \text{ NO} + 9 \text{ H}_2\text{O}$$
$$6 \text{ NO} + \text{P}_4 \rightarrow \text{P}_4\text{O}_6 + 3 \text{ N}_2$$
$$\text{P}_4\text{O}_6 + 6 \text{ H}_2\text{O} \rightarrow 4 \text{ H}_3\text{PO}_3$$
$$4 \text{ H}_3\text{PO}_3 \rightarrow \text{PH}_3 + 3 \text{ H}_3\text{PO}_4$$
Therefore, 6 mol NH$_3$ produces 1 mol PH$_3$.

$$1.00 \text{ g NH}_3 \times \frac{1 \text{ mol NH}_3}{17.04 \text{ g NH}_3} \times \frac{1 \text{ mol PH}_3}{6 \text{ mol NH}_3} \times \frac{34.00 \text{ g PH}_3}{1 \text{ mol PH}_3} = 0.333 \text{ g PH}_3$$

Check: The answer is g PH$_3$, which is correct. The magnitude is reasonable. Even though the molar mass of PH$_3$ is greater than NH$_3$, 6 mol of NH$_3$ are required to produce 1 mol PH$_3$.

4.123   **Given:** 10.0 kg mixture; 30.35% hexane; 15.85% heptane; 53.80% octane   **Find:** total mass CO$_2$
Conceptual Plan: kg hexane $\rightarrow$ kmol hexane $\rightarrow$ kmol CO$_2$ $\rightarrow$ kg CO$_2$

$$\frac{1 \text{ kmol C}_6\text{H}_{14}}{86.20 \text{ kg C}_6\text{H}_{14}} \qquad \frac{12 \text{ kmol CO}_2}{2 \text{ kmol C}_6\text{H}_{14}} \qquad \frac{44.01 \text{ kg CO}_2}{1 \text{ kmol CO}_2}$$

kg heptane $\rightarrow$ kmol heptane $\rightarrow$ kmol CO$_2$ $\rightarrow$ kg CO$_2$

$$\frac{1 \text{ kmol C}_7\text{H}_{16}}{100.23 \text{ kg C}_7\text{H}_{16}} \qquad \frac{7 \text{ kmol CO}_2}{1 \text{ kmol C}_7\text{H}_{16}} \qquad \frac{44.01 \text{ kg CO}_2}{1 \text{ kmol CO}_2}$$

kg octane $\rightarrow$ kmol octane $\rightarrow$ kmol CO$_2$ $\rightarrow$ kg CO$_2$

$$\frac{1 \text{ kmol C}_8\text{H}_{18}}{114.26 \text{ kg C}_8\text{H}_{18}} \qquad \frac{16 \text{ kmol CO}_2}{2 \text{ kmol C}_8\text{H}_{18}} \qquad \frac{44.01 \text{ kg CO}_2}{1 \text{ kmol CO}_2}$$

Solution: Balanced Reactions:

$$2\,C_6H_{14}(l) + 19\,O_2(g) \rightarrow 12\,CO_2(g) + 14\,H_2O(l)$$
$$C_7H_{16}(l) + 11\,O_2(g) \rightarrow 7\,CO_2(g) + 8\,H_2O(l)$$
$$2\,C_8H_{18}(l) + 25\,O_2(g) \rightarrow 16\,CO_2(g) + 18\,H_2O(l)$$

$$10.0\ \text{kg mix} \times \frac{30.35\ \text{kg C}_6\text{H}_{14}}{100.0\ \text{kg mix}} \times \frac{1\ \text{kmol C}_6\text{H}_{14}}{86.20\ \text{kg C}_6\text{H}_{14}} \times \frac{12\ \text{kmol CO}_2}{2\ \text{kmol C}_6\text{H}_{14}} \times \frac{44.01\ \text{kg CO}_2}{1\ \text{kmol CO}_2} = 9.29\underline{7}\ \text{kg CO}_2$$

$$10.0\ \text{kg mix} \times \frac{15.85\ \text{kg C}_7\text{H}_{16}}{100.0\ \text{kg mix}} \times \frac{1\ \text{kmol C}_7\text{H}_{16}}{100.23\ \text{kg C}_7\text{H}_{16}} \times \frac{7\ \text{kmol CO}_2}{1\ \text{kmol C}_7\text{H}_{16}} \times \frac{44.01\ \text{kg CO}_2}{1\ \text{kmol CO}_2} = 4.87\underline{2}\ \text{kg CO}_2$$

$$10.0\ \text{kg mix} \times \frac{53.80\ \text{kg C}_8\text{H}_{18}}{100.0\ \text{kg mix}} \times \frac{1\ \text{kmol C}_8\text{H}_{18}}{114.26\ \text{kg C}_8\text{H}_{18}} \times \frac{16\ \text{kmol CO}_2}{2\ \text{kmol C}_8\text{H}_{18}} \times \frac{44.01\ \text{kg CO}_2}{1\ \text{kmol CO}_2} = 16.\underline{5}78\ \text{kg CO}_2$$

Total CO_2 = 9.30 kg + 4.87 kg + 16.6 kg = 30.8 kg CO_2

Check: The units of the answer (kg CO_2) are correct. The magnitude of the answer is reasonable because a large amount of CO_2 is produced per mole of hydrocarbon.

Challenge Problems

4.125   **Given:** g C_3H_8 + C_2H_2 = 2.0 g; mol CO_2 = 1.5 mol H_2O **Find:** original g C_2H_2
Conceptual Plan: mol $C_3H_8 \rightarrow$ mol CO_2 and mol H_2O and mol $C_2H_2 \rightarrow$ mol CO_2 and mol H_2O
Solution: Let a = mol C_3H_8 and b = mol C_2H_2

$$\begin{array}{llll} C_3H_8 + 5\,O_2 \rightarrow 3\,CO_2 + 4\,H_2O & & C_2H_2 + 3/2\,O_2 \rightarrow 2\,CO_2 + H_2O \\ a \qquad\qquad\quad 3a \quad\ \ 4a & & b \qquad\qquad\quad 2b \quad\ b \end{array}$$

Total mol CO_2 = $3a + 2b$ and total mol H_2O = $4a + b$

mol CO_2 = 1.5(mol H_2O)

So $3a + 2b = 1.5(4a + b)$

And $\left(a\ \text{mol C}_3\text{H}_8 \times \dfrac{44.11\ C_3H_8}{1\ \text{mol C}_3\text{H}_8} \right) + \left(b\ \text{mol C}_2\text{H}_2 \times \dfrac{26.01\ C_2H_2}{\text{mol C}_2\text{H}_2} \right) = 2.0\ \text{g}$

Solve the simultaneous equations: a = 9.98×10^{-3} mol C_3H_8 and b = 0.059 mol C_2H_2.
Substitute for b and solve for grams C_2H_2.

$$0.06\underline{0}\ \text{mol C}_2\text{H}_2 \times \frac{26.01\ C_2H_2}{\text{mol C}_2\text{H}_2} = 1.\underline{5}6\ \text{g C}_2\text{H}_2 = 1.6\ \text{g C}_2\text{H}_2$$

Check: The units of the answer (g C_2H_2) are correct. The magnitude is reasonable because it is less than the total mass.

4.127   **Given:** 0.100 L, 1.22 M NaI; total mass = 28.1 g   **Find:** g AgI
Conceptual Plan: vol, M $\rightarrow$ mol NaI $\rightarrow$ mol I$^-$; total mol I$^- \rightarrow$ mol AgI and HgI$_2$

Solution: $0.100\ \text{L soln} \times \dfrac{1.22\ \text{mol NaI}}{\text{L soln}} \times \dfrac{1\ \text{mol I}^-}{\text{mol NaI}} = 0.122\ \text{mol I}^-$

Let x = mol AgI   and y = mol HgI$_2$

$x + 2y$ = 0.122 mol I$^-$ so y = 0.061 − 0.5x

$\left(x\ \text{mol AgI} \times \dfrac{234.77\ \text{g AgI}}{1\ \text{mol AgI}} \right) + \left(y\ \text{mol HgI}_2 \times \dfrac{454.39\ \text{g HgI}_2}{\text{mol HgI}_2} \right) = 28.1\ \text{g}$

Solve the simultaneous equations and x = 0.0504 mol AgI.

$$0.0504\ \text{mol AgI} \times \frac{234.77\ \text{g AgI}}{1\ \text{mol AgI}} = 11.8\ \text{g AgI}$$

Check: The units of the answer (g AgI) are correct. The magnitude is reasonable because it is less than the total mass.

4.129   **Given:** 3.5×10^{-3} M Ca^{2+}; 1.1×10^{-3} M Mg^{2+}; 19.5 gal H_2O; 0.65 kg detergent/load   **Find:** % by mass Na$_2$CO$_3$
Conceptual Plan: gal $H_2O \rightarrow$ L H_2O then $V, M \rightarrow$ mol Ca^{2+} and $V, M \rightarrow$ mol Mg^{2+}

$$\frac{3.785\ \text{L}}{1\ \text{gal}} \qquad\qquad \text{vol} \times \text{M} = \text{mol} \qquad\qquad \text{vol} \times \text{M} = \text{mol}$$

then total moles ions → mol CO$_3^{2-}$ → mol Na$_2$CO$_3$ → g Na$_2$CO$_3$ → kg Na$_2$CO$_3$ → % Na$_2$CO$_3$

$$\frac{1 \text{ mol CO}_3^{2-}}{1 \text{ mol ion}} \quad \frac{1 \text{ mol Na}_2\text{CO}_3}{1 \text{ mol CO}_3^{2-}} \quad \frac{106.01 \text{ g Na}_2\text{CO}_3}{1 \text{ mol Na}_2\text{CO}_3} \quad \frac{\text{kg}}{1000 \text{ g}} \quad \frac{\text{kg Na}_2\text{CO}_3}{\text{kg detergent}} \times 100\%$$

Solution: $19.5 \text{ gal} \times \dfrac{3.785 \text{ L}}{1 \text{ gal}} \times \dfrac{3.5 \times 10^{-3} \text{ mol Ca}^{2+}}{\text{L}} = 0.258 \text{ mol Ca}^{2+}$

$19.5 \text{ gal} \times \dfrac{3.785 \text{ L}}{1 \text{ gal}} \times \dfrac{1.1 \times 10^{-3} \text{ mol Mg}^{2+}}{\text{L}} = 0.08119 \text{ mol Mg}^{2+}$

$0.3392 \text{ mol ions} \times \dfrac{1 \text{ mol CO}_3^{2-}}{\text{mol ions}} \times \dfrac{1 \text{ mol Na}_2\text{CO}_3}{1 \text{ mol CO}_3^{2-}} \times \dfrac{106.01 \text{ g Na}_2\text{CO}_3}{1 \text{ mol Na}_2\text{CO}_3} \times \dfrac{\text{kg Na}_2\text{CO}_3}{1000 \text{ g Na}_2\text{CO}_3} = 0.03596 \text{ kg Na}_2\text{CO}_3$

$\dfrac{0.03596 \text{ kg Na}_2\text{CO}_3}{0.65 \text{ kg detergent}} \times 100\% = 5.5\% \text{ Na}_2\text{CO}_3$

Check: The units of the answer ($\%$ Na$_2$CO$_3$) are correct. The magnitude of the answer is reasonable. The percent is less than 100%.

4.131   In designing the unit, you would need to consider the theoretical yield and % yield of the reaction, how changing the limiting reactant would affect the reaction, and the stoichiometry between KO$_2$ and O$_2$ to determine the mass of KO$_2$ required to produce enough O$_2$ for 10 minutes. You might also consider the speed of the reaction and whether the reaction produced heat. In addition, because your body does not use 100% of the oxygen taken in with each breath, the apparatus would need to replenish only the oxygen used. The percentage of oxygen in air is about 20%, and the percentage in exhaled air is about 16%; so we will assume that 4% of the air would need to be replenished with oxygen. (Note: The problem can also be solved by finding the amount of KO$_2$ that would be required to react with all of the exhaled CO$_2$.)

Given: air = 4% O$_2$; volume = 5 − 8 L/min; 1 mol gas = 22.4 L gas   **Find:** O$_2$ for 10 min breathing time
Conceptual Plan: 10 min → vol air → vol O$_2$ → mol O$_2$ → mol KO$_2$ → g KO$_2$

$$\frac{8 \text{ L air}}{1 \text{ min}} \quad \frac{4 \text{ L O}_2}{100 \text{ L air}} \quad \frac{1 \text{ mol O}_2}{22.4 \text{ L O}_2} \quad \frac{4 \text{ mol KO}_2}{3 \text{ mol O}_2} \quad \frac{71.10 \text{ g KO}_2}{1 \text{ mol KO}_2}$$

Solution: $10 \text{ min} \times \dfrac{8 \text{ L air}}{\text{min}} \times \dfrac{4 \text{ L O}_2}{100 \text{ L air}} \times \dfrac{1 \text{ mol O}_2}{22.4 \text{ L O}_2} \times \dfrac{4 \text{ mol KO}_2}{3 \text{ mol O}_2} \times \dfrac{71.10 \text{ g KO}_2}{1 \text{ mol KO}_2} = 14 \text{ g KO}_2$

Check: The units of the answer (g KO$_2$) are correct. The magnitude of the answer is reasonable because it is an amount that could be carried in a portable device.

4.133   **Given:** 151 g Na$_2$B$_4$O$_7$ **Find:** g B$_5$H$_9$
Conceptual Plan: g Na$_2$B$_4$O$_7$ → mol Na$_2$B$_4$O$_7$ → mol B$_5$H$_9$ → g B$_5$H$_9$

$$\frac{\text{mol Na}_2\text{B}_4\text{O}_7}{201.22 \text{ g Na}_2\text{B}_4\text{O}_7} \quad \frac{4 \text{ mol B}_5\text{H}_9}{5 \text{ mol Na}_2\text{B}_4\text{O}_7} \quad \frac{63.13 \text{ g B}_5\text{H}_9}{\text{mol B}_5\text{H}_9}$$

Solution: All of the B in B$_5$H$_9$ goes to the Na$_2$B$_4$O$_7$, so the mole ratio between the two can be used.

$$151 \text{ g Na}_2\text{B}_4\text{O}_7 \times \frac{1 \text{ mol Na}_2\text{B}_4\text{O}_7}{201.22 \text{ g Na}_2\text{B}_4\text{O}_7} \times \frac{4 \text{ mol B}_5\text{H}_9}{5 \text{ mol Na}_2\text{B}_4\text{O}_7} \times \frac{63.13 \text{ g B}_5\text{H}_9}{1 \text{ mol B}_5\text{H}_9} = 37.9 \text{ g B}_5\text{H}_9$$

Check: The units of the answer (g B$_5$H$_9$) are correct. The magnitude of the answer is reasonable because the molar mass of B$_5$H$_9$ is less than the molar mass of Na$_2$B$_4$O$_7$.

Conceptual Problems

4.135   **Given:** 5 mol NO; 10 mol H$_2$ **Find:** conditions of product mixture
Conceptual Plan: mol H$_2$ → mol NO and mol H$_2$ → mol NH$_3$ and mol H$_2$ → mol H$_2$O
Solution: The correct answer is (a). Because the mol ratio of H$_2$ to NO is 5:2, the 10 mol of H$_2$ will require 4 mol NO and H$_2$ is the limiting reactant. This eliminates answers (b) and (c). Because there is excess NO, this eliminates (d), leaving answer (a).

4.137   **Given:** 6 molecules N_2H_4; 4 molecules N_2O_4; a) contains 9 molecules N_2, 12 molecules H_2O, and 1 molecule N_2H_4; solution (b) contains 12 molecules N_2, 16 molecules H_2O, and 2 molecules N_2H_4; solution (c) contains 9 molecules N_2 and 12 molecules H_2O **Find:** theoretical yield N_2, H_2O

Conceptual Plan: molecules $N_2H_4 \rightarrow$ molecules N_2

$$\frac{3 \text{ molecules } N_2}{2 \text{ molecules } N_2H_4}$$ $\rightarrow$ **smaller molecules amount determines limiting reactant**

molecules $N_2O_4 \rightarrow$ molecules N_2

$$\frac{3 \text{ molecules } N_2}{1 \text{ molecules } N_2O_4}$$

molecules $N_2H_4 \rightarrow$ molecules H_2O

$$\frac{4 \text{ molecules } H_2O}{2 \text{ molecules } N_2H_4}$$

molecules $N_2H_4 \rightarrow$ molecules N_2O_4

$$\frac{1 \text{ molecule } N_2O_4}{2 \text{ molecules } N_2H_4}$$

Solution: $6 \text{ molecules } N_2H_4 \times \dfrac{3 \text{ molecules } N_2}{2 \text{ molecules } N_2H_4} = 9$ molecules N_2

$6 \text{ molecules } N_2O_4 \times \dfrac{3 \text{ molecules } N_2}{1 \text{ molecules } N_2O_4} = 18$ molecules N_2

Limiting reactant $= N_2H_4$ because it produced the least molecules of N_2.

$6 \text{ molecules } N_2H_4 \times \dfrac{4 \text{ molecules } H_2O}{2 \text{ molecules } N_2H_4} = 12$ molecules H_2O

$6 \text{ molecules } N_2H_4 \times \dfrac{1 \text{ molecule } N_2O_4}{2 \text{ molecules } N_2H_4} = 3$ molecules N_2O_4 used

Reaction mixture should contain 9 molecules N_2, 12 molecules H_2O, and 1 molecule N_2O_4; this is best represented by (a).

5 Gases

Review Questions

5.1   Pressure is the force exerted per unit area by gas molecules as they strike the surfaces around them. Pressure is caused by collisions of gas molecules with surfaces or other gas molecules.

5.3   When you exhale, you reverse the process of inhalation. The chest cavity muscles relax, which decreases the lung volume, increasing the pressure within the lungs and forcing the air out of the lungs.

5.5   A manometer is a U-shaped tube containing a dense liquid, usually mercury. In an open-ended manometer, one end of the tube is open to atmospheric pressure and the other is attached to a flask containing the gas sample. If the pressure of the gas sample is exactly equal to atmospheric pressure, the mercury levels on both sides of the tube are the same. If the pressure of the sample is greater than atmospheric pressure, the mercury level on the sample side of the tube is lower than on the side open to the atmosphere. If the pressure of the sample is less than atmospheric pressure, the mercury level on the sample side is higher than on the side open to the atmosphere. This type of manometer always measures the pressure of the gas sample relative to atmospheric pressure. The difference in height between the two levels is equal to the pressure difference from atmospheric pressure.

5.7   This pain is caused by air-containing cavities within your ear. When you ascend a mountain, the external pressure (the pressure that surrounds you) drops, while the pressure within your ear cavities (the internal pressure) remains the same. This creates an imbalance—the greater internal pressure forces your eardrum to bulge outward, causing pain. With time, and the help of a yawn or two, the excess air within your ear cavities escapes, equalizing the internal and external pressure and relieving the pain.

5.9   When we breathe, we expand the volume of our chest cavity, reducing the pressure on the outer surface of the lungs to less than 1 atm (Boyle's law). Because of this pressure differential, the lungs expand, the pressure in them falls, and air from outside our lungs then flows into them. Extra-long snorkels do not work because of the pressure exerted by water at an increased depth. A diver at 10 m experiences an external pressure of 2 atm. This is more than the muscles of the chest cavity can overcome—the chest cavity and lungs are compressed, resulting in an air pressure within them of more than 1 atm. If the diver had a snorkel that went to the surface—where the air pressure is 1 atm—air would flow out of his lungs, not into them. It would be impossible to breathe.

5.11   The ideal gas law ($PV = nRT$) combines all of the relationships between the four variables relevant to gases [pressure, volume, number of moles, and temperature (in kelvin's)] in one simple expression.

5.13   The molar volume of an ideal gas is the volume occupied by one mole of gas at STP which is $T = 0\,°C$ (273 K) and $P = 1.00$ atm. Substituting these values into the ideal gas law, one can calculate this value as 22.414 L.

5.15   The pressure due to any individual component in a gas mixture is called the partial pressure (P_n) of that component and can be calculated from the ideal gas law by assuming that each gas component acts independently. The sum of the partial pressures of the components in a gas mixture must equal the total pressure: $P_{total} = P_a + P_b + P_c + \ldots$ where P_{total} is the total pressure and $P_a, P_b, P_c \ldots$ are the partial pressures of the components.

5.17   No, when collecting a gas over water, it will contain some water molecules. The vapor pressure of water can be found in Table 5.4. Therefore, $P_{Gas} = P_{Total} - P_{H_2O}$.

5.19   The basic postulates of kinetic molecular theory are as follows: (1) The size of a particle is negligibly small, (2) the average kinetic energy of a particle is proportional to the temperature in kelvins, and (3) the collision of one particle with another (or with the walls) is completely elastic. Pressure is defined as force divided by area. According to kinetic molecular theory, a gas is a collection of particles in constant motion. The motion results in collisions between the particles and the surfaces around them. As each particle collides with a surface, it exerts a force upon that surface. The result of many particles in a gas sample exerting forces on the surfaces around them is constant pressure.

5.21   Postulate 2 of kinetic molecular theory states that the average kinetic energy is proportional to the temperature in kelvins. The root mean that square velocity of a collection of gas particles is inversely proportional to the square root of the molar mass of the particles in kilograms per mole.

5.23   The process by which gas molecules spread out in response to a concentration gradient is called diffusion. Effusion is the process by which a gas escapes from a container into a vacuum through a small hole. The rate of effusion is inversely proportional to the square root of the molar mass of the gas.

Problems by Topic

Converting between Pressure Units

5.25   (a)   **Given:** 24.9 in Hg   **Find:** atm
              Conceptual Plan: in Hg → atm
$$\frac{1\ atm}{29.92\ in\ Hg}$$
              Solution: $24.9\ \text{in Hg} \times \dfrac{1\ atm}{29.92\ \text{in Hg}} = 0.832\ atm$

              Check: The units (atm) are correct. The magnitude of the answer (< 1) makes physical sense because we started with less than 29.92 in Hg.

        (b)   **Given:** 24.9 in Hg   **Find:** mmHg
              Conceptual Plan: Use answer from part (a) then convert atm → mmHg
$$\frac{760\ mmHg}{1\ atm}$$
              Solution: $0.832\ \text{atm} \times \dfrac{760\ mmHg}{1\ \text{atm}} = 632\ mmHg$

              Check: The units (mmHg) are correct. The magnitude of the answer $(< 760\ mmHg)$ makes physical sense because we started with less than 1 atm.

        (c)   **Given:** 24.9 in Hg   **Find:** psi
              Conceptual Plan: Use answer from part (a) then convert atm → psi
$$\frac{14.7\ psi}{1\ atm}$$
              Solution: $0.832\ \text{atm} \times \dfrac{14.7\ psi}{1\ \text{atm}} = 12.2\ psi$

              Check: The units (psi) are correct. The magnitude of the answer $(< 14.7\ psi)$ makes physical sense because we started with less than 1 atm.

        (d)   **Given:** 24.9 in Hg   **Find:** Pa
              Conceptual Plan: Use answer from part (a) then convert atm → Pa
$$\frac{101,325\ Pa}{1\ atm}$$
              Solution: $0.832\ \text{atm} \times \dfrac{101,325\ Pa}{1\ \text{atm}} = 8.43 \times 10^4\ Pa$

              Check: The units (Pa) are correct. The magnitude of the answer $(< 101,325\ Pa)$ makes physical sense because we started with less than 1 atm.

5.27   (a)   **Given:** 31.85 in Hg   **Find:** mmHg
Conceptual Plan: in Hg → mmHg

$$\frac{25.4 \text{ mmHg}}{1 \text{ in Hg}}$$

Solution: $31.85 \text{ in Hg} \times \dfrac{25.4 \text{ mmHg}}{1 \text{ in Hg}} = 809.0 \text{ mmHg}$

Check: The units (mmHg) are correct. The magnitude of the answer (809) makes physical sense because inches are larger than millimeters.

(b)   **Given:** 31.85 in Hg   **Find:** atm
Conceptual Plan: Use answer from part (a) then convert mmHg → atm

$$\frac{1 \text{ atm}}{760 \text{ mmHg}}$$

Solution: $809.0 \text{ mmHg} \times \dfrac{1 \text{ atm}}{760 \text{ mmHg}} = 1.064 \text{ atm}$

Check: The units (atm) are correct. The magnitude of the answer (>1) makes physical sense because we started with more than 760 mmHg.

(c)   **Given:** 31.85 in Hg   **Find:** torr
Conceptual Plan: Use answer from part (a) then convert mmHg → torr

$$\frac{1 \text{ torr}}{1 \text{ mmHg}}$$

Solution: $809.0 \text{ mmHg} \times \dfrac{1 \text{ torr}}{1 \text{ mmHg}} = 809.0 \text{ torr}$

Check: The units (torr) are correct. The magnitude of the answer (809) makes physical sense because both units are of the same size.

(d)   **Given:** 31.85 in Hg   **Find:** kPa
Conceptual Plan: Use answer from part (b) then convert atm → Pa → kPa

$$\frac{101,325 \text{ Pa}}{1 \text{ atm}} \quad \frac{1 \text{ kPa}}{1000 \text{ Pa}}$$

Solution: $1.064 \text{ atm} \times \dfrac{101,325 \text{ Pa}}{1 \text{ atm}} \times \dfrac{1 \text{ kPa}}{1000 \text{ Pa}} = 107.8 \text{ kPa}$

Check: The units (kPa) are correct. The magnitude of the answer (108) makes physical sense because we started with more than 1 atm and there are ~ 101 kPa in an atm.

5.29   (a)   **Given:** $P_{bar} = 762.4$ mmHg and figure   **Find:** P_{gas}
Conceptual Plan: Measure height difference then convert cm Hg → mmHg → mmHg

$$\frac{10 \text{ mmHg}}{1 \text{ cm Hg}} \quad P_{gas} = h + P_{bar}$$

Solution:

$$h = 7.0 \text{ cm Hg} \times \frac{10 \text{ mmHg}}{1 \text{ cm Hg}} = 70. \text{ mmHg} \quad P_{gas} = 70. \text{ mmHg} + 762.4 \text{ mmHg} = 832 \text{ mmHg}$$

Check: The units (mmHg) are correct. The magnitude of the answer (832 mmHg) makes physical sense because the mercury column is higher on the right, indicating that the pressure is above barometric pressure. No significant figures to the right of the decimal point can be reported because the mercury height is known only to the ones place.

(b)   **Given:** $P_{bar} = 762.4$ mmHg and figure   **Find:** P_{gas}
Conceptual Plan: Measure height difference then convert cm Hg → mmHg → mmHg

$$\frac{10 \text{ mmHg}}{1 \text{ cm Hg}} \quad P_{gas} = h + P_{bar}$$

Solution:

$$h = -4.4 \text{ cm Hg} \times \frac{10 \text{ mmHg}}{1 \text{ cm Hg}} = -44 \text{ mmHg} \quad P_{gas} = -44 \text{ mmHg} + 762.4 \text{ mmHg} = 718 \text{ mmHg}$$

Check: The units (mmHg) are correct. The magnitude of the answer (718 mmHg) makes physical sense because the mercury column is higher on the left, indicating that the pressure is below barometric pressure. No significant figures to the right of the decimal point can be reported because the mercury height is known only to the ones place.

Simple Gas Laws

5.31   **Given:** $V_1 = 5.6$ L, $P_1 = 735$ mmHg, and $V_2 = 9.4$ L   **Find:** P_2
        Conceptual Plan: $V_1, P_1, V_2 \rightarrow P_2$

$$P_1 V_1 = P_2 V_2$$

        Solution:

$P_1 V_1 = P_2 V_2$ Rearrange to solve for P_2.

$$P_2 = P_1 \frac{V_1}{V_2} = 735 \text{ mmHg} \times \frac{5.6 \, \cancel{L}}{9.4 \, \cancel{L}} = 4\underline{3}7.872 \text{ mmHg} = 4.4 \times 10^2 \text{ mmHg}$$

Check: The units (mmHg) are correct. The magnitude of the answer (440 mmHg) makes physical sense because Boyle's law indicates that as the volume increases, the pressure decreases.

5.33   **Given:** $V_1 = 48.3$ mL, $T_1 = 22$ °C, and $T_2 = 87$ °C   **Find:** V_2
        Conceptual Plan: °C $\rightarrow$ K then $V_1, T_1, T_2 \rightarrow V_2$

$$K = °C + 273.15 \qquad \frac{V_1}{T_1} = \frac{V_2}{T_2}$$

        Solution: $T_1 = 22$ °C $+ 273.15 = 295$ K and $T_2 = 87$ °C $+ 273.15 = 360.$ K

$$\frac{V_1}{T_1} = \frac{V_2}{T_2} \quad \text{Rearrange to solve for } V_2. \quad V_2 = V_1 \frac{T_2}{T_1} = 48.3 \text{ mL} \times \frac{360 \, K}{295 \, K} = 58.9 \text{ mL}$$

Check: The units (mL) are correct. The magnitude of the answer (59 mL) makes physical sense because Charles's law indicates that as the volume increases, the temperature increases.

5.35   **Given:** $V_1 = 2.46$ L, $n_1 = 0.158$ mol, and $\Delta n = 0.113$ mol   **Find:** V_2
        Conceptual Plan: $n_1 \rightarrow n_2$ then $V_1, n_1, n_2 \rightarrow V_2$

$$n_1 + \Delta n = n_2 \qquad \frac{V_1}{n_1} = \frac{V_2}{n_2}$$

        Solution: $n_2 = 0.158$ mol $+ 0.113$ mol $= 0.271$ mol

$$\frac{V_1}{n_1} = \frac{V_2}{n_2} \quad \text{Rearrange to solve for } V_2. \quad V_2 = V_1 \frac{n_2}{n_1} = 2.46 \text{ L} \times \frac{0.271 \, \cancel{mol}}{0.158 \, \cancel{mol}} = 4.2\underline{1}937 \text{ L} = 4.22 \text{ L}$$

Check: The units (L) are correct. The magnitude of the answer (4 L) makes physical sense because Avogadro's law indicates that as the number of moles increases, the volume increases.

Ideal Gas Law

5.37   **Given:** $n = 0.118$ mol, $P = 0.97$ atm, and $T = 305$ K   **Find:** V
        Conceptual Plan: $n, P, T \rightarrow V$

$$PV = nRT$$

        Solution: $PV = nRT$ Rearrange to solve for V. $V = \dfrac{nRT}{P} = \dfrac{0.118 \, \cancel{mol} \times 0.08206 \dfrac{L \cdot \cancel{atm}}{\cancel{mol} \cdot \cancel{K}} \times 305 \, \cancel{K}}{0.97 \, \cancel{atm}} = 3.0 \text{ L}$

The volume would be the same for argon gas because the ideal gas law does not care about the mass of the gas, only the number of moles of gas.

Check: The units (L) are correct. The magnitude of the answer (3 L) makes sense because, as you will see in the next section, one mole of an ideal gas under standard conditions (273 K and 1 atm) occupies 22.4 L. Although these are not standard conditions, they are close enough for a ballpark check of the answer. Because this gas sample contains 0.118 moles, a volume of 3 L is reasonable.

5.39   **Given:** $V = 10.0$ L, $n = 0.448$ mol, and $T = 315$ K   **Find:** P
        Conceptual Plan: $n, V, T \rightarrow P$

$$PV = nRT$$

Solution:

$$PV = nRT \quad \text{Rearrange to solve for } P. \; P = \frac{nRT}{V} = \frac{0.448 \; \cancel{mol} \times 0.08206 \dfrac{\cancel{L} \cdot atm}{\cancel{mol} \cdot \cancel{K}} \times 315 \; \cancel{K}}{10.0 \; \cancel{L}} = 1.16 \; atm$$

Check: The units (atm) are correct. The magnitude of the answer (~ 1 atm) makes sense because, as you will see in the next section, one mole of an ideal gas under standard conditions (273 K and 1 atm) occupies 22.4 L. Although these are not standard conditions, they are close enough for a ballpark check of the answer. Because this gas sample contains 0.448 moles in a volume of 10 L, a pressure of 1 atm is reasonable.

5.41   **Given:** $V = 28.5 \; L, P = 1.8 \; atm,$ and $T = 298 \; K$ **Find:** n
           Conceptual Plan: $V, P, T \rightarrow n$

$$PV = nRT$$

           Solution: $PV = nRT$ Rearrange to solve for n. $n = \dfrac{PV}{RT} = \dfrac{1.8 \; \cancel{atm} \times 28.5 \; \cancel{L}}{0.08206 \dfrac{\cancel{L} \cdot \cancel{atm}}{mol \cdot \cancel{K}} \times 298 \; \cancel{K}} = 2.1 \; mol$

           Check: The units (mol) are correct. The magnitude of the answer (2 mol) makes sense because, as you will see in the next section, one mole of an ideal gas under standard conditions (273 K and 1 atm) occupies 22.4 L. Although these are not standard conditions, they are close enough for a ballpark check of the answer. Because this gas sample has a volume of 28.5 L and a pressure of 1.8 atm, ~ 2 mol is reasonable.

5.43   **Given:** $P_1 = 36.0 \; psi \; (\text{gauge } P), V_1 = 11.8 \; L, T_1 = 12.0 \, °C, V_2 = 12.2 \; L,$ and $T_2 = 65.0 \, °C$
           Find: P_2 and compare to $P_{max} = 38.0 \; psi \; (\text{gauge P})$
           Conceptual Plan: $°C \rightarrow K$ and gauge $P \rightarrow psi \rightarrow atm$ then $P_1, V_1, T_1, V_2, T_2 \rightarrow P_2$

$$K = °C + 273.15 \qquad psi = \text{gauge P} + 14.7 \; \frac{1 \; atm}{14.7 \; psi} \qquad \frac{P_1 V_1}{T_1} = \frac{P_2 V_2}{T_2}$$

           Solution: $T_1 = 12.0 \, °C + 273.15 = 285.2 \; K$ and $T_2 = 65.0 \, °C + 273.15 = 338.2 \; K$

$$P_1 = 36.0 \; psi \; (\text{gauge P}) + 14.7 = 50.7 \; \cancel{psi} \times \frac{1 \; atm}{14.7 \; \cancel{psi}} = 3.4\underline{4}898 \; atm$$

$$P_{max} = 38.0 \; psi \; (\text{gauge P}) + 14.7 = 52.7 \; \cancel{psi} \times \frac{1 \; atm}{14.7 \; \cancel{psi}} = 3.59 \; atm$$

$$\frac{P_1 V_1}{T_1} = \frac{P_2 V_2}{T_2} \quad \text{Rearrange to solve for } P_2. \; P_2 = P_1 \frac{V_1}{V_2} \frac{T_2}{T_1} = 3.4\underline{4}898 \; atm \times \frac{11.8 \; \cancel{L}}{12.2 \; \cancel{L}} \times \frac{338.2 \; \cancel{K}}{285.2 \; \cancel{K}} = 3.96 \; atm$$

This exceeds the maximum tire rating of 3.59 atm or 38.0 psi (gauge P).

           Check: The units (atm) are correct. The magnitude of the answer (3.96 atm) makes physical sense because the relative increase in T is greater than the relative increase in V; so P should increase.

5.45   **Given:** $m \, (CO_2) = 28.8 \; g, P = 742 \; mmHg,$ and $T = 22 \, °C$ **Find:** V
           Conceptual Plan: $°C \rightarrow K$ and $mmHg \rightarrow atm$ and $g \rightarrow mol$ then $n, P, T \rightarrow V$

$$K = °C + 273.15 \qquad \frac{1 \; atm}{760 \; mmHg} \qquad \frac{1 \; mol}{44.01 \; g} \qquad PV = nRT$$

           Solution: $T_1 = 22 \, °C + 273.15 = 295 \; K, P = 742 \; \cancel{mmHg} \times \dfrac{1 \; atm}{760 \; \cancel{mmHg}} = 0.97\underline{6}316 \; atm$

$$n = 28.8 \; \cancel{g} \times \frac{1 \; mol}{44.01 \; \cancel{g}} = 0.65\underline{4}397 \; mol \quad PV = nRT \quad \text{Rearrange to solve for } V.$$

$$V = \frac{nRT}{P} = \frac{0.65\underline{4}397 \; \cancel{mol} \times 0.08206 \dfrac{L \cdot \cancel{atm}}{\cancel{mol} \cdot \cancel{K}} \times 295 \; \cancel{K}}{0.97\underline{6}316 \; \cancel{atm}} = 16.2 \; L$$

           Check: The units (L) are correct. The magnitude of the answer (16 L) makes sense because one mole of an ideal gas under standard conditions (273 K and 1 atm) occupies 22.4 L. Although these are not standard conditions, they are close enough for a ballpark check of the answer. Because this gas sample contains 0.65 mole, a volume of 16 L is reasonable.

5.47   **Given:** 26.0 g argon, $V = 55.0$ mL, and $T = 295$ K   **Find:** P
Conceptual Plan: g $\rightarrow$ n and mL $\rightarrow$ L then $n, V, T \rightarrow P$

$$\frac{1 \text{ mol}}{39.95 \text{ g}} \qquad \frac{1 \text{ L}}{1000 \text{ mL}} \qquad PV = nRT$$

Solution: $26.0 \text{ g Ar} \times \dfrac{1 \text{ mol Ar}}{39.95 \text{ g Ar}} = 0.65\underline{0}8135 \text{ mol Ar} \quad 55.0 \text{ mL} \times \dfrac{1 \text{ L}}{1000 \text{ mL}} = 0.0550 \text{ L}$

$PV = nRT$ Rearrange to solve for P.

$$P = \frac{nRT}{V} = \frac{0.65\underline{0}8135 \text{ mol} \times 0.08206 \dfrac{\text{L} \cdot \text{atm}}{\text{mol} \cdot \text{K}} \times 295 \text{ K}}{0.0550 \text{ L}} = 286.44906 \text{ atm} = 286 \text{ atm}$$

Check: The units (atm) are correct. The magnitude of the answer (300 atm) makes sense because, as you will see in the next section, one mole of an ideal gas under standard conditions (273 K and 1 atm) occupies 22.4 L. Although these are not standard conditions, they can be used for a ballpark check of the answer. Because the volume is ~$1/400^{\text{th}}$ the molar volume and we have ~2/3 of a mole, the resulting pressure should be $(400)(2/3) = 270$ atm.

Given: $V_1 = 55.0$ mL, $P_1 = 286$ atm, and $P_2 = 1.20$ atm   **Find:** V_2 (number of 750 mL bottles)
Conceptual Plan: $V_1, P_1, P_2 \rightarrow V_2$

$$P_1 V_1 = P_2 V_2$$

Solution: $P_1 V_1 = P_2 V_2$ Rearrange to solve for V_2.

$$V_2 = V_1 \frac{P_1}{P_2} = 55.0 \text{ mL} \times \frac{286.44906 \text{ atm}}{1.20 \text{ atm}} = 1.3\underline{1}289 \times 10^4 \text{ mL} \times \frac{1 \text{ bottle}}{750.0 \text{ mL}} = 17.5 \text{ bottles}$$

Check: The units (bottles) are correct. The magnitude of the answer (18 bottles) makes physical sense because Boyle's law indicates that as the volume decreases, the pressure increases. The pressure is decreasing by a factor of ~250, so the volume should increase by this factor.

5.49   **Given:** sample a = 5 gas particles, sample b = 10 gas particles, and sample c = 8 gas particles, with all temperatures and volumes the same   **Find:** sample with largest P
Conceptual Plan: $n, V, T \rightarrow P$

$$PV = nRT$$

Solution: $PV = nRT$ Because V and T are constant, $P \, \alpha \, n$. The sample with the largest number of gas particles will have the highest P. $P_b > P_c > P_a$

5.51   **Given:** $P_1 = 755$ mmHg, $T_1 = 25 \,°C$, and $T_2 = 1155 \,°C$ **Find:** P_2
Conceptual Plan: $°C \rightarrow K$ and mmHg $\rightarrow$ atm then $P_1, T_1, T_2 \rightarrow P_2$

$$K = °C + 273.15 \qquad \frac{1 \text{ atm}}{760 \text{ mmHg}} \qquad \frac{P_1}{T_1} = \frac{P_2}{T_2}$$

Solution: $T_1 = 25 \,°C + 273.15 = 298$ K and $T_2 = 1155 \,°C + 273.15 = 1428$ K

$$P = 755 \text{ mmHg} \times \frac{1 \text{ atm}}{760 \text{ mmHg}} = 0.99\underline{3}421 \text{ atm} \quad \frac{P_1}{T_1} = \frac{P_2}{T_2} \quad \text{Rearrange to solve for } P_2.$$

$$P_2 = P_1 \frac{T_2}{T_1} = 0.99\underline{3}421 \text{ atm} \times \frac{1428 \text{ K}}{298 \text{ K}} = 4.76 \text{ atm}$$

Check: The units (atm) are correct. The magnitude of the answer (5 atm) makes physical sense because there is a significant increase in T, which will increase P significantly.

Molar Volume, Density, and Molar Mass of a Gas

5.53   **Given:** STP and m (Ne) $= 33.6$ g   **Find:** V
Conceptual Plan: g $\rightarrow$ mol $\rightarrow$ V

$$\frac{1 \text{ mol}}{20.18 \text{ g}} \qquad \frac{22.414 \text{ L}}{1 \text{ mol}}$$

Solution: $33.6 \text{ g} \times \dfrac{1 \text{ mol}}{20.18 \text{ g}} \times \dfrac{22.414 \text{ L}}{1 \text{ mol}} = 37.3 \text{ L}$

Check: The units (L) are correct. The magnitude of the answer (37 L) makes sense because one mole of an ideal gas under standard conditions (273 K and 1 atm) occupies 22.4 L and we have about 1.7 mol.

5.55 **Given:** H_2, $P = 1655$ psi, and $T = 20.0\,°C$ **Find:** d
Conceptual Plan: $°C \rightarrow K$ and psi $\rightarrow$ atm then $P, T, \mathcal{M} \rightarrow d$

$$K = °C + 273.15 \qquad \frac{1\ atm}{14.70\ psi} \qquad d = \frac{P\mathcal{M}}{RT}$$

Solution: $T = 20.0\,°C + 273.15 = 293.2\ K$ $P = 1655\ psi \times \dfrac{1\ atm}{14.70\ psi} = 112.\underline{5}85\ atm$

$$d = \frac{P\mathcal{M}}{RT} = \frac{112.\underline{5}85\ atm \times 2.016\ \dfrac{g}{mol}}{0.08206\dfrac{L \cdot atm}{K \cdot mol} \times 293.2\ K} = 9.434\ \frac{g}{L}$$

Check: The units (g/L) are correct. The magnitude of the answer (9 g/L) makes physical sense because this is a high pressure; so the gas density will be on the high side.

5.57 **Given:** $V = 248$ mL, $m = 0.433$ g, $P = 745$ mmHg, and $T = 28\,°C$ **Find:** $\mathcal{M}$
Conceptual Plan: $°C \rightarrow K$ mmHg $\rightarrow$ atm mL $\rightarrow$ L then $V, m \rightarrow d$ then $d, P, T \rightarrow \mathcal{M}$

$$K = °C + 273.15 \qquad \frac{1\ atm}{760\ mmHg} \qquad \frac{1\ L}{1000\ mL} \qquad d = \frac{m}{V} \qquad d = \frac{PM}{RT}$$

Solution: $T = 28\,°C + 273.15 = 301\ K$ $P = 745\ mmHg \times \dfrac{1\ atm}{760\ mmHg} = 0.980263\ atm$

$$V = 248\ mL \times \frac{1\ L}{1000\ mL} = 0.248\ L \quad d = \frac{m}{V} = \frac{0.433\ g}{0.248\ L} = 1.7\underline{4}597\ g/L \quad d = \frac{P\mathcal{M}}{RT} \quad \text{Rearrange to solve for } \mathcal{M}.$$

$$\mathcal{M} = \frac{dRT}{P} = \frac{1.7\underline{4}597\ \dfrac{g}{L} \times 0.08206\ \dfrac{L \cdot atm}{K \cdot mol} \times 301\ K}{0.980263\ atm} = 44.0\ g/mol$$

Check: The units (g/mol) are correct. The magnitude of the answer (44 g/mol) makes physical sense because this is a reasonable number for a molecular weight of a gas.

5.59 **Given:** $m = 38.8$ mg, $V = 224$ mL, $T = 55\,°C$, and $P = 886$ torr   **Find:** $\mathcal{M}$
Conceptual Plan: mg $\rightarrow$ g mL $\rightarrow$ L $°C \rightarrow K$ torr $\rightarrow$ atm then $V, m \rightarrow d$ then $d, P, T \rightarrow \mathcal{M}$

$$\frac{1\ g}{1000\ mg} \qquad \frac{1\ L}{1000\ mL} \qquad K = °C + 273.15 \qquad \frac{1\ atm}{760\ torr} \qquad d = \frac{m}{V} \qquad d = \frac{P\mathcal{M}}{RT}$$

Solution:

$$m = 38.8\ mg \times \frac{1\ g}{1000\ mg} = 0.0388\ g \quad V = 224\ mL \times \frac{1\ L}{1000\ mL} = 0.224\ L \quad T = 55\,°C + 273.15 = 328\ K$$

$$P = 886\ torr \times \frac{1\ atm}{760\ torr} = 1.1\underline{6}5789\ atm \quad d = \frac{m}{V} = \frac{0.0388\ g}{0.224\ L} = 0.17\underline{3}214\ g/L \quad d = \frac{P\mathcal{M}}{RT}$$

Rearrange to solve for $\mathcal{M}$. $\mathcal{M} = \dfrac{dRT}{P} = \dfrac{0.17\underline{3}214\ \dfrac{g}{L} \times 0.08206\ \dfrac{L \cdot atm}{K \cdot mol} \times 328\ K}{1.1\underline{6}5789\ atm} = 4.00\ g/mol$

Check: The units (g/mol) are correct. The magnitude of the answer (4 g/mol) makes physical sense because this is a reasonable number for a molecular weight of a gas, especially because the density is on the low side.

Partial Pressure

5.61 **Given:** $P_{N_2} = 215$ torr, $P_{O_2} = 102$ torr, $P_{He} = 117$ torr, $V = 1.35$ L, and $T = 25.0\,°C$ **Find:** $P_{Total}, m_{N_2}, m_{O_2}, m_{He}$
Conceptual Plan: $°C \rightarrow K$ and torr $\rightarrow$ atm and $P, V, T \rightarrow n$ then mol $\rightarrow$ g

$$K = °C + 273.15 \qquad \frac{1\ atm}{760\ torr} \qquad PV = nRT \qquad \mathcal{M}$$

and $P_{N_2}, P_{O_2}, P_{He} \rightarrow P_{Total}$

$$P_{Total} = P_{N_2} + P_{O_2} + P_{He}$$

Solution: $T_1 = 25.0\,°C + 273.15 = 298.2\,K$, $PV = nRT$ Rearrange to solve for n.

$$n = \frac{PV}{RT} \quad P_{N_2} = 215 \text{ torr} \times \frac{1 \text{ atm}}{760 \text{ torr}} = 0.2828947 \text{ atm} \quad n_{N_2} = \frac{0.2828947 \text{ atm} \times 1.35 \text{ L}}{0.08206 \frac{\text{L} \cdot \text{atm}}{\text{mol} \cdot \text{K}} \times 298.2 \text{ K}} = 0.01560700 \text{ mol}$$

$$0.01560700 \text{ mol} \times \frac{28.02 \text{ mol}}{1 \text{ mol}} = 0.437 \text{ g N}_2$$

$$P_{O_2} = 102 \text{ torr} \times \frac{1 \text{ atm}}{760 \text{ torr}} = 0.1342105 \text{ atm} \quad n_{O_2} = \frac{0.1342105 \text{ atm} \times 1.35 \text{ L}}{0.08206 \frac{\text{L} \cdot \text{atm}}{\text{mol} \cdot \text{K}} \times 298.2 \text{ K}} = 0.007404252 \text{ mol}$$

$$0.007404252 \text{ mol} \times \frac{32.00 \text{ mol}}{1 \text{ mol}} = 0.237 \text{ g O}_2$$

$$P_{He} = 117 \text{ torr} \times \frac{1 \text{ atm}}{760 \text{ torr}} = 0.1539474 \text{ atm} \quad n_{He} = \frac{0.1539474 \text{ atm} \times 1.35 \text{ L}}{0.08206 \frac{\text{L} \cdot \text{atm}}{\text{mol} \cdot \text{K}} \times 298.2 \text{ K}} = 0.008493113 \text{ mol}$$

$$0.008493113 \text{ mol} \times \frac{4.003 \text{ mol}}{1 \text{ mol}} = 0.0340 \text{ g He and}$$

$P_{Total} = P_{N_2} + P_{O_2} + P_{He} = 0.283 \text{ atm} + 0.134 \text{ atm} + 0.154 \text{ atm} = 0.571 \text{ atm or}$
$P_{Total} = P_{N_2} + P_{O_2} + P_{He} = 215 \text{ torr} + 102 \text{ torr} + 117 \text{ torr} = 434 \text{ torr}$

Check: The units (g and atm) are correct. The magnitude of the answer (1 g) makes sense because gases are not very dense and these pressures are <1 atm. Because all of the pressures are small, the total is <1 atm.

5.63   **Given:** $m\,(CO_2) = 1.20$ g, $V = 755$ mL, $P_{N_2} = 725$ mmHg, and $T = 25.0\,°C$ **Find:** P_{Total}
Conceptual Plan: mL → L and °C → K and g → mol and $n, P, T → V$ then atm → mmHg

$$\frac{1 \text{ L}}{1000 \text{ mL}} \quad K = °C + 273.15 \quad \frac{1 \text{ mol}}{44.01 \text{ g}} \quad PV = nRT \quad \frac{760 \text{ mmHg}}{1 \text{ atm}}$$

finally $P_{CO_2}, P_{N_2} → P_{Total}$

$$P_{Total} = P_{CO_2} + P_{N_2}$$

Solution: $V = 755 \text{ mL} \times \dfrac{1 \text{ L}}{1000 \text{ mL}} = 0.755 \text{ L} \qquad T = 25.0\,°C + 273.15 = 298.2 \text{ K}$

$$n = 1.20 \text{ g} \times \frac{1 \text{ mol}}{44.01 \text{ g}} = 0.0272665 \text{ mol}, PV = nRT \quad \text{Rearrange to solve for } P.$$

$$P = \frac{nRT}{V} = \frac{0.0272665 \text{ mol} \times 0.08206 \frac{\text{L} \cdot \text{atm}}{\text{mol} \cdot \text{K}} \times 298.2 \text{ K}}{0.755 \text{ L}} = 0.883734 \text{ atm}$$

$$P_{CO_2} = 0.883734 \text{ atm} \times \frac{760 \text{ mmHg}}{1 \text{ atm}} = 672 \text{ mmHg}$$

$$P_{Total} = P_{CO_2} + P_{N_2} = 672 \text{ mmHg} + 725 \text{ mmHg} = 1397 \text{ mmHg or } 1397 \text{ torr} \times \frac{1 \text{ atm}}{760 \text{ torr}} = 1.84 \text{ atm}$$

Check: The units (mmHg) are correct. The magnitude of the answer (1400 mmHg) makes sense because it must be greater than 725 mmHg.

5.65   **Given:** $m\,(N_2) = 1.25$ g, $m\,(O_2) = 0.85$ g, $V = 1.55$ L, and $T = 18\,°C$ **Find:** $\chi_{N_2}, \chi_{O_2}, P_{N_2}, P_{O_2}$
Conceptual Plan: g → mol °C → K then $n_{N_2}, n_{O_2} → \chi_{N_2}$ and $n_{N_2}, n_{O_2} → \chi_{O_2}$

$$\mathcal{M} \quad K = °C + 273.15 \quad \chi_{N_2} = \frac{n_{N_2}}{n_{N_2} + n_{O_2}} \quad \chi_{O_2} = \frac{n_{O_2}}{n_{N_2} + n_{O_2}}$$

then $n, V, T → P$

$$PV = nRT$$

Solution: $n_{N_2} = 1.25 \text{ g} \times \dfrac{1 \text{ mol}}{28.02 \text{ g}} = 0.0446110 \text{ mol}, n_{O_2} = 0.85 \text{ g} \times \dfrac{1 \text{ mol}}{32.00 \text{ g}} = 0.026563 \text{ mol},$

$$T = 18\,°C + 273.15 = 291 \text{ K}, \chi_{N_2} = \frac{n_{N_2}}{n_{N_2} + n_{O_2}} = \frac{0.0446110 \text{ mol}}{0.0446110 \text{ mol} + 0.026563 \text{ mol}} = 0.626792 = 0.627,$$

$$\chi_{O_2} = \frac{n_{O_2}}{n_{N_2} + n_{O_2}} = \frac{0.026563 \text{ mol}}{0.0446110 \text{ mol} + 0.026563 \text{ mol}} = 0.373212 \quad \text{We can also calculate this as}$$

$$\chi_{O_2} = 1 - \chi_{N_2} = 1 - 0.626792 = 0.373208 = 0.373 \quad PV = nRT \quad \text{Rearrange to solve for } P. \ P = \frac{nRT}{V}$$

$$P_{N_2} = \frac{0.044611 \text{ mol} \times 0.08206 \frac{L \cdot atm}{mol \cdot K} \times 291 \text{ K}}{1.55 \text{ L}} = 0.687 \text{ atm}$$

$$P_{O_2} = \frac{0.026563 \text{ mol} \times 0.08206 \frac{L \cdot atm}{mol \cdot K} \times 291 \text{ K}}{1.55 \text{ L}} = 0.409 \text{ atm}$$

Check: The units (none and atm) are correct. The magnitude of the answers makes sense because the mole fractions should total 1, and because the weight of N_2 is greater than O_2, its mole fraction is larger. The number of moles is $<< 1$, so we expect the pressures to be < 1 atm, given the V (1.55 L).

5.67   **Given:** $T = 30.0\,°C$, $P_{Total} = 732$ mmHg, and $V = 722$ mL   **Find:** P_{H_2} and m_{H_2}
Conceptual Plan: $T \rightarrow P_{H_2O}$ then $P_{Total}, P_{H_2O} \rightarrow P_{H_2}$ then mmHg $\rightarrow$ atm and mL $\rightarrow$ L

$$\text{Table 5.4} \qquad P_{Total} = P_{H_2O} + P_{H_2} \qquad \frac{1 \text{ atm}}{760 \text{ mmHg}} \qquad \frac{1 \text{ L}}{1000 \text{ mL}}$$

and $°C \rightarrow K$ $P, V, T \rightarrow n$ then mol $\rightarrow$ g

$$K = °C + 273.15 \quad PV = nRT \qquad \frac{2.016 \text{ g}}{1 \text{ mol}}$$

Solution: Table 5.4 states that at 30 °C, $P_{H_2O} = 31.86$ mmHg $P_{Total} = P_{H_2O} + P_{H_2}$ Rearrange to solve for P_{H_2}. $P_{H_2} = P_{Total} - P_{H_2O} = 732$ mmHg $- 31.86$ mmHg $= 700.$ mmHg

$$P_{H_2} = 700. \text{ mmHg} \times \frac{1 \text{ atm}}{760 \text{ mmHg}} = 0.921052 \text{ atm} \quad V = 722 \text{ mL} \times \frac{1 \text{ L}}{1000 \text{ mL}} = 0.722 \text{ L}$$

$$T = 30.0\,°C + 273.15 = 303.2 \text{ K}, PV = nRT \quad \text{Rearrange to solve for } n. \ n = \frac{PV}{RT}$$

$$n_{H_2} = \frac{0.921052 \text{ atm} \times 0.722 \text{ L}}{0.08206 \frac{L \cdot atm}{mol \cdot K} \times 303.2 \text{ K}} = 0.0267276 \text{ mol then } 0.0267276 \text{ mol} \times \frac{2.016 \text{ g}}{1 \text{ mol}} = 0.0539 \text{ g } H_2$$

Check: The units (g) are correct. The magnitude of the answer ($<< 1$ g) makes sense because gases are not very dense, hydrogen is light, the volume is small, and the pressure is ~ 1 atm.

5.69   **Given:** $T = 25\,°C$, $P_{Total} = 748$ mmHg, and $V = 0.951$ L   **Find:** P_{H_2} and m_{H_2}
Conceptual Plan: $T \rightarrow P_{H_2O}$ then $P_{Total}, P_{H_2O} \rightarrow P_{H_2}$ then mmHg $\rightarrow$ atm and mL $\rightarrow$ L

$$\text{Table 5.4} \qquad P_{Total} = P_{H_2O} + P_{H_2} \qquad \frac{1 \text{ atm}}{760 \text{ mmHg}} \qquad \frac{1 \text{ L}}{1000 \text{ mL}}$$

and $°C \rightarrow K$ $P, V, T \rightarrow n$ then mol $\rightarrow$ g

$$K = °C + 273.15 \quad PV = nRT \qquad \frac{2.016 \text{ g}}{1 \text{ mol}}$$

Solution: Table 5.4 states that at 25 °C, $P_{H_2O} = 23.78$ mmHg $P_{Total} = P_{H_2O} + P_{H_2}$
Rearrange to solve for P_{H_2}. $P_{H_2} = P_{Total} - P_{H_2O} = 748$ mmHg $- 23.78$ mmHg $= 724$ mmHg

$$P_{H_2} = 724 \text{ mmHg} \times \frac{1 \text{ atm}}{760 \text{ mmHg}} = 0.952632 \text{ atm} \quad T = 25\,°C + 273.15 = 298 \text{ K} \quad PV = nRT$$

$$\text{Rearrange to solve for } n. \ n_{H_2} = \frac{PV}{RT} = \frac{0.952632 \text{ atm} \times 0.951 \text{ L}}{0.08206 \frac{L \cdot atm}{mol \cdot K} \times 298 \text{ K}} = 0.0370474 \text{ mol}$$

$$0.0370474 \text{ mol} \times \frac{2.016 \text{ g}}{1 \text{ mol}} = 0.0747 \text{ g } H_2$$

Check: The units (g) are correct. The magnitude of the answer ($<< 1$ g) makes sense because gases are not very dense, hydrogen is light, the volume is small, and the pressure is ~ 1 atm.

Reaction Stoichiometry Involving Gases

5.71   **Given:** $m(C) = 15.7$ g, $P = 1.0$ atm, and $T = 355$ K   **Find:** V
Conceptual Plan: g C → mol C → mol H_2 then n (mol H_2), P, T → V

$$\frac{1 \text{ mol}}{12.01 \text{ g C}} \qquad \frac{1 \text{ mol } H_2}{1 \text{ mol C}} \qquad\qquad PV = nRT$$

Solution: $15.7 \text{ g C} \times \dfrac{1 \text{ mol C}}{12.01 \text{ g C}} \times \dfrac{1 \text{ mol } H_2}{1 \text{ mol C}} = 1.30724 \text{ mol } H_2$ $PV = nRT$ Rearrange to solve for V.

$$V = \frac{nRT}{P} = \frac{1.30724 \text{ mol} \times 0.08206 \dfrac{L \cdot atm}{mol \cdot K} \times 355 \text{ K}}{1.0 \text{ atm}} = 38 \text{ L}$$

Check: The units (L) are correct. The magnitude of the answer (38 L) makes sense because we have more than one mole of gas; so we expect more than 22 L.

5.73   **Given:** $P = 748$ mmHg, $T = 86\,°C$, and $m(CH_3OH) = 25.8$ g   **Find:** V_{H_2} and V_{CO}
Conceptual Plan: g CH_3OH → mol CH_3OH → mol H_2 and mmHg → atm and °C → K

$$\frac{1 \text{ mol } CH_3OH}{32.04 \text{ g } CH_3OH} \qquad \frac{2 \text{ mol } H_2}{1 \text{ mol } CH_3OH} \qquad \frac{1 \text{ atm}}{760 \text{ mmHg}} \qquad K = °C + 273.15$$

then n (mol H_2), P, T → V and mol H_2 → mol CO then n (mol CO), P, T → V

$$PV = nRT \qquad\qquad \frac{1 \text{ mol CO}}{2 \text{ mol } H_2} \qquad\qquad PV = nRT$$

Solution: $25.8 \text{ g } CH_3OH \times \dfrac{1 \text{ mol } CH_3OH}{32.04 \text{ g } CH_3OH} \times \dfrac{2 \text{ mol } H_2}{1 \text{ mol } CH_3OH} = 1.61049 \text{ mol } H_2$

$$P_{H_2} = 748 \text{ mmHg} \times \frac{1 \text{ atm}}{760 \text{ mmHg}} = 0.984211 \text{ atm}, \; T = 86\,°C + 273.15 = 359 \text{ K}, PV = nRT$$

Rearrange to solve for V. $V = \dfrac{nRT}{P}$ $V_{H_2} = \dfrac{1.61049 \text{ mol} \times 0.08206 \dfrac{L \cdot atm}{mol \cdot K} \times 359 \text{ K}}{0.984211 \text{ atm}} = 48.2 \text{ L } H_2$

$1.61049 \text{ mol } H_2 \times \dfrac{1 \text{ mol CO}}{2 \text{ mol } H_2} = 0.80525 \text{ mol CO}, \; V_{CO} = \dfrac{0.80525 \text{ mol} \times 0.08206 \dfrac{L \cdot atm}{mol \cdot K} \times 359 \text{ K}}{0.984211 \text{ atm}} = 24.1 \text{ L CO}$

Check: The units (L) are correct. The magnitude of the answers (48 L and 24 L) makes sense because we have more than one mole of hydrogen gas and half that of CO; so we expect significantly more than 22 L for hydrogen and half that for CO.

5.75   **Given:** $V = 11.8$ L at STP   **Find:** $m(NaN_3)$
Conceptual Plan: V_{N_2} → mol N_2 → mol NaN_3 → g NaN_3

$$\frac{1 \text{ mol } N_2}{22.414 \text{ L } N_2} \qquad \frac{2 \text{ mol } NaN_3}{3 \text{ mol } N_2} \qquad \frac{65.02 \text{ g } NaN_3}{1 \text{ mol } NaN_3}$$

Solution: $11.8 \text{ L } N_2 \times \dfrac{1 \text{ mol } N_2}{22.414 \text{ L } N_2} \times \dfrac{2 \text{ mol } NaN_3}{3 \text{ mol } N_2} \times \dfrac{65.02 \text{ g } NaN_3}{1 \text{ mol } NaN_3} = 22.8 \text{ g } NaN_3$

Check: The units (g) are correct. The magnitude of the answer (23 g) makes sense because we have about half a mole of nitrogen gas, which translates to even fewer moles of NaN_3; so we expect significantly less than 65 g.

5.77   **Given:** $V_{CH_4} = 25.5$ L, $P_{CH_4} = 732$ torr, and $T = 25\,°C$; mixed with $V_{H_2O} = 22.8$ L, $P_{H_2O} = 702$ torr, and $T = 125\,°C$; forms $P_{H_2} = 26.2$ L at STP   **Find:** % yield
Conceptual Plan: CH_4 : torr → atm and °C → K and P, V, T → n_{CH_4} → n_{H_2}

$$\frac{1 \text{ atm}}{760 \text{ torr}} \qquad K = °C + 273.15 \qquad PV = nRT \qquad \frac{3 \text{ mol } H_2}{1 \text{ mol } CH_4}$$

H_2O : torr → atm and °C → K and P, V, T → n_{H_2O} → n_{H_2}

$$\frac{1 \text{ atm}}{760 \text{ torr}} \qquad K = °C + 273.15 \qquad PV = nRT \qquad \frac{3 \text{ mol } H_2}{1 \text{ mol } H_2O}$$

Copyright © 2014 Pearson Education, Inc.

Select smaller n_{H_2} as theoretical yield,
then $L_{H_2} \to$ mol H_2 (actual yield) finally actual yield, theoretical yield $\to$ % yield

$$\frac{1 \text{ mol } H_2}{22.414 \text{ L } H_2}$$

$$\% \text{ yield} = \frac{\text{actual yield}}{\text{theoretical yield}} \times 100\%$$

Solution: CH_4: $P_{CH_4} = 732 \text{ torr} \times \frac{1 \text{ atm}}{760 \text{ torr}} = 0.963158 \text{ atm}, T = 25\,°C + 273.15 = 298 \text{ K}, PV = nRT$

Rearrange to solve for n. $n = \frac{PV}{RT}$ $\quad n_{CH_4} = \frac{0.963158 \text{ atm} \times 25.5 \text{ L}}{0.08206 \frac{L \cdot atm}{mol \cdot K} \times 298 \text{ K}} = 1.00436 \text{ mol } CH_4$

$1.00436 \text{ mol } CH_4 \times \frac{3 \text{ mol } H_2}{1 \text{ mol } CH_4} = 3.01308 \text{ mol } H_2$

H_2O: $P_{H_2O} = 702 \text{ torr} \times \frac{1 \text{ atm}}{760 \text{ torr}} = 0.923684 \text{ atm}, T = 125\,°C + 273.15 = 398 \text{ K}, n = \frac{PV}{RT}$

$n_{H_2O} = \frac{0.923684 \text{ atm} \times 22.8 \text{ L}}{0.08206 \frac{L \cdot atm}{mol \cdot K} \times 398 \text{ K}} = 0.644828 \text{ mol}$ $\quad H_2O \; 0.644828 \text{ mol } H_2O \times \frac{3 \text{ mol } H_2}{1 \text{ mol } H_2O} = 1.93448 \text{ mol } H_2$

Water is the limiting reagent because the moles of hydrogen generated is lower.
theoretical yield $= 1.93448 \text{ mol } H_2$

$26.2 \text{ L } H_2 \times \frac{1 \text{ mol } H_2}{22.414 \text{ L } H_2} = 1.16891 \text{ mol } H_2 = \text{actual yield}$

$\% \text{ yield} = \frac{\text{actual yield}}{\text{theoretical yield}} \times 100\% = \frac{1.16891 \text{ mol } H_2}{1.93448 \text{ mol } H_2} \times 100\% = 60.4\%$

Check: The units (%) are correct. The magnitude of the answer (60%) makes sense because it is between 0 and 100%.

5.79   **Given:** $V = 2.00 \text{ L}, P_{Cl_2} = 337 \text{ mmHg}, P_{F_2} = 729 \text{ mmHg}, T = 298 \text{ K}$ **Find:** limiting reactant and m_{ClF_3}
Conceptual Plan: Determine limiting reactant by comparing the pressures of each reactant then

$$\frac{3 \text{ mmHg } F_2}{1 \text{ mmHg } Cl_2}$$

mmHg limiting reactant $\to$ mmHg ClF_3 and mmHg $\to$ atm then

$$\frac{2 \text{ mmHg } ClF_3}{1 \text{ mmHg } Cl_2} \text{ or } \frac{2 \text{ mmHg } ClF_3}{3 \text{ mmHg } F_2} \qquad \frac{1 \text{ atm}}{760 \text{ mmHg}}$$

$P, V, T \to n \to m$

$$PV = nRT \quad \frac{92.45 \text{ g } ClF_3}{1 \text{ mol } ClF_3}$$

Solution: To determine the limiting reactant, calculate the pressure of fluorine needed to react all of the chlorine and compare to the pressure of fluorine available.

$337 \text{ mmHg } Cl_2 \times \frac{3 \text{ mmHg } F_2}{1 \text{ mmHg } Cl_2} = 1011 \text{ mmHg } F_2 \text{ needed.}$

Because only 729 mmHg F_2 is available, F_2 is the limiting reactant; then

$P_{ClF_3} = 729 \text{ mmHg } F_2 \times \frac{2 \text{ mmHg } ClF_3}{3 \text{ mmHg } F_2} \times \frac{1 \text{ atm } ClF_3}{760 \text{ mmHg } ClF_3} = 0.6394737 \text{ atm } ClF_3 \; PV = nRT$ Rearrange to

solve for n. $n = \frac{PV}{RT}$ $\quad n_{ClF_3} = \frac{0.6394737 \text{ atm } ClF_3 \times 2.00 \text{ L}}{0.08206 \frac{L \cdot atm}{mol \cdot K} \times 298 \text{ K}} = 0.052300387 \text{ mol } ClF_3$

Finally, $0.052300387 \text{ mol } ClF_3 \times \frac{92.45 \text{ g } ClF_3}{1 \text{ mol } ClF_3} = 4.8351708 \text{ g } ClF_3 = 4.84 \text{ g } ClF_3$

Check: The units (g) are correct. The magnitude of the answer (5 g) makes sense because we have much less than a mole of fluorine; so we expect a mass much less than the molar mass of ClF_3.

Kinetic Molecular Theory

5.81   (a)   Yes, because the average kinetic energy of a particle is proportional to the temperature in kelvins and the two gases are at the same temperature, they have the same average kinetic energy.

(b)   No, because the helium atoms are lighter, they must move faster to have the same kinetic energy as argon atoms.

(c)   No, because the Ar atoms are moving slower to compensate for their larger mass, they will exert the same pressure on the walls of the container.

(d)   Because He is lighter, it has the faster rate of effusion.

5.83   **Given:** F_2, Cl_2, Br_2, and $T = 298$ K   **Find:** u_{rms} and KE_{avg} for each gas and relative rates of effusion

Conceptual Plan: $\mathcal{M}, T \rightarrow u_{rms} \rightarrow KE_{avg}$

$$u_{rms} = \sqrt{\frac{3RT}{\mathcal{M}}} \qquad KE_{avg} = \tfrac{1}{2}N_A m u_{rms}^2 = \tfrac{3}{2}RT$$

Solution:

$$F_2: \mathcal{M} = \frac{38.00 \text{ g}}{1 \text{ mol}} \times \frac{1 \text{ kg}}{1000 \text{ g}} = 0.03800 \text{ kg/mol}, \, u_{rms} = \sqrt{\frac{3RT}{\mathcal{M}}} = \sqrt{\frac{3 \times 8.314\frac{J}{K \cdot mol} \times 298 \text{ K}}{0.03800 \frac{kg}{mol}}} = 442 \text{ m/s}$$

$$Cl_2: \mathcal{M} = \frac{70.90 \text{ g}}{1 \text{ mol}} \times \frac{1 \text{ kg}}{1000 \text{ g}} = 0.07090 \text{ kg/mol}, \, u_{rms} = \sqrt{\frac{3RT}{\mathcal{M}}} = \sqrt{\frac{3 \times 8.314\frac{J}{K \cdot mol} \times 298 \text{ K}}{0.07090 \frac{kg}{mol}}} = 324 \text{ m/s}$$

$$Br_2: \mathcal{M} = \frac{159.80 \text{ g}}{1 \text{ mol}} \times \frac{1 \text{ kg}}{1000 \text{ g}} = 0.15980 \text{ kg/mol}, \, u_{rms} = \sqrt{\frac{3RT}{\mathcal{M}}} = \sqrt{\frac{3 \times 8.314\frac{J}{K \cdot mol} \times 298 \text{ K}}{0.15980 \frac{kg}{mol}}} = 216 \text{ m/s}$$

All molecules have the same kinetic energy:

$$KE_{avg} = \frac{3}{2}RT = \frac{3}{2} \times 8.314 \frac{J}{K \cdot mol} \times 298 \text{ K} = 3.72 \times 10^3 \text{ J/mol}$$

Because rate of effusion is proportional to $\sqrt{\dfrac{1}{\mathcal{M}}}$, F_2 will have the fastest rate and Br_2 will have the slowest rate.

Check: The units (m/s) are correct. The magnitude of the answer $(200-450 \text{ m/s})$ makes sense because it is consistent with what was shown in the text and the heavier the molecule, the slower the molecule.

5.85   **Given:** $^{238}UF_6$ and $^{235}UF_6$ U-235 = 235.054 amu, U-238 = 238.051 amu

Find: ratio of effusion rates $^{238}UF_6 / \, ^{235}UF_6$

Conceptual Plan: $\mathcal{M}_{238_{UF_6}}, \mathcal{M}_{235_{UF_6}} \rightarrow Rate_{238_{UF_6}} / Rate_{235_{UF_6}}$

$$\frac{Rate_{238_{UF_6}}}{Rate_{235_{UF_6}}} = \sqrt{\frac{\mathcal{M}_{235_{UF_6}}}{\mathcal{M}_{238_{UF_6}}}}$$

Solution: $^{238}UF_6: \mathcal{M} = \dfrac{352.05 \text{ g}}{1 \text{ mol}} \times \dfrac{1 \text{ kg}}{1000 \text{ g}} = 0.35205 \text{ kg/mol}$

$^{235}UF_6: \mathcal{M} = \dfrac{349.05 \text{ g}}{1 \text{ mol}} \times \dfrac{1 \text{ kg}}{1000 \text{ g}} = 0.34905 \text{ kg/mol}$

$$\frac{Rate_{238_{UF_6}}}{Rate_{235_{UF_6}}} = \sqrt{\frac{\mathcal{M}_{235_{UF_6}}}{\mathcal{M}_{238_{UF_6}}}} = \sqrt{\frac{0.34905 \text{ kg/mol}}{0.35205 \text{ kg/mol}}} = 0.99573$$

Check: The units (none) are correct. The magnitude of the answer (< 1) makes sense because the heavier molecule has the lower effusion rate because it moves more slowly.

5.87   **Given:** Ne and unknown gas; Ne effusion in 76 s and unknown in 155 s   **Find:** identify unknown gas

Conceptual Plan: $\mathcal{M}_{Ne}, Rate_{Ne}, Rate_{Unk} \rightarrow \mathcal{M}_{Kr}$

$$\frac{Rate_{Ne}}{Rate_{Unk}} = \sqrt{\frac{\mathcal{M}_{Unk}}{\mathcal{M}_{Ne}}}$$

Solution: Ne: $\mathcal{M} = \dfrac{20.18 \text{ g}}{1 \text{ mol}} \times \dfrac{1 \text{ kg}}{1000 \text{ g}} = 0.02018 \text{ kg/mol}, \quad \dfrac{Rate_{Ne}}{Rate_{Unk}} = \sqrt{\dfrac{\mathcal{M}_{unk}}{\mathcal{M}_{Ne}}}$

Rearrange to solve for $\mathcal{M}_{Unk}$. $\mathcal{M}_{Unk} = \mathcal{M}_{Ne}\left(\dfrac{Rate_{Ne}}{Rate_{Unk}}\right)^2$ Because Rate α 1/(effusion time),

$$\mathcal{M}_{Unk} = \mathcal{M}_{Ne}\left(\frac{Time_{Unk}}{Time_{Ne}}\right)^2 = 0.02018 \frac{\text{kg}}{\text{mol}} \times \left(\frac{155 \text{ s}}{76 \text{ s}}\right)^2 = 0.084 \frac{\text{kg}}{\text{mol}} \times \frac{1000 \text{ g}}{1 \text{ kg}} = 84 \text{ g/mol or Kr}$$

Check: The units (g/mol) are correct. The magnitude of the answer ($>$ Ne) makes sense because Ne effused faster; so it must be lighter.

5.89   Gas A has the higher molar mass because it has the slower average velocity. Gas B will have the higher effusion rate because it has the higher velocity.

Real Gases

5.91   The postulate that the volume of the gas particles is small compared to the space between them breaks down at high pressure. At high pressures, the number of molecules per unit volume increases; so the volume of the gas particles becomes more significant. Because the spacing between the particles is reduced, the molecules themselves occupy a significant portion of the volume.

5.93   **Given:** Ne, $n = 1.000$ mol, $P = 500.0$ atm, and $T = 355.0$ K   **Find:** V(ideal) and V(van der Waals)

Conceptual Plan: $n, P, T \rightarrow V$ and $n, P, T \rightarrow V$

$$PV = nRT \qquad \left(P + \frac{an^2}{V^2}\right)(V - nb) = nRT$$

Solution: $PV = nRT$ Rearrange to solve for V.

$$V = \frac{nRT}{P} = \frac{1.000 \text{ mol} \times 0.08206 \dfrac{\text{L} \cdot \text{atm}}{\text{mol} \cdot \text{K}} \times 355.0 \text{ K}}{500.0 \text{ atm}} = 0.05826 \text{ L}$$

$$\left(P + \frac{an^2}{V^2}\right)(V - nb) = nRT \quad \text{Rearrange to solve for } V = \frac{nRT}{\left(P + \dfrac{an^2}{V^2}\right)} + nb$$

Using a = 0.211 L^2 atm/mol^2 and b = 0.0171 L/mol from Table 5.5 and the V from the ideal gas law calculation above, solve for V by successive approximations.

$$V = \frac{1.000 \text{ mol} \times 0.08206 \dfrac{\text{L} \cdot \text{atm}}{\text{mol} \cdot \text{K}} \times 355.0 \text{ K}}{500.0 \text{ atm} + \dfrac{0.211 \dfrac{\text{L}^2 \cdot \text{atm}}{\text{mol}^2} \times (1.000 \text{ mol})^2}{(0.05826 \text{ L})^2}} + \left(1.000 \text{ mol} \times 0.0171 \dfrac{\text{L}}{\text{mol}}\right) = 0.068920 \text{ L}$$

Plug in this new value.

$$V = \frac{1.000 \text{ mol} \times 0.08206 \dfrac{\text{L} \cdot \text{atm}}{\text{mol} \cdot \text{K}} \times 355.0 \text{ K}}{500.0 \text{ atm} + \dfrac{0.211 \dfrac{\text{L}^2 \cdot \text{atm}}{\text{mol}^2} \times (1.000 \text{ mol})^2}{(0.068920 \text{ L})^2}} + \left(1.000 \text{ mol} \times 0.0171 \dfrac{\text{L}}{\text{mol}}\right) = 0.070609 \text{ L}$$

Plug in this new value.

$$V = \frac{1.000 \ \cancel{mol} \times 0.08206 \dfrac{L \cdot \cancel{atm}}{\cancel{mol} \cdot \cancel{K}} \times 355.0 \ \cancel{K}}{500.0 \ \cancel{atm} + \dfrac{0.211 \dfrac{\cancel{L^2} \cdot \cancel{atm}}{\cancel{mol^2}} \times (1.000 \ \cancel{mol})^2}{(0.070609 \ \cancel{L})^2}} + \left(1.000 \ \cancel{mol} \times 0.0171 \dfrac{L}{\cancel{mol}}\right) = 0.07081\underline{6} \ L$$

Plug in this new value.

$$V = \frac{1.000 \ \cancel{mol} \times 0.08206 \dfrac{L \cdot \cancel{atm}}{\cancel{mol} \cdot \cancel{K}} \times 355.0 \ \cancel{K}}{500.0 \ \cancel{atm} + \dfrac{0.211 \dfrac{\cancel{L^2} \cdot \cancel{atm}}{\cancel{mol^2}} \times (1.000 \ \cancel{mol})^2}{(0.070816 \ \cancel{L})^2}} + \left(1.000 \ \cancel{mol} \times 0.0171 \dfrac{L}{\cancel{mol}}\right) = 0.07084\underline{0} \ L = 0.0708 \ L$$

The two values are different because we are at very high pressures. The pressure is corrected from 500.0 atm to 542.1 atm, and the final volume correction is 0.0171 L.

Check: The units (L) are correct. The magnitude of the answer (~ 0.07 L) makes sense because we are at such a high pressure and have one mole of gas.

Cumulative Problems

5.95 **Given:** m (penny) $= 2.482$ g, $T = 25\,°C$, $V = 0.899$ L, and $P_{Total} = 791$ mmHg   **Find:** % Zn in penny
Conceptual Plan: $T \rightarrow P_{H_2O}$ then $P_{Total}, P_{H_2O} \rightarrow P_{H_2}$ then mmHg $\rightarrow$ atm and $°C \rightarrow K$

$$\text{Table 5.4} \qquad\qquad P_{Total} = P_{H_2O} + P_{H_2} \qquad\qquad \frac{1 \ atm}{760 \ mmHg} \qquad K = °C + 273.15$$

and $P, V, T \quad \rightarrow \quad n_{H_2} \quad \rightarrow \quad n_{Zn} \quad \rightarrow \quad g_{Zn} \quad \rightarrow \quad$ **% Zn**

$$PV = nRT \qquad \frac{1 \ mol \ Zn}{1 \ mol \ H_2} \qquad \frac{65.39 \ g \ Zn}{1 \ mol \ Zn} \qquad \% \ Zn = \frac{g_{Zn}}{g_{penny}} \times 100\%$$

Solution: Table 5.4 states that $P_{H_2O} = 23.78$ mmHg at $25\,°C$. $P_{Total} = P_{H_2O} + P_{H_2}$ Rearrange to solve for P_{H_2}.

$$P_{H_2} = P_{Total} - P_{H_2O} = 791 \ \text{mmHg} - 23.78 \ \text{mmHg} = 767 \ \text{mmHg} \quad P_{H_2} = 767 \ \cancel{mmHg} \times \frac{1 \ atm}{760 \ \cancel{mmHg}} = 1.00\underline{9}2 \ atm$$

then $T = 25\,°C + 273.15 = 298$ K, $PV = nRT$

Rearrange to solve for n. $n_{H_2} = \dfrac{PV}{RT} = \dfrac{1.0092 \ \cancel{atm} \times 0.899 \ \cancel{L}}{0.08206 \dfrac{\cancel{L} \cdot \cancel{atm}}{mol \cdot \cancel{K}} \times 298 \ \cancel{K}} = 0.037\underline{1}013 \ mol$

$$0.037\underline{1}013 \ \cancel{mol \ H_2} \times \frac{1 \ \cancel{mol \ Zn}}{1 \ \cancel{mol \ H_2}} \times \frac{65.39 \ g \ Zn}{1 \ \cancel{mol \ Zn}} = 2.4\underline{2}605 \ g \ Zn$$

$$\% \ Zn = \frac{g_{Zn}}{g_{penny}} \times 100\% = \frac{2.4\underline{2}605 \ \cancel{g}}{2.482 \ \cancel{g}} \times 100\% = 97.7\% \ Zn$$

Check: The units (% Zn) are correct. The magnitude of the answer (98%) makes sense because it should be between 0 and 100%. We expect about 1/22 a mole of gas because our conditions are close to STP and we have ~ 1 L of gas.

5.97 **Given:** $V = 255$ mL, m(flask) $= 143.187$ g, m(flask + gas) $= 143.289$ g, $P = 267$ torr, and $T = 25\,°C$ **Find:** $\mathcal{M}$
Conceptual Plan: $°C \rightarrow K$ torr $\rightarrow$ atm mL $\rightarrow$ L m(flask), m(flask + gas) $\rightarrow m$(gas)

$$K = °C + 273.15 \quad \frac{1 \ atm}{760 \ torr} \qquad \frac{1 \ L}{1000 \ mL} \qquad m(gas) = m(flask + gas) - m(flask)$$

then $V, m \rightarrow d$ **then** $d, P, T, \rightarrow \mathcal{M}$

$$d = \frac{m}{V} \qquad\qquad d = \frac{P \mathcal{M}}{RT}$$

Solution: $T = 25\,°C + 273.15 = 298$ K, $P = 267 \ \cancel{torr} \times \dfrac{1 \ atm}{760 \ \cancel{torr}} = 0.35131\underline{6} \ atm$

$V = 255 \ \cancel{mL} \times \dfrac{1 \ L}{1000 \ \cancel{mL}} = 0.255 \ L$

$$m(gas) = m(flask + gas) - m(flask) = 143.289 \text{ g} - 143.187 \text{ g} = 0.102 \text{ g}$$

$$d = \frac{m}{V} = \frac{0.102 \text{ g}}{0.255 \text{ L}} = 0.400 \text{ g/L}, \quad d = \frac{P\mathcal{M}}{RT} \quad \text{Rearrange to solve for } \mathcal{M}.$$

$$\mathcal{M} = \frac{dRT}{P} = \frac{0.400 \frac{\text{g}}{\text{L}} \times 0.08206 \frac{\text{L} \cdot \text{atm}}{\text{K} \cdot \text{mol}} \times 298 \text{ K}}{0.351316 \text{ atm}} = 27.8 \text{ g/mol}$$

An alternative method to solve this problem would be to calculate the moles of gas using the ideal gas law (0.00366 moles) and then taking the mass (0.102 g) and dividing it by the number of moles.

Check: The units (g/mol) are correct. The magnitude of the answer (28 g/mol) makes physical sense because this is a reasonable number for a molecular weight of a gas.

5.99   **Given:** $V = 158 \text{ mL}, m(gas) = 0.275 \text{ g}, P = 556 \text{ mmHg}, T = 25 \,°\text{C}, \text{gas} = 82.66\% \text{ C and } 17.34\% \text{ H}$
          Find: molecular formula
          Conceptual Plan: $°\text{C} \to \text{K} \quad \text{mmHg} \to \text{atm} \quad \text{mL} \to \text{L then } V, m \to d$

$$\text{K} = °\text{C} + 273.15 \qquad \frac{1 \text{ atm}}{760 \text{ mmHg}} \qquad \frac{1 \text{ L}}{1000 \text{ mL}} \qquad d = \frac{m}{V}$$

then $d, P, T, \to \mathcal{M}$ **then** $\% \text{ C}, \% \text{ H}, \mathcal{M} \to$ **formula**

$$d = \frac{P\mathcal{M}}{RT} \qquad \#\text{C} = \frac{0.8266 \text{ g C}}{12.01 \frac{\text{g C}}{\text{mol C}}} \qquad \#\text{H} = \frac{\mathcal{M} \, 0.1734 \text{ g H}}{1.008 \frac{\text{g H}}{\text{mol H}}}$$

Solution: $T = 25 \,°\text{C} + 273.15 = 298 \text{ K}, P = 556 \text{ mmHg} \times \dfrac{1 \text{ atm}}{760 \text{ mmHg}} = 0.731579 \text{ atm}$

$$V = 158 \text{ mL} \times \frac{1 \text{ L}}{1000 \text{ mL}} = 0.158 \text{ L}, d = \frac{m}{V} = \frac{0.275 \text{ g}}{0.158 \text{ L}} = 1.74051 \text{ g/L}, d = \frac{P\mathcal{M}}{RT} \quad \text{Rearrange to solve for } \mathcal{M}.$$

$$\mathcal{M} = \frac{dRT}{P} = \frac{1.74051 \frac{\text{g}}{\text{L}} \times 0.08206 \frac{\text{L} \cdot \text{atm}}{\text{K} \cdot \text{mol}} \times 298 \text{ K}}{0.731579 \text{ atm}} = 58.2 \text{ g/mol}$$

$$\#\text{C} = \frac{\mathcal{M} \times 0.8266 \text{ g C}}{12.01 \frac{\text{g C}}{\text{mol C}}} = \frac{58.2 \frac{\text{g HC}}{\text{mol HC}} \times \frac{0.8266 \text{ g C}}{1 \text{ g HC}}}{12.01 \frac{\text{g C}}{\text{mol C}}} = 4.00 \frac{\text{mol C}}{\text{mol HC}}$$

$$\#\text{H} = \frac{\mathcal{M} \times 0.1734 \text{ g H}}{1.0008 \frac{\text{g H}}{\text{mol H}}} = \frac{58.2 \frac{\text{g HC}}{\text{mol HC}} \times \frac{0.1734 \text{ g H}}{1 \text{ g HC}}}{1.008 \frac{\text{g H}}{\text{mol H}}} = 10.0 \frac{\text{mol H}}{\text{mol HC}} \quad \text{Formula is } C_4H_{10} \text{ or butane.}$$

An alternative method would be to calculate the empirical formula (C_2H_5) and then use the ideal gas law to calculate the number of moles of gas (0.00473 moles). The molar mass can be calculated by dividing the mass (0.275 g) by the number of moles. Divide the molar mass by the molar mass (58.2 g/mol) of the empirical formula (29.07 g/mol), giving a multiplier of 2 for the empirical formula.

Check: The answer came up with integer number of C and H atoms in the formula and a molecular weight (58 g/mol) that is reasonable for a gas.

5.101   **Given:** $m \, (\text{NiO}) = 24.78 \text{ g}, T = 40.0 \,°\text{C}, \text{and } P_{\text{Total}} = 745 \text{ mmHg}$ **Find:** V_{O_2}
          Conceptual Plan: $T \to P_{\text{H}_2\text{O}} \text{ then } P_{\text{Total}}, P_{\text{H}_2\text{O}} \to P_{\text{O}_2} \text{ then } \text{mmHg} \to \text{atm and } °\text{C} \to \text{K}$

$$\text{Table 5.4} \qquad P_{\text{Total}} = P_{\text{H}_2\text{O}} + P_{\text{O}_2} \qquad \frac{1 \text{ atm}}{760 \text{ mmHg}} \qquad \text{K} = °\text{C} + 273.15$$

and $g_{\text{NiO}} \to n_{\text{NiO}} \to n_{\text{O}_2} \text{ then } P, V, T \to n_{\text{O}_2}$

$$\frac{1 \text{ mol NiO}}{74.69 \text{ g NiO}} \quad \frac{1 \text{ mol O}_2}{2 \text{ mol NiO}} \qquad PV = nRT$$

Solution: Table 5.4 states that $P_{H_2O} = 55.40$ mmHg at $40\,°C$ $P_{Total} = P_{H_2O} + P_{O_2}$ Rearrange to solve for P_{O_2}.

$P_{O_2} = P_{Total} - P_{H_2O} = 745$ mmHg $- 55.40$ mmHg $= 689.\underline{6}$ mmHg

$P_{O_2} = 689.\underline{6}$ mmHg $\times \dfrac{1\ atm}{760\ mmHg} = 0.907368$ atm $T = 40.0\,°C + 273.15 = 313.2$ K

24.78 g NiO $\times \dfrac{1\ mol\ NiO}{74.69\ mol\ NiO} \times \dfrac{1\ mol\ O_2}{2\ mol\ NiO} = 0.165\underline{8}857$ mol O_2 $PV = nRT$

Rearrange to solve for V. $V_{O_2} = \dfrac{nRT}{P} = \dfrac{0.165\underline{8}857\ mol \times 0.08206\ \dfrac{L \cdot atm}{mol \cdot K} \times 313.2\ K}{0.907\underline{3}68\ atm} = 4.70$ L

Check: The units (L) are correct. The magnitude of the answer (5 L) makes sense because we have much less than 0.5 mole of NiO; so we get less than a mole of oxygen. Thus, we expect a volume much less than 22 L.

5.103   **Given:** HCl, K_2S to H_2S, $V_{H_2S} = 42.9$ mL, $P_{H_2S} = 752$ mmHg, and $T = 25.8\,°C$ **Find:** $m(K_2S)$
Conceptual Plan: Read description of reaction and convert words to equation then $°C \rightarrow K$

$$K = °C + 273.15$$

and mmHg $\rightarrow$ atm and mL $\rightarrow$ L then $P, V, T \rightarrow n_{H_2S} \rightarrow n_{K_2S} \rightarrow m_{K_2S}$

$$\dfrac{1\ atm}{760\ mmHg} \qquad \dfrac{1\ L}{1000\ mL} \qquad PV = nRT \quad \dfrac{1\ mol\ K_2S}{1\ mol\ H_2S}\ \dfrac{1\ mol\ K_2S}{110.27\ g\ K_2S}$$

Solution: $2\ HCl(aq) + K_2S(s) \rightarrow H_2S(g) + 2\ KCl(aq)$

$T = 25.8\,°C + 273.15 = 299.0$ K, $P_{H_2S} = 752$ mmHg $\times \dfrac{1\ atm}{760\ mmHg} = 0.98\underline{9}474$ atm

$V_{H_2S} = 42.9$ mL $\times \dfrac{1\ L}{1000\ mL} = 0.0429$ L $PV = nRT$ Rearrange to solve for n_{H_2S}.

$n_{H_2S} = \dfrac{PV}{RT} = \dfrac{0.98\underline{9}474\ atm \times 0.0429\ L}{0.08206\ \dfrac{L \cdot atm}{mol \cdot K} \times 299.0\ K} = 0.0017\underline{3}005$ mol

$0.0017\underline{3}005$ mol H_2S $\times \dfrac{1\ mol\ K_2S}{1\ mol\ H_2S} \times \dfrac{110.27\ g\ K_2S}{1\ mol\ K_2S} = 0.191$ g K_2S

Check: The units (g) are correct. The magnitude of the answer (0.2 g) makes sense because such a small volume of gas is generated.

5.105   **Given:** $T = 22\,°C$, $P = 1.02$ atm, and $m = 11.83$ g   **Find:** V_{Total}
Conceptual Plan: $°C \rightarrow K$ and $g_{(NH_4)_2CO_3} \rightarrow n_{(NH_4)_2CO_3} \rightarrow n_{Gas}$ then $p, n, T \rightarrow V$

$$K = °C + 273.15 \qquad \dfrac{1\ mol\ (NH_4)_2CO_3}{96.09\ g\ (NH_4)_2CO_3}\ \dfrac{(2 + 1 + 1 = 4)\ mol\ gas}{1\ mol\ NH_4CO_3} \qquad PV = nRT$$

Solution: $T = 22\,°C + 273.15 = 295$ K

11.83 g NH_4CO_3 $\times \dfrac{1\ mol\ (NH_4)_2CO_3}{96.09\ g\ (NH_4)_2CO_3} \times \dfrac{4\ mol\ gas}{1\ mol\ (NH_4)_2CO_3} = 0.49\underline{2}455$ mol gas

$PV = nRT$ Rearrange to solve for V_{Gas}.

$V_{Gas} = \dfrac{nRT}{P} = \dfrac{0.49\underline{2}455\ mol\ gas \times 0.08206\ \dfrac{L \cdot atm}{mol \cdot K} \times 295\ K}{1.02\ atm} = 11.7$ L

Check: The units (L) are correct. The magnitude of the answer (12 L) makes sense because about half a mole of gas is generated.

5.107   **Given:** He and air; $V = 855$ mL, $P = 125$ psi, $T = 25\,°C$, $\mathcal{M}(\text{air}) = 28.8$ g/mol   **Find:** $\Delta = m(\text{air}) - m(\text{He})$

Conceptual Plan: mL $\rightarrow$ L and psi $\rightarrow$ atm and $°C \rightarrow$ K then $P, T, \mathcal{M} \rightarrow d$

$$\frac{1\,\text{L}}{1000\,\text{mL}} \qquad \frac{1\,\text{atm}}{14.7\,\text{psi}} \qquad K = °C + 273.15 \qquad d = \frac{P\mathcal{M}}{RT}$$

then $d, V \rightarrow m$ then $m(\text{air}), m(\text{He}) \rightarrow \Delta$

$$d = \frac{m}{V} \qquad\qquad \Delta = m(air) - m(He)$$

Solution: $V = 855\ \text{mL} \times \dfrac{1\,\text{L}}{1000\,\text{mL}} = 0.855\ \text{L}$, $P = 125\ \text{psi} \times \dfrac{1\,\text{atm}}{14.7\,\text{psi}} = 8.50\underline{3}40$ atm

$T = 25\,°C + 273.15 = 298$ K, $d_{\text{air}} = \dfrac{P\mathcal{M}}{RT} = \dfrac{8.50\underline{3}40\ \text{atm} \times 28.8\ \dfrac{\text{g air}}{\text{mol air}}}{0.08206\ \dfrac{\text{L}\cdot\text{atm}}{\text{K}\cdot\text{mol}} \times 298\ \text{K}} = 10.0\underline{1}47\dfrac{\text{g air}}{\text{L}}$, $d = \dfrac{m}{V}$

Rearrange to solve for m. $m = dV$

$m_{\text{air}} = 10.0\underline{1}47\dfrac{\text{g air}}{\text{L}} \times 0.855\ \text{L} = 8.5\underline{6}657$ g air

$d_{\text{He}} = \dfrac{P\mathcal{M}}{RT} = \dfrac{8.50\underline{3}40\ \text{atm} \times 4.03\ \dfrac{\text{g He}}{\text{mol He}}}{0.08206\ \dfrac{\text{L}\cdot\text{atm}}{\text{K}\cdot\text{mol}} \times 298\ \text{K}} = 1.40\underline{1}36\dfrac{\text{g He}}{\text{L}}$

$m_{\text{He}} = 1.40\underline{1}36\dfrac{\text{g He}}{\text{L}} \times 0.855\ \text{L} = 1.1\underline{9}816$ g He

$\Delta = m_{\text{air}} - m_{\text{He}} = 8.5\underline{6}657$ g air $- 1.1\underline{9}816$ g He $= 7.37$ g

Check: The units (g) are correct. We expect the difference to be less than the difference in the molecular weights because we have less than a mole of gas.

5.109   **Given:** flow $= 335$ L/s, $P_{\text{NO}} = 22.4$ torr, $T_{\text{NO}} = 955$ K, $P_{\text{NH}_3} = 755$ torr, $T_{\text{NO}} = 298$ K, and NH_3 purity $= 65.2\%$
Find: flow$_{\text{NH}_3}$

Conceptual Plan: torr $\rightarrow$ atm then $P_{\text{NO}}, V_{\text{NO}}/\text{s}, T_{\text{NO}} \rightarrow n_{\text{NO}}/\text{s} \rightarrow n_{\text{NH}_3}/\text{s (pure)}$

$$\frac{1\,\text{atm}}{760\,\text{torr}} \qquad\qquad PV = nRT \quad \frac{4\,\text{mol NH}_3}{4\,\text{mol NO}}$$

then $n_{\text{NH}_3}/\text{s (pure)} \rightarrow n_{\text{NH}_3}/\text{s (impure)}$ then $n_{\text{NH}_3}/\text{s (impure)}, P_{\text{NH}_3}, T_{\text{NH}_3} \rightarrow V_{\text{NH}_3}/\text{s}$

$$\frac{100\,\text{mol NH}_3\ \text{impure}}{65.2\,\text{mol NH}_3\ \text{pure}} \qquad\qquad\qquad PV = nRT$$

Solution: $P_{\text{NO}} = 22.4\ \text{torr} \times \dfrac{1\,\text{atm}}{760\,\text{torr}} = 0.029\underline{4}737$ atm, $P_{\text{NH}_3} = 755\ \text{torr} \times \dfrac{1\,\text{atm}}{760\,\text{torr}} = 0.99\underline{3}421$ atm

$PV = nRT$ Rearrange to solve for n_{NO}. Note that we can substitute V/s for V and get n/s as a result.

$\dfrac{n_{\text{NO}}}{\text{s}} = \dfrac{PV}{RT} = \dfrac{0.029\underline{4}737\ \text{atm} \times 335\ \text{L/s}}{0.08206\ \dfrac{\text{L}\cdot\text{atm}}{\text{mol}\cdot\text{K}} \times 955\ \text{K}} = 0.12\underline{5}992\dfrac{\text{mol NO}}{\text{s}}$

$0.12\underline{5}992\dfrac{\text{mol NO}}{\text{s}} \times \dfrac{4\,\text{mol NH}_3}{4\,\text{mol NO}} \times \dfrac{100\,\text{mol NH}_3\ \text{impure}}{65.2\,\text{mol NH}_3\ \text{pure}} = 0.19\underline{3}240\dfrac{\text{mol NH}_3\ \text{impure}}{\text{s}}$ $PV = nRT$

Rearrange to solve for V_{NH_3}. Note that we can substitute n/s for n and get V/s as a result.

$\dfrac{V_{\text{NH}_3}}{\text{s}} = \dfrac{nRT}{P} = \dfrac{0.19\underline{3}240\dfrac{\text{mol NH}_3\ \text{impure}}{\text{s}} \times 0.08206 \times \dfrac{\text{L}\cdot\text{atm}}{\text{mol}\cdot\text{K}} \times 298\text{K}}{0.99\underline{3}421\ \text{atm}} = 4.76\dfrac{\text{L}}{\text{s}}$ impure NH_3

Check: The units (L) are correct. The magnitude of the answer (5 L/s) makes sense because we expect it to be less than that for the NO. The NO is at a very low concentration and a high temperature. When this converts to a much higher pressure and lower temperature, it goes down significantly even though the ammonia is impure. From a practical standpoint, you would like a low flow rate to make it economical.

5.111   **Given:** l = 30.0 cm, w = 20.0 cm, h = 15.0 cm, 14.7 psi   **Find:** force (lb)

Conceptual Plan: *l, w, h* → **Surface Area, SA(cm²)** → **Surface Area(in²)** → **Force**

$$SA = 2(lh) + 2(wh) + 2(lw) \qquad \frac{(1\text{ in})^2}{(2.54\text{ cm})^2} \qquad \frac{14.7\text{ lb}}{1\text{ in}^2}$$

Solution: $SA = 2(lh) + 2(wh) + 2(lw) = 2(30.0\text{ cm} \times 15.0\text{ cm}) + 2(20.0\text{ cm} \times 15.0\text{ cm})$
$+ 2(30.0\text{ cm} \times 20.0\text{ cm}) = 2700\text{ cm}^2$

$$2700\text{ cm}^2 \times \frac{(1\text{ in})^2}{(2.54\text{ cm})^2} = 418.50\text{ in}^2, 418.50\text{ in}^2 \times \frac{14.7\text{ lb}}{1\text{ in}^2} = 6150\text{ lb. The can would be crushed.}$$

Check: The units (lb) are correct. The magnitude of the answer (6150 lb) is not unreasonable because there is a large surface area.

5.113   **Given:** $V_1 = 160.0$ L, $P_1 = 1855$ psi, 3.5 L/balloon, $P_2 = 1.0$ atm = 14.7 psi, and $T = 298$ K   **Find:** # balloons
Conceptual Plan: $V_1, P_1, P_2 \rightarrow V_2$ **then L → # balloons**

$$P_1V_1 = P_2V_2 \qquad \frac{1\text{ balloon}}{3.5\text{ L}}$$

Solution: $P_1V_1 = P_2V_2$ Rearrange to solve for V_2. $V_2 = \dfrac{P_1}{P_2}V_1 = \dfrac{1855\text{ psi}}{14.7\text{ psi}} \times 160.0\text{ L} = 20190.5\text{ L}$

$$20190.5\text{ L} \times \frac{1\text{ balloon}}{3.5\text{ L}} = 5768\text{ balloons} = 5800\text{ balloons}$$

Check: The units (balloons) are correct. The magnitude of the answer (5800) is reasonable because a store does not want to buy a new helium tank very often.

5.115   **Given:** $r_1 = 2.5$ cm, $P_1 = 4.00$ atm, $T = 298$ K, and $P_2 = 1.00$ atm   **Find:** r_2
Conceptual Plan: $r_1 \rightarrow V_1 V_1, P_1, P_2 \rightarrow V_2$ **then** $V_2 \rightarrow r_2$

$$V = \frac{4}{3}\pi r^3 \qquad P_1V_1 = P_2V_2 \qquad V = \frac{4}{3}\pi r^3$$

Solution: $V = \dfrac{4}{3}\pi r^3 = \dfrac{4}{3} \times \pi \times (2.5\text{ cm})^3 = 65.450\text{ cm}^3$ $P_1 V_1 = P_2V_2$ Rearrange to solve for V_2.

$$V_2 = \frac{P_1}{P_2}V_1 = \frac{4.00\text{ atm}}{1.00\text{ atm}} \times 65.450\text{ cm}^3 = 261.80\text{ cm}^3, V = \frac{4}{3}\pi r^3$$

Rearrange to solve for r. $r = \sqrt[3]{\dfrac{3V}{4\pi}} = \sqrt[3]{\dfrac{3 \times 261.80\text{ cm}^3}{4 \times \pi}} = 4.0\text{ cm}$

Check: The units (cm) are correct. The magnitude of the answer (4 cm) is reasonable because the bubble will expand as the pressure is decreased.

5.117   **Given:** 2.0 mol CO : 1.0 mol O_2, $V = 2.45$ L, $P_1 = 745$ torr, $P_2 = 552$ torr, and $T = 552$ °C
Find: % reacted
Conceptual Plan: From $PV = nRT$**, we know that** $P \, \alpha \, n$**; looking at the chemical reaction, we see that 2 + 1 = 3 moles of gas gets converted to 2 moles of gas. If all of the gas reacts,** $P_2 = 2/3\,P_1$**. Calculate** $-\Delta P$ **for 100% reacted and for actual case. Then calculate % reacted.**

$$-\Delta P_{100\%\text{ reacted}} = P_1 - \frac{2}{3}P_1 \quad -\Delta P_{\text{actual}} = P_1 - P_2 \quad \%\text{ reacted} = \frac{\Delta P_{\text{actual}}}{\Delta P_{100\%\text{ reacted}}} \times 100\%$$

Solution: $-\Delta P_{100\%\text{ reacted}} = P_1 - \dfrac{2}{3}P_1 = 745\text{ torr} - \dfrac{2}{3}745\text{ torr} = 248.333\text{ torr}$

$-\Delta P_{\text{actual}} = P_1 - P_2 = 745\text{ torr} - 552\text{ torr} = 193\text{ torr}$

$\%\text{ reacted} = \dfrac{\Delta P_{\text{actual}}}{\Delta P_{100\%\text{ reacted}}} \times 100\% = \dfrac{193\text{ torr}}{248.333\text{ torr}} \times 100\% = 77.7\%$

Check: The units (%) are correct. The magnitude of the answer (78%) makes sense because the pressure dropped most of the way to the pressure if all of the reactants had reacted (2/3(745 torr) = 497 torr). **Note: This problem can be solved many ways, including calculating the moles of reactants and products using** $PV = nRT$**.**

5.119 **Given:** $P(\text{Total})_1 = 2.2 \text{ atm} = CO + O_2$, $P(\text{Total})_2 = 1.9 \text{ atm} = CO + O_2 + CO_2$, $V = 1.0 \text{ L}$, and $T = 1.0 \times 10^3 \text{ K}$
Find: mass CO_2 made
Conceptual Plan: $P(\text{Total})_1 = 2.2 \text{ atm} = P(CO)_1 + P(O_2)_1$, $P(\text{Total})_2 = 1.9 \text{ atm} = P(CO)_2 + P(O_2)_2 + P(CO_2)_2$.
Let x = amount of $P(O_2)$ reacted. From stoichiometry: $P(CO)_2 = P(CO)_1 - 2x$, $P(O2)_2 = P(O2)_1 - x$,
$P(CO_2)_2 = 2x$. Thus, $P(\text{Total})_2 = 1.9 \text{ atm} = P(CO)_1 - 2x + P(O2)_1 - x + 2x = P(\text{Total})_1 - x$. Using
the initial conditions: $1.9 \text{ atm} = 2.2 \text{ atm} - x$. So $x = 0.3 \text{ atm}$, and because $2x = P(CO_2)_2 = 0.6 \text{ atm}$,
$P, V, T \rightarrow n \rightarrow g$.

$$PV = nRT \quad \frac{44.01 \text{ g}}{1 \text{ mol}}$$

Solution: $PV = nRT$ Rearrange to solve for n.

$$n = \frac{PV}{RT} = \frac{0.6 \text{ atm} \times 1.0 \text{ L}}{0.08206 \dfrac{\text{L} \cdot \text{atm}}{\text{mol} \cdot \text{K}} \times 1000 \text{ K}} = 0.0073117 \text{ mol}$$

$$0.0073117 \text{ mol} \times \frac{44.01 \text{ g}}{1 \text{ mol}} = 0.321788 \text{ g } CO_2 = 0.3 \text{ g } CO_2$$

Check: The units (g) are correct. The magnitude of the answer (0.3 g) makes sense because we have such a small volume at a very high temperature and such a small pressure. This leads us to expect a very small number of moles.

5.121 **Given:** $h_1 = 22.6 \text{ m}$, $T_1 = 22 \,^\circ\text{C}$, and $h_2 = 23.8 \text{ m}$ **Find:** T_2
Conceptual Plan: $^\circ\text{C} \rightarrow \text{K}$ because $V_{\text{cylinder}} \, \alpha \, h$ we do not need to know r to use $V_1, T_1, T_2 \rightarrow V_2$

$$K = {}^\circ\text{C} + 273.15 \qquad V = \pi r^2 h \qquad\qquad \frac{V_1}{T_1} = \frac{V_2}{T_2}$$

Solution: $T_1 = 22 \,^\circ\text{C} + 273.15 = 295\text{K}$, $\dfrac{V_1}{T_1} = \dfrac{V_2}{T_2}$ Rearrange to solve for T_2.

$$T_2 = T_1 \times \frac{V_2}{V_1} = T_1 \times \frac{\pi r^2 t_2}{\pi r^2 t_1} = 295 \text{ K} \times \frac{23.8 \text{ m}}{22.6 \text{ m}} = 311 \text{ K}$$

Check: The units (K) are correct. We expect the temperature to increase because the volume increased.

5.123 **Given:** He, $V = 0.35 \text{ L}$, $P_{\text{max}} = 88 \text{ atm}$, and $T = 299 \text{ K}$ **Find:** m_{He}
Conceptual Plan: $P, V, T \rightarrow n$ then mol $\rightarrow$ g

$$PV = nRT \qquad\qquad \mathcal{M}$$

Solution: $PV = nRT$ Rearrange to solve for n.

$$n_{\text{He}} = \frac{PV}{RT} = \frac{88 \text{ atm} \times 0.35 \text{ L}}{0.08206 \dfrac{\text{L} \cdot \text{atm}}{\text{mol} \cdot \text{K}} \times 299 \text{ K}} = 1.2553 \text{ mol}, \; 1.2553 \text{ mol} \times \frac{4.003 \text{ g}}{1 \text{ mol}} = 5.0 \text{ g He}$$

Check: The units (g) are correct. The magnitude of the answer (5 g) makes sense because the high pressure and the low volume cancel out (remember 22 L/mol at STP). So we expect ~1 mol; so ~4 g.

5.125 **Given:** 15.0 mL HBr in 1.0 minute and 20.3 mL unknown hydrocarbon gas in 1.0 minute   **Find:** formula of unknown gas
Conceptual Plan: Because these gases are under the same conditions, $V \, \alpha \, n$, V, time $\rightarrow$ Rate then

$$\text{Rate} = \frac{V}{time}$$

$$\mathcal{M}_{\text{HBr}}, \text{Rate}_{\text{HBr}}, \text{Rate}_{\text{Unk}} \rightarrow \mathcal{M}_{\text{Unk}}$$

$$\frac{\text{Rate}_{\text{HBr}}}{\text{Rate}_{\text{Unk}}} = \sqrt{\frac{\mathcal{M}_{\text{Unk}}}{\mathcal{M}_{\text{HBr}}}}$$

Solution: $\text{Rate}_{\text{HBr}} = \dfrac{V}{time} = \dfrac{15.0 \text{ mL}}{1.0 \text{ min}} = 15.0 \dfrac{\text{mL}}{\text{min}}$, $\text{Rate}_{\text{Unk}} = \dfrac{V}{time} = \dfrac{20.3 \text{ mL}}{1.0 \text{ min}} = 20.3 \dfrac{\text{mL}}{\text{min}}$,

$$\frac{\text{Rate}_{\text{HBr}}}{\text{Rate}_{\text{Unk}}} = \sqrt{\frac{\mathcal{M}_{\text{Unk}}}{\mathcal{M}_{\text{HBr}}}} \qquad \text{Rearrange to solve for } \mathcal{M}_{\text{Unk}}.$$

$$\mathcal{M}_{\text{Unk}} = \mathcal{M}_{\text{HBr}} \left(\frac{\text{Rate}_{\text{HBr}}}{\text{Rate}_{\text{Unk}}} \right)^2 = 80.91 \frac{\text{g}}{\text{mol}} \times \left(\frac{15.0 \dfrac{\text{mL}}{\text{min}}}{20.3 \dfrac{\text{mL}}{\text{min}}} \right)^2 = 44.2 \frac{\text{g}}{\text{mol}} \text{ The formula is } C_3H_8, \text{ or propane.}$$

Check: The units (g/mol) are correct. The magnitude of the answer ($< $ HBr) makes sense because the unknown diffused faster and so must be lighter.

5.127   **Given:** 0.583 g neon, $V = 8.00 \times 10^2$ cm^3, $P_{Total} = 1.17$ atm, and $T = 295$ K   **Find:** g argon
Conceptual Plan: g $\rightarrow$ n and mL $\rightarrow$ L then n, V, T $\rightarrow$ P_{Ne} then P_{Ne}, P_{Total} $\rightarrow$ P_{Ar} then

$$\frac{1 \text{ mol}}{20.18 \text{ g}} \qquad \frac{1 \text{ L}}{1000 \text{ mL}} \qquad PV = nRT \qquad P_{Total} = P_{Ne} + P_{Ar}$$

$P_{Ar}, V, T \rightarrow n \rightarrow g$

$$PV = nRT \quad \frac{39.95 \text{ g}}{1 \text{ mol}}$$

Solution: $0.583 \text{ g Ne} \times \dfrac{1 \text{ mol Ne}}{20.18 \text{ g Ne}} = 0.02888999 \text{ mol Ne} \quad 8.00 \times 10^2 \text{ mL} \times \dfrac{1 \text{ L}}{1000 \text{ mL}} = 0.800 \text{ L}$

$PV = nRT$ Rearrange to solve for P.

$$P = \frac{nRT}{V} = \frac{0.02888999 \text{ mol} \times 0.08206 \dfrac{\text{L} \cdot \text{atm}}{\text{mol} \cdot \text{K}} \times 295 \text{ K}}{0.800 \text{ L}} = 0.874200 \text{ atm Ne}$$

$P_{Total} = P_{Ne} + P_{Ar}$ Rearrange to solve for P_{Ar}. $P_{Ar} = P_{Total} - P_{Ne} = 1.17 \text{ atm} - 0.874200 \text{ atm} = 0.295800 \text{ atm}$

$PV = nRT$ Rearrange to solve for n. $n_{Ar} = \dfrac{PV}{RT} = \dfrac{0.295800 \text{ atm} \times 0.800 \text{ L}}{0.08206 \dfrac{\text{L} \cdot \text{atm}}{\text{mol} \cdot \text{K}} \times 295 \text{ K}} = 0.00977539 \text{ mol Ar}$

$0.00977539 \text{ mol Ar} \times \dfrac{39.95 \text{ g Ar}}{1 \text{ mol Ar}} = 0.390527 \text{ g Ar} = 0.39 \text{ g Ar}$

Check: The units (g) are correct. The magnitude of the answer (0.4 g) makes sense because the pressure and volume are small.

5.129   **Given:** 75.2% by mass nitrogen + 24.8% by mass krypton and $P_{Total} = 745$ mmHg   **Find:** P_{Kr}
Solution: Assume that 100 g total, so we have 75.2 g N$_2$ and 24.8 g Kr. Converting these masses to moles,

$$75.2 \text{ g N}_2 \times \frac{1 \text{ mol N}_2}{28.02 \text{ g N}_2} = 2.683797 \text{ mol N}_2 \text{ and } 24.8 \text{ g Kr} \times \frac{1 \text{ mol Kr}}{83.80 \text{ g Kr}} = 0.2959427 \text{ mol Kr}$$

$$P_{Kr} = \chi_{Kr} P_{Total} = \frac{0.2959427 \text{ mol Kr}}{2.683797 \text{ mol N}_2 + 0.2959427 \text{ mol Kr}} 745 \text{ mmHg Kr} = 74.0 \text{ mmHg Kr}$$

Check: The units (mmHg) are correct. The magnitude of the answer (74 mmHg) makes sense because the mixture is mostly nitrogen by mass and this dominance is magnified because the molar mass of krypton is larger than the molar mass of nitrogen.

Challenge Problems

5.131   **Given:** $2 \text{ NH}_3(g) \rightarrow \text{N}_2(g) + 3 \text{ H}_2(g)$; $\text{N}_2\text{H}_4(g) \rightarrow \text{N}_2(g) + 2 \text{ H}_2(g)$; initially $P = 0.50$ atm, $T = 300$ K; finally $P = 4.5$ atm, $T = 1200$ K   **Find:** N$_2$H$_4$ percent initially
Conceptual Plan: $P_{initial}, T_{initial}, T_{final} \rightarrow P_{final}$ then determine change in moles of gas

$$\frac{P_{initial}}{T_{initial}} = \frac{P_{final}}{T_{final}} \qquad \frac{2 \text{ atm added gas}}{2 \text{ atm NH}_3 \text{ reacted}} \text{ and } \frac{2 \text{ atm added gas}}{1 \text{ atm N}_2\text{H}_4 \text{ reacted}}$$

$P_1, P_2 \rightarrow \Delta P$ write expression for ΔP then solve for $P_{1, \text{N}_2\text{H}_4}$ and P_{1, NH_3} finally $P_{1, \text{N}_2\text{H}_4}, P_{1, \text{NH}_3} \rightarrow$ % N$_2$H$_4$

$$\Delta P = P_2 - P_1 \quad \Delta P = P_{1, \text{NH}_3} \frac{3 \text{ atm added gas}}{2 \text{ atm reacted}} + P_{1, \text{N}_2\text{H}_4} \frac{2 \text{ atm added gas}}{1 \text{ atm reacted}} \text{ where } P_{1,1200 \text{ K}} = P_{1, \text{NH}_3} + P_{1, \text{N}_2\text{H}_4} \quad \% \text{ N}_2\text{H}_4 = \frac{P_{\text{N}_2\text{H}_4}}{P_{\text{N}_2\text{H}_4} + P_{\text{NH}_3}} \times 100\%$$

Solution: $\dfrac{P_{initial}}{T_{initial}} = \dfrac{P_{final}}{T_{final}}$ Rearrange to solve for P_{final}. $P_2 = P_1 \times \dfrac{T_2}{T_1} = 0.50 \text{ atm} \times \dfrac{1200 \text{ K}}{300 \text{ K}} = 2.0 \text{ atm}$ if no reaction occurred.

$\Delta P = P_{final} - P_{initial} = 4.5 \text{ atm} - 2.0 \text{ atm} = 2.5 \text{ atm}$, and $P_{1,1200 \text{ K}} = 2.0 \text{ atm} = P_{1, \text{NH}_3} + P_{1, \text{N}_2\text{H}_4}$ or $P_{1, \text{NH}_3} = 2.0 \text{ atm} - P_{1, \text{N}_2\text{H}_4}$

Substitute this into $\Delta P = P_{1, \text{NH}_3} \dfrac{2 \text{ atm added gas}}{2 \text{ atm reacted}} + P_{1, \text{N}_2\text{H}_4} \dfrac{2 \text{ atm added gas}}{1 \text{ atm reacted}}$ and solve for $P_{1, \text{N}_2\text{H}_4}$.

$\Delta P = 2.5 \text{ atm} = (2.0 \text{ atm} - P_{1, \text{N}_2\text{H}_4}) \dfrac{2 \text{ atm added gas}}{2 \text{ atm reacted}} + P_{1, \text{N}_2\text{H}_4} \dfrac{2 \text{ atm added gas}}{1 \text{ atm reacted}} \rightarrow$

$P_{1, N_2H_4} = 2.5 \text{ atm} - 2.0 \text{ atm} = 0.5 \text{ atm}$ and $P_{NH_3} = 2.0 \text{ atm} - 0.5 \text{ atm} = 1.5 \text{ atm}$; finally

$$\% \ N_2H_4 = \frac{P_{N_2H_4}}{P_{N_2H_4} + P_{NH_3}} \times 100\% = \frac{0.5 \text{ atm}}{0.5 \text{ atm} + 1.5 \text{ atm}} \times 100\% = \underline{2}5\% \ N_2H_4 = 30\% \ N_2H_4$$

Check: The units (%) are correct. The magnitude of the answer (30%) makes sense because if it were all N_2H_4, the final pressure would have been 6 atm. Because we are closer to the initial pressure than this maximum pressure, less than half of the gas is N_2H_4.

5.133   **Given:** $2 \ CO_2(g) \rightarrow 2 \ CO(g) + O_2(g)$; initially $P = 10.0 \text{ atm}$, $T = 701 \text{ K}$; finally $P = 22.5 \text{ atm}$, $T = 1401 \text{ K}$
          Find: mole percent decomposed
          Conceptual Plan: $P_{initial}, T_{initial}, T_{final} \rightarrow P_{final}$ then determine change in moles of gas

$$\frac{P_{initial}}{T_{initial}} = \frac{P_{final}}{T_{final}} \qquad\qquad\qquad \frac{1 \text{ atm added gas}}{2 \text{ atm } CO_2 \text{ reacted}}$$

          $P_1, P_2 \rightarrow \Delta P$ **write expression for** ΔP **then solve for** $P_{CO_2 \text{reacted}}$ **finally**

$$\Delta P = P_2 - P_1 \qquad \Delta P = P_{CO_2 \text{ reacted}} \frac{1 \text{ atm added gas}}{2 \text{ atm } CO_2 \text{ reacted}}$$

          $P_{final}, P_{CO_2 \text{ reacted}} \rightarrow \%\ CO_2$ **decomposed**

$$\% \ CO_2 \text{ decomposed} = \frac{P_{CO_2 \text{ reacted}}}{P_{final}} \times 100\%$$

          Solution: $\dfrac{P_{initial}}{T_{initial}} = \dfrac{P_{final}}{T_{final}}$ Rearrange to solve for P_{final}. $P_2 = P_1 \times \dfrac{T_2}{T_1} = 10.0 \text{ atm} \times \dfrac{1401 \text{ K}}{701 \text{ K}} = 19.985735 \text{ atm}$

$$\Delta P = P_{final} - P_{initial} = 22.5 \text{ atm} - 19.985735 \text{ atm} = 2.\underline{5}14265 \text{ atm}$$

$\Delta P = P_{CO_2 \text{ reacted}} \dfrac{1 \text{ atm added gas}}{2 \text{ atm } CO_2 \text{ reacted}}$ or the pressure increases 1 atm for each 2 atm of gas decomposed, so

5.0285 3 atm decomposes and then

$$\% \ CO_2 \text{ decomposed} = \frac{P_{CO_2 \text{ reacted}}}{P_{final}} \times 100\% = \frac{5.0\underline{2}853 \text{ atm}}{19.985735 \text{ atm}} \times 100\% = 2\underline{5}.1606\% CO_2 \text{ decomposed} =$$

25% CO_2 decomposed

Check: The units (%) are correct. The magnitude of the answer (11%) makes sense because if all of the gas had decomposed, the final pressure would have been 40 atm. Because we are much closer to the initial pressure than this maximum pressure, much less than half of the gas had decomposed.

5.135   **Given:** CH_4: $V = 155 \text{ mL}$ at STP; O_2: $V = 885 \text{ mL}$ at STP; NO: $V = 55.5 \text{ mL}$ at STP; mixed in a flask: $V = 2.0 \text{ L}$, $T = 275 \text{ K}$, and 90.0% of limiting reagent used.   **Find:** Ps of all components and P_{Total}
          Conceptual Plan: CH_4: $mL \rightarrow L \rightarrow mol_{CO_4} \rightarrow mol_{CO_2}$ and

$$\frac{1 \text{ L}}{1000 \text{ mL}} \quad \frac{1 \text{ mol}}{22.414 \text{ L}} \qquad \frac{1 \text{ mol } CO_2}{5 \text{ mol NO}}$$

          O_2: $mL \rightarrow L \rightarrow mol_{O_2} \rightarrow mol_{CO_2}$ and NO: $mL \rightarrow L \rightarrow mol_{NO} \rightarrow mol_{CO_2}$

$$\frac{1 \text{ L}}{1000 \text{ mL}} \quad \frac{1 \text{ mol}}{22.414 \text{ L}} \quad \frac{1 \text{ mol } CO_2}{5 \text{ mol } O_2} \qquad\qquad \frac{1 \text{ L}}{1000 \text{ mL}} \quad \frac{1 \text{ mol}}{22.414 \text{ L}} \quad \frac{1 \text{ mol } CO_2}{5 \text{ mol NO}}$$

          the smallest yield determines the limiting reagent then initial $mol_{NO} \rightarrow$ **reacted** $mol_{NO} \rightarrow$ **final** mol_{NO}

$$\text{NO is the limiting reagent} \qquad\qquad 90.0\% \qquad\qquad 0.100 \times \text{initial } mol_{NO}$$

          reacted $mol_{NO} \rightarrow$ **reacted** mol_{CH_4} **then initial** mol_{CH_4}, **reacted** $mol_{CH_4} \rightarrow$ **final** mol_{CH_4} **then**

$$\frac{1 \text{ mol } CH_4}{5 \text{ mol NO}} \qquad\qquad\qquad \text{initial } mol_{CH_4} - \text{reacted } mol_{CH_4} = \text{final } mol_{CH_4}$$

          final mol_{CH_4}, $V, T \rightarrow$ **final** P_{CH_4} **and reacted** $mol_{NO} \rightarrow$ **reacted** mol_{O_2} **then**

$$PV = nRT \qquad\qquad\qquad \frac{5 \text{ mol } O_2}{5 \text{ mol NO}}$$

          initial mol_{O_2}, **reacted** $mol_{O_2} \rightarrow$ **final** mol_{O_2} **then final** mol_{O_2}, $V, T \rightarrow$ **final** P_{O_2} **and**

$$\text{initial } mol_{O_2} - \text{reacted } mol_{O_2} = \text{final } mol_{O_2} \qquad\qquad PV = nRT$$

          final mol_{NO}, $V, T \rightarrow$ **final** P_{NO} **and theoretical** mol_{CO_2} **from NO** $\rightarrow$ **final** mol_{CO_2}

$$PV = nRT \qquad\qquad\qquad 90.0\%$$

          final mol_{CO_2}, $V, T \rightarrow P_{CO_2}$ **then final** $mol_{CO_2} \rightarrow mol_{H_2O}$, $V, T \rightarrow P_{H_2O}$ **and**

$$PV = nRT \qquad\qquad \frac{1 \text{ mol } H_2O}{1 \text{ mol } CO_2} \qquad\qquad PV = nRT$$

final $mol_{CO_2} \rightarrow mol_{NO_2}$ **then** $mol_{NO_2}, V, T \rightarrow P_{NO_2}$ **and final** $mol_{CO_2} \rightarrow mol_{OH}$ **then**

$$\frac{1 \text{ mol } NO_2}{1 \text{ mol } CO_2} \qquad\qquad PV = nRT \qquad\qquad \frac{2 \text{ mol } OH}{1 \text{ mol } CO_2}$$

$mol_{OH}, V, T \rightarrow P_{OH}$ **finally** $P_{CH_4}, P_{O_2}, P_{NO}, P_{CO_2}, P_{H_2O}, P_{NO_2}, P_{OH} \rightarrow P_{T \text{ total}}$

$$PV = nRT \qquad\qquad\qquad\qquad\qquad P_{Total} = \Sigma P$$

Solution: CH_4: $155 \text{ mL} \times \dfrac{1 \text{ L}}{1000 \text{ mL}} \times \dfrac{1 \text{ mol } CH_4}{22.414 \text{ L}} \times \dfrac{1 \text{ mol } CO_2}{1 \text{ mol } CH_4} = 0.00691532 \text{ mol } CO_2$

O_2: $885 \text{ mL} \times \dfrac{1 \text{ L}}{1000 \text{ mL}} \times \dfrac{1 \text{ mol } O_2}{22.414 \text{ L}} = 0.0394843 \text{ mol } O_2 \times \dfrac{1 \text{ mol } CO_2}{5 \text{ mol } O_2} = 0.00789686 \text{ mol } CO_2$

NO: $55.5 \text{ mL} \times \dfrac{1 \text{ L}}{1000 \text{ mL}} \times \dfrac{1 \text{ mol } NO}{22.414 \text{ L}} \times \dfrac{1 \text{ mol } CO_2}{5 \text{ mol } NO} = 0.000495226 \text{ mol } CO_2$

$0.000495226 \text{ mol } CO_2$ is the smallest yield, so NO is the limiting reagent.

$55.5 \text{ mL} \times \dfrac{1 \text{ L}}{1000 \text{ mL}} \times \dfrac{1 \text{ mol } NO}{22.414 \text{ L}} = 0.00247613 \text{ mol } NO$

reacted mol NO $= 0.900 \times$ mol NO $= 0.900 \times 0.00247613$ mol NO $= 0.00222852$ mol NO

unreacted mol NO $= 0.100 \times$ mol NO $= 0.100 \times 0.00247613$ mol NO $= 0.000247613$ mol NO

$$P = \frac{nRT}{V} = \frac{0.000247613 \text{ mol} \times 0.08206 \dfrac{L \cdot atm}{mol \cdot K} \times 275 \text{ K}}{2.0 \text{ L}} = 0.00279 \text{ atm NO remaining}$$

$0.00222852 \text{ mol } NO \times \dfrac{1 \text{ mol } CH_4}{5 \text{ mol } NO} = 0.000445704 \text{ mol } CH_4$ reacted

$0.00691532 \text{ mol } CH_4 - 0.000445704 \text{ mol } CH_4$ reacted $= 0.00646962 \text{ mol } CH_4$ unreacted then $PV = nRT$

Rearrange to solve for P. $P = \dfrac{nRT}{V} = \dfrac{0.00646962 \text{ mol} \times 0.08206 \dfrac{L \cdot atm}{mol \cdot K} \times 275 \text{ K}}{2.0 \text{ L}} = 0.0730 \text{ atm } CH_4$ remaining

$0.00222852 \text{ mol } NO \times \dfrac{5 \text{ mol } O_2}{5 \text{ mol } NO} = 0.00222852 \text{ mol } O_2$ reacted

$0.0394843 \text{ mol } O_2 - 0.00222852 \text{ mol } O_2$ reacted $= 0.0372558 \text{ mol } O_2$ unreacted

$$P = \frac{nRT}{V} = \frac{0.0372558 \text{ mol} \times 0.08206 \dfrac{L \cdot atm}{mol \cdot K} \times 275 \text{ K}}{2.0 \text{ L}} = 0.420 \text{ atm } O_2 \text{ remaining}$$

$0.00222852 \text{ mol } NO \times \dfrac{1 \text{ mol } CO_2}{5 \text{ mol } NO} = 0.000445704 \text{ mol } CO_2$

$$P = \frac{nRT}{V} = \frac{0.000445704 \text{ mol} \times 0.08206 \dfrac{L \cdot atm}{mol \cdot K} \times 275 \text{ K}}{2.0 \text{ L}} = 0.00503 \text{ atm } CO_2 \text{ produced}$$

$0.00222852 \text{ mol } NO \times \dfrac{1 \text{ mol } H_2O}{5 \text{ mol } NO} = 0.000445704 \text{ mol } H_2O$

$$P = \frac{nRT}{V} = \frac{0.000445704 \text{ mol} \times 0.08206 \dfrac{L \cdot atm}{mol \cdot K} \times 275 \text{ K}}{2.0 \text{ L}} = 0.00503 \text{ atm } H_2O \text{ produced}$$

$0.00222852 \text{ mol } NO \times \dfrac{5 \text{ mol } NO_2}{5 \text{ mol } NO} = 0.00222852 \text{ mol } NO_2$

$$P = \frac{nRT}{V} = \frac{0.00222852 \text{ mol} \times 0.08206 \dfrac{L \cdot atm}{mol \cdot K} \times 275 \text{ K}}{2.0 \text{ L}} = 0.0251 \text{ atm } NO_2 \text{ produced}$$

$0.00222852 \text{ mol } NO \times \dfrac{2 \text{ mol } OH}{5 \text{ mol } NO} = 0.000891408 \text{ mol } OH$

$$P = \frac{nRT}{V} = \frac{0.000891408 \text{ mol} \times 0.08206 \dfrac{L \cdot atm}{mol \cdot K} \times 275 \text{ K}}{2.0 \text{ L}} = 0.0101 \text{ atm } OH \text{ produced}$$

$P_{Total} = \displaystyle\sum P$

$= 0.0730 \text{ atm} + 0.420 \text{ atm} + 0.00279 \text{ atm} + 0.00503 \text{ atm} + 0.00503 \text{ atm} + 0.0251 \text{ atm} + 0.0101 \text{ atm}$

$= 0.541 \text{ atm}$

Check: The units (atm) are correct. The magnitude of the answers is reasonable. The limiting reagent has the lowest pressure. The product pressures are in line with the ratios of the stoichiometric coefficients.

5.137   **Given:** $P_{CH_4} + P_{C_2H_4} = 0.53$ atm and $P_{CO_2} + P_{H_2O} = 2.2$ atm   **Find:** χ_{CH_4}
Conceptual Plan: Write balanced reactions to determine change in moles of gas for CH$_4$ and C$_2$H$_6$.

$$2\,CH_4(g) + 4\,O_2(g) \rightarrow 4\,H_2O(g) + 2\,CO_2(g) \text{ and } 2\,C_2H_6(g) + 7\,O_2(g) \rightarrow 6\,H_2O(g) + 4\,CO_2(g); \text{ thus,}\quad \frac{6\text{ mol gases}}{2\text{ mol }CH_4}\quad \frac{10\text{ mol gases}}{2\text{ mol }C_2H_6}$$

Write expression for final pressure, substituting in data given $\rightarrow \chi_{CH_4}$.

$$\chi_{CH_4} = \frac{n_{CH_4}}{n_{CH_4} + n_{C_2H_6}} \text{ and } \chi_{C_2H_6} = 1 - \chi_{CH_4}$$

$$P_{CH_4} = \chi_{CH_4}P_{Total} \qquad P_{C_2H_6} = \chi_{C_2H_6}P_{Total} \qquad P_{Final} = \left(\chi_{CH_4}P_{Total} \times \frac{6\text{ mol gases}}{2\text{ mol }CH_4}\right) + \left((1 - \chi_{CH_4})P_{Total} \times \frac{10\text{ mol gases}}{2\text{ mol }C_2H_6}\right)$$

Solution:

$$P_{Final} = \left(\chi_{CH_4} \times 0.53\text{ atm} \times \frac{6\text{ mol gases}}{2\text{ mol }CH_4}\right) + \left((1 - \chi_{CH_4}) \times 0.53\text{ atm} \times \frac{10\text{ mol gases}}{2\text{ mol }C_2H_6}\right) = 2.2\text{ atm}$$

Substitute as above for $\chi_{C_2H_6}$, then solve for $\chi_{CH_4} = 0.42$.

Check: The units (none) are correct. The magnitude of the answer (0.42) makes sense because if it had been all methane, the final pressure would have been 1.59 atm and if it had been all ethane, the final pressure would have been 2.65 atm. Because we are closer to the latter pressure, we expect the mole fraction of methane to be less than 0.5.

Conceptual Problems

5.139   Because the passengers have more mass than the balloon, they have more momentum than the balloon. The passengers will continue to travel in their original direction longer. The car is slowing, so the relative position of the passengers is to move forward and the balloon to move backward. The opposite happens upon acceleration.

5.141   B is the limiting reactant (2.0 L of B requires 1.0 L A to completely react). The final container will have 0.5 L A and 2.0 L C, so the final volume will be 2.5 L. The change will be $[(2.5\text{ L}/3.5\text{ L}) \times 100\%] - 100\% = -29\%$.

5.143   (a)   False—All gases have the same average kinetic energy at the same temperature.
(b)   False—The gases will have the same partial pressures because we have the same number of moles of each.
(c)   False—The average velocity of the B molecules will be less than that of the A molecules because the Bs are heavier.
(d)   True—Because B molecules are heavier, they will contribute more to the density $(d = m/V)$.

5.145   When the volume of a gas is cut in half, the pressure doubles. When the temperature of a gas in kelvins doubles, the pressure doubles. The next effect is that the pressure increases by a factor of four.

5.147   Because the velocity is inversely proportional to the molar mass, the tails on the helium are $\sqrt{20/4} = \sim 2.2$ times as long as those for neon and $\sqrt{84/4} = \sim 4.6$ times as long as those for krypton.

smallest particles = He

medium sized particles = Ne

largest = Kr

6 Thermochemistry

Review Questions

6.1 Thermochemistry is the study of the relationship between chemistry and energy. It is important because energy and its uses are critical to society. It is important to understand how much energy is required or released in a process.

6.3 Kinetic energy is associated with the motion of an object. Potential energy is associated with the position or composition of an object. Examples of kinetic energy are a moving billiard ball, movement of gas molecules, and a raging river. Examples of potential energy are a billiard ball raised above the surface of a billiard table, a compressed spring, and bonds in molecules.

6.5 The SI unit of energy is $kg \cdot m^2/s^2$, defined as the joule (J), named after the English scientist James Joule. Other units of energy are the kilojoule (kJ), the calorie (cal), the Calorie (Cal), and the kilowatt-hour (kWh).

6.7 According to the first law, a device that would continually produce energy with no energy input, sometimes known as a perpetual motion machine, cannot exist because the best we can do with energy is break even.

6.9 The internal energy (E) of a system is the sum of the kinetic and potential energies of all of the particles that compose the system. Internal energy is a state function.

6.11 If the reactants have a lower internal energy than the products, ΔE_{sys} is positive and energy flows into the system from the surroundings.

6.13 The internal energy (E) of a system is the sum of the kinetic and potential energies of all of the particles that compose the system. The change in the internal energy of the system (ΔE) must be the sum of the heat transferred (q) and the work done (w): $\Delta E = q + w$.

6.15 The heat capacity of a system is usually defined as the quantity of heat required to change its temperature by $1\,°C$. Heat capacity (C) is a measure of the system's ability to hold thermal energy without undergoing a large change in temperature. The difference between heat capacity (C) and specific heat capacity (C_s) is that the specific heat capacity is the amount of heat required to raise the temperature of *1 gram* of the substance by $1\,°C$.

6.17 When two objects of different temperatures come in direct contact, heat flows from the higher temperature object to the lower temperature object. The amount of heat lost by the warmer object is equal to the amount of heat gained by the cooler object. The warmer object's temperature will drop and the cooler object's temperature will rise until they reach the same temperature. The magnitude of these temperature changes depends on the mass and heat capacities of the two objects.

6.19 In calorimetry, the thermal energy exchanged between the reaction (defined as the system) and the surroundings is measured by observing the change in temperature of the surroundings. A bomb calorimeter is used to measure the ΔE_{rxn} for combustion reactions. The calorimeter includes a tight-fitting, sealed container that forces the reaction to occur at constant volume. A coffee-cup calorimeter is used to measure ΔH_{rxn} for many aqueous reactions. The calorimeter consists of two Styrofoam® coffee cups, one inserted into the other, to provide insulation from the laboratory environment. Because the reaction happens under conditions of constant pressure (open to the atmosphere), $q_{rxn} = q_p = \Delta H_{rxn}$.

6.21   An endothermic reaction has a positive ΔH and absorbs heat from the surroundings. An endothermic reaction feels cold to the touch. An exothermic reaction has a negative ΔH and gives off heat to the surroundings. An exothermic reaction feels warm to the touch.

6.23   The internal energy of a chemical system is the sum of its kinetic energy and its potential energy. It is this potential energy that absorbs the energy in an endothermic chemical reaction. In an endothermic reaction, as some bonds break and others form, the protons and electrons go from an arrangement of lower potential energy to one of higher potential energy, absorbing thermal energy in the process. This absorption of thermal energy reduces the kinetic energy of the system. This is detected as a drop in temperature.

6.25   (a) If a reaction is multiplied by a factor, the ΔH is multiplied by the same factor.
       (b) If a reaction is reversed, the sign of ΔH is reversed.
       The relationships hold because H is a state function. Twice as much energy is contained in twice the quantity of reactants or products. If the reaction is reversed, the final and initial states have been switched and the direction of heat flow is reversed.

6.27   The standard state is defined as follows: for a gas, the pure gas at a pressure of exactly 1 atm; for a liquid or solid, the pure substance in its most stable form at a pressure of 1 atm and the temperature of interest (often taken to be 25 °C); and for a substance in solution, a concentration of exactly 1 M. The standard enthalpy change $(\Delta H°)$ is the change in enthalpy for a process when all reactants and products are in their standard states. The superscript degree sign indicates standard states.

6.29   To calculate $\Delta H°_{rxn}$, subtract the heats of formations of the reactants multiplied by their stoichiometric coefficients from the heats of formation of the products multiplied by their stoichiometric coefficients. In the form of an equation:

$$\Delta H°_{rxn} = \sum n_p \Delta H°_f(\text{products}) - \sum n_R \Delta H°_f(\text{reactants})$$

6.31   One of the main problems associated with the burning of fossil fuels is that, even though they are abundant in the Earth's crust, they are a finite and nonrenewable energy source. The other major problems associated with fossil fuel use are related to the products of combustion. Three major environmental problems associated with the emissions of fossil fuel combustion are air pollution, acid rain, and global warming. One of the main products of fossil fuel combustion is carbon dioxide (CO_2), which is a greenhouse gas.

Problems by Topic

Energy Units

6.33   (a)   **Given:** 534 kWh   **Find:** J
             Conceptual Plan: kWh $\rightarrow$ J
             $$\frac{3.60 \times 10^6 \text{ J}}{1 \text{ kWh}}$$
             Solution: $534 \text{ kWh} \times \dfrac{3.60 \times 10^6 \text{ J}}{1 \text{ kWh}} = 1.92 \times 10^9 \text{ J}$

             Check: The units (J) are correct. The magnitude of the answer (10^9) makes physical sense because a kWh is much larger than a Joule; so the answer increases.

       (b)   **Given:** 215 kJ   **Find:** Cal
             Conceptual Plan: kJ $\rightarrow$ J $\rightarrow$ Cal
             $$\frac{1000 \text{ J}}{1 \text{ kJ}} \quad \frac{1 \text{ Cal}}{4184 \text{ J}}$$
             Solution: $215 \text{ kJ} \times \dfrac{1000 \text{ J}}{1 \text{ kJ}} \times \dfrac{1 \text{ Cal}}{4184 \text{ J}} = 51.4 \text{ Cal}$

             Check: The units (Cal) are correct. The magnitude of the answer (51) makes physical sense because a Calorie is about $\frac{1}{4}$ of a kJ; so the answer decreases by a factor of about four.

(c)   **Given:** 567 Cal   **Find:** J

     Conceptual Plan: Cal → J

$$\frac{4184 \text{ J}}{1 \text{ Cal}}$$

     Solution: $567 \cancel{\text{ Cal}} \times \dfrac{4184 \text{ J}}{1 \cancel{\text{ Cal}}} = 2.37 \times 10^6 \text{ J}$

     Check: The units (J) are correct. The magnitude of the answer (10^6) makes physical sense because a Calorie is much larger than a Joule; so the answer increases.

(d)   **Given:** 2.85×10^3 J   **Find:** cal

     Conceptual Plan: J → cal

$$\frac{1 \text{ cal}}{4.184 \text{ J}}$$

     Solution: $2.85 \times 10^3 \cancel{\text{ J}} \times \dfrac{1 \text{ cal}}{4.184 \cancel{\text{ J}}} = 681 \text{ cal}$

     Check: The units (cal) are correct. The magnitude of the answer (680) makes physical sense because a J is about $\frac{1}{4}$ the size of a calorie; so the answer decreases by a factor of about four.

6.35   (a)   **Given:** 2387 Cal   **Find:** J

            Conceptual Plan: Cal → J

$$\frac{4184 \text{ J}}{1 \text{ Cal}}$$

            Solution: $2387 \cancel{\text{ Cal}} \times \dfrac{4184 \text{ J}}{1 \cancel{\text{ Cal}}} = 9.987 \times 10^6 \text{ J}$

            Check: The units (J) are correct. The magnitude of the answer (10^7) makes physical sense because a Calorie is much larger than a Joule; so the answer increases.

     (b)   **Given:** 2387 Cal   **Find:** kJ

            Conceptual Plan: Cal → J → kWh

$$\frac{4184 \text{ J}}{1 \text{ Cal}} \quad \frac{1 \text{ kJ}}{1000 \text{ J}}$$

            Solution: $2387 \cancel{\text{ Cal}} \times \dfrac{4184 \cancel{\text{ J}}}{1 \cancel{\text{ Cal}}} \times \dfrac{1 \text{ kJ}}{1000 \cancel{\text{ J}}} = 9.987 \times 10^3 \text{ kJ}$

            Check: The units (kJ) are correct. The magnitude of the answer (10^4) makes physical sense because a Calorie is larger than a kJ; so the answer increases.

     (c)   **Given:** 2387 Cal   **Find:** kWh

            Conceptual Plan: Cal → J → kWh

$$\frac{4184 \text{ J}}{1 \text{ Cal}} \quad \frac{1 \text{ kWh}}{3.60 \times 10^6 \text{ J}}$$

            Solution: $2387 \cancel{\text{ Cal}} \times \dfrac{4184 \cancel{\text{ J}}}{1 \cancel{\text{ Cal}}} \times \dfrac{1 \text{ kWh}}{3.60 \times 10^6 \cancel{\text{ J}}} = 2.774 \text{ kWh}$

            Check: The units (kWh) are correct. The magnitude of the answer (3) makes physical sense because a Calorie is much smaller than a kWh; so the answer decreases.

Internal Energy, Heat, and Work

6.37   (d) $\Delta E_{\text{sys}} = -\Delta E_{\text{surr}}$ If energy change of the system is negative, energy is being transferred from the system to the surroundings, decreasing the energy of the system and increasing the energy of the surroundings. The amount of energy lost by the system must go somewhere; so the amount gained by the surroundings is equal and opposite to that lost by the system.

6.39   (a)   The energy exchange is primarily heat because the skin (part of the surroundings) is cooled. There is a small expansion (work) because water is being converted from a liquid to a gas. The sign of ΔE_{sys} is positive because the surroundings cool.

     (b)   The energy exchange is primarily work. The sign of ΔE_{sys} is negative because the system is expanding (doing work on the surroundings).

     (c)   The energy exchange is primarily heat. The sign of ΔE_{sys} is positive because the system is being heated by the flame.

6.41   **Given:** 622 kJ heat released; 105 kJ work done on surroundings   **Find:** ΔE_{sys}
        Conceptual Plan: Interpret language to determine the sign of the two terms then $q, w \rightarrow \Delta E_{sys}$

$$\Delta E = q + w$$

        Solution: Because heat is released from the system to the surroundings, $q = -622$ kJ; because the system is doing work on the surroundings, $w = -105$ kJ. $\Delta E = q + w = -622$ kJ $- 105$ kJ $= -727$ kJ $= -7.27 \times 10^2$ kJ.

        Check: The units (kJ) are correct. The magnitude of the answer (-730) makes physical sense because both terms are negative.

6.43   **Given:** 655 J heat absorbed; 344 J work done on surroundings   **Find:** ΔE_{sys}
        Conceptual Plan: Interpret language to determine the sign of the two terms then $q, w \rightarrow \Delta E_{sys}$

$$\Delta E = q + w$$

        Solution: Because heat is absorbed by the system, $q = +655$ J; because the system is doing work on the surroundings, $w = -344$ J. $\Delta E = q + w = 655$ J $- 344$ J $= 311$ J.

        Check: The units (J) are correct. The magnitude of the answer $(+300)$ makes physical sense because the heat term dominates over the work term.

Heat, Heat Capacity, and Work

6.45   Cooler A had more ice after 3 hours because most of the ice in cooler B was melted to cool the soft drinks that started at room temperature. In cooler A, the drinks were already cold; so the ice only needed to maintain this cool temperature.

6.47   **Given:** 1.50 L water, $T_i = 25.0\,°C$, $T_f = 100.0\,°C$, $d = 1.0$ g/mL   **Find:** q
        Conceptual Plan: L $\rightarrow$ mL $\rightarrow$ g and pull C_s from Table 6.4 and $T_i, T_f \rightarrow \Delta T$ then $m, C_s, \Delta T \rightarrow q$

$$\frac{1000\ mL}{1\ L} \qquad \frac{1.0\ g}{1.0\ mL} \qquad 4.18\frac{J}{g \cdot °C} \qquad\qquad\qquad \Delta T = T_f - T_i \qquad\qquad q = mC_s\Delta T$$

        Solution: $1.50\ \cancel{L} \times \dfrac{1000\ \cancel{mL}}{1\ \cancel{L}} \times \dfrac{1.0\ g}{1.0\ \cancel{mL}} = 1500$ g and $\Delta T = T_f - T_i = 100.0\,°C - 25.0\,°C = 75.0\,°C$

        then $q = mC_s\Delta T = 1500\ \cancel{g} \times 4.18\dfrac{J}{\cancel{g} \cdot \cancel{°C}} \times 75.0\,\cancel{°C} = 4.7 \times 10^5$ J

        Check: The units (J) are correct. The magnitude of the answer (10^6) makes physical sense because there is such a large mass, a significant temperature change, and a high specific heat capacity material.

6.49   **(a)**   **Given:** 25 g gold; $T_i = 27.0\,°C$; $q = 2.35$ kJ   **Find:** T_f
              Conceptual Plan: kJ $\rightarrow$ J and pull C_s from Table 6.4 then $m, C_s, q \rightarrow \Delta T$ then $T_i, \Delta T \rightarrow T_f$

$$\frac{1000\ J}{1\ kJ} \qquad\qquad 0.128\frac{J}{g \cdot °C} \qquad\quad q = mC_s\Delta T \qquad\quad \Delta T = T_f - T_i$$

              Solution: $2.35\ \cancel{kJ} \times \dfrac{1000\ J}{1\ \cancel{kJ}} = 2350$ J then $q = mC_s\Delta T$.   Rearrange to solve for ΔT.

              $\Delta T = \dfrac{q}{mC_s} = \dfrac{2350\ J}{25\ \cancel{g} \times 0.128\dfrac{\cancel{J}}{\cancel{g} \cdot °C}} = 734.375\,°C$ finally $\Delta T = T_f - T_i$.   Rearrange to solve for T_f.

              $T_f = \Delta T + T_i = 734.375\,°C + 27.0\,°C = 760\,°C$

              Check: The units (°C) are correct. The magnitude of the answer (760) makes physical sense because such a large amount of heat is absorbed and there are a small mass and specific heat capacity. The temperature change should be very large.

       (b)   **Given:** 25 g silver; $T_i = 27.0\,°C$; $q = 2.35$ kJ   **Find:** T_f
              Conceptual Plan: kJ $\rightarrow$ J and pull C_s from Table 6.4 then $m, C_s, q \rightarrow \Delta T$ then $T_i, \Delta T \rightarrow T_f$

$$\frac{1000\ J}{1\ kJ} \qquad\qquad 0.235\frac{J}{g \cdot °C} \qquad\quad q = mC_s\Delta T \qquad\quad \Delta T = T_f - T_i$$

              Solution: $2.35\ \cancel{kJ} \times \dfrac{1000\ J}{1\ \cancel{kJ}} = 2350$ J then $q = mC_s\Delta T$.   Rearrange to solve for ΔT.

$$\Delta T = \frac{q}{mC_s} = \frac{2350\ \text{J}}{25\ \cancel{g} \times 0.235\dfrac{\text{J}}{\cancel{g} \cdot {}^\circ\text{C}}} = 4\underline{0}0\ {}^\circ\text{C} \quad \text{finally } \Delta T = T_f - T_i. \quad \text{Rearrange to solve for } T_f.$$

$$T_f = \Delta T + T_i = 4\underline{0}0\ {}^\circ\text{C} + 27.0\ {}^\circ\text{C} = 430\ {}^\circ\text{C}$$

Check: The units (°C) are correct. The magnitude of the answer (430) makes physical sense because such a large amount of heat is absorbed and there are a small mass and specific heat capacity. The temperature change should be very large. The temperature change should be less than that of the gold because the specific heat capacity is greater.

(c) **Given:** 25 g aluminum; $T_i = 27.0\ {}^\circ\text{C}$; $q = 2.35$ kJ   **Find:** T_f
 Conceptual Plan: kJ → J and pull C_s from Table 6.4 then $m, C_s, q \to \Delta T$ then $T_i, \Delta T \to T_f$

$$\frac{1000\ \text{J}}{1\ \text{kJ}} \qquad\qquad 0.903\ \frac{\text{J}}{\text{g} \cdot {}^\circ\text{C}} \qquad\qquad q = mC_s\Delta T \qquad\qquad \Delta T = T_f - T_i$$

Solution: $2.35\ \text{kJ} \times \dfrac{1000\ \text{J}}{1\ \text{kJ}} = 2350\ \text{J}$ then $q = mC_s\Delta T$.   Rearrange to solve for ΔT.

$$\Delta T = \frac{q}{mC_s} = \frac{2350\ \text{J}}{25\ \cancel{g} \times 0.903\dfrac{\text{J}}{\cancel{g} \cdot {}^\circ\text{C}}} = 10\underline{4}.10\ {}^\circ\text{C} \quad \text{finally } \Delta T = T_f - T_i. \quad \text{Rearrange to solve for } T_f.$$

$$T_f = \Delta T + T_i = 10\underline{4}.10\ {}^\circ\text{C} + 27.0\ {}^\circ\text{C} = 130\ {}^\circ\text{C}$$

Check: The units (°C) are correct. The magnitude of the answer (130) makes physical sense because such a large amount of heat is absorbed and there is such a small mass. The temperature change should be less than that of the silver because the specific heat capacity is greater.

(d) **Given:** 25 g water; $T_i = 27.0\ {}^\circ\text{C}$; $q = 2.35$ kJ   **Find:** T_f
 Conceptual Plan: kJ → J and pull C_s from Table 6.4 then $m, C_s, q \to \Delta T$ then $T_i, \Delta T \to T_f$

$$\frac{1000\ \text{J}}{1\ \text{kJ}} \qquad\qquad 4.18\ \frac{\text{J}}{\text{g} \cdot {}^\circ\text{C}} \qquad\qquad q = mC_s\Delta T \qquad\qquad \Delta T = T_f - T_i$$

Solution: $2.35\ \text{kJ} \times \dfrac{1000\ \text{J}}{1\ \text{kJ}} = 2350\ \text{J}$ then $q = mC_s\Delta T$.   Rearrange to solve for ΔT.

$$\Delta T = \frac{q}{mC_s} = \frac{2350\ \text{J}}{25\ \cancel{g} \times 4.18\dfrac{\text{J}}{\cancel{g} \cdot {}^\circ\text{C}}} = 2\underline{2}.488\ {}^\circ\text{C} \quad \text{finally } \Delta T = T_f - T_i. \quad \text{Rearrange to solve for } T_f.$$

$$T_f = \Delta T + T_i = 2\underline{2}.488\ {}^\circ\text{C} + 27.0\ {}^\circ\text{C} = 49\ {}^\circ\text{C}$$

Check: The units (°C) are correct. The magnitude of the answer (49) makes physical sense because such a large amount of heat is absorbed and there is such a small mass. The temperature change should be less than that of the aluminum because the specific heat capacity is greater.

6.51   **Given:** $V_i = 0.0$ L; $V_f = 2.5$ L; $P = 1.1$ atm   **Find:** w (J)
 Conceptual Plan: $V_i, V_f \to \Delta V$ then $P, \Delta V \to w$ (L atm) $\to w$ (J)

$$\Delta V = V_f - V_i \qquad\qquad w = -P\Delta V \qquad\qquad \frac{101.3\ \text{J}}{1\ \text{L} \cdot \text{atm}}$$

Solution: $\Delta V = V_f - V_i = 2.5\ \text{L} - 0.0\ \text{L} = 2.5\ \text{L}$ then

$$w = -P\Delta V = -1.1\ \cancel{\text{atm}} \times 2.5\ \cancel{\text{L}} \times \frac{101.3\ \text{J}}{1\ \cancel{\text{L}} \cdot \cancel{\text{atm}}} = -280\ \text{J} \ \text{or} \ -2.8 \times 10^2\ \text{J}$$

Check: The units (J) are correct. The magnitude of the answer (-280) makes physical sense because this is an expansion (negative work) and we have atmospheric pressure and a small volume of expansion.

6.53   **Given:** $q = 565$ J absorbed; $V_i = 0.10$ L; $V_f = 0.85$ L; $P = 1.0$ atm   **Find:** ΔE_{sys}
 Conceptual Plan: $V_i, V_f \to \Delta V$ and interpret language to determine the sign of the heat

$$\Delta V = V_f - V_i \qquad\qquad\qquad q = +565\ \text{J}$$

then $P, \Delta V \to w$ (L atm) $\to w$ (J) finally $q, w \to \Delta E_{sys}$

$$w = -P\Delta V \qquad \frac{101.3\ \text{J}}{1\ \text{L} \cdot \text{atm}} \qquad\qquad \Delta E = q + w$$

Solution: $\Delta V = V_f - V_i = 0.85\,L - 0.10\,L = 0.75\,L$ then

$$w = -P\Delta V = -1.0\,\text{atm} \times 0.75\,\cancel{L} \times \frac{101.3\,J}{1\,\cancel{L} \cdot \text{atm}} = -75.975\,J \qquad\qquad \Delta E = q + w = +565\,J - 75.975\,J = 489\,J$$

Check: The units (J) are correct. The magnitude of the answer (500) makes physical sense because the heat absorbed dominated the small expansion work (negative work).

Enthalpy and Thermochemical Stoichiometry

6.55   **Given:** 1 mol fuel, 3452 kJ heat produced; 11 kJ work done on surroundings   **Find:** ΔE_{sys}, ΔH
              Conceptual Plan: Interpret language to determine the sign of the two terms then $q \rightarrow \Delta H$ and $q, w \rightarrow \Delta E_{sys}$
                                                            $\Delta H = q_p$ $\Delta E = q + w$

             Solution: Because heat is produced by the system to the surroundings, $q = -3452\,kJ$; because the system is doing work on the surroundings, $w = -11\,kJ$. $\Delta H = q_p = -3452\,kJ$ and
             $\Delta E = q + w = -3452\,kJ - 11\,kJ = -3463\,kJ$

             Check: The units (kJ) are correct. The magnitude of the answer (-3500) makes physical sense because both terms are negative. We expect significant amounts of energy from fuels.

6.57   (a)   Combustion is an exothermic process; ΔH is negative.
           (b)   Evaporation requires an input of energy, so it is endothermic; ΔH is positive.
           (c)   Condensation is the reverse of evaporation, so it is exothermic; ΔH is negative.

6.59   **Given:** 177 mL acetone (C_3H_6O), $\Delta H°_{rxn} = -1790\,kJ$; $d = 0.788\,g/mL$ **Find:** q
             Conceptual Plan: mL acetone $\rightarrow$ g acetone $\rightarrow$ mol acetone $\rightarrow$ q

                                $\dfrac{0.788\,g}{1\,mL}$ $\dfrac{1\,mol}{58.08\,g}$ $\dfrac{-1790\,kJ}{1\,mol}$

             Solution: $177\,\cancel{mL} \times \dfrac{0.788\,\cancel{g}}{1\,\cancel{mL}} \times \dfrac{1\,\cancel{mol}}{58.08\,\cancel{g}} \times \dfrac{-1790\,kJ}{1\,\cancel{mol}} = -4.30 \times 10^3\,kJ$ or $4.30 \times 10^3\,kJ$ released

             Check: The units (kJ) are correct. The magnitude of the answer (-10^3) makes physical sense because the enthalpy change is negative and we have more than a mole of acetone. We expect more than 1790 kJ to be released.

6.61   **Given:** 5.56 kg nitromethane (CH_3NO_2); $\Delta H°_{rxn} = -1418\,kJ/2\,mol$ nitromethane   **Find:** q
             Conceptual Plan: kg nitromethane $\rightarrow$ g nitromethane $\rightarrow$ mol nitromethane $\rightarrow$ q

                             $\dfrac{1000\,g}{1\,kg}$ $\dfrac{1\,mol}{61.04\,g}$ $\dfrac{-1418\,kJ}{2\,mol}$

             Solution: $5.56\,\cancel{kg} \times \dfrac{1000\,\cancel{g}}{1\,\cancel{kg}} \times \dfrac{1\,\cancel{mol}}{61.04\,\cancel{g}} \times \dfrac{-1418\,kJ}{2\,\cancel{mol}} = -6.46 \times 10^4\,kJ$ or $6.46 \times 10^4\,kJ$ released

             Check: The units (kJ) are correct. The magnitude of the answer (-10^4) makes physical sense because the enthalpy change is negative and we have more than a mole of acetone. We expect more than 1418 kJ to be released.

6.63   **Given:** pork roast, $\Delta H°_{rxn} = -2217\,kJ$; q needed $= 1.6 \times 10^3\,kJ$, 10% efficiency   **Find:** $m(CO_2)$
             Conceptual Plan: q used $\rightarrow$ q generated $\rightarrow$ mol CO_2 $\rightarrow$ g CO_2

                         $\dfrac{100\,kJ\ generated}{10\,kJ\ used}$ $\dfrac{3\,mol}{2217\,kJ}$ $\dfrac{44.01\,g}{1\,mol}$

             Solution: $1.6 \times 10^3\,\cancel{kJ} \times \dfrac{100\,\cancel{kJ\ generated}}{10\,\cancel{kJ\ used}} \times \dfrac{3\,mol\,\cancel{CO_2}}{2217\,\cancel{kJ}} \times \dfrac{44.01\,g\,CO_2}{1\,mol\,\cancel{CO_2}} = 950\,g\,CO_2$

             Check: The units (g) are correct. The magnitude of the answer (~1000) makes physical sense because the process is not very efficient and a great deal of energy is needed.

Thermal Energy Transfer

6.65   **Given:** silver block; $T_{Ag, i} = 58.5\ °C$; 100.0 g water; $T_{H_2O, i} = 24.8\ °C$; $T_f = 26.2\ °C$ **Find:** mass of silver block
Conceptual Plan: Pull C_s values from table then H_2O: $m, C_s, T_i, T_f \rightarrow q$ and $Ag : C_s, T_i, T_f \rightarrow m$

$$\text{Ag: } 0.235\ \frac{J}{g \cdot °C} \quad H_2O: 4.18 \frac{J}{g \cdot °C} \quad q = mC_s(T_f - T_i) \text{ then set } q_{Ag} = -q_{H_2O}$$

Solution: $q = mC_s(T_f - T_i)$ substitute in values and set $q_{Ag} = -q_{H_2O}$.

$$q_{Ag} = m_{Ag}C_{Ag}(T_f - T_{Ag, i}) = m_{Ag} \times 0.235\ \frac{J}{g \cdot °C} \times (26.2\ °C - 58.5\ °C) =$$

$$-q_{H_2O} = -m_{H_2O}C_{H_2O}(T_f - T_{H_2O, i}) = -100.0\ g \times 4.18\ \frac{J}{g \cdot °C} \times (26.2\ °C - 24.8\ °C)$$

Rearrange to solve for m_{Ag}.

$$m_{Ag} \times \left(-7.5\underline{9}05\ \frac{J}{g}\right) = -58\underline{5}.2\ J \rightarrow m_{Ag} = \frac{-585.2\ J}{-7.5905\ \dfrac{J}{g}} = 77.\underline{0}964\ g\ Ag = 77.1\ g\ Ag$$

Check: The units (g) are correct. The magnitude of the answer (77 g) makes physical sense because the heat capacity of water is much greater than the heat capacity of silver.

6.67   **Given:** 31.1 g gold; $T_{Au, i} = 69.3\ °C$; 64.2 g water; $T_{H_2O, i} = 27.8\ °C$ **Find:** T_f
Conceptual Plan: Pull C_s values from table then $m, C_s, T_i \rightarrow T_f$

$$\text{Au: } 0.128\ \frac{J}{g \cdot °C} \quad H_2O: 4.18 \frac{J}{g \cdot °C} \quad q = mC_s(T_f - T_i) \text{ then set } q_{Au} = -q_{H_2O}$$

Solution: $q = mC_s(T_f - T_i)$ substitute in values and set $q_{Au} = -q_{H_2O}$.

$$q_{Au} = m_{Au}C_{Au}(T_f - T_{Au, i}) = 31.1\ g \times 0.128\ \frac{J}{g \cdot °C} \times (T_f - 69.3\ °C) =$$

$$-q_{H_2O} = -m_{H_2O}C_{H_2O}(T_f - T_{H_2O, i}) = -64.2\ g \times 4.18\ \frac{J}{g \cdot °C} \times (T_f - 27.8\ °C)$$

Rearrange to solve for T_f.

$$3.9\underline{8}08\ \frac{J}{°C} \times (T_f - 69.3\ °C) = -26\underline{8}.356\ \frac{J}{°C} \times (T_f - 27.8\ °C) \rightarrow$$

$$3.9\underline{8}08\ \frac{J}{°C}\ T_f - 27\underline{5}.8694\ J = -26\underline{8}.356\ \frac{J}{°C}\ T_f + 74\underline{6}0.2968\ J \rightarrow$$

$$26\underline{8}.356\ \frac{J}{°C}\ T_f + 3.9\underline{8}08\ \frac{J}{°C}\ T_f = 27\underline{5}.8694\ J + 74\underline{6}0.2968\ J \rightarrow 27\underline{2}.3368\ \frac{J}{°C}\ T_f = 77\underline{3}6.1662\ J \rightarrow$$

$$T_f = \frac{77\underline{3}6.1662\ J}{27\underline{2}.3368\ \dfrac{J}{°C}} = 28.4\ °C$$

Check: The units (°C) are correct. The magnitude of the answer (28) makes physical sense because the heat transfer is dominated by the water (larger mass and larger specific heat capacity). The final temperature should be closer to the initial temperature of water than of gold.

6.69   **Given:** 6.15 g substance A; $T_{A, i} = 20.5\ °C$; 25.2 g substance B; $T_{B, i} = 52.7\ °C$; $C_s = 1.17\ J/g \cdot °C$; $T_f = 46.7\ °C$
Find: specific heat capacity of substance A
Conceptual Plan: A: $m, T_i, T_f \rightarrow q$ B: $m, C_s, T_i, T_f \rightarrow q$ and solve for C

$$q = mC_s(T_f - T_i) \qquad \text{then set } q_A = -q_B$$

Solution: $q = mC_s(T_f - T_i)$ substitute in values and set $q_A = -q_B$.

$$q_A = m_A C_A(T_f - T_{A, i}) = 6.15\ g \times C_A \times (46.7\ °C - 20.5\ °C) =$$

$$-q_B = -m_B C_B(T_f - T_{B, i}) = -25.2\ g \times 1.17\ \frac{J}{g \cdot °C} \times (46.7\ °C - 52.7\ °C)$$

Rearrange to solve for C_A.

$$C_A \times (161.13 \text{ g} \cdot {}^\circ\text{C}) = 176.904 \text{ J} \rightarrow C_A = \frac{176.904 \text{ J}}{161.13 \text{ g} \cdot {}^\circ\text{C}} = 1.097896 \frac{\text{J}}{\text{g} \cdot {}^\circ\text{C}} = 1.10 \frac{\text{J}}{\text{g} \cdot {}^\circ\text{C}}$$

Check: The units $(\text{J}/\text{g} \cdot {}^\circ\text{C})$ are correct. The magnitude of the answer $(1 \text{ J}/\text{g} \cdot {}^\circ\text{C})$ makes physical sense because the mass of substance B is greater than the mass of substance A by a factor of ~ 4.1 and the temperature change for substance A is greater than the temperature change of substance B by a factor of ~ 4.4; so the heat capacity of substance A will be slightly smaller.

Calorimetry

6.71 $\Delta H_{rxn} = q_p$ and $\Delta E_{rxn} = q_V = \Delta H - P\Delta V$. Because combustions always involve expansions, expansions do work and therefore have a negative value. Combustions are always exothermic and therefore have a negative value. This means that ΔE_{rxn} is more negative than ΔH°_{rxn}; so A (-25.9 kJ) is the constant volume process, and B (-23.3 kJ) is the constant pressure process.

6.73   **Given:** 0.514 g biphenyl$(C_{12}H_{10})$; bomb calorimeter; $T_i = 25.8 \,{}^\circ\text{C}$; $T_f = 29.4 \,{}^\circ\text{C}$; $C_{cal} = 5.86$ kJ/°C   **Find:** ΔE_{rxn}
      Conceptual Plan: $T_i, T_f \rightarrow \Delta T$ then $\Delta T, C_{cal} \rightarrow q_{cal} \rightarrow q_{rxn}$ then g $C_{12}H_{10} \rightarrow$ mol $C_{12}H_{10}$

$$\Delta T = T_f - T_i \qquad q_{cal} = C_{cal}\Delta T \quad q_{cal} = -q_{rxn} \qquad \frac{1 \text{ mol}}{154.20 \text{ g}}$$

      then q_{rxn}, mol $C_{12}H_{10} \rightarrow \Delta E_{rxn}$

$$\Delta E_{rxn} = \frac{q_V}{\text{mol } C_{12}H_{10}}$$

      Solution: $\Delta T = T_f - T_i = 29.4 \,{}^\circ\text{C} - 25.8 \,{}^\circ\text{C} = 3.6 \,{}^\circ\text{C}$ then $q_{cal} = C_{cal}\Delta T = 5.86 \frac{\text{kJ}}{{}^\circ\text{C}} \times 3.6 \,{}^\circ\text{C} = 21.096 \text{ kJ}$

      then $q_{cal} = -q_{rxn} = -21.096 \text{ kJ}$ and $0.514 \text{ g } C_{12}H_{10} \times \frac{1 \text{ mol } C_{12}H_{10}}{154.20 \text{ g } C_{12}H_{10}} = 0.00333333 \text{ mol } C_{12}H_{10}$ then

$$\Delta E_{rxn} = \frac{q_V}{\text{mol } C_{12}H_{10}} = \frac{-21.096 \text{ kJ}}{0.00333333 \text{ mol } C_{12}H_{10}} = -6.3 \times 10^3 \text{ kJ/mol}$$

      Check: The units (kJ/mol) are correct. The magnitude of the answer (-6000) makes physical sense because such a large amount of heat is generated from a very small amount of biphenyl.

6.75   **Given:** 0.103 g zinc; coffee-cup calorimeter; $T_i = 22.5 \,{}^\circ\text{C}$; $T_f = 23.7 \,{}^\circ\text{C}$; 50.0 mL solution; $d(\text{solution}) = 1.0$ g/mL; $C_{soln} = 4.18$ kJ/g $\cdot$ °C   **Find:** ΔH_{rxn}
      Conceptual Plan: $T_i, T_f \rightarrow \Delta T$ and mL soln $\rightarrow$ g soln then $\Delta T, C_{soln} \rightarrow q_{cal} \rightarrow q_{rxn}$ then

$$\Delta T = T_f - T_i \qquad \frac{1.0 \text{ g}}{1.0 \text{ mL}} \qquad q_{soln} = mC_{soln}\Delta T \quad q_{soln} = -q_{rxn}$$

      g Zn $\rightarrow$ mol Zn then q_{rxn}, mol Zn $\rightarrow \Delta H_{rxn}$

$$\frac{1 \text{ mol}}{65.37 \text{ g}} \qquad \Delta H_{rxn} = \frac{q_p}{\text{mol Zn}}$$

      Solution: $\Delta T = T_f - T_i = 23.7 \,{}^\circ\text{C} - 22.5 \,{}^\circ\text{C} = 1.2 \,{}^\circ\text{C}$ and $50.0 \text{ mL} \times \frac{1.0 \text{ g}}{1.0 \text{ mL}} = 50.0 \text{ g}$

      then $q_{soln} = mC_{soln}\Delta T = 50.0 \text{ g} \times 4.18 \frac{\text{J}}{\text{g} \cdot {}^\circ\text{C}} \times 1.2 \,{}^\circ\text{C} = 250.8 \text{ J}$ then

      $q_{soln} = -q_{rxn} = -250.8 \text{ J}$ and $0.103 \text{ g Zn} \times \frac{1 \text{ mol Zn}}{65.37 \text{ g Zn}} = 0.00157565 \text{ mol Zn}$ then

$$\Delta H_{rxn} = \frac{q_p}{\text{mol Zn}} = \frac{-250.8 \text{ J}}{0.00157565 \text{ mol Zn}} = -1.6 \times 10^5 \text{ J/mol} = -1.6 \times 10^2 \text{ kJ/mol}$$

      Check: The units (kJ/mol) are correct. The magnitude of the answer (-160) makes physical sense because such a large amount of heat is generated from a very small amount of zinc.

Quantitative Relationships Involving ΔH and Hess's Law

6.77   (a)   Because $A + B \rightarrow 2C$ has ΔH_1, $2C \rightarrow A + B$ will have a $\Delta H_2 = -\Delta H_1$. When the reaction direction is reversed, it changes from exothermic to endothermic (or vice versa); so the sign of ΔH changes.

        (b)   Because $A + \frac{1}{2}B \rightarrow C$ has ΔH_1, $2A + B \rightarrow 2C$ will have a $\Delta H_2 = 2\,\Delta H_1$. When the reaction amount doubles, the amount of heat (or ΔH) doubles.

        (c)   Because $A \rightarrow B + 2C$ has ΔH_1, $\frac{1}{2}A \rightarrow \frac{1}{2}B + C$ will have a $\Delta H_1' = \frac{1}{2}\Delta H_1$. When the reaction amount is cut in half, the amount of heat (or ΔH) is cut in half. Then $\frac{1}{2}B + C \rightarrow \frac{1}{2}A$ will have a $\Delta H_2 = -H_1' = -\frac{1}{2}\Delta H_1$. When the reaction direction is reversed, it changes from exothermic to endothermic (or vice versa); so the sign of ΔH changes.

6.79   Because the first reaction has Fe_2O_3 as a product and the reaction of interest has it as a reactant, we need to reverse the first reaction. When the reaction direction is reversed, ΔH changes.

     $Fe_2O_3(s) \rightarrow 2\,Fe(s) + 3/2\,O_2(g)$ $\Delta H = +824.2$ kJ

     Because the second reaction has 1 mole CO as a reactant and the reaction of interest has 3 moles of CO as a reactant, we need to multiply the second reaction and the ΔH by 3.

     $3[\,CO(g) + 1/2\,O_2(g) \rightarrow CO_2(g)\,]$ $\Delta H = 3(-282.7\text{ kJ}) = -848.1$ kJ

     Hess's law states that the ΔH of the net reaction is the sum of the ΔH of the steps.

     The rewritten reactions are as follows:

     $Fe_2O_3(s) \rightarrow 2\,Fe(s) + \cancel{3/2\,O_2(g)}$ $\Delta H = +824.2$ kJ

     $3\,CO(g) + \cancel{3/2\,O_2(g)} \rightarrow 3\,CO_2(g)$ $\Delta H = -848.1$ kJ

     ───

     $Fe_2O_3(s) + 3\,CO(g) \rightarrow 2\,Fe(s) + 3\,CO_2(g)$ $\Delta H_{rxn} = -23.9$ kJ

6.81   Because the first reaction has C_5H_{12} as a reactant and the reaction of interest has it as a product, we need to reverse the first reaction. When the reaction direction is reversed, ΔH changes.

     $5\,CO_2(g) + 6\,H_2O(g) \rightarrow C_5H_{12}(l) + 8\,O_2(g)$ $\Delta H = +3244.8$ kJ

     Because the second reaction has 1 mole C as a reactant and the reaction of interest has 5 moles of C as a reactant, we need to multiply the second reaction and the ΔH by 5.

     $5[\,C(s) + O_2(g) \rightarrow CO_2(g)\,]$ $\Delta H = 5(-393.5\text{ kJ}) = -1967.5$ kJ

     Because the third reaction has 2 moles H_2 as a reactant and the reaction of interest has 6 moles of H_2 as a reactant, we need to multiply the third reaction and the ΔH by 3.

     $3[\,2\,H_2(g) + O_2(g) \rightarrow 2\,H_2O(g)\,]$ $\Delta H = 3(-483.5\text{ kJ}) = -1450.5$ kJ

     Hess's law states that the ΔH of the net reaction is the sum of the ΔH of the steps. The rewritten reactions are as follows:

     $\cancel{5\,CO_2(g)} + \cancel{6\,H_2O(g)} \rightarrow C_5H_{12}(l) + \cancel{8\,O_2(g)}$ $\Delta H = +3244.8$ kJ

     $5\,C(s) + \cancel{5\,O_2(g)} \rightarrow \cancel{5\,CO_2(g)}$ $\Delta H = -1967.5$ kJ

     $6\,H_2(g) + \cancel{3\,O_2(g)} \rightarrow \cancel{6\,H_2O(g)}$ $\Delta H = -1450.5$ kJ

     ───

     $5\,C(s) + 6\,H_2(g) \rightarrow C_5H_{12}(l)$ $\Delta H_{rxn} = -173.2$ kJ

Enthalpies of Formation and ΔH

6.83   (a) $\dfrac{1}{2}N_2(g) + \dfrac{3}{2}H_2(g) \rightarrow NH_3(g)$ $\Delta H_f^\circ = -45.9$ kJ/mol

        (b) $C(s) + O_2(g) \rightarrow CO_2(g)$ $\Delta H_f^\circ = -393.5$ kJ/mol

        (c) $2\,Fe(s) + \dfrac{3}{2}O_2(g) \rightarrow Fe_2O_3(s)$ $\Delta H_f^\circ = -824.2$ kJ/mol

        (d) $C(s) + 2\,H_2(g) \rightarrow CH_4(g)$ $\Delta H_f^\circ = -74.6$ kJ/mol

6.85   **Given:** $N_2H_4(l) + N_2O_4(g) \rightarrow 2\,N_2O(g) + 2\,H_2O(g)$ **Find:** ΔH_{rxn}°

     Conceptual Plan: $\Delta H_{rxn}^\circ = \sum n_P \Delta H_f^\circ(\text{products}) - \sum n_R \Delta H_f^\circ(\text{reactants})$

     Solution:

Reactant/Product	ΔH_f° (kJ/mol from Appendix IIB)
$N_2H_4(l)$	50.6
$N_2O_4(g)$	9.16
$N_2O(g)$	81.6
$H_2O(g)$	−241.8

Be sure to pull data for the correct formula and phase.

$$\Delta H^{\circ}_{rxn} = \sum n_P \Delta H^{\circ}_f(products) - \sum n_R \Delta H^{\circ}_f(reactants)$$

$$= [2(\Delta H^{\circ}_f(N_2O(g))) + 2(\Delta H^{\circ}_f(H_2O(g)))] - [1(\Delta H^{\circ}_f(N_2H_4(l))) + 1(\Delta H^{\circ}_f(N_2O_4(g)))]$$

$$= [2(81.6\,kJ) + 2(-241.8\,kJ)] - [1(50.6\,kJ) + 1(9.16\,kJ)]$$

$$= [-320.4\,kJ] - [59.8\,kJ]$$

$$= -380.2\,kJ$$

Check: The units (kJ) are correct. The answer is negative, which means that the reaction is exothermic. The answer is dominated by the negative heat of formation of water.

6.87   (a)   **Given:** $C_2H_4(g) + H_2(g) \rightarrow C_2H_6(g)$ **Find:** ΔH°_{rxn}

              Conceptual Plan: $\Delta H^{\circ}_{rxn} = \sum n_P \Delta H^{\circ}_f(products) - \sum n_R \Delta H^{\circ}_f(reactants)$

              Solution:

Reactant/Product	ΔH°_f (kJ/mol from Appendix IIB)
$C_2H_4(g)$	52.4
$H_2(g)$	0.0
$C_2H_6(g)$	−84.68

Be sure to pull data for the correct formula and phase.

$$\Delta H^{\circ}_{rxn} = \sum n_P \Delta H^{\circ}_f(products) - \sum n_R \Delta H^{\circ}_f(reactants)$$

$$= [1(\Delta H^{\circ}_f(C_2H_6(g)))] - [1(\Delta H^{\circ}_f(C_2H_4(g))) + 1(\Delta H^{\circ}_f(H_2(g)))]$$

$$= [1(-84.68\,kJ)] - [1(52.4\,kJ) + 1(0.0\,kJ)]$$

$$= [-84.68\,kJ] - [52.4\,kJ]$$

$$= -137.1\,kJ$$

Check: The units (kJ) are correct. The answer is negative, which means that the reaction is exothermic. Both hydrocarbon terms are negative, so the final answer is negative.

     (b)   **Given:** $CO(g) + H_2O(g) \rightarrow H_2(g) + CO_2(g)$ **Find:** ΔH°_{rxn}

              Conceptual Plan: $\Delta H^{\circ}_{rxn} = \sum n_P \Delta H^{\circ}_f(products) - \sum n_R \Delta H^{\circ}_f(reactants)$

              Solution:

Reactant/Product	ΔH°_f (kJ/mol from Appendix IIB)
$CO(g)$	−110.5
$H_2O(g)$	−241.8
$H_2(g)$	0.0
$CO_2(g)$	−393.5

Be sure to pull data for the correct formula and phase.

$$\Delta H^{\circ}_{rxn} = \sum n_P \Delta H^{\circ}_f(products) - \sum n_R \Delta H^{\circ}_f(reactants)$$

$$= [1(\Delta H^{\circ}_f(H_2(g))) + 1(\Delta H^{\circ}_f(CO_2(g)))] - [1(\Delta H^{\circ}_f(CO(g))) + 1(\Delta H^{\circ}_f(H_2O(g)))]$$

$$= [1(0.0\,kJ) + 1(-393.5\,kJ)] - [1(-110.5\,kJ) + 1(-241.8\,kJ)]$$

$$= [-393.5\,kJ] - [-352.3\,kJ]$$

$$= -41.2\,kJ$$

Check: The units (kJ) are correct. The answer is negative, which means that the reaction is exothermic.

     (c)   **Given:** $3\,NO_2(g) + H_2O(l) \rightarrow 2\,HNO_3(aq) + NO(g)$ **Find:** ΔH°_{rxn}

              Conceptual Plan: $\Delta H^{\circ}_{rxn} = \sum n_P \Delta H^{\circ}_f(products) - \sum n_R \Delta H^{\circ}_f(reactants)$

              Solution:

Reactant/Product	ΔH°_f (kJ/mol from Appendix IIB)
$NO_2(g)$	33.2
$H_2O(l)$	−285.8
$HNO_3(aq)$	−207
$NO(g)$	91.3

Be sure to pull data for the correct formula and phase.

$$\Delta H^\circ_{rxn} = \sum n_P \Delta H^\circ_f(products) - \sum n_R \Delta H^\circ_f(reactants)$$
$$= [2(\Delta H^\circ_f(HNO_3(aq))) + 1(\Delta H^\circ_f(NO(g)))] - [3(\Delta H^\circ_f(NO_2(g))) + 1(\Delta H^\circ_f(H_2O(l)))]$$
$$= [2(-207\,kJ) + 1(91.3\,kJ)] - [3(33.2\,kJ) + 1(-285.8\,kJ)]$$
$$= [-32\underline{2}.7\,kJ] - [-186.2\,kJ]$$
$$= -137\,kJ$$

Check: The units (kJ) are correct. The answer is negative, which means that the reaction is exothermic.

(d) **Given:** $Cr_2O_3(s) + 3\,CO(g) \rightarrow 2\,Cr(s) + 3\,CO_2(g)$ **Find:** ΔH°_{rxn}
Conceptual Plan: $\Delta H^\circ_{rxn} = \sum n_P \Delta H^\circ_f(products) - \sum n_R \Delta H^\circ_f(reactants)$
Solution:

Reactant/Product	ΔH°_f (kJ/mol from Appendix IIB)
$Cr_2O_3(s)$	-1139.7
$CO(g)$	-110.5
$Cr(s)$	0.0
$CO_2(g)$	-393.5

Be sure to pull data for the correct formula and phase.

$$\Delta H^\circ_{rxn} = \sum n_P \Delta H^\circ_f(products) - \sum n_R \Delta H^\circ_f(reactants)$$
$$= [2(\Delta H^\circ_f(Cr(s))) + 3(\Delta H^\circ_f(CO_2(g)))] - [1(\Delta H^\circ_f(Cr_2O_3(s))) + 3(\Delta H^\circ_f(CO(g)))]$$
$$= [2(0.0\,kJ) + 3(-393.5\,kJ)] - [1(-1139.7\,kJ) + 3(-110.5\,kJ)]$$
$$= [-1180.5\,kJ] - [-1471.2\,kJ]$$
$$= 290.7\,kJ$$

Check: The units (kJ) are correct. The answer is positive, which means that the reaction is endothermic.

6.89 **Given:** form glucose $(C_6H_{12}O_6)$ and oxygen from sunlight, carbon dioxide, and water **Find:** ΔH°_{rxn}
Conceptual Plan: Write balanced reaction then $\Delta H^\circ_{rxn} = \sum n_P \Delta H^\circ_f(products) - \sum n_R \Delta H^\circ_f(reactants)$
Solution: $6\,CO_2(g) + 6\,H_2O(l) \rightarrow C_6H_{12}O_6(s) + 6\,O_2(g)$

Reactant/Product	ΔH°_f (kJ/mol from Appendix IIB)
$CO_2(g)$	-393.5
$H_2O(l)$	-285.8
$C_6H_{12}O(s, glucose)$	-1273.3
$O_2(g)$	0.0

Be sure to pull data for the correct formula and phase.

$$\Delta H^\circ_{rxn} = \sum n_P \Delta H^\circ_f(products) - \sum n_R \Delta H^\circ_f(reactants)$$
$$= [1(\Delta H^\circ_f(C_6H_{12}O(s))) + 6(\Delta H^\circ_f(O_2(g)))] - [6(\Delta H^\circ_f(CO_2(g))) + 6(\Delta H^\circ_f(H_2O(l)))]$$
$$= [1(-1273.3\,kJ) + 6(0.0\,kJ)] - [6(-393.5\,kJ) + 6(-285.8\,kJ)]$$
$$= [-1273.3\,kJ] - [-4075.8\,kJ]$$
$$= +2802.5\,kJ$$

Check: The units (kJ) are correct. The answer is positive, which means that the reaction is endothermic. The reaction requires the input of light energy, so we expect that this will be an endothermic reaction.

6.91 **Given:** $2\,CH_3NO_2(l) + 3/2\,O_2(g) \rightarrow 2\,CO_2(g) + 3\,H_2O(l) + N_2(g)$ and $\Delta H^\circ_{rxn} = -1418.4\,kJ$
Find: $\Delta H^\circ_f(CH_3NO_2(l))$
Conceptual Plan: Fill known values into $\Delta H^\circ_{rxn} = \sum n_P \Delta H^\circ_f(products) - \sum n_R \Delta H^\circ_f(reactants)$ **and rearrange to solve for** $\Delta H^\circ_f(CH_3NO_2(l))$.
Solution:

Reactant/Product	ΔH°_f (kJ/mol from Appendix IIB)
$O_2(g)$	0.0
$CO_2(g)$	-393.5
$H_2O(l)$	-285.8
$N_2(g)$	0.0

Be sure to pull data for the correct formula and phase.

$$\Delta H^\circ_{rxn} = \sum n_P \Delta H^\circ_f(products) - \sum n_R \Delta H^\circ_f(reactants)$$
$$= [2(\Delta H^\circ_f(CO_2(g))) + 3(\Delta H^\circ_f(H_2O(g))) + 1(\Delta H^\circ_f(N_2(g)))] - [2(\Delta H^\circ_f(CH_3NO_2(l))) + 3/2(\Delta H^\circ_f(O_2(g)))]$$
$$(-1418.4 \text{ kJ}) = [2(-393.5 \text{ kJ}) + 3(-285.8 \text{ kJ}) + 1(0.0 \text{ kJ})] - [2(\Delta H^\circ_f(CH_3NO_2(l)) + 3/2(0.0 \text{ kJ})]$$
$$- 1418.4 \text{ kJ} = [-1644.4 \text{ kJ}] - [2(\Delta H^\circ_f(CH_3NO_2(l)))]$$
$$\Delta H^\circ_f(CH_3NO_2(l)) = -113.0 \text{ kJ/mol}$$

Check: The units (kJ/mol) are correct. The answer is negative (but not as negative as water and carbon dioxide), which is consistent with an exothermic combustion reaction.

Energy Use and the Environment

6.93   (a)   **Given:** methane, $\Delta H^\circ_{rxn} = -802.3$ kJ; $q = 1.00 \times 10^2$ kJ   **Find:** $m(CO_2)$
Conceptual Plan: $q \rightarrow$ **mol CO$_2$** $\rightarrow$ **g CO$_2$**

$$\frac{1 \text{ mol}}{-802.3 \text{ kJ}} \quad \frac{44.01 \text{ g}}{1 \text{ mol}}$$

Solution: $-1.00 \times 10^2 \text{ kJ} \times \dfrac{1 \text{ mol CO}_2}{-802.3 \text{ kJ}} \times \dfrac{44.01 \text{ g CO}_2}{1 \text{ mol CO}_2} = 5.49 \text{ g CO}_2$

Check: The units (g) are correct. The magnitude of the answer (~ 5) makes physical sense because less than a mole of fuel is used.

(b)   **Given:** propane, $\Delta H^\circ_{rxn} = -2217$ kJ; $q = 1.00 \times 10^2$ kJ   **Find:** $m(CO_2)$
Conceptual Plan: $q \rightarrow$ **mol CO$_2$** $\rightarrow$ **g CO$_2$**

$$\frac{3 \text{ mol}}{-2217 \text{ kJ}} \quad \frac{44.01 \text{ g}}{1 \text{ mol}}$$

Solution: $-1.00 \times 10^2 \text{ kJ} \times \dfrac{3 \text{ mol CO}_2}{-2217 \text{ kJ}} \times \dfrac{44.01 \text{ g CO}_2}{1 \text{ mol CO}_2} = 5.96 \text{ g CO}_2$

Check: The units (g) are correct. The magnitude of the answer (~ 6) makes physical sense because less than a mole of fuel is used.

(c)   **Given:** octane, $\Delta H^\circ_{rxn} = -5074.1$ kJ; $q = 1.00 \times 10^2$ kJ   **Find:** $m(CO_2)$
Conceptual Plan: $q \rightarrow$ **mol CO$_2$** $\rightarrow$ **g CO$_2$**

$$\frac{8 \text{ mol}}{-5074.1 \text{ kJ}} \quad \frac{44.01 \text{ g}}{1 \text{ mol}}$$

Solution: $-1.00 \times 10^2 \text{ kJ} \times \dfrac{8 \text{ mol CO}_2}{-5074.1 \text{ kJ}} \times \dfrac{44.01 \text{ g CO}_2}{1 \text{ mol CO}_2} = 6.94 \text{ g CO}_2$

Check: The units (g) are correct. The magnitude of the answer (~ 7) makes physical sense because less than a mole of fuel is used.
The methane generated the least carbon dioxide for the given amount of heat, while the octane generated the most carbon dioxide for the given amount of heat.

6.95   **Given:** 7×10^{12} kg/yr octane (C_8H_{18}), $\Delta H^\circ_{rxn} = -5074.1$ kJ (using Problem 6.93), 3×10^{15} kg CO$_2$ in the atmosphere   **Find:** $m(CO_2)$ in kg and time to double atmospheric CO$_2$
Conceptual Plan: Use reaction from Problem 6.93 $C_8H_{18}(l) + 25/2 \, O_2(g) \rightarrow 8 \, CO_2(g) + 9 \, H_2O(g)$
kg (C_8H_{18}) $\rightarrow$ **g (C_8H_{18})** $\rightarrow$ **mol (C_8H_{18})** $\rightarrow$ **mol CO$_2$** $\rightarrow$ **g CO$_2$** $\rightarrow$ **kg CO$_2$ produced**

$$\frac{1000 \text{ g}}{1 \text{ kg}} \quad \frac{1 \text{ mol C}_8H_{18}}{114.22 \text{ g}} \quad \frac{8 \text{ mol CO}_2}{1 \text{ mol C}_8H_{18}} \quad \frac{44.01 \text{ g}}{1 \text{ mol}} \quad \frac{1 \text{ kg}}{1000 \text{ g}}$$

then kg CO$_2$ $\rightarrow$ **yr**

$$\frac{3 \times 10^{15} \text{ kg CO}_2}{\left(\dfrac{\text{kg CO}_2 \text{ produced}}{\text{yr}}\right)}$$

Solution:

$$7 \times 10^{12} \text{ kg} \times \frac{1000 \text{ g}}{1 \text{ kg}} \times \frac{1 \text{ mol C}_8H_{18}}{114.22 \text{ g}} \times \frac{8 \text{ mol CO}_2}{1 \text{ mol C}_8H_{18}} \times \frac{44.01 \text{ g CO}_2}{1 \text{ mol CO}_2} \times \frac{1 \text{ kg}}{1000 \text{ g}} = \underline{2}.158 \times 10^{13} \text{ kg CO}_2$$

$$= 2 \times 10^{13} \text{ kg CO}_2/\text{yr}$$

$$\frac{3 \times 10^{15}\ \text{kg}\ \cancel{CO_2}}{\left(\dfrac{2.158 \times 10^{13}\ \text{kg}\ \cancel{CO_2\ \text{produced}}}{\text{yr}}\right)} = 1\underline{3}9\ \text{yr} = 100\ \text{yr}$$

Check: The units (kg and yr) are correct. The magnitude of the answer $(10^{13}\ \text{kg})$ makes physical sense because the mass of CO_2 is larger than the hydrocarbon mass in a combustion (because O is much heavier than H). The time is consistent with earlier Chapter 5 problems where ~1% of the atmospheric carbon dioxide is generated each year.

Cumulative Problems

6.97　　**Given:**

billiard ball$_A$ = system: $m_A = 0.17$ kg, $v_{A1} = 4.5$ m/s slows to $v_{A2} = 3.8$ m/s and $v_{A3} = 0$; ball$_B$: $m_B = 0.17$ kg, $v_{B1} = 0$ and $v_{B2} = 3.8$ m/s, and $KE = \frac{1}{2}mv^2$ **Find:** $w, q, \Delta E_{sys}$

Conceptual Plan: $m, v \rightarrow KE$ then $KE_{A3}, KE_{A1} \rightarrow \Delta E_{sys}$ and $KE_{A2}, KE_{A1} \rightarrow q$ and $KE_{B2}, KE_{B1} \rightarrow w_B$

$$KE = \frac{1}{2}mv^2 \qquad \Delta E_{sys} = KE_{A3} - KE_{A1} \qquad q = KE_{A2} - KE_{A1} \qquad w_B = KE_{B2} - KE_{B1}$$

$\Delta E_{sys}, q \rightarrow w_A$ **verify that** $w_A = -w_B$ **so that no heat is transferred to ball$_B$**

$$\Delta E = q + w$$

Solution: $KE = \dfrac{1}{2}mv^2$ because m is in kg and v is in m/s; KE will be in $\text{kg} \cdot \text{m}^2/\text{s}^2$, which is the definition of a joule.

$$KE_{A1} = \frac{1}{2}(0.17\ \text{kg})\left(4.5\ \frac{\text{m}}{\text{s}}\right)^2 = 1.\underline{7}213\ \frac{\text{kg} \cdot \text{m}^2}{\text{s}^2} = 1.\underline{7}213\ \text{J}$$

$$KE_{A2} = \frac{1}{2}(0.17\ \text{kg})\left(3.8\ \frac{\text{m}}{\text{s}}\right)^2 = 1.\underline{2}274\ \frac{\text{kg} \cdot \text{m}^2}{\text{s}^2} = 1.\underline{2}274\ \text{J}$$

$$KE_{A3} = \frac{1}{2}(0.17\ \text{kg})\left(0\ \frac{\text{m}}{\text{s}}\right)^2 = 0\ \frac{\text{kg} \cdot \text{m}^2}{\text{s}^2} = 0\ \text{J},$$

$$KE_{B1} = \frac{1}{2}(0.17\ \text{kg})\left(0\ \frac{\text{m}}{\text{s}}\right)^2 = 0\ \frac{\text{kg} \cdot \text{m}^2}{\text{s}^2} = 0\ \text{J}$$

$$KE_{B2} = \frac{1}{2}(0.17\ \text{kg})\left(3.8\ \frac{\text{m}}{\text{s}}\right)^2 = 1.\underline{2}274\ \frac{\text{kg} \cdot \text{m}^2}{\text{s}^2} = 1.\underline{2}274\ \text{J}$$

$\Delta E_{sys} = KE_{A3} - KE_{A1} = 0\ \text{J} - 1.\underline{7}213\ \text{J} = -1.\underline{7}213\ \text{J} = -1.7\ \text{J}$

$q = KE_{A2} - KE_{A1} = 1.\underline{2}274\ \text{J} - 1.\underline{7}213\ \text{J} = -0.\underline{4}939\ \text{J} = -0.5\ \text{J}$

$w_B = KE_{B2} - KE_{B1} = 1.\underline{2}274\ \text{J} - 0\ \text{J} = 1.\underline{2}274\ \text{J}$

$w = \Delta E - q = -1.\underline{7}213\ \text{J} - (-0.\underline{4}939\ \text{J}) = -1.\underline{2}274\ \text{J} = -1.2\ \text{J}$

Because $w_A = -w_B$, no heat is transferred to ball$_B$.

Check: The units (J) are correct. Because the ball is initially moving and is stopped at the end, it has lost energy (negative ΔE_{sys}). As the ball slows due to friction, it is releasing heat (negative q). The kinetic energy is transferred to a second ball, so it does work (w negative).

6.99　　**Given:** $H_2O(l) \rightarrow H_2O(g)$ $\Delta H^\circ_{rxn} = +44.01$ kJ/mol; $\Delta T_{body} = -0.50\,°\text{C}$, $m_{body} = 95$ kg, $C_{body} = 4.0$ J/g$°$C
Find: m_{H_2O}

Conceptual Plan: $\text{kg} \rightarrow \text{g}$ then $m_{body}, \Delta T, C_{body} \rightarrow q_{body} \rightarrow q_{rxn}(\text{J}) \rightarrow q_{rxn}(\text{kJ}) \rightarrow \text{mol}\ H_2O \rightarrow \text{g}\ H_2O$

$$\frac{1000\ \text{g}}{1\ \text{kg}} \qquad q_{body} = m_{body}C_{body}\Delta T_{body} \qquad q_{rxn} = -q_{body} \qquad \frac{1\ \text{kJ}}{1000\ \text{J}} \qquad \frac{1\ \text{mol}}{44.01\ \text{kJ}} \qquad \frac{18.02\ \text{g}}{1\ \text{mol}}$$

Solution: $95\ \cancel{\text{kg}} \times \dfrac{1000\ \text{g}}{1\ \cancel{\text{kg}}} = 95000\ \text{g}$ then

$$q_{body} = m_{body}C_{body}\Delta T_{body} = 95000 \ \cancel{g} \times 4.0 \ \frac{J}{\cancel{g} \cdot \cancel{°C}} \times (-0.50 \ \cancel{°C}) = -190000 \ J \text{ then}$$

$$q_{rxn} = -q_{body} = 190000 \ \cancel{J} \times \frac{1 \ \cancel{kJ}}{1000 \ \cancel{J}} \times \frac{1 \ \cancel{mol}}{44.01 \ \cancel{kJ}} \times \frac{18.02 \ g}{1 \ \cancel{mol}} = 78 \ g \ H_2O$$

Check: The units (g) are correct. The magnitude of the answer (78) makes physical sense because a person can sweat this much on a hot day.

6.101   **Given:** $H_2O(s) \rightarrow H_2O(l)$ $\Delta H_f^\circ(H_2O(s)) = -291.8 \ kJ/mol$; 355 mL beverage $T_{Bev, \ i} = 25.0 \ °C$, $T_{Bev, \ f} = 0.0 \ °C$, $C_{Bev} = 4.184 \ J/g \ °C$, $d_{Bev} = 1.0 \ g/mL$ **Find:** ΔH_{rxn}° (ice melting) and m_{ice}

        Conceptual Plan: $\Delta H_{rxn}^\circ = \sum n_P \Delta H_f^\circ(products) - \sum n_R \Delta H_f^\circ(reactants)$ **mL $\rightarrow$ g and $T_i, T_f \rightarrow \Delta T$ then**

$$\frac{1.0 \ g}{1.0 \ mL} \qquad\qquad \Delta T = T_f - T_i$$

$$m_{H_2O}, \ \Delta T_{H_2O}, \ C_{H_2O} \rightarrow q_{H_2O} \rightarrow q_{rxn}(J) \rightarrow q_{rxn}\ (kJ) \rightarrow \text{mol ice} \rightarrow \text{g ice}$$

$$q_{Bev} = m_{Bev}C_{Bev}\Delta T_{Bev} \qquad q_{rxn} = -q_{Bev} \qquad \frac{1 \ kJ}{1000 \ J} \quad \frac{1 \ mol}{\Delta H_{rxn}^\circ} \quad \frac{18.02 \ g}{1 \ mol}$$

Solution:

Reactant/Product	ΔH_f° (kJ/mol from Appendix IIB)
$H_2O(s)$	-291.8
$H_2O(l)$	-285.8

Be sure to pull data for the correct formula and phase.

$$\begin{aligned} \Delta H_{rxn}^\circ &= \sum n_P \Delta H_f^\circ(products) - \sum n_R \Delta H_f^\circ(reactants) \\ &= [1(\Delta H_f^\circ(H_2O(l)))] - [1(\Delta H_f^\circ(H_2O(s)))] \\ &= [1(-285.8 \ kJ)] - [1(-291.8 \ kJ)] \\ &= +6.0 \ kJ \end{aligned}$$

$$355 \ \cancel{mL} \times \frac{1.0 \ g}{1.0 \ \cancel{mL}} = 355 \ g \text{ and } \Delta T = T_f - T_i = 0.0 \ °C - 25.0 \ °C = -25.0 \ °C \text{ then}$$

$$q_{Bev} = m_{Bev}C_{Bev}\Delta T_{Bev} = 355 \ \cancel{g} \times 4.184 \ \frac{J}{\cancel{g} \cdot \cancel{°C}} \times (-25.0 \ \cancel{°C}) = -37133 \ J \text{ then}$$

$$q_{rxn} = -q_{Bev} = 37133 \ \cancel{J} \times \frac{1 \ \cancel{kJ}}{1000 \ \cancel{J}} \times \frac{1 \ \cancel{mol}}{6.0 \ \cancel{kJ}} \times \frac{18.02 \ g}{1 \ \cancel{mol}} = 110 \ g \ \text{ice}$$

Check: The units (kJ and g) are correct. The answer is positive, which means that the reaction is endothermic. We expect an endothermic reaction because we know that heat must be added to melt ice. The magnitude of the answer (110 g) makes physical sense because it is much smaller than the weight of the beverage and it would fit in a glass with the beverage.

6.103   **Given:** 25.5 g aluminum; $T_{Al, \ i} = 65.4 \ °C$; 55.2 g water; $T_{H_2O, \ i} = 22.2 \ °C$ **Find:** T_f

        Conceptual Plan: Pull C_s values from table then $m, C_s, T_i \rightarrow T_f$

$$\text{Al: } 0.903 \ \frac{J}{g \cdot °C} \quad H_2O: 4.18 \ \frac{J}{g \cdot °C} \qquad q = mC_s(T_f - T) \text{ then set } q_{Al} = -q_{H_2O}$$

Solution: $q = mC_s(T_f - T_i)$ substitute in values and set $q_{Al} = -q_{H_2O}$.

$$q_{Al} = m_{Al}C_{Al}(T_f - T_{Al, \ i}) = 25.5 \ \cancel{g} \times 0.903 \ \frac{J}{\cancel{g} \cdot °C} \times (T_f - 65.4 \ °C) =$$

$$-q_{H_2O} = -m_{H_2O}C_{H_2O}(T_f - T_{H_2O, \ i}) = -55.2 \ \cancel{g} \times 4.18 \ \frac{J}{\cancel{g} \cdot °C} \times (T_f - 22.2 \ °C)$$

Rearrange to solve for T_f.

$$23.0265 \ \frac{J}{°C} \times (T_f - 65.4 \ °C) = -230.736 \ \frac{J}{°C} \times (T_f - 22.2 \ °C) \rightarrow$$

$$23.0265 \ \frac{J}{°C}T_f - 1505.93 \ J = -230.736 \ \frac{J}{°C}T_f + 5122.34 \ J \rightarrow$$

$$-5122.34 \ J - 1505.93 \ J = -230.736 \ \frac{J}{°C}T_f - 23.0265 \ \frac{J}{°C}T_f \rightarrow$$

$$6\underline{6}28.27 \text{ J} = 253.\underline{7}625 \frac{\text{J}}{°\text{C}} T_{\text{f}} \rightarrow$$

$$T_{\text{f}} = \frac{6628.27 \cancel{\text{J}}}{253.\underline{7}625 \dfrac{\cancel{\text{J}}}{°\text{C}}} = 26.1 °\text{C}$$

Check: The units (°C) are correct. The magnitude of the answer (26) makes physical sense because the heat transfer is dominated by the water (larger mass and larger specific heat capacity). The final temperature should be closer to the initial temperature of water than of aluminum.

6.105   **Given:** palmitic acid ($C_{16}H_{32}O_2$) combustion $\Delta H_{\text{f}}^{\circ}(C_{16}H_{32}O_2(s)) = -208 \text{ kJ/mol}$; sucrose ($C_{12}H_{22}O_{11}$) combustion $\Delta H_{\text{f}}^{\circ}(C_{12}H_{22}O_{11}(s)) = -2226.1 \text{ kJ/mol}$ **Find:** $\Delta H_{\text{rxn}}^{\circ}$ in kJ/mol and Cal/g

     Conceptual Plan: Write balanced reaction then $\Delta H_{\text{rxn}}^{\circ} = \sum n_{\text{P}} \Delta H_{\text{f}}^{\circ}(products) - \sum n_{\text{R}} \Delta H_{\text{f}}^{\circ}(reactants)$ **then**

     kJ/mol $\rightarrow$ J/mol $\rightarrow$ Cal/mol $\rightarrow$ Cal/g

         $\dfrac{1000 \text{ J}}{1 \text{ kJ}}$ $\dfrac{1 \text{ Cal}}{4184 \text{ J}}$ PA: $\dfrac{1 \text{ mol}}{256.42 \text{ g}}$ S: $\dfrac{1 \text{ mol}}{342.30 \text{ g}}$

     Solution: Combustion is the combination with oxygen to form carbon dioxide and water (*l*):

     $C_{16}H_{32}O_2(s) + 23 \, O_2(g) \rightarrow 16 \, CO_2(g) + 16 \, H_2O(l)$

Reactant/Product	$\Delta H_{\text{f}}^{\circ}$ (kJ/mol from Appendix IIB)
$C_{16}H_{32}O_2(s)$	-208
$O_2(g)$	0.0
$CO_2(g)$	-393.5
$H_2O(l)$	-285.8

     Be sure to pull data for the correct formula and phase.

$$\begin{aligned}
\Delta H_{\text{rxn}}^{\circ} &= \sum n_{\text{P}} \Delta H_{\text{f}}^{\circ}(products) - \sum n_{\text{R}} \Delta H_{\text{f}}^{\circ}(reactants) \\
&= [16(\Delta H_{\text{f}}^{\circ}(CO_2(g))) + 16(\Delta H_{\text{f}}^{\circ}(H_2O(l)))] - [1(\Delta H_{\text{f}}^{\circ}(C_{16}H_{32}O_2(s))) + 23(\Delta H_{\text{f}}^{\circ}(O_2(g)))] \\
&= [16(-393.5 \text{ kJ}) + 16(-285.8 \text{ kJ})] - [1(-208 \text{ kJ}) + 23(0.0 \text{ kJ})] \\
&= [-10868.8 \text{ kJ}] - [-208 \text{ kJ}] \\
&= -10660.8 \text{ kJ/mol} = -10661 \text{ kJ/mol}
\end{aligned}$$

$$-1066\underline{0}.8 \, \frac{\cancel{\text{kJ}}}{\cancel{\text{mol}}} \times \frac{1000 \cancel{\text{J}}}{1 \cancel{\text{kJ}}} \times \frac{1 \text{ Cal}}{4184 \cancel{\text{J}}} \times \frac{1 \cancel{\text{mol}}}{256.42 \text{ g}} = -9.9378 \text{ Cal/g}$$

     $C_{12}H_{22}O_{11}(s) + 12 \, O_2(g) \rightarrow 12 \, CO_2(g) + 11 \, H_2O(l)$

Reactant/Product	$\Delta H_{\text{f}}^{\circ}$ (kJ/mol from Appendix IIB)
$C_{12}H_{22}O_{11}(s)$	-2226.1
$O_2(g)$	0.0
$CO_2(g)$	-393.5
$H_2O(l)$	-285.8

     Be sure to pull data for the correct formula and phase.

$$\begin{aligned}
\Delta H_{\text{rxn}}^{\circ} &= \sum n_{\text{P}} \Delta H_{\text{f}}^{\circ}(products) - \sum n_{\text{R}} \Delta H_{\text{f}}^{\circ}(reactants) \\
&= [12(\Delta H_{\text{f}}^{\circ}(CO_2(g))) + 11(\Delta H_{\text{f}}^{\circ}(H_2O(l)))] - [1(\Delta H_{\text{f}}^{\circ}(C_{12}H_{22}O_{11}(s))) + 12(\Delta H_{\text{f}}^{\circ}(O_2(g)))] \\
&= [12(-393.5 \text{ kJ}) + 11(-285.8 \text{ kJ})] - [1(-2226.1 \text{ kJ}) + 12(0.0 \text{ kJ})] \\
&= [-7865.8 \text{ kJ}] - [-2226.1 \text{ kJ}] \\
&= -5639.7 \text{ kJ/mol}
\end{aligned}$$

$$-5639.7 \frac{\cancel{\text{kJ}}}{\cancel{\text{mol}}} \times \frac{1000 \cancel{\text{J}}}{1 \cancel{\text{kJ}}} \times \frac{1 \text{ Cal}}{4184 \cancel{\text{J}}} \times \frac{1 \cancel{\text{mol}}}{342.30 \text{ g}} = -3.9378 \text{ Cal/g}$$

     Check: The units (kJ/mol and Cal/g) are correct. The magnitude of the answers are consistent with the food labels we see every day. Palmitic acid gives more Cal/g than does sucrose.

6.107   At constant pressure, $\Delta H_{\text{rxn}} = q_{\text{P}}$, and at constant volume, $\Delta E_{\text{rxn}} = q_{\text{V}} = \Delta H_{\text{rxn}} - P\Delta V$. Recall that $PV = nRT$. Because the conditions are at constant P and a constant number of moles of gas, as T changes, the only variable that can change is V. This means that $P\Delta V = nR\Delta T$. Substituting into the equation for ΔE_{rxn}, we get $\Delta E_{\text{rxn}} = \Delta H_{\text{rxn}} - nR\Delta T$ or $\Delta H_{\text{rxn}} = \Delta E_{\text{rxn}} + nR\Delta T$.

6.109   **Given:** 16 g peanut butter; bomb calorimeter; $T_i = 22.2\,°C$; $T_f = 25.4\,°C$; $C_{cal} = 120.0\,kJ/°C$ **Find:** calories in peanut butter
Conceptual Plan: $T_i, T_f \rightarrow \Delta T$ then $\Delta T, C_{cal} \rightarrow q_{cal} \rightarrow q_{rxn}\,(kJ) \rightarrow q_{rxn}\,(kJ) \rightarrow q_{rxn}\,(Cal)$

$$\Delta T = T_f - T_i \qquad q_{cal} = C_{cal}\Delta T \qquad q_{rxn} = -q_{cal} \qquad \frac{1000\,J}{1\,kJ} \qquad \frac{1\,Cal}{4184\,J}$$

then $q_{rxn}\,(Cal) \rightarrow Cal/g$

$$\div\,16\,g\ peanut\ butter$$

Solution: $\Delta T = T_f - T_i = 25.4\,°C - 22.2\,°C = 3.2\,°C$ then $q_{cal} = C_{cal}\Delta T = 120.0\,\dfrac{kJ}{°C} \times 3.2\,°C = 3\underline{8}4\,kJ$

then $q_{rxn} = -q_{cal} = -3\underline{8}4\,kJ \times \dfrac{1000\,J}{1\,kJ} \times \dfrac{1\,Cal}{4184\,J} = 91.778\,Cal$ then $\dfrac{91.778\,Cal}{16\,g} = 5.7\,Cal/g$

Check: The units (Cal/g) are correct. The magnitude of the answer (6) makes physical sense because there is a significant percentage of fat and sugar in peanut butter. The answer is in line with the answers in Problem 6.105.

6.111   **Given:** $V_1 = 20.0\,L$ at $P_1 = 3.0\,atm$; $P_2 = 1.5\,atm$; let it expand at constant T **Find:** $w, q, \Delta E_{sys}$
Conceptual Plan: $V_1, P_1, P_2 \rightarrow V_2$ then $V_1, V_2 \rightarrow \Delta V$ then $P, \Delta V \rightarrow w\,(L\,atm) \rightarrow w\,(J)$

$$P_1V_1 = P_2V_2 \qquad \Delta V = V_2 - V_1 \qquad w = -P\Delta V \qquad \frac{101.3\,J}{1\,L\cdot atm}$$

for an ideal gas $\Delta E_{sys}\,\alpha\,T$; **so because this is a constant temperature process,** $\Delta E_{sys} = 0$ **finally** $\Delta E_{sys}, w \rightarrow q$

$$\Delta E = q + w$$

Solution: $P_1V_1 = P_2V_2$.   Rearrange to solve for V_2. $V_2 = V_1\dfrac{P_1}{P_2} = (20.0\,L) \times \dfrac{3.0\,atm}{1.5\,atm} = 40.\,L$ and

$\Delta V = V_2 - V_1 = 40.\,L - 20.0\,L = 20.\,L$ then

$w = -P\Delta V = -1.5\,atm \times 20.\,L \times \dfrac{101.3\,J}{1\,L\cdot atm} = -30\underline{3}9\,J = -3.0 \times 10^3\,J \qquad \Delta E = q + w$

Rearrange to solve for q. $q = \Delta E_{sys} - w = +0\,J - (-30\underline{3}9\,J) = 3.0 \times 10^3\,J$

Check: The units (J) are correct. Because there is no temperature change, we expect no energy change ($\Delta E_{sys} = 0$). The piston expands as it does work (negative work), so heat is absorbed (positive q).

6.113   The oxidation of $S(g)$ to SO_3 can be written as follows:
$S(g) + 3/2\,O_2(g) \rightarrow SO_3(g)$ $\Delta H = -204\,kJ$
The oxidation of $SO_2(g)$ to SO_3 can be written as follows:
$SO_2(g) + 1/2\,O_2(g) \rightarrow SO_3(g)$ $\Delta H = +89.5\,kJ$
The enthalpy of formation reaction for $SO_2(g)$ under these conditions can be written as follows:
$S(g) + O_2(g) \rightarrow SO_2(g)$ $\Delta H = ??$
Because the second reaction has 1 mole SO_2 as a reactant and the reaction of interest has 1 mole of SO_2 as a product, we need to reverse the second reaction. When the reaction direction is reversed, ΔH changes sign.
$SO_3(g) \rightarrow SO_2(g) + 1/2\,O_2(g)$ $\Delta H = -89.5\,kJ$
Hess's law states that the ΔH of the net reaction is the sum of the ΔH of the steps.
The rewritten reactions are as follows:
$S(g) + 3\!/\!2\,O_2(g) \rightarrow \cancel{SO_3(g)}$ $\Delta H = -204\,kJ$
$\cancel{SO_3(g)} \rightarrow SO_2(g) + \cancel{1\!/\!2\,O_2(g)}$ $\Delta H = -89.5\,kJ$

$S(g) + O_2(g) \rightarrow SO_2(g)$ $\Delta H = -294\,kJ = \Delta H_f$
Note that this is not under standard conditions because S is not a solid.

6.115   **Given:** 25.3% methane (CH_4), 38.2% ethane (C_2H_6), and the rest propane (C_3H_8) by volume; $V = 1.55\,L$ tank, $P = 755\,mmHg$, and $T = 298\,K$ **Find:** heat for combustion
Conceptual Plan: Percent composition $\rightarrow$ **mmHg** $\rightarrow$ **atm then** $P, V, T \rightarrow n$ **then**

$$\text{Dalton's Law of Partial Pressures} \qquad \frac{1\,atm}{760\,mmHg} \qquad PV = nRT$$

use data in Problem 6.93 for methane and propane and calculate heat of combustion for ethane

$\Delta H°_{rxn}(CH_4) = -802.3\,kJ;$ $\Delta H°_{rxn}(C_3H_8) = -2043\,kJ$ write balanced reaction then $\Delta H°_{rxn} = \sum n_P\Delta H°_f(products) - \sum n_R\Delta H°_f(reactants)$

then $n, \Delta H \rightarrow q$

Solution: $P_{CH_4} = \dfrac{25.3 \text{ mmHg } CH_4}{100 \text{ mmHg gas}} \times 755 \text{ mmHg gas} \times \dfrac{1 \text{ atm } CH_4}{760 \text{ mmHg}} = 0.25\underline{1}3355 \text{ atm } CH_4$

$P_{C_2H_6} = \dfrac{38.2 \text{ mmHg } C_2H_6}{100 \text{ mmHg gas}} \times 755 \text{ mmHg gas} \times \dfrac{1 \text{ atm } C_2H_6}{760 \text{ mmHg}} = 0.37\underline{9}48684 \text{ atm } C_2H_6$

$P_{C_3H_8} = \dfrac{100 - (25.3 + 38.2) \text{ mmHg } C_3H_8}{100 \text{ mmHg gas}} \times 755 \text{ mmHg gas} \times \dfrac{1 \text{ atm } C_3H_8}{760 \text{ mmHg}} = 0.36\underline{2}59868 \text{ atm } C_3H_8$

$PV = nRT$.   Rearrange to solve for n.

$n_{CH_4} = \dfrac{PV}{RT} = \dfrac{0.25\underline{1}3355 \text{ atm } CH_4 \times 1.55 \text{ L}}{0.08206 \dfrac{\text{L} \cdot \text{atm}}{\text{mol} \cdot \text{K}} \times 298 \text{ K}} = 0.015\underline{9}3081 \text{ mol } CH_4$

$n_{C_2H_6} = \dfrac{PV}{RT} = \dfrac{0.37\underline{9}48684 \text{ atm } C_2H_6 \times 1.55 \text{ L}}{0.08206 \dfrac{\text{L} \cdot \text{atm}}{\text{mol} \cdot \text{K}} \times 298 \text{ K}} = 0.024\underline{0}5363 \text{ mol } C_2H_6$

$n_{C_2H_6} = \dfrac{PV}{RT} = \dfrac{0.36\underline{2}59868 \text{ atm } C_3H_8 \times 1.55 \text{ L}}{0.08206 \dfrac{\text{L} \cdot \text{atm}}{\text{mol} \cdot \text{K}} \times 298 \text{ K}} = 0.022\underline{9}83181 \text{ mol } C_3H_8$

$C_2H_6(g) + 7/2\,O_2(g) \rightarrow 2\,CO_2(g) + 3\,H_2O(g)$

Reactant/Product	ΔH_f° (kJ/mol from Appendix IIB)
$C_2H_6(g)$	-84.68
$O_2(g)$	0.0
$CO_2(g)$	-393.5
$H_2O(g)$	-241.8

Be sure to pull data for the correct formula and phase.

$\Delta H_{rxn}^\circ = \sum n_P \Delta H_f^\circ (products) - \sum n_R \Delta H_f^\circ (reactants)$

$= [2(\Delta H_f^\circ(CO_2(g))) + 3(\Delta H_f^\circ(H_2O(g)))] - [1(\Delta H_f^\circ(C_2H_6(g))) + 7/2(\Delta H_f^\circ(O_2(g)))]$

$= [2(-393.5 \text{ kJ}) + 3(-241.8 \text{ kJ})] - [1(-84.68 \text{ kJ}) + 7/2(0.0 \text{ kJ})]$

$= [-1512.4 \text{ kJ}] - [-84.68 \text{ kJ}]$

$= -1427.7 \text{ kJ}$

$0.015\underline{9}3081 \text{ mol } CH_4 \times \dfrac{-802.3 \text{ kJ}}{1 \text{ mol } CH_4} = -12.\underline{7}81289 \text{ kJ}$

$0.024\underline{0}5363 \text{ mol } C_2H_6 \times \dfrac{-1427.7 \text{ kJ}}{1 \text{ mol } C_2H_6} = -34.\underline{3}4137 \text{ kJ}$

$0.022\underline{9}83181 \text{ mol } C_3H_8 \times \dfrac{-2043 \text{ kJ}}{1 \text{ mol } C_3H_8} = -46.\underline{9}546 \text{ kJ}$

The total heat is $-12.\underline{7}81289 \text{ kJ} - 34.\underline{3}4137 \text{ kJ} - 46.\underline{9}546 \text{ kJ} = -94.\underline{0}7726 \text{ kJ} = -94.1 \text{ kJ}$

Check: The units (kJ) are correct. The magnitude of the answer (-100 kJ) makes sense because heats of combustion are typically large and negative.

6.117   **Given:** 1.55 cm copper cube and 1.62 cm aluminum cube, $T_{Metals, i} = 55.0\,^\circ C$, 100.0 mL water, $T_{H_2O, i} = 22.2\,^\circ C$
Other: density (water) $= 0.998 \text{ g/mL}$ **Find:** T_f
Conceptual Plan: Pull d values from Table 1.4 then edge length $\rightarrow V \rightarrow m$ then

   Cu: 8.96 g/mL Al: 2.70 g/mL $V = l^3 \quad d = m/V$

pull C_s values from Table 6.4 then $m, C_s, T_i \rightarrow T_f$

Cu: $0.385 \dfrac{J}{g \cdot {}^\circ C}$ Al: $0.903 \dfrac{J}{g \cdot {}^\circ C}$ H$_2$O: $4.18 \dfrac{J}{g \cdot {}^\circ C}$ $q = mC_s(T_f - T_i)$ then set $q_{Cu} + q_{Al} = -q_{H_2O}$

Solution: $V_{Cu} = l^3 = (1.55 \text{ cm})^3 = 3.7\underline{2}3875 \text{ cm}^3 = 3.7\underline{2}3875 \text{ mL}$ and

$V_{Al} = l^3 = (1.62 \text{ cm})^3 = 4.2\underline{5}1528 \text{ cm}^3 = 4.2\underline{5}1528 \text{ mL}$ then $d = m/V$.   Rearrange to solve for m. $m = dV$

$m_{Cu} = 8.96 \dfrac{g}{mL} \times 3.7\underline{2}3875 \text{ mL} = 33.3\underline{6}592 \text{ g Cu}$

$$m_{Al} = 2.70 \frac{g}{mL} \times 4.2\underline{5}1528 \text{ mL} = 11.4791256 \text{ g Al and } m_{H_2O} = 0.998 \frac{g}{mL} \times 100.0 \text{ mL} = 99.8 \text{ g H}_2\text{O}$$

$q = mC_s(T_f - T_i)$ substitute in values and set $q_{Cu} + q_{Al} = -q_{H_2O}$

$$q_{Cu} + q_{Al} = m_{Cu}C_{Cu}(T_f - T_{Cu,i}) + m_{Al}C_{Al}(T_f - T_{Al,i}) =$$

$$33.\underline{3}6592 \text{ g} \times 0.385 \frac{J}{g \cdot {}^\circ C} \times (T_f - 55.0\,{}^\circ C) + 11.\underline{4}791256 \text{ g} \times 0.903 \frac{J}{g \cdot {}^\circ C} \times (T_f - 55.0\,{}^\circ C) =$$

$$-q_{H_2O} = -m_{H_2O}C_{H_2O}(T_f - T_{H_2O,i}) = -99.8 \text{ g} \times 4.18 \frac{J}{g \cdot {}^\circ C} \times (T_f - 22.2\,{}^\circ C)$$

Rearrange to solve for T_f.

$$12.\underline{8}4588 \frac{J}{{}^\circ C} \times (T_f - 55.0\,{}^\circ C) + 10.36565 \frac{J}{{}^\circ C} \times (T_f - 55.0\,{}^\circ C) = -417.\underline{1}64 \frac{J}{{}^\circ C} \times (T_f - 22.2\,{}^\circ C) \rightarrow$$

$$12.\underline{8}4588 \frac{J}{{}^\circ C}T_f - 706.5234 \text{ J} + 10.\underline{3}6565 \frac{J}{{}^\circ C}T_f - 570.1108 \text{ J} = -417.\underline{1}64 \frac{J}{{}^\circ C}T_f + 92\underline{6}1.041 \text{ J} \rightarrow$$

$$12.\underline{8}4588 \frac{J}{{}^\circ C}T_f + 10.\underline{3}6565 \frac{J}{{}^\circ C}T_f + 417.\underline{1}64 \frac{J}{{}^\circ C}T_f = +706.5234 \text{ J} + 570.1108 \text{ J} + 92\underline{6}1.041 \text{ J} \rightarrow$$

$$44\underline{0}.3755 \frac{J}{{}^\circ C}T_f = 105\underline{3}7.675 \text{ J} \rightarrow T_f = \frac{105\underline{3}7.675 \text{ J}}{44\underline{0}.3755 \frac{J}{{}^\circ C}} = 23.\underline{9}2884\,{}^\circ C = 23.9\,{}^\circ C$$

Check: The units (°C) are correct. The magnitude of the answer (24) makes physical sense because the heat transfer is dominated by the water (larger mass and larger specific heat capacity). The final temperature should be closer to the initial temperature of water than of copper and aluminum.

Challenge Problems

6.119   **Given:** 655 kWh/yr; coal is 3.2% S; remainder is C; S emitted as $SO_2(g)$ and gets converted to H_2SO_4 when reacting with water   **Find:** $m(H_2SO_4)/\text{yr}$

    Conceptual Plan: Write balanced reaction then $\Delta H^\circ_{rxn} = \sum n_P \Delta H^\circ_f(products) - \sum n_R \Delta H^\circ_f(reactants)$

    (because the form of sulfur is not given, assume that all heat is from combustion of only carbon) then

    kWh $\rightarrow$ J $\rightarrow$ kJ $\rightarrow$ mol (C) $\rightarrow$ g (C) $\rightarrow$ g (S) $\rightarrow$ mol (H_2SO_4) $\rightarrow$ mol (H_2SO_4) $\rightarrow$ g (H_2SO_4)

$$\frac{3.60 \times 10^6 \text{J}}{1 \text{ kWh}} \quad \frac{1 \text{ kJ}}{1000 \text{ J}} \quad \frac{\text{mol C}}{\Delta H^\circ_f(CO_2(g))} \quad \frac{12.01 \text{ g}}{1 \text{ mol}} \quad \frac{3.2 \text{ g S}}{(100.0 - 3.2) \text{ g C}} \quad \frac{1 \text{ mol}}{32.07 \text{ g}} \quad \frac{1 \text{ mol H}_2SO_4}{1 \text{ mol S}} \quad \frac{98.09 \text{ g}}{1 \text{ mol}}$$

    Solution: $C(s) + O_2(g) \rightarrow CO_2(g)$. This reaction is the heat of formation of $CO_2(g)$, so

    $\Delta H^\circ_{rxn} = \Delta H^\circ_f(CO_2(g)) = -393.5 \text{ kJ/mol}$ then

$$655 \text{ kWh} \times \frac{3.60 \times 10^6 \text{ J}}{1 \text{ kWh}} \times \frac{1 \text{ kJ}}{1000 \text{ J}} \times \frac{\text{mol C}}{393.5 \text{ kJ}} \times \frac{12.01 \text{ g C}}{1 \text{ mol C}} \times \frac{3.2 \text{ g S}}{(100.0 - 3.2) \text{ g C}} \times \frac{1 \text{ mol S}}{32.07 \text{ g S}}$$

$$\times \frac{1 \text{ mol H}_2SO_4}{1 \text{ mol S}} \times \frac{98.09 \text{ g H}_2SO_4}{1 \text{ mol H}_2SO_4} = 7.3 \times 10^3 \text{ g H}_2SO_4$$

    Check: The units (g) are correct. The magnitude (7300) is reasonable, considering this calculation is for only one home.

6.121   **Given:** methane combustion, 100% efficiency; $\Delta T = 10.0\,{}^\circ C$; house $= 30.0 \text{ m} \times 30.0 \text{ m} \times 3.0 \text{ m}$; C_s (air) $= 30 \text{ J/K} \cdot \text{mol}$; 1.00 mol air $= 22.4 \text{ L}$ **Find:** m (CH_4)

    Conceptual Plan: $l, w, h \rightarrow V(\text{m}^3) \rightarrow V(\text{cm}^3) \rightarrow V(\text{L}) \rightarrow$ mol (air) then $m, C_s, \Delta T \rightarrow q_{air}$ (J)

$$V = lwh \qquad \frac{(100 \text{ cm})^3}{(1 \text{ m})^3} \qquad \frac{1 \text{ L}}{1000 \text{ cm}^3} \qquad \frac{1 \text{ mol air}}{22.4 \text{ L}} \qquad q = mC_s\Delta T$$

    then q_{air} **(J)** $\rightarrow q_{rxn}$ **(J)** $\rightarrow q$ **(kJ) then write balanced reaction for methane combustion**

$$q_{rxn} = -q_{air} \qquad \frac{1 \text{ kJ}}{1000 \text{ J}}$$

    then $\Delta H^\circ_{rxn} = \sum n_P \Delta H^\circ_f(products) - \sum n_R \Delta H^\circ_f(reactants)$, **and then** q **(kJ)** $\rightarrow$ **mol (CH_4)** $\rightarrow$ **g (CH_4)**

$$\Delta H^\circ_{rxn} \qquad \frac{16.04 \text{ g}}{1 \text{ mol}}$$

Solution: $V = lwh = 30.0 \text{ m} \times 30.0 \text{ m} \times 3.0 \text{ m} = 2\underline{7}00 \text{ m}^3$, then

$$2\underline{7}00 \text{ m}^3 \times \frac{(100 \text{ cm})^3}{(1 \text{ m}^3)} \times \frac{1 \text{ L}}{1000 \text{ cm}^3} \times \frac{1 \text{ mol air}}{22.4 \text{ L}} = 1.\underline{2}0536 \times 10^5 \text{ mol air, and then}$$

$$q = mC_s\Delta T = 1.\underline{2}0536 \times 10^5 \text{ mol} \times 30 \frac{J}{\text{mol} \cdot {}^\circ C} \times 10.0 {}^\circ C = 3.\underline{6}161 \times 10^7 \text{ J} \times \frac{1 \text{ kJ}}{1000 \text{ J}} = 3.\underline{6}161 \times 10^4 \text{ J needed}$$

$$CH_4(g) + 2\,O_2(g) \rightarrow CO_2(g) + 2\,H_2O(g)$$

Reactant/Product	ΔH_f° (kJ/mol from Appendix IIB)
$CH_4(g)$	-74.6
$O_2(g)$	0.0
$CO_2(g)$	-393.5
$H_2O(g)$	-241.8

Be sure to pull data for the correct formula and phase.

$$\Delta H_{rxn}^\circ = \sum n_P \Delta H_f^\circ(products) - \sum n_R \Delta H_f^\circ(reactants)$$

$$= [1(\Delta H_f^\circ(CO_2(g))) + 2(\Delta H_f^\circ(H_2O(g)))] - [1(\Delta H_f^\circ(CH_4(g))) + 2(\Delta H_f^\circ(O_2(g)))]$$

$$= [(-393.5 \text{ kJ}) + 2(-241.8 \text{ kJ})] - [1(-74.6 \text{ kJ}) + 2(0.0 \text{ kJ})]$$

$$= [-877.1 \text{ kJ}] - [-74.6 \text{ kJ}]$$

$$= -802.5 \text{ kJ}$$

$$q_{rxn} = -q_{air} = -3.\underline{6}161 \times 10^4 \text{ kJ} \times \frac{1 \text{ mol } CH_4}{-802.5 \text{ kJ}} \times \frac{16.04 \text{ g } CH_4}{1 \text{ mol } CH_4} = 7\underline{2}2.8 \text{ g } CH_4 = 700 \text{ g } CH_4$$

Check: The units (g) are correct. The magnitude (700) is not surprising because the volume of a house is large.

6.123　　**Given:** $m(\text{ice}) = 9.0 \text{ g}$; coffee: $T_1 = 90.0\,^\circ C$, $m = 120.0 \text{ g}$, $C_s = C_{H_2O}$, $\Delta H_{fus}^\circ = 6.0 \text{ kJ/mol}$　**Find:** T_f of coffee
　　Conceptual Plan: $q_{ice} = -q_{coffee}$ so g (ice) → mol (ice) → q_{fus} (kJ) → q_{fus} (J) → q_{coffee} (J) then

$$\frac{1 \text{ mol}}{18.01 \text{ g}} \qquad \frac{6.0 \text{ kJ}}{1 \text{ mol}} \qquad \frac{1000 \text{ J}}{1 \text{ kJ}} \qquad q_{coffee} = -q_{ice}$$

$q, m, C_s \rightarrow \Delta T$ then T_i, $\Delta T \rightarrow T_2$; now we have slightly cooled coffee in contact with 0.0 °C water

$$q = mC_s\Delta T \qquad \Delta T = T_2 - T_i$$

so $q_{ice} = -q_{coffee}$ with m, C_s, $T_i \rightarrow T_f$

$$q = mC_s(T_f - T_i) \text{ then set } q_{H_2O} = -q_{coffee}$$

Solution: $9.0 \text{ g} \times \dfrac{1 \text{ mol}}{18.01 \text{ g}} \times \dfrac{6.0 \text{ kJ}}{1 \text{ mol}} \times \dfrac{1000 \text{ J}}{1 \text{ kJ}} = 2.\underline{9}983 \times 10^3 \text{ J}$, $q_{coffee} = -q_{ice} = -2.\underline{9}983 \times 10^3 \text{ J}$

$q = mC_s\Delta T$.　Rearrange to solve for ΔT. $\Delta T = \dfrac{q}{mC_s} = \dfrac{-2.\underline{9}983 \times 10^3 \text{ J}}{120.0 \text{ g} \times 4.18 \dfrac{J}{\text{g} \cdot {}^\circ C}} = -5.\underline{9}775\,^\circ C$ then

$\Delta T = T_2 - T_i$.　Rearrange to solve for T_2. $T_2 = \Delta T + T_i = -5.\underline{9}775\,^\circ C + 90.0\,^\circ C = 84.\underline{0}225\,^\circ C$

$q = mC_s(T_f - T_i)$ substitute in values and set $q_{H_2O} = -q_{coffee}$

$$q_{H_2O} = m_{H_2O}C_{H_2O}(T_f - T_{H_2O,i}) = 9.0 \text{ g} \times 4.18 \frac{J}{\text{g} \cdot {}^\circ C} \times (T_f - 0.0\,^\circ C) =$$

$$-q_{coffee} = -m_{coffee}C_{coffee}(T_f - T_{coffee2}) = -120.0 \text{ g} \times 4.18 \frac{J}{\text{g} \cdot {}^\circ C} \times (T_f - 84.\underline{0}225\,^\circ C)$$

Rearrange to solve for T_f.

$9.0 \text{ g } T_f = -120.0 \text{ g }(T_f - 84.\underline{0}225\,^\circ C) \rightarrow 9.0 \text{ g } T_f = -120.0 \text{ g } T_f + 10082.7 \text{ g} \rightarrow$

$$-10082.7 \text{ g} = -129.0 \frac{\text{g}}{{}^\circ C} T_f \rightarrow T_f = \frac{-10082.7 \text{ g}}{-129.0 \dfrac{\text{g}}{{}^\circ C}} = 78.2\,^\circ C$$

Check: The units (°C) are correct. The temperature is closer to the original coffee temperature because the mass of coffee is so much larger than the ice mass.

6.125 $KE = \frac{1}{2}mv^2$. For an ideal gas, $v = u_{rms} = \sqrt{\frac{3RT}{\mathcal{M}}}$; so $KE_{avg} = \frac{1}{2}N_A m u_{rms}^2 = \frac{3}{2}RT$ then

$\Delta E_{sys} = KE_2 - KE_1 = \frac{3}{2}RT_2 - \frac{3}{2}RT_1 = \frac{3}{2}R\Delta T$. At constant V, $\Delta E_{sys} = C_V \Delta T$; so $C_V = \frac{3}{2}R$.

At constant P, $\Delta E_{sys} = q + w = q_P - P\Delta V = \Delta H - P\Delta V$. Because $PV = nRT$, for one mole of an

ideal gas at constant P, $P\Delta V = R\Delta T$; so $\Delta E_{sys} = q + w = q_P - P\Delta V = \Delta H - P\Delta V = \Delta H - R\Delta T$. This gives

$\frac{3}{2}R\Delta T = \Delta H - R\Delta T$ or $\Delta H = \frac{5}{2}R\Delta T = C_P \Delta T$, so $C_P = \frac{5}{2}R$.

6.127 $q = \Delta H = 454 \text{ g} \times \frac{1 \text{ mol}}{18.02 \text{ g}} \times \frac{40.7 \text{ kJ}}{1 \text{ mol}} = 1025.405 \text{ kJ} = 1030 \text{ kJ}$ and $w = -P\Delta V$. Assume that $P = 1 \text{ atm}$

(exactly) and $\Delta V = V_G - V_L$, where $V_L = 454 \text{ g} \times \frac{1 \text{ mL}}{0.9998 \text{ g}} \times \frac{1 \text{ L}}{1000 \text{ mL}} = 0.4540908 \text{ L}$ and $PV = nRT$.

Rearrange to solve for V_G. $V_G = \frac{nRT}{P} = \dfrac{454 \text{ g} \times \dfrac{1 \text{ mol}}{18.02 \text{ g}} \times 0.08206 \dfrac{L \cdot atm}{mol \cdot K} \times 373 \text{ K}}{1 \text{ atm}} = 771.1545 \text{ L}$

$\Delta V = V_G - V_L = 771.1545 \text{ L} - 0.4540908 \text{ L} = 770.7004 \text{ L}$ and so

$w = -P\Delta V = -1.0 \text{ atm} \times 770.7004 \text{ L} \times \frac{101.3 \text{ J}}{1 \text{ L} \cdot atm} = -78071.9515 \text{ J} = -7.81 \times 10^4 \text{ J} = -78.1 \text{ kJ}$

Finally, $\Delta E = q + w = 1025.405 \text{ kJ} - 78.0719515 \text{ kJ} = 947.333 \text{ kJ} = 950 \text{ kJ}$.

6.129 $C_8H_{18}(l) + 25/2 \, O_2(g) \rightarrow 8 \, CO_2(g) + 9 \, H_2O(l)$; $q = \Delta H = -1303 \text{ kJ/mol}$ and $w = -P\Delta V$. Assume that $P = 1 \text{ atm}$ (exactly) and $\Delta V = \Delta V_G$ because the gas volumes are so much larger than the liquid volumes. The

change in the number of moles of gas $\Delta n_G = 8 \text{ mol} - \frac{25}{2} \text{ mol} = -4.5 \text{ mol}$ and $P\Delta V = \Delta nRT$.   Rearrange to

solve for ΔV_G. $\Delta V_G = \frac{\Delta nRT}{P} = \dfrac{-4.5 \text{ mol} \times 0.08206 \dfrac{L \cdot atm}{mol \cdot K} \times 298 \text{ K}}{1 \text{ atm}} = -110.0425 \text{ L}$ and so

$w = -P\Delta V = -1.0 \text{ atm} \times (-110.0425 \text{ L}) \times \frac{101.3 \text{ J}}{1 \text{ L} \cdot atm} = +11147.30 \text{ J} = +11000 \text{ J} = +11 \text{ kJ}$

Finally, $\Delta E = q + w = -1303 \text{ kJ} + 11.14730 \text{ kJ} = -1291.8527 \text{ kJ} = -1292 \text{ kJ}$.

Conceptual Problems

6.131   (d)   Only one answer is possible: $\Delta E_{sys} = -\Delta E_{surr}$.

6.133   (a)   At constant P, $\Delta E_{sys} = q + w = q_P + w = \Delta H + w$; so $\Delta E_{sys} - w = \Delta H = q$.

6.135   The aluminum cylinder will be cooler after 1 hour because it has a lower heat capacity than does water (less heat needs to be pulled out for every °C temperature change).

6.137   **Given:** 2418 J heat produced; 5 J work done on surroundings at constant P **Find:** ΔE, ΔH, q, and w
          Conceptual Plan: Interpret language to determine the sign of the two terms then $q, w \rightarrow \Delta E_{sys}$
                                                                        $\Delta E = q + w$

          Solution: Because heat is released from the system to the surroundings, $q = -2418 \text{ J}$;
          because the system is doing work on the surroundings, $w = -5 \text{ kJ}$. At constant P, $\Delta H = q = -2.418 \text{ kJ}$;
          $\Delta E = q + w = -2418 \text{ J} - 5 \text{ J} = -2423 \text{ J} = -2 \text{ kJ}$.

          Check: The units (kJ) are correct. The magnitude of the answer (-2) makes physical sense because both terms are negative and the amount of work done is negligibly small.

6.139   (b)   If ΔV is positive, then $w = -P\Delta V < 0$. Because $\Delta E_{sys} = q + w = q_P + w = \Delta H + w$; if w is negative, then $\Delta H > \Delta E_{sys}$.

7 The Quantum-Mechanical Model of the Atom

Review Questions

7.1   The quantum-mechanical model of the atom is important because it explains how electrons exist in atoms and how those electrons determine the chemical and physical properties of elements.

7.3   The wavelength (λ) of the wave is the distance in space between adjacent crests and is measured in units of distance. The amplitude of the wave is the vertical height of a crest. The more closely spaced the waves (i.e., the shorter the wavelength), the more energy there is. The amplitude of the electric and magnetic field waves in light determine the intensity or brightness of the light. The higher the amplitude is the more energy the wave has.

7.5   For visible light, wavelength determines the color. Red light has a wavelength of 750 nm, the longest wavelength of visible light, and blue has a wavelength of 500 nm.

7.7   (a)   Gamma rays (γ)—the wavelength range is 10^{-11} to 10^{-15} m. Gamma rays are produced by the sun, other stars, and certain unstable atomic nuclei on Earth. Human exposure to gamma rays is dangerous because the high energy of gamma rays can damage biological molecules.

          (b)   X-rays—the wavelength range is 10^{-8} to 10^{-11} m. X-rays are used in medicine. X-rays pass through many substances that block visible light and are therefore used to image bones and internal organs. X-rays are sufficiently energetic to damage biological molecules, so while several yearly exposures to X-rays are harmless, excessive exposure increases cancer risk.

          (c)   Ultraviolet radiation (UV)—the wavelength range is 0.4×10^{-6} to 10^{-8} m. Ultraviolet radiation is most familiar as the component of sunlight that produces a sunburn or suntan. While not as energetic as gamma rays or X-rays, ultraviolet light still carries enough energy to damage biological molecules. Excessive exposure to ultraviolet light increases the risk of skin cancer and cataracts and causes premature wrinkling of the skin.

          (d)   Visible light—the wavelength range is 0.75×10^{-6} to 0.4×10^{-6} m (750 nm to 400 nm). Visible light, as long as the intensity is not too high, does not carry enough energy to damage biological molecules. It does, however, cause certain molecules in our eyes to change their shape, sending a signal to brains that results in vision.

          (e)   Infrared radiation (IR)—the wavelength range is 0.75×10^{-6} to 10^{-3} m. The heat you feel when you place your hand near a hot object is infrared radiation. All warm objects, including human bodies, emit infrared light. Although infrared light is invisible to our eyes, infrared sensors can detect it and are often used in night vision technology to "see" in the dark.

          (f)   Microwave radiation—the wavelength range is 10^{-3} to 10^{-1} m. Microwave radiation is used in radar and in microwave ovens. Microwave radiation is efficiently absorbed by water and can therefore heat substances that contain water.

          (g)   Radio waves—the wavelength range is 10^{-1} to 10^{5} m. Radio waves are used to transmit the signals responsible for AM and FM radio, cell telephones, television, and other forms of communication.

7.9   Diffraction occurs when a wave encounters an obstacle or a slit that is comparable in size to its wavelength. The wave bends around the slit. The diffraction of light through two slits separated by a distance comparable to the wavelength of the light results in an interference pattern. Each slit acts as a new wave source, and the two new waves interfere with each other. This results in a pattern of bright and dark lines.

Interference from Two Slits

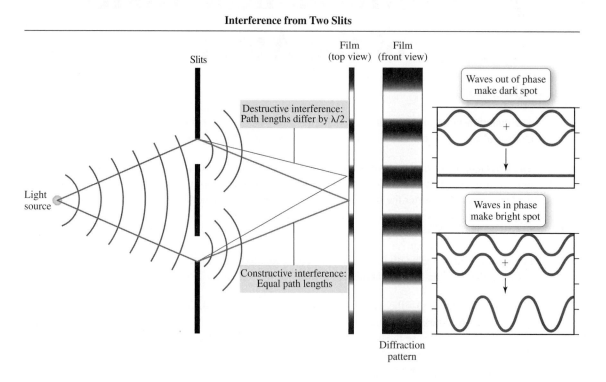

7.11   Because of the results of the experiments with the photoelectric effect, Einstein proposed that light energy must come in packets. The amount of energy in a light packet depends on its frequency (wavelength). The emission of electrons depends on whether a single photon has sufficient energy to dislodge a single electron.

7.13   An emission spectrum occurs when an atom absorbs energy and re-emits that energy as light. The light emitted contains distinct wavelengths for each element. The emission spectrum of a particular element is always the same and can be used to identify the element. A white light spectrum is continuous, meaning that there are no sudden interruptions in the intensity of the light as a function of wavelengths. It consists of all wavelengths. Emission spectra are not continuous. They consist of bright lines at specific wavelengths, with complete darkness in between.

7.15   Electron diffraction occurs when an electron beam is aimed at two closely spaced slits and a series of detectors is arranged to detect the electrons after they pass through the slits. An interference pattern similar to that observed for light is recorded behind the slits. Electron diffraction is evidence of the wave nature of electrons.

7.17   Complementary properties exclude one another. The more you know about one, the less you know about the other. Which of two complementary properties you observe depends on the experiment you perform. In electron diffraction, when you try to observe which hole the electron goes through (particle nature), you lose the interference pattern (wave nature). When you try to observe the interference pattern, you cannot determine which hole the electron goes through.

7.19   A trajectory is a path that is determined by the particle's velocity (the speed and direction of travel), its position, and the forces acting on it. Both position and velocity are required to predict a trajectory.

7.21   Deterministic means that the present determines the future. That means that under the identical condition, identical results will occur.

7.23   A probability distribution map is a statistical map that shows where an electron is likely to be found under a given set of conditions.

7.25   An orbital is a probability distribution map showing where the electron is likely to be found.

7.27 The principal quantum number (n) is an integer that has possible values of 1, 2, 3, etc. The principal quantum number determines the overall size and energy of an orbital.

7.29 The magnetic quantum number (m_l) is an integer ranging from $-l$ to $+l$. For example, if $l = 1, m_l = -1, 0, +1$. The magnetic quantum number specifies the orientation of the orbital.

7.31 The probability density is the probability per unit volume of finding the electron at a point in space. The radial distribution function represents the total probability of finding the electron within a thin spherical shell at a distance r from the nucleus. In contrast to probability density, which has a maximum at the nucleus for an s orbital, the radial distribution function has a value of zero at the nucleus. It increases to a maximum and then decreases again with increasing r.

7.33 The sublevels are s ($l = 0$), which can hold a maximum of 2 electrons; p ($l = 1$), which can hold a maximum of 6 electrons; d ($l = 2$), which can hold a maximum of 10 electrons; and f ($l = 3$), which can hold a maximum of 14 electrons.

Problems by Topic

Electromagnetic Radiation

7.35 **Given:** distance to sun $= 1.496 \times 10^8$ km   **Find:** time for light to travel from sun to Earth
Conceptual Plan: distance km → distance m → time
$$\frac{1000 \text{ m}}{\text{km}} \qquad \text{time} = \frac{\text{distance}}{3.00 \times 10^8 \text{ m/s}}$$
Solution: $1.496 \times 10^8 \text{ km} \times \dfrac{1000 \text{ m}}{\text{km}} \times \dfrac{\text{s}}{3.00 \times 10^8 \text{ m}} = 499 \text{ s}$

Check: The units of the answer (s) are correct. The magnitude of the answer is reasonable because it corresponds to about 8 minutes.

7.37 (i) By increasing wavelength, the order is (d) ultraviolet < (c) infrared < (b) microwave < (a) radio waves.
(ii) By increasing energy, the order is (a) radio waves < (b) microwaves < (c) infrared < (d) ultraviolet.

7.39 (a) **Given:** $\lambda = 632.8$ nm   **Find:** frequency (ν)
Conceptual Plan: nm → m → ν
$$\frac{1 \text{ m}}{10^9 \text{ nm}} \qquad \nu = \frac{c}{\lambda}$$
Solution:
$$632.8 \text{ nm} \times \frac{1 \text{ m}}{10^9 \text{ nm}} = 6.328 \times 10^{-7} \text{ m}; \ \nu = \frac{3.00 \times 10^8 \text{ m}}{\text{s}} \times \frac{1}{6.328 \times 10^{-7} \text{ m}} = 4.74 \times 10^{14} \text{ s}^{-1}$$
Check: The units of the answer (s^{-1}) are correct. The magnitude of the answer seems reasonable because wavelength and frequency are inversely proportional.

     (b) **Given:** $\lambda = 503$ nm   **Find:** frequency (ν)
Conceptual Plan: nm → m → ν
$$\frac{1 \text{ m}}{10^9 \text{ nm}} \qquad \nu = \frac{c}{\lambda}$$
Solution: $503 \text{ nm} \times \dfrac{1 \text{ m}}{10^9 \text{ nm}} = 5.03 \times 10^{-7} \text{ m}; \ \nu = \dfrac{3.00 \times 10^8 \text{ m}}{\text{s}} \times \dfrac{1}{5.03 \times 10^{27} \text{ m}} = 5.96 \times 10^{14} \text{ s}^{-1}$
Check: The units of the answer (s^{-1}) are correct. The magnitude of the answer seems reasonable because wavelength and frequency are inversely proportional.

     (c) **Given:** $\lambda = 0.052$ nm   **Find:** frequency (ν)
Conceptual Plan: nm → m → ν
$$\frac{1 \text{ m}}{10^9 \text{ nm}} \qquad \nu = \frac{c}{\lambda}$$
Solution: $0.052 \text{ nm} \times \dfrac{1 \text{ m}}{10^9 \text{ nm}} = 5.2 \times 10^{-11} \text{ m}; \ \nu = \dfrac{3.00 \times 10^8 \text{ m}}{\text{s}} \times \dfrac{1}{5.2 \times 10^{-11} \text{ m}} = 5.8 \times 10^{18} \text{ s}^{-1}$
Check: The units of the answer (s^{-1}) are correct. The magnitude of the answer seems reasonable because wavelength and frequency are inversely proportional.

7.41   (a)   **Given:** frequency (ν) from Problem 7.39a $= 4.74 \times 10^{14}\,\text{s}^{-1}$ **Find:** Energy
                Conceptual Plan: $v \rightarrow E$

$$E = h\nu \quad h = 6.626 \times 10^{-34}\,\text{J}\cdot\text{s}$$

                Solution: $6.626 \times 10^{-34}\,\text{J}\cdot\cancel{\text{s}} \times \dfrac{4.74 \times 10^{14}}{\cancel{\text{s}}} = 3.14 \times 10^{-19}\,\text{J}$

                Check: The units of the answer (J) are correct. The magnitude of the answer is reasonable because we are talking about the energy of one photon.

       (b)   **Given:** frequency (ν) from Problem 7.39b $= 5.96 \times 10^{14}\,\text{s}^{-1}$ **Find:** Energy
                Conceptual Plan: $v \rightarrow E$

$$E = h\nu \quad h = 6.626 \times 10^{-34}\,\text{J}\cdot\text{s}$$

                Solution: $6.626 \times 10^{-34}\,\text{J}\cdot\cancel{\text{s}} \times \dfrac{5.96 \times 10^{14}}{\cancel{\text{s}}} = 3.95 \times 10^{-19}\,\text{J}$

                Check: The units of the answer (J) are correct. The magnitude of the answer is reasonable because we are talking about the energy of one photon.

       (c)   **Given:** frequency (ν) from Problem 7.39c $= 5.8 \times 10^{18}\,\text{s}^{-1}$ **Find:** Energy
                Conceptual Plan: $v \rightarrow E$

$$E = h\nu \quad h = 6.626 \times 10^{-34}\,\text{J}\cdot\text{s}$$

                Solution: $6.626 \times 10^{234}\,\text{J}\cdot\cancel{\text{s}} \times \dfrac{5.8 \times 10^{18}}{\cancel{\text{s}}} = 3.8 \times 10^{-15}\,\text{J}$

                Check: The units of the answer (J) are correct. The magnitude of the answer is reasonable because we are talking about the energy of one photon.

7.43   **Given:** $\lambda = 532\,\text{nm}$ and $E_{\text{pulse}} = 3.85\,\text{mJ}$ **Find:** number of photons
       Conceptual Plan: $\text{nm} \rightarrow \text{m} \rightarrow E_{\text{photon}} \rightarrow \text{number of photons}$

$$\frac{\text{m}}{10^9\,\text{nm}} \quad E = \frac{hc}{\lambda}; h = 6.626 \times 10^{-34}\,\text{J}\cdot\text{s} \quad \frac{E_{\text{pulse}}}{E_{\text{photon}}}$$

       Solution:

$$532\,\cancel{\text{nm}} \times \frac{\text{m}}{10^9\,\cancel{\text{nm}}} = 5.32 \times 10^{-7}\,\text{m}; \quad E = \frac{6.626 \times 10^{-34}\,\text{J}\cdot\cancel{\text{s}} \times \dfrac{3.00 \times 10^8\,\cancel{\text{m}}}{\cancel{\text{s}}}}{5.32 \times 10^{-7}\,\cancel{\text{m}}} = 3.7\underline{3}65 \times 10^{-19}\,\text{J/photon}$$

$$3.85\,\cancel{\text{mJ}} \times \frac{\cancel{\text{J}}}{1000\,\cancel{\text{mJ}}} \times \frac{1\,\text{photon}}{3.7\underline{3}65 \times 10^{-19}\,\cancel{\text{J}}} = 1.03 \times 10^{16}\,\text{photons}$$

       Check: The units of the answer (number of photons) are correct. The magnitude of the answer is reasonable for the amount of energy involved.

7.45   (a)   **Given:** $\lambda = 1500\,\text{nm}$ **Find:** E for 1 mol photons
                Conceptual Plan: $\text{nm} \rightarrow \text{m} \rightarrow E_{\text{photon}} \rightarrow E(\text{J})_{\text{mol}} \rightarrow E(\text{kJ})_{\text{mol}}$

$$\frac{\text{m}}{10^9\,\text{nm}} \quad E = \frac{hc}{\lambda}; h = 6.626 \times 10^{-34}\,\text{J}\cdot\text{s} \quad \frac{6.022 \times 10^{23}\,\text{photons}}{\text{mol}} \quad \frac{\text{kJ}}{1000\,\text{J}}$$

       Solution:

$$1500\,\cancel{\text{nm}} \times \frac{\text{m}}{10^9\,\cancel{\text{nm}}} = 1.500 \times 10^{-6}\,\text{m} \quad E = \frac{6.626 \times 10^{-34}\,\text{J}\cdot\cancel{\text{s}} \times \dfrac{3.00 \times 10^8\,\cancel{\text{m}}}{\cancel{\text{s}}}}{1.500 \times 10^{-6}\,\cancel{\text{m}}} = 1.3\underline{2}52 \times 10^{-19}\,\text{J/photon}$$

$$\frac{1.3\underline{2}52 \times 10^{-19}\,\cancel{\text{J}}}{\cancel{\text{photon}}} \times \frac{6.022 \times 10^{23}\,\cancel{\text{photons}}}{\text{mol}} \times \frac{\text{kJ}}{1000\,\cancel{\text{J}}} = 79.8\,\text{kJ/mol}$$

       Check: The units of the answer (kJ/mol) are correct. The magnitude of the answer is reasonable for a wavelength in the infrared region.

(b) **Given:** $\lambda = 500$ nm   **Find:** E for 1 mol photons
Conceptual Plan: nm $\rightarrow$ m $\rightarrow E_{photon} \rightarrow E_{mol} \rightarrow E(kJ)_{mol}$

$$\frac{m}{10^9 \text{ nm}} \quad E = \frac{hc}{\lambda}; h = 6.626 \times 10^{-34} \text{ J} \cdot \text{s} \quad \frac{6.022 \times 10^{23} \text{ photons}}{\text{mol}} \quad \frac{\text{kJ}}{1000 \text{ J}}$$

Solution:

$$500 \text{ nm} \times \frac{m}{10^9 \text{ nm}} = 5.00 \times 10^{27} \text{ m}; \quad E = \frac{6.626 \times 10^{234} \text{ J} \cdot \text{s} \times \frac{3.00 \times 10^8 \text{ m}}{\text{s}}}{5.00 \times 10^{27} \text{ m}} = 3.9756 \times 10^{-19} \text{ J/photon}$$

$$\frac{3.9756 \times 10^{-19} \text{ J}}{\text{photon}} \times \frac{6.022 \times 10^{23} \text{ photons}}{\text{mol}} \times \frac{\text{kJ}}{1000 \text{ J}} = 239 \text{ kJ/mol}$$

Check: The units of the answer (kJ/mol) are correct. The magnitude of the answer is reasonable for a wavelength in the visible region.

(c) **Given:** $\lambda = 150$ nm   **Find:** E for 1 mol photons
Conceptual Plan: nm $\rightarrow$ m $\rightarrow E_{photon} \rightarrow E_{mol} \rightarrow E(kJ)_{mol}$

$$\frac{m}{10^9 \text{ nm}} \quad E = \frac{hc}{\lambda}; h = 6.626 \times 10^{-34} \text{ J} \cdot \text{s} \quad \frac{6.022 \times 10^{23} \text{ photons}}{\text{mol}} \quad \frac{\text{kJ}}{1000 \text{ J}}$$

Solution:

$$1.50 \text{ nm} \times \frac{m}{10^9 \text{ nm}} = 1.50 \times 10^{-7} \text{ m}; \quad E = \frac{6.626 \times 10^{-34} \text{ J} \cdot \text{s} \times \frac{3.00 \times 10^8 \text{ m}}{\text{s}}}{1.50 \times 10^{-7} \text{ m}} = 1.3252 \times 10^{-18} \text{ J/photon}$$

$$\frac{1.3252 \times 10^{-18} \text{ J}}{\text{photon}} \times \frac{6.022 \times 10^{23} \text{ photons}}{\text{mol}} \times \frac{\text{kJ}}{1000 \text{ J}} = 798 \text{ kJ/mol}$$

Check: The units of the answer (kJ/mol) are correct. The magnitude of the answer is reasonable for a wavelength in the ultraviolet region. Note: The energy increases from the IR to the Vis to the UV as expected.

The Wave Nature of Matter and the Uncertainty Principle

7.47   The interference pattern would be a series of light and dark lines.

7.49   **Given:** $m = 9.109 \times 10^{-31}$ kg, $\lambda = 0.20$ nm   **Find:** v
Conceptual Plan: $m, \lambda \rightarrow v$

$$v = \frac{h}{m\lambda}$$

Solution: $\dfrac{6.626 \times 10^{-34} \frac{\text{kg} \cdot \text{m}^2}{\text{s}^2} \cdot \text{s}}{(9.109 \times 10^{-31} \text{ kg})(0.20 \text{ nm})\left(\dfrac{1 \text{ m}}{1 \times 10^9 \text{ nm}}\right)} = 3.6 \times 10^6 \text{ m/s}$

Check: The units of the answer (m/s) are correct. The magnitude of the answer is large, as would be expected for the speed of the electron.

7.51   **Given:** $m = 9.109 \times 10^{-31}$ kg; $v = 1.35 \times 10^5$ m/s   **Find:** λ

Conceptual Plan: $m, v \rightarrow \lambda$

$$\lambda = \frac{h}{mv}$$

Solution: $\dfrac{6.626 \times 10^{-34} \dfrac{\text{kg} \cdot \text{m}^2}{\text{s}^2} \cdot \text{s}}{(9.109 \times 10^{-31} \text{ kg})\left(\dfrac{1.35 \times 10^5 \text{ m}}{\text{s}}\right)} = 5.39 \times 10^{-9} \text{ m} = 5.39 \text{ nm}$$

Check: The units of the answer (m) are correct. The magnitude is reasonable because we are looking at an electron.

7.53   **Given:** $m = 143$ g; $v = 95$ mph   **Find:** λ

Conceptual Plan: $m, v \rightarrow \lambda$

$$\lambda = \frac{h}{mv}$$

Solution: $\dfrac{6.626 \times 10^{-34} \dfrac{\text{kg} \cdot \text{m}^2}{\text{s}^2} \cdot \text{s}}{(143 \text{ g})\left(\dfrac{\text{kg}}{1000 \text{ g}}\right)\left(\dfrac{95 \text{ mi}}{\text{hr}}\right)\left(\dfrac{1.609 \text{ km}}{\text{mi}}\right)\left(\dfrac{1000 \text{ m}}{\text{km}}\right)\left(\dfrac{\text{hr}}{3600 \text{ s}}\right)} = 1.1 \times 10^{-34} \text{ m}$$

The value of the wavelength $(1.1 \times 10^{-34}$ m$)$ is so small that it will not have an effect on the trajectory of the baseball.

Check: The units of the answer (m) are correct. The magnitude of the answer is very small, as would be expected for the de Broglie wavelength of a baseball.

7.55   **Given:** $\Delta x = 552$ pm, $m = 9.109 \times 10^{-31}$ kg   **Find:** Δv

Conceptual Plan: $\Delta x, m \rightarrow \Delta v$

$$\Delta x \times m\Delta v \geq \frac{h}{4\pi}$$

Solution: $\dfrac{6.626 \times 10^{-34} \dfrac{\text{kg} \cdot \text{m}^2}{\text{s}^2} \cdot \text{s}}{4(3.141)(9.109 \times 10^{-31} \text{ kg})(552 \text{ pm})\left(\dfrac{\text{m}}{1 \times 10^{12} \text{ pm}}\right)} = 1.05 \times 10^5 \text{ m/s}$$

Check: The units of the answer (m/s) are correct. The magnitude is reasonable for the uncertainty in the speed of an electron.

Orbitals and Quantum Numbers

7.57   Because the size of the orbital is determined by the n quantum, with the size increasing with increasing n, an electron in a $2s$ orbital is closer, on average, to the nucleus than is an electron in a $3s$ orbital.

7.59   The value of l is an integer that lies between 0 and $n - 1$.

   (a)   When $n = 1$, l can only be $l = 0$.
   (b)   When $n = 2$, l can be $l = 0$ or $l = 1$.
   (c)   When $n = 3$, l can be $l = 0$, $l = 1$, or $l = 2$.
   (d)   When $n = 4$, l can be $l = 0$, $l = 1$, $l = 2$, or $l = 3$.

7.61   Set c cannot occur together as a set of quantum numbers to specify an orbital. l must lie between 0 and $n - 1$; so for $n = 3$, l can only be as high as 2.

7.63   The 2s orbital would be the same shape as the 1s orbital but would be larger in size, and the 3p orbitals would have the same shape as the 2p orbitals but would be larger in size. Also, the 2s and 3p orbitals would have more nodes.

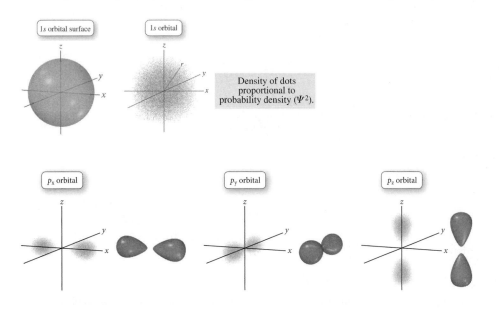

Atomic Spectroscopy

7.65   When the atom emits the photon of energy that was needed to raise the electron to the $n = 2$ level, the photon has the same energy as the energy absorbed to move the electron to the excited state. Therefore, the electron has to be in $n = 1$ (the ground state) following the emission of the photon.

7.67   According to the quantum-mechanical model, the higher the n level, the higher the energy. So the transition from $3p \rightarrow 1s$ would be a greater energy difference than a transition from $2p \rightarrow 1s$. The lower energy transition would have the longer wavelength. Therefore, the $2p \rightarrow 1s$ transition would produce a longer wavelength.

7.69   (a)   **Given:** $n = 2 \rightarrow n = 1$ **Find:** λ

        Conceptual Plan: $n = 1, n = 2 \rightarrow \Delta E_{atom} \rightarrow \Delta E_{photon} \rightarrow \lambda$

$$\Delta E_{atom} = E_1 - E_2 \quad \Delta E_{atom} \rightarrow -\Delta E_{photom} \quad E = \frac{hc}{\lambda}$$

        Solution:

$$\Delta E = E_1 - E_2$$

$$= -2.18 \times 10^{-18} \, J \left(\frac{1}{1^2} \right) - \left[-2.18 \times 10^{-18} \, J \left(\frac{1}{2^2} \right) \right] = -2.18 \times 10^{-18} \, J \left[\left(\frac{1}{1^2} \right) - \left(\frac{1}{2^2} \right) \right] = -1.6\underline{3}5 \times 10^{-18} \, J$$

$$\Delta E_{photon} = -\Delta E_{atom} = 1.6\underline{3}5 \times 10^{-18} \, J; \lambda = \frac{hc}{E} = \frac{(6.626 \times 10^{234} \, J \cdot s)(3.00 \times 10^8 \, m/s)}{1.6\underline{3}5 \times 10^{-18} \, J} = 1.22 \times 10^{-7} \, m$$

        This transition would produce a wavelength in the UV region.

        Check: The units of the answer (m) are correct. The magnitude of the answer is reasonable because it is in the region of UV radiation.

    (b)   **Given:** $n = 3 \rightarrow n = 1$ **Find:** λ

        Conceptual Plan: $n = 1, n = 3 \rightarrow \Delta E_{atom} \rightarrow \Delta E_{photon} \rightarrow \lambda$

$$\Delta E_{atom} = E_1 - E_3 \quad \Delta E_{atom} \rightarrow -\Delta E_{photom} \quad E = \frac{hc}{\lambda}$$

        Solution:

$$\Delta E = E_1 - E_3$$

$$= -2.18 \times 10^{-18} \text{ J}\left(\frac{1}{1^2}\right) - \left[-2.18 \times 10^{-18} \text{ J}\left(\frac{1}{3^2}\right)\right] = -2.18 \times 10^{-18} \text{ J}\left[\left(\frac{1}{1^2}\right) - \left(\frac{1}{3^2}\right)\right] = -1.9\underline{3}8 \times 10^{-18} \text{ J}$$

$$\Delta E_{\text{photon}} = -\Delta E_{\text{atom}} = 1.9\underline{3}8 \times 10^{-18} \text{ J}; \ \lambda = \frac{hc}{E} = \frac{(6.626 \times 10^{-34} \text{ J} \cdot \text{s})(3.00 \times 10^8 \text{ m/s})}{1.9\underline{3}8 \times 10^{-18} \text{ J}} = 1.03 \times 10^{-7} \text{ m}$$

This transition would produce a wavelength in the UV region.

Check: The units of the answer (m) are correct. The magnitude of the answer is reasonable because it is in the region of UV radiation.

(c)   **Given:** $n = 4 \rightarrow n = 2$ **Find:** λ
   Conceptual Plan: $n = 2, n = 4 \rightarrow \Delta E_{\text{atom}} \rightarrow \Delta E_{\text{photon}} \rightarrow \lambda$

$$\Delta E_{\text{atom}} = E_2 - E_4 \quad \Delta E_{\text{atom}} \rightarrow -\Delta E_{\text{photom}} \quad E = \frac{hc}{\lambda}$$

Solution:
$$\Delta E = E_2 - E_4$$

$$= 2.18 \times 10^{-18} \text{ J}\left(\frac{1}{2^2}\right) - \left[-2.18 \times 10^{-18} \text{ J}\left(\frac{1}{4^2}\right)\right] = -2.18 \times 10^{-18} \text{ J}\left[\left(\frac{1}{2^2}\right) - \left(\frac{1}{4^2}\right)\right] = -4.0\underline{8}8 \times 10^{-19} \text{ J}$$

$$\Delta E_{\text{photon}} = -\Delta E_{\text{atom}} = 4.0\underline{8}8 \times 10^{-19} \text{ J} \quad \lambda = \frac{hc}{E} = \frac{(6.626 \times 10^{-34} \text{ J} \cdot \text{s})(3.00 \times 10^8 \text{ m/s})}{4.0\underline{8}8 \times 10^{-19} \text{ J}} = 4.86 \times 10^{-7} \text{ m}$$

This transition would produce a wavelength in the visible region.

Check: The units of the answer (m) are correct. The magnitude of the answer is reasonable because it is in the region of visible light.

(d)   **Given:** $n = 5 \rightarrow n = 2$ **Find:** λ
   Conceptual Plan: $n = 2, n = 5 \rightarrow \Delta E_{\text{atom}} \rightarrow \Delta E_{\text{photon}} \rightarrow \lambda$

$$\Delta E_{\text{atom}} = E_2 - E_5 \quad \Delta E_{\text{atom}} \rightarrow -\Delta E_{\text{photom}} \quad E = \frac{hc}{\lambda}$$

Solution:
$$\Delta E = E_2 - E_5$$

$$= -2.18 \times 10^{-18} \text{ J}\left(\frac{1}{2^2}\right) - \left[-2.18 \times 10^{-18} \text{ J}\left(\frac{1}{5^2}\right)\right] = -2.18 \times 10^{-18} \text{ J}\left[\left(\frac{1}{2^2}\right) - \left(\frac{1}{5^2}\right)\right] = -4.5\underline{7}8 \times 10^{-19} \text{ J}$$

$$\Delta E_{\text{photon}} = -\Delta E_{\text{atom}} = 4.5\underline{7}8 \times 10^{-19} \text{ J} \quad \lambda = \frac{hc}{E} = \frac{(6.626 \times 10^{-34} \text{ J} \cdot \text{s})(3.00 \times 10^8 \text{ m/s})}{4.5\underline{7}8 \times 10^{-19} \text{ J}} = 4.34 \times 10^{-7} \text{ m}$$

This transition would produce a wavelength in the visible region.

Check: The units of the answer (m) are correct. The magnitude of the answer is reasonable because it is in the region of visible light.

7.71   **Given:** $n(\text{initial}) = 7; \lambda = 397 \text{ nm}$ **Find:** $n(\text{final})$
   Conceptual Plan: $\lambda \rightarrow \Delta E_{\text{photon}} \rightarrow \Delta E_{\text{atom}} \rightarrow n = x, n = 7$

$$E = \frac{hc}{\lambda} \quad \Delta E_{\text{photon}} \rightarrow -\Delta E_{\text{atom}} \quad \Delta E_{\text{atom}} = E_x - E_7$$

Solution: $E = \dfrac{hc}{\lambda} = \dfrac{(6.626 \times 10^{-34} \text{ J} \cdot \text{s})(3.00 \times 10^8 \text{ m/s})}{(397 \text{ nm})\left(\dfrac{\text{m}}{10^9 \text{ nm}}\right)} = 5.0\underline{0}7 \times 10^{-19} \text{ J}$

$$\Delta E_{\text{atom}} = -\Delta E_{\text{photon}} = -5.0\underline{0}7 \times 10^{-19} \text{ J}$$

$$\Delta E = E_x - E_7 = -5.0\underline{0}7 \times 10^{-19} = -2.18 \times 10^{-18} \text{ J}\left(\frac{1}{x^2}\right) - \left[-2.18 \times 10^{-18} \text{ J}\left(\frac{1}{7^2}\right)\right] = -2.18 \times 10^{-18} \text{ J}\left[\left(\frac{1}{x^2}\right) - \left(\frac{1}{7^2}\right)\right]$$

$$0.2297 = \left(\frac{1}{x^2}\right) - \left(\frac{1}{7^2}\right) \quad 0.25011 = \left(\frac{1}{x^2}\right) \quad x^2 = 3.998 \quad x = 2$$

Check: The answer is reasonable because it is an integer less than the initial value of 7.

Cumulative Problems

7.73   **Given:** 348 kJ/mol   **Find:** λ
Conceptual Plan: kJ/mol $\rightarrow$ kJ/molec $\rightarrow$ kJ/molec $\rightarrow$ λ

$$\frac{\text{mol C} - \text{C bonds}}{6.022 \times 10^{23}\,\text{C} - \text{C bonds}} \qquad \frac{1000\,\text{J}}{\text{kJ}} \qquad E = \frac{hc}{\lambda}$$

Solution: $\dfrac{348\,\cancel{\text{kJ}}}{\cancel{\text{mol C} - \text{C bonds}}} \times \dfrac{\cancel{\text{mol C} - \text{C bonds}}}{6.022 \times 10^{23}\,\cancel{\text{C} - \text{C bonds}}} \times \dfrac{1000\,\text{J}}{\cancel{\text{kJ}}} = 5.7\underline{7}9 \times 10^{-19}\,\text{J}$

$$\lambda = \frac{(6.626 \times 10^{-34}\,\cancel{\text{J}} \cdot \cancel{\text{s}})(3.00 \times 10^8\,\text{m}/\cancel{\text{s}})}{5.7\underline{7}9 \times 10^{-19}\,\cancel{\text{J}}} = 3.44 \times 10^{-7}\,\text{m} = 344\,\text{nm}$$

Check: The units of the answer (m or nm) are correct. The magnitude of the answer is reasonable because this wavelength is in the UV region.

7.75   **Given:** $E_{\text{pulse}} = 5.0$ watts; d $= 5.5$ mm; hole $= 1.2$ mm; $\lambda = 532$ nm   **Find:** photons/s
Conceptual Plan: fraction of beam through hole $\rightarrow$ fraction of power and then $E_{\text{photon}} \rightarrow$ number photons/s

$$\frac{\text{area hole}}{\text{area beam}} \qquad \text{fraction} \times \text{power} \qquad E = \frac{hc}{\lambda} \qquad \frac{\text{power/s}}{E/\text{photon}}$$

Solution: $A = \pi r^2 \quad \dfrac{\pi(0.60\,\cancel{\text{mm}})^2}{\pi(2.75\,\cancel{\text{mm}})^2} = 0.0476 \qquad 0.0476 \times 5.0\,\cancel{\text{watts}} \times \dfrac{\text{J/s}}{\cancel{\text{watt}}} = 0.2\underline{3}8\,\text{J/s}$

$$E_{\text{photon}} = \frac{(6.626 \times 10^{-34}\,\text{J} \cdot \cancel{\text{s}})(3.00 \times 10^8\,\cancel{\text{m}}/\cancel{\text{s}})}{(532\,\cancel{\text{nm}})\left(\dfrac{\cancel{\text{m}}}{10^9\,\cancel{\text{nm}}}\right)} = 3.7\underline{3}6 \times 10^{-19}\,\text{J/photon}$$

$$\frac{0.2\underline{3}8\,\cancel{\text{J}}/\text{s}}{3.7\underline{3}6 \times 10^{-19}\,\cancel{\text{J}}/\text{photon}} = 6.4 \times 10^{17}\,\text{photons/s}$$

Check: The units of the answer (number of photons/s) are correct. The magnitude of the answer is reasonable.

7.77   **Given:** KE $= 506$ eV   **Find:** λ
Conceptual Plan: $\text{KE}_{\text{ev}} \rightarrow \text{KE}_{\text{J}} \rightarrow v \rightarrow \lambda$

$$\frac{1.602 \times 10^{-19}\,\text{J}}{\text{eV}} \qquad KE = 1/2\,mv^2 \qquad \lambda = \frac{h}{mv}$$

Solution:

$$506\,\text{eV}\left(\frac{1.602 \times 10^{-19}\,\text{J}}{\text{eV}}\right)\left(\frac{\text{kg} \cdot \text{m}^2}{\dfrac{\text{s}^2}{\text{J}}}\right) = \frac{1}{2}(9.11 \times 10^{-31}\,\text{kg})\,v^2$$

$$v^2 = \frac{506\,\cancel{\text{eV}}\left(\dfrac{1.602 \times 10^{-19}\,\cancel{\text{J}}}{\cancel{\text{eV}}}\right)\left(\dfrac{\text{kg} \cdot \text{m}^2}{\dfrac{\text{s}^2}{\cancel{\text{J}}}}\right)}{\dfrac{1}{2}(9.11 \times 10^{-31}\,\cancel{\text{kg}})} = 1.7796 \times 10^{14}\,\frac{\text{m}^2}{\text{s}^2}$$

$$v = 1.33 \times 10^7\,\text{m/s} \qquad \lambda = \frac{h}{mv} = \frac{6.626 \times 10^{-34}\,\dfrac{\cancel{\text{kg}} \cdot \text{m}^2}{\cancel{\text{s}^2}} \cdot \cancel{\text{s}}}{(9.11 \times 10^{-31}\,\cancel{\text{kg}})(1.33 \times 10^7\,\cancel{\text{m}}/\cancel{\text{s}})} = 5.47 \times 10^{-11}\,\text{m} = 0.0547\,\text{nm}$$

Check: The units of the answer (m or nm) are correct. The magnitude of the answer is reasonable because a de Broglie wavelength is usually a very small number.

7.79   **Given:** $n = 1 \rightarrow n = \infty$ **Find:** E; λ
Conceptual Plan: $n = \infty, n = 1 \rightarrow \Delta E_{\text{atom}} \rightarrow \Delta E_{\text{photon}} \rightarrow \lambda$

$$\Delta E_{\text{atom}} = E_\infty - E_1 \qquad \Delta E_{\text{atom}} = \Delta E_{\text{photon}} \qquad E = \frac{hc}{\lambda}$$

Solution: $\Delta E = E_\infty - E_1 = 0 - \left[-2.18 \times 10^{-18} \text{J} \left(\frac{1}{1^2} \right) \right] = +2.18 \times 10^{-18}$ J

$$\Delta E_{\text{photon}} = -\Delta E_{\text{atom}} = +2.18 \times 10^{-18} \text{ J}$$

$$\lambda = \frac{hc}{E} = \frac{(6.626 \times 10^{-34} \text{ J} \cdot \text{s})(3.00 \times 10^8 \text{ m/s})}{2.18 \times 10^{-18} \text{ J}} = 9.12 \times 10^{-8} \text{ m} = 91.2 \text{ nm}$$

Check: The units of the answers (J for E and m or nm for the first part) are correct. The magnitude of the answer is reasonable because it would require more energy to completely remove the electron than just moving it to a higher n level. This results in a shorter wavelength.

7.81 (a) **Given:** $n = 1$ **Find:** number of orbitals if $l = 0 \rightarrow n$
      Conceptual Plan: value $n \rightarrow$ **values** $l \rightarrow$ **values** $m_l \rightarrow$ **number of orbitals**

$l = 0 \rightarrow n$ $m_l = -1 \rightarrow +1$ total m_l

Solution: $n =$ 1
$l =$ 0   1
$m_l =$ 0 $-1, 0, +1$
total 4 orbitals

Check: The total orbitals will be equal to the (number of l sublevels)^2.

(b) **Given:** $n = 2$ **Find:** number of orbitals if $l = 0 \rightarrow n$
     Conceptual Plan: value n $\rightarrow$ **values** $l \rightarrow$ **values** $m_l \rightarrow$ **number of orbitals**

$l = 0 \rightarrow n$ $m_l = -1 \rightarrow +1$ total m_l

Solution: $n =$ 2
$l =$ 0   1   2
$m_l =$ 0 $-1, 0, +1$ $-2, -1, 0, 1, 2$
total 9 orbitals

Check: The total orbitals will be equal to the (number of l sublevels)^2.

(c) **Given:** $n = 3$ **Find:** number of orbitals if $l = 0 \rightarrow n$
     Conceptual Plan: value $n \rightarrow$ **values** $l \rightarrow$ **values** $m_l \rightarrow$ **number of orbitals**

$l = 0 \rightarrow n$ $m_l = -1 \rightarrow +1$ total m_l

Solution:

$n =$ 3
$l =$ 0   1   2   3
$m_l =$ 0 $-1, 0, +1$ $-2, -1, 0, 1, 2$ $-3, -2, -1, 0, 1, 2, 3$
total 16 orbitals

Check: The total orbitals will be equal to the (number of l sublevels)^2.

7.83 **Given:** $\lambda = 1875$ nm; 1282 nm; 1093 nm   **Find:** equivalent transitions
      Conceptual Plan: $\lambda \rightarrow E_{\text{photon}} \rightarrow E_{\text{atom}} \rightarrow n$

$E = \frac{hc}{\lambda}$ $E_{\text{photon}} = -E_{\text{atom}}$ $E = -2.18 \times 10^{-18} \text{ J} \left(\frac{1}{n_f^2} - \frac{1}{n_i^2} \right)$

Solution: Because the wavelength of the transitions are longer wavelengths than those obtained in the visual region, the electron must relax to a higher n level. Therefore, we can assume that the electron returns to the $n = 3$ level.

For $\lambda = 1875$ nm: $E = \dfrac{(6.626 \times 10^{-34} \text{ J} \cdot \text{s})(3.00 \times 10^8 \text{ m/s})}{1875 \text{ nm} \left(\dfrac{\text{m}}{10^9 \text{ nm}} \right)} = 1.060 \times 10^{-19}$ J 1.060×10^{-19} J $= -1.060 \times 10^{-19}$ J

$-1.060 \times 10^{-19} \text{ J} = -2.18 \times 10^{-18} \text{ J} \left(\dfrac{1}{3^2} - \dfrac{1}{n^2} \right); n = 4$

For $\lambda = 1282$ nm: $E = \dfrac{(6.626 \times 10^{-34} \text{ J} \cdot \text{s})(3.00 \times 10^8 \text{ m/s})}{1282 \text{ nm} \left(\dfrac{\text{m}}{10^9 \text{ nm}} \right)} = 1.551 \times 10^{-19}$ J 1.551×10^{-19} J $= -1.551 \times 10^{-19}$ J

$-1.551 \times 10^{-19} \text{ J} = -2.18 \times 10^{-18} \text{ J} \left(\dfrac{1}{3^2} - \dfrac{1}{n^2} \right); n = 5$

For $\lambda = 1093$ nm: $E = \dfrac{(6.626 \times 10^{-34}\,\text{J}\cdot\text{s})(3.00 \times 10^8\,\text{m/s})}{1093\,\text{nm}\left(\dfrac{\text{m}}{10^9\,\text{nm}}\right)} = 1.819 \times 10^{-19}\,\text{J};\quad 1.819 \times 10^{-19}\,\text{J} = -1.819 \times 10^{-19}\,\text{J}$

$$-1.819 \times 10^{-19}\,\text{J} = -2.18 \times 10^{-18}\,\text{J}\left(\dfrac{1}{3^2} - \dfrac{1}{n^2}\right); n = 6$$

Check: The values obtained are all integers, which is correct. The values of n (4, 5, and 6) are reasonable. The values of n increase as the wavelength decreases because the two n levels involved are farther apart and more energy is released as the electron relaxes to the $n = 3$ level.

7.85 **Given:** $\Phi = 193$ kJ/mol **Find:** threshold frequency (ν)

Conceptual Plan: Φ **KJ/mol** $\rightarrow \Phi$ **KJ/atom** $\rightarrow \Phi$ **J/atom** $\rightarrow \nu$

$$\dfrac{\text{mol}}{6.022 \times 10^{23}\,\text{atoms}} \qquad \dfrac{1000\,\text{J}}{\text{kJ}} \qquad \Phi = h\nu$$

Solution: $\nu = \dfrac{\Phi}{h} = \dfrac{\left(\dfrac{193\,\text{kJ}}{\text{mol}}\right)\left(\dfrac{\text{mol}}{6.022 \times 10^{23}\,\text{atoms}}\right)\left(\dfrac{1000\,\text{J}}{\text{kJ}}\right)}{6.626 \times 10^{-34}\,\text{J}\cdot\text{s}} = 4.84 \times 10^{14}\,\text{s}^{-1}$

Check: The units of the answer (s^{-1}) are correct. The magnitude of the answer puts the frequency in the infrared range and is a reasonable answer.

7.87 **Given:** $\nu_{\text{low}} = 30\text{s}^{-1}$ $\nu_{\text{hi}} = 1.5 \times 10^4\,\text{s}^{-1}$; speed $= 344$ m/s **Find:** $\lambda_{\text{low}} - \lambda_{\text{hi}}$

Conceptual Plan: $\nu_{\text{low}} \rightarrow \lambda_{\text{low}}$ **and** $\nu_{\text{hi}} = \lambda_{\text{hi}}$ **then** $\lambda_{\text{low}} - \lambda_{\text{hi}}$

$$\lambda\nu = \text{speed}$$

Solution: $\lambda = \dfrac{\text{speed}}{\nu}$; $\lambda_{\text{low}} = \dfrac{344\,\text{m/s}}{30\,\text{s}^{-1}} = 11$ m; $\lambda_{\text{hi}} = \dfrac{344\,\text{m/s}}{1.5 \times 10^4\,\text{s}^{-1}} = 0.023$ m; 11 m $- 0.023$ m $= 11$ m

Check: The units of the answer (m) are correct. The magnitude is reasonable because the value is only determined by the low-frequency value because of significant figures.

7.89 **Given:** $\lambda = 792$ nm; $V = 100.0$ mL; $P = 55.7$ mtorr; $T = 25\,°\text{C}$ **Find:** E to dissociate 15.0%

Conceptual Plan: $\lambda \rightarrow E/\text{molecule}$ **and then** $P,V,T \rightarrow n \rightarrow$ **molecules**

$$E = \dfrac{hc}{\lambda} \qquad\qquad n = \dfrac{PV}{RT} \quad \dfrac{6.022 \times 10^{23}\,\text{molecules}}{\text{mole}}$$

Solution: $E = \dfrac{(6.626 \times 10^{-34}\,\text{J}\cdot\text{s})(3.00 \times 10^8\,\text{m/s})}{792\,\text{nm}\left(\dfrac{\text{m}}{10^9\,\text{nm}}\right)} = 2.51 \times 10^{-19}\,\text{J/molecule}$

$PV = nRT$. Rearrange to solve for n, so $n = PV/RT =$

$$\dfrac{(55.7\,\text{mtorr})\left(\dfrac{1\,\text{torr}}{1000\,\text{mtorr}}\right)\left(\dfrac{1\,\text{atm}}{760\,\text{torr}}\right)(100.0\,\text{mL})\left(\dfrac{\text{L}}{1000\,\text{mL}}\right)}{\left(\dfrac{0.0821\,\text{L}\cdot\text{atm}}{\text{mol}\cdot\text{K}}\right)(298\,\text{K})} = 2.7\underline{5}503 \times 10^{-4}\,\text{mol}$$

$2.7\underline{5}503 \times 10^{-4}\,\text{mol} \times \left(\dfrac{6.022 \times 10^{23}\,\text{molecules}}{\text{mol}}\right) = 1.80 \times 10^{17}\,\text{molecules}$

$(1.80 \times 10^{17}\,\text{molecules})(0.15\%) = 2.70 \times 10^{16}\,\text{molecules dissociated}$

$(2.51 \times 10^{-19}\,\text{J/molecule})(2.70 \times 10^{16}\,\text{molecules}) = 6.777 \times 10^{-3}\,\text{J} = 6.78 \times 10^{-3}\,\text{J}$

Check: The units of the answer (J) are correct. The magnitude is reasonable because it is for a part of a mole of molecules.

7.91 **Given:** 20.0 mW; 1.00 hr.; 2.29×10^{20} photons **Find:** λ

Conceptual Plan: $\text{mW} \rightarrow \text{W} \rightarrow \text{J} \rightarrow \text{J/photon} \rightarrow \lambda$

$$\dfrac{\text{W}}{1000\,\text{mW}} \quad E = \text{W} \times \text{s} \quad \dfrac{E}{\text{number of photons}} \quad \lambda = \dfrac{hc}{E}$$

Solution: $(20.0 \, \text{mW})\left(\dfrac{1 \, \text{W}}{1000 \, \text{mW}}\right)\left(\dfrac{\frac{\text{J}}{\text{s}}}{\text{W}}\right)\left(\dfrac{3600 \, \text{s}}{2.29 \times 10^{20} \, \text{photons}}\right) = 3.14 \times 10^{-19} \, \text{J/photon}$

$$\dfrac{(6.626 \times 10^{-34} \, \text{J} \cdot \text{s})(3.00 \times 10^{8} \, \text{m/s})\left(\dfrac{10^{9} \, \text{nm}}{\text{m}}\right)}{3.14 \times 10^{-19} \, \text{J}} = 632 \, \text{nm}$$

Check: The units of the answer (nm) are correct. The magnitude is reasonable because it is in the red range.

Challenge Problems

7.93   (a)   **Given:** $n = 1; n = 2; n = 3; \text{L} = 155 \, \text{pm}$ **Find:** E_1, E_2, E_3
            Conceptual Plan: $n \rightarrow E$

$$E_n = \dfrac{n^2 h^2}{8 \, m \, \text{L}^2}$$

Solution:

$$E_1 = \dfrac{1^2 (6.626 \times 10^{-34} \, \text{J} \cdot \text{s})^2}{8(9.11 \times 10^{-31} \, \text{kg})(155 \, \text{pm})^2\left(\dfrac{\text{m}}{10^{12} \, \text{pm}}\right)^2} = \dfrac{1(6.626 \times 10^{-34})^2 \, \text{J}^2 \text{s}^2}{8(9.11 \times 10^{-31} \, \text{kg})(155 \times 10^{-12})^2 \, \text{m}^2}$$

$$= \dfrac{1(6.626 \times 10^{-34})^2 \left(\dfrac{\text{kg} \cdot \text{m}^2}{\text{s}^2}\right) \text{J} \, \text{s}^2}{8(9.11 \times 10^{-31} \, \text{kg})(155 \times 10^{-12})^2 \, \text{m}^2} = 2.51 \times 10^{-18} \, \text{J}$$

$$E_2 = \dfrac{2^2 (6.626 \times 10^{-34} \, \text{J} \cdot \text{s})^2}{8(9.11 \times 10^{-31} \, \text{kg})(155 \, \text{pm})^2\left(\dfrac{\text{m}}{10^{12} \, \text{pm}}\right)^2} = \dfrac{4(6.626 \times 10^{-34})^2 \, \text{J}^2 \, \text{s}^2}{8(9.11 \times 10^{-31} \, \text{kg})(155 \times 10^{-12})^2 \, \text{m}^2}$$

$$= \dfrac{4(6.626 \times 10^{-34})^2 \left(\dfrac{\text{kg} \cdot \text{m}^2}{\text{s}^2}\right) \text{J} \, \text{s}^2}{8(9.11 \times 10^{-31} \, \text{kg})(155 \times 10^{-12})^2 \, \text{m}^2} = 1.00 \times 10^{-17} \, \text{J}$$

$$E_3 = \dfrac{3^2 (6.626 \times 10^{-34} \, \text{J} \cdot \text{s})^2}{8(9.11 \times 10^{-31} \, \text{kg})(155 \, \text{pm})^2\left(\dfrac{\text{m}}{10^{12} \, \text{pm}}\right)^2} = \dfrac{9(6.626 \times 10^{-34})^2 \, \text{J}^2 \, \text{s}^2}{8(9.11 \times 10^{-31} \, \text{kg})(155 \times 10^{-12})^2 \, \text{m}^2}$$

$$= \dfrac{9(6.626 \times 10^{-34})^2 \left(\dfrac{\text{kg} \cdot \text{m}^2}{\text{s}^2}\right) \text{J} \, \text{s}^2}{8(9.11 \times 10^{-31} \, \text{kg})(155 \times 10^{-12})^2 \, \text{m}^2} = 2.26 \times 10^{-17} \, \text{J}$$

Check: The units of the answers (J) are correct. The answers seem reasonable because the energy is increasing with increasing n level.

(b)   **Given:** $n = 1 \rightarrow n = 2$ and $n = 2 \rightarrow n = 3$ **Find:** λ
        Conceptual Plan: $n = 1, n = 2 \rightarrow \Delta E_{\text{atom}} \rightarrow \Delta E_{\text{photon}} \rightarrow \lambda$

$$\Delta E_{\text{atom}} = E_2 - E_1 \qquad \Delta E_{\text{atom}} \rightarrow -\Delta E_{\text{photon}} \qquad E = \dfrac{hc}{\lambda}$$

Solution: Use the energies calculated in Part a.

$$E_2 - E_1 = (1.00 \times 10^{-17} \, \text{J} - 2.51 \times 10^{-18} \, \text{J}) = 7.49 \times 10^{-18} \, \text{J}$$

$$\lambda = \dfrac{(6.626 \times 10^{-34} \, \text{J} \cdot \text{s})(3.00 \times 10^{8} \, \text{m/s})}{7.49 \times 10^{-18} \, \text{J}} = 2.65 \times 10^{-8} \, \text{m} = 26.5 \, \text{nm}$$

$$E_3 - E_2 = (2.26 \times 10^{-17} \, \text{J} - 1.00 \times 10^{-17} \, \text{J}) = 1.26 \times 10^{-17} \, \text{J}$$

$$\lambda = \dfrac{(6.626 \times 10^{-34} \, \text{J} \cdot \text{s})(3.00 \times 10^{8} \, \text{m/s})}{1.26 \times 10^{-17} \, \text{J}} = 1.58 \times 10^{-8} \, \text{m} = 15.8 \, \text{nm}$$

These wavelengths would lie in the UV region.

Check: The units of the answers (m) are correct. The magnitude of the answers is reasonable based on the energies obtained for the levels.

7.95   For the 1s orbital in the Excel(R) spreadsheet, call the columns as follows: column A as r and column B as $\Psi(1s)$. Make the values for r column A as follows: 0–200. In column B, put the equation for the wave function written as follows: = (POWER(1/3.1415,1/2))*(1/POWER(53,3/2))*(EXP(-A2/53)). Go to make chart, choose *xy* scatter.

e.g., sample values	
r	$\Psi(1s)$
0	7.000146224
1	7.000143491
2	7.000140809
3	7.000138177
4	7.000135594
5	7.00013306
6	7.000130573

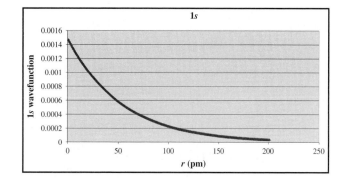

For the 2s orbital in the same Excel spreadsheet, call the columns as follows: column A as r and column C as $\Psi(2s)$. Use the same values for r in column A: 0–200. In column C, put the equation for the wave function written as follows: = (POWER(1/((32)*(3.1415)),1/2))*(1/POWER(53,3/2))*(2-(A2/53))*(EXP(-A2/53)). Go to make chart; choose *xy* scatter.

e.g., sample values	
r	$\Psi(2s)$
0	7. 0000516979
1	7. 000050253
2	7. 0000488441
3	7. 0000474702
4	7. 0000461307
5	7. 0000448247
6	7. 0000435513

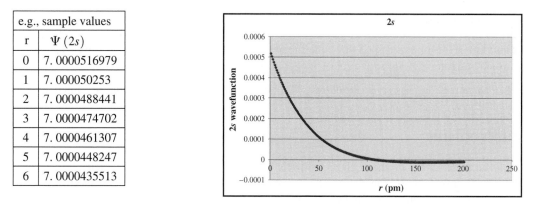

Note: The plot for the 2s orbital extends below the *x*-axis. The *x*-intercept represents the radial node of the orbital.

7.97   **Given:** threshold frequency = 2.25×10^{14} s^{-1}; $\lambda = 5.00 \times 10^{-7}$ m   **Find:** v of electron
Conceptual Plan: $v \rightarrow \Phi$ **and then** $\lambda \rightarrow E$ **and then** $\rightarrow$ **KE** $\rightarrow v$

$$\Phi = hv \qquad E = \frac{hc}{\lambda} \qquad KE = E - \Phi \quad KE = 1/2\, mv^2$$

Solution:

$$\Phi = (6.626 \times 10^{-34}\,\text{J} \cdot \text{s})(2.25 \times 10^{14}\,\text{s}^{-1}) = 1.4\underline{9}1 \times 10^{-19}\,\text{J}; \quad E = \frac{(6.626 \times 10^{-34}\,\text{J} \cdot \text{s})(3.00 \times 10^{8}\,\text{m/s})}{5.00 \times 10^{-7}\,\text{m}} = 3.9\underline{7}6 \times 10^{-19}\,\text{J}$$

$$KE = 3.9\underline{7}6 \times 10^{-19}\,\text{J} - 1.4\underline{9}1 \times 10^{-19}\,\text{J} = 2.485 \times 10^{-19}\,\text{J}; \quad KE = 1/2mv^2; \text{ Substituting and rearranging gives}$$

$$v^2 = \frac{2.485 \times 10^{-19}\,\dfrac{\text{kg} \cdot \text{m}^2}{\text{s}^2}}{\dfrac{1}{2}(9.11 \times 10^{-31}\,\text{kg})} = 5.455 \times 10^{11}\,\frac{\text{m}^2}{\text{s}^2}$$

$$v = 7.39 \times 10^{5}\,\text{m/s}$$

Check: The units of the answer (m/s) are correct. The magnitude of the answer is reasonable for the speed of an electron.

7.99   **Given:** $t = 5.0 \text{ fs}$; $\lambda_{\text{low}} = 722 \text{ nm}$ **Find:** ΔE and λ_{high}

Conceptual Plan: $t \rightarrow \Delta E$ and then $\lambda_{\text{low}} \rightarrow E_{\text{high}} \rightarrow E_{\text{low}} \rightarrow \lambda_{\text{high}}$

$$\Delta t \times \Delta E \geq \frac{h}{4\pi} \qquad E = \frac{hc}{\lambda} \quad E - \Delta E \quad \lambda = \frac{hc}{E}$$

Solution: $\dfrac{6.626 \times 10^{-34} \text{ J} \cdot \cancel{s}}{4(3.141)(5.0 \cancel{\text{ fs}})\left(\dfrac{\cancel{s}}{1 \times 10^{15} \cancel{\text{ fs}}}\right)} = 1.0\underline{5}5 \times 10^{-20} \text{ J}$

$$E = \frac{(6.626 \times 10^{-34} \text{ J} \cdot \cancel{s})(3.00 \times 10^8 \cancel{m}/\cancel{s})}{722 \cancel{\text{ nm}}\left(\dfrac{\cancel{m}}{10^9 \cancel{\text{ nm}}}\right)} = 2.7\underline{5} \times 10^{-19} \text{ J}$$

$2.75 \times 10^{-19} \text{ J} - 1.06 \times 10^{-20} \text{ J} = 2.\underline{6}4 \times 10^{-19} \text{ J}$

$$\frac{(6.626 \times 10^{-34} \text{ J} \cdot \cancel{s})(3.00 \times 10^8 \cancel{m}/\cancel{s})\left(\dfrac{10^9 \text{ nm}}{\cancel{m}}\right)}{(2.64 \times 10^{-19} \cancel{J})} = 751.8 \text{ nm} = 7.5 \times 10^2 \text{ nm}$$

Check: The units of the answer (nm) are correct. The magnitude of the answer is reasonable because it is a longer wavelength but it is close to the original wavelength.

7.101   **Given:** $r = 1.8 \text{ m}$ **Find:** λ

Conceptual Plan: $r \rightarrow C$

$$C = 2\pi r$$

Solution: $(2)(3.141)(1.8\text{m}) = 1\underline{1}.3 \text{ m} =$ the circumference of the orbit. So the largest wavelength that would fit the orbit would be 11 m.

Check: The units of the answer (m) are correct. The magnitude of the wave is about the circumference of the orbit.

Conceptual Problems

7.103   In the Bohr model of the atom, the electron travels in a circular orbit around the nucleus. It is a two-dimensional model. The electron is constrained to move only from one orbit to another orbit. But the electron is treated as a particle that behaves according to the laws of classical physics.

The quantum-mechanical model of the atom is three-dimensional. In this model, we treat the electron, an absolutely small particle, differently than we treat particles with classical physics. The electron is in an orbital, which gives us the probability of finding the electron within a volume of space.

Because the electron in the Bohr model is constrained to a circular orbit, it would theoretically be possible to know both the position and the velocity of the electron simultaneously. This contradicts the Heisenberg uncertainty principle, which states that position and velocity are complementary terms that cannot both be known with precision. So in the quantum mechanical model we can only know the position or the velocity of an electron at any given time, never both.

7.105   (a)   Because the interference pattern is caused by single electrons interfering with themselves, the pattern remains the same even when the rate of the electrons passing through the slits is one electron per minute. It will simply take longer for the full pattern to develop.

(b)   When a light is placed behind the slits, it flashes to indicate which hole the electron passed through, but the interference pattern is now absent. With the laser on, the electrons hit positions directly behind each slit, as if they were ordinary particles.

(c)   Diffraction occurs when a wave encounters an obstacle of a slit that is comparable in size to its wavelength. The wave bends around the slit. The diffraction of light through two slits separated by a distance comparable to the wavelength of the light results in an interference pattern. Each slit acts as a new wave source, and the two new waves interfere with each other, which results in a pattern of bright and dark lines.

(d)   Because the mass of the bullets and their particle size are not absolutely small, the bullets will not produce an interference pattern when they pass through the slits. The de Broglie wavelength produced by the bullets will not be sufficiently large enough to interfere with the bullet trajectory, and no interference pattern will be observed.

8 Periodic Properties of the Elements

Review Questions

8.1   A periodic property is predictable based on the element's position within the periodic table.

8.3   The first attempt to organize the elements according to similarities in their properties was made by the German chemist Johann Döbereiner. He grouped elements into triads: three elements with similar properties. A more complex approach was attempted by the English chemist John Newlands. He organized elements into octaves, analogous to musical notes. When arranged this way, the properties of every eighth element were similar.

8.5   Meyer proposed an organization of the known elements based on some periodic properties. Moseley listed elements according to the atomic number rather than the atomic mass. This resolved the problems in Mendeleev's table where an increase in atomic mass did not correlate with similar properties.

8.7   An electron configuration shows the particular orbitals that are occupied by electrons in an atom. Some examples are $H = 1s^1$, $He = 1s^2$, and $Li = 1s^2 2s^1$.

8.9   Shielding or screening occurs when one electron is blocked from the full effects of the nuclear charge so that the electron experiences only a part of the nuclear charge. It is the inner (core) electrons that shield the outer electrons from the full nuclear charge.

8.11   The sublevels within a principle level split in multielectron atoms because of penetration of the outer electrons into the region of the core electrons. The sublevels in hydrogen are not split because they are empty in the ground state.

8.13   The Pauli exclusion principle states the following: No two electrons in an atom can have the same four quantum numbers. Because two electrons occupying the same orbital have three identical quantum numbers (n, l, m_l), they must have different spin quantum numbers. The Pauli exclusion principle implies that each orbital can have a maximum of only two electrons, with opposing spins.

8.15   In order of increasing energy the orbitals are $1s < 2s < 2p < 3s < 3p < 4s < 3d < 4p < 5s$. The $4s$ orbital fills before the $3d$ and the $5s$ fills before the $4d$. They are lower in energy because of greater penetration of the $4s$ and $5s$ orbitals.

8.17

Groups

Legend:
- s-block elements
- d-block elements
- p-block elements
- f-block elements

Main table (Periods 1–7):

Period	1 / 1A	2 / 2A	3 / 3B	4 / 4B	5 / 5B	6 / 6B	7 / 7B	8 (8B)	9 (8B)	10 (8B)	11 / 1B	12 / 2B	13 / 3A	14 / 4A	15 / 5A	16 / 6A	17 / 7A	18 / 8A
1	1 H $1s^1$																	2 He $1s^2$
2	3 Li $2s^1$	4 Be $2s^2$											5 B $2s^2 2p^1$	6 C $2s^2 2p^2$	7 N $2s^2 2p^3$	8 O $2s^2 2p^4$	9 F $2s^2 2p^5$	10 Ne $2s^2 2p^6$
3	11 Na $3s^1$	12 Mg $3s^2$											13 Al $3s^2 3p^1$	14 Si $3s^2 3p^2$	15 P $3s^2 3p^3$	16 S $3s^2 3p^4$	17 Cl $3s^2 3p^5$	18 Ar $3s^2 3p^6$
4	19 K $4s^1$	20 Ca $4s^2$	21 Sc $4s^2 3d^1$	22 Ti $4s^2 3d^2$	23 V $4s^2 3d^3$	24 Cr $4s^1 3d^5$	25 Mn $4s^2 3d^5$	26 Fe $4s^2 3d^6$	27 Co $4s^2 3d^7$	28 Ni $4s^2 3d^8$	29 Cu $4s^1 3d^{10}$	30 Zn $4s^2 3d^{10}$	31 Ga $4s^2 4p^1$	32 Ge $4s^2 4p^2$	33 As $4s^2 4p^3$	34 Se $4s^2 4p^4$	35 Br $4s^2 4p^5$	36 Kr $4s^2 4p^6$
5	37 Rb $5s^1$	38 Sr $5s^2$	39 Y $5s^2 4d^1$	40 Zr $5s^2 4d^2$	41 Nb $5s^1 4d^4$	42 Mo $5s^1 4d^5$	43 Tc $5s^2 4d^5$	44 Ru $5s^1 4d^7$	45 Rh $5s^1 4d^8$	46 Pd $4d^{10}$	47 Ag $5s^1 4d^{10}$	48 Cd $5s^2 4d^{10}$	49 In $5s^2 5p^1$	50 Sn $5s^2 5p^2$	51 Sb $5s^2 5p^3$	52 Te $5s^2 5p^4$	53 I $5s^2 5p^5$	54 Xe $5s^2 5p^6$
6	55 Cs $6s^1$	56 Ba $6s^2$	57 La $6s^2 5d^1$	72 Hf $6s^2 5d^2$	73 Ta $6s^2 5d^3$	74 W $6s^2 5d^4$	75 Re $6s^2 5d^5$	76 Os $6s^2 5d^6$	77 Ir $6s^2 5d^7$	78 Pt $6s^1 5d^9$	79 Au $6s^1 5d^{10}$	80 Hg $6s^2 5d^{10}$	81 Tl $6s^2 6p^1$	82 Pb $6s^2 6p^2$	83 Bi $6s^2 6p^3$	84 Po $6s^2 6p^4$	85 At $6s^2 6p^5$	86 Rn $6s^2 6p^6$
7	87 Fr $7s^1$	88 Ra $7s^2$	89 Ac $7s^2 6d^1$	104 Rf $7s^2 6d^2$	105 Db $7s^2 6d^3$	106 Sg $7s^2 6d^4$	107 Bh	108 Hs	109 Mt	110 Ds	111 Rg	112	113	114	115	116		

Lanthanides:

58 Ce $6s^2 4f^1 5d^1$	59 Pr $6s^2 4f^3$	60 Nd $6s^2 4f^4$	61 Pm $6s^2 4f^5$	62 Sm $6s^2 4f^6$	63 Eu $6s^2 4f^7$	64 Gd $6s^2 4f^7 5d^1$	65 Tb $6s^2 4f^9$	66 Dy $6s^2 4f^{10}$	67 Ho $6s^2 4f^{11}$	68 Er $6s^2 4f^{12}$	69 Tm $6s^2 4f^{13}$	70 Yb $6s^2 4f^{14}$	71 Lu $6s^2 4f^{14} 6d^1$

Actinides:

90 Th $7s^2 6d^2$	91 Pa $7s^2 5f^2 6d^1$	92 U $7s^2 5f^3 6d^1$	93 Np $7s^2 5f^4 6d^1$	94 Pu $7s^2 5f^6$	95 Am $7s^2 5f^7$	96 Cm $7s^2 5f^7 6d^1$	97 Bk $7s^2 5f^9$	98 Cf $7s^2 5f^{10}$	99 Es $7s^2 5f^{11}$	100 Fm $7s^2 5f^{12}$	101 Md $7s^2 5f^{13}$	102 No $7s^2 5f^{14}$	103 Lr $7s^2 5f^{14} 6d^1$

8.19   The rows in the periodic table grow progressively longer because you are adding sublevels as the n level increases.

8.21   The row number of a main-group element is equal to the highest principal quantum number of that element. However, the principal quantum number of the d orbital being filled across each row in the transition series is equal to the row number minus one. For the inner transition elements, the principal quantum number of the f orbital being filled across each row is the row number minus two.

8.23   To use the periodic table to write the electron configuration, find the noble gas that precedes the element. The element has the inner electron configuration of that noble gas. Place the symbol for the noble gas in [ ]. Obtain the outer electron configuration by tracing the element across the period and assigning electrons in the appropriate orbitals.

8.25   (a)   The alkali metals (group 1A) have one valence electron, and are among the most reactive metals because their outer electron configuration (ns^1) is one electron beyond a noble gas configuration. They react to lose the ns^1 electron, obtaining a noble gas configuration. This is why the group 1A metals tend to form 1+ cations.

(b)   The alkaline earth metals (group 2A) have two valence electrons, have an outer electron configuration of ns^2, and also tend to be reactive metals. They lose their ns^2 electrons to form 2+ cations.

(c)   The halogens (group 7A) have seven valence electrons and have an outer electron configuration of $ns^2 np^5$. They are among the most reactive nonmetals. They are only one electron short of a noble gas configuration and tend to react to gain that one electron, forming 1− anions.

(d)   The oxygen family (group 6A) has six valence electrons and has an outer electron configuration of $ns^2 np^4$. They are two electrons short of a noble gas configuration and tend to react to gain those two electrons, forming 2− anions.

8.27   The effective nuclear charge (Z_{eff}) is the average or net charge from the nucleus experienced by the electrons in the outermost levels. Shielding is the blocking of nuclear charge from the outermost electrons. The shielding is primarily due to the inner (core) electrons, although there is some interaction and shielding from the electron repulsions of the outer electrons with each other.

8.29   (a)   The radii of transition elements stay roughly constant across each row instead of decreasing in size as in the main-group elements. The difference is that across a row of transition elements, the number of electrons in the outermost principal energy level is nearly constant. As another proton is added to the nucleus with each successive element, another electron is added as well, but the electron goes into an $n_{highest} - 1$ orbital. The number of outermost electrons stays constant, and they experience a roughly constant effective nuclear charge, keeping the radius approximately constant.

(b)   As you go down the first two rows of a column within the transition metals, the elements follow the same general trend in atomic radii and the main-group elements; that is the radii get larger because you are adding outermost electrons into higher n levels.

8.31   An important exception to simply subtracting the number of electrons occurs for transition metal cations. When writing the electron configuration of a transition metal cation, remove the electrons in the highest n-value orbitals first, even if this does not correspond to the reverse order of filling. Normally, even though the d orbital electrons add after the s orbital electrons, the s orbital electrons are lost first. This is because (1) the ns and $(n-1)d$ orbitals are extremely close in energy and, depending on the exact configuration, can vary in relative energy ordering and (2) as the $(n-1)d$ orbitals begin to fill in the first transition series, the increasing nuclear charge stabilizes the $(n-1)d$ orbitals relative to the ns orbitals. This happens because the $(n-1)d$ orbitals are not outermost orbitals and therefore are not effectively shielded from the increasing nuclear charge by the ns orbitals.

8.33   The ionization energy (IE) of an atom or ion is the energy required to remove an electron from the atom or ion in the gaseous state. The ionization energy is always positive because removing an electron takes energy. The energy required to remove the first electron is called the first ionization energy (IE_1). The energy required to remove the second electron is called the second ionization energy (IE_2). The second IE is always greater than the first IE.

8.35   Exceptions occur with elements Be, Mg, and Ca in group 2A having a higher first ionization energy than elements B, Al, and Ga in group 3A. This exception is caused by the change in going from the s block to the p block. The result is that the electrons in the s orbital shield the electron in the p orbital from nuclear charge, making it easier to remove.

Another exception occurs with N, P, and As in group 5A having a higher first ionization energy than O, S, and Se in group 6A. This exception is caused by the repulsion between electrons when they must occupy the same orbital. Group 5A has $3p$ electrons, while group 6A has $4p$ electrons. In the group 5A elements, the p orbitals are half-filled, which makes the configuration particularly stable. The fourth group 6A electron must pair with another electron, making it easier to remove.

8.37   The electron affinity (EA) of an atom or ion is the energy change associated with the gaining of an electron by the atom in the gaseous state. The electron affinity is usually—although not always—negative because an atom or ion usually releases energy when it gains an electron. The trends in electron affinity are not as regular as trends in other properties. For main-group elements, electron affinity generally becomes more negative as you move to the right across a row in the periodic table. There is no corresponding trend in electron affinity going down a column with the exception of group IA which becomes more positive as you go down the column.

8.39   (a)   The reactions of the alkali metals with halogens result in the formation of metal halides.
$$2\,M(s) + X_2 \rightarrow 2\,MX(s)$$
(b)   Alkali metals react with water to form the dissolved alkali metal ion, the hydroxide ion, and hydrogen gas.
$$2\,M(s) + 2\,H_2O(l) \rightarrow 2\,M^+(aq) + 2\,OH^-(aq) + H_2(g)$$

Problems by Topic

Electron Configurations

8.41   (a)   Si   Silicon has 14 electrons. Distribute two of these into the $1s$ orbital, two into the $2s$ orbital, six into the $2p$ orbital, two into the $3s$ orbital, and two into the $3p$ orbital. $1s^2 2s^2 2p^6 3s^2 3p^2$

(b)   O   Oxygen has 8 electrons. Distribute two of these into the $1s$ orbital, two into the $2s$ orbital, and four into the $2p$ orbital. $1s^2 2s^2 2p^4$

(c)   K   Potassium has 19 electrons. Distribute two of these into the $1s$ orbital, two into the $2s$ orbital, six into the $2p$ orbital, two into the $3s$ orbital, six into the $3p$ orbital, and one into the $4s$ orbital. $1s^2 2s^2 2p^6 3s^2 3p^6 4s^1$

(d)   Ne   Neon has 10 electrons. Distribute two of these into the $1s$ orbital, two into the $2s$ orbital, and six into the $2p$ orbital. $1s^2 2s^2 2p^6$

8.43   (a)   N   Nitrogen has seven electrons and has the electron configuration $1s^2 2s^2 2p^3$. Draw a box for each orbital, putting the lowest energy orbital ($1s$) on the far left and proceeding to orbitals of higher energy to the right. Distribute the seven electrons into the boxes representing the orbitals, allowing a maximum of two electrons per orbital and remembering Hund's rule. You can see from the diagram that nitrogen has three unpaired electrons.

        (b)   F   Fluorine has nine electrons and has the electron configuration $1s^2 2s^2 2p^5$. Draw a box for each orbital, putting the lowest energy orbital ($1s$) on the far left and proceeding to orbitals of higher energy to the right. Distribute the nine electrons into the boxes representing the orbitals, allowing a maximum of two electrons per orbital and remembering Hund's rule. You can see from the diagram that fluorine has one unpaired electron.

        (c)   Mg   Magnesium has 12 electrons and has the electron configuration $1s^2 2s^2 2p^6 3s^2$. Draw a box for each orbital, putting the lowest energy orbital ($1s$) on the far left and proceeding to orbitals of higher energy to the right. Distribute the 12 electrons into the boxes representing the orbitals, allowing a maximum of two electrons per orbital and remembering Hund's rule. You can see from the diagram that magnesium has no unpaired electrons.

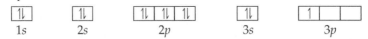

        (d)   Al   Aluminum has 13 electrons and has the electron configuration $1s^2 2s^2 2p^6 3s^2 3p^1$. Draw a box for each orbital, putting the lowest energy orbital ($1s$) on the far left and proceeding to orbitals of higher energy to the right. Distribute the 13 electrons into the boxes representing the orbitals, allowing a maximum of two electrons per orbital and remembering Hund's rule. You can see from the diagram that aluminum has one unpaired electron.

8.45   (a)   P   The atomic number of P is 15. The noble gas that precedes P in the periodic table is neon, so the inner electron configuration is [Ne]. Obtain the outer electron configuration by tracing the elements between Ne and P and assigning electrons to the appropriate orbitals. Begin with [Ne]. Because P is in row 3, add two $3s$ electrons. Next, add three $3p$ electrons as you trace across the p block to P, which is in the third column of the p block.

                P $[Ne]3s^2 3p^3$

        (b)   Ge   The atomic number of Ge is 32. The noble gas that precedes Ge in the periodic table is argon, so the inner electron configuration is [Ar]. Obtain the outer electron configuration by tracing the elements between Ar and Ge and assigning electrons to the appropriate orbitals. Begin with [Ar]. Because Ge is in row 4, add two $4s$ electrons. Next, add ten $3d$ electrons as you trace across the d block. Finally, add two $4p$ electrons as you trace across the p block to Ge, which is in the second column of the p block.

                Ge $[Ar]4s^2 3d^{10} 4p^2$

        (c)   Zr   The atomic number of Zr is 40. The noble gas that precedes Zr in the periodic table is krypton, so the inner electron configuration is [Kr]. Obtain the outer electron configuration by tracing the elements between Kr and Zr and assigning electrons to the appropriate orbitals. Begin with [Kr]. Because Zr is in row 5, add two $5s$ electrons. Next, add two $4d$ electrons as you trace across the d block to Zr, which is in the second column.

                Zr $[Kr]5s^2 4d^2$

        (d)   I   The atomic number of I is 53. The noble gas that precedes I in the periodic table is krypton, so the inner electron configuration is [Kr]. Obtain the outer electron configuration by tracing the elements between Kr and I and assigning electrons to the appropriate orbitals. Begin with [Kr]. Because I is in row 5, add two $5s$ electrons. Next, add ten $4d$ electrons as you trace across the d block. Finally, add five $5p$ electrons as you trace across the p block to I, which is in the fifth column of the p block.

                I $[Kr]5s^2 4d^{10} 5p^5$

8.47 (a) Li is in period 2 and the first column in the *s* block, so Li has one 2*s* electron.
  (b) Cu is in period 4 and the ninth column in the *d* block $(n - 1)$, so Cu should have nine 3*d* electrons. However, it is one of our exceptions; so it has ten 3*d* electrons.
  (c) Br is in period 4 and the fifth column of the *p* block, so Br has five 4*p* electrons.
  (d) Zr is in period 5 and the second column of the *d* block $(n - 1)$, so Zr has two 4*d* electrons.

8.49 (a) In period 4, an element with five valence electrons could be V or As.
  (b) In period 4, an element with four 4*p* electrons would be in the fourth column of the *p* block and is Se.
  (c) In period 4, an element with three 3*d* electrons would be in the third column of the *d* block $(n - 1)$ and is V.
  (d) In period 4, an element with a complete outer shell would be in the sixth column of the *p* block and is Kr.

Valence Electrons and Simple Chemical Behavior from the Periodic Table

8.51 (a) Ba is in column 2A, so it has two valence electrons.
  (b) Cs is in column 1A, so it has one valence electron.
  (c) Ni is in column 8 of the *d* block, so it has ten valence electrons (eight from the *d* block and two from the *s* block).
  (d) S is in column 6A, so it has six valence electrons.

8.53 (a) The outer electron configuration ns^2 would belong to a reactive metal in the alkaline earth family.
  (b) The outer electron configuration ns^2np^6 would belong to an unreactive nonmetal in the noble gas family.
  (c) The outer electron configuration ns^2np^5 would belong to a reactive nonmetal in the halogen family.
  (d) The outer electron configuration ns^2np^2 would belong to an element in the carbon family. If $n = 2$, the element is a nonmetal; if $n = 3$ or 4, the element is a metalloid; and if $n = 5$ or 6, the element is a metal.

Coulomb's Law and Effective Nuclear Charge

8.55 Coulomb's law states that the potential energy (E) of two charge particles depends on their charges $(q_1$ and $q_2)$ and on their separation (r). $E = \dfrac{1}{4\pi\varepsilon_o} \dfrac{q_1 q_2}{r}$. The potential energy is positive for charges of the same sign and negative for charges of opposite signs. The magnitude of the potential energy depends inversely on the separation between the charged particles.

(a) **Given:** $q_1 = 1-$; $q_2 = 2+$; $r = 150$ pm   **Find:** E
Conceptual Plan: Magnitude of Potential Energy depends on the charge and the separation.
$$E = \frac{1}{4\pi\varepsilon_o} \frac{q_1 q_2}{r}$$
Solution: $E = \dfrac{1}{4\pi\varepsilon_o} \dfrac{(1-)(2+)}{150} = \dfrac{-0.0133}{4\pi\varepsilon_o}$

(b) **Given:** $q_1 = 1-$; $q_2 = 1+$; $r = 150$ pm   **Find:** E
Conceptual Plan: Magnitude of Potential Energy depends on the charge and the separation.
$$E = \frac{1}{4\pi\varepsilon_o} \frac{q_1 q_2}{r}$$
Solution: $E = \dfrac{1}{4\pi\varepsilon_o} \dfrac{(1-)(1+)}{150} = \dfrac{-0.00667}{4\pi\varepsilon_o}$

(c) **Given:** $q_1 = 1-$; $q_2 = 3+$; $r = 150$ pm   **Find:** E
Conceptual Plan: Magnitude of Potential Energy depends on the charge and the separation.
$$E = \frac{1}{4\pi\varepsilon_o} \frac{q_1 q_2}{r}$$
Solution: $E = \dfrac{1}{4\pi\varepsilon_o} \dfrac{(1-)(3+)}{100} = \dfrac{-0.0300}{4\pi\varepsilon_o}$
c is most negative and will have the lowest potential energy.

8.57   The valence electrons in nitrogen would experience a greater effective nuclear charge. Be has four protons, and N has seven protons. Both atoms have two core electrons that predominately contribute to the shielding, while the valence electrons will contribute a slight shielding effect. So Be has an effective nuclear charge of slightly more than 2+, and N has an effective nuclear charge of slightly more than 5+.

8.59   (a) $K(19) = [Ar]4s^1$ $Z_{eff} = Z - \text{core electrons} = 19 - 18 = 1+$
        (b) $Ca(20) = [Ar]4s^2$ $Z_{eff} = Z - \text{core electrons} = 20 - 18 = 2+$
        (c) $O(8) = [He]2s^22p^4$ $Z_{eff} = Z - \text{core electrons} = 8 - 2 = 6+$
        (d) $C(6) = [He]2s^22p^2$ $Z_{eff} = Z - \text{core electrons} = 6 - 2 = 4+$

Atomic Radius

8.61   (a)   Al or In   In atoms are larger than Al atoms because as you trace the path between Al and In on the periodic table, you move down a column. Atomic size increases as you move down a column because the outermost electrons occupy orbitals with a higher principal quantum number that are larger, resulting in a larger atom.

        (b)   Si or N   Si atoms are larger than N atoms because as you trace the path between N and Si on the periodic table, you move down a column (atomic size increases) and then to the left across a period (atomic size increases). These effects add together for an overall increase.

        (c)   P or Pb   Pb atoms are larger than P atoms because as you trace the path between P and Pb on the periodic table, you move down a column (atomic size increases) and then to the left across a period (atomic size increases). These effects add together for an overall increase.

        (d)   C or F   C atoms are larger than F atoms because as you trace the path between C and F on the periodic table, you move to the right within the same period. As you move to the right across a period, the effective nuclear charge experienced by the outermost electrons increases, which results in a smaller size.

8.63   Ca, Rb, S, Si, Ge, F   F is above and to the right of the other elements, so you start with F as the smallest atom. As you trace a path from F to S, you move to the left (size increases) and down (size increases). Next, you move left from S to Si (size increases), then down to Ge (size increases), then to the left to Ca (size increases), and then to the left and down to Rb (size increases). So in order of increasing atomic radii, F < S < Si < Ge < Ca < Rb.

Ionic Electron Configurations, Ionic Radii, Magnetic Properties, and Ionization Energy

8.65   (a) O^{2-} Begin by writing the electron configuration of the neutral atom.
                O $1s^22s^22p^4$
                Because this ion has a 2– charge, add two electrons to write the electron configuration of the ion.
                O^{2-} $1s^22s^22p^6$ This is isoelectronic with Ar.

        (b) Br^- Begin by writing the electron configuration of the neutral atom.
                Br $[Ar]4s^23d^{10}4p^5$
                Because this ion has a 1– charge, add one electron to write the electron configuration of the ion.
                Br^- $[Ar]4s^23d^{10}4p^6$ This is isoelectronic with Kr.

        (c) Sr^{2+} Begin by writing the electron configuration of the neutral atom.
                Sr $[Kr]5s^2$
                Because this ion has a 2+ charge, remove two electrons to write the electron configuration of the ion.
                Sr^{2+} $[Kr]$

        (d) Co^{3+} Begin by writing the electron configuration of the neutral atom.
                Co $[Ar]4s^23d^7$
                Because this ion has a 3+ charge, remove three electrons to write the electron configuration of the ion. Because it is a transition metal, remove the electrons from the $4s$ orbital before removing electrons from the $3d$ orbitals.
                Co^{3+} $[Ar]4s^03d^6$ or $[Ar]3d^6$

(e)　Cu^{2+}　Begin by writing the electron configuration of the neutral atom. Remember, Cu is one of our exceptions.

Cu　　　$[Ar]4s^1 3d^{10}$

Because this ion has a 2+ charge, remove two electrons to write the electron configuration of the ion. Because it is a transition metal, remove the electrons from the $4s$ orbital before removing electrons from the $3d$ orbitals.

Cu^{2+}　　$[Ar]4s^0 3d^9$ or $[Ar]3d^9$

8.67　(a)　V^{5+}　Begin by writing the electron configuration of the neutral atom.

V　　　　$[Ar]4s^2 3d^3$

Because this ion has a 5+ charge, remove five electrons to write the electron configuration of the ion. Because it is a transition metal, remove the electrons from the $4s$ orbital before removing electrons from the $3d$ orbitals.

V^{5+}　　　$[Ar]4s^0 3d^0 = [Ne]3s^2 3p^6$

$[Ne]$　　$\boxed{\uparrow\downarrow}$　　$\boxed{\uparrow\downarrow}\,\boxed{\uparrow\downarrow}\,\boxed{\uparrow\downarrow}$

　　　　　$3s$　　　　　$3p$

V^{5+} is diamagnetic.

(b)　Cr^{3+}　Begin by writing the electron configuration of the neutral atom. Remember, Cr is one of our exceptions.

Cr　　　$[Ar]4s^1 3d^5$

Because this ion has a 3+ charge, remove three electrons to write the electron configuration of the ion. Because it is a transition metal, remove the electrons from the $4s$ orbital before removing electrons from the $3d$ orbitals.

Cr^{3+}　　$[Ar]4s^0 3d^3$

$[Ar]$　　$\boxed{}$　　$\boxed{\uparrow}\,\boxed{\uparrow}\,\boxed{\uparrow}\,\boxed{}\,\boxed{}$

　　　　　$4s$　　　　　$3d$

Cr^{3+} is paramagnetic.

(c)　Ni^{2+}　Begin by writing the electron configuration of the neutral atom.

Ni　　　$[Ar]4s^2 3d^8$

Because this ion has a 2+ charge, remove two electrons to write the electron configuration of the ion. Because it is a transition metal, remove the electrons from the $4s$ orbital before removing electrons from the $3d$ orbitals.

Ni^{2+}　　$[Ar]4s^2 3d^8$

$[Ar]$　　$\boxed{}$　　$\boxed{\uparrow\downarrow}\,\boxed{\uparrow\downarrow}\,\boxed{\uparrow\downarrow}\,\boxed{\uparrow}\,\boxed{\uparrow}$

　　　　　$4s$　　　　　$3d$

Ni^{2+} is paramagnetic.

(d)　Fe^{3+}　Begin by writing the electron configuration of the neutral atom.

Fe　　　$[Ar]4s^2 3d^6$

Because this ion has a 3+ charge, remove three electrons to write the electron configuration of the ion. Because it is a transition metal, remove the electrons from the $4s$ orbital before removing electrons from the $3d$ orbitals.

Fe^{3+}　　$[Ar]4s^0 3d^5$

$[Ar]$　　$\boxed{}$　　$\boxed{\uparrow}\,\boxed{\uparrow}\,\boxed{\uparrow}\,\boxed{\uparrow}\,\boxed{\uparrow}$

　　　　　$4s$　　　　　$3d$

Fe^{3+} is paramagnetic.

8.69　(a)　Li or Li^+　A Li atom is larger than Li^+ because cations are smaller than the atoms from which they are formed.

(b)　I^- or Cs^+　An I^- ion is larger than a Cs^+ ion because, although they are isoelectronic, I^- has two fewer protons than Cs^+, resulting in a lesser pull on the electrons and therefore a larger radius.

(c)　Cr or Cr^{3+}　A Cr atom is larger than Cr^{3+} because cations are smaller than the atoms from which they are formed.

(d)　O or O^{2-}　An O^{2-} ion is larger than an O atom because anions are larger than the atoms from which they are formed.

8.71   Because all of the species are isoelectronic, the radius will depend on the number of protons in each species. The fewer the protons, the larger the radius.

F: $Z = 9$; Ne: $Z = 10$; O: $Z = 8$; Mg: $Z = 12$; Na: $Z = 11$

So $O^{2-} > F^- > Ne > Na^+ > Mg^{2+}$

8.73   (a)   Br or Bi   Br has a higher ionization energy than Bi because as you trace the path between Br and Bi on the periodic table, you move down a column (ionization energy decreases) and then to the left across a period (ionization energy decreases). These effects sum together for an overall decrease.

(b)   Na or Rb   Na has a higher ionization energy than Rb because as you trace a path between Na and Rb on the periodic table, you move down a column. Ionization energy decreases as you go down a column because of the increasing size of orbitals with increasing n.

(c)   As or At   Based on periodic trends alone, it is impossible to tell which one has a higher ionization energy because as you trace the path between As and At, you move to the right across a period (ionization energy increases) and then down a column (ionization energy decreases). These effects tend to oppose each other, and it is not obvious which one will dominate.

(d)   P or Sn   P has a higher ionization energy than Sn because as you trace the path between P and Sn on the periodic table, you move down a column (ionization energy decreases) and then to the left across a period (ionization energy decreases). These effects sum together for an overall decrease.

8.75   Because ionization energy increases as you move to the right across a period and increases as you move up a column, the element with the smallest first ionization energy would be the element farthest to the left and lowest on the periodic table. So In has the smallest ionization energy. As you trace a path to the right and up on the periodic table, the next element you reach is Si; continuing up and to the right, you reach N; and continuing to the right, you reach F. So in the order of increasing first ionization energy, the elements are In $<$ Si $<$ N $<$ F.

8.77   The jump in ionization energy occurs when you change from removing a valence electron to removing a core electron. To determine where this jump occurs, you need to look at the electron configuration of the atom.

(a)   Be $1s^2 2s^2$ The first and second ionization energies involve removing $2s$ electrons, while the third ionization energy removes a core electron; so the jump will occur between the second and third ionization energies.

(b)   N $1s^2 2s^2 2p^3$ The first five ionization energies involve removing the $2p$ and $2s$ electrons, while the sixth ionization energy removes a core electron; so the jump will occur between the fifth and sixth ionization energies.

(c)   O $1s^2 2s^2 2p^4$ The first six ionization energies involve removing the $2p$ and $2s$ electrons, while the seventh ionization energy removes a core electron; so the jump will occur between the sixth and seventh ionization energies.

(d)   Li $1s^2 2s^1$ The first ionization energy involves removing a $2s$ electron, while the second ionization energy removes a core electron; so the jump will occur between the first and second ionization energies.

Electron Affinities and Metallic Character

8.79   (a)   Na or Rb   Na has a more negative electron affinity than Rb. In column 1A, electron affinity becomes less negative as you go down the column.

(b)   B or S   S has a more negative electron affinity than B. As you trace from B to S in the periodic table, you move to the right, which shows the value of the electron affinity becoming more negative. Also, as you move from period 2 to period 3, the value of the electron affinity becomes more negative. Both of these trends sum together for the value of the electron affinity to become more negative.

(c)   C or N   C has the more negative electron affinity. As you trace from C to N across the periodic table, you normally expect N to have the more negative electron affinity. However, N has a half-filled p sublevel, which lends it extra stability; therefore, it is harder to add an electron.

(d)   Li or F   F has the more negative electron affinity. As you trace from Li to F on the periodic table, you move to the right in the period. As you move to the right across a period, the value of the electron affinity generally becomes more negative.

8.81   (a)   Sr or Sb   Sr is more metallic than Sb because as we trace the path between Sr and Sb on the periodic table, we move to the right within the same period. Metallic character decreases as we move to the right.

      (b)   As or Bi   Bi is more metallic because as we trace a path between As and Bi on the periodic table, we move down a column in the same family (metallic character increases).

      (c)   Cl or O   Based on periodic trends alone, we cannot tell which is more metallic because as we trace the path between O and Cl, we move to the right across a period (metallic character decreases) and then down a column (metallic character increases). These effects tend to oppose each other, and it is not easy to tell which will predominate.

      (d)   S or As   As is more metallic than S because as we trace the path between S and As on the periodic table, we move down a column (metallic character increases) and then to the left across a period (metallic character increases). These effects add together for an overall increase.

8.83   The order of increasing metallic character is S < Se < Sb < In < Ba < Fr.

Metallic character decreases as you move left to right across a period and decreases as you move up a column; therefore, the element with the least metallic character will be to the top right of the periodic table. So of these elements, S has the least metallic character. As you move down the column, the next element is Se. As you continue down and to the left, you reach Sb; continuing to the left, you reach In; moving down the column and to the left, you come to Ba; and down the column and to the left is Fr.

Chemical Behavior of the Alkali Metals and Halogens

8.85   Alkaline earth metals react with halogens to form metal halides. Write the formulas for the reactants and the metal halide product.

$$Sr(s) + I_2(g) \rightarrow SrI_2(s)$$

8.87   Alkali metals react with water to form the dissolved metal ion, the hydroxide ion, and hydrogen gas. Write the skeletal equation including each of these and then balance it.

$$Li(s) + H_2O(l) \rightarrow Li^+(aq) + OH^-(aq) + H_2(g)$$
$$2\,Li(s) + 2\,H_2O(l) \rightarrow 2\,Li^+(aq) + 2\,OH^-(aq) + H_2(g)$$

8.89   The halogens react with hydrogen to form hydrogen halides. Write the skeletal reaction with each of the halogens and hydrogens as the reactants and the hydrogen halide compound as the product and balance the equation.

$$H_2(g) + Br_2(g) \rightarrow HBr(g)$$
$$H_2(g) + Br_2(g) \rightarrow 2\,HBr(g)$$

Cumulative Problems

8.91   Br: $1s^2\,2s^2\,2p^6\,3s^2\,3p^6\,4s^2\,3d^{10}\,4p^5$
Kr: $1s^2\,2s^2\,2p^6\,3s^2\,3p^6\,4s^2\,3d^{10}\,4p^6$

Krypton has a completely filled p sublevel, giving it chemical stability. Bromine needs one electron to achieve a completely filled p sublevel and thus has a highly negative electron affinity. Therefore, it easily takes on an electron and is reduced to the bromide ion, giving it the added stability of the filled p sublevel.

8.93   Write the electron configuration of vanadium.
V: $[\,Ar\,]4s^2 3d^3$

Because this ion has a 3+ charge, remove three electrons to write the electron configuration of the ion. Because it is a transition metal, remove the electrons from the $4s$ orbital before removing electrons from the $3d$ orbitals.
V^{3+}: $[\,Ar\,]4s^0 3d^2$ or $[Ar]3d^2$

Both vanadium and the V^{3+} ion have unpaired electrons and are paramagnetic.

8.95   Because K^+ has a 1+ charge, we would need a cation with a similar size and a 1+ charge. Looking at the ions in the same family, Na^+ would be too small and Rb^+ would be too large. If we consider Ar^+ and Ca^+, we would have ions of similar size and charge. Between these two, Ca^+ would be easier to achieve because the first ionization energy of Ca is similar to that of K, while the first ionization energy of Ar is much larger. However, the second ionization energy of Ca is relatively low, making it easy to lose the second electron.

8.97   C has an outer shell electron configuration of ns^2np^2; based on this, we would expect Si and Ge, which are in the same family, to be most like carbon. Ionization energies for both Si and Ge are similar and tend to be slightly lower than that of C, but all are intermediate in the range of first ionization energies. The electron affinities of Si and Ge are close to that of C.

8.99   (a)   N: $[\text{He}]2s^22p^3$ Mg: $[\text{Ne}]3s^2$ O: $[\text{He}]2s^22p^4$
                 F: $[\text{He}]2s^22p^5$ Al: $[\text{Ne}]3s^23p^1$

         (b)   Mg > Al > N > O > F

         (c)   Al < Mg < O < N < F   (from the table)

         (d)   Mg and Al would have the largest radius because they are in period $n = 3$; Al is smaller than Mg because radius decreases as you move to the right across the period. F is smaller than O and O is smaller than N because as you move to the right across the period, radius decreases.

                The first ionization energy of Al is smaller than the first ionization energy of Mg because Al loses the electron from the $3p$ orbital, which is shielded by the electrons in the $3s$ orbital. Mg loses the electron from the filled $3s$ orbital, which has added stability because it is a filled orbital. The first ionization energy of O is lower than the first ionization energy of N because N has a half-filled $2p$ orbital, which adds extra stability, thus making it harder to remove the electron. The fourth electron in the O $2p$ orbitals experiences added electron-electron repulsion because it must pair with another electron in the same $2p$ orbital, thus making it easier to remove.

8.101   As you move to the right across a row in the periodic table for the main-group elements, the effective nuclear charge (Z_{eff}) experienced by the electrons in the outermost principal energy level increases, resulting in a stronger attraction between the outermost electrons and the nucleus and therefore a smaller atomic radii.

         Across the row of transition elements, the number of electrons in the outermost principal energy level (highest n value) is nearly constant. As another proton is added to the nucleus with each successive element, another electron is added, but that electron goes into an $n_{\text{highest}} - 1$ orbital (a core level). The number of outermost electrons stays constant, and they experience a roughly constant effective nuclear charge, keeping the radius approximately constant after the first couple of elements in the series.

8.103   All of the noble gases have a filled outer quantum level, very high first ionization energies, and positive values for the electron affinity; thus, the noble gases are particularly unreactive. The lighter noble gases will not form any compounds because the ionization energies of both He and Ne are over 2000 kJ/mol. Because ionization energy decreases as you move down a column, you find that the heavier noble gases (Ar, Kr, and Xe) do form some compounds. They have ionization energies that are close to the ionization energy of H and thus can be forced to lose an electron.

8.105   Group 6A: ns^2np^4 Group 7A: ns^2np^5
         The electron affinity of the group 7A elements is more negative than that of the group 6A elements in the same period because group 7A requires only one electron to achieve the noble gas configuration ns^2np^6, while the group 6A elements require two electrons. Adding one electron to the group 6A element will not give them any added stability and leads to extra electron-electron repulsions; so the value of the electron affinity is less negative than that for group 7A.

8.107   Br(35) = $[\text{Ar}]4s^23d^{10}4p^5$ I(53) = $[\text{Kr}]5s^24d^{10}5p^5$
         Br and I are both halogens with an outermost electron configuration of ns^2np^5; the next element with the same outermost electron configuration is 85, At.

8.109   (a) $10 - 2 = 8 \rightarrow \text{O}$; $12 - 2 = 10 \rightarrow \text{Ne}$; $58 - 5 = 53 \rightarrow \text{I}$; $11 - 2 = 9 \rightarrow \text{F}$; $7 - 2 = 5 \rightarrow \text{B}$; $44 - 5 = 39$
                $\rightarrow \text{Y}$; $63 - 6 = 57 \rightarrow \text{La}$; $66 - 6 = 60 \rightarrow \text{Nd}$ One If by Land

         (b) $9 - 2 = 7 \rightarrow \text{N}$; $99 - 7 = 92 \rightarrow \text{U}$; $30 - 4 = 26 \rightarrow \text{Fe}$; $95 - 7 = 88 \rightarrow \text{Ra}$; $19 - 3 = 16 \rightarrow \text{S}$ $47 - 5$
                $= 42 \rightarrow \text{Mo}$; $79 - 6 = 73 \rightarrow \text{Ta}$ (backwards) Atoms Are Fun

8.111   **Given:** $r = 100.00$ pm, $q_{\text{proton}} = 1.60218 \times 10^{-19}$ C, and $q_{\text{electron}} = -1.60218 \times 10^{-19}$ C
         Find: IE in kJ/mol and λ of ionization
         Conceptual Plan: $r, q_{\text{proton}}, q_{\text{electron}}, \rightarrow E_{\text{atom}} \rightarrow E_{\text{mol}}$ **and then** $E_{\text{atom}} \rightarrow \lambda$

$$E = \frac{1}{4\pi\epsilon_0}\frac{q_pq_e}{r} \quad \frac{\text{kJ}}{(1000\,\text{J})} \quad \frac{6.022 \times 10^{23}\,\text{atom}}{\text{mol}} \qquad \lambda = \frac{hc}{E}$$

Solution:

$$E = \frac{1}{(4)(3.141)\left(8.85 \times 10^{-12}\,\frac{C^2}{J\,m}\right)} \times \frac{(1.60218 \times 10^{-19}\,C)(-1.60218 \times 10^{-19}\,C)}{(100.00\,pm)\left(\frac{1\,m}{1 \times 10^{12}\,pm}\right)} = -2.3\underline{0}89 \times 10^{-18}\,J/atom$$

$$-2.3\underline{0}9 \times 10^{-18}\,J/atom \times \frac{6.022 \times 10^{23}\,atom}{mol} \times \frac{kJ}{(1000\,J)} = -1.39 \times 10^3\,kJ/mol$$

$$IE = 0 - (-1.39 \times 10^3\,kJ/mol) = 1.39 \times 10^3\,kJ/mol$$

$$\lambda = \frac{(6.626 \times 10^{-34}\,J \cdot s)(3.00 \times 10^8\,m \cdot s)\left(\frac{1 \times 10^9\,nm}{m}\right)}{(2.3\underline{0}8 \times 10^{-18}\,J)} = 86.1\,nm$$

Check: The units of the answer (kJ/mol) are correct. The magnitude of the answer is reasonable because the value is positive and energy must be added to the atom to remove the electron. The units of the wavelength (nm) are correct and the magnitude is reasonable based on the ionization energy.

Challenge Problems

8.113   (a)   Using Excel, make a table of radius, atomic number, and density. Using *xy* scatter, make a chart of radius versus density. With an exponential trendline, estimate the density of argon and xenon. Also make a chart of atomic number versus density. With a linear trendline, estimate the density of argon and xenon.

Element	Atomic Radius (pm)	Atomic Number	Density (g/mL)
He	32	2	0.18
Ne	70	10	0.90
Ar	98	18	–
Kr	112	36	3.75
Xe	130	54	–
Rn	–	86	9.73

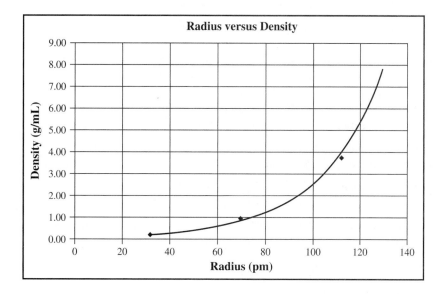

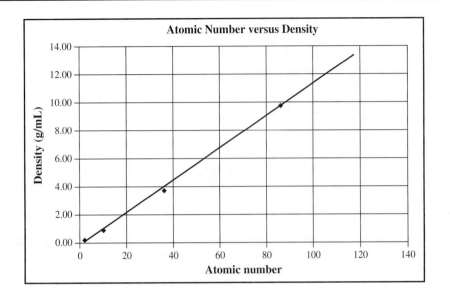

From the radius versus density chart, Ar has a density of ~2 g/L and Xe has a density of ~7.7 g/L. From the atomic number versus density chart, Ar has a density of ~1.8 g/L and Xe has a density of ~6 g/L.

(b) Using the chart of atomic number versus density, element 118 would be predicted to have a density of ~13 g/L.

(c) **Given:** Ne: $M = 20.18$ g/mol; $r = 70$ pm **Find:** mass of neon; d neon

Conceptual Plan: $M \rightarrow m_{atom}$ and then $r \rightarrow vol_{atom}$ and then $\rightarrow d$

$$\frac{mol}{6.022 \times 10^{23} \text{ atoms}} \qquad V = \frac{4}{3}\pi r^3 \qquad d = \frac{m}{v}$$

Solution: $\dfrac{20.18 \text{ g}}{mol} \times \dfrac{mol}{6.022 \times 10^{23} \text{ atoms}} = 3.35 \times 10^{-23}$ g/atom

$$V = \frac{4}{3} \times 3.14 \times (70 \text{ pm})^3 \times \left(\frac{m}{10^{12} \text{ pm}}\right)^3 \times \frac{L}{0.0010 \text{ m}^3} = 1.\underline{4}4 \times 10^{-27} \text{ L}$$

$$d = \frac{3.35 \times 10^{-23} \text{ g}}{1.\underline{4}4 \times 10^{-27} \text{ L}} = 2.\underline{3}3 \times 10^4 = 2.3 \times 10^4 \text{ g/L}$$

Check: The units of the answer (g/L) are correct. This density is significantly larger than the actual density of neon gas. This suggests that a liter of neon is composed primarily of empty space.

(d) **Given:** Ne: $M = 20.18$ g/mol, $d = 0.90$ g/L; Kr: $M = 83.30$ g/mol, $d = 3.75$ g/L; Ar: $M = 39.95$ g/mol
Find: d of argon in g/L

Conceptual Plan: $d \rightarrow$ mol/L $\rightarrow$ atoms/L for Kr and Ne and then atoms/L $\rightarrow$ mol/L $\rightarrow d$ for Ar

$$\text{mol} = \frac{\text{mass}}{\text{molar mass}} \quad \frac{6.022 \times 10^{23} \text{ atoms}}{\text{mol}} \qquad\qquad \frac{\text{mol}}{6.022 \times 10^{23} \text{ atoms}} \quad \frac{39.95 \text{ g}}{\text{mol}}$$

Solution: for Ne: $\dfrac{0.90 \text{ g}}{L} \times \dfrac{mol}{20.18 \text{ g}} \times \dfrac{6.022 \times 10^{23} \text{ atoms}}{mol} = 2.69 \times 10^{22}$ atoms/L

for Kr: $\dfrac{3.75 \text{ g}}{L} \times \dfrac{mol}{83.80 \text{ g}} \times \dfrac{6.022 \times 10^{23} \text{ atoms}}{mol} = 2.69 \times 10^{22}$ atoms/L

for Ar: $\dfrac{2.69 \times 10^{22} \text{ atoms}}{L} \times \dfrac{mol}{6.022 \times 10^{23} \text{ atoms}} \times \dfrac{39.95 \text{ g}}{mol} = 1.78$ g/L

This value is similar to the value calculated in part a. The value of the density calculated from the radius was 2 g/L, and the value of the density calculated from the atomic number was 1.8 g/L.

Check: The units of the answer (g/L) are correct. The value of the answer agrees with the published value.

8.115 The density increases as you move to the right across the first transition series. For the first transition series, the mass increases as you move to the right across the periodic table. However, the radius of the transition series elements stays nearly constant as you move to the right across the periodic table; thus, the volume will remain nearly constant. Because density is mass/volume, the density of the elements increases.

8.117   The longest wavelength would be associated with the lowest energy state next to the ground state of carbon, which has two unpaired electrons.

Ground state of carbon:

Longest wavelength: One of the p electrons flipped in its orbital, which requires the least amount of energy.

The next wavelength would be associated with the pairing of the two p electrons in the same orbital because this requires energy and raises the energy.

$$\boxed{\uparrow\downarrow}\quad\boxed{\uparrow\downarrow}\quad\boxed{\uparrow\downarrow\;|\;|\;}$$
$$1s\qquad 2s\qquad 2p$$

The next wavelength would be associated with the energy needed to promote one of the s electrons to a p orbital.

$$\boxed{\uparrow\downarrow}\quad\boxed{\uparrow\;}\quad\boxed{\uparrow\;|\uparrow\;|\uparrow\;|\uparrow}$$
$$1s\qquad 2s\qquad 2p$$

8.119   The element that would fill the $8s$ and $8p$ orbitals would have atomic number 168. The element is in the noble gas family and would have the properties of noble gases. It would have the electron configuration of $[118]8s^2 5g^{18}6f^{14}7d^{10}8p^6$. The outer shell electron (highest n level) configuration would be $8s^2 8p^6$. The element would be relatively inert, have a first ionization energy less than 1037 kJ/mol (the first ionization energy of Rn), and have a positive electron affinity. It would be difficult to form compounds with most elements but would be able to form compounds with fluorine.

8.121   When you move down the column from Al to Ga, the size of the atom actually decreases because not much shielding is contributed by the $3d$ electrons in the Ga atom, while there is a large increase in the nuclear charge. Therefore, the effective nuclear charge is greater for Ga than for Al; so the ionization energy does not decrease. As you go from In to Tl, the ionization energy actually increases because the $4f$ electrons do not contribute to the shielding of the outermost electrons and there is a large increase in the effective nuclear charge.

8.123   The second electron is added to an ion with a $1-$ charge, so a large repulsive force must be overcome to add the second electron. Thus, it will require energy to add the second electron, and the second electron affinity will have a positive value.

8.125   **Given:** Ra, $Z = 88$ **Find:** Z for next two alkaline earth metals
Solution: The next element would lie in period 8, column 2A. The largest currently known element is 116 in period 7, column 6A. To reach period 8 column 2A, you need to add four protons and would have $Z = 120$.

The alkaline earth metal following 120 would lie in period 9, column 2A. To reach this column, you need to add 18 g-block element protons, 10 d block element protons, 14 f block element protons, 6 p block element protons, and 2 s block element protons. This would give $Z = 170$.

8.127   Francium would have an electron configuration of $[Rn]7s^1$; the atomic radius would be > 265 pm (atomic radius of Cs); the first ionization energy would be less than 376 kJ/mol; the density would be greater than 1.879 g/cm^3; and the melting would be less than 29 °C.
(a) $2\,Fr + 2\,H_2O \rightarrow 2\,Fr^+(aq) + 2\,OH^-(aq) + H_2(g)$
(b) $4\,Fr + O_2(g) \rightarrow 2\,Fr_2O(s)$
(c) $2\,Fr + Cl_2(g) \rightarrow 2\,FrCl(s)$

Conceptual Problems

8.129   If six electrons rather than eight electrons led to a stable configuration, the electron configuration of the stable configuration would be $ns^2 np^4$.
(a)   A noble gas would have the electron configuration $ns^2 np^4$. This could correspond to the O atom.
(b)   A reactive nonmetal would have one less electron than the stable configuration. This would have the electron configuration $ns^2 np^3$. This could correspond to the N atom.
(c)   A reactive metal would have one more electron than the stable configuration. This would have the electron configuration of ns^1. This could correspond to the Li atom.

8.131 (a) True: An electron in a 3*s* orbital is more shielded than an electron in a 2*s* orbital. This is true because there are more core electrons below a 3*s* orbital.

(b) True: An electron in a 3*s* orbital penetrates the region occupied by the core electrons more than electrons in a 3*p* orbital. Examine Figure 8.5, the radial distribution functions for the 3*s*, 3*p*, and 3*d* orbitals. You will see that the 3*s* electrons penetrate more deeply than the 3*p* electrons and more than the 3*d* electrons.

(c) False: An electron in an orbital that penetrates closer to the nucleus will experience *less* shielding than an electron in an orbital that does not penetrate as far.

(d) True: An electron in an orbital that penetrates close to the nucleus will tend to experience a higher effective nuclear charge than one that does not. Because the orbital penetrates closer to the nucleus, the electron will experience less shielding and therefore a higher effective nuclear charge.

8.133   The 4*s* electrons in calcium have relatively low ionization energies (IE_1 = 590 kJ/mol; IE_2 = 1145 kJ/mol) because they are valence electrons. The energetic cost for calcium to lose a third electron is extraordinarily high because the next electron to be lost is a core electron. Similarly, the electron affinity of fluorine to gain one electron (-328 kJ/mol) is highly exothermic because the added electron completes fluoride's valence shell. The gain of a second electron by the negatively charged fluoride anion would not be favorable. Therefore, we would expect calcium and fluoride to combine in a 1:2 ratio.

9 Chemical Bonding I: The Lewis Theory

Review Questions

9.1 Bonding theories are central to chemistry because they explain how atoms bond together to form molecules. Bonding theories explain why some combinations of atoms are stable and others are not.

9.3 The three types of bonds are ionic bonds, which occur between metals and nonmetals and are characterized by the transfer of electrons; covalent bonds, which occur between nonmetals and are characterized by the sharing of electrons; and metallic bonds, which occur between metals and are characterized by electrons being pooled.

9.5 Bonds are formed when atoms attain a stable electron configuration. Because the stable configuration usually has eight electrons in the outermost shell, this is known as the octet rule.

9.7 In Lewis theory, we represent ionic bonding by moving electron dots from the metal to the nonmetal and then allowing the resultant ions to form a crystalline lattice composed of alternating cations and anions. The cation loses its valence electron(s) and is left with an octet in the previous principal energy level; the anion gains electron(s) to form an octet. The Lewis structure of the anion is usually written within brackets with the charge in the upper right-hand corner, outside the brackets. The positive and negative charges attract one another, resulting in the compound.

9.9 Lattice energy is associated with forming a crystalline lattice of alternating cations and anions from the gaseous ions. Because the cations are positively charged and the anions are negatively charged, there is a lowering of potential—as described by Coulomb's law—when the ions come together to form a lattice. That energy is emitted as heat when the lattice forms.

9.11 The Born–Haber cycle is a hypothetical series of steps that represents the formation of an ionic compound from its constituent elements. The steps are chosen so that the change in enthalpy of each step is known except for the last one, which is the lattice energy. In terms of the formation of NaCl, the steps are as follows:
Step 1: The formation of gaseous sodium from solid sodium (heat of sublimation of sodium)
Step 2: The formation of a chlorine atom from a chlorine molecule (bond energy of chlorine)
Step 3: The ionization of gaseous sodium (ionization energy of sodium)
Step 4: The addition of an electron to gaseous chlorine (the electron affinity of chlorine)
Step 5: The formation of the crystalline solid from the gaseous ions (the lattice energy)

The overall reaction is the formation of NaCl(s), so we can use Hess's law to determine the lattice energy.
$$\Delta H_f^\circ = \Delta H_{step\,1} + \Delta H_{step\,2} + \Delta H_{step\,3} + \Delta H_{step\,4} + \Delta H_{step\,5}$$
ΔH_f° = heat of sublimation + ½ bond energy + ionization energy + electron affinity + lattice energy
Because all of the terms are known except the lattice energy, we can calculate the lattice energy.

9.13 We modeled ionic solids as a lattice of individual ions held together by coulombic forces, which are equal in all directions. To melt the solid, these forces must be overcome, which requires a significant amount of heat. Therefore, the model accounts for the high melting points of ionic solids.

9.15   A pair of electrons that is shared between two atoms is called a bonding pair, while a pair of electrons that is associated with only one atom—and therefore, not involved in bonding—is called a lone pair.

9.17   Generally, combinations of atoms that can satisfy the octet rule on each atom are stable, while those combinations that do not satisfy the octet rule are not stable.

9.19   Electronegativity is the ability of an atom to attract electrons to itself in a chemical bond. This results in a polar bond. Electronegativity generally increases across a period in the periodic table. And electronegativity generally decreases down a column (group) in the periodic table. The most electronegative element is fluorine.

9.21   Percent ionic character is defined as the ratio of a bond's actual dipole moment to the dipole moment it would have if the electron were completely transferred from one atom to the other, multiplied by 100.

          A bond in which an electron is completely transferred from one atom to another would have 100% ionic character. However, no bond is 100% ionic. Percent ionic character generally increases as the electronegativity difference increases. In general, bonds with greater than 50% ionic character are referred to as ionic bonds.

9.23   To calculate the dipole moment, we use $\mu = qr$:

          For 100 pm: $\mu = 1.6 \times 10^{-19}\ \cancel{C} \times 100\ \cancel{pm} \times \dfrac{\cancel{m}}{10^{12}\ \cancel{pm}} \times \dfrac{D}{3.34 \times 10^{-30}\ \cancel{C} \cdot \cancel{m}} = 4.8\ D$

          For 200 pm: $\mu = 1.6 \times 10^{-19}\ \cancel{C} \times 200\ \cancel{pm} \times \dfrac{\cancel{m}}{10^{12}\ \cancel{pm}} \times \dfrac{D}{3.34 \times 10^{-30}\ \cancel{C} \cdot \cancel{m}} = 9.6\ D$

9.25   The total number of electrons for a Lewis structure of a molecule is the sum of the valence electrons of each atom in the molecule.

          The total number of electrons for the Lewis structure of an ion is found by summing the number of valence electrons for each atom and then subtracting one electron for each positive charge or adding one electron for each negative charge.

9.27   In some cases, we can write resonance structures that are not equivalent. One possible resonance structure may be somewhat better than another. In such cases, the true structure may still be represented as an average of the resonance structures, but with the better resonance structure contributing more to the true structure. Multiple nonequivalent resonance structures may be weighted differently in their contributions to the true overall structure of a molecule.

9.29   The octet rule has some exceptions because not all atoms have eight electrons surrounding them. The three major categories are (1) odd octets—electron species, molecules, or ions with an odd number of electron (e.g., NO); (2) incomplete octets—molecules or ions with fewer than eight electrons around an atom (e.g., BF_3); and (3) expanded octets—molecules or ions with more than eight electrons around an atom (e.g., AsF_5).

9.31   The bond energy of a chemical bond is the energy required to break 1 mole of the bond in the gas phase. Because breaking bonds is endothermic and forming bonds is exothermic, we can calculate the overall enthalpy change as a sum of the enthalpy changes associated with breaking the required bonds in the reactants and forming the required bonds in the products.

9.33   When metal atoms bond together to form a solid, each metal atom donates one or more electrons to an electron sea.

Problems by Topic

Valence Electrons and Dot Structures

9.35   N: $1s^2 2s^2 2p^3$ $\cdot\overset{\displaystyle .}{\underset{\displaystyle .}{N}}:$ The electrons included in the Lewis structure are $2s^2 2p^3$.

9.37   (a)   Al: $1s^2 2s^2 2p^6 3s^2 3p^1$

                                   $\cdot\overset{\displaystyle .}{Al}\cdot$

          (b) Na^+: $1s^2 2s^2 2p^6$

                                   Na^+

(c)   Cl: $1s^2 2s^2 2p^6 3s^2 3p^5$

$:\ddot{\underset{..}{Cl}}\cdot$

(d)   Cl⁻: $1s^2 2s^2 2p^6 3s^2 3p^6$

$:\ddot{\underset{..}{Cl}}:$

Ionic Lewis Structures and Lattice Energy

9.39   (a)   NaF: Draw the Lewis structures for Na and F based on their valence electrons.

Na: $3s^1$ F: $2s^2 2p^5$

Na· $:\ddot{F}\cdot$

Sodium must lose one electron and be left with the octet from the previous shell, while fluorine needs to gain one electron to get an octet.

Na⁺ $\left[:\ddot{\underset{..}{F}}:\right]^-$

(b)   CaO: Draw the Lewis structures for Ca and O based on their valence electrons. Ca: $4s^2$ O: $2s^2 2p^4$

Ca: $:\ddot{O}\cdot$

Calcium must lose two electrons and be left with two $1s$ electrons from the previous shell, while oxygen needs to gain two electrons to get an octet.

Ca²⁺ $\left[:\ddot{\underset{..}{O}}:\right]^{2-}$

(c)   SrBr₂: Draw the Lewis structures for Sr and Br based on their valence electrons. Sr: $5s^2$ Br: $4s^2 4p^5$

Sr: $:\ddot{\underset{.}{Br}}\cdot$

Strontium must lose two electrons and be left with the octet from the previous shell, while bromine needs to gain one electron to get an octet.

Sr²⁺ 2$\left[:\ddot{\underset{..}{Br}}:\right]^-$

(d)   K₂O: Draw the Lewis structures for K and O based on their valence electrons. K: $4s^1$ O: $2s^2 2p^4$

K· $:\ddot{O}\cdot$

Potassium must lose one electron and be left with the octet from the previous shell, while oxygen needs to gain two electrons to get an octet.

2K⁺ $\left[:\ddot{\underset{..}{O}}:\right]^{2-}$

9.41   (a)   Sr and Se: Draw the Lewis structures for Sr and Se based on their valence electrons.

Sr: $5s^2$ Se: $4s^2 4p^4$

Sr: $:\ddot{Se}\cdot$

Strontium must lose two electrons and be left with the octet from the previous shell, while selenium needs to gain two electrons to get an octet.

Sr²⁺ $\left[:\ddot{\underset{..}{Se}}:\right]^{2-}$

Thus, we need one Sr²⁺ and one Sr²⁻. Write the formula with subscripts (if necessary) to indicate the number of atoms.

SrSe

(b)   Ba and Cl: Draw the Lewis structures for Ba and Cl based on their valence electrons.

Ba: $6s^2$ Cl: $3s^2 3p^5$

Ba: $:\ddot{\underset{..}{Cl}}\cdot$

Barium must lose two electrons and be left with the octet from the previous shell, while chlorine needs to gain one electron to get an octet.

$$Ba^{2+} \quad 2\left[:\ddot{\underset{..}{Cl}}:\right]^{-}$$

Thus, we need one Ba^{2+} and two Cl^-. Write the formula with subscripts (if necessary) to indicate the number of atoms.

$BaCl_2$

(c)   Na and S: Draw the Lewis structures for Na and S based on their valence electrons.

Na: $3s^1$ S: $3s^2 3p^4$

$Na\cdot \quad :\ddot{\underset{..}{S}}\cdot$

Sodium must lose one electron and be left with the octet from the previous shell, while sulfur needs to gain two electrons to get an octet.

$$2\,Na^{+} \quad \left[:\ddot{\underset{..}{S}}:\right]^{2-}$$

Thus, we need two Na^+ and one S^{2-}. Write the formula with subscripts (if necessary) to indicate the number of atoms.

Na_2S

(d)   Al and O: Draw the Lewis structures for Al and O based on their valence electrons.

Al: $3s^2 3p^1$ O: $2s^2 2p^4$

$\ddot{Al}\cdot \quad :\ddot{\underset{..}{O}}\cdot$

Aluminum must lose three electrons and be left with the octet from the previous shell, while oxygen needs to gain two electrons to get an octet.

$$2\,Al^{3+} \quad 3\left[:\ddot{\underset{..}{O}}:\right]^{2-}$$

Thus, we need two Al^{3+} and three O^{2-} to lose and gain the same number of electrons. Write the formula with subscripts (if necessary) to indicate the number of atoms.

Al_2O_3

9.43   As the size of the alkaline metal ions increases down the column, so does the distance between the metal cation and the oxide anion. Therefore, the magnitude of the lattice energy of the oxides decreases, making the formation of the oxides less exothermic and the compounds less stable. Because the ions cannot get as close to each other, they do not release as much energy.

9.45   Cesium is slightly larger than barium, but oxygen is slightly larger than fluorine; so we cannot use size to explain the difference in the lattice energy. However, the charge on the cesium ion is 1+ and the charge on the fluoride ion is 1−, while the charge on the barium ion is 2+ and the charge on the oxide ion is 2−. The coulombic equation states that the magnitude of the potential also depends on the product of the charges. Because the product of the charges for CsF = 1− and the product of the charges for BaO = 4−, the stabilization for BaO relative to CsF should be about four times greater, which is what we see in its more exothermic lattice energy.

9.47   **Given:** $\Delta H_f^\circ KCl = -436.5 \text{ kJ/mol}$; $IE_1(K) = 419 \text{ kJ/mol}$; $\Delta H_{sub}(K) = 89.0 \text{ kJ/mol}$; $Cl_2(g)$ bond energy $= 243 \text{ kJ/mol}$; $EA(Cl) = -349 \text{ kJ/mol}$ **Find:** lattice energy

Conceptual Plan:

$$K(s) + 1/2\,Cl_2(g) \xrightarrow{\Delta H_{sub}} K(g) + 1/2\,Cl_2(g) \xrightarrow{IE_1} K^+(g) + 1/2\,Cl_2(g) \xrightarrow{\text{bond energy}} K^+(g) + Cl(g) \xrightarrow{EA} K^+(g) + Cl^-(g) \xrightarrow{\text{lattice energy}} KCl(s)$$

ΔH_f°

Solution: $\Delta H_f^\circ = \Delta H_{sub} + IE_1 + 1/2 \text{ bond energy} + EA + \text{lattice energy}$

$$-436.5\,\frac{kJ}{mol} = +89.0\,\frac{kJ}{mol} + 419\,\frac{kJ}{mol} + \frac{1}{2}\,(243)\,\frac{kJ}{mol} + (-349)\,\frac{kJ}{mol} + \text{lattice energy}$$

lattice energy $= -717 \text{ kJ/mol}$

Simple Covalent Lewis Structures, Electronegativity, and Bond Polarity

9.49   (a)   Hydrogen: Write the Lewis structure of each atom based on the number of valence electrons.

H· ·H

When the two hydrogen atoms share their electrons, they each get a duet, which is a stable configuration for hydrogen.

H—H

(b)   The halogens: Write the Lewis structure of each atom based on the number of valence electrons.

:Ẍ· ·Ẍ:

If the two halogens pair together, each can achieve an octet, which is a stable configuration. So the halogens are predicted to exist as diatomic molecules.

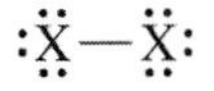

:Ẍ—Ẍ:

(c)   Oxygen: Write the Lewis structure of each atom based on the number of valence electrons.

:Ö· ·Ö:

To achieve a stable octet on each oxygen, the oxygen atoms will need to share two electron pairs. So oxygen is predicted to exist as a diatomic molecule with a double bond.

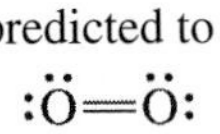

:Ö=Ö:

(d)   Nitrogen: Write the Lewis structure of each atom based on the number of valence electrons.

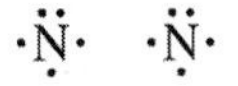

·N̈· ·N̈·

To achieve a stable octet on each nitrogen, the nitrogen atoms will need to share three electron pairs. So nitrogen is predicted to exist as a diatomic molecule with a triple bond.

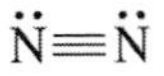

N̈≡N̈

9.51   (a) PH_3:   Write the Lewis structure for each atom based on the number of valence electrons.

·P̈· ·H

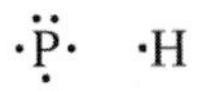

Phosphorus will share an electron pair with each hydrogen to achieve a stable octet.

H—P̈—H
     |
     H

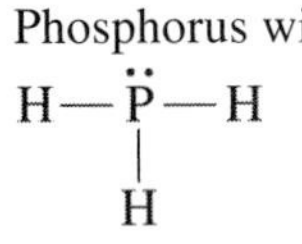

(b) SCl_2:   Write the Lewis structure for each atom based on the number of valence electrons.

:S̈· :C̈l·

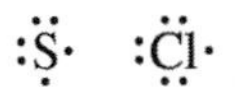

The sulfur will share an electron pair with each chlorine to achieve a stable octet.

:S̈—C̈l:
 |
:C̈l:

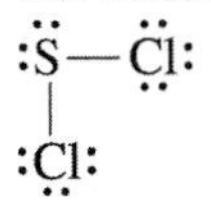

(c)   HI: Write the Lewis structure for each atom based on the number of valence electrons.

H· ·Ï:

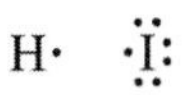

The iodine will share an electron pair with hydrogen to achieve a stable octet.

H—Ï:

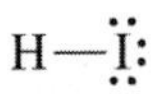

(d) CH_4:   Write the Lewis structure for each atom based on the number of valence electrons.

·C̈· H·

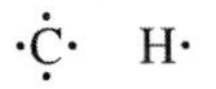

The carbon will share an electron pair with each hydrogen to achieve a stable octet.

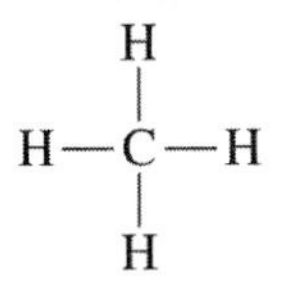

    H
    |
H—C—H
    |
    H

9.53　(a)　SF_2:　Write the Lewis structure for each atom based on the number of valence electrons.

$$: \overset{..}{\underset{.}{S}} \cdot \quad : \overset{..}{\underset{..}{F}} \cdot$$

The sulfur will share an electron pair with each fluorine to achieve a stable octet.

$$: \overset{..}{S} - \overset{..}{\underset{..}{F}} :$$
$$|$$
$$: \overset{..}{\underset{..}{F}} :$$

(b)　SiH_4:　Write the Lewis structure for each atom based on the number of valence electrons.

$$H \cdot \quad \cdot \overset{.}{\underset{.}{Si}} \cdot$$

The silicon will share an electron pair with each hydrogen to achieve a stable octet.

$$\begin{array}{c} H \\ | \\ H - Si - H \\ | \\ H \end{array}$$

(c)　HCOOH　Write the Lewis structure for each atom based on the number of valence electrons.

$$H \cdot \quad \cdot \overset{.}{\underset{.}{C}} \cdot \quad : \overset{..}{\underset{.}{O}} \cdot$$

The carbon will share an electron pair with hydrogen, an electron pair with the interior oxygen, and two electron pairs with the terminal oxygen to achieve a stable octet. The terminal oxygen will share two electron pairs with carbon to achieve a stable octet. The interior oxygen will share an electron pair with carbon and an electron pair with hydrogen to achieve a stable octet.

$$\begin{array}{c} : O : \\ \| \\ H - C - \overset{..}{\underset{..}{O}} - H \end{array}$$

(d)　CH_3SH　Write the Lewis structure for each atom based on the number of valence electrons.

$$H \cdot \quad \cdot \overset{.}{\underset{.}{C}} \cdot \quad : \overset{..}{\underset{.}{S}} \cdot$$

The carbon will share an electron pair with each hydrogen and an electron pair with sulfur to achieve a stable octet. The sulfur will share an electron pair with carbon and an electron pair with hydrogen to achieve a stable octet.

$$\begin{array}{c} H \\ | \\ H - C - \overset{..}{\underset{..}{S}} - H \\ | \\ H \end{array}$$

9.55　(a)　Br and Br:　pure covalent　From Figure 9.8, we find that the electronegativity of Br is 2.5. Because both atoms are the same, the electronegativity difference (Δ EN) = 0, and using Table 9.1, we classify this bond as pure covalent.

(b)　C and Cl:　polar covalent　From Figure 9.8, we find that the electronegativity of C is 2.5 and Cl is 3.0. The electronegativity difference (Δ EN) is Δ EN = 3.0 − 2.5 = 0.5. Using Table 9.1, we classify this bond as polar covalent.

(c)　C and S:　pure covalent　From Figure 9.8, we find that the electronegativity of C is 2.5 and S is 2.5. The electronegativity difference (Δ EN) is Δ EN = 2.5 − 2.5 = 0. Using Table 9.1, we classify this bond as pure covalent.

(d)　Sr and O:　ionic　From Figure 9.8, we find that the electronegativity of Sr is 1.0 and O is 3.5. The electronegativity difference (Δ EN) is Δ EN = 3.5 − 1.0 = 2.5. Using Table 9.1, we classify this bond as ionic.

9.57　CO: Write the Lewis structure for each atom based on the number of valence electrons.

$$\cdot \overset{..}{\underset{.}{C}} \cdot \quad : \overset{..}{\underset{.}{O}} \cdot$$

The carbon will share three electron pairs with oxygen to achieve a stable octet.

The oxygen atom is more electronegative than the carbon atom; so the oxygen will have a partial negative charge, and the carbon will have a partial positive charge.

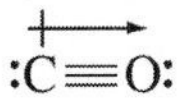

To estimate the percent ionic character, determine the difference in electronegativity between carbon and oxygen. From Figure 9.8, we find that the electronegativity of C is 2.5 and O is 3.5. The electronegativity difference (Δ EN) is Δ EN $= 3.5 - 2.5 = 1.0$. From Figure 9.10, we can estimate a percent ionic character of 25%.

Covalent Lewis Structures, Resonance, and Formal Charge

9.59   (a) CI_4:   Write the correct skeletal structure for the molecule.

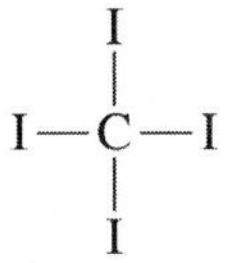

$$\begin{array}{c} I \\ | \\ I-C-I \\ | \\ I \end{array}$$

Calculate the total number of electrons for the Lewis structure by summing the number of valence electrons of each atom in the molecule.

(number of valence e$^-$ for C) $+$ 4(number of valence e$^-$ for I) $= 4 + 4(7) = 32$

Distribute the electrons among the atoms, giving octets to as many atoms as possible. Begin with the bonding electrons; then proceed to lone pairs on terminal atoms and finally to lone pairs on the central atom.

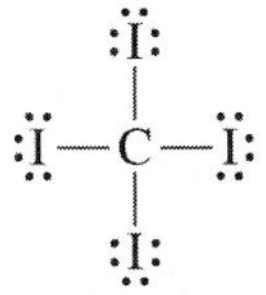

All 32 valence electrons are used.

If any atom lacks an octet, form double or triple bonds as necessary to give them octets.

All atoms have octets; the structure is complete.

(b) N_2O: Write the correct skeletal structure for the molecule.

N is less electronegative, so it is central.

N—N—O

Calculate the total number of electrons for the Lewis structure by summing the number of valence electrons of each atom in the molecule.

2(number of valence e$^-$ for N) $+$ (number of valence e$^-$ for O) $= 2(5) + 6 = 16$

Distribute the electrons among the atoms, giving octets to as many atoms as possible. Begin with the bonding electrons; then proceed to lone pairs on terminal atoms and finally to lone pairs on the central atom.

:N̈—N—Ö:

All 16 valence electrons are used.

If any atom lacks an octet, form double or triple bonds as necessary.

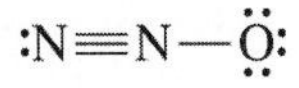

:N≡N—Ö:

All atoms have octets; the structure is complete. The formal charges are 0, +1, and −1 for the N, N, and O, respectively.

An alternate structure with two double bonds is not preferred because the formal charges would be −1, +1, and 0, and the most negative formal charges should be on the O atom.

(c) SiH_4:   Write the correct skeletal structure for the molecule.
H is always terminal, so Si is the central atom.

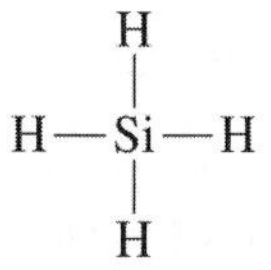

$$
\begin{array}{c}
H \\
| \\
H-Si-H \\
| \\
H
\end{array}
$$

Calculate the total number of electrons for the Lewis structure by summing the number of valence electrons of each atom in the molecule   .

(number of valence e⁻ for Si) + 4(number of valence e⁻ for H) = 4 + 4(1) = 8

Distribute the electrons among the atoms, giving octets (or duets for H) to as many atoms as possible. Begin with the bonding electrons; then proceed to lone pairs on terminal atoms and finally to lone pairs on the central atom.

All eight valence electrons are used.
If any atom lacks an octet, form double or triple bonds as necessary to give them octets.
All atoms have octets; the structure is complete.

(d) Cl_2CO: Write the correct skeletal structure for the molecule.
C is least electronegative, so it is the central atom.

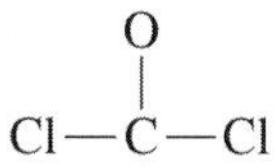

$$
\begin{array}{c}
O \\
| \\
Cl-C-Cl
\end{array}
$$

Calculate the total number of electrons for the Lewis structure by summing the number of valence electrons of each atom in the molecule.

(number of valence e⁻ for C) + 2(number of valence e⁻ for Cl) + (number of valence e⁻ for O) = 4 + 2(7) + 6 = 24

Distribute the electrons among the atoms, giving octets to as many atoms as possible. Begin with the bonding electrons; then proceed to lone pairs on terminal atoms and finally to lone pairs on the central atom.

$$
\begin{array}{c}
:\ddot{O}: \\
| \\
:\ddot{C}l-C-\ddot{C}l:
\end{array}
$$

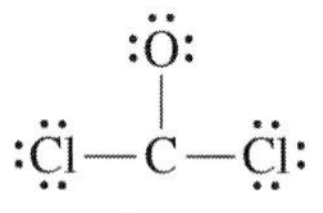

All 24 valence electrons are used.
If any atom lacks an octet, form double or triple bonds as necessary.

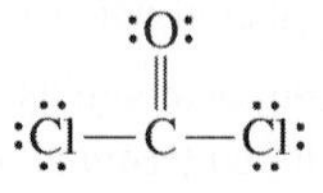

$$
\begin{array}{c}
:O: \\
\| \\
:\ddot{C}l-C-\ddot{C}l:
\end{array}
$$

All atoms have octets; the structure is complete.

9.61   (a) N_2H_2:   Write the correct skeletal structure for the molecule.

H—N—N—H

Calculate the total number of electrons for the Lewis structure by summing the number of valence electrons of each atom in the molecule.

2(number of valence e⁻ for N) + 2(number of valence e⁻ for H) = 2(5) + 2(1) = 12

Distribute the electrons among the atoms, giving octets (or duets for H) to as many atoms as possible. Begin with the bonding electrons; then proceed to lone pairs on terminal atoms and finally to lone pairs on the central atom.

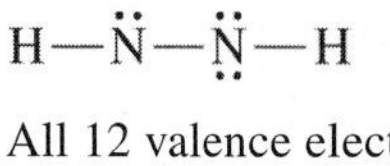

All 12 valence electrons are used.

If any atom lacks an octet, form double or triple bonds as necessary.

$$H—\ddot{N}=\ddot{N}—H$$

All atoms have octets (duets for H); the structure is complete.

(b) N_2H_4:   Write the correct skeletal structure for the molecule.

$$\begin{matrix} H\diagdown & & \diagup H \\ & N—N & \\ H\diagup & & \diagdown H \end{matrix}$$

Calculate the total number of electrons for the Lewis structure by summing the valence electrons of each atom in the molecule.

$$2(\text{number of valence e}^- \text{ for N}) + 4(\text{number of valence e}^- \text{ for H}) = 2(5) + 4(1) = 14$$

Distribute the electrons among the atoms, giving octets (or duets for H) to as many atoms as possible. Begin with the bonding electrons; then proceed to lone pairs on terminal atoms and finally to lone pairs on the central atom.

$$\begin{matrix} H\diagdown & \ddot{}\ddot{} & \diagup H \\ & \ddot{N}—\ddot{N} & \\ H\diagup & & \diagdown H \end{matrix}$$

All 14 valence electrons are used.

If any atom lacks an octet, form double or triple bonds as necessary to give them octets.

All atoms have octets (duets for H); the structure is complete.

(c) C_2H_2:   Write the correct skeletal structure for the molecule.

$$H—C—C—H$$

Calculate the total number of electrons for the Lewis structure by summing the number of valence electrons of each atom in the molecule.

$$2(\text{number of valence e}^- \text{ for C}) + 2(\text{number of valence e}^- \text{ for H}) = 2(4) + 2(1) = 10$$

Distribute the electrons among the atoms, giving octets (or duets for H) to as many atoms as possible. Begin with the bonding electrons; then proceed to lone pairs on terminal atoms and finally to lone pairs on the central atom.

$$H—C—\ddot{\underset{..}{C}}—H$$

All ten valence electrons are used.

If any atom lacks an octet, form double or triple bonds as necessary.

$$H—C\equiv C—H$$

All atoms have octets (duets for H); the structure is complete.

(d) C_2H_4:   Write the correct skeletal structure for the molecule.

$$\begin{matrix} H\diagdown & & \diagup H \\ & C—C & \\ H\diagup & & \diagdown H \end{matrix}$$

Calculate the total number of electrons for the Lewis structure by summing the number of valence electrons of each atom in the molecule.

$$2(\text{number of valence e}^- \text{ for C}) + 4(\text{number of valence e}^- \text{ for H}) = 2(4) + 4(1) = 12$$

Distribute the electrons among the atoms, giving octets (or duets for H) to as many atoms as possible. Begin with the bonding electrons; then proceed to lone pairs on terminal atoms and finally to lone pairs on the central atom.

$$\begin{matrix} H\diagdown & \ddot{} & \diagup H \\ & \ddot{C}—C & \\ H\diagup & & \diagdown H \end{matrix}$$

All 12 valence electrons are used.

If any atom lacks an octet, form double or triple bonds as necessary.

$$\begin{matrix} H\diagdown & & \diagup H \\ & C=C & \\ H\diagup & & \diagdown H \end{matrix}$$

All atoms have octets (duets for H); the structure is complete.

9.63   (a) SeO_2:   Write the correct skeletal structure for the molecule.

Se is less electronegative, so it is central.

$$O—Se—O$$

Calculate the total number of electrons for the Lewis structure by summing the number of valence electrons of each atom in the molecule.

$$(\text{number of valence } e^- \text{ for Se}) + 2(\text{number of valence } e^- \text{ for O}) = 6 + 2(6) = 18$$

Distribute the electrons among the atoms, giving octets to as many atoms as possible. Begin with the bonding electrons; then proceed to lone pairs on terminal atoms and finally to lone pairs on the central atom.

$$:\overset{..}{\underset{..}{\text{O}}}-\overset{..}{\text{Se}}-\overset{..}{\underset{..}{\text{O}}}:$$

All 18 valence electrons are used.

If any atom lacks an octet, form double or triple bonds as necessary.

$$:\overset{..}{\underset{..}{\text{O}}}-\overset{..}{\text{Se}}=\overset{..}{\underset{..}{\text{O}}}:$$

All atoms have octets; the structure is complete. However, the double bond can form from either oxygen atom, so there are two resonance forms.

$$:\overset{..}{\underset{..}{\text{O}}}-\overset{..}{\text{Se}}=\overset{..}{\underset{..}{\text{O}}}: \longleftrightarrow :\overset{..}{\underset{..}{\text{O}}}=\overset{..}{\text{Se}}-\overset{..}{\underset{..}{\text{O}}}:$$

Calculate the formal charge on each atom by finding the number of valence electrons and subtracting the number of lone pair electrons and one-half the number of bonding electrons.

$$:\overset{..}{\underset{..}{\text{O}}}-\overset{..}{\text{Se}}=\overset{..}{\underset{..}{\text{O}}}: \longleftrightarrow :\overset{..}{\underset{..}{\text{O}}}=\overset{..}{\text{Se}}-\overset{..}{\underset{..}{\text{O}}}:$$

number of valence electrons	6	6	6	6	6	6
− number of lone pair electrons	6	2	4	4	2	6
− 1/2(number of bonding electrons)	1	3	2	2	3	1
Formal charge	−1	+1	0	0	+1	−1

(b) CO_3^{2-}:   Write the correct skeletal structure for the ion.

$$\begin{array}{c} \text{O} \\ | \\ \text{O}-\text{C}-\text{O} \end{array}$$

Calculate the total number of electrons for the Lewis structure by summing the number of valence electrons of each atom in the ion and adding 2 for the 2− charge.

$$3(\text{number of valence } e^- \text{ for O}) + (\text{number of valence } e^- \text{ for C}) + 2 = 3(6) + 4 + 2 = 24$$

Distribute the electrons among the atoms, giving octets to as many atoms as possible. Begin with the bonding electrons; then proceed to lone pairs on terminal atoms and finally to lone pairs on the central atom.

All 24 valence electrons are used.

If any atom lacks an octet, form double or triple bonds as necessary.

Finally, write the Lewis structure in brackets with the charge of the ion in the upper right-hand corner.

$$\left[\begin{array}{c} :\overset{..}{\text{O}}: \\ | \\ :\overset{..}{\underset{..}{\text{O}}}-\text{C}=\overset{..}{\underset{..}{\text{O}}}: \end{array} \right]^{2-}$$

All atoms have octets; the structure is complete. However, the double bond can form from any oxygen atom, so there are three resonance forms.

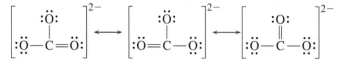

Calculate the formal charge on each atom by finding the number of valence electrons and subtracting the number of lone pair electrons and one-half the number of bonding electrons.

$$\left[\begin{array}{c} :\ddot{O}: \\ | \\ :\ddot{O}-C=\ddot{O}: \end{array}\right]^{2-}$$

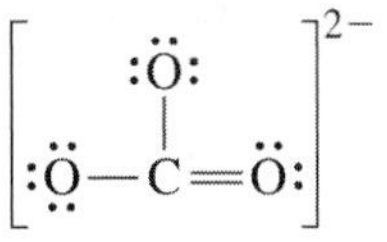

	O_{left}	O_{top}	O_{right}	C
number of valence electrons	6	6	6	4
− number of lone pair electrons	6	6	4	0
− 1/2(number of bonding electrons)	1	1	2	4
Formal charge	−1	−1	0	0

The sum of the formal charges is −2, which is the overall charge of the ion. The other resonance forms would have the same values for the single- and double-bonded oxygen atoms.

(c) ClO^-:   Write the correct skeletal structure for the ion.

Cl—O

Calculate the total number of electrons for the Lewis structure by summing the number of valence electrons of each atom in the ion and adding 1 for the 1− charge.

(number of valence e⁻ for O) + (number of valence e⁻ for Cl) + 1 = 6 + 7 + 1 = 14

Distribute the electrons among the atoms, giving octets to as many atoms as possible. Begin with the bonding electrons; then proceed to lone pairs on terminal atoms and finally to lone pairs on the central atom.

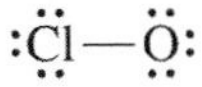

:$\ddot{\ddot{Cl}}$—$\ddot{\ddot{O}}$:

All 14 valence electrons are used.

If any atom lacks an octet, form double or triple bonds as necessary to give them octets.

Finally, write the Lewis structure in brackets with the charge of the ion in the upper right-hand corner.

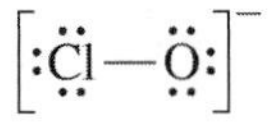

$$\left[:\ddot{\ddot{Cl}}-\ddot{\ddot{O}}:\right]^-$$

All atoms have octets; the structure is complete.

Calculate the formal charge on each atom by finding the number of valence electrons and subtracting the number of lone pair electrons and one-half the number of bonding electrons.

	Cl	O
number of valence electrons	7	6
− number of lone pair electrons	6	6
− 1/2(number of bonding electrons)	1	1
Formal charge	0	−1

The sum of the formal charges is −1, which is the overall charge of the ion.

(d) NO_2^-:   Write the correct skeletal structure for the ion.

O—N—O

Calculate the total number of electrons for the Lewis structure by summing the number of valence electrons of each atom in the ion and adding 1 for the 1− charge.

2(number of valence e⁻ for O) + (number of valence e⁻ for N) + 1 = 2(6) + 5 + 1 = 18

Distribute the electrons among the atoms, giving octets to as many atoms as possible. Begin with the bonding electrons; then proceed to lone pairs on terminal atoms and finally to lone pairs on the central atom.

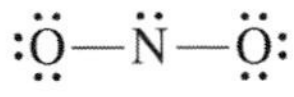

:$\ddot{\ddot{O}}$—$\ddot{N}$—$\ddot{\ddot{O}}$:

All 14 valence electrons are used.

If any atom lacks an octet, form double or triple bonds as necessary.

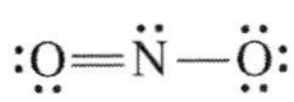

:$\ddot{O}$=$\ddot{N}$—$\ddot{\ddot{O}}$:

Finally, write the Lewis structure in brackets with the charge of the ion in the upper right-hand corner.

$$\left[:\ddot{O}\!=\!\ddot{N}\!-\!\ddot{O}: \right]^{-}$$

All atoms have octets; the structure is complete. However, the double bond can form from either oxygen atom, so there are two resonance forms.

$$\left[:\ddot{O}\!=\!\ddot{N}\!-\!\ddot{O}: \right]^{-} \longleftrightarrow \left[:\ddot{O}\!-\!\ddot{N}\!=\!\ddot{O}: \right]^{-}$$

Calculate the formal charge on each atom by finding the number of valence electrons and subtracting the number of lone pair electrons and one-half the number of bonding electrons. Using the left-side structure:

	O	N	O
number of valence electrons	6	5	6
− number of lone pair electrons	4	2	6
− 1/2(number of bonding electrons)	2	3	1
Formal charge	0	0	−1

The sum of the formal charges is −1, which is the overall charge of the ion.

9.65

$$\underset{I}{\overset{\overset{\displaystyle H}{|}}{H\!-\!C\!=\!\ddot{S}}} \qquad \underset{II}{\overset{\overset{\displaystyle H}{|}}{H\!-\!S\!=\!\ddot{C}}}$$

Calculate the formal charge on each atom in structure I by finding the number of valence electrons and subtracting the number of lone pair electrons and one-half the number of bonding electrons.

	H_{left}	H_{top}	C	S
number of valence electrons	1	1	4	6
− number of lone pair electrons	0	0	0	4
− 1/2(number of bonding electrons)	1	1	4	2
Formal charge	0	0	0	0

The sum of the formal charges is 0, which is the overall charge of the molecule.
Calculate the formal charge on each atom in structure II by finding the number of valence electrons and subtracting the number of lone pair electrons and one-half the number of bonding electrons.

	H_{left}	H_{top}	C	S
number of valence electrons	1	1	6	4
− number of lone pair electrons	0	0	0	4
− 1/2(number of bonding electrons)	1	1	4	2
Formal charge	0	0	+2	−2

The sum of the formal charges is 0, which is the overall charge of the molecule.
Structure I is the better Lewis structure because it has the least amount of formal charge on each atom.

9.67 $:O\!\equiv\!C\!-\!\ddot{O}:$ does not provide a significant contribution to the resonance hybrid as it has a +1 formal charge on a very electronegative oxygen.

	O_{left}	O_{right}	C
number of valence electrons	6	6	4
− number of lone pair electrons	2	6	0
− 1/2(number of bonding electrons)	3	1	4
Formal charge	+1	−1	0

9.69   CH₃COO⁻:   Write the correct skeletal structure for the ion.

Calculate the total number of electrons for the Lewis structure by summing the valence electrons of each atom in the ion and adding 1 for the 1− charge.
3(number of valence e⁻ for H) + 2(number of valence e⁻ for C) + 2(number of valence e⁻ for O) + 1 = 3(1) + 2(4) + 2(6) + 1 = 24
Distribute the electrons among the atoms, giving octets (or duets for H) to as many atoms as possible. Begin with the bonding electrons; then proceed to lone pairs on terminal atoms and finally to lone pairs of the central atom.

All 24 valence electrons are used.
If any atom lacks an octet, form double or triple bonds as necessary to give them octets.

Finally write the Lewis structure in brackets with the charge of the ion in the upper right-hand corner.

All atoms have octets (duets for H); the structure is complete. However, the double bond can form from any oxygen atom, so there are two resonance forms.

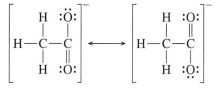

Calculate the formal charge on each atom by finding the number of valence electrons and subtracting the number of lone pair electrons and one-half the number of bonding electrons.

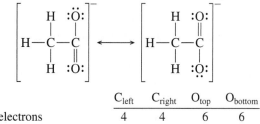

	C_{left}	C_{right}	O_{top}	O_{bottom}		C_{left}	C_{right}	O_{top}	O_{bottom}
number of valence electrons	4	4	6	6		4	4	6	6
− number of lone pair electrons	0	0	6	4		0	0	4	6
− 1/2(number of bonding electrons)	4	4	1	2		4	4	2	1
Formal charge	0	0	−1	0		0	0	0	−1

The sum of the formal charges is −1, which is the overall charge of the ion.

9.71

	N	O
number of valence electrons	5	6
− number of lone pair electrons	0	6
− 1/2(number of bonding electrons)	4	1
Formal charge	+1	−1

Odd-Electron Species, Incomplete Octets, and Expanded Octets

9.73   (a) BCl_3:   Write the correct skeletal structure for the molecule.

B is less electronegative, so it is central.

$$\begin{array}{c} \text{Cl} \\ | \\ \text{Cl}-\text{B}-\text{Cl} \end{array}$$

Calculate the total number of electrons for the Lewis structure by summing the number of valence electrons of each atom in the molecule.

(number of valence e⁻ for B) + 3(number of valence e⁻ for Cl) = 3 + 3(7) = 24

Distribute the electrons among the atoms, giving octets to as many atoms as possible. Begin with the bonding electrons; then proceed to lone pairs on terminal atoms and finally to lone pairs on the central atom.

$$\begin{array}{c} :\ddot{\text{Cl}}: \\ | \\ :\ddot{\text{Cl}}-\text{B}-\ddot{\text{Cl}}: \end{array}$$

All 24 valence electrons are used.

B has an incomplete octet. If we complete the octet, there is a formal charge of −1 on the B, which is less electronegative than Cl.

    (b) NO_2:   Write the correct skeletal structure for the molecule.

N is less electronegative, so it is central.

$$\text{O}-\text{N}-\text{O}$$

Calculate the total number of electrons for the Lewis structure by summing the number of valence electrons of each atom in the molecule.

(number of valence e⁻ for N) + 2(number of valence e⁻ for O) = 5 + 2(6) = 17

Distribute the electrons among the atoms, giving octets to as many atoms as possible. Begin with the bonding electrons; then proceed to lone pairs on terminal atoms and finally to lone pairs on the central atom.

$$\ddot{\text{O}}=\dot{\text{N}}-\ddot{\text{O}}: \longleftrightarrow :\ddot{\text{O}}-\dot{\text{N}}=\ddot{\text{O}}$$

All 17 valence electrons are used.

N has an incomplete octet. It has seven electrons because we have an odd number of valence electrons.

    (c) BH_3:   Write the correct skeletal structure for the molecule.

B is less electronegative, so it is central.

$$\begin{array}{c} \text{H} \\ | \\ \text{H}-\text{B}-\text{H} \end{array}$$

Calculate the total number of electrons for the Lewis structure by summing the number of valence electrons of each atom in the molecule.

(number of valence e⁻ for B) + 3(number of valence e⁻ for H) = 3 + 3(1) = 6

Distribute the electrons among the atoms, giving octets (or duets for H) to as many atoms as possible. Begin with the bonding electrons; then proceed to lone pairs on terminal atoms and finally to lone pairs on the central atom.

$$\begin{array}{c} H \\ | \\ H-B-H \end{array}$$

All six valence electrons are used.

B has an incomplete octet. H cannot double-bond, so it is not possible to complete the octet on B with a double-bond.

9.75   (a) PO_4^{3-}:   Write the correct skeletal structure for the ion.

$$\begin{array}{c} O \\ | \\ O-P-O \\ | \\ O \end{array}$$

Calculate the total number of electrons for the Lewis structure by summing the number of valence electrons of each atom in the ion and adding 3 for the 3− charge.

$$4(\text{number of valence } e^- \text{ for O}) + (\text{number of valence } e^- \text{ for P}) + 3 = 4(6) + 5 + 3 = 32$$

Distribute the electrons among the atoms, giving octets to as many atoms as possible. Begin with the bonding electrons; then proceed to lone pairs on terminal atoms and finally to lone pairs on the central atom.

All 32 valence electrons are used.

Finally, write the Lewis structure in brackets with the charge of the ion in the upper right-hand corner.

$$\left[\begin{array}{c} \ddot{:O:} \\ | \\ :\ddot{O}-P-\ddot{O}: \\ | \\ :\ddot{O}: \end{array} \right]^{3-}$$

All atoms have octets (duets for H); the structure is complete.

Calculate the formal charge on each atom by finding the number of valence electrons and subtracting the number of lone pair electrons and one-half the number of bonding electrons.

	O_{left}	O_{top}	O_{right}	O_{bottom}	P
number of valence electrons	6	6	6	6	5
− number of lone pair electrons	6	6	6	6	0
− 1/2(number of bonding electrons)	1	1	1	1	4
Formal charge	−1	−1	−1	−1	+1

The sum of the formal charges is -3, which is the overall charge of the ion. However, we can write a resonance structure with a double bond to an oxygen because P can expand its octet. This leads to lower formal charges on P and O.

Using the leftmost structure, calculate the formal charge on each atom by finding the number of valence electrons and subtracting the number of lone pair electrons and one-half the number of bonding electrons.

	O_{left}	O_{top}	O_{right}	O_{bottom}	P
number of valence electrons	6	6	6	6	5
− number of lone pair electrons	4	6	6	6	0
− 1/2(number of bonding electrons)	2	1	1	1	5
Formal charge	0	−1	−1	−1	0

The sum of the formal charges is -3, which is the overall charge of the ion. These resonance forms would all have the lower formal charges associated with the double-bonded O and P.

(b)   CN⁻:   Write the correct skeletal structure for the ion.

C—N

Calculate the total number of electrons for the Lewis structure by summing the number of valence electrons of each atom in the ion and adding 1 for the 1− charge.

(number of valence e⁻ for C) + (number of valence e⁻ for N) + 1 = 4 + 5 + 1 = 10

Distribute the electrons among the atoms, giving octets to as many atoms as possible. Begin with the bonding electrons; then proceed to lone pairs on terminal atoms and finally to lone pairs on the central atom.

:C—N̈:

All ten valence electrons are used.

If any atom lacks an octet, form double or triple bonds as necessary.

:C≡N:

Finally, write the Lewis structure in brackets with the charge of the ion in the upper right-hand corner.

$$\left[:\!C\!\!\equiv\!\!N\!: \right]^{-}$$

All atoms have octets; the structure is complete.

Calculate the formal charge on each atom by finding the number of valence electrons and subtracting the number of lone pair electrons and one-half the number of bonding electrons.

$$\left[:\!C\!\!\equiv\!\!N\!: \right]^{-}$$

	C	N
number of valence electrons	4	5
− number of lone pair electrons	2	2
− 1/2(number of bonding electrons)	3	3
Formal charge	−1	0

The sum of the formal charges is -1, which is the overall charge of the ion.

(c) SO_3^{2-}:   Write the correct skeletal structure for the ion.

O
|
O—S—O

Calculate the total number of electrons for the Lewis structure by summing the valence electrons of each atom in the ion and adding 2 for the 2− charge.

3(number of valence e⁻ for O) + (number of valence e⁻ for S) + 2 = 3(6) + 6 + 2 = 26

Distribute the electrons among the atoms, giving octets to as many atoms as possible. Begin with the bonding electrons; then proceed to lone pairs on terminal atoms and finally to lone pairs on the central atom.

$$:\ddot{O}:$$
$$|$$
$$:\ddot{O}-\ddot{S}-\ddot{O}:$$

All 26 valence electrons are used.

Finally, write the Lewis structure in brackets with the charge of the ion in the upper right-hand corner.

$$\left[\begin{array}{c} :\ddot{O}: \\ | \\ :\ddot{O}-\ddot{S}-\ddot{O}: \end{array}\right]^{2-}$$

Calculate the formal charge on each atom by finding the number of valence electrons and subtracting the number of lone pair electrons and one-half the number of bonding electrons.

$$\left[\begin{array}{c} :\ddot{O}: \\ | \\ :\ddot{O}-\ddot{S}-\ddot{O}: \end{array}\right]^{2-}$$

	O_{left}	O_{top}	O_{right}	S
number of valence electrons	6	6	6	6
− number of lone pair electrons	6	6	6	2
− 1/2(number of bonding electrons)	1	1	1	3
Formal charge	−1	−1	−1	+1

The sum of the formal charges is −2, which is the overall charge of the ion. However, we can write a resonance structure with a double bond to an oxygen because S can expand its octet. This leads to a lower formal charge.

$$\left[\begin{array}{c} :\ddot{O}: \\ | \\ \ddot{O}=\ddot{S}-\ddot{O}: \end{array}\right]^{2-} \longleftrightarrow \left[\begin{array}{c} :O: \\ || \\ :\ddot{O}-\ddot{S}-\ddot{O}: \end{array}\right]^{2-} \longleftrightarrow \left[\begin{array}{c} :\ddot{O}: \\ | \\ :\ddot{O}-\ddot{S}=\ddot{O} \end{array}\right]^{2-}$$

Using the leftmost resonance form, calculate the formal charge on each atom by finding the number of valence electrons and subtracting the number of lone pair electrons and one-half the number of bonding electrons.

	O_{left}	O_{top}	O_{right}	S
number of valence electrons	6	6	6	6
− number of lone pair electrons	4	6	6	2
− 1/2(number of bonding electrons)	2	1	1	4
Formal charge	0	−1	−1	0

The sum of the formal charges is −2, which is the overall charge of the ion. These resonance forms would all have the lower formal charge on the double-bonded O and S.

(d) ClO_2^-:   Write the correct skeletal structure for the ion.

$$O-Cl-O$$

Calculate the total number of electrons for the Lewis structure by summing the number of valence electrons of each atom in the ion and adding 1 for the 1− charge.

$$2(\text{number of valence e}^- \text{ for O}) + (\text{number of valence e}^- \text{ for Cl}) + 1 = 2(6) + 7 + 1 = 20$$

Distribute the electrons among the atoms, giving octets (or duets for H) to as many atoms as possible. Begin with the bonding electrons; then proceed to lone pairs on terminal atoms and finally to lone pairs on the central atom.

$$:\ddot{O}-\ddot{Cl}-\ddot{O}:$$

All 20 valence electrons are used.

Finally, write the Lewis structure in brackets with the charge of the ion in the upper right-hand corner.

$$\left[:\ddot{O}-\ddot{C}l-\ddot{O}:\right]^-$$

All atoms have octets; the structure is complete.

Calculate the formal charge on each atom by finding the number of valence electrons and subtracting the number of lone pair electrons and one-half the number of bonding electrons.

$$\left[:\ddot{O}-\ddot{C}l-\ddot{O}:\right]^-$$

	O_{left}	O_{right}	Cl
number of valence electrons	6	6	7
− number of lone pair electrons	6	6	4
− 1/2(number of bonding electrons)	1	1	2
Formal charge	−1	−1	+1

The sum of the formal charges is −1, which is the overall charge of the ion. However, we can write a resonance structure with a double bond to an oxygen because Cl can expand its octet. This leads to a lower formal charge.

$$\left[\ddot{O}=\ddot{C}l-\ddot{O}:\right]^- \longleftrightarrow \left[:\ddot{O}-\ddot{C}l=\ddot{O}\right]^-$$

Using the leftmost resonance form, calculate the formal charge on each atom by finding the number of valence electrons and subtracting the number of lone pair electrons and one-half the number of bonding electrons.

	O_{left}	O_{right}	Cl
number of valence electrons	6	6	7
− number of lone pair electrons	4	6	4
− 1/2(number of bonding electrons)	2	1	3
Formal charge	0	−1	0

The sum of the formal charges is −1, which is the overall charge of the ion. These resonance forms would all have the lower formal charge on the double-bonded O and Cl.

9.77   (a) PF_5:   Write the correct skeletal structure for the molecule.

Calculate the total number of electrons for the Lewis structure by summing the number of valence electrons of each atom in the molecule.

(number of valence e^- for P) + 5(number of valence e^- for F) 5 + 5(7) = 40

Distribute the electrons among the atoms, giving octets to as many atoms as possible. Begin with the bonding electrons; then proceed to lone pairs on terminal atoms and finally to lone pairs on the central atom. Arrange additional electrons around the central atom, giving it an expanded octet of up to 12 electrons.

(b) I_3^-:   Write the correct skeletal structure for the ion.

I—I—I

Calculate the total number of electrons for the Lewis structure by summing the number of valence electrons of each atom in the ion and adding 1 for the 1− charge.

3(number of valence e^- for I) + 1 = 3(7) + 1 = 22

Distribute the electrons among the atoms, giving octets to as many atoms as possible. Begin with the bonding electrons; then proceed to lone pairs on terminal atoms and finally to lone pairs on the central atom. Arrange additional electrons around the central atom, giving it an expanded octet of up to 12 electrons.

$$:\ddot{I}-\ddot{I}-\ddot{I}:$$

Finally, write the Lewis structure in brackets with the charge of the ion in the upper right-hand corner.

$$\left[:\ddot{I}-\ddot{I}-\ddot{I}:\right]^{-}$$

(c) SF_4:   Write the correct skeletal structure for the molecule.

$$
\begin{array}{c}
F \\
| \\
F-S-F \\
| \\
F
\end{array}
$$

Calculate the total number of electrons for the Lewis structure by summing the number of valence electrons of each atom in the molecule.

(number of valence e⁻ for S) + 4(number of valence e⁻ for F) = 6 + 4(7) = 34

Distribute the electrons among the atoms, giving octets (or duets for H) to as many atoms as possible. Begin with the bonding electrons; then proceed to lone pairs on terminal atoms and finally to lone pairs on the central atom. Arrange additional electrons around the central atom, giving it an expanded octet of up to 12 electrons.

$$
\begin{array}{c}
:\ddot{F}: \\
| \\
:\ddot{F}-\ddot{S}-\ddot{F}: \\
| \\
:\ddot{F}:
\end{array}
$$

(d) GeF_4:   Write the correct skeletal structure for the molecule.

$$
\begin{array}{c}
F \\
| \\
F-Ge-F \\
| \\
F
\end{array}
$$

Calculate the total number of electrons for the Lewis structure by summing the number of valence electrons of each atom in the molecule.

(number of valence e⁻ for Ge) + 4(number of valence e⁻ for F) = 4 + 4(7) = 32

Distribute the electrons among the atoms, giving octets to as many atoms as possible. Begin with the bonding electrons; then proceed to lone pairs on terminal atoms and finally to lone pairs on the central atom.

$$
\begin{array}{c}
:\ddot{F}: \\
| \\
:\ddot{F}-Ge-\ddot{F}: \\
| \\
:\ddot{F}:
\end{array}
$$

Bond Energies and Bond Lengths

9.79   Bond strength: $H_3CCH_3 < H_2CCH_2 < HCCH$
Bond length: $H_3CCH_3 > H_2CCH_2 > HCCH$
Write the Lewis structures for the three compounds. Compare the C—C bonds. Triple bonds are stronger than double bonds; double bonds are stronger than single bonds. Also, single bonds are longer than double bonds, which are longer than triple bonds.

HCCH (10 e⁻) H_2CCH_2 (12 e⁻) H_3CCH_3 (14 e⁻)

$$H-C \equiv C-H$$

9.81   Rewrite the reaction using the Lewis structures of the molecules involved.

Determine which bonds are broken in the reaction and sum the bond energies of the following:

$$\Sigma(\Delta H\text{'s bonds broken})$$
$$= 4\,\text{mol}(C-H) + 1\,\text{mol}(C=C) + 1\,\text{mol}(H-H)$$
$$= 4\,\text{mol}(414\,\text{kJ/mol}) + 1\,\text{mol}(611\,\text{kJ/mol}) + 1\,\text{mol}(436\,\text{kJ/mol})$$
$$= 2703\,\text{kJ/mol}$$

Determine which bonds are formed in the reaction and sum the negatives of the bond energies of the following:

$$\Sigma(-\Delta H\text{'s bonds formed})$$
$$= -6\,\text{mol}(C-H) - 1\,\text{mol}(C-C)$$
$$= -6\,\text{mol}(414\,\text{kJ/mol}) - 1\,\text{mol}(347\,\text{kJ/mol})$$
$$= -2831\,\text{kJ/mol}$$

Find ΔH_{rxn} by summing the results of the two steps.

$$\Delta H_{rxn} = \Sigma(\Delta H\text{'s bonds broken}) + \Sigma(\Delta H - \Delta H\text{'s bonds formed})$$
$$= 2703\,\text{kJ/mol} - 2831\,\text{kJ/mol}$$
$$= -128\,\text{kJ/mol}$$

9.83   Rewrite the reaction using the Lewis structures of the molecules involved.

$$C + 2H-\overset{..}{\underset{..}{O}}-H \longrightarrow 2H-H + \overset{..}{\underset{..}{O}}=C=\overset{..}{\underset{..}{O}}$$

Determine which bonds are broken in the reaction and sum the bond energies of the following:

$$\Sigma(\Delta H\text{'s bonds broken})$$
$$= 4(O-H)$$
$$= 4(464\,\text{kJ/mol})$$
$$= 1856\,\text{kJ/mol}$$

Determine which bonds are formed in the reaction and sum the negatives of the bond energies of the following:

$$\Sigma(-\Delta H\text{'s bonds formed})$$
$$= -2(C=O) - 2(H-H)$$
$$= -2(799\,\text{kJ/mol}) - 2(436\,\text{kJ/mol})$$
$$= -2470\,\text{kJ/mol}$$

Find ΔH_{rxn} by summing the results of the two steps.

$$\Delta H_{rxn} = \Sigma(\Delta H\text{'s bonds broken}) + \Sigma(-\Delta H\text{'s bonds formed})$$
$$= 1856\,\text{kJ/mol} - 2470\,\text{kJ/mol}$$
$$= -614\,\text{kJ/mol}$$

Cumulative Problems

9.85   (a) BI_3: This is a covalent compound between two nonmetals.

Write the correct skeletal structure for the molecule.

$$\begin{array}{c} I \\ | \\ I-B-I \end{array}$$

Calculate the total number of electrons for the Lewis structure by summing the number of valence electrons of each atom in the molecule.

(number of valence e^- for B) + (number of valence e^- for I) = 3 + 3(7) = 24

Distribute the electrons among the atoms, giving octets to as many atoms as possible. Begin with the bonding electrons; then proceed to lone pairs on terminal atoms and finally to lone pairs on the central atom.

$$\begin{array}{c} :\overset{..}{I}: \\ | \\ :\overset{..}{\underset{..}{I}}-B-\overset{..}{\underset{..}{I}}: \end{array}$$

(b) K_2S: This is an ionic compound between a metal and nonmetal.

Draw the Lewis structures for K and S based on their valence electrons. K: $4s^1$ S: $3s^2 3p^4$

$$K\cdot \quad \cdot \ddot{\underset{\cdot\cdot}{S}}:$$

Potassium must lose one electron and be left with the octet from the previous shell, while sulfur needs to gain two electrons to get an octet.

$$2K^+ \quad \left[:\ddot{\underset{\cdot\cdot}{S}}:\right]^{2-}$$

(c)   HCFO: This is a covalent compound between nonmetals.

Write the correct skeletal structure for the molecule.

$$\begin{array}{c} O \\ | \\ H-C-F \end{array}$$

Calculate the total number of electrons for the Lewis structure by summing the number of valence electrons of each atom in the molecule.

(number of valence e^- for H) + (number of valence e^- for C) + (number of valence e^- for F) + (number of valence e^- for O) = 1 + 4 + 7 + 6 = 18

Distribute the electrons among the atoms, giving octets to as many atoms as possible. Begin with the bonding electrons; then proceed to lone pairs on terminal atoms and finally to lone pairs on the central atom.

$$\begin{array}{c} :\ddot{O}: \\ | \\ H-C-\ddot{\underset{\cdot\cdot}{F}}: \end{array}$$

If any atom lacks an octet, form double or triple bonds as necessary to give them octets.

$$\begin{array}{c} :O: \\ \| \\ H-C-\ddot{\underset{\cdot\cdot}{F}}: \end{array}$$

(d) PBr_3: This is a covalent compound between two nonmetals.

Write the correct skeletal structure for the molecule.

$$\begin{array}{c} Br \\ | \\ Br-P-Br \end{array}$$

Calculate the total number of electrons for the Lewis structure by summing the number of valence electrons of each atom in the molecule.

(number of valence e^- for P) + 3(number of valence e^- for Br) = 5 + 3(7) = 26

Distribute the electrons among the atoms, giving octets to as many atoms as possible. Begin with the bonding electrons; then proceed to lone pairs on terminal atoms and finally to lone pairs on the central atom.

$$\begin{array}{c} :\ddot{Br}: \\ | \\ :\ddot{Br}-\ddot{P}-\ddot{Br}: \end{array}$$

9.87   (a) $BaCO_3$: Ba^{2+}

$$\left[\begin{array}{c} :\ddot{O}: \\ | \\ \ddot{\underset{\cdot\cdot}{O}}=C-\ddot{\underset{\cdot\cdot}{O}}: \end{array} \right]^{2-}$$

Determine the cation and anion.

$$Ba^{2+} \quad CO_3^{2-}$$

Write the Lewis structure for the barium cation based on the valence electrons.

$$\begin{array}{ll} Ba \quad 5s^2 & Ba^{2+} \quad 5s^0 \\ Ba: & Ba^{2+} \end{array}$$

Ba must lose two electrons and be left with the octet from the previous shell.

Write the Lewis structure for the covalent anion.

Write the correct skeletal structure for the ion.

$$O$$
$$|$$
$$O—C—O$$

Calculate the total number of electrons for the Lewis structure by summing the number of valence electrons of each atom in the ion and adding two for the 2− charge.

(number of valence e⁻ for C) + 3(number of valence e⁻ for O) = 4 + 3(6) + 2 = 24

Distribute the electrons among the atoms, giving octets to as many atoms as possible. Begin with the bonding electrons; then proceed to lone pairs on terminal atoms and finally to lone pairs on the central atom.

$$:\ddot{O}:$$
$$|$$
$$:\ddot{O}—C—\ddot{O}:$$

If any atom lacks an octet, form double or triple bonds as necessary.

$$:\ddot{O}:$$
$$|$$
$$\ddot{O}=C—\ddot{O}:$$

Finally, write the Lewis structure in brackets with the charge of the ion in the upper right-hand corner.

$$Ba^{2+} \left[\begin{array}{c} :\ddot{O}: \\ | \\ \ddot{O}=C—\ddot{O}: \end{array} \right]^{2-}$$

The double bond can be between the C and any of the oxygen atoms, so there are resonance structures.

$$Ba^{2+} \left[\begin{array}{c} :\ddot{O}: \\ | \\ \ddot{O}=C—\ddot{O}: \end{array} \right]^{2-} \longleftrightarrow Ba^{2+} \left[\begin{array}{c} :O: \\ \| \\ :\ddot{O}—C—\ddot{O}: \end{array} \right]^{2-} \longleftrightarrow Ba^{2+} \left[\begin{array}{c} :\ddot{O}: \\ | \\ :\ddot{O}—C=\ddot{O} \end{array} \right]^{2-}$$

(b) $Ca(OH)_2$: Ca^{2+}

$$2 \left[:\ddot{O}—H \right]^{-}$$

Determine the cation and anion.

$$Ca^{2+} \qquad OH^{-}$$

Write the Lewis structure for the calcium cation based on the valence electrons.

Ca $4s^2$ Ca^{2+} $4s^0$

Ca: Ca^{2+}

Ca must lose two electrons and be left with the octet from the previous shell.

Write the Lewis structure for the covalent anion.

Write the correct skeletal structure for the ion.

$$O—H$$

Calculate the total number of electrons for the Lewis structure by summing the valence electrons of each atom in the ion and adding one for the 1− charge.

(number of valence e⁻ for H) + (number of valence e⁻ for O) + 1 = 1 + 6 + 1 = 8

Distribute the electrons among the atoms, giving octets (or duets for H) to as many atoms as possible. Begin with the bonding electrons; then proceed to lone pairs on terminal atoms and finally to lone pairs on the central atom.

$$:\ddot{O}—H$$

Finally, write the Lewis structure in brackets with the charge of the ion in the upper right-hand corner.

$$Ca^{2+} \left[:\ddot{O}—H \right]^{-}$$

(c) KNO_3: K^{+}

Determine the cation and anion.

$$K^+ \qquad NO_3^-$$

Write the Lewis structure for the potassium cation based on the valence electrons.

$$K\, 4s^1 \qquad K^+4s^0$$
$$K\cdot \qquad K^+$$

K must lose one electron and be left with the octet from the previous shell.

Write the Lewis structure for the covalent anion.

Write the correct skeletal structure for the ion.

$$
\begin{array}{c}
O \\
| \\
O{-}N{-}O
\end{array}
$$

Calculate the total number of electrons for the Lewis structure by summing the valence electrons of each atom in the ion and adding one for the 1− charge.

$$(\text{number of valence e}^-\text{ for N}) + (\text{number of valence e}^-\text{ for O}) = 5 + 3(6) + 1 = 24$$

Distribute the electrons among the atoms, giving octets to as many atoms as possible. Begin with the bonding electrons; then proceed to lone pairs on terminal atoms and finally to lone pairs on the central atom.

$$
\begin{array}{c}
:\ddot{O}: \\
| \\
:\ddot{O}{-}N{-}\ddot{O}:
\end{array}
$$

If any atom lacks an octet, form double or triple bonds as necessary.

$$
\begin{array}{c}
:\ddot{O}: \\
| \\
\ddot{O}{=}N{-}\ddot{O}:
\end{array}
$$

Finally, write the Lewis structure in brackets with the charge of the ion in the upper right-hand corner.

$$
K^+ \left[\begin{array}{c}
:\ddot{O}: \\
| \\
\ddot{O}{=}N{-}\ddot{O}:
\end{array}\right]^-
$$

The double bond can be between the N and any of the oxygen atoms, so there are resonance structures.

$$
K^+ \left[\begin{array}{c}
:\ddot{O}: \\
| \\
\ddot{O}{=}N{-}\ddot{O}:
\end{array}\right]^- \longleftrightarrow
K^+ \left[\begin{array}{c}
:O: \\
\| \\
:\ddot{O}{-}N{-}\ddot{O}:
\end{array}\right]^- \longleftrightarrow
K^+ \left[\begin{array}{c}
:\ddot{O}: \\
| \\
:\ddot{O}{-}N{=}\ddot{O}
\end{array}\right]^-
$$

(d) LiIO: Li⁺

$$\left[:\ddot{I}{-}\ddot{O}:\right]^-$$

Determine the cation and anion.

$$Li^+ \qquad IO^-$$

Write the Lewis structure for the lithium cation based on the valence electrons.

$$Li\, 2s^1 \qquad Li^+2s^0$$
$$Li\cdot \qquad Li^+$$

Li must lose one electron and be left with the octet from the previous shell.

Write the Lewis structure for the covalent anion.

Write the correct skeletal structure for the ion.

$$I{-}O$$

Calculate the total number of electrons for the Lewis structure by summing the number of valence electrons of each atom in the ion and adding one for the 1− charge.

$$(\text{number of valence e}^-\text{ for I}) + (\text{number of valence e}^-\text{ for O}) = 7 + 6 + 1 = 14$$

Distribute the electrons among the atoms, giving octets to as many atoms as possible. Begin with the bonding electrons; then proceed to lone pairs on terminal atoms and finally to lone pairs on the central atom.

$$:\ddot{I}{-}\ddot{O}:$$

Finally, write the Lewis structure in brackets with the charge of the ion in the upper right-hand corner.

$$Li^+ \left[:\ddot{I}{-}\ddot{O}:\right]^-$$

9.89   (a) C_4H_8:   Write the correct skeletal structure for the molecule.

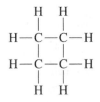

Calculate the total number of electrons for the Lewis structure by summing the number of valence electrons of each atom in the molecule.

4(number of valence e⁻ for C) + 8(number of valence e⁻ for H) = 4(4) + 8(1) = 24

Distribute the electrons among the atoms, giving octets (or duets for H) to as many atoms as possible.

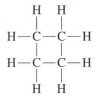

All atoms have octets or duets for H.

   (b) C_4H_4:   Write the correct skeletal structure for the molecule.

H—C—C—H
     | |
H—C—C—H

Calculate the total number of electrons for the Lewis structure by summing the number of valence electrons of each atom in the molecule.

4(number of valence e⁻ for C) + 4(number of valence e⁻ for H) = 4(4) + 4(1) = 20

Distribute the electrons among the atoms, giving octets (or duets for H) to as many atoms as possible.

H—C̈—C̈—H
     | |
H—C—C—H

Complete octets by forming double bonds on alternating carbons; draw resonance structures.

H—C—C—H   H—C=C—H
     ‖   ‖ ⟷ | |
H—C—C—H   H—C=C—H

   (c) C_6H_{12}: Write the correct skeletal structure for the molecule.

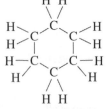

Calculate the total number of electrons for the Lewis structure by summing the valence electrons of each atom in the molecule.

6(number of valence e⁻ for C) + 12(number of valence e⁻ for H) = 6(4) + 12(1) = 36

Distribute the electrons among the atoms, giving octets (or duets for H) to as many atoms as possible. Begin with the bonding.

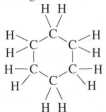

All 36 electrons are used, and all atoms have octets or duets for H.

(d) C_6H_6:   Write the correct skeletal structure for the molecule.

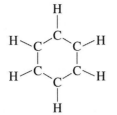

Calculate the total number of electrons for the Lewis structure by summing the number of valence electrons of each atom in the molecule.

6(number of valence e⁻ for C) + 6(number of valence e⁻ for H) = 6(4) + 6(1) = 30

Distribute the electrons among the atoms, giving octets (or duets for H) to as many atoms as possible.

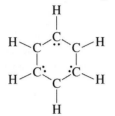

Complete octets by forming double bonds on alternating carbons; draw resonance structures.

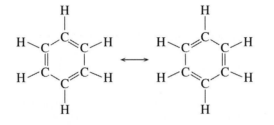

9.91   **Given:** 26.01% C; 4.38% H; 69.52% O; molar mass = 46.02 g/mol
Find: molecular formula and Lewis structure
Conceptual Plan: Convert mass to mol of each element → pseudoformula → empirical formula

$$\frac{1\ mol\ C}{12.01\ g\ C} \qquad \frac{1\ mol\ H}{1.008\ g\ H} \qquad \frac{1\ mol\ O}{16.00\ g\ O} \qquad\qquad \text{divide by smallest number}$$

→ molecular formula → Lewis structure

empirical formula × n

Solution: $26.01\ \cancel{g\ C} \times \dfrac{1\ mol\ C}{12.01\ \cancel{g\ C}} = 2.166\ mol\ C$

$4.38\ \cancel{g\ H} \times \dfrac{1\ mol\ H}{1.008\ \cancel{g\ H}} = 4.3\underline{4}5\ mol\ H$

$69.52\ \cancel{g\ O} \times \dfrac{1\ mol\ O}{16.00\ \cancel{g\ O}} = 4.345\ mol\ O$

$C_{2.166}H_{4.345}O_{4.345}$

$C_{\underset{2.166}{2.166}}H_{\underset{2.166}{4.345}}O_{\underset{2.166}{4.345}} \rightarrow CH_2O_2$

The correct empirical formula is CH_2O_2.

empirical formula mass = (12.01 g/mol) + 2(1.008 g/mol) + 2(16.00 g/mol) = 46.03 g/mol

$$n = \frac{\text{molar mass}}{\text{formula molar mass}} = \frac{46.02\ \cancel{g/mol}}{46.03\ \cancel{g/mol}} = 1$$

molecular formula   = $CH_2O_2 \times 1$

= CH_2O_2

Write the correct skeletal structure for the molecule.

$$\overset{\displaystyle O}{\underset{\displaystyle}{H-C-O-H}}$$

Calculate the total number of electrons for the Lewis structure by summing the number of valence electrons of each atom in the molecule.

(number of valence e^- for C) + 2(number of valence e^- for O) + 2(number of valence e^- for H)
= 4 + 2(6) + 2(1) = 18

Distribute the electrons among the atoms, giving octets (or duets for H) to as many atoms as possible. Begin with the bonding electrons; then proceed to lone pairs on terminal atoms and finally to lone pairs on the central atoms.

$$\begin{array}{c} :\ddot{O}: \\ | \\ H-C-\ddot{O}-H \end{array}$$

Complete the octet on C by forming a double bond.

$$\begin{array}{c} :O: \\ \| \\ H-C-\ddot{O}-H \\ \quad\ \ \ \ddot{} \end{array}$$

9.93   To determine the values of the lattice energy, it is necessary to look them up online. The lattice energy of Al_2O_3 is −15,916 kJ/mol; the value for Fe_2O_3 is −14,774 kJ/mol. The thermite reaction is exothermic due to the energy released when the Al_2O_3 lattice forms. The lattice energy of Al_2O_3 is more negative than the lattice energy of Fe_2O_3.

9.95 HNO_3 Write the correct skeletal structure for the molecule.

$$\begin{array}{c} O \\ | \\ H-O-N-O \end{array}$$

Calculate the total number of electrons for the Lewis structure by summing the number of valence electrons of each atom in the molecule.

3(number of valence e^- for O) + (number of valence e^- for N) + (number of valence e^- for H) = 3(6) + 5 + 1 = 24

Distribute the electrons among the atoms, giving octets (or duets for H) to as many atoms as possible. Begin with the bonding electrons; then proceed to lone pairs on terminal atoms and finally to lone pairs on the central atom.

$$\begin{array}{c} :\ddot{O}: \\ | \\ H-\ddot{O}-N-\ddot{O}: \end{array}$$

All 24 valence electrons are used.

If any atoms lack an octet, form double or triple bonds as necessary. The double bond can be formed to any of the three oxygen atoms, so there are three resonance forms.

$$\underset{\text{I}}{\begin{array}{c} :O: \\ \| \\ H-\ddot{O}-N-\ddot{O}: \end{array}} \longleftrightarrow \underset{\text{II}}{\begin{array}{c} :\ddot{O}: \\ | \\ H-\ddot{O}-N=\ddot{O} \end{array}} \longleftrightarrow \underset{\text{III}}{\begin{array}{c} :\ddot{O}: \\ | \\ H-\ddot{O}=N-\ddot{O}: \end{array}}$$

All atoms have octets (duets for H); the structure is complete.

To determine which resonance hybrid(s) is most important, calculate the formal charge on each atom in each structure by finding the number of valence electrons and subtracting the number of lone pair electrons and one-half the number of bonding electrons.

	Structure I					Structure II				
	O_{left}	O_{top}	O_{right}	N	H	O_{left}	O_{top}	O_{right}	N	H
number of valence electrons	6	6	6	5	1	6	6	6	5	1
− number of lone pair electrons	4	4	6	0	0	4	6	4	0	0
− 1/2(number of bonding electrons)	2	2	1	4	1	2	1	2	4	1
Formal charge	0	0	−1	+1	0	0	−1	0	+1	0

Structure III

	O_{left}	O_{top}	O_{right}	N	H
number of valence electrons	6	6	6	5	1
− number of lone pair electrons	2	6	6	0	0
− 1/2(number of bonding electrons)	3	1	1	4	1
Formal charge	+1	−1	−1	+1	0

The sum of the formal charges is 0 for each structure, which is the overall charge of the molecule. However, in structures I and II, the individual formal charges are lower. These two forms would contribute equally to the structure of HNO_3. Structure III would be less important because the individual formal charges are higher. In addition, oxygen is more electronegative than N and in this structure N has the positive formal charge. So structure III will be unacceptable.

9.97 CNO^- Write the skeletal structure.

C—N—O

Determine the number of valence electrons.

(valence e^- from C) + (valence e^- from N) + (valence e^- from O) + 1 (from the negative charge)

4 + 5 + 6 + 1 = 16

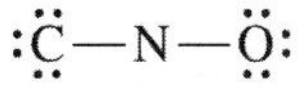

:C̈—N—Ö:

Distribute the electrons to complete octets if possible.

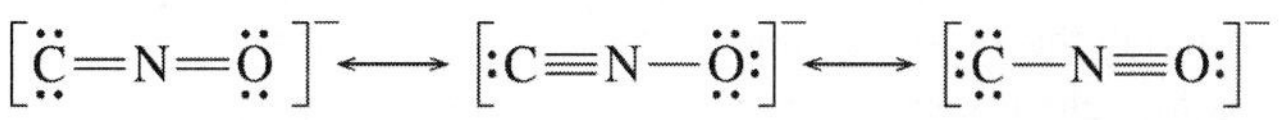

$$\left[\ddot{C}{=}N{=}\ddot{O}\right]^- \longleftrightarrow \left[:C{\equiv}N{-}\ddot{O}:\right]^- \longleftrightarrow \left[:\ddot{C}{-}N{\equiv}O:\right]^-$$

I   II   III

Determine the formal charge on each atom for each structure.

	Structure I			Structure II		
	C	N	O	C	N	O
number of valence electrons	4	5	6	4	5	6
− number of lone pair electrons	4	0	4	2	0	6
− 1/2(number of bonding electrons)	2	4	2	3	4	1
Formal charge	−2	+1	0	−1	+1	−1

	Structure III		
	C	N	O
number of valence electrons	4	5	6
− number of lone pair electrons	6	0	2
− 1/2(number of bonding electrons)	1	4	3
Formal charge	−3	+1	+1

Structures I, II, and III all follow the octet rule but have varying degrees of negative formal charge on carbon, which is the least electronegative atom. Also, the amount of formal charge is very high in all three resonance forms. Although structure II is the best of the resonance forms, it has a −1 charge on the least electronegative atom, C, and a +1 charge on the more electronegative atom, N. Therefore, none of these resonance forms contributes strongly to the stability of the fulminate ion and the ion is not very stable.

9.99   (a) C_2H_4:   Write the correct skeletal structure for the molecule.

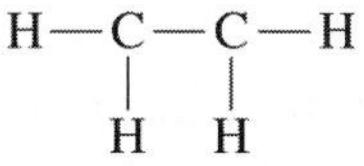

H—C—C—H
   | |
   H H

Calculate the total number of electrons for the Lewis structure by summing the valence electrons of each atom in the molecule.

2(number of valence e^- for C) + 4(number of valence e^- for H) = 2(4) + 4(1) = 12

Distribute the electrons among the atoms, giving octets (or duets for H) to as many atoms as possible. Begin with the bonding electrons; then proceed to lone pairs on terminal atoms and finally to lone pairs on the central atom.

$$\begin{array}{c} \overset{\displaystyle \cdot\cdot}{\text{C}}\text{C} \\ \text{H}-\overset{|}{\underset{|}{\text{C}}}-\overset{|}{\underset{|}{\text{C}}}-\text{H} \\ \text{H}\quad \text{H} \end{array}$$

All 12 valence electrons are used.

If any atom lacks an octet, form double or triple bonds to give them octets.

$$\begin{array}{c} \text{H}-\text{C}=\text{C}-\text{H} \\ \ \ |\quad\ | \\ \ \ \text{H}\quad \text{H} \end{array}$$

All atoms have octets (duets for hydrogen); the structure is complete.

(b) CH_3NH_2:   Write the correct skeletal structure for the molecule.

$$\begin{array}{c} \text{H} \\ | \\ \text{H}-\text{C}-\text{N}-\text{H} \\ |\quad\ | \\ \text{H}\quad \text{H} \end{array}$$

Calculate the total number of electrons for the Lewis structure by summing the valence electrons of each atom in the molecule.

number of valence e^- for C + 5(number of valence e^- for H) + number of valence electrons for N = 4 + 5(1) + 5 = 14

Distribute the electrons among the atoms, giving octets (or duets for H) to as many atoms as possible. Begin with the bonding electrons; then proceed to lone pairs on terminal atoms and finally to lone pairs on the central atom.

$$\begin{array}{c} \text{H} \\ | \\ \text{H}-\text{C}-\overset{\displaystyle \cdot\cdot}{\text{N}}-\text{H} \\ |\quad\ | \\ \text{H}\quad \text{H} \end{array}$$

All 14 valence electrons are used. All atoms have octets (duets for hydrogen); the structure is complete.

(c)   HCHO: Write the correct skeletal structure for the molecule.

$$\begin{array}{c} \text{O} \\ | \\ \text{H}-\text{C}-\text{H} \end{array}$$

Calculate the total number of electrons for the Lewis structure by summing the valence electrons of each atom in the molecule.

number of valence e^- for C + 2(number of valence e^- for H) + number of valence electrons for O = 4 + 2(1) + 6 = 12

Distribute the electrons among the atoms, giving octets (or duets for H) to as many atoms as possible. Begin with the bonding electrons; then proceed to lone pairs on terminal atoms and finally to lone pairs on the central atom.

$$\begin{array}{c} :\overset{\displaystyle \cdot\cdot}{\text{O}}: \\ | \\ \text{H}-\text{C}-\text{H} \end{array}$$

All 12 valence electrons are used.

If any atom lacks an octet, form double or triple bonds to give them octets.

$$\begin{array}{c} :\text{O}: \\ \| \\ \text{H}-\text{C}-\text{H} \end{array}$$

All atoms have octets (duets for hydrogen); the structure is complete.

(d) CH_3CH_2OH: Write the correct skeletal structure for the molecule.

$$\begin{array}{c} \text{H}\quad \text{H} \\ |\quad\ | \\ \text{H}-\text{C}-\text{C}-\text{O}-\text{H} \\ |\quad\ | \\ \text{H}\quad \text{H} \end{array}$$

Calculate the total number of electrons for the Lewis structure by summing the valence electrons of each atom in the molecule.

$$2(\text{number of valence e}^- \text{ for C}) + 6(\text{number of valence e}^- \text{ for H}) + \text{number of electrons}$$
$$\text{for O} = 2(4) + 6(1) + 6 = 20$$

Distribute the electrons among the atoms, giving octets (or duets for H) to as many atoms as possible. Begin with the bonding electrons; then proceed to lone pairs on terminal atoms and finally to lone pairs on the central atom.

All 20 valence electrons are used. All atoms have octets (duets for hydrogen); the structure is complete.

(e)   HCOOH: Write the correct skeletal structure for the molecule.

Calculate the total number of electrons for the Lewis structure by summing the valence electrons of each atom in the molecule.

$$\text{number of valence e}^- \text{ for C} + 2(\text{number of valence e}^- \text{ for H}) + 2(\text{number of electrons}$$
$$\text{for O}) = 4 + 2(1) + 2(6) = 18$$

Distribute the electrons among the atoms, giving octets (or duets for H) to as many atoms as possible. Begin with the bonding electrons; then proceed to lone pairs on terminal atoms and finally to lone pairs on the central atom.

All 18 valence electrons are used.

If any atom lacks an octet, form double or triple bonds to give them octets.

All atoms have octets (duets for hydrogen); the structure is complete.

9.101   HCSNH$_2$:   Write the correct skeletal structure for the molecule.

Calculate the total number of electrons for the Lewis structure by summing the number of valence electrons of each atom in the molecule.

(number of valence e$^-$ for N) + (number of valence e$^-$ for S) + (number of valence e$^-$ for C) + 3(number of valence e$^-$ for H) = 5 + 6 + 4 + 3(1) = 18

Distribute the electrons among the atoms, giving octets (or duets for H) to as many atoms as possible. Begin with the bonding electrons; then proceed to lone pairs on terminal atoms and finally to lone pairs on the central atoms.

Complete the octet on C by forming a double bond.

9.103   (a) O_2^-:   Write the correct skeletal structure for the radical.

O—O

Calculate the total number of electrons for the Lewis structure by summing the number of valence electrons of each atom in the radical and adding 1 for the 1− charge.

2(number of valence e^- for O) + 1 = 2(6) + 1 = 13

Distribute the electrons among the atoms, giving octets to as many atoms as possible. Begin with the bonding electrons; then proceed to lone pairs on terminal atoms and finally to lone pairs on the central atom.

$$\left[\cdot \ddot{\text{O}} - \ddot{\text{O}} \colon \right]^-$$

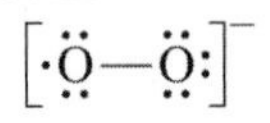

All 13 valence electrons are used.

O has an incomplete octet. It has seven electrons because we have an odd number of valence electrons.

(b) O^-:   Write the Lewis structure based on the valence electrons $2s^2 2p^5$.

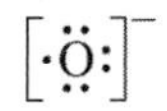

$$\left[\cdot \ddot{\text{O}} \colon \right]^-$$

(c)   OH: Write the correct skeletal structure for the molecule.

H—O

Calculate the total number of electrons for the Lewis structure by summing the number of valence electrons of each atom in the molecule.

(number of valence e^- for O) + (number of valence e^- for H) = 6 + 1 = 7

Distribute the electrons among the atoms, giving octets (or duets for H) to as many atoms as possible. Begin with the bonding electrons; then proceed to lone pairs on terminal atoms and finally to lone pairs on the central atom.

$$\text{H} - \ddot{\text{O}} \cdot$$

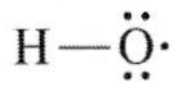

All seven valence electrons are used.

O has an incomplete octet. It has seven electrons because we have an odd number of valence electrons.

(d) CH_3OO: Write the correct skeletal structure for the radical.

C is the least electronegative atom, so it is central.

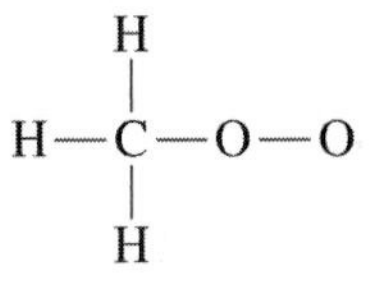

$$\begin{array}{c} \text{H} \\ | \\ \text{H} - \text{C} - \text{O} - \text{O} \\ | \\ \text{H} \end{array}$$

Calculate the total number of electrons for the Lewis structure by summing the number of valence electrons of each atom in the molecule.

3(number of valence e^- for H) + (number of valence e^- for C) + 2(number of valence e^- for O) = 3(1) + 4 + 2(6) = 19

Distribute the electrons among the atoms, giving octets (or duets for H) to as many atoms as possible. Begin with the bonding electrons; then proceed to lone pairs on terminal atoms and finally to lone pairs on the central atoms.

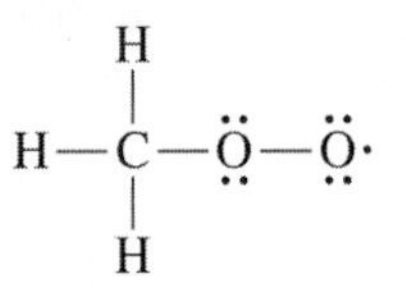

$$\begin{array}{c} \text{H} \\ | \\ \text{H} - \text{C} - \ddot{\text{O}} - \ddot{\text{O}} \cdot \\ | \\ \text{H} \end{array}$$

All 19 valence electrons are used.

O has an incomplete octet. It has seven electrons because we have an odd number of valence electrons.

9.105   Rewrite the reaction using the Lewis structures of the molecules involved.

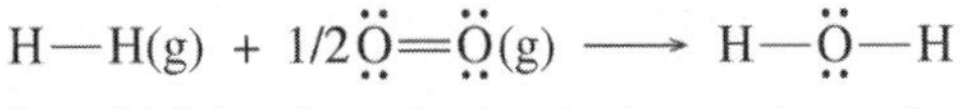

$$\text{H} - \text{H}(g) \ + \ 1/2 \, \ddot{\text{O}} = \ddot{\text{O}}(g) \ \longrightarrow \ \text{H} - \ddot{\text{O}} - \text{H}$$

Determine which bonds are broken in the reaction and sum the bond energies of the following:

$\sum (\Delta H\text{'s bonds broken})$

= 1 mol(H—H) + 1/2 mol(O=O)

$$= 1 \text{ mol}(436 \text{ kJ/mol}) + 1/2 \text{ mol}(498)$$
$$= 685 \text{ kJ/mol}$$

Determine which bonds are formed in the reaction and sum the negatives of the bond energies of the following:

$$\Sigma(-\Delta H\text{'s bonds formed})$$
$$= -2\text{mol}(O-H)$$
$$= -2\text{mol}(464 \text{ kJ/mol})$$
$$= -928 \text{ kJ/mol}$$

Find ΔH_{rxn} by summing the results of the two steps.

$$\Delta H_{rxn} = \Sigma(\Delta H\text{'s bonds broken}) + \Sigma(-\Delta H\text{'s bonds formed})$$
$$= 685 \text{ kJ/mol} - 928 \text{ kJ/mol}$$
$$= -243 \text{ kJ/mol}$$

$$CH_4(g) + 2\,O_2(g) \rightarrow CO_2(g) + 2\,H_2O(g)$$

Rewrite the reaction using the Lewis structures of the molecules involved.

Determine which bonds are broken in the reaction and sum the bond energies of the following:

$$\Sigma(\Delta H\text{'s bonds broken})$$
$$= 4 \text{ mol}(C-H) + 2 \text{ mol}(O=O)$$
$$= 4 \text{ mol}(414 \text{ kJ/mol}) + 2 \text{ mol}(498)$$
$$= 2652 \text{ kJ/mol}$$

Determine which bonds are formed in the reaction and sum the negatives of the bond energies of the following:

$$\Sigma(-\Delta H\text{'s bonds formed})$$
$$= -2 \text{ mol}(C=O) - 4 \text{ mol}(O-H)$$
$$= -2 \text{ mol}(799 \text{ kJ/mol}) - 4 \text{ mol}(464 \text{ kJ/mol})$$
$$= -3454 \text{ kJ/mol}$$

Find ΔH_{rxn} by summing the results of the two steps.

$$\Delta H_{rxn} = \Sigma(\Delta H\text{'s bonds broken}) + \Sigma(-\Delta H\text{'s bonds formed})$$
$$= 2652 \text{ kJ/mol} - 3454 \text{ kJ/mol}$$
$$= -802 \text{ kJ/mol}$$

Compare the following:

	kJ/mol	kJ/g
H_2	-243	-121 (using a molar mass of 2.016)
CH_4	-802	-50.0 (using a molar mass of 16.04)

$$\frac{-243 \text{ kJ}}{\text{mol } H_2} \times \frac{\text{mol } H_2}{2.016 \text{ g } H_2} = \frac{-121 \text{ kJ}}{\text{g } H_2} \qquad \frac{-802 \text{ kJ}}{\text{mol } CH_4} \times \frac{\text{mol } CH_4}{16.04 \text{ g } CH_4} = \frac{-50.0 \text{ kJ}}{\text{g } CH_4}$$

So methane yields more energy per mole, but hydrogen yields more energy per gram.

9.107   (a) Cl_2O_7:   Write the correct skeletal structure for the molecule.

Calculate the total number of electrons for the Lewis structure by summing the valence electrons of each atom in the molecule.

2(number of valence e⁻ for Cl) + 7(number of valence e⁻ for O) = 2(7) + 7(6) = 56

Distribute the electrons among the atoms, giving octets (or duets for H) to as many atoms as possible. Begin with the bonding electrons; then proceed to lone pairs on terminal atoms and finally to lone pairs on the central atom.

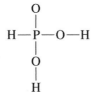

Form double bonds to minimize formal charge.

(b) H_3PO_3:   Write the correct skeletal structure for the molecule.

$$
\begin{array}{c}
\text{O} \\
\| \\
\text{H}-\text{P}-\text{O}-\text{H} \\
| \\
\text{O} \\
| \\
\text{H}
\end{array}
$$

Calculate the total number of electrons for the Lewis structure by summing the valence electrons of each atom in the molecule.

(number of valence e⁻ for P) + 3(number of valence e⁻ for O) + 3(number of valence e⁻ for H) = 5 + 3(6) + 3(1) = 26

Distribute the electrons among the atoms, giving octets (or duets for H) to as many atoms as possible. Begin with the bonding electrons; then proceed to lone pairs on terminal atoms and finally to lone pairs on central atoms.

Form double bonds to minimize formal charge.

(c) H_3AsO_4:   Write the correct skeletal structure for the molecule.

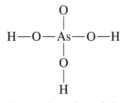

Calculate the total number of electrons for the Lewis structure by summing the valence electrons of each atom in the molecule.

(number of valence e⁻ for As) + 4(number of valence e⁻ for O) + 3(number of valence e⁻ for H) = 5 + 4(6) + 3(1) = 32

Distribute the electrons among the atoms, giving octets (or duets for H) to as many atoms as possible. Begin with the bonding electrons; then proceed to lone pairs on terminal atoms and finally to lone pairs on the central atom.

$$
\begin{array}{c}
\ddot{\text{O}}\!: \\
| \\
\text{H}-\ddot{\text{O}}-\text{As}-\ddot{\text{O}}-\text{H} \\
| \\
\!:\!\text{O}\!: \\
| \\
\text{H}
\end{array}
$$

Form double bonds to minimize formal charge.

$$
\begin{array}{c}
\!:\!\text{O}\!: \\
\| \\
\text{H}-\ddot{\text{O}}-\text{As}-\ddot{\text{O}}-\text{H} \\
| \\
\!:\!\text{O}\!: \\
| \\
\text{H}
\end{array}
$$

9.109 $Na^+F^- < Na^+O^{2-} < Mg^{2+}F^- < Mg^{2+}O^{2-} < Al^{3+}O^{2-}$

The lattice energy is proportional to the magnitude of the charge and inversely proportional to the distance between the atoms. Na^+F^- would have the smallest lattice energy because the magnitude of the charges on Na and F are the smallest. $Mg^{2+}F^-$ and Na^+O^{2-} both have the same magnitude formal charge, the O^{2-} is larger than F^- in size, and Na^+ is larger than Mg^{2+}; so Na^+O^{2-} should be less than $Mg^{2+}F^-$. The magnitude of the charge makes $Mg^{2+}O^{2-} < Al^{3+}O^{2-}$.

9.111   **Given:** heat atomization CH_4 = 1660 kJ/mol, CH_2Cl_2 = 1495 kJ/mol   **Find:** bond energy C—Cl Write the reaction using the Lewis structure.

$$
\begin{array}{c}
\text{H} \\
| \\
\text{H}-\text{C}-\text{H}(g) \longrightarrow \text{C}(g) + 4\,\text{H}(g) \\
| \\
\text{H}
\end{array}
$$

Determine the number and kinds of bonds broken and then ΔH atomization $= \Sigma$ bonds broken.
ΔH atomization $= \Sigma\,4\,(\text{C—H})$ bonds broken

$$
\frac{1660\ \text{kJ}}{1\ \text{mol CH}_4} \times \frac{1\ \text{mol CH}_4}{4\ \text{C—H bonds}} = 415\ \text{kJ/C—H bond}
$$

Write the reaction using the Lewis structure.

$$
\begin{array}{c}
\text{H} \\
| \\
\!:\!\ddot{\text{C}}\text{l}-\text{C}-\text{H}(g) \longrightarrow \text{C}(g) + 2\,\text{H}(g) + 2\,\text{Cl}(g) \\
| \\
\!:\!\ddot{\text{C}}\text{l}\!:
\end{array}
$$

Determine the number and kinds of bonds broken and ΔH atomization $= \Sigma$ bonds broken.
ΔH atomization $= \Sigma\,2\,(\text{C—H})$ bonds broken $+ 2(\text{C—Cl})$ bonds broken
$1495\ \text{kJ/mol} = 2(415\ \text{kJ/mol}) + 2(x)$ $x = 333\ \text{kJ/mol}$ for the C—Cl bond energy

Check: The bond energy found (333 kJ/mol) is very close to the table value of 339 kJ/mol.

9.113   **Given:** 7.743% H   **Find:** Lewis structure
Conceptual Plan: **%H → %C → mass C, H → mol C, H → pseudoformula → empirical formula**

$$
100\% - \%\text{H} \qquad \text{Assume 100 g sample} \quad \frac{1\ \text{mol C}}{12.01\ \text{g}}\ \frac{1\ \text{mol N}}{1.008\ \text{g}} \qquad \text{divide by smallest number}
$$

Solution: $\%\text{C} = 100\% - 7.743\% = 92.568\%\ \text{C}$

In a 100.00 g sample: 7.743 g H, 92.568 g C

$$
7.743\ \text{g H} \times \frac{1\ \text{mol H}}{1.008\ \text{g H}} = 7.682\ \text{mol}
$$

$$
92.568\ \text{g C} \times \frac{1\ \text{mol C}}{12.01\ \text{g C}} = 7.708\ \text{mol C}
$$

$C_{7.708}H_{7.682}$

$\dfrac{C_{7.708}H_{7.682}}{7.682 \quad 7.682} \rightarrow CH$

The smallest molecular formula would be C_2H_2.

Write the correct skeletal structure for the molecule.

H—C—C—H

Calculate the total number of electrons for the Lewis structure by summing the valence electrons of each atom in the molecule.

2(number of valence e^- for C) + 2(number of valence e^- for H) = 2(4) + 2(1) = 10

Distribute the electrons among the atoms, giving octets (or duets for H) to as many atoms as possible. Begin with the bonding electrons; then proceed to lone pairs on terminal atoms and finally to lone pairs on the central atom.

H—C̤—C̤—H

Complete the octet on C by forming a triple bond.

H—C≡C—H

Challenge Problems

9.115

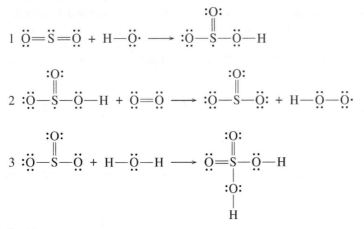

Step 1:
　　Bonds broken: $2\,\text{mol}(S{=}O) + 1\,\text{mol}(H{-}O) = 2\,\text{mol}(523\,\text{kJ/mol}) + 1\,\text{mol}(464\,\text{kJ/mol}) = 1510\,\text{kJ/mol}$
　　Bonds formed: $-2\,\text{mol}(S{-}O) - 1\,\text{mol}(S{=}O) - 1\,\text{mol}(O{-}H) =$
　　　　$-2\,\text{mol}(265\,\text{kJ/mol}) - 1\,\text{mol}(523\,\text{kJ/mol}) - 1\,\text{mol}(464\,\text{kJ/mol}) = -1517\,\text{kJ/mol}$
　　　　　　　　$\Delta H_{\text{step}} = -7\,\text{kJ/mol}$

Step 2:
　　Bonds broken: $2\,\text{mol}(S{-}O) + 1\,\text{mol}(S{=}O) + 1\,\text{mol}(O{-}H) + 1\,\text{mol}(O{=}O) =$
　　　　$2\,\text{mol}(265\,\text{kJ/mol}) + 1\,\text{mol}(523\,\text{kJ/mol}) + 1\,\text{mol}(464\,\text{kJ/mol}) + 1\,\text{mol}(498\,\text{kJ/mol}) =$
　　　　　　　　$2015\,\text{kJ/mol}$
　　Bonds formed: $-2\,\text{mol}(S{-}O) - 1\,\text{mol}(S{=}O) - 1\,\text{mol}(O{-}H) - 1\,\text{mol}(O{=}O) =$
　　　　$-2\,\text{mol}(265\,\text{kJ/mol}) - 1\,\text{mol}(523\,\text{kJ/mol}) - 1\,\text{mol}(464\,\text{kJ/mol}) - 1\,\text{mol}(142\,\text{kJ/mol}) =$
　　　　　　　　　　$-1659\,\text{kJ/mol}$
　　　　　　　　$\Delta H_{\text{step}} = +356\,\text{kJ/mol}$

Step 3:
　　Bonds broken: $2\,\text{mol}(S{-}O) + 1\,\text{mol}(S{=}O) + 2\,\text{mol}(O{-}H) =$
　　　　$2\,\text{mol}(265\,\text{kJ/mol}) + 1\,\text{mol}(523\,\text{kJ/mol}) + 2\,\text{mol}(464\,\text{kJ/mol}) = 1981\,\text{kJ/mol}$

Bonds formed: $-2\,\text{mol}(S\!\!-\!\!O) - 2\,\text{mol}(S\!\!=\!\!O) - 2\,\text{mol}(O\!\!-\!\!H) =$
$-2\,\text{mol}(265\,\text{kJ/mol}) + -2\,\text{mol}(523\,\text{kJ/mol}) + -2\,\text{mol}(464\,\text{kJ/mol}) = -2504\,\text{kJ/mol}$
$\Delta H_{\text{step}} = -523\,\text{kJ/mol}$

Hess's law states that ΔH for the reaction is the sum of ΔH of the steps:
$\Delta H_{\text{rxn}} = (-7\,\text{kJ/mol}) + (+356\,\text{kJ/mol}) + (-523\,\text{kJ/mol}) = -174\,\text{kJ/mol}$

9.117 **Given:** $\mu = 1.08$ D HCl, 20% ionic and $\mu = 1.82$ D HF, 45% ionic **Find:** r
Conceptual Plan: $\boldsymbol{\mu \rightarrow \mu_{\text{calc}} \rightarrow r}$

% ionic character $= \dfrac{\mu}{\mu_{\text{calc}}} \times 100\%$

Solution: For HCl $\qquad \mu_{\text{calc}} = \dfrac{1.08}{0.20} = 5.4\,\text{D}$ $\qquad \dfrac{5.4\,\cancel{D} \times \dfrac{3.34 \times 10^{-30}\,\cancel{C} \cdot \cancel{m}}{\cancel{D}} \times \dfrac{10^{12}\,\text{pm}}{\cancel{m}}}{1.6 \times 10^{-19}\,\cancel{C}} = 113\,\text{pm}$

For HF $\qquad \mu_{\text{calc}} = \dfrac{1.82}{0.45} = 4.0\underline{4}\,\text{D}$ $\qquad \dfrac{4.04\,\cancel{D} \times \dfrac{3.34 \times 10^{-30}\,\cancel{C} \cdot \cancel{m}}{\cancel{D}} \times \dfrac{10^{12}\,\text{pm}}{\cancel{m}}}{1.6 \times 10^{-19}\,\cancel{C}} = 84\,\text{pm}$

From Table 9.4, the bond length of HCl = 127 pm, and HF = 92 pm. Both of these values are slightly higher than the calculated values.

9.119 For the four P atoms to be equivalent, they must all be in the same electronic environment. That is, they must all see the same number of bonds and lone pair electrons. The only way to achieve this is with a tetrahedral configuration where the P atoms are at the four points of the tetrahedron.

9.121 **Given:** $\Delta H_{\text{f}}^{\circ}\,\text{PI}_3(s) = -24.7\,\text{kJ/mol}$; $\text{P}\!\!-\!\!\text{I} = 184\,\text{kJ/mol}$; $\text{I}\!\!-\!\!\text{I} = 151\,\text{kJ/mol}$; $\Delta H_{\text{f}}^{\circ}\,\text{P}(g) = 334\,\text{kJ/mol}$; $\Delta H_{\text{f}}^{\circ}\text{I}_2(g) = 62\,\text{kJ/mol}$ **Find:** $\Delta H_{\text{sub}}\text{PI}_3(s)$
Conceptual Plan: $\text{PI}_3(s) \rightarrow \text{PI}_3(g)$; use Hess's law
Solution:

Reaction	$\Delta H(\text{kJ/mol})$	
$\text{PI}_3(s) \rightarrow \cancel{\text{P}(s)} + \cancel{3/2\,\text{I}_2(s)}$	$+24.7$	(this is the reverse of the formation reaction)
$\cancel{\text{P}(s)} \rightarrow \cancel{\text{P}(g)}$	$+334$	(formation of $\text{P}(g)$)
$\cancel{3/2\,\text{I}_2(s)} \rightarrow \cancel{3/2\,\text{I}_2(g)}$	$3/2(62)$	(formation of $\text{I}_2(g)$)
$\cancel{3/2\,\text{I}_2(g)} \rightarrow \cancel{3\,\text{I}(g)}$	$3/2(151)$	(breaking I—I bond)
$\cancel{\text{P}(g)} + \cancel{3\,\text{I}(g)} \rightarrow \text{PI}_3(g)$	$-3(184)$	(forming P—I bond)
$\text{PI}_3(s) \rightarrow \text{PI}_3(g)$	$+126$	(sublimation of $\text{PI}_3(s)$)

9.123 **Given:** H_2S_4 linear **Find:** oxidation number of each S
Write the correct skeletal structure for the molecule.

$\text{H}\!\!-\!\!\text{S}\!\!-\!\!\text{S}\!\!-\!\!\text{S}\!\!-\!\!\text{S}\!\!-\!\!\text{H}$

Calculate the total number of electrons for the Lewis structure by summing the number of valence electrons of each atom in the molecule.

4(number of valence e^- for S) + 2(number of valence e^- for H) = 4(6) + 2(1) = 26

Distribute the electrons among the atoms, giving octets (or duets for H) to as many atoms as possible.

$\text{H}\!\!-\!\!\overset{..}{\underset{..}{\text{S}}}\!\!-\!\!\overset{..}{\underset{..}{\text{S}}}\!\!-\!\!\overset{..}{\underset{..}{\text{S}}}\!\!-\!\!\overset{..}{\underset{..}{\text{S}}}\!\!-\!\!\text{H}$

Determine the oxidation number on each atom. EN(H) < EN(S); so the electrons in the H—S bond belong to the S atom, while the electrons in the S—S bonds split between the two S atoms.

O. N. = valence electrons − electrons that belong to the atom

$\text{H}\!\!-\!\!\overset{..}{\underset{..}{\text{S}}}_{\!A}\!\!-\!\!\overset{..}{\underset{..}{\text{S}}}_{\!B}\!\!-\!\!\overset{..}{\underset{..}{\text{S}}}_{\!C}\!\!-\!\!\overset{..}{\underset{..}{\text{S}}}_{\!D}\!\!-\!\!\text{H}$

H = 1 − 0 = +1 for each H
S_A = 6 − 7 = −1
S_B = 6 − 6 = 0
S_C = 6 − 6 = 0
S_D = 6 − 7 = −1

9.125   **Given:** $\Delta H_f(SO_2) = -296.8$ kJ/mol, $\Delta H(S(g)) = 277.2$ kJ/mol, break $O{=}O$ bond 498 kJ
          Find: $S{=}O$ bond energy
          Conceptual Plan: Use ΔH_f for SO_2 and $S(g)$ and the bond energy of O_2 to determine heat of atomization of SO_2.
          Solution:

Reaction		ΔH(kJ/mol)
$SO_2(g)$	$\rightarrow \cancel{S(s, \text{rhombic})} + \cancel{O_2(g)}$	+296.8
$\cancel{S(s, \text{rhombic})}$	$\rightarrow S(g)$	+277.2
$\cancel{O{=}O(g)}$	$\rightarrow 2\,O(g)$	+498
$SO_2(g)$	$\rightarrow S(g) + 2\,O(g)$	+1072

Write the reaction using the Lewis structure.

$\ddot{\underset{\cdot\cdot}{O}}{=}\ddot{S}{=}\ddot{\underset{\cdot\cdot}{O}}(g) \longrightarrow S(g) + 2O(g)$

Determine the number and kinds of bonds broken and then ΔH atomization = bonds broken.
ΔH atomization = $2 \sum(S{=}O)$ bonds broken.
1072 kJ/mol = 2 (S=O) bonds broken.
S=O bond energy = 536 kJ/mol.

Check: The S=O bond energy is close to the table value of 523 kJ/mol

Conceptual Problems

9.127   When we say that a compound is "energy rich," we mean that it gives off a great amount of energy when it reacts. It means that a lot of energy is stored in the compound. This energy is released when the weak bonds in the compound break and much stronger bonds are formed in the product, thereby releasing energy.

9.129   Lewis theory is successful because it allows us to understand and predict many chemical observations. We can use it to determine the formulae of ionic compounds and to account for the low melting points and boiling points of molecular compounds compared to ionic compounds. Lewis theory allows us to predict what molecules or ions will be stable, which will be more reactive, and which will not exist. Lewis theory, however, does not tell us anything about how the bonds in the molecules and ions form. It does not give us a way to account for the paramagnetism of oxygen. And by itself, Lewis theory does not tell us anything about the shape of the molecule or ion.

10 Chemical Bonding II: Molecular Shapes, Valence Bond Theory, and Molecular Orbital Theory

Review Questions

10.1   The properties of molecules are directly related to their shape. The sensation of taste, immune response, the sense of smell, and many types of drug action all depend on shape-specific interactions between molecules and proteins.

10.3   The five basic electron geometries are as follows:
1.   Linear, which has two electron groups
2.   Trigonal planar, which has three electron groups
3.   Tetrahedral, which has four electron groups
4.   Trigonal bipyramidal, which has five electron groups
5.   Octahedral, which has six electron groups
An electron group is defined as a lone pair of electrons, a single bond, a multiple bond, or even a single electron.

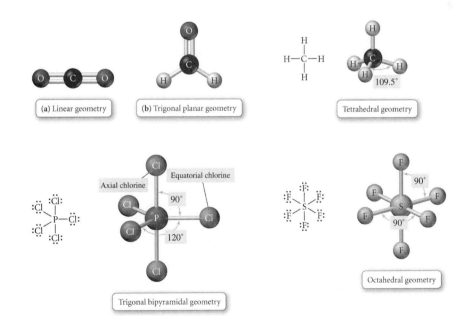

(a) Linear geometry   (b) Trigonal planar geometry   Tetrahedral geometry

Axial chlorine   Equatorial chlorine

Trigonal bipyramidal geometry   Octahedral geometry

10.5   (a)   Four electron groups give tetrahedral electron geometry, while three bonding groups and one lone pair give a trigonal pyramidal molecular geometry.

         (b)   Four electron groups give a tetrahedral electron geometry, while two bonding groups and two lone pairs give a bent molecular geometry.

         (c)   Five electron groups give a trigonal bipyramidal electron geometry, while four bonding groups and one lone pair give a seesaw molecular geometry.

         (d)   Five electron groups give a trigonal bipyramidal electron geometry, while three bonding groups and two lone pairs give a T-shaped molecular geometry.

         (e)   Five electron groups gives a trigonal bipyramidal electron geometry, while two bonding groups and three lone pairs give a linear geometry.

         (f)   Six electron groups give an octahedral electron geometry, while five bonding groups and one lone pair give a square pyramidal molecular geometry.

         (g)   Six electron groups give an octahedral electron geometry, while four bonding groups and two lone pairs give a square planar molecular geometry.

10.7   To determine whether a molecule is polar, do the following:
         1.   Draw the Lewis structure for the molecule and determine the molecular geometry.
         2.   Determine whether the molecule contains polar bonds.
         3.   Determine whether the polar bonds add together to form a net dipole moment.
         Polarity is important because polar and nonpolar molecules have different properties. Polar molecules interact strongly with other polar molecules but do not interact with nonpolar molecules, and vice versa.

10.9   According to valence bond theory, the shape of the molecule is determined by the geometry of the overlapping orbitals.

10.11   Hybridization is a mathematical procedure in which the standard atomic orbitals are combined to form new atomic orbitals called hybrid orbitals. Hybrid orbitals are still localized on individual atoms, but they have different shapes and energies from those of standard atomic orbitals. They are necessary in valence bond theory because they correspond more closely to the actual distribution of electrons in chemically bonded atoms.

10.13   The number of standard atomic orbitals added together always equals the number of hybrid orbitals formed. The total number of orbitals is conserved.

10.15   The double bond in Lewis theory is simply two pairs of electrons that are shared between the same two atoms. However, in valence bond theory, we see that the double bond is made up of two different kinds of bonds. The double bond in valence bond theory consists of one σ bond and one π bond. Valence bond theory shows us that rotation about a double bond is severely restricted. Because of the side-by-side overlap of the p orbitals, the π bond must essentially break for rotation to occur. The σ bond consists of end-to-end overlap. Because the overlap is linear, rotation is not restricted.

10.17   In molecular orbital theory, atoms will bond when the electrons in the atoms can lower their energy by occupying the molecular orbitals of the resultant molecule.

10.19   A bonding molecular orbital is lower in energy than the atomic orbitals from which it is formed. There is an increased electron density in the internuclear region.

10.21   The electrons in orbitals behave like waves. The bonding molecular orbital arises from the constructive interference between the atomic orbitals and is lower in energy than the atomic orbitals. The antibonding molecular orbital arises from the destructive interference between the atomic orbitals and is higher in energy than the atomic orbitals.

10.23   Molecular orbitals can be approximated by a linear combination of atomic orbitals (AOs). The total number of MOs formed from a particular set of AOs will always equal the number of AOs used.

10.25

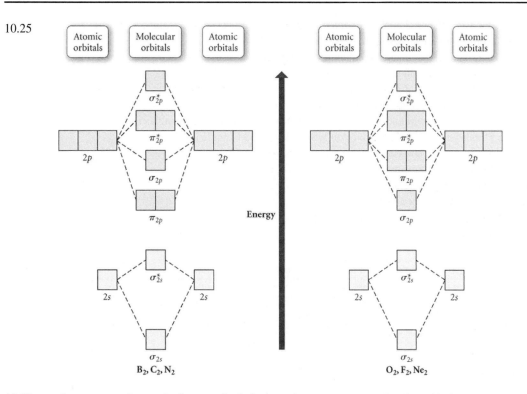

B₂, C₂, N₂   **O₂, F₂, Ne₂**

10.27   A paramagnetic species has unpaired electrons in one or more molecular orbitals. A paramagnetic species is attracted
         to a magnetic field. The magnetic property is a direct result of the unpaired electron(s). The spin and angular momen-
         tum of the electrons generate tiny magnetic fields. A diamagnetic species has all of its electrons paired. The magnetic
         fields caused by the electron spin, and orbital angular momentum tend to cancel each other. A diamagnetic species is
         not attracted to a magnetic field, and is, in fact, slightly repelled.

10.29   Nonbonding orbitals are atomic orbitals not involved in a bond that remain localized on the atom.

Problems by Topic

VSEPR Theory and Molecular Geometry

10.31   Four electron groups: A trigonal pyramidal molecular geometry has three bonding groups and one lone pair of elec-
         trons, so there are four electron groups on atom A.

10.33   (a)   4 total electron groups, 4 bonding groups, 0 lone pairs
                A tetrahedral molecular geometry has four bonding groups and no lone pairs. So there are four total electron
                groups, four bonding groups, and no lone pairs.
         (b)   5 total electron groups, 3 bonding groups, 2 lone pairs
                A T-shaped molecular geometry has three bonding groups and two lone pairs. So there are five total electron
                groups, three bonding groups, and two lone pairs.
         (c)   6 total electron groups, 5 bonding groups, 1 lone pair
                A square pyramidal molecular geometry has five bonding groups and one lone pair. So there are six total elec-
                tron groups, five bonding groups, and one lone pair.

10.35   (a) PF_3:   Electron geometry—tetrahedral; molecular geometry—trigonal pyramidal; bond angle = 109.5°
                          Because of the lone pair, the bond angle will be less than 109.5°.
                          Draw a Lewis structure for the molecule:
                          PF_3 has 26 valence electrons.

$$:\ddot{F}:$$
$$|$$
$$:\ddot{F}-\underset{..}{\overset{..}{P}}-\ddot{F}:$$

Determine the total number of electron groups around the central atom:

There are four electron groups on P.

Determine the number of bonding groups and the number of lone pairs around the central atom:

There are three bonding groups and one lone pair.

Use Table 10.1 to determine the electron geometry, molecular geometry, and bond angles:

Four electron groups give a tetrahedral electron geometry; three bonding groups and one lone pair give a trigonal pyramidal molecular geometry; the idealized bond angles for tetrahedral geometry are 109.5°. The lone pair will make the bond angle less than idealized.

(b)   SBr$_2$:   Electron geometry—tetrahedral; molecular geometry—bent; bond angle = 109.5°

Because of the lone pairs, the bond angle will be less than 109.5°.

Draw a Lewis structure for the molecule:

SBr$_2$ has 20 valence electrons.

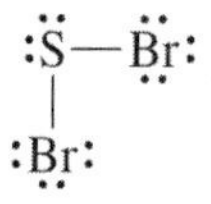

$$:\ddot{S}—\ddot{B}r:$$
$$\quad\;\;|$$
$$:\ddot{B}r:$$

Determine the total number of electron groups around the central atom:

There are four electron groups on S.

Determine the number of bonding groups and the number of lone pairs around the central atom:

There are two bonding groups and two lone pairs.

Use Table 10.1 to determine the electron geometry, molecular geometry, and bond angles:

Four electron groups give a tetrahedral electron geometry; two bonding groups and two lone pairs give a bent molecular geometry; the idealized bond angles for tetrahedral geometry are 109.5°. The lone pairs will make the bond angle less than idealized.

(c)   CHCl$_3$:   Electron geometry—tetrahedral; molecular geometry—tetrahedral; bond angle = 109.5°

Because there are no lone pairs, the bond angle will be 109.5°.

Draw a Lewis structure for the molecule:

CHCl$_3$ has 26 valence electrons.

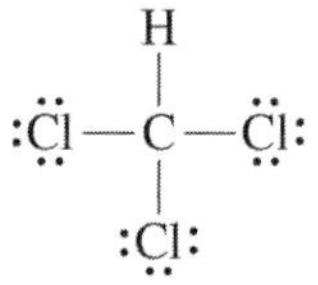

$$\begin{array}{c} H \\ | \\ :\ddot{C}l—C—\ddot{C}l: \\ | \\ :\ddot{C}l: \end{array}$$

Determine the total number of electron groups around the central atom:

There are four electron groups on C.

Determine the number of bonding groups and the number of lone pairs around the central atom:

There are four bonding groups and no lone pairs.

Use Table 10.1 to determine the electron geometry, molecular geometry, and bond angles:

Four electron groups give a tetrahedral electron geometry; four bonding groups and no lone pairs give a tetrahedral molecular geometry; the idealized bond angles for tetrahedral geometry are 109.5°. However, because the attached atoms have different electronegativities, the bond angles are less than idealized.

(d)   CS$_2$:   Electron geometry—linear; molecular geometry—linear; bond angle = 180°

Because there are no lone pairs, the bond angle will be 180°.

Draw a Lewis structure for the molecule:

CS$_2$ has 16 valence electrons.

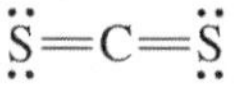

$$\ddot{S}=C=\ddot{S}$$

Determine the total number of electron groups around the central atom:

There are two electron groups on C.

Determine the number of bonding groups and the number of lone pairs around the central atom:

There are two bonding groups and no lone pairs.

Use Table 10.1 to determine the electron geometry, molecular geometry, and bond angles:

Two electron groups give a linear geometry; two bonding groups and no lone pairs give a linear molecular geometry; the idealized bond angle is 180°. The molecule will not deviate from this.

10.37   H₂O will have the smaller bond angle because lone pair–lone pair repulsions are greater than lone pair–bonding pair repulsions.

Draw the Lewis structures for both structures:

H_3O^+ has eight valence electrons. H_2O has eight valence electrons.

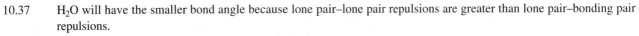

There are three bonding groups and one lone pair.

There are two bonding groups and two lone pairs.

Both have four electron groups, but the two lone pairs in H_2O will cause the bond angle to be smaller because of the lone pair–lone pair repulsions.

10.39   (a) SF_4 Draw a Lewis structure for the molecule:

SF_4 has 34 valence electrons.

Determine the total number of electron groups around the central atom:
There are five electron groups on S.
Determine the number of bonding groups and the number of lone pairs around the central atom:
There are four bonding groups and one lone pair.
Use Table 10.1 to determine the electron geometry and molecular geometry:
The electron geometry is trigonal bipyramidal, so the molecular geometry is seesaw.
Sketch the molecule:

(b) ClF_3 Draw a Lewis structure for the molecule:

ClF_3 has 28 valence electrons.

Determine the total number of electron groups around the central atom:
There are five electron groups on Cl.
Determine the number of bonding groups and the number of lone pairs around the central atom:
There are three bonding groups and two lone pairs.
Use Table 10.1 to determine the electron geometry and molecular geometry:
The electron geometry is trigonal bipyramidal, so the molecular geometry is T-shaped.
Sketch the molecule:

(c) IF_2^- Draw a Lewis structure for the ion:

IF_2^- has 22 valence electrons.

Determine the total number of electron groups around the central atom:
There are five electron groups on I.
Determine the number of bonding groups and the number of lone pairs around the central atom:
There are two bonding groups and three lone pairs.

Use Table 10.1 to determine the electron geometry and molecular geometry:
The electron geometry is trigonal bipyramidal, so the molecular geometry is linear.
Sketch the ion:

$$\left[\text{F—I—F} \right]^-$$

(d) IBr_4^- Draw a Lewis structure for the ion:
IBr_4^- has 36 valence electrons.

$$\left[\begin{array}{c} :\ddot{B}r: \\ | \\ :\ddot{B}r\!-\!\dot{\overset{\cdot}{I}}\!-\!\ddot{B}r: \\ | \\ :\ddot{B}r: \end{array} \right]^-$$

Determine the total number of electron groups around the central atom:
There are six electron groups on I.
Determine the number of bonding groups and the number of lone pairs around the central atom:
There are four bonding groups and two lone pairs.
Use Table 10.1 to determine the electron geometry and molecular geometry:
The electron geometry is octahedral, so the molecular geometry is square planar.
Sketch the ion:

$$\left[\begin{array}{cc} Br & Br \\ \diagdown\!I\!\diagup \\ Br & Br \end{array} \right]^-$$

10.41   (a) C_2H_2 Draw the Lewis structure:

$$\text{H—C}\equiv\text{C—H}$$

Atom	Number of Electron Groups	Number of Lone Pairs	Molecular Geometry
Left C	2	0	Linear
Right C	2	0	Linear

Sketch the molecule:

$$\text{H—C}\equiv\text{C—H}$$

(b) C_2H_4 Draw the Lewis structure:

$$\begin{array}{cc} \text{H}\diagdown & \diagup\text{H} \\ \text{C} & =\text{C} \\ \text{H}\diagup & \diagdown\text{H} \end{array}$$

Atom	Number of Electron Groups	Number of Lone Pairs	Molecular Geometry
Left C	3	0	Trigonal planar
Right C	3	0	Trigonal planar

Sketch the molecule:

$$\begin{array}{cc} \text{H}\diagdown & \diagup\text{H} \\ \text{C}=\text{C} \\ \text{H}\diagup & \diagdown\text{H} \end{array}$$

(c) C_2H_6 Draw the Lewis structure:

Atom	Number of Electron Groups	Number of Lone Pairs	Molecular Geometry
Left C	4	0	Tetrahedral
Right C	4	0	Tetrahedral

Sketch the molecule:

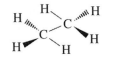

10.43   (a)   Four pairs of electrons give a tetrahedral electron geometry. The lone pair would cause lone pair–bonded pair repulsions and would have a trigonal pyramidal molecular geometry.

   (b)   Five pairs of electrons give a trigonal bipyramidal electron geometry. The lone pair occupies an equatorial position to minimize lone pair–bonded pair repulsions, and the molecule would have a seesaw molecular geometry.

   (c)   Six pairs of electrons give an octahedral electron geometry. The two lone pairs would occupy opposite positions to minimize lone pair–lone pair repulsions. The molecular geometry would be square planar.

10.45   (a)   CH₃OH   Draw the Lewis structure and determine the geometry about each interior atom:

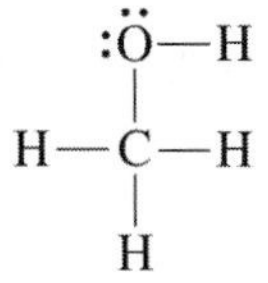

$$:\ddot{O}-H$$
$$|$$
$$H-C-H$$
$$|$$
$$H$$

Atom	Number of Electron Groups	Number of Lone Pairs	Molecular Geometry
C	4	0	Tetrahedral
O	4	2	Bent

Sketch the molecule:

   (b)   CH₃OCH₃ Draw the Lewis structure and determine the geometry about each interior atom:

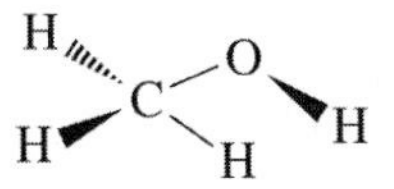

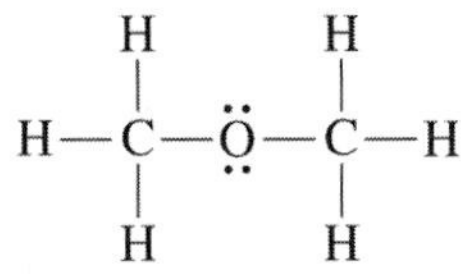

$$H \qquad H$$
$$| \qquad\quad |$$
$$H-C-\ddot{O}-C-H$$
$$| \qquad\quad |$$
$$H \qquad H$$

Atom	Number of Electron Groups	Number of Lone Pairs	Molecular Geometry
C	4	0	Tetrahedral
O	4	2	Bent
C	4	0	Tetrahedral

Sketch the molecule:

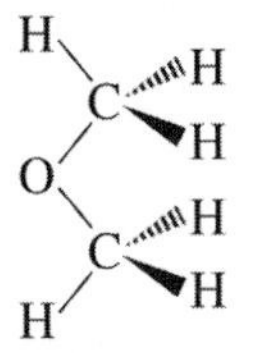

   (c)   H₂O₂   Draw the Lewis structure and determine the geometry about each interior atom:

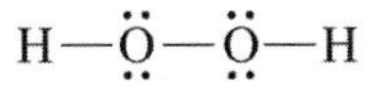

$$H-\ddot{O}-\ddot{O}-H$$

Atom	Number of Electron Groups	Number of Lone Pairs	Molecular Geometry
O	4	2	Bent
O	4	2	Bent

Sketch the molecule:

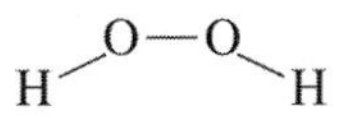

Molecular Shape and Polarity

10.47   Draw the Lewis structure for CO_2 and CCl_4; determine the molecular geometry and the polarity.

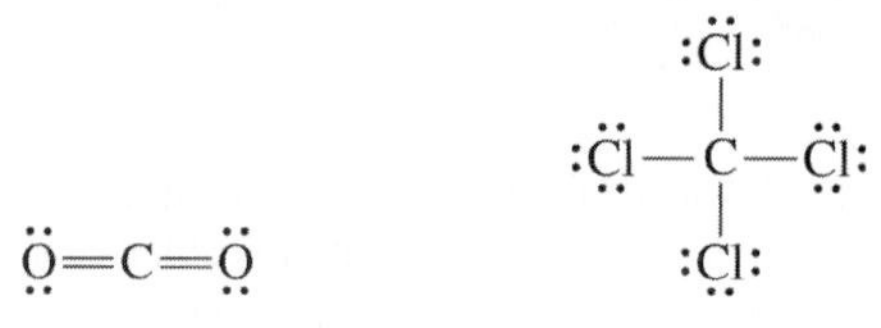

$$:\ddot{Cl}:$$
$$|$$
$$:\ddot{Cl}-C-\ddot{Cl}:$$
$$|$$
$$:\ddot{Cl}:$$

$$\ddot{O}=C=\ddot{O}$$

Number of electron groups on C	2	4
Number of lone pairs	0	0
Molecular geometry	linear	tetrahedral

Even though each molecule contains polar bonds, the sum of the bond dipoles gives a net dipole of zero for each molecule.

The linear molecular geometry of CO_2 will have bond vectors that are equal and opposite. ← —→

The tetrahedral molecular geometry of CCl_4 will have bond vectors that are equal and have a net dipole of zero.

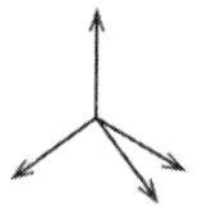

10.49   (a) PF_3—polar

Use the Lewis structure and molecular geometry determined in Problem 10.35, which is trigonal pyramidal.

Determine whether the molecule contains polar bonds:

The electronegativity of P = 2.1 and F = 4. Therefore, the bonds are polar.

Determine whether the polar bonds add together to form a net dipole:

Because the molecule is trigonal pyramidal, the three dipole moments sum to a nonzero net dipole moment. The molecule is polar. See Table 10.2 to see how dipole moments add to determine polarity.

(b) SBr_2—nonpolar

Use the Lewis structure and molecular geometry determined in Problem 10.35, which is bent.

Determine whether the molecule contains polar bonds:

The electronegativity of S = 2.5 and Br = 2.8. Therefore, the bonds are nonpolar.

Even though the molecule is bent, because the bonds are nonpolar, the molecule is nonpolar.

(c) $CHCl_3$—polar

Use the Lewis structure and molecular geometry determined in Problem 10.35, which is tetrahedral.

Determine whether the molecule contains polar bonds:

The electronegativity of C = 2.5, H = 2.1, and Cl = 3.0. Therefore, the bonds are polar.

Determine whether the polar bonds add together to form a net dipole:

Because the bonds have different dipole moments due to the different atoms involved, the four dipole moments sum to a nonzero net dipole moment. The molecule is polar. See Table 10.2 to see how dipole moments add to determine polarity.

(d) CS_2—nonpolar

Use the Lewis structure and molecular geometry determined in Problem 10.35, which is linear.

Determine whether the molecule contains polar bonds:

The electronegativity of C = 2.5 and S = 2.5. Therefore, the bonds are nonpolar. Also, the molecule is linear, which would result in a zero net dipole even if the bonds were polar.

The molecule is nonpolar. See Table 10.2 to see how dipole moments add to determine polarity.

10.51   (a) ClO_3^-—polar

Draw the Lewis structure and determine the molecular geometry:

$$\left[\begin{array}{c} :\ddot{O}: \\ | \\ :\ddot{O}-Cl-\ddot{O}: \end{array}\right]^-$$

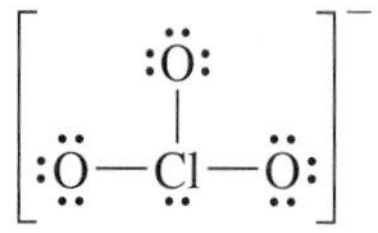

Four electron pairs with one lone pair give a trigonal pyramidal molecular geometry.

Determine whether the molecule contains polar bonds:

The electronegativities of Cl = 3.0 and O = 3.5. Therefore, the bonds are polar.

Determine whether the polar bonds add together to form a net dipole:

Because the molecular geometry is trigonal pyramidal, the three dipole moments sum to a nonzero net dipole moment. The molecule is polar. Refer to Table 10.2 to see how dipole moments add to determine polarity.

(b) SCl_2—polar

Draw the Lewis structure and determine the molecular geometry:

$$:\ddot{S}—\ddot{Cl}:$$
$$|$$
$$:\ddot{Cl}:$$

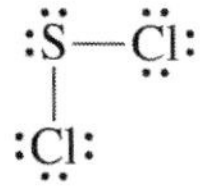

Four electron pairs with two lone pairs give a bent molecular geometry.

Determine whether the molecule contains polar bonds:

The electronegativity of S = 2.5 and Cl = 3.0. Therefore, the bonds are polar.

Determine whether the polar bonds add together to form a net dipole:

Because the molecular geometry is bent, the two dipole moments sum to a nonzero net dipole moment. The molecule is polar. See Table 10.2 to see how dipole moments add to determine polarity.

(c) SCl_4—polar

Draw the Lewis structure and determine the molecular geometry:

$$:\ddot{Cl}:$$
$$|$$
$$:\ddot{Cl}—\ddot{S}—\ddot{Cl}:$$
$$|$$
$$:\ddot{Cl}:$$

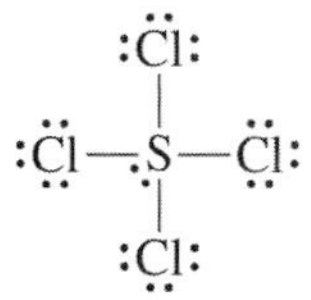

Five electron pairs with one lone pair give a seesaw molecular geometry.

Determine whether the molecule contains polar bonds:

The electronegativity of S = 2.5 and Cl = 3.0. Therefore, the bonds are polar.

Determine whether the polar bonds add together to form a net dipole:

Because the molecular geometry is seesaw, the four equal dipole moments sum to a nonzero net dipole moment. The molecule is polar.

The seesaw molecular geometry will not have offsetting bond vectors.

(d) $BrCl_5$—nonpolar

Draw the Lewis structure and determine the molecular geometry.

$$:\ddot{Cl}:$$
$$:\ddot{Cl}\diagdown\ |\ \diagup\ddot{Cl}:$$
$$Br$$
$$:\ddot{Cl}\diagup\ \ddot{}\ \diagdown\ddot{Cl}:$$

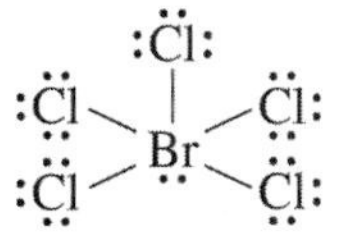

Six electron pairs with one lone pair give square pyramidal molecular geometry.

Determine whether the molecule contains polar bonds:

The electronegativity of Br = 2.8 and Cl = 3.0. The difference is only 0.2; therefore, the bonds are nonpolar. Even though the molecular geometry is square pyramidal, the five bonds are nonpolar; so there is no net dipole. The molecule is nonpolar.

Valence Bond Theory

10.53   (a)   Be $2s^2$ No bonds can form. Beryllium contains no unpaired electrons, so no bonds can form without hybridization.

(b)   P $3s^23p^3$ Three bonds can form. Phosphorus contains three unpaired electrons, so three bonds can form without hybridization.

(c)   F $2s^22p^5$ One bond can form. Fluorine contains one unpaired electron, so one bond can form without hybridization.

10.55 PH$_3$

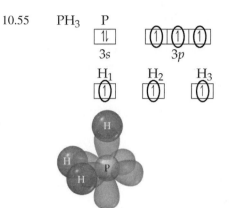

The unhybridized bond angles should be 90°. So without hybridization, there is good agreement between valence bond theory and the actual bond angle of 93.3°.

10.57 C 2s^{2}2p^2

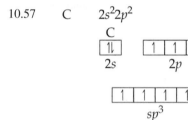

10.59 sp^2 Only sp^2 hybridization of this set of orbitals has a remaining p orbital to form a π bond.
sp^3 hybridization utilizes all $3p$ orbitals.
sp^3d^2 hybridization utilizes all $3p$ orbitals and $2d$ orbitals.

10.61 (a) CCl$_4$ Write the Lewis structure for the molecule:

$$:\ddot{C}l:$$
$$|$$
$$:\ddot{C}l - C - \ddot{C}l:$$
$$|$$
$$:\ddot{C}l:$$

Use VSEPR to predict the electron geometry:
Four electron groups around the central atom give a tetrahedral electron geometry.
Select the correct hybridization for the central atom based on the electron geometry:
Tetrahedral electron geometry has sp^3 hybridization.
Sketch the molecule and label the bonds:

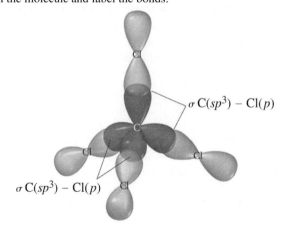

(b)   NH₃   Write the Lewis structure for the molecule:

$$\overset{\displaystyle H}{\underset{\displaystyle \ddot{} }{H-N-H}}$$

Use VSEPR to predict the electron geometry:
Four electron groups around the central atom give a tetrahedral electron geometry.
Select the correct hybridization for the central atom based on the electron geometry:
Tetrahedral electron geometry has sp^3 hybridization.
Sketch the molecule and label the bonds:

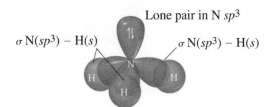

Lone pair in N $sp3$

σ N($sp3$) – H(s) σ N($sp3$) – H(s)

(c)   OF₂   Write the Lewis structure for the molecule:

$$:\ddot{F}-\ddot{O}-\ddot{F}:$$

Use VSEPR to predict the electron geometry:
Four electron groups around the central atom give a tetrahedral electron geometry.
Select the correct hybridization for the central atom based on the electron geometry:
Tetrahedral electron geometry has sp^3 hybridization.
Sketch the molecule and label the bonds:

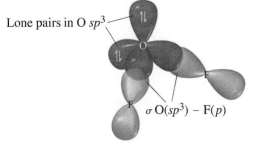

Lone pairs in O $sp3$

σ O($sp3$) – F(p)

(d)   CO₂   Write the Lewis structure for the molecule:

$$\ddot{O}=C=\ddot{O}$$

Use VSEPR to predict the electron geometry:
Two electron groups around the central atom give a linear electron geometry.
Select the correct hybridization for the central atom based on the electron geometry:
Linear electron geometry has sp hybridization.
Sketch the molecule and label the bonds:

π C(p_y) – O(p_y) π C(p_z) – O(p_z)

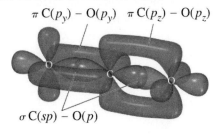

σ C(sp) – O(p)

10.63   (a)   COCl₂   Write the Lewis structure for the molecule:

$$\overset{\displaystyle :O:}{\underset{\displaystyle }{:\ddot{C}l-C-\ddot{C}l:}}$$

Use VSEPR to predict the electron geometry:

Three electron groups around the central atom give a trigonal planar electron geometry.
Select the correct hybridization for the central atom based on the electron geometry:
Trigonal planar electron geometry has sp^2 hybridization.
Sketch the molecule and label the bonds:

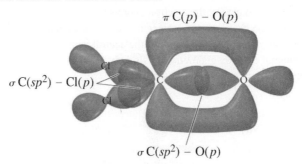

$\pi\ C(p) - O(p)$

$\sigma\ C(sp^2) - Cl(p)$

$\sigma\ C(sp^2) - O(p)$

(b) BrF_5 Write the Lewis structure for the molecule:

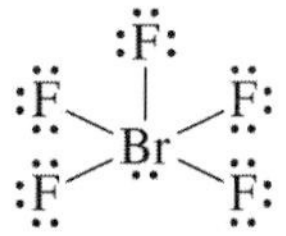

Use VSEPR to predict the electron geometry:
Six electron pairs around the central atoms give an octahedral electron geometry.
Select the correct hybridization for the central atom based on the electron geometry:
Octahedral electron geometry has sp^3d^2 hybridization.
Sketch the molecule and label the bonds:

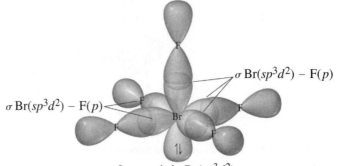

$\sigma\ Br(sp^3d^2) - F(p)$

$\sigma\ Br(sp^3d^2) - F(p)$

Lone pair in $Br(sp^3d^2)$

(c) XeF_2 Write the Lewis structure for the molecule:

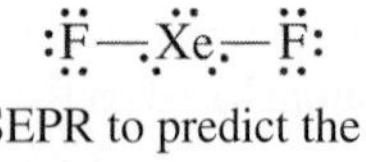

$:\!\ddot{F}\!-\!\dot{\ddot{X}}\!e\!-\!\ddot{F}\!:$

Use VSEPR to predict the electron geometry:
Five electron groups around the central atom give a trigonal bipyramidal geometry.
Select the correct hybridization for the central atom based on the electron geometry:
Trigonal bipyramidal geometry has sp^3d hybridization.

Sketch the molecule and label the bonds:

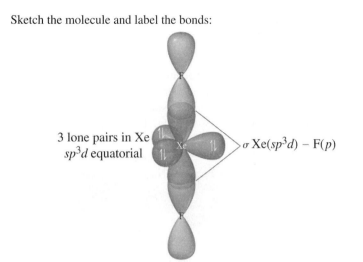

3 lone pairs in Xe
sp^3d equatorial $\sigma\, Xe(sp^3d) - F(p)$

(d) I_3^- Write the Lewis structure for the molecule:

$$\left[\ddot{\underset{\cdot\cdot}{I}}-\ddot{\underset{\cdot\cdot}{I}}-\ddot{\underset{\cdot\cdot}{I}}\,\right]^-$$

Use VSEPR to predict the electron geometry:
Five electron groups around the central atom give a trigonal bipyramidal geometry.
Select the correct hybridization for the central atom based on the electron geometry:
Trigonal bipyramidal geometry has sp^3d hybridization.
Sketch the molecule and label the bonds:

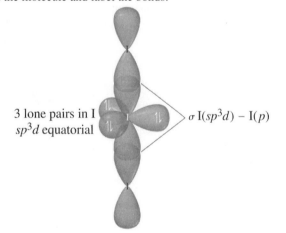

3 lone pairs in I
sp^3d equatorial $\sigma\, I(sp^3d) - I(p)$

10.65   (a) N_2H_2 Write the Lewis structure for the molecule:

$$H-\ddot{N}=\ddot{N}-H$$

Use VSEPR to predict the electron geometry:
Three electron groups around each interior atom give a trigonal planar electron geometry.
Select the correct hybridization for the central atoms based on the electron geometry:
Trigonal planar electron geometry has sp^2 hybridization.
Sketch the molecule and label the bonds:

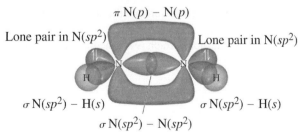

$\pi\, N(p) - N(p)$

Lone pair in $N(sp^2)$ Lone pair in $N(sp^2)$

$\sigma\, N(sp^2) - H(s)$ $\sigma\, N(sp^2) - H(s)$

$\sigma\, N(sp^2) - N(sp^2)$

(b)   N₂H₄   Write the Lewis structure for the molecule:

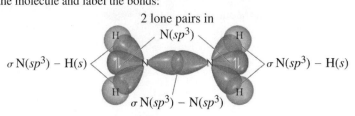

Use VSEPR to predict the electron geometry:
Four electron groups around each interior atom give tetrahedral electron geometry.
Select the correct hybridization for the central atoms based on the electron geometry:
Tetrahedral electron geometry has sp^3 hybridization.
Sketch the molecule and label the bonds:

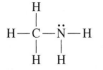
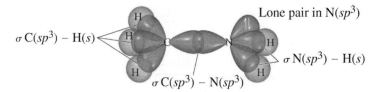

(c)   CH₃NH₂ Write the Lewis structure for the molecule:

$$H - C - \ddot{N} - H$$

Use VSEPR to predict the electron geometry:
Four electron groups around the C give a tetrahedral electron geometry around the C atom, and four electron groups around the N give a tetrahedral geometry around the N atom.
Select the correct hybridization for the central atoms based on the electron geometry:
Tetrahedral electron geometry has sp^3 hybridization of both C and N.
Sketch the molecule and label the bonds:

10.67

The Cs in positions 1 and 2 have four electron groups around the atom, which is tetrahedral electron geometry. Tetrahedral electron geometry is sp^3 hybridization.
C in position 3 has three electron groups around the atom, which is trigonal planar electron geometry. Trigonal planar electron geometry is sp^2 hybridization.
O (bonded to C and H) has four electron groups around the atom, which is tetrahedral electron geometry. Tetrahedral electron geometry is sp^3 hybridization.
N has four electron groups around the atom, which is tetrahedral electron geometry. Tetrahedral electron geometry is sp^3 hybridization.

Molecular Orbital Theory

10.69 $1s + 1s$ constructive interference results in a bonding orbital:

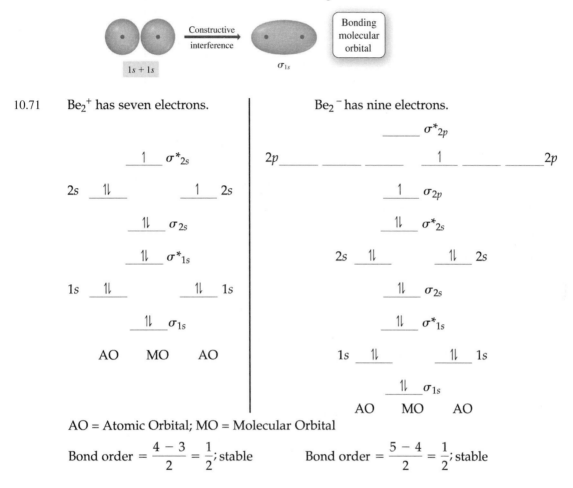

10.71 Be_2^+ has seven electrons. Be_2^- has nine electrons.

AO = Atomic Orbital; MO = Molecular Orbital

Bond order $= \dfrac{4-3}{2} = \dfrac{1}{2}$; stable   Bond order $= \dfrac{5-4}{2} = \dfrac{1}{2}$; stable

10.73   The bonding and antibonding molecular orbitals from the combination of p_x and p_x atomic orbitals lie along the internuclear axis.

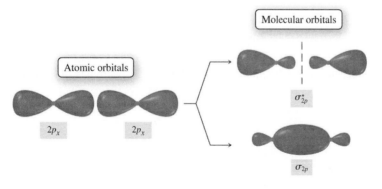

10.75

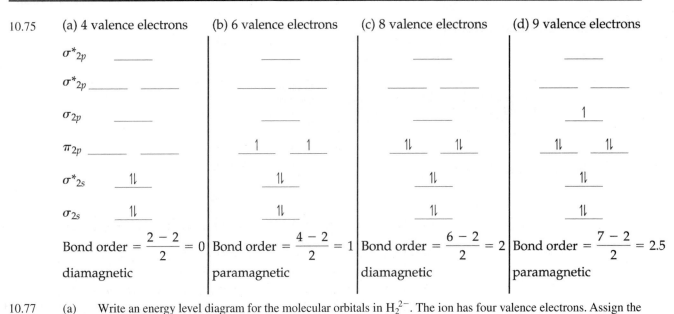

(a) 4 valence electrons

σ^*_{2p}

σ^*_{2p}

σ_{2p}

π_{2p}

σ^*_{2s} ⇅

σ_{2s} ⇅

Bond order $= \dfrac{2-2}{2} = 0$

diamagnetic

(b) 6 valence electrons

π_{2p} ↿ ↿

σ^*_{2s} ⇅

σ_{2s} ⇅

Bond order $= \dfrac{4-2}{2} = 1$

paramagnetic

(c) 8 valence electrons

π_{2p} ⇅ ⇅

σ^*_{2s} ⇅

σ_{2s} ⇅

Bond order $= \dfrac{6-2}{2} = 2$

diamagnetic

(d) 9 valence electrons

σ_{2p} ↿

π_{2p} ⇅ ⇅

σ^*_{2s} ⇅

σ_{2s} ⇅

Bond order $= \dfrac{7-2}{2} = 2.5$

paramagnetic

10.77   (a)   Write an energy level diagram for the molecular orbitals in H_2^{2-}. The ion has four valence electrons. Assign the electrons to the molecular orbitals beginning with the lowest energy orbitals and following Hund's rule.

σ^*_{1s} ⇅

σ_{1s} ⇅   Bond order $= \dfrac{2-2}{2} = 0$. With a bond order of 0, the ion will not exist.

(b)   Write an energy level diagram for the molecular orbitals in Ne_2. The molecule has 16 valence electrons. Assign the electrons to the molecular orbitals beginning with the lowest energy orbitals and following Hund's rule.

σ^*_{2p} ⇅

π^*_{2p} ⇅   ⇅

π_{2p} ⇅   ⇅

σ_{2p} ⇅

σ^*_{2s} ⇅

σ_{2s} ⇅

Bond order $= \dfrac{8-8}{2} = 0$. With a bond order of 0, the molecule will not exist.

(c)   Write an energy level diagram for the molecular orbitals in He_2^{2+}. The ion has two valence electrons. Assign the electrons to the molecular orbitals beginning with the lowest energy orbitals and following Hund's rule.

σ^*_{1s} ⎯⎯

σ_{1s} ⇅

Bond order $= \dfrac{2-0}{2} = 1$. With a bond order of 1, the ion will not exist.

(d)   Write an energy level diagram for the molecular orbitals in F_2^{2-}. The molecule has 16 valence electrons. Assign the electrons to the molecular orbitals beginning with the lowest energy orbitals and following Hund's rule.

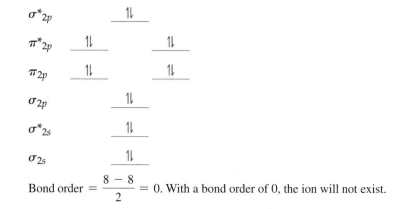

σ^*_{2p} ⇅

π^*_{2p} ⇅   ⇅

π_{2p} ⇅   ⇅

σ_{2p} ⇅

σ^*_{2s} ⇅

σ_{2s} ⇅

Bond order $= \dfrac{8-8}{2} = 0$. With a bond order of 0, the ion will not exist.

10.79 C_2^- has the highest bond order, the highest bond energy, and the shortest bond.
Write an energy level diagram for the molecular orbitals in each of the C_2 species.
Assign the electrons to the molecular orbitals beginning with the lowest energy orbitals and following Hund's rule for each of the species.

C_2 (8 valence electrons) C_2^+ (7 valence electrons) C_2^- (9 valence electrons)

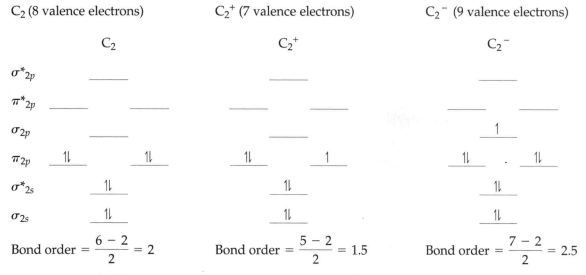

	C_2	C_2^+	C_2^-
σ^*_{2p}			
π^*_{2p}			
σ_{2p}			↑
π_{2p}	⇅ ⇅	⇅ ↑	⇅ ⇅
σ^*_{2s}	⇅	⇅	⇅
σ_{2s}	⇅	⇅	⇅

Bond order $= \dfrac{6-2}{2} = 2$ Bond order $= \dfrac{5-2}{2} = 1.5$ Bond order $= \dfrac{7-2}{2} = 2.5$

C_2^- has the highest bond order at 2.5. Bond order is directly related to bond energy, so C_2^- has the largest bond energy. Bond order is inversely related to bond length, so C_2^- has the shortest bond length.

10.81   Write an energy level diagram for the molecular orbitals in CO using O_2 energy ordering.
Assign the electrons to the molecular orbitals beginning with the lowest energy orbitals and following Hund's rule. CO has 10 valence electrons.

σ^*_{2p} ___

π^*_{2p} ___   ___

π_{2p} ⇅   ⇅

σ_{2p} ⇅

σ^*_{2s} ⇅

σ_{2s} ⇅

Bond order $= \dfrac{8-2}{2} = 3$

The electron density is toward the O atom because it is more electronegative.

Cumulative Problems

10.83   (a)   COF₂   Write the Lewis structure for the molecule:

$$:\ddot{O}:$$
$$\|$$
$$:\ddot{F}-C-\ddot{F}:$$

Use VSEPR to predict the electron geometry:
Three electron groups around the central atom give a trigonal planar electron geometry. Three bonding pairs of electrons give a trigonal planar molecular geometry.
Determine whether the molecule contains polar bonds:
The electronegativity of C = 2.5, O = 3.5, and F = 4.0. Therefore, the bonds are polar.
Determine whether the polar bonds add together to form a net dipole:
Even though a trigonal planar molecular geometry normally is nonpolar, because the bonds have different dipole moments, the sum of the dipole moments is not zero. The molecule is polar. See Table 10.2 to see how dipole moments add to determine polarity.
Select the correct hybridization for the central atom based on the electron geometry:
Trigonal planar geometry has sp^2 hybridization.
Sketch the molecule and label the bonds:

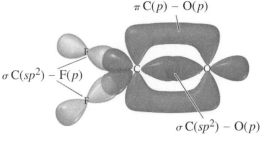

$\pi \, C(p) - O(p)$

$\sigma \, C(sp^2) - F(p)$

$\sigma \, C(sp^2) - O(p)$

(b)   S₂Cl₂   Write the Lewis structure for the molecule:

$$:\ddot{C}l-\ddot{S}-\ddot{S}-\ddot{C}l:$$

Use VSEPR to predict the electron geometry:
Four electron groups around the central atom give a tetrahedral electron geometry. Two bonding pairs and two lone pairs of electrons give a bent molecular geometry.
Determine whether the molecule contains polar bonds:
The electronegativity of S = 2.5 and Cl = 3.0. Therefore, the bonds are polar.
Determine whether the polar bonds add together to form a net dipole:
In a bent molecular geometry, the sum of the dipole moments is not zero. The molecule is polar. See Table 10.2 to see how dipole moments add to determine polarity.
Select the correct hybridization for the central atom based on the electron geometry:
Tetrahedral geometry has sp^3 hybridization.
Sketch the molecule and label the bonds:

$\sigma \, S(sp^3) - S(sp^3)$

2 lone pairs in sp^3 orbital on S

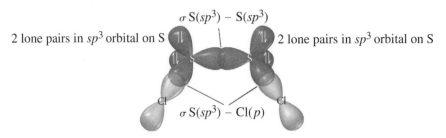

2 lone pairs in sp^3 orbital on S

$\sigma \, S(sp^3) - Cl(p)$

(c)   SF₄   Write the Lewis structure for the molecule:

Use VSEPR to predict the electron geometry:

Five electron groups around the central atom give a trigonal bipyramidal electron geometry.

Four bonding pairs and one lone pair of electrons give a seesaw molecular geometry.

Determine whether the molecule contains polar bonds:

The electronegativity of S = 2.5 and F = 4.0. Therefore, the bonds are polar.

Determine whether the polar bonds add together to form a net dipole:

In a seesaw molecular geometry, the sum of the dipole moments is not zero. The molecule is polar.

See Table 10.2 to see how dipole moments add to determine polarity.

Select the correct hybridization for the central atom based on the electron geometry:

Trigonal bipyramidal electron geometry has sp^3d hybridization.

Sketch the molecule and label the bonds:

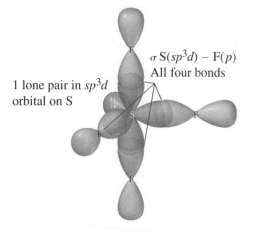

σ S(sp^3d) – F(p)
All four bonds

1 lone pair in sp^3d
orbital on S

10.85   (a)   serine

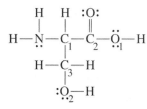

The Cs in positions 1 and 3 have four electron groups around the atom. Four electron groups give a tetrahedral electron geometry; tetrahedral electron geometry has sp^3 hybridization. Four bonding groups and zero lone pairs give a tetrahedral molecular geometry.

C in position 2 has three electron groups around the atom. Three electron groups give a trigonal planar geometry; trigonal planar geometry has sp^2 hybridization. Three bonding groups and zero lone pairs give a trigonal planar molecular geometry.

N has four electron groups around the atom. Four electron groups give a tetrahedral electron geometry; tetrahedral electron geometry has sp^3 hybridization. Three bonding groups and one lone pair give a trigonal pyramidal molecular geometry.

The Os in positions 1 and 2 each have four electron groups around the atom. Four electron groups give a tetrahedral electron geometry; tetrahedral electron geometry has sp^3 hybridization. Two bonding groups and two lone pairs give a bent molecular geometry.

(b)   asparagine

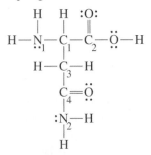

The Cs in positions 1 and 3 each have four electron groups around the atom. Four electron groups give a tetrahedral electron geometry; tetrahedral electron geometry has sp^3 hybridization. Four bonding groups and zero lone pairs give a tetrahedral molecular geometry.

The Cs in positions 2 and 4 each have three electron groups around the atom. Three electron groups give a trigonal planar geometry; trigonal planar geometry has sp^2 hybridization. Three bonding groups and zero lone groups give a trigonal planar molecular geometry.

The Ns in positions 1 and 2 each have four electron groups around the atom. Four electron groups give a tetrahedral electron geometry; tetrahedral electron geometry has sp^3 hybridization. Three bonding groups and one lone pair give a trigonal pyramidal molecular geometry.

O has four electron groups around the atom. Four electron groups give a tetrahedral electron geometry; tetrahedral electron geometry has sp^3 hybridization. Two bonding groups and two lone pairs give a bent molecular geometry.

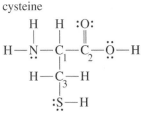

(c)   cysteine

The Cs in positions 1 and 3 each have four electron groups around the atom. Four electron groups give a tetrahedral electron geometry; tetrahedral electron geometry has sp^3 hybridization. Four bonding groups and zero lone pairs give a tetrahedral molecular geometry.

C in position 2 has three electron groups around the atom. Three electron groups give a trigonal planar geometry; trigonal planar geometry has sp^2 hybridization. Three bonding groups and zero lone pairs give a trigonal planar molecular geometry.

N has four electron groups around the atom. Four electron groups give a tetrahedral electron geometry; tetrahedral electron geometry has sp^3 hybridization. Three bonding groups and one lone pair give a trigonal pyramidal molecular geometry.

O and S have four electron groups around the atom. Four electron groups give a tetrahedral electron geometry; tetrahedral electron geometry has sp^3 hybridization. Two bonding groups and two lone pairs give bent molecular geometry.

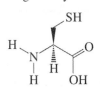

10.87 4π bonds, 25σ bonds; the lone pair on the Os and N in position 2 occupies sp^2 orbitals; the lone pairs on N in position 1, N in position 3 and N in position 4 occupy sp^3 orbitals.

caffeine

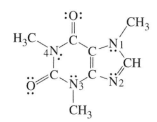

10.89   (a)   Water-soluble: The 4 C—OH bonds, the C=O bond, and the C—O bonds in the ring make the molecule polar. Because of the large electronegativity difference between the C and O, each of the bonds will have a dipole moment. The sum of the dipole moments does *not* give a net zero dipole moment, so the molecule is polar. Because it is polar, it will be water-soluble.

   (b)   Fat-soluble: There is only one C—O bond in the molecule. The dipole moment from this bond is not enough to make the molecule polar because of all of the nonpolar components of the molecule. The C—H bonds in the structure lead to a net dipole of zero for most of the sites in the molecule. Because the molecule is nonpolar, it is fat-soluble.

   (c)   Water-soluble: The carboxylic acid function (COOH group) along with the N atom in the ring make the molecule polar. Because of the electronegativity difference between the C and O and the C and N atoms, the bonds will have a dipole moment and the net dipole moment of the molecule is *not* zero; so the molecule is polar. Because the molecule is polar, it is water-soluble.

   (d)   Fat-soluble: The two O atoms in the structure contribute a very small amount to the net dipole moment of this molecule. The majority of the molecule is nonpolar because there is no net dipole moment around the interior C atoms. Because the molecule is nonpolar, it is fat-soluble.

10.91   ClF has 14 valence electrons. Assign the electrons to the lowest energy MOs first and then follow Hund's rule. The MOs are formed from the $2s$ and $2p$ orbitals on F and the $3s$ and $3p$ orbitals on Cl.

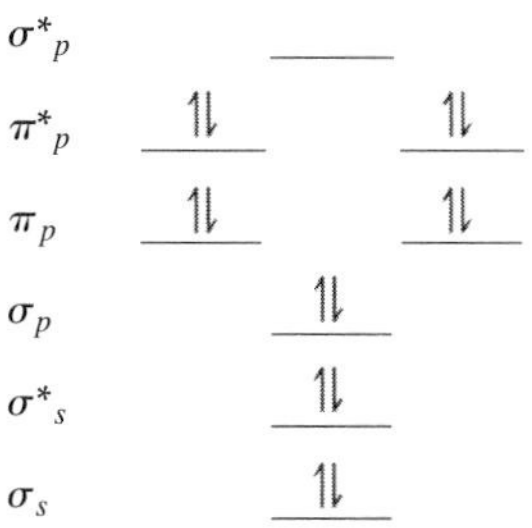

σ^*_p ___

π^*_p ⇅   ⇅

π_p ⇅   ⇅

σ_p ⇅

σ^*_s ⇅

σ_s ⇅

Bond order $= \dfrac{8-6}{2} = 1$

10.93   BrF (14 valence electrons)

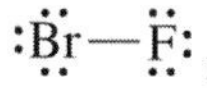

   $:\ddot{B}r\!-\!\ddot{F}:$ no central atom, no hybridization, no electron structure.

   BrF_2^- (22 valence electrons)

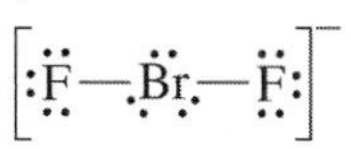

   $\left[:\ddot{F}\!-\!\ddot{B}r\!-\!\ddot{F}:\right]^-$ There are five electron groups on the central atom, so the electron geometry is trigonal bipyramidal. The two bonding groups and three lone pairs give a linear molecular geometry. An electron geometry of trigonal bipyramidal has sp^3d hybridization.

   BrF_3 (28 valence electrons)

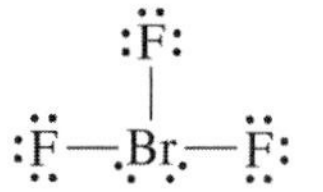

   $:\ddot{F}:$
   |
   $:\ddot{F}\!-\!\ddot{B}r\!-\!\ddot{F}:$ There are five electron groups on the central atom, so the electron geometry is trigonal bipyramidal. The three bonding groups and two lone pairs give a T-shaped molecular geometry. An electron geometry of trigonal bipyramidal has sp^3d hybridization.

BrF$_4^-$ (36 valence electrons)

There are six electron groups on the central atom, so the electron geometry is octahedral. The four bonding groups and two lone pairs give a square planar molecular geometry. An electron geometry of octahedral has sp^3d^2 hybridization.

BrF$_5$ (42 valence electrons)

There are six electron groups on the central atom, so the electron geometry is octahedral. The five bonding groups and one lone pair give a square pyramidal molecular geometry. An electron geometry of octahedral has sp^3d^2 hybridization.

10.95   Draw the Lewis structure: C$_4$H$_6$Cl$_2$ (36 valence electrons)

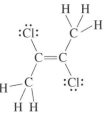

Even though the C—Cl bonds are polar, the net dipole will be zero because the C—Cl bonds and the C—CH$_3$ bonds are on opposite sides of the double bond. This will result in bond vectors that cancel each other.

10.97   (a)   N$_2$O$_5$ Draw the Lewis structure: (40 valence electrons)

Each N has a trigonal planar electron geometry, so there are 3 sp^2 hybrid orbitals on each N. The central O has tetrahedral electron geometry, so there are 4 sp^3 hybrid orbitals. There are a total of 10 hybrid orbitals.

(b)   C$_2$H$_5$NO Draw the Lewis structure: (24 valence electrons)

C$_A$ has a trigonal planar electron geometry, so there are 3 sp^2 hybrid orbitals.
N has a trigonal planar electron geometry, so there are 3 sp^2 hybrid orbitals.
C$_B$ has a tetrahedral electron geometry, so there are 4 sp^3 hybrid orbitals.
O has a tetrahedral electron geometry, so there are 4 sp^3 hybrid orbitals.
There are a total of 14 hybrid orbitals.

(c)   BrCN   Draw the Lewis structure: (16 valence electrons)

The C has linear electron geometry, so there are 2 sp hybrid orbitals.

Challenge Problems

10.99   According to valence bond theory, CH_4, NH_3, and H_2O are all sp^3 hybridized. This hybridization results in a tetrahedral electron group configuration with a 109.5° bond angle. NH_3 and H_2O deviate from this idealized bond angle because their lone electron pairs exist in their own sp^3 orbitals. The presence of lone pairs lowers the tendency for the central atom's orbitals to hybridize. As a result, as lone pairs are added, the bond angle moves from the 109.5° hybrid angle toward the 90° unhybridized angle.

10.101   Using the MO diagram for NH_3, assign the eight valence electrons to the molecular orbitals. Start with the lowest energy orbital first and follow Hund's rule.

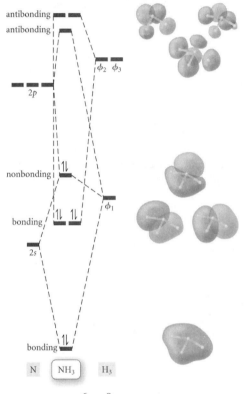

$$\text{Bond order} = \frac{6 - 0}{2} = 3$$

With a bond order of 3, the molecule is stable.

10.103   For each write the Lewis structure:
Determine electron pair geometry around each central atom.
Determine the molecular geometry, determine idealized bond angles, and predict actual bond angles.
NO_2

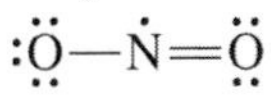

:Ö—Ṅ=Ö

Two bonding groups and a lone electron give a trigonal planar electron geometry; the molecular geometry will be bent. Trigonal planar electron geometry has idealized bond angles of 120°. The bond angle is expected to be slightly less than 120° because of the lone electron occupying the third sp^2 orbital.
NO_2^+

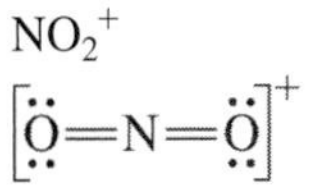

[Ö=N=Ö]$^+$

Two bonding groups of electrons and no lone pairs give a linear electron geometry and molecular geometry. Linear electron geometry has a bond angle of 180°.

NO_2^-

$$\left[:\ddot{O}\!\!-\!\!\ddot{N}\!\!=\!\!\ddot{O} \right]^-$$

Two bonding groups of electrons and one lone pair give a trigonal planar electron geometry; the molecular geometry will be bent.

Trigonal planar electron geometry has idealized bond angles of 120°. The bond angle is expected to be less than 120° because of the lone pair electrons occupying the third sp^2 orbital. Further, the bond angle should be less than the bond angle in NO_2 because the presence of lone pairs lowers the tendency for the central atom's orbitals to hybridize. As a result, as lone pairs are added, the bond angle moves further from the 120° hybrid angle to the 90° unhybridized angle, and the two electrons will increase this tendency.

10.105 CH_5^+ Draw the Lewis structure: (nine valence electrons)

$$\left[\begin{array}{c} H \\ | \\ H\!\!-\!\!C \!\!\begin{array}{c} \diagup H \\ \diagdown H \end{array} \\ | \\ H \end{array} \right]^+$$

To accommodate the five σ bonds, you need five equal energy hybrid orbitals. So you need to combine five atomic orbitals. The valence electrons on C are in the $2s$ and $2p$ orbitals, so you can combine the s and $3p$ orbitals, but this would only give you four hybrid orbitals. The next lowest energy orbital available on C is the $3s$, so this would be the next atomic orbital added in. This gives a hybridization of s^2p^3. VSEPR theory would predict that the geometry is trigonal bipyramidal to accommodate the five bonds.

10.107 CH_3CONH_2 Draw the Lewis structure: (24 valence electrons)

$$\begin{array}{ccc} H & :O: & \\ | & \| & \\ H\!\!-\!\!C\!\!-\!\!C\!\!-\!\!\ddot{N}\!\!-\!\!H \\ |_1 & |_2 & | \\ H & & H \end{array}$$

Structure I

In structure I: C_1 has 4σ bonds and no lone pairs. The electron geometry would be tetrahedral.

C_2 has 3σ bonds, 1π bond, and no lone pairs. The electron geometry would be trigonal planar.

N has 3σ bonds and 1 lone pair. The electron geometry would be tetrahedral.

This structure would have a trigonal pyramidal molecular geometry around the nitrogen and would not be planar. A second resonance form can be drawn.

$$\begin{array}{ccc} H & :\ddot{O}: & \\ | & | & \\ H\!\!-\!\!C\!\!-\!\!C\!\!=\!\!N\!\!-\!\!H \\ |_1 & |_2 & | \\ H & & H \end{array}$$

Structure II

In structure II: C_1 has 4σ bonds and no lone pairs. The electron geometry would be tetrahedral.

C_2 has 3σ bonds, 1π bond, and no lone pairs. The electron geometry would be trigonal planar.

N has 3σ bonds, 1π bond, and no lone pairs. The electron geometry would be trigonal planar.

This resonance form would account for a planar configuration around the N.

Conceptual Problems

10.109 Statement a is the best statement.

Statement b neglects the lowering of potential energy that arises from the interaction of the lone pair electrons with the bonding electrons.

Statement c neglects the interaction of the electrons altogether. The molecular geometries are determined by the number and types of electron groups around the central atom.

10.111   In Lewis theory, a covalent bond comes from the sharing of electrons.

> A single bond shares two electrons (one pair).
> A double bond shares four electrons (two pairs).
> A triple bond shares six electrons (three pairs).

In valence bond theory, a covalent bond forms when orbitals overlap. The orbitals can be unhybridized or hybridized orbitals.

> A single bond forms when a σ bond is formed from the overlap of an s orbital with an s orbital, an s orbital with a p orbital, or a p orbital and a p orbital overlapping end to end. A σ can also form from the overlap of a hybridized orbital on the central atom with an s orbital or with a p orbital overlapping end to end.
> A double bond is a combination of a σ bond and a π bond. The π bond forms from the sideways overlap of a p orbital on each of the atoms involved in the bond. The p orbitals must have the same orientation.
> A triple bond is a combination of a σ bond and 2π bonds. The π bonds form from the sideways overlap of a p orbital on each of the atoms involved in the bond. The p orbitals must have the same orientation, so each π bond is formed from a different set of p orbitals.

In molecular orbital theory, molecular orbitals form. These are combinations of the atomic orbitals of the atoms involved in the bond. The bonds form when the valence electrons occupy more bonding molecular orbitals than anti-bonding molecular orbitals. This is calculated by the bond order.

> A single bond has a bond order of 1.
> A double bond has a bond order of 2.
> A triple bond has a bond order of 3.

All three models show the formation of bonds between two atoms. All three models show the formation of the same number of bonds between the atoms involved. Lewis theory tells us only about the number of bonds formed and, combined with VSEPR theory, allows us to predict the shape of the molecule. It does not, however, tell us anything about how the bonds are formed. Valence bond theory addresses the formation of the different types of bonds, sigma and pi. In valence bond theory, the bonds form from the overlap of atomic orbitals on the individual atoms involved in the bonds and the atoms are localized between the two atoms involved in the bond. Molecular orbital theory approaches the formation of bonds by looking at the entire molecule. The electrons are not restricted to any two individual atoms, but are treated as belonging to the whole molecule. The electrons reside in molecular orbitals that are part of the entire molecule rather than being restricted to individual atoms. Each model gives us information about the molecule. The amount and type of information we need determines the model we choose to use.

11 Liquids, Solids, and Intermolecular Forces

Review Questions

11.1 The key to the gecko's sticky feet lies in the millions of microhairs, called setae, that line its toes. Each seta is between 30 and 130 μm long and branches out to end in several hundred flattened tips called spatulae. This unique structure allows the gecko's toes to have unusually close contact with the surfaces it climbs. The close contact allows intermolecular forces—which are significant only at short distances—to hold the gecko to the wall.

11.3 The main properties of liquids are that liquids have much higher densities in comparison to gases and generally have lower densities in comparison to solids, liquids have an indefinite shape and assume the shape of their container, liquids have a definite volume, and liquids are not easily compressed.

11.5 Solids may be crystalline, in which case the atoms or molecules that compose them are arranged in a well-ordered three-dimensional array, or they may be amorphous, in which case the atoms or molecules that compose them have no long-range order.

11.7 Because the most molecular motion occurs in the gas phase and the least molecular motion occurs in the solid phase (atoms are pushed closer together), a substance will be converted from a solid to a liquid and then to a gas as the temperature increases. The strength of the intermolecular interactions is least in the gas phase because there are large distances between particles and they are moving very fast. Intermolecular forces are stronger in liquids and solids, where molecules are "touching" one another. The strength of the interactions in the condensed phases determines at what temperature the substance will melt and boil.

11.9 Intermolecular forces, even the strongest ones, are generally much weaker than bonding forces. The reason for the relative weakness of intermolecular forces compared to bonding forces is related to Coulomb's law $\left(E = \dfrac{1}{4\pi\varepsilon_{\mathrm{o}}} \dfrac{q_1 q_2}{r} \right)$. Bonding forces are the result of large charges (the charges on protons and electrons, q_1 and q_2) interacting at very close distances (r). Intermolecular forces are the result of smaller charges (as we will see in the following discussion) interacting at greater distances.

11.11 The dipole–dipole force exists in all molecules that are polar. Polar molecules have permanent dipoles that interact with the permanent dipoles of neighboring molecules. The positive end of one permanent dipole is attracted to the negative end of another; this attraction is the dipole–dipole force.

11.13 The hydrogen bond is a sort of super dipole–dipole force. Polar molecules containing hydrogen atoms bonded directly to fluorine, oxygen, or nitrogen exhibit an intermolecular force called hydrogen bonding. The large electronegativity difference between hydrogen and these electronegative elements means that the H atoms will have fairly large partial positive charges ($\delta+$), while the F, O, or N atoms will have fairly large partial negative charges ($\delta-$). In addition, because these atoms are all quite small, they can approach one another very closely. The result is a strong attraction between the hydrogen in each of these molecules and the F, O, or N on its neighbors, an attraction called a hydrogen bond.

11.15   Surface tension is the tendency of liquids to minimize their surface area. Molecules at the surface have relatively fewer neighbors with which to interact because there are no molecules above the surface. Consequently, molecules at the surface are inherently less stable—they have higher potential energy—than those in the interior. To increase the surface area of the liquid, some molecules from the interior must be moved to the surface, a process requiring energy. The surface tension of a liquid is the energy required to increase the surface area by a unit amount. Surface tension decreases with decreasing intermolecular forces.

11.17   Capillary action is the ability of a liquid to flow against gravity up a narrow tube. Capillary action results from a combination of two forces: the attraction between molecules in a liquid, called cohesive forces, and the attraction between these molecules and the surface of the tube, called adhesive forces. The adhesive forces cause the liquid to spread out over the surface of the tube, while the cohesive forces cause the liquid to stay together. If the adhesive forces are greater than the cohesive forces (as is the case for water in a glass tube), the attraction to the surface draws the liquid up the tube while the cohesive forces pull along those molecules that are not in direct contact with the tube walls. The liquid rises up the tube until the force of gravity balances the capillary action—the thinner the tube, the higher the rise. If the adhesive forces are smaller than the cohesive forces (as is the case for liquid mercury), the liquid does not rise up the tube at all (and, in fact, will drop to a level below the level of the surrounding liquid).

11.19   The molecules that leave the liquid are at the high end of the energy curve—the most energetic. If no additional heat enters the liquid, the average energy of the entire collection of molecules goes down—much as the class average on an exam goes down when the highest-scoring students are eliminated. So vaporization is an endothermic process; it takes energy to vaporize the molecules in a liquid. Also, vaporization requires overcoming the intermolecular forces that hold liquids together. Because energy must be absorbed to pull the molecules apart, the process is endothermic. Condensation is the opposite process, so it must be exothermic. Also, gas particles have more energy than those in the liquid. It is the least energetic of these that condense, adding energy to the liquid.

11.21   The heat of vaporization (ΔH_{vap}) is the amount of heat required to vaporize 1 mole of a liquid to a gas. The heat of vaporization of a liquid can be used to calculate the amount of heat energy required to vaporize a given mass of the liquid (or the amount of heat given off by the condensation of a given mass of liquid) and can be used to compare the volatility of two substances.

11.23   When a system in dynamic equilibrium is disturbed, the system responds so as to minimize the disturbance and return to a state of equilibrium.

11.25   The boiling point of a liquid is the temperature at which its vapor pressure equals the external pressure. The normal boiling point of a liquid is the temperature at which its vapor pressure equals 1 atm.

11.27   As the temperature rises, more liquid vaporizes and the pressure within the container increases. As more and more gas is forced into the same amount of space, the density of the gas becomes higher and higher. At the same time, the increasing temperature causes the density of the liquid to become lower and lower. At the critical temperature, the meniscus between the liquid and gas disappears and the gas and liquid phases commingle to form a supercritical fluid.

11.29   Fusion, or melting, is the phase transition from solid to liquid. The term fusion is used for melting because if you heat several crystals of a solid, they will fuse into a continuous liquid upon melting. Fusion is endothermic because solids have less kinetic energy than liquids, so energy must be added to a solid to get it to melt.

11.31   There are two horizontal lines (i.e., heat is added, but the temperature stays constant) in the heating curve because there are two endothermic phase changes. The heat that is added is used to change the phase from solid to liquid or from liquid to gas and therefore there is no rise in temperature.

11.33   A phase diagram is a map of the phase of a substance as a function of pressure (on the *y*-axis) and temperature (on the *x*-axis).

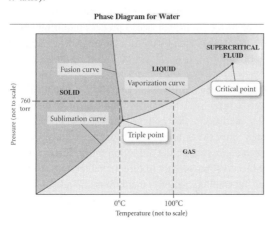

Phase Diagram for Water

11.35   Water has a low molar mass (18.01 g/mol), yet it is a liquid at room temperature. Water's high boiling point for its molar mass can be understood by examining the structure of the water molecule. The bent geometry of the water molecule and the highly polar nature of the O—H bonds result in a molecule with a significant dipole moment. Water's two O—H bonds (hydrogen directly bonded to oxygen) allow a water molecule to form strong hydrogen bonds with four other water molecules, resulting in a relatively high boiling point. Water's high polarity also allows it to dissolve many other polar and ionic compounds, and even a number of nonpolar gases such as oxygen and carbon dioxide (by inducing a dipole moment in their molecules). Water has an exceptionally high specific heat capacity. One significant difference between the phase diagram of water and that of other substances is that the fusion curve for water has a negative slope. The fusion curve within the phase diagrams for most substances has a positive slope because increasing pressure favors the denser phase, which for most substances is the solid phase. This negative slope means that ice is less dense than liquid water and so ice floats. The solid phase sinks in the liquid of most other substances.

11.37   A crystalline lattice is the regular arrangements of atoms within a crystalline solid. The crystalline lattice can be represented by a small collection of atoms, ions, or molecules—a fundamental building block called the unit cell. When the unit cell is repeated over and over—like tiles in a floor or the pattern in a wallpaper design, but in three dimensions—the entire lattice can be reproduced.

11.39   Atoms in a simple cubic cell structure have a coordination number of 6, an edge length of $2r$, and one atom in the unit cell. Atoms in a body-centered cubic cell structure have a coordination number of 8, an edge length of $4r/\sqrt{3}$, and two atoms in the unit cell. Atoms in a face-centered cubic cell structure have a coordination number of 12, an edge length of $2\sqrt{2}r$, and four atoms in the unit cell.

11.41   The three types of solids are molecular solids, ionic solids, and atomic solids. Molecular solids are those solids whose composite units are molecules. The lattice sites in a crystalline molecular solid are therefore occupied by molecules. Ice (solid H_2O) and dry ice (solid CO_2) are examples of molecular solids. Molecular solids are held together by the kinds of intermolecular forces—dispersion forces, dipole–dipole forces, and hydrogen bonding. Ionic solids are those solids whose composite units are ions. Table salt (NaCl) and calcium fluoride (CaF_2) are good examples of ionic solids. Ionic solids are held together by the coulombic interactions that occur between the cations and anions occupying the lattice sites in the crystal, which is an ionic bond. Atomic solids are those solids whose composite units are individual atoms. Atomic solids can themselves be divided into three categories—nonbonding atomic solids, metallic atomic solids, and network covalent atomic solids—each held together by a different kind of force. Nonbonding atomic solids, which include only the noble gases in their solid form, are held together by relatively weak dispersion forces. Metallic atomic solids, such as iron and gold, are held together by metallic bonds, which in the simplest model are represented by the interaction of metal cations with a sea of electrons that surround them. Network covalent atomic solids, such as diamond, graphite, and silicon dioxide, are held together by covalent bonds.

11.43   Cesium chloride (CsCl) is a good example of an ionic compound containing cations and anions of similar size (Cs^+ radius = 167 pm; Cl^- radius = 181 pm). In the cesium chloride structure, the chloride ions occupy the lattice sites of a simple cubic cell and one cesium ion lies in the very center of the cell, as shown in Figure 11.51. Notice that the cesium chloride unit cell contains one chloride anion ($8 \times 1/8 = 1$) and one cesium cation (the cesium ion in the middle belongs entirely to the unit cell) for a ratio of Cs to Cl of 1:1, as in the formula for the compound.

The crystal structure of sodium chloride must accommodate the more disproportionate sizes of Na^+ (radius = 97 pm) and Cl^- (radius = 181 pm). The larger chloride anion could theoretically fit many of the smaller sodium cations around it, but charge neutrality requires that each sodium cation be surrounded by an equal number of chloride anions. The structure that minimizes the energy is shown in Figure 11.52 and has a coordination number of 6 (each chloride anion is surrounded by six sodium cations, and vice versa). You can visualize this structure, called the rock salt structure, as chloride anions occupying the lattice sites of a face-centered cubic structure with the smaller sodium cations occupying the holes between the anions. (Alternatively, you can visualize this structure as the sodium cations occupying the lattice sites of a face-centered cubic structure with the larger chloride anions occupying the spaces between the cations.) Each unit cell contains four chloride anions $[(8 \times 1/8) + (6 \times \frac{1}{2}) = 4]$ and four sodium cations $(12 \times \frac{1}{4})$, resulting in a ratio of 1:1, as in the formula of the compound.

You can visualize this structure (shown in Figure 11.53), called the zinc blende structure, as sulfide anions occupying the lattice sites of a face-centered cubic structure with the smaller zinc cations occupying four of the eight tetrahedral holes located directly beneath each corner atom. A tetrahedral hole is the empty space that lies in the center of a tetrahedral arrangement of four atoms. Each unit cell contains four sulfide anions $[(8 \times 1/8) + (6 \times \frac{1}{2}) = 4]$ and four zinc cations (each of the four zinc cations is completely contained within the unit cell), resulting in a ratio of 1:1, as in the formula of the compound.

11.45   Atomic solids can be divided into three categories—nonbonding atomic solids, metallic atomic solids, and network covalent atomic solids. Nonbonding atomic solids, which include only the noble gases in their solid form, are held together by relatively weak dispersion forces. Metallic atomic solids, such as iron and gold, are held together by metallic bonds, which in the simplest model are represented by the interaction of metal cations with a sea of electrons around them. Network covalent atomic solids, such as diamond, graphite, and silicon dioxide, are held together by covalent bonds.

11.47   The band gap is an energy gap that exists between the valence band and conduction band. In metals, the valence band and conduction band are energetically continuous—the energy difference between the top of the valence band and the bottom of the conduction band is infinitesimally small. In semiconductors, the band gap is small, allowing some electrons to be promoted at ordinary temperatures, resulting in limited conductivity. In insulators, the band gap is large, and electrons are not promoted into the conduction band at ordinary temperatures, resulting in no electrical conductivity.

Problems by Topic

Intermolecular Forces

11.49   (a)   dispersion forces
         (b)   dispersion forces, dipole–dipole forces, and hydrogen bonding
         (c)   dispersion forces and dipole–dipole forces
         (d)   dispersion forces

11.51   (a)   dispersion forces and dipole–dipole forces
         (b)   dispersion forces, dipole–dipole forces, and hydrogen bonding
         (c)   dispersion forces
         (d)   dispersion forces

11.53   (a) CH_4 < (b) CH_3CH_3 < (c) CH_3CH_2Cl < (d) CH_3CH_2OH. The first two molecules only exhibit dispersion forces, so the boiling point increases with increasing molar mass. The third molecule also exhibits dipole–dipole forces, which are stronger than dispersion forces. The last molecule exhibits hydrogen bonding. Because these are by far the strongest intermolecular forces in this group, the last molecule has the highest boiling point.

11.55   (a) CH_3OH has the higher boiling point because it exhibits hydrogen bonding.
         (b) CH_3CH_2OH has the higher boiling point because it exhibits hydrogen bonding.
         (c) CH_3CH_3 has the higher boiling point because it has the larger molar mass.

11.57   (a) Br_2 has the higher vapor pressure because it has the smaller molar mass.
         (b) H_2S has the higher vapor pressure because it does not exhibit hydrogen bonding.
         (c) PH_3 has the higher vapor pressure because it does not exhibit hydrogen bonding.

11.59   (a)   This will not form a homogeneous solution because one is polar and one is nonpolar.

(b)   This will form a homogeneous solution. There will be ion–dipole interactions between the K^+ and Cl^- ions and the water molecules. There will also be dispersion forces, dipole–dipole forces, and hydrogen bonding between the water molecules.

(c)   This will form a homogeneous solution. Dispersion forces will be present among all of the molecules.

(d)   This will form a homogeneous solution. There will be dispersion forces, dipole–dipole forces, and hydrogen bonding among all of the molecules.

Surface Tension, Viscosity, and Capillary Action

11.61   Water will have the higher surface tension because it exhibits hydrogen bonding, a strong intermolecular force. Acetone cannot form hydrogen bonds.

11.63   Compound A will have the higher viscosity because it can interact with other molecules along the entire molecule. The more branched isomer has a smaller surface area, allowing for fewer interactions. Also, the molecule is very flexible, and the molecules can get tangled with each other.

11.65   In a clean glass tube, the water can generate strong adhesive interactions with the glass (due to the dipoles at the surface of the glass). Water experiences adhesive forces with glass that are stronger than its cohesive forces, causing it to climb the surface of a glass tube. When grease or oil coats the glass, this interferes with the formation of these adhesive interactions with the glass because oils are nonpolar and cannot interact strongly with the dipoles in the water. Without experiencing these strong intermolecular forces with oil, the water's cohesive forces will be greater and water will be drawn away from the surface of the tube.

Vaporization and Vapor Pressure

11.67   The water in the 12 cm diameter beaker will evaporate more quickly because there is more surface area for the molecules to evaporate from. The vapor pressure will be the same in the two containers because the vapor pressure is the pressure of the gas when it is in dynamic equilibrium with the liquid (evaporation rate = condensation rate). The vapor pressure is dependent only on the substance and the temperature. The 12 cm diameter container will reach this dynamic equilibrium faster.

11.69   The boiling point and higher heat of vaporization of oil are much higher than that of water, so it will not vaporize as quickly as the water. The evaporation of water cools your skin because evaporation is an endothermic process.

11.71   **Given:** 915 kJ from candy bar, water $d = 1.00$ g/mL   **Find:** $L(H_2O)$ vaporized at 100.0 °C
Other: $\Delta H^\circ_{vap} = 40.7$ kJ/mol
Conceptual Plan: $q \rightarrow$ mol $H_2O \rightarrow$ g $H_2O \rightarrow$ mL $H_2O \rightarrow$ L H_2O

$$\frac{1 \text{ mol}}{40.7 \text{ kJ}} \qquad \frac{18.01 \text{ g}}{1 \text{ mol}} \qquad \frac{1.00 \text{ mL}}{1.00 \text{ g}} \qquad \frac{1 \text{ L}}{1000 \text{ mL}}$$

Solution: $915 \text{ kJ} \times \dfrac{1 \text{ mol}}{40.7 \text{ kJ}} \times \dfrac{18.02 \text{ g}}{1 \text{ mol}} \times \dfrac{1.00 \text{ mL}}{1 \text{ g}} \times \dfrac{1 \text{ L}}{1000 \text{ mL}} = 0.405 \text{ L } H_2O$

Check: The units (L) are correct. The magnitude of the answer (< 1 L) makes physical sense because we are vaporizing about 22 moles of water.

11.73   **Given:** 0.95 g water condenses on iron block 75.0 g at $T_i = 22$ °C   **Find:** T_f (iron block)
Other: $\Delta H^\circ_{vap} = 44.0$ kJ/mol; $C_{Fe} = 0.449$ J/g · °C from text
Conceptual Plan: g $H_2O \rightarrow$ mol $H_2O \rightarrow q_{H_2O}$ (kJ) $\rightarrow q_{H_2O}$ (J) $\rightarrow q_{Fe}$ then $q_{Fe}, m_{Fe}, T_i \rightarrow T_f$

$$\frac{1 \text{ mol}}{18.01 \text{ g}} \qquad \frac{-44.0 \text{ kJ}}{1 \text{ mol}} \qquad \frac{1000 \text{ J}}{1 \text{ kJ}} \qquad -q_{H_2O} = q_{Fe} \qquad q = mC_s(T_f - T_i)$$

Solution: $0.95 \text{ g} \times \dfrac{1 \text{ mol}}{18.02 \text{ g}} \times \dfrac{-44.0 \text{ kJ}}{1 \text{ mol}} \times \dfrac{1000 \text{ J}}{1 \text{ kJ}} = -2\underline{3}19.64 \text{ J}$ then $-q_{H_2O} = q_{Fe} = 2\underline{3}19.64 \text{ J}$ then

$q = mC_s(T_f - T_i)$.   Rearrange to solve for T_f.

$$T_f = \frac{mC_s T_i + q}{mC_s} = \frac{\left(75.0 \text{ g} \times 0.449 \frac{J}{g \cdot °C} \times 22 \text{ °C}\right) + 2\underline{3}19.64 \text{ J}}{75.0 \text{ g} \times 0.449 \frac{J}{g \cdot °C}} = 91 \text{ °C}$$

Check: The units (°C) are correct. The temperature rose, which is consistent with heat being added to the block. The magnitude of the answer (91 °C) makes physical sense because even though we have $\sim\frac{1}{20}$ of a mole, the energy involved in condensation is very large.

11.75 **Given:**

Temperature (K)	Vapor Pressure (torr)
200	65.3
210	134.3
220	255.7
230	456.0
235	597.0

Find: $\Delta H_{vap}^°(NH_3)$ and normal boiling point

Conceptual Plan: To find the heat of vaporization, use Excel or similar software to make a plot of the natural log of vapor pressure (ln P) as a function of the inverse of the temperature in K (1/T). Then fit the points to a line and determine the slope of the line. Because the *slope* $= -\Delta H_{vap}/R$, we find the heat of vaporization as follows:
slope $= -\Delta H_{vap}/R \rightarrow \Delta H_{vap} = -$ *slope* $\times R$ then $J \rightarrow kJ$

$$\frac{1 \text{ kJ}}{1000 \text{ J}}$$

For the normal boiling point, use the equation of the best-fit line, substitute 760 torr for the pressure, and calculate the temperature.

Solution: Data was plotted in Excel.
The slope of the best-fitting line is $-29\underline{6}9.9$ K.

$$\Delta H_{vap} = -\text{slope} \times R = -(-29\underline{6}9.9 \text{ K}) \times \frac{8.314 \text{ J}}{K \text{ mol}} = \frac{2.4\underline{6}917 \times 10^4 \text{ J}}{\text{mol}} \times \frac{1 \text{ kJ}}{1000 \text{ J}} = 24.7 \frac{\text{kJ}}{\text{mol}}$$

$$\ln P = -29\underline{6}9.9 \text{ K}\left(\frac{1}{T}\right) + 19.03\underline{6} \rightarrow$$

$$\ln 760 = -29\underline{6}9.9 \text{ K}\left(\frac{1}{T}\right) + 19.03\underline{6} \rightarrow$$

$$29\underline{6}9.9 \text{ K}\left(\frac{1}{T}\right) = 19.03\underline{6} - 6.63\underline{3}32 \rightarrow$$

$$T = \frac{29\underline{6}9.9 \text{ K}}{12.4026\underline{8}} = 239 \text{ K}$$

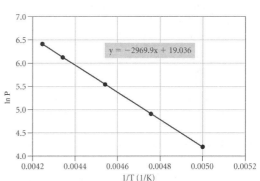
y = −2969.9x + 19.036

Check: The units (kJ/mol) are correct. The magnitude of the answer (25 kJ/mol) is consistent with other values in the text.

11.77 **Given:** ethanol, $\Delta H_{vap}^° = 38.56$ kJ/mol; normal boiling point $= 78.4$ °C **Find:** $P_{Ethanol}$ at 15 °C
Conceptual Plan: °C $\rightarrow$ K and kJ $\rightarrow$ J then $\Delta H_{vap}^°, T_1, P_1, T_2 \rightarrow P_2$

$$K = °C + 273.15 \quad \frac{1000 \text{ J}}{1 \text{ kJ}} \qquad \ln\frac{P_2}{P_1} = \frac{-\Delta H_{vap}}{R}\left(\frac{1}{T_2} - \frac{1}{T_1}\right)$$

Solution: $T_1 = 78.4 \text{ °C} + 273.15 = 351.6 \text{ K} \quad T_2 = 15 \text{ °C} + 273.15 = 288 \text{ K}$

$$\frac{38.56 \text{ kJ}}{\text{mol}} \times \frac{1000 \text{ J}}{1 \text{ kJ}} = 3.856 \times 10^4 \frac{J}{\text{mol}} \quad P_1 = 760 \text{ torr} \quad \ln\frac{P_2}{P_1} = \frac{-\Delta H_{vap}}{R}\left(\frac{1}{T_2} - \frac{1}{T_1}\right) \text{ Substitute values in}$$

equation. $\ln\dfrac{P_2}{760 \text{ torr}} = \dfrac{-3.856 \times 10^4 \frac{J}{\text{mol}}}{8.314\frac{J}{K \cdot \text{mol}}}\left(\dfrac{1}{288 \text{ K}} - \dfrac{1}{351.6 \text{ K}}\right) = -2.9\underline{1}302 \rightarrow$

$$\frac{P_2}{760 \text{ torr}} = e^{-2.9\underline{1}302} = 0.05\underline{4}311 \rightarrow P_2 = 0.05\underline{4}311 \times 760 \text{ torr} = 41 \text{ torr}$$

Check: The units (torr) are correct. Because 15 °C is significantly below the boiling point, we expect the answer to be much less than 760 torr.

Sublimation and Fusion

11.79   **Given:** 65.8 g water freezes   **Find:** energy released   **Other:** $\Delta H_{fus}^{\circ} = 6.02$ kJ/mol from text
Conceptual Plan: $gH_2O \rightarrow$ mol $H_2O \rightarrow q_{H_2O}$ (kJ) $\rightarrow q_{H_2O}$ (J)

$$\frac{1 \text{ mol}}{18.02 \text{ g}} \qquad \frac{-6.02 \text{ kJ}}{1 \text{ mol}} \qquad \frac{1000 \text{ J}}{1 \text{ kJ}}$$

Solution: $65.8 \text{ g} \times \dfrac{1 \text{ mol}}{18.02 \text{ g}} \times \dfrac{-6.02 \text{ kJ}}{1 \text{ mol}} \times \dfrac{1000 \text{ J}}{1 \text{ kJ}} = -21{,}982 \text{ J} = 2.20 \times 10^4 \text{ J so } 2.20 \times 10^4 \text{ J released}$

or 22.0 kJ released

Check: The units (J) are correct. The magnitude (22,000 J) makes sense because we are freezing about 3 moles of water. Freezing is exothermic, so heat is released.

11.81   **Given:** 8.5 g ice; 255 g water   **Find:** ΔT of water   **Other:** $\Delta H_{fus}^{\circ} = 6.0$ kJ/mol; $C_{H_2O} = 4.18$ J/g $\cdot$ °C from text
Conceptual Plan: The first step is to calculate how much heat is removed from the water to melt the ice.
$q_{ice} = -q_{water}$ so g(ice) $\rightarrow$ mol(ice) $\rightarrow q_{fus}$(kJ) $\rightarrow q_{fus}$(J) $\rightarrow q_{water}$(J) then $q, m, C_s \rightarrow \Delta T_1$

$$\frac{1 \text{ mol}}{18.02 \text{ g}} \qquad \frac{6.0 \text{ kJ}}{1 \text{ mol}} \qquad \frac{1000 \text{ J}}{1 \text{ kJ}} \qquad q_{water} = -q_{ice} \qquad q = mC_s\Delta T_1$$

Now we have slightly cooled water (at a temperature of T_1) in contact with 0.0 °C water, and we can calculate a second temperature drop of the water due to mixing of the water that was ice with the water that was initially room temperature; so $q_{ice} = -q_{water}$ with $m, C_s \rightarrow \Delta T_2$ with $\Delta T_1 \Delta T_2 \rightarrow \Delta T_{Total}$.

$$q = mC_s\Delta T_2 \text{ then set } q_{ice} = -q_{water} \quad \Delta T_{Total} = \Delta T_1 + \Delta T_2$$

Solution: $8.5 \text{ g} \times \dfrac{1 \text{ mol}}{18.02 \text{ g}} \times \dfrac{6.0 \text{ kJ}}{1 \text{ mol}} \times \dfrac{1000 \text{ J}}{1 \text{ kJ}} = 2.83176 \times 10^3 \text{ J}, q_{water} = -q_{ice} = -2.83176 \times 10^3 \text{ J}$

$q = mC_s\Delta T$ Rearrange to solve for ΔT. $\Delta T_1 = \dfrac{q}{mC_s} = \dfrac{-2.83176 \times 10^3 \text{ J}}{255 \text{ g} \times 4.18 \dfrac{\text{J}}{\text{g} \cdot °C}} = -2.6567 °C$

$q = mC_s\Delta T$ Substitute values and set $q_{ice} = -q_{H_2O}$.

$q_{ice} = m_{ice}C_{ice}(T_f - T_{icei}) = 8.5 \text{ g} \times 2.09 \dfrac{\text{J}}{\text{g} \cdot °C} \times (T_f - 0.0 °C) =$

$-q_{water} = -m_{water}C_{water}\Delta T_{water2} = -255 \text{ g} \times 4.18 \dfrac{\text{J}}{\text{g} \cdot °C} \times \Delta T_{water2} \rightarrow$

$17.765 \, T_f = -1065.9\Delta T_{water2} = -1065.9(T_f - T_{f1})$ Rearrange to solve for T_f. $17.765 \, T_f + 1065.9 \, T_f = 1065.9 \, T_{f1} \rightarrow$
$1083.665 \, T_f = 1065.9 \, T_{f1} \rightarrow T_f = 0.983607 \, T_{f1}$ but $\Delta T_1 = (T_{f1} - T_{i1}) = -2.6567 °C$, which says that
$T_{f1} = T_{i1} - 2.6567 °C$ and $\Delta T_{Total} = (T_f - T_{i1})$; so
$\Delta T_{Total} = 0.983607 \, T_{f1} - T_{i1} = 0.983607(T_{i1} - 2.6567 °C) - T_{i1} = -2.6131 °C - 0.016393 \, T_{i1}$.

This implies that the larger the initial temperature of the water, the larger the temperature drop. If the initial temperature was 90 °C, the temperature drop would be 4.1 °C. If the initial temperature was 25 °C, the temperature drop would be 3.0 °C. If the initial temperature was 5 °C, the temperature drop would be 2.7 °C. This makes physical sense because the lower the initial temperature of the water, the less kinetic energy it initially has and the smaller the heat transfer from the water to the melted ice will be.

Check: The units (°C) are correct. The temperature drop from the melting of the ice is only 2.7 °C because the mass of the water is so much larger than that of the ice.

11.83   **Given:** 10.0 g ice $T_i = -10.0 °C$ to steam at $T_f = 110.0 °C$ **Find:** heat required (kJ)
Other: $\Delta H_{fus}^{\circ} = 6.02$ kJ/mol; $\Delta H_{vap}^{\circ} = 40.7$ kJ/mol; $C_{ice} = 2.09$ J/g $\cdot$ °C; $C_{water} = 4.18$ J/g $\cdot$ °C;
$C_{steam} = 2.01$ J/g $\cdot$ °C
Conceptual Plan: Follow the heating curve in Figure 11.24. $q_{Total} = q_1 + q_2 + q_3 + q_4 + q_5$ where q_1, q_3, and q_5 are heating of a single phase then J $\rightarrow$ kJ and q_2 and q_4 are phase transitions.

$$q = mC_s(T_f - T_i) \qquad\qquad \frac{1 \text{ kJ}}{1000 \text{ J}} \qquad q = m \times \frac{1 \text{ mol}}{18.02 \text{ g}} \times \frac{\Delta H}{1 \text{ mol}}$$

Solution:

Heating ice from $-10.0\,°C$ to $0\,°C$:

$$q_1 = m_{ice}C_{ice}(T_{icef} - T_{icei}) = 10.0\ \text{g} \times 2.09\ \frac{\text{J}}{\text{g}\cdot°C} \times (0.0\,°C - (-10.0\,°C)) = 209\ \text{J} \times \frac{1\ \text{kJ}}{1000\ \text{J}} = 0.209\ \text{kJ}$$

Melting the ice—phase change:

$$q_2 = m \times \frac{1\ \text{mol}}{18.02\ \text{g}} \times \frac{\Delta H_{fus}}{1\ \text{mol}} = 10.0\ \text{g} \times \frac{1\ \text{mol}}{18.02\ \text{g}} \times \frac{6.02\ \text{kJ}}{1\ \text{mol}} = 3.3\underline{4}1\ \text{kJ}$$

Heating water to $100\,°C$:

$$q_3 = m_{water}C_{water}(T_{waterf} - T_{wateri}) = 10.0\ \text{g} \times 4.18\ \frac{\text{J}}{\text{g}\cdot°C} \times (100.0\,°C - 0.0\,°C) = 4180\ \text{J} \times \frac{1\ \text{kJ}}{1000\ \text{J}} = 4.18\ \text{kJ}$$

Converting water to steam—phase change:

$$q_4 = m \times \frac{1\ \text{mol}}{18.02\ \text{g}} \times \frac{\Delta H_{vap}}{1\ \text{mol}} = 10.0\ \text{g} \times \frac{1\ \text{mol}}{18.02\ \text{g}} \times \frac{40.7\ \text{kJ}}{1\ \text{mol}} = 22.5\underline{8}6\ \text{kJ}$$

Heating steam to $110\,°C$:

$$q_5 = m_{steam}C_{steam}(T_{steamf} - T_{steami}) = 10.0\ \text{g} \times 2.01\ \frac{\text{J}}{\text{g}\cdot°C} \times (110.0\,°C - 100.0\,°C)$$

$$= 201\ \text{J} \times \frac{1\ \text{kJ}}{1000\ \text{J}} = 0.201\ \text{kJ}$$

$$q_{Total} = q_1 + q_2 + q_3 + q_4 + q_5 = 0.209\ \text{kJ} + 3.3\underline{4}1\ \text{kJ} + 4.18\ \text{kJ} + 22.5\underline{8}6\ \text{kJ} + 0.201\ \text{kJ} = 30.5\ \text{kJ}$$

Check: The units (kJ) are correct. The total amount of heat is dominated by the vaporization step. Because we have less than 1 mole, we expect less than 41 kJ.

Phase Diagrams

11.85 (a) solid
(b) liquid
(c) gas
(d) supercritical fluid
(e) solid/liquid equilibrium or fusion curve
(f) liquid/gas equilibrium or vaporization curve
(g) solid/liquid/gas equilibrium or triple point

11.87 **Given:** nitrogen; normal boiling point $= 77.3$ K; normal melting point $= 63.1$ K; critical temperature $= 126.2$ K; critical pressure $= 2.55 \times 10^4$ torr; triple point at 63.1 K and 94.0 torr
Find: Sketch phase diagram. Does nitrogen have a stable liquid phase at 1 atm?

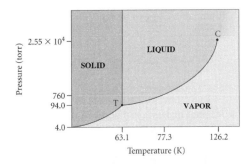

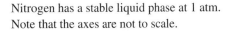
Nitrogen has a stable liquid phase at 1 atm.
Note that the axes are not to scale.

11.89 (a) 0.027 mmHg, the higher of the two triple points
(b) The rhombic phase is denser because if we start in the monoclinic phase at $100\,°C$ and increase the pressure, we will cross into the rhombic phase.

The Uniqueness of Water

11.91   Water has a low molar mass (18.02 g/mol), yet it is a liquid at room temperature. Water's high boiling point for its molar mass can be understood by examining the structure of the water molecule. The bent geometry of the water molecule and the highly polar nature of the O—H bonds result in a molecule with a significant dipole moment. Water's two O—H bonds (hydrogen directly bonded to oxygen) allow a water molecule to form very strong hydrogen bonds with four other water molecules, resulting in a relatively high boiling point.

11.93   Water has an exceptionally high specific heat capacity, which has a moderating effect on the climate of coastal cities. Also, its high ΔH_{vap} causes water evaporation and condensation to have a strong effect on temperature. A tremendous amount of heat can be stored in large bodies of water. Heat will be absorbed or released from large bodies of water preferentially over the land around it. In some cities, such as San Francisco, the daily fluctuation in temperature can be less than 10 °C. This same moderating effect occurs over the entire planet, two-thirds of which is covered by water. In other words, without water, the daily temperature fluctuations on our planet might be more like those on Mars, where temperature fluctuations of 63 °C (113 °F) have been measured between early morning and midday.

Types of Solids and Their Structures

11.95   **Given:** X-ray with $\lambda = 154$ pm, maximum reflection angle of $\theta = 28.3°$; assume $n = 1$
        Find: distance between layers
        Conceptual Plan: $\lambda, \theta, n \rightarrow d$

$$n\lambda = 2\,d\sin\theta$$

Solution: $n\lambda = 2\,d\sin\theta$. Rearrange to solve for d. $d = \dfrac{n\lambda}{2\sin\theta} = \dfrac{1 \times 154\text{ pm}}{2\sin 28.3°} = 162$ pm

Check: The units (pm) are correct. The magnitude (162 pm) makes sense because $n = 1$ and the sin is always <1. The number is consistent with interatomic distances.

11.97   (a)   8 corner atoms $\times$ (1/8 atom/unit cell) = 1 atom/unit cell
        (b)   8 corner atoms $\times$ (1/8 atom/unit cell) + 1 atom in center = $(1 + 1)$ atoms/unit cell = 2 atoms/unit cell
        (c)   8 corner atoms $\times$ (1/8 atom/unit cell) + 6 face-centered atoms $\times$ (1/2 atom/unit cell)
              $= (1 + 3)$ atoms/unit cell $= 4$ atoms/unit cell

11.99   **Given:** platinum; face-centered cubic structure; $r = 139$ pm   **Find:** edge length of unit cell and density (g/cm^3)
        Conceptual Plan: $r \rightarrow l$ and $l \rightarrow V(\text{pm}^3) \rightarrow V(\text{cm}^3)$ and $\mathcal{M}$, FCC structure $\rightarrow m$ then $m, V \rightarrow d$

$$l = 2\sqrt{2}\,r \qquad V = l^3 \qquad \frac{(1\text{ cm})^3}{(10^{10}\text{ pm})^3} \qquad m = \frac{4\text{ atoms}}{\text{unit cell}} \times \frac{\mathcal{M}}{N_A} \qquad d = m/V$$

Solution: $l = 2\sqrt{2}\,r = 2\sqrt{2} \times 139$ pm $= 39\underline{3}.151$ pm $= 393$ pm and

$$V = l^3 = (39\underline{3}.151\text{ pm})^3 \times \frac{(1\text{ cm})^3}{(10^{10}\text{ pm})^3} = 6.0\underline{7}684 \times 10^{-23}\text{ cm}^3 \text{ and}$$

$$m = \frac{4\text{ atoms}}{\text{unit cell}} \times \frac{\mathcal{M}}{N_A} = \frac{4\text{ atoms}}{\text{unit cell}} \times \frac{195.09\text{ g}}{1\text{ mol}} \times \frac{1\text{ mol}}{6.022 \times 10^{23}\text{ atoms}} = 1.29\underline{5}849 \times 10^{-21}\frac{\text{g}}{\text{unit cell}} \text{ then}$$

$$d = \frac{m}{V} = \frac{1.29\underline{5}849 \times 10^{-21}\frac{\text{g}}{\text{unit cell}}}{6.0\underline{7}684 \times 10^{-23}\frac{\text{cm}^3}{\text{unit cell}}} = 21.3\frac{\text{g}}{\text{cm}^3}$$

Check: The units (pm and g/cm^3) are correct. The magnitude (393 pm) makes sense because it must be larger than the radius of an atom. The magnitude (21 g/cm^3) is consistent for Pt from Chapter 1.

11.101   **Given:** rhodium; face-centered cubic structure; $d = 12.41$ g/cm^3 **Find:** r(Rh)
         Conceptual Plan: $\mathcal{M}$, FCC structure $\rightarrow m$ then $m, V \rightarrow d$ then $V(\text{cm}^3) \rightarrow l(\text{cm}) \rightarrow l(\text{pm})$ then $l \rightarrow r$

$$m = \frac{4\text{ atoms}}{\text{unit cell}} \times \frac{\mathcal{M}}{N_A} \qquad d = m/V \qquad V = l^3 \qquad \frac{10^{10}\text{ pm}}{1\text{ cm}} \qquad l = 2\sqrt{2}\,r$$

Solution: $m = \dfrac{4\text{ atoms}}{\text{unit cell}} \times \dfrac{\mathcal{M}}{N_A} = \dfrac{4\text{ atoms}}{\text{unit cell}} \times \dfrac{102.905\text{ g}}{1\text{ mol}} \times \dfrac{1\text{ mol}}{6.022 \times 10^{23}\text{ atoms}} = 6.83\underline{5}271 \times 10^{-22}\dfrac{\text{g}}{\text{unit cell}}$

then $d = \dfrac{m}{V}$ Rearrange to solve for V. $V = \dfrac{m}{d} = \dfrac{6.83\underline{5}271 \times 10^{-22} \dfrac{g}{\text{unit cell}}}{12.41\dfrac{g}{\text{cm}^3}} = 5.50\underline{7}873 \times 10^{-23} \dfrac{\text{cm}^3}{\text{unit cell}}$

then $V = l^3$ Rearrange to solve for l.

$l = \sqrt[3]{V} = \sqrt[3]{5.50\underline{7}873 \times 10^{-23} \text{ cm}^3} = 3.80\underline{4}831 \times 10^{-8} \text{ cm} \times \dfrac{10^{10} \text{ pm}}{1 \text{ cm}} = 380.\underline{4}831 \text{ pm}$ then $l = 2\sqrt{2}r$

Rearrange to solve for r. $r = \dfrac{l}{2\sqrt{2}} = \dfrac{380.\underline{4}831 \text{ pm}}{2\sqrt{2}} = 134.5 \text{ pm}$

Check: The units (pm) are correct. The magnitude (135 pm) is consistent with an atomic diameter.

11.103   **Given:** polonium, simple cubic structure $d = 9.3$ g/cm^3; $r = 167$ pm; $\mathcal{M} = 209$ g/mol   **Find:** estimate N_A
Conceptual Plan: $r \rightarrow l$ and $l \rightarrow V(\text{pm}^3) \rightarrow V(\text{cm}^3)$ then $d, V \rightarrow m$ then $\mathcal{M}$, SC structure $\rightarrow m$

$$l = 2r \qquad V = l^3 \qquad \dfrac{(1 \text{ cm})^3}{(10^{10} \text{ pm})^3} \qquad d = m/V \qquad m = \dfrac{1 \text{ atom}}{\text{unit cell}} \times \dfrac{\mathcal{M}}{N_A}$$

Solution: $l = 2r = 2 \times 167 \text{ pm} = 334 \text{ pm}$ and $V = l^3 = (334 \text{ pm})^3 \times \dfrac{(1 \text{ cm})^3}{(10^{10} \text{ pm})^3} = 3.7\underline{2}597 \times 10^{-23} \text{ cm}^3$ then

$d = \dfrac{m}{V}$ Rearrange to solve for m. $m = dV = 9.3 \dfrac{g}{\text{cm}^3} \times \dfrac{3.7\underline{2}597 \times 10^{-23} \text{ cm}^3}{\text{unit cell}} = 3.\underline{4}6515 \times 10^{-22} \dfrac{g}{\text{unit cell}}$

then $m = \dfrac{1 \text{ atom}}{\text{unit cell}} \times \dfrac{\mathcal{M}}{N_A}$ Rearrange to solve for N_A.

$N_A = \dfrac{1 \text{ atom}}{\text{unit cell}} \times \dfrac{\mathcal{M}}{m} = \dfrac{1 \text{ atom}}{\text{unit cell}} \times \dfrac{209 \text{ g}}{1 \text{ mol}} \times \dfrac{1 \text{ unit cell}}{3.\underline{4}6515 \times 10^{-22} \text{ g}} = 6.0 \times 10^{23} \dfrac{\text{atoms}}{\text{mol}}$

Check: The units (atoms/mol) are correct. The magnitude (6×10^{23} atoms/mol) is consistent with Avogadro's number.

11.105   (a)   atomic because argon (Ar) is an atom
(b)   molecular because water (H_2O) is a molecule
(c)   ionic because potassium oxide (K_2O) is an ionic solid
(d)   atomic because iron (Fe) is an atom

11.107   LiCl has the highest melting point because it is the only ionic solid in the group. The other three solids are held together by intermolecular forces, while LiCl is held together by stronger coulombic interactions between the cations and anions of the crystal lattice.

11.109   (a) TiO_2 because it is an ionic solid
(b) $SiCl_4$ because it has a higher molar mass and therefore has stronger dispersion forces
(c)   Xe because it has a higher molar mass and therefore has stronger dispersion forces
(d)   CaO because the ions have greater charge and therefore stronger dipole–dipole interactions

11.111   The Ti atoms occupy the corner positions and the center of the unit cell: 8 corner atoms $\times$ (1/8 atom/unit cell) $+1$ atom in center $= (1 + 1)$ Ti atoms/unit cell $= 2$ Ti atoms/unit cell. The O atoms occupy four positions on the top and bottom faces and two positions inside the unit cell: 4 face-centered atoms $\times$ (1/2 atom/unit cell) $+2$ atoms in the interior $= (2 + 2)$ O atoms/unit cell $= 4$ O atoms/unit cell. Therefore, there are 2 Ti atoms/ unit cell and 4 O atoms/unit cell; so the ratio Ti:O is 2:4, or 1:2. The formula for the compound is TiO_2.

11.113   In CsCl: The Cs atoms occupy the center of the unit cell: 1 atom in center $= 1$ Cs atom/unit cell. The Cl atoms occupy corner positions of the unit cell: 8 corner atoms $\times$ (1/8 atom/unit cell) $= 1$ Cl atom/unit cell. Therefore, there are 1 Cl atom/unit cell and 1 Cl atom/unit cell, so the ratio Cs:Cl is 1:1. The formula for the compound is CsCl, as expected.

In BaCl$_2$: The Ba atoms occupy the corner positions and the face-centered positions of the unit cell: 8 corner atoms $\times$ (1/8 atom/unit cell) $+$ 6 face-centered atoms $\times$ (1/2 atom/unit cell) $= (1 + 3)$ Ba atoms/unit cell $= 4$ Ba atoms/unit cell. The Cl atoms occupy eight positions inside the unit cell: 8 Cl atoms/unit cell. Therefore, there are 4 Ba atoms/unit cell and 8 Cl atoms/unit cell so the ratio Ba:Cl is 4:8, or 1:2. The formula for the compound is BaCl$_2$, as expected.

Band Theory

11.115   (a)   Zn should have little or no band gap because it is the only metal in the group.

11.117   (a)   p-type semiconductor: Ge is Group 4A, and Ga is Group 3A; so the Ga will generate electron "holes."

         (b)   n-type semiconductor: Si is Group 4A, and As is Group 5A; so the As will add electrons to the conduction band.

Cumulative Problems

11.119   The general trend is that melting point increases with increasing molar mass. This is because the electrons of the larger molecules are held more loosely and a stronger dipole moment can be induced more easily. HF is the exception to the rule. It has a relatively high melting point due to strong intermolecular forces due to hydrogen bonding.

11.121   **Given:** $P_{H_2O} = 23.76$ torr at 25 °C; 1.25 g water in 1.5 L container   **Find:** m (H_2O) as liquid

       Conceptual Plan: °C $\rightarrow$ K and torr $\rightarrow$ atm then $P, V, T \rightarrow$ mol $(g) \rightarrow$ g (g) then g (g), g $(l)_i \rightarrow$ g $(l)_f$

$$K = °C + 273.15 \qquad \frac{1 \text{ atm}}{760 \text{ torr}} \qquad PV = nRT \qquad \frac{18.01 \text{ g}}{1 \text{ mol}} \qquad g(l)_f = g(l)_i - g(g)$$

       Solution: $T = 25\,°C + 273.15 = 298$ K $23.76 \text{ torr} \times \dfrac{1 \text{ atm}}{760 \text{ torr}} = 0.0312\underline{6}32$ atm then $PV = nRT$

       Rearrange to solve for n. $n = \dfrac{PV}{RT} = \dfrac{0.0312632 \text{ atm} \times 1.5 \text{ L}}{0.08206 \dfrac{\text{L} \cdot \text{atm}}{\text{K} \cdot \text{mol}} \times 298 \text{ K}} = 0.00191768$ mol in the gas phase then

       $0.00191768 \text{ mol} \times \dfrac{18.02 \text{ g}}{1 \text{ mol}} = 0.034\underline{5}566$ g in the gas phase then

       g $(l)_f$ = g $(l)_i$ − g (g) = 1.25 g − 0.034\underline{5}566 g = 1.22 g remaining as liquid. Yes, there is 1.22 g of liquid.

       Check: The units (g) are correct. The magnitude (1.2 g) is expected because very little material is expected to be in the gas phase.

11.123   Because we are starting at a temperature that is higher and a pressure that is lower than the triple point, the phase transitions will be gas $\rightarrow$ liquid $\rightarrow$ solid, or condensation followed by freezing.

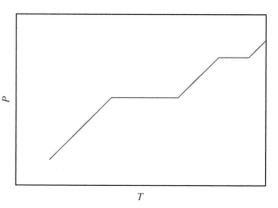

11.125   **Given:** ice: $T_1 = 0\,°C$ exactly, $m = 53.5$ g; water: $T_1 = 75\,°C$, $m = 115$ g   **Find:** T_f

       Other: $\Delta H°_{fus} = 6.0$ kJ/mol; $C_{water} = 4.18$ J/g·°C

       Conceptual Plan: $q_{ice} = -q_{water}$ so g(ice) $\rightarrow$ mol(ice) $\rightarrow q_{fus}$(kJ) $\rightarrow q_{fus}$(J) $\rightarrow q_{water}$(J) then

$$\frac{1 \text{ mol}}{18.02 \text{ g}} \qquad \frac{6.02 \text{ kJ}}{1 \text{ mol}} \qquad \frac{1000 \text{ J}}{1 \text{ kJ}} \qquad q_{water} = -q_{ice}$$

       $q, m, C_s \rightarrow \Delta T$ then $T_i, \Delta T \rightarrow T_2$ now we have slightly cooled water in contact with 0.0 °C water

$$q = mC_s \Delta T \qquad \Delta T = T_2 - T_i$$

       so $q_{ice} = -q_{water}$ with $m, C_s, T_i \rightarrow T_f$

$$q = mC_s(T_f - T_i) \text{ then set } q_{ice} = -q_{water}$$

       Solution: $53.5 \text{ g} \times \dfrac{1 \text{ mol}}{18.02 \text{ g}} \times \dfrac{6.02 \text{ kJ}}{1 \text{ mol}} \times \dfrac{1000 \text{ J}}{1 \text{ kJ}} = 1.78828 \times 10^4$ J $q_{water} = -q_{ice} = -1.78828 \times 10^4$ J

       $q_{water} = mC_s \Delta T$ Rearrange to solve for ΔT. $\Delta T = \dfrac{q}{mC_s} = \dfrac{-1.78828 \times 10^4 \text{ J}}{115 \text{ g} \times 4.18 \dfrac{\text{J}}{\text{g} \cdot °C}} = -37.\underline{2}016\,°C$ then

$\Delta T = T_2 - T_i$ Rearrange to solve for T_2. $T_2 = \Delta T + T_i = -37.\underline{2}016\,°C + 75\,°C = 37.\underline{7}98\,°C$

$q = mC_s(T_f - T_i)$ Substitute values and set $q_{ice} = -q_{water}$.

$q_{ice} = m_{ice}C_{ice}(T_f - T_{icei}) = 53.5\,\cancel{g} \times 4.18\,\dfrac{J}{\cancel{g} \cdot °C} \times (T_f - 0.0\,°C) =$

$-q_{water} = -m_{water}\,C_{water}(T_f - T_{water2}) = -115\,\cancel{g} \times 4.18\,\dfrac{J}{\cancel{g} \cdot °C} \times (T_f - 37.\underline{7}98\,°C)$

Rearrange to solve for T_f.

$53.5T_f = -115(T_f - 37.\underline{7}98\,°C) \rightarrow 53.5\,T_f = -115\,T_f + 4\underline{3}46.8\,°C \rightarrow -4\underline{3}46.8\,°C = -168.5\,T_f$

$\rightarrow T_f = \dfrac{-4\underline{3}46.8\,°C}{-168.\underline{5}} = 25.\underline{8}\,°C = 26\,°C$

Check: The units (°C) are correct. The temperature is between the two initial temperatures. Because the ice mass is about half the water mass, we are not surprised that the temperature is closer to the original ice temperature.

11.127   **Given:** 1 mole of methanol   **Find:** Draw a heating curve beginning at 170 K and ending at 350 K
         Other: melting point = 176 K, boiling point = 338 K, $\Delta H_{fus} = 2.2$ kJ/mol, $\Delta H_{vap} = 35.2$ kJ/mol,
         $C_{s,solid} = 105$ J/mol $\cdot$ K, $C_{s,liquid} = 81.3$ J/mol $\cdot$ K, $C_{s,gas} = 48$ J/mol $\cdot$ K
         Conceptual Plan:
         Calculate the temperature range of each phase:

$T_{Starting}, T_{Melting} \rightarrow \Delta T_{Solid}$ and $T_{Melting}, T_{Boiling} \rightarrow \Delta T_{Liquid}$ and $T_{Boiling}, T_{Ending} \rightarrow \Delta T_{Gas}$

      $\Delta T_{Solid} = T_{Melting} - T_{Starting}$ $\Delta T_{Liquid} = T_{Boiling} - T_{Melting}$ $\Delta T_{Gas} = T_{Ending} - T_{Boiling}$

         Because there is 1 mole of methanol, the heat for each phase change is simply the ΔH for that phase change.
         Calculate the heat required for each step:

$\Delta T_{Solid}, C_{s,solid} \rightarrow q_{Solid}$ and $\Delta T_{Liquid}, C_{s,liquid} \rightarrow q_{Liquid}$ and $\Delta T_{Gas}, C_{s,gas} \rightarrow q_{Gas}$

        $q = C_s \times \Delta T$ $q = C_s \times \Delta T$ $q = C_s \times \Delta T$

         Finally, plot each of the segments.
         Solution:

$\Delta T_{Solid} = T_{Melting} - T_{Starting} = 176\,K - 170\,K = 6\,K$ $\Delta T_{Liquid} = T_{Boiling} - T_{Melting} = 338\,K - 176\,K = 162\,K$

and $\Delta T_{Gas} = T_{Ending} - T_{Boiling} = 350\,K - 338\,K = 12\,K$

$q_{fus} = \Delta H_{fus} = 2.2$ kJ/mol and $q_{vap} = \Delta H_{vap} = 35.2$ kJ/mol

$q_{Solid} = C_{s,Solid} \times \Delta T_{Solid} = 105\dfrac{J}{mol \cdot \cancel{K}} \times 6\,\cancel{K} = \underline{6}30\,J/mol = 0.\underline{6}3\,kJ/mol$

$q_{Liquid} = C_{s,Liquid} \times \Delta T_{Liquid} = 81.3\dfrac{J}{mol \cdot \cancel{K}} \times 162\,\cancel{K} = 13\underline{1}70.6\,J/mol = 13.\underline{1}706\,kJ/mol$ and

$q_{Gas} = C_{s,Gas} \times \Delta T_{Gas} = 48\dfrac{J}{mol \cdot \cancel{K}} \times 12\,\cancel{K} = 5\underline{7}6\,J/mol = 0.5\underline{7}6\,kJ/mol$

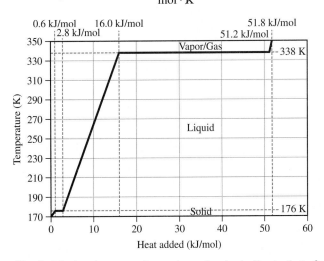

Check: The heating curve has a shape that is similar to that of water.

11.129 **Given:** home: $6.0 \text{ m} \times 10.0 \text{ m} \times 2.2 \text{ m}$; $T = 30\,°C$, $P_{H_2O} = 85\%$ of $P^\circ_{H_2O}$ **Find:** $m(H_2O)$ removed
Other: $P^\circ_{H_2O} = 31.8$ mmHg from text
Conceptual Plan: $l, w, h \rightarrow V(m^3) \rightarrow V(cm^3) \rightarrow V(L)$ and $P^\circ_{H_2O} \rightarrow P_{H_2O}(mmHg) \rightarrow P_{H_2O}(atm)$ and

$$V = l\,w\,h \qquad \frac{(100\ cm)^3}{(1\ m)^3} \qquad \frac{1\ L}{1000\ cm^3} \qquad\qquad P_{H_2O} = 0.85\,P^\circ_{H_2O} \qquad \frac{1\ atm}{760\ mmHg}$$

$°C \rightarrow K$ then $P, V, T \rightarrow mol(H_2O) \rightarrow g(H_2O)$

$$K = °C + 273.15 \qquad PV = nRT \qquad \frac{18.02\ g}{1\ mol}$$

Solution: $V = l\,w\,h = 6.0 \text{ m} \times 10.0 \text{ m} \times 2.2 \text{ m} = 132\ m^3 \times \dfrac{(100\ cm)^3}{(1\ m)^3} \times \dfrac{1\ L}{1000\ cm^3} = 1.32 \times 10^5\ L$

$P_{H_2O} = 0.85\,P^\circ_{H_2O} = 0.85 \times 31.8\ mmHg \times \dfrac{1\ atm}{760\ mmHg} = 0.035566\ atm \quad T = 30\,°C + 273.15 = 303\ K$

then $PV = nRT$ Rearrange to solve for n.

$$n = \frac{PV}{RT} = \frac{0.035566\ atm \times 1.32 \times 10^5\ L}{0.08206\ \dfrac{L \cdot atm}{K \cdot mol} \times 303\ K} = 188.81\ mol \text{ then } 188.81\ mol \times \frac{18.02\ g}{1\ mol} = 3400\ g \text{ to remove}$$

Check: The units (g) are correct. The magnitude of the answer (3400 g) makes sense because the volume of the house is so large. We are removing almost 200 moles of water.

11.131 CsCl has a higher melting point than AgI because of its higher coordination number. In CsCl, one anion bonds to eight cations (and vice versa), while in AgI, one anion bonds only to four cations.

11.133 (a) Atoms are connected across the face diagonal (c), so $c = 4r$.
(b) From the Pythagorean Theorem, $c^2 = a^2 + b^2$; from part (a), $c = 4r$; and for a cubic structure, $a = l, b = l$. So $(4r)^2 = l^2 + l^2 \rightarrow 16r^2 = 2l^2 \rightarrow 8r^2 = l^2 \rightarrow l = \sqrt{8r^2} \rightarrow l = 2\sqrt{2}r$.

11.135 **Given:** diamond, $V(\text{unit cell}) = 0.0454\ nm^3$; $d = 3.52\ g/cm^3$ **Find:** number of carbon atoms/unit cell
Conceptual Plan: $V(nm^3) \rightarrow V(cm^3)$ then $d, V \rightarrow m \rightarrow mol \rightarrow atoms$

$$\frac{(1\ cm)^3}{(10^7\ nm)^3} \qquad\qquad d = m/V \quad \frac{1\ mol}{12.01\ g} \quad \frac{6.022 \times 10^{23}\ atoms}{1\ mol}$$

Solution: $0.0454\ nm^3 \times \dfrac{(1\ cm)^3}{(10^7\ nm)^3} = 4.54 \times 10^{-23}\ cm^3$ then $d = \dfrac{m}{V}$ Rearrange to solve for m.

$m = dV = 3.52\ \dfrac{g}{cm^3} \times 4.54 \times 10^{-23}\ cm^3 = 1.59808 \times 10^{-22}\ g$ then

$$\frac{1.59808 \times 10^{-22}\ g}{\text{unit cell}} \times \frac{1\ mol}{12.01\ g} \times \frac{6.022 \times 10^{23}\ atoms}{1\ mol} = 8.01\ \frac{C\ atoms}{\text{unit cell}} = 8\ \frac{C\ atoms}{\text{unit cell}}$$

Check: The units (atoms) are correct. The magnitude (8 atoms) makes sense because it is a fairly small number and our answer is within calculation error of an integer.

11.137 (a) $CO_2(s) \rightarrow CO_2(g)$ at 194.7 K
(b) $CO_2(s) \rightarrow$ triple point at 216.5 K $\rightarrow CO_2(g)$ just above 216.5 K
(c) $CO_2(s) \rightarrow CO_2(l)$ at somewhat above 216 K $\rightarrow CO_2(g)$ at around 250 K
(d) $CO_2(s) \rightarrow CO_2$ above the critical point where there is no distinction between liquid and gas. This change occurs at about 300 K.

11.139 **Given:** metal, $d = 7.8748\ g/cm^3$; $l = 0.28664$ nm, body-centered cubic lattice **Find:** $\mathcal{M}$
Conceptual Plan: $l \rightarrow V(nm^3) \rightarrow V(cm^3)$ then $d, V \rightarrow m$ then m, FCC structure $\rightarrow \mathcal{M}$

$$V = l^3 \qquad \frac{(1\ cm)^3}{(10^7\ nm)^3} \qquad\qquad d = m/V \qquad\qquad m = \frac{2\ atoms}{\text{unit cell}} \times \frac{\mathcal{M}}{N_A}$$

Solution: $V = l^3 = (0.28664\ nm)^3 = 0.02355105602\ nm^3 \times \dfrac{(1\ cm)^3}{(10^7\ nm)^3} = 2.355105602 \times 10^{-23}\ cm^3$ then $d = \dfrac{m}{V}$

Rearrange to solve for m. $m = dV = 7.8748 \frac{g}{cm^3} \times 2.355\underline{1}05602 \times 10^{-23} \, cm^3 = 1.854\underline{5}98559 \times 10^{-22} \, g$

then $m = \frac{2 \text{ atoms}}{\text{unit cell}} \times \frac{\mathcal{M}}{N_A}$ Rearrange to solve for $\mathcal{M}$.

$\mathcal{M} = \frac{\text{unit cell}}{2 \text{ atoms}} \times N_A \times m = \frac{\text{unit cell}}{2 \text{ atoms}} \times \frac{6.022 \times 10^{23} \text{ atoms}}{1 \text{ mol}} \times \frac{1.854\underline{5}98559 \times 10^{-22} \, g}{\text{unit cell}} = 55.84\underline{2} \frac{g}{mol}$

$= 55.84 \frac{g}{mol} \text{ iron}$

Check: The units (g/mol) are correct. The magnitude (55.8 g/mol) makes sense because it is a reasonable atomic mass for a metal and it is close to iron.

Challenge Problems

11.141   **Given:** KCl, rock salt structure   **Find:** density (g/cm^3) **Other:** $r(K^+) = 133$ pm; $r(Cl^-) = 181$ pm from Chapter 8
Conceptual Plan: Rock salt structure is a face-centered cubic structure with anions at the lattice points and cations in the holes between lattice sites $\rightarrow$ **assume** $r = r(Cl^-)$**, but** $\mathcal{M} = \mathcal{M}(KCL)$
$r(K^+), r(Cl^-) \rightarrow l$ **and** $l \rightarrow V(pm^3) \rightarrow V(cm^3)$**, and FCC structure** $\rightarrow m$ **then** $m, V \rightarrow d$

from Figure 11.43 $l = 2r(Cl^-) + 2r(K^+)$ $V = l^3 \quad \frac{(1 \, cm)^3}{(10^{10} \, pm)^3}$ $m = \frac{4 \text{ formula units}}{\text{unit cell}} \times \frac{\mathcal{M}}{N_A}$ $d = m/V$

Solution: $l = 2r(Cl^-) + 2r(K^+) = 2(181 \text{ pm}) + 2(133 \text{ pm}) = 628 \text{ pm}$ and

$V = l^3 = (628 \text{ pm})^3 \times \frac{(1 \, cm)^3}{(10^{10} \, pm)^3} = 2.47\underline{6}73 \times 10^{-22} \, cm^3$ and

$m = \frac{4 \text{ formula units}}{\text{unit cell}} \times \frac{\mathcal{M}}{N_A} = \frac{4 \text{ formula units}}{\text{unit cell}} \times \frac{74.55 \, g}{1 \text{ mol}} \times \frac{1 \text{ mol}}{6.022 \times 10^{23} \text{ formula units}}$

$= 4.951\underline{8}43 \times 10^{-22} \frac{g}{\text{unit cell}}$

then $d = \frac{m}{V} = \frac{4.951\underline{8}43 \times 10^{-22} \dfrac{g}{\text{unit cell}}}{2.47\underline{6}73 \times 10^{-22} \dfrac{cm^3}{\text{unit cell}}} = 1.99\underline{9}35 \frac{g}{cm^3} = 2.00 \frac{g}{cm^3}$

Check: The units (g/cm^3) are correct. The magnitude $(2 \, g/cm^3)$ is reasonable for a salt density. The published value is 1.98 g/cm^3. This method of estimating the density gives a value that is close to the experimentally measured density.

11.143   Decreasing the pressure will decrease the temperature of liquid nitrogen. Because the nitrogen is boiling, its temperature must be constant at a given pressure. As the pressure decreases, the boiling point decreases, as does the temperature. Remember that vaporization is an endothermic process; so as the nitrogen vaporizes, it removes heat from the liquid, dropping its temperature. If the pressure drops below the pressure of the triple point, the phase change will shift from vaporization to sublimation and the liquid nitrogen will become solid.

11.145   **Given:** cubic closest packing structure = cube with touching spheres of radius $= r$ on alternating corners of a cube
Find: body diagonal of cube and radius of tetrahedral hole
Solution: The cell edge length $= l$ and $l^2 + l^2 = (2r)^2 \rightarrow 2l^2 = 4r^2 \rightarrow l^2 = 2r^2$. So, $l = \sqrt{2}r$. Because body diagonal $= BD$ is the hypotenuse of the right triangle formed by the face diagonal and the cell edge, we have $(BD)^2 = l^2 + (2r)^2 = 2r^2 + 4r^2 = 6r^2 \rightarrow BD = \sqrt{6}r$. The radius of the tetrahedral hole $= r_T$ is half the body diagonal minus the radius of the sphere, or

$r_T = \frac{BD}{2} - r = \frac{\sqrt{6}r}{2} - r = \left(\frac{\sqrt{6}}{2} - 1\right)r = \left(\frac{\sqrt{6} - 2}{2}\right)r = \left(\frac{\sqrt{3}\sqrt{2} - \sqrt{2}\sqrt{2}}{\sqrt{2}\sqrt{2}}\right)r$

$= \left(\frac{\sqrt{3} - \sqrt{2}}{\sqrt{2}}\right)r \approx 0.22474r$

11.147 **Given:** 1.00 L water, $T_i = 298$ K, $T_f = 373$ K-vapor; $P_{CH_4} = 1.00$ atm   **Find:** $V(CH_4)$
Other: $\Delta H^\circ_{comb}(CH_4) = 890.4$ kJ/mol; $C_{water} = 75.2$ J/mol·K; $\Delta H^\circ_{vap}(H_2O) = 40.7$ kJ/mol, $d = 1.00$ g/mL
Conceptual Plan: $L \rightarrow mL \rightarrow g \rightarrow mol$ then heat liquid water: $n, C_s, T_i, T_f \rightarrow q_{1water}(J)$

$$\frac{1000\ mL}{1\ L} \quad \frac{1.00\ g}{1.00\ mL} \quad \frac{1\ mol}{18.02\ g} \qquad q = mC_s(T_f - T_i)$$

vaporize water: $n_{water}, \Delta H^\circ_{vap} \rightarrow q_{2water}(J)$ then calculate total heat $q_{1water}, q_{2water} \rightarrow q_{water}(J)$ then

$$q = n\Delta H \qquad\qquad\qquad\qquad q_{1water} + q_{2water} = q_{water}$$

$q_{water}(J) \rightarrow -q_{CH_4comb}(J) \rightarrow n_{CH_4}$ finally $n_{CH_4}, P,\ T \rightarrow V$

$$q_{water} = -q_{CH_4comb} \qquad q = n\Delta H \qquad\qquad PV = nRT$$

Solution: $1.00\ L \times \dfrac{1000\ mL}{1\ L} \times \dfrac{1.00\ g}{1.00\ mL} \times \dfrac{1\ mol}{18.02\ g} = 55.5247\ mol\ H_2O$

$q_{1water} = n_{water}C_{water}(T_f - T_i) = 55.5247\ mol \times 75.2\ \dfrac{J}{mol \cdot K} \times (373\ K - 298\ K) = 3.1315931 \times 10^5\ J$

$= 313.15931\ kJ$

$q_{2water} = n\Delta H = 55.5247\ mol \times 40.7\dfrac{kJ}{mol} = 2.259855 \times 10^3\ kJ$

$q_{1water} + q_{2water} = q_{water} = 313.15931\ kJ + 2.259855 \times 10^3\ kJ = 2.57301431 \times 10^3\ kJ$

$q_{water} = -q_{CH_4comb} = 2.57301431 \times 10^3\ kJ$ then $q_{CH_4comb} = n\Delta H$ Rearrange to solve for n.

$n_{CH_4} = \dfrac{q_{CH_4}}{\Delta H_{CH_4comb}} = \dfrac{-2.57301431 \times 10^3\ kJ}{-890.4\dfrac{kJ}{mol}} = 2.8897286\ mol$ then $PV = nRT$

Rearrange to solve for V. $V = \dfrac{nRT}{P} = \dfrac{2.8897286\ mol \times 0.08206\dfrac{L \cdot atm}{mol \cdot K} \times 298\ K}{1.00\ atm} = 70.665076\ L = 70.7\ L$

Check: The units (L) are correct. The volume (71 L) is reasonable because we are using about 3 moles of methane.

11.149 $P_{Total} = P_{N_2} + P_{H_2O} + P_{ethanol}$
P_{N_2}: Use Boyle's law to calculate $P_1V_1 = P_2V_2$ Rearrange to solve for P_2.
$P_2 = P_1\dfrac{V_1}{V_2} = 1.0\ atm \times \dfrac{1.0\ L}{3.0\ L} \times \dfrac{760\ mmHg}{1\ atm} = 253.333\ mmHg$

For water and ethanol, we need to calculate the pressure if all of the liquid were to vaporize in the 3.0 L apparatus. $PV = nRT$. Rearrange to solve for P.

$P = \dfrac{nRT}{V} = \dfrac{2.0\ g \times \dfrac{1\ mol}{18.02\ g} \times 0.08206\dfrac{L \cdot atm}{mol \cdot K} \times \dfrac{760\ mmHg}{1\ atm} \times 308\ K}{3.00\ L} = 710.640\ mmHg\ H_2O.$ Because this

pressure is greater than the vapor pressure of water at this temperature, $P_{H_2O} = 42\ mmHg$.

$P = \dfrac{nRT}{V} = \dfrac{0.50\ g \times \dfrac{1\ mol}{46.07\ g} \times 0.08206\dfrac{L \cdot atm}{mol \cdot K} \times \dfrac{760\ mmHg}{1\ atm} \times 308\ K}{3.00\ L} = 69.4937\ mmHg\ ethanol.$ Because this

pressure is less than the vapor pressure of ethanol at this temperature (102 mmHg), all of the liquid will vaporize and $P_{ethanol} = 69.4937\ mmHg$.

Finally, the total pressure is
$P_{Total} = P_{N_2} + P_{H_2O} + P_{ethanol} = 253.333\ mmHg + 42\ mmHg + 69.4937\ mmHg = 364.827\ mmHg = 360\ mmHg.$

Conceptual Problems

11.151 The water; a container with a larger surface area will evaporate more quickly because there is more surface area for the molecules to evaporate from. Vapor pressure is the pressure of the gas when it is in dynamic equilibrium with the liquid (evaporation rate = condensation rate). The vapor pressure is dependent only on the substance and the temperature. The larger the surface area, the more quickly it will reach this equilibrium state.

11.153   The triple point will be at a lower temperature because the fusion equilibrium line has a positive slope. This means that we will be increasing both temperature and pressure as we travel from the triple point to the normal melting point.

11.155   The liquid segment will have the least steep slope because it takes the most kJ/mol to raise the temperature of the phase.

11.157   The heat of fusion of a substance is always smaller than the heat of vaporization because the number of interactions between particles that are broken is less in fusion than in vaporization. When we melt a solid, the particles have increased mobility but are still strongly interacting with other liquid particles. In vaporization, all of the interactions between particles must be broken (gas particles have essentially no intermolecular interactions) and the particles must absorb enough energy to move much more rapidly.

11.159   The oxygen atoms would be attracted to the —CH_2 groups through dipole–dipole forces.

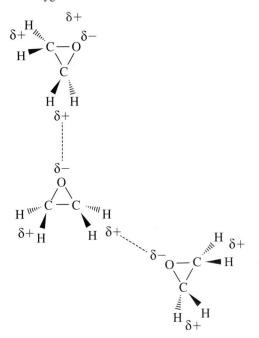

12 Solutions

Review Questions

12.1   As seawater moves through the intestine, it flows past cells that line the digestive tract, which consist of largely fluid interiors surrounded by membranes. Although cellular fluids themselves contain dissolved ions, including sodium and chloride, the fluids are more dilute than seawater. Nature's tendency toward mixing (which tends to produce solutions of uniform concentration), together with the selective permeability of the cell membranes (which allow water to flow in and out but restrict the flow of dissolved solids), cause a flow of solvent out of the body's cells and into the seawater.

12.3   A substance is soluble in another substance if they can form a homogeneous mixture. The solubility of a substance is the amount of the substance that will dissolve in a given amount of solvent. Many different units can be used to express solubility, including grams of solute per 100 grams of solvent, grams of solute per liter of solvent, moles of solute per liter of solution, and moles of solute per kilogram of solvent.

12.5   Entropy is a measure of energy randomization or energy dispersal in a system. When two substances mix to form a solution, there is an increase in randomness because the components are no longer segregated to separate regions. This makes the formation of a solution energetically favorable, even when it is endothermic.

12.7   A solution always forms if the solvent–solute interactions are comparable to, or stronger than, the solvent–solvent interactions and the solute–solute interactions.

12.9   Step 1: Separate the solute into its constituent particles. This step is always endothermic (positive ΔH) because energy is required to overcome the forces that hold the solute together.

Step 2: Separate the solvent particles from each other to make room for the solute particles. This step is also endothermic because energy is required to overcome the intermolecular forces among the solvent particles.

Step 3: Mix the solute particles with the solvent particles. This step is exothermic because energy is released as the solute particles interact with the solvent particles through the various types of intermolecular forces.

12.11   In any solution formation, the initial rate of dissolution far exceeds the rate of deposition. But as the concentration of dissolved solute increases, the rate of deposition also increases. Eventually, the rate of dissolution and deposition become equal—dynamic equilibrium has been reached.

A saturated solution is a solution in which the dissolved solute is in dynamic equilibrium with the solid (or undissolved) solute. If you add additional solute to a saturated solution, it will not dissolve.

An unsaturated solution is a solution containing less than the equilibrium amount of solute. If you add additional solute to an unsaturated solution, it will dissolve.

A supersaturated solution is a solution containing more than the equilibrium amount of solute. Such solutions are unstable and the excess solute normally precipitates out of the solution. However, in some cases, if left undisturbed, a supersaturated solution can exist for an extended period of time.

12.13   The solubility of gases in liquids decreases with increasing temperature. The decreasing solubility of gases with increasing temperature results in a lower oxygen concentration available for fish and other aquatic life in warm waters.

12.15   Henry's law quantifies the solubility of gases with increasing pressure as follows: $S_{gas} = k_H P_{gas}$, where S_{gas} is the solubility of the gas; k_H is a constant of proportionality (called the Henry's law constant) that depends on the specific solute, solvent, and temperature; and P_{gas} is the partial pressure of the gas. The equation simply shows that the solubility of a gas in a liquid is directly proportional to the pressure of the gas above the liquid. If the solubility of a gas is known at a certain temperature, the solubility at another pressure at this temperature can be calculated.

12.17   Parts by mass and parts by volume are ratios of masses and volume, respectively. A parts by mass concentration is the ratio of the mass of the solute to the mass of the solution, all multiplied by a multiplication factor, where percent by mass (%) is the desired unit, the factor = 100; where parts per million by mass (ppm) is the desired unit, the factor = 10^6; and for parts per billion by mass (ppb), the factor = 10^9. The size of the multiplication factor depends on the concentration of the solution. For example, in percent by mass, the multiplication factor is 100%; so

$$\text{percent by mass} = \frac{\text{mass solute}}{\text{mass solution}} \times 100\%.$$ A solution with a concentration of 28% by mass contains 28 g of solute per 100 g of solution.

12.19   Raoult's law quantifies the relationship between the vapor pressure of a solution and its concentration as $P_{solution} = \chi_{solvent} P^\circ_{solvent}$, where $P_{solution}$ is the vapor pressure of the solution, $\chi_{solvent}$ is the mole fraction of the solvent, and $P^\circ_{solvent}$ is the vapor pressure of the pure solvent. This equation allows you to calculate the vapor pressure of a solution or to calculate the concentration of a solution, given the vapor pressure of the solution.

12.21   If the solute–solvent interactions are particularly strong (stronger than solvent–solvent interactions), the solute tends to prevent the solvent from vaporizing as easily as it would otherwise and the vapor pressure of the solution will be less than that predicted by Raoult's law. If the solute–solvent interactions are particularly weak (weaker than solvent–solvent interactions), the solute tends to allow more vaporization than would occur with just the solvent and the vapor pressure of the solution will be greater than predicted by Raoult's law.

12.23   Colligative properties depend on the amount of solute and not the type of solute. Examples of colligative properties are vapor pressure lowering, freezing point depression, boiling point elevation, and osmotic pressure.

12.25   The van't Hoff factor (i) is the ratio of moles of particles in solution to moles of formula units dissolved: $$i = \frac{\text{moles of particles}}{\text{moles of formula units dissolved}}.$$ The van't Hoff factor often does not match its theoretical value because the ionic solute is not completely dissolved into the expected number of ions, leaving ion pairs in solution. The result is that the number of particles in the solution is not as high as theoretically expected.

12.27   The Tyndall effect is the scattering of light by a colloidal dispersion. The Tyndall effect is often used as a test to determine whether a mixture is a solution or a colloid because solutions contain completely dissolved solute molecules that are too small to scatter light.

Problems by Topic

Solubility

12.29   (a)   hexane, toluene, or CCl_4; dispersion forces
         (b)   water, methanol, acetone; dispersion, dipole–dipole, hydrogen bonding (except for acetone)
         (c)   hexane, toluene, or CCl_4; dispersion forces
         (d)   water, acetone, methanol, ethanol; dispersion, ion–dipole

12.31 $HOCH_2CH_2CH_2OH$ would be more soluble in water because it has —OH groups on both ends of the molecule; so it can hydrogen bond on both ends.

12.33   (a)   water; dispersion, dipole–dipole, hydrogen bonding
         (b)   hexane; dispersion forces
         (c)   water; dispersion, dipole–dipole
         (d)   water; dispersion, dipole–dipole, hydrogen bonding

Energetics of Solution Formation

12.35   (a)   endothermic

          (b)   The lattice energy is greater in magnitude than is the heat of hydration.

          (c)

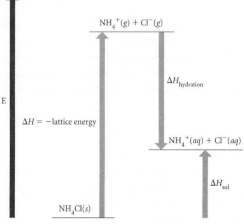

          (d)   The solution forms because chemical systems tend toward greater entropy.

12.37   **Given:** $AgNO_3$: lattice energy $= -820.$ kJ/mol, $\Delta H_{soln} = +22.6$ kJ/mol   **Find:** $\Delta H_{hydration}$
          Conceptual Plan: lattice energy, $\Delta H_{soln} \rightarrow \Delta H_{hydration}$
$$\Delta H_{soln} = \Delta H_{solute} + \Delta H_{hydration} \; where \; \Delta H_{solute} = -\Delta H_{lattice}$$
          Solution: $\Delta H_{soln} = \Delta H_{solute} + \Delta H_{hydration}$ where $\Delta H_{solute} = -\Delta H_{lattice}$ so $\Delta H_{hydration} = \Delta H_{soln} + \Delta H_{lattice}$
          $\Delta H_{hydration} = 22.6$ kJ/mol $- 820.$ kJ/mol $= -797$ kJ/mol

          Check: The units (kJ/mol) are correct. The magnitude of the answer (-800 kJ/mol) makes physical sense because
          the lattice energy is so negative; thus, it dominates the calculation.

12.39   **Given:** LiI: lattice energy $= -7.3 \times 10^2$ kJ/mol, $\Delta H_{hydration} = -793$ kJ/mol; 15.0 g LiI
          Find: ΔH_{soln} and heat evolved
          Conceptual Plan: lattice energy, $\Delta H_{hydration} \rightarrow \Delta H_{soln}$ and g $\rightarrow$ mol then mol, $\Delta H_{soln} \rightarrow q$
$$\Delta H_{soln} = \Delta H_{solute} + \Delta H_{hydration} \; where \; \Delta H_{solute} = -\Delta H_{lattice} \quad \frac{1 \; mol}{133.84 \; g} \qquad q = n\Delta H_{soln}$$
          Solution: $\Delta H_{soln} = \Delta H_{solute} + \Delta H_{hydration}$ where $\Delta H_{solute} = -\Delta H_{lattice}$ so $\Delta H_{soln} = \Delta H_{hydration} - \Delta H_{lattice}$

          $\Delta H_{soln} = -793$ kJ/mol $- (-7\underline{3}0$ kJ/mol$) = -\underline{6}3$ kJ/mol $= -6.3 \times 10^1$ kJ/mol and

          $15.0 \; g \times \dfrac{1 \; mol}{133.84 \; g} = 0.11\underline{2}074$ mol then

          $q = n\Delta H_{soln} = 0.11\underline{2}074$ mol $\times -6.3 \times 10^1 \dfrac{kJ}{mol} = -7.1$ kJ or 7.0 kJ released

          Check: The units (kJ/mol and kJ) are correct. The magnitude of the answer (-60 kJ/mol) makes physical sense
          because the lattice energy and the heat of hydration are about the same. The magnitude of the heat (7 kJ) makes
          physical sense because 15 g is much less than a mole; thus, the amount of heat released is going to be small.

Solution Equilibrium and Factors Affecting Solubility

12.41   The solution is unsaturated because we are dissolving 25 g of NaCl per 100 g of water and the solubility from the
          figure is $\sim$35 g NaCl per 100 g of water at 25 °C.

12.43   At 40 °C, the solution has 45 g of KNO_3 per 100 g of water, and it can contain up to 63 g of KNO_3 per 100 g of
          water. At 0 °C, the solubility from the figure is $\sim$14 g KNO_3 per 100 g of water; so $\sim$31 g KNO_3 per 100 g of water will
          precipitate out of solution.

12.45 Because the solubility of gases decreases as the temperature increases, boiling will cause dissolved oxygen to be removed from the solution.

12.47 Henry's law says that as pressure increases, nitrogen will more easily dissolve in blood. To reverse this process, divers should ascend to lower pressures.

12.49 **Given:** room temperature, 80.0 L aquarium, $P_{Total} = 1.0$ atm; $\chi_{N_2} = 0.78$ **Find:** $m(N_2)$
Other: $k_H(N_2) = 6.1 \times 10^{-4}$ M/L at 25 °C
Conceptual Plan: $P_{Total}, \chi_{N_2} \rightarrow P_{N_2}$ then $P_{N_2}, k_H(N_2) \rightarrow S_{N_2}$ then $L \rightarrow mol \rightarrow g$

$$P_{N_2} = \chi_{N_2} P_{Total} \qquad S_{N_2} = k_H(N_2)P_{N_2} \qquad M = \frac{\text{amount solute (moles)}}{\text{volume solution (L)}} \frac{28.01 \text{ g } N_2}{1 \text{ mol } N_2}$$

Solution: $P_{N_2} = \chi_{N_2} P_{Total} = 0.78 \times 1.0$ atm $= 0.78$ atm then

$$S_{N_2} = k_H(N_2)P_{N_2} = 6.1 \times 10^{-4} \frac{M}{\text{atm}} \times 0.78 \text{ atm} = 4.\underline{7}58 \times 10^{-4} \text{ M then}$$

$$80.0 \text{ L} \times 4.\underline{7}58 \times 10^{-4} \frac{\text{mol}}{\text{L}} \times \frac{28.01 \text{ g}}{1 \text{ mol}} = 1.1 \text{ g}$$

Check: The units (g) are correct. The magnitude of the answer (1 g) seems reasonable because we have 80 L of water and expect much less than a mole of nitrogen.

Concentrations of Solutions

12.51 **Given:** NaCl and water; 112 g NaCl in 1.00 L solution **Find:** M, m, and mass percent **Other:** $d = 1.08$ g/mL
Conceptual Plan: $g_{NaCl} \rightarrow$ mol and $L \rightarrow mL \rightarrow g_{soln}$ and $g_{soln}, g_{NaCl} \rightarrow g_{H_2O} \rightarrow kg_{H_2O}$ then

$$\frac{1 \text{ mol NaCl}}{58.44 \text{ g NaCl}} \qquad \frac{1000 \text{ mL}}{1 \text{ L}} \quad \frac{1.08 \text{ g}}{1 \text{ mL}} \qquad g_{H_2O} = g_{soln} - g_{NaCl} \quad \frac{1 \text{ kg}}{1000 \text{ g}}$$

mol, $V \rightarrow$ M and mol, $kg_{H_2O} \rightarrow m$ and $g_{soln}, g_{NaCl} \rightarrow$ mass percent

$$M = \frac{\text{amount solute (moles)}}{\text{volume solution (L)}} \quad m = \frac{\text{amount solute (moles)}}{\text{mass solvent (kg)}} \quad \text{mass percent} = \frac{\text{mass solute}}{\text{mass solution}} \times 100\%$$

Solution: $112 \text{ g NaCl} \times \frac{1 \text{ mol NaCl}}{58.44 \text{ g NaCl}} = 1.9164956 \text{ mol NaCl and}$

$$1.00 \text{ L} \times \frac{1000 \text{ mL}}{1 \text{ L}} \times \frac{1.08 \text{ g}}{1 \text{ mL}} = 10\underline{8}0 \text{ g soln and}$$

$$g_{H_2O} = g_{soln} - g_{NaCl} = 10\underline{8}0 \text{ g} - 112 \text{ g} = 9\underline{6}8 \text{ g } H_2O \times \frac{1 \text{ kg}}{1000 \text{ g}} = 0.9\underline{6}8 \text{ kg } H_2O \text{ then}$$

$$M = \frac{\text{amount solute (moles)}}{\text{volume solution (L)}} = \frac{1.9\underline{1}64956 \text{ mol NaCl}}{1.00 \text{ L soln}} = 1.92 \text{ M and}$$

$$m = \frac{\text{amount solute (moles)}}{\text{mass solvent (kg)}} = \frac{1.9\underline{1}64956 \text{ mol NaCl}}{0.9\underline{6}8 \text{ kg } H_2O} = 2.0 \text{ } m \text{ and}$$

$$\text{mass percent} = \frac{\text{mass solute}}{\text{mass solution}} \times 100\% = \frac{112 \text{ g NaCl}}{10\underline{8}0 \text{ g soln}} \times 100\% = 10.4\% \text{ by mass}$$

Check: The units (M, m, and percent by mass) are correct. The magnitude of the answer (2 M) seems reasonable because we have 112 g NaCl, which is a couple of moles and we have 1 L. The magnitude of the answer (2 m) seems reasonable because it is a little higher than the molarity, which we expect because we only use the solvent weight in the denominator. The magnitude of the answer (10%) seems reasonable because we have 112 g NaCl and just over 1000 g of solution.

12.53 **Given:** initial solution: 50.0 mL of 5.00 M KI; final solution contains 3.05 g KI in 25.0 mL
Find: final volume of diluted solution
Conceptual Plan: final solution: $g_{KI} \rightarrow$ mol and mL $\rightarrow$ L then mol, $V \rightarrow M_2$ then $M_1, V_1, M_2 \rightarrow V_2$

$$\frac{1 \text{ mol KI}}{166.006 \text{ g KI}} \qquad \frac{1 \text{ L}}{1000 \text{ mL}} \qquad M = \frac{\text{amount solute (moles)}}{\text{volume solution (L)}} \qquad M_1 V_1 = M_2 V_2$$

Solution: $3.05 \text{ g KI} \times \dfrac{1 \text{ mol KI}}{166.006 \text{ g KI}} = 0.01\underline{8}37283 \text{ mol KI}$ and $25.0 \text{ mL} \times \dfrac{1 \text{ L}}{1000 \text{ mL}} = 0.0250 \text{ L}$

then $M = \dfrac{\text{amount solute (moles)}}{\text{volume solution (L)}} = \dfrac{0.01\underline{8}37283 \text{ mol KI}}{0.0250 \text{L soln}} = 0.73\underline{4}9132 \text{ M}$ then $M_1 V_1 = M_2 V_2$

Rearrange to solve for V_2. $V_2 = \dfrac{M_1}{M_2} \times V_1 = \dfrac{5.00 \text{ M}}{0.73\underline{4}9132 \text{ M}} \times 50.0 \text{ mL} = 340. \text{ mL diluted volume}$

Check: The units (mL) are correct. The magnitude of the answer (340 mL) seems reasonable because we are starting with a concentration of 5 M and ending with a concentration of less than 1 M.

12.55   **Given:** $AgNO_3$ and water; 3.4% Ag by mass, 4.8 L solution   **Find:** m (Ag)   **Other:** $d = 1.01 \text{ g/mL}$
          Conceptual Plan: $\mathbf{L \to mL \to g_{soln} \to g_{Ag}}$

$$\dfrac{1000 \text{ mL}}{1 \text{ L}} \quad \dfrac{1.01 \text{ g}}{1 \text{ mL}} \quad \dfrac{3.4 \text{ g Ag}}{100 \text{ g soln}}$$

Solution: $4.8 \text{ L} \times \dfrac{1000 \text{ mL}}{1 \text{ L}} \times \dfrac{1.01 \text{ g}}{1 \text{ mL}} = 4\underline{8}48 \text{ g soln}$ then

$4\underline{8}48 \text{ g soln} \times \dfrac{3.4 \text{ g Ag}}{100 \text{ g soln}} = 160 \text{ g Ag} = 1.6 \times 10^2 \text{ g Ag}$

Check: The units (g) are correct. The magnitude of the answer (160 g) seems reasonable because we have almost 5000 g of solution.

12.57   **Given:** Ca^{2+} and water; 0.0085% Ca^{2+} by mass, 1.2 g Ca   **Find:** m (water)
          Conceptual Plan: $\mathbf{g_{Ca} \to g_{soln} \to g_{H_2O}}$

$$\dfrac{100 \text{ g soln}}{0.0085 \text{ g Ca}} \quad g_{H_2O} = g_{soln} - g_{Ca}$$

Solution: $1.2 \text{ g Ca} \times \dfrac{100 \text{ g soln}}{0.0085 \text{ g Ca}} = 1\underline{4}118 \text{ g soln}$ then

$g_{H_2O} = g_{soln} - g_{Ca} = 1\underline{4}118 \text{ g} - 1.2 \text{ g} = 1.4 \times 10^4 \text{ g water}$

Check: The units (g) are correct. The magnitude of the answer (10^4 g) seems reasonable because we have such a low concentration of Ca.

12.59   **Given:** concentrated HNO_3: 70.3% HNO_3 by mass, $d = 1.41 \text{ g/mL}$; final solution: 1.15 L of 0.100 M HNO_3
          Find: describe final solution preparation
          Conceptual Plan: $\mathbf{M_2, V_2 \to mol_{HNO_3} \to g_{HNO_3} \to g_{conc\ acid} \to mL_{conc\ acid}}$ **then describe method**

$$mol = MV \quad \dfrac{63.02 \text{ g HNO}_3}{1 \text{ mol HNO}_3} \quad \dfrac{100 \text{ g conc acid}}{70.3 \text{ g HNO}_3} \quad \dfrac{1 \text{ mL}}{1.41 \text{ g}}$$

Solution: $mol = MV = 0.100 \dfrac{\text{mol HNO}_3}{1 \text{ L soln}} \times 1.15 \text{ L soln} = 0.115 \text{ mol HNO}_3$ then

$0.115 \text{ mol HNO}_3 \times \dfrac{63.02 \text{ g HNO}_3}{1 \text{ mol HNO}_3} \times \dfrac{100 \text{ g conc acid}}{70.3 \text{ g HNO}_3} \times \dfrac{1 \text{ mL conc acid}}{1.41 \text{ g conc acid}} = 7.31 \text{ mL conc acid}$

Prepare the solution by putting about 1.00 L of distilled water in a container. Carefully pour in the 7.31 mL of the concentrated acid, mix the solution, and allow it to cool. Finally, add enough water to generate the total volume of solution (1.15 L). It is important to add acid to water, not the reverse, because such a large amount of heat is released upon mixing.

Check: The units (mL) are correct. The magnitude of the answer (7 mL) seems reasonable because we are starting with such a concentrated solution and diluting it to a low concentration.

12.61   (a)   **Given:** 1.00×10^2 mL of 0.500 M KCl   **Find:** describe final solution preparation
                  Conceptual Plan: $\mathbf{mL \to L}$ **then** $\mathbf{M, V \to mol_{KCl} \to g_{KCl}}$ **then describe method**

$$\dfrac{1 \text{ L}}{1000 \text{ mL}} \qquad mol = MV \quad \dfrac{74.55 \text{ g KCl}}{1 \text{ mol KCl}}$$

Solution: $1.00 \times 10^2 \text{ mL} \times \dfrac{1 \text{ L}}{1000 \text{ mL}} = 0.100 \text{ L}$

$mol = MV = 0.500 \dfrac{\text{mol KCl}}{1 \text{ L soln}} \times 0.100 \text{ L soln} = 0.0500 \text{ mol KCl}$

then $0.0500 \text{ mol KCl} \times \dfrac{74.55 \text{ g KCl}}{1 \text{ mol KCl}} = 3.73 \text{ g KCl}$

Prepare the solution by carefully adding 3.73 g KCl to a 100 mL volumetric flask. Add ~75 mL of distilled water and agitate the solution until the salt dissolves completely. Finally, add enough water to generate a total volume of solution (add water to the mark on the flask).

Check: The units (g) are correct. The magnitude of the answer (4 g) seems reasonable because we are making a small volume of solution and the formula weight of KCl is ~75 g/mol.

(b)　**Given:** 1.00×10^2 g of 0.500 *m* KCl　**Find:** describe final solution preparation

Conceptual Plan:

$m \rightarrow \text{mol}_{\text{KCl}}/1 \text{ kg solvent} \rightarrow g_{\text{KCl}}/1 \text{ kg solvent then } g_{\text{KCl}}/1 \text{ kg solvent}, g_{\text{soln}} \rightarrow g_{\text{KCl}}, g_{\text{H}_2\text{O}}$

$$m = \dfrac{\text{amount solute (moles)}}{\text{mass solvent (kg)}} \qquad \dfrac{74.55 \text{ g KCl}}{1 \text{ mol KCl}} \qquad\qquad g_{\text{soln}} = g_{\text{KCl}} + g_{\text{H}_2\text{O}}$$

then describe method

Solution: $m = \dfrac{\text{amount solute (moles)}}{\text{mass solvent (kg)}}$ so $0.500 \, m = \dfrac{0.500 \text{ mol KCl}}{1 \text{ kg H}_2\text{O}}$ so

$\dfrac{0.500 \text{ mol KCl}}{1 \text{ kg H}_2\text{O}} \times \dfrac{74.55 \text{ g KCl}}{1 \text{ mol KCl}} = \dfrac{37.\underline{2}8 \text{ g KCl}}{1000 \text{ g H}_2\text{O}}$　$g_{\text{soln}} = g_{\text{KCl}} + g_{\text{H}_2\text{O}}$ so $g_{\text{soln}} - g_{\text{KCl}} = g_{\text{H}_2\text{O}}$

Substitute into ratio. $\dfrac{37.\underline{2}8 \text{ g KCl}}{1037.28 \text{ g solution}} = \dfrac{x \text{ g KCl}}{100 \text{ g solution}}$. Multiply both sides by 100 g solution.

$\dfrac{37.\underline{2}8 \text{ g KCl} \, (100 \text{ g soln})}{1037.28 \text{ g soln}} = 3.59 \text{ g}$

$g_{\text{H}_2\text{O}} = g_{\text{soln}} - g_{\text{KCl}} = 100. \text{ g} - 3.59 \text{ g} = 96.41 \text{ g H}_2\text{O}$

Prepare the solution by carefully adding 3.59 g KCl to a container with 96.41 g of distilled water and agitate the solution until the salt dissolves completely.

Check: The units (g) are correct. The magnitude of the answer (3.6 g) seems reasonable because we are making a small volume of solution and the formula weight of KCl is ~75 g/mol.

(c)　**Given:** 1.00×10^2 g of 5.0% KCl by mass　**Find:** describe final solution preparation

Conceptual Plan: $g_{\text{soln}} \rightarrow g_{\text{KCl}} \text{ then } g_{\text{KCl}}, g_{\text{soln}} \rightarrow g_{\text{H}_2\text{O}}$

$$\dfrac{5.0 \text{ g KCl}}{100 \text{ g soln}} \qquad\qquad g_{\text{soln}} = g_{\text{KCl}} + g_{\text{H}_2\text{O}}$$

then describe method

Solution: $1.00 \times 10^2 \text{ g soln} \times \dfrac{5.0 \text{ g KCl}}{100 \text{ g soln}} = 5.0 \text{ g KCl}$ then $g_{\text{soln}} = g_{\text{KCl}} + g_{\text{H}_2\text{O}}$

So $g_{\text{H}_2\text{O}} = g_{\text{soln}} - g_{\text{KCl}} = 100. \text{ g} - 5.0 \text{ g} = 95 \text{ g H}_2\text{O}$

Prepare the solution by carefully adding 5.0 g KCl to a container with 95 g of distilled water and agitate the solution until the salt dissolves completely.

Check: The units (g) are correct. The magnitude of the answer (5 g) seems reasonable because we are making a small volume of solution and the solution is 5% by mass KCl.

12.63　(a)　**Given:** 28.4 g of glucose $(C_6H_{12}O_6)$ in 355 g water; final volume = 378 mL　**Find:** molarity

Conceptual Plan: $mL \rightarrow L$ and $g_{C_6H_{12}O_6} \rightarrow \text{mol}_{C_6H_{12}O_6}$ then $\text{mol}_{C_6H_{12}O_6}, V \rightarrow M$

$$\dfrac{1 \text{ L}}{1000 \text{ mL}} \qquad \dfrac{1 \text{ mol } C_6H_{12}O_6}{180.16 \text{ g } C_6H_{12}O_6} \qquad M = \dfrac{\text{amount solute (moles)}}{\text{volume solution (L)}}$$

Solution: $378 \text{ mL} \times \dfrac{1 \text{ L}}{1000 \text{ mL}} = 0.378 \text{ L}$ and

$28.4 \text{ g } C_6H_{12}O_6 \times \dfrac{1 \text{ mol } C_6H_{12}O_6}{180.16 \text{ g } C_6H_{12}O_6} = 0.15\underline{7}638 \text{ mol } C_6H_{12}O_6$

$M = \dfrac{\text{amount solute (moles)}}{\text{volume solution (L)}} = \dfrac{0.15\underline{7}638 \text{ mol } C_6H_{12}O_6}{0.378 \text{ L}} = 0.417 \text{ M}$

Check: The units (M) are correct. The magnitude of the answer (0.4 M) seems reasonable because we have 1/8 mole in about 1/3 L.

(b)   **Given:** 28.4 g of glucose ($C_6H_{12}O_6$) in 355 g water; final volume = 378 mL   **Find:** molality
Conceptual Plan: $g_{H_2O} \rightarrow kg_{H_2O}$ and $g_{C_6H_{12}O_6} \rightarrow mol_{C_6H_{12}O_6}$ then $mol_{C_6H_{12}O_6}$, $kg_{H_2O} \rightarrow m$

$$\frac{1 \text{ kg}}{1000 \text{ g}} \qquad \frac{1 \text{ mol } C_6H_{12}O_6}{180.16 \text{ g } C_6H_{12}O_6} \qquad m = \frac{\text{amount solute (moles)}}{\text{mass solvent (kg)}}$$

Solution: $355 \text{ g} \times \dfrac{1 \text{ kg}}{1000 \text{ g}} = 0.355 \text{ kg}$ and

$28.4 \text{ g } C_6H_{12}O_6 \times \dfrac{1 \text{ mol } C_6H_{12}O_6}{180.16 \text{ g } C_6H_{12}O_6} = 0.15\underline{7}638 \text{ mol } C_6H_{12}O_6$

$m = \dfrac{\text{amount solute (moles)}}{\text{mass solvent (kg)}} = \dfrac{0.15\underline{7}638 \text{ mol } C_6H_{12}O_6}{0.355 \text{ kg}} = 0.444 \ m$

Check: The units (m) are correct. The magnitude of the answer (0.4 m) seems reasonable because we have 1/8 mole in about 1/3 kg.

(c)   **Given:** 28.4 g of glucose ($C_6H_{12}O_6$) in 355 g water; final volume = 378 mL   **Find:** percent by mass
Conceptual Plan: $g_{C_6H_{12}O_6}$, $g_{H_2O} \rightarrow g_{soln}$ then $g_{C_6H_{12}O_6}$, $g_{soln} \rightarrow$ **percent by mass**

$$g_{soln} = g_{C_6H_{12}O_6} + g_{H_2O} \qquad \text{mass percent} = \frac{\text{mass solute}}{\text{mass solution}} \times 100\%$$

Solution: $g_{soln} = g_{C_6H_{12}O_6} + g_{H_2O} = 28.4 \text{ g} + 355 \text{ g} = 38\underline{3}.4 \text{ g soln}$ then

$$\text{mass percent} = \frac{\text{mass solute}}{\text{mass solution}} \times 100\% = \frac{28.4 \text{ g } C_6H_{12}O_6}{38\underline{3}.4 \text{ g soln}} \times 100\% = 7.41\% \text{ by mass}$$

Check: The units (percent by mass) are correct. The magnitude of the answer (7%) seems reasonable because we are dissolving 28 g in 355 g.

(d)   **Given:** 28.4 g of glucose ($C_6H_{12}O_6$) in 355 g water; final volume = 378 mL   **Find:** mole fraction
Conceptual Plan: $g_{C_6H_{12}O_6} \rightarrow mol_{C_6H_{12}O_6}$ and $g_{H_2O} \rightarrow mol_{H_2O}$ then $mol_{C_6H_{12}O_6}$, $mol_{H_2O} \rightarrow \chi_{C_6H_{12}O_6}$

$$\frac{1 \text{ mol } C_6H_{12}O_6}{180.16 \text{ g } C_6H_{12}O_6} \qquad \frac{1 \text{ mol } H_2O}{18.02 \text{ g } H_2O} \qquad \chi = \frac{\text{amount solute (in moles)}}{\text{total amount of solute and solvent (in moles)}}$$

Solution: $28.4 \text{ g } C_6H_{12}O_6 \times \dfrac{1 \text{ mol } C_6H_{12}O_6}{180.16 \text{ g } C_6H_{12}O_6} = 0.15\underline{7}638 \text{ mol } C_6H_{12}O_6$ and

$355 \text{ g } H_2O \times \dfrac{1 \text{ mol } H_2O}{18.02 \text{ g } H_2O} = 19.\underline{7}003 \text{ mol } H_2O$ then

$\chi = \dfrac{\text{amount solute (in moles)}}{\text{total amount of solute and solvent (in moles)}} = \dfrac{0.15\underline{7}638 \text{ mol}}{0.15\underline{7}638 \text{ mol} + 19.\underline{7}003 \text{ mol}} = 0.00794$

Check: The units (none) are correct. The magnitude of the answer (0.008) seems reasonable because we have many more grams of water and water has a much lower molecular weight.

(e)   **Given:** 28.4 g of glucose ($C_6H_{12}O_6$) in 355 g water; final volume = 378 mL   **Find:** mole percent
Conceptual Plan: Use answer from part (d) then $\chi_{C_6H_{12}O_6} \rightarrow$ **mole percent**

$$\chi \times 100\%$$

Solution: mole percent $= \chi \times 100\% = 0.00794 \times 100\% = 0.794$ mole percent

Check: The units (mol %) are correct. The magnitude of the answer (0.8) seems reasonable because we have many more grams of water, water has a much lower molecular weight than glucose, and we are increasing the answer from part (d) by a factor of 100.

12.65   **Given:** 3.0% H_2O_2 by mass, $d = 1.01$ g/mL   **Find:** molarity
Conceptual Plan:
Assume exactly 100 g of solution; $g_{solution} \rightarrow g_{H_2O_2} \rightarrow mol_{H_2O_2}$ and $g_{solution} \rightarrow mL_{solution} \rightarrow L_{solution}$

$$\frac{3.0 \text{ g } H_2O_2}{100 \text{ g solution}} \quad \frac{1 \text{ mol } H_2O_2}{34.02 \text{ g } H_2O_2} \qquad \frac{1 \text{ mL}}{1.01 \text{ g}} \quad \frac{1 \text{ L}}{1000 \text{ mL}}$$

then $mol_{H_2O_2}$, $L_{solution} \rightarrow M$

$$M = \frac{\text{amount solute (moles)}}{\text{volume solution (L)}}$$

Solution: $100 \text{ g solution} \times \dfrac{3.0 \text{ g } H_2O_2}{100 \text{ g solution}} \times \dfrac{1 \text{ mol } H_2O_2}{34.02 \text{ g } H_2O_2} = 0.08\underline{8}1834 \text{ mol } H_2O_2$ and

$$100 \text{ g solution} \times \frac{1 \text{ mL solution}}{1.01 \text{ g solution}} \times \frac{1 \text{ L solution}}{1000 \text{ mL solution}} = 0.0990099 \text{ L solution then}$$

$$M = \frac{\text{amount solute (moles)}}{\text{volume solution (L)}} = \frac{0.0881834 \text{ mol H}_2\text{O}_2}{0.0990099 \text{ L solution}} = 0.89 \text{ M H}_2\text{O}_2$$

Check: The units (M) are correct. The magnitude of the answer (1) seems reasonable because we are starting with a low concentration solution and pure water is ~55.5 M.

12.67   **Given:** 36% HCl by mass   **Find:** molality and mole fraction

     Conceptual Plan: Assume exactly 100 g of solution; $g_{solution} \rightarrow g_{HCl} \rightarrow mol_{HCl}$ **and** $g_{HCl}, g_{solution} \rightarrow$

$$\frac{36 \text{ g HCl}}{100 \text{ g solution}} \qquad \frac{1 \text{ mol HCl}}{36.46 \text{ g HCl}} \qquad g_{soln} = g_{HCl} + g_{H_2O}$$

     $g_{solvent} \rightarrow kg_{solvent}$ **then** $mol_{HCl}, kg_{solvent} \rightarrow m$ **and** $g_{solvent} \rightarrow mol_{solvent}$ **then** $mol_{HCl}, mol_{solvent} \rightarrow \chi_{HCl}$

$$\frac{1 \text{ kg}}{1000 \text{ g}} \qquad m = \frac{\text{amount solute (moles)}}{\text{mass solvent (kg)}} \qquad \frac{1 \text{ mol H}_2\text{O}}{18.02 \text{ g H}_2\text{O}} \qquad \chi = \frac{\text{amount solute (in moles)}}{\text{total amount of solute and solvent (in moles)}}$$

     Solution: $100 \text{ g solution} \times \dfrac{36 \text{ g HCl}}{100 \text{ g solution}} = 36 \text{ g HCl} \times \dfrac{1 \text{ mol HCl}}{36.46 \text{ g HCl}} = 0.987383 \text{ mol HCl and}$

     $g_{soln} = g_{HCl} + g_{H_2O}$ Rearrange to solve for $g_{solvent}$. $g_{H_2O} = g_{soln} - g_{HCl} = 100 \text{ g} - 36 \text{ g} = 64 \text{ g H}_2\text{O}$

$$64 \text{ g H}_2\text{O} \times \frac{1 \text{ kg H}_2\text{O}}{1000 \text{ g H}_2\text{O}} = 0.064 \text{ kg H}_2\text{O then}$$

$$m = \frac{\text{amount solute (moles)}}{\text{mass solvent (kg)}} = \frac{0.987383 \text{ mol HCl}}{0.064 \text{ kg}} = 15 \text{ } m \text{ HCl and}$$

$$64 \text{ g H}_2\text{O} \times \frac{1 \text{ mol H}_2\text{O}}{18.02 \text{ g H}_2\text{O}} = 3.55161 \text{ mol H}_2\text{O then}$$

$$\chi = \frac{\text{amount solute (in moles)}}{\text{total amount of solute and solvent (in moles)}} = \frac{0.987383 \text{ mol}}{0.987383 \text{ mol} + 3.55161 \text{ mol}} = 0.22$$

Check: The units (m and unitless) are correct. The magnitude of the answers (15 and 0.2) seems reasonable because we are starting with a high concentration solution and the molar mass of water is much less than that of HCl.

Vapor Pressure of Solutions

12.69   The level has decreased more in the beaker filled with pure water. The dissolved salt in the seawater decreases the vapor pressure and subsequently lowers the rate of vaporization.

12.71   **Given:** 24.5 g of glycerin $(\text{C}_3\text{H}_8\text{O}_3)$ in 135 mL water at 30 °C; $P^\circ_{\text{H}_2\text{O}} = 31.8$ torr   **Find:** $P_{\text{H}_2\text{O}}$

     Other: $d\,(\text{H}_2\text{O}) = 1.00$ g/mL; glycerin is not an ionic solid

     Conceptual Plan: $g_{\text{C}_3\text{H}_8\text{O}_3} \rightarrow mol_{\text{C}_3\text{H}_8\text{O}_3}$ **and** $mL \text{ H}_2\text{O} \rightarrow g\text{H}_2\text{O} \rightarrow mol_{\text{H}_2\text{O}}$ **then** $mol_{\text{C}_3\text{H}_8\text{O}_3}, mol_{\text{H}_2\text{O}} \rightarrow \chi_{\text{H}_2\text{O}}$

$$\frac{1 \text{ mol C}_3\text{H}_8\text{O}_3}{92.09 \text{ g C}_3\text{H}_8\text{O}_3} \qquad \frac{1.00 \text{ g}}{1 \text{ mL}} \quad \frac{1 \text{ mol H}_2\text{O}}{18.01 \text{ g H}_2\text{O}} \qquad \chi = \frac{\text{amount solvent (in moles)}}{\text{total amount of solute and solvent (in moles)}}$$

     then $\chi_{\text{H}_2\text{O}}, P^\circ_{\text{H}_2\text{O}} \rightarrow P_{\text{H}_2\text{O}}$

$$P_{\text{solution}} = \chi_{\text{solvent}} P^\circ_{\text{solvent}}$$

     Solution: $24.5 \text{ g C}_3\text{H}_8\text{O}_3 \times \dfrac{1 \text{ mol C}_3\text{H}_8\text{O}_3}{92.09 \text{ g C}_3\text{H}_8\text{O}_3} = 0.2660441 \text{ mol C}_3\text{H}_8\text{O}_3 \text{ and}$

$$135 \text{ mL} \times \frac{1.00 \text{ g}}{1 \text{ mL}} \times \frac{1 \text{ mol H}_2\text{O}}{18.01 \text{ g H}_2\text{O}} = 7.495836 \text{ mol H}_2\text{O then}$$

$$\chi = \frac{\text{amount solvent (in moles)}}{\text{total amount of solute and solvent (in moles)}} = \frac{7.495836 \text{ mol}}{0.2660441 \text{ mol} + 7.495836 \text{ mol}} = 0.9657243 \text{ then}$$

$$P_{\text{solution}} = \chi_{\text{solvent}} P^\circ_{\text{solvent}} = 0.9657243 \times 31.8 \text{ torr} = 30.7 \text{ torr}$$

Check: The units (torr) are correct. The magnitude of the answer (31 torr) seems reasonable because it is a drop from the pure vapor pressure. Very few moles of glycerin are added, so the pressure will not drop much.

12.73   **Given:** 50.0 g of heptane (C_7H_{16}) and 50.0 g of octane (C_8H_{18}) at 25 °C; $P^{\circ}_{C_7H_{16}} = 45.8$ torr; $P^{\circ}_{C_8H_{18}} = 10.9$ torr

(a)   **Find:** $P_{C_7H_{16}}, P_{C_8H_{18}}$
Conceptual Plan: $g_{C_7H_{16}} \rightarrow mol_{C_7H_{16}}$ and $g_{C_8H_{18}} \rightarrow mol_{C_8H_{18}}$ then $mol_{C_7H_{16}}$,

$$\frac{1 \text{ mol } C_7H_{16}}{100.20 \text{ g } C_7H_{16}} \quad \frac{1 \text{ mol } C_8H_{18}}{114.22 \text{ g } C_8H_{18}} \quad \chi_{C_7H_{16}} = \frac{\text{amount } C_7H_{16} \text{ (in moles)}}{\text{total amount (in moles)}}$$

$mol_{C_8H_{18}} \rightarrow \chi_{C_7H_{16}}, \chi_{C_8H_{18}}$ then $\chi_{C_7H_{16}}, P^{\circ}_{C_7H_{16}} \rightarrow P_{C_7H_{16}}$ and $\chi_{C_8H_{18}}, P^{\circ}_{C_8H_{18}} \rightarrow P_{C_8H_{18}}$

$$\chi_{C_8H_{18}} = 1 - \chi_{C_7H_{16}} \qquad\qquad P_{C_7H_{16}} = \chi_{C_7H_{16}} P^{\circ}_{C_7H_{16}} \qquad\qquad P_{C_8H_{18}} = \chi_{C_8H_{18}} P^{\circ}_{C_8H_{18}}$$

Solution: $50.0 \text{ g } C_7H_{16} \times \dfrac{1 \text{ mol } C_7H_{16}}{100.20 \text{ g } C_7H_{16}} = 0.499002 \text{ mol } C_7H_{16}$ and

$50.0 \text{ g } C_8H_{18} \times \dfrac{1 \text{ mol } C_8H_{18}}{114.22 \text{ g } C_8H_{18}} = 0.437752 \text{ mol } C_8H_{18}$ then

$$\chi_{C_7H_{16}} = \frac{\text{amount } C_7H_{16} \text{ (in moles)}}{\text{total amount (in moles)}} = \frac{0.499002 \text{ mol}}{0.499002 \text{ mol} + 0.437752 \text{ mol}} = 0.532693$$ and

$\chi_{C_8H_{18}} = 1 - \chi_{C_7H_{16}} = 1 - 0.532693 = 0.467307$ then

$P_{C_7H_{16}} = \chi_{C_7H_{16}} P^{\circ}_{C_7H_{16}} = 0.532693 \times 45.8 \text{ torr} = 24.4 \text{ torr}$ and

$P_{C_8H_{18}} = \chi_{C_8H_{18}} P^{\circ}_{C_8H_{18}} = 0.467307 \times 10.9 \text{ torr} = 5.09 \text{ torr}$

Check: The units (torr) are correct. The magnitude of the answers (24 and 5 torr) seems reasonable because we expect a drop in half from the pure vapor pressures because we have roughly a 50:50 mole ratio of the two components.

(b)   **Find:** P_{Total}
Conceptual Plan: $P_{C_7H_{16}}, P_{C_8H_{18}} \rightarrow P_{\text{Total}}$

$$P_{\text{Total}} = P_{C_7H_{16}} + P_{C_8H_{18}}$$

Solution: $P_{\text{Total}} = P_{C_7H_{16}} + P_{C_8H_{18}} = 24.4 \text{ torr} + 5.09 \text{ torr} = 29.5 \text{ torr}$

Check: The units (torr) are correct. The magnitude of the answer (30 torr) seems reasonable considering the two pressures.

(c)   **Find:** mass percent composition of the gas phase
Conceptual Plan: Because $n \, \alpha \, P$ and we are calculating a mass percent, which is a ratio of masses, we can simply convert 1 torr to 1 mole so
$P_{C_7H_{16}}, P_{C_8H_{18}} \rightarrow n_{C_7H_{16}}, n_{C_8H_{18}}$ then $mol_{C_7H_{16}} \rightarrow g_{C_7H_{16}}$ and $mol_{C_8H_{18}} \rightarrow g_{C_8H_{18}}$

$$\frac{100.20 \text{ g } C_7H_{16}}{1 \text{ mol } C_7H_{16}} \qquad\qquad \frac{114.22 \text{ g } C_8H_{18}}{1 \text{ mol } C_8H_{18}}$$

then $g_{C_7H_{16}}, g_{C_8H_{18}} \rightarrow$ mass percents

$$\text{mass percent} = \frac{\text{mass solute}}{\text{mass solution}} \times 100\%$$

Solution: so $n_{C_7H_{16}} = 24.4$ mol and $n_{C_8H_{18}} = 5.09$ mol then

$24.4 \text{ mol } C_7H_{16} \times \dfrac{100.20 \text{ g } C_7H_{16}}{1 \text{ mol } C_7H_{16}} = 2444.88 \text{ g } C_7H_{16}$ and

$5.09 \text{ mol } C_8H_{18} \times \dfrac{114.22 \text{ g } C_8H_{18}}{1 \text{ mol } C_8H_{18}} = 581.380 \text{ g } C_8H_{18}$ then

$$\text{mass percent} = \frac{\text{mass solute}}{\text{mass solution}} \times 100\% = \frac{2444.88 \text{ g } C_7H_{16}}{2444.88 \text{ g } C_7H_{16} + 581.380 \text{ g } C_8H_{18}} \times 100\% =$$

80.8% by mass C_7H_{16}
then $100\% - 80.8\% = 19.2\%$ by mass C_8H_{18}

Check: The units (mol %) are correct. The magnitude of the answers (81% and 19%) seems reasonable considering the two pressures.

(d)   The two mass percents are different because the vapor is richer in the more volatile component (the lighter molecule).

12.75 **Given:** 4.08 g of chloroform $(CHCl_3)$ and 9.29 g of acetone (CH_3COCH_3); at 35 °C, $P°_{CHCl_3} = 295$ torr; $P°_{CH_3COCH_3} = 332$ torr; assume ideal behavior; $P_{Total\ measured} = 312$ torr
Find: P_{CHCl_3}, $P_{CH_3COCH_3}$, P_{Total}, and if the solution is ideal.
Conceptual Plan: $g_{CHCl_3} \rightarrow mol_{CHCl_3}$ and $g_{CH_3COCH_3} \rightarrow mol_{CH_3COCH_3}$ then

$$\frac{1\ mol\ CHCl_3}{119.38\ g\ CHCl_3} \qquad \frac{1\ mol\ CH_3COCH_3}{58.08\ g\ CH_3COCH_3}$$

mol_{CHCl_3}, $mol_{CH_3COCH_3} \rightarrow \chi_{CHCl_3}$, $\chi_{CH_3COCH_3}$ then χ_{CHCl_3}, $P°_{CHCl_3} \rightarrow P_{CHCl_3}$ and

$$\chi_{CHCl_3} = \frac{amount\ CHCl_3\ (in\ moles)}{total\ amount\ (in\ moles)} \qquad \chi_{CH_3COCH_3} = 1 - \chi_{CHCl_3} \qquad P_{CHCl_3} = \chi_{CHCl_3}P°_{CHCl_3}$$

$\chi_{CH_3COCH_3}$, $P°_{CH_3COCH_3} \rightarrow P_{CH_3COCH_3}$ then P_{CHCl_3}, $P_{CH_3COCH_3} \rightarrow P_{Total}$ then compare values

$$P_{CH_3COCH_3} = \chi_{CH_3COCH_3}P°_{CH_3COCH_3} \qquad\qquad P_{Total} = P_{CHCl_3} + P_{CH_3COCH_3}$$

Solution:

$$4.08\ g\ CHCl_3 \times \frac{1\ mol\ CHCl_3}{119.38\ g\ CHCl_3} = 0.034\underline{1}766\ mol\ CHCl_3\ and$$

$$9.29\ g\ CH_3COCH_3 \times \frac{1\ mol\ CH_3COCH_3}{58.08\ g\ CH_3COCH_3} = 0.15\underline{9}952\ mol\ CH_3COCH_3\ then$$

$$\chi_{CHCl_3} = \frac{amount\ CHCl_3\ (in\ moles)}{total\ amount\ (in\ moles)} = \frac{0.034\underline{1}766\ mol}{0.034\underline{1}766\ mol + 0.15\underline{9}952\ mol} = 0.17\underline{6}052\ and$$

$\chi_{CH_3COCH_3} = 1 - \chi_{CHCl_3} = 1 - 0.17\underline{6}052 = 0.82\underline{3}948$ then
$P_{CHCl_3} = \chi_{CHCl_3}P°_{CHCl_3} = 0.17\underline{6}052 \times 295$ torr $= 51.9$ torr and
$P_{CH_3COCH_3} = \chi_{CH_3COCH_3}P°_{CH_3COCH_3} = 0.82\underline{3}948 \times 332$ torr $= 274$ torr then
$P_{Total} = P_{CHCl_3} + P_{CH_3COCH_3} = 51.9$ torr $+ 274$ torr $= 326$ torr
Because 326 torr $\neq$ 312 torr, the solution is not behaving ideally. The chloroform–acetone interactions are stronger than the chloroform–chloroform and acetone–acetone interactions.

Check: The units (torr) are correct. The magnitude of the answer (326) seems reasonable because each is a fraction of the pure vapor pressure. We are not surprised that the solution is not ideal because the types of bonds in the two molecules are very different.

Freezing Point Depression, Boiling Point Elevation, and Osmosis

12.77 **Given:** 55.8 g of glucose $(C_6H_{12}O_6)$ in 455 g water **Find:** T_f and T_b **Other:** $K_f = 1.86\ °C/m$; $K_b = 0.512\ °C/m$;
Conceptual Plan: $g_{H_2O} \rightarrow kg_{H_2O}$ and $g_{C_6H_{12}O_6} \rightarrow mol_{C_6H_{12}O_6}$ then $mol_{C_6H_{12}O_6}$, $kg_{H_2O} \rightarrow m$

$$\frac{1\ kg}{1000\ g} \qquad \frac{1\ mol\ C_6H_{12}O_6}{180.16\ g\ C_6H_{12}O_6} \qquad m = \frac{amount\ solute\ (moles)}{mass\ solvent\ (kg)}$$

m, $K_f \rightarrow \Delta T_f \rightarrow T_f$ and m, $K_b \rightarrow \Delta T_b \rightarrow T_b$

$$\Delta T_f = K_f \times m \qquad T_f = T°_f - \Delta T_f \qquad \Delta T_b = K_b \times m \qquad \Delta T_b = T_b - T°_b$$

Solution: $455\ g \times \dfrac{1\ kg}{1000\ g} = 0.455\ kg$ and $55.8\ g\ C_6H_{12}O_6 \times \dfrac{1\ mol\ C_6H_{12}O_6}{180.16\ g\ C_6H_{12}O_6} = 0.30\underline{9}725\ mol\ C_6H_{12}O_6$ then

$$m = \frac{amount\ solute\ (moles)}{mass\ solvent\ (kg)} = \frac{0.30\underline{9}725\ mol\ C_6H_{12}O_6}{0.455\ kg} = 0.68\underline{0}714\ m\ then$$

$$\Delta T_f = K_f \times m = 1.86\ \frac{°C}{m} \times 0.68\underline{0}714\ m = 1.27\ °C\ then\ T_f = T°_f - \Delta T_f = 0.00\ °C - 1.27\ °C = -1.27\ °C\ and$$

$$\Delta T_b = K_b \times m = 0.512\ \frac{°C}{m} \times 0.68\underline{0}714\ m = 0.349\ °C\ and\ \Delta T_b = T_b - T°_b\ so$$

$$T_b = T°_b + \Delta T_b = 100.000\ °C + 0.349\ °C = 100.349\ °C$$

Check: The units (°C) are correct. The magnitude of the answers (100.3) seems reasonable because the molality is ~2/3. The shift in boiling point is less than the shift in freezing point because the constant for boiling is smaller than the constant for freezing.

12.79   **Given:** 10.0 g of naphthalene $(C_{10}H_8)$ in 100.0 mL benzene (C_6H_6) **Find:** T_f and T_b
Other: $d(\text{benzene}) = 0.877 \text{ g/cm}^3$; $K_f = 5.12 \text{ °C/}m$; $K_b = 2.53 \text{ °C/}m$; $T_f^\circ = 5.5 \text{ °C}$; $T_b^\circ = 80.1 \text{ °C}$
Conceptual Plan: $mL_{C_6H_6} \rightarrow g_{C_6H_6} \rightarrow kg_{C_6H_6}$ and $g_{C_{10}H_8} \rightarrow mol_{C_{10}H_8}$ then $mol_{C_{10}H_8}, kg_{C_6H_6} \rightarrow m$

$$\frac{0.877 \text{ g}}{1 \text{ cm}^3} \quad \frac{1 \text{ kg}}{1000 \text{ g}} \qquad \frac{1 \text{ mol } C_{10}H_8}{128.2 \text{ g } C_{10}H_8} \qquad m = \frac{\text{amount solute (moles)}}{\text{mass solvent (kg)}}$$

$m, K_f \rightarrow \Delta T_f \rightarrow T_f$ and $m, K_b \rightarrow \Delta T_b \rightarrow T_b$

$$\Delta T_f = K_f \times m \quad T_f = T_f^\circ - \Delta T_f \quad \Delta T_b = K_b \times m \quad \Delta T_b = T_b - T_b^\circ$$

Solution: $100.0 \text{ cm}^3 \times \dfrac{0.877 \text{ g}}{1 \text{ cm}^3} \times \dfrac{1 \text{ kg}}{1000 \text{ g}} = 0.0877 \text{ kg}$ and

$10.0 \text{ g } C_{10}H_8 \times \dfrac{1 \text{ mol } C_{10}H_8}{128.2 \text{ g } C_{10}H_8} = 0.07802503 \text{ mol } C_{10}H_8$ then

$m = \dfrac{\text{amount solute (moles)}}{\text{mass solvent (kg)}} = \dfrac{0.07802503 \text{ mol } C_{10}H_8}{0.0877 \text{ kg}} = 0.8896811 \, m$ then

$\Delta T_f = K_f \times m = 5.12 \dfrac{\text{°C}}{m} \times 0.8896811 \, m = 4.56 \text{ °C}$ then $T_f = T_f^\circ - \Delta T_f = 5.5 \text{ °C} - 4.56 \text{ °C} = 0.94 \text{ °C} = 0.9 \text{ °C}$

and $\Delta T_b = K_b \times m = 2.53 \dfrac{\text{°C}}{m} \times 0.8896811 \, m = 2.25 \text{ °C}$ and $\Delta T_b = T_b - T_b^\circ$ so

$T_b = T_b^\circ + \Delta T_b = 80.1 \text{ °C} + 2.25 \text{ °C} = 82.35 \text{ °C} = 82.4 \text{ °C}$

Check: The units (°C) are correct. The magnitude of the answers (82.4) seems reasonable because the molality is almost 1. Because the constants are larger for benzene than for water, we expect larger temperature shifts. The shift in boiling point is less than the shift in freezing point because the constant is smaller for freezing.

12.81   **Given:** 17.5 g of unknown nonelectrolyte in 100.0 g water, $T_f = -1.8 \text{ °C}$ **Find:** $\mathcal{M}$
Other: $K_f = 1.86 \text{ °C/}m$
Conceptual Plan: $g_{H_2O} \rightarrow kg_{H_2O}$ and $T_f \rightarrow \Delta T_f$ then $\Delta T_f, K_f \rightarrow m$ then $m, kg_{H_2O} \rightarrow mol_{Unk}$

$$\frac{1 \text{ kg}}{1000 \text{ g}} \qquad T_f = T_f^\circ - \Delta T_f \qquad \Delta T_f = K_f \times m \qquad m = \frac{\text{amount solute (moles)}}{\text{mass solvent (kg)}}$$

then $g_{Unk}, mol_{Unk} \rightarrow \mathcal{M}$

$$\mathcal{M} = \frac{g_{Unk}}{mol_{Unk}}$$

Solution: $100.0 \text{ g} \times \dfrac{1 \text{ kg}}{1000 \text{ g}} = 0.1000 \text{ kg}$ and $T_f = T_f^\circ - \Delta T_f$ so

$\Delta T_f = T_f^\circ - T_f = 0.00 \text{ °C} - (-1.8 \text{ °C}) = +1.8 \text{ °C} \; \Delta T_f = K_f \times m$ Rearrange to solve for m.

$m = \dfrac{\Delta T_f}{K_f} = \dfrac{1.8 \text{ °C}}{1.86 \dfrac{\text{°C}}{m}} = 0.96774 \, m$ then $m = \dfrac{\text{amount solute (moles)}}{\text{mass solvent (kg)}}$ so

$mol_{Unk} = m_{Unk} \times kg_{H_2O} = 0.96774 \dfrac{\text{mol Unk}}{\text{kg}} \times 0.1000 \text{ kg} = 0.096774 \text{ mol Unk}$ then

$\mathcal{M} = \dfrac{g_{Unk}}{mol_{Unk}} = \dfrac{17.5 \text{ g}}{0.096774 \text{ mol}} = 180 \dfrac{\text{g}}{\text{mol}} = 1.8 \times 10^2 \dfrac{\text{g}}{\text{mol}}$

Check: The units (g/mol) are correct. The magnitude of the answer (180 g/mol) seems reasonable because the molality is ~0.1 and we have ~18 g. It is a reasonable molecular weight for a solid or liquid.

12.83   **Given:** 24.6 g of glycerin $(C_3H_8O_3)$ in 250.0 mL of solution at 298 K   **Find:** Π
Conceptual Plan: $mL_{soln} \rightarrow L_{soln}$ and $g_{C_3H_8O_3} \rightarrow mol_{C_3H_8O_3}$ then $mol_{C_3H_8O_3}, L_{soln} \rightarrow M$ then

$$\frac{1 \text{ L}}{1000 \text{ mL}} \qquad \frac{1 \text{ mol } C_3H_8O_3}{92.09 \text{ g } C_3H_8O_3} \qquad M = \frac{\text{amount solute (moles)}}{\text{volume solution (L)}}$$

$M, T \rightarrow \Pi$

$$\Pi = MRT$$

Solution:

$250.0 \text{ mL} \times \dfrac{1 \text{ L}}{1000 \text{ mL}} = 0.2500 \text{ L}$ and $24.6 \text{ g } C_3H_8O_3 \times \dfrac{1 \text{ mol } C_3H_8O_3}{92.09 \text{ g } C_3H_8O_3} = 0.267130 \text{ mol } C_3H_8O_3$ then

$$M = \frac{\text{amount solute (moles)}}{\text{volume solution (L)}} = \frac{0.26\underline{7}130 \text{ mol } C_3H_8O_3}{0.2500 \text{ L}} = 1.0\underline{6}852 \text{ M then}$$

$$\Pi = MRT = 1.0\underline{6}852\frac{\text{mol}}{\cancel{L}} \times 0.08206\frac{\cancel{L} \cdot \text{atm}}{\cancel{K} \cdot \cancel{\text{mol}}} \times 298 \cancel{K} = 26.1 \text{ atm}$$

Check: The units (atm) are correct. The magnitude of the answer (26 atm) seems reasonable because the molarity is ~ 1.

12.85 **Given:** 27.55 mg unknown protein in 25.0 mL solution; $\Pi = 3.22$ torr at 25 °C **Find:** $\mathcal{M}_{\text{unknown protein}}$

Conceptual Plan: °C $\rightarrow$ K and torr $\rightarrow$ atm then $\Pi, T \rightarrow$ M then mL$_{\text{soln}} \rightarrow$ L$_{\text{soln}}$ then

$$K = °C + 273.15 \qquad \frac{1 \text{ atm}}{760 \text{ torr}} \qquad \Pi = MRT \qquad \frac{1 \text{ L}}{1000 \text{ mL}}$$

L$_{\text{soln}}$, M $\rightarrow$ mol$_{\text{unknown protein}}$ and mg $\rightarrow$ g then g$_{\text{unknown protein}}$, mol$_{\text{unknown protein}} \rightarrow \mathcal{M}_{\text{unknown protein}}$

$$M = \frac{\text{amount solute (moles)}}{\text{volume solution (L)}} \qquad \frac{1 \text{ g}}{1000 \text{ mg}} \qquad \mathcal{M} = \frac{\text{g}_{\text{unknown protein}}}{\text{mol}_{\text{unknown protein}}}$$

Solution: $25 °C + 273.15 = 298$ K and $3.22 \text{ torr} \times \frac{1 \text{ atm}}{760 \text{ torr}} = 0.0042\underline{3}684$ atm then solve $\Pi = MRT$ for M.

$$M = \frac{\Pi}{RT} = \frac{0.0042\underline{3}684 \text{ atm}}{0.08206\dfrac{\text{L} \cdot \text{atm}}{K \cdot \text{mol}} \times 298 \, K} = 1.7\underline{3}258 \times 10^{-4}\frac{\text{mol}}{\text{L}} \text{ then } 25.0 \text{ mL} \times \frac{1 \text{ L}}{1000 \text{ mL}} = 0.0250 \text{ L then}$$

$$M = \frac{\text{amount solute (moles)}}{\text{volume solution (L)}} \quad \text{Rearrange to solve for mol}_{\text{unknown protein}}.$$

$$\text{mol}_{\text{unknown protein}} = M \times L = 1.7\underline{3}258 \times 10^{-4}\frac{\text{mol}}{\cancel{L}} \times 0.0250 \, \cancel{L} = 4.3\underline{3}146 \times 10^{-6} \text{ mol and}$$

$$27.55 \text{ mg} \times \frac{1 \text{ g}}{1000 \text{ mg}} = 0.02755 \text{ g then } \mathcal{M} = \frac{\text{g}_{\text{unknown protein}}}{\text{mol}_{\text{unknown protein}}} = \frac{0.02755 \text{ g}}{4.3\underline{3}146 \times 10^{-6} \text{ mol}} = 6.36 \times 10^3\frac{\text{g}}{\text{mol}}$$

Check: The units (g/mol) are correct. The magnitude of the answer (6400 g/mol) seems reasonable for a large biological molecule. A small amount of material is put into 0.025 L; so the concentration is very small, and the molecular weight is large.

12.87 (a) **Given:** 0.100 m of K_2S, completely dissociated **Find:** T_f, T_b

Other: $K_f = 1.86 °C/m$; $K_b = 0.512 °C/m$

Conceptual Plan: $m, i, K_f \rightarrow \Delta T_f$ then $\Delta T_f \rightarrow T_f$ and $m, i, K_b \rightarrow \Delta T_b$ then $\Delta T_b \rightarrow T_b$

$$\Delta T_f = K_f \times i \times m_{i=3} \qquad T_f = T_f° - \Delta T_f \qquad \Delta T_b = K_b \times i \times m_{i=3} \qquad T_b = T_b° + \Delta T_b$$

Solution: $\Delta T_f = K_f \times i \times m = 1.86 \dfrac{°C}{\cancel{m}} \times 3 \times 0.100 \, \cancel{m} = 0.558 °C$ then

$T_f = T_f° - \Delta T_f = 0.000 °C - 0.558 °C = -0.558 °C$ and

$\Delta T_b = K_b \times i \times m = 0.512 \dfrac{°C}{\cancel{m}} \times 3 \times 0.100 \, \cancel{m} = 0.154 °C$ then

$T_b = T_b° + \Delta T_b = 100.000 °C + 0.154 °C = 100.154 °C$

Check: The units (°C) are correct. The magnitude of the answers ($-0.6 °C$ and $100.2 °C$) seems reasonable because the molality of the particles is 0.3. The shift in boiling point is less than the shift in freezing point because the constant for boiling is larger than the constant for freezing.

(b) **Given:** 21.5 g $CuCl_2$ in 4.50×10^2 g water, completely dissociated **Find:** T_f, T_b

Other: $K_f = 1.86 °C/m$; $K_b = 0.512 °C/m$

Conceptual Plan: $g_{H_2O} \rightarrow kg_{H_2O}$ and $g_{CuCl_2} \rightarrow mol_{CuCl_2}$ then $mol_{CuCl_2}, kg_{H_2O} \rightarrow m$

$$\frac{1 \text{ kg}}{1000 \text{ g}} \qquad \frac{1 \text{ mol } CuCl_2}{134.45 \text{ g } CuCl_2} \qquad m = \frac{\text{amount solute (moles)}}{\text{mass solvent (kg)}}$$

$m, i, K_f \rightarrow \Delta T_f \rightarrow T_f$ and $m, i, K_b \rightarrow \Delta T_b \rightarrow T_b$

$$\Delta T_f = K_f \times i \times m_{i=3} \quad T_f = T_f° - \Delta T_f \quad \Delta T_b = K_b \times i \times m_{i=3} \quad T_b = T_b° + \Delta T_b$$

Solution:

$$4.5 \times 10^2 \, \cancel{g} \times \frac{1 \text{ kg}}{1000 \, \cancel{g}} = 0.450 \text{ kg and } 21.5 \text{ g } \cancel{CuCl_2} \times \frac{1 \text{ mol } CuCl_2}{134.45 \text{ g } \cancel{CuCl_2}} = 0.15\underline{9}911 \text{ mol } CuCl_2 \text{ then}$$

$$m = \frac{\text{amount solute (moles)}}{\text{mass solvent (kg)}} = \frac{0.159911 \text{ mol CuCl}_2}{0.450 \text{ kg}} = 0.355358 \, m \text{ then}$$

$$\Delta T_f = K_f \times i \times m = 1.86\frac{^{\circ}\text{C}}{m} \times 3 \times 0.355358 \, m = 1.98 \,^{\circ}\text{C then}$$

$$T_f = T_f^{\circ} - \Delta T_f = 0.000 \,^{\circ}\text{C} - 1.98 \,^{\circ}\text{C} = -1.98 \,^{\circ}\text{C and}$$

$$\Delta T_b = K_b \times i \times m = 0.512\frac{^{\circ}\text{C}}{m} \times 3 \times 0.355358 \, m = 0.546 \,^{\circ}\text{C then}$$

$$T_b = T_b^{\circ} + \Delta T_b = 100.000 \,^{\circ}\text{C} + 0.546 \,^{\circ}\text{C} = 100.546 \,^{\circ}\text{C}$$

Check: The units ($^{\circ}$C) are correct. The magnitude of the answers ($-2\,^{\circ}$C and $100.5\,^{\circ}$C) seems reasonable because the molality of the particles is $\sim$1. The shift in boiling point is less than the shift in freezing point because the constant for boiling is larger than the constant for freezing.

(c) **Given:** 5.5% by mass NaNO$_3$, completely dissociated   **Find:** T_f, T_b
Other: $K_f = 1.86 \,^{\circ}\text{C}/m$; $K_b = 0.512 \,^{\circ}\text{C}/m$
Conceptual Plan: Percent by mass $\rightarrow$ g$_{\text{NaNO}_3}$, g$_{\text{H}_2\text{O}}$ then g$_{\text{H}_2\text{O}} \rightarrow$ kg$_{\text{H}_2\text{O}}$ and g$_{\text{NaNO}_3} \rightarrow$ mol$_{\text{NaNO}_3}$

$$\text{mass percent} = \frac{\text{mass solute}}{\text{mass solution}} \times 100\% \qquad \frac{1 \text{ kg}}{1000 \text{ g}} \qquad \frac{1 \text{ mol NaNO}_3}{85.00 \text{ g NaNO}_3}$$

then mol$_{\text{NaNO}_3}$**, kg**$_{\text{H}_2\text{O}}$ $\rightarrow$ m **then** m, i, K_f $\rightarrow$ $\Delta T_f \rightarrow T_f$ **and** m, i, K_b $\rightarrow$ ΔT_b $\rightarrow$ T_b

$$m = \frac{\text{amount solute (moles)}}{\text{mass solvent (kg)}} \qquad \Delta T_f = K_f \times i \times m_{i=2} \quad T_f = T_f^{\circ} - \Delta T_f \qquad \Delta T_b = K_b \times i \times m_{i=2} \, T_b = T_b^{\circ} + \Delta T_b$$

Solution: mass percent $= \dfrac{\text{mass solute}}{\text{mass solution}} \times 100\%$, so 5.5% by mass NaNO$_3$ means 5.5 g NaNO$_3$ and

$$100.0 \text{ g} - 5.5 \text{ g} = 94.5 \text{ g water. Then } 94.5 \text{ g} \times \frac{1 \text{ kg}}{1000 \text{ g}} = 0.0945 \text{ kg and}$$

$$5.5 \text{ g NaNO}_3 \times \frac{1 \text{ mol NaNO}_3}{85.00 \text{ g NaNO}_3} = 0.064706 \text{ mol NaNO}_3 \text{ then}$$

$$m = \frac{\text{amount solute (moles)}}{\text{mass solvent (kg)}} = \frac{0.064706 \text{ mol NaNO}_3}{0.0945 \text{ kg}} = 0.68472 \, m \text{ then}$$

$$\Delta T_f = K_f \times i \times m = 1.86\frac{^{\circ}\text{C}}{m} \times 2 \times 0.68472 \, m = 2.5 \,^{\circ}\text{C then}$$

$$T_f = T_f^{\circ} - \Delta T_f = 0.000 \,^{\circ}\text{C} - 2.5 \,^{\circ}\text{C} = -2.5 \,^{\circ}\text{C and}$$

$$\Delta T_b = K_b \times i \times m = 0.512\frac{^{\circ}\text{C}}{m} \times 2 \times 0.68472 \, m = 0.70 \,^{\circ}\text{C then}$$

$$T_b = T_b^{\circ} + \Delta T_b = 100.000 \,^{\circ}\text{C} + 0.70 \,^{\circ}\text{C} = 100.70 \,^{\circ}\text{C}$$

Check: The units ($^{\circ}$C) are correct. The magnitude of the answers ($-2.5\,^{\circ}$C and $100.7\,^{\circ}$C) seems reasonable because the molality of the particles is $\sim$1. The shift in boiling point is less than the shift in freezing point because the constant for boiling is larger than the constant for freezing.

12.89 **Given:** NaCl complete dissociation; 1.0 L water and $T_f = -10.0 \,^{\circ}\text{C}$ **Find:** mass of NaCl
Other: $K_f = 1.86 \,^{\circ}\text{C}/m$; $d(\text{water}) = 1.0 \text{ g/mL}$
Conceptual Plan:
$T_f \rightarrow \Delta T_f$ **then** $\Delta T_f, i, K_f \rightarrow m$ **then** $\text{L}_{\text{H}_2\text{O}} \rightarrow \text{mL}_{\text{H}_2\text{O}} \rightarrow \text{g}_{\text{H}_2\text{O}} \rightarrow \text{kg}_{\text{H}_2\text{O}}$ **then** $m, \text{kg}_{\text{H}_2\text{O}} \rightarrow \text{mol}_{\text{NaCl}}$

$$T_f = T_f^{\circ} - \Delta T_f \qquad \Delta T_f = K_f \times i \times m \qquad \frac{1000 \text{ mL}}{1 \text{ L}} \quad \frac{1.0 \text{ g}}{1 \text{ mL}} \quad \frac{1 \text{ kg}}{1000 \text{ g}} \qquad m = \frac{\text{amount solute (moles)}}{\text{mass solvent (kg)}}$$

then mol$_{\text{NaCl}} \rightarrow$ **g**$_{\text{NaCl}}$

$$\frac{58.44 \text{ g NaCl}}{1 \text{ mol NaCl}}$$

Solution: $T_f = T_f^{\circ} - \Delta T_f$ so $\Delta T_f = T_f^{\circ} - T_f = 0.0 \,^{\circ}\text{C} - 10.0 \,^{\circ}\text{C} = -10.0 \,^{\circ}\text{C}$ then $\Delta T_f = K_f \times i \times m$

Rearrange to solve for m. Because the salt completely dissolves, $i = 2$.

$$m = \frac{\Delta T_f}{K_f \times i} = \frac{10.0 \,^{\circ}\text{C}}{1.86\dfrac{^{\circ}\text{C}}{m} \times 2} = 2.688172 \, m \text{ NaCl then } 1.0 \text{ L} \times \frac{1000 \text{ mL}}{1 \text{ L}} \times \frac{1.0 \text{ g}}{1 \text{ mL}} \times \frac{1 \text{ kg}}{1000 \text{ g}} = 1.0 \text{ kg then}$$

$$m = \frac{\text{amount solute (moles)}}{\text{mass solvent (kg)}} \text{ so } \text{mol}_{NaCl} = m \times \text{kg}_{H_2O} = 2.6\underline{8}8172 \frac{\text{mol NaCl}}{\text{kg}_{H_2O}} \times 1.0 \text{ kg}_{H_2O} = 2.6\underline{8}8172 \text{ mol NaCl}$$

$$\text{then } 2.6\underline{8}8172 \text{ mol NaCl} \times \frac{58.44 \text{ g NaCl}}{1 \text{ mol NaCl}} = 1\underline{5}7.097 \text{ g NaCl} = 160 \text{ g NaCl}$$

Check: The units (g) are correct. The magnitude of the answer (160 g) seems reasonable because the temperature change is moderate and NaCl has a low formula mass.

12.91   (a)   **Given:** 0.100 m of $FeCl_3$ **Find:** T_f **Other:** $K_f = 1.86 \,°C/m$; $i_{measured} = 3.4$
Conceptual Plan: $m, i, K_f \rightarrow \Delta T_f$ then $\Delta T_f \rightarrow T_f$

$$\Delta T_f = K_f \times i \times m \quad T_f = T_f° - \Delta T_f$$

Solution: $\Delta T_f = K_f \times i \times m = 1.86 \dfrac{°C}{m} \times 3.4 \times 0.100 \, m = 0.632 \,°C$ then

$$T_f = T_f° - \Delta T_f = 0.000 \,°C - 0.632 \,°C = -0.632 \,°C$$

Check: The units (°C) are correct. The magnitude of the answer ($-0.6 \,°C$) seems reasonable because the theoretical molality of the particles is 0.4.

(b)   **Given:** 0.085 M of K_2SO_4 at 298 K   **Find:** Π **Other:** $i_{measured} = 2.6$
Conceptual Plan: $M, i, T \rightarrow \Pi$

$$\Pi = i \times MRT$$

Solution: $\Pi = i \times MRT = 2.6 \times 0.085 \dfrac{\text{mol}}{L} \times 0.08206 \dfrac{L \cdot \text{atm}}{K \cdot \text{mol}} \times 298 \, K = 5.4 \text{ atm}$

Check: The units (atm) are correct. The magnitude of the answer (5 atm) seems reasonable because the molarity of particles is $\sim 0.2 \, m$.

(c)   **Given:** 1.22% by mass $MgCl_2$ **Find:** T_b **Other:** $K_b = 0.512 \,°C/m$; $i_{measured} = 2.7$
Conceptual Plan: percent by mass $\rightarrow g_{MgCl_2}, g_{H_2O}$ **then** $g_{H_2O} \rightarrow kg_{H_2O}$ **and** $g_{MgCl_2} \rightarrow mol_{MgCl_2}$ **then**

$$\text{mass percent} = \frac{\text{mass solute}}{\text{mass solution}} \times 100\% \qquad \frac{1 \text{ kg}}{1000 \text{ g}} \qquad \frac{1 \text{ mol } MgCl_2}{95.21 \text{ g } MgCl_2}$$

$mol_{MgCl_2}, kg_{H_2O} \rightarrow m$ **then** $m, i, K_b \rightarrow \Delta T_b \rightarrow T_b$

$$m = \frac{\text{amount solute (moles)}}{\text{mass solvent (kg)}} \qquad \Delta T_b = K_b \times i \times m \ T_b = T_b° + \Delta T_b$$

Solution: $\text{mass percent} = \dfrac{\text{mass solute}}{\text{mass solution}} \times 100\%$ so 1.22% by mass $MgCl_2$ means 1.22 g $MgCl_2$ and

$$100.00 \text{ g} - 1.22 \text{ g} = 98.78 \text{ g water then } 98.78 \text{ g} \times \frac{1 \text{ kg}}{1000 \text{ g}} = 0.09878 \text{ kg and}$$

$$1.22 \text{ g } MgCl_2 \times \frac{1 \text{ mol } MgCl_2}{95.21 \text{ g } MgCl_2} = 0.012\underline{8}138 \text{ mol } MgCl_2 \text{ then}$$

$$m = \frac{\text{amount solute (moles)}}{\text{mass solvent (kg)}} = \frac{0.012\underline{8}138 \text{ mol } MgCl_2}{0.09878 \text{ kg}} = 0.12\underline{9}721 \, m \text{ then}$$

$$\Delta T_b = K_b \times i \times m = 0.512 \frac{°C}{m} \times 2.7 \times 0.12\underline{9}721 \, m = 0.18 \,°C \text{ then}$$

$$T_b = T_b° - \Delta T_b = 100.000 \,°C + 0.18 \,°C = 100.18 \,°C$$

Check: The units (°C) are correct. The magnitude of the answer (100.2 °C) seems reasonable because the molality of the particles is $\sim 1/3$.

12.93   **Given:** 1.2 m MX_2 in water and $T_b = 101.4 \,°C$ **Find:** i
Other: $K_b = 0.512 \,°C/m$
Conceptual Plan: $T_b \rightarrow \Delta T_b$ **then** $m, \Delta T_b, K_b \rightarrow i$

$$\Delta T_b = T_b - T_b° \qquad \Delta T_b = K_b \times i \times m$$

Solution: $\Delta T_b = T_b - T_b° = 101.4 \,°C - 100.0 \,°C = 1.4 \,°C$ then $\Delta T_b = K_b \times i \times m$

Rearrange to solve for i. $i = \dfrac{\Delta T_b}{K_b \times m} = \dfrac{1.4\,°C}{0.512\dfrac{°C}{m} \times 1.2\,m} = 2.2\underline{7}865 = 2.3$

Check: The units (none) are correct. The magnitude of the answer (2.3) seems reasonable because the formula of the salt is MX_2, where $i = 3$ if it completely dissociated.

12.95   **Given:** 0.100 M of ionic solution, $\Pi = 8.3$ atm at 25 °C   **Find:** $i_{measured}$
Conceptual Plan: °C → K then $\Pi, M, T → i$

$$K = °C + 273.15 \qquad \Pi = i \times MRT$$

Solution: 25 °C + 273.15 = 298 K then $\Pi = i \times MRT$. Rearrange to solve for i.

$$i = \frac{\Pi}{MRT} = \frac{8.3\ \text{atm}}{0.100\dfrac{mol}{L} \times 0.08206\dfrac{L \cdot atm}{K \cdot mol} \times 298\ K} = 3.4$$

Check: The units (none) are correct. The magnitude of the answer (3) seems reasonable for an ionic solution with a high osmotic pressure.

12.97   **Given:** 5.50 % NaCl by mass in water at 25 °C   **Find:** P_{H_2O} **Other:** $P°_{H_2O} = 23.78$ torr
Conceptual Plan: % NaCl by mass → g_{NaCl}, g_{H_2O} then $g_{NaCl} → mol_{NaCl}$ and

$$\frac{5.50\ \text{g NaCl}}{100\ \text{g (NaCl} + H_2O)} \qquad \frac{1\ \text{mol NaCl}}{58.44\ \text{g NaCl}}$$

$g_{H_2O} → mol_{H_2O}$ then $mol_{NaCl}, mol_{H_2O} → \chi_{H_2O}$ then $\chi_{H_2O}, P°_{H_2O} → P_{H_2O}$

$$\frac{1\ \text{mol } H_2O}{18.01\ \text{g } H_2O} \qquad \chi = \frac{\text{amount solvent (in moles)}}{\text{total amount of solute and solvent particles (in moles)}} \qquad P_{solution} = \chi_{solvent} P°_{solvent}$$

Solution: $\dfrac{5.50\ \text{g NaCl}}{100\ \text{g (NaCl} + H_2O)}$ means 5.50 g NaCl and $(100\ \text{g} - 5.50\ \text{g}) = 94.5$ g H_2O then

$5.50\ \text{g NaCl} \times \dfrac{1\ \text{mol NaCl}}{58.44\ \text{g NaCl}} = 0.094\underline{1}136$ mol NaCl and $94.5\ \text{g } H_2O \times \dfrac{1\ \text{mol } H_2O}{18.01\ \text{g } H_2O} = 5.2\underline{4}708$ mol H_2O then

number of moles of solute $= i_{NaCl} \times n_{NaCl}$ so $\chi_{solv} = \dfrac{\text{amount solvent (in moles)}}{\text{total amount solute and solvent particles (in moles)}} =$

$\dfrac{5.2\underline{4}708\ \text{mol}}{5.2\underline{4}708\ \text{mol} + 2(0.094\underline{1}136\ \text{mol})} = 0.96\underline{5}36$ then $P_{soln} = \chi_{solv} P°_{solv} = 0.96\underline{5}36 \times 23.78$ torr $= 23.0$ torr

Check: The units (torr) are correct. The magnitude of the answer (23 torr) seems reasonable because it is a drop from the pure vapor pressure. Only a fraction of a mole of NaCl is added, so the pressure will not drop much.

Cumulative Problems

12.99   Chloroform is polar and has stronger solute–solvent interactions than does nonpolar carbon tetrachloride.

12.101   **Given:** $KClO_4$: lattice energy $= -599$ kJ/mol, $\Delta H_{hydration} = -548$ kJ/mol; 10.0 g $KClO_4$ in 100.00 mL solution
Find: ΔH_{soln} and ΔT **Other:** $C_s = 4.05$ J/g °C; $d = 1.05$ g/mL
Conceptual Plan: Lattice energy, $\Delta H_{hydration} → \Delta H_{soln}$ and g → mol then mol, $\Delta H_{soln} → q(kJ) → q(J)$

$$\Delta H_{soln} = \Delta H_{solute} + \Delta H_{hydration}\ where\ \Delta H_{solute} = -\Delta H_{lattice} \qquad \frac{1\ \text{mol}}{138.55\ \text{g}} \qquad q = n\,\Delta H_{soln} \qquad \frac{1000\ \text{J}}{1\ \text{kJ}}$$

then $mL_{soln} → g_{soln}$ then $q, g_{soln}, C_s → \Delta T$

$$\frac{1.05\ \text{g}}{1\ \text{mL}} \qquad q = mC_s\Delta T$$

Solution: $\Delta H_{soln} = \Delta H_{solute} + \Delta H_{hydration}$ where $\Delta H_{solute} = -\Delta H_{lattice}$ so $\Delta H_{soln} = \Delta H_{hydration} - \Delta H_{lattice}$

$\Delta H_{soln} = -548$ kJ/mol $- (-599$ kJ/mol$) = +51$ kJ/mol and $10.0\ \text{g} \times \dfrac{1\ \text{mol}}{138.55\ \text{g}} = 0.072\underline{1}761$ mol then

$q = n\,\Delta H_{soln} = 0.072\underline{1}761\ \text{mol} \times 51\dfrac{\text{kJ}}{\text{mol}} = +3.\underline{6}810\ \text{kJ} \times \dfrac{1000\ \text{J}}{1\ \text{kJ}} = +3\underline{6}81.0$ J absorbed then

$100.0 \; \cancel{mL} \times \dfrac{1.05 \text{ g}}{1 \; \cancel{mL}} = 105$ g. Because heat is absorbed when $KClO_4$ dissolves, the temperature will drop or

$q = -3681.0$ J and $q = mC_s\Delta T$. Rearrange to solve for ΔT.

$$\Delta T = \dfrac{q}{mC_s} = \dfrac{-3681.0 \; \cancel{J}}{105 \; \cancel{g} \times 4.05 \; \dfrac{\cancel{J}}{\cancel{g} \cdot {}^\circ C}} = -8.7 \; {}^\circ C$$

Check: The units (kJ/mol and °C) are correct. The magnitude of the answer (51 kJ/mol) makes physical sense because the lattice energy is larger than the heat of hydration. The magnitude of the temperature change $(-9 \; {}^\circ C)$ makes physical sense because heat is absorbed and the heat of solution is fairly small.

12.103 **Given:** argon, 0.0537 L; 25 °C, $P_{Ar} = 1.0$ atm to make 1.0 L saturated solution **Find:** $k_H(Ar)$
Conceptual Plan: °C → K and P_{Ar}, V, T → mol_{Ar} then $mol_{Ar}, V_{soln}, P_{Ar}$ → $k_H(Ar)$

$$K = {}^\circ C + 273.15 \qquad\qquad PV = nRT \qquad\qquad S_{Ar} = k_H(Ar) \, P_{Ar} \text{ with } S_{Ar} = \dfrac{mol_{Ar}}{L_{soln}}$$

Solution: $25 \; {}^\circ C + 273.15 = 298$ K and $PV = nRT$ Rearrange to solve for n.

$$n = \dfrac{PV}{RT} = \dfrac{1.0 \; \cancel{atm} \times 0.0537 \; \cancel{L}}{0.08206 \dfrac{\cancel{L} \cdot \cancel{atm}}{\cancel{K} \cdot mol} \times 298 \; \cancel{K}} = 0.00219597 \text{ mol then } S_{Ar} = k_H(Ar)P_{Ar} \text{ with } S_{Ar} = \dfrac{mol_{Ar}}{L_{soln}}$$

Substitute in values and rearrange to solve for k_H.

$$k_H(Ar) = \dfrac{mol_{Ar}}{L_{soln} \, P_{Ar}} = \dfrac{0.00219597 \text{ mol}}{1.0 \; L_{soln} \times 1.0 \text{ atm}} = 2.2 \times 10^{-3} \dfrac{M}{atm}$$

Check: The units (M/atm) are correct. The magnitude of the answer (10^{-3}) seems reasonable because it is consistent with other values in the text.

12.105 **Given:** 0.0020 ppm by mass Hg = legal limit; 0.0040 ppm by mass Hg = contaminated water; 50.0 mg Hg ingested **Find:** volume of contaminated water
Conceptual Plan: mg_{Hg} → g_{Hg} → g_{H_2O} → mL_{H_2O} → L_{H_2O}

$$\dfrac{1 \text{ g}}{1000 \text{ mg}} \qquad \dfrac{10^6 \text{ g water}}{0.0040 \text{ g Hg}} \qquad \dfrac{1 \text{ mL}}{1.00 \text{ g}} \qquad \dfrac{1 \text{ L}}{1000 \text{ mL}}$$

Solution: $50.0 \; \cancel{mg \; Hg} \times \dfrac{1 \; \cancel{g \; Hg}}{1000 \; \cancel{mg \; Hg}} \times \dfrac{10^6 \; \cancel{g \; water}}{0.0040 \; \cancel{g \; Hg}} \times \dfrac{1 \; \cancel{mL \; water}}{1.00 \; \cancel{g \; water}} \times \dfrac{1 \text{ L water}}{1000 \; \cancel{mL \; water}} = 1.3 \times 10^4$ L water

Check: The units (L) are correct. The magnitude of the answer (10^4 L) seems reasonable because the concentration is so low.

12.107 **Given:** 12.5% NaCl by mass in water at 55 °C; 2.5 L vapor **Find:** g_{H_2O} in vapor
Other: $P^\circ_{H_2O} = 118$ torr, $i_{NaCl} = 2.0$ (complete dissociation)
Conceptual Plan: % NaCl by mass → g_{NaCl}, g_{H_2O} then g_{NaCl} → mol_{NaCl} and g_{H_2O} → mol_{H_2O}

$$\dfrac{12.5 \text{ g NaCl}}{100 \text{ g (NaCl} + H_2O)} \qquad\qquad \dfrac{1 \text{ mol NaCl}}{58.44 \text{ g NaCl}} \qquad\qquad \dfrac{1 \text{ mol } H_2O}{18.02 \text{ g } H_2O}$$

then mol_{NaCl}, mol_{H_2O} → χ_{NaCl} → χ_{H_2O} then $\chi_{H_2O}, P^\circ_{H_2O}$ → P_{H_2O}

$$\chi = \dfrac{\text{amount solute (in moles)}}{\text{total amount of solute and solvent (in moles)}} \quad \chi_{H_2O} = 1 - i_{NaCl} \, \chi_{NaCl} \quad P_{solution} = \chi_{solvent} P^\circ_{solvent}$$

then torr → atm and °C → K P, V, T → mol_{H_2O} → g_{H_2O}

$$\dfrac{1 \text{ atm}}{760 \text{ torr}} \qquad K = {}^\circ C + 273.15 \qquad PV = nRT \qquad \dfrac{18.02 \text{ g } H_2O}{1 \text{ mol } H_2O}$$

Solution: $\dfrac{12.5 \text{ g NaCl}}{100 \text{ g (NaCl} + H_2O)}$ means 12.5 g NaCl and $(100 \text{ g} - 12.5 \text{ g}) = 87.5$ g H_2O then

$12.5 \; \cancel{g \; NaCl} \times \dfrac{1 \text{ mol NaCl}}{58.44 \; \cancel{g \; NaCl}} = 0.213895$ mol NaCl and $87.5 \; \cancel{g \; H_2O} \times \dfrac{1 \text{ mol } H_2O}{18.02 \; \cancel{g \; H_2O}} = 4.85841$ mol H_2O

then $\chi = \dfrac{\text{amount solute (in moles)}}{\text{total amount of solute and solvent (in moles)}} = \dfrac{0.213895 \text{ mol}}{0.213895 \text{ mol} + 4.85841 \text{ mol}} = 0.0421692$

then $\chi_{H_2O} = 1 - i_{NaCl}\chi_{NaCl} = 1 - (2.0 \times 0.0421492) = 0.915662$ then

$P_{solution} = \chi_{solvent}P^{\circ}_{solvent} = 0.915662 \times 118 \text{ torr} = 108.048 \text{ torr } H_2O$ then

$108.048 \text{ torr } H_2O \times \dfrac{1 \text{ atm}}{760 \text{ torr}} = 0.142168 \text{ atm}$

and $55\,^{\circ}C + 273.15 = 328 \text{ K}$ then $PV = nRT$. Rearrange to solve for n.

$n = \dfrac{PV}{RT} = \dfrac{0.142168 \text{ atm} \times 2.5 \text{ L}}{0.08206\dfrac{\text{L} \cdot \text{atm}}{\text{K} \cdot \text{mol}} \times 328 \text{ K}} = 0.013205 \text{ mol}$ then

$0.013205 \text{ mol } H_2O \times \dfrac{18.02 \text{ g } H_2O}{1 \text{ mol } H_2O} = 0.24 \text{ g } H_2O$

Check: The units (g) are correct. The magnitude of the answer (0.2 g) seems reasonable because there is very little mass in a vapor.

12.109   **Given:** $T_b = 106.5\,^{\circ}C$ aqueous solution   **Find:** T_f **Other:** $K_f = 1.86\,^{\circ}C/m; K_b = 0.512\,^{\circ}C/m$
           Conceptual Plan: $T_b \rightarrow \Delta T_b$ then $\Delta T_b, K_b \rightarrow m$ then $m, K_f \rightarrow \Delta T_f \rightarrow T_f$

$\qquad\qquad\qquad T_b = T_b^{\circ} + \Delta T_b \qquad\qquad \Delta T_b = K_b \times m \quad \Delta T_f = K_f \times m \quad T_f = T_f^{\circ} - \Delta T_f$

Solution: $T_b = T_b^{\circ} + \Delta T_b$ so $\Delta T_b = T_b - T_b^{\circ} = 106.5\,^{\circ}C - 100.0\,^{\circ}C = 6.5\,^{\circ}C$ then $\Delta T_b = K_b \times m$

Rearrange to solve for m. $m = \dfrac{\Delta T_b}{K_b} = \dfrac{6.5\,^{\circ}C}{0.512\dfrac{^{\circ}C}{m}} = 12.695\ m$ then

$\Delta T_f = K_f \times m = 1.86\dfrac{^{\circ}C}{m} \times 12.695\ m = 23.6\,^{\circ}C$ then $T_f = T_f^{\circ} - \Delta T_f = 0.000\,^{\circ}C - 23.6\,^{\circ}C = -24\,^{\circ}C$

Check: The units $(^{\circ}C)$ are correct. The magnitude of the answer $(-24\,^{\circ}C)$ seems reasonable because the shift in boiling point is less than the shift in freezing point because the constant for boiling is smaller than the constant for freezing.

12.111   (a)   **Given:** 0.90% NaCl by mass per volume; isotonic aqueous solution at $25\,^{\circ}C$; KCl; $i = 1.9$
                   Find: % KCl by mass per volume
                   Conceptual Plan: Isotonic solutions will have the same number of particles. Because i is the same,

$\qquad\qquad\qquad\qquad\qquad\qquad\qquad\qquad\qquad\qquad\qquad\qquad\qquad\qquad \dfrac{1 \text{ mol KCl}}{1 \text{ mol NaCl}}$

the new % mass per volume will be the mass ratio of the two salts.

$\qquad$ percent by mass per volume $= \dfrac{\text{mass solute}}{V} \times 100\% \quad \dfrac{1 \text{ mol NaCl}}{58.44 \text{ g NaCl}}$ and $\dfrac{74.55 \text{ g KCl}}{1 \text{ mol KCl}}$

Solution: percent by mass per volume $= \dfrac{\text{mass solute}}{V} \times 100\%$

$= \dfrac{0.0090 \text{ g NaCl}}{V} \times \dfrac{1 \text{ mol NaCl}}{58.44 \text{ g NaCl}} \times \dfrac{1 \text{ mol KCl}}{1 \text{ mol NaCl}} \times \dfrac{74.55 \text{ g KCl}}{1 \text{ mol KCl}} \times 100\%$

$= 1.1\% \text{ KCl by mass per volume}$

Check: The units (% KCl by mass per volume) are correct. The magnitude of the answer (1.1%) seems reasonable because the molar mass of KCl is larger than the molar mass of NaCl.

           (b)   **Given:** 0.90% NaCl by mass per volume; isotonic aqueous solution at $25\,^{\circ}C$; NaBr; $i = 1.9$
                   Find: % NaBr by mass per volume
                   Conceptual Plan: Isotonic solutions will have the same number of particles. Because i is the same,

$\qquad\qquad\qquad\qquad\qquad\qquad\qquad\qquad\qquad\qquad\qquad\qquad\qquad\qquad \dfrac{1 \text{ mol NaBr}}{1 \text{ mol NaCl}}$

the new % mass per volume will be the mass ratio of the two salts.

$\qquad$ percent by mass per volume $= \dfrac{\text{mass solute}}{V} \times 100\% \quad \dfrac{1 \text{ mol NaCl}}{58.44 \text{ g NaCl}}$ and $\dfrac{102.89 \text{ g NaBr}}{1 \text{ mol NaBr}}$

Solution: percent by mass per volume $= \dfrac{\text{mass solute}}{V} \times 100\%$

$= \dfrac{0.0090 \text{ g NaCl}}{V} \times \dfrac{1 \text{ mol NaCl}}{58.44 \text{ g NaCl}} \times \dfrac{1 \text{ mol NaBr}}{1 \text{ mol NaCl}} \times \dfrac{102.89 \text{ g NaBr}}{1 \text{ mol NaBr}} \times 100\%$

$= 1.6\%$ NaBr by mass per volume

Check: The units (% NaBr by mass per volume) are correct. The magnitude of the answer (1.6%) seems reasonable because the molar mass of NaBr is larger than the molar mass of NaCl.

(c) **Given:** 0.90% NaCl by mass per volume; isotonic aqueous solution at 25 °C; glucose ($C_6H_{12}O_6$); $i = 1.9$
Find: % glucose by mass per volume
Conceptual Plan: Isotonic solutions will have the same number of particles. Because glucose is a nonelectrolyte, the i is not the same; then use the mass ratio of the two compounds.

$\dfrac{1.9 \text{ mol } C_6H_{12}O_6}{1 \text{ mol NaCl}}$ percent by mass per volume $= \dfrac{\text{mass solute}}{V} \times 100\%$ $\dfrac{1 \text{ mol NaCl}}{58.44 \text{ g NaCl}}$ and $\dfrac{180.16 \text{ g } C_6H_{12}O_6}{1 \text{ mol } C_6H_{12}O_6}$

Solution: percent by mass per volume $= \dfrac{\text{mass solute}}{V} \times 100\% =$

$\dfrac{0.0090 \text{ g NaCl}}{V} \times \dfrac{1 \text{ mol NaCl}}{58.44 \text{ g NaCl}} \times \dfrac{1.9 \text{ mol } C_6H_{12}O_6}{1 \text{ mol NaCl}} \times \dfrac{180.16 \text{ g } C_6H_{12}O_6}{1 \text{ mol } C_6H_{12}O_6} \times 100\% =$

$5.3 \, C_6H_{12}O_6$ by mass per volume

Check: The units (% $C_6H_{12}O_6$ by mass per volume) are correct. The magnitude of the answer (5.3%) seems reasonable because the molar mass of $C_6H_{12}O_6$ is larger than the molar mass of NaCl and we need more moles of $C_6H_{12}O_6$ because it is a nonelectrolyte.

12.113 **Given:** 4.5701 g of $MgCl_2$ and 43.238 g water, $P_{\text{soln}} = 0.3624$ atm, $P^\circ_{\text{soln}} = 0.3804$ atm at 348.0 K
Find: i_{measured}
Conceptual Plan: $g_{MgCl_2} \rightarrow \text{mol}_{MgCl_2}$ and $g_{H_2O} \rightarrow \text{mol}_{H_2O}$ then $P_{\text{soln}}, P^\circ_{\text{soln}}, \rightarrow \chi_{MgCl_2}$

$\dfrac{1 \text{ mol } MgCl_2}{95.21 \text{ g } MgCl_2}$ $\dfrac{1 \text{ mol } H_2O}{18.02 \text{ g } H_2O}$ $P_{\text{Soln}} = (1 - \chi_{MgCl_2})P^\circ_{H_2O}$

then $\text{mol}_{MgCl_2}, \text{mol}_{H_2O}, \chi_{MgCl_2} \rightarrow i$

$\chi_{MgCl_2} = \dfrac{i(\text{moles } MgCl_2)}{\text{moles } H_2O + i(\text{moles } MgCl_2)}$

Solution: $4.5701 \text{ g } MgCl_2 \times \dfrac{1 \text{ mol } MgCl_2}{95.21 \text{ g } MgCl_2} = 0.04800021 \text{ mol } MgCl_2$ and

$43.238 \text{ g } H_2O \times \dfrac{1 \text{ mol } H_2O}{18.015 \text{ g } H_2O} = 2.4001110 \text{ mol } H_2O$ then $P_{\text{soln}} = (1 - \chi_{MgCl_2})P^\circ_{H_2O}$ so

$\chi_{MgCl_2} = 1 - \dfrac{P_{\text{soln}}}{P^\circ_{H_2O}} = 1 - \dfrac{0.3624 \text{ atm}}{0.3804 \text{ atm}} = 0.04731861$. Solve for i.

$i(0.04800021) = 0.04731861(2.4001110 + i(0.04800021)) \rightarrow$

$i(0.04800021 - 0.002271303) = 0.1135699 \rightarrow i = \dfrac{0.1135699}{0.04572891} = 2.484$

Check: The units (none) are correct. The magnitude of the answer (2.5) seems reasonable for $MgCl_2$ because we expect i to be 3 if it completely dissociates. Because Mg is small and doubly charged, we expect a significant drop from 3.

12.115 **Given:** $T_b = 375.3$ K aqueous solution   **Find:** P_{H_2O} **Other:** $P^\circ_{H_2O} = 0.2467$ atm; $K_b = 0.512 \,°C/m$
Conceptual Plan: $T_b \rightarrow \Delta T_b$ then $\Delta T_b, K_b \rightarrow m$ assume 1 kg water $kg_{H_2O} \rightarrow \text{mol}_{H_2O}$ then

$T_b = T^\circ_b + \Delta T_b$ $\Delta T_b = K_b \times m$ $\dfrac{1 \text{ mol } H_2O}{18.02 \text{ g } H_2O}$

$m \rightarrow \text{mol}_{\text{solute}}$ then $\text{mol}_{H_2O}, \text{mol}_{\text{solute}} \rightarrow \chi_{H_2O}$ then $\chi_{H_2O}, P^\circ_{H_2O} \rightarrow P_{H_2O}$

$m = \dfrac{\text{amount solute (moles)}}{\text{mass solvent (kg)}}$ $\chi_{H_2O} = \dfrac{\text{moles } H_2O}{\text{moles } H_2O + \text{moles solute}}$ $P_{H_2O} = \chi_{H_2O}P^\circ_{H_2O}$

Solution: $T_b = T_b^{\circ} + \Delta T_b$ so $\Delta T_b = T_b - T_b^{\circ} = 375.3 \text{ K} - 373.15 \text{ K} = 2.2 \text{ K} = 2.2 \,^{\circ}\text{C}$ then

$\Delta T_b = K_b \times m$ Rearrange to solve for m. $m = \dfrac{\Delta T_b}{K_b} = \dfrac{2.2 \,^{\circ}\text{C}}{0.512 \dfrac{^{\circ}\text{C}}{m}} = 4.2\underline{9}6875 \, m$ then

$$1000 \text{ g } H_2O \times \frac{1 \text{ mol } H_2O}{18.02 \text{ g } H_2O} = 55.4\underline{9}390 \text{ mol } H_2O \text{ then}$$

$$m = \frac{\text{amount solute (moles)}}{\text{mass solvent (kg)}} = \frac{x \text{ mol}}{1 \text{ kg}} = 4.2\underline{9}6875 \, m \text{ so } x = 4.2\underline{9}6875 \text{ mol then}$$

$$\chi_{H_2O} = \frac{\text{moles } H_2O}{\text{moles } H_2O + \text{ moles solute}} = \frac{55.4\underline{9}390 \text{ mol}}{55.4\underline{9}390 \text{ mol} + 4.2\underline{9}6875 \text{ mol}} = 0.92\underline{8}1348$$

then $P_{H_2O} = \chi_{H_2O} P_{H_2O}^{\circ} = 0.92\underline{8}1348 \times 0.2467 \text{ atm} = 0.229 \text{ atm}$

Check: The units (atm) are correct. The magnitude of the answer (0.229 atm) seems reasonable because the mole fraction is lowered by $\sim 7\%$.

12.117 **Given:** equal masses of carbon tetrachloride (CCl_4) and chloroform ($CHCl_3$) at 316 K; $P_{CCl_4}^{\circ} = 0.354$ atm; $P_{CHCl_3}^{\circ} = 0.526$ atm **Find:** χ_{CCl_4}, χ_{CHCl_3} in vapor and P_{CHCl_3} in flask of condensed vapor

Conceptual Plan: Assume 100 grams of each $g_{CCl_4} \rightarrow mol_{CCl_4}$ **and** $g_{CHCl_3} \rightarrow mol_{CHCl_3}$ **then**

$$\frac{1 \text{ mol } CCl_4}{153.81 \text{ g } CCl_4} \qquad \frac{1 \text{ mol } CHCl_3}{119.37 \text{ g } CHCl_3}$$

$mol_{CCl_4}, mol_{CHCl_3} \rightarrow \chi_{CCl_4}, \chi_{CHCl_3}$ **then** $\chi_{CCl_4}, P_{CCl_4}^{\circ} \rightarrow P_{CCl_4}$ **and** $\chi_{CHCl_3}, P_{CHCl_3}^{\circ} \rightarrow P_{CHCl_3}$ **then**

$$\chi_{CCl_4} = \frac{\text{amount } CCl_4 \text{ (in moles)}}{\text{total amount (in moles)}} \quad \chi_{CHCl_3} = 1 - \chi_{CCl_4} \qquad P_{CCl_4} = \chi_{CCl_4} P_{CCl_4}^{\circ} \qquad P_{CHCl_3} = \chi_{CHCl_3} P_{CHCl_3}^{\circ}$$

$P_{CCl_4}^{\circ}, P_{CHCl_3} \rightarrow P_{Total}$ **then because** $n \, \alpha \, P$ **and we are calculating a mass percent, which is a ratio of masses,**

$$P_{Total} = P_{CCl_4} + P_{CHCl_3}$$

we can simply convert 1 atm to 1 mole so $P_{CCl_4}, P_{CHCl_3} \rightarrow n_{CHCl_4}, n_{CHCl_3}$ **then**

$$\chi_{CCl_4} = \frac{\text{amount } CCl_4 \text{ (in moles)}}{\text{total amount (in moles)}}$$

$mol_{CCl_4}, mol_{CHCl_3} \rightarrow \chi_{CCl_4}, \chi_{CHCl_3}$ **then for the second vapor** $\chi_{CHCl_3}, P_{CHCl_3}^{\circ} \rightarrow P_{CHCl_3}$

$$\chi_{CHCl_3} = 1 - \chi_{CCl_4} \qquad\qquad P_{CHCl_3} = \chi_{CHCl_3} P_{CHCl_3}^{\circ}$$

Solution: $100.00 \text{ g } CCl_4 \times \dfrac{1 \text{ mol } CCl_4}{153.81 \text{ g } CCl_4} = 0.6501\underline{5}279 \text{ mol } CCl_4$ and

$100.00 \text{ g } CHCl_3 \times \dfrac{1 \text{ mol } CHCl_3}{119.37 \text{ g } CHCl_3} = 0.8377\underline{3}142 \text{ mol } CHCl_3$ then

$\chi_{CCl_4} = \dfrac{\text{amount } CCl_4 \text{ (in moles)}}{\text{total amount (in moles)}} = \dfrac{0.6501\underline{5}279 \text{ mol}}{0.6501\underline{5}279 \text{ mol} + 0.8377\underline{3}142 \text{ mol}} = 0.43\underline{6}96464$ and

$\chi_{CHCl_3} = 1 - \chi_{CCl_4} = 1 - 0.43\underline{6}96464 = 0.563\underline{0}3536$ then

$P_{CCl_4} = \chi_{CCl_4} P_{CCl_4}^{\circ} = 0.43\underline{6}96464 \times 0.354 \text{ atm} = 0.15\underline{4}685 \text{ atm}$ and

$P_{CHCl_3} = \chi_{CHCl_3} P_{CHCl_3}^{\circ} = 0.563\underline{0}3536 \times 0.526 \text{ atm} = 0.29\underline{6}157 \text{ atm}$ then

$P_{Total} = P_{CCl_4} + P_{CHCl_3} = 0.15\underline{4}685 \text{ atm} + 0.29\underline{6}157 \text{ atm} = 0.45\underline{0}841 \text{ atm}$ then

$mol_{CCl_4} = 0.15\underline{4}687 \text{ mol}$ and $mol_{CHCl_3} = 0.29\underline{6}157 \text{ mol}$ then

$\chi_{CCl_4} = \dfrac{\text{amount } CCl_4 \text{ (in moles)}}{\text{total amount (in moles)}} = \dfrac{0.15\underline{4}685 \text{ mol}}{0.15\underline{4}685 \text{ mol} + 0.29\underline{6}157 \text{ mol}} = 0.34\underline{3}102 = 0.343$ in the first vapor and

$\chi_{CHCl_3} = 1 - \chi_{CCl_4} = 1 - 0.34\underline{3}102 = 0.65\underline{6}898 = 0.657$ in the first vapor; then in the second vapor,

$P_{CHCl_3} = \chi_{CHCl_3} P_{CHCl_3}^{\circ} = 0.65\underline{6}898 \times 0.526 \text{ atm} = 0.34\underline{5}528 \text{ atm} = 0.346 \text{ atm}$

Check: The units (none and atm) are correct. The magnitude of the answers seems reasonable because we expect the lighter component to be found preferentially in the vapor phase. This effect is magnified in the second vapor.

12.119 **Given:** 49.0% H_2SO_4 by mass, $d = 1.39 \text{ g/cm}^3$, 25.0 mL diluted to 99.8 cm^3 **Find:** molarity

Conceptual Plan: Initial $mL_{solution} \rightarrow g_{solution} \rightarrow g_{H_2SO_4} \rightarrow mol_{H_2SO_4}$ **and final** $mL_{solution} \rightarrow L_{solution}$

$$\frac{1.39 \text{ g}}{1 \text{ mL}} \quad \frac{49.0 \text{ g } H_2SO_4}{100 \text{ g solution}} \quad \frac{1 \text{ mol } H_2SO_4}{98.09 \text{ g } H_2SO_4} \qquad\qquad \frac{1 \text{ L}}{1000 \text{ mL}}$$

then $mol_{H_2SO_4}, L_{solution} \rightarrow M$

$$M = \frac{\text{amount solute (moles)}}{\text{volume solution (L)}}$$

Solution:

$$25.0 \ \cancel{\text{mL solution}} \times \frac{1.39 \ \cancel{\text{g solution}}}{1 \ \cancel{\text{mL solution}}} \times \frac{49.0 \ \cancel{\text{g H}_2\text{SO}_4}}{100 \ \cancel{\text{g solution}}} \times \frac{1 \ \text{mol H}_2\text{SO}_4}{98.09 \ \cancel{\text{g H}_2\text{SO}_4}} = 0.17\underline{3}5906 \ \text{mol H}_2\text{SO}_4 \ \text{and}$$

$$99.8 \ \cancel{\text{mL solution}} \times \frac{1 \ \text{L solution}}{1000 \ \cancel{\text{mL solution}}} = 0.0998 \ \text{L solution then}$$

$$M = \frac{\text{amount solute (moles)}}{\text{volume solution (L)}} = \frac{0.17\underline{3}5906 \ \text{mol H}_2\text{SO}_4}{0.0998 \ \text{L solution}} = 1.74 \ \text{M H}_2\text{SO}_4$$

Check: The units (M) are correct. The magnitude of the answer (1.74 M) seems reasonable because the solutions is ~1/6 sulfuric acid.

12.121 **Given:** 10.05 g of unknown compound in 50.0 g water, $T_f = -3.16 \,^\circ\text{C}$, mass percent composition of the compound is 60.97% C and 11.94% H; the rest is O **Find:** molecular formula
Other: $K_f = 1.86 \,^\circ\text{C}/m$; $d = 1.00 \ \text{g/mL}$
Conceptual Plan: $g_{\text{H}_2\text{O}} \rightarrow kg_{\text{H}_2\text{O}}$ and $T_f \rightarrow \Delta T_f$ then $\Delta T_f, K_f \rightarrow m$ then $m, kg_{\text{H}_2\text{O}} \rightarrow \text{mol}_{\text{Unk}}$

$$\frac{1 \ \text{kg}}{1000 \ \text{g}} \qquad T_f = T_f^\circ - \Delta T_f \qquad \Delta T_f = K_f \times m \qquad m = \frac{\text{amount solute (moles)}}{\text{mass solvent (kg)}}$$

then $g_{\text{Unk}}, \text{mol}_{\text{Unk}} \rightarrow \mathcal{M} \rightarrow g_{\text{C}}, g_{\text{H}}, g_{\text{O}} \rightarrow \text{mol}_{\text{C}}, \text{mol}_{\text{H}}, \text{mol}_{\text{O}} \rightarrow$ **molecular formula**

$$\mathcal{M} = \frac{g_{\text{Unk}}}{\text{mol}_{\text{Unk}}} \quad \text{mass percents} \qquad \frac{1 \ \text{mol C}}{12.01 \ \text{g C}} \quad \frac{1 \ \text{mol H}}{1.008 \ \text{g H}} \quad \frac{1 \ \text{mol O}}{16.00 \ \text{g O}}$$

Solution: $50.0 \ \cancel{g} \times \dfrac{1 \ \text{kg}}{1000 \ \cancel{g}} = 0.0500 \ \text{kg}$ and $T_f = T_f^\circ - \Delta T_f$ so

$$\Delta T_f = T_f^\circ - T_f = 0.00 \,^\circ\text{C} - (-3.16 \,^\circ\text{C}) = +3.16 \,^\circ\text{C} \quad \Delta T_f = K_f \times m. \quad \text{Rearrange to solve for } m.$$

$$m = \frac{\Delta T_f}{K_f} = \frac{3.16 \,\cancel{^\circ\text{C}}}{1.86 \dfrac{\cancel{^\circ\text{C}}}{m}} = 1.6\underline{9}892 \ m \ \text{then} \ m = \frac{\text{amount solute (moles)}}{\text{mass solvent (kg)}} \ \text{so}$$

$$\text{mol}_{\text{Unk}} = m_{\text{Unk}} \times kg_{\text{H}_2\text{O}} = 1.6\underline{9}892 \ \frac{\text{mol Unk}}{\cancel{kg}} \times 0.0500 \ \cancel{kg} = 0.08\underline{4}94600 \ \text{mol Unk then}$$

$$\mathcal{M} = \frac{g_{\text{Unk}}}{\text{mol}_{\text{Unk}}} = \frac{10.05 \ \text{g}}{0.08\underline{4}94600 \ \text{mol}} = 11\underline{8}.3105 \frac{\text{g}}{\text{mol}} \ \text{then}$$

$$\frac{11\underline{8}.3105 \ \cancel{\text{g Unk}}}{1 \ \text{mol Unk}} \times \frac{60.97 \ \cancel{\text{g C}}}{100 \ \cancel{\text{g Unk}}} \times \frac{1 \ \text{mol C}}{12.01 \ \cancel{\text{g C}}} = \frac{6.01 \ \text{mol C}}{1 \ \text{mol Unk}}$$

$$\frac{11\underline{8}.3105 \ \cancel{\text{g Unk}}}{1 \ \text{mol Unk}} \times \frac{11.94 \ \cancel{\text{g H}}}{100 \ \cancel{\text{g Unk}}} \times \frac{1 \ \text{mol H}}{1.008 \ \cancel{\text{g H}}} = \frac{14.0 \ \text{mol H}}{1 \ \text{mol Unk}} \ \text{and}$$

$$\frac{11\underline{8}.3105 \ \cancel{\text{g Unk}}}{1 \ \text{mol Unk}} \times \frac{(100 - (60.97 + 11.94)) \ \cancel{\text{g O}}}{100 \ \cancel{\text{g Unk}}} \times \frac{1 \ \text{mol O}}{16.00 \ \cancel{\text{g O}}} = \frac{2.00 \ \text{mol O}}{1 \ \text{mol Unk}}$$

So the molecular formula is $C_6H_{12}O_2$.

Check: The units (formula) are correct. The magnitude of the answer (formula with ~118 g/mol) seems reasonable because the molality is ~1.7 and we have ~10 g. It is a reasonable molecular weight for a solid or liquid. The formula does have the correct molar mass.

12.123 **Given:** 100.0 mL solution 13.5% by mass NaCl, $d = 1.12 \ \text{g/mL}$; $T_b = 104.4 \,^\circ\text{C}$ **Find:** g NaCl or water to add
Other: $K_b = 0.512 \,^\circ\text{C}/m$; $i_{\text{measured}} = 1.8$
Conceptual Plan: $T_b \rightarrow \Delta T_b$ then $\Delta T_b, i, K_b \rightarrow m$ then $\text{mL}_{\text{solution}} \rightarrow g_{\text{solution}} \rightarrow g_{\text{NaCl}} \rightarrow \text{mol}_{\text{NaCl}}$ then

$$\Delta T_b = T_b - T_b^\circ \qquad \Delta T_b = K_b \times i \times m \qquad \frac{1.12 \ \text{g solution}}{1 \ \text{mL solution}} \quad \frac{13.5 \ \text{g NaCl}}{100 \ \text{g solution}} \quad \frac{1 \ \text{mol NaCl}}{58.44 \ \text{g NaCl}}$$

$m, \text{mol}_{\text{NaCl}} \rightarrow kg_{\text{H}_2\text{O}} \rightarrow g_{\text{H}_2\text{O}}$ and $g_{\text{solution}}, g_{\text{NaCl}} \rightarrow g_{\text{H}_2\text{O}}$ **then compare the initial and final** $g_{\text{H}_2\text{O}}$ **then**

$$m = \frac{\text{amount solute (moles)}}{\text{mass solvent (kg)}} \quad \frac{1000 \ \text{g}}{1 \ \text{kg}} \qquad g_{\text{solution}} = g_{\text{NaCl}} + g_{\text{H}_2\text{O}}$$

calculate the total NaCl in final solution by scaling up the amount from the initial solution. Then calculate the difference between the needed and starting amounts of NaCl.

Solution: $\Delta T_b = T_b - T_b^\circ = 104.4\,^\circ C - 100.0\,^\circ C = 4.4\,^\circ C$ then $\Delta T_b = K_b \times i \times m$

Rearrange to solve for m. $m = \dfrac{\Delta T_b}{K_b\, i} = \dfrac{4.4\,^\circ\!C}{0.512\dfrac{^\circ\!C}{m} \times 1.8} = 4.\underline{7}74306\ m$ NaCl then

$100.0\ \text{mL solution} \times \dfrac{1.12\ \text{g solution}}{1\ \text{mL solution}} = 112\ \text{g solution}$ $\qquad$ $112\ \text{g solution} \times \dfrac{13.5\ \text{g NaCl}}{100\ \text{g solution}} = 15.\underline{1}2\ \text{g NaCl}$

$15.\underline{1}2\ \text{g NaCl} \times \dfrac{1\ \text{mol NaCl}}{58.44\ \text{g NaCl}} = 0.258\underline{7}269\ \text{mol NaCl}$

then $m = \dfrac{\text{amount solute (moles)}}{\text{mass solvent (kg)}}$ $\quad$ Rearrange to solve for kg_{H_2O}.

$\text{kg}_{H_2O} = \dfrac{mol_{NaCl}}{m} = \dfrac{0.258\underline{7}269\ \text{mol NaCl}}{\dfrac{4.\underline{7}74306\ \text{mol NaCl}}{1\ \text{kg}_{H_2O}}} = 0.05\underline{4}1915\ \text{kg}_{H_2O} \times \dfrac{1000\ \text{g}_{H_2O}}{1\ \text{kg}_{H_2O}} = 54.\underline{1}915\ \text{g}_{H_2O}$ in final solution

then $g_{\text{solution}} = g_{NaCl} + g_{H_2O} = 112\ \text{g solution} - 15.\underline{1}2\ \text{g NaCl} = 96.88\ \text{g H}_2\text{O}$ in initial solution. Comparing the initial and final solutions, there is a lot more water in the initial solution; so NaCl needs to be added.

In the solution with a boiling point of 104.4 °C, $\dfrac{15.\underline{1}2\ \text{g NaCl}}{54.\underline{1}915\ \text{g H}_2\text{O}} = \dfrac{x\ \text{g NaCl}}{96.88\ \text{g H}_2\text{O}}$. Solve for x g NaCl.

$x\ \text{g NaCl} = \dfrac{15.\underline{1}2\ \text{g NaCl}}{54.\underline{1}915\ \text{g H}_2\text{O}} \times 96.88\ \text{g H}_2\text{O} = 27.031\ \text{g NaCl}$; so the amount to be added is

$27.031\ \text{g NaCl} - 15.\underline{1}2\ \text{g NaCl} = 1\underline{1}.911\ \text{g NaCl} = 12\ \text{g NaCl}$

Check: The units (g) are correct. The magnitude of the answer (12 g) seems reasonable because there is approximately twice as much water in the initial solution as is desired; so the NaCl amount needs to be approximately doubled.

Challenge Problems

12.125 $\quad$ **Given:** N_2: $k_H(N_2) = 6.1 \times 10^{-4}$ M/L at 25 °C; 14.6 mg/L at 50 °C and 1.00 atm $P_{N_2} = 0.78$ atm
O_2: $k_H(O_2) = 1.3 \times 10^{-3}$ M/L at 25 °C; 27.8 mg/L at 50 °C and 1.00 atm; $P_{O_2} = 0.21$ atm; and 1.5 L water
Find: $V(N_2)$ and $V(O_2)$
Conceptual Plan: at 25 °C : $P_{\text{Total}}, \chi_{N_2} \to P_{N_2}$ then $P_{N_2}, k_H(N_2) \to S_{N_2}$ then $L \to \text{mol}$

$\qquad\qquad\qquad\qquad\qquad P_{N_2} = \chi_{N_2} P_{\text{Total}} \qquad\qquad S_{N_2} = k_H(N_2)P_{N_2} \qquad S_{N_2}$

at 50 °C : $L \to mL \to mg \to g \to \text{mol}$ then $\text{mol}_{25\,°C}, \text{mol}_{25\,°C} \to \text{mol}_{\text{removed}}$ then $°C \to K$

$\qquad \dfrac{1000\ \text{mL}}{1\ \text{L}}\ \dfrac{14.6\ \text{mg}}{1\ \text{L}}\ \dfrac{1\ \text{g}}{1000\ \text{mg}}\ \dfrac{1\ \text{mol}}{28.01\ \text{g}} \qquad \text{mol}_{\text{removed}} = \text{mol}_{25\,°C} - \text{mol}_{50\,°C} \qquad K = °C + 273.15$

then $P, n, T \to V$

$\qquad PV = nRT$

at 25 °C : $P_{\text{Total}}, \chi_{O_2} \to P_{O_2}$ then $P_{O_2}, k_H(O_2) \to S_{O_2}$ then $L \to \text{mol}$

$\qquad\qquad\qquad\qquad\quad P_{O_2} = \chi_{O_2} P_{\text{Total}} \qquad\qquad S_{O_2} = k_H(O_2)P_{O_2} \qquad S_{O_2}$

at 50 °C : $L \to mL \to mg \to g \to \text{mol}$ then $\text{mol}_{25\,°C}, \text{mol}_{25\,°C} \to \text{mol}_{\text{removed}}$ then $°C \to K$

$\qquad \dfrac{1000\ \text{mL}}{1\ \text{L}}\ \dfrac{27.8\ \text{mg}}{1\ \text{L}}\ \dfrac{1\ \text{g}}{1000\ \text{mg}}\ \dfrac{1\ \text{mol}}{32.00\ \text{g}} \qquad \text{mol}_{\text{removed}} = \text{mol}_{25\,°C} - \text{mol}_{50\,°C} \qquad K = °C + 273.15$

then $P, n, T \to V$

$\qquad PV = nRT$

Solution: at 25 °C: $P_{N_2} = \chi_{N_2} P_{\text{Total}} = 0.78 \times 1.0\ \text{atm} = 0.78\ \text{atm}$ then

$S_{N_2} = k_H(N_2)P_{N_2} = 6.1 \times 10^{-4}\dfrac{\text{M}}{\text{atm}} \times 0.78\ \text{atm} = 4.\underline{7}58 \times 10^{-4}\ \text{M}$ then

$1.5\ \text{L} \times 4.\underline{7}58 \times 10^{-4}\dfrac{\text{mol}}{\text{L}} = 0.0007\underline{1}370\ \text{mol}$

at 50 °C: $1.5\ \text{L} \times \dfrac{14.6\ \text{mg}}{1\ \text{L} \cdot \text{atm}} \times 0.78\ \text{atm} \times \dfrac{1\ \text{g}}{1000\ \text{mg}} \times \dfrac{1\ \text{mol}}{28.01\ \text{g}} = 0.0006\underline{0}985\ \text{mol}$ then

$\text{mol}_{\text{removed}} = \text{mol}_{25\,°C} - \text{mol}_{50\,°C} = 0.0007\underline{1}370\ \text{mol} - 0.0006\underline{0}985\ \text{mol} = 1.0\underline{3}9 \times 10^{-4}\ \text{mol N}_2$

then $50\,°C + 273.15 = 323\,K$ then $PV = nRT$. Rearrange to solve for V.

$$V = \frac{nRT}{P} = \frac{1.039 \times 10^{-4}\,\text{mol} \times 0.08206 \frac{L \cdot atm}{K \cdot mol} \times 323\,K}{1.00\,atm} = 0.0027\underline{5}39\,L\;N_2$$

at $25\,°C$: $P_{O_2} = \chi_{O_2} P_{Total} = 0.21 \times 1.0\,atm = 0.21\,atm$ then

$$S_{O_2} = k_H(O_2)P_{O_2} = 1.3 \times 10^{-3}\frac{M}{atm} \times 0.21\,atm = 2.7\underline{3} \times 10^{-4}\,M \text{ then}$$

$$1.5\,L \times 2.7\underline{3} \times 10^{-4}\frac{mol}{L} = 0.00040\underline{9}5\,mol$$

at $50\,°C$: $1.5\,L \times \dfrac{27.8\,mg}{1\,L \cdot atm} \times 0.21\,atm \times \dfrac{1\,g}{1000\,mg} \times \dfrac{1\,mol}{32.00\,g} = 0.000273\underline{6}6\,mol$ then

$mol_{removed} = mol_{25\,°C} - mol_{50\,°C} = 0.00040\underline{9}5\,mol - 0.000273\underline{6}6\,mol = 1.\underline{3}58 \times 10^{-4}\,mol\;O_2$

then $50\,°C + 273.15 = 323\,K$ then $PV = nRT$. Rearrange to solve for V.

$$V = \frac{nRT}{P} = \frac{1.\underline{3}58 \times 10^{-4}\,\text{mol} \times 0.08206\frac{L \cdot atm}{K \cdot mol} \times 323\,K}{1.00\,atm} = 0.003\underline{5}994\,L\;O_2 \text{ finally}$$

$$V_{Total} = V_{N_2} + V_{O_2} = 0.0027\underline{5}26\,L + 0.003\underline{5}994\,L = 0.0064\,L$$

Check: The units (L) are correct. The magnitude of the answer (0.006 L) seems reasonable because we have so little dissolved gas at room temperature and most is still soluble at $50\,°C$.

12.127   **Given:** 1.10 g glucose ($C_6H_{12}O_6$) and sucrose ($C_{12}H_{22}O_{11}$) mixture in 25.0 mL solution and $\Pi = 3.78\,atm$ at 298 K
Find: percent composition of mixture
Conceptual Plan: $\Pi, T \rightarrow M$ then $mL_{soln} \rightarrow L_{soln}$ then $L_{soln}, M \rightarrow mol_{mixture}$ then

$$\Pi = MRT \qquad\qquad \frac{1\,L}{1000\,mL} \qquad\qquad M = \frac{\text{amount solute (moles)}}{\text{volume solution (L)}}$$

$mol_{mixture}, g_{mixture} \rightarrow mol_{C_6H_{12}O_6}, mol_{C_{12}H_{22}O_{11}}$ then

$g_{mixture} = mol\;C_6H_{12}O_6 \times \dfrac{180.16\,g\;C_6H_{12}O_6}{1\,mol\;C_6H_{12}O_6} + mol\;C_{12}H_{22}O_{11} \times \dfrac{342.30\,g\;C_{12}H_{22}O_{11}}{1\,mol\;C_{12}H_{22}O_{11}}$ with $mol_{mixture} = mol_{C_6H_{12}O_6} + mol_{C_{12}H_{22}O_{11}}$

$mol_{C_6H_{12}O_6} \rightarrow g_{C_6H_{12}O_6}$ and $mol_{C_{12}H_{22}O_{11}} \rightarrow g_{C_{12}H_{22}O_{11}}$ and $g_{C_6H_{12}O_6}, g_{C_{12}H_{22}O_{11}} \rightarrow$ **mass percents**

$$\frac{180.16\,g\;C_6H_{12}O_6}{1\,mol\;C_6H_{12}O_6} \qquad\qquad \frac{342.30\,g\;C_{12}H_{22}O_{11}}{1\,mol\;C_{12}H_{22}O_{11}} \qquad\qquad \text{mass percent} = \frac{\text{mass solute}}{\text{mass solution}} \times 100\%$$

Solution: $\Pi = MRT$. Rearrange to solve for M.

$$M = \frac{\Pi}{RT} = \frac{3.78\,atm}{0.08206\frac{L \cdot atm}{K \cdot mol} \times 298\,K} = 0.15\underline{4}577\frac{\text{mol mixture}}{L} \text{ then } 25.0\,mL \times \frac{1\,L}{1000\,mL} = 0.0250\,L$$

then $M = \dfrac{\text{amount solute (moles)}}{\text{volume solution (L)}}$ so

$$mol_{mixture} = M \times L_{soln} = 0.15\underline{4}577\frac{\text{mol mixture}}{L} \times 0.0250\,L = 0.0038\underline{6}442\,\text{mol mixture then}$$

$$g_{mixture} = mol\;C_6H_{12}O_6 \times \frac{180.16\,g\;C_6H_{12}O_6}{1\,mol\;C_6H_{12}O_6} + mol\;C_{12}H_{22}O_{11} \times \frac{342.30\,g\;C_{12}H_{22}O_{11}}{1\,mol\;C_{12}H_{22}O_{11}} \text{ with}$$

$mol_{mixture} = mol_{C_6H_{12}O_6} + mol_{C_{12}H_{22}O_{11}}$ so

$$1.10\,g = mol\;C_6H_{12}O_6 \times \frac{180.16\,g\;C_6H_{12}O_6}{1\,mol\;C_6H_{12}O_6} + (0.0038\underline{6}442\,mol - mol\;C_6H_{12}O_6) \times \frac{342.30\,g\;C_{12}H_{22}O_{11}}{1\,mol\;C_{12}H_{22}O_{11}} \rightarrow$$

$1.10 = 180.16 \times mol\;C_6H_{12}O_6 + 1.3\underline{2}228 - 342.30 \times mol\;C_6H_{12}O_6 \rightarrow 162.14\,x\,mol\;C_6H_{12}O_6 = 0.2\underline{2}228 \rightarrow$

$$x\,mol\;C_6H_{12}O_6 = \frac{0.2\underline{2}228}{162.14} = 0.001\underline{3}7091\,mol\;C_6H_{12}O_6 \text{ then}$$

$mol_{C_{12}H_{22}O_{11}} = mol_{mixture} - mol_{C_6H_{12}O_6} = 0.0038\underline{6}442\,mol - 0.001\underline{3}7091\,mol$

$= 0.002\underline{4}935\,mol\;C_{12}H_{22}O_{11}$ then

$$0.00137091 \text{ mol } C_6H_{12}O_6 \times \frac{180.16 \text{ g } C_6H_{12}O_6}{1 \text{ mol } C_6H_{12}O_6} = 0.24698 \text{ g } C_6H_{12}O_6 \text{ and}$$

$$0.0024935 \text{ mol } C_{12}H_{22}O_{11} \times \frac{342.30 \text{ g } C_{12}H_{22}O_{11}}{1 \text{ mol } C_{12}H_{22}O_{11}} = 0.85353 \text{ g } C_{12}H_{22}O_{11} \text{ finally}$$

$$\text{mass percent} = \frac{\text{mass solute}}{\text{mass solution}} \times 100\% = \frac{0.24698 \text{ g } C_6H_{12}O_6}{0.24698 \text{ g } C_6H_{12}O_6 + 0.85353 \text{ g } C_{12}H_{22}O_{11}} \times 100\%$$

$$= 22.44\% \text{ } C_6H_{12}O_6 \text{ by mass and } 100.00\% - 22.44\% = 77.56\% \text{ } C_{12}H_{22}O_{11} \text{ by mass}$$

Check: The units (% by mass) are correct. We expect the percent by $C_6H_{12}O_6$ to be larger than that for $C_{12}H_{22}O_{11}$ because the $g_{\text{mixture}}/\text{mol}_{\text{mixture}} = 285$ g/mol, which is closer to $C_{12}H_{22}O_{11}$ than $C_6H_{12}O_6$ and the molar mass of $C_{12}H_{22}O_{11}$ is larger than the molar mass of $C_6H_{12}O_6$. In addition, and most definitively, the masses obtained for sucrose and glucose sum to 1.1 g, the initial amount of solid dissolved.

12.129   **Given:** isopropyl alcohol $((CH_3)_2CHOH)$ and propyl alcohol $(CH_3CH_2CH_2OH)$ at 313 K; solution 2/3 by mass isopropyl alcohol $P_{2/3} = 0.110$ atm; solution 1/3 by mass isopropyl alcohol $P_{1/3} = 0.089$ atm
Find: $P°_{\text{iso}}$ and $P°_{\text{pro}}$ and explain why they are different
Conceptual Plan: Because these are isomers, they have the same molar mass and so the fraction by mass is the same as the mole fraction so mole fractions, $P_{\text{soln}}\text{s} \rightarrow P°\text{s}$

$$\chi_{\text{iso}} = \frac{\text{amount iso (in moles)}}{\text{total amount (in moles)}} \qquad \chi_{\text{pro}} = 1 - \chi_{\text{iso}} \qquad P_{\text{iso}} = \chi_{\text{iso}}P°_{\text{iso}} \qquad P_{\text{pro}} = \chi_{\text{pro}}P°_{\text{pro}} \text{ and } P_{\text{soln}} = P_{\text{iso}} + P_{\text{pro}}$$

Solution:
Solution 1: $\chi_{\text{iso}} = 2/3$ and $\chi_{\text{iso}} = 1/3 \; P_{\text{soln}} = P_{\text{iso}} + P_{\text{pro}}$ so $0.110 \text{ atm} = 2/3 P°_{\text{iso}} + 1/3 P°_{\text{pro}}$
Solution 2: $\chi_{\text{iso}} = 1/3$ and $\chi_{\text{iso}} = 2/3 \; P_{\text{soln}} = P_{\text{iso}} + P_{\text{pro}}$ so $0.089 \text{ atm} = 1/3 P°_{\text{iso}} + 2/3 P°_{\text{pro}}$ We now have two equations and two unknowns and a number of ways to solve this. One way is to rearrange the first equation for $P°_{\text{iso}}$ and substitute into the other equation. Thus, $P°_{\text{iso}} = 3/2(0.110 \text{ atm} - 1/3 P°_{\text{pro}})$ and

$$0.089 \text{ atm} = \frac{1\cancel{3}}{\cancel{3}2}(0.110 \text{ atm} - 1/3 \, P°_{\text{pro}}) + \frac{2}{3}P°_{\text{pro}} \rightarrow 0.089 \text{ atm} = 0.0550 \text{ atm} - \frac{1}{6}\,P°_{\text{pro}} + \frac{2}{3}\,P°_{\text{pro}} \rightarrow$$

$$\frac{1}{2}P°_{\text{pro}} = 0.0340 \text{ atm} \rightarrow P°_{\text{pro}} = 0.0680 \text{ atm} = 0.068 \text{ atm and then}$$

$$P°_{\text{iso}} = 3/2(0.110 \text{ atm} - 1/3 P°_{\text{pro}}) = 3/2(0.110 \text{ atm} - 1/3(0.0680 \text{ atm})) = 0.131 \text{ atm}$$

The major intermolecular attractions are between the OH groups. The OH group at the end of the chain in propyl alcohol is more accessible than the one in the middle of the chain in isopropyl alcohol. In addition, the molecular shape of propyl alcohol is a straight chain of carbon atoms, while that of isopropyl alcohol has a branched chain and is more like a ball. The contact area between two ball-like objects is smaller than that of two chain-like objects. The smaller contact area in isopropyl alcohol means that that the molecules do not attract each other as strongly as do those of propyl alcohol. As a result of both of these factors, the vapor pressure of isopropyl alcohol is higher.

Check: The units (atm) are correct. The magnitude of the answers seems reasonable because both solution partial pressures are ~ 0.1 atm.

12.131   **Given:** 0.1000 m H_2SO_4 solution; complete dissociation to H^+ and HSO_4^-; limited dissociation to SO_4^{2-}; $T_f = 272.76$ K   **Find:** $m(SO_4^{2-})$ **Other:** $K_f = 1.86$ K/m
Conceptual Plan: $T_f \rightarrow \Delta T_f$ then $m, \Delta T_f, K_f \rightarrow i \rightarrow m(SO_4^{2-})$

$$T_f = T°_f - \Delta T_f \qquad \Delta T_f = K_f \, im \quad m(SO_4^{2-}) = m \, H_2SO_4 \, (i - 2.0)/2$$

Solution: $T_f = T°_f - \Delta T_f$ Rearrange to solve for ΔT_f. So
$\Delta T_f = T°_f - T_f = 273.15 \text{ K} - 272.76 \text{ K} = -0.39$ K then $\Delta T_f = K_f \, im$. Rearrange to solve for i.

$$i = \frac{\Delta T_f}{K_f \, m} = \frac{0.39 \text{ K}}{1.86 \frac{\text{K}}{m} \times 0.1000 \, m} = 2.0968.$$

Remember that when H_2SO_4 completely dissociates, two particles are formed (H^+ and HSO_4^-); so $i = 2$. When HSO_4^- dissociates, two particles are generated (SO_4^{2-} and H^+).

$$m(SO_4^{2-}) = m \, H_2SO_4 \frac{i - 2}{2} = (0.1000 \, m) \frac{2.0968 - 2}{2} = 0.04839 \, m = 0.4 \, m \, SO_4^{2-}$$

Check: The units (m) are correct. The magnitude of the answer (0.4 m) seems reasonable because not much of the HSO_4^- dissociates.

12.133   **Given:** $Na_2CO_3 + NaHCO_3 = 11.60$ g in 1.00 L; treat 300.0 cm^3 of solution with HNO_3 and collect 0.940 L CO_2 at 298 K and 0.972 atm   **Find:** $M(Na_2CO_3)$ and $M(NaHCO_3)$

Conceptual Plan: *P, V, T* $\rightarrow$ *n* in 300.0 cm^3 then *n* in 300.0 cm^3 $\rightarrow$ *n* in 1.00 L then

$$PV = nRT \qquad\qquad \text{take ratio of volumes}$$

set up equations for the total mass and the total moles and solve. Then calculate concentrations.

$$g_{Na_2CO_3} + g_{NaHCO_3} = 11.60\,g \quad \frac{1\,mol\,Na_2CO_3}{105.99\,g\,Na_2CO_3} \quad \frac{1\,mol\,NaHCO_3}{84.01\,g\,NaHCO_3} \quad n_{Na_2CO_3} + n_{NaHCO_3} = n$$

Solution: $PV = nRT$. Rearrange to solve for *n*. $n = \dfrac{PV}{RT}$

$$n_{CO_2} = \frac{0.972\,\text{atm} \times 0.940\,L}{0.08206\dfrac{L \cdot \text{atm}}{mol \cdot K} \times 298\,K} = 0.03736340\,mol\,CO_2 \text{ in } 300.0\,cm^3 \text{ then take ratio of moles to volume to get the}$$

moles in 1.00 L

$$\frac{0.03736340\,mol\,CO_2}{300.0\,cm^3} \times 1000\,cm^3 = 0.1245447\,mol\,CO_2 \text{ in } 1.00\,L \text{ because 1 mole of } CO_2 \text{ is generated for each}$$

mole of carbonate. So $n = n_{Na_2CO_3} + n_{NaHCO_3} = 0.1245447$ mol and $g_{Na_2CO_3} + g_{NaHCO_3} = 11.60$ g or

$g_{Na_2CO_3} = 11.60$ g $- g_{NaHCO_3}$ using molar masses and substituting

$$n_{Na_2CO_3} + n_{NaHCO_3} = 0.1245447\,mol = g_{Na_2CO_3} \times \frac{1\,mol\,Na_2CO_3}{105.99\,g\,Na_2CO_3} + g_{NaHCO_3} \times \frac{1\,mol\,NaHCO_3}{84.01\,g\,NaHCO_3} =$$

$$(11.60\,g - g_{NaHCO_3}) \times \frac{1\,mol\,Na_2CO_3}{105.99\,g\,Na_2CO_3} + g_{NaHCO_3} \times \frac{1\,mol\,NaHCO_3}{84.01\,g\,NaHCO_3}$$

$$\rightarrow 0.1245447\,mol = 0.10944429\,mol - g_{NaHCO_3} \times 0.009434852\frac{mol}{g} + g_{NaHCO_3}\,0.011903345\frac{mol}{g}$$

$$\rightarrow 0.01510041\,mol = g_{NaHCO_3} \times 0.002468493\frac{mol}{g} \rightarrow$$

$$g_{NaHCO_3} = \frac{0.01510041\,\text{mol}}{0.002468493\dfrac{\text{mol}}{g}} = 6.117259\,g\,NaHCO_3 = 6.1\,g\,NaHCO_3 \text{ and then } g_{Na_2CO_3} =$$

11.60 g $- g_{NaHCO_3} = 11.60$ g $- 6.117259$ g $NaHCO_3 = 5.48274$ g $Na_2CO_3 = 5.5$ g Na_2CO_3 then M $=$ mol/L

$$\frac{6.117259\,g\,NaHCO_3 \times \dfrac{1\,mol\,NaHCO_3}{84.01\,g\,NaHCO_3}}{1.00\,L} = 0.073\,M\,NaHCO_3 \text{ and}$$

$$\frac{5.48274\,g\,Na_2CO_3 \times \dfrac{1\,mol\,Na_2CO_3}{105.99\,g\,Na_2CO_3}}{1.00\,L} = 0.052\,M\,Na_2CO_3$$

Check: The units (M and M) are correct. The magnitude of the answers (0.073 M and 0.052 M) makes sense because the number of moles of CO_2 is small (0.12 M). The balance of the two components makes sense because if it were all Na_2CO_3, the number of moles would have been 0.109 moles (11.6/105.99) and if it were all $NaHCO_3$, the number of moles would have been 0.138 moles (11.6/84.01)—the actual number of moles is roughly in the middle of these two values.

Conceptual Problems

12.135   The warm coolant water should not be put directly in the river without being cooled because it will raise the temperature of the water. When water is warmed, there is less dissolved oxygen in the water; this will be detrimental to aquatic life that depends on the dissolved oxygen.

12.137   (e)   NaCl. If all of the substances have the same cost per kilogram, we need to determine which substance will generate the largest number of particles per kilogram (or gram). $HOCH_2CH_2OH$ generates 1 mol particle/62.07 g, NaCl generates 2 mol particles/58.44 g, KCl generates 2 mol particles/74.56 g, $MgCl_2$ generates 3 mol particles/95.22 g, and $SrCl_2$ generates 3 particles/158.53 g. So NaCl will generate 1 mole of particles for each 29 g.

13 Chemical Kinetics

Review Questions

13.1 Unlike mammals, which actively regulate their body temperature through metabolic activity, lizards are ectotherms—their body temperature depends on their surroundings. When splashed with cold water, a lizard's body gets colder. The drop in body temperature immobilizes the lizard because its movement depends on chemical reactions that occur within its muscles, and the rates of those reactions—how fast they occur—are highly sensitive to temperature. In other words, when the temperature drops, the reactions that produce movement occur more slowly; therefore, the movement itself slows down. Cold reptiles are lethargic, unable to move very quickly. For this reason, reptiles try to maintain their body temperature in a narrow range by moving between sun and shade.

13.3 The rate of a chemical reaction is measured as a change in the amounts of reactants or products (usually in terms of concentration) divided by the change in time. Typical units are molarity per second (M/s), molarity per minute (M/min), and molarity per year (M/yr), depending on how fast the reaction proceeds.

13.5 The average rate of the reaction can be calculated for any time interval as

$$\text{Rate} = -\frac{1}{a}\frac{[A]_{t_2} - [A]_{t_1}}{t_2 - t_1} = -\frac{1}{b}\frac{[B]_{t_2} - [B]_{t_1}}{t_2 - t_1} = \frac{1}{c}\frac{[C]_{t_2} - [C]_{t_1}}{t_2 - t_1} = \frac{1}{d}\frac{[D]_{t_2} - [D]_{t_1}}{t_2 - t_1} \quad \text{for the}$$

chemical reaction $aA + bB \rightarrow cC + dD$. The instantaneous rate of the reaction is the rate at any one point in time, represented by the instantaneous slope of the plot of concentration versus time at that point. We can obtain the instantaneous rate from the slope of the tangent to this curve at the point of interest.

13.7 The reaction order cannot be determined by the stoichiometry of the reaction. It can only be determined by running controlled experiments where the concentrations of the reactants are varied and the reaction rates are measured and analyzed.

13.9 The rate law shows the relationship between the rate of a reaction and the concentrations of the reactants. The integrated rate law for a chemical reaction is a relationship between the concentration of a reactant and time.

13.11 The half-life $(t_{1/2})$ of a reaction is the time required for the concentration of a reactant to fall to one-half of its initial value. For a zero-order reaction, $t_{1/2} = \dfrac{[A]_0}{2k}$. For a first-order reaction, $t_{1/2} = \dfrac{0.693}{k}$. For a second-order reaction, $t_{1/2} = \dfrac{1}{k[A]_0}$.

13.13 The modern form of the Arrhenius equation, which relates the rate constant (k) and the temperature in kelvin (T), is as follows: $k = A\,e^{-E_a/RT}$, where R is the gas constant ($8.314\,\text{J/mol} \cdot \text{K}$), A is a constant called the frequency factor (or the pre-exponential factor), and E_a is called the activation energy (or activation barrier). The frequency factor is the number of times the reactants approach the activation barrier per unit time. The exponential factor ($-E_a/RT$) is the fraction of approaches that are successful in surmounting the activation barrier and forming products. The exponential factor increases with increasing temperature, but decreases with an increasing value for the activation energy. As the temperature increases, the number of collisions increases and the number of molecules having enough thermal energy to surmount the activation barrier increases. At any given temperature, a sample of molecules will have a distribution of energies, as shown in Figure 13.14. Under common circumstances, only a

252

small fraction of the molecules have enough energy to make it over the activation barrier. Because of the shape of the energy distribution curve, however, a small change in temperature results in a large difference in the number of molecules having enough energy to surmount the activation barrier.

13.15 In the collision model, a chemical reaction occurs after a sufficiently energetic collision between the two reactant molecules. In collision theory, therefore, each approach to the activation barrier is a collision between the reactant molecules. The value of the frequency factor should simply be the number of collisions that occur per second. In the collision model, $k = p \, z \, e^{-E_a/RT}$, where the frequency factor (A) has been separated into two separate parts—p is called the orientation factor, and z is the collision frequency. The collision frequency is simply the number of collisions that occur per unit time. The orientation factor says that if two molecules are to react with each other, they must collide in such a way that allows the necessary bonds to break and form. The small orientation factor indicates that the orientational requirements for this reaction are fairly stringent—the molecules must be aligned in a very specific way for the reaction to occur. When two molecules with sufficient energy and the correct orientation collide, something unique happens: The electrons on one of the atoms or molecules are attracted to the nuclei of the other; some bonds begin to weaken, and other bonds begin to form. If all goes well, the reactants go through the transition state and are transformed into the products and a chemical reaction occurs.

13.17 An elementary step is a single step in a reaction mechanism. Elementary steps cannot be broken down into simpler steps—they occur as they are written. Elementary steps are characterized by their molecularity, the number of reactant particles involved in the step. The molecularity of the three most common types of elementary steps are as follows: unimolecular—$A \rightarrow$ products and Rate $= k[A]$; bimolecular—$A + A \rightarrow$ products and Rate $= k[A]^2$; and bimolecular—$A + B \rightarrow$ products and Rate $= k[A][B]$. Elementary steps in which three reactant particles collide, called termolecular steps, are very rare because the probability of three particles simultaneously colliding is small.

13.19 Reaction intermediates are species that are formed in one step of a mechanism and consumed in another step. An intermediate is not found in the balanced equation for the overall reaction, but plays a key role in the mechanism.

13.21 In homogeneous catalysis, the catalyst exists in the same phase as the reactants. In heterogeneous catalysis, the catalyst exists in a phase different from that of the reactants. The most common type of heterogeneous catalyst is a solid catalyst.

13.23 Enzymes are biological catalysts that increase the rates of biochemical reactions. Enzymes are large protein molecules with complex three-dimensional structures. Within that structure is a specific area called the active site. The properties and shape of the active site are just right to bind the reactant molecule, usually called the substrate. The substrate fits into the active site in a manner that is analogous to a key fitting into a lock.

Problems by Topic

Reaction Rates

13.25 (a) $$\text{Rate} = -\frac{1}{2}\frac{\Delta[\text{HBr}]}{\Delta t} = \frac{\Delta[\text{H}_2]}{\Delta t} = \frac{\Delta[\text{Br}_2]}{\Delta t}$$

(b) **Given:** first 25.0 s; 0.600 M to 0.512 M   **Find:** average rate
Conceptual Plan: $t_1, t_2, [\text{HBr}]_1, [\text{HBr}]_2 \rightarrow$ **average rate**
$$\text{Rate} = -\frac{1}{2}\frac{\Delta[\text{HBr}]}{\Delta t}$$

Solution:
$$\text{Rate} = -\frac{1}{2}\frac{[\text{HBr}]_{t_2} - [\text{HBr}]_{t_1}}{t_2 - t_1} = -\frac{1}{2}\frac{0.512 \text{ M} - 0.600 \text{ M}}{25.0 \text{ s} - 0.0 \text{ s}} = 1.\underline{7}6 \times 10^{-3} \text{ M} \cdot \text{s}^{-1} = 1.8 \times 10^{-3} \text{ M} \cdot \text{s}^{-1}$$

Check: The units ($\text{M} \cdot \text{s}^{-1}$) are correct. The magnitude of the answer ($10^{-3} \text{ M} \cdot \text{s}^{-1}$) makes physical sense because rates are always positive and we are not changing the concentration much in 25 s.

(c) **Given:** 1.50 L vessel, first 15.0 s of reaction, and part (b) data   **Find:** mol_{Br_2} formed
Conceptual Plan: Average rate, $t_1, t_2, \rightarrow \Delta[\text{Br}_2]$ **then** $\Delta[\text{Br}_2]$, L $\rightarrow$ mol_{Br_2} **formed**
$$\text{Rate} = \frac{\Delta[\text{Br}_2]}{\Delta t} \qquad\qquad M = \frac{\text{mol}_{\text{Br}_2}}{\text{L}}$$

Solution: $\text{Rate} = 1.\underline{7}6 \times 10^{-3}\,\text{M}\cdot\text{s}^{-1} = \dfrac{\Delta[Br_2]}{\Delta t} = \dfrac{\Delta[Br_2]}{15.0\,\text{s} - 0.0\,\text{s}}$ Rearrange to solve for $\Delta[Br_2]$.

$\Delta[Br_2] = 1.\underline{7}6 \times 10^{-3}\,\dfrac{M}{\cancel{s}} \times 15.0\,\cancel{s} = 0.02\underline{6}4\,\text{M}$ then $M = \dfrac{\text{mol}_{Br_2}}{L}$ Rearrange to solve for mol_{Br_2}.

$0.02\underline{6}4\dfrac{\text{mol Br}_2}{L} \times 1.50\,\text{L} = 0.040\,\text{mol Br}_2$

Check: The units (mol) are correct. The magnitude of the answer (0.04 mol) makes physical sense because the time is shorter than in part (b) and we need to divide by 2 and multiply by 1.5 because of the stoichiometric coefficient difference and the volume of the vessel, respectively.

13.27   (a) $\text{Rate} = -\dfrac{1}{2}\dfrac{\Delta[A]}{\Delta t} = -\dfrac{\Delta[B]}{\Delta t} = \dfrac{1}{3}\dfrac{\Delta[C]}{\Delta t}$

(b)   **Given:** $\dfrac{\Delta[A]}{\Delta t} = -0.100\,\text{M/s}$ **Find:** $\dfrac{\Delta[B]}{\Delta t}$ and $\dfrac{\Delta[C]}{\Delta t}$

Conceptual Plan: $\dfrac{\Delta[A]}{\Delta t} \rightarrow \dfrac{\Delta[B]}{\Delta t}$ and $\dfrac{\Delta[C]}{\Delta t}$

$\text{Rate} = -\dfrac{1}{2}\dfrac{\Delta[A]}{\Delta t} = -\dfrac{\Delta[B]}{\Delta t} = \dfrac{1}{3}\dfrac{\Delta[C]}{\Delta t}$

Solution: $\text{Rate} = -\dfrac{1}{2}\dfrac{\Delta[A]}{\Delta t} = -\dfrac{\Delta[B]}{\Delta t} = \dfrac{1}{3}\dfrac{\Delta[C]}{\Delta t}$. Substitute value and solve for the two desired

values. $-\dfrac{1}{2}\dfrac{-0.100\,M}{s} = -\dfrac{\Delta[B]}{\Delta t}$ so $\dfrac{\Delta[B]}{\Delta t} = -0.0500\,\text{M}\cdot\text{s}^{-1}$ and $-\dfrac{1}{2}\dfrac{-0.100\,M}{s} = \dfrac{1}{3}\dfrac{\Delta[C]}{\Delta t}$

so $\dfrac{\Delta[C]}{\Delta t} = 0.150\,\text{M}\cdot\text{s}^{-1}$

Check: The units $(M\cdot s^{-1})$ are correct. The magnitude of the answer $(-0.05\,M\cdot s^{-1})$ makes physical sense because fewer moles of B are reacting for every mole of A and the change in concentration with time is negative because this is a reactant. The magnitude of the answer $(0.15\,M\cdot s^{-1})$ makes physical sense because more moles of C are being formed for every mole of A reacting and the change in concentration with time is positive because this is a product.

13.29   **Given:** $Cl_2(g) + 3\,F_2(g) \rightarrow 2\,ClF_3(g)$; $\Delta[Cl_2]/\Delta t = -0.012\,\text{M/s}$ **Find:** $\Delta[F_2]/\Delta t$; $\Delta[ClF_3]/\Delta t$; and Rate
Conceptual Plan: Write the expression for the rate with respect to each species then

$\text{Rate} = -\dfrac{\Delta[Cl_2]}{\Delta t} = -\dfrac{1}{3}\dfrac{\Delta[F_2]}{\Delta t} = \dfrac{1}{2}\dfrac{\Delta[ClF_3]}{\Delta t}$

rate expression, $\dfrac{\Delta[Cl_2]}{\Delta t} \rightarrow \dfrac{\Delta[F_2]}{\Delta t}; \dfrac{\Delta[ClF_3]}{\Delta t};$ **Rate**

$\text{Rate} = -\dfrac{\Delta[Cl_2]}{\Delta t} = -\dfrac{1}{3}\dfrac{\Delta[F_2]}{\Delta t} = \dfrac{1}{2}\dfrac{\Delta[ClF_3]}{\Delta t}$

Solution: $\text{Rate} = -\dfrac{\Delta[Cl_2]}{\Delta t} = -\dfrac{1}{3}\dfrac{\Delta[F_2]}{\Delta t} = \dfrac{1}{2}\dfrac{\Delta[ClF_3]}{\Delta t}$ so

$-\dfrac{\Delta[Cl_2]}{\Delta t} = -\dfrac{1}{3}\dfrac{\Delta[F_2]}{\Delta t}$ Rearrange to solve for $\dfrac{\Delta[F_2]}{\Delta t}$. $\dfrac{\Delta[F_2]}{\Delta t} = 3\dfrac{\Delta[Cl_2]}{\Delta t} = 3\,(-0.012\,\text{M}\cdot\text{s}^{-1}) = -0.036\,\text{M}\cdot\text{s}^{-1}$

and $-\dfrac{\Delta[Cl_2]}{\Delta t} = \dfrac{1}{2}\dfrac{\Delta[ClF_3]}{\Delta t}$ Rearrange to solve for $\dfrac{\Delta[ClF_3]}{\Delta t}$.

$\dfrac{\Delta[ClF_3]}{\Delta t} = -2\dfrac{\Delta[Cl_2]}{\Delta t} = -2\,(-0.012\,\text{M}\cdot\text{s}^{-1}) = 0.024\,\text{M}\cdot\text{s}^{-1}$

Check: The units $(M\cdot s^{-1})$ are correct. The magnitude of the answers $(-0.036\,M\cdot s^{-1}$ and $0.024\,M\cdot s^{-1})$ makes physical sense because F_2 is being used at three times the rate of Cl_2, ClF_3 is being formed at two times the rate of Cl_2 disappearance, and Cl_2 has a stoichiometric coefficient of 1.

13.31   (a)   **Given:** $[C_4H_8]$ versus time data   **Find:** average rate between 0 and 10 s and between 40 and 50 s
Conceptual Plan: $t_1, t_2, [C_4H_8]_1, [C_4H_8]_2 \rightarrow$ **average rate**

$\text{Rate} = -\dfrac{\Delta[C_4H_8]}{\Delta t}$

Solution: For 0 to 10 s, Rate $= -\dfrac{[C_4H_8]_{t_2} - [C_4H_8]_{t_1}}{t_2 - t_1} = -\dfrac{0.913\text{ M} - 1.000\text{ M}}{10.\text{ s} - 0.\text{ s}} = 8.7 \times 10^{-3}\text{ M} \cdot \text{s}^{-1}$ and

for 40 to 50 s, Rate $= -\dfrac{[C_4H_8]_{t_2} - [C_4H_8]_{t_1}}{t_2 - t_1} = -\dfrac{0.637\text{ M} - 0.697\text{ M}}{50.\text{ s} - 40.\text{ s}} = 6.0 \times 10^{-3}\text{ M} \cdot \text{s}^{-1}$

Check: The units $(\text{M} \cdot \text{s}^{-1})$ are correct. The magnitude of the answers $(10^{-3}\text{ M} \cdot \text{s}^{-1})$ makes physical sense because rates are always positive and we are not changing the concentration much in 10 s. Also, reactions slow as they proceed because the concentration of the reactants is decreasing.

(b)   **Given:** $[C_4H_8]$ versus time data   **Find:** $\dfrac{\Delta[C_2H_4]}{\Delta t}$ between 20 and 30 s

Conceptual Plan: $t_1, t_2, [C_4H_8]_1, [C_4H_8]_2 \rightarrow \dfrac{\Delta[C_2H_4]}{\Delta t}$

$$\text{Rate} = -\frac{\Delta[C_4H_8]}{\Delta t} = \frac{1}{2}\frac{\Delta[C_2H_4]}{\Delta t}$$

Solution: Rate $= -\dfrac{[C_4H_8]_{t_2} - [C_4H_8]_{t_1}}{t_2 - t_1} = -\dfrac{0.763\text{ M} - 0.835\text{ M}}{30.\text{ s} - 20.\text{ s}} = 7.2 \times 10^{-3}\text{ M} \cdot \text{s}^{-1} = \dfrac{1}{2}\dfrac{\Delta[C_2H_4]}{\Delta t}$

Rearrange to solve for $\dfrac{\Delta[C_2H_4]}{\Delta t}$. So $\dfrac{\Delta[C_2H_4]}{\Delta t} = 2(7.2 \times 10^{-3}\text{ M} \cdot \text{s}^{-1}) = 1.4 \times 10^{-2}\text{ M} \cdot \text{s}^{-1}$

Check: The units $(\text{M} \cdot \text{s}^{-1})$ are correct. The magnitude of the answer $(10^{-2}\text{ M} \cdot \text{s}^{-1})$ makes physical sense because rate of product formation is always positive and we are not changing the concentration much in 10 s. The rate of change of the product is faster than the decline of the reactant because of the stoichiometric coefficients.

13.33   (a)   **Given:** $[Br_2]$ versus time plot
Find: (i) average rate between 0 and 25 s; (ii) instantaneous rate at 25 s; (iii) instantaneous rate of HBr formation at 50 s
Conceptual Plan: (i) $t_1, t_2, [Br_2]_1, [Br_2]_2 \rightarrow$ **average rate then**

$$\text{Rate} = -\frac{\Delta[Br_2]}{\Delta t}$$

(ii) draw tangent at 25 s and determine slope $\rightarrow$ **instantaneous rate then**

$$\text{Rate} = -\frac{\Delta[Br_2]}{\Delta t}$$

(iii) draw tangent at 50 s and determine slope $\rightarrow$ **instantaneous rate** $\rightarrow \dfrac{\Delta[HBr]}{\Delta t}$

$$\text{Rate} = -\frac{\Delta[Br_2]}{\Delta t} \qquad \text{Rate} = \frac{1}{2}\frac{\Delta[HBr]}{\Delta t}$$

Solution:

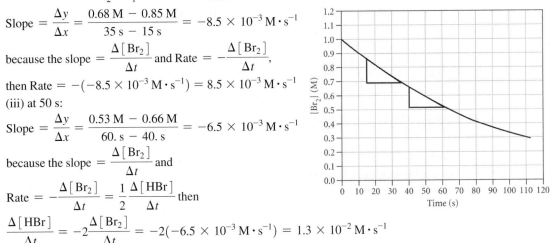

(i) Rate $= -\dfrac{[Br_2]_{t_2} - [Br_2]_{t_1}}{t_2 - t_1} = -\dfrac{0.75\text{ M} - 1.00\text{ M}}{25\text{ s} - 0.\text{ s}} = 1.0 \times 10^{-2}\text{ M} \cdot \text{s}^{-1}$ and (ii) at 25 s:

Slope $= \dfrac{\Delta y}{\Delta x} = \dfrac{0.68\text{ M} - 0.85\text{ M}}{35\text{ s} - 15\text{ s}} = -8.5 \times 10^{-3}\text{ M} \cdot \text{s}^{-1}$

because the slope $= \dfrac{\Delta[Br_2]}{\Delta t}$ and Rate $= -\dfrac{\Delta[Br_2]}{\Delta t}$,

then Rate $= -(-8.5 \times 10^{-3}\text{ M} \cdot \text{s}^{-1}) = 8.5 \times 10^{-3}\text{ M} \cdot \text{s}^{-1}$

(iii) at 50 s:

Slope $= \dfrac{\Delta y}{\Delta x} = \dfrac{0.53\text{ M} - 0.66\text{ M}}{60.\text{ s} - 40.\text{ s}} = -6.5 \times 10^{-3}\text{ M} \cdot \text{s}^{-1}$

because the slope $= \dfrac{\Delta[Br_2]}{\Delta t}$ and

Rate $= -\dfrac{\Delta[Br_2]}{\Delta t} = \dfrac{1}{2}\dfrac{\Delta[HBr]}{\Delta t}$ then

$\dfrac{\Delta[HBr]}{\Delta t} = -2\dfrac{\Delta[Br_2]}{\Delta t} = -2(-6.5 \times 10^{-3}\text{ M} \cdot \text{s}^{-1}) = 1.3 \times 10^{-2}\text{ M} \cdot \text{s}^{-1}$

Check: The units $(M \cdot s^{-1})$ are correct. The magnitude of the first answer is larger than that of the second answer because the rate is slowing down and the first answer includes the initial portion of the data. The magnitude of the answers $(10^{-3} M \cdot s^{-1})$ makes physical sense because rates are always positive and we are not changing the concentration much.

(b)   **Given:** $[Br_2]$ versus time data and $[HBr]_0 = 0 M$ **Find:** plot [HBr] with time

Conceptual Plan: Because Rate $= -\dfrac{\Delta[Br_2]}{\Delta t} = \dfrac{1}{2}\dfrac{\Delta[HBr]}{\Delta t}$. **The rate of change of [HBr] will be twice that of $[Br_2]$. The plot will start at the origin.**
Solution:

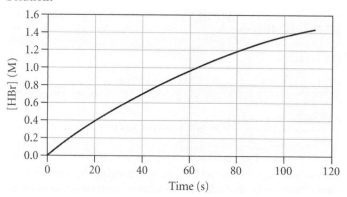

Check: The units (M versus s) are correct. The plot makes sense because it has the same general shape of the original plot except that we are increasing instead of decreasing our concentration axis by a factor of two (to account for the difference in stoichiometric coefficients).

The Rate Law and Reaction Orders

13.35   (a)   **Given:** Rate versus [A] plot   **Find:** reaction order
Conceptual Plan: Look at shape of plot and match to possibilities.
Solution: The plot is a linear plot, so Rate α [A] or the reaction is first-order.
Check: The order of the reaction is a common reaction order.

(b)   **Given:** part (a)   **Find:** sketch plot of [A] versus time
Conceptual Plan: Using the result from part (a), shape plot of [A] versus time should be curved with [A] decreasing. Use 1.0 M as initial concentration.
Solution:

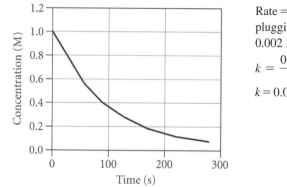

Rate $= k$ [A]
plugging in any of the values from the graph
$0.002 M \cdot s^{-1} = k$ [0.2 M]
$k = \dfrac{0.002 M \cdot s^{-1}}{0.2 M}$
$k = 0.01 s^{-1}$

Check: The plot has a shape that matches the one in the text for first-order plots.

(c) **Given:** part (a) **Find:** write a rate law and estimate k

Conceptual Plan: Using result from part (a), the slope of the plot is the rate constant.

Solution: Slope $= \dfrac{\Delta y}{\Delta x} = \dfrac{0.010\frac{M}{s} - 0.00\frac{M}{s}}{1.0\,M - 0.0\,M} = 0.010\ s^{-1}$ so Rate $= k[A]^1$ or Rate $= k[A]$ or

Rate $= 0.010\ s^{-1}\,[A]$

Check: The units (s^{-1}) are correct. The magnitude of the answer $(10^{-2}\ s^{-1})$ makes physical sense because of the rate and concentration data. Remember that concentration is in units of M; so plugging the rate constant into the equation has the units of the rate as $M \cdot s^{-1}$, which is correct.

13.37 **Given:** reaction order: (a) first-order, (b) second-order, and (c) zero-order **Find:** units of k

Conceptual Plan: Using rate law, rearrange to solve for k.

$$\text{Rate} = k[A]^n, \text{where } n = \text{reaction order}$$

Solution: For all cases, rate has units of $M \cdot s^{-1}$ and [A] has units of M.

(a) Rate $= k[A]^1 = k[A]$ so $k = \dfrac{\text{Rate}}{[A]} = \dfrac{\frac{M}{s}}{M} = s^{-1}$

(b) Rate $= k[A]^2$ so $k = \dfrac{\text{Rate}}{[A]^2} = \dfrac{\frac{M}{s}}{M \cdot M} = M^{-1} \cdot s^{-1}$

(c) Rate $= k[A]^0 = k = M \cdot s^{-1}$

Check: The units $(s^{-1}, M^{-1} \cdot s^{-1}, \text{and } M \cdot s^{-1})$ are correct. The units for k change with the reaction order so that the units on the rate remain as $M \cdot s^{-1}$.

13.39 **Given:** A, B, and C react to form products. Reaction is first-order in A, second-order in B, and zero-order in C.
Find: (a) rate law; (b) overall order of reaction; (c) factor change in rate if [A] doubled; (d) factor change in rate if [B] doubled; (e) factor change in rate if [C] doubled; and (f) factor change in rate if [A], [B], and [C] doubled

Conceptual Plan:

(a) **Using general rate law form, substitute values for orders.**

$$\text{Rate} = k[A]^m[B]^n[C]^p, \text{where } m, n, \text{and } p = \text{reaction orders}$$

(b) **Using rate law in part (a), add up all reaction orders.**

$$\textit{overall reaction order} = m + n + p$$

(c) **Through (f), using rate law from part (a), substitute concentration changes.**

$$\frac{\text{Rate 2}}{\text{Rate 1}} = \frac{k[A]_2^1[B]_2^2}{k[A]_1^1[B]_1^2}$$

Solution:

(a) $m = 1, n = 2,$ and $p = 0$ so Rate $= k[A]^1[B]^2[C]^0$ or Rate $= k[A][B]^2$

(b) *overall reaction order* $= m + n + p = 1 + 2 + 0 = 3$, so it is a third-order reaction overall

(c) $\dfrac{\text{Rate 2}}{\text{Rate 1}} = \dfrac{k[A]_2^1\,[B]_2^2}{k[A]_1^1\,[B]_1^2}$ and $[A]_2 = 2[A]_1, [B]_2 = [B]_1, [C]_2 = [C]_1,$ so $\dfrac{\text{Rate 2}}{\text{Rate 1}} = \dfrac{k(2[A]_1)^1\,[B]_1^2}{k[A]_1^1\,[B]_1^2} = 2$

so the reaction rate doubles (factor of 2)

(d) $\dfrac{\text{Rate 2}}{\text{Rate 1}} = \dfrac{k[A]_2^1\,[B]_2^2}{k[A]_1^1\,[B]_1^2}$ and $[A]_2 = [A]_1, [B]_2 = 2[B]_1, [C]_2 = [C]_1,$ so $\dfrac{\text{Rate 2}}{\text{Rate 1}} = \dfrac{k[A]_1^1(2\,[B]_1)^2}{k[A]_1^1\,[B]_1^2} = 2^2 = 4$

so the reaction rate quadruples (factor of 4)

(e) $\dfrac{\text{Rate 2}}{\text{Rate 1}} = \dfrac{k[A]_2^1\,[B]_2^2}{k[A]_1^1\,[B]_1^2}$ and $[A]_2 = [A]_1, [B]_2 = [B]_1, [C]_2 = 2[C]_1,$ so $\dfrac{\text{Rate 2}}{\text{Rate 1}} = \dfrac{k[A]_1^1\,[B]_1^2}{k[A]_1^1\,[B]_1^2} = 1$

so the reaction rate is unchanged (factor of 1)

(f) $\dfrac{\text{Rate 2}}{\text{Rate 1}} = \dfrac{k[A]_2^1\,[B]_2^2}{k[A]_1^1\,[B]_1^2}$ and $[A]_2 = 2[A]_1, [B]_2 = 2[B]_1, [C]_2 = 2[C]_1,$ so

$\dfrac{\text{Rate 2}}{\text{Rate 1}} = \dfrac{k(2[A]_1)^1\,(2[B]_1)^2}{k[A]_1^1\,[B]_1^2} = 2 \times 2^2 = 8$ so the reaction rate goes up by a factor of 8

Check: The units (none) are correct. The rate law is consistent with the orders given, and the overall order is larger than any of the individual orders. The factors are consistent with the reaction orders. The larger the order, the larger the factor. When all concentrations are changed, the rate changes the most. If a reactant is not in the rate law, then changing its concentration has no effect on the reaction rate.

13.41 **Given:** table of [A] versus initial rate   **Find:** rate law and k
Conceptual Plan: Using general rate law form, compare rate ratios to determine reaction order.

$$\frac{\text{Rate } 2}{\text{Rate } 1} = \frac{k[A]_2^n}{k[A]_1^n}$$

Then use one of the concentration/initial rate pairs to determine k.

$$\text{Rate} = k[A]^n$$

Solution: $\dfrac{\text{Rate } 2}{\text{Rate } 1} = \dfrac{k[A]_2^n}{k[A]_1^n}$ Comparing the first two sets of data, $\dfrac{0.210 \, \cancel{\text{M}}/\text{s}}{0.053 \, \cancel{\text{M}}/\text{s}} = \dfrac{k(0.200 \, \text{M})^n}{k(0.100 \, \text{M})^n}$ and $3.9\underline{6}23 = 2^n$

so $n = 2$ If we compare the first and the last data sets, $\dfrac{0.473 \, \cancel{\text{M}}/\text{s}}{0.053 \, \cancel{\text{M}}/\text{s}} = \dfrac{k(0.300 \, \text{M})^n}{k(0.100 \, \text{M})^n}$ and $8.9\underline{2}45 = 3^n$ so $n = 2$

This second comparison is not necessary, but it increases our confidence in the reaction order. So $\text{Rate} = k[A]^2$. Selecting the second data set and rearranging the rate equation,

$$k = \frac{\text{Rate}}{[A]^2} = \frac{0.210 \dfrac{\text{M}}{\text{s}}}{(0.200 \, \text{M})^2} = 5.25 \, \text{M}^{-1} \cdot \text{s}^{-1}, \text{ so Rate} = 5.25 \, \text{M}^{-1} \cdot \text{s}^{-1}[A]^2$$

Check: The units (none and $\text{M}^{-1} \cdot \text{s}^{-1}$) are correct. The rate law is a common form. The rate is changing more rapidly than the concentration, so second order is consistent. The rate constant is consistent with the units necessary to get rate as M/s, and the magnitude is reasonable because we have a second-order reaction.

13.43 **Given:** table of $[NO_2]$ and $[F_2]$ versus initial rate   **Find:** rate law, k, and overall order
Conceptual Plan: Using general rate law form, compare rate ratios to determine reaction order of each reactant. Be sure to choose data that changes only one concentration at a time.

$$\frac{\text{Rate } 2}{\text{Rate } 1} = \frac{k[NO_2]_2^m [F_2]_2^n}{k[NO_2]_1^m [F_2]_1^n}$$

Then use one of the concentration/initial rate pairs to determine k.

$$\text{Rate} = k[NO_2]^m [F_2]^n$$

Solution: $\dfrac{\text{Rate } 2}{\text{Rate } 1} = \dfrac{k[NO_2]_2^m [F_2]_2^n}{k[NO_2]_1^m [F_2]_1^n}$ Comparing the first two sets of data,

$\dfrac{0.051 \, \cancel{\text{M}}/\text{s}}{0.026 \, \cancel{\text{M}}/\text{s}} = \dfrac{k(0.200 \, \text{M})^m (\cancel{0.100 \, \text{M}})^n}{k(0.100 \, \text{M})^m (\cancel{0.100 \, \text{M}})^n}$ and $1.9\underline{6}15 = 2^m$; so $m = 1$. If we compare the second and third

data sets, $\dfrac{0.103 \, \cancel{\text{M}}/\text{s}}{0.051 \, \cancel{\text{M}}/\text{s}} = \dfrac{k(\cancel{0.200 \, \text{M}})^m (0.200 \, \text{M})^n}{k(\cancel{0.200 \, \text{M}})^m (0.100 \, \text{M})^n}$ and $2 = 2^n$, $n = 1$. Other comparisons can be made, but they are

not necessary. They should reinforce these values of the reaction orders. So $\text{Rate} = k[NO_2][F_2]$. Selecting the last

data set and rearranging the rate equation, $k = \dfrac{\text{Rate}}{[NO_2][F_2]} = \dfrac{0.411 \dfrac{\text{M}}{\text{s}}}{(0.400 \, \text{M})(0.400 \, \text{M})} = 2.57 \, \text{M}^{-1} \cdot \text{s}^{-1}$; so

$\text{Rate} = 2.57 \, \text{M}^{-1} \cdot \text{s}^{-1} [NO_2][F_2]$, and the reaction is second-order overall.

Check: The units (none and $\text{M}^{-1} \cdot \text{s}^{-1}$) are correct. The rate law is a common form. The rate is changing as rapidly as each concentration is changing, which is consistent with first order in each reactant. The rate constant is consistent with the units necessary to get rate as M/s, and the magnitude is reasonable because we have a second-order reaction.

The Integrated Rate Law and Half-Life

13.45 (a) The reaction is zero order. Because the slope of the plot is independent of the concentration, there is no dependence of the concentration of the reactant in the rate law.

(b) The reaction is first-order. The expression for the half-life of a first-order reaction is $t_{1/2} = \dfrac{0.693}{k}$, which is independent of the reactant concentration.

(c) The reaction is second-order. The integrated rate expression for a second-order reaction is $\dfrac{1}{[A]_t} = kt + \dfrac{1}{[A]_0}$, which is linear when the inverse of the concentration is plotted versus time.

13.47 **Given:** table of [AB] versus time **Find:** reaction order, k, and [AB] at 25 s

Conceptual Plan: Look at the data and see if any common reaction orders can be eliminated. If the data does not show an equal concentration drop with time, zero order can be eliminated. Look for changes in the half-life (compare time for concentration to drop to one-half of any value). If the half-life is not constant, the first order can be eliminated. If the half-life is getting longer as the concentration drops, this might suggest second order. Plot the data as indicated by the appropriate rate law. Determine k from the slope of the plot. Finally, calculate the [AB] at 25 s by using the appropriate integrated rate expression.

Solution: By the preceding logic, we can eliminate both the zero-order and first-order reactions. (Alternatively, you could make all three plots and only one should be linear.) This suggests that we should have a second-order reaction. Plot 1/[AB] versus time.

Because $\dfrac{1}{[AB]_t} = kt + \dfrac{1}{[AB]_0}$, the slope will be the rate constant. The slope can be determined by measuring $\Delta y / \Delta x$ on the plot or by using functions such as "add trendline" in Excel. Thus, the rate constant is 0.0225 $M^{-1} \cdot s^{-1}$, and the rate law is Rate = 0.0225 $M^{-1} \cdot s^{-1}[AB]^2$.

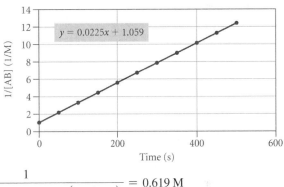

Finally, use $\dfrac{1}{[AB]_t} = kt + \dfrac{1}{[AB]_0}$; substitute the values of $[AB]_0$, 25 s, and k; and rearrange to solve

for [AB] at 25 s. $[AB]_t = \dfrac{1}{kt + \dfrac{1}{[AB]_0}} = \dfrac{1}{(0.0225\ M^{-1} \cdot s^{-1})(25\ s) + \left(\dfrac{1}{0.950\ M}\right)} = 0.619\ M$

Check: The units (none, $M^{-1} \cdot s^{-1}$, and M) are correct. The rate law is a common form. The plot was extremely linear, confirming second-order kinetics. The rate constant is consistent with the units necessary to get rate as M/s, and the magnitude is reasonable because we have a second-order reaction. The [AB] at 25 s is between the values at 0 s and 50 s.

13.49 **Given:** table of $[C_4H_8]$ versus time **Find:** reaction order, k, and reaction rate when $[C_4H_8] = 0.25$ M

Conceptual Plan: Look at the data and see if any common reaction orders can be eliminated. If the data does not show an equal concentration drop with time, zero order can be eliminated. Look for changes in half-life (compare time for concentration to drop to one-half of any value). If the half-life is not constant, the first order can be eliminated. If the half-life is getting longer as the concentration drops, this might suggest second order. Plot the data as indicated by the appropriate rate law. Determine k from the slope of the plot. Finally, calculate the reaction rate when $[C_4H_8] = 0.25$ M by using the rate law.

Solution: By the preceding logic, we can see that the reaction is most likely first-order. It takes about 60 s for the concentration to be cut in half for any concentration. Plot $\ln[C_4H_8]$ versus time. Because $\ln[A]_t = -kt + \ln[A]_0$, the negative of the slope will be the rate constant. The slope can be determined by measuring $\Delta y / \Delta x$ on the plot or by using functions such as "add trendline" in Excel. Thus, the rate constant is 0.0112 s^{-1}, and the rate law is Rate = 0.0112 $s^{-1}[C_4H_8]$. Finally, use Rate = 0.0112 $s^{-1}[C_4H_8]$ and substitute the values of $[C_4H_8]$: Rate = 0.0112 $s^{-1}[0.25\ M] = 2.8 \times 10^{-3}\ M \cdot s^{-1}$

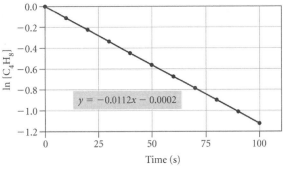

Check: The units (none, s^{-1}, and $M \cdot s^{-1}$) are correct. The rate law is a common form. The plot was extremely linear, confirming first-order kinetics. The rate constant is consistent with the units necessary to get rate as M/s, and the magnitude is reasonable because we have a first-order reaction. The rate when $[C_4H_8] = 0.25$ M is consistent with the average rate using 90 s and 100 s.

13.51 **Given:** plot of ln[A] versus time has slope $= -0.0045/\text{s}$; $[A]_0 = 0.250$ M **Find:** (a) k, (b) rate law, (c) t_{12}, and (d) [A] after 225 s

Conceptual Plan:

(a) A plot of ln[A] versus time is linear for a first-order reaction. Using $\ln[A]_t = -kt + \ln[A]_0$, the rate constant is the negative of the slope.

(b) Rate law is first-order. Add rate constant from part (a).

(c) For a first-order reaction, $t_{1/2} = \dfrac{0.693}{k}$. Substitute k from part (a).

(d) Use the integrated rate law, $\ln[A]_t = -kt + \ln[A]_0$, and substitute k and the initial concentration.

Solution:

(a) Because the rate constant is the negative of the slope, $k = 4.5 \times 10^{-3} \, s^{-1}$.

(b) Because the reaction is first-order, Rate $= k[A] = 4.5 \times 10^{-3} \, s^{-1}[A]$.

(c) $t_{1/2} = \dfrac{0.693}{k} = \dfrac{0.693}{0.0045/\text{s}} = 1.5 \times 10^2$ s

(d) $\ln[A]_t = -kt + \ln[A]_0$ and substitute k and the initial concentration. So
$\ln[A]_t = -(0.0045/\cancel{s})(225 \, \cancel{s}) + \ln 0.250 \, M = -2.39879$ and $[A]_{250\,s} = e^{-2.39879} = 0.0908$ M

Check: The units (s^{-1}, none, s, and M) are correct. The rate law is a common form. The rate constant is consistent with value of the slope. The half-life is consistent with a small value of k. The concentration at 225 s is consistent with being between one and two half-lives.

13.53 **Given:** decomposition of SO_2Cl_2, first order; $k = 1.42 \times 10^{-4} \, s^{-1}$
Find: (a) $t_{1/2}$, (b) t to decrease to 25% of $[SO_2Cl_2]_0$, (c) t to 0.78 M when $[SO_2Cl_2]_0 = 1.00$ M, and
(d) $[SO_2Cl_2]$ after 2.00×10^2 s and 5.00×10^2 s when $[SO_2Cl_2]_0 = 0.150$ M
Conceptual Plan:

(a) $k \rightarrow t_{1/2}$

$t_{1/2} = \dfrac{0.693}{k}$

(b) $[SO_2Cl_2]_0, 25\% \text{ of } [SO_2Cl_2]_0, k \rightarrow t$

$\ln[A]_t = -kt + \ln[A]_0$

(c) $[SO_2Cl_2]_0, [SO_2Cl_2]_t, k \rightarrow t$

$\ln[A]_t = -kt + \ln[A]_0$

(d) $[SO_2Cl_2]_0, t, k \rightarrow [SO_2Cl_2]_t$

$\ln[A]_t = -kt + \ln[A]_0$

Solution:

(a) $t_{1/2} = \dfrac{0.693}{k} = \dfrac{0.693}{1.42 \times 10^{-4} \, s^{-1}} = 4.88 \times 10^3$ s

(b) $[SO_2Cl_2]_t = 0.25[SO_2Cl_2]_0$ Because $\ln[SO_2Cl_2]_t = -kt + \ln[SO_2Cl_2]_0$, rearrange to solve for t.

$t = -\dfrac{1}{k}\ln\dfrac{[SO_2Cl_2]_t}{[SO_2Cl_2]_0} = -\dfrac{1}{1.42 \times 10^{-4} \, s^{-1}}\ln\dfrac{0.25\,\cancel{[SO_2Cl_2]_0}}{\cancel{[SO_2Cl_2]_0}} = 9.8 \times 10^3$ s

(c) $[SO_2Cl_2]_t = 0.78$ M; $[SO_2Cl_2]_0 = 1.00$ M Because $\ln[SO_2Cl_2]_t = -kt + \ln[SO_2Cl_2]_0$, rearrange to solve

for t. $t = -\dfrac{1}{k}\ln\dfrac{[SO_2Cl_2]_t}{[SO_2Cl_2]_0} = -\dfrac{1}{1.42 \times 10^{-4} \, s^{-1}}\ln\dfrac{0.78 \, \cancel{M}}{1.00 \, \cancel{M}} = 1.7 \times 10^3$ s

(d) $[SO_2Cl_2]_0 = 0.150$ M and 2.00×10^2 s in

$\ln[SO_2Cl_2]_t = -(1.42 \times 10^{-4} \, \cancel{s^{-1}})(2.00 \times 10^2 \, \cancel{s}) + \ln 0.150 \, M = -1.92552 \rightarrow$
$[SO_2Cl_2]_t = e^{-1.92552} = 0.146$ M
$[SO_2Cl_2]_0 = 0.150$ M and 5.00×10^2 s in

$$\ln[SO_2Cl_2]_t = -(1.42 \times 10^{-4}\,s^{-1})(5.00 \times 10^2\,s) + \ln 0.150\,M = -1.9\underline{6}812 \rightarrow$$
$$[SO_2Cl_2]_t = e^{-1.9\underline{6}812} = 0.140\,M$$

Check: The units (s, s, s, and M) are correct. The rate law is a common form. The half-life is consistent with a small value of *k*. The time to 25% is consistent with two half-lives. The time to 0.78 M is consistent with being less than one half-life. The final concentrations are consistent with the time being less than one half-life.

13.55   **Given:** $t_{1/2}$ for radioactive decay of U-238 = 4.5 billion years and independent of $[U\text{-}238]_0$
          Find: *t* to decrease by 10%; number U-238 atoms today, when 1.5×10^8 atoms formed 13.8 billion years ago
          Conceptual Plan: $t_{1/2}$ **independent of concentration implies first-order kinetics,** $t_{1/2} \rightarrow k$ **then**

$$t_{1/2} = \frac{0.693}{k}$$

90% of $[U\text{-}238]_0, k \rightarrow t$ and $[U\text{-}238]_0, t, k \rightarrow [U\text{-}238]_t$

$$\ln[A]_t = -kt + \ln[A]_0 \qquad \ln[A]_t = -kt + \ln[A]_0$$

Solution: $t_{1/2} = \dfrac{0.693}{k}$ Rearrange to solve for *k*. $k = \dfrac{0.693}{t_{1/2}} = \dfrac{0.693}{4.5 \times 10^9\,yr} = 1.\underline{5}4 \times 10^{-10}\,yr^{-1}$ then

$[U\text{-}238]_t = 0.10\,[U\text{-}238]_0$ Because $\ln[U\text{-}238]_t = -kt + \ln[U\text{-}238]_0$, rearrange to solve for *t*.

$$t = -\frac{1}{k}\ln\frac{[U\text{-}238]_t}{[U\text{-}238]_0} = -\frac{1}{1.\underline{5}4 \times 10^{-10}\,yr^{-1}}\ln\frac{0.90\,[U\text{-}238]_0}{[U\text{-}238]_0} = 6.8 \times 10^8\,yr$$

and $[U\text{-}238]_0 = 1.15 \times 10^{18}$ atoms; $t = 13.8 \times 10^9\,yr$

$$\ln[U\text{-}238]_t = -kt + \ln[U\text{-}238]_0 = -(1.\underline{5}4 \times 10^{-10}\,yr^{-1})(13.8 \times 10^9\,yr)$$
$$+\ln(1.5 \times 10^{18}\,atoms) = 39.7\underline{2}6797 \rightarrow$$
$$[U\text{-}238]_t = e^{39.7\underline{2}6797} = 1.8 \times 10^{17}\,atoms$$

Check: The units (yr and atoms) are correct. The time to 10% decay is consistent with less than one half-life. The final concentration is consistent with the time being about three half-lives.

The Effect of Temperature and the Collision Model

13.57

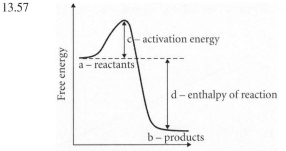

13.59   **Given:** activation energy = 56.8 kJ/mol, frequency factor = 1.5×10^{11}/s, 25 °C   **Find:** rate constant
          Conceptual Plan: °C → K and kJ/mol → J/mol then $E_a, T, A \rightarrow k$

$$K = °C + 273.15 \qquad \frac{1000\,J}{1\,kJ} \qquad k = Ae^{-E_a/RT}$$

Solution: $T = 25\,°C + 273.15 = 298\,K$ and $\dfrac{56.8\,kJ}{mol} \times \dfrac{1000\,J}{1\,kJ} = 5.68 \times 10^4\,\dfrac{J}{mol}$ then

$$k = Ae^{-E_a/RT} = (1.5 \times 10^{11}\,s^{-1})e^{\dfrac{-5.68\times10^4\,\frac{J}{mol}}{\left(8.314\,\frac{J}{K\cdot mol}\right)298\,K}} = 17\,s^{-1}$$

Check: The units (s^{-1}) are correct. The rate constant is consistent with a large activation energy and a large frequency factor.

13.61 **Given:** plot of ln k versus $1/T$ (in K) is linear with a slope of -7445 K **Find:** E_a

Conceptual Plan: Because $\ln k = \dfrac{-E_a}{R}\left(\dfrac{1}{T}\right) + \ln A$ plot of ln k versus $1/T$ will have a *slope* $= -E_a/R$.

Solution: Because the *slope* $= -7445$ K $= -E_a/R$,

$$E_a = -(slope)R = -(-7445\ \text{K})\left(8.314\dfrac{J}{\text{K} \cdot \text{mol}}\right)\left(\dfrac{1\,\text{kJ}}{1000\ J}\right) = 61.90\dfrac{\text{kJ}}{\text{mol}}$$

Check: The units (kJ/mol) are correct. The activation energy is typical for many reactions.

13.63 **Given:** table of rate constant versus T **Find:** E_a and A

Conceptual Plan: Because $\ln k = \dfrac{-E_a}{R}\left(\dfrac{1}{T}\right) + \ln A$, a plot of ln k versus $1/T$ will have a *slope* $= -E_a/R$

and an *intercept* $= \ln A$.

Solution: The slope can be determined by measuring $\Delta y/\Delta x$ on the plot or by using functions such as "add trendline" in Excel. Because the *slope* $= -30\underline{1}89$ K $= -E_a/R$, $E_a = -(slope)R =$

$$-(-30\underline{1}89\ \text{K})\left(8.314\dfrac{J}{\text{K} \cdot \text{mol}}\right)\left(\dfrac{1\,\text{kJ}}{1000\ J}\right)$$

$= 251\dfrac{\text{kJ}}{\text{mol}}$ and *intercept* $= 27.3\underline{9}9 = \ln A$ then

$A = e^{intercept} = e^{27.3\underline{9}9} = 7.93 \times 10^{11}\text{s}^{-1}.$

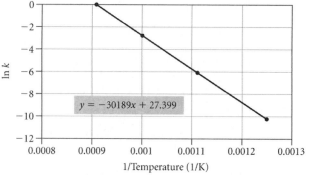

Check: The units (kJ/mol and s^{-1}) are correct. The plot was extremely linear, confirming Arrhenius behavior. The activation and frequency factor are typical for many reactions.

13.65 **Given:** table of rate constant versus T **Find:** E_a and A

Conceptual Plan: Because $\ln k = \dfrac{-E_a}{R}\left(\dfrac{1}{T}\right) + \ln A$, a plot of ln k versus $1/T$ will have a *slope* $= -E_a/R$ and

an *intercept* $= \ln A$.

Solution: The slope can be determined by measuring $\Delta y/\Delta x$ on the plot or by using functions such as "add trendline" in Excel. Because the *slope* $= -27\underline{6}7.2$ K $= -E_a/R$, $E_a = -(slope)R$

$$= -(-27\underline{6}7.2\ \text{K})\left(8.314\dfrac{J}{\text{K} \cdot \text{mol}}\right)\left(\dfrac{1\,\text{kJ}}{1000\ J}\right)$$

$= 23.0\dfrac{\text{kJ}}{\text{mol}}$ and *intercept* $= 25.1\underline{1}2 = \ln A$ then

$A = e^{intercept} = e^{25.1\underline{1}2} = 8.05 \times 10^{10}\ \text{s}^{-1}.$

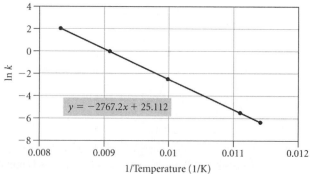

Check: The units (kJ/mol and s^{-1}) are correct. The plot was extremely linear, confirming Arrhenius behavior. The activation and frequency factor are typical for many reactions.

13.67 **Given:** rate constant $= 0.0117/\text{s}$ at 400.0 K and 0.689/s at 450.0 K **Find:** (a) E_a and (b) rate constant at 425 K

Conceptual Plan:

(a) $k_1, T_1, k_2, T_2, \rightarrow E_a$ then J/mol $\rightarrow$ kJ/mol

$$\ln\left(\dfrac{k_2}{k_1}\right) = \dfrac{E_a}{R}\left(\dfrac{1}{T_1} - \dfrac{1}{T_2}\right) \qquad \dfrac{1\,\text{kJ}}{1000\,\text{J}}$$

(b) $E_a, k_1, T_1, T_2 \rightarrow k_2$

$$\ln\left(\dfrac{k_2}{k_1}\right) = \dfrac{E_a}{R}\left(\dfrac{1}{T_1} - \dfrac{1}{T_2}\right)$$

Solution:

(a) $\ln\left(\dfrac{k_2}{k_1}\right) = \dfrac{E_a}{R}\left(\dfrac{1}{T_1} - \dfrac{1}{T_2}\right)$ $\quad$ Rearrange to solve for E_a.

$$E_a = \dfrac{R\ln\left(\dfrac{k_2}{k_1}\right)}{\left(\dfrac{1}{T_1} - \dfrac{1}{T_2}\right)} = \dfrac{8.314\,\dfrac{J}{K\cdot mol}\ln\left(\dfrac{0.689\,s^{-1}}{0.0117\,s^{-1}}\right)}{\left(\dfrac{1}{400.0\cdot K} - \dfrac{1}{450.0\cdot K}\right)} = 1.22 \times 10^5\,\dfrac{J}{mol} \times \dfrac{1\,kJ}{1000\,J} = 122\,\dfrac{kJ}{mol}$$

(b) $\ln\left(\dfrac{k_2}{k_1}\right) = \dfrac{E_a}{R}\left(\dfrac{1}{T_1} - \dfrac{1}{T_2}\right)$ with $k_1 = 0.0117/s$, $T_1 = 400.0\,K$, $T_2 = 425\,K$ $\quad$ Rearrange to solve for k_2.

$$\ln k_2 = \dfrac{E_a}{R}\left(\dfrac{1}{T_1} - \dfrac{1}{T_2}\right) + \ln k_1 = \dfrac{1.22 \times 10^5\,\dfrac{J}{mol}}{8.314\dfrac{J}{K\cdot mol}}\left(\dfrac{1}{400.0\cdot K} - \dfrac{1}{425\cdot K}\right) + \ln 0.0117\,s^{-1} = -2.2902 \rightarrow$$

$$k_2 = e^{-2.2902} = 0.101\,s^{-1}$$

Check: The units $(kJ/mol$ and $s^{-1})$ are correct. The activation energy is typical for a reaction. The rate constant at 425 K is between the values given at 400 K and 450 K.

13.69 **Given:** rate constant doubles from $10.0\,°C$ to $20.0\,°C$ $\quad$ **Find:** E_a
Conceptual Plan: $°C \rightarrow K$ then $k_1, T_1, k_2, T_2 \rightarrow E_a$ then $J/mol \rightarrow kJ/mol$

$$K = °C + 273.15 \qquad \ln\left(\dfrac{k_2}{k_1}\right) = \dfrac{E_a}{R}\left(\dfrac{1}{T_1} - \dfrac{1}{T_2}\right) \qquad \dfrac{1\,kJ}{1000\,J}$$

Solution: $T_1 = 10.0\,°C + 273.15 = 283.2\,K$ and $T_2 = 20.0\,°C + 273.15 = 293.2\,K$ and $k_2 = 2\,k_1$ then

$\ln\left(\dfrac{k_2}{k_1}\right) = \dfrac{E_a}{R}\left(\dfrac{1}{T_1} - \dfrac{1}{T_2}\right).$ $\quad$ Rearrange to solve for E_a.

$$E_a = \dfrac{R\ln\left(\dfrac{k_2}{k_1}\right)}{\left(\dfrac{1}{T_1} - \dfrac{1}{T_2}\right)} = \dfrac{8.314\,\dfrac{J}{K\cdot mol}\ln\left(\dfrac{2\,k_1}{k_1}\right)}{\left(\dfrac{1}{283.2\,K} - \dfrac{1}{293.2\,K}\right)} = 4.7851 \times 10^4\,\dfrac{J}{mol} \times \dfrac{1\,kJ}{1000\,J} = 47.85\,\dfrac{kJ}{mol}$$

Check: The units (kJ/mol) are correct. The activation energy is typical for a reaction.

13.71 Reaction (a) would have the faster rate because the orientation factor, p, would be larger for this reaction because the reactants are symmetrical.

Reaction Mechanisms

13.73 The overall reaction is the sum of the elementary steps in the proposed mechanism. Because the first reaction is the slow step, it is the rate determining step. Using this first step to determine the rate law, Rate $= k_1[AB]^2$. Because this is the observed rate law, this mechanism is consistent with the experimental data.

13.75 (a) The overall reaction is the sum of the steps in the mechanism:

$$Cl_2(g) \underset{k_2}{\overset{k_1}{\rightleftharpoons}} 2\,\cancel{Cl(g)}$$

$$\cancel{Cl(g)} + CHCl_3(g) \xrightarrow{k_3} HCl(g) + \cancel{CCl_3(g)}$$

$$\cancel{Cl(g)} + \cancel{CCl_3(g)} \xrightarrow{k_4} CCl_4(g)$$

$$\overline{Cl_2(g) + CHCl_3(g) \longrightarrow HCl(g) + CCl_4(g)}$$

(b) The intermediates are the species that are generated by one step and consumed by other steps. These are a $Cl(g)$ and $CCl_3(g)$.

(c)   Because the second step is the rate determining step, Rate $= k_3[Cl][CHCl_3]$. Because Cl is an intermediate, its concentration cannot appear in the rate law. Using the fast equilibrium in the first step, we see that $k_1[Cl_2] = k_2[Cl]^2$ or $[Cl] = \sqrt{\dfrac{k_1}{k_2}}[Cl_2]$. Substituting this into the first rate expression, we get

Rate $= k_3\sqrt{\dfrac{k_1}{k_2}}[Cl_2]^{1/2}[CHCl_3]$. Simplifying this expression, we see that Rate $= k[Cl_2]^{1/2}[CHCl_3]$.

Catalysis

13.77   Heterogeneous catalysts require a large surface area because catalysis can only happen at the active sites on the surface. A greater surface area means greater opportunity for the substrate to react, which results in a speedier reaction.

13.79   Assume rate ratio $\propto k$ ratio (because concentration terms will cancel each other) and $k = Ae^{-E_a/RT}$. $T = 25\,°C + 273.15 = 298\ K$, $E_{a_1} = 1.25 \times 10^5\ J/mol$, and $E_{a_2} = 5.5 \times 10^4\ J/mol$. Ratio of rates will be

$$\frac{k_2}{k_1} = \frac{\cancel{A}e^{-E_{a_2}/RT}}{\cancel{A}e^{-E_{a_1}/RT}} = \frac{e^{\frac{-5.5\times10^4\,\frac{\cancel{J}}{\cancel{mol}}}{\left(8.314\,\frac{\cancel{J}}{\cancel{K}\cdot\cancel{mol}}\right)298\,\cancel{K}}}}{e^{\frac{-1.25\times10^5\,\frac{\cancel{J}}{\cancel{mol}}}{\left(8.314\,\frac{\cancel{J}}{\cancel{K}\cdot\cancel{mol}}\right)298\,\cancel{K}}}} = \frac{e^{-22.199}}{e^{-50.\underline{4}53}} = 10^{12}$$ Note that no additional significant figures are quoted since the exponent has its last significant digit in the 1's place.

Cumulative Problems

13.81   **Given:** table of $[CH_3CN]$ versus time   **Find:** (a) reaction order, k; (b) $t_{1/2}$; and (c) t for 90% conversion
Conceptual Plan: (a) and (b) Look at the data and see if any common reaction orders can be eliminated. If the data does not show an equal concentration drop with time, zero order can be eliminated. Look for changes in the half-life (compare time for concentration to drop to one half of any value). If the half-life is not constant, the first order can be eliminated. If the half-life is getting longer as the concentration drops, this might suggest second order. Plot the data as indicated by the appropriate rate law, or if it is first-order and there is an obvious half-life in the data, a plot is not necessary. Determine k from the slope of the plot (or using the half-life equation for first order). (c) Finally, calculate the time to 90% conversion using the appropriate integrated rate equation.
Solution: (a) and (b) By the preceding logic, we can see that the reaction is first-order. It takes 15.0 h for the concentration to be cut in half for any concentration (1.000 M to 0.501 M, 0.794 M to 0.398 M, and 0.631 M to 0.316 M), so $t_{1/2} = 15.0$ h. Then use $t_{1/2} = \dfrac{0.693}{k}$ and rearrange to solve for k.

$k = \dfrac{0.693}{t_{1/2}} = \dfrac{0.693}{15.0\ h} = 0.0462\ h^{-1}$

(c) $[CH_3CN]_t = 0.10\ [CH_3CN]_0$ Because $\ln[CH_3CN]_t = -kt + \ln[CH_3CN]_0$, rearrange to solve for t.

$t = -\dfrac{1}{k}\ln\dfrac{[CH_3CN]_t}{[CH_3CN]_0} = -\dfrac{1}{0.0462\ h^{-1}}\ln\dfrac{0.10\,\cancel{[CH_3CN]_0}}{\cancel{[CH_3CN]_0}} = 49.8\ h$

Check: The units (none, h^{-1}, h, and h) are correct. The rate law is a common form. The data showed a constant half-life very clearly. The rate constant is consistent with the units necessary to get rate as M/s, and the magnitude is reasonable because we have a first-order reaction. The time to 90% conversion is consistent with a time between three and four half-lives.

13.83 **Given:** Rate $= k \dfrac{[A][C]^2}{[B]^{1/2}} = 0.0115$ M/s at certain initial concentrations of A, B, and C; double A and C concentration and triple B concentration   **Find:** reaction rate

Conceptual Plan: $[A]_1, [B]_1, [C]_1, \text{Rate 1}, [A]_2, [B]_2, [C]_2 \rightarrow \text{Rate 2}$

$$\frac{\text{Rate 2}}{\text{Rate 1}} = \frac{k\dfrac{[A]_2[C]_2^2}{[B]_2^{1/2}}}{k\dfrac{[A]_1[C]_1^2}{[B]_1^{1/2}}}$$

Solution: $\dfrac{\text{Rate 2}}{\text{Rate 1}} = \dfrac{k\dfrac{[A]_2[C]_2^2}{[B]_2^{1/2}}}{k\dfrac{[A]_1[C]_1^2}{[B]_1^{1/2}}}$ Rearrange to solve for Rate 2. Rate 2 $= \dfrac{k\dfrac{[A]_2[C]_2^2}{[B]_2^{1/2}}}{k\dfrac{[A]_1[C]_1^2}{[B]_1^{1/2}}}$

Rate 1 $[A]_2 = 2[A]_1$, $[B]_2 = 3[B]_1$, $[C]_2 = 2[C]_1$, and Rate 1 $= 0.0115$ M/s so

$$\text{Rate 2} = \frac{k\dfrac{2[A]_1(2[C]_1)^2}{(3[B]_1)^{1/2}}}{k\dfrac{[A]_1[C]_1^2}{[B]_1^{1/2}}} \, 0.0115 \frac{M}{s} = \frac{2^3}{3^{1/2}} 0.0115 \frac{M}{s} = 0.0531 \frac{M}{s}$$

Check: The units $(M \cdot s^{-1})$ are correct. They should increase because we have a factor of eight (2^3) divided by the square root of three (1.73).

13.85 **Given:** table of P_{Total} versus time   **Find:** rate law, k, and P_{Total} at 2.00×10^4 s

Conceptual Plan: Because two moles of gas are generated for each mole of CH_3CHO decomposed, this $P_{CH_3CHO} = P_{Total}^\circ - (P_{Total} - P_{Total}^\circ)$. Look at the data and see if any common reaction orders can be eliminated. If the data does not show an equal P_{Total} rise (or P_{CH_3CHO} drop) with time, zero order can be eliminated. There is not enough data to look for changes in the half-life (compare time for P_{CH_3CHO} to drop to one-half of any value). It does appear that the half-life is getting longer, so the first order can be eliminated. Plot the data as indicated by the appropriate rate law. Determine k from the slope of the plot. Finally, calculate the P_{CH_3CHO} at 2.00×10^4 s using the appropriate integrated rate expression and convert this to P_{Total} using the reaction stoichiometry.

Solution: Calculate $P_{CH_3CHO} = P_{Total}^\circ - (P_{Total} - P_{Total}^\circ)$.

Time(s)	P_{Total}(atm)	P_{CH_3CHO}(atm)
0	0.22	0.22
1000	0.24	0.20
3000	0.27	0.17
7000	0.31	0.13

By the preceding logic, we can eliminate both the zero-order and the first-order reactions. (Alternatively, you could make all three plots, and only one should be linear.) This suggests that we should have a second-order reaction. Plot $1/P_{CH_3CHO}$ versus time. Because $\dfrac{1}{P_{CH_3CHO}} = kt + \dfrac{1}{P_{CH_3CHO}^\circ}$, the slope will be the rate constant. The slope can be determined by measuring $\Delta y/\Delta x$ on the plot or by using functions such as "add trendline" in Excel. Thus, the rate constant is 4.5×10^{-4} atm$^{-1} \cdot$ s^{-1}, and the rate law is Rate $= 4.5 \times 10^{-4}$ atm$^{-1} \cdot$ s$^{-1} P_{CH_3CHO}$.

Finally, use $\dfrac{1}{P_{CH_3CHO}} = kt + \dfrac{1}{P_{CH_3CHO}^\circ}$; substitute the values of $P_{CH_3CHO}^\circ$, 2.00×10^4 s, and k and rearrange to solve for $P_{CH_3CHO}^\circ$ at 2.00×10^4 s.

$$P_{CH_3CHO} = \frac{1}{kt + \dfrac{1}{P_{CH_3CHO}^\circ}} =$$

$$\frac{1}{(4.5 \times 10^{-4}\,\text{atm}^{-1} \cdot \text{s}^{-1})(2.00 \times 10^{4}\,\text{s}) + \left(\dfrac{1}{0.22\,\text{atm}}\right)} = 0.07\underline{3}8255\,\text{atm} = 0.074\,\text{atm}$$

Finally, from the first equation in the solution, $P_{\text{Total}} = 2P^{\circ}_{\text{Total}} - P_{\text{CH}_3\text{CHO}} = 2(0.22\,\text{atm}) - 0.07\underline{3}8255\,\text{atm} =$ 0.3$\underline{6}$6175 atm = 0.37 atm.

Check: The units (none, $\text{atm}^{-1} \cdot \text{s}^{-1}$, and atm) are correct. The rate law is a common form. The plot was extremely linear, confirming second-order kinetics. The rate constant is consistent with the units necessary to get rate as atm/s, and the magnitude is reasonable because we have a second-order reaction. The P_{Total} at 2.00×10^{4} s is consistent with the changes we see in the data table through 7000 s.

13.87   **Given:** N_2O_5 decomposes to NO_2 and O_2, first order in $[N_2O_5]$; $t_{1/2} = 2.81$ h at 25 °C; $V = 1.5$ L, $P^{\circ}_{N_2O_5} = 745$ torr
Find: P_{O_2} after 215 minutes
Conceptual Plan: Write a balanced reaction. Then $t_{1/2} \rightarrow k$ **then** °C $\rightarrow$ K **and torr** $\rightarrow$ **atm then**

$$N_2O_5 \rightarrow 2\,NO_2 + \tfrac{1}{2}O_2 \qquad t_{1/2} = \frac{0.693}{k} \qquad K = {}^{\circ}C + 273.15 \qquad \frac{1\,\text{atm}}{760\,\text{torr}}$$

$P_{N_2O_5}, V, T \rightarrow n/V$ **then** min $\rightarrow$ h **then** $[N_2O_5]_0$, $t, k \rightarrow [N_2O_5]_t$ **then** $[N_2O_5]_0$, $[N_2O_5]_t \rightarrow [O_2]_t$

$$PV = nRT \qquad \frac{1\,\text{h}}{60\,\text{min}} \qquad \ln[A]_t = -kt + \ln[A]_0 \qquad [O_2]_t = ([N_2O_5]_0 - [N_2O_5]_t) \times \frac{1/2\,\text{mol}\,O_2}{1\,\text{mol}\,N_2O_5}$$

then $[O_2]_t$, $V, T \rightarrow P^{\circ}_{O_2}$ **and finally atm** $\rightarrow$ **torr**

$$PV = nRT \qquad \frac{760\,\text{torr}}{1\,\text{atm}}$$

Solution: $t_{1/2} = \dfrac{0.693}{k}$ and rearrange to solve for k. $k = \dfrac{0.693}{t_{1/2}} = \dfrac{0.693}{2.81\,\text{h}} = 0.24\underline{6}619\,\text{h}^{-1}$ then

$T = 25\,{}^{\circ}C + 273.15 = 298$ K. $745\,\text{torr} \times \dfrac{1\,\text{atm}}{760\,\text{torr}} = 0.98\underline{0}263$ atm then $PV = nRT$ Rearrange to solve for n/V.

$$\frac{n}{V} = \frac{P}{RT} = \frac{0.98\underline{0}263\,\text{atm}}{0.08206\dfrac{\text{L} \cdot \text{atm}}{\text{K} \cdot \text{mol}} \times 298\,\text{K}} = 0.040\underline{0}862\,\text{M} \text{ then } 215\,\text{min} \times \frac{1\,\text{h}}{60\,\text{min}} = 3.5\underline{8}333\,\text{h}$$

Because $\ln[N_2O_5]_t = -kt + \ln[N_2O_5]_0 = -(0.24\underline{6}619\,\text{h}^{-1})(3.5\underline{8}333\,\text{h}) + \ln(0.040\underline{0}862\,\text{M}) = -4.1\underline{0}044 \rightarrow$
$[N_2O_5]_t = e^{-4.10044} = 0.016\underline{5}654$ M then

$$[O_2]_t = ([N_2O_5]_0 - [N_2O_5]_t) \times \frac{1/2\,\text{mol}\,O_2}{1\,\text{mol}\,N_2O_5} = \left(0.040\underline{0}862\frac{\text{mol}\,N_2O_5}{\text{L}} - 0.016\underline{5}654\frac{\text{mol}\,N_2O_5}{\text{L}}\right) \times \frac{1/2\,\text{mol}\,O_2}{1\,\text{mol}\,N_2O_5}$$

$= 0.011\underline{7}604$ M O_2
then finally $PV = nRT$ and rearrange to solve for P

$$P = \frac{n}{V}RT = 0.011\underline{7}604\frac{\text{mol}}{\text{L}} \times 0.08206\frac{\text{L} \cdot \text{atm}}{\text{K} \cdot \text{mol}} \times 298\,\text{K} = 0.287587\,\text{atm} \times \frac{760\,\text{torr}}{1\,\text{atm}} = 219\,\text{torr}$$

Check: The units (torr) are correct. The pressure is reasonable because it must be less than one-half of the original pressure.

13.89   **Given:** I_2 formation from I atoms, second order in I; $k = 1.5 \times 10^{10}\,\text{M}^{-1} \cdot \text{s}^{-1}$, $[I]_0 = 0.0100$ M
Find: t to decrease by 95%
Conceptual Plan: $[I]_0$, $[I]_t$, $k \rightarrow t$

$$\frac{1}{[A]_t} = kt + \frac{1}{[A]_0}$$

Solution: $[I]_t = 0.05\,[I]_0 = 0.05 \times 0.0100$ M $= 0.0005$ M $\dfrac{1}{[I]_t} = kt + \dfrac{1}{[I]_0}$ Rearrange to solve for t.

$$t = \frac{1}{k}\left(\frac{1}{[I]_t} - \frac{1}{[I]_0}\right) = \frac{1}{(1.5 \times 10^{10}\,\text{M}^{-1} \cdot \text{s}^{-1})}\left(\frac{1}{0.0005\,\text{M}} - \frac{1}{0.0100\,\text{M}}\right) = \underline{1}.267 \times 10^{-7}\,\text{s} = 1 \times 10^{-7}\,\text{s}$$

Check: The units (s) are correct. We expect the time to be extremely small because the rate constant is so large.

13.91   **Given:** $AB(aq) \rightarrow A(g) + B(g)$; $k = 0.0118\,\text{M}^{-1} \cdot \text{s}^{-1}$; 250.0 mL of 0.100 M AB; collect gas over water
$T = 25.0\,{}^{\circ}C$, $P_{\text{Total}} = 755.1$ mmHg, and $V = 200.0$ mL; $P^{\circ}_{H_2O} = 23.8$ mmHg   **Find:** t
Conceptual Plan: $P_{\text{Total}}, P_{H_2O} \rightarrow P_A + P_B$ **then** mmHg $\rightarrow$ atm **and** mL $\rightarrow$ L

$$P_{\text{Total}} = P_{H_2O} + P_A + P_B \qquad \frac{1\,\text{atm}}{760\,\text{mmHg}} \qquad \frac{1\,\text{L}}{1000\,\text{mL}}$$

and $°C \to K$ $P, V, T \to n_{A+B} \to \Delta n_{AB}$ then $[AB]_0, V_{AB}, \Delta n_{AB} \to [AB]$ then $k, [AB] \to t$

$$K = °C + 273.15 \quad PV = nRT \quad \Delta n_{AB} = \tfrac{1}{2}n_{A+B} \qquad [AB] = [AB]_0 - \dfrac{\Delta n_{AB}}{V_{AB} \times \dfrac{1\,L}{1000\,mL}} \qquad \dfrac{1}{[AB]_t} = kt + \dfrac{1}{[AB]_0}$$

Solution: $P_{Total} = P_{H_2O} + P_A + P_B$ Rearrange to solve for $P_A + P_B$. $P_A + P_B = P_{Total} - P_{H_2O} = 755.1\ mmHg$ $- 23.8\ mmHg = 731.3\ mmHg$

$$P_A + P_B = 731.3\ \cancel{mmHg} \times \dfrac{1\ atm}{760\ \cancel{mmHg}} = 0.96223684\ atm \quad V = 200.0\ \cancel{mL} \times \dfrac{1\ L}{1000\ \cancel{mL}} = 0.2000\ L$$

$$T = 25.0\,°C + 273.15 = 298.2\ K \quad PV = nRT \quad \text{Rearrange to solve for } n. \ n = \dfrac{PV}{RT}$$

$$n_{A+B} = \dfrac{0.96223684\ \cancel{atm} \times 0.2000\ \cancel{L}}{0.08206\dfrac{\cancel{L} \cdot \cancel{atm}}{mol \cdot \cancel{K}} \times 298.2\ \cancel{K}} = 0.0078645309\ mol\ A + B \quad \text{Because one mole each of A and B is generated}$$

for each mole of AB reacting, $\Delta n_{AB} = \tfrac{1}{2}n_{A+B} = \tfrac{1}{2}(0.0078645309\ mol\ A + B) = 0.0039322655\ mol\ AB$ then

$$[AB] = [AB]_0 - \dfrac{\Delta n_{AB}}{V_{AB} \times \dfrac{1\ L}{1000\ mL}} = 0.100\ M - \dfrac{0.0039322655\ mol\ AB}{250.0\ \cancel{mL} \times \dfrac{1\ L}{1000\ \cancel{mL}}} = 0.100\ M - 0.015729062\ M$$

$$= 0.0842709\ M$$

Because $\dfrac{1}{[AB]_t} = kt + \dfrac{1}{[AB]_0}$, rearrange to solve for t.

$$t = \dfrac{1}{k}\left(\dfrac{1}{[XY]_t} - \dfrac{1}{[XY]_0}\right) = \dfrac{1}{(0.0118\ M^{-1} \cdot s^{-1})}\left(\dfrac{1}{0.0842709\ M} - \dfrac{1}{0.100\ M}\right) = 158.1773\ s = 160\ s$$

Check: The units (s) are correct. The magnitude of the answer (160 s) makes sense because the rate constant is $0.0118\ M^{-1}\,s^{-1}$ and a small volume of gas is generated.

13.93   (a)   There are two elementary steps in the reaction mechanism because there are two peaks in the reaction progress diagram.

      (b)

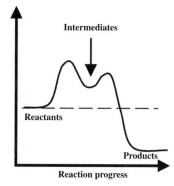

      (c)   The first step is the rate limiting step because it has the higher activation energy.

      (d)   The overall reaction is exothermic because the products are at a lower energy than the reactants.

13.95   **Given:** n-butane desorption from single crystal aluminum oxide, first order; $k = 0.128\ s^{-1}$ at 150 K; initially completely covered

      Find: (a) $t_{1/2}$; (b) t for 25% and for 50% to desorb; (c) fraction remaining after 10 s and 20 s

      Conceptual Plan: (a) $k \to t_{1/2}$ (b) $[C_4H_{10}]_0, [C_4H_{10}]_t, k \to t$ (c) $[C_4H_{10}]_0, t, k \to [C_4H_{10}]_t$

$$t_{1/2} = \dfrac{0.693}{k} \qquad\qquad \ln[A]_t = -kt + \ln[A]_0 \qquad \ln[A]_t = -kt + \ln[A]_0$$

      Solution:

      (a) $t_{1/2} = \dfrac{0.693}{k} = \dfrac{0.693}{0.128\ s^{-1}} = 5.41\ s$

(b) $\ln[C_4H_{10}]_t = -kt + \ln[C_4H_{10}]_0$ Rearrange to solve for t. For 25% desorbed, $[C_4H_{10}]_t = 0.75[C_4H_{10}]_0$

and $t = -\dfrac{1}{k}\ln\dfrac{[C_4H_{10}]_t}{[C_4H_{10}]_0} = -\dfrac{1}{0.128\text{ s}^{-1}}\ln\dfrac{0.75[C_4H_{10}]_0}{[C_4H_{10}]_0} = 2.2\text{ s}$. For 50% desorbed,

$[C_4H_{10}]_t = 0.50[C_4H_{10}]_0$ and $t = -\dfrac{1}{k}\ln\dfrac{[C_4H_{10}]_t}{[C_4H_{10}]_0} = -\dfrac{1}{0.128\text{ s}^{-1}}\ln\dfrac{0.50[C_4H_{10}]_0}{[C_4H_{10}]_0} = 5.4\text{ s}$

(c)   For 10 s, $\ln[C_4H_{10}]_t = -kt + \ln[C_4H_{10}]_0 = -(0.128\text{ s}^{-1})(10\text{ s}) + \ln(1.00) = -1.28 \rightarrow$

$[C_4H_{10}]_t = e^{-1.28} = 0.28 = $ fraction covered

For 20 s, $\ln[C_4H_{10}]_t = -kt + \ln[C_4H_{10}]_0 = -(0.128\text{ s}^{-1})(20\text{ s}) + \ln(1.00) = -2.56 \rightarrow$

$[C_4H_{10}]_t = e^{-2.56} = 0.077 = $ fraction covered

Check: The units (s, s, s, none, and none) are correct. The half-life is reasonable considering the size of the rate constant. The time to 25% desorbed is less than one half-life. The time to 50% desorbed is the half-life. The fraction at 10 s is consistent with about two half-lives. The fraction covered at 20 s is consistent with about four half-lives.

13.97   (a)   **Given:** table of rate constant versus T **Find:** E_a and A

Conceptual Plan: First, convert temperature data into kelvin (°C + 273.15 = K). Because

$\ln k = \dfrac{-E_a}{R}\left(\dfrac{1}{T}\right) + \ln A$, **a plot of** $\ln k$ **versus** $1/T$ **will have a** *slope* $= -E_a/R$ **and an** *intercept* $= \ln A$.

Solution: The slope can be determined by measuring $\Delta y/\Delta x$ on the plot or by using functions such as "add trendline" in Excel. Because the *slope* $= -10759\text{ K} = -E_a/R$,
$E_a = -(slope)R$

$= -(-10759\text{ K})\left(8.314\dfrac{J}{K\cdot mol}\right)\left(\dfrac{1\text{ kJ}}{1000\text{ J}}\right)$

$= 89.5\dfrac{kJ}{mol}$

and *intercept* $= 26.769 = \ln A$ then
$A = e^{intercept} = e^{26.769} = 4.22 \times 10^{11}\text{ s}^{-1}$

Check: The units (kJ/mol and s^{-1}) are correct. The plot was extremely linear, confirming Arrhenius behavior. The activation and frequency factor are typical for many reactions.

$y = -10759x + 26.769$

(b)   **Given:** part (a) results   **Find:** k at 15 °C
Conceptual Plan: °C $\rightarrow$ K then $T, E_a, A \rightarrow k$

$$°C + 273.15 = K \qquad \ln k = \dfrac{-E_a}{R}\left(\dfrac{1}{T}\right) + \ln A$$

Solution: 15 °C + 273.15 = 288 K then

$$\ln k = \dfrac{-E_a}{R}\left(\dfrac{1}{T}\right) + \ln A = \dfrac{-89.5\dfrac{kJ}{mol} \times \dfrac{1000\text{ J}}{1\text{ kJ}}}{8.314\dfrac{J}{K\cdot mol}}\left(\dfrac{1}{288\text{ K}}\right) + \ln(4.22 \times 10^{11}\text{ s}^{-1}) = -10.610 \rightarrow$$

$k = e^{-10.610} = 2.5 \times 10^{-5}\text{ M}^{-1}\cdot\text{s}^{-1}$

Check: The units (M$^{-1}\cdot$s^{-1}) are correct. The value of the rate constant is less than the value at 25 °C.

(c)   **Given:** part (a) results, 0.155 M C_2H_5Br and 0.250 M OH^- at 75 °C   **Find:** initial reaction rate
Conceptual Plan: °C $\rightarrow$ K then $T, E_a, A \rightarrow k$ then $k, [C_2H_5Br], [OH^-] \rightarrow$ initial reaction rate

$$°C + 273.15 = K \qquad \ln k = \dfrac{-E_a}{R}\left(\dfrac{1}{T}\right) + A \qquad\qquad Rate = k[C_2H_5Br][OH^-]$$

Solution: 75 °C + 273.15 = 348 K then

$$\ln k = \frac{-E_a}{R}\left(\frac{1}{T}\right) + \ln A = \frac{-89.5\frac{kJ}{mol} \times \frac{1000\,J}{1\,kJ}}{8.314\frac{J}{K \cdot mol}}\left(\frac{1}{348\,K}\right) + \ln(4.22 \times 10^{11}\,s^{-1}) = -4.\underline{1}656 \rightarrow$$

$$k = e^{-4.\underline{1}656} = 1.\underline{5}521 \times 10^{-2}\,M^{-1} \cdot s^{-1}$$

Rate $= k[C_2H_5Br][OH^-] = (1.\underline{5}521 \times 10^{-2}\,M^{-1} \cdot s^{-1})(0.155\,M)(0.250\,M) = 6.0 \times 10^{-4}\,M \cdot s^{-1}$

Check: The units $(M \cdot s^{-1})$ are correct. The value of the rate is reasonable considering the value of the rate constant (larger than in the table) and the fact that the concentrations are less than 1 M.

13.99   (a)   No, because the activation energy is zero. This means that the rate constant $(k = Ae^{-E_a/RT})$ will be independent of temperature.

(b)   No bond is broken, and the two radicals (CH_3) attract each other.

(c)   Formation of diatomic gases from atomic gases

13.101   **Given:** $t_{1/2}$ for radioactive decay of C-14 = 5730 years; bone has 19.5% C-14 in living bone
Find: age of bone
Conceptual Plan: Radioactive decay implies first-order kinetics, $t_{1/2} \rightarrow k$ **then 19.5% of** $[\text{C-14}]_0, k \rightarrow t$

$$t_{1/2} = \frac{0.693}{k} \qquad \ln[A]_t = -kt + \ln[A]_0$$

Solution: $t_{1/2} = \frac{0.693}{k}$ Rearrange to solve for k. $k = \frac{0.693}{t_{1/2}} = \frac{0.693}{5730\,yr} = 1.20\underline{9}42 \times 10^{-4}\,yr^{-1}$ then

$[\text{C-14}]_t = 0.195\,[\text{C-14}]_0$ Because $\ln[\text{C-14}]_t = -kt + \ln[\text{C-14}]_0$, rearrange to solve for t.

$$t = -\frac{1}{k}\ln\frac{[\text{C-14}]_t}{[\text{C-14}]_0} = -\frac{1}{1.20\underline{9}42 \times 10^{-4}\,yr^{-1}}\ln\frac{0.195\,[\text{C-14}]_0}{[\text{C-14}]_0} = 1.35 \times 10^4\,yr$$

Check: The units (yr) are correct. The time to 19.5% decay is consistent with the time being between two and three half-lives.

13.103   (a)   For each, check that all steps sum to overall reaction and that the predicted rate law is consistent with experimental data (Rate $= k[H_2][I_2]$).
For the first mechanism, the single step is the overall reaction. The rate law is determined by the stoichiometry; so Rate $= k[H_2][I_2]$, and the mechanism is valid.
For the second mechanism, the overall reaction is the sum of the steps in the mechanism:

$I_2(g) \underset{k_2}{\overset{k_1}{\rightleftharpoons}} 2I(g)$
$H_2(g) + 2I(g) \xrightarrow{k_3} 2HI(g)$ So the sum matches the overall reaction.

$H_2(g) + I_2(g) \longrightarrow 2HI(g)$

Because the second step is the rate determining step, Rate $= k_3[H_2][I]^2$. Because I is an intermediate, its concentration cannot appear in the rate law. Using the fast equilibrium in the first step, we see that

$k_1[I_2] = k_2[I]^2$ or $[I]^2 = \frac{k_1}{k_2}[I_2]$. Substituting this into the first rate expression, we get

Rate $= k_3\frac{k_1}{k_2}[H_2][I_2]$, and the mechanism is valid.

(b)   To distinguish between mechanisms, you could look for the buildup of $I(g)$, the intermediate in the second mechanism, or for the buildup of $I(g)$ and $H_2I(g)$, the intermediates in the third mechanism.

13.105   The steps in the mechanism are as follows:
$Br_2(g) \underset{k_{-1}}{\overset{k_1}{\rightleftharpoons}} 2Br(g)$
$Br(g) + H_2(g) \xrightarrow{k_2} HBr(g) + H(g)$
$H(g) + Br_2(g) \xrightarrow{k_3} HBr(g) + Br(g)$

Because the second step is the rate determining step, Rate $= k_2[H_2][Br]$. Because Br is an intermediate, its concentration cannot appear in the rate law. Using the fast equilibrium in the first step, we see that

$k_1[Br_2] = k_{-1}[Br]^2$ or $[Br] = \sqrt{\dfrac{k_1}{k_{-1}}}[Br_2]^{1/2}$. Substituting this into the first rate expression, we get

Rate $= k_2\sqrt{\dfrac{k_1}{k_{-1}}}[H_2][Br_2]^{1/2}$. The rate law is 3/2 order overall.

13.107   (a)   For a zero-order reaction, the rate is independent of the concentration. If the first half goes in the first 100 minutes, the second half will go in the second 100 minutes. This means that none, or 0%, will be left at 200 minutes.

      (b)   For a first-order reaction, the half-life is independent of concentration. This means that if half of the reactant decomposes in the first 100 minutes, half of this (or another 25% of the original amount) will decompose in the second 100 minutes. This means that at 200 minutes, 50% + 25% = 75% has decomposed or 25% remains.

      (c)   For a second-order reaction, $t_{1/2} = \dfrac{1}{k[A]_0} = 100$ min and the integrated rate expression is $\dfrac{1}{[A]_t} = kt + \dfrac{1}{[A]_0}$.

We can rearrange the first expression to solve for k as $k = \dfrac{1}{100\ \text{min}\ [A]_0}$. Substituting this and 200 minutes

into the integrated rate expression, we get $\dfrac{1}{[A]_t} = \dfrac{200\ \cancel{\text{min}}}{100\ \cancel{\text{min}}\ [A]_0} + \dfrac{1}{[A]_0} \rightarrow \dfrac{1}{[A]_t} = \dfrac{3}{[A]_0} \rightarrow \dfrac{[A]_t}{[A]_0} = \dfrac{1}{3}$
or 33% remains.

13.109   Using the energy diagram shown and using Hess's law, we can see that the activation energy for the decomposition is equal to the activation energy for the formation reaction plus the heat of formation of 2 moles of HI, or $E_{a\ \text{formation}} = E_{a\ \text{decomposition}} + 2\Delta H_f^{\circ}(HI)$. So $E_{a\ \text{formation}} = 185\ \text{kJ} + 2\ \text{mol}(-5.65\ \text{kJ/mol}) = 174\ \text{kJ}$.

Check: Because the reaction is endothermic, we expect the activation energy in the reverse direction to be less than that in the forward direction.

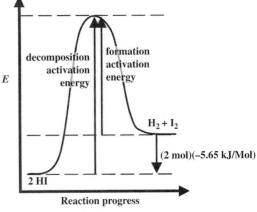

Note: Energy axis is not to scale.

Challenge Problems

13.111   (a)   Because the rate determining step involves the collision of two molecules, the expected reaction order would be second-order.

      (b)   The proposed mechanism is

$CH_3NC + CH_3NC \underset{k_2}{\overset{k_1}{\rightleftharpoons}} CH_3NC^* + CH_3NC$ (fast)

$CH_3NC^* \xrightarrow{k_3} CH_3CN$ (slow)

$\overline{CH_3NC \longrightarrow CH_3CN}$ So the sum matches the overall reaction.

CH_3NC^* is the activated molecule. Because the second step is the rate determining step, Rate $= k_3[CH_3NC^*]$. Because CH_3NC^* is an intermediate, its concentration cannot appear in the rate law. Using the fast equilibrium

in the first step, we see that $k_1[CH_3NC]^2 = k_2[CH_3NC^*][CH_3NC]$ or $[CH_3NC^*] = \dfrac{k_1}{k_2}[CH_3NC]$. Substi-

tuting this into the first rate expression, we get Rate $= k_3\dfrac{k_1}{k_2}[CH_3NC]$, which simplifies to Rate $= k[CH_3NC]$.

This matches the experimental observation of first order, and the mechanism is valid.

13.113   Rate $= k[A]^2$ and Rate $= -\dfrac{d[A]}{dt}$, so $\dfrac{d[A]}{dt} = -k[A]^2$. Moving the A terms to the left and the t and constants to

the right, we have $\dfrac{d[A]}{[A]^2} = -k\,dt$. Integrating, we get $\displaystyle\int_{[A]_0}^{[A]} \dfrac{d[A]}{[A]^2} = -\int_0^t k\,dt$. When we evaluate this integral,

$-[A]^{-1}\Big|_{[A]_0}^{[A]} = -kt\Big|_0^t \rightarrow -[A]_t^{-1} - (-[A]_0^{-1}) = -kt \rightarrow [A]_t^{-1} = kt + [A]_0^{-1}$ or $\dfrac{1}{[A]_t} = kt + \dfrac{1}{[A]_0}$, which is

the desired integrated rate law.

13.115   For this mechanism, the overall reaction is the sum of the steps in the mechanism:

$$Cl_2(g) \underset{k_2}{\overset{k_1}{\rightleftharpoons}} 2\,Cl(g)$$

$$\cancel{Cl(g)} + CO(g) \underset{k_4}{\overset{k_3}{\rightleftharpoons}} \cancel{ClCO(g)}$$

$$\cancel{ClCO(g)} + Cl_2(g) \xrightarrow{k_5} \cancel{Cl_2CO(g)} + Cl(g)$$

$$\overline{CO(g) + 2\,Cl_2(g) \rightarrow Cl_2CO(g) + 2\,Cl(g)}$$

No overall reaction is given. Because the third step is the rate determining step, $Rate = k_5[ClCO][Cl_2]$. Because ClCO is an intermediate, its concentration cannot appear in the rate law. Using the fast equilibrium in the second step, we see that $k_3[Cl][CO] = k_4[ClCO]$ or $[ClCO] = \dfrac{k_3}{k_4}[Cl][CO]$. Substituting this expression into the first rate expression, we get $Rate = k_5\dfrac{k_3}{k_4}[Cl][CO][Cl_2]$. Because Cl is an intermediate, its concentration cannot appear in the rate law. Using the fast equilibrium in the first step, we see that $k_1[Cl_2] = k_2[Cl]^2$ or $[Cl] = \sqrt{\dfrac{k_1}{k_2}[Cl_2]}$. Substituting this expression into last first rate expression, we get $Rate = k_5\dfrac{k_3}{k_4}\sqrt{\dfrac{k_1}{k_2}[Cl_2]}[CO][Cl_2] = k_5\dfrac{k_3}{k_4}\sqrt{\dfrac{k_1}{k_2}}[CO][Cl_2]^{3/2}$.

Simplifying this expression, we see that $Rate = k\,[CO][Cl_2]^{3/2}$.

13.117   For the elementary reaction $2\,NOCl(g) \underset{k_{-1}}{\overset{k_1}{\rightleftharpoons}} 2\,NO(g) + Cl_2(g)$, we see that $k_1[NOCl]^2 = k_{-1}[NO]^2[Cl_2]$. For each mole of NOCl that reacts, 1 mole of NO and 0.5 mole of Cl_2 are generated. Because before any reaction only NOCl is present, $[NO] = 2[Cl_2]$. Substituting this into the first expression, we get $k_1[NOCl]^2 = k_{-1}(2[Cl_2])^2[Cl_2] \rightarrow k_1[NOCl]^2 = 4k_{-1}[Cl_2]^3$. Rearranging and substituting the specific values into this expression, we get

$$[Cl_2] = \sqrt[3]{\dfrac{k_1}{4k_{-1}}}[NOCl]^{2/3} = \sqrt[3]{\dfrac{7.8 \times 10^{-2}\,\dfrac{L^2}{mol^2 \cdot s}}{4\left(4.7 \times 10^2\dfrac{L^2}{mol \cdot s}\right)}}\left(0.12\,\dfrac{mol}{L}\right)^{2/3} = 0.008\underline{4}2235\,\dfrac{mol}{L}\,Cl_2$$

$$= 0.0084\ M\ Cl_2 \text{ and } [Cl_2] = 2[NO] = 2\left(0.008\underline{4}2235\dfrac{mol}{L}\right) = 0.01\underline{6}8445\dfrac{mol}{L}NO = 0.017\ M\ NO$$

Conceptual Problems

13.119   Reactant concentrations drop more quickly for first-order reactions, than for second-order reactions, so reaction A must be second-order. A plot of 1/[A] versus time will be linear $\left(\dfrac{1}{[A]_t} = kt + \dfrac{1}{[A]_0}\right)$. Reaction B is first-order. A plot of ln[A] versus time will be linear $(\ln[A]_t = -kt + \ln[A]_0)$.

14 Chemical Equilibrium

Review Questions

14.1   Like adult hemoglobin, fetal hemoglobin is in equilibrium with oxygen. However, the equilibrium constant for fetal hemoglobin is larger than the equilibrium constant for adult hemoglobin, meaning that the reaction tends to go farther in the direction of the product. Consequently, fetal hemoglobin loads oxygen at a lower oxygen concentration than does adult hemoglobin. In the placenta, fetal blood flows in close proximity to maternal blood, without the two ever mixing. Because of the different equilibrium constants, the maternal hemoglobin unloads oxygen that the fetal hemoglobin then binds and carries into its own circulatory system. Nature has thus evolved a chemical system through which the mother's hemoglobin can in effect hand off oxygen to the hemoglobin of the fetus.

14.3   The general expression of the equilibrium constant is $K = \frac{[C]^c[D]^d}{[A]^a[B]^b}$.

14.5   If you reverse the equation, invert the equilibrium constant. $K_{reverse} = \dfrac{1}{K_{forward}}$

If you multiply the coefficients in the equation by a factor, raise the equilibrium constant to the same factor. That is, if you multiply the reaction by n, raise the equilibrium constant K to the n. $K' = K^n$

14.7 K_c is the equilibrium constant with respect to concentration in molarity. K_p is the equilibrium constant with respect to partial pressures in atmospheres. The two can be related as follows: $K_p = K_c(RT)^{\Delta n}$

14.9   The concentration of a solid does not change because a solid does not expand to fill its container. Its concentration, therefore, depends only on its density, which is constant as long as some solid is present. Consequently, solids are not included in the equilibrium expression. Similarly, the concentration of a pure liquid does not change, so liquids are also excluded from the equilibrium expression.

14.11   When we know the initial concentrations of the reactants and products and the equilibrium concentration of one reactant or product, the other equilibrium concentrations can be deduced from the stoichiometry of the reaction. From the initial and equilibrium concentration of the one reactant, we can find the change in concentration for that reactant. Using the stoichiometry of the reaction, we can determine the equilibrium concentrations of the other reactants and products. We generally use an ICE table (I = Initial C = Change E = Equilibrium) to keep track of the changes.

14.13   Standard states are defined as 1 M concentration, or 1 atm pressure. So the value of K_c and K_p when the reactants and products are in their standard states is 1.

14.15   To solve a problem given initial concentrations and the equilibrium constant, you would prepare an ICE table, calculate Q, compare Q and K_c, predict the direction of the reaction, represent the change with x, sum the table, determine the equilibrium values, put the equilibrium values in the equilibrium expression, and solve for x. Determine the reactant and product concentration.

14.17   According to Le Châtelier's principle, when a chemical system at equilibrium is disturbed, the system shifts in a direction that minimizes the disturbance.

14.19   If a chemical system is at equilibrium:
- decreasing the volume causes the reaction to shift in the direction that has the fewer moles of gas particles.
- increasing the volume causes the reaction to shift in the direction that has the greater number of moles of gas particles.
- if a reaction has an equal number of moles of gas on both sides of the chemical equation, then a change in volume produces no effect on the equilibrium.
- adding an inert gas to the mixture at a fixed volume has no effect on the equilibrium.

Problems by Topic

Equilibrium and the Equilibrium Constant

14.21   The equilibrium constant is defined as the concentrations of the products raised to their stoichiometric coefficients divided by the concentrations of the reactants raised to their stoichiometric coefficients.

(a) $K = \dfrac{[SbCl_3][Cl_2]}{[SbCl_5]}$

(b) $K = \dfrac{[NO]^2[Br_2]}{[BrNO]^2}$

(c) $K = \dfrac{[CS_2][H_2]^4}{[CH_4][H_2S]^2}$

(d) $K = \dfrac{[CO_2]^2}{[CO]^2[O_2]}$

14.23   With an equilibrium constant of 1.4×10^{-5}, the value of the equilibrium constant is small; therefore, the concentration of reactants will be greater than the concentration of products. This is independent of the initial concentration of the reactants and products.

14.25   (i)   has 10 H_2 and 10 I_2
(ii)   has 7 H_2 and 7 I_2 and 6 HI
(iii)   has 5 H_2 and 5 I_2 and 10 HI
(iv)   has 4 H_2 and 4 I_2 and 12 HI
(v)   has 3 H_2 and 3 I_2 and 14 HI
(vi)   has 3 H_2 and 3 I_2 and 14 HI

(a)   Concentration of (v) and (vi) are the same, so the system reached equilibrium at (v).
(b)   If a catalyst was added to the system, the system would reach the conditions at (v) sooner because a catalyst speeds up the reaction but does not change the equilibrium conditions.
(c)   The final figure (vi) would have the same amount of reactants and products because a catalyst speeds up the reaction but does not change the equilibrium concentrations.

14.27   (a)   If you reverse the reaction, invert the equilibrium constant. So $K' = \dfrac{1}{K_p} = \dfrac{1}{2.26 \times 10^4} = 4.42 \times 10^{-5}$.

The reactants will be favored.
(b)   If you multiply the coefficients in the equation by a factor, raise the equilibrium constant to the same factor. So $K' = (K_p)^{1/2} = (2.26 \times 10^4)^{1/2} = 1.50 \times 10^2$. The products will be favored.
(c)   Begin with the reverse of the reaction and invert the equilibrium constant.

$K_{reverse} = \dfrac{1}{K_p} = \dfrac{1}{2.26 \times 10^4} = 4.42 \times 10^{-5}$

Then multiply the reaction by 2 and raise the value of $K_{reverse}$ to the second power.

$K' = (K_{reverse})^2 = (4.42 \times 10^{-5})^2 = 1.95 \times 10^{-9}$. The reactants will be favored.

14.29 To find the equilibrium constant for reaction 3, you need to combine reactions 1 and 2 to get reaction 3. Begin by reversing reaction 2; then multiply reaction 1 by 2 and add the two new reactions. When you add reactions, you multiply the values of K.

$$N_2(g) + O_2(g) \rightleftharpoons 2\cancel{NO}(g) \qquad\qquad K_1 = \frac{1}{K_p} = \frac{1}{2.1 \times 10^{30}} = 4.\underline{7}6 \times 10^{-31}$$

$$\underline{2\cancel{NO}(g) + Br_2(g) \rightleftharpoons 2\,NOBr(g)} \qquad\qquad K_2 = (K_p)^2 = (5.3)^2 = 28.09$$

$$N_2(g) + O_2(g) + Br_2(g) \rightleftharpoons 2\,NOBr(g) \qquad K_3 = K_1 K_2 = (4.\underline{7}6 \times 10^{-31})(2\underline{8}.09) = 1.3 \times 10^{-29}$$

K_p, K_c, and Heterogeneous Equilibria

14.31 (a) **Given:** $K_p = 6.26 \times 10^{-22}$ $T = 298\,\text{K}$ **Find:** K_c
Conceptual Plan: $K_p \rightarrow K_c$

$$K_p = K_c(RT)^{\Delta n}$$

Solution: $\Delta n = \text{mol product gas} - \text{mol reactant gas} = 2 - 1 = 1$

$$K_c = \frac{K_p}{(RT)^{\Delta n}} = \frac{6.26 \times 10^{-22}}{\left(0.08206\dfrac{\text{L} \cdot \text{atm}}{\text{mol} \cdot \text{K}} \times 298\,\text{K}\right)^1} = 2.56 \times 10^{-23}$$

Check: Substitute into the equation and confirm that you get the original value of K_p.

$$K_p = K_c(RT)^{\Delta n} = (2.56 \times 10^{-23})\left(0.08206\dfrac{\text{L} \cdot \text{atm}}{\text{mol} \cdot \text{K}} \times 298\right)^1 = 6.26 \times 10^{-22}$$

(b) **Given:** $K_p = 7.7 \times 10^{24}$ $T = 298\,\text{K}$ **Find:** K_c
Conceptual Plan: $K_p \rightarrow K_c$

$$K_p = K_c(RT)^{\Delta n}$$

Solution: $\Delta n = \text{mol product gas} - \text{mol reactant gas} = 4 - 2 = 2$

$$K_c = \frac{K_p}{(RT)^{\Delta n}} = \frac{7.7 \times 10^{24}}{\left(0.08206\dfrac{\text{L} \cdot \text{atm}}{\text{mol} \cdot \text{K}} \times 298\,\text{K}\right)^2} = 1.3 \times 10^{22}$$

Check: Substitute into the equation and confirm that you get the original value of K_p.

$$K_p = K_c(RT)^{\Delta n} = (1.3 \times 10^{22})\left(0.08206\dfrac{\text{L} \cdot \text{atm}}{\text{mol} \cdot \text{K}} \times 298\right)^2 = 7.7 \times 10^{24}$$

(c) **Given:** $K_p = 81.9$ $T = 298\,\text{K}$ **Find:** K_c
Conceptual Plan: $K_p \rightarrow K_c$

$$K_p = K_c(RT)^{\Delta n}$$

Solution: $\Delta n = \text{mol product gas} - \text{mol reactant gas} = 2 - 2 = 0$

$$K_c = \frac{K_p}{(RT)^{\Delta n}} = \frac{81.9}{\left(0.08206\dfrac{\text{L} \cdot \text{atm}}{\text{mol} \cdot \text{K}} \times 298\,\text{K}\right)^0} = 81.9$$

Check: Substitute into the equation and confirm that you get the original value of K_p.

$$K_p = K_c(RT)^{\Delta n} = (81.9)\left(0.08206\dfrac{\text{L} \cdot \text{atm}}{\text{mol} \cdot \text{K}} \times 298\right)^0 = 81.9$$

14.33 (a) Because H_2O is a liquid, it is omitted from the equilibrium expression. $K_{eq} = \dfrac{[HCO_3^-][OH^-]}{[CO_3^{2-}]}$

(b) Because $KClO_3$ and KCl are both solids, they are omitted from the equilibrium expression. $K_{eq} = [O_2]^3$

(c) Because H_2O is a liquid, it is omitted from the equilibrium expression. $K_{eq} = \dfrac{[H_3O^+][F^-]}{[HF]}$

(d) Because H_2O is a liquid, it is omitted from the equilibrium expression. $K_{eq} = \dfrac{[NH_4^+][OH^-]}{[NH_3]}$

Relating the Equilibrium Constant to Equilibrium Concentrations and Equilibrium Partial Pressures

14.35   **Given:** at equilibrium: $[CO] = 0.105$ M, $[H_2] = 0.114$ M, $[CH_3OH] = 0.185$ M   **Find:** K_c
Conceptual Plan: Balanced reaction → equilibrium expression → K_c

Solution: $K_c = \dfrac{[CH_3OH]}{[CO][H_2]^2} = \dfrac{(0.185)}{(0.105)(0.114)^2} = 136$

Check: The answer is reasonable because the concentration of products is greater than the concentration of reactants and the equilibrium constant should be greater than 1.

14.37   At 500 K: **Given:** at equilibrium: $[N_2] = 0.115$ M, $[H_2] = 0.105$ M, and $[NH_3] = 0.439$ M   **Find:** K_c
Conceptual Plan: Balanced reaction → equilibrium expression → K_c

Solution: $K_c = \dfrac{[NH_3]^2}{[N_2][H_2]^3} = \dfrac{(0.439)^2}{(0.115)(0.105)^3} = 1.45 \times 10^3$

Check: The value is reasonable because the concentration of products is greater than the concentration of reactants.

At 575 K: **Given:** at equilibrium: $[N_2] = 0.110$ M, $[NH_3] = 0.128$ M, $K_c = 9.6$ **Find:** $[H_2]$
Conceptual Plan: Balanced reaction → equilibrium expression → $[H_2]$

Solution: $K_c = \dfrac{[NH_3]^2}{[N_2][H_2]^3}$ $9.6 = \dfrac{(0.128)^2}{(0.110)(x)^3}$ $x = 0.249$

Check: Plug the value for x back into the equilibrium expression and check the value.

$9.6 = \dfrac{(0.128)^2}{(0.110)(0.249)^3}$

At 775 K: **Given:** at equilibrium: $[N_2] = 0.120$ M, $[H_2] = 0.140$ M, $K_c = 0.0584$ **Find:** $[NH_3]$
Conceptual Plan: Balanced reaction → equilibrium expression → $[NH_3]$

Solution: $K_c = \dfrac{[NH_3]^2}{[N_2][H_2]^3}$ $0.0584 = \dfrac{(x)^2}{(0.120)(0.140)^3}$ $x = 0.00439$

Check: Plug the value for x back into the equilibrium expression and check the value.

$0.0584 = \dfrac{(0.00439)^2}{(0.120)(0.140)^3}$

14.39   **Given:** $P_{NO} = 108$ torr; $P_{Br_2} = 126$ torr, $K_p = 28.4$ **Find:** P_{NOBr}
Conceptual Plan: torr → atm and then balanced reaction → equilibrium expression → P_{NOBr}

$\dfrac{1\ atm}{760\ torr}$

Solution: $P_{NO} = 108\ \text{torr} \times \dfrac{1\ atm}{760\ \text{torr}} = 0.142\underline{1}$ atm $P_{Br_2} = 126\ \text{torr} \times \dfrac{1\ atm}{760\ \text{torr}} = 0.165\underline{8}$ atm

$K_p = \dfrac{P_{NOBr}^2}{P_{NO}^2 P_{Br_2}}$ $28.4 = \dfrac{x^2}{(0.142\underline{1})^2(0.165\underline{8})}$ $x = 0.308$ atm $= 234$ torr

Check: Plug the value for x back into the equilibrium expression and check the value.

$28.3 = \dfrac{(0.308)^2}{(0.142\underline{1})^2(0.165\underline{8})}$

14.41   **Given:** $P_{A,\ initial} = 1.32$ atm; $P_{A,\ eq} = 0.25$ atm   **Find:** K_p
Conceptual Plan: 1. **Prepare ICE table.**
2. **Calculate pressure change for known value.**
3. **Calculate pressure changes for other reactants/products.**
4. **Determine equilibrium pressures.**
5. **Write the equilibrium expression and determine K_p.**

Solution: $A(g) \rightleftharpoons 2B(g)$

	P_A	P_B
Initial	1.32 atm	0.00
Change	$-x$	$+2x$
Equil	0.25 atm	$2x$

$$x = 1.32 - 0.25 = 1.07 \text{ atm}$$
$$2x = 2(1.07 \text{ atm}) = 2.14 \text{ atm} = P_B$$
$$K_p = \frac{P_B^2}{P_A} = \frac{(2.14)^2}{(0.25)} = 1\underline{8}.32 = 18$$

14.43   **Given:** $[Fe^{3+}]_{initial} = 1.0 \times 10^{-3}$ M; $[SCN^-]_{initial} = 8.0 \times 10^{-4}$ M; $[FeSCN^{2+}]_{eq} = 1.7 \times 10^{-4}$ M   **Find:** K_c
Conceptual Plan: 1. Prepare ICE table.
　　　　　　　　　　　2. Calculate concentration change for known value.
　　　　　　　　　　　3. Calculate concentration changes for other reactants/products.
　　　　　　　　　　　4. Determine equilibrium concentration.
　　　　　　　　　　　5. Write the equilibrium expression and determine K_c.

Solution: $Fe^{3+}(aq) \quad + \quad SCN^-(aq) \rightleftharpoons FeSCN^{3+}(aq)$

	$[Fe^{3+}]$	$[SCN^-]$	$[FeSCN^{3+}]$
Initial	1.0×10^{-3}	8.0×10^{-4}	0.00
Change	-1.7×10^{-4}	-1.7×10^{-4}	$+1.7 \times 10^{-4}$
Equil	8.3×10^{-4}	6.3×10^{-4}	1.7×10^{-4}

$$K_c = \frac{[FeSCN^{2+}]}{[Fe^{3+}][SCN^-]} = \frac{(1.7 \times 10^{-4})}{(8.3 \times 10^{-4})(6.3 \times 10^{-4})} = 3.3 \times 10^2$$

14.45   **Given:** 3.67 L flask, 0.763 g H_2 initial, 96.9 g I_2 initial, 90.4 g HI equilibrium   **Find:** K_c
Conceptual Plan: g → mol → M and then

$$n = \frac{g}{\text{molar mass}} \quad M = \frac{n}{V}$$

　　　　　　　　　　　1. Prepare ICE table.
　　　　　　　　　　　2. Calculate concentration change for known value.
　　　　　　　　　　　3. Calculate concentration changes for other reactants/products.
　　　　　　　　　　　4. Determine equilibrium concentration.
　　　　　　　　　　　5. Write the equilibrium expression and determine K_c.

Solution: $0.763 \text{ g } H_2 \times \dfrac{1 \text{ mol } H_2}{2.016 \text{ g } H_2} = 0.37\underline{8}5 \text{ mol } H_2$ then $\dfrac{0.37\underline{8}5 \text{ mol } H_2}{3.67 \text{ L}} = 0.103 \text{ M}$ This is an initial concentration.

$96.9 \text{ g } I_2 \times \dfrac{1 \text{ mol } I_2}{253.8 \text{ g } I_2} = 0.38\underline{1}8 \text{ mol } I_2$ then $\dfrac{0.38\underline{1}8 \text{ mol } I_2}{3.67 \text{ L}} = 0.104 \text{ M}$ This is an initial concentration.

$90.4 \text{ g } HI \times \dfrac{1 \text{ mol } HI}{127.9 \text{ g } I_2} = 0.70\underline{6}8 \text{ mol } HI$ then $\dfrac{0.70\underline{6}8 \text{ mol } HI}{3.67 \text{ L}} = 0.193 \text{ M}$ This is an equilibrium concentration.

$H_2(g) \quad + \quad I_2(g) \rightleftharpoons 2 HI(g)$

	$[H_2]$	$[I_2]$	$[HI]$
Initial	0.103	0.104	0.00
Change	-0.0965	-0.0965	$+0.193$
Equil	0.0065	0.0075	0.193

Because HI gained 0.193 M, H_2 and I_2 had to lose $0.193/2 = 0.0965$ M from the stoichiometry of the balanced reaction.

$$K_c = \frac{[HI]^2}{[H_2][I_2]} = \frac{(0.193)^2}{(0.0065)(0.0075)} = 764$$

The Reaction Quotient and Reaction Direction

14.47 **Given:** $K_c = 8.5 \times 10^{-3}$; $[NH_3] = 0.166$ M; $[H_2S] = 0.166$ M **Find:** whether solid will form or decompose
Conceptual Plan: Calculate $Q \rightarrow$ **compare** $Q \rightarrow$ **and** K_c
Solution: $Q = [NH_3][H_2S] = (0.166)(0.166) = 0.0276$
$Q = 0.0276$ and $K_c = 8.5 \times 10^{-3}$ so $Q > K_c$ and the reaction will shift to the left; so more solid will form.

14.49 **Given:** 6.55 g Ag_2SO_4, 1.5 L solution, $K_c = 1.1 \times 10^{-5}$ **Find:** whether more solid will dissolve
Conceptual Plan: g $Ag_2SO_4 \rightarrow$ **mol** $Ag_2SO_4 \rightarrow [Ag_2SO_4] \rightarrow [Ag^+], [SO_4^{-2}] \rightarrow$ **calculate** Q **and compare to** K_c

$$\frac{1 \text{ mol } Ag_2SO_4}{311.81 \text{ g}} \qquad [\,] = \frac{\text{mol } Ag_2SO_4}{\text{vol solution}} \qquad\qquad Q = [Ag^+]^2[SO_4^{-2}]$$

Solution: $6.55 \text{ g } Ag_2SO_4 \left(\dfrac{1 \text{ mol } Ag_2SO_4}{311.81 \text{ g } Ag_2SO_4} \right) = 0.0210 \text{ mol } Ag_2SO_4$

$$\frac{0.0210 \text{ mol } Ag_2SO_4}{1.5 \text{ L solution}} = 0.01\underline{4}0 \text{ M } Ag_2SO_4$$

$[Ag^+] = 2[Ag_2SO_4] = 2(0.01\underline{4}0 \text{ M}) = 0.02\underline{8}0 \text{ M}$; $[SO_4^{2-}] = [Ag_2SO_4] = 0.01\underline{4}0 \text{ M}$
$Q = [Ag^+]^2[SO_4^{2-}] = (0.02\underline{8}0)^2(0.01\underline{4}0) = 1.1 \times 10^{-5}$
$Q = K_c$, so the system is at equilibrium and is a saturated solution. Therefore, if more solid is added, it will not dissolve.

Finding Equilibrium Concentrations from Initial Concentrations and the Equilibrium Constant

14.51 (a) **Given:** $[A] = 1.0$ M, $[B] = 0.0$, $K_c = 4.0$; $a = 1, b = 1$ **Find:** $[A], [B]$ at equilibrium
Conceptual Plan: Prepare an ICE table, calculate Q, **compare** Q **and** K_c, **predict the direction of the reaction, represent the change with** x, **sum the table, determine the equilibrium values, put the equilibrium values in the equilibrium expression, and solve for** x. **Determine [A] and [B].**
Solution: $A(g) \rightleftharpoons B(g)$

	$[A]$	$[B]$
Initial	1.0	0.00
Change	$-x$	$+x$
Equil	$1.0 - x$	x

$Q = \dfrac{[B]}{[A]} = \dfrac{0}{1.0} = 0$ $Q < K$ Therefore, the reaction will proceed to the right by x.

$K_c = \dfrac{[B]}{[A]} = \dfrac{(x)}{(1.0 - x)} = 4.0$; $x = 0.80$

$[A] = 1 - 0.80 = 0.20$ M; $[B] = 0.80$ M

Check: Plug the values into the equilibrium expression: $K_c = \dfrac{0.80}{0.20} = 4.0$

(b) **Given:** $[A] = 1.0$ M, $[B] = 0.0$, $K_c = 4.0$; $a = 2, b = 2$ **Find:** $[A], [B]$ at equilibrium
Conceptual Plan: Prepare an ICE table, calculate Q, **compare** Q **and** K_c, **predict the direction of the reaction, represent the change with** x, **sum the table, determine the equilibrium values, put the equilibrium values in the equilibrium expression, and solve for** x. **Determine [A] and [B].**
Solution: $2 A(g) \rightleftharpoons 2 B(g)$

	$[A]$	$[B]$
Initial	1.0	0.00
Change	$-2x$	$+2x$
Equil	$1.0 - 2x$	$2x$

$$Q = \frac{[B]^2}{[A]^2} = \frac{0}{1.0} = 0 \quad Q < K \quad \text{Therefore, the reaction will proceed to the right by } x.$$

$$K_c = \frac{[B]^2}{[A]^2} = \frac{(2x)^2}{(1.0 - 2x)^2} = 4.0$$

$$\sqrt{\frac{(2x)^2}{(1.0 - 2x)^2}} = \sqrt{4.0} \quad \text{Take the square root of this expression. } x = 0.33$$

$$[A] = 1 - 2(0.33) = 0.34 \text{ M}; \quad [B] = 2(0.33) = 0.66 \text{ M}$$

Check: Plug the values into the equilibrium expression: $K_c = \dfrac{(0.66)^2}{(0.34)^2} = 3.8.$

(c) **Given:** $[A] = 1.0 \text{ M}, [B] = 0.0, K_c = 4.0; a = 1, b = 2$ **Find:** $[A], [B]$ at equilibrium
Conceptual Plan: Prepare an ICE table, calculate Q, compare Q and K_c, predict the direction of the reaction, represent the change with x, sum the table, determine the equilibrium values, put the equilibrium values in the equilibrium expression, and solve for x. Determine [A] and [B].
Solution: $\quad A(g) \rightleftharpoons 2 B(g)$

	$[A]$	$[B]$
Initial	1.0	0.00
Change	$-x$	$+2x$
Equil	$1.0 - x$	$2x$

$$Q = \frac{[B]^2}{[A]} = \frac{0}{1.0} = 0 \quad Q < K \quad \text{Therefore, the reaction will proceed to the right by } x.$$

$$K_c = \frac{[B]^2}{[A]} = \frac{(2x)^2}{(1.0 - x)} = 4.0 \qquad 4x^2 + 4x - 4 = 0 \quad \text{Solve using the quadratic equation, found in Appendix I.}$$

$$x = -1.6 \quad \text{or } x = 0.62; \text{ therefore, } x = 0.62.$$

$$[A] = 1.0 - 0.62 = 0.38 \text{ M} \quad [B] = 2x = 2(0.62) = 1.2 \text{ M}$$

Check: Plug the values into the equilibrium expression: $K_c = \dfrac{(1.2)^2}{0.38} = 3.8.$

14.53 **Given:** $[N_2O_4] = 0.0500 \text{ M}, [NO_2] = 0.0, K_c = 0.513$ **Find:** $[N_2O_4], [NO_2]$ at equilibrium
Conceptual Plan: Prepare an ICE table, calculate Q, compare Q and K_c, predict the direction of the reaction, represent the change with x, sum the table, determine the equilibrium values, put the equilibrium values in the equilibrium expression, and solve for x. Determine $[N_2O_4]$ and $[NO_2]$.
Solution:
$$N_2O_4(g) \rightleftharpoons 2 NO_2(g)$$

	$[N_2O_4]$	$[NO_2]$
Initial	0.0500	0.00
Change	$-x$	$+2x$
Equil	$0.0500 - x$	$2x$

$$Q = \frac{[NO_2]^2}{[N_2O_4]} = \frac{0}{0.0500} = 0 \qquad Q < K \quad \text{Therefore, the reaction will proceed to the right by } x.$$

$$K_c = \frac{[NO_2]^2}{[N_2O_4]} = \frac{(2x)^2}{(0.0500 - x)} = 0.513 \quad 4x^2 + 0.513x - 0.02565 = 0$$

$$\frac{-b \pm \sqrt{b^2 - 4ac}}{2a} = \frac{-0.513 \pm \sqrt{(0.513)^2 - 4(4)(-0.02565)}}{2(4)} = \frac{-0.513 \pm \sqrt{0.6735}}{2(4)}$$

$$x = -0.1667 \text{ or } x = 0.0385 \quad \text{Therefore, } x = 0.0385.$$

$$[N_2O_4] = 0.0500 - 0.0385 = 0.0115 \text{ M} \quad [NO_2] = 2x = 2(0.0385) = 0.0770 \text{ M}$$

Check: Plug the values into the equilibrium expression: $K_c = \dfrac{(0.0770)^2}{0.0115} = 0.516.$

14.55 **Given:** $[CO] = 0.20 \, M$, $[CO_2] = 0.0$, $K_c = 4.0 \times 10^3$ **Find:** $[CO_2]$ at equilibrium
Conceptual Plan: Prepare an ICE table, calculate Q, compare Q and K_c, predict the direction of the reaction, represent the change with x, sum the table, determine the equilibrium values, put the equilibrium values in the equilibrium expression, and solve for x. Determine $[CO]$, $[Cl_2]$, and $[COCl_2]$.
Solution:

$$NiO(s) + CO(g) \rightleftharpoons Ni(s) + CO_2(g)$$

	$[CO]$	$[Cl_2]$
Initial	0.20	0.0
Change	$-x$	$+x$
Equil	$0.20 - x$	x

$$Q = \frac{[CO_2]}{[CO]} = \frac{0}{(0.20)} = 0 \qquad Q < K \quad \text{Therefore, the reaction will proceed to the right by } x.$$

$$K_c = \frac{[CO_2]}{[CO]} = \frac{x}{(0.20 - x)} = 4.0 \times 10^3 \qquad 4.0 \times 10^3 (0.20 - x) = x$$

$$x = 0.199$$

$$[CO_2] = 0.199 \, M$$

Check: Because the equilibrium constant is so large, the reaction goes essentially to completion; therefore, it is reasonable that the concentration of the product is 0.199 M.

14.57 **Given:** $[HC_2H_3O_2] = 0.210 \, M$, $[H_3O^+] = 0.0$, $[C_2H_3O_2^-] = 0.0$, $K_c = 1.8 \times 10^{-5}$
Find: $[HC_2H_3O_2]$, $[H_2O^+]$, $[C_2H_3O_2^-]$ at equilibrium
Conceptual Plan: Prepare an ICE table, calculate Q, compare Q and K_c, predict the direction of the reaction, represent the change with x, sum the table, determine the equilibrium values, put the equilibrium values in the equilibrium expression, and solve for x. Determine $[CO]$, $[Cl_2]$, and $[COCl_2]$.
Solution:

$$HC_2H_3O_2(aq) + H_2O(l) \rightleftharpoons H_3O^+(aq) + C_2H_3O_2^-(aq)$$

	$HC_2H_3O_2$	$[H_2O]$	$[H_3O^+]$	$[C_2H_3O_2^-]$
Initial	0.210		0.0	0.0
Change	$-x$		$+x$	$+x$
Equil	$0.210 - x$		x	x

$$Q = \frac{[H_3O^+][C_2H_3O_2^-]}{[HC_2H_3O_2]} = \frac{0}{(0.210)} = 0 \qquad Q < K \quad \text{Therefore, the reaction will proceed to the right by } x.$$

$$K_c = \frac{[H_3O^+][C_2H_3O_2^-]}{[HC_2H_3O_2]} = \frac{(x)(x)}{(0.210 - x)} = 1.8 \times 10^{-5}$$

Assume that x is small compared to 0.210.
$$x^2 = 0.210(1.8 \times 10^{-5})$$

$$x = 0.00194 \quad \text{Check assumption: } \frac{0.00194}{0.210} \times 100 = 0.92\%; \text{assumption is valid.}$$

$$[H_3O^+] = [C_2H_3O_2^-] = 0.00194 \, M$$
$$[HC_2H_3O_2] = 0.210 - 0.00194 = 0.2081 = 0.208 \, M$$

Check: Plug the values into the equilibrium expression: $K_c = \dfrac{(0.00194)(0.00194)}{(0.208)} = 1.81 \times 10^{-5}$.

The answer is the same to two significant figures with the true value, so the answers are valid.

14.59   **Given:** $P_{Br_2} = 755$ torr, $P_{Cl_2} = 735$ torr, $P_{BrCl} = 0.0$, $K_p = 1.11 \times 10^{-4}$ **Find:** P_{BrCl} at equilibrium
    Conceptual Plan: Torr → **atm and then prepare an ICE table, calculate Q, compare Q and K_c, predict the direction of the reaction, represent the change with x, sum the table, determine the equilibrium values, put the equilibrium values in the equilibrium expression, and solve for x. Determine P_{BrCl}.**

    Solution: $P_{Br_2} = 755 \text{ torr} \times \dfrac{1 \text{ atm}}{760 \text{ torr}} = 0.99\underline{3}4 \text{ atm}$ $P_{Cl_2} = 735 \text{ torr} \times \dfrac{1 \text{ atm}}{760 \text{ torr}} = 0.96\underline{7}1 \text{ atm}$

	$Br_2(g)$	$+$	$Cl_2(g)$	$\rightleftharpoons$	$2\,BrCl(g)$
	P_{Br_2}		P_{Cl_2}		P_{BrCl}
Initial	0.9934		0.9671		0.0
Change	$-x$		$-x$		$+2x$
Equil	$0.9934 - x$		$0.9671 - x$		$2x$

$$Q = \frac{P_{BrCl}^2}{P_{Br_2}P_{Cl_2}} = \frac{0}{(0.9934)(0.9671)} = 0 \qquad\qquad Q < K \quad \text{Therefore, the reaction will proceed to the right by } x.$$

$$K_p = \frac{P_{BrCl}^2}{P_{Br_2}P_{Cl_2}} = \frac{(2x)^2}{(0.9934 - x)(0.9671 - x)} = 1.11 \times 10^{-4}$$

Assume that x is small compared to 0.9934 and 0.9671.

$$\frac{(2x)^2}{(0.9934)(0.9671)} = 1.11 \times 10^{-4} \qquad 4x^2 = 1.066 \times 10^{-4}$$

$x = 0.00516$ atm $= 3.92$ torr

$P_{BrCl} = 2x = 2(3.92 \text{ torr}) = 7.84$ torr

Check: Plug the values into the equilibrium expression:

$$K_c = \frac{(2(0.00516))^2}{(0.9934 - 0.00516)(0.9671 - 0.00516)} = 1.120 \times 10^{-4} = 1.12 \times 10^{-4}.$$

This is within 0.01×10^{-4} of the true value; therefore, the answers are valid.

14.61   (a)   **Given:** $[A] = 1.0$ M, $[B] = [C] = 0.0$, $K_c = 1.0$ **Find:** $[A], [B], [C]$ at equilibrium
               Conceptual Plan: Prepare an ICE table, calculate Q, compare Q and K_c, predict the direction of the reaction, represent the change with x, sum the table, determine the equilibrium values, put the equilibrium values in the equilibrium expression, and solve for x. Determine $[A], [B]$, and $[C]$.
               Solution:

	$A(g)$	$\rightleftharpoons$	$B(g)$	$+$	$C(g)$
	$[A]$		$[B]$		$[C]$
Initial	1.0		0.0		0.0
Change	$-x$		$+x$		$+x$
Equil	$1.0 - x$		x		x

$$Q = \frac{[B][C]}{[A]} = \frac{0}{(1.0)} = 0 \qquad\qquad Q < K \quad \text{Therefore, the reaction will proceed to the right by } x.$$

$$K_c = \frac{[B][C]}{[A]} = \frac{(x)(x)}{(1.0 - x)} = 1.0$$

$x^2 = 1.0(1.0 - x)$

$x^2 + x - 1 = 0$

$$\frac{-b \pm \sqrt{b^2 - 4ac}}{2a} = \frac{-1 \pm \sqrt{1^2 - 4(1)(-1)}}{2(1)}$$

$x = 0.6\underline{1}8$ or $x = -1.618$ Therefore, $x = 0.6\underline{1}8$.

$[B] = [C] = x = 0.6\underline{1}8 = 0.62\,M$

$[A] = 1.0 - 0.6\underline{1}8 = 0.382 = 0.38\,M$

Check: Plug the values into the equilibrium expression:

$K_c = \dfrac{(0.62)(0.62)}{(0.38)} = 1.01 = 1.0$, which is the equilibrium constant; so the values are correct.

(b) **Given:** $[A] = 1.0\,M$, $[B] = [C] = 0.0$, $K_c = 0.010$ **Find:** $[A], [B], [C]$ at equilibrium
Conceptual Plan: Prepare an ICE table, calculate Q, compare Q and K_c, predict the direction of the reaction, represent the change with x, sum the table, determine the equilibrium values, put the equilibrium values in the equilibrium expression, and solve for x. Determine $[A], [B]$, and $[C]$.

Solution:

$$A(g) \rightleftharpoons B(g) + C(g)$$

	$[A]$	$[B]$	$[C]$
Initial	1.0	0.0	0.0
Change	$-x$	$+x$	$+x$
Equil	$1.0 - x$	x	x

$Q = \dfrac{[B][C]}{[A]} = \dfrac{0}{(1.0)} = 0$ $Q < K$ Therefore, the reaction will proceed to the right by x.

$K_c = \dfrac{[B][C]}{[A]} = \dfrac{(x)(x)}{(1.0 - x)} = 0.010$

$x^2 = 0.010(1.0 - x)$

$x^2 + 0.010x - 0.010 = 0$

$\dfrac{-b \pm \sqrt{b^2 - 4ac}}{2a} = \dfrac{-(0.010) \pm \sqrt{(0.010)^2 - 4(1)(-0.010)}}{2(1)}$

$x = 0.09\underline{5}12$ or $x = -0.1051$ Therefore, $x = 0.09\underline{5}12$.

$[B] = [C] = x = 0.09\underline{5}12 = 0.095\,M$

$[A] = 1.0 - 0.09\underline{5}12 = 0.90488 = 0.90\,M$

Check: Plug the values into the equilibrium expression:

$K_c = \dfrac{(0.095)(0.095)}{(0.90)} = 0.01003 = 0.010$, which is the equilibrium constant; so the values are correct.

(c) **Given:** $[A] = 1.0\,M$, $[B] = [C] = 0.0$, $K_c = 1.0 \times 10^{-5}$ **Find:** $[A], [B], [C]$ at equilibrium
Conceptual Plan: Prepare an ICE table, calculate Q, compare Q and K_c, predict the direction of the reaction, represent the change with x, sum the table, determine the equilibrium values, put the equilibrium values in the equilibrium expression, and solve for x. Determine $[A], [B]$, and $[C]$.

Solution:

$$A(g) \rightleftharpoons B(g) + C(g)$$

	$[A]$	$[B]$	$[C]$
Initial	1.0	0.0	0.0
Change	$-x$	$+x$	$+x$
Equil	$1.0 - x$	x	x

$Q = \dfrac{[B][C]}{[A]} = \dfrac{0}{(1.0)} = 0$ $Q < K$ Therefore, the reaction will proceed to the right by x.

$K_c = \dfrac{[B][C]}{[A]} = \dfrac{(x)(x)}{(1.0 - x)} = 1.0 \times 10^{-5}$

Assume that x is small compared to 1.0.

$x^2 = 1.0(1.0 \times 10^{-5})$

$x = 0.003\underline{1}6$ Check assumption: $\dfrac{0.003\underline{1}6}{1.0} \times 100\% = 0.32\%$; assumption is valid.

$[B] = [C] = x = 0.003\underline{16} = 0.0032\,M$

$[A] = 1.0 - 0.003\underline{16} = 0.9968 = 1.0\,M$

Check: Plug the values into the equilibrium expression:

$$K_c = \frac{(0.0032)(0.0032)}{(1.0)} = 1.024 \times 10^{-5} = 1.0 \times 10^{-5},$$

which is the equilibrium constant; so the values are correct

Le Châtelier's Principle

14.63   **Given:** $CO(g) + Cl_2(g) \rightleftharpoons COCl_2(g)$ at equilibrium   **Find:** What is the effect of each of the following?

(a) $COCl_2$ is added to the reaction mixture. Adding $COCl_2$ increases the concentration of $COCl_2$ and causes the reaction to shift to the left.

(b) Cl_2 is added to the reaction mixture. Adding Cl_2 increases the concentration of Cl_2 and causes the reaction to shift to the right.

(c) $COCl_2$ is removed from the reaction mixture. Removing the $COCl_2$ decreases the concentration of $COCl_2$ and causes the reaction to shift to the right.

14.65   **Given:** $2\,KClO_3(s) \rightleftharpoons 2\,KCl(s) + 3\,O_2(g)$ at equilibrium   **Find:** What is the effect of each of the following?

(a) O_2 is removed from the reaction mixture. Removing the O_2 decreases the concentration of O_2 and causes the reaction to shift to the right.

(b)   KCl is added to the reaction mixture. Adding KCl does not cause any change in the reaction. KCl is a solid, and the concentration remains constant; so the addition of more solid does not change the equilibrium concentration.

(c) $KClO_3$ is added to the reaction mixture. Adding $KClO_3$ does not cause any change in the reaction. $KClO_3$ is a solid, and the concentration remains constant; so the addition of more solid does not change the equilibrium concentration.

(d) O_2 is added to the reaction mixture. Adding O_2 increases the concentration of O_2 and causes the reaction to shift to the left.

14.67   (a)   **Given:** $I_2(g) \rightleftharpoons 2\,I(g)$ at equilibrium   **Find:** the effect of increasing the volume

The chemical equation has 2 moles of gas on the right and 1 mole of gas on the left. Increasing the volume of the reaction mixture decreases the pressure and causes the reaction to shift to the right (toward the side with more moles of gas particles).

(b)   **Given:** $2\,H_2S(g) \rightleftharpoons 2\,H_2(g) + S_2(g)$ **Find:** the effect of decreasing the volume

The chemical equation has 3 moles of gas on the right and 2 moles of gas on the left. Decreasing the volume of the reaction mixture increases the pressure and causes the reaction to shift to the left (toward the side with fewer moles of gas particles).

(c)   **Given:** $I_2(g) + Cl_2(g) \rightleftharpoons 2\,ICl(g)$ **Find:** the effect of decreasing the volume

The chemical equation has 2 moles of gas on the right and 2 moles of gas on the left. Decreasing the volume of the reaction mixture increases the pressure but causes no shift in the reaction because the moles are equal on both sides.

14.69   **Given:** $C(s) + CO_2(g) \rightleftharpoons 2\,CO(g)$ is endothermic.   **Find:** the effect of increasing the temperature

Because the reaction is endothermic, we can think of the heat as a reactant. Increasing the temperature is equivalent to adding a reactant, causing the reaction to shift to the right. This will cause an increase in the concentration of products and a decrease in the concentration of reactants; therefore, the value of K will increase.

Find: the effect of decreasing the temperature

Because the reaction is endothermic, we can think of the heat as a reactant. Decreasing the temperature is equivalent to removing a reactant, causing the reaction to shift to the left. This will cause a decrease in the concentration of products and an increase in the concentration of reactants; therefore, the value of K will decrease.

14.71   **Given:** $C(s) + 2\,H_2(g) \rightleftharpoons CH_4(g)$ is exothermic.   **Find:** Determine which will favor CH_4.

(a)   Adding more C to the reaction mixture does *not* favor CH_4. Adding C does not cause any change in the reaction. C is a solid, and the concentration remains constant; so the addition of more solid does not change the equilibrium concentration.

(b)   Adding more H_2 to the reaction mixture favors CH_4. Adding H_2 increases the concentration of H_2, causing the reaction to shift to the right.

(c)   Raising the temperature of the reaction mixture does *not* favor CH_4. Because the reaction is exothermic, we can think of heat as a product. Raising the temperature is equivalent to adding a product, causing the reaction to shift to the left.

(d)   Lowering the volume of the reaction mixture favors CH_4. The chemical equation has 1 mole of gas on the right and 2 moles of gas on the left. Decreasing the volume of the reaction mixture increases the pressure and causes the reaction to shift to the right (toward the side with fewer moles of gas particles).

(e)   Adding a catalyst to the reaction mixture does *not* favor CH_4. A catalyst added to the reaction mixture only speeds up the reaction; it does not change the equilibrium concentration.

(f)   Adding neon gas to the reaction mixture does *not* favor CH_4. Adding an inert gas to a reaction mixture at a fixed volume has no effect on the equilibrium.

Cumulative Problems

14.73   (a)   To find the value of K for the new equation, combine the two given equations to yield the new equation. Reverse equation 1 and use $1/K_1$, then add to equation 2. To find K for equation 3, use $(1/K_1)(K_2)$.

$$HbO_2(aq) \rightleftharpoons \cancel{Hb(aq)} + O_2(aq) \qquad K_1 = 1/1.8$$

$$\cancel{Hb(aq)} + CO(aq) \rightleftharpoons HbCO(aq) \qquad K_2 = 306$$

$$\overline{HbO_2(aq) + CO(aq) \rightleftharpoons HbCO(aq) + O_2(aq) \qquad K_3 = K_1 K_2 = (1/1.8)(306) = 170}$$

(b)   **Given:** $O_2 = 20\%, CO = 0.10\%$ **Find:** the ratio $\dfrac{[HbCO]}{[HbO_2]}$

Conceptual Plan: Determine the equilibrium expression and then determine $\dfrac{[HbCO]}{[HbO_2]}$**.**

Solution: $K = \dfrac{[HbCO][O_2]}{[HbO_2][CO]}$ $170 = \dfrac{[HbCO](20.0)}{[HbO_2](0.10)}$ then $\dfrac{[HbCO]}{[HbO_2]} = 170\left(\dfrac{0.10}{20.0}\right) = \dfrac{0.85}{1.0}$

Because the ratio is almost 1:1, 0.10% CO will replace about 50% of the O_2 in the blood. The CO blocks the uptake of O_2 by the blood and is therefore highly toxic.

14.75   (a)   **Given:** 4.45 g CO_2, 10.0 L, 1200 K, 2.00 g C, $K_p = 5.78$ **Find:** total pressure

Conceptual Plan: g $CO_2 \rightarrow$ mol CO_2 and g C $\rightarrow$ mol C and then determine limiting reactant

$$\frac{1\ mol\ CO_2}{44.01\ g\ CO_2} \qquad \frac{1\ mol\ C}{12.01\ g\ C}$$

and then mol $CO_2 \rightarrow P\ CO_2$. Prepare an ICE table, represent the change with x, sum the table,

$$PV = nRT$$

determine the equilibrium values, put the equilibrium values in the equilibrium expression, and solve for x.

Solution: $CO_2(g) + C(s) \rightleftharpoons 2\,CO(g)$

$$4.45\ \cancel{g\ CO_2} \times \frac{1\ mol}{44.01\ \cancel{g\ CO_2}} = 0.10\underline{1}1\ mol\ CO_2 \quad 2.00\ \cancel{g\ C} \times \frac{1\ mol}{12.01\ \cancel{g\ C}} = 0.16\underline{6}5\ mol\ C$$

Because the stoichiometry is 1:1, the CO_2 is the limiting reactant.

$$P_{CO_2} = \frac{(0.10\underline{1}1\ \cancel{mol})\left(\dfrac{0.08206\ \cancel{L} \cdot atm}{\cancel{mol} \cdot \cancel{K}}\right)(1200\ \cancel{K})}{10.0\ \cancel{L}} = 0.99\underline{5}6\ atm$$

$$CO_2(g) + C(s) \rightleftharpoons 2\,CO(g)$$

	P_{CO_2}	P_{CO}
Initial	0.996	0.0
Change	$-x$	$+2x$
Equil	$0.996 - x$	$2x$

The reaction will proceed to the right by x.

$$K_p = \frac{P_{CO}^2}{P_{CO_2}} = \frac{(2x)^2}{(0.996 - x)} = 5.78. \text{ Solve using the quadratic equation, found in Appendix I.}$$

$$x = 0.678 \text{ atm}$$

$$P_{CO_2} = 0.996 \text{ atm} - 0.678 \text{ atm} = 0.318 \text{ atm} \quad P_{CO} = 2(0.678 \text{ atm}) = 1.3\underline{5}6 \text{ atm}$$

$$P \text{ total} = 1.67 \text{ atm}$$

Check: Plug the values for the partial pressure into the equilibrium expression. $\dfrac{(1.356)^2}{0.318} = 5.78$, which is the value of the equilibrium constant.

(b)   **Given:** 4.45 g CO_2, 10.0 L, 1200 K, 0.50 g C, $K_p = 5.78$ **Find:** total pressure

Conceptual Plan: g $CO_2 \rightarrow$ mol CO_2 and g C $\rightarrow$ mol C and then determine limiting reactant

$$\frac{1 \text{ mol } CO_2}{44.01 \text{ g } CO_2} \qquad\qquad \frac{1 \text{ mol C}}{12.01 \text{ g C}}$$

and then mol $CO_2 \rightarrow P_{CO_2}$ and mol C $\rightarrow$ mol CO $\rightarrow P_{CO}$

$$PV = nRT \qquad\qquad\qquad\qquad PV = nRT$$

Solution: $CO_2(g) + C(s) \rightleftharpoons 2\,CO(g)$

$$4.45 \text{ g } CO_2 \times \frac{1 \text{ mol}}{44.01 \text{ g } CO_2} = 0.10\underline{1}1 \text{ mol } CO_2 \qquad 0.50 \text{ g C} \times \frac{1 \text{ mol}}{12.01 \text{ g C}} = 0.04\underline{1}6 \text{ mol C}$$

Because the stoichiometry is 1:1, the C is the limiting reactant; therefore, the moles of CO formed will be determined from the reaction, not the equilibrium.

$$0.04\underline{1}6 \text{ mol C} \times \frac{2 \text{ mol CO}}{1 \text{ mol C}} = 0.08\underline{3}2 \text{ mol CO}$$

	$CO_2(g)$	$+$	$C(s)$	$\rightleftharpoons$	$2\,CO(g)$
Initial	0.1011		0.04\underline{1}6		0.0
Change	$-0.04\underline{1}6$		$-0.04\underline{1}6$		$+2(0.04\underline{1}6)$
Equil	0.05\underline{9}5		0		0.08\underline{3}2

$$P_{CO_2} = \frac{(0.05\underline{9}5 \text{ mol})\left(\dfrac{0.08206 \text{ L} \cdot \text{atm}}{\text{mol} \cdot \text{K}}\right)(1200 \text{ K})}{10.0 \text{ L}} = 0.5\underline{8}6 \text{ atm}$$

$$P_{CO} = \frac{(0.08\underline{3}2 \text{ mol})\left(\dfrac{0.08206 \text{ L} \cdot \text{atm}}{\text{mol} \cdot \text{K}}\right)(1200 \text{ K})}{10.0 \text{ L}} = 0.8\underline{1}9 \text{ atm}$$

$$P_{total} = 0.5\underline{8}6 + 0.8\underline{1}9 = 1.405 = 1.41 \text{ atm}$$

Check: The pressure is less than the equilibrium pressure, which is reasonable because the C was the limiting reactant.

14.77   **Given:** $V = 10.0$ L, $T = 650$ K, 1.0 g MgO, $P_{CO_2} = 0.026$ atm, $K_p = 0.0260$
Find: mass $MgCO_3$ when volume is 0.100 L
Conceptual Plan: $P(10.0 \text{ L}) \rightarrow P(0.100 \text{ L})$. **Prepare an ICE table, represent the change with x, sum the table, determine the equilibrium value, put the equilibrium values in the equilibrium expression, and solve for x. Then determine moles CO_2, the limiting reactant, and the mass of $MgCO_3$ formed.**

$$P_1V_1 = P_2V_2 \quad PV = nRT$$

Solution:

$$P_1V_1 = P_2V_2 \qquad (0.0260 \text{ atm})(10.0 \text{ L}) = (x)(0.100 \text{ L}) \qquad x = 2.60 \text{ atm}$$

$$MgCO_3(s) \rightleftharpoons MgO_3(s) + CO_2(g)$$

Initial	2.60
Change	$-x$
Equil	$2.60 - x$

$$K_p = P_{CO_2} = 0.026 = 2.60 - x \qquad x = 2.5\underline{7}9 \text{ atm}$$

$$n_{CO_2} = \frac{(2.5\underline{7}9 \text{ atm})10.0 \text{ L}}{\left(\dfrac{0.08206 \text{ L} \cdot \text{atm}}{\text{mol} \cdot \text{K}}\right)(650 \text{ K})} = 0.48\underline{3}5 \text{ mol } CO_2 \quad 1.0 \text{ g MgO} \times \frac{1 \text{ mol}}{40.30 \text{ g MgO}} = 0.024\underline{8} \text{ mol MgO}$$

Therefore, MgO is the limiting reactant and produces 0.0248 mol $MgCO_3$.

$$0.0248 \text{ mol MgCO}_3 \times \frac{84.31 \text{ g MgCO}_3}{1 \text{ mol MgCO}_3} = 2.09 \text{ g MgCO}_3$$

14.79 **Given:** $C_2H_4(g) + Cl_2(g) \rightleftharpoons C_2H_4Cl_2(g)$ is exothermic **Find:** Which of the following will maximize $C_2H_4Cl_2$?

(a) Increasing the reaction volume will not maximize $C_2H_4Cl_2$. The chemical equation has 1 mole of gas on the right and 2 moles of gas on the left. Increasing the volume of the reaction mixture decreases the pressure and causes the reaction to shift to the left (toward the side with more moles of gas particles).

(b) Removing $C_2H_4Cl_2$ as it forms will maximize $C_2H_4Cl_2$. Removing the $C_2H_4Cl_2$ will decrease the concentration of $C_2H_4Cl_2$ and will cause the reaction to shift to the right, producing more $C_2H_4Cl_2$.

(c) Lowering the reaction temperature will maximize $C_2H_4Cl_2$. The reaction is exothermic, so we can think of heat as a product. Lowering the temperature will cause the reaction to shift to the right, producing more $C_2H_4Cl_2$.

(d) Adding Cl_2 will maximize $C_2H_4Cl_2$. Adding Cl_2 increases the concentration of Cl_2, so the reaction shifts to the right, which will produce more $C_2H_4Cl_2$.

14.81 **Given:** reaction 1 at equilibrium: $P_{H_2} = 0.958$ atm, $P_{I_2} = 0.877$ atm, $P_{HI} = 0.020$ atm; reaction 2: $P_{H_2} = P_{I_2} = 0.621$ atm, $P_{HI} = 0.101$ atm **Find:** Is reaction 2 at equilibrium? If not, what is the P_{HI} at equilibrium? **Conceptual Plan: Use equilibrium partial pressures to determine K_p. Use K_p to determine whether reaction 2 is at equilibrium. Prepare an ICE table, calculate Q, compare Q and K_p, predict the direction of the reaction, represent the change with x, sum the table, determine the equilibrium values, put the equilibrium values in the equilibrium expression, and solve for x. Determine P_{HI}.**

Solution:

$$H_2(g) + I_2(g) \rightleftharpoons 2 HI(g)$$

Reaction 1:	P_{H_2}	P_{I_2}	P_{HI}
Equil	0.958	0.877	0.020

$$K_p = \frac{P_{HI}^2}{P_{H_2}P_{I_2}} = \frac{(0.020)^2}{(0.958)(0.877)} = 4.7\underline{6}10 \times 10^{-4}$$

$$Q = \frac{P_{HI}^2}{P_{H_2}P_{I_2}} = \frac{(0.101)^2}{(0.621)(0.621)} = 0.0264 \qquad Q > K, \text{ so the reaction shifts to the left.}$$

$$H_2(g) + I_2(g) \rightleftharpoons 2 HI(g)$$

Reaction 2:	P_{H_2}	P_{I_2}	P_{HI}
Initial	0.621	0.621	0.101
Change	$+x$	$+x$	$-2x$
Equil	$0.621 + x$	$0.621 + x$	$0.101 - 2x$

$$K_p = \frac{P_{HI}^2}{P_{H_2}P_{I_2}} = \frac{(0.101 - 2x)^2}{(0.621 + x)(0.621 + x)} = 4.7\underline{6}10 \times 10^{-4}$$

$$\sqrt{\frac{(0.101 - 2x)^2}{(0.621 + x)(0.621 + x)}} = \sqrt{4.7\underline{6}10 \times 10^{-4}}$$

$$\frac{(0.101 - 2x)}{(0.621 + x)} = 2.1\underline{8}2 \times 10^{-2}$$

$$x = 0.04325 = 0.0433$$

$$P_{H_2} = P_{I_2} = 0.621 + x = 0.621 + 0.0433 = 0.664 \text{ atm}$$

$$P_{HI} = 0.101 - 2x = 0.101 - 2(0.0433) = 0.0144 \text{ atm}$$

Check: Plug the values into the equilibrium expression:

$$K_p = \frac{(0.0144)^2}{(0.664)^2} = 4.703 \times 10^{-4} = 4.70 \times 10^{-4}; \text{ this value is close to the original equilibrium constant.}$$

14.83   **Given:** 200.0 L container; 1.27 kg N_2; 0.310 kg H_2; 725 K; $K_p = 5.3 \times 10^{-5}$ **Find:** mass in g of NH_3 and % yield
Conceptual Plan: $K_p \to K_c$ and then kg $\to$ g $\to$ mol $\to$ M and then prepare an ICE table. Represent

$$K_p = K_c(RT)^{\Delta n} \qquad \frac{1000 \text{ g}}{\text{kg}} \qquad \frac{\text{g}}{\text{molar mass}} \qquad \frac{\text{mol}}{\text{vol}}$$

the change with x, sum the table, determine the equilibrium values, put the equilibrium values in the equilibrium expression, and solve for x. Determine [NH_3]. Then M $\to$ mol $\to$ g and then determine

$$\text{M} \times \text{vol} \quad \text{mol} \times \text{molar mass}$$

theoretical yield $NH_3 \to$ % yield.

determine limiting reactant   % yield $= \dfrac{\text{actual yield}}{\text{theoretical yield}} \times 100\%$

Solution: $K_p = K_c(RT)^{\Delta n}$ $K_c = \dfrac{K_p}{(RT)^{\Delta n}} = \dfrac{5.3 \times 10^{-5}}{\left(\left(0.08206 \dfrac{\text{L} \cdot \text{atm}}{\text{mol} \cdot \text{K}}\right)(725 \text{ K})\right)^{-2}} = 0.1\underline{8}76$

$$n_{N_2} = 1.27 \text{ kg N}_2 \times \frac{1000 \text{ g}}{\text{kg}} \times \frac{1 \text{ mol N}_2}{28.02 \text{ g N}_2} = 45.\underline{3}25 \text{ mol N}_2 \quad [N_2] = \frac{45.\underline{3}25 \text{ mol}}{200.0 \text{ L}} = 0.22\underline{6}63 \text{ M}$$

$$n_{H_2} = 0.310 \text{ kg H}_2 \times \frac{1000 \text{ g}}{\text{kg}} \times \frac{1 \text{ mol H}_2}{2.016 \text{ g H}_2} = 153.\underline{7}7 \text{ mol H}_2 \quad [H_2] = \frac{153.\underline{7}7 \text{ mol}}{200.0 \text{ L}} = 0.76\underline{8}85 \text{ M}$$

$$N_2(g) \quad + \quad 3 H_2(g) \rightleftharpoons 2 NH_3(g)$$

Reaction 1:	$[N_2]$	$[H_2]$	$[NH_3]$
Initial	0.2266	0.7689	0.0
Change	$-x$	$-3x$	$+2x$
Equil	$0.2266 - x$	$0.7689 - 3x$	$2x$

Reaction shifts to the right.

$$K_c = \frac{[NH_3]^2}{[N_2][H_2]^3} = \frac{(2x)^2}{(0.2266 - x)(0.7689 - 3x)^3} = 0.1876$$

Assume that x is small compared to 0.2268 and $3x$ is small compared to 0.7689.

$$\frac{(2x)^2}{(0.2266)(0.7689)^3} = 0.1877 \qquad x = 0.06951$$

Check assumptions: $\dfrac{0.06951}{0.2268} \times 100\% = 30.6$, which is not valid, and $\dfrac{3(0.06951)}{0.7689} \times 100\% = 27.1\%$.

Use method of successive substitution to solve for x. This yields $x = 0.0460$.
$[NH_3] = 2x = 2(0.0460) = 0.0920 \text{ M}$

Check: Plug the values into the equilibrium expression:

$$K_c = \frac{(0.0920)^2}{(0.2266 - 0.0460)((0.7689 - 3(0.0460))^3} = 0.1866;$$

this value is close to the original equilibrium constant.

Determine grams NH_3: $\dfrac{0.0922 \text{ mol } NH_3}{L} \times 200.0 \text{ L} \times \dfrac{17.02 \text{ g } NH_3}{\text{mol } NH_3} = 3\underline{1}3.8 \text{ g} = 3.1 \times 10^2 \text{ g}$

Determine the theoretical yield and the limiting reactant:

$1.27 \text{ kg } N_2 \times \dfrac{1000 \text{ g}}{\text{kg}} \times \dfrac{1 \text{ mol } N_2}{28.0 \text{ g } N_2} \times \dfrac{2 \text{ mol } NH_3}{1 \text{ mol } N_2} \times \dfrac{17.02 \text{ g } NH_3}{\text{mol } NH_3} = 15\underline{4}4 \text{ g } NH_3$

$0.310 \text{ kg } H_2 \times \dfrac{1000 \text{ g}}{\text{kg}} \times \dfrac{1 \text{ mol } H_2}{2.016 \text{ g } H_2} \times \dfrac{2 \text{ mol } NH_3}{3 \text{ mol } H_2} \times \dfrac{17.02 \text{ g } NH_3}{\text{mol } NH_3} = 1745 \text{ g } NH_3$

N_2 produces the least amount of NH_3; therefore, it is the limiting reactant, and the theoretical yield is $1.54 \times 10^3 \text{ g } NH_3$.

% yield $= \dfrac{3.1 \times 10^2 \text{ g}}{1.54 \times 10^3 \text{ g}} \times 100\% = 20.\%$

14.85   **Given:** at equilibrium: $P_{CO} = 0.30$ atm; $P_{Cl_2} = 0.10$ atm; $P_{COCl_2} = 0.60$ atm, add 0.40 atm Cl_2

Find: P_{CO} when system returns to equilibrium

Conceptual Plan: Use equilibrium partial pressures to determine K_p. For the new conditions, prepare an ICE table, represent the change with x, sum the table, determine the equilibrium values, put the equilibrium values in the equilibrium expression, and solve for x. Determine P_{CO}.

Solution: $CO(g) + Cl_2(g) \rightleftharpoons COCl_2(g)$

Condition 1:	P_{CO}	P_{Cl_2}	P_{COCl_2}
	0.30	0.10	0.60

$K_p = \dfrac{P_{COCl_2}}{P_{CO}P_{Cl_2}} = \dfrac{(0.60)}{(0.30)(0.10)} = 20$

$CO(g) \quad + \quad Cl_2(g) \rightleftharpoons COCl_2(g)$

Condition 2:	P_{CO}	P_{Cl_2}	P_{COCl_2}
Initial	0.30	0.10 + 0.40	0.60
Change	$-x$	$-x$	$+x$
Equil	$0.30 - x$	$0.50 - x$	$0.60 + x$

Reaction shifts to the right because the concentration of Cl_2 was increased.

$K_p = \dfrac{P_{COCl_2}}{P_{CO}P_{Cl_2}} = \dfrac{(0.60 + x)}{(0.30 - x)(0.50 - x)} = 20$

$20x^2 - 17x + 2.4 = 0$

$\dfrac{-b \pm \sqrt{b^2 - 4ac}}{2a} = \dfrac{-(-17) \pm \sqrt{(-17)^2 - 4(20)(2.4)}}{2(20)}$

$x = 0.67$ or 0.18 so $x = 0.18$

$P_{CO} = 0.30 - 0.18 = 0.12$ atm; $P_{Cl_2} = 0.50 - 0.18 = 0.32$; $P_{COCl_2} = 0.60 + 0.18 = 0.78$ atm

Check: Plug the values into the equilibrium expression:

$K_p = \dfrac{(0.78)}{(0.12)(0.32)} = 20.3 = 20.$; this is the same as the original equilibrium constant.

14.87   **Given:** $K_p = 0.76$; P_{total} at equilibrium $= 1.0$ atm   **Find:** $P_{initial}$ CCl_4

Conceptual Plan: Prepare an ICE table, represent the P_{CCl_4} with A and the change with x, sum the table, determine the equilibrium values, use the total pressure, and solve for A in terms of x. Determine partial pressure of each at equilibrium, use the equilibrium expression to determine x, and determine A.

Solution: $CCl_4(g) \rightleftharpoons C(s) + 2 Cl_2(g)$

	P_{CCl_4}	P_C	P_{Cl_2}
Initial	A	constant	0.00
Change	$-x$		$+2x$
Equil	$A - x$		$2x$

$$P_{total} = P_{CCl_4} + P_{Cl_2} \quad 1.0 = A - x + 2x \quad A = 1.0 - x$$
$$P_{CCl_4} = (A - x) = (1.0 - x) - x = 1.0 - 2x; P_{Cl_2} = 2x$$

$$K_p = \frac{P_{Cl_2}^2}{P_{CCl_4}} = \frac{(2x)^2}{(1.0 - 2x)} = 0.76$$

$4x^2 + 1.52x - 0.76 = 0$ Solve quadratic equation. $x = 0.285$ or -0.665 so $x = 0.28\underline{5}$
$A = 1.0 - x = 1.0 - 0.285 = 0.715 = 0.72$ atm

Check: Plug the values into the equilibrium expression:

$$K_p = \frac{P_{Cl_2}^2}{P_{CCl_4}} = \frac{(2x)^2}{(A - x)} = \frac{(2(0.285))^2}{(0.715 - 0.285)} = 0.756 = 0.76; \text{ the original equilibrium is constant.}$$

14.89   **Given:** $V = 0.654$ L, $T = 1000$ K, $K_p = 3.9 \times 10^{-2}$ **Find:** mass CaO as equilibrium
Conceptual Plan: $K_p \rightarrow P_{CO_2} \rightarrow n_{(CO_2)} \rightarrow n_{(CaO)} \rightarrow g$
$\qquad\qquad\qquad PV = nRT \qquad\quad \text{stoichiometry} \qquad g = n(\text{molar mass})$
Solution: Because $CaCO_3$ and CaO are solids, they are not included in the equilibrium expression.

$$K_p = P_{CO_2} = 3.9 \times 10^{-2}; \quad n = \frac{PV}{RT} = \frac{(3.9 \times 10^{-2} \text{ atm})(0.654 \text{ L})}{\left(0.08206 \dfrac{\text{L} \cdot \text{atm}}{\text{mol} \cdot \text{K}}\right)(1000 \text{ K})} = 3.\underline{1}08 \times 10^{-4} \text{ mol } CO_2$$

$$3.\underline{1}08 \times 10^{-4} \text{ mol } CO_2 \times \frac{1 \text{ mol CaO}}{1 \text{ mol } CO_2} \times \frac{56.1 \text{ g CaO}}{1 \text{ mol CaO}} = 0.0174 \text{ g} = 0.0174 \text{ g CaO}$$

Check: The small value of K would give a small amount of products, so we would not expect a large mass of CaO to form.

14.91   **Given:** $K_p = 3.10$, initial $P_{CO} = 215$ torr, $P_{Cl_2} = 245$ torr   **Find:** mole fraction $COCl_2$
Conceptual Plan: *P* in torr $\rightarrow$ *P* in atm. **Prepare an ICE table, represent the change with *x*, sum the table,**
$\qquad\qquad\qquad\qquad \dfrac{1 \text{ atm}}{760 \text{ torr}}$
determine the equilibrium values, use the total pressure, and solve for mole fraction.
$\qquad\qquad\qquad\qquad\qquad\qquad\qquad\qquad\qquad\qquad\qquad\qquad\qquad \dfrac{P_{COCl_2}}{P_{total}}$

Solution: $P_{CO} = (215 \text{ torr})\left(\dfrac{1 \text{ atm}}{760 \text{ torr}}\right) = 0.28\underline{2}9 \text{ atm} \qquad P_{Cl_2} = (245 \text{ torr})\left(\dfrac{1 \text{ atm}}{760 \text{ torr}}\right) = 0.32\underline{2}4 \text{ atm}$

	$CO(g)$	$+$	$Cl_2(g)$	$\rightleftharpoons$	$COCl_2(g)$	
Initial	0.28$\underline{2}$9		0.32$\underline{2}$4		0	$Q < K$, so the reaction shifts to the right.
Change	$-x$		$-x$		$+x$	
Equil	0.28$\underline{2}$9 $- x$		0.32$\underline{2}$4 $- x$		x	

$$K_p = \frac{P_{COCl_2}}{P_{CO}P_{Cl_2}} = \frac{x}{(0.28\underline{2}9 - x)(0.32\underline{2}4 - x)} = 3.10 \text{ so } 3.10x^2 - 2.87634x + 0.2827 = 0 \text{ Solve using the quadratic}$$

equation, found in Appendix I. $x = 0.81\underline{6}1$ or $0.11\underline{1}7$ so $x = 0.1117$ *P* is proportional to *n* under like conditions so the mole fraction can be determined using the pressure values.
$P_{CO} = 0.28\underline{2}9 - 0.11\underline{1}7 = 0.17\underline{1}2 \qquad P_{Cl_2} = 0.32\underline{2}4 - 0.11\underline{1}7 = 0.21\underline{0}7 \qquad P_{COCl_2} = 0.11\underline{1}7$

$$\text{mole fraction } COCl_2 = \frac{P_{COCl_2}}{P_{CO} + P_{Cl_2} + P_{COCl_2}} = \frac{0.11\underline{1}7}{0.17\underline{1}2 + 0.21\underline{0}7 + 0.1117} = 0.22\underline{6}3 = 0.226$$

Check: Plug the equilibrium pressures into the equilibrium expression:

$$K_p = \frac{0.11\underline{1}7}{(0.17\underline{1}2)(0.21\underline{0}7)} = 3.0966 = 3.10, \text{ which is the equilibrium constant; so the answer is reasonable.}$$

Challenge Problems

14.93 (a)   **Given:** $P_{NO} = 522$ torr, $P_{O_2} = 421$ torr; at equilibrium, $P_{total} = 748$ torr   **Find:** K_p

Conceptual Plan: Prepare an ICE table, represent the change with x, sum the table, determine the equilibrium values, use the total pressure, and solve for x. torr → atm → K_p

Solution: $2 NO(g) + O_2(g) \rightleftharpoons 2 NO_2(g)$

	P_{NO}	P_{O_2}	P_{NO_2}
Initial	522 torr	421 torr	0.00
Change	$-2x$	$-x$	$+2x$
Equil	$522 - 2x$	$421 - x$	$2x$

$P_{total} = P_{NO} + P_{O_2} + P_{NO_2}$ $748 = 522 - 2x + (421 - x) + 2x$
$x = 195$ torr $P_{NO} = (522 - 2(195)) = 132$ torr
$P_{O_2} = (421 - 195) = 226$ torr $P_{NO_2} = 2(195) = 390$ torr

$P_{NO} = 132 \text{ torr} \times \dfrac{1 \text{ atm}}{760 \text{ torr}} = 0.17\underline{3}7 \text{ atm};$ $P_{O_2} = 226 \text{ torr} \times \dfrac{1 \text{ atm}}{760 \text{ torr}} = 0.29\underline{7}4 \text{ atm}$

$P_{NO_2} = 390 \text{ torr} \times \dfrac{1 \text{ atm}}{760 \text{ torr}} = 0.51\underline{3}2 \text{ atm}$

$K_p = \dfrac{P_{NO_2}^2}{P_{NO}^2 P_{O_2}} = \dfrac{(0.51\underline{3}2)^2}{(0.17\underline{3}7)^2(0.29\underline{7}4)} = 29.35 = 29.3$

(b)   **Given:** $P_{NO} = 255$ torr, $P_{O_2} = 185$ torr, $K_p = 29.3$ **Find:** equilibrium P_{NO_2}

Conceptual Plan:

torr → atm and then prepare an ICE table. Represent the change with x, sum the table,

$\dfrac{\text{atm}}{760 \text{ torr}}$

determine the equilibrium values, put the equilibrium values in the equilibrium expression, and solve for x. Determine P_{NO_2}.

Solution: $P_{NO} = 255 \text{ torr} \times \dfrac{1 \text{ atm}}{760 \text{ torr}} = 0.33\underline{5}5 \text{ atm}$ $P_{O_2} = 185 \text{ torr} \times \dfrac{1 \text{ atm}}{760 \text{ torr}} = 0.24\underline{3}4 \text{ atm}$

$2 NO(g) + O_2(g) \rightleftharpoons 2 NO_2(g)$

	P_{NO}	P_{O_2}	P_{NO_2}
Initial	0.3355	0.2434	0.00
Change	$-2x$	$-x$	$+2x$
Equil	$0.3355 - 2x$	$0.2434 - x$	$2x$

$K_p = \dfrac{P_{NO_2}^2}{P_{NO}^2 P_{O_2}} = \dfrac{(2x)^2}{(0.3355 - 2x)^2(0.2434 - x)} = 29.3$

$-117.2x^3 + 63.847x^2 - 12.867x + 0.80282 = 0.$ Solve using successive approximations or a cubic equation calculator found on the Internet.
$x = 0.11\underline{1}3$ $P_{NO_2} = 2x = 2(0.11\underline{1}3) = 0.22\underline{2}6$ atm
$P_{NO} = (0.3355 - 2(0.1113)) = 0.1129$ $P_{O_2} = (0.2434 - 0.1113) = 0.1321$

$0.22\underline{2}6 \text{ atm} \times \dfrac{760 \text{ torr}}{1 \text{ atm}} = 169.2 \text{ torr} = 169 \text{ torr}$

Check: Plug the values into the equilibrium expression:

$K_p = \dfrac{(0.22\underline{2}6)^2}{(0.1129)^2(0.1321)} = 29.428 = 29.4$; this is within 0.1 of the original equilibrium constant.

14.95   **Given:** P_{NOCl} at equilibrium $= 115$ torr; $K_p = 0.27$, $T = 700$ K   **Find:** initial pressure NO, Cl_2
Conceptual Plan: torr → **atm and then prepare an ICE table. Represent the change with x, sum the table,**

$$\frac{atm}{760 \, torr}$$

determine the equilibrium values, put the equilibrium values in the equilibrium expression, and determine initial pressure.

Solution: $115 \, \cancel{torr} \times \dfrac{1 \, atm}{760 \, \cancel{torr}} = 0.151\underline{3} \, atm$

$$2 \, NO(g) \quad + \quad Cl_2(g) \rightleftharpoons 2 \, NOCl(g)$$

	P_{NO}	P_{Cl_2}	P_{NOCl}
Initial	A	A	0.00
Change	$-2x$	$-x$	$+2x$
Equil	$A - 2x$	$A - x$	0.151
	$A - 0.151$	$A - 0.0756$	

Let $A =$ initial pressure of NO and Cl_2.

$2x = 0.151;$ $x = 0.0756$

$$K_p = \frac{P_{NOCl}^2}{P_{NO}^2 \, P_{Cl_2}} = \frac{(0.151)^2}{(A - 0.151)^2(A - 0.0756)} = 0.27$$

$0.27A^3 - 0.1019A^2 + 0.01231A - 0.023266 = 0$. Solve using successive approximations or a cubic equation calculator found on the Internet.
$A = 0.566$
$P_{NO} = P_{Cl_2} = A = 0.566 \, atm = 430 \, torr$

Check: Plug the values into the equilibrium expression:

$$K_p = \frac{P_{NOCl}^2}{P_{NO}^2 \, P_{Cl_2}} = \frac{(0.151)^2}{(0.566 - 0.151)^2(0.566 - 0.0756)} = 0.2699 = 0.27;$$

this is the same as the original equilibrium constant.

14.97   **Given:** $P = 0.750$ atm, density $= 0.520$ g/L, $T = 337\,^{\circ}C$ **Find:** K_c
Conceptual Plan: Prepare an ICE table, represent the P_{CCl_4} with A and the change with x, sum the table, determine the equilibrium values, use the total pressure, and solve for A in terms of x. Determine partial pressure of each at equilibrium in terms of x, use the density to determine the apparent molar mass, and use

$$d = \frac{P\mathcal{M}}{RT}$$

the mole fraction (in terms of P) and the molar mass of each gas to determine x.

$$\chi_A = \frac{P_A}{P_{total}}$$

Solution: $2 \, NO_2(g) \rightleftharpoons 2 \, NO(g) + O_2(g)$

	P_{NO_2}	P_{NO}	P_{O_2}
Initial	A	0.00	0.00
Change	$-2x$	$+2x$	$+x$
Equil	$A - 2x$	$2x$	x

$P_{total} = P_{NO_2} + P_{NO} + P_{O_2}$ $0.750 = A - 2x + 2x + x$ $A = 0.750 - x$
$P_{NO_2} = (A - 2x) = (0.750 - x) - 2x = (0.750 - 3x); P_{NO} = (2x); P_{O_2} = x$

$$d = \frac{P\mathcal{M}}{RT} \qquad \mathcal{M} = \frac{dRT}{P} = \frac{\left(0.520\dfrac{g}{\cancel{L}}\right)\left(0.08206\dfrac{\cancel{L} \cdot \cancel{atm}}{mol \cdot \cancel{K}}\right)(610 \, \cancel{K})}{0.750 \, \cancel{atm}} = 34.71 \, g/mol$$

$$\mathcal{M} = \chi_{NO_2}\mathcal{M}_{NO_2} + \chi_{NO}\mathcal{M}_{NO} + \chi_{O_2}\mathcal{M}_{O_2} = \frac{P_{NO_2}}{P_{total}}\mathcal{M}_{NO_2} + \frac{P_{NO}}{P_{total}}\mathcal{M}_{NO} + \frac{P_{O_2}}{P_{total}}\mathcal{M}_{O_2}$$

$$P_{total}\mathcal{M} = P_{NO_2}\mathcal{M}_{NO_2} + P_{NO}\mathcal{M}_{NO} + P_{O_2}\mathcal{M}_{O_2}$$

$(0.750)(34.7) = (0.750 - 3x)(46.0) + 2x(30.0) + x(32.0)$
$x = 0.184$

$$K_p = \frac{P_{NO}^2 P_{O_2}}{P_{NO_2}^2} = \frac{(2x)^2(x)}{(0.750 - 3x)} = \frac{(2(0.184))^2(0.184)}{(0.750 - 3(0.184))^2} = 0.63\underline{5}6$$

$$K_p = K_c(RT)^{\Delta n} \quad 0.63\underline{5}6 = K_c\left(\left(0.08206\frac{\text{L} \cdot \text{atm}}{\text{mol} \cdot \text{K}}\right)(610\,\text{K})\right)^1$$

$$K_c = 1.27 \times 10^{-2}$$

14.99   **Given:** $P_{\text{total}} = 3.0$ atm, mole fraction $O_2 = 0.12$, $T = 600$ K   **Find:** K_p

Conceptual Plan: mole fraction $\rightarrow P_{O_2} \rightarrow P_{SO_2} \rightarrow P_{SO_3} \rightarrow K_p$

$$\text{mole fraction} = \frac{P_{O_2}}{P_{\text{total}}} \qquad \frac{2P_{SO_2}}{P_{O_2}} \qquad P_{SO_3} = P_{\text{total}} - P_{SO_2} - P_{O_2} \qquad K_p = \frac{P_{SO_2}^2 P_{O_2}}{P_{SO_3}^2}$$

Solution: $P_{O_2} = (0.12)(3.0) = 0.36$ atm $P_{SO_2} = 2P_{O_2} = 2(0.36\text{ atm}) = 0.72$ atm

$$P_{SO_3} = 3.0 - 0.36 - 0.72 = 1.\underline{9}2 \quad K_p = \frac{P_{SO_2}^2 P_{O_2}}{P_{SO_3}^2} = \frac{(0.72)^2(0.36)}{(1.\underline{9}2)^2} = 0.050\underline{6} = 5.1 \times 10^{-2}$$

Conceptual Problems

14.101   Yes, the direction will depend on the volume. If the initial moles of A and B are equal, the initial concentrations of A
and B are equal regardless of the volume. Because $K_c = \dfrac{[\text{B}]^2}{[\text{A}]} = 1$, if the volume is such that the $[\text{A}] = [\text{B}] < 1.0$,
then $Q < K$ and the reaction goes to the right to reach equilibrium. However, if the volume is such that the
$[\text{A}] = [\text{B}] > 1.0$, then $Q > K$ and the reaction goes to the left to reach equilibrium.

14.103   An examination of the data shows that when $P_A = 1.0$, $P_B = 1.0$; therefore, $K_p = \dfrac{P_B^b}{P_A^a} = \dfrac{(1.0)^b}{(1.0)^a} = 1.0$. Therefore, the
value of the numerator and denominator must be equal. We see from the data that $P_B = \sqrt{P_A}$, so $P_B^2 = P_A$. Because
the stoichiometric coefficients become exponents in the equilibrium expression, $a = 1$ and $b = 2$.

15 Acids and Bases

Review Questions

15.1 The pain of heartburn is caused by hydrochloric acid that is excreted in the stomach to kill microorganisms and activate enzymes that break down food. The hydrochloric acid can sometimes back up out of the stomach into the esophagus, a phenomenon known as acid reflux. When hydrochloric acid comes in contact with the lining of the esophagus, the H^+ ions irritate the esophageal tissues, resulting in a burning sensation. The simplest way to relieve mild heartburn is to swallow repeatedly. Saliva contains bicarbonate ion (HCO_3^-) that acts as a base and, when swallowed, neutralizes some of the acid in the esophagus. Heartburn can also be treated with antacids such as Tums, milk of magnesia, and Mylanta. Because these over-the-counter medications contain more base than does saliva, they do a better job of neutralizing the esophageal acid.

15.3 A carboxylic acid is an organic acid that contains the following group of atoms: $H-O-\overset{\overset{\displaystyle O}{\|}}{C}-$. Carboxylic acids are often found in substances derived from living organisms. Examples include citric acid, malic acid, and acetic acid.

15.5 The hydronium ion is H_3O^+. In water, H^+ ions always associate with H_2O molecules to form hydronium ions and other associated species with the general formula $H(H_2O)_n^+$.

15.7 According to Huheey, "The differences between the various acid–base concepts are not concerned with which is right, but which is most convenient to use in a particular situation." There is no single correct definition; we use the definition that is best for a particular situation.

15.9 A conjugate acid–base pair consists of two substances related to each other by the transfer of a proton. In the reaction:

$NH_3(aq) + H_2O(l) \rightleftharpoons NH_4^+(aq) + OH^-(aq)$, the conjugate pairs are NH_3/NH_4^+ and H_2O/OH^-.

15.11 A diprotic acid contains two ionizable protons. An example is H_2SO_4.

A triprotic acid contains three ionizable protons. An example is H_3PO_4.

15.13 The autoionization of water is $H_2O(l) + H_2O(l) \rightleftharpoons H_3O^+(aq) + OH^-(aq)$ and has the ion product of water $K_w = [H_3O^+][OH^-]$.

It can also be written $H_2O(l) \rightleftharpoons H^+(aq) + OH^-(aq)$, and the ion product of water is $K_w = [H^+][OH^-]$. At 25 °C, $K_w = 1.0 \times 10^{-14}$.

15.15 We define pH as $pH = -\log[H_3O^+]$. An acidic solution has a pH < 7, a basic solution has a pH > 7, and a neutral solution has a pH $= 7$.

15.17 We can neglect the contribution of the autoionization of water to the H_3O^+ concentration in a solution of a strong or weak acid because it is negligible compared to the concentration of H_3O^+ ions generated from the dissociation of the acid. Even the weakest of the weak acids has an ionization constant that is four orders of magnitude larger than that of water.

15.19   The percent ionization is defined as percent ionization $= \dfrac{\text{concentration of ionized acid}}{\text{initial concentration of acid}} \times 100$. The percent ionization of a weak acid decreases with the increasing concentration of the acid.

15.21 $B(aq) + H_2O(l) \rightleftharpoons BH^+(aq) + OH^-(aq)$

15.23   At 25 °C, the product of K_a for an acid and K_b for its conjugate base is $K_a \times K_b = K_w = 1.0 \times 10^{-14}$.

15.25   For most polyprotic acids, K_{a_1} is much larger than K_{a_2}. Therefore, the amount of H_3O^+ contributed by the first ionization step is much larger than that contributed by the second or third ionization step. In addition, the production of H_3O^+ by the first step inhibits additional production of H_3O^+ by the second step because of Le Châtelier's principle.

15.27   For an H—Y binary acid, the factors affecting the ease with which this hydrogen will be donated are the polarity of the bond and the strength of the bond.

15.29   A Lewis acid is an electron pair acceptor. A Lewis base is an electron pair donor.

15.31   The combustion of fossil fuels produces oxides of sulfur and nitrogen, which react with oxygen and water to form sulfuric and nitric acids. These acids then combine with rain to form acid rain. Acid rain is a significant problem in the northeastern United States.

Problems by Topic

The Nature and Definitions of Acids and Bases

15.33   (a)   acid $HNO_3(aq) \rightarrow H^+(aq) + NO_3^-(aq)$
         (b)   acid $NH_4^+(aq) \rightarrow H^+(aq) + NH_3(aq)$
         (c)   base $KOH(aq) \rightarrow K^+(aq) + OH^-(aq)$
         (d)   acid $HC_2H_3O_2(aq) \rightarrow H^+(aq) + C_2H_3O_2^-(aq)$

15.35   (a)   Because H_2CO_3 donates a proton to H_2O, it is the acid. After H_2CO_3 donates the proton, it becomes HCO_3^-, the conjugate base. Because H_2O accepts a proton, it is the base. After H_2O accepts the proton, it becomes H_3O^+, the conjugate acid.
         (b)   Because H_2O donates a proton to NH_3, it is the acid. After H_2O donates the proton, it becomes OH^-, the conjugate base. Because NH_3 accepts a proton, it is the base. After NH_3 accepts the proton, it becomes NH_4^+, the conjugate acid.
         (c)   Because HNO_3 donates a proton to H_2O, it is the acid. After HNO_3 donates the proton, it becomes NO_3^-, the conjugate base. Because H_2O accepts a proton, it is the base. After H_2O accepts the proton, it becomes H_3O^+, the conjugate acid.
         (d)   Because H_2O donates a proton to C_5H_5N, it is the acid. After H_2O donates the proton, it becomes OH^-, the conjugate base. Because C_5H_5N accepts a proton, it is the base. After C_5H_5N accepts the proton, it becomes $C_5H_5NH^+$, the conjugate acid.

15.37   (a) Cl^- $HCl(aq) + H_2O(l) \rightarrow H_3O^+(aq) + Cl^-(aq)$
         (b) HSO_3^- $H_2SO_3(aq) + H_2O(l) \rightleftharpoons H_3O^+(aq) + HSO_3^-(aq)$
         (c) CHO_2^- $HCHO_2(aq) + H_2O(l) \rightleftharpoons H_3O^+(aq) + CHO_2^-(aq)$
         (d) F^- $HF(aq) + H_2O(l) \rightleftharpoons H_3O^+(aq) + F^-(aq)$

15.39 $H_2PO_4^-(aq) + H_2O(l) \rightleftharpoons H_3O^+(aq) + HPO_4^{2-}(aq)$
         $H_2PO_4^-(aq) + H_2O(l) \rightleftharpoons H_3PO_4(aq) + OH^-(aq)$

Acid Strength and K_a

15.41   (a) HNO_3 is a strong acid.
        (b)   HCl is a strong acid.
        (c)   HBr is a strong acid.
        (d) H_2SO_3 is a weak acid. $H_2SO_3(aq) + H_2O(l) \rightleftharpoons H_3O^+(aq) + HSO_3^-(aq)$

$$K_a = \frac{[H_3O^+][HSO_3^-]}{[H_2SO_3]}$$

15.43   The strength of an acid is determined by how much it ionizes and how much undissociated acid is left over. The larger the fraction of the acid that ionized, the stronger the acid.
        (a)   contains no HA, 10 H^+, and 10 A^-
        (b)   contains 3 HA, 3 H^+, and 7 A^-
        (c)   contains 9 HA, 1 H^+, and 1 A^-
        So solution a > solution b > solution c.

15.45   (a) F^- is a stronger base than is Cl^-.
              F^- is the conjugate base of HF (a weak acid); Cl^- is the conjugate base of HCl (a strong acid); the weaker the acid, the stronger the conjugate base.
        (b) NO_2^- is a stronger base than is NO_3^-.
              NO_2^- is the conjugate base of HNO_2 (a weak acid); NO_3^- is the conjugate base of HNO_3 (a strong acid); the weaker the acid, the stronger the conjugate base.
        (c) ClO^- is a stronger base than is F^-.
              F^- is the conjugate base of HF ($K_a = 3.5 \times 10^{-4}$); ClO^- is the conjugate base of HClO ($K_a = 2.9 \times 10^{-8}$); HClO is the weaker acid so ClO^- will be the stronger base—the weaker the acid, the stronger the conjugate base.

Autoionization of Water and pH

15.47   (a)   **Given:** $K_w = 1.0 \times 10^{-14}$, $[H_3O^+] = 1.2 \times 10^{-8}$ M   **Find:** $[OH^-]$
              Conceptual Plan: $[\mathbf{H_3O^+}] \rightarrow [\mathbf{OH^-}]$
                    $K_w = 1.0 \times 10^{-14} = [H_3O^+][OH^-]$
              Solution:
              $K_w = 1.0 \times 10^{-14} = (1.2 \times 10^{-8})[OH^-]$
              $[OH^-] = 8.3 \times 10^{-7}$ M
              $[OH^-] > [H_3O^+]$, so the solution is basic.
        (b)   **Given:** $K_w = 1.0 \times 10^{-14}$, $[H_3O^+] = 8.5 \times 10^{-5}$ M   **Find:** $[OH^-]$
              Conceptual Plan: $[\mathbf{H_3O^+}] \rightarrow [\mathbf{OH^-}]$
                    $K_w = 1.0 \times 10^{-14} = [H_3O^+][OH^-]$
              Solution:
              $K_w = 1.0 \times 10^{-14} = (8.5 \times 10^{-5})[OH^-]$
              $[OH^-] = 1.2 \times 10^{-10}$ M
              $[H_3O^+] > [OH^-]$, so the solution is acidic.
        (c)   **Given:** $K_w = 1.0 \times 10^{-14}$, $[H_3O^+] = 3.5 \times 10^{-2}$ M   **Find:** $[OH^-]$
              Conceptual Plan: $[\mathbf{H_3O^+}] \rightarrow [\mathbf{OH^-}]$
                    $K_w = 1.0 \times 10^{-14} = [H_3O^+][OH^-]$
              Solution:
              $K_w = 1.0 \times 10^{-14} = (3.5 \times 10^{-2})[OH^-]$
              $[OH^-] = 2.9 \times 10^{-13}$ M
              $[H_3O^+] > [OH^-]$, so the solution is acidic.

15.49 (a) **Given:** $[H_3O^+] = 1.7 \times 10^{-8}\,M$ **Find:** pH and pOH
Conceptual Plan: $[H_3O^+] \to pH \to pOH$

$$pH = -\log[H_3O^+] \quad pH + pOH = 14$$

Solution: $pH = -\log(1.7 \times 10^{-8}) = 7.77$ $pOH = 14.00 - 7.77 = 6.23$
$pH > 7$, so the solution is basic.

(b) **Given:** $[H_3O^+] = 1. \times 10^{-7}\,M$ **Find:** pH and pOH
Conceptual Plan: $[H_3O^+] \to pH \to pOH$

$$pH = -\log[H_3O^+] \quad pH + pOH = 14$$

Solution: $pH = -\log(1.0 \times 10^{-7}) = 7.00$ $pOH = 14.00 - 7.00 = 7.00$
$pH = 7$, so the solution is neutral.

(c) **Given:** $[H_3O^+] = 2.2 \times 10^{-6}\,M$ **Find:** pH and pOH
Conceptual Plan: $[H_3O^+] \to pH \to pOH$

$$pH = -\log[H_3O^+] \quad pH + pOH = 14$$

Solution: $pH = -\log(2.2 \times 10^{-6}) = 5.66$ $pOH = 14.00 - 5.66 = 8.34$
$pH < 7$, so the solution is acidic.

15.51 $pH = -\log[H_3O^+] \quad K_w = 1.0 \times 10^{-14} = [H_3O^+][OH^-]$

$[H_3O^+]$	$[OH^-]$	pH	Acidic or basic
7.1×10^{-4}	1.4×10^{-11}	**3.15**	acidic
3.7×10^{-9}	2.7×10^{-6}	8.43	basic
8×10^{-12}	1×10^{-3}	**11.1**	basic
6.3×10^{-4}	**1.6×10^{-11}**	3.20	acidic

$$[H_3O^+] = 10^{-3.15} = 7.1 \times 10^{-4} \qquad [OH^-] = \frac{1.0 \times 10^{-14}}{7.1 \times 10^{-4}} = 1.4 \times 10^{-11}$$

$$[OH^-] = \frac{1.0 \times 10^{-14}}{3.7 \times 10^{-9}} = 2.7 \times 10^{-6} \qquad pH = -\log(3.7 \times 10^{-9}) = 8.43$$

$$[H_3O^+] = 10^{-11.1} = 8 \times 10^{-12} \qquad [OH^-] = \frac{1.0 \times 10^{-14}}{8 \times 10^{-12}} = 1 \times 10^{-3}$$

$$[H_3O^+] = \frac{1.0 \times 10^{-14}}{1.6 \times 10^{-11}} = 6.3 \times 10^{-4} \qquad pH = -\log(6.2 \times 10^{-4}) = 3.21$$

15.53 **Given:** $K_w = 2.4 \times 10^{-14}$ at $37\,°C$ **Find:** $[H_3O^+]$, pH
Conceptual Plan: $K_w \to [H_3O^+] \to pH$

$$K_w = [H_3O^+][OH^-] \quad pH = -\log[H_3O^+]$$

Solution: $H_2O(l) + H_2O(l) \rightleftharpoons H_3O^+(aq) + OH^-(aq)$

$$K_w = [H_3O^+][OH^-]$$
$$[H_3O^+] = [OH^-] = \sqrt{K_w} = \sqrt{2.4 \times 10^{-14}} = 1.5 \times 10^{-7}$$
$$pH = -\log[H_3O^+] = -\log(1.5 \times 10^{-7}) = 6.82$$

Check: The value of K_w increased, indicating more products formed; so the H_3O^+ increases and the pH decreases from the values at $25\,°C$.

15.55 (a) **Given:** $[H_3O^+] = 0.044\,M$ **Find:** pH
Conceptual Plan: $[H_3O^+] \to pH$

$$pH = -\log[H_3O^+]$$

Solution: $pH = -\log(0.044) = 1.3\underline{5}7 = 1.36$

(b) **Given:** $[H_3O^+] = 0.045\,M$ **Find:** pH
Conceptual Plan: $[H_3O^+] \to pH$

$$pH = -\log[H_3O^+]$$

Solution: $pH = -\log(0.045) = 1.3\underline{4}7 = 1.35$

(c)   **Given:** $[H_3O^+] = 0.046\,M$ **Find:** pH
Conceptual Plan: $[H_3O^+] \rightarrow pH$

$$pH = -\log[H_3O^+]$$

Solution: $pH = -\log(0.046) = 1.3\underline{3}7 = 1.34$

If the pH of the solution did not carry as many decimal places as the significant digits to the right of the decimal point, you would not be able to distinguish a difference in the pH of solutions (b) and (c). Also, solution (a) would have a pH that was too high.

Acid Solutions

15.57   (a)   **Given:** 0.25 M HCl (strong acid)   **Find:** $[H_3O^+], [OH^-], pH$
Conceptual Plan: $[HCl] \rightarrow [H_3O^+] \rightarrow pH$ and then $[H_3O^+] \rightarrow [OH^-]$

$$[HCl] \rightarrow [H_3O^+] \quad pH = -\log[H_3O^+] \qquad [H_3O^+][OH^-] = 1.0 \times 10^{-14}$$

Solution: $0.25\,M\,HCl = 0.25\,M\,H_3O^+ \quad pH = -\log(0.25) = 0.60$
$$[OH^-] = 1.0 \times 10^{-14}/0.25\,M = 4.0 \times 10^{-14}$$

Check: HCl is a strong acid with a relatively high concentration, so we expect the pH to be low and the $[OH^-]$ to be small.

(b)   **Given:** 0.015 M HNO_3 (strong acid)   **Find:** $[H_3O^+], [OH^-], pH$
Conceptual Plan: $[HNO_3] \rightarrow [H_3O^+] \rightarrow pH$ and then $[H_3O^+] \rightarrow [OH^-]$

$$[HNO_3] \rightarrow [H_3O^+] \quad pH = -\log[H_3O^+] \qquad [H_3O^+][OH^-] = 1.0 \times 10^{-14}$$

Solution: $0.015\,M\,HNO_3 = 0.015\,M\,H_3O^+ \quad pH = -\log(0.015) = 1.82$
$$[OH^-] = 1.0 \times 10^{-14}/0.015\,M = 6.7 \times 10^{-13}$$

Check: HNO_3 is a strong acid, so we expect the pH to be low and the $[OH^-]$ to be small.

(c)   **Given:** 0.052 M HBr and 0.015 M HNO_3 (strong acids)   **Find:** $[H_3O^+], [OH^-], pH$
Conceptual Plan: $[HBr] + [HNO_3] \rightarrow [H_3O^+] \rightarrow pH$ and then $[H_3O^+] \rightarrow [OH^-]$

$$[HBr]+[HNO_3] \rightarrow [H_3O^+] \quad pH = -\log[H_3O^+] \qquad [H_3O^+][OH^-] = 1.0 \times 10^{-14}$$

Solution: $0.052\,M\,HBr = 0.052\,M\,H_3O^+$ and $0.020\,M\,HNO_3 = 0.020\,M\,H_3O^+$
Total $H_3O^+ = 0.072\,M + 0.020\,M = 0.072\,M \quad pH = -\log(0.072) = 1.14$
$$[OH^-] = 1.0 \times 10^{-14}/0.072\,M = 1.4 \times 10^{-13}$$

Check: HBr and HNO_3 are both strong acids that completely dissociate. This gives a relatively high concentration, so we expect the pH to be low and the $[OH^-]$ to be small.

(d)   **Given:** $HNO_3 = 0.655\%$ by mass, $d_{solution} = 1.01\,g/mL$ **Find:** $[H_3O^+], [OH^-], pH$
Conceptual Plan:
% mass $HNO_3 \rightarrow$ g $HNO_3 \rightarrow$ mol HNO_3 and then g soln $\rightarrow$ mL soln $\rightarrow$ L soln $\rightarrow$ M HNO_3

$$\frac{\%}{100} \qquad \frac{mol\,HNO_3}{63.018\,g\,HNO_3} \qquad\qquad \frac{1.01\,g\,soln}{mL\,soln} \quad \frac{1000\,mL\,soln}{L\,soln} \quad \frac{mol\,HNO_3}{L\,soln}$$

$\rightarrow$ M $H_3O^+ \rightarrow$ pH and then $[H_3O^+] \rightarrow [OH^-]$

$$[HNO_3] \rightarrow [H_3O^+] \qquad pH = -\log[H_3O^+] \quad [H_3O^+][OH^-] = 1.0 \times 10^{-14}$$

Solution: $\dfrac{0.655\,\cancel{g\,HNO_3}}{100\,\cancel{g\,soln}} \times \dfrac{1\,mol\,HNO_3}{63.018\,\cancel{g\,HNO_3}} \times \dfrac{1.01\,\cancel{g\,soln}}{\cancel{mL\,soln}} \times \dfrac{1000\,\cancel{mL\,soln}}{L\,soln} = 0.105\,M\,HNO_3$

$0.105\,M\,HNO_3 = 0.105\,M\,H_3O^+ \quad pH = -\log(0.105) = 0.979$
$$[OH^-] = 1.00 \times 10^{-14}/0.105\,M = 9.52 \times 10^{-14}$$

Check: HNO_3 is a strong acid that completely dissociates. This gives a relatively high concentration, so we expect the pH to be low and the $[OH^-]$ to be small.

15.59   (a)   **Given:** pH $= 1.25, 0.250$ L   **Find:** g HI

         Conceptual Plan: pH $\rightarrow [H_3O^+] \rightarrow [HI] \rightarrow$ mol HI $\rightarrow$ g HI

$$\text{pH} = -\log[H_3O^+] \quad [H_3O^+] \rightarrow [HI] \quad \text{mol} = MV \quad g = \frac{127.9 \text{ g HI}}{\text{mol HI}}$$

         Solution: $[H_3O^+] = 10^{-1.25} = 0.056$ M $= [HI] \dfrac{0.056 \text{ mol HI}}{L} \times 0.250 \text{ L} \times \dfrac{127.9 \text{ g HI}}{\text{mol HI}} = 1.8$ g HI

   (b)   **Given:** pH $= 1.75, 0.250$ L   **Find:** g HI

         Conceptual Plan: pH $\rightarrow [H_3O^+] \rightarrow [HI] \rightarrow$ mol HI $\rightarrow$ g HI

$$\text{pH} = -\log[H_3O^+] \quad [H_3O^+] \rightarrow [HI] \quad \text{mol} = MV \quad g = \frac{127.9 \text{ g HI}}{\text{mol HI}}$$

         Solution: $[H_3O^+] = 10^{-1.75} = 0.017\underline{8}$ M $= [HI] \dfrac{0.017\underline{8} \text{ mol HI}}{L} \times 0.250 \text{ L} \times \dfrac{127.9 \text{ g HI}}{\text{mol HI}} = 0.57$ g HI

   (c)   **Given:** pH $= 2.85, 0.250$ L   **Find:** g HI

         Conceptual Plan: pH $\rightarrow [H_3O^+] \rightarrow [HI] \rightarrow$ mol HI $\rightarrow$ g HI

$$\text{pH} = -\log[H_3O^+] \quad [H_3O^+] \rightarrow [HI] \quad \text{mol} = MV \quad g = \frac{127.9 \text{ g HI}}{\text{mol HI}}$$

         Solution: $[H_3O^+] = 10^{-2.85} = 0.0014$ M $= [HI] \dfrac{0.0014 \text{ mol HI}}{L} \times 0.250 \text{ L} \times \dfrac{127.9 \text{ g HI}}{\text{mol HI}} = 0.045$ g HI

15.61   **Given:** 224 mL HCl, 27.2 °C, 1.02 atm, 1.5 L solution   **Find:** pH

        Conceptual Plan: vol HCl $\rightarrow$ mol HCl $\rightarrow [HCl] \rightarrow [H_3O^+] \rightarrow$ pH

$$PV = nRT \qquad M = \frac{\text{mol HCl}}{\text{L soln}} \qquad [HCl] = [H_3O^+] \quad \text{pH} = -\log[H_3O^+]$$

        Solution: $n = \dfrac{(1.02 \text{ atm})(224 \text{ mL})\left(\dfrac{L}{1000 \text{ mL}}\right)}{\left(\dfrac{0.08206 \text{ L atm}}{\text{mol K}}\right)((27.2 + 273.15) \text{ K})} = 0.00927$ mol

$$[HCl] = \frac{0.00927 \text{ mol}}{1.5 \text{ L}} = 0.006\underline{18} \text{ M} = [H_3O^+] \quad \text{pH} = -\log(0.006\underline{18}) = 2.21$$

15.63   **Given:** 0.100 M benzoic acid, $K_a = 6.5 \times 10^{-5}$ **Find:** $[H_3O^+]$, pH

        Conceptual Plan: Write a balanced reaction. Prepare an ICE table, represent the change with x, sum the table, determine the equilibrium values, put the equilibrium values in the equilibrium expression, and solve for x. Determine $[H_3O^+]$ and pH.

        Solution: $HC_7H_5O_2(aq) + H_2O(l) \rightleftharpoons H_3O^+(aq) + C_7H_5O_2^-(aq)$

Initial	0.100	0.0	0.0
Change	$-x$	$+x$	$+x$
Equil	$0.100 - x$	x	x

$$K_a = \frac{[H_3O^+][C_7H_5O_2^-]}{[HC_7H_5O_2]} = \frac{(x)(x)}{(0.100 - x)} = 6.5 \times 10^{-5}$$

        Assume that x is small compared to 0.100.

        $x^2 = (6.5 \times 10^{-5})(0.100); \quad x = 2.5 \times 10^{-3}$ M $= [H_3O^+]$

        Check assumption: $\dfrac{2.5 \times 10^{-3}}{0.100} \times 100\% = 2.5\%$; assumption is valid.

        pH $= -\log(2.5 \times 10^{-3}) = 2.60$

15.65   (a)   **Given:** 0.500 M $HNO_2, K_a = 4.6 \times 10^{-4}$ **Find:** pH

         Conceptual Plan: Write a balanced reaction. Prepare an ICE table, represent the change with x, sum the table, determine the equilibrium values, put the equilibrium values in the equilibrium expression, and solve for x. Determine $[H_3O^+]$ and pH.

Solution: $HNO_2(aq) + H_2O(l) \rightleftharpoons H_3O^+(aq) + NO_2^-(aq)$

Initial	0.500	0.0	0.0
Change	$-x$	$+x$	$+x$
Equil	$0.500 - x$	x	x

$$K_a = \frac{[H_3O^+][NO_2^-]}{[HNO_2]} = \frac{(x)(x)}{(0.500 - x)} = 4.6 \times 10^{-4}$$

Assume that x is small compared to 0.500.

$$x^2 = (4.6 \times 10^{-4})(0.500); \quad x = 0.015 \, M = [H_3O^+]$$

Check assumption: $\dfrac{0.015}{0.500} \times 100\% = 3.0\%$; assumption is valid.

$$pH = -\log(0.015) = 1.82$$

(b)   **Given:** 0.100 M HNO_2, $K_a = 4.6 \times 10^{-4}$ **Find:** pH
Conceptual Plan: Write a balanced reaction. Prepare an ICE table, represent the change with x, sum the table, determine the equilibrium values, put the equilibrium values in the equilibrium expression, and solve for x. Determine $[H_3O^+]$ and pH.

Solution: $HNO_2(aq) + H_2O(l) \rightleftharpoons H_3O^+(aq) + NO_2^-(aq)$

Initial	0.100	0.0	0.0
Change	$-x$	$+x$	$+x$
Equil	$0.100 - x$	x	x

$$K_a = \frac{[H_3O^+][NO_2^-]}{[HNO_2]} = \frac{(x)(x)}{(0.100 - x)} = 4.6 \times 10^{-4}$$

Assume that x is small compared to 0.100.

$$x^2 = (4.6 \times 10^{-4})(0.100) \quad x = 0.0068 \, M = [H_3O^+]$$

Check assumption: $\dfrac{0.0068}{0.100} \times 100\% = 6.8\%$; assumption is not valid; solve using the quadratic equation.

$$x^2 = (4.6 \times 10^{-4})(0.100 - x); \quad x^2 + 4.6 \times 10^{-4}x - 4.6 \times 10^{-5} = 0$$

$$x = 0.00656$$

$$pH = -\log(0.00656) = 2.18$$

(c)   **Given:** 0.0100 M HNO_2, $K_a = 4.6 \times 10^{-4}$ **Find:** pH
Conceptual Plan: Write a balanced reaction. Prepare an ICE table, represent the change with x, sum the table, determine the equilibrium values, put the equilibrium values in the equilibrium expression, and solve for x. Determine $[H_3O^+]$ and pH.

Solution: $HNO_2(aq) + H_2O(l) \rightleftharpoons H_3O^+(aq) + NO_2^-(aq)$

Initial	0.0100	0.0	0.0
Change	$-x$	$+x$	$+x$
Equil	$0.0100 - x$	x	x

$$K_a = \frac{[H_3O^+][NO_2^-]}{[HNO_2]} = \frac{(x)(x)}{(0.0100 - x)} = 4.6 \times 10^{-4}$$

Assume that x is small compared to 0.100.

$$x^2 = (4.6 \times 10^{-4})(0.0100); \quad x = 0.0021 \, M = [H_3O^+]$$

Check assumption: $\dfrac{0.0021}{0.0100} \times 100\% = 21\%$; assumption is not valid; solve using the quadratic equation.

$$x^2 = (4.6 \times 10^{-4})(0.0100 - x) \quad x^2 + 4.6 \times 10^{-4}x - 4.6 \times 10^{-6} = 0$$

$$x = 0.0019$$

$$pH = -\log(0.0019) = 2.72$$

15.67 **Given:** 15.0 mL glacial acetic, $d = 1.05$ g/mL, dilute to 1.50 L, $K_a = 1.8 \times 10^{-5}$ **Find:** pH

Conceptual Plan: mL acetic acid → g acetic acid → mol acetic acid → M and then write a balanced reaction.

$$\frac{1.05 \text{ g}}{\text{mL}} \qquad \frac{\text{mol acetic acid}}{60.05 \text{ g}} \qquad M = \frac{\text{mol}}{L}$$

Prepare an ICE table, represent the change with x, sum the table, determine the equilibrium values, put the equilibrium values in the equilibrium expression, and solve for x. Determine $[H_3O^+]$ and pH.

Solution: $15.0 \text{ mL} \times \dfrac{1.05 \text{ g}}{\text{mL}} \times \dfrac{1 \text{ mol}}{60.05 \text{ g}} \times \dfrac{1}{1.50 \text{ L}} = 0.174\underline{9}$ M

$$HC_2H_3O_2(aq) + H_2O(l) \rightleftharpoons H_3O^+(aq) + C_2H_3O_2^-(aq)$$

Initial	0.1749	0.0	0.0
Change	$-x$	$+x$	$+x$
Equil	$0.1749 - x$	x	x

$$K_a = \frac{[H_3O^+][C_2H_3O_2^-]}{[HC_2H_3O_2]} = \frac{(x)(x)}{(0.1749 - x)} = 1.8 \times 10^{-5}$$

Assume that x is small compared to 0.1749.

$$x^2 = (1.8 \times 10^{-5})(0.1749); \quad x = 0.001\underline{77} \text{ M} = [H_3O^+]$$

Check assumption: $\dfrac{0.001\underline{77}}{0.1749} \times 100\% = 1.0\%$; assumption is valid.

$$pH = -\log(0.001\underline{77}) = 2.75$$

15.69 **Given:** 0.185 M HA, pH $= 2.95$ **Find:** K_a

Conceptual Plan: pH → $[H_3O^+]$ and then write a balanced reaction. Prepare an ICE table, calculate equilibrium concentrations, and plug into the equilibrium expression to solve for K_a.

Solution: $[H_3O^+] = 10^{-2.95} = 0.00112 \text{ M} = [A^-]$

$$HA(aq) + H_2O(l) \rightleftharpoons H_3O^+(aq) + A^-(aq)$$

Initial	0.185	0.0	0.0
Change	$-x$	$+x$	$+x$
Equil	$0.185 - 0.00112$	0.00112	0.00112

$$K_a = \frac{[H_3O^+][A^-]}{[HA]} = \frac{(0.00112)(0.00112)}{(0.185 - 0.00112)} = 6.82 \times 10^{-6}$$

15.71 **Given:** 0.125 M HCN, $K_a = 4.9 \times 10^{-10}$ **Find:** % ionization

Conceptual Plan: Write a balanced reaction. Prepare an ICE table, represent the change with x, sum the table, determine the equilibrium values, put the equilibrium values in the equilibrium expression, solve for x, and then $x \rightarrow$ % ionization.

$$\% \text{ ionization} = \frac{x}{[HCN]_{\text{original}}} \times 100\%$$

Solution: $\quad HCN(aq) + H_2O(l) \rightleftharpoons H_3O^+(aq) + CN^-(aq)$

Initial	0.125	0.0	0.0
Change	$-x$	$+x$	$+x$
Equil	$0.125 - x$	x	x

$$K_a = \frac{[H_3O^+][CN^-]}{[HCN]} = \frac{(x)(x)}{(0.125 - x)} = 4.9 \times 10^{-10}$$

Assume that x is small compared to 0.125.

$$x^2 = (4.9 \times 10^{-10})(0.125); \quad x = 7.8\underline{3} \times 10^{-6}$$

$$\% \text{ ionization} = \frac{7.8\underline{3} \times 10^{-6}}{0.125} \times 100\% = 0.0063\% \text{ ionized}$$

15.73   (a)   **Given:** 1.00 M $HC_2H_3O_2$, $K_a = 1.8 \times 10^{-5}$ **Find:** % ionization
Conceptual Plan: Write a balanced reaction. Prepare an ICE table, represent the change with x, sum the table, determine the equilibrium values, put the equilibrium values in the equilibrium expression, solve for x, and then $x \rightarrow$ % ionization.

$$\% \text{ ionization} = \frac{x}{[HC_2H_3O_2]_{\text{original}}} \times 100\%$$

Solution: $HC_2H_3O_2(aq) + H_2O(l) \rightleftharpoons H_3O^+(aq) + C_2H_3O_2^-(aq)$

	$HC_2H_3O_2$	H_3O^+	$C_2H_3O_2^-$
Initial	1.00	0.0	0.0
Change	$-x$	$+x$	$+x$
Equil	$1.00 - x$	x	x

$$K_a = \frac{[H_3O^+][C_2H_3O_2^-]}{[HC_2H_3O_2]} = \frac{(x)(x)}{(1.00 - x)} = 1.8 \times 10^{-5}$$

Assume that x is small compared to 1.00.

$x^2 = (1.8 \times 10^{-5})(1.00); \quad x = 0.00424$

$$\% \text{ ionization} = \frac{0.00424}{1.00} \times 100\% = 0.42\% \text{ ionized}$$

(b)   **Given:** 0.500 M $HC_2H_3O_2$, $K_a = 1.8 \times 10^{-5}$ **Find:** % ionization
Conceptual Plan: Write a balanced reaction. Prepare an ICE table, represent the change with x, sum the table, determine the equilibrium values, put the equilibrium values in the equilibrium expression, solve for x, and then $x \rightarrow$ % ionization.

$$\% \text{ ionization} = \frac{x}{[HC_2H_3O_2]_{\text{original}}} \times 100\%$$

Solution: $HC_2H_3O_2(aq) + H_2O(l) \rightleftharpoons H_3O^+(aq) + C_2H_3O_2^-(aq)$

	$HC_2H_3O_2$	H_3O^+	$C_2H_3O_2^-$
Initial	0.500	0.0	0.0
Change	$-x$	$+x$	$+x$
Equil	$0.500 - x$	x	x

$$K_a = \frac{[H_3O^+][C_2H_3O_2^-]}{[HC_2H_3O_2]} = \frac{(x)(x)}{(0.500 - x)} = 1.8 \times 10^{-5}$$

Assume that x is small compared to 0.500.
$x^2 = (1.8 \times 10^{-5})(0.500) \quad x = 0.00300$

$$\% \text{ ionization} = \frac{0.00300}{0.500} \times 100\% = 0.60\% \text{ ionized}$$

(c)   **Given:** 0.100 M $HC_2H_3O_2$, $K_a = 1.8 \times 10^{-5}$ **Find:** % ionization
Conceptual Plan: Write a balanced reaction. Prepare an ICE table, represent the change with x, sum the table, determine the equilibrium values, put the equilibrium values in the equilibrium expression, solve for x, and then $x \rightarrow$ % ionization.

$$\% \text{ ionization} = \frac{x}{[HC_2H_3O_2]_{\text{original}}} \times 100\%$$

Solution: $HC_2H_3O_2(aq) + H_2O(l) \rightleftharpoons H_3O^+(aq) + C_2H_3O_2^-(aq)$

	$HC_2H_3O_2$	H_3O^+	$C_2H_3O_2^-$
Initial	0.100	0.0	0.0
Change	$-x$	$+x$	$+x$
Equil	$0.100 - x$	x	x

$$K_a = \frac{[H_3O^+][C_2H_3O_2^-]}{[HC_2H_3O_2]} = \frac{(x)(x)}{(0.100 - x)} = 1.8 \times 10^{-5}$$

Assume that x is small compared to 0.100.
$x^2 = (1.8 \times 10^{-5})(0.100) \quad x = 0.00134$

$$\% \text{ ionization} = \frac{0.00134}{0.100} \times 100\% = 1.3\% \text{ ionized}$$

(d) **Given:** 0.0500 M $HC_2H_3O_2$, $K_a = 1.8 \times 10^{-5}$ **Find:** % ionization

Conceptual Plan: Write a balanced reaction. Prepare an ICE table, represent the change with x, sum the table, determine the equilibrium values, put the equilibrium values in the equilibrium expression, solve for x, and then $x \rightarrow$ % ionization.

$$\% \text{ ionization} = \frac{x}{[HC_2H_3O_2]_{original}} \times 100\%$$

Solution: $HC_2H_3O_2(aq) + H_2O(l) \rightleftharpoons H_3O^+(aq) + C_2H_3O_2^-(aq)$

Initial	0.0500	0.0	0.0
Change	$-x$	$+x$	$+x$
Equil	$0.0500 - x$	x	x

$$K_a = \frac{[H_3O^+][C_2H_3O_2^-]}{[HC_2H_3O_2]} = \frac{(x)(x)}{(0.0500 - x)} = 1.8 \times 10^{-5}$$

Assume that x is small compared to 0.0500.

$$x^2 = (1.8 \times 10^{-5})(0.0500); \quad x = 9.\underline{49} \times 10^{-4}$$

$$\% \text{ ionization} = \frac{9.\underline{49} \times 10^{-4}}{0.0500} \times 100\% = 1.9\% \text{ ionized}$$

15.75   Given: 0.148 M HA, 1.55% dissociation   **Find:** K_a

Conceptual Plan: $M \rightarrow [H_3O^+] \rightarrow K_a$ **and then write a balanced reaction, determine equilibrium concentration, and plug into the equilibrium expression.**

Solution: $(0.148 \text{ M HA})(0.0155) = 0.0022\underline{94}; \quad [H_3O^+] = [A^-]$

	$HA(aq) + H_2O(l) \rightleftharpoons H_3O^+(aq) + A^-(aq)$		
Initial	0.148	0.0	0.0
Change	$-x$	$+x$	$+x$
Equil	$0.148 - 0.0022\underline{94}$	$0.0022\underline{94}$	$0.0022\underline{94}$

$$K_a = \frac{[H_3O^+][A^-]}{[HA]} = \frac{(0.0022\underline{94})(0.0022\underline{94})}{(0.148 - 0.0022\underline{94})} = 3.61 \times 10^{-5}$$

15.77   (a) **Given:** 0.250 M HF, $K_a = 3.5 \times 10^{-4}$ **Find:** pH, % dissociation

Conceptual Plan: Write a balanced reaction. Prepare an ICE table, represent the change with x, sum the table, determine the equilibrium values, put the equilibrium values in the equilibrium expression, solve for x, and then $x \rightarrow$ % ionization.

$$\% \text{ ionization} = \frac{x}{[HF]_{original}} \times 100\%$$

Solution: $HF(aq) + H_2O(l) \rightleftharpoons H_3O^+(aq) + F^-(aq)$

Initial	0.250	0.0	0.0
Change	$-x$	$+x$	$+x$
Equil	$0.250 - x$	x	x

$$K_a = \frac{[H_3O^+][F^-]}{[HF]} = \frac{(x)(x)}{(0.250 - x)} = 3.5 \times 10^{-4}$$

Assume that x is small compared to 0.250.

$$x^2 = (3.5 \times 10^{-4})(0.250) \quad x = 0.009\underline{35} \text{ M} = [H_3O^+]$$

$$\frac{0.009\underline{35}}{0.250} \times 100\% = 3.7\%$$

$$pH = -\log(0.009\underline{35}) = 2.03$$

(b) **Given:** 0.100 M HF, $K_a = 3.5 \times 10^{-4}$ **Find:** pH, % dissociation

Conceptual Plan: Write a balanced reaction. Prepare an ICE table, represent the change with x, sum the table, determine the equilibrium values, put the equilibrium values in the equilibrium expression, solve for x, and then $x \rightarrow$ % ionization.

$$\% \text{ ionization} = \frac{x}{[HF]_{original}} \times 100\%$$

Solution: $HF(aq) + H_2O(l) \rightleftharpoons H_3O^+(aq) + F^-(aq)$

Initial	0.1000	0.0	0.0
Change	$-x$	$+x$	$+x$
Equil	$0.100 - x$	x	x

$$K_a = \frac{[H_3O^+][F^-]}{[HF]} = \frac{(x)(x)}{(0.100 - x)} = 3.5 \times 10^{-4}$$

Assume that x is small compared to 0.100.

$x^2 = (3.5 \times 10^{-4})(0.100); \quad x = 0.00592 \text{ M} = [H_3O^+]$

$$\frac{0.00592}{0.100} \times 100\% = 5.9\%$$

$x > 5.0\%$ Therefore, assumption is invalid; solve using the quadratic equation.

$x^2 + 3.5 \times 10^{-4}x - 3.5 \times 10^{-5} = 0$

$x = 0.00574$ or -0.00609

$\text{pH} = -\log(0.00574) = 2.24$

$$\% \text{ dissociation} = \frac{0.00574}{0.100} \times 100\% = 5.7\%$$

(c)   **Given:** 0.050 M HF, $K_a = 3.5 \times 10^{-4}$ **Find:** pH, % dissociation

Conceptual Plan: Write a balanced reaction. Prepare an ICE table, represent the change with x, sum the table, determine the equilibrium values, put the equilibrium values in the equilibrium expression, solve for x, and then $x \rightarrow$ % ionization.

$$\% \text{ ionization} = \frac{x}{[HF]_{original}} \times 100\%$$

Solution: $HF(aq) + H_2O(l) \rightleftharpoons H_3O^+(aq) + F^-(aq)$

Initial	0.050	0.0	0.0
Change	$-x$	$+x$	$+x$
Equil	$0.050 - x$	x	x

$$K_a = \frac{[H_3O^+][F^-]}{[HF]} = \frac{(x)(x)}{(0.050 - x)} = 3.5 \times 10^{-4}$$

Assume that x is small compared to 0.050.

$x^2 = (3.5 \times 10^{-4})(0.050); \quad x = 0.00418 \text{ M} = [H_3O^+]$

$$\frac{0.00418}{0.050} \times 100\% = 8.4\%$$

Assumption is invalid; solve using the quadratic equation.

$x^2 + 3.5 \times 10^{-4}x - 1.75 \times 10^{-5} = 0$

$x = 0.00401$ or -0.00436

$\text{pH} = -\log(0.00401) = 2.40$

$$\% \text{ dissociation} = \frac{0.00401}{0.050} \times 100\% = 8.0\%$$

15.79   (a)   **Given:** 0.115 M HBr (strong acid), 0.125 M HCHO$_2$ (weak acid)   **Find:** pH

Conceptual Plan: Because the mixture is a strong acid and a weak acid, the strong acid will dominate. Use the concentration of the strong acid to determine $[H_3O^+]$ and then pH.

Solution: 0.115 M HBr = 0.115 M $[H_3O^+]$ pH $= -\log(0.115) = 0.939$

(b)   **Given:** 0.150 M HNO$_2$ (weak acid), 0.085 M HNO$_3$ (strong acid)   **Find:** pH

Conceptual Plan: Because the mixture is a strong acid and a weak acid, the strong acid will dominate. Use the concentration of the strong acid to determine $[H_3O^+]$ and then pH.

Solution: 0.085 M HNO$_3$ = 0.085 M $[H_3O^+]$ pH $= -\log(0.085) = 1.07$

(c)   **Given:** 0.185 M HCHO$_2$, $K_a = 1.8 \times 10^{-4}$; 0.225 M HC$_2$H$_3$O$_2$, $K_a = 1.8 \times 10^{-5}$ **Find:** pH

Conceptual Plan: Because the mixture is a weak acid and a weak acid and the K values are only 10^1 apart, you need to find $[H_3O^+]$ from each reaction. Write a balanced reaction. Prepare an ICE table, represent the change with x, sum the table, determine the equilibrium values, put the equilibrium values in the equilibrium expression, and solve for x.

Solution: $HCHO_2(aq) + H_2O(l) \rightleftharpoons H_3O^+(aq) + CHO_2^-(aq)$

Initial	0.185	0.0	0.0
Change	$-x$	$+x$	$+x$
Equil	$0.185 - x$	x	x

$K_a = \dfrac{[H_3O^+][CHO_2^-]}{[HCHO_2]} = \dfrac{(x)(x)}{(0.185 - x)} = 1.8 \times 10^{-4}$

Assume that x is small compared to 0.100.

$x^2 = (1.8 \times 10^{-4})(0.185); \quad x = 0.005\underline{7}706$

$HC_2H_3O_2(aq) + H_2O(l) \rightleftharpoons H_3O^+(aq) + C_2H_3O_2^-(aq)$

Initial	0.225	0.005\underline{7}706	0.0
Change	$-x$	$+x$	$+x$
Equil	$0.225 - x$	$0.005\underline{7}706 + x$	x

$K_a = \dfrac{[H_3O^+][C_2H_3O_2^-]}{[HC_2H_3O_2]} = \dfrac{(0.005\underline{7}706 + x)(x)}{(0.225 - x)} = 1.8 \times 10^{-5}$

Assume that x is small compared to 0.225.

$x^2 + 0.005\underline{7}706x - 4.05 \times 10^{-6} = 0; \quad x = 0.006\underline{3}250$

$[H_3O^+] = 0.005\underline{7}706 + x = 0.005\underline{7}706 + 0.006\underline{3}250 = 0.006\underline{4}031 \text{ M}$

$pH = -\log(0.006\underline{4}031) = 2.19$

(d)   **Given:** 0.050 M $HC_2H_3O_2$, $K_a = 1.8 \times 10^{-5}$; 0.050 M HCN, $K_a = 4.9 \times 10^{-10}$ **Find:** pH

Conceptual Plan: Because the values of K are more than 10^1 apart, the acid with the larger K will dominate the reaction. Write a balanced reaction. Prepare an ICE table, represent the change with x, sum the table, determine the equilibrium values, put the equilibrium values in the equilibrium expression, and solve for x.

Solution: $HC_2H_3O_2(aq) + H_2O(l) \rightleftharpoons H_3O^+(aq) + C_2H_3O_2^-(aq)$

Initial	0.0500	0.0	0.0
Change	$-x$	$+x$	$+x$
Equil	$0.0500 - x$	x	x

$K_a = \dfrac{[H_3O^+][C_2H_3O_2^-]}{[HC_2H_3O_2]} = \dfrac{(x)(x)}{(0.0500 - x)} = 1.8 \times 10^{-5}$

Assume that x is small compared to 0.0500.

$x^2 = (1.8 \times 10^{-5})(0.0500) \; x = 9.\underline{4}9 \times 10^{-4}$

$pH = -\log(9.\underline{4}9 \times 10^{-4}) = 3.02$

Base Solutions

15.81   (a)   **Given:** 0.15 M NaOH   **Find:** $[OH^-]$, $[H_3O^+]$, pH, pOH

Conceptual Plan: $[NaOH] \rightarrow [OH^-] \rightarrow [H_3O^+] \rightarrow pH \rightarrow pOH$

$\qquad\qquad K_w = [H_3O^+][OH^-] \quad pH = -\log[H_3O^+] \quad pH + pOH = 14$

Solution: $[OH^-] = [NaOH] = 0.15 \text{ M}$

$[H_3O^+] = \dfrac{K_w}{[OH^-]} = \dfrac{1.0 \times 10^{-14}}{0.15 \text{ M}} = 6.7 \times 10^{-14} \text{ M}$

$pH = -\log(6.7 \times 10^{-14}) = 13.17$

$pOH = 14.00 - 13.17 = 0.83$

(b)   **Given:** 1.5×10^{-3} M $Ca(OH)_2$ **Find:** $[OH^-]$, $[H_3O^+]$, pH, pOH

Conceptual Plan: $[Ca(OH)_2] \rightarrow [OH^-] \rightarrow [H_3O^+] \rightarrow pH \rightarrow pOH$

$\qquad\qquad K_w = [H_3O^+][OH^-] \quad pH = -\log[H_3O^+] \quad pH + pOH = 14$

Solution: $[OH^-] = 2[Ca(OH)_2] = 2(1.5 \times 10^{-3}) = 0.0030 \text{ M}$

$[H_3O^+] = \dfrac{K_w}{[OH^-]} = \dfrac{1.0 \times 10^{-14}}{0.0030 \text{ M}} = 3.\underline{3}3 \times 10^{-12} \text{ M}$

$pH = -\log(3.\underline{3}3 \times 10^{-12}) = 11.48$

$pOH = 14.00 - 11.48 = 2.52$

(c)   **Given:** 4.8×10^{-4} M $Sr(OH)_2$ **Find:** $[OH^-]$, $[H_3O^+]$, pH, pOH

Conceptual Plan: $[Sr(OH)_2] \rightarrow [OH^-] \rightarrow [H_3O^+] \rightarrow pH \rightarrow pOH$

$$K_w = [H_3O^+][OH^-] \quad pH = -\log[H_3O^+] \quad pH + pOH = 14$$

Solution: $[OH^+] = [Sr(OH)_2] = 2(4.8 \times 10^{-4}) = 9.6 \times 10^{-4}$ M

$$[H_3O^-] = \frac{K_w}{[OH^-]} = \frac{1.0 \times 10^{-14}}{9.6 \times 10^{-4} \text{ M}} = 1.0\underline{4} \times 10^{-11} \text{ M}$$

$pH = -\log(1.0\underline{4} \times 10^{-11}) = 10.98$
$pOH = 14.00 - 10.98 = 3.02$

(d)   **Given:** 8.7×10^{-5} M KOH   **Find:** $[OH^-]$, $[H_3O^+]$, pH, pOH

Conceptual Plan: $[KOH] \rightarrow [OH^-] \rightarrow [H_3O^+] \rightarrow pH \rightarrow pOH$

$$K_w = [H_3O^+][OH^-] \quad pH = -\log[H_3O^+] \quad pH + pOH = 14$$

Solution: $[OH^-] = [KOH] = 8.7 \times 10^{-5}$ M

$$[H_3O^+] = \frac{K_w}{[OH^-]} = \frac{1.0 \times 10^{-14}}{8.7 \times 10^{-5} \text{ M}} = 1.\underline{1} \times 10^{-10} \text{ M}$$

$pH = -\log(1.\underline{1} \times 10^{-10}) = 9.96$
$pOH = 14.00 - 9.96 = 4.04$

15.83   **Given:** 3.85% KOH by mass, $d = 1.01$ g/mL   **Find:** pH

Conceptual Plan:

% mass $\rightarrow$ g KOH $\rightarrow$ mol KOH and mass soln $\rightarrow$ mL soln $\rightarrow$ L soln $\rightarrow$ M KOH $\rightarrow [OH^+]$

$$\frac{1 \text{ mol KOH}}{56.11 \text{ g KOH}} \qquad \frac{1.01 \text{ g soln}}{\text{mL soln}} \quad \frac{1000 \text{ mL soln}}{\text{L soln}} \quad \frac{\text{mol KOH}}{\text{L soln}}$$

$\rightarrow$ **pOH** $\rightarrow$ **pH**

$$pOH = -\log[OH^-] \quad pH + pOH = 14$$

Solution: $\dfrac{3.85 \text{ g KOH}}{100.0 \text{ g soln}} \times \dfrac{1 \text{ mol KOH}}{56.11 \text{ g KOH}} \times \dfrac{1.01 \text{ g soln}}{\text{mL soln}} \times \dfrac{1000 \text{ mL soln}}{\text{L soln}} = 0.69\underline{3}0$ M KOH

$[OH^-] = [KOH] = 0.69\underline{3}0$ M   pOH $= -\log(0.69\underline{3}0) = 0.159$ pH $= 14.000 - 0.159 = 13.841$

15.85   **Given:** 3.55 L, pH $= 12.4$; 0.855 M KOH   **Find:** Vol

Conceptual Plan: pH $\rightarrow [H_3O^+] \rightarrow [OH^-]$ and then $V_1M_1 = V_2M_2$

$$[H_3O^+] = 10^{-pH} \quad 1.0 \times 10^{-14} = [H_3O^+][OH^-] \quad V_1M_1 = V_2M_2$$

Solution: $[H_3O^+] = 10^{-12.4} = 3.98 \times 10^{-13}$ $1.0 \times 10^{-14} = 3.98 \times 10^{-13}[OH^-]$

$[OH^-] = 0.025\underline{1}3$

$V_1M_1 = V_2M_2$ $V_1(0.855 \text{ M}) = (3.55 \text{ L})(0.0251 \text{ M})$ $V_1 = 0.104$ L

15.87

(a) $NH_3(aq) + H_2O(l) \rightleftharpoons NH_4^+(aq) + OH^-(aq)$ $K_b = \dfrac{[NH_4^+][OH^-]}{[NH_3]}$

(b) $HCO_3^-(aq) + H_2O(l) \rightleftharpoons H_2CO_3(aq) + OH^-(aq)$ $K_b = \dfrac{[H_2CO_3][OH^-]}{[HCO_3^-]}$

(c) $CH_3NH_2(aq) + H_2O(l) \rightleftharpoons CH_3NH_3^+(aq) + OH^-(aq)$ $K_b = \dfrac{[CH_3NH_3^+][OH^-]}{[CH_3NH_2]}$

15.89   **Given:** 0.15 M NH_3, $K_b = 1.76 \times 10^{-5}$ **Find:** $[OH^-]$, pH, pOH

Conceptual Plan: Write a balanced reaction. Prepare an ICE table, represent the change with x, sum the table, determine the equilibrium values, put the equilibrium values in the equilibrium expression, and solve for x.

$x = [OH^-] \rightarrow [pOH] \rightarrow pH$

$$pOH = -\log[OH^-] \quad pH + pOH = 14$$

Solution: $NH_3(aq) + H_2O(l) \rightleftharpoons NH_4^+(aq) + OH^-(aq)$

	NH_3	NH_4^+	OH^-
Initial	0.15	0.0	0.0
Change	$-x$	$+x$	$+x$
Equil	$0.15 - x$	x	x

$$K_b = \frac{[NH_4^+][OH^-]}{[NH_3]} = \frac{(x)(x)}{(0.15 - x)} = 1.76 \times 10^{-5}$$

Assume that x is small.

$x^2 = (1.76 \times 10^{-5})(0.15)$ $x = [OH^-] = 0.0016\underline{2}$ M

$pOH = -\log(0.0016\underline{2}) = 2.79$

$pH = 14.00 - 2.79 = 11.21$

15.91 **Given:** $pK_b = 10.4$, 455 mg/L caffeine   **Find:** pH

Conceptual Plan: $pK_b \rightarrow K_b$ and then mg/L $\rightarrow$ g/L $\rightarrow$ mol/L and then write a balanced reaction. Prepare an ICE table, represent the change with x, sum the table, determine the equilibrium values, put the equilibrium values in the equilibrium expression, and solve for x.

$x = [OH^-] \rightarrow [pOH] \rightarrow pH$

$pOH = -\log[OH^-] \quad pH + pOH = 14$

Solution: $K_b = 10^{-10.4} = \underline{3}.98 \times 10^{-11}$

$$\frac{455 \text{ mg caffeine}}{\text{L soln}} \times \frac{\text{g caffeine}}{1000 \text{ mg caffeine}} \times \frac{1 \text{ mol caffeine}}{194.20 \text{ g}} = 0.00234\underline{3} \text{ M caffeine}$$

$$C_8H_{10}N_4O_2(aq) + H_2O(l) \rightleftharpoons HC_8H_{10}N_4O_2^+(aq) + OH^-(aq)$$

Initial	0.00234\underline{3}	0.0	0.0
Change	$-x$	$+x$	$+x$
Equil	$0.00234\underline{3} - x$	x	x

$$K_b = \frac{[HC_8H_{10}N_4O_3^+][OH^-]}{[C_8H_{10}N_4O_2]} = \frac{(x)(x)}{(0.00234\underline{3} - x)} = \underline{3}.98 \times 10^{-11}$$

Assume that x is small.

$x^2 = (\underline{3}.98 \times 10^{-11})(0.00234\underline{3})$ $x = [OH^-] = \underline{3}.05 \times 10^{-7}$ M

$$\frac{\underline{3}.05 \times 10^{-7} \text{ M}}{0.00234\underline{3}} \times 100\% = 0.013\%; \text{ assumption is valid.}$$

$pOH = -\log(\underline{3}.05 \times 10^{-7}) = 6.5$

$pH = 14.00 - 6.5 = 7.5$

15.93 **Given:** 0.150 M morphine, pH = 10.5   **Find:** K_b

Conceptual Plan:

$pH \rightarrow pOH \rightarrow [OH^-]$ and then write a balanced equation, prepare an ICE table, and determine

$pH + pOH = 14 \qquad pOH = -\log[OH^-]$

equilibrium concentrations $\rightarrow K_b$.

Solution: $pOH = 14.0 - 10.5 = 3.5;$ $[OH^-] = 10^{-3.5} = \underline{3}.16 \times 10^{-4} = [Hmorphine^+]$

$$morphine(aq) + H_2O(l) \rightleftharpoons Hmorphine^+(aq) + OH^-(aq)$$

Initial	0.150	0.0	0.0
Change	$-x$	$+x$	$+x$
Equil	$0.150 - x$	3.16×10^{-4}	3.16×10^{-4}

$$K_b = \frac{[Hmorphine^+][OH^-]}{[morphine]} = \frac{(3.16 \times 10^{-4})(3.16 \times 10^{-4})}{(0.150 - 3.16 \times 10^{-4})} = \underline{6}.67 \times 10^{-7} = 7 \times 10^{-7}$$

Acid–Base Properties of Ions and Salts

15.95 (a) pH-neutral: Br^- is the conjugate base of a strong acid; therefore, it is pH-neutral.

(b) weak base: ClO^- is the conjugate base of a weak acid; therefore, it is a weak base.

$ClO^-(aq) + H_2O(l) \rightleftharpoons HClO(aq) + OH^-(aq)$

(c) weak base: CN^- is the conjugate base of a weak acid; therefore, it is a weak base.

$CN^-(aq) + H_2O(l) \rightleftharpoons HCN(aq) + OH^-(aq)$

(d) pH-neutral: Cl^- is the conjugate base of a strong acid; therefore, it is pH-neutral.

15.97   **Given:** $[F^-] = 0.140$ M, $K_a(HF) = 3.5 \times 10^{-4}$ **Find:** $[OH^-]$, pOH

Conceptual Plan: Determine K_b. Write a balanced reaction. Prepare an ICE table, represent the change

$$K_b = \frac{K_w}{K_a}$$

with x, sum the table, determine the equilibrium values, put the equilibrium values in the equilibrium expression, and solve for x. Determine $[OH^-] \rightarrow pOH \rightarrow pH$.

$$pOH = -\log[OH^-] \qquad pH + pOH = 14$$

Solution: $F^-(aq) + H_2O(l) \rightleftharpoons HF(aq) + OH^-(aq)$

Initial	0.140	0.0	0.0
Change	$-x$	$+x$	$+x$
Equil	$0.140 - x$	x	x

$$K_b = \frac{K_w}{K_a} = \frac{1 \times 10^{-14}}{3.5 \times 10^{-4}} = 2.\underline{8}6 \times 10^{-11} = \frac{(x)(x)}{(0.140 - x)}$$

Assume that x is small.

$x = 2.0 \times 10^{-6} = [OH^-]$ $pOH = -\log(2.0 \times 10^{-6}) = 5.70$

$pH = 14.00 - 5.70 = 8.30$

15.99   (a)   weak acid: NH_4^+ is the conjugate acid of a weak base; therefore, it is a weak acid.

$$NH_4^+(aq) + H_2O(l) \rightleftharpoons H_3O^+(aq) + NH_3(aq)$$

   (b)   pH-neutral: Na^+ is the counterion of a strong base; therefore, it is pH-neutral.

   (c)   weak acid: The Co^{3+} cation is a small, highly charged metal cation; therefore, it is a weak acid.

$$Co(H_2O)_6^{3+}(aq) + H_2O(l) \rightleftharpoons Co(H_2O)_5(OH)^{2+}(aq) + H_3O^+(aq)$$

   (d)   weak acid: $CH_2NH_3^+$ is the conjugate acid of a weak base; therefore, it is a weak acid.

$$CH_2NH_3^+(aq) + H_2O(l) \rightleftharpoons H_3O^+(aq) + CH_2NH_2(aq)$$

15.101   (a)   acidic: $FeCl_3$ Fe^{3+} is a small, highly charged metal cation; therefore, it is acidic. Cl^- is the conjugate base of a strong acid; therefore, it is pH-neutral.

   (b)   basic: NaF Na^+ is the counterion of a strong base; therefore, it is pH-neutral. F^- is the conjugate base of a weak acid; therefore, it is basic.

   (c)   pH-neutral: $CaBr_2$ Ca^{2+} is the counterion of a strong base; therefore, it is pH-neutral. Br^- is the conjugate base of a strong acid; therefore, it is pH-neutral.

   (d)   acidic: NH_4Br NH_4^+ is the conjugate acid of a weak base; therefore, it is acidic. Br^- is the conjugate base of a strong acid; therefore, it is pH-neutral.

   (e)   acidic: $C_6H_5NH_3NO_2$ $C_6H_5NH_3^+$ is the conjugate acid of a weak base; therefore, it is acidic. NO_2^- is the conjugate base of a weak acid; therefore, it is basic. To determine pH, compare K values.

$$K_a(C_6H_5NH_3^+) = \frac{1.0 \times 10^{-14}}{3.9 \times 10^{-10}} = 2.6 \times 10^{-5} \quad K_b(NO_2^-) = \frac{1.0 \times 10^{-14}}{4.6 \times 10^{-4}} = 2.2 \times 10^{-11}$$

$K_a > K_b$; therefore, the solution is acidic.

15.103   **Conceptual Plan:** Identify each species and determine whether it is acidic, basic, or neutral.

$NaCl$ pH-neutral: Na^+ is the counterion of a strong base; therefore, it is pH-neutral. Cl^- is the conjugate base of a strong acid; therefore, it is pH-neutral.

NH_4Cl acidic: NH_4^+ is the conjugate acid of a weak base; therefore, it is acidic. Cl^- is the conjugate base of a strong acid; therefore, it is pH-neutral.

$NaHCO_3$ basic: Na^+ is the counterion of a strong base; therefore, it is pH-neutral. HCO_3^- is the conjugate base of a weak acid; therefore, it is basic.

NH_4ClO_2 acidic: NH_4^+ is the conjugate acid of a weak base; therefore, it is acidic. ClO_2^- is the conjugate base of a weak acid; therefore, it is basic. $K_a(NH_4^+) = 5.6 \times 10^{-10}$ $K_b(ClO_2^-) = 9.1 \times 10^{-13}$

$K_a > K_b$; therefore, the solution is acidic.

$NaOH$ strong base

Increasing acidity: $NaOH < NaHCO_3 < NaCl < NH_4ClO_2 < NH_4Cl$

15.105   (a)   **Given:** 0.10 M NH_4Cl **Find:** pH
Conceptual Plan: Identify each species and determine which will contribute to pH. Write a balanced re-action. Prepare an ICE table, represent the change with *x*, sum the table, determine the equilibrium values, put the equilibrium values in the equilibrium expression, and solve for *x*. Determine $[H_3O^+] \rightarrow$ **pH.**
Solution: NH_4^+ is the conjugate acid of a weak base; therefore, it is acidic. Cl^- is the conjugate base of a strong acid; therefore, it is pH-neutral.

$$NH_4^+(aq) + H_2O(l) \rightleftharpoons NH_3(aq) + H_3O^+(aq)$$

Initial	0.10	0.0	0.0
Change	$-x$	$+x$	$+x$
Equil	$0.10 - x$	x	x

$$K_a = \frac{K_w}{K_b} = \frac{1.0 \times 10^{-14}}{1.76 \times 10^{-5}} = 5.\underline{6}8 \times 10^{-10} = \frac{(x)(x)}{(0.10 - x)}$$

Assume that *x* is small.
$$x = 7.\underline{5}4 \times 10^{-6} = [H_3O^+] \quad pH = -\log(7.\underline{5}4 \times 10^{-6}) = 5.12$$

   (b)   **Given:** 0.10 M $NaC_2H_3O_2$ **Find:** pH
Conceptual Plan: Identify each species and determine which will contribute to pH. Write a balanced reaction. Prepare an ICE table, represent the change with *x*, sum the table, determine the equilibrium values, put the equilibrium values in the equilibrium expression, and solve for *x*. Determine $[OH^-] \rightarrow pOH \rightarrow$ **pH.**
$pOH = -\log[OH^-] \quad pH + pOH = 14$
Solution: Na^+ is the counterion of a strong base; therefore, it is pH-neutral. $C_2H_3O_2^-$ is the conjugate base of a weak acid; therefore, it is basic.

$$C_2H_3O_2^-(aq) + H_2O(l) \rightleftharpoons HC_2H_3O_2(aq) + OH^-(aq)$$

Initial	0.10	0.0	0.0
Change	$-x$	$+x$	$+x$
Equil	$0.10 - x$	x	x

$$K_b = \frac{K_w}{K_a} = \frac{1.0 \times 10^{-14}}{1.8 \times 10^{-5}} = 5.\underline{5}6 \times 10^{-10} = \frac{(x)(x)}{(0.10 - x)}$$

Assume that *x* is small.
$$x = 7.\underline{4}6 \times 10^{-6} = [OH^-] \quad pOH = -\log(7.\underline{4}6 \times 10^{-6}) = 5.13$$
$$pH = 14.00 - 5.13 = 8.87$$

   (c)   **Given:** 0.10 M NaCl   **Find:** pH
Conceptual Plan: Identify each species and determine which will contribute to pH.
Solution: Na^+ is the counterion of a strong base; therefore, it is pH-neutral. Cl^- is the conjugate base of a strong acid; therefore, it is pH-neutral.
$$pH = 7.0$$

15.107   **Given:** 0.15 M KF   **Find:** concentration of all species
Conceptual Plan: Identify each species and determine which will contribute to pH. Write a balanced reaction. Prepare an ICE table, represent the change with *x*, sum the table, determine the equilibrium values, put the equilibrium values in the equilibrium expression, and solve for *x*. Then $[OH^-] \rightarrow [H_3O^+]$.
$$K_w = [H_3O^+][OH^-]$$
Solution: K^+ is the counterion of a strong base; therefore, it is pH-neutral. F^- is the conjugate base of a weak acid; therefore, it is basic.

$$F^-(aq) + H_2O(l) \rightleftharpoons HF(aq) + OH^-(aq)$$

Initial	0.15	0.0	0.0
Change	$-x$	$+x$	$+x$
Equil	$0.15 - x$	x	x

$$K_b = \frac{K_w}{K_a} = \frac{1.0 \times 10^{-14}}{3.5 \times 10^{-4}} = \frac{(x)(x)}{(0.15 - x)}$$

Assume that x is small.

$$x = 2.1 \times 10^{-6} = [OH^-] = [HF]; \qquad [H_3O^+] = \frac{K_w}{[OH^-]} = \frac{1 \times 10^{-14}}{2.1 \times 10^{-6}} = 4.8 \times 10^{-9}$$

$[K^+] = 0.15 \, M$
$[F^-] = (0.15 - 2.1 \times 10^{-6}) = 0.15 \, M$
$[HF] = 2.1 \times 10^{-6}$
$[OH^-] = 2.1 \times 10^{-6}$
$[H_3O^+] = 4.8 \times 10^{-9}$

Polyprotic Acids

15.109 $H_3PO_4(aq) + H_2O(l) \rightleftharpoons H_3O^+(aq) + H_2PO_4^-(aq)$ $K_{a_1} = \dfrac{[H_3O^+][H_2PO_4^-]}{[H_3PO_4]}$

$H_2PO_4^-(aq) + H_2O(l) \rightleftharpoons H_3O^+(aq) + HPO_4^{2-}(aq)$ $K_{a_2} = \dfrac{[H_3O^+][HPO_4^{2-}]}{[H_2PO_4^-]}$

$HPO_4^{2-}(aq) + H_2O(l) \rightleftharpoons H_3O^+(aq) + PO_4^{3-}(aq)$ $K_{a_3} = \dfrac{[H_3O^+][PO_4^{3-}]}{[HPO_4^{2-}]}$

15.111   (a)   **Given:** 0.350 M H_3PO_4 $K_{a_1} = 7.5 \times 10^{-3}, K_{a_2} = 6.2 \times 10^{-8}$ **Find:** $[H_3O^+]$, pH
Conceptual Plan: K_{a_1} **is much larger than** K_{a_2}**, so use** K_{a_1} **to calculate** $[H_3O^+]$**. Write a balanced reaction. Prepare an ICE table, represent the change with x, sum the table, determine the equilibrium values, put the equilibrium values in the equilibrium expression, and solve for x. Repeat this process by preparing a second ICE table using these values and representing the change with y.**
Solution: $H_3PO_4(aq) + H_2O(l) \rightleftharpoons H_3O^+(aq) + H_2PO_4^-(aq)$

		H_3O^+	$H_2PO_4^-$
Initial	0.350	0.0	0.0
Change	$-x$	$+x$	$+x$
Equil	$0.350 - x$	x	x

$$K_{a_1} = \frac{[H_3O^+][H_2PO_4^-]}{[H_3PO_4]} = \frac{(x)(x)}{(0.350 - x)} = 7.5 \times 10^{-3}$$

Assume that x is small compared to 0.350.
$x^2 = (7.5 \times 10^{-3})(0.350)$ $x = 0.0505 \, M = [H_3O^+]$

Check assumption: $\dfrac{0.0505}{0.350} \times 100\% = 14.4\%$ assumption is not valid, solve using the quadratic equation.

$x^2 + 7.5 \times 10^{-3}x - 0.002625 = 0; \quad x = 0.04\underline{7}62 = [H_3O^+]$
$pH = -\log(0.04\underline{7}62) = 1.32$

(b)   **Given:** 0.350 M $H_2C_2O_4$ $K_{a_1} = 6.0 \times 10^{-2}, K_{a_2} = 6.0 \times 10^{-5}$ **Find:** $[H_3O^+]$, pH
Conceptual Plan: K_{a_1} **is much larger than** K_{a_2}**, so use** K_{a_1} **to calculate** $[H_3O^+]$**. Write a balanced reaction. Prepare an ICE table, represent the change with x, sum the table, determine the equilibrium values, put the equilibrium values in the equilibrium expression, and solve for x.**
Solution: $H_2C_2O_4(aq) + H_2O(l) \rightleftharpoons H_3O^+(aq) + HC_2O_4^-(aq)$

		H_3O^+	$HC_2O_4^-$
Initial	0.350	0.0	0.0
Change	$-x$	$+x$	$+x$
Equil	$0.350 - x$	x	x

$$K_{a_1} = \frac{[H_3O^+][HC_2O_4^-]}{[H_2C_2O_4]} = \frac{(x)(x)}{(0.350 - x)} = 6.0 \times 10^{-2}$$

$x^2 + 6.0 \times 10^{-2}x - 0.021 = 0$ $x = 0.1\underline{1}79 = 0.12 \, M \, [H_3O^+]$
$pH = -\log(0.1\underline{1}79) = 0.93$

15.113　**Given:** 0.500 M H_2SO_3　$K_{a_1} = 1.6 \times 10^{-2}, K_{a_2} = 6.4 \times 10^{-8}$　**Find:** concentration of all species
Conceptual Plan: K_{a_1} **is much larger than** K_{a_2}**, so use** K_{a_1} **to calculate** $[H_3O^+]$**. Write a balanced reaction.**
Prepare an ICE table, represent the change with x**, sum the table, determine the equilibrium values, put the**
equilibrium values in the equilibrium expression, and solve for x**. Repeat this process by preparing a second**
ICE table using these values and representing the change with y**.**
Solution:　$H_2SO_3(aq) + H_2O(l) \rightleftharpoons H_3O^+(aq) + HSO_3^-(aq)$

Initial	0.500	0.0	0.0
Change	$-x$	$+x$	$+x$
Equil	$0.500 - x$	x	x

$$K_{a_1} = \frac{[H_3O^+][HSO_3^-]}{[H_2SO_3]} = \frac{(x)(x)}{(0.500 - x)} = 1.6 \times 10^{-2}$$

$x^2 + 1.6 \times 10^{-2}x - 0.0080 = 0$　$x = 0.08\underline{1}8 = 0.082$ M $[H_3O^+] = [HSO_3^-]$

Use the values from reaction 1 in reaction 2.

$$HSO_3^-(aq) + H_2O(l) \rightleftharpoons H_3O^+(aq) + SO_3^{2-}(aq)$$

Initial	0.0818	0.0818	0.0
Change	$-y$	$+y$	$+y$
Equil	$0.0818 - y$	$0.0818 + y$	y

$$K_{a_1} = \frac{[H_3O^+][SO_3^{2-}]}{[HSO_3^-]} = \frac{(0.0818 + y)(y)}{(0.0818 - y)} = 6.4 \times 10^{-8}$$

Assume that y is small.　$y = 6.4 \times 10^{-8}$
$[H_2SO_3] = 0.500 - 0.0818 = 0.418$ M
$[HSO_3^-] = x = 0.0818 = 0.082$ M
$[SO_3^{2-}] = y = 6.4 \times 10^{-8}$ M
$[H_3O^+] = x + y = 0.0818$ M $+ 6.4 \times 10^{-8}$ M $= 0.082$ M

$$[OH^-] = \frac{K_w}{[H_3O^+]} = \frac{1.0 \times 10^{-14}}{0.0818} = 1.2 \times 10^{-13} \text{ M}$$

15.115　(a)　**Given:** $[H_2SO_4] = 0.50$ M　$K_{a_2} = 0.012$　**Find:** $[H_3O^+]$, pH
Conceptual Plan: The first ionization step is strong. Use K_{a_2} **and reaction 2. Write a balanced reaction.**
Prepare an ICE table, represent the change with x**, sum the table, determine the equilibrium values, put**
the equilibrium values in the equilibrium expression, and solve for x**.**
Solution: $H_2SO_4(aq) + H_2O(l) \rightarrow H_3O^+(aq) + HSO_4^-(aq)$ strong
$\qquad\qquad$ 0.50 M
$\qquad\qquad [H_3O^+] = [HSO_4^-] = 0.50$ M
$\qquad\qquad HSO_4^-(aq) + H_2O(l) \rightleftharpoons H_3O^+(aq) + SO_4^{2-}(aq)$

Initial	0.50	0.50	0.0
Change	$-x$	$+x$	$+x$
Equil	$0.500 - x$	$0.50 + x$	x

$$K_{a_2} = \frac{[H_3O^+][SO_4^{2-}]}{[HSO_4^-]} = \frac{(0.50 + x)(x)}{(0.50 - x)} = 0.012$$

$x^2 + 0.512x - 0.006 = 0$　$x = 0.01\underline{1}5 = [H_3O^+]$ from second ionization step
$[H_3O^+] = 0.50 + 0.012 = 0.51$ M
pH $= -\log(0.51) = 0.29$

(b)　**Given:** $[H_2SO_4] = 0.10$ M　$K_{a_2} = 0.012$　**Find:** $[H_3O^+]$, pH
Conceptual Plan: The first ionization step is strong. Use K_{a_2} **and reaction 2. Write a balanced reaction.**
Prepare an ICE table, represent the change with x**, sum the table, determine the equilibrium values, put**
the equilibrium values in the equilibrium expression, and solve for x**.**

Solution: $H_2SO_4(aq) + H_2O(l) \rightarrow H_3O^+(aq) + HSO_4^-(aq)$ strong

$\quad$ 0.10 M

$\quad [H_3O^+] = [HSO_4^-] = 0.10\,M$

$\quad HSO_4^-(aq) + H_2O(l) \rightleftharpoons H_3O^+(aq) + SO_4^{2-}(aq)$

Initial	0.10	0.10	0.0
Change	$-x$	$+x$	$+x$
Equil	$0.10 - x$	$0.10 + x$	x

$$K_{a_2} = \frac{[H_3O^+][SO_4^{2-}]}{[HSO_4^-]} = \frac{(0.10 + x)(x)}{(0.10 - x)} = 0.012$$

$x^2 + 0.112x - 0.0012 = 0; \quad x = 0.009\underline{8}48$

$\dfrac{0.009\underline{8}48}{0.10} \times 100\% = 9.8\%$ contribution of second ionizations step is not negligible.

$[H_3O^+] = 0.10 + 0.009\underline{8}48 = 0.1\underline{0}9848\,M = 0.11\,M$

$pH = -\log(0.11) = 0.96$

(c) $\quad$ **Given:** $[H_2SO_4] = 0.050\,M \quad K_{a_2} = 0.012$ $\quad$ **Find:** $[H_3O^+]$, pH

$\quad$ **Conceptual Plan: The first ionization step is strong. Use K_{a_2} and reaction 2. Write a balanced reaction. Prepare an ICE table, represent the change with x, sum the table, determine the equilibrium values, put the equilibrium values in the equilibrium expression, and solve for x.**

$\quad$ **Solution:** $H_2SO_4(aq) + H_2O(l) \rightarrow H_3O^+(aq) + HSO_4^-(aq)$ strong

$\quad\quad$ 0.050 M

$\quad [H_3O^+] = [HSO_4^-] = 0.050\,M$

$\quad\quad HSO_4^-(aq) + H_2O(l) \rightleftharpoons H_3O^+(aq) + SO_4^{2-}(aq)$

Initial	0.050	0.050	0.0
Change	$-x$	$+x$	$+x$
Equil	$0.050 - x$	$0.050 + x$	x

$$K_{a_2} = \frac{[H_3O^+][SO_4^{2-}]}{[HSO_4^-]} = \frac{(0.050 + x)(x)}{(0.050 - x)} = 0.012$$

$x^2 + 0.062\,x - 0.0006 = 0 \quad x = 0.008\underline{5}09$

$\dfrac{0.008\underline{5}09}{0.05} \times 100\% = 17\%$ contribution of second ionizations step is not negligible.

$[H_3O^+] = 0.050 + 0.0085 = 0.0585 = 0.059\,M \quad\quad pH = -\log(0.059) = 1.23$

Molecular Structure and Acid Strength

15.117 $\quad$ (a) $\quad$ HCl is the stronger acid. HCl is the weaker bond; therefore, it is more acidic.

$\quad\quad$ (b) $\quad$ HF is the stronger acid. F is more electronegative than is O, so the bond is more polar and more acidic.

$\quad\quad$ (c) $\quad$ H_2Se is the stronger acid. The H—Se bond is weaker; therefore, it is more acidic.

15.119 $\quad$ (a) $\quad$ H_2SO_4 is the stronger acid because it has more oxygen atoms.

$\quad\quad$ (b) $\quad$ $HClO_2$ is the stronger acid because it has more oxygen atoms.

$\quad\quad$ (c) $\quad$ HClO is the stronger acid because Cl is more electronegative than is Br.

$\quad\quad$ (d) $\quad$ CCl_3COOH is the stronger acid because Cl is more electronegative than is H.

15.121 $\quad$ S^{2-} is the stronger base. Base strength is determined from the corresponding acid. The weaker the acid, the stronger the base. H_2S is the weaker acid because it has a stronger bond.

Lewis Acids and Bases

15.123 $\quad$ (a) $\quad$ Lewis acid: Fe^{3+} has an empty d orbital and can accept lone pair electrons.

$\quad\quad$ (b) $\quad$ Lewis acid: BH_3 has an empty p orbital to accept a lone pair of electrons.

$\quad\quad$ (c) $\quad$ Lewis base: NH_3 has a lone pair of electrons to donate.

$\quad\quad$ (d) $\quad$ Lewis base: F^- has lone pair electrons to donate.

15.125 (a) Fe^{3+} accepts an electron pair from H_2O; so Fe^{3+} is the Lewis acid, and H_2O is the Lewis base.

(b) Zn^{2+} accepts an electron pair from NH_3; so Zn^{2+} is the Lewis acid, and NH_3 is the Lewis base.

(c) The empty p orbital on B accepts an electron pair from $(CH_3)_3N$; so BF_3 is the Lewis acid, and $(CH_3)_3N$ is the Lewis base.

Cumulative Problems

15.127 (a) weak acid: The beaker contains 10 HF molecule, 2 H_3O^+ ions, and 2 F^- ions. Because both the molecule and the ions exist in solution, the acid is a weak acid.

(b) strong acid: The beaker contains 12 H_3O^+ ions and 12 I^- ions. Because the molecule is completely ionized in solution, the acid is a strong acid.

(c) weak acid: The beaker contains 10 $HCHO_2$ molecules, 2 H_3O^+ ions, and 2 CHO_2^- ions. Because both the molecule and the ions exist in solution, the acid is a weak acid.

(d) strong acid: The beaker contains 12 H_3O^+ ions and 12 NO_3^- ions. Because the molecule is completely ionized in solution, the acid is a strong acid.

15.129 $HbH^+(aq) + O_2(aq) \rightleftharpoons HbO_2(aq) + H^+(aq)$

Using Le Châtelier's principle, if the $[H^+]$ increases, the reaction will shift left and if the $[H^+]$ decreases, the reaction will shift right. So if the pH of blood is too acidic (low pH; $[H^+]$ increased), the reaction will shift to the left. This will cause less of the HbO_2 in the blood and decrease the oxygen-carrying capacity of the hemoglobin in the blood.

15.131 **Given:** 4.00×10^2 mg $Mg(OH)_2$, 2.00×10^2 mL HCl solution, pH = 1.3   **Find:** volume neutralized, % neutralized
Conceptual Plan:
mg $Mg(OH)_2 \rightarrow$ g $Mg(OH)_2 \rightarrow$ mol $Mg(OH)_2$ and then pH $\rightarrow [H_3O^+]$ and then mol $Mg(OH)_2$

$$\frac{\text{g } Mg(OH)_2}{1000 \text{ mg}} \qquad \frac{\text{mol } Mg(OH)_2}{58.326 \text{ g}} \qquad\qquad pH = -\log[H_3O^+]$$

mol $OH^- \rightarrow$ mol $H_3O^+ \rightarrow$ vol $H_3O^+ \rightarrow$ % neutralized

$$\frac{2 \text{ } OH^-}{Mg(OH)_2} \quad \frac{H_3O^+}{OH^-} \quad \frac{\text{mol } H_3O^+}{M(H_3O^+)} \quad \frac{\text{vol HCl neutralized}}{\text{total vol HCl}} \times 100$$

Solution: $[H_3O^+] = 10^{-1.3} = 0.05012 \text{ M} = 0.05 \text{ M}$

$$4.00 \times 10^2 \text{ mg } \cancel{Mg(OH)_2} \times \frac{\text{g } \cancel{Mg(OH)_2}}{1000 \text{ mg } \cancel{Mg(OH)_2}} \times \frac{1 \text{ mol } \cancel{Mg(OH)_2}}{58.326 \text{ g } \cancel{Mg(OH)_2}} \times \frac{2 \text{ mol } \cancel{OH^-}}{1 \text{ mol } \cancel{Mg(OH)_2}}$$

$$\times \frac{1 \text{ mol } \cancel{H_3O^+}}{1 \text{ mol } \cancel{OH^-}} \times \frac{1 \cancel{L}}{0.0501 \text{ mol } \cancel{H_3O^+}} \times \frac{1000 \text{ mL}}{\cancel{L}} = 273.8 \text{ mL} = 274 \text{ mL neutralized}$$

The stomach contains 2.00×10^2 mL HCl at pH = 1.3, and 4.00×10^2 mg will neutralize 274 mL of pH 1.3 HCl; so all of the stomach acid will be neutralized.

15.133 **Given:** pH of Great Lakes acid rain = 4.5, West Coast = 5.4   **Find:** $[H_3O^+]$ and ratio of Great Lakes/West Coast
Conceptual Plan: pH $\rightarrow [H_3O^+]$ and then ratio of $[H_3O^+]$ Great Lakes to West Coast

$$pH = -\log[H_3O^+]$$

Solution: Great Lakes: $[H_3O^+] = 10^{-4.5} = 3.16 \times 10^{-5} \text{ M}$ West Coast: $[H_3O^+] = 10^{-5.4} = 3.98 \times 10^{-6} \text{ M}$

$$\frac{\text{Great Lakes}}{\text{West Coast}} = \frac{3.16 \times 10^{-5} \text{ M}}{3.98 \times 10^{-6} \text{ M}} = 7.94 = 8 \text{ times more acidic}$$

15.135 **Given:** 6.5×10^2 mg aspirin, 8 oz water, $pK_a = 3.5$ **Find:** pH of solution
Conceptual Plan:
mg aspirin $\rightarrow$ g aspirin $\rightarrow$ mol aspirin and ounces $\rightarrow$ quart $\rightarrow$ L and then [aspirin] and pK_a

$$\frac{\text{g aspirin}}{1000 \text{ mg}} \quad \frac{\text{mol aspirin}}{180.15 \text{ g}} \qquad\qquad \frac{32 \text{ oz}}{qt} \quad \frac{1.0567 \text{ qt}}{1 \text{ L}} \quad \frac{\frac{\text{mol aspirin}}{\text{L soln}}}{} \quad pK_a = -\log K_a$$

$\rightarrow K_a$. Write a balanced reaction. Prepare an ICE table, represent the change with x, sum the table, determine the equilibrium values, put the equilibrium values in the equilibrium expression, and solve for x. Determine $[H_3O^+] \rightarrow$ pH.

$$pH = -\log[H_3O^+]$$

Solution: $\dfrac{6.5 \times 10^2 \text{ mg-aspirin}}{8 \text{ oz}} \times \dfrac{1 \text{ g aspirin}}{1000 \text{ mg-aspirin}} \times \dfrac{\text{mol aspirin}}{180.15 \text{ g}} \times \dfrac{32 \text{ oz}}{\text{qt}} \times \dfrac{1.0567 \text{ qt}}{1 \text{ L}} = 0.01\underline{5}3 \text{ M}$

$$K_a = 10^{-3.5} = 3.\underline{1}6 \times 10^{-4}$$

$$\text{aspirin}(aq) + H_2O(l) \rightleftharpoons H_3O^+(aq) + \text{aspirin}^-(aq)$$

Initial	0.01$\underline{5}$3	0.0	0.0
Change	$-x$	$+x$	$+x$
Equil	0.01$\underline{5}$3 $- x$	x	x

$$K_a = \dfrac{[H_3O^+][\text{aspirin}^-]}{[\text{aspirin}]} = \dfrac{(x)(x)}{(0.0153 - x)} = 3.\underline{1}6 \times 10^{-4}$$

$x^2 + 3.16 \times 10^{-4}x - 4.83 \times 10^{-6} = 0$; Solve the quadric equation. $x = 2.\underline{0}4 \times 10^{-3} \text{ M} = [H_3O^+]$

$\text{pH} = -\log(2.\underline{0}4 \times 10^{-3}) = 2.69 = 2.7$

15.137   **(a)**   **Given:** 0.0100 M HClO$_4$ **Find:** pH

           Conceptual Plan: $[HClO_4] \rightarrow [H_3O^+] \rightarrow pH$

                     $[HClO_4] \rightarrow [H_3O^+] \quad pH = -\log[H_3O^+]$

           Solution: HClO$_4$ is a strong acid, so 0.0100 M HClO$_4$ = 0.0100 M H$_3$O$^+$ pH = $-\log(0.0100)$ = 2.000

  (b)   **Given:** 0.115 M HClO$_2$, $K_a = 1.1 \times 10^{-2}$ **Find:** pH

           Conceptual Plan: Write a balanced reaction. Prepare an ICE table, represent the change with x, sum the table, determine the equilibrium values, put the equilibrium values in the equilibrium expression, and solve for x. Determine $[H_3O^+]$ and pH.

           Solution: $HClO_2(aq) + H_2O(l) \rightleftharpoons H_3O^+(aq) + ClO_2^-(aq)$

Initial	0.115	0.0	0.0
Change	$-x$	$+x$	$+x$
Equil	0.115 $- x$	x	x

$$K_a = \dfrac{[H_3O^+][ClO_2^-]}{[HClO_2]} = \dfrac{(x)(x)}{(0.115 - x)} = 1.1 \times 10^{-2}$$

           Assume that x is small compared to 0.115.

$$x^2 = (1.1 \times 10^{-2})(0.115); \quad x = 0.03\underline{5}6 \text{ M} = [H_3O^+]$$

           Check assumption: $\dfrac{0.03\underline{5}6}{0.115} \times 100\% = 31.0\%$; assumption is not valid; solve using the quadratic equation.

$x^2 + 1.1 \times 10^{-2}x - 0.001265 = 0$;   Solve the quadric equation.

$x = 0.030\underline{4}9$ or -0.0415

$\text{pH} = -\log(0.030\underline{4}9) = 1.52$

  (c)   **Given:** 0.045 M Sr(OH)$_2$ **Find:** pH

           Conceptual Plan: $[Sr(OH)_2] \rightarrow [OH^-] \rightarrow [H_3O^+] \rightarrow pH$

                  $K_w = [H_3O^+][OH^-] \quad pH = -\log[H_3O^+] \quad pH + pOH = 14$

           Solution: Sr(OH)$_2$ is a strong base, so $[OH^-] = 2[Sr(OH)_2] = 2(0.045) = 0.090 \text{ M}$

$$[H_3O^+] = \dfrac{K_w}{[OH^-]} = \dfrac{1.0 \times 10^{-14}}{0.090 \text{ M}} = 1.\underline{1}1 \times 10^{-13} \text{ M}$$

$\text{pH} = -\log(1.\underline{1}1 \times 10^{-13}) = 12.95$

  (d)   **Given:** 0.0852 KCN, $K_a(\text{HCN}) = 4.9 \times 10^{-10}$ **Find:** pH

           Conceptual Plan: Identify each species and determine which will contribute to pH. Write a balanced reaction. Prepare an ICE table, represent the change with x, sum the table, determine the equilibrium values, put the equilibrium values in the equilibrium expression, and solve for x. Determine $[OH^-] \rightarrow pOH \rightarrow pH$.

           $pOH = -\log[OH^-] \quad pH + pOH = 14$

Solution: K^+ is the counterion of a strong base; therefore, it is pH-neutral. CN^- is the conjugate base of a weak acid; therefore, it is basic.

$$CN^-(aq) + H_2O(l) \rightleftharpoons HCN(aq) + OH^-(aq)$$

	$CN^-(aq)$	$HCN(aq)$	$OH^-(aq)$
Initial	0.0852	0.0	0.0
Change	$-x$	$+x$	$+x$
Equil	$0.0852 - x$	x	x

$$K_b = \frac{K_w}{K_a} = \frac{1.0 \times 10^{-14}}{4.9 \times 10^{-10}} = 2.0\underline{4} \times 10^{-5} = \frac{(x)(x)}{(0.0852 - x)}$$

Assume that x is small.

$x = 1.\underline{32} \times 10^{-3}\,M = [OH^-];\quad pOH = -\log(1.\underline{32} \times 10^{-3}) = 2.88$
$pH = 14.00 - 2.88 = 11.12$

(e) **Given:** 0.155 NH_4Cl, $K_b\,(NH_3) = 1.76 \times 10^{-5}$ **Find:** pH
Conceptual Plan: Identify each species and determine which will contribute to pH. Write a balanced reaction. Prepare an ICE table, represent the change with x, sum the table, determine the equilibrium values, put the equilibrium values in the equilibrium expression, and solve for x. Determine $[H_3O^+] \rightarrow$ pH.
Solution: NH_4^+ is the conjugate acid of a weak base; therefore, it is acidic. Cl^- is the conjugate base of a strong acid; therefore, it is pH-neutral.

$$NH_4^+(aq) + H_2O(l) \rightleftharpoons NH_3(aq) + H_3O^+(aq)$$

	$NH_4^+(aq)$	$NH_3(aq)$	$H_3O^+(aq)$
Initial	0.155	0.0	0.0
Change	$-x$	$+x$	$+x$
Equil	$0.155 - x$	x	x

$$K_a = \frac{K_w}{K_b} = \frac{1 \times 10^{-14}}{1.76 \times 10^{-5}} = 5.6\underline{82} \times 10^{-10} = \frac{(x)(x)}{(0.155 - x)}$$

Assume that x is small.

$x = 9.3\underline{85} \times 10^{-6}\,M = [H_3O^+];\quad pH = -\log(9.3\underline{85} \times 10^{-6}) = 5.028$

15.139 (a) **Given:** 0.0550M HI (strong acid), 0.00850M HF (weak acid)   **Find:** pH
Conceptual Plan: Because the mixture is a strong acid and a weak acid, the strong acid will dominate. Use the concentration of the strong acid to determine $[H_3O^+]$ and then pH.
Solution: 0.0550M HI = 0.0550 M $[H_3O^+]$ pH = $-\log(0.0550) = 1.260$

(b) **Given:** 0.112 M NaCl (salt), 0.0953 M KF (salt)   **Find:** pH
Conceptual Plan: Identify each species and determine which will contribute to pH. Write a balanced reaction. Prepare an ICE table, represent the change with x, sum the table, determine the equilibrium values, put the equilibrium values in the equilibrium expression, and solve for x. Determine $[H_3O^+] \rightarrow$ pH.
Solution: Na^+ is the counterion of a strong base; therefore, it is pH-neutral. Cl^- is the conjugate base of a strong acid; therefore, it is pH-neutral. K^+ is the counterion of a strong base; therefore, it is pH-neutral. F^- is the conjugate base of a weak acid. Therefore, it will produce a basic solution.

$$F^-(aq) + H_2O(l) \rightleftharpoons HF(aq) + OH^-(aq)$$

	$F^-(aq)$	$HF(aq)$	$OH^-(aq)$
Initial	0.0953	0.0	0.0
Change	$-x$	$+x$	$+x$
Equil	$0.0953 - x$	x	x

$$K_b = \frac{K_w}{K_a} = \frac{1.0 \times 10^{-14}}{3.5 \times 10^{-4}} = \frac{(x)(x)}{(0.0953 - x)}$$

Assume that x is small.

$x = 1.\underline{650} \times 10^{-6} = [OH^-]\quad pOH = -\log(1.\underline{650} \times 10^{-6}) = 5.78$
$pH = 14.00 - 5.78 = 8.22$

(c) **Given:** 0.132 M NH_4Cl (salt), 0.150 M HNO_3 (strong acid)   **Find:** pH
Conceptual Plan: Because the mixture is a strong acid and a salt, the strong acid will dominate. Use the concentration of the strong acid to determine $[H_3O^+]$ and then pH.
Solution: 0.150 M HNO_3 = 0.150 M $[H_3O^+]$ pH = $-\log(0.150) = 0.824$

(d)   **Given:** 0.0887 M $NaC_7H_5O_2$ (salt) 0.225 M KBr (salt)
Conceptual Plan: Identify each species and determine which will contribute to pH. Write a balanced reaction. Prepare an ICE table, represent the change with x, sum the table, determine the equilibrium values, put the equilibrium values in the equilibrium expression, and solve for x. Determine $[H_3O^+] \rightarrow$ pH.
Solution: Na^+ is the counterion of a strong base; therefore, it is pH-neutral. $C_7H_5O_2^-$ is the conjugate base of a weak acid. Therefore, it will produce a basic solution. K^+ is the counterion of a strong base; therefore, it is pH-neutral. Cl^- is the conjugate base of a strong acid; therefore, it is pH-neutral.

$$C_7H_5O_2^-(aq) + H_2O(l) \rightleftharpoons HC_7H_5O_2(aq) + OH^-(aq)$$

	$C_7H_5O_2^-$	$HC_7H_5O_2$	OH^-
Initial	0.0887	0.0	0.0
Change	$-x$	$+x$	$+x$
Equil	$0.0887 - x$	x	x

$$K_b = \frac{K_w}{K_a} = \frac{1.0 \times 10^{-14}}{6.5 \times 10^{-5}} = \frac{(x)(x)}{(0.0887 - x)}$$

Assume that x is small.

$$x = 3.\underline{6}94 \times 10^{-6} = [OH^-] \quad pOH = -\log(3.\underline{6}94 \times 10^{-6}) = 5.43$$

$$pH = 14.00 - 5.43 = 8.57$$

(e)   **Given:** 0.0450 M HCl (strong acid), 0.0225 M HNO_3 (strong acid)   **Find:** pH
Conceptual Plan: Because the mixture is a strong acid and a strong acid, $[H_3O^+]$ is the sum of the concentration of both acids, and then determine pH.
Solution: 0.0450 M HCl = 0.0450 $[H_3O^+]$, 0.0225 M HNO_3 = 0.0225 M $[H_3O^+]$
$[H_3O^+]$ = 0.0450 + 0.0225 = 0.0675 M pH = $-\log(0.0675)$ = 1.171

15.141   (a)   sodium cyanide = NaCN nitric acid = HNO_3
$$H^+(aq) + CN^-(aq) \rightleftharpoons HCN(aq)$$
(b)   ammonium chloride = NH_4Cl sodium hydroxide = NaOH
$$NH_4^+(aq) + OH^-(aq) \rightleftharpoons NH_3(aq) + H_2O(l)$$
(c)   sodium cyanide = NaCN ammonium bromide = NH_4Br
$$NH_4^+(aq) + CN^-(aq) \rightleftharpoons NH_3(aq) + HCN(aq)$$
(d)   potassium hydrogen sulfate = $KHSO_4$ lithium acetate = $LiC_2H_3O_2$
$$HSO_4^-(aq) + C_2H_3O_2^-(aq) \rightleftharpoons SO_4^{2-}(aq) + HC_2H_3O_2(aq)$$
(e)   sodium hypochlorite = NaClO ammonia = NH_3
No reaction; both are bases.

15.143   **Given:** 1.0 M urea, pH = 7.050   **Find:** K_a $Hurea^+$
Conceptual Plan: pH $\rightarrow$ pOH $\rightarrow [OH^-] \rightarrow K_b$(urea) $\rightarrow K_a(Hurea^+)$. Write a balanced reaction.
$$pH + pOH = 14 \quad pOH = -\log[OH^-]$$
Prepare an ICE table, represent the change with x, sum the table, determine the equilibrium values, put the equilibrium values in the equilibrium expression, and determine K_b.
Solution: pOH = 14.000 $-$ 7.050 = 6.950 $[OH^-]$ = $10^{-6.950}$ = $1.1\underline{2}2 \times 10^{-7}$ M

$$urea(aq) + H_2O(l) \rightleftharpoons Hurea^+(aq) + OH^-(aq)$$

	urea	$Hurea^+$	OH^-
Initial	1.0	0.0	0.0
Change	$-x$	$+x$	$+x$
Equil	$1.0 - 1.1\underline{2}2 \times 10^{-7}$	$1.1\underline{2}2 \times 10^{-7}$	$1.1\underline{2}2 \times 10^{-7}$

$$K_b = \frac{[Hurea^+][OH^-]}{[urea]} = \frac{(1.1\underline{2}2 \times 10^{-7})(1.1\underline{2}2 \times 10^{-7})}{(1.0 - 1.1\underline{2}2 \times 10^{-7})} = 1.2\underline{5}89 \times 10^{-14}$$

$$K_a = \frac{K_w}{K_b} = \frac{1.00 \times 10^{-14}}{1.2\underline{5}89 \times 10^{-14}} = 0.79\underline{4}3 = 0.794$$

15.145   **Given:** $Ca(Lact)_2$, $[Ca^{2+}] = 0.26$ M, pH $= 8.40$ **Find:** K_a lactic acid
           Conceptual Plan: $[Ca^{2+}] \rightarrow [Lact^-]$; determine K_b lactate ion. **Prepare an ICE table, represent the**

$$\frac{2 \text{ mol lactate ion}}{1 \text{ mol } Ca^{2+}}$$

change with x, sum the table, and determine the equilibrium constant. $K_b \rightarrow K_a$

$$K_a = \frac{K_w}{K_b}$$

Solution: $0.26 \text{ M } Ca^{2+} \left(\dfrac{2 \text{ mol Lact}^-}{1 \text{ mol } Ca^{2+}} \right) = 0.52 \text{ M Lact}^-$

$$Lact^-(aq) + H_2O(l) \rightleftharpoons H\,Lact(aq) + OH^-(aq)$$

	Lact⁻		H Lact	OH⁻
Initial	0.52		0	0
Change	$-x$		$+x$	$+x$
Equil	$0.52 - x$		x	x

$$pH = 8.40 \quad [H_3O^+] = 10^{-8.40} = 4.0 \times 10^{-9} \quad [OH^-] = \frac{1.0 \times 10^{-14}}{4.0 \times 10^{-9}} = 2.5 \times 10^{-6} = x$$

$$K_b = \frac{[H\,Lact][OH^-]}{[Lact^-]} = \frac{(2.5 \times 10^{-6})(2.5 \times 10^{-6})}{(0.52 - 2.5 \times 10^{-6})} = 1.2 \times 10^{-11}$$

$$K_a = \frac{K_w}{K_b} = \frac{1.0 \times 10^{-14}}{1.2 \times 10^{-11}} = 8.3 \times 10^{-4}$$

Challenge Problems

15.147   The calculation is incorrect because it neglects the contribution from the autoionization of water.
           HI is a strong acid, so $[H_3O^+]$ from HI $= 1.0 \times 10^{-7}$.

$$H_2O(l) + H_2O(l) \rightleftharpoons H_3O^+(aq) + OH^-(aq)$$

		H₃O⁺	OH⁻
Initial		1×10^{-7}	0.0
Change	$-x$	$+x$	$+x$
Equil		$1 \times 10^{-7} + x$	x

$$K_w = [H_3O^+][OH^-] = 1.0 \times 10^{-14}$$
$$(1 \times 10^{-7} + x)(x) = 1.0 \times 10^{-14} \quad x^2 + 1 \times 10^{-7}x - 1.0 \times 10^{-14} = 0$$
$$x = 6.18 \times 10^{-8}$$
$$[H_3O^+] = (1 \times 10^{-7} + x) = (1 \times 10^{-7} + 6.18 \times 10^{-8}) = 1.618 \times 10^{-7}$$
$$pH = -\log(1.618 \times 10^{-7}) = 6.79$$

15.149   **Given:** 0.00115 M HCl, 0.01000 M $HClO_2$, $K_a = 1.1 \times 10^{-2}$ **Find:** pH
           Conceptual Plan: Use HCl to determine $[H_3O^+]$. Use $[H_3O^+]$ and $HClO_2$ to determine dissociation of $HClO_2$.
           Write a balanced reaction, prepare an ICE table, calculate equilibrium concentrations, and plug into the equilibrium expression.
           Solution: 0.00115 M HCl $= 0.00115$ M H_3O^+

$$HClO_2(aq) + H_2O(l) \rightleftharpoons H_3O^+(aq) + ClO_2^-(aq)$$

	HClO₂		H₃O⁺	ClO₂⁻
Initial	0.0100		0.00115	0.0
Change	$-x$		$+x$	$+x$
Equil	$0.01000 - x$		$0.00115 + x$	x

$$K_a = \frac{[H_3O^+][ClO_2^-]}{[HClO_2]} = \frac{(0.00115 + x)(x)}{(0.0100 - x)} = 1.1 \times 10^{-2}$$
$$x^2 + 0.01215x - 1.1 \times 10^{-4} = 0 \quad x = 0.006045$$
$$[H_3O^+] = 0.00115 + 0.006045 = 0.007195$$
$$pH = -\log(0.007195) = 2.14$$

15.151   **Given:** 1.0 M HA, $K_a = 1.0 \times 10^{-8}$, $K = 4.0$ for reaction 2   **Find:** $[H^+]$, $[A^-]$, $[HA_2^-]$

Conceptual Plan: Combine reaction 1 and reaction 2; then determine the equilibrium expression and the value of K. Prepare an ICE table, calculate equilibrium concentrations, and plug into the equilibrium expression.

Solution:

$$
\begin{array}{lll}
HA(aq) & \rightleftharpoons H^+(aq) + A^-(aq) & K = 1.0 \times 10^{-8} \\
HA(aq) + A^-(aq) \rightleftharpoons HA_2^-(aq) & & K = 4.0 \\
\hline
2\,HA(aq) & \rightleftharpoons H^+(aq) + HA_2^-(aq) & K = 4.0 \times 10^{-8}
\end{array}
$$

Initial	1.0	0.0	0.0
Change	$-2x$	$+x$	$+x$
Equil	$1.0 - 2x$	x	x

$$K = \frac{[H^+][HA_2^-]}{[HA]^2} = 4.0 \times 10^{-8} = \frac{(x)(x)}{(1.0 - 2x)^2}$$

Take the square root of both sides of the equation. $x = 2.0 \times 10^{-4}\,M = [H^+]$

$$
\begin{array}{ccccc}
& HA(aq) & \rightleftharpoons & H^+(aq) & + & A^-(aq) & K = 1.0 \times 10^{-8}
\end{array}
$$

Initial	1.0	2.0×10^{-4}	0.0
Change	$-y$	$+y$	$+y$
Equil	$1.0 - y$	$2.0 \times 10^{-4} + y$	y

$$K = \frac{[H^+][A^-]}{[HA]} = 1.0 \times 10^{-8} = \frac{(2.0 \times 10^{-4} + y)(y)}{(1.0 - y)}$$

Assume that y is small compared to 1.0.

$y^2 + 2.0 \times 10^{-4}y - 1.0 \times 10^{-8} = 0;$ $y = 4.\underline{1}4 \times 10^{-5}$

$[H^+] = x + y = 2.0 \times 10^{-4}\,M + 4.\underline{1}4 \times 10^{-5}\,M = 2.1 \times 10^{-4}\,M$

$[A^-] = y = 4.1 \times 10^{-5}\,M$

$[HA_2^-] = x = 2.0 \times 10^{-4}\,M$

15.153   **Given:** 0.200 mol NH₄CN, 1.00 L, $K_b(NH_3) = 1.76 \times 10^{-5}$, $K_a(HCN) = 4.9 \times 10^{-10}$ **Find:** pH

Conceptual Plan: $K_b\,NH_3 \rightarrow K_a\,NH_4^+$; $K_a\,HCN \rightarrow K_b\,CN^-$ Prepare an ICE table, represent the

$$K_a K_b = K_w$$

change with x, sum the table, and determine the equilibrium conditions.

Solution: $K_a(NH_4^+) = \dfrac{1.0 \times 10^{-14}}{1.76 \times 10^{-5}} = 5.68 \times 10^{-10}$ $K_b(CN^-) = \dfrac{1.0 \times 10^{-14}}{4.9 \times 10^{-10}} = 2.0 \times 10^{-5}$

0.200 M NH₄CN = 0.200 M NH₄⁺ and 0.200 M CN⁻

Because the value for K for CN^- is greater than K for NH_4^+, the CN^- reaction will be larger and the solution will be basic.

$$CN^-(aq) + H_2O(l) \rightleftharpoons HCN(aq) + OH^-(aq)$$

Initial	0.200	0	0
Change	$-x$	$+x$	$+x$
Equil	$0.200 - x$	x	x

$K_b = \dfrac{[HCN][OH^-]}{[CN^-]}$ $2.0 \times 10^{-5} = \dfrac{(x)(x)}{(0.200 - x)}$ Solve using the quadratic equation.

$x = 0.0020 = [OH^-]$

$[H_3O^+] = \dfrac{1.0 \times 10^{-14}}{0.0020} = 5.0 \times 10^{-12}$ pH $= -\log(5.0 \times 10^{-12}) = 11.30$

15.155   **Given:** mixture Na₂CO₃ and NaHCO₃ = 82.2 g, 1.0 L, pH = 9.95, $K_a(H_2CO_3) = K_{a_1} = 4.3 \times 10^{-7}$, $K_{a_2} = 5.6 \times 10^{-11}$ **Find:** mass NaHCO₃

Conceptual Plan: Let x = g NaHCO₃, y = g Na₂CO₃ → mol NaHCO₃, Na₂CO₃ → [NaHCO₃], [Na₂CO₃]

$$\frac{1\ mol\ NaHCO_3}{84.01\ g} \qquad \frac{1\ mol\ Na_2CO_3}{105.99\ g} \qquad\qquad M = \frac{mol}{L}$$

$\rightarrow [HCO_3^-], [CO_3^{2-}], K_a(HCO_3^-) \rightarrow K_b(CO_3^{2-})$**. Prepare an ICE table, represent the change with** z**,**

$$\frac{HCO_3^-}{NaHCO_3} \quad \frac{CO_3^{-2}}{Na_2CO_3} \qquad K_b = \frac{K_w}{K_a}$$

sum the table, and determine the equilibrium conditions.

Solution: Let $x = $ g $NaHCO_3$ and $y = $ g Na_2CO_3 $x + y = 82.2$, so $y = 82.2 - x$

$$\text{mol } NaHCO_3 = x \text{ g } NaHCO_3\left(\frac{1 \text{ mol}}{84.01 \text{ g } NaHCO_3}\right) = \frac{x}{84.01}$$

$$[HCO_3^-] = \left(\frac{\frac{x}{84.01} \text{ mol } NaHCO_3}{1 \text{ L}}\right)\frac{1 \text{ mol } HCO_3^-}{1 \text{ mol } NaHCO_3} = \frac{x}{84.01} M \text{ } HCO_3^-$$

$$\text{mol } Na_2CO_3 = 82.2 - x \text{ g } Na_2CO_3\left(\frac{1 \text{ mol}}{105.99 \text{ g } Na_2CO_3}\right) = \frac{82.2 - x}{105.99} Na_2CO_3$$

$$[CO_3^{-2}] = \left(\frac{\frac{82.2 - x}{105.99} \text{ mol } Na_2CO_3}{1 \text{ L}}\right)\frac{1 \text{ mol } CO_3^{2-}}{1 \text{ mol } Na_2CO_3} = \frac{82.2 - x}{105.99} M \text{ } CO_3^{2-}$$

Because the pH of the solution is basic, it is the hydrolysis of CO_3^{2-} that dominates in the solution.

$$K_b(CO_3^{2-}) = \frac{1.0 \times 10^{-14}}{5.6 \times 10^{-11}} = 1.79 \times 10^{-4} \text{ and } [H_3O^+] = 10^{-9.95} = 1.12 \times 10^{-10} [OH^-] = \frac{1.0 \times 10^{-14}}{1.12 \times 10^{-10}}$$

$$= 8.93 \times 10^{-5}$$

$$CO^{2-}(aq) + H_2O(l) \rightleftharpoons HCO_3^-(aq) + OH^-(aq)$$

Initial	$\dfrac{82.2 - x}{105.99}$	$\dfrac{x}{84.01}$
Change	$-z$	$+z$
Equil	$\left(\dfrac{82.2 - x}{105.99}\right) - z$	$\left(\dfrac{x}{84.01}\right) + z \qquad 8.93 \times 10^{-5}$

$$K_b(CO_3^{2-}) = 1.79 \times 10^{-4} = \frac{[HCO_3^-][OH^-]}{[CO_3^{2-}]} = \frac{\left(\left(\dfrac{x}{84.01}\right) - z\right)(8.93 \times 10^{-5})}{\left(\dfrac{82.2 - x}{105.99}\right) - z} \qquad \text{Assume that } z \text{ is small.}$$

$$\frac{1.79 \times 10^{-4}}{8.93 \times 10^{-5}} = \frac{\left(\dfrac{x}{84.01}\right)}{\left(\dfrac{82.2 - x}{105.99}\right)}; \quad x = 50.\underline{38} = 50.4 \text{ g } NaHCO_3$$

Conceptual Problems

15.157 Solution (b) would be most acidic.

(a) 0.0100 M HCl (strong acid) and 0.0100 M KOH (strong base) because the concentrations are equal, the acid and base will completely neutralize each other, and the resulting solution will be pH-neutral.

(b) 0.0100 M HF (weak acid) and 0.0100 M KBr (salt). $K_a(HF) = 3.5 \times 10^{-4}$. The weak acid will produce an acidic solution. K^+ is the counterion of a strong base and is pH-neutral. Br^- is the conjugate base of a strong acid and is pH-neutral.

(c)   0.0100 M NH$_4$Cl (salt) and 0.100 M CH$_3$NH$_3$Br. K_b(NH$_3$) = 1.8 × 10^{-5}, K_b(CH$_3$NH$_2$) = 4.4 × 10^{-4}. NH$_4^+$ is the conjugate acid of a weak base, and CH$_3$NH$_3^+$ is the conjugate acid of a weak base. Cl$^-$ and Br$^-$ are the conjugate bases of strong acids and will be pH-neutral. K_a of NH$_4^+$ = 5.68 × 10^{-10}, K_a of CH$_3$NH$_3^+$ = 2.27 × 10^{-11}. Since NH$_4^+$ has a higher K_a, it will have a higher concentration of [H$_3$O$^+$] ions and therefore will be more acidic. The solution will be acidic. However, because the K_a for the conjugate acids in this solution is smaller K_a for HF, the solution will be acidic, but not as acidic as HF.

(d)   0.100 M NaCN (salt) and 0.100 M CaCl$_2$. Na$^+$ and Ca^{2+} ion are the counterion of a strong base; therefore, they are pH-neutral. Cl$^-$ is the conjugate base of a strong acid and is pH-neutral. CN$^-$ is the conjugate base of a weak acid and will produce a basic solution.

15.159   CH$_3$COOH < CH$_2$ClCOOH < CHCl$_2$COOH < CCl$_3$COOH
Because Cl is more electronegative than is H, as you add Cl, you increase the number of electronegative atoms, which pulls the electron density away from the O—H group, polarizing the O—H bond, making it more acidic.

16 Aqueous Ionic Equilibrium

Review Questions

16.1 The pH range of human blood is between 7.36 and 7.42. This nearly constant blood pH is maintained by buffers that are chemical systems that resist pH changes, neutralizing an added acid or base. An important buffer system in blood is a mixture of carbonic acid (H_2CO_3) and bicarbonate ion (HCO_3^-).

16.3 The common ion effect occurs when a solution contains two substances ($HC_2H_3O_2$ and $NaC_2H_3O_2$) that share a common ion ($C_2H_3O_2^-$). The presence of the $C_2H_3O_2^-(aq)$ ion causes the acid to ionize even less than it normally would, resulting in a less acidic solution (higher pH). This effect is an example of Le Châtelier's principle shifting an equilibrium because of the addition (or removal) of the common ion from the solution.

16.5 When the concentration of the conjugate acid and base components of a buffer system are equal, the pH is equal to the pK_a of the weak acid of the buffer system. When more of the acid component is present, the pH becomes more acidic (pH drops). When more of the base component is present, the pH becomes more basic (pH rises). The pH in both cases can be calculated using the Henderson–Hasselbalch equation.

16.7 To find the pH of this solution, determine which component is the acid, determine which component is the base, and substitute their concentrations into the Henderson–Hasselbalch equation. The pK_a is the negative of the log of the equilibrium constant of the acid dissociation reaction where the weak acid component and water are the reactants and the conjugate base and H_3O^+ are the products. At the end, confirm that the "x is small" approximation is valid by calculating the $[H_3O^+]$ from the pH. Because H_3O^+ is formed by ionization of the acid, the calculated $[H_3O^+]$ has to be less than 0.05 (or 5%) of the initial concentration of the acid for the "x is small" approximation to be valid.

16.9 The relative concentrations of acid and conjugate base should not differ by more than a factor of 10 for a buffer to be reasonably effective. Using the Henderson–Hasselbalch equation, this means that the pH should be within one pH unit of the weak acid's pK_a.

16.11 The titration of weak acid by a strong base always has a basic equivalence point because at the equivalence point, all of the acid has been converted into its conjugate base, resulting in a weakly basic solution.

16.13 (a) The initial pH of the solution is simply the pH of the strong acid. Because strong acids completely dissociate, the concentration of H_3O^+ is the concentration of the strong acid and $pH = -\log[H_3O^+]$.

(b) Before the equivalence point, H_3O^+ is in excess. Calculate the $[H_3O^+]$ by subtracting the number of moles of added OH^- from the initial number of moles of H_3O^+ and dividing by the *total* volume. Then convert to pH using $-\log[H_3O^+]$.

(c) At the equivalence point, neither reactant is in excess and the $pH = 7.00$.

(d) Beyond the equivalence point, OH^- is in excess. Calculate the $[OH^-]$ by subtracting the initial number of moles of H_3O^+ from the number of moles of added OH^- and dividing by the *total* volume. Then convert to pH using $-\log[H_3O^+]$.

16.15 When a polyprotic acid is titrated with a strong base and if K_{a_1} and K_{a_2} are sufficiently different, the pH curve will have two equivalence points because the two acidic protons will be titrated sequentially. The titration of the first acidic proton will be completed before the titration of the second acidic proton.

16.17 The end point is the point when the indicator changes color in an acid–base titration. The equivalence point is when stoichiometrically equivalent amounts of acid and base have reacted. With the correct indicator, the end point of the titration will occur at the equivalence point.

16.19 The solubility product constant (K_{sp}) is the equilibrium expression for a chemical equation representing the dissolution of an ionic compound. The expression of the solubility product constant of A_mX_n is $K_{sp} = [A^{n+}]^m [X^{m-}]^n$.

16.21 In accordance with Le Châtelier's principle, the presence of a common ion in solution causes the equilibrium to shift to the left (compared to its position with pure water as the solvent), which means that less of the ionic compound dissolves. Thus, the solubility of an ionic compound is lower in a solution containing a common ion than in pure water. The exact value of the solubility can be calculated by working an equilibrium problem in which the concentration of the common ion is accounted for in the initial conditions. The molar solubility of a compound, A_mX_n, in a solution with an initial concentration of $A^{n+} = [A^{n+}]_0$ and an initial concentration of $X^{m-} = [X^{m-}]_0$ can be computed directly from K_{sp} by solving for S in the expression $K_{sp} = ([A^{n+}]_0 + mS)^m ([X^{m-}]_0 + nS)^n$.

16.23 Q is the reaction quotient, the product of the concentrations of the ionic components raised to their stoichiometric coefficients, and K is the product of the concentrations of the ionic components raised to their stoichiometric coefficients at equilibrium. A solution contains an ionic compound: if $Q < K_{sp}$, the solution is unsaturated. More of the solid ionic compound can dissolve in the solution: if $Q = K_{sp}$, the solution is saturated, the solution is holding the equilibrium amount of the dissolved ions, and additional solid will not dissolve in the solution. If $Q > K_{sp}$, the solution is supersaturated and under most circumstances, the excess solid will precipitate out.

16.25 Qualitative analysis is a systematic way to determine the metal ions present in an unknown solution by the selective precipitation of the ions. The word qualitative means "involving quality or kind." So qualitative analysis involves finding the kind of ions present in the solution. Quantitative analysis is concerned with quantity, or the amounts of substances in a solution or mixture.

Problems by Topic

The Common Ion Effect and Buffers

16.27 The only solution in which HNO_2 will ionize less is (d) 0.10 M $NaNO_2$. It is the only solution that generates a common ion NO_2^- with nitrous acid.

16.29 (a) **Given:** 0.20 M $HCHO_2$ and 0.15 M $NaCHO_2$ **Find:** pH **Other:** $K_a (HCHO_2) = 1.8 \times 10^{-4}$
Conceptual Plan: M $NaCHO_2 \rightarrow$ M CHO_2^- then M $HCHO_2$, M $CHO_2^- \rightarrow [H_3O^+] \rightarrow$ pH

$\quad\quad\quad\quad NaCHO_2(aq) \rightarrow Na^+(aq) + CHO_2^-(aq) \quad\quad\quad\quad\quad$ ICE table $\quad\quad$ pH $= -\log[H_3O^+]$

Solution: Because 1 CHO_2^- ion is generated for each $NaCHO_2$, $[CHO_2^-] = 0.15$ M CHO_2^-.

$$HCHO_2(aq) + H_2O(l) \rightleftharpoons H_3O^+(aq) + CHO_2^-(aq)$$

	$[HCHO_2]$	$[H_3O^+]$	$[CHO_2^-]$
Initial	0.20	≈ 0.00	0.15
Change	$-x$	$+x$	$+x$
Equil	$0.20 - x$	$+x$	$0.15 + x$

$K_a = \dfrac{[H_3O^+][CHO_2^-]}{[HCHO_2]} = 1.8 \times 10^{-4} = \dfrac{x(0.15 + x)}{0.20 - x}$ Assume that x is small ($x << 0.15 < 0.20$), so

$\dfrac{x(0.15 + \cancel{x})}{0.20 - \cancel{x}} = 1.8 \times 10^{-4} = \dfrac{x(0.15)}{0.20}$ and $x = 2.4 \times 10^{-4}$ M $= [H_3O^+]$. Confirm that the more stringent assumption is valid.

$\dfrac{2.4 \times 10^{-4}}{0.15} \times 100\% = 0.16\%$, so assumption is valid. Finally,

pH $= -\log[H_3O^+] = -\log(2.4 \times 10^{-4}) = 3.62$.

Check: The units (none) are correct. The magnitude of the answer makes physical sense because pH should be greater than $-\log(0.20) = 0.70$ because this is a weak acid and there is a common ion effect.

(b)   **Given:** 0.16 M NH_3 and 0.22 M NH_4Cl **Find:** pH   **Other:** $K_b(NH_3) = 1.79 \times 10^{-5}$

Conceptual Plan: $M\ NH_4Cl \rightarrow M\ NH_4^+$ then $M\ NH_3, M\ NH_4^+ \rightarrow [OH^-] \rightarrow [H_3O^+] \rightarrow pH$

$$NH_4Cl(aq) \rightarrow NH_4^+(aq) + Cl^-(aq) \quad\quad \text{ICE table} \quad K_w = [H_3O^+][OH^-] \quad pH = -\log[H_3O^+]$$

Solution: Because 1 NH_4^+ ion is generated for each NH_4Cl, $[NH_4^+] = 0.22$ M NH_4^+.

$$NH_3(aq) + H_2O(l) \rightleftharpoons NH_4^+(aq) + OH^-(aq)$$

	$[NH_3]$	$[NH_4^+]$	$[OH^-]$
Initial	0.16	0.22	≈ 0.00
Change	$-x$	$+x$	$+x$
Equil	$0.16 - x$	$0.22 + x$	$+x$

$$K_b = \frac{[NH_4^+][OH^-]}{[NH_3]} = 1.79 \times 10^{-5} = \frac{(0.22 + x)x}{0.16 - x}$$

Assume that x is small ($x << 0.16 < 0.22$), so $\dfrac{(0.22 + \cancel{x})x}{0.16 - \cancel{x}} = 1.79 \times 10^{-5} = \dfrac{(0.22)x}{0.16}$ and

$x = 1.3\underline{0}182 \times 10^{-5}$ M $= [OH^-]$. Confirm that the more stringent assumption is valid.

$$\frac{1.3\underline{0}182 \times 10^{-5}}{0.16} \times 100\% = 8.1 \times 10^{-3}\%, \text{ so assumption is valid.}$$

$$K_w = [H_3O^+][OH^-] \text{ so } [H_3O^+] = \frac{K_w}{[OH^-]} = \frac{1.0 \times 10^{-14}}{1.30182 \times 10^{-5}} = 7.\underline{6}816 \times 10^{-10} \text{ M}$$

Finally, pH $= -\log[H_3O^+] = -\log(7.\underline{6}816 \times 10^{-10}) = 9.11$.

Check: The units (none) are correct. The magnitude of the answer makes physical sense because pH should be less than $14 + \log(0.16) = 13.2$ because this is a weak base and there is a common ion effect.

16.31   **Given:** 0.15 M $HC_7H_5O_2$ in pure water and in 0.10 M $NaC_7H_5O_2$

Find: % ionization in both solutions   **Other:** $K_a(HC_7H_5O_2) = 6.5 \times 10^{-5}$

Conceptual Plan: Pure water: $M\ HC_7H_5O_2 \rightarrow [H_3O^+] \rightarrow$ % ionization then in $NaC_7H_5O_2$ solution:

$$\text{ICE table} \quad \text{\% ionization} = \frac{[H_3O^+]_{equil}}{[HC_7H_5O_2]_0} \times 100\%$$

$M\ NaC_7H_5O_2 \rightarrow M\ C_7H_5O_2^-$ then $M\ HC_7H_5O_2, M\ C_7H_5O_2^- \rightarrow [H_3O^+] \rightarrow$ % ionization

$$NaC_7H_5O_2(aq) \rightarrow Na^+(aq) + C_7H_5O_2^-(aq) \quad\quad \text{ICE table} \quad \text{\% ionization} = \frac{[H_3O^-]_{equil}}{[HC_7H_5O_2]_0} \times 100\%$$

Solution: in pure water:

$$HC_7H_5O_2(aq) + H_2O(l) \rightleftharpoons H_3O^+(aq) + C_7H_5O_2^-(aq)$$

	$[HC_7H_5O_2]$	$[H_3O^+]$	$[C_7H_5O_2^-]$
Initial	0.15	≈ 0.00	0.00
Change	$-x$	$+x$	$+x$
Equil	$0.15 - x$	$+x$	$+x$

$$K_a = \frac{[H_3O^+][C_7H_5O_2^-]}{[HC_7H_5O_2]} = 6.5 \times 10^{-5} = \frac{x^2}{0.15 - x}$$

Assume that x is small ($x << 0.10$), so $\dfrac{x^2}{0.15 - \cancel{x}} = 6.5 \times 10^{-5} = \dfrac{x^2}{0.15}$ and $x = 3.\underline{1}225 \times 10^{-3}$ M $= [H_3O^+]$. Then

% ionization $= \dfrac{[H_3O^+]_{equil}}{[HC_7H_5O_2]_0} \times 100\% = \dfrac{3.\underline{1}225 \times 10^{-3}}{0.15} \times 100\% = 2.1\%$, which also confirms that the

assumption is valid (because it is less than 5%). In $NaC_7H_5O_2$ solution: because one $C_7H_5O_2^-$ ion is generated for each $NaC_7H_5O_2$, $[C_7H_5O_2^-] = 0.10$ M $C_7H_5O_2^-$.

$$HC_7H_5O_2(aq) + H_2O(l) \rightleftharpoons H_3O^+(aq) + C_7H_5O_2^-(aq)$$

	$[HC_7H_5O_2]$	$[H_3O^+]$	$[C_7H_5O_2^-]$
Initial	0.15	≈ 0.00	0.10
Change	$-x$	$+x$	$+x$
Equil	$0.15 - x$	$+x$	$0.10 + x$

$K_a = \dfrac{[H_3O^+][C_7H_5O_2^-]}{[HC_7H_5O_2]} = 6.5 \times 10^{-5} = \dfrac{x(0.10 + x)}{0.15 - x}$ Assume that x is small ($x \ll 0.10 < 0.15$), so

$\dfrac{x(0.10 + \cancel{x})}{0.15 - \cancel{x}} = 6.5 \times 10^{-5} = \dfrac{x(0.10)}{0.15}$ and $x = 9.7\underline{5} \times 10^{-5}$ M $= [H_3O^+]$. Then

% ionization $= \dfrac{[H_3O^+]_{equil}}{[HC_7H_5O_2]_0} \times 100\% = \dfrac{9.7\underline{5} \times 10^{-5}}{0.15} \times 100\% = 0.065\%$, which also confirms that the

assumption is valid (because it is less than 5%). The percent ionization in the sodium benzoate solution is less than in pure water because of the common ion effect. An increase in one of the products (benzoate ion) shifts the equilibrium to the left, so less acid dissociates.

Check: The units (%) are correct. The magnitude of the answer makes physical sense because the acid is weak; so the percent ionization is low. With a common ion present, the percent ionization decreases.

16.33   (a)   **Given:** 0.15 M HF   **Find:** pH   **Other:** $K_a(\text{HF}) = 3.5 \times 10^{-4}$
Conceptual Plan: M HF $\rightarrow [H_3O^+] \rightarrow$ pH
ICE table   pH $= -\log[H_3O^+]$
Solution:

$$HF(aq) + H_2O(l) \rightleftharpoons H_3O^+(aq) + F^-(aq)$$

	$[HF]$	$[H_3O^+]$	$[F^-]$
Initial	0.15	≈ 0.00	0.00
Change	$-x$	$+x$	$+x$
Equil	$0.15 - x$	$+x$	$+x$

$K_a = \dfrac{[H_3O^+][F^-]}{[HF]} = 3.5 \times 10^{-4} = \dfrac{x^2}{0.15 - x}$ Assume that x is small ($x \ll 0.15$), so

$\dfrac{x^2}{0.15 - \cancel{x}} = 3.5 \times 10^{-4} = \dfrac{x^2}{0.15}$ and $x = 7.2\underline{4}57 \times 10^{-3}$ M $= [H_3O^+]$. Confirm that

assumption is valid. $\dfrac{7.2\underline{4}57 \times 10^{-3}}{0.15} \times 100\% = 4.8\% < 5\%$, so assumption is valid.

Finally, pH $= -\log[H_3O^+] = -\log(7.2\underline{4}57 \times 10^{-3}) = 2.14$.

Check: The units (none) are correct. The magnitude of the answer makes physical sense because the pH should be greater than $-\log(0.15) = 0.82$ because this is a weak acid.

(b)   **Given:** 0.15 M NaF   **Find:** pH   **Other:** $K_a(\text{HF}) = 3.5 \times 10^{-4}$
Conceptual Plan: M NaF $\rightarrow$ M F$^-$ and $K_a \rightarrow K_b$ then M F$^- \rightarrow [OH^-] \rightarrow [H_3O^+] \rightarrow$ pH
NaF$(aq) \rightarrow$ Na$^+(aq)$ + F$^-(aq)$ $K_w = K_a K_b$ ICE table $K_w = [H_3O^+][OH^-]$ pH $= -\log[H_3O^+]$
Solution:
Because one F$^-$ ion is generated for each NaF, $[F^-] = 0.15$ M F$^-$. Because $K_w = K_a K_b$, rearrange to solve for K_b.

$K_b = \dfrac{K_w}{K_a} = \dfrac{1.0 \times 10^{-14}}{3.5 \times 10^{-4}} = 2.\underline{8}571 \times 10^{-11}$

$$F^-(aq) + H_2O(l) \rightleftharpoons HF(aq) + OH^-(aq)$$

	$[F^-]$	$[HF]$	$[OH^-]$
Initial	0.15	0.00	≈ 0.00
Change	$-x$	$+x$	$+x$
Equil	$0.15 - x$	$+x$	$+x$

$$K_b = \frac{[HF][OH^-]}{[F^-]} = 2.\underline{8}571 \times 10^{-11} = \frac{x^2}{0.15 - x}$$

Assume that x is small ($x \ll 0.15$), so $\dfrac{x^2}{0.15 - \cancel{x}} = 2.\underline{8}571 \times 10^{-11} = \dfrac{x^2}{0.15}$ and

$x = 2.\underline{0}702 \times 10^{-6}\,M = [OH^-]$.

Confirm that assumption is valid. $\dfrac{2.\underline{0}702 \times 10^{-6}}{0.15} \times 100\% = 0.0014\% < 5\%$, so assumption is valid.

$K_w = [H_3O^+][OH^-]$, so $[H_3O^+] = \dfrac{K_w}{[OH^-]} = \dfrac{1.0 \times 10^{-14}}{2.\underline{0}702 \times 10^{-6}} = 4.\underline{8}305 \times 10^{-9}\,M$.

Finally, $pH = -\log[H_3O^+] = -\log(4.\underline{8}305 \times 10^{-9}) = 8.32$.

Check: The units (none) are correct. The magnitude of the answer makes physical sense because the pH should be slightly basic because the fluoride ion is a very weak base.

(c) **Given:** 0.15 M HF and 0.15 M NaF **Find:** pH **Other:** $K_a(HF) = 3.5 \times 10^{-4}$
Conceptual Plan: M NaF $\rightarrow$ M F$^-$ then M HF, M F$^-$ $\rightarrow$ $[H_3O^+]$ $\rightarrow$ pH

$$NaF(aq) \rightarrow Na^+(aq) + F^-(aq) \qquad\qquad \text{ICE table} \qquad pH = -\log[H_3O^+]$$

Solution: Because one F$^-$ ion is generated for each NaF, $[F^-] = 0.15\,M\,F^-$.

$$HF(aq) + H_2O(l) \rightleftharpoons H_3O^+(aq) + F^-(aq)$$

	$[HF]$	$[H_3O^+]$	$[F^-]$
Initial	0.15	≈ 0.00	0.15
Change	$-x$	$+x$	$+x$
Equil	$0.15 - x$	$+x$	$0.15 + x$

$$K_a = \frac{[H_3O^+][F^-]}{[HF]} = 3.5 \times 10^{-4} = \frac{x(0.15 + x)}{0.15 - x}$$

Assume that x is small ($x \ll 0.15$), so $\dfrac{x(0.15 + \cancel{x})}{0.15 - \cancel{x}} = 3.5 \times 10^{-4} = \dfrac{x(\cancel{0.15})}{\cancel{0.15}}$ and

$x = 3.5 \times 10^{-4}\,M = [H_3O^+]$.

Confirm that assumption is valid. $\dfrac{3.5 \times 10^{-4}}{0.15} \times 100\% = 0.23\% < 5\%$, so assumption is valid.

Finally, $pH = -\log[H_3O^+] = -\log(3.5 \times 10^{-4}) = 3.46$.

Check: The units (none) are correct. The magnitude of the answer makes physical sense because the pH should be greater than that in part (a) (2.14) because of the common ion effect suppressing the dissociation of the weak acid.

16.35 When an acid (such as HCl) is added, it reacts with the conjugate base of the buffer system as follows: HCl + NaC$_2$H$_3$O$_2$ $\rightarrow$ HC$_2$H$_3$O$_2$ + NaCl. When a base (such as NaOH) is added, it reacts with the weak acid of the buffer system as follows: NaOH + HC$_2$H$_3$O$_2$ $\rightarrow$ H$_2$O + NaC$_2$H$_3$O$_2$. The reaction generates the other buffer system component.

16.37 (a) **Given:** 0.15 M HCHO$_2$ and 0.10 M NaCHO$_2$ **Find:** pH **Other:** $K_a(HCHO_2) = 1.8 \times 10^{-4}$
Conceptual Plan: Identify acid and base components then M NaCHO$_2$ $\rightarrow$ M CHO$_2^-$ then

$$\text{acid} = HCHO_2 \text{ base} = CHO_2^- \qquad\qquad NaCHO_2(aq) \rightarrow Na^+(aq) + CHO_2^-(aq)$$

K_a, M HCHO$_2$, M CHO$_2^-$ $\rightarrow$ pH

$$pH = pK_a + \log\frac{[\text{base}]}{[\text{acid}]}$$

Solution: Acid $= HCHO_2$, so $[\text{acid}] = [HCHO_2] = 0.15$ M. Base $= CHO_2^-$. Because one CHO_2^- ion is generated for each $NaCHO_2$, $[CHO_2^-] = 0.10$ M $CHO_2^- = [\text{base}]$. Then

$$pH = pK_a + \log \frac{[\text{base}]}{[\text{acid}]} = -\log(1.8 \times 10^{-4}) + \log \frac{0.10\,M}{0.15\,M} = 3.57.$$

Note that to use the Henderson–Hasselbalch equation, the assumption that x is small must be valid. This was confirmed in Problem 16.29.

Check: The units (none) are correct. The magnitude of the answer makes physical sense because the pH should be less than the pK_a of the acid because there is more acid than base. The answer agrees with Problem 16.29.

(b)   **Given:** 0.12 M NH_3 and 0.18 M NH_4Cl **Find:** pH   **Other:** $K_b(NH_3) = 1.76 \times 10^{-5}$
     Conceptual Plan: Identify acid and base components then M $NH_4Cl \rightarrow$ M NH_4^+ and $K_b \rightarrow pK_b \rightarrow pK_a$

$$\text{acid} = NH_4^+ \text{ base} = NH_3 \quad NH_4Cl(aq) \rightarrow NH_4^+(aq) + Cl^-(aq) \quad pK_b = -\log K_b \quad 14 = pK_a + pK_b$$

     then pK_a, M NH_3, M $NH_4^+ \rightarrow$ pH

$$pH = pK_a + \log \frac{[\text{base}]}{[\text{acid}]}$$

Solution: Base $= NH_3$, $[\text{base}] = [NH_3] = 0.12$ M. Acid $= NH_4^+$. Because one NH_4^+ ion is generated for each NH_4Cl, $[NH_4^+] = 0.18$ M $NH_4^+ = [\text{acid}]$.

Because $K_b(NH_3) = 1.76 \times 10^{-5}$, $pK_b = -\log K_b = -\log(1.76 \times 10^{-5}) = 4.75$. Because $14 = pK_a + pK_b$,

$$pK_a = 14 - pK_b = 14 - 4.75 = 9.25. \text{ Then } pH = pK_a + \log \frac{[\text{base}]}{[\text{acid}]} = 9.25 + \log \frac{0.12\,M}{0.18\,M} = 9.07.$$

Note that to use the Henderson–Hasselbalch equation, the assumption that x is small must be valid. This was confirmed in Problem 16.29.

Check: The units (none) are correct. The magnitude of the answer makes physical sense because the pH should be less than the pK_a of the acid because there is more acid than base. The answer agrees with Problem 16.29 within the error of the value.

16.39   (a)   **Given:** 0.135 M $HClO$ and 0.155 M $KClO$ **Find:** pH   **Other:** $K_a(HClO) = 2.9 \times 10^{-8}$
     Conceptual Plan: Identify acid and base components then M $KClO \rightarrow$ M ClO^- then

$$\text{acid} = HClO, \text{ base} = ClO^- \quad\quad\quad KClO(aq) \rightarrow K^+(aq) + ClO^-(aq)$$

     M $HClO$, M $ClO^- \rightarrow$ pH

$$pH = pK_a + \log \frac{[\text{base}]}{[\text{acid}]}$$

Solution: Acid $= HClO$, so $[\text{acid}] = [HClO] = 0.135$ M. Base $= ClO^-$. Because 1 ClO^- ion is generated for each $KClO$, $[ClO^-] = 0.155$ M $ClO^- = [\text{base}]$. Then

$$pH = pK_a + \log \frac{[\text{base}]}{[\text{acid}]} = -\log(2.9 \times 10^{-8}) + \log \frac{0.155\,M}{0.135\,M} = 7.60.$$

Check: The units (none) are correct. The magnitude of the answer makes physical sense because pH should be greater than the pK_a of the acid because there is more base than acid.

(b)   **Given:** 1.05% by mass $C_2H_5NH_2$ and 1.10% by mass $C_2H_5NH_3Br$ **Find:** pH
     Other: $K_b(C_2H_5NH_2) = 5.6 \times 10^{-4}$
     Conceptual Plan: Assume exactly 100 g of solution. Because both components are in the same solution (i.e., the same final volume of solution), the ratio of the moles of each component is the same as the ratio of the molarity of these components and only the relative number of moles needs to be calculated.

$$g_{\text{solution}} \rightarrow g_{C_2H_5NH_2} \rightarrow \text{mol}_{C_2H_5NH_2} \text{ and } g_{\text{solution}} \rightarrow g_{C_2H_5NH_3Br} \rightarrow \text{mol}_{C_2H_5NH_3Br}$$

$$\frac{1.05\text{ g } C_2H_5NH_2}{100\text{ g solution}} \quad \frac{1\text{ mol } C_2H_5NH_2}{45.09\text{ g } C_2H_5NH_2} \quad\quad\quad \frac{1.10\text{ g } C_2H_5NH_3Br}{100\text{ g solution}} \quad \frac{1\text{ mol } C_2H_5NH_3Br}{125.99\text{ g } C_2H_5NH_3Br}$$

     identify acid and base components then M $C_2H_5NH_3Br \rightarrow$ M $C_2H_5NH_3^+$ and

$$\text{acid} = C_2H_5NH_3^+, \text{ base} = C_2H_5NH_2 \quad C_2H_5NH_3Br(aq) \rightarrow C_2H_5NH_3^+(aq) + Br^-(aq)$$

     $K_b \rightarrow pK_b \rightarrow pK_a$ then pK_a, M $C_2H_5NH_2$, M $C_2H_5NH_3^+ \rightarrow$ pH

$$pK_b = -\log K_b \quad 14 = pK_a + pK_b \quad\quad\quad\quad pH = pK_a + \log \frac{[\text{base}]}{[\text{acid}]}$$

Solution:

$$100 \ \text{g solution} \times \frac{1.05 \ \text{g } C_2H_5NH_2}{100 \ \text{g solution}} \times \frac{1 \ \text{mol } C_2H_5NH_2}{45.09 \ \text{g } C_2H_5NH_2} = 0.02328676 \ \text{mol } C_2H_5NH_2 \ \text{ and}$$

$$100 \ \text{g solution} \times \frac{1.10 \ \text{g } C_2H_5NH_3Br}{100 \ \text{g solution}} \times \frac{1 \ \text{mol } C_2H_5NH_3Br}{125.99 \ \text{g } C_2H_5NH_3Br} = 0.008730852 \ \text{mol } C_2H_5NH_3Br$$

Then base $= C_2H_5NH_2$, moles of base $= \text{mol } C_2H_5NH_2 = 0.02328676 \ \text{mol}$; acid $= C_2H_5NH_3^+$.
Because 1 $C_2H_5NH_3^+$ ion is generated for each $C_2H_5NH_3Br$, $\text{mol } C_2H_5NH_3^+ = 0.008730852 \ \text{mol}$.
Because $K_b \ (C_2H_5NH_2) = 5.6 \times 10^{-4}$, $pK_b = -\log K_b = -\log(5.6 \times 10^{-4}) = 3.25$.
Because $14 = pK_a + pK_b$, $pK_a = 14 - pK_b = 14 - 3.25 = 10.75$, then

$$pH = pK_a + \log \frac{\text{mol base}}{\text{mol acid}} = 10.75 + \log \frac{0.02328676 \ \text{mol}}{0.008730852 \ \text{mol}} = 11.18.$$

Check: The units (none) are correct. The magnitude of the answer makes physical sense because pH should be greater than the pK_a of the acid because there is more base than acid.

(c)   **Given:** 10.0 g $HC_2H_3O_2$ and 10.0 g $NaC_2H_3O_2$ in 150.0 mL solution   **Find:** pH
Other: $K_a \ (HC_2H_3O_2) = 1.8 \times 10^{-5}$
Conceptual Plan: Identify acid and base components then mL $\rightarrow$ L and g $HC_2H_3O_2 \rightarrow$ mol $HC_2H_3O_2$

$$\text{acid} = HC_2H_3O_2, \text{base} = C_2H_3O_2^- \qquad \frac{1 \ L}{1000 \ \text{mL}} \qquad \frac{1 \ \text{mol } HC_2H_3O_2}{60.05 \ \text{g } HC_2H_3O_2}$$

then mol $HC_2H_3O_2$, L $\rightarrow$ M $HC_2H_3O_2$ and g $NaC_2H_3O_2 \rightarrow$ mol $NaC_2H_3O_2$ then

$$M = \frac{\text{mol}}{L} \qquad \frac{1 \ \text{mol } NaC_2H_3O_2}{82.04 \ \text{g } NaC_2H_3O_2}$$

mol $NaC_2H_3O_2$, L $\rightarrow$ M $NaC_2H_3O_2 \rightarrow$ M $C_2H_3O_2^-$ then M $HC_2H_3O_2$, M $C_2H_3O_2^- \rightarrow$ pH

$$M = \frac{\text{mol}}{L} \quad NaC_2H_3O_2(aq) \rightarrow Na^+(aq) + C_2H_3O_2^-(aq) \qquad pH = pK_a + \log \frac{[\text{base}]}{[\text{acid}]}$$

Solution: $150.0 \ \text{mL} \times \dfrac{1 \ L}{1000 \ \text{mL}} = 0.1500 \ L$ and

$$10.0 \ \text{g } HC_2H_3O_2 \times \frac{1 \ \text{mol } HC_2H_3O_2}{60.05 \ \text{g } HC_2H_3O_2} = 0.166528 \ \text{mol } HC_2H_3O_2$$

$$\text{then } M = \frac{\text{mol}}{L} = \frac{0.166528 \ \text{mol } HC_2H_3O_2}{0.1500 \ L} = 1.11019 \ M \ HC_2H_3O_2 \ \text{ and}$$

$$10.0 \ \text{g } NaC_2H_3O_2 \times \frac{1 \ \text{mol } NaC_2H_3O_2}{82.04 \ \text{g } NaC_2H_3O_2} = 0.121892 \ \text{mol } NaC_2H_3O_2 \ \text{then}$$

$$M = \frac{\text{mol}}{L} = \frac{0.121892 \ \text{mol } NaC_2H_3O_2}{0.1500 \ L} = 0.812612 \ M \ NaC_2H_3O_2$$

Acid $= HC_2H_3O_2$, so $[\text{acid}] = [HC_2H_3O_2] = 1.11019 \ M$ and base $= C_2H_3O_2^-$. Because 1 $C_2H_3O_2^-$ ion is generated for each $NaC_2H_3O_2$, $[C_2H_3O_2^-] = 0.812612 \ M \ C_2H_3O_2^- = [\text{base}]$. Then

$$pH = pK_a + \log \frac{[\text{base}]}{[\text{acid}]} = -\log(1.8 \times 10^{-5}) + \log \frac{0.812612 \ M}{1.11019 \ M} = 4.61.$$

Check: The units (none) are correct. The magnitude of the answer makes physical sense because pH should be less than the pK_a of the acid because there is more acid than base.

16.41   (a)   **Given:** 50.0 mL of 0.15 M $HCHO_2$ and 75.0 mL of 0.13 M $NaCHO_2$ **Find:** pH
Other: $K_a(HCHO_2) = 1.8 \times 10^{-4}$
Conceptual Plan: Identify acid and base components then mL $HCHO_2$, mL $NaCHO_2 \rightarrow$ total mL then

$$\text{acid} = HCHO_2, \text{base} = CHO_2^- \qquad\qquad \text{total mL} = \text{mL } HCHO_2 + \text{mL } NaCHO_2$$

mL $HCHO_2$, M $HCHO_2$, total mL $\rightarrow$ buffer M $HCHO_2$ and

$$M_1V_1 = M_2V_2$$

mL $NaCHO_2$, M $NaCHO_2$, total mL $\rightarrow$ buffer M $NaCHO_2 \rightarrow$ buffer M CHO_2^- then

$$M_1V_1 = M_2V_2 \qquad NaCHO_2(aq) \rightarrow Na^+(aq) + CHO_2^-(aq)$$

K_a, M HCHO$_2$, M CHO$_2{}^-$ → pH

$$pH = pK_a + \log \frac{[\text{base}]}{[\text{acid}]}$$

Solution: total mL = mL HCHO$_2$ + mL NaCHO$_2$ = 50.0 mL + 75.0 mL = 125.0 mL
Then because $M_1 V_1 = M_2 V_2$, rearrange to solve for M$_2$.

$$M_2 = \frac{M_1 V_1}{V_2} = \frac{(0.15 \text{ M})(50.0 \text{ mL})}{125.0 \text{ mL}} = 0.060 \text{ M HCHO}_2 \text{ and}$$

$$M_2 = \frac{M_1 V_1}{V_2} = \frac{(0.13 \text{ M})(75.0 \text{ mL})}{125.0 \text{ mL}} = 0.078 \text{ M NaCHO}_2. \text{ Acid} = \text{HCHO}_2, \text{ so } [\text{acid}] = [\text{HCHO}_2] =$$

0.060 M. Base = CHO$_2{}^-$. Because one CHO$_2{}^-$ ion is generated for each NaCHO$_2$, $[\text{CHO}_2{}^-] =$

0.078 M CHO$_2{}^-$ = $[\text{base}]$. Then pH = $pK_a + \log \dfrac{[\text{base}]}{[\text{acid}]} = -\log(1.8 \times 10^{-4}) + \log \dfrac{0.078 \text{ M}}{0.060 \text{ M}} = 3.86.$

Check: The units (none) are correct. The magnitude of the answer makes physical sense because the pH should be greater than the pK_a of the acid because there is more base than acid.

(b)   **Given:** 125.0 mL of 0.10 M NH$_3$ and 250.0 mL of 0.10 M NH$_4$Cl   **Find:** pH
Other: $K_b(\text{NH}_3) = 1.76 \times 10^{-5}$
Conceptual Plan: Identify acid and base components then mL NH$_3$, mL NH$_4$Cl → total mL then

$$\text{acid} = \text{NH}_4{}^+ \text{ base} = \text{NH}_3 \qquad\qquad \text{total mL} = \text{mL NH}_3 + \text{mL NH}_4\text{Cl}$$

mL NH$_3$, M NH$_3$, total mL → buffer M NH$_3$ and

$$M_1 V_1 = M_2 V_2$$

mL NH$_4$Cl, M NH$_4$Cl, total mL → buffer M NH$_4$Cl → buffer M NH$_4{}^+$ and K_b → pK_b → pK_a then

$$M_1 V_1 = M_2 V_2 \qquad \text{NH}_4\text{Cl}(aq) \rightarrow \text{NH}_4{}^+(aq) + \text{Cl}^-(aq) \qquad pK_b = -\log K_b \qquad 14 = pK_a + pK_b$$

pK_a, M NH$_3$, M NH$_4{}^+$ → pH

$$pH = pK_a + \log \frac{[\text{base}]}{[\text{acid}]}$$

Solution: total mL = mL NH$_3$ + mL NH$_4$Cl = 125.0 mL + 250.0 mL = 375.0 mL
Then because $M_1 V_1 = M_2 V_2$, rearrange to solve for M_2.

$$M_2 = \frac{M_1 V_1}{V_2} = \frac{(0.10 \text{ M})(125.0 \text{ mL})}{375.0 \text{ mL}} = 0.033333 \text{ M NH}_3 \text{ and}$$

$$M_2 = \frac{M_1 V_1}{V_2} = \frac{(0.10 \text{ M})(250.0 \text{ mL})}{375.0 \text{ mL}} = 0.066667 \text{ M NH}_4\text{Cl. Base} = \text{NH}_3, [\text{base}] = [\text{NH}_3] =$$

0.033333 M acid = NH$_4{}^+$. Because one NH$_4{}^+$ ion is generated for each NH$_4$Cl, $[\text{NH}_4{}^+] =$
0.066667 M NH$_4{}^+$ = $[\text{acid}]$. Because $K_b(\text{NH}_3) = 1.79 \times 10^{-5}$,
p$K_b = -\log K_b = -\log(1.76 \times 10^{-5}) = 4.75$. Because $14 = pK_a + pK_b$,

$$pK_a = 14 - pK_b = 14 - 4.75 = 9.25. \text{ Then } pH = pK_a + \log \frac{[\text{base}]}{[\text{acid}]} = 9.25 + \log \frac{0.033333 \text{ M}}{0.066667 \text{ M}} = 8.95.$$

Check: The units (none) are correct. The magnitude of the answer makes physical sense because the pH should be less than the pK_a of the acid because there is more acid than base.

16.43   **Given:** NaF/HF buffer at pH = 4.00   **Find:** [NaF]/[HF]   **Other:** $K_a(\text{HF}) = 3.5 \times 10^{-4}$
Conceptual Plan: Identify acid and base components then pH, K_a → [NaF]/[HF]

$$\text{acid} = \text{HF base} = \text{F}^- \qquad\qquad pH = pK_a + \log \frac{[\text{base}]}{[\text{acid}]}$$

Solution: pH = $pK_a + \log \dfrac{[\text{base}]}{[\text{acid}]} = -\log(3.5 \times 10^{-4}) + \log \dfrac{[\text{NaF}]}{[\text{HF}]} = 4.00.$ Solve for [NaF]/[HF].

$$\log \frac{[\text{NaF}]}{[\text{HF}]} = 4.00 - 3.46 = 0.54 \rightarrow \frac{[\text{NaF}]}{[\text{HF}]} = 10^{0.54} = 3.5.$$

Check: The units (none) are correct. The magnitude of the answer makes physical sense because the pH is greater than the pK_a of the acid; so there needs to be more base than acid.

16.45 **Given:** 150.0 mL buffer of 0.15 M benzoic acid at pH = 4.25 **Find:** mass sodium benzoate
Other: $K_a(HC_7H_5O_2) = 6.5 \times 10^{-5}$
Conceptual Plan: Identify acid and base components then pH, K_a, $[HC_7H_5O_2] \rightarrow [NaC_7H_5O_2]$

$$acid = HC_7H_5O_2 \text{ base} = C_7H_5O_2^-$$

$$pH = pK_a + \log \frac{[base]}{[acid]}$$

mL $\rightarrow$ L then $[NaC_7H_5O_2]$, L $\rightarrow$ mol $NaC_7H_5O_2 \rightarrow$ g $NaC_7H_5O_2$

$$\frac{1 \text{ L}}{1000 \text{ mL}} \qquad M = \frac{mol}{L} \qquad \frac{144.11 \text{ g } NaC_7H_5O_2}{1 \text{ mol } NaC_7H_5O_2}$$

Solution: $pH = pK_a + \log \dfrac{[base]}{[acid]} = -\log(6.5 \times 10^{-5}) + \log \dfrac{[NaC_7H_5O_2]}{0.15 \text{ M}} = 4.25.$ Solve for $[NaC_7H_5O_2]$.

$$\log \frac{[NaC_7H_5O_2]}{0.15 \text{ M}} = 4.25 - 4.18\underline{7}0866 = 0.0\underline{6}291 \rightarrow \frac{[NaC_7H_5O_2]}{0.15 \text{ M}} = 10^{0.06291} = 1.1\underline{5}59 \rightarrow$$

$$[NaC_7H_5O_2] = 0.17\underline{3}38 \text{ M}$$

Convert to moles using $M = \dfrac{mol}{L}$.

$$\frac{0.17\underline{3}38 \text{ mol } NaC_7H_5O_2}{1 \text{ L}} \times 0.150 \text{ L} = 0.02\underline{6}007 \text{ mol } NaC_7H_5O_2$$

$$0.02\underline{6}007 \text{ mol } NaC_7H_5O_2 \times \frac{144.11 \text{ g } NaC_7H_5O_2}{1 \text{ mol } NaC_7H_5O_2} = 3.7 \text{ g } NaC_7H_5O_2$$

Check: The units (g) are correct. The magnitude of the answer makes physical sense because the volume of solution is small and the concentration is low; so much less than a mole is needed.

16.47 **(a)** **Given:** 250.0 mL buffer of 0.250 M $HC_2H_3O_2$ and 0.250 M $NaC_2H_3O_2$ **Find:** initial pH
Other: $K_a(HC_2H_3O_2) = 1.8 \times 10^{-5}$
Conceptual Plan: Identify acid and base components then M $NaC_2H_3O_2 \rightarrow$ M $C_2H_3O_2^-$ then

$$acid = HC_2H_3O_2 \text{ base} = C_2H_3O_2^- \qquad NaC_2H_3O_2(aq) \rightarrow Na^+(aq) + C_2H_3O_2^-(aq)$$

M $HC_2H_3O_2$, M $C_2H_3O_2^- \rightarrow$ pH

$$pH = pK_a + \log \frac{[base]}{[acid]}$$

Solution: Acid = $HC_2H_3O_2$, so $[acid] = [HC_2H_3O_2] = 0.250$ M. Base = $C_2H_3O_2^-$. Because one $C_2H_3O_2^-$ ion is generated for each $NaC_2H_3O_2$, $[C_2H_3O_2^-] = 0.250$ M $C_2H_3O_2^- = [base]$. Then

$$pH = pK_a + \log \frac{[base]}{[acid]} = -\log(1.8 \times 10^{-5}) + \log \frac{0.250 \text{ M}}{0.250 \text{ M}} = 4.74.$$

Check: The units (none) are correct. The magnitude of the answer makes physical sense because the pH is equal to the pK_a of the acid because there are equal amounts of acid and base.

(b) **Given:** 250.0 mL buffer of 0.250 M $HC_2H_3O_2$ and 0.250 M $NaC_2H_3O_2$, add 0.0050 mol HCl **Find:** pH
Other: $K_a(HC_2H_3O_2) = 1.8 \times 10^{-5}$
Conceptual Plan: Part I: Stoichiometry:
mL $\rightarrow$ L then $[NaC_2H_3O_2]$, L $\rightarrow$ mol $NaC_2H_3O_2$ and $[HC_2H_3O_2]$, L $\rightarrow$ mol $HC_2H_3O_2$

$$\frac{1 \text{ L}}{1000 \text{ mL}} \qquad M = \frac{mol}{L} \qquad M = \frac{mol}{L}$$

write balanced equation then

$$HCl + NaC_2H_3O_2 \rightarrow HC_2H_3O_2 + NaCl$$

mol $NaC_2H_3O_2$, mol $HC_2H_3O_2$, mol HCl $\rightarrow$ mol $NaC_2H_3O_2$, mol $HC_2H_3O_2$ then

set up stoichiometry table

Part II: Equilibrium:
mol $NaC_2H_3O_2$, mol $HC_2H_3O_2$, L, $K_a \rightarrow$ pH

$$pH = pK_a + \log \frac{[base]}{[acid]}$$

Solution: $250.0 \text{ mL} \times \dfrac{1 \text{ L}}{1000 \text{ mL}} = 0.2500$ L then

$$\frac{0.250 \text{ mol } HC_2H_3O_2}{1 \text{ L}} \times 0.250 \text{ L} = 0.0625 \text{ mol } HC_2H_3O_2 \text{ and}$$

$$\frac{0.250 \text{ mol NaC}_2\text{H}_3\text{O}_2}{1 \text{ L}} \times 0.250 \text{ L} = 0.0625 \text{ mol NaC}_2\text{H}_3\text{O}_2.$$ Set up a table to track changes:

	HCl(aq)	+	NaC$_2$H$_3$O$_2$(aq)	→	HC$_2$H$_3$O$_2$(aq)	+	NaCl(aq)
Before addition	≈0.00 mol		0.0625 mol		0.0625 mol		0.00 mol
Addition	0.0050 mol		—		—		—
After addition	≈0.00 mol		0.0575 mol		0.0675 mol		0.0050 mol

Because the amount of HCl is small, there are still significant amounts of both buffer components; so the Henderson–Hasselbalch equation can be used to calculate the new pH.

$$\text{pH} = \text{p}K_a + \log \frac{[\text{base}]}{[\text{acid}]} = -\log(1.8 \times 10^{-5}) + \log \frac{\dfrac{0.0575 \text{ mol}}{0.250 \text{ L}}}{\dfrac{0.0675 \text{ mol}}{0.250 \text{ L}}} = 4.68$$

Check: The units (none) are correct. The magnitude of the answer makes physical sense because the pH dropped slightly when acid was added.

(c) **Given:** 250.0 mL buffer of 0.250 M HC$_2$H$_3$O$_2$ and 0.250 M NaC$_2$H$_3$O$_2$, add 0.0050 mol NaOH
Find: pH   **Other:** $K_a(\text{HC}_2\text{H}_3\text{O}_2) = 1.8 \times 10^{-5}$
Conceptual Plan: Part I: Stoichiometry:
mL → L then [NaC$_2$H$_3$O$_2$], L → mol NaC$_2$H$_3$O$_2$ and [HC$_2$H$_3$O$_2$], L → mol HC$_2$H$_3$O$_2$

$$\frac{1 \text{ L}}{1000 \text{ mL}} \qquad\qquad M = \frac{\text{mol}}{\text{L}} \qquad\qquad M = \frac{\text{mol}}{\text{L}}$$

write balanced equation then

NaOH + HC$_2$H$_3$O$_2$ → H$_2$O + NaC$_2$H$_3$O$_2$

mol NaC$_2$H$_3$O$_2$, mol HC$_2$H$_3$O$_2$, mol NaOH → mol NaC$_2$H$_3$O$_2$, mol HC$_2$H$_3$O$_2$ then

set up stoichiometry table

Part II: Equilibrium:
mol NaC$_2$H$_3$O$_2$, mol HC$_2$H$_3$O$_2$, L, K_a → pH

$$\text{pH} = \text{p}K_a + \log \frac{[\text{base}]}{[\text{acid}]}$$

Solution: $250.0 \text{ mL} \times \dfrac{1 \text{ L}}{1000 \text{ mL}} = 0.2500 \text{ L}$ then

$$\frac{0.250 \text{ mol HC}_2\text{H}_3\text{O}_2}{1 \text{ L}} \times 0.2500 \text{ L} = 0.0625 \text{ mol HC}_2\text{H}_3\text{O}_2 \text{ and}$$

$$\frac{0.250 \text{ mol NaC}_2\text{H}_3\text{O}_2}{1 \text{ L}} \times 0.2500 \text{ L} = 0.0625 \text{ mol NaC}_2\text{H}_3\text{O}_2.$$ Set up a table to track changes:

	NaOH(aq)	+	HC$_2$H$_3$O$_2$(aq)	→	NaC$_2$H$_3$O$_2$(aq)	+	H$_2$O(l)
Before addition	≈0.00 mol		0.0625 mol		0.0625 mol		—
Addition	0.0050 mol		—		—		—
After addition	≈0.00 mol		0.0575 mol		0.0675 mol		—

Because the amount of NaOH is small, there are still significant amounts of both buffer components; so the Henderson–Hasselbalch equation can be used to calculate the new pH.

$$\text{pH} = \text{p}K_a + \log \frac{[\text{base}]}{[\text{acid}]} = -\log(1.8 \times 10^{-5}) + \log \frac{\dfrac{0.0675 \text{ mol}}{0.2500 \text{ L}}}{\dfrac{0.0575 \text{ mol}}{0.2500 \text{ L}}} = 4.81$$

Check: The units (none) are correct. The magnitude of the answer makes physical sense because the pH rose slightly when base was added.

16.49 (a) **Given:** 500.0 mL pure water   **Find:** initial pH and pH after adding 0.010 mol HCl
Conceptual Plan: Pure water has a pH of 7.00 then mL → L then mol HCl, L → [H$_3$O$^+$] → pH

$$\frac{1 \text{ L}}{1000 \text{ mL}} \qquad\qquad M = \frac{\text{mol}}{\text{L}} \quad \text{pH} = -\log[\text{H}_3\text{O}^+]$$

Solution: Pure water has a pH of 7.00, so initial pH $= 7.00$, $500.0 \text{ mL} \times \dfrac{1 \text{ L}}{1000 \text{ mL}} = 0.5000 \text{ L}$,

then $M = \dfrac{\text{mol}}{\text{L}} = \dfrac{0.010 \text{ mol HCl}}{0.5000 \text{ L}} = 0.020 \text{ M HCl}$. Because HCl is a strong acid, it dissociates completely; so

$\text{pH} = -\log \left[H_3O^- \right] = -\log (0.020) = 1.70$.

Check: The units (none) are correct. The magnitude of the answers makes physical sense because the pH starts neutral and then drops significantly when acid is added and no buffer is present.

(b)   **Given:** 500.0 mL buffer of 0.125 M $HC_2H_3O_2$ and 0.115 M $NaC_2H_3O_2$
Find: initial pH and pH after adding 0.010 mol HCl   **Other:** $K_a(HC_2H_3O_2) = 1.8 \times 10^{-5}$
Conceptual Plan: Initial pH:
Identify acid and base components then M $NaC_2H_3O_2 \rightarrow$ M $C_2H_3O_2^-$ then

$$\text{acid} = HC_2H_3O_2, \text{base} = C_2H_3O_2^- \quad NaC_2H_3O_2(aq) \rightarrow Na^+(aq) + C_2H_3O_2^-(aq)$$

M $HC_2H_3O_2$, M $C_2H_3O_2^- \rightarrow$ pH

$$\text{pH} = pK_a + \log \frac{[\text{base}]}{[\text{acid}]}$$

pH after HCl addition: Part I: Stoichiometry:
mL $\rightarrow$ L then $[NaC_2H_3O_2]$, L $\rightarrow$ mol $NaC_2H_3O_2$ and $[HC_2H_3O_2]$, L $\rightarrow$ mol $HC_2H_3O_2$

$$\dfrac{1 \text{ L}}{1000 \text{ mL}} \qquad\qquad M = \dfrac{\text{mol}}{\text{L}} \qquad\qquad M = \dfrac{\text{mol}}{\text{L}}$$

write balanced equation then

$$HCl + NaC_2H_3O_2 \rightarrow HC_2H_3O_2 + NaCl$$

mol $NaC_2H_3O_2$, mol $HC_2H_3O_2$, mol HCl $\rightarrow$ mol $NaC_2H_3O_2$, mol $HC_2H_3O_2$ then

set up stoichiometry table

Part II: Equilibrium:
mol $NaC_2H_3O_2$, mol $HC_2H_3O_2$, L, $K_a \rightarrow$ pH

$$\text{pH} = pK_a + \log \frac{[\text{base}]}{[\text{acid}]}$$

Solution: Initial pH: Acid $= HC_2H_3O_2$, so $[\text{acid}] = [HC_2H_3O_2] = 0.125 \text{ M}$. Base $= C_2H_3O_2^-$. Because one $C_2H_3O_2^-$ ion is generated for each $NaC_2H_3O_2$, $[C_2H_3O_2^-] = 0.115 \text{ M } C_2H_3O_2^- = [\text{base}]$. Then

$$\text{pH} = pK_a + \log \frac{[\text{base}]}{[\text{acid}]} = -\log(1.8 \times 10^{-5}) + \log \frac{0.115 \text{ M}}{0.125 \text{ M}} = 4.71.$$

pH after HCl addition:

$$500.0 \text{ mL} \times \frac{1 \text{ L}}{1000 \text{ mL}} = 0.5000 \text{ L then} \quad \frac{0.125 \text{ mol } HC_2H_3O_2}{1 \text{ L}} \times 0.5000 \text{ L} = 0.0625 \text{ mol } HC_2H_3O_2 \text{ and}$$

$$\frac{0.115 \text{ mol } NaC_2H_3O_2}{1 \text{ L}} \times 0.5000 \text{ L} = 0.0575 \text{ mol } NaC_2H_3O_2. \text{ Set up a table to track changes:}$$

	$HCl(aq)$	$+$	$NaC_2H_3O_2(aq)$	$\rightarrow$	$HC_2H_3O_2(aq)$	$+$	$NaCl(aq)$
Before addition	≈ 0.00 mol		0.0575 mol		0.0625 mol		0.00 mol
Addition	0.010 mol		—		—		—
After addition	≈ 0.00 mol		0.04$\underline{7}$5 mol		0.07$\underline{2}$5 mol		0.10 mol

Because the amount of HCl is small, there are still significant amounts of both buffer components; so the Henderson–Hasselbalch equation can be used to calculate the new pH.

$$\text{pH} = pK_a + \log \frac{[\text{base}]}{[\text{acid}]} = -\log(1.8 \times 10^{-5}) + \log \frac{\dfrac{0.04\underline{7}5 \text{ mol}}{0.5000 \text{ L}}}{\dfrac{0.07\underline{2}5 \text{ mol}}{0.5000 \text{ L}}} = 4.56$$

Check: The units (none) are correct. The magnitude of the answers makes physical sense because the pH started below the pK_a of the acid and it dropped slightly when acid was added.

(c) **Given:** 500.0 mL buffer of 0.155 M $C_2H_5NH_2$ and 0.145 M $C_2H_5NH_3Cl$
Find: initial pH and pH after adding 0.010 mol HCl   **Other:** $K_b(C_2H_5NH_2) = 5.6 \times 10^{-4}$
Conceptual Plan: Initial pH:
Identify acid and base components then M $C_2H_5NH_3Cl \rightarrow$ M $C_2H_5NH_3^+$

$$\text{acid} = C_2H_5NH_3^+ \text{ base} = C_2H_5NH_2 \quad C_2H_5NH_3Cl(aq) \rightarrow C_2H_5NH_3^+(aq) + Cl^-(aq)$$

and $K_b \rightarrow pK_b \rightarrow pK_a$ **then** pK_a, M $C_2H_5NH_2$, M $C_2H_5NH_3^+ \rightarrow$ **pH**

$$pK_b = -\log K_b \quad 14 = pK_a + pK_b \qquad\qquad pH = pK_a + \log\frac{[\text{base}]}{[\text{acid}]}$$

pH after HCl addition: Part I: Stoichiometry:
mL $\rightarrow$ L then $[C_2H_5NH_2]$, L $\rightarrow$ mol $C_2H_5NH_2$ and

$$\frac{1\text{ L}}{1000\text{ mL}} \qquad\qquad M = \frac{\text{mol}}{L}$$

$[C_2H_5NH_3Cl]$, L $\rightarrow$ mol $C_2H_5NH_3Cl$ write balanced equation then

$$M = \frac{\text{mol}}{L} \qquad HCl + C_2H_5NH_2 \rightarrow C_2H_5NH_3Cl$$

mol $C_2H_5NH_2$, mol $C_2H_5NH_3Cl$, mol HCl $\rightarrow$ mol $C_2H_5NH_2$, mol $C_2H_5NH_3Cl$ then

set up stoichiometry table

Part II: Equilibrium:
mol $C_2H_5NH_2$, mol $C_2H_5NH_3Cl$, L, $K_a \rightarrow$ pH

$$pH = pK_a + \log\frac{[\text{base}]}{[\text{acid}]}$$

Solution: Base = $C_2H_5NH_2$, so $[\text{base}] = [C_2H_5NH_2] = 0.155$ M. Acid = $C_2H_5NH_3^+$. Because one $C_2H_5NH_3^+$ ion is generated for each $C_2H_5NH_3Cl$, $[C_2H_5NH_3^+] = 0.145$ M $C_2H_5NH_3^+ = [\text{acid}]$. Because $K_b(C_2H_5NH_2) = 5.6 \times 10^{-4}$, $pK_b = -\log K_b = -\log(5.6 \times 10^{-4}) = 3.25$. Because $14 = pK_a + pK_b$, $pK_a = 14 - pK_b = 14 - 3.25 = 10.75$ then

$$pH = pK_a + \log\frac{[\text{base}]}{[\text{acid}]} = 10.75 + \log\frac{0.155\text{ M}}{0.145\text{ M}} = 10.78.$$

pH after HCl addition: $500.0\text{ mL} \times \dfrac{1\text{ L}}{1000\text{ mL}} = 0.5000$ L then

$$\frac{0.155\text{ mol } C_2H_5NH_2}{1\text{ L}} \times 0.5000\text{ L} = 0.0775\text{ mol } C_2H_5NH_2 \text{ and}$$

$$\frac{0.145\text{ mol } C_2H_5NH_3Cl}{1\text{ L}} \times 0.5000\text{ L} = 0.0725\text{ mol } C_2H_5NH_3Cl.$$ Set up a table to track changes:

	HCl(aq)	+	$C_2H_5NH_2(aq)$	$\rightarrow$	$C_2H_5NH_3Cl(aq)$
Before addition	≈0.00 mol		0.0775 mol		0.0725 mol
Addition	0.010 mol		—		—
After addition	≈0.00 mol		0.0675 mol		0.0825 mol

Because the amount of HCl is small, there are still significant amounts of both buffer components; so the Henderson–Hasselbalch equation can be used to calculate the new pH.

$$pH = pK_a + \log\frac{[\text{base}]}{[\text{acid}]} = 10.75 + \log\frac{\frac{0.0675\text{ mol}}{0.5000\text{ L}}}{\frac{0.0825\text{ mol}}{0.5000\text{ L}}} = 10.66$$

Check: The units (none) are correct. The magnitude of the answers makes physical sense because the initial pH should be greater than the pK_a of the acid because there is more base than acid and the pH drops slightly when acid is added.

16.51 **Given:** 350.00 mL 0.150 M HF and 0.150 M NaF buffer
Find: mass NaOH to raise pH to 4.00 and mass NaOH to raise pH to 4.00 with buffer concentrations raised to 0.350 M
Other: $K_a(HF) = 3.5 \times 10^{-4}$
Conceptual Plan: Identify acid and base components. Because $[NaF] = [HF]$, then initial pH $= pK_a$

$$\text{acid} = HF, \text{ base} = F^- \qquad\qquad pH = pK_a$$

final pH, pK_a → [NaF]/[HF] and mL → L then [HF], L → mol HF and [NaF], L → mol NaF

$$pH = pK_a + \log \frac{[\text{base}]}{[\text{acid}]} \qquad \frac{1\,\text{L}}{1000\,\text{mL}} \qquad M = \frac{\text{mol}}{\text{L}} \qquad M = \frac{\text{mol}}{\text{L}}$$

then write balanced equation then

$$NaOH + HF \rightarrow NaF + H_2O$$

mol HF, mol NaF, [NaF]/[HF] → mol NaOH → g NaOH.

$$\text{set up stoichiometry table} \quad \frac{40.00\text{ g NaOH}}{1\text{ mol NaOH}}$$

Finally, when the buffer concentrations are raised to 0.350 M, simply multiply the g NaOH by the ratio of concentrations (0.350 M/0.150 M).

Solution: initial pH = $pK_a = -\log(3.5 \times 10^{-4}) = 3.46$ then

$$pH = pK_a + \log \frac{[\text{base}]}{[\text{acid}]} = -\log(3.5 \times 10^{-4}) + \log \frac{[\text{NaF}]}{[\text{HF}]} = 4.00. \text{ Solve for [NaF]/[HF].}$$

$$\log \frac{[\text{NaF}]}{[\text{HF}]} = 4.00 - 3.46 = 0.54 \rightarrow \frac{[\text{NaF}]}{[\text{HF}]} = 10^{0.54} = 3.5. \; 350.0\text{ mL} \times \frac{1\text{ L}}{1000\text{ mL}} = 0.3500\text{ L then}$$

$$\frac{0.150\text{ mol HF}}{1\text{ L}} \times 0.3500\text{ L} = 0.0525\text{ mol HF and } \frac{0.150\text{ mol NaF}}{1\text{ L}} \times 0.3500\text{ L} = 0.0525\text{ mol NaF}$$

Set up a table to track changes:

	NaOH(aq)	+	HF(aq)	→	NaF(aq)	+	H$_2$O(aq)
Before addition	≈0.00 mol		0.0525 mol		0.0525 mol		—
Addition	x		—		—		—
After addition	≈0.00 mol		(0.0525 − x) mol		(0.0525 + x) mol		—

Because $\dfrac{[\text{NaF}]}{[\text{HF}]} = 3.5 = \dfrac{(0.0525 + x)\text{ mol}}{(0.0525 - x)\text{ mol}}$, solve for x. Note that the ratio of moles is the same as the ratio of concentrations because the volume for both terms is the same. $3.5(0.0525 - x) = (0.0525 + x) \rightarrow$ $0.18375 - 3.5x = 0.0525 + x \rightarrow 0.13125 = 4.5x \rightarrow x = 0.029167$ mol NaOH then

$$0.029167\text{ mol NaOH} \times \frac{40.00\text{ g NaOH}}{1\text{ mol NaOH}} = 1.1667\text{ g NaOH} = 1.2\text{ g NaOH}$$

To scale the amount of NaOH up to a 0.350 M HF and NaF solution, multiply the NaOH mass by the ratio of concentrations. $1.1667\text{ g NaOH} \times \dfrac{0.350\text{ M}}{0.150\text{ M}} = 2.7\text{ g NaOH}$

Check: The units (g) are correct. The magnitude of the answers makes physical sense because there is much less than a mole of each of the buffer components; so there must be much less than a mole of NaOH. The higher the buffer concentrations, the higher the buffer capacity and the mass of NaOH it can neutralize.

16.53   (a)   Yes, this will be a buffer because NH$_3$ is a weak base and NH$_4^+$ is its conjugate acid. The ratio of base to acid is $0.10/0.15 = 0.67$, so the pH will be within 1 pH unit of the pK_a.

        (b)   No, this will not be a buffer solution because HCl is a strong acid and NaOH is a strong base.

        (c)   Yes, this will be a buffer because HF is a weak acid and the NaOH will convert $(20.0/50.0) \times 100\% = 40\%$ of the acid to its conjugate base.

        (d)   No, this will not be a buffer solution because both components are bases.

        (e)   No, this will not be a buffer solution because both components are bases.

16.55   (a)   **Given:** blood buffer 0.024 M HCO$_3^-$ and 0.0012 M H$_2$CO$_3$, pK_a = 6.1   **Find:** initial pH

            Conceptual Plan: Identify acid and base components then M HCO$_3^-$, M H$_2$CO$_3$ → pH

$$\text{acid} = H_2CO_3 \text{ base} = HCO_3^- \qquad\qquad pH = pK_a + \log \frac{[\text{base}]}{[\text{acid}]}$$

            Solution: Acid = H$_2$CO$_3$, so $[\text{acid}] = [H_2CO_3] = 0.0012$ M. Base = HCO$_3^-$, so $[\text{base}] = [HCO_3^-] = $

            0.024 M HCO$_3^-$. Then $pH = pK_a + \log \dfrac{[\text{base}]}{[\text{acid}]} = 6.1 + \log \dfrac{0.024\text{ M}}{0.0012\text{ M}} = 7.4$.

            Check: The units (none) are correct. The magnitude of the answer makes physical sense because the pH is greater than the pK_a of the acid because there is more base than acid.

(b) **Given:** 5.0 L of blood buffer   **Find:** mass HCl to lower pH to 7.0

Conceptual Plan: Final pH, $pK_a \rightarrow [HCO_3^-]/[H_2CO_3]$ then $[HCO_3^-]$, L → mol HCO_3^- and

$$pH = pK_a + \log \frac{[\text{base}]}{[\text{acid}]} \qquad\qquad M = \frac{mol}{L}$$

$[H_2CO_3]$, L → mol H_2CO_3 then write balanced equation then

$$M = \frac{mol}{L} \qquad\qquad H^+ + HCO_3^- \rightarrow H_2CO_3$$

mol HCO_3^-, mol H_2CO_3, $[HCO_3^-]/[H_2CO_3]$ → mol HCl → g HCl

$$\text{set up stoichiometry table} \quad \frac{36.46 \text{ g HCl}}{1 \text{ mol HCl}}$$

Solution: $pH = pK_a + \log \dfrac{[\text{base}]}{[\text{acid}]} = 6.1 + \log \dfrac{[HCO_3^-]}{[H_2CO_3]} = 7.0$. Solve for $[HCO_3^-]/[H_2CO_3]$.

$\log \dfrac{[HCO_3^-]}{[H_2CO_3]} = 7.0 - 6.1 = 0.9 \rightarrow \dfrac{[HCO_3^-]}{[H_2CO_3]} = 10^{0.9} = \underline{7}.9433$. Then

$\dfrac{0.024 \text{ mol } HCO_3^-}{1\,\cancel{L}} \times 5.0\,\cancel{L} = 0.12 \text{ mol } HCO_3^-$ and $\dfrac{0.0012 \text{ mol } H_2CO_3}{1\,\cancel{L}} \times 5.0\,\cancel{L} = 0.0060 \text{ mol } H_2CO_3$.

Because HCl is a strong acid, $[HCl] = [H^+]$. Set up a table to track changes:

	$H^+(aq)$	+	$HCO_3^-(aq)$	→	$H_2CO_3(aq)$
Before addition	≈ 0.00 mol		0.12 mol		0.0060 mol
Addition	x		—		—
After addition	≈ 0.00 mol		$(0.12 - x)$ mol		$(0.0060 + x)$ mol

Because $\dfrac{[HCO_3^-]}{[H_2CO_3]} = \underline{7}.9433 = \dfrac{(0.12 - x)\,\cancel{mol}}{(0.0060 + x)\,\cancel{mol}}$, solve for x. Note that the ratio of moles is the same as the ratio of concentrations because the volume for both terms is the same.

$\underline{7}.9433(0.0060 + x) = (0.12 - x) \rightarrow 0.04\underline{7}6598 + \underline{7}.9433x = 0.12 - x \rightarrow \underline{8}.9433x = 0.0\underline{7}234 \rightarrow$

$x = 0.00\underline{8}0888$ mol HCl then $0.00\underline{8}0888 \,\cancel{\text{mol HCl}} \times \dfrac{36.46 \text{ g HCl}}{1\,\cancel{\text{mol HCl}}} = 0.2\underline{9}492$ g HCl $= 0.3$ g HCl

Check: The units (g) are correct. The amount of acid needed is small because the concentrations of the buffer components are very low and the buffer starts only 0.4 pH unit above the final pH.

(c) **Given:** 5.0 L of blood buffer   **Find:** mass NaOH to raise pH to 7.8

Conceptual Plan: Final pH, $pK_a \rightarrow [HCO_3^-]/[H_2CO_3]$ then $[HCO_3^-]$, L → mol HCO_3^- and

$$pH = pK_a + \log \frac{[\text{base}]}{[\text{acid}]} \qquad\qquad M = \frac{mol}{L}$$

$[H_2CO_3]$, L → mol H_2CO_3 then write balanced equation then

$$M = \frac{mol}{L} \qquad\qquad OH^- + H_2CO_3 \rightarrow HCO_3^- + H_2O$$

mol HCO_3^-, mol H_2CO_3, $[HCO_3^-]/[H_2CO_3]$ → mol NaOH → g NaOH

$$\text{set up stoichiometry table} \quad \frac{40.00 \text{ g NaOH}}{1 \text{ mol NaOH}}$$

Solution: $pH = pK_a + \log \dfrac{[\text{base}]}{[\text{acid}]} = 6.1 + \log \dfrac{[HCO_3^-]}{[H_2CO_3]} = 7.8$

Solve for $[HCO_3^-]/[H_2CO_3]$. $\log \dfrac{[HCO_3^-]}{[H_2CO_3]} = 7.8 - 6.1 = 1.7 \rightarrow \dfrac{[HCO_3^-]}{[H_2CO_3]} = 10^{1.7} = 50.\underline{1}1872$. Then

$\dfrac{0.024 \text{ mol } HCO_3^-}{1\,\cancel{L}} \times 5.0\,\cancel{L} = 0.12 \text{ mol } HCO_3^-$ and $\dfrac{0.0012 \text{ mol } H_2CO_3}{1\,\cancel{L}} \times 5.0\,\cancel{L} = 0.0060 \text{ mol } H_2CO_3$.

Because NaOH is a strong base, $[NaOH] = [OH^-]$. Set up a table to track changes:

	$OH^-(aq)$	+	$H_2CO_3(aq)$	→	$HCO_3^-(aq)$	+	$H_2O(l)$
Before addition	≈ 0.00 mol		0.0060 mol		0.12 mol		—
Addition	x		—		—		
After addition	≈ 0.00 mol		$(0.0060 - x)$ mol		$(0.12 + x)$ mol		

Because $\dfrac{[\text{HCO}_3^-]}{[\text{H}_2\text{CO}_3]} = 50.\underline{1}1872 = \dfrac{(0.12 + x)\ \text{mol}}{(0.0060 - x)\ \text{mol}}$, solve for x. Note that the ratio of moles is the same as the ratio of concentrations because the volume for both terms is the same.

$50.\underline{1}1872(0.0060 - x) = (0.12 + x) \rightarrow 0.3\underline{0}071 - 50.11872x = 0.12 + x \rightarrow 51.11872x = 0.1\underline{8}071$
$\rightarrow x = 0.003\underline{5}351$ mol NaOH then

$0.003\underline{5}351\ \text{mol NaOH} \times \dfrac{40.00\ \text{g NaOH}}{1\ \text{mol NaOH}} = 0.1\underline{4}140\ \text{g NaOH} = 0.14\ \text{g NaOH}$

Check: The units (g) are correct. The amount of base needed is small because the concentrations of the buffer components are very low.

16.57 **Given:** $\text{HC}_2\text{H}_3\text{O}_2/\text{KC}_2\text{H}_3\text{O}_2$, $\text{HClO}_2/\text{KClO}_2$, $\text{NH}_3/\text{NH}_4\text{Cl}$, and HClO/KClO potential buffer systems to create buffer at pH = 7.20 **Find:** best buffer system and ratio of component masses
Other: $K_a(\text{HC}_2\text{H}_3\text{O}_2) = 1.8 \times 10^{-5}$, $K_a(\text{HClO}_2) = 1.8 \times 10^{-4}$, $K_b(\text{NH}_3) = 1.76 \times 10^{-5}$, $K_a(\text{HClO}) = 2.9 \times 10^{-8}$
Conceptual Plan: Calculate pK_a of all potential buffer acids for the base $K_b \rightarrow pK_b \rightarrow pK_a$ and

$$pK_a = -\log K_a \qquad\qquad pK_b = -\log K_b \quad 14 = pK_a + pK_b$$

choose the pK_a that is closest to 7.20 then pH, $K_a \rightarrow$ [base]/[acid] $\rightarrow$ mass base/mass acid

$$pH = pK_a + \log \dfrac{[\text{base}]}{[\text{acid}]} \qquad \dfrac{\mathcal{M}\,(\text{base})}{\mathcal{M}\,(\text{acid})}$$

Solution: For $\text{HC}_2\text{H}_3\text{O}_2/\text{KC}_2\text{H}_3\text{O}_2$: $pK_a = -\log K_a = -\log(1.8 \times 10^{-5}) = 4.74$
For $\text{HClO}_2/\text{KClO}_2$: $pK_a = -\log K_a = -\log(1.8 \times 10^{-4}) = 3.74$
For $\text{NH}_3/\text{NH}_4\text{Cl}$: $pK_b = -\log K_b = -\log(1.76 \times 10^{-5}) = 4.75$
Because $14 = pK_a + pK_b$, $pK_a = 14 - pK_b = 14 - 4.75 = 9.25$.
And for HClO/KClO: $pK_a = -\log K_a = -\log(2.9 \times 10^{-8}) = 7.54$ So the HClO/KClO buffer system has the pK_a
that is closest to 7.20. So pH $= pK_a + \log \dfrac{[\text{base}]}{[\text{acid}]} = 7.54 + \log \dfrac{[\text{KClO}]}{[\text{HClO}]} = 7.20$. Solve for $[\text{KClO}]/[\text{HClO}]$.

$\log \dfrac{[\text{KClO}]}{[\text{HClO}]} = 7.20 - 7.54 = -0.34 \rightarrow \dfrac{[\text{KClO}]}{[\text{HClO}]} = 10^{-0.34} = 0.4\underline{5}7088$ Then convert to mass ratio using

$$\dfrac{\mathcal{M}\,(\text{base})}{\mathcal{M}\,(\text{acid})}, \quad 0.4\underline{5}7088 \dfrac{\dfrac{\text{KClO mol}}{L}}{\dfrac{\text{HClO mol}}{L}} \times \dfrac{\dfrac{90.55\ \text{g KClO}}{\text{mol KClO}}}{\dfrac{52.46\ \text{g HClO}}{\text{mol HClO}}} = 0.79 \dfrac{\text{g KClO}}{\text{g HClO}}.$$

Check: The units (none and g base/g acid) are correct. The buffer system with the K_a closest to 10^{-7} is the best choice. The magnitude of the answer makes physical sense because the buffer needs more acid than base (and this fact is not overcome by the heavier molar mass of the base).

16.59 **Given:** 500.0 mL of 0.100 M HNO_2/0.150 M KNO_2 buffer and (a) 250 mg NaOH, (b) 350 mg KOH, (c) 1.25 g HBr, and (d) 1.35 g HI **Find:** whether buffer capacity is exceeded
Conceptual Plan: mL $\rightarrow$ L then $[\text{HNO}_2]$, L $\rightarrow$ mol HNO_2 and $[\text{KNO}_2]$, L $\rightarrow$ mol KNO_2

$$\dfrac{1\,L}{1000\ \text{mL}} \qquad\qquad M = \dfrac{\text{mol}}{L} \qquad\qquad M = \dfrac{\text{mol}}{L}$$

then calculate moles of acid or base to be added to the buffer mg $\rightarrow$ g $\rightarrow$ mol then

$$\dfrac{1\,g}{1000\ \text{mg}} \quad \mathcal{M}$$

compare the added amount to the buffer amount of the opposite component. Ratio of base/acid must be between 0.1 and 10 to maintain the buffer integrity.

Solution: $500.00\ \text{mL} \times \dfrac{1\,L}{1000\ \text{mL}} = 0.5000\ L$ then $\dfrac{0.100\ \text{mol HNO}_2}{1\,L} \times 0.5000\,L = 0.0500\ \text{mol HNO}_2$ and

$\dfrac{0.150\ \text{mol KNO}_2}{1\,L} \times 0.5000\,L = 0.0750\ \text{mol KNO}_2$

(a) For NaOH: $250 \text{ mg NaOH} \times \dfrac{1 \text{ g NaOH}}{1000 \text{ mg NaOH}} \times \dfrac{1 \text{ mol NaOH}}{40.00 \text{ g NaOH}} = 0.0062\underline{5} \text{ mol NaOH}$. Because the buffer contains 0.0500 mol acid, the amount of acid is reduced by $0.00625/0.0500 = 0.125$ and the ratio of base/acid is still between 0.1 and 10. The buffer capacity is not exceeded.

(b) For KOH: $350 \text{ mg KOH} \times \dfrac{1 \text{ g KOH}}{1000 \text{ mg KOH}} \times \dfrac{1 \text{ mol KOH}}{56.11 \text{ g KOH}} = 0.0062\underline{4} \text{ mol KOH}$. Because the buffer contains 0.0500 mol acid, the amount of acid is reduced by $0.00624/0.0500 = 0.125$ and the ratio of base/acid is still between 0.1 and 10. The buffer capacity is not exceeded.

(c) For HBr: $1.25 \text{ g HBr} \times \dfrac{1 \text{ mol HBr}}{80.91 \text{ g HBr}} = 0.015\underline{4}493 \text{ mol HBr}$. Because the buffer contains 0.0750 mol base, the amount of acid is reduced by $0.0154/0.0750 = 0.206$ and the ratio of base/acid is still between 0.1 and 10. The buffer capacity is not exceeded.

(d) For HI: $1.35 \text{ g HI} \times \dfrac{1 \text{ mol HI}}{127.91 \text{ g HI}} = 0.010\underline{5}543 \text{ mol HI}$. Because the buffer contains 0.0750 mol base, the amount of acid is reduced by $0.0106/0.0750 = 0.141$ and the ratio of base/acid is still between 0.1 and 10. The buffer capacity is not exceeded.

Titrations, pH Curves, and Indicators

16.61 (i) The equivalence point of a titration is where the pH rises sharply as base is added. The pH at the equivalence point is the midpoint of the sharp rise at ~50 mL added base. For (a), the pH = ~8, and for (b), the pH = ~7.

(ii) Graph (a) represents a weak acid, and graph (b) represents a strong acid. A strong acid titration starts at a lower pH, has a flatter initial region, and has a sharper rise at the equivalence point than does a weak acid. The pH at the equivalence point of a strong acid is neutral, while the pH at the equivalence point of a weak acid is basic.

16.63 **Given:** 20.0 mL 0.200 M KOH and 0.200 M CH_3NH_2 titrated with 0.100 M HI

(a) **Find:** volume of base to reach equivalence point

Conceptual Plan: The answer for both titrations will be the same because the initial concentration and volumes of the bases are the same. Write balanced equation then mL → L then

$$HI + KOH \rightarrow KI + H_2O \text{ and } HI + CH_3NH_2 \rightarrow CH_3NH_3I \qquad \dfrac{1 \text{ L}}{1000 \text{ mL}}$$

[base], L → mol base then set mol base = mol acid and [HI], mol HI → L HI → mL HI

$$M = \dfrac{\text{mol}}{\text{L}} \qquad \text{balanced equation has 1:1 stoichiometry} \qquad M = \dfrac{\text{mol}}{\text{L}} \quad \dfrac{1000 \text{ mL}}{1 \text{ L}}$$

Solution: $20.0 \text{ mL base} \times \dfrac{1 \text{ L}}{1000 \text{ mL}} = 0.0200 \text{ L base}$ then

$\dfrac{0.200 \text{ mol base}}{1 \text{ L}} \times 0.0200 \text{ L} = 0.00400 \text{ mol base}$. So mol base $= 0.00400$ mol $=$ mol HI then

$0.00400 \text{ mol HI} \times \dfrac{1 \text{ L HI}}{0.100 \text{ mol HI}} = 0.0400 \text{ L HI} \times \dfrac{1000 \text{ mL}}{1 \text{ L}} = 40.0 \text{ mL HI for both titrations}$

Check: The units (mL) are correct. The volume of acid is twice the volume of bases because the concentration of the base is twice that of the acid in each case. The answer for both titrations is the same because the stoichiometry is the same for both titration reactions.

(b) The pH at the equivalence point will be neutral for KOH (because it is a strong base), and it will be acidic for CH_3NH_2 (because it is a weak base and will produce a conjugate acid when titrated).

(c) The initial pH will be lower for CH_3NH_2 (because it is a weak base and will only partially dissociate) and not raise the pH as high as KOH will (because it is a strong base and so dissociates completely) at the same base concentration.

(d)   The titration curves will look like the following:

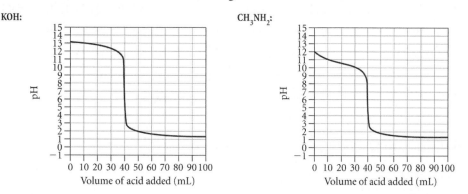

KOH: CH_3NH_2:

Important features to include are a high initial pH (pH is over 13 for a strong base and is lower for a weak base), flat initial region (very flat for strong base, not as flat for weak base where pH halfway to equivalence point is the pK_b of the base), sharp drop at equivalence point, pH at equivalence point (neutral for strong base and lower for weak base), and then flattening out at low pH.

16.65   (a)   The equivalence point of a titration is where the pH rises sharply as base is added. The volume at the equivalence point is ~30 mL. The pH at the equivalence point is the midpoint of the sharp rise at ~30 mL added base, which is a pH = ~9.

(b)   At 0 mL, the pH is calculated by doing an equilibrium calculation of a weak acid in water (as done in Chapter 15).

(c)   The pH halfway to the equivalence point is equal to the pK_a of the acid, or ~15 mL.

(d)   The pH at the equivalence point, or ~30 mL, is calculated by doing an equilibrium problem with the K_b of the acid. At the equivalence point, all of the acid has been converted to its conjugate base.

(e)   Beyond the equivalence point (30 mL), there is excess base. All of the acid has been converted to its conjugate base, so the pH is calculated by focusing on this excess base concentration.

16.67   **Given:** 35.0 mL of 0.175 M HBr titrated with 0.200 M KOH

(a)   **Find:** initial pH
Conceptual Plan: Because HBr is a strong acid, it will dissociate completely; so initial
$$pH = -\log[H_3O^+] = -\log[HBr].$$
Solution: $pH = -\log[HBr] = -\log 0.175 = 0.757$

Check: The units (none) are correct. The pH is reasonable because the concentration is greater than 0.1 M, and when the acid dissociates completely, the pH becomes less than 1.

(b)   **Find:** volume of base to reach equivalence point
Conceptual Plan: Write balanced equation then mL → L then [HBr], L → mol HBr then

$$HBr + KOH \rightarrow KBr + H_2O \qquad \frac{1\ L}{1000\ mL} \qquad M = \frac{mol}{L}$$

set mol acid (HBr) = mol base (KOH) and [KOH], mol KOH → L KOH → mL KOH

balanced equation has 1:1 stoichiometry $M = \frac{mol}{L} \quad \frac{1000\ mL}{1\ L}$

Solution: $35.0\ \text{mL HBr} \times \dfrac{1\ L}{1000\ \text{mL}} = 0.0350\ L\ HBr$ then

$\dfrac{0.175\ mol\ HBr}{1\ L} \times 0.0350\ L = 0.006125\ mol\ HBr$

So mol acid = mol HBr = 0.006125 mol = mol KOH then

$0.006125\ \text{mol KOH} \times \dfrac{1\ L}{0.200\ \text{mol KOH}} = 0.030625\ \text{L KOH} \times \dfrac{1000\ mL}{1\ L} = 30.6\ mL\ KOH$

Check: The units (mL) are correct. The volume of base is a little less than the volume of acid because the concentration of the base is a little greater than that of the acid.

(c) **Find:** pH after adding 10.0 mL of base

Conceptual Plan: Use calculations from part (b). Then mL → L then [KOH], L → mol KOH then

$$\frac{1\ L}{1000\ mL} \qquad M = \frac{mol}{L}$$

mol HBr, mol KOH → mol excess HBr and L HBr, L KOH → total L then

set up stoichiometry table   L HBr + L KOH = total L

mol excess HBr, L → [HBr] → pH.

$$M = \frac{mol}{L} \qquad pH = -\log[\,HBr\,]$$

Solution: $10.0\ \cancel{mL\ KOH} \times \dfrac{1\ L}{1000\ \cancel{mL}} = 0.0100\ L\ KOH$ then

$$\frac{0.200\ mol\ KOH}{1\ \cancel{L}} \times 0.0100\ \cancel{L} = 0.00200\ mol\ KOH$$

Because KOH is a strong base, $[\,KOH\,] = [\,OH^-\,]$. Set up a table to track changes:

	KOH(*aq*)	+	HBr(*aq*)	→	KBr(*aq*)	+	H₂O(*l*)
Before addition	≈0.00 mol		0.006125 mol		0.00 mol		—
Addition	0.00200 mol		—		—		—
After addition	≈0.00 mol		0.004125 mol		0.00200 mol		—

Then $0.0350\ L\ HBr + 0.0100\ L\ KOH = 0.0450\ L$ total volume.

So mol excess acid $=$ mol HBr $= 0.004125$ mol in 0.0450 L so

$$[\,HBr\,] = \frac{0.004125\ mol\ HBr}{0.0450\ L} = 0.0916667\ M \text{ and}$$

$$pH = -\log[\,HBr\,] = -\log 0.0916667 = 1.038$$

Check: The units (none) are correct. The pH is a little higher than the initial pH, which is expected because this is a strong acid.

(d) **Find:** pH at equivalence point

Solution: Because this is a strong acid–strong base titration, the pH at the equivalence point is neutral, or 7.

(e) **Find:** pH after adding 5.0 mL of base beyond the equivalence point

Conceptual Plan: Use calculations from parts (b) and (c). Then the pH is only dependent on the amount of excess base and the total solution volumes.

mL excess → L excess then [KOH], L excess → mol KOH excess

$$\frac{1\ L}{1000\ mL} \qquad\qquad M = \frac{mol}{L}$$

then L HBr, L KOH to equivalence point, L KOH excess → total L then

L HBr + L KOH to equivalence point + L KOH excess = total L

mol excess KOH, total L → [KOH] = [OH⁻] → [H₃O⁺] → pH

$$M = \frac{mol}{L} \qquad K_w = [\,H_3O^+\,][\,OH^-\,] \qquad pH = -\log[\,H_3O^+\,]$$

Solution: $5.0\ \cancel{mL\ KOH} \times \dfrac{1\ L}{1000\ \cancel{mL}} = 0.0050\ L\ KOH$ excess then

$\dfrac{0.200\ mol\ KOH}{1\ \cancel{L}} \times 0.0050\ \cancel{L} = 0.0010\ mol\ KOH$ excess. Then $0.0350\ L\ HBr + 0.0306\ L\ KOH + 0.0050\ L$

KOH $= 0.0706$ L total volume. $[\,KOH\ excess\,] = \dfrac{0.0010\ mol\ KOH\ excess}{0.0706\ L} = 0.014164\ M$ KOH excess.

Because KOH is a strong base, $[\,KOH\,]$ excess $= [\,OH^-\,]$. $K_w = [\,H_3O^+\,][\,OH^-\,]$, so

$$[\,H_3O^+\,] = \frac{K_w}{[\,OH^-\,]} = \frac{1.0 \times 10^{-14}}{0.014164} = 7.06 \times 10^{-13}\ M.$$

Finally, $pH = -\log[\,H_3O^+\,] = -\log(7.06 \times 10^{-13}) = 12.15$.

Check: The units (none) are correct. The pH is rising sharply at the equivalence point, so the pH after 5 mL past the equivalence point should be quite basic.

16.69   **Given:** 25.0 mL of 0.115 M RbOH titrated with 0.100 M HCl

(a)   **Find:** initial pH

Conceptual Plan: Because RbOH is a strong base, it will dissociate completely, so

$$[RbOH] = [OH^-] \rightarrow [H_3O^+] \rightarrow pH$$

$$K_w = [H_3O^+][OH^-] \quad pH = -\log[H_3O^+]$$

Solution: Because RbOH is a strong base, $[RbOH]$ excess $= [OH^-]$. $K_w = [H_3O^+][OH^-]$, so

$$[H_3O^+] = \frac{K_w}{[OH^-]} = \frac{1.0 \times 10^{-14}}{0.115} = 8.\underline{6}9565 \times 10^{-14} \, M \text{ and}$$

$$pH = -\log[H_3O^+] = -\log(8.\underline{6}9565 \times 10^{-14}) = 13.06.$$

Check: The units (none) are correct. The pH is reasonable because the concentration is greater than 0.1 M, and when the base dissociates completely, the pH becomes greater than 13.

(b)   **Find:** volume of acid to reach equivalence point

Conceptual Plan: Write balanced equation then mL $\rightarrow$ L then [RbOH], L $\rightarrow$ mol RbOH then

$$HCl + RbOH \rightarrow RbCl + H_2O \qquad \frac{1 \, L}{1000 \, mL} \qquad M = \frac{mol}{L}$$

set mol base (RbOH) = mol acid (HCl) and [HCl], mol HCl $\rightarrow$ L HCl $\rightarrow$ mL HCl

balanced equation has 1:1 stoichiometry $\qquad M = \frac{mol}{L} \quad \frac{1000 \, mL}{1 \, L}$

Solution: $25.0 \, \text{mL RbOH} \times \dfrac{1 \, L}{1000 \, \text{mL}} = 0.0250 \, L \, RbOH$ then

$$\frac{0.115 \, mol \, RbOH}{1 \, L} \times 0.0250 \, L = 0.002875 \, mol \, RbOH. \text{ So mol base} = mol \, RbOH = 0.002\underline{8}75 \, mol = mol$$

HCl then $0.002\underline{8}75 \, \text{mol HCl} \times \dfrac{1 \, L}{0.100 \, \text{mol HCl}} = 0.02875 \, \text{L HCl} \times \dfrac{1000 \, mL}{1 \, L} = 28.8 \, mL \, HCl.$

Check: The units (mL) are correct. The volume of acid is greater than the volume of base because the concentration of the base is a little greater than that of the acid.

(c)   **Find:** pH after adding 5.0 mL of acid

Conceptual Plan: Use calculations from part (b). Then mL $\rightarrow$ L then [HCl], L $\rightarrow$ mol HCl then

$$\frac{1 \, L}{1000 \, mL} \qquad M = \frac{mol}{L}$$

mol RbOH, mol HCl $\rightarrow$ mol excess RbOH and L RbOH, L HCl $\rightarrow$ total L then

set up stoichiometry table $\qquad\qquad$ L RbOH + L HCl = total L

mol excess RbOH, L $\rightarrow$ [RbOH] = [OH$^-$] $\rightarrow$ [H$_3$O$^+$] $\rightarrow$ pH.

$$M = \frac{mol}{L} \qquad K_w = [H_3O^+][OH^-] \quad pH = -\log[H_3O^+]$$

Solution: $5.0 \, \text{mL HCl} \times \dfrac{1 \, L}{1000 \, \text{mL}} = 0.0050 \, L \, HCl$ then $\dfrac{0.100 \, mol \, HCl}{1 \, L} \times 0.0050 \, L = 0.00050 \, mol \, HCl$

Because HCl is a strong acid, $[HCl] = [H_3O^+]$. Set up a table to track changes:

	HCl(aq)	+	RbOH(aq)	$\rightarrow$	RbCl(aq)	+	H$_2$O(l)
Before addition	0.00 mol		0.002875 mol		0.00 mol		—
Addition	0.00050 mol		—		—		—
After addition	≈0.00 mol		0.002375 mol		0.00050 mol		—

Then $0.0250 \, L \, RbOH + 0.0050 \, L \, HCl = 0.0300 \, L$ total volume. So mol excess base = mol RbOH =

$0.002\underline{3}75 \, mol$ in $0.0300 \, L$, so $[RbOH] = \dfrac{0.002375 \, mol \, RbOH}{0.0300 \, L} = 0.079\underline{1}667 \, M.$ Because RbOH is a strong

base, $[RbOH]$ excess $= [OH^-]$. $K_w = [H_3O^+][OH^-]$ so $[H_3O^+] = \dfrac{K_w}{[OH^-]} = \dfrac{1.0 \times 10^{-14}}{0.079\underline{1}667}$

$= 1.\underline{2}6316 \times 10^{-13}$ M and pH $= -\log [H_3O^+] = -\log (1.\underline{2}6316 \times 10^{-13}) = 12.90$

Check: The units (none) are correct. The pH is a little lower than the initial pH, which is expected because this is a strong base.

(d)   **Find:** pH at equivalence point
Solution: Because this is a strong acid–strong base titration, the pH at the equivalence point is neutral, or 7.

(e)   **Find:** pH after adding 5.0 mL of acid beyond the equivalence point
Conceptual Plan: Use calculations from parts (b) and (c). Then the pH is only dependent on the amount of excess acid and the total solution volumes. Then mL excess → L excess then [HCl], L excess → mol HCl excess

$$\frac{1\,L}{1000\,mL} \qquad\qquad M = \frac{mol}{L}$$

then L RbOH, L HCl to equivalence point, L HCl excess → total L then

L RbOH + L HCl to equivalence point + L HCl excess = total L

mol excess HCl, total L → [HCl] $= [H_3O^+] \rightarrow$ pH.

$$M = \frac{mol}{L} \qquad\qquad pH = -\log[H_3O^+]$$

Solution: $5.0 \text{ mL HCl} \times \dfrac{1\,L}{1000\,mL} = 0.0050$ L HCl excess then

$\dfrac{0.100 \text{ mol HCl}}{1\,L} \times 0.0050\,L = 0.00050$ mol HCl excess. Then 0.0250 L RbOH $+ 0.0288$ L HCl $+ 0.0050$ L

HCl $= 0.0588$ L total volume. $[\text{HCl excess}] = \dfrac{0.00050 \text{ mol HCl excess}}{0.0588\,L} = 0.008\underline{5}034$ M HCl excess

Because HCl is a strong acid, $[\text{HCl}]$ excess $= [H_3O^+]$.

Finally, pH $= -\log[H_3O^+] = -\log (0.008\underline{5}034) = 2.07$.

Check: The units (none) are correct. The pH is dropping sharply at the equivalence point, so the pH after 5 mL past the equivalence point should be quite acidic.

16.71   **Given:** 20.0 mL of 0.105 M $HC_2H_3O_2$ titrated with 0.125 M NaOH   **Other:** $K_a(HC_2H_3O_2) = 1.8 \times 10^{-5}$

(a)   **Find:** initial pH
**Conceptual Plan: Because $HC_2H_3O_2$ is a weak acid, set up an equilibrium problem using the initial concentration.
So M $HC_2H_3O_2 \rightarrow [H_3O^+] \rightarrow$ pH**

       ICE table   pH $= -\log [H_3O^+]$

Solution:

$$HC_3H_3O_2(aq) + H_2O(l) \rightleftharpoons H_3O^+(aq) + C_2H_3O_2^-(aq)$$

	$[HC_2H_3O_2]$		$[H_3O^+]$	$[C_2H_3O_2^-]$
Initial	0.105		≈ 0.00	0.00
Change	$-x$		$+x$	$+x$
Equil	$0.105 - x$		$+x$	$+x$

$K_a = \dfrac{[H_3O^+][C_2H_3O_2^-]}{[HC_2H_3O_2]} = 1.8 \times 10^{-5} = \dfrac{x^2}{0.105 - x}$ Assume that x is small $(x << 0.105)$, so

$\dfrac{x^2}{0.105 - \cancel{x}} = 1.8 \times 10^{-5} = \dfrac{x^2}{0.105}$ and $x = 1.\underline{3}748 \times 10^{-3}$ M $= [H_3O^+]$. Confirm that the assumption is valid.

$\dfrac{1.\underline{3}748 \times 10^{-3}}{0.105} \times 100\% = 1.3\% < 5\%$, so the assumption is valid. Finally,

pH $= -\log [H_3O^+] = -\log (1.\underline{3}748 \times 10^{-3}) = 2.86$.

Check: The units (none) are correct. The magnitude of the answer makes physical sense because the pH should be greater than $-\log (0.105) = 0.98$ because this is a weak acid.

(b)   **Find:** volume of base to reach equivalence point
      Conceptual Plan: Write a balanced equation then mL → L then [HC₂H₃O₂], L → mol HC₂H₃O₂ then

$$HC_2H_3O_2 + NaOH \rightarrow NaC_2H_3O_2 + H_2O \qquad \frac{1\ L}{1000\ mL} \qquad\qquad M = \frac{mol}{L}$$

set mol acid(HC₂H₃O₂) = mol base(NaOH) and [NaOH], mol NaOH → L NaOH → mL NaOH

$$\text{balanced equation has 1:1 stoichiometry} \qquad\qquad M = \frac{mol}{L} \qquad \frac{1000\ mL}{1\ L}$$

Solution: $20.0\ \cancel{mL\ HC_2H_3O_2} \times \dfrac{1\ L}{1000\ \cancel{mL}} = 0.0200\ L\ HC_2H_3O_2$ then

$$\frac{0.105\ mol\ HC_2H_3O_2}{1\ \cancel{L}} \times 0.0200\ \cancel{L} = 0.00210\ mol\ HC_2H_3O_2$$

So mol acid = mol HC₂H₃O₂ = 0.00210 mol = mol NaOH then

$$0.00210\ \cancel{mol\ NaOH} \times \frac{1\ L}{0.125\ \cancel{mol\ NaOH}} = 0.0168\ \cancel{L\ NaOH} \times \frac{1000\ mL}{1\ \cancel{L}} = 16.8\ mL\ NaOH$$

Check: The units (mL) are correct. The volume of base is a little less than the volume of acid because the concentration of the base is a little greater than that of the acid.

(c)   **Find:** pH after adding 5.0 mL of base
      Conceptual Plan: Use calculations from part (b). Then mL → L then [NaOH], L → mol NaOH then

$$\frac{1\ L}{1000\ mL} \qquad\qquad M = \frac{mol}{L}$$

mol HC₂H₃O₂, mol NaOH → mol excess HC₂H₃O₂, mol C₂H₃O₂⁻ and

$$\text{set up stoichiometry table}$$

L HC₂H₃O₂, L NaOH → total L then

$$L\ HC_2H_3O_2 + L\ NaOH = total\ L$$

mol excess HC₂H₃O₂, L → [HC₂H₃O₂] and mol excess C₂H₃O₂⁻, L → [C₂H₃O₂⁻] then

$$M = \frac{mol}{L} \qquad\qquad\qquad M = \frac{mol}{L}$$

M HC₂H₃O₂, M C₂H₃O₂⁻ → [H₃O⁺] → pH.

$$\text{ICE table} \qquad pH = -\log\left[H_3O^+\right]$$

Solution: $5.0\ \cancel{mL\ NaOH} \times \dfrac{1\ L}{1000\ \cancel{mL}} = 0.0050\ L\ NaOH$ then

$$\frac{0.125\ mol\ NaOH}{1\ \cancel{L}} \times 0.0050\ \cancel{L} = 0.000\underline{6}25\ mol\ NaOH.\ \text{Set up a table to track changes:}$$

	NaOH(*aq*)	+	HC₂H₃O₂(*aq*)	→	NaC₂H₃O₂(*aq*)	+	H₂O(*l*)
Before addition	0.00 mol		0.00210 mol		0.00 mol		—
Addition	0.000\underline{6}25 mol		—		—		—
After addition	≈0.00 mol		0.0014\underline{7}5 mol		0.000625 mol		—

Then 0.0200 L HC₂H₃O₂ + 0.0050 L NaOH = 0.0250 L total volume. Then

$$\left[HC_2H_3O_2\right] = \frac{0.0014\underline{7}5\ mol\ HC_2H_3O_2}{0.0250\ L} = 0.0590\ M\ \text{and}$$

$$\left[NaC_2H_3O_2\right] = \frac{0.000\underline{6}25\ mol\ C_2H_3O_2^-}{0.0250\ L} = 0.025\ M.$$

Because one C₂H₃O₂⁻ ion is generated for each NaC₂H₃O₂, [C₂H₃O₂⁻] = 0.025 M C₂H₃O₂⁻.

$$HC_2H_3O_2(aq) + H_2O(l) \rightleftharpoons H_3O^+(aq) + C_2H_3O_2^-(aq)$$

	[HC₂H₃O₂]	[H₃O⁺]	[C₂H₃O₂⁻]
Initial	0.0590	≈0.00	0.025
Change	−*x*	+*x*	+*x*
Equil	0.0590 − *x*	+*x*	0.025 + *x*

$$K_a = \frac{[H_3O^+][C_2H_3O_2^-]}{[HC_2H_3O_2]} = 1.8 \times 10^{-5} = \frac{x(0.025 + x)}{0.0590 - x}$$ Assume that x is small ($x << 0.025 < 0.0590$), so

$$\frac{x(0.025 + \cancel{x})}{0.0590 - \cancel{x}} = 1.8 \times 10^{-5} = \frac{x(0.025)}{0.0590}$$ and $x = 4.\underline{2}48 \times 10^{-5}$ M $= [H_3O^+]$. Confirm that the assumption is valid.

$$\frac{4.\underline{2}48 \times 10^{-5}}{0.025} \times 100\% = 0.17\% < 5\%,$$ so the assumption is valid.

Finally, pH $= -\log[H_3O^+] = -\log(4.\underline{2}48 \times 10^{-5}) = 4.37$.

Check: The units (none) are correct. The pH is a little higher than the initial pH, which is expected because some of the acid has been neutralized.

(d)   **Find:** pH at one-half the equivalence point
Conceptual Plan: Because this is a weak acid–strong base titration, the pH at one-half the equivalence point is the pK_a of the weak acid.
Solution: pH $=$ p$K_a = -\log K_a = -\log(1.8 \times 10^{-5}) = 4.74$

Check: The units (none) are correct. Because this is a weak acid–strong base titration, the pH at one-half the equivalence point is the pK_a of the weak acid; so it should be a little below 5.

(e)   **Find:** pH at equivalence point
Conceptual Plan: Use calculations from part (b). Then because all of the weak acid has been converted to its conjugate base, the pH is only dependent on the hydrolysis reaction of the conjugate base. The mol $C_2H_3O_2^-$ = initial mol $HC_2H_3O_2$ and L $HC_2H_3O_2$,
L NaOH to equivalence point $\rightarrow$ total L then

$$\text{L } HC_2H_3O_2 + \text{L NaOH} = \text{total L}$$

mol excess $C_2H_3O_2^-$, L $\rightarrow [C_2H_3O_2^-]$ and $K_a \rightarrow K_b$ then do an equilibrium calculation:

$$M = \frac{\text{mol}}{L} \qquad\qquad K_w = K_a K_b$$

$$[C_2H_3O_2^-], K_b \rightarrow [OH^-] \rightarrow [H_3O^+] \rightarrow \text{pH}$$
set up ICE table $K_w = [H_3O^+][OH^-]$ pH $= -\log[H_3O^+]$
Solution: mol $C_2H_3O_2^-$ = initial mol $HC_2H_3O_2$ = 0.00210 mol and total volume $=$ L $HC_2H_3O_2$ +
L NaOH $= 0.020$ L $+ 0.0168$ L $= 0.0368$ L then

$$[C_2H_3O_2^-] = \frac{0.00210 \text{ mol } C_2H_3O_2^-}{0.0368 \text{ L}} = 0.057\underline{0}652 \text{ M and } K_w = K_a K_b. \text{ Rearrange to solve for } K_b.$$

$$K_b = \frac{K_w}{K_a} = \frac{1.0 \times 10^{-14}}{1.8 \times 10^{-5}} = 5.\underline{5}556 \times 10^{-10}. \text{ Set up an ICE table:}$$

$$C_2H_3O_2^-(aq) + H_2O(l) \rightleftharpoons HC_2H_3O_2(aq) + OH^-(aq)$$

	$[C_2H_3O_2^-]$	$[HC_2H_3O_2]$	$[OH^-]$
Initial	0.057$\underline{0}$652	≈ 0.00	≈ 0.00
Change	$-x$	$+x$	$+x$
Equil	0.057$\underline{0}$652 $- x$	$+x$	$+x$

$$K_b = \frac{[HC_2H_3O_2][OH^-]}{[C_2H_3O_2^-]} = 5.\underline{5}556 \times 10^{-10} = \frac{x^2}{0.057\underline{0}652 - x}.$$ Assume that x is small ($x << 0.057$), so

$$\frac{x^2}{0.057\underline{0}652 - \cancel{x}} = 5.\underline{5}556 \times 10^{-10} = \frac{x^2}{0.057\underline{0}652}$$ and $x = 5.6306 \times 10^{-6}$ M $= [OH^-]$.

Confirm that the assumption is valid. $\dfrac{5.6306 \times 10^{-6}}{0.057\underline{0}652} \times 100\% = 0.0099\% < 5\%,$ so the assumption is valid.

$$K_w = [H_3O^+][OH^-], \text{ so } [H_3O^+] = \frac{K_w}{[OH^-]} = \frac{1.0 \times 10^{-14}}{5.\underline{6}305 \times 10^{-6}} = 1.\underline{7}760 \times 10^{-9} \text{ M}.$$

Finally, pH $= -\log[H_3O^+] = -\log(1.\underline{7}760 \times 10^{-9}) = 8.75$.

Check: The units (none) are correct. Because this is a weak acid–strong base titration, the pH at the equivalence point is basic.

(f) **Find:** pH after adding 5.0 mL of base beyond the equivalence point

Conceptual Plan: Use calculations from parts (b) and (c). Then the pH is only dependent on the amount of excess base and the total solution volumes.

mL excess → L excess then [NaOH], L excess → mol NaOH excess

$$\frac{1\text{ L}}{1000\text{ mL}} \qquad M = \frac{mol}{L}$$

then L HC$_2$H$_3$O$_2$, L NaOH to equivalence point, L NaOH excess → total L then

L HC$_2$H$_3$O$_2$ + L NaOH to equivalence point + L NaOH excess = total L

mol excess NaOH, total L → [NaOH] = [OH$^-$] → [H$_3$O$^+$] → pH

$$M = \frac{mol}{L} \qquad K_w = [H_3O^+][OH^-] \quad pH = -\log[H_3O^+]$$

Solution: 5.0 ~~mL NaOH~~ $\times \dfrac{1\text{ L}}{1000\text{ mL}} = 0.0050$ L NaOH excess then

$\dfrac{0.125\text{ mol NaOH}}{1\text{ L}} \times 0.0050\text{ L} = 0.000625$ mol NaOH excess. Then 0.0200 L HC$_2$H$_3$O$_2$ + 0.0168 L NaOH

+ 0.0050 L NaOH = 0.0418 L total volume.

$[\text{NaOH excess}] = \dfrac{0.000625\text{ mol NaOH excess}}{0.0418\text{ L}} = 0.0149522$ M NaOH excess. Because NaOH is a strong

base, [NaOH] excess = [OH$^-$]. The strong base overwhelms the weak base, which is insignificant in the

calculation. $K_w = [H_3O^+][OH^-]$, so $[H_3O^+] = \dfrac{K_w}{[OH^-]} = \dfrac{1.0 \times 10^{-14}}{0.0149522} = 6.\underline{6}88 \times 10^{-13}$ M.

Finally, pH $= -\log[H_3O^+] = -\log(6.\underline{6}88 \times 10^{-13}) = 12.17$.

Check: The units (none) are correct. The pH is rising sharply at the equivalence point, so the pH after 5 mL past the equivalence point should be quite basic.

16.73 **Given:** 25.0 mL of 0.175 M CH$_3$NH$_2$ titrated with 0.150 M HBr **Other:** K_b(CH$_3$NH$_2$) = 4.4 × 10^{-4}

(a) **Find:** initial pH

Conceptual Plan: Because CH$_3$NH$_2$ is a weak base, set up an equilibrium problem using the initial concentration, so M CH$_3$NH$_2$ → [OH$^-$] → [H$_3$O$^+$] → pH

ICE table $\quad K_w = [H_3O^+][OH^-] \quad pH = -\log[H_3O^+]$

Solution:

$$CH_3NH_2(aq) + H_2O(l) \rightleftharpoons CH_3NH_3^+(aq) + OH^-(aq)$$

	[CH$_3$NH$_2$]	[CH$_3$NH$_3^+$]	[OH$^-$]
Initial	0.175	0.00	≈0.00
Change	−x	+x	+x
Equil	0.175 − x	+x	+x

$K_b = \dfrac{[CH_3NH_3^+][OH^-]}{[CH_3NH_2]} = 4.4 \times 10^{-4} = \dfrac{x^2}{0.175 - x}$

Assume that x is small ($x \ll 0.175$), so $\dfrac{x^2}{0.175 - x} = 4.4 \times 10^{-4} = \dfrac{x^2}{0.175}$ and $x = 8.\underline{7}750 \times 10^{-3}$ M = [OH$^-$].

Confirm that the assumption is valid. $\dfrac{8.\underline{7}750 \times 10^{-3}}{0.175} \times 100\% = 5.0\%$, so the assumption is valid.

$K_w = [H_3O^+][OH^-]$, so $[H_3O^+] = \dfrac{K_w}{[OH^-]} = \dfrac{1.0 \times 10^{-14}}{8.\underline{7}750 \times 10^{-3}} = 1.\underline{1}396 \times 10^{-12}$ M.

Finally, pH $= -\log[H_3O^+] = -\log(1.\underline{1}396 \times 10^{-12}) = 11.94$.

Check: The units (none) are correct. The magnitude of the answer makes physical sense because the pH should be less than $14 + \log(0.175) = 13.2$ because this is a weak base.

(b) **Find:** volume of acid to reach equivalence point

Conceptual Plan: Write a balanced equation, then mL → L then [CH₃NH₂], L → mol CH₃NH₂

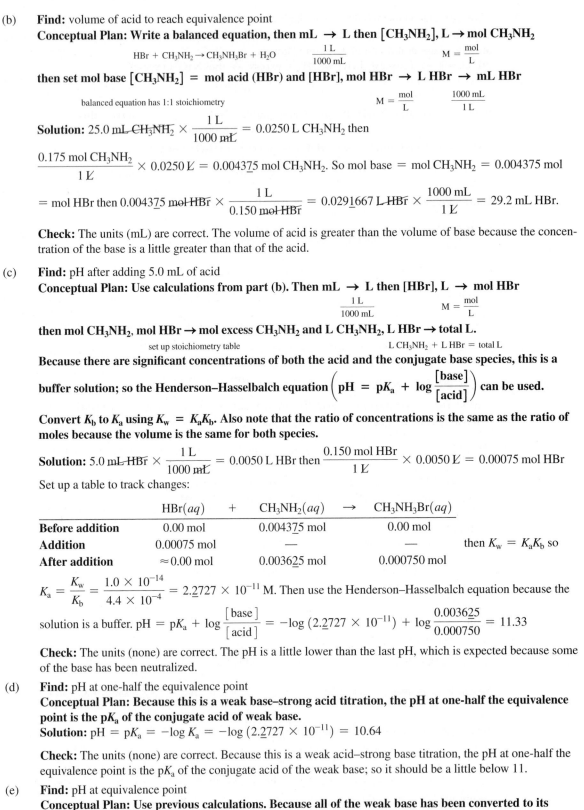

$$HBr + CH_3NH_2 \rightarrow CH_3NH_3Br + H_2O \qquad \frac{1\ L}{1000\ mL} \qquad M = \frac{mol}{L}$$

then set mol base [CH₃NH₂] = mol acid (HBr) and [HBr], mol HBr → L HBr → mL HBr

balanced equation has 1:1 stoichiometry $\qquad M = \dfrac{mol}{L} \qquad \dfrac{1000\ mL}{1\ L}$

Solution: $25.0\ \text{mL CH}_3\text{NH}_2 \times \dfrac{1\ L}{1000\ mL} = 0.0250\ L\ CH_3NH_2$ then

$\dfrac{0.175\ mol\ CH_3NH_2}{1\ L} \times 0.0250\ L = 0.004375\ mol\ CH_3NH_2.$ So mol base = mol CH₃NH₂ = 0.004375 mol

$= \text{mol HBr then } 0.004375\ \text{mol HBr} \times \dfrac{1\ L}{0.150\ \text{mol HBr}} = 0.0291667\ \text{L HBr} \times \dfrac{1000\ mL}{1\ L} = 29.2\ mL\ HBr.$

Check: The units (mL) are correct. The volume of acid is greater than the volume of base because the concentration of the base is a little greater than that of the acid.

(c) **Find:** pH after adding 5.0 mL of acid

Conceptual Plan: Use calculations from part (b). Then mL → L then [HBr], L → mol HBr

$$\frac{1\ L}{1000\ mL} \qquad M = \frac{mol}{L}$$

then mol CH₃NH₂, mol HBr → mol excess CH₃NH₂ and L CH₃NH₂, L HBr → total L.

set up stoichiometry table $\qquad$ L CH₃NH₂ + L HBr = total L

Because there are significant concentrations of both the acid and the conjugate base species, this is a buffer solution; so the Henderson–Hasselbalch equation $\left(pH = pK_a + \log \dfrac{[\text{base}]}{[\text{acid}]} \right)$ can be used.

Convert K_b to K_a using $K_w = K_a K_b$. Also note that the ratio of concentrations is the same as the ratio of moles because the volume is the same for both species.

Solution: $5.0\ \text{mL HBr} \times \dfrac{1\ L}{1000\ mL} = 0.0050\ L\ HBr$ then $\dfrac{0.150\ mol\ HBr}{1\ L} \times 0.0050\ L = 0.00075\ mol\ HBr$

Set up a table to track changes:

	HBr(*aq*)	+	CH₃NH₂(*aq*)	→	CH₃NH₃Br(*aq*)	
Before addition	0.00 mol		0.004375 mol		0.00 mol	
Addition	0.00075 mol		—		—	then $K_w = K_a K_b$ so
After addition	≈0.00 mol		0.003625 mol		0.000750 mol	

$K_a = \dfrac{K_w}{K_b} = \dfrac{1.0 \times 10^{-14}}{4.4 \times 10^{-4}} = 2.2727 \times 10^{-11}$ M. Then use the Henderson–Hasselbalch equation because the

solution is a buffer. $pH = pK_a + \log \dfrac{[\text{base}]}{[\text{acid}]} = -\log (2.2727 \times 10^{-11}) + \log \dfrac{0.003625}{0.000750} = 11.33$

Check: The units (none) are correct. The pH is a little lower than the last pH, which is expected because some of the base has been neutralized.

(d) **Find:** pH at one-half the equivalence point

Conceptual Plan: Because this is a weak base–strong acid titration, the pH at one-half the equivalence point is the pK_a of the conjugate acid of weak base.

Solution: $pH = pK_a = -\log K_a = -\log (2.2727 \times 10^{-11}) = 10.64$

Check: The units (none) are correct. Because this is a weak acid–strong base titration, the pH at one-half the equivalence point is the pK_a of the conjugate acid of the weak base; so it should be a little below 11.

(e) **Find:** pH at equivalence point

Conceptual Plan: Use previous calculations. Because all of the weak base has been converted to its conjugate acid, the pH is only dependent on the hydrolysis reaction of the conjugate acid. The mol CH₃NH₃⁺ = initial mol CH₃NH₂ and

L CH₃NH₂, L HBr to equivalence point → total L then mol CH₃NH₃⁺, L → [CH₃NH₃⁺]

L CH₃NH₂ + L HBr = total L $\qquad M = \dfrac{mol}{L}$

then do an equilibrium calculation: $[CH_3NH_3^+], K_a \rightarrow [H_3O^+] \rightarrow pH$

set up ICE table $pH = -\log[H_3O^+]$

Solution: mol base $=$ mol acid $=$ mol $CH_3NH_3^+ = 0.004375$ mol. Then total volume $=$ L $CH_3NH2 +$

L HBr $= 0.0250$ L $+ 0.0292$ L $= 0.0542$ then $[CH_3NH_3^+] = \dfrac{0.004375 \text{ mol } CH_3NH_3^+}{0.0542 \text{ L}} = 0.0807196$ M. Set

up an ICE table:

$$CH_3NH_3^+(aq) + H_2O(l) \rightleftharpoons CH_3NH_2(aq) + H_3O^+(aq)$$

	$[CH_3NH_3^+]$	$[CH_3NH_2]$	$[H_3O^+]$
Initial	0.0807196	≈ 0.00	≈ 0.00
Change	$-x$	$+x$	$+x$
Equil	$0.0807196 - x$	$+x$	$+x$

$K_a = \dfrac{[CH_3NH_2][H_3O^+]}{[CH_3NH_3^+]} = 2.2727 \times 10^{-11} = \dfrac{x^2}{0.0807196 - x}$ Assume that x is small ($x << 0.0807$), so

$\dfrac{x^2}{0.0807196 - \cancel{x}} = 2.2727 \times 10^{-11} = \dfrac{x^2}{0.0807196}$ and $x = 1.3544 \times 10^{-6} = [H_3O^+]$.

Confirm that the assumption is valid. $\dfrac{1.3544 \times 10^{-6}}{0.0807106} \times 100\% = 0.0017\% < 5\%$, so the assumption is valid.

Finally, $pH = -\log[H_3O^+] = -\log(1.3544 \times 10^{-6}) = 5.87$.

Check: The units (none) are correct. Because this is a weak base–strong acid titration, the pH at the equivalence point is acidic.

(f)   **Find:** pH after adding 5.0 mL of acid beyond the equivalence point

Conceptual Plan: Use calculations from parts (b) and (c). Then the pH is only dependent on the amount of excess acid and the total solution volumes.

mL excess $\rightarrow$ L excess then [HBr], L excess $\rightarrow$ mol HBr excess

$\frac{1 \text{ L}}{1000 \text{ mL}}$ $M = \frac{mol}{L}$

then L CH_3NH_2, L HBr to equivalence point, L HBr excess $\rightarrow$ total L then

L CH_3NH_2 + L HBr to equivalence point + L HBr excess = total L

mol excess HBr, total L $\rightarrow [HBr] = [H_3O^+] \rightarrow pH$

$M = \frac{mol}{L}$ $pH = -\log[H_3O^+]$

Solution: $5.0 \cancel{\text{ mL HBr}} \times \dfrac{1 \text{ L}}{1000 \cancel{\text{ mL}}} = 0.0050$ L HBr excess then

$\dfrac{0.150 \text{ mol HBr}}{1 \cancel{L}} \times 0.0050 \cancel{L} = 0.00075$ mol HBr excess. Then 0.0250 L $CH_3NH_2 + 0.0292$ L HBr $+$

0.0050 L HBr $= 0.0592$ L total volume.

$[HBr \text{ excess}] = \dfrac{0.00075 \text{ mol HBr excess}}{0.0592 \text{ L}} = 0.012669$ M HBr excess

Because HBr is a strong acid, $[HBr]$ excess $= [H_3O^+]$. The strong acid overwhelms the weak acid which is insignificant in the calculation. Finally, $pH = -\log[H_3O^+] = -\log(0.012669) = 1.90$.

Check: The units (none) are correct. The pH is dropping sharply at the equivalence point, so the pH after 5 mL past the equivalence point should be quite acidic.

16.75   (i)   Acid a is more concentrated because the equivalence point (where sharp pH rise occurs) is at a higher volume of added base.

       (ii)   Acid b has the larger K_a because the pH at a volume of added base equal to half the equivalence point volume is lower.

16.77 **Given:** 0.229 g unknown monoprotic acid titrated with 0.112 M NaOH and curve
Find: molar mass and pK_a of acid
Conceptual Plan: The equivalence point is where sharp pH rise occurs. The pK_a is the pH at a volume of added base equal to half the equivalence point volume. Then mL NaOH → L NaOH

$$\frac{1\,L}{1000\,mL}$$

then [NaOH], L NaOH → mol NaOH = mol acid then mol acid, g acid → molar mass.

$$M = \frac{mol}{L}$$
$$\frac{g\ acid}{mol\ acid}$$

Solution: The equivalence point is at 25 mL NaOH. The pH at 0.5 × 25 mL = 13 mL is ~3 = pK_a. Then

$$25\ \text{mL NaOH} \times \frac{1\,L}{1000\,mL} = 0.025\ \text{L NaOH then}$$

$$\frac{0.112\ \text{mol NaOH}}{1\,L} \times 0.025\,L = 0.0028\ \text{mol NaOH} = 0.0028\ \text{mol acid then}$$

$$\text{molar mass} = \frac{0.229\ \text{g acid}}{0.0028\ \text{mol acid}} = 82\ \text{g/mol}$$

Check: The units (none and g/mol) are correct. The pK_a is consistent with a weak acid. The molar mass is reasonable for an acid (> 1 g/mol).

16.79 **Given:** 20.0 mL of 0.115 M sulfurous acid (H_2SO_3) titrated with 0.1014 M KOH **Find:** volume of base added
Conceptual Plan: Because this is a diprotic acid, each proton is titrated sequentially. Write balanced equations.

$$H_2SO_3 + OH^- \rightarrow HSO_3^- + H_2O \text{ and } HSO_3^- + OH^- \rightarrow SO_3^{2-} + H_2O$$

Then mL → L then [H_2SO_3], L → mol H_2SO_3 then set mol base (H_2SO_3) = mol acid (KOH) and

$$\frac{1\,L}{1000\,mL}$$
$$M = \frac{mol}{L}$$
balanced equation has 1:1 stoichiometry (1st equivalence point)

[KOH], mol KOH → L KOH → mL KOH the volume to the second equivalence point will be

$$M = \frac{mol}{L} \qquad \frac{1000\,mL}{1\,L}$$

double the volume to the first equivalence point.
Solution: $20.0\ \text{mL } H_2SO_3 \times \dfrac{1\,L}{1000\,mL} = 0.0200\ \text{L } H_2SO_3$ then

$$\frac{0.115\ \text{mol } H_2SO_3}{1\,L} \times 0.0200\,L = 0.00230\ \text{mol } H_2SO_3. \text{ So mol base} = \text{mol } H_2SO_3 = 0.00230\ \text{mol} = \text{mol KOH}$$

then $0.00230\ \text{mol KOH} \times \dfrac{1\,L}{0.1014\ \text{mol KOH}} = 0.02268245\ \text{L KOH} \times \dfrac{1000\,mL}{1\,L} = 22.7\ \text{mL KOH to first}$

equivalence point. The volume to the second equivalence point is simply twice this amount, or 45.4 mL, to the second equivalence point.

Check: The units (mL) are correct. The volume of base is a greater than the volume of acid because the concentration of the acid is a little greater that of the base. The volume to the second equivalence point is twice the volume to the first equivalence point.

16.81 The indicator will be in its acid form at an acidic pH, so the color in the HCl sample will be red. The color change will occur over the pH range from pH = pK_a − 1.0 to pH = pK_a + 1.0, so the color will start to change at pH = 5.0 − 1.0 = 4.0 and finish changing by pH = 5.0 + 1.0 = 6.0.

16.83 Because the exact conditions of the titration are not given, a rough calculation will suffice. Recall that at the equivalence point, the moles of acid and base are equal. If it is assumed that the concentrations of the acid and the base are equal, the total volume of the solution will have doubled. Assuming an acid and base concentration of 0.1 M, the conjugate based formed must have a concentration of ~0.05 M. From earlier calculations, it can be seen that

the $K_b = \dfrac{K_w}{K_a} = \dfrac{[OH^-]^2}{0.05}$; thus,

$$[OH^-] = \sqrt{\frac{0.05\,K_w}{K_a}} = \sqrt{\frac{5 \times 10^{-16}}{K_a}} \text{ and the pH} = 14 + \log\sqrt{\frac{5 \times 10^{-16}}{K_a}}.$$

(a) For HF, the $K_a = 3.5 \times 10^{-4}$; so the above equation approximates the pH at the equivalence point of ~ 8.0. Looking at Table 16.1, phenol red or m-nitrophenol will change at the appropriate pH range.

(b) For HCl, the pH at the equivalence point is 7 because HCl is a strong acid. Looking at Table 16.1, alizarin, bromthymol blue, m-nitrophenol, or phenol red will change at the appropriate pH range.

(c) For HCN, the $K_a = 4.9 \times 10^{-10}$; so the preceding equation approximates the pH at the equivalence point of ~ 11.0. Looking at Table 16.1, alizarin yellow R will change at the appropriate pH range.

Solubility Equilibria

16.85   For the dissolution reaction, start with the ionic compound as a solid and put it in equilibrium with the appropriate cation and anion, making sure to include the appropriate stoichiometric coefficients. The K_{sp} expression is the product of the concentrations of the cation and anion concentrations raised to their stoichiometric coefficients.

(a) $BaSO_4(s) \rightleftharpoons Ba^{2+}(aq) + SO_4^{2-}(aq)$ and $K_{sp} = [Ba^{2+}][SO_4^{2-}]$

(b) $PbBr_2(s) \rightleftharpoons Pb^{2+}(aq) + 2\,Br^-(aq)$ and $K_{sp} = [Pb^{2+}][Br^-]^2$

(c) $Ag_2CrO_4(s) \rightleftharpoons 2\,Ag^+(aq) + CrO_4^{2-}(aq)$ and $K_{sp} = [Ag^+]^2[CrO_4^{2-}]$

16.87   **Given:** ionic compound formula and Table 16.2 of K_{sp} values   **Find:** molar solubility (S)
Conceptual Plan: Because the balanced chemical equation for the dissolution of A_mX_n is $A_mX_n(s) \rightleftharpoons mA^{n+}(aq) + nX^{m-}(aq)$, the expression of the solubility product constant of A_mX_n is $K_{sp} = [A^{n+}]^m[X^{m-}]^n$. The molar solubility of a compound, A_mX_n, can be computed directly from K_{sp} by solving for S in the expression $K_{sp} = (mS)^m(nS)^n = m^m n^n S^{m+n}$.
Solution:

(a) For AgBr, $K_{sp} = 5.35 \times 10^{-13}$, $A = Ag^+$, $m = 1$, $X = Br^-$, and $n = 1$; so $K_{sp} = 5.35 \times 10^{-13} = S^2$.

Rearrange to solve for S. $S = \sqrt{5.35 \times 10^{-13}} = 7.31 \times 10^{-7}\,M$

(b) For $Mg(OH)_2$, $K_{sp} = 2.06 \times 10^{-13}$, $A = Mg^{2+}$, $m = 1$, $X = OH^-$, and $n = 2$; so $K_{sp} = 2.06 \times 10^{-13} = 2^2 S^3$.

Rearrange to solve for S. $S = \sqrt[3]{\dfrac{2.06 \times 10^{-13}}{4}} = 3.72 \times 10^{-5}\,M$

(c) For CaF_2, $K_{sp} = 1.46 \times 10^{-10}$, $A = Ca^{2+}$, $m = 1$, $X = F^-$, and $n = 2$; so $K_{sp} = 1.46 \times 10^{-10} = 2^2 S^3$.

Rearrange to solve for S. $S = \sqrt[3]{\dfrac{1.46 \times 10^{-10}}{4}} = 3.32 \times 10^{-4}\,M$

Check: The units (M) are correct. The molar solubilities are much less than 1 and dependent not only on the value of the K_{sp}, but also on the stoichiometry of the ionic compound. The more ions generated, the greater the molar solubility for the same value of the K_{sp}.

16.89   **Given:** ionic compound formula and molar solubility (S)   **Find:** K_{sp}
Conceptual Plan: The expression of the solubility product constant of A_mX_n is $K_{sp} = [A^{n+}]^m[X^{m-}]^n$. The molar solubility of a compound, A_mX_n, can be computed directly from K_{sp} by solving for S in the expression $K_{sp} = (mS)^m(nS)^n = m^m n^n S^{m+n}$.

Solution:

(a) For MX, $S = 3.27 \times 10^{-11}\,M$, $A = M^+$, $m = 1$, $X = X^-$, and $n = 1$; so $K_{sp} = S^2 = (3.27 \times 10^{-11})^2 = 1.07 \times 10^{-21}$.

(b) For PbF_2, $S = 5.63 \times 10^{-3}\,M$, $A = Pb^{2+}$, $m = 1$, $X = F^-$, and $n = 2$; so $K_{sp} = 2^2 S^3 = 2^2 (5.63 \times 10^{-3})^3 = 7.14 \times 10^{-7}$.

(d) For MgF_2, $S = 2.65 \times 10^{-4}\,M$, $A = Mg^{2+}$, $m = 1$, $X = F^-$, and $n = 2$; so $K_{sp} = 2^2 S^3 = 2^2 (2.65 \times 10^{-4})^3 = 7.44 \times 10^{-11}$.

Check: The units (none) are correct. The K_{sp} values are much less than 1 and dependent not only on the value of the solubility, but also on the stoichiometry of the ionic compound. The more ions generated, the smaller the K_{sp} for the same value of the S.

16.91 **Given:** ionic compound formulas AX and AX_2 and $K_{sp} = 1.5 \times 10^{-5}$ **Find:** higher molar solubility (S)
Conceptual Plan: The expression of the solubility product constant of A_mX_n is $K_{sp} = [A^{n+}]^m[X^{m-}]^n$.
The molar solubility of a compound, A_mX_n, can be computed directly from K_{sp} by solving for S in the
expression $K_{sp} = (mS)^m(nS)^n = m^m n^n S^{m+n}$.

Solution: For AX, $K_{sp} = 1.5 \times 10^{-5}$, $m = 1$, and $n = 1$; so $K_{sp} = 1.5 \times 10^{-5} = S^2$. Rearrange to solve for S.

$S = \sqrt{1.5 \times 10^{-5}} = 3.9 \times 10^{-3}$ M. For AX_2, $K_{sp} = 1.5 \times 10^{-5}$, $m = 1$, and $n = 2$; so $K_{sp} = 1.5 \times 10^{-5} = 2^2 S^3$.

Rearrange to solve for S. $S = \sqrt[3]{\dfrac{1.5 \times 10^{-5}}{4}} = 1.6 \times 10^{-2}$ M. Because 10^{-2} M $> 10^{-3}$ M, AX_2 has a higher molar

solubility.

Check: The units (M) are correct. The more ions generated, the greater the molar solubility for the same value of
the K_{sp}.

16.93 **Given:** $Fe(OH)_2$ in 100.0 mL solution **Find:** grams of $Fe(OH)_2$ **Other:** $K_{sp} = 4.87 \times 10^{-17}$
Conceptual Plan: The expression of the solubility product constant of A_mX_n is $K_{sp} = [A^{ns}]^m[X^{m-}]^n$. The
molar solubility of a compound, A_mX_n, can be computed directly from K_{sp} by solving for S in the expression
$K_{sp} = (mS)^m(nS)^n = m^m n^n S^{m+n}$. Then solve for S, then mL $\rightarrow$ L then

$$\frac{1\,L}{1000\,mL}$$

S, L $\rightarrow$ mol $Fe(OH)_2$ $\rightarrow$ g $Fe(OH)_2$.

$$M = \frac{mol}{L} \quad \frac{89.87\,g\,Fe(OH)_2}{1\,mol\,Fe(OH)_2}$$

Solution: For $Fe(OH)_2$, $K_{sp} = 4.87 \times 10^{-17}$, A $= Fe^{2+}$, $m = 1$, X $= OH^-$, and $n = 2$; so $K_{sp} = 4.87 \times 10^{-17} = 2^2 S^3$.

Rearrange to solve for S. $S = \sqrt[3]{\dfrac{4.87 \times 10^{-17}}{4}} = 2.3\underline{0}050 \times 10^{-6}$ M. Then $100.0\,\cancel{mL} \times \dfrac{1\,L}{1000\,\cancel{mL}} = 0.1000$ L

then $\dfrac{2.3\underline{0}050 \times 10^{-6}\,mol\,Fe(OH)_2}{1\,\cancel{L}} \times 0.1000\,\cancel{L} = 2.3\underline{0}050 \times 10^{-7}\,\cancel{mol\,Fe(OH)_2} \times \dfrac{89.87\,g\,Fe(OH)_2}{1\,\cancel{mol\,Fe(OH)_2}}$

$= 2.07 \times 10^{-5}$ g $Fe(OH)_2$.

Check: The units (g) are correct. The solubility rules from Chapter 4 (most hydroxides are insoluble) suggest that
very little $Fe(OH)_2$ will dissolve; so the magnitude of the answer is not surprising.

16.95 (a) **Given:** BaF_2 **Find:** molar solubility (S) in pure water **Other:** $K_{sp}(BaF_2) = 2.45 \times 10^{-5}$
Conceptual Plan: The expression of the solubility product constant of A_mX_n is $K_{sp} = [A^{n+}]^m[X^{m-}]^n$. The
molar solubility of a compound, A_mX_n, can be computed directly from K_{sp} by solving for S in the expression $K_{sp} = (mS)^m(nS)^n = m^m n^n S^{m+n}$.

Solution: For BaF_2, $K_{sp} = 2.45 \times 10^{-5}$, A $= Ba^{2+}$, $m = 1$, X $= F^-$, and $n = 2$; so $K_{sp} = 2.45 \times 10^{-5} = 2^2 S^3$.

Rearrange to solve for S. $S = \sqrt[3]{\dfrac{2.45 \times 10^{-5}}{4}} = 1.83 \times 10^{-2}$ M

(b) **Given:** BaF_2 **Find:** molar solubility (S) in 0.10 M $Ba(NO_3)_2$ **Other:** $K_{sp}(BaF_2) = 2.45 \times 10^{-5}$
Conceptual Plan: M $Ba(NO_3)_2 \rightarrow$ M Ba^{2+} then M Ba^{2+}, $K_{sp} \rightarrow$ S

$$Ba(NO_3)_2(s) \rightarrow Ba^{2+}(aq) + 2\,NO_3^-(aq) \qquad \text{ICE table}$$

Solution: Because one Ba^{2+} ion is generated for each $Ba(NO_3)_2$, $[Ba^{2+}] = 0.10$ M.

$BaF_2(s) \rightleftharpoons$	$Ba^{2+}(aq)$	$+\ 2\,F^-(aq)$
Initial	0.10	0.00
Change	S	$2S$
Equil	$0.10 + S$	$2S$

$K_{sp}(BaF_2) = [Ba^{2+}][F^-]^2 = 2.45 \times 10^{-5} = (0.10 + S)(2S)^2$

Assume that $S << 0.10$, $2.45 \times 10^{-5} = (0.10)(2S)^2$, and $S = 7.83 \times 10^{-3}$ M. Confirm that the assumption is valid. $\dfrac{7.83 \times 10^{-3}}{0.10} \times 100\% = 7.8\% > 5\%$, so the assumption is not valid. Because expanding the expression will give a third-order polynomial, that is not easily solved directly. Solve by successive approximations. Substitute $S = 7.83 \times 10^{-3}$ M for the S term that is part of a sum [i.e., the one in $(0.10 + S)$]. Thus, $2.45 \times 10^{-5} = (0.10 + 7.83 \times 10^{-3})(2S)^2$ and $S = 7.53 \times 10^{-3}$ M. Substitute this new S value again. Thus, $2.45 \times 10^{-5} = (0.10 + 7.53 \times 10^{-3})(2S)^2$ and $S = 7.55 \times 10^{-3}$ M. Substitute this new S value again. Thus, $2.45 \times 10^{-5} = (0.10 + 7.55 \times 10^{-3})(2S)^2$ and $S = 7.55 \times 10^{-3}$ M. So the solution has converged and $S = 7.55 \times 10^{-3}$ M.

(c) **Given:** BaF_2 **Find:** molar solubility (S) in 0.15 M NaF **Other:** $K_{sp}(BaF_2) = 2.45 \times 10^{-5}$
Conceptual Plan: M NaF $\rightarrow$ **M F⁻** then **M F⁻**, $K_{sp} \rightarrow S$

$$NaF(s) \rightarrow Na^+(aq) + F^-(aq) \qquad \text{ICE table}$$

Solution: Because one F^- ion is generated for each NaF, $[F^-] = 0.15$ M.

$BaF_2(s) \rightleftharpoons$	$Ba^{2+}(aq)$	$+\ 2\,F^-(aq)$
Initial	0.00	0.15
Change	S	$2S$
Equil	S	$0.15 + 2S$

$$K_{sp}(BaF_2) = [Ba^{2+}][F^-]^2 = 2.45 \times 10^{-5} = (S)(0.15 + 2S)^2$$

Because $2S << 0.15$, $2.45 \times 10^{-5} = (S)(0.15)^2$ and $S = 1.09 \times 10^{-3}$ M. Confirm that the assumption is valid. $\dfrac{2(1.09 \times 10^{-3})}{0.15} \times 100\% = 1.5\% < 5\%$, so the assumption is valid.

Check: The units (M) are correct. The solubility of the BaF_2 decreases in the presence of a common ion. The effect of the anion is greater because the K_{sp} expression has the anion concentration squared.

16.97 **Given:** $Ca(OH)_2$ **Find:** molar solubility (S) in buffers at (a) pH = 4, (b) pH = 7, and (c) pH = 9
Other: $K_{sp}(Ca(OH)_2) = 4.68 \times 10^{-6}$
Conceptual Plan: pH $\rightarrow$ **[H₃O⁺]** $\rightarrow$ **[OH⁻]** then **M OH⁻**, $K_{sp} \rightarrow S$

$$[H_3O^+] = 10^{-pH} \qquad K_w = [H_3O^+][OH^-] \qquad \text{set up ICE table}$$

Solution:
(a) pH = 4, so $[H_3O^+] = 10^{-pH} = 10^{-4} = 1 \times 10^{-4}$ M then $K_w = [H_3O^+][OH^-]$ so

$$[OH^-] = \frac{K_w}{[H_3O^+]} = \frac{1.0 \times 10^{-14}}{1 \times 10^{-4}} = 1 \times 10^{-10}\ \text{M then}$$

$Ca(OH)_2(s) \rightleftharpoons$	$Ca^{2+}(aq)$	$+\ 2\,OH^-(aq)$
Initial	0.00	1×10^{-10}
Change	S	—
Equil	S	1×10^{-10}

$$K_{sp}(Ca(OH)_2) = [Ca^{2+}][OH^-]^2 = 4.68 \times 10^{-6} = S(1 \times 10^{-10})^2 \text{ and } S = 5 \times 10^{14}\ \text{M}$$

(b) pH = 7, so $[H_3O^+] = 10^{-pH} = 10^{-7} = 1 \times 10^{-7}$ M then $K_w = [H_3O^+][OH^-]$ so

$$[OH^-] = \frac{K_w}{[H_3O^+]} = \frac{1.0 \times 10^{-14}}{1 \times 10^{-7}} = 1 \times 10^{-7}\ \text{M then}$$

$Ca(OH)_2(s) \rightleftharpoons$	$Ca^{2+}(aq)$	$+\ 2\,OH^-(aq)$
Initial	0.00	1×10^{-7}
Change	S	—
Equil	S	1×10^{-7}

$$K_{sp}(Ca(OH)_2) = [Ca^{2+}][OH^-]^2 = 4.68 \times 10^{-6} = S(1 \times 10^{-7})^2 \text{ and } S = 5 \times 10^{8}\ \text{M}$$

(c)   pH = 9, so $[H_3O^+] = 10^{-pH} = 10^{-9} = 1 \times 10^{-9}$ M then $K_w = [H_3O^+][OH^-]$ so

$$[OH^-] = \frac{K_w}{[H_3O^+]} = \frac{1.0 \times 10^{-14}}{1 \times 10^{-9}} = 1 \times 10^{-5} \text{ M then}$$

$$Ca(OH)_2(s) \rightleftharpoons Ca^{2+}(aq) + 2\,OH^-(aq)$$

Initial	0.00	1×10^{-5}
Change	S	—
Equil	S	1×10^{-5}

$$K_{sp}(Ca(OH)_2) = [Ca^{2+}][OH^-]^2 = 4.68 \times 10^{-6} = S(1 \times 10^{-5})^2 \text{ and } S = 5 \times 10^4 \text{ M}$$

Check: The units (M) are correct. The solubility of the $Ca(OH)_2$ decreases as the pH increases (and the hydroxide ion concentration increases). These molar solubilities are not achievable because the saturation point of pure $Ca(OH)_2$ is ~30 M. The bottom line is that as long as the hydroxide concentration can be controlled with a buffer, the $Ca(OH)_2$ will be very soluble.

16.99   (a) $BaCO_3$ will be more soluble in acidic solutions because CO_3^{2-} is basic. In acidic solutions, it can be converted to HCO_3^- and H_2CO_3. These species are not CO_3^{2-}, so they do not appear in the K_{sp} expression.

(b) CuS will be more soluble in acidic solutions because S^{2-} is basic. In acidic solutions, it can be converted to HS^- and H_2S. These species are not S^{2-}, so they do not appear in the K_{sp} expression.

(c) $AgCl$ will not be more soluble in acidic solutions because Cl^- will not react with acidic solutions because HCl is a strong acid.

(d) PbI_2 will not be more soluble in acidic solutions because I^- will not react with acidic solutions because HI is a strong acid.

Precipitation and Qualitative Analysis

16.101   **Given:** 0.015 M NaF and 0.010 M $Ca(NO_3)_2$ **Find:** Will a precipitate form? If so, identify it.
Other: $K_{sp}(CaF_2) = 1.46 \times 10^{-10}$
Conceptual Plan: Look at all possible combinations and consider the solubility rules from Chapter 4. Salts of alkali metals (Na) are very soluble, so NaF and $NaNO_3$ will be very soluble. Nitrate compounds are very soluble, so $NaNO_3$ will be very soluble. The only possibility for a precipitate is CaF_2. Determine whether a precipitate will form by determining the concentration of the Ca^{2+} and F^- in solution. Then compute the reaction quotient, Q. If $Q > K_{sp}$, a precipitate will form.
Solution: Because the only possible precipitate is CaF_2, calculate the concentrations of Ca^{2+} and F^-. $NaF(s) \rightarrow Na^+(aq) + F^-(aq)$. Because one F^- ion is generated for each NaF, $[F^-] = 0.015$ M.
$Ca(NO_3)_2(s) \rightarrow Ca^{2+}(aq) + 2\,NO_3^-(aq)$. Because one Ca^{2+} ion is generated for each $Ca(NO_3)_2$, $[Ca^{2+}] = 0.010$ M.
Then calculate $Q(CaF_2)$, $A = Ca^{2+}$, $m = 1$, $X = F^-$, and $n = 2$. Because $Q = [A^{n+}]^m[X^{m-}]^n$,
$Q(CaF_2) = [Ca^{2+}][F^-]^2 = (0.010)(0.015)^2 = 2.3 \times 10^{-6} > 1.46 \times 10^{-10} = K_{sp}(CaF_2)$; so a precipitate will form.

Check: The units (none) are correct. The solubility of the CaF_2 is low, and the concentration of ions is extremely large compared to the K_{sp}; so a precipitate will form.

16.103   **Given:** 75.0 mL of NaOH with pOH = 2.58 and 125.0 mL of 0.0018 M $MgCl_2$
Find: Will a precipitate form? If so, identify it.   **Other:** $K_{sp}(Mg(OH)_2) = 2.06 \times 10^{-13}$
Conceptual Plan: Look at all possible combinations and consider the solubility rules from Chapter 4. Salts of alkali metals (Na) are very soluble, so NaOH and NaCl will be very soluble. Chloride compounds are generally very soluble, so $MgCl_2$ and NaCl will be very soluble. The only possibility for a precipitate is $Mg(OH)_2$. Determine whether a precipitate will form by determining the concentration of the Mg^{2+} and OH^- in solution. Because pOH, not NaOH concentration, is given, pOH → $[OH^-]$ then

$$[OH^-] = 10^{-pOH}$$

mix solutions and calculate diluted concentrations mL NaOH, mL $MgCl_2$ → mL total then

$$\text{mL NaOH} + \text{mL MgCl}_2 = \text{total mL}$$

mL, initial M → final M then compute the reaction quotient, Q.

$$M_1V_1 = M_2V_2$$

If $Q > K_{sp}$, a precipitate will form.

Solution: Because the only possible precipitate is $Mg(OH)_2$, calculate the concentrations of Mg^{2+} and OH^-. For NaOH at pOH = 2.58, so $[OH^-] = 10^{-pOH} = 10^{-2.58} = 2.\underline{6}3027 \times 10^{-3}$ M and $MgCl_2(s) \rightarrow Mg^{2+}(aq) + 2\,Cl^-(aq)$. Because one Mg^{2+} ion is generated for each $MgCl_2$, $[Mg^{2+}] = 0.0018$ M. Then total mL = mL NaOH + mL $MgCl_2$ = 75.0 mL + 125.0 mL = 200.0 mL. Then $M_1V_1 = M_2V_2$; rearrange to solve for M_2. $M_2 = M_1\dfrac{V_1}{V_2} = 2.\underline{6}3027 \times 10^{-3}$ M $OH^- \times \dfrac{75.0\ \text{mL}}{200.0\ \text{mL}} = 9.\underline{8}635 \times 10^{-4}$ M OH^- and

$M_2 = M_1\dfrac{V_1}{V_2} = 0.0018$ M $Mg^+ \times \dfrac{125.0\ \text{mL}}{200.0\ \text{mL}} = 1.\underline{1}25 \times 10^{-3}$ M Mg^{2+}. Calculate $Q(Mg(OH)_2)$, A = Mg^{2+},

$m = 1$, X = OH^-, and $n = 2$. Because $Q = [A^{n+}]^m\,[X^{m-}]^n$, $Q(Mg(OH)_2) = [Mg^{2+}][OH^-]^2 =$

$(1.\underline{1}25 \times 10^{-3})(9.\underline{8}635 \times 10^{-4})^2 = 1.1 \times 10^{-9} > 2.06 \times 10^{-13} = K_{sp}(Mg(OH)_2)$; so a precipitate will form.

Check: The units (none) are correct. The solubility of the $Mg(OH)_2$ is low, and the NaOH (a base) is high enough that the product of the concentration of ions is large compared to the K_{sp}; so a precipitate will form.

16.105 **Given:** KOH as precipitation agent in (a) 0.015 M $CaCl_2$, (b) 0.0025 M $Fe(NO_3)_2$, and (c) 0.0018 M $MgBr_2$
Find: concentration of KOH necessary to form a precipitate
Other: $K_{sp}(Ca(OH)_2) = 4.68 \times 10^{-6}$, $K_{sp}(Fe(OH)_2) = 4.87 \times 10^{-17}$, $K_{sp}(Mg(OH)_2) = 2.06 \times 10^{-13}$
Conceptual Plan: The solubility rules from Chapter 4 state that most hydroxides are insoluble, so all precipitates will be hydroxides. Determine the concentration of the cation in solution. Because all metals have an oxidation

state of $+2$ and $[OH^-] = [KOH]$, all of the $K_{sp} = [\text{cation}][\text{KOH}]^2$; so $[\text{KOH}] = \sqrt{\dfrac{K_{sp}}{[\text{cation}]}}$.

Solution:
(a) $CaCl_2(s) \rightarrow Ca^{2+}(aq) + 2\,Cl^-(aq)$. Because one Ca^{2+} ion is generated for each $CaCl_2$, $[Ca^{2+}] = 0.015$ M.

Then $[KOH] = \sqrt{\dfrac{K_{sp}}{[\text{cation}]}} = \sqrt{\dfrac{4.68 \times 10^{-6}}{0.015}} = 0.018$ M KOH.

(b) $Fe(NO_3)_2(s) \rightarrow Fe^{2+}(aq) + 2\,NO_3^-(aq)$. Because one Fe^{2+} ion is generated for each $Fe(NO_3)_2$, $[Fe^{2+}] = 0.0025$ M.

Then $[KOH] = \sqrt{\dfrac{K_{sp}}{[\text{cation}]}} = \sqrt{\dfrac{4.87 \times 10^{-17}}{0.0025}} = 1.4 \times 10^{-7}$ M KOH.

(c) $MgBr_2(s) \rightarrow Mg^{2+}(aq) + 2\,Br^-(aq)$. Because one Mg^{2+} ion is generated for each $MgBr_2$, $[Mg^{2+}] = 0.0018$ M.

Then $[KOH] = \sqrt{\dfrac{K_{sp}}{[\text{cation}]}} = \sqrt{\dfrac{2.06 \times 10^{-13}}{0.0018}} = 1.1 \times 10^{-5}$ M KOH.

Check: The units (none) are correct. Because all cations have an oxidation state of $+2$, it can be seen that the [KOH] needed to precipitate the hydroxide is lower the smaller the K_{sp}.

16.107 **Given:** solution with 0.010 M Ba^{2+} and 0.020 M Ca^{2+}, add Na_2SO_4 to form precipitates
Find: (a) which ion precipitates first and minimum $[Na_2SO_4]$ needed and (b) [first cation] when second cation precipitates **Other:** $K_{sp}(BaSO_4) = 1.07 \times 10^{-10}$, $K_{sp}(CaSO_4) = 7.10 \times 10^{-5}$
Conceptual Plan: (a) The precipitates that will form are $BaSO_4$ and $CaSO_4$.
Use the equation derived in Problem 16.19 to define K_{sp}. Substitute concentration of cation and solve for

for ionic compound, A_mX_n, $K_{sp} = [A^{n+}]^m[X^{m-}]^n$

concentration of anion to form precipitate. The cation with the lower anion concentration will precipitate first.
(b) Substitute the higher anion concentration into the K_{sp} expression for the first cation to precipitate and calculate the amount of this first cation to remain in solution.
Solution:
(a) Derive expression for $K_{sp}(BaSO_4)$, A = Ba^{2+}, $m = 1$, X = SO_4^{2-}, and $n = 1$ Because $K_{sp} = [Ba^{2+}][SO_4^{2-}]$,
$K_{sp}(BaSO_4) = 1.07 \times 10^{-10} = 0.010\,[SO_4^{2-}]$. Solve for $[SO_4^{2-}]$. $[SO_4^{2-}] = 1.07 \times 10^{-8}$ M SO_4^{2-}.
Because $Na_2SO_4(s) \rightarrow 2\,Na^+(aq) + SO_4^{2-}(aq)$. Because one SO_4^{2-} ion is generated for each
Na_2SO_4, $[Na_2SO_4] = 1.1 \times 10^{-8}$ M Na_2SO_4 to precipitate $BaSO_4$. Derive expression for
$K_{sp}(CaSO_4)$, A = Ca^{2+}, $m = 1$, X = SO_4^{2-}, and $n = 1$ Because $K_{sp} = [Ca^{2+}][SO_4^{2-}]$,

K_{sp} (CaSO$_4$) $= 7.10 \times 10^{-5} = 0.020\,[SO_4^{2-}]$. Solve for $[SO_4^{2-}]$. $[SO_4^{2-}] = 0.0036$ M SO$_4^{2-} =$ 0.0036 M Na$_2$SO$_4$ $= [Na_2SO_4]$ to precipitate CaSO$_4$. Because 1.1×10^{-8} M Na$_2$SO$_4$ $\ll$ 0.0036 M Na$_2$SO$_4$, the Ba^{2+} will precipitate first.

(b)   Because Ca^{2+} will not precipitate until $[Na_2SO_4] = 0.00355$ M Na$_2$SO$_4$, substitute this value into the K_{sp} expression for BaSO$_4$. So K_{sp} (BaSO$_4$) $= [Ba^{2+}][SO_4^{2-}] = 1.07 \times 10^{-10} = [Ba^{2+}]\,0.0035$. Solve for $[Ba^{2+}] = 3.0 \times 10^{-8}$ M Ba^{2+}.

Check: The units (none, M, and M) are correct. Comparing the two K_{sp} values, it can be seen that the Ba^{2+} will precipitate first because the solubility product is so much lower. Because the K_{sp} value is so low, the concentration of precipitating agent is very low. Because the CaSO$_4$ K_{sp} value is so much higher, the higher $[SO_4^{2-}]$ to precipitate Ca will force the concentration of Ba^{2+} to very low levels.

Complex Ion Equilibria

16.109   **Given:** solution with 1.1×10^{-3} M Zn(NO$_3$)$_2$ and 0.150 M NH$_3$ **Find:** $[Zn^{2+}]$ at equilibrium
Other: $K_f(\text{Zn}(\text{NH}_3)_4{}^{2+}) = 2.8 \times 10^9$
Conceptual Plan: Write a balanced equation and expression for K_f. Use initial concentrations to set up an ICE table. Because the K_f is so large, assume that reaction essentially goes to completion. Solve for $[\mathbf{Zn^{2+}}]$ at equilibrium.
Solution: Zn(NO$_3$)$_2$(s) $\rightarrow$ Zn^{2+}(aq) $+$ 2 NO$_3^-$(aq). Because one Zn^{2+} ion is generated for each Zn(NO$_3$)$_2$, $[Zn^{2+}] = 1.1 \times 10^{-3}$ M. Balanced equation is:

$$\text{Zn}^{2+}(aq) \;+\; 4\,\text{NH}_3(aq) \;\rightleftharpoons\; \text{Zn}(\text{NH}_3)_4{}^{2+}(aq)$$

	$[Zn^{2+}]$	$[NH_3]$	$[\text{Zn}(\text{NH}_3)_4{}^{2+}]$
Initial	1.1×10^{-3}	0.150	0.00
Change	$\approx 1.1 \times 10^{-3}$	$\approx -4(1.1 \times 10^{-3})$	$\approx 1.1 \times 10^{-3}$
Equil	x	0.14$\underline{5}$6	1.1×10^{-3}

Set up an ICE table with initial concentrations. Because K_f is so large and because initially $[NH_3] > 4\,[Zn^{2+}]$, the reaction essentially goes to completion.

Write equilibrium expression and solve for x.

$$K_f = \frac{[\text{Zn}(\text{NH}_3)_4{}^{2+}]}{[Zn^{2+}][NH_3]^4} = 2.8 \times 10^9 = \frac{1.1 \times 10^{-3}}{x\,(0.14\underline{5}6)^4}. \text{ So } x = 8.7 \times 10^{-10} \text{ M Zn}^{2+}. \text{ Because } x \text{ is insignificant compared}$$

to the initial concentration, the assumption is valid.

Check: The units (M) are correct. Because K_f is so large, the reaction essentially goes to completion and $[Zn^{2+}]$ is extremely small.

16.111   **Given:** FeS(s) $+$ 6 CN$^-$(aq) $\rightleftharpoons$ Fe(CN)$_6{}^{4-}$(aq) $+$ S^{2-}(aq) use K_{sp} and K_f values   **Find:** K
Other: K_f (Fe(CN)$_6{}^{4-}$) $= 1.5 \times 10^{35}$, K_{sp} (FeS) $= 3.72 \times 10^{-19}$
Conceptual Plan: Identify the appropriate solid and complex ion. Write balanced equations for dissolving the solid and forming the complex ion. Add these two reactions to get the desired overall reaction. Using the rules from Chapter 14, multiply the individual reaction Ks to get the overall K for the sum of these reactions.
Solution: Identify the solid as FeS and the complex ion as Fe(CN)$_6{}^{4-}$. Write the individual reactions and add them together.

FeS(s) $\rightleftharpoons$ ~~Fe^{2+}(aq)~~ $+$ S^{2-}(aq)	$K_{sp} = 1.5 \times 10^{-19}$
~~Fe^{2+}(aq)~~ $+$ 6 CN$^-$(aq) $\rightleftharpoons$ Fe(CN)$_6{}^{4-}$(aq)	$K_f = 3.72 \times 10^{35}$
FeS(s) $+$ 6 CN$^-$(aq) $\rightleftharpoons$ Fe(CN)$_6{}^{4-}$(aq) $+$ S^{2-}(aq)	

Because the overall reaction is the simple sum of the two reactions, the overall reaction $K = K_f \times K_{sp} =$ $(1.5 \times 10^{35}) \times (3.72 \times 10^{-19}) = 5.6 \times 10^{16}$.

Check: The units (none) are correct. Because K_f is so large, it overwhelms the K_{sp} and the overall reaction is very spontaneous.

Cumulative Problems

16.113 **Given:** 150.0 mL solution of 2.05 g sodium benzoate and 2.47 g benzoic acid   **Find:** pH

Other: $K_a(HC_7H_5O_2) = 6.5 \times 10^{-5}$

Conceptual Plan: g $NaC_7H_5O_2$ → mol $NaC_7H_5O_2$ and g $HC_7H_5O_2$ → mol $HC_7H_5O_2$

$$\frac{1 \text{ mol } NaC_7H_5O_2}{144.10 \text{ g } NaC_7H_5O_2} \qquad \frac{1 \text{ mol } HC_7H_5O_2}{122.12 \text{ g } HC_7H_5O_2}$$

Because the two components are in the same solution, the ratio of [base]/[acid] = (mol base)/(mol acid).
Then K_a, mol $NaC_7H_5O_2$, mol $HC_7H_5O_2$ → pH.

$$pH = pK_a + \log \frac{[\text{base}]}{[\text{acid}]}$$

Solution: $2.05 \text{ g } NaC_7H_5O_2 \times \dfrac{1 \text{ mol } NaC_7H_5O_2}{144.10 \text{ g } NaC_7H_5O_2} = 0.014\underline{2}262 \text{ mol } NaC_7H_5O_2$ and

$2.47 \text{ g } HC_7H_5O_2 \times \dfrac{1 \text{ mol } HC_7H_5O_2}{122.12 \text{ g } HC_7H_5O_2} = 0.020\underline{2}260 \text{ mol } HC_7H_5O_2$ then

$pH = pK_a + \log \dfrac{[\text{base}]}{[\text{acid}]} = pK_a + \log \dfrac{\text{mol base}}{\text{mol acid}} = -\log(6.5 \times 10^{-5}) + \log \dfrac{0.014\underline{2}262 \text{ mol}}{0.020\underline{2}260 \text{ mol}} = 4.03$

Check: The units (none) are correct. The magnitude of the answer makes physical sense because the pH is a little lower than the pK_a of the acid because there is more acid than base in the buffer solution.

16.115 **Given:** 150.0 mL of 0.25 M $HCHO_2$ and 75.0 ml of 0.20 M NaOH   **Find:** pH

Other: $K_a(HCHO_2) = 1.8 \times 10^{-4}$

Conceptual Plan: In this buffer, the base is generated by converting some of the formic acid to the formate ion.
Part I: Stoichiometry:
mL → L then L, initial $HCHO_2$ M → mol $HCHO_2$ then mL → L then

$$\frac{1 \text{ L}}{1000 \text{ mL}} \qquad M = \frac{\text{mol}}{L} \qquad \frac{1 \text{ L}}{1000 \text{ mL}}$$

L, initial NaOH M → mol NaOH then write a balanced equation then

$$M = \frac{\text{mol}}{L} \qquad NaOH + HCHO_2 \rightarrow H_2O + NaCHO_2$$

mol $HCHO_2$, mol NaOH → mol $NaCHO_2$, mol $HCHO_2$ then

set up stoichiometry table

Part II: Equilibrium:
Because the two components are in the same solution, the ratio of [base]/[acid] = (mol base)/(mol acid).
Then K_a, mol $NaCHO_2$, mol $HCHO_2$ → pH.

$$pH = pK_a + \log \frac{[\text{base}]}{[\text{acid}]}$$

Solution: $150.0 \text{ mL} \times \dfrac{1 \text{ L}}{1000 \text{ mL}} = 0.1500 \text{ L}$ then

$0.1500 \text{ L } HCHO_2 \times \dfrac{0.25 \text{ mol } HCHO_2}{1 \text{ L } HCHO_2} = 0.037\underline{5} \text{ mol } HCHO_2$

Then $75.0 \text{ mL} \times \dfrac{1 \text{ L}}{1000 \text{ mL}} = 0.0750 \text{ L}$ then $0.0750 \text{ L } NaOH \times \dfrac{0.20 \text{ mol } NaOH}{1 \text{ L } NaOH} = 0.015 \text{ mol } NaOH$

Set up a table to track changes:

	NaOH(*aq*)	+	$HCHO_2$(*aq*)	→	$NaCHO_2$(*aq*)	+	H_2O(*l*)
Before addition	0.00 mol		0.037$\underline{5}$ mol		0.00 mol		—
Addition	0.015 mol		—		—		—
After addition	≈0.00 mol		0.0225 mol		0.015 mol		—

Because the amount of NaOH is small, there are significant amounts of both buffer components; so the Henderson–Hasselbalch equation can be used to calculate the pH.

$pH = pK_a + \log \dfrac{[\text{base}]}{[\text{acid}]} = pK_a + \log \dfrac{\text{mol base}}{\text{mol acid}} = -\log(1.8 \times 10^{-4}) + \log \dfrac{0.015 \text{ mol}}{0.022\underline{5} \text{ mol}} = 3.57$

Check: The units (none) are correct. The magnitude of the answer makes physical sense because the pH is a little lower than the pK_a of the acid because there is more acid than base in the buffer solution.

16.117 **Given:** 1.0 L of buffer of 0.25 mol NH_3 and 0.25 mol NH_4Cl; adjust to pH $= 8.75$
Find: mass NaOH or HCl   **Other:** $K_b(NH_3) = 1.76 \times 10^{-5}$
Conceptual Plan: To decide which reagent needs to be added to adjust pH, calculate the initial pH. Because the mol NH_3 = mol NH_4Cl, the pH = pK_a so $K_b \rightarrow pK_b \rightarrow pK_a$ then

$$\text{acid} = NH_4^+ \quad \text{base} = NH_3 \qquad pK_b = -\log K_b \quad 14 = pK_a + pK_b$$

final pH, $pK_a \rightarrow [NH_3]/[NH_4^+]$ then $[NH_3], L \rightarrow$ mol $[NH_3]$ and $[NH_4^+], L \rightarrow$ mol $[NH_4^+]$

$$pH = pK_a + \log \frac{[\text{base}]}{[\text{acid}]} \qquad M = \frac{\text{mol}}{L} \qquad M = \frac{\text{mol}}{L}$$

then write a balanced equation then

$$H^+ + NH_3 \rightarrow NH_4^+$$

mol NH_3, mol NH_4^+, $[NH_3]/[NH_4^+] \rightarrow$ mol HCl $\rightarrow$ g HCl.

$$\text{set up stoichiometry table} \qquad \frac{36.46 \text{ g HCl}}{1 \text{ mol HCl}}$$

Solution: Because $K_b(NH_3) = 1.76 \times 10^{-5}$, $pK_b = -\log K_b = -\log(1.76 \times 10^{-5}) = 4.75$. Because $14 = pK_a + pK_b$, $pK_a = 14 - pK_b = 14 - 4.75 = 9.25$. Because the desired pH is lower (8.75), HCl (a strong acid) needs to be added. Then $pH = pK_a + \log \dfrac{[\text{base}]}{[\text{acid}]} = 9.25 + \log \dfrac{[NH_3]}{[NH_4^+]} = 8.75$. Solve for $\dfrac{[NH_3]}{[NH_4^+]}$.

$$\log \frac{[NH_3]}{[NH_4^+]} = 8.75 - 9.25 = -0.50 \rightarrow \frac{[NH_3]}{[NH_4^+]} = 10^{-0.50} = 0.3\underline{1}623 \text{ then}$$

$$\frac{0.25 \text{ mol } NH_3}{1 \text{ L}} \times 1.0 \text{ L} = 0.25 \text{ mol } NH_3 \text{ and}$$

$$\frac{0.25 \text{ mol } NH_4Cl}{1 \text{ L}} \times 1.0 \text{ L} = 0.25 \text{ mol } NH_4Cl = 0.25 \text{ mol } NH_4^+. \text{ Because HCl is a strong acid, } [HCl] = [H^+].$$

Set up a table to track changes:

	$H^+(aq)$	$+$	$NH_3(aq)$	$\rightarrow$	$NH_4^+(aq)$
Before addition	≈ 0.00 mol		0.25 mol		0.25 mol
Addition	x		—		—
After addition	≈ 0.00 mol		$(0.25 - x)$ mol		$(0.25 + x)$ mol

Because $\dfrac{[NH_3]}{[NH_4^+]} = 0.3\underline{1}623 = \dfrac{(0.25 - x) \text{ mol}}{(0.25 + x) \text{ mol}}$, solve for x. Note that the ratio of moles is the same as the ratio of concentrations because the volume for both terms is the same. $0.3\underline{1}623(0.25 + x) = (0.25 - x) \rightarrow 0.0\underline{7}90575 + 0.3\underline{1}623x = 0.25 - x \rightarrow 1.3\underline{1}623x = 0.1\underline{7}094 \rightarrow x = 0.1\underline{2}987$ mol HCl then

$$0.1\underline{2}987 \text{ mol HCl} \times \frac{36.46 \text{ g HCl}}{1 \text{ mol HCl}} = 4.7 \text{ g HCl}$$

Check: The units (g) are correct. The magnitude of the answer makes physical sense because there is much less than a mole of each of the buffer components; so there must be much less than a mole of HCl.

16.119 (a) **Given:** potassium hydrogen phthalate = KHP = $KHC_8H_4O_4$ titration with NaOH   **Find:** balanced equation
Conceptual Plan: The reaction will be a titration of the acid proton, leaving the phthalate ion intact. The K will not be titrated because it is basic.
Solution: $NaOH(aq) + KHC_8H_4O_4(aq) \rightarrow Na^+(aq) + K^+(aq) + C_8H_4O_4^{2-}(aq) + H_2O(l)$

Check: An acid–base reaction generates a salt (soluble here) and water. There is only one acidic proton in KHP.

(b) **Given:** 0.5527 g KHP titrated with 25.87 mL of NaOH solution   **Find:** [NaOH]
Conceptual Plan:
g KHP $\rightarrow$ mol KHP $\rightarrow$ mol NaOH and mL $\rightarrow$ L then mol NaOH and mL $\rightarrow$ M NaOH

$$\frac{1 \text{ mol KHP}}{204.22 \text{ g KHP}} \quad \text{1:1 from balanced equation} \qquad \frac{1 \text{ L}}{1000 \text{ mL}} \qquad M = \frac{\text{mol}}{L}$$

Solution: $0.5527 \, \text{g KHP} \times \dfrac{1 \, \text{mol KHP}}{204.22 \, \text{g KHP}} = 0.002706\underline{3}95 \, \text{mol KHP}$; mol KHP = mol acid = mol base =

$0.002706\underline{3}95 \, \text{mol NaOH}$ then $25.87 \, \text{mL} \times \dfrac{1 \, \text{L}}{1000 \, \text{mL}} = 0.02587 \, \text{L}$ then

$$[\text{NaOH}] = \dfrac{0.002706\underline{3}95 \, \text{mol NaOH}}{0.02587 \, \text{L}} = 0.1046 \, \text{M NaOH}$$

Check: The units (M) are correct. The magnitude of the answer makes physical sense because there is much less than a mole of acid. The magnitude of the moles of acid and base are smaller than the volume of base in liters.

16.121   **Given:** 0.25 mol weak acid with 10.0 mL of 3.00 M KOH diluted to 1.5000 L has pH = 3.85   **Find:** pK_a of acid
Conceptual Plan: mL → L then M KOH, L → mol KOH then write a balanced reaction

$$\dfrac{1 \, \text{L}}{1000 \, \text{mL}} \qquad\qquad M = \dfrac{\text{mol}}{\text{L}} \qquad\qquad \text{KOH} + \text{HA} \rightarrow \text{NaA} + \text{H}_2\text{O}$$

added mol KOH, initial mol acid → equil. mol KOH, equil. mol acid then

set up stoichiometry table

equil. mol KOH, equil. mol acid, pH → pK_a

$$\text{pH} = pK_a + \log \dfrac{[\text{base}]}{[\text{acid}]}$$

Solution: $10.00 \, \text{mL} \times \dfrac{1 \, \text{L}}{1000 \, \text{mL}} = 0.01000 \, \text{L}$ then $0.01000 \, \text{L KOH} \times \dfrac{3.00 \, \text{mol KOH}}{1 \, \text{L KOH}} = 0.0300 \, \text{mol KOH}$

Because KOH is a strong base, $[\text{KOH}] = [\text{OH}^-]$. Set up a table to track changes:

	KOH(aq)	+	HA(aq)	→	KA(aq)	+	H$_2$O(l)
Before addition	≈0.00 mol		0.25 mol		0.00 mol		—
Addition	0.0300 mol		—		—		—
After addition	≈0.00 mol		0.22 mol		0.0300 mol		—

Because the ratio of base to acid is between 0.1 and 10, it is a buffer solution. Note that the ratio of moles is the same as the ratio of concentrations because the volume for both terms is the same.

$$\text{pH} = pK_a + \log \dfrac{[\text{base}]}{[\text{acid}]} = pK_a + \log \dfrac{0.0300 \, \text{mol}}{0.22 \, \text{mol}} = 3.85 \text{ Solve for } pK_a.$$

$$pK_a = 3.85 - \log \dfrac{0.0300 \, \text{mol}}{0.22 \, \text{mol}} = 4.72$$

Check: The units (none) are correct. The magnitude of the answer makes physical sense because there is more acid than base at equilibrium; so the pK_a is higher than the pH of the solution.

16.123   **Given:** 0.552 g ascorbic acid dissolved in 20.00 mL and titrated with 28.42 mL of 0.1103 M KOH solution; pH = 3.72 when 10.0 mL KOH added   **Find:** molar mass and K_a of acid
Conceptual Plan: mL → L then M KOH, L → mol KOH → mol acid then mol acid, g acid → $\mathcal{M}$

$$\dfrac{1 \, \text{L}}{1000 \, \text{mL}} \qquad M = \dfrac{\text{mol}}{\text{L}} \quad \text{1:1 for monoprotic acid} \qquad\qquad \mathcal{M} = \dfrac{\text{g acid}}{\text{mol acid}}$$

For the second part of the problem:
mL → L then M KOH, L → mol KOH then write a balanced reaction

$$\dfrac{1 \, \text{L}}{1000 \, \text{mL}} \qquad\qquad M = \dfrac{\text{mol}}{\text{L}} \qquad\qquad \text{KOH} + \text{HA} \rightarrow \text{NaA} + \text{H}_2\text{O}$$

added mol KOH, initial mol acid → equil. mol KOH, equil. mol acid then

set up stoichiometry table

equil. mol KOH, equil. mol acid, pH → pK_a → K_a.

$$\text{pH} = pK_a + \log \dfrac{[\text{base}]}{[\text{acid}]} \quad pK_a = -\log K_a$$

Solution: $28.42 \, \text{mL} \times \dfrac{1 \, \text{L}}{1000 \, \text{mL}} = 0.02842 \, \text{L}$ then

$0.02842 \, \text{L KOH} \times \dfrac{0.1103 \, \text{mol KOH}}{1 \, \text{L KOH}} = 0.003134\underline{7}26 \, \text{mol KOH} = \text{mol base} = \text{mol acid} = 0.003134\underline{7}26 \, \text{mol}$

ascorbic acid then $M = \dfrac{\text{g acid}}{\text{mol acid}} = \dfrac{0.552 \, \text{g acid}}{0.003134\underline{7}26 \, \text{mol acid}} = 176 \, \text{g/mol}$. For the second part of the problem,

$$10.00 \text{ mL} \times \frac{1 \text{ L}}{1000 \text{ mL}} = 0.01000 \text{ L then } 0.01000 \text{ L KOH} \times \frac{0.1103 \text{ mol KOH}}{1 \text{ L KOH}} = 0.001103 \text{ mol KOH}$$

Because KOH is a strong base, $[\text{KOH}] = [\text{OH}^-]$. Set up a table to track changes:

	KOH(aq)	+	HA(aq)	→	KA(aq)	+	H$_2$O(l)
Before addition	≈0.00 mol		0.003134726 mol		0.00 mol		—
Addition	0.001103 mol		—		—		—
After addition	≈0.00 mol		0.002031726 mol		0.001103 mol		—

Because the ratio of base to acid is between 0.1 and 10, it is a buffer solution. Note that the ratio of moles is the same as the ratio of concentrations because the volume for both terms is the same.

$$\text{pH} = \text{p}K_a + \log \frac{[\text{base}]}{[\text{acid}]} = \text{p}K_a + \log \frac{0.001103 \text{ mol}}{0.002031726 \text{ mol}} = 3.72. \text{ Solve for p}K_a.$$

$$\text{p}K_a = 3.72 - \log \frac{0.001103}{0.002031726} = 3.98529 \text{ and so p}K_a = -\log K_a$$

or $K_a = 10^{-\text{p}K_a} = 10^{-3.98529} = 1.0 \times 10^{-4}$.

Check: The units (g/mol and none) are correct. The magnitude of the answer makes physical sense because there is much less than a mole of acid and about half a gram of acid; so the molar mass will be high. The number is reasonable for an acid (must be >20 g/mol—lightest acid is HF). The K_a is reasonable because the pK_a is within 1 unit of the pH when the titration solution is behaving as a buffer.

16.125   **Given:** saturated CaCO$_3$ solution; precipitate 1.00×10^2 mg CaCO$_3$ **Find:** volume of solution evaporated
     Other: $K_{sp}(\text{CaCO}_3) = 4.96 \times 10^{-9}$
     Conceptual Plan: mg CaCO$_3$ → g CaCO$_3$ → mol CaCO$_3$

$$\frac{1 \text{ g CaCO}_3}{1000 \text{ mg CaCO}_3} \qquad \frac{1 \text{ mol CaCO}_3}{100.09 \text{ g CaCO}_3}$$

     The expression of the solubility product constant of A$_m$X$_n$ is $K_{sp} = [\text{A}^{n+}]^m [\text{X}^{m-}]^n$. The molar solubility of a compound, A$_m$X$_n$, can be computed directly from K_{sp} by solving for S in the expression $K_{sp} = (mS)^m (nS)^n = m^n n^n S^{m+n}$. Then mol CaCO$_3$, S → L.

$$M = \frac{\text{mol}}{\text{L}}$$

     Solution: $1.00 \times 10^2 \text{ mg CaCO}_3 \times \dfrac{1 \text{ g CaCO}_3}{1000 \text{ mg CaCO}_3} \times \dfrac{1 \text{ mol CaCO}_3}{100.09 \text{ g CaCO}_3} = 9.99101 \times 10^{-4} \text{ mol CaCO}_3$ then

$K_{sp} = 4.96 \times 10^{-9}, \text{A} = \text{Ca}^{2+}, m = 1, \text{X} = \text{CO}_3^{2-}, \text{and } n = 1;$ so $K_{sp} = 4.96 \times 10^{-9} = S^2$. Rearrange to solve for S. $S = \sqrt{4.96 \times 10^{-9}} = 7.04273 \times 10^{-5}$ M. Finally,

$$9.99101 \times 10^{-4} \text{ mol CaCO}_3 \times \frac{1 \text{ L}}{7.04273 \times 10^{-5} \text{ mol CaCO}_3} = 14.2 \text{ L}.$$

     Check: The units (L) are correct. The volume should be large because the solubility is low.

16.127   **Given:** $[\text{Ca}^{2+}] = 9.2$ mg/dL and $K_{sp}(\text{Ca}_2\text{P}_2\text{O}_7) = 8.64 \times 10^{-13}$ **Find:** $[\text{P}_2\text{O}_7^{4-}]$ to form precipitate
     Conceptual Plan: mg Ca^{2+}/dL → g Ca^{2+}/dL → mol Ca^{2+}/dL → mol Ca^{2+}/L then

$$\frac{1 \text{ g Ca}^{2+}}{1000 \text{ mg Ca}^{2+}} \qquad \frac{1 \text{ mol Ca}^{2+}}{40.08 \text{ g Ca}^{2+}} \qquad \frac{10 \text{ dL}}{1 \text{ L}}$$

     write a balanced equation and expression for K_{sp}. Then $[\text{Ca}^{2+}]$, K_{sp} → $[\text{P}_2\text{O}_7^{4-}]$.

     Solution: $9.2 \dfrac{\text{mg Ca}^{2+}}{\text{dL}} \times \dfrac{1 \text{ g Ca}^{2+}}{1000 \text{ mg Ca}^{2+}} \times \dfrac{1 \text{ mol Ca}^{2+}}{40.08 \text{ g Ca}^{2+}} \times \dfrac{10 \text{ dL}}{1 \text{ L}} = 2.29541 \times 10^{-3} \text{ M Ca}^{2+}$ then write

equation $\text{Ca}_2\text{P}_2\text{O}_7(s) \rightarrow 2 \text{ Ca}^{2+}(aq) + \text{P}_2\text{O}_7^{4-}(aq)$. So $K_{sp} = [\text{Ca}^{2+}]^2 [\text{P}_2\text{O}_7^{4-}] = 8.64 \times 10^{-13} = (2.29541 \times 10^{-3})^2 [\text{P}_2\text{O}_7^{4-}]$. Solve for $[\text{P}_2\text{O}_7^{4-}]$ then $[\text{P}_2\text{O}_7^{4-}] = 1.6 \times 10^{-7}$ M.

     Check: The units (M) are correct. Because K_{sp} is so small and the calcium concentration is relatively high, the diphosphate concentration required is at a very low level.

16.129   **Given:** CuX in 0.150 M NaCN   **Find:** molar solubility (S)
Other: $K_f(Cu(CN)_4{}^{2-}) = 1.0 \times 10^{25}$, $K_{sp}(CuX) = [Cu^{2+}][X^{2-}] = 1.27 \times 10^{-36}$
Conceptual Plan: Identify the appropriate solid and complex ion. Write balanced equations for dissolving the solid and forming the complex ion. Add these two reactions to get the desired overall reaction. Using the rules from Chapter 14, multiply the individual reaction Ks to get the overall K for the sum of these reactions. Then M NaCN, $K \rightarrow S$.
         ICE table

Solution: Identify the solid as MX and the complex ion as $M(CN)_4{}^{2-}$. Write the individual reactions and add them together.

$$CuX(s) \rightleftharpoons \cancel{Cu^+(aq)} + X^{2-}(aq) \qquad\qquad K_{sp} = 1.27 \times 10^{-36}$$
$$\cancel{Cu^{2+}(aq)} + 4\,CN^-(aq) \rightleftharpoons Cu(CN)_4{}^{2-}(aq) \qquad K_f = 1.0 \times 10^{25}$$

$$\overline{Cu(s) + 4\,CN^-(aq) \rightleftharpoons Cu(CN)_4{}^{2-}(aq) + X^{2-}(aq)}$$

Because the overall reaction is the simple sum of the two reactions, the overall reaction $K = K_f K_{sp} = (1.0 \times 10^{25}) \times (1.27 \times 10^{-36}) = 1.\underline{2}7 \times 10^{-11}$. $NaCN(s) \rightarrow Na^+(aq) + CN^-(aq)$. Because one CN^- ion is generated for each NaCN, $[CN^-] = 0.150$ M. Set up an ICE table:

	$Cu(s) \; + $	$4\,CN^-(aq)$	$\rightleftharpoons$	$Cu(CN)_4{}^{2-}(aq)$	$+$	$X^{2-}(aq)$
		$[CN^-]$		$[Cu(CN)_4{}^{2-}]$		$[X^{2-}]$
Initial		0.150		0.00		0.00
Change		$-4S$		$+S$		$+S$
Equil		$0.150 - 4S$		$+S$		$+S$

$$K = \frac{[Cu(CN)_4{}^{2-}][X^{2-}]}{[CN^-]^4} = 1.\underline{2}7 \times 10^{-11} = \frac{S^2}{(0.150 - 4S)^4}$$

Assume that S is small ($4S << 0.150$), so $\dfrac{S^2}{(0.150 - \cancel{4S})^4} = 1.\underline{2}7 \times 10^{-11} = \dfrac{S^2}{(0.150)^4}$ and $S = 8.\underline{0}183 \times 10^{-8} =$

8.0×10^{-8} M. Confirm that the assumption is valid. $\dfrac{4(8.\underline{0}183 \times 10^{-8})}{0.150} \times 100\% = 0.00021\% << 5\%$, so the assumption is valid.

Check: The units (M) are correct. Because K_f is large, the overall K is larger than the original K_{sp} and the solubility of MX increases over that of pure water ($\sqrt{1.27 \times 10^{-36}} = 1.13 \times 10^{-18}$ M).

16.131   **Given:** 100.0 mL of 0.36 M NH_2OH and 50.0 mL of 0.26 M HCl and $K_b(NH_2OH) = 1.10 \times 10^{-8}$ **Find:** pH
Conceptual Plan: Identify acid and base components mL $\rightarrow$ L then $[NH_2OH]$, L $\rightarrow$ mol NH_2OH

$$\text{acid} = NH_3OH^+ \; \text{base} = NH_2OH \qquad \frac{1\,L}{1000\,mL} \qquad M = \frac{mol}{L}$$

then mL $\rightarrow$ L then [HCl], L $\rightarrow$ mol HCl then write balanced equation then

$$\frac{1\,L}{1000\,mL} \qquad M = \frac{mol}{L} \qquad HCl + NH_2OH \rightarrow NH_3OHCl$$

mol NH_2OH, mol HCL $\rightarrow$ mol excess NH_2OH, mol NH_3OH^+.
       set up stoichiometry table

Because there are significant amounts of both the acid and the conjugate base species, this is a buffer solution; so the Henderson–Hasselbalch equation $\left(pH = pK_a + \log \dfrac{[\text{base}]}{[\text{acid}]} \right)$ **can be used. Convert K_b to K_a using $K_w = K_a K_b$. Also note that the ratio of concentrations is the same as the ratio of moles because the volume is the same for both species.**

Solution: $100 \; \cancel{mL\,NH_2OH} \times \dfrac{1\,L}{1000 \; \cancel{mL}} = 0.1\,L\,NH_2OH$ then

$\dfrac{0.36\,mol\,NH_2OH}{1 \; \cancel{L}} \times 0.1000 \; \cancel{L} = 0.036\,mol\,NH_2OH$. $50.0 \; \cancel{mL\,HCl} \times \dfrac{1\,L}{1000 \; \cancel{mL}} = 0.0500\,L\,HCl$ then

$\dfrac{0.26\,mol\,HCl}{1 \; \cancel{L}} \times 0.0500 \; \cancel{L} = 0.013\,mol\,HCl$. Set up a table to track changes:

	$HCl(aq)$	$+$	$NH_2OH(aq)$	$\rightarrow$	$NH_3OHCl(aq)$
Before addition	0.00 mol		0.036 mol		0.00 mol
Addition	0.013 mol		—		—
After addition	≈ 0.00 mol		0.023 mol		0.013 mol

Then $K_w = K_aK_b$, so $K_a = \dfrac{K_w}{K_b} = \dfrac{1.0 \times 10^{-14}}{1.10 \times 10^{-8}} = 9.\underline{0}909 \times 10^{-7}$ M. Then use the Henderson–Hasselbalch equation because the solution is a buffer. Note that the ratio of moles is the same as the ratio of concentrations because the volume for both terms is the same.

$$pH = pK_a + \log \frac{[\,base\,]}{[\,acid\,]} = -\log(9.\underline{0}909 \times 10^{-7}) + \log \frac{0.023 \; \text{mol}}{0.013 \; \text{mol}} = 6.2\underline{8}918 = 6.29$$

Check: The units (none) are correct. The magnitude of the answer makes physical sense because the pH should be more than the pK_a of the acid because there is more base than acid.

16.133 **Given:** 25.0 mL of NaOH titrated with 19.6 mL of 0.189 M HCl solution; 10.0 mL of H_3PO_4 titrated with 34.9 mL NaOH **Find:** concentration of H_3PO_4 solution

Conceptual Plan: Write the first balanced reaction then mL $\rightarrow$ L then M HCl, L $\rightarrow$ mol HCl $\rightarrow$ mol NaOH

$$HCl + NaOH \rightarrow NaCl + H_2O \qquad \frac{1\,L}{1000\,mL} \qquad M = \frac{mol}{L} \qquad 1{:}1$$

then mL $\rightarrow$ L then mol NaOH, L $\rightarrow$ M NaOH then write 2nd balanced reaction then mL $\rightarrow$ L

$$\frac{1\,L}{1000\,mL} \qquad H_3PO_4 + 3\,NaOH \rightarrow Na_3PO_4 + 3\,H_2O \qquad \frac{1\,L}{1000\,mL}$$

then M NaOH, L $\rightarrow$ mol NaOH $\rightarrow$ mol H_3PO_4 then mL $\rightarrow$ L then mol H_3PO_4, L $\rightarrow$ M H_3PO_4.

$$M = \frac{mol}{L} \qquad 3{:}1 \qquad \frac{1\,L}{1000\,mL} \qquad M = \frac{mol}{L}$$

Solution: In the first titration, $19.6 \; \text{mL} \times \dfrac{1\,L}{1000\,mL} = 0.0196$ L then

$$0.0196 \; \text{L HCl} \times \frac{0.189 \; \text{mol HCl}}{1 \; \text{L HCl}} = 0.003\underline{7}044 \; \text{mol HCl} \times \frac{1 \; \text{mol NaOH}}{1 \; \text{mol HCl}} = 0.003\underline{7}044 \; \text{mol NaOH then}$$

$$25.0 \; \text{mL} \times \frac{1\,L}{1000\,mL} = 0.0250 \; \text{L then} \quad \frac{0.003\underline{7}044 \; \text{mol NaOH}}{0.0250 \; \text{L NaOH}} = 0.14\underline{8}176 \; \text{M NaOH}$$

In the second titration, $34.9 \; \text{mL} \times \dfrac{1\,L}{1000\,mL} = 0.0349$ L then

$$0.0349 \; \text{L NaOH} \times \frac{0.14\underline{8}176 \; \text{mol NaOH}}{1 \; \text{L NaOH}} = 0.0051\underline{7}134 \; \text{mol NaOH} \times \frac{1 \; \text{mol } H_3PO_4}{3 \; \text{mol NaOH}} = 0.0017\underline{2}378 \; \text{mol } H_3PO_4$$

$$\text{then } 10.0 \; \text{mL} \times \frac{1\,L}{1000\,mL} = 0.0100 \; \text{L then} \quad \frac{0.0017\underline{2}378 \; \text{mol } H_3PO_4}{0.0100 \; \text{L } H_3PO_4} = 0.172 \; \text{M } H_3PO_4$$

Check: The units (M) are correct. The magnitude of the answer makes physical sense because the concentration of NaOH is a little lower than the HCl (because the volume of NaOH is greater than HCl) and the concentration of H_3PO_4 is more than the NaOH (because the ratio of the volume of NaOH to volume of H_3PO_4 is just over 3 and H_3PO_4 is a triprotic acid).

16.135 **Given:** $(CH_3)_2NH/(CH_3)_2NH_2Cl$ buffer at pH $= 10.43$ **Find:** relative masses of $(CH_3)_2NH$ and $(CH_3)_2NH_2Cl$

Other: $K_b((CH_3)_2NH) = 5.4 \times 10^{-4}$

Conceptual Plan: $K_b \rightarrow pK_b \rightarrow pK_a$ **then** pH, $pK_a \rightarrow [(CH_3)_2NH]/[(CH_3)_2NH_2^+]$ **then**

$$pK_b = -\log K_b \qquad 14 = pK_a + pK_b \qquad acid = (CH_3)_2NH_2^+ \; base = (CH_3)_2NH \qquad pH = pK_a + \log \frac{[\,base\,]}{[\,acid\,]}$$

$[(CH_3)_2NH]/[(CH_3)_2NH_2^+] \rightarrow g(CH_3)_2NH/g(CH_3)_2NH_2^+$

$$\frac{45.09 \; g\,(CH_3)_2NH}{1 \; mol\,(CH_3)_2NH} \qquad \frac{1 \; mol\,(CH_3)_2NH_2Cl}{81.54 \; g\,(CH_3)_2NH_2Cl}$$

Solution: Because $K_b((CH_3)_2NH) = 5.4 \times 10^{-4}$, $pK_b = -\log K_b = -\log(5.4 \times 10^{-4}) = 3.27$. Because $14 = pK_a + pK_b$, $pK_a = 14 - pK_b = 14 - 3.27 = 10.73$. Because $[\,acid\,] = [(CH_3)_2NH_2^+] = [(CH_3)_2NH_2Cl]$

and $[\,base\,] = [(CH_3)_2NH]$, $pH = pK_a + \log \dfrac{[\,base\,]}{[\,acid\,]} = 10.73 + \log \dfrac{[(CH_3)_2NH]}{[(CH_3)_2NH_2Cl]} = 10.43$. Solve for

$$\frac{[(CH_3)_2NH]}{[(CH_3)_2NH_2Cl]}. \quad \log \frac{[(CH_3)_2NH]}{[(CH_3)_2NH_2Cl]} = 10.43 - 10.73 = -0.30 \rightarrow \frac{[(CH_3)_2NH]}{[(CH_3)_2NH_2Cl]} = 10^{-0.30} = 0.5\underline{0}1187.$$

Then $\dfrac{0.5\underline{0}1187 \text{ mol(CH}_3)_2\text{NH}}{1 \text{ mol(CH}_3)_2\text{NH}_2\text{Cl}} \times \dfrac{45.09 \text{ g(CH}_3)_2\text{NH}}{1 \text{ mol(CH}_3)_2\text{NH}} \times \dfrac{1 \text{ mol(CH}_3)_2\text{NH}_2\text{Cl}}{81.54 \text{ g(CH}_3)_2\text{NH}_2\text{Cl}} = \dfrac{0.2\underline{7}7146 \text{ g(CH}_3)_2\text{NH}}{\text{g(CH}_3)_2\text{NH}_2\text{Cl}}$

$= \dfrac{0.28 \text{ g (CH}_3)_2\text{NH}}{\text{g (CH}_3)_2\text{NH}_2\text{Cl}}$ or $\dfrac{3.6 \text{ g (CH}_3)_2\text{NH}_2\text{Cl}}{\text{g (CH}_3)_2\text{NH}}$

Check: The units (g/g) are correct. The magnitude of the answer makes physical sense because there are more moles of acid than base in the buffer (pH < pK_a) and the molar mass of the acid is greater than the molar mass of the base. Thus, the ratio of the mass of the base to the mass of the acid is expected to be less than 1.

16.137 **Given:** $HC_7H_5O_2/C_7H_5O_2Na$ buffer at pH = 4.55, complete dissociation of $C_7H_5O_2Na$, $d = 1.01$ g/mL; $T_f = -2.0 \,°C$

Find: $[HC_7H_5O_2]$ and $[C_7H_5O_2Na]$ **Other:** $K_a(HC_7H_5O_2) = 6.5 \times 10^{-5}$, $K_f = 1.86 \,°C/m$

Conceptual Plan: $K_a \rightarrow pK_a$ then pH, p$K_a \rightarrow [C_7H_5O_2Na]/[HC_7H_5O_2]$ and $T_f \rightarrow \Delta T_f$ then

$$pK_a = -\log K_a \quad \text{acid} = HC_7H_5O_2 \, \text{base} = C_7H_5O_2^- \quad pH = pK_a + \log\dfrac{[\text{base}]}{[\text{acid}]} \quad T_f = T_f^\circ - \Delta T_f$$

$\Delta T_f, i, K_f \rightarrow m$ then assume 1 kg water (or 1000 g water)

$$\Delta T_f = K_f \times m \quad m = \dfrac{\text{amount solute (moles)}}{\text{mass solvent (kg)}}$$

Assume that $HC_7H_5O_2$ does not dissociate and that $C_7H_5O_2Na$ completely dissociates, then mol particles = mol $HC_7H_5O_2$ + 2(mol $C_7H_5O_2Na$). Use $[C_7H_5O_2Na]/[HC_7H_5O_2]$ and total mol particles = mol $[HC_7H_5O_2]$ + 2(mol $C_7H_5O_2Na$) to solve for mol $HC_7H_5O_2$ and mol $C_7H_5O_2Na$. Then mol $HC_7H_5O_2 \rightarrow$ g $HC_7H_5O_2$ and mol $C_7H_5O_2$ Na $\rightarrow$ g $C_7H_5O_2$ Na then

$$\dfrac{122.12 \text{ g } HC_7H_5O_2}{1 \text{ mol } HC_7H_5O_2} \qquad \dfrac{144.10 \text{ g } C_7H_5O_2Na}{1 \text{ mol } HC_7H_5O_2Na}$$

g $HC_7H_5O_2$, g $C_7H_5O_2Na$, g water $\rightarrow$ g solution $\rightarrow$ mL solution $\rightarrow$ L solution then

$$\text{g } HC_7H_5O_2 + \text{g } C_7H_5O_2Na + \text{g water} = \text{g solution} \qquad d = \dfrac{1 \text{ mL}}{1.01 \text{ g}} \qquad \dfrac{1 \text{ L}}{1000 \text{ mL}}$$

mol $HC_7H_5O_2$, L $\rightarrow$ M $HC_7H_5O_2$ and mol $C_7H_5O_2Na$, L $\rightarrow$ M $C_7H_5O_2Na$.

$$M = \dfrac{\text{mol}}{L} \qquad\qquad M = \dfrac{\text{mol}}{L}$$

Solution: Because $K_a(HC_7H_5O_2) = 6.5 \times 10^{-5}$, p$K_a = -\log K_a = -\log(6.5 \times 10^{-5}) = 4.19$. Because $[\text{acid}] = [HC_7H_5O_2]$ and $[\text{base}] = [C_7H_5O_2^-] = [C_7H_5O_2Na]$,

$$pH = pK_a + \log\dfrac{[\text{base}]}{[\text{acid}]} = 4.19 + \log\dfrac{[C_7H_5O_2Na]}{[HC_7H_5O_2]} = 4.55. \text{ Solve for } \dfrac{[C_7H_5O_2Na]}{[HC_7H_5O_2]}.$$

$$\log\dfrac{[C_7H_5O_2Na]}{[HC_7H_5O_2]} = 4.55 - 4.19 = 0.36 \rightarrow \dfrac{[C_7H_5O_2Na]}{[HC_7H_5O_2]} = 10^{0.36} = 2.2\underline{9}087. \text{ Then } T_f = T_f^\circ - \Delta T_f, \text{ so}$$

$\Delta T_f = T_f^\circ - T_f = 0.0 \,°C - (-2.0 \,°C) = 2.0 \,°C$. Then $\Delta T_f = K_f \times m$. Rearrange to solve for m.

$$m = \dfrac{\Delta T_f}{K_f} = \dfrac{2.0 \,°C}{1.86 \dfrac{°C}{m}} = 1.\underline{0}7527 \, \dfrac{\text{mol particles}}{\text{kg solvent}}. \text{ Assume 1 kg water, so we have } 1.\underline{0}7527 \text{ mol particles.}$$

Assume that $HC_7H_5O_2$ does not dissociate and that $C_7H_5O_2Na$ completely dissociates; then 1.$\underline{0}$7527 mol particles = mol $HC_7H_5O_2$ + 2(mol $C_7H_5O_2Na$). Use $[C_7H_5O_2Na]/[HC_7H_5O_2] = 2.2\underline{9}087$ and 1.$\underline{0}$7527 mol particles = mol $[HC_7H_5O_2]$ + 2(mol $C_7H_5O_2Na$) to solve for mol $HC_7H_5O_2$ and mol $C_7H_5O_2Na$. So mol $C_7H_5O_2Na$ = 2.2$\underline{9}$087(mol $HC_7H_5O_2$) $\rightarrow$ 1.$\underline{0}$7527 mol particles = mol $HC_7H_5O_2$ + 2(2.29087 mol $HC_7H_5O_2$) $\rightarrow$ 1.$\underline{0}$7527 mol particles = 5.$\underline{5}$8174 mol $HC_7H_5O_2 \rightarrow$ mol $HC_7H_5O_2$ = 0.19$\underline{2}$641 mol and mol $C_7H_5O_2Na$ = 2.2$\underline{9}$087 × mol $HC_7H_5O_2$ = 2.2$\underline{9}$087 × 0.19$\underline{2}$641 mol $HC_7H_5O_2$ = 0.44$\underline{1}$315 mol $C_7H_5O_2Na$.

$$0.19\underline{2}641 \text{ mol } HC_7H_5O_2 \times \dfrac{122.12 \text{ g } HC_7H_5O_2}{1 \text{ mol } HC_7H_5O_2} = 2\underline{3}.5253 \text{ g } HC_7H_5O_2 \text{ and}$$

$$0.44\underline{1}315 \text{ mol } C_7H_5O_2Na \times \dfrac{144.10 \text{ g } C_7H_5O_2Na}{1 \text{ mol } C_7H_5O_2Na} = 6\underline{3}.5935 \text{ g } C_7H_5O_2Na \text{ then}$$

2$\underline{3}$.5253 g $HC_7H_5O_2$ + 6$\underline{3}$.5935 g $C_7H_5O_2Na$ + 1000 g water = 108$\underline{7}$.119 g solution then

$1087.119 \text{ g solution} \times \dfrac{1 \text{ mL}}{1.01 \text{ g}} \times \dfrac{1 \text{ L}}{1000 \text{ mL}} = 1.076355$ L. Finally,

$\dfrac{0.192641 \text{ mol HC}_7\text{H}_5\text{O}_2}{1.076355 \text{ L}} = 0.178975 \text{ M HC}_7\text{H}_5\text{O}_2 = 0.18 \text{ M HC}_7\text{H}_5\text{O}_2$ and

$\dfrac{0.441315 \text{ mol C}_7\text{H}_5\text{O}_2\text{Na}}{1.076355 \text{ L}} = 0.410009 \text{ M C}_7\text{H}_5\text{O}_2\text{Na} = 0.41 \text{ M C}_7\text{H}_5\text{O}_2\text{Na}.$

Check: The units (M and M) are correct. The magnitude of the answer makes physical sense because there are more moles of base than acid in the buffer (pH $>$ pK_a).

Challenge Problems

16.139   **Given:** 10.0 L of 75 ppm $CaCO_3$ and 55 ppm $MgCO_3$ (by mass)

Find: mass Na_2CO_3 to precipitate 90.0% of ions

Other: $K_{sp}(CaCO_3) = 4.96 \times 10^{-9}$ and $K_{sp}(MgCO_3) = 6.82 \times 10^{-6}$

Conceptual Plan: Assume that the density of water is 1.00 g/mL L water $\rightarrow$ mL water $\rightarrow$ g water then

$$\dfrac{1000 \text{ mL}}{1 \text{ L}} \qquad \dfrac{1.00 \text{ g water}}{1 \text{ mL}}$$

g water $\rightarrow$ g $CaCO_3$ $\rightarrow$ mol $CaCO_3$ $\rightarrow$ mol Ca^{2+} and g water $\rightarrow$ g $MgCO_3$ $\rightarrow$ mol $MgCO_3$ then

$$\dfrac{75 \text{ g CaCO}_3}{10^6 \text{ g water}} \quad \dfrac{1 \text{ mol CaCO}_3}{100.09 \text{ g CaCO}_3} \quad \dfrac{1 \text{ mol Ca}^{2+}}{1 \text{ mol CaCO}_3} \qquad \dfrac{55 \text{ g MgCO}_3}{10^6 \text{ g water}} \quad \dfrac{1 \text{ mol MgCO}_3}{84.32 \text{ g MgCO}_3}$$

mol $MgCO_3$ $\rightarrow$ mol Mg^{2+} then comparing the two K_{sp} values, essentially all of the Ca^{2+} will

$$\dfrac{1 \text{ mol Mg}^{2+}}{1 \text{ mol MgCO}_3}$$

precipitate before the Mg^{2+} will begin to precipitate. Because 90.0% of the ions are to be precipitates, 10.0% of the ions will be left in solution (all will be Mg^{2+}).

$$(0.100)(\text{mol Ca}^{2+} + \text{mol Mg}^{2+})$$

Calculate the moles of ions remaining in solution. Then mol Mg^{2+}, L $\rightarrow$ M Mg^{2+}.

$$M = \dfrac{\text{mol}}{\text{L}}$$

The solubility product constant (K_{sp}) is the equilibrium expression for a chemical equation representing the dissolution of an ionic compound. The expression of the solubility product constant of A_mX_n is $K_{sp} = [A^{n+}]^m[X^{m-}]^n$. Use this equation to M Mg^{2+}, K_{sp} $\rightarrow$ M CO_3^{2-} then M CO_3^{2-}, L $\rightarrow$ mol CO_3^{2-}

$$\text{for ionic compound, } A_mX_n, K_{sp} = [A^{n+}]^m[X^{m-}]^n \qquad M = \dfrac{\text{mol}}{\text{L}}$$

then mol CO_3^{2-} $\rightarrow$ mol Na_2CO_3 $\rightarrow$ g Na_2CO_3.

$$\dfrac{1 \text{ mol CO}_3^{2-}}{1 \text{ mol Na}_2\text{CO}_3} \qquad \dfrac{105.99 \text{ g Na}_2\text{CO}_3}{1 \text{ mol Na}_2\text{CO}_3}$$

Solution: $10.0 \text{ L} \times \dfrac{1000 \text{ mL}}{1 \text{ L}} \times \dfrac{1.00 \text{ g water}}{1 \text{ mL}} = 1.00 \times 10^4$ g water then

$1.00 \times 10^4 \text{ g water} \times \dfrac{75 \text{ g CaCO}_3}{10^6 \text{ g water}} \times \dfrac{1 \text{ mol CaCO}_3}{100.09 \text{ g CaCO}_3} \times \dfrac{1 \text{ mol Ca}^{2+}}{1 \text{ mol CaCO}_3} = 0.0074933 \text{ mol Ca}^{2+}$ and

$1.00 \times 10^4 \text{ g water} \times \dfrac{55 \text{ g MgCO}_3}{10^6 \text{ g water}} \times \dfrac{1 \text{ mol MgCO}_3}{84.32 \text{ g MgCO}_3} \times \dfrac{1 \text{ mol Mg}^{2+}}{1 \text{ mol MgCO}_3} = 0.0065228 \text{ mol Mg}^{2+}$ so the ions

remaining in solution after 90.0% precipitate out $= (0.100)(\text{mol Ca}^{2+} + \text{mol Mg}^{2+})$

$= (0.100)(0.0074933 \text{ mol Ca}^{2+} + 0.0065228 \text{ mol Mg}^{2+}) = 0.00140161 \text{ mol ions}$

so $\dfrac{0.00140161 \text{ mol Mg}^{2+}}{10.0 \text{ L}} = 0.000140161 \text{ M Mg}^{2+}$. Then $K_{sp} = 6.82 \times 10^{-6}$, A $= Mg^{2+}$, $m = 1$, X $= CO_3^{2-}$, and

$n = 1$, so $K_{sp} = 6.82 \times 10^{-6} = [\text{Mg}^{2+}][\text{CO}_3^{2-}] = (0.00140161)[\text{CO}_3^{2-}]$. Rearrange to solve for $[\text{CO}_3^{2-}]$. So

$[\text{CO}_3^{2-}] = 0.0486583$ M. Then

$\dfrac{0.0486583 \text{ mol CO}_3^{2-}}{1 \text{ L}} \times 10.0 \text{ L} \times \dfrac{1 \text{ mol Na}_2\text{CO}_3}{1 \text{ mol CO}_3^{2-}} \times \dfrac{105.99 \text{ g Na}_2\text{CO}_3}{1 \text{ mol Na}_2\text{CO}_3} = 51.6 \text{ g Na}_2\text{CO}_3.$

Check: The units (g) are correct. The mass is reasonable to put in a washing machine load.

16.141   **Given:** excess $Mg(OH)_2$ in 1.00 L of 1.0 M NH_4Cl has pH = 9.00   **Find:** $K_{sp}(Mg(OH)_2)$
         Other: $K_b(NH_3) = 1.76 \times 10^{-5}$
         Conceptual Plan: M $NH_4Cl \rightarrow$ M NH_4^+ and $K_b \rightarrow K_a$ then final pH $\rightarrow [H_3O^+]$ then

$$NH_4Cl(aq) \rightarrow NH_4^+(aq) + Cl^-(aq) \quad K_w = K_a K_b \qquad [H_3O^+] = 10^{-pH}$$

M NH_4^+, M H_3O^+, $K_a \rightarrow x$. Because x is significant compared to initial M NH_4^+, this is a buffer solution.

                    ICE table

The NH_4^+ is neutralized with $Mg(OH)_2$. Because $Mg(OH)_2(s) \rightleftharpoons Mg^{2+}(aq) + 2\,OH^-(aq)$, two moles of OH^- are generated for each mole of $Mg(OH)_2$ dissolved. Thus, $\frac{1}{2}(x \text{ mol } OH^-) = $ mol $Mg(OH)_2$ dissolved in 1.00 L of solution. Because there is 1.00 L solution, mol $Mg(OH)_2 = [Mg^{2+}]$ and $[H_3O^+] \rightarrow [OH^-]$.

                                                        $K_w = [H_3O^+][OH^-]$

Finally, write an expression for $K_{sp}(Mg(OH)_2)$ and substitute values for $[Mg^{2+}]$ and $[OH^-]$.
         Solution: Because one NH_4^+ ion is generated for each NH_4Cl, $[NH_4^+] = 1.0$ M NH_4^+. Because

$$K_w = K_a K_b, \text{ rearrange to solve for } K_a. \; K_a = \frac{K_w}{K_b} = \frac{1.0 \times 10^{-14}}{1.76 \times 10^{-5}} = 5.\underline{6}818 \times 10^{-10}. \text{ Final pH = 9.00, so}$$

$$[H_3O^+] = 10^{-pH} = 10^{-9.00} = 1.0 \times 10^{-9} \text{ M. Set up an ICE table:}$$

$$NH_4^+(aq) + H_2O(l) \rightleftharpoons H_3O^+(aq) + NH_3(aq)$$

	$[NH_4^+]$	$[H_3O^+]$	$[NH_3]$
Initial	1.0	≈ 0.00	0.00
Change	$-x$	$+x$	$+x$
Equil	$1.0 - x$	1×10^{-9}	x

$$K_a = \frac{[H_3O^+][NH_3]}{[NH_4^+]} = 5.\underline{6}818 \times 10^{-10} = \frac{(1.0 \times 10^{-9})x}{1.0 - x}$$

Solve for x. $5.\underline{6}818 \times 10^{-10}(1.0 - x) = (1.0 \times 10^{-9})x \rightarrow 5.\underline{6}818 \times 10^{-10} = (1.0 \times 10^{-9} + 5.\underline{6}818 \times 10^{-10})x \rightarrow x = 0.3\underline{6}2318$, so this is a buffer solution. Because there is 1.00 L of solution, 0.3\underline{6}2318 mol of NH_4^+ is neutralized with $Mg(OH)_2$. Because $Mg(OH)_2(s) \rightleftharpoons Mg^{2+}(aq) + 2\,OH^-(aq)$, two moles of OH^- are generated for each mole of $Mg(OH)_2$ dissolved. Thus, $\frac{1}{2}(0.3\underline{6}2318 \text{ mol } OH^-) = 0.1\underline{8}1159$ mol $Mg(OH)_2$ was dissolved in 1.00 L of solution. Thus, the $[Mg^{2+}] = 0.1\underline{8}1159$ M. Because $K_w = [H_3O^+][OH^-]$,

$$[OH^-] = \frac{K_w}{[H_3O^+]} = \frac{1.0 \times 10^{-14}}{1.0 \times 10^{-9}} = 1.0 \times 10^{-5} \text{ M then}$$

$$K_{sp}(Mg(OH)_2) = [Mg^{2+}][OH^-]^2 = (0.1\underline{8}1159)(1.0 \times 10^{-5})^2 = 1.8 \times 10^{-11}.$$

         Check: The units (none) are correct. The magnitude of the answer makes physical sense because the concentration of NH_4Cl is high; so it took a significant amount of $Mg(OH)_2$ to raise the pH to 9.00. Note that this number disagrees with the accepted value for the $K_{sp}(Mg(OH)_2)$. This is most likely due to errors in the measurements in this experiment.

16.143   (a)   **Given:** $Au(OH)_3$ in pure water   **Find:** molar solubility (S)   **Other:** $K_{sp} = 5.5 \times 10^{-46}$
               Conceptual Plan: Use equations derived in Problems 16.19 and 16.20 and solve for S. Then

                     for ionic compound, A_mX_n, $K_{sp} = [A^{n+}]^m[X^{m-}]^n = m^m n^n S^{m+n}$

               check answer for validity.
               Solution: For $Au(OH)_3$, $K_{sp} = 5.5 \times 10^{-46}$, $A = Au^{3+}$, $m = 1$, $X = OH^-$, and $n = 3$; so $K_{sp} = [Au^{3+}][OH^-]^3$

$$= 5.5 \times 10^{-46} = 3^3 S^4. \text{ Rearrange to solve for } S. \; S \; \sqrt[4]{\frac{5.5 \times 10^{-46}}{27}} = 2.1 \times 10^{-12} \text{ M. This answer}$$

suggests that the $[OH^-] = 3(2.1 \times 10^{-12} \text{ M}) = 6.3 \times 10^{-12}$ M. This result is lower than what is found in pure water (1.0×10^{-7} M), so substitute this value for $[OH^-]$ and solve for $S = [Au^{3+}]$. So $K_{sp} = [Au^{3+}][OH^-]^3 = 5.5 \times 10^{-46} = S(1.0 \times 10^{-7} \text{ M})^3$ and solving for S gives $S = 5.5 \times 10^{-25}$ M.

               Check: The units (M) are correct. Because K_{sp} is so small, the autoionization of water must be considered and the solubility is smaller than what is normally anticipated.

    (b)   **Given:** $Au(OH)_3$ in 1.0 M HNO_3 **Find:** molar solubility (S)   **Other:** $K_{sp} = 5.5 \times 10^{-46}$
               Conceptual Plan: Because HNO_3 is a strong acid, it will neutralize the gold(III) hydroxide (through the reaction of H^+ with OH^- to form water (the reverse of the autoionization of water equilibrium). Write balanced equations for dissolving the solid and for the neutralization reaction. Add these two reactions

to get the desired overall reaction. Using the rules from Chapter 14, multiply the individual reaction Ks to get the overall K for the sum of these reactions. Then M $HNO_3, K \rightarrow S$.

ICE table

Solution: Identify the solid as $Au(OH)_3$. Write the individual reactions and add them together.

$$Au(OH)_3(s) \rightleftharpoons Au^{3+}(aq) + 3\,\cancel{OH^-(aq)} \qquad\qquad K_{sp} = 5.5 \times 10^{-46}$$

$$3\,H^+(aq) + 3\,\cancel{OH^-(aq)} \rightleftharpoons 3\,H_2O(l) \qquad \left(\frac{1}{K_w}\right)^3 = \left(\frac{1}{1.0 \times 10^{-14}}\right)^3$$

$$Au(OH)_3(s) + 3\,H^+(aq) \rightleftharpoons Au^{3+}(aq) + 3\,H_2O(l)$$

Because the overall reaction is the sum of the dissolution reaction and three times the reverse of the autoionization of water reaction, the overall reaction

$$K = K_{sp}\left(\frac{1}{K_w}\right)^3 = (5.5 \times 10^{-46})\left(\frac{1}{1.0 \times 10^{-14}}\right)^3 = 5.5 \times 10^{-4} = \frac{[Au^{3+}]}{[H^+]^3}; \text{ then because } HNO_3 \text{ is a strong}$$

acid, it will completely dissociate to H^+ and NO_3^-. Set up an ICE table:

$$Au(OH)_3(s) + 3\,H^+(aq) \rightleftharpoons Au^{3+}(aq) + 3\,H_2O(l)$$

	$[H^+]$	$[Au^{3+}]$
Initial	1.0	0.00
Change	$-3S$	$+S$
Equil	$1.0 - 3S$	$+S$

$$K = \frac{[Au^{3+}]}{[H^+]^3} = 5.5 \times 10^{-4} = \frac{S}{(1.0 - 3S)^3}.$$

Assume that S is small ($3S \ll 1.0$), so $\dfrac{S}{(1.0 - \cancel{3S})^3} = 5.5 \times 10^{-4}\,\text{M} = \dfrac{S}{(1.0)^3} = S$. Confirm that the

assumption is valid. $\dfrac{3(5.5 \times 10^{-4})}{1.0} \times 100\% = 0.017\% \ll 5\%$, so the assumption is valid.

Check: The units (M) are correct. K is much larger than the original K_{sp}, so the solubility of $Au(OH)_3$ increases over that of pure water.

16.145   **Given:** 1.00 L of 0.100 M $MgCO_3$ **Find:** volume of 0.100 M Na_2CO_3 to precipitate 99% of Mg^{2+} ions
Other: $K_{sp}(MgCO_3) = 6.82 \times 10^{-6}$
Conceptual Plan: Because 99% of the Mg^{2+} ions are to be precipitated, 1% of the ions will be left in solution.

$$(0.01)(0.100\,\text{M } Mg^{2+})$$

Let x = required volume (in L). Calculate the amount of CO_3^{2-} added and the amount of Mg^{2+} that does not precipitate and remains in solution. Use these to calculate the $[Mg^{2+}]$ and $[CO_3^{2-}]$. The solubility product constant (K_{sp}) is the equilibrium expression for a chemical equation representing the dissolution of an ionic compound. The expression of the solubility product constant of A_mX_n is $K_{sp} = [A^{n+}]^m[X^{m-}]^n$. Substitute these expressions in this equation to $[Mg^{2+}], [CO_3^{2-}], K_{sp} \rightarrow x$.

for ionic compound, $A_mX_n, K_{sp} = [A^{n+}]^m[X^{m-}]^n$

Solution: Because 99% of the Mg^{2+} ions are to be precipitated, 1% of the ions will be left in solution, or $(0.01)(0.100\,\text{M } Mg^{2+}) = 0.001\,\text{M } Mg^{2+}$. Let x = required volume (in L). The volume of the solution after precipitation is $(1.00 + x)$. The amount of CO_3^{2-} added $= (0.100\text{M})(x\,\text{L}) = 0.100x$ mol CO_3^{2-}. The amount of Mg^{2+} that does not precipitate and remains in solution is $(0.100\,\text{M})(1.00\,\text{L})(0.01) = 1.00 \times 10^{-3}$ mol, and the amount that precipitates $= 0.099$ mol, which is also equal to the amount of CO_3^{2-} used. The amount of CO_3^{2-} remaining in solution is $(0.10x - 0.099)$. Thus, $[Mg^{2+}] = 1.00 \times 10^{-3}\text{mol}/(1.00 + x)$ L and $[CO_3^{2-}] = (0.10x - 0.099)$ mol/$(1.00 + x)$ L. Then $K_{sp} = 6.82 \times 10^{-6}$, A $= Mg^{2+}, m = 1, X = CO_3^{2-}$, and

$n = 1$; so $K_{sp} = 6.82 \times 10^{-6} = [Mg^{2+}][CO_3^{2-}] = \dfrac{(1.00 \times 10^{-3})(0.10x - 0.099)}{(1.00 + x)^2}$. Rearrange to solve for x.

$1.00 + 2.00x + x^2 = \dfrac{1.0 \times 10^{-4}x - 9.9 \times 10^{-5}}{6.82 \times 10^{-6}} \rightarrow 0 = x^2 - 12.\underline{6}628x + 15.\underline{5}161$. Using quadratic equation,

$x = 1.\underline{3}75\,\text{L} = 1.4\,\text{L}$.

Check: The units (L) are correct. The necessary concentration is very low, so the volume is fairly large.

16.147   **Given:** 1.0 L solution with 0.10 M $Ba(OH)_2$ and excess $Zn(OH)_2$ **Find:** pH
Other: $K_{sp}(Zn(OH)_2) = 3 \times 10^{-15}$, $K_f(Zn(OH)_4{}^{2-}) = 2 \times 10^{15}$
Conceptual Plan: Because $[Ba(OH)_2] = 0.10$ M, $[OH^-] = 0.20$ M. Write balanced equations for the solubility of $Zn(OH)_2$ and reaction with excess OH^- and expressions for K_{sp} and K_f. Use initial concentrations to set up an ICE table. Solve for $[OH^-]$ at equilibrium. Then $[OH^-] \rightarrow [H_3O^+] \rightarrow$ pH.

$$K_w = [H_3O^+][OH^-] \quad pH = -\log[H_3O^+]$$

Solution: Write two reactions and combine.

$Zn(OH)_2(s) \rightleftharpoons Zn^{2+}\cancel{(aq)} + \cancel{2}\,OH^-(aq)$ with $K_{sp} = [Zn^{2+}][OH^-]^2 = 3 \times 10^{-15}$

$\cancel{Zn^{2+}(aq)} + \cancel{2}4\,OH^-(aq) \rightleftharpoons Zn(OH)_4{}^{2-}(aq)$ with $K_f = \dfrac{[Zn(OH)_4{}^{2-}]}{[Zn^{2+}][OH^-]^4} = 2 \times 10^{15}$

$Zn(OH)_2(s) + 2\,OH^-(aq) \rightleftharpoons Zn(OH)_4{}^{2-}(aq)$

with $K = K_{sp}K_f = \cancel{[Zn^{2+}]}[OH^-]^2\dfrac{[Zn(OH)_4{}^{2-}]}{\cancel{[Zn^{2+}]}[OH^-]^{\cancel{4}\,2}} = (3 \times 10^{-15})(2 \times 10^{15})$

$K = \dfrac{[Zn(OH)_4{}^{2-}]}{[OH^-]^2} = 6$. Set up an ICE table with initial concentration and solve for x.

$$Zn(OH)_2(s) + 2\,OH^-(aq) \rightleftharpoons Zn(OH)_4{}^{2-}(aq)$$

	$[OH^-]$	$[Zn(OH)_4{}^{2-}]$
Initial	0.20	0.00
Change	$-2x$	x
Equil	$0.20 - 2x$	x

$K = \dfrac{[Zn(OH)_4{}^{2-}]}{[OH^-]^2} = \dfrac{x}{(0.20 - 2x)^2} = 6$

$x = 6(0.20 - 2x)^2 \rightarrow x = 6(4x^2 - 0.80x + 0.040) \rightarrow x = 24x^2 - 4.8x + 0.24 \rightarrow 0 = 24x^2 - 5.8x + 0.24$. Using quadratic equation, $x = 0.0530049 \rightarrow [OH^-] = 0.20 - 2x = 0.20 - 2(0.0530049) = 0.093990$ M OH^-. Then $K_w = [H_3O^+][OH^-]$, so

$$[H_3O^+] = \frac{K_w}{[OH^-]} = \frac{1.0 \times 10^{-14}}{0.093990} = 1.06394 \times 10^{-13} \text{ M. Finally,}$$

$$pH = -\log[H_3O^+] = -\log(1.06394 \times 10^{-13}) = 12.97.$$

Check: The units (none) are correct. Because the pH of the solution before the addition of the $Zn(OH)_2$ is 13.30, the reaction decreases the $[OH^-]$ and the pH drops.

Conceptual Problems

16.149   If the concentration of the acid is greater than the concentration of the base, the pH will be less than the pK_a. If the concentration of the acid is equal to the concentration of the base, the pH will be equal to the pK_a. If the concentration of the acid is less than the concentration of the base, the pH will be greater than the pK_a.
   (a) $pH < pK_a$
   (b) $pH > pK_a$
   (c) $pH = pK_a$; the OH^- will convert half of the acid to base.
   (d) $pH > pK_a$; the OH^- will convert more than half of the acid to base.

16.151   Only (a) is the same for both solutions. The volume to the first equivalence point will be the same because the number of moles of acid is the same. The pH profiles of the two titrations will be different.

16.153   (a)   The solubility will be unchanged because the pH is constant and no common ions are added.
   (b)   The solubility will be less because extra fluoride ions are added, suppressing the solubility of the fluoride ionic compound.
   (c)   The solubility will increase because some of the fluoride ion will be converted to HF; so more of the ionic compound can be dissolved.

17 Free Energy and Thermodynamics

Review Questions

17.1 The first law of thermodynamics states that energy is conserved in chemical processes. When we burn gasoline to run a car, for example, the amount of energy produced by the chemical reaction does not vanish, nor does any new energy appear that was not present before the combustion. Some of the energy from the combustion reaction goes toward driving the car forward (about 20%), and the rest is dissipated into the surroundings as heat (just feel the engine). However, the total energy given off by the combustion reaction exactly equals the sum of the amount of energy driving the car forward and the amount dissipated as heat—energy is conserved. In other words, when it comes to energy, you can't win; you cannot create energy that was not there to begin with.

17.3 A perpetual motion machine perpetually moves without any energy input. This machine is not possible because if the machine is to be in motion, it must pay the heat tax with each cycle of its motion—over time, then, it will run down and stop moving.

17.5 A spontaneous process occurs without ongoing outside intervention (such as the performance of work by some external force). For example, when you drop a book in a gravitational field, the book spontaneously drops to the floor.

17.7 Entropy (S) is a thermodynamic function that is proportional to the number of energetically equivalent ways to arrange the components of a system to achieve a particular state. Entropy, like enthalpy, is a state function—its value depends only on the state of the system, not on how the system got to that state. Therefore, for any process, the change in entropy is just the entropy of the final state minus the entropy of the initial state, or $\Delta S = S_{final} - S_{initial}$.

17.9 Microstates are the number of internal arrangements. These microstates give rise to the same external arrangement, or macrostate. If there are three gas particles (A, B, and C) and two containers, the fact that the first container has two particles and the other container has one particle is a macrostate. The fact that particles A and C are in the first container and particle B is in the other container is a microstate. There are at least as many microstates as there are macrostates.

17.11 The second law of thermodynamics states that for any spontaneous process, the entropy of the universe increases ($\Delta S_{univ} > 0$). The criterion for spontaneity is the entropy of the universe. Processes that increase the entropy of the universe—those that result in greater dispersal or randomization of energy—occur spontaneously. Processes that decrease the entropy of the universe do not occur spontaneously. Heat travels from a substance at higher temperature to one at lower temperature because the process disperses thermal energy. The cooler object has less thermal energy, so transferring heat from the warmer object results in greater energy randomization—the energy that was concentrated in the hot substance becomes dispersed between the two substances.

17.13 When water freezes at temperatures below 0 °C, the entropy of the water decreases, yet the process is spontaneous because the entropy of the universe increases ($\Delta S_{univ} > 0$). The entropy of the system can decrease ($\Delta S_{sys} < 0$) as long as the entropy of the surroundings increases by a greater amount ($\Delta S_{surr} > -\Delta S_{sys}$) so that the overall entropy of the universe undergoes a net increase. For liquid water freezing, the change in entropy for the system (ΔS_{sys}) is negative because the water becomes more ordered. For ΔS_{univ} to be positive, ΔS_{surr} must be positive and greater in absolute value (or magnitude) than ΔS_{sys}. We learned in Chapter 6 that freezing is an exothermic process: it gives off heat to the surroundings. If we think of entropy as the dispersal or randomization of energy, the release of heat energy by the system disperses that energy into the surroundings, increasing the entropy of the surroundings. The freezing of water below 0 °C increases the entropy of the universe because the heat given off to the surroundings increases the entropy of the surroundings to a sufficient degree to overcome the entropy decrease in the water. The freezing of water becomes nonspontaneous above 0 °C because the magnitude of the increase in the entropy of the surroundings due to the dispersal of energy into the surroundings is temperature-dependent. The higher the temperature, the smaller the percent increase in entropy for a given rise in temperature. Therefore, the impact of the heat released to the surroundings by the freezing of water depends on the temperature of the surroundings—the higher the temperature, the smaller the impact.

17.15   The change in Gibbs free energy for a process is proportional to the negative of ΔS_{univ}. Because ΔS_{univ} is a criterion for spontaneity, ΔG is also a criterion for spontaneity (although opposite in sign).

17.17   The third law of thermodynamics states that the entropy of a perfect crystal at absolute zero (0 K) is zero. For enthalpy, we defined a standard state so that we could define a "zero" for the scale. This is not necessary for entropy because there is an absolute zero.

17.19   The larger the molar mass, the greater the entropy at 25 °C. For a given state of matter, entropy generally increases with increasing molecular complexity.

17.21   The three ways of calculating the ΔG_{rxn} are:

- Use tabulated values of standard enthalpies of formation to calculate $\Delta H_{rxn}°$ and use tabulated values of standard entropies to calculate $\Delta S_{rxn}°$; then use the values of $\Delta H_{rxn}°$ and $\Delta S_{rxn}°$ calculated in these ways to calculate the standard free energy change for a reaction by using the equation $\Delta G_{rxn}° = \Delta H_{rxn}° - T\Delta S_{rxn}°$.
- Use tabulated values of the standard free energies of formation to calculate $\Delta G_{rxn}°$ using an equation similar to that used for standard enthalpy of a reaction $\Delta G_{rxn}° = \sum n_p \Delta G_f°(\text{products}) - \sum n_r \Delta G_f°(\text{reactants})$.
- Use a reaction pathway or stepwise reaction to sum the changes in free energy for each of the steps in a manner similar to that used in Chapter 6 for enthalpy of stepwise reactions.

The method to calculate the free energy of a reaction at temperatures other than at 25 °C is the first method. The second method is only applicable at 25 °C. The third method is only applicable at the temperature of the individual reactions, generally 25 °C.

17.23   The standard free energy change for a reaction $(\Delta G_{rxn}°)$ applies only to standard conditions. For a gas, standard conditions are those in which the pure gas is present at a partial pressure of 1 atm. For nonstandard conditions, we need to calculate ΔG_{rxn} (not $\Delta G_{rxn}°$) to predict spontaneity.

17.25   The free energy of reaction under nonstandard conditions (ΔG_{rxn}) can be calculated from $\Delta G_{rxn}°$ using the relationship $\Delta G_{rxn} = \Delta G_{rxn}° + RT \ln Q$, where Q is the reaction quotient (defined in Section 14.7), T is the temperature in K, and R is the gas constant in the appropriate units (8.314 J/mol K).

Problems by Topic

Entropy, the Second Law of Thermodynamics, and the Direction of Spontaneous Change

17.27   (a) and (c) are spontaneous processes.

17.29   System B has the greatest entropy. There is only one energetically equivalent arrangement for System A. However, the particles of System B may exchange positions for a second energetically equivalent arrangement.

17.31   (a) $\Delta S > 0$ because a gas is being generated.
              (b) $\Delta S < 0$ because 2 moles of gas are being converted to 1 mole of gas.
              (c) $\Delta S < 0$ because a gas is being converted to a solid.
              (d) $\Delta S < 0$ because 4 moles of gas are being converted to 2 moles of gas.

17.33   (a) $\Delta S_{sys} > 0$ because 6 moles of gas are being converted to 7 moles of gas. Because $\Delta H < 0$ and $\Delta S_{surr} > 0$, the reaction is spontaneous at all temperatures.

(b) $\Delta S_{sys} < 0$ because 2 moles of different gases are being converted to 2 moles of one gas. Because $\Delta H > 0$ and $\Delta S_{surr} < 0$, the reaction is nonspontaneous at all temperatures.

(c) $\Delta S_{sys} < 0$ because 3 moles of gas are being converted to 2 moles of gas. Because $\Delta H > 0$ and $\Delta S_{surr} < 0$, the reaction is nonspontaneous at all temperatures.

(d) $\Delta S_{sys} > 0$ because 9 moles of gas are being converted to 10 moles of gas. Because $\Delta H < 0$ and $\Delta S_{surr} > 0$, the reaction is spontaneous at all temperatures.

17.35   (a)   **Given:** $\Delta H^{\circ}_{rxn} = -385$ kJ, $T = 298$ K   **Find:** ΔS_{surr}
Conceptual Plan: kJ $\rightarrow$ J then $\Delta H^{\circ}_{rxn}, T \rightarrow \Delta S_{surr}$

$$\frac{1000 \text{ J}}{1 \text{ kJ}} \qquad \Delta S_{surr} = \frac{-\Delta H_{sys}}{T}$$

Solution: $-385 \text{ kJ} \times \dfrac{1000 \text{ J}}{1 \text{ kJ}} = -385{,}000$ J then

$$\Delta S_{surr} = \frac{-\Delta H_{sys}}{T} = \frac{-(-385{,}000 \text{ J})}{298 \text{ K}} = 1290 \text{ J/K} = 1.29 \times 10^3 \text{ J/K}$$

Check: The units (J/K) are correct. The magnitude of the answer (10^3 J/KJ) makes sense because the kJ and the temperature started with very similar values and then a factor of 10^3 was applied.

(b)   **Given:** $\Delta H^{\circ}_{rxn} = -385$ kJ, $T = 77$ K   **Find:** ΔS_{surr}
Conceptual Plan: kJ $\rightarrow$ J then $\Delta H^{\circ}_{rxn}, T \rightarrow \Delta S_{surr}$

$$\frac{1000 \text{ J}}{1 \text{ kJ}} \qquad \Delta S_{surr} = \frac{-\Delta H_{sys}}{T}$$

Solution: $-385 \text{ kJ} \times \dfrac{1000 \text{ J}}{1 \text{ kJ}} = -385{,}000$ J then $\Delta S_{surr} = \dfrac{-\Delta H_{sys}}{T} = \dfrac{-(-385{,}000 \text{ J})}{77 \text{ K}} = 5.00 \times 10^3 \text{ J/K}$

Check: The units (J/K) are correct. The magnitude of the answer (5×10^3 J/K) makes sense because the temperature is much lower than in part (a); so the answer should increase.

(c)   **Given:** $\Delta H^{\circ}_{rxn} = +114$ kJ, $T = 298$ K   **Find:** ΔS_{surr}
Conceptual Plan: kJ $\rightarrow$ J then $\Delta H^{\circ}_{rxn}, T \rightarrow \Delta S_{surr}$

$$\frac{1000 \text{ J}}{1 \text{ kJ}} \qquad \Delta S_{surr} = \frac{-\Delta H_{sys}}{T}$$

Solution: $+114 \text{ kJ} \times \dfrac{1000 \text{ J}}{1 \text{ kJ}} = +114{,}000$ J then $\Delta S_{surr} = \dfrac{-\Delta H_{sys}}{T} = \dfrac{-114{,}000 \text{ J}}{298 \text{ K}} = -383 \text{ J/K}$

Check: The units (J/K) are correct. The magnitude of the answer (-400 J/K) makes sense because the kJ are less and of the opposite sign than part (a); so the answer should decrease.

(d)   **Given:** $\Delta H^{\circ}_{rxn} = +114$ kJ, $T = 77$ K   **Find:** ΔS_{surr}
Conceptual Plan: kJ $\rightarrow$ J then $\Delta H^{\circ}_{rxn}, T \rightarrow \Delta S_{surr}$

$$\frac{1000 \text{ J}}{1 \text{ kJ}} \qquad \Delta S_{surr} = \frac{-\Delta H_{sys}}{T}$$

Solution: $+114 \text{ kJ} \times \dfrac{1000 \text{ J}}{1 \text{ kJ}} = +114{,}000$ J then

$$\Delta S_{surr} = \frac{-\Delta H_{sys}}{T} = \frac{-114{,}000 \text{ J}}{77 \text{ K}} = -1480 \frac{\text{J}}{\text{K}} = -1.48 \times 10^3 \text{ J/K}$$

Check: The units (J/K) are correct. The magnitude of the answer (-2×10^3 J/K) makes sense because the temperature is much lower than in part (c); so the answer should increase.

17.37   (a)   **Given:** $\Delta H^{\circ}_{rxn} = +115$ kJ, $\Delta S_{rxn} = -263$ J/K, $T = 298$ K   **Find:** ΔS_{univ} and spontaneity
Conceptual Plan: kJ $\rightarrow$ J then $\Delta H^{\circ}_{rxn}, T \rightarrow \Delta S_{surr}$ then $\Delta S_{rxn}, \Delta S_{surr} \rightarrow \Delta S_{univ}$

$$\frac{1000 \text{ J}}{1 \text{ kJ}} \qquad \Delta S_{surr} = \frac{-\Delta H_{sys}}{T} \qquad \Delta S_{univ} = \Delta S_{sys} + \Delta S_{surr}$$

Solution: $115 \text{ kJ} \times \dfrac{1000 \text{ J}}{1 \text{ kJ}} = 115{,}000$ J then $\Delta S_{surr} = \dfrac{-\Delta H_{sys}}{T} = \dfrac{-(115{,}000 \text{ J})}{298 \text{ K}} = -385.906 \text{ J/K}$ then

$\Delta S_{univ} = \Delta S_{sys} + \Delta S_{surr} = -263 \text{ J/K} - 385.906 \text{ J/K} = -649 \text{ J/K}$; so the reaction is nonspontaneous.

Check: The units (J/K) are correct. The magnitude of the answer (-650 J/K) makes sense because both terms are negative; so the reaction is nonspontaneous.

(b)   **Given:** $\Delta H°_{rxn} = -115$ kJ, $\Delta S_{rxn} = +263$ J/K, $T = 298$ K   **Find:** ΔS_{univ} and spontaneity
Conceptual Plan: kJ → J then $\Delta H°_{rxn}, T → \Delta S_{surr}$ then $\Delta S_{rxn}, \Delta S_{surr} → \Delta S_{univ}$

$$\frac{1000 \text{ J}}{1 \text{ kJ}} \qquad \Delta S_{surr} = \frac{-\Delta H_{sys}}{T} \qquad \Delta S_{univ} = \Delta S_{sys} + \Delta S_{surr}$$

Solution: $-115 \text{ kJ} \times \dfrac{1000 \text{ J}}{1 \text{ kJ}} = -115,000 \text{ J}$ then $\Delta S_{surr} = \dfrac{-\Delta H_{sys}}{T} = \dfrac{-(-115,000 \text{ J})}{298 \text{ K}} = 385.906 \text{ J/K}$

then $\Delta S_{univ} = \Delta S_{sys} + \Delta S_{surr} = +263 \text{ J/K} + 385.906 \text{ J/K} = +649 \text{ J/K}$; so the reaction is spontaneous.

Check: The units (J/K) are correct. The magnitude of the answer ($+650$ J/K) makes sense because both terms were positive; so the reaction is spontaneous.

(c)   **Given:** $\Delta H°_{rxn} = -115$ kJ, $\Delta S_{rxn} = -263$ J/K, $T = 298$ K   **Find:** ΔS_{univ} and spontaneity
Conceptual Plan: kJ → J then $\Delta H°_{rxn}, T → \Delta S_{surr}$ then $\Delta S_{rxn}, \Delta S_{surr} → \Delta S_{univ}$

$$\frac{1000 \text{ J}}{1 \text{ kJ}} \qquad \Delta S_{surr} = \frac{-\Delta H_{sys}}{T} \qquad \Delta S_{univ} = \Delta S_{sys} + \Delta S_{surr}$$

Solution: $-115 \text{ kJ} \times \dfrac{1000 \text{ J}}{1 \text{ kJ}} = -115,000 \text{ J}$ then $\Delta S_{surr} = \dfrac{-\Delta H_{sys}}{T} = \dfrac{-(-115,000 \text{ J})}{298 \text{ K}} = +385.906 \text{ J/K}$

then $\Delta S_{univ} = \Delta S_{sys} + \Delta S_{surr} = -263 \text{ J/K} + 385.906 \text{ J/K} = +123 \text{ J/K}$; so the reaction is spontaneous.

Check: The units (J/K) are correct. The magnitude of the answer (120 J/K) makes sense because the larger term was positive; so the reaction is spontaneous.

(d)   **Given:** $\Delta H°_{rxn} = -115$ kJ, $\Delta S_{rxn} = -263$ J/K, $T = 615$ K   **Find:** ΔS_{univ} and spontaneity
Conceptual Plan: kJ → J then $\Delta H°_{rxn}, T → \Delta S_{surr}$ then $\Delta S_{rxn}, \Delta S_{surr} → \Delta S_{univ}$

$$\frac{1000 \text{ J}}{1 \text{ kJ}} \qquad \Delta S_{surr} = \frac{-\Delta H_{sys}}{T} \qquad \Delta S_{univ} = \Delta S_{sys} + \Delta S_{surr}$$

Solution: $-115 \text{ kJ} \times \dfrac{1000 \text{ J}}{1 \text{ kJ}} = -115,000 \text{ J}$ then $\Delta S_{surr} = \dfrac{-\Delta H_{sys}}{T} = \dfrac{-(-115,000 \text{ J})}{615 \text{ K}} = +186.992 \text{ J/K}$

then $\Delta S_{univ} = \Delta S_{sys} + \Delta S_{surr} = -263 \text{ J/K} + 186.992 \text{ J/K} = -76 \text{ J/K}$; so the reaction is nonspontaneous.

Check: The units (J/K) are correct. The magnitude of the answer (-80 J/K) makes sense because the larger term was negative, so the reaction is nonspontaneous.

Standard Entropy Changes and Gibbs Free Energy

17.39   (a)   **Given:** $\Delta H°_{rxn} = +115$ kJ, $\Delta S_{rxn} = -263$ J/K, $T = 298$ K   **Find:** ΔG and spontaneity
Conceptual Plan: J/K → kJ/K then $\Delta H°_{rxn}, \Delta S_{rxn}, T → \Delta G$

$$\frac{1000 \text{ J}}{1 \text{ kJ}} \qquad\qquad \Delta G = \Delta H_{rxn} - T\Delta S_{rxn}$$

Solution: $-263 \dfrac{\text{J}}{\text{K}} \times \dfrac{1 \text{ kJ}}{1000 \text{ J}} = -0.263 \text{ kJ/K}$ then

$\Delta G = \Delta H_{rxn} - T\Delta S_{rxn} = +115 \text{ kJ} - \left((298 \text{ K})\left(-0.263 \dfrac{\text{kJ}}{\text{K}}\right)\right) = +1.93 \times 10^2 \text{ kJ} = +1.93 \times 10^5 \text{ J}$; so the reaction is nonspontaneous.

Check: The units (kJ) are correct. The magnitude of the answer ($+190$ kJ) makes sense because both terms are positive; so the reaction is nonspontaneous.

(b)   **Given:** $\Delta H°_{rxn} = -115$ kJ, $\Delta S_{rxn} = +263$ J/K, $T = 298$ K   **Find:** ΔG and spontaneity
Conceptual Plan: J/K → kJ/K then $\Delta H°_{rxn}, \Delta S_{rxn}, T → \Delta G$

$$\frac{1000 \text{ J}}{1 \text{ kJ}} \qquad\qquad \Delta G = \Delta H_{rxn} - T\Delta S_{rxn}$$

Solution: $+263 \dfrac{J}{K} \times \dfrac{1 \text{ kJ}}{1000 \text{ } J} = +0.263 \text{ kJ/K}$ then

$\Delta G = \Delta H_{rxn} - T\Delta S_{rxn} = -115 \text{ kJ} - \left((298 \text{ } K)\left(0.263 \dfrac{\text{kJ}}{K} \right) \right) = -193 \text{ kJ} = -1.93 \times 10^2 \text{ kJ} = -1.93 \times 10^5 \text{ J};$
so the reaction is spontaneous.

Check: The units (kJ) are correct. The magnitude of the answer (-190 kJ) makes sense because both terms are negative; so the reaction is spontaneous.

(c)   **Given:** $\Delta H^{\circ}_{rxn} = -115 \text{ kJ}, \Delta S_{rxn} = -263 \text{ J/K}, T = 298 \text{ K}$ **Find:** ΔG and spontaneity
Conceptual Plan: $\text{J/K} \rightarrow \text{kJ/K}$ then $\Delta H^{\circ}_{rxn}, \Delta S_{rxn}, T \rightarrow \Delta G$

$$\dfrac{1000 \text{ J}}{1 \text{ kJ}} \qquad\qquad \Delta G = \Delta H_{rxn} - T\Delta S_{rxn}$$

Solution: $-263 \dfrac{J}{K} \times \dfrac{1 \text{ kJ}}{1000 \text{ } J} = -0.263 \text{ kJ/K}$ then $\Delta G = \Delta H_{rxn} - T\Delta S_{rxn} =$

$-115 \text{ kJ} - \left((298 \text{ } K)\left(-0.263 \dfrac{\text{kJ}}{K} \right) \right) = -3\underline{6}.626 \text{ kJ} = -3.7 \times 10^1 \text{ kJ} = -3.7 \times 10^4 \text{ J};$ so the reaction is
spontaneous.

Check: The units (kJ) are correct. The magnitude of the answer (-40 kJ) makes sense because the larger term was negative; so the reaction is spontaneous.

(d)   **Given:** $\Delta H^{\circ}_{rxn} = -115 \text{ kJ}, \Delta S_{rxn} = -263 \text{ J/K}, T = 615 \text{ K}$ **Find:** ΔG and spontaneity
Conceptual Plan: $\text{J/K} \rightarrow \text{kJ/K}$ then $\Delta H^{\circ}_{rxn}, \Delta S_{rxn}, T \rightarrow \Delta G$

$$\dfrac{1000 \text{ J}}{1 \text{ kJ}} \qquad\qquad \Delta G = \Delta H_{rxn} - T\Delta S_{rxn}$$

Solution: $-263 \dfrac{J}{K} \times \dfrac{1 \text{ kJ}}{1000 \text{ } J} = -0.263 \text{ kJ/K}$ then

$\Delta G = \Delta H_{rxn} - T\Delta S_{rxn} = -115 \text{ kJ} - \left((615 \text{ } K)\left(-0.263 \dfrac{\text{kJ}}{K} \right) \right) = +47 \text{ kJ} = +4.7 \times 10^4 \text{ J};$ so the reaction
is nonspontaneous.

Check: The units (J/K) are correct. The magnitude of the answer $(+47 \text{ kJ})$ makes sense because the larger term was positive; so the reaction is nonspontaneous.

17.41   **Given:** $\Delta H^{\circ}_{rxn} = -2217 \text{ kJ}, \Delta S_{rxn} = +101.1 \text{ J/K}, T = 25\,^{\circ}\text{C}$ **Find:** ΔG and spontaneity
Conceptual Plan: $^{\circ}\text{C} \rightarrow \text{K}$ then $\text{J/K} \rightarrow \text{kJ/K}$ then $\Delta H^{\circ}_{rxn}, \Delta S_{rxn}, T \rightarrow \Delta G$

$$K = 273.15 + \,^{\circ}\text{C} \qquad \dfrac{1 \text{ kJ}}{1000 \text{ J}} \qquad\qquad \Delta G = \Delta H_{rxn} - T\Delta S_{rxn}$$

Solution: $T = 273.15 + 25\,^{\circ}\text{C} = 298 \text{ K}$ then $+101.1 \dfrac{J}{K} \times \dfrac{1 \text{ kJ}}{1000 \text{ } J} = +0.1011 \dfrac{\text{kJ}}{K}$ then

$\Delta G = \Delta H_{rxn} - T\Delta S_{rxn} = -2217 \text{ kJ} - \left((298 \text{ } K)\left(0.1011 \dfrac{\text{kJ}}{K} \right) \right) = -2247 \text{ kJ} = -2.247 \times 10^6 \text{ J};$ so the reaction is
spontaneous.

Check: The units (kJ) are correct. The magnitude of the answer (-2250 kJ) makes sense because both terms are negative; so the reaction is spontaneous.

17.43

ΔH	ΔS	ΔG	Low Temp.	High Temp.
−	+	−	Spontaneous	Spontaneous
−	−	Temp.-dependent	Spontaneous	Nonspontaneous
+	+	Temp.-dependent	Nonspontaneous	Spontaneous
+	−	+	Nonspontaneous	Nonspontaneous

17.45   The molar entropy of a substance increases with increasing temperatures. The kinetic energy and the molecular motion increase. The substance will have access to an increased number of energy levels.

17.47   (a) $CO_2(g)$ because it has greater molar mass/complexity

      (b) $CH_3OH(g)$ because it is in the gas phase

      (c) $CO_2(g)$ because it has greater molar mass/complexity

      (d) $SiH_4(g)$ because it has greater molar mass

      (e) $CH_3CH_2CH_3(g)$ because it has greater molar mass/complexity

      (f) $NaBr(aq)$ because a solution has more entropy than does a solid crystal

17.49   (a) $He(g) < Ne(g) < SO_2(g) < NH_3(g) < CH_3CH_2OH(g)$. All are in the gas phase. From He to Ne, there is an increase in molar mass; beyond that, the molecules increase in complexity.

      (b) $H_2O(s) < H_2O(l) < H_2O(g)$. Entropy increases as we go from a solid to a liquid to a gas.

      (c) $CH_4(g) < CF_4(g) < CCl_4(g)$. Entropy increases as the molar mass increases.

17.51   (a)   **Given:** $C_2H_4(g) + H_2(g) \rightarrow C_2H_6(g)$ **Find:** ΔS°_{rxn}
           Conceptual Plan: $\Delta S^\circ_{rxn} = \sum n_p S^\circ(\text{products}) - \sum n_r S^\circ(\text{reactants})$
           Solution:

Reactant/Product	S°(J/mol K from Appendix IIB)
$C_2H_4(g)$	219.3
$H_2(g)$	130.7
$C_2H_6(g)$	229.2

           Be sure to pull data for the correct formula and phase.

$$\begin{aligned}
\Delta S^\circ_{rxn} &= \sum n_p S^\circ(\text{products}) - \sum n_r S^\circ(\text{reactants}) \\
&= [1(S^\circ(C_2H_6(g)))] - [1(S^\circ(C_2H_4(g))) + 1(S^\circ(H_2(g)))] \\
&= [1(229.2\,\text{J/K})] - [1(219.3\,\text{J/K}) + 1(130.7\,\text{J/K})] \\
&= [229.2\,\text{J/K}] - [350.0\,\text{J/K}] \\
&= -120.8\,\text{J/K} \quad \text{The moles of gas are decreasing.}
\end{aligned}$$

           Check: The units (J/K) are correct. The answer is negative, which is consistent with 2 moles of gas going to 1 mole of gas.

      (b)   **Given:** $C(s) + H_2O(g) \rightarrow CO(g) + H_2(g)$ **Find:** ΔS°_{rxn}
           Conceptual Plan: $\Delta S^\circ_{rxn} = \sum n_p S^\circ(\text{products}) - \sum n_r S^\circ(\text{reactants})$
           Solution:

Reactant/Product	S°(J/mol K from Appendix IIB)
$C(s)$	5.7
$H_2O(g)$	188.8
$CO(g)$	197.7
$H_2(g)$	130.7

           Be sure to pull data for the correct formula and phase.

$$\begin{aligned}
\Delta S^\circ_{rxn} &= \sum n_p S^\circ(\text{products}) - \sum n_r S^\circ(\text{reactants}) \\
&= [1(S^\circ(CO(g))) + 1(S^\circ(H_2(g)))] - [1(S^\circ(C(s))) + 1(S^\circ(H_2O(g)))] \\
&= [1(197.7\,\text{J/K}) + 1(130.7\,\text{J/K})] - [1(5.7\,\text{J/K}) + 1(188.8\,\text{J/K})] \\
&= [328.4\,\text{J/K}] - [194.5\,\text{J/K}] \\
&= +133.9\,\text{J/K} \quad \text{The moles of gas are increasing.}
\end{aligned}$$

           Check: The units (J/K) are correct. The answer is positive, which is consistent with 1 mole of gas going to 2 moles of gas.

      (c)   **Given:** $CO(g) + H_2O(g) \rightarrow H_2(g) + CO_2(g)$ **Find:** ΔS°_{rxn}
           Conceptual Plan: $\Delta S^\circ_{rxn} = \sum n_p S^\circ(\text{products}) - \sum n_r S^\circ(\text{reactants})$
           Solution:

Reactant/Product	S°(J/mol K from Appendix IIB)
$CO(g)$	197.7
$H_2O(g)$	188.8
$H_2(g)$	130.7
$CO_2(g)$	213.8

Be sure to pull data for the correct formula and phase.

$$\Delta S_{rxn}^\circ = \sum n_p S^\circ (\text{products}) - \sum n_r S^\circ (\text{reactants})$$
$$= [1(S^\circ(H_2(g))) + 1(S^\circ(CO_2(g)))] - [1(S^\circ(CO(g))) + 1(S^\circ(H_2O(g)))]$$
$$= [1(130.7\,\text{J/K}) + 1(213.8\,\text{J/K})] - [1(197.7\,\text{J/K}) + 1(188.8\,\text{J/K})]$$
$$= [344.5\,\text{J/K}] - [386.5\,\text{J/K}]$$
$$= -42.0\,\text{J/K}$$

The change is small because the number of moles of gas is constant.

Check: The units (J/K) are correct. The answer is small and negative, which is consistent with a constant number of moles of gas. Water molecules are bent, and carbon dioxide molecules are linear; so the water has more complexity. Also, carbon monoxide is more complex than is hydrogen gas.

(d)   **Given:** $2\,H_2S(g) + 3\,O_2(g) \rightarrow 2\,H_2O(l) + 2\,SO_2(g)$ **Find:** ΔS_{rxn}°

Conceptual Plan: $\Delta S_{rxn}^\circ = \sum n_p S^\circ (\text{products}) - \sum n_r S^\circ (\text{reactants})$
Solution:

Reactant/Product	S°(J/mol K from Appendix IIB)
$H_2S(g)$	205.8
$O_2(g)$	205.2
$H_2O(l)$	70.0
$SO_2(g)$	248.2

Be sure to pull data for the correct formula and phase.

$$\Delta S_{rxn}^\circ = \sum n_p S^\circ (\text{products}) - \sum n_r S^\circ (\text{reactants})$$
$$= [2(S^\circ(H_2O(l))) + 2(S^\circ(SO_2(g)))] - [2(S^\circ(H_2S(g))) + 3(S^\circ(O_2(g)))]$$
$$= [2(70.0\,\text{J/K}) + 2(248.2\,\text{J/K})] - [2(205.8\,\text{J/K}) + 3(205.2\,\text{J/K})]$$
$$= [636.4\,\text{J/K}] - [1027.2\,\text{J/K}]$$
$$= -390.8\,\text{J/K}$$

The number of moles of gas is decreasing.

Check: The units (J/K) are correct. The answer is negative, which is consistent with a decrease in the number of moles of gas.

17.53   **Given:** $CH_2Cl_2(g)$ formed from elements in standard states   **Find:** ΔS° and rationalize sign
Conceptual Plan: Write a balanced reaction, then $\Delta S_{rxn}^\circ = \sum n_p S^\circ (\text{products}) - \sum n_r S^\circ (\text{reactants}).$
Solution: $C(s) + H_2(g) + Cl_2(g) \rightarrow CH_2Cl_2(g)$

Reactant/Product	S°(J/mol K from Appendix IIB)
$C(s)$	5.7
$H_2(g)$	130.7
$Cl_2(g)$	223.1
$CH_2Cl_2(g)$	270.2

Be sure to pull data for the correct formula and phase.

$$\Delta S_{rxn}^\circ = \sum n_p S^\circ (\text{products}) - \sum n_r S^\circ (\text{reactants})$$
$$= [1(S^\circ(CH_2Cl_2(g)))] - [1(S^\circ(C(s))) + 1(S^\circ(H_2(g))) + 1(S^\circ(Cl_2(g)))]$$
$$= [1(270.2\,\text{J/K})] - [1(5.7\,\text{J/K}) + 1(130.7\,\text{J/K}) + 1(223.1\,\text{J/K})]$$
$$= [270.2\,\text{J/K}] - [359.5\,\text{J/K}]$$
$$= -89.3\,\text{J/K}$$

The moles of gas are decreasing.

Check: The units (J/K) are correct. The answer is negative, which is consistent with 2 moles of gas going to 1 mole of gas.

17.55   **Given:** methanol (CH_3OH) combustion at 25 °C   **Find:** ΔH_{rxn}°, ΔS_{rxn}°, ΔG_{rxn}°, and sponteneity
Conceptual Plan: Write a balanced reaction then $\Delta H_{rxn}^\circ = \sum n_p H_f^\circ (\text{products}) - \sum n_r H_f^\circ (\text{reactants})$ **then**
$\Delta S_{rxn}^\circ = \sum n_p S^\circ (\text{products}) - \sum n_r S^\circ (\text{reactants})$ **then** °C → K **then** J/K → kJ/K **then**

$$K = 273.15 + \,°C \qquad \frac{1\,\text{kJ}}{1000\,\text{J}}$$

ΔH_{rxn}°, ΔS_{rxn}°, $T \rightarrow \Delta G^\circ$.

$$\Delta G = \Delta H_{rxn} - T\Delta S_{rxn}$$

Solution: Combustion is combined with oxygen to form carbon dioxide and water.

$$2 CH_3OH(l) + 3 O_2(g) \rightarrow 2 CO_2(g) + 4 H_2O(g)$$

Reactant/Product	ΔH_f°(kJ/mol from Appendix IIB)
$CH_3OH(l)$	−238.6
$O_2(g)$	0.0
$CO_2(g)$	−393.5
$H_2O(g)$	−241.8

Be sure to pull data for the correct formula and phase.

$$\begin{aligned}
\Delta H_{rxn}^\circ &= \sum n_p \Delta H_f^\circ(products) - \sum n_r \Delta H_f^\circ(reactants) \\
&= [2(\Delta H_f^\circ(CO_2(g))) + 4(\Delta H_f^\circ(H_2O(g)))] - [2(\Delta H_f^\circ(CH_3OH(l))) + 3(\Delta H_f^\circ(O_2(g)))] \\
&= [2(-393.5 \text{ kJ}) + 4(-241.8 \text{ kJ})] - [2(-238.6 \text{ kJ}) + 3(0.0 \text{ kJ})] \\
&= [-1754.2 \text{ kJ}] - [-477.2 \text{ kJ}] \\
&= -1277 \text{ kJ then}
\end{aligned}$$

Reactant/Product	S°(J/mol K from Appendix IIB)
$CH_3OH(l)$	126.8
$O_2(g)$	205.2
$CO_2(g)$	213.8
$H_2O(g)$	188.8

Be sure to pull data for the correct formula and phase.

$$\begin{aligned}
\Delta S_{rxn}^\circ &= \sum n_p S^\circ(products) - \sum n_r S^\circ(reactants) \\
&= [2(S^\circ(CO_2(g))) + 4(S^\circ(H_2O(g)))] - [2(S^\circ(CH_3OH(l))) + 3(S^\circ(O_2(g)))] \\
&= [2(213.8 \text{ J/K}) + 4(188.8 \text{ J/K})] - [2(126.8 \text{ J/K}) + 3(205.2 \text{ J/K})] \\
&= [1182.8 \text{ J/K}] - [869.2 \text{ J/K}] \\
&= 313.6 \text{ J/K}
\end{aligned}$$

then $T = 273.15 + 25\,^\circ\text{C} = 298$ K then $+313.6\dfrac{J}{K} \times \dfrac{1 \text{ kJ}}{1000 \text{ J}} = +0.3136$ kJ/K then

$$\Delta G = \Delta H_{rxn} - T\Delta S_{rxn} = -1277 \text{ kJ} - \left((298 \text{ K}) \left(+0.3136 \dfrac{kJ}{K} \right) \right) = -1370. \text{ kJ} = -1.370 \times 10^6 \text{ J; so the reaction}$$
is spontaneous.

Check: The units (kJ, J/K, and kJ) are correct. Combustion reactions are exothermic, and we see a large negative enthalpy. We expect a large positive entropy because we have an increase in the number of moles of gas. The free energy is the sum of two negative terms; so we expect a large negative free energy, and the reaction is spontaneous.

17.57 (a) **Given:** $N_2O_4(g) \rightarrow 2 NO_2(g)$ at 25 °C

Find: ΔH_{rxn}°, ΔS_{rxn}°, ΔG_{rxn}°, and spontaneity. Can temperature be changed to make it spontaneous?

Conceptual Plan: $\Delta H_{rxn}^\circ = \sum n_p H_f^\circ(products) - \sum n_r H_f^\circ(reactants)$ then

$\Delta S_{rxn}^\circ = \sum n_p S^\circ(products) - \sum n_r S^\circ(reactants)$ then °C → K then J/K → kJ/K then

$$K = 273.15 + ^\circ C \qquad \dfrac{1 \text{ kJ}}{1000 \text{ J}}$$

$\Delta H_{rxn}^\circ, \Delta S_{rxn}^\circ, T \rightarrow \Delta G^\circ$

$$\Delta G = \Delta H_{rxn} - T\Delta S_{rxn}$$

Solution:

Reactant/Product	ΔH_f° (kJ/mol from Appendix IIB)
$N_2O_4(g)$	9.16
$NO_2(g)$	33.2

Be sure to pull data for the correct formula and phase.

$$\begin{aligned}
\Delta H_{rxn}^\circ &= \sum n_p \Delta H_f^\circ(products) - \sum n_r \Delta H_f^\circ(reactants) \\
&= [2(\Delta H_f^\circ(NO_2(g)))] - [1(\Delta H_f^\circ(N_2O_4(g)))] \\
&= [2(33.2 \text{ kJ})] - [1(9.16 \text{ kJ})]
\end{aligned}$$

$$= [66.4 \text{ kJ}] - [9.16 \text{ kJ}])]$$
$$= +57.2\underline{4} \text{ kJ then}$$

Reactant/Product	$S°$(J/mol K from Appendix IIB)
$N_2O_4(g)$	304.4
$NO_2(g)$	240.1

Be sure to pull data for the correct formula and phase.

$$\Delta S°_{\text{rxn}} = \sum n_p S°(\text{products}) - \sum n_r S°(\text{reactants})$$
$$= [2(S°(NO_2(g)))] - [1(S°(N_2O_4(g)))]$$
$$= [2(240.1 \text{ J/K})] - [1(304.4 \text{ J/K})]$$
$$= [480.2 \text{ J/K}] - [304.4 \text{ J/K}]$$
$$= +175.8 \text{ J/K then } T = 273.15 + 25 °C = 298 \text{ K then}$$

$$+175.8\frac{J}{K} \times \frac{1 \text{ kJ}}{1000 \text{ } J} = +0.1758 \text{ kJ then}$$

$$\Delta G° = \Delta H°_{\text{rxn}} - T\Delta S°_{\text{rxn}} = 57.24 - \left((298 \text{ K})\left(+0.1758\frac{\text{kJ}}{\text{K}}\right)\right) = +4.8\underline{5}16 \text{ kJ} = +4.9 \text{ kJ} = +4.9 \times 10^3 \text{ J};$$

so the reaction is nonspontaneous. It can be made spontaneous by raising the temperature.

Check: The units (kJ, J/K, and kJ) are correct. The reaction requires the breaking of a bond, so we expect that this will be an endothermic reaction. We expect a positive entropy change because we are increasing the number of moles of gas. Because the positive enthalpy term dominates at room temperature, the reaction is nonspontaneous. The second term can dominate if we raise the temperature high enough.

(b)   **Given:** $NH_4Cl(s) \rightarrow HCl(g) + NH_3(g)$ at 25 °C
Find: $\Delta H°_{\text{rxn}}$, $\Delta S°_{\text{rxn}}$, $\Delta G°_{\text{rxn}}$, and spontaneity. Can temperature be changed to make it spontaneous?
Conceptual Plan: $\Delta H°_{\text{rxn}} = \sum n_p H°_f(\text{products}) - \sum n_r H°_f(\text{reactants})$ then

$$\Delta S°_{\text{rxn}} = \sum n_p S°(\text{products}) - \sum n_r S°(\text{reactants}) \text{ then } °C \rightarrow K \text{ then } J/K \rightarrow kJ/K \text{ then}$$

$$K = 273.15 + °C \qquad \frac{1 \text{ kJ}}{1000 \text{ J}}$$

$$\Delta H°_{\text{rxn}}, \Delta S°_{\text{rxn}}, T \rightarrow \Delta G°$$
$$\Delta G = \Delta H_{\text{rxn}} - T\Delta S_{\text{rxn}}$$

Solution:

Reactant/Product	$\Delta H°_f$ (kJ/mol from Appendix IIB)
$NH_4Cl(s)$	−314.4
$HCl(g)$	−92.3
$NH_3(g)$	−45.9

Be sure to pull data for the correct formula and phase.

$$\Delta H°_{\text{rxn}} = \sum n_p \Delta H°_f(\text{products}) - \sum n_r \Delta H°_f(\text{reactants})$$
$$= [1(\Delta H°_f(HCl(g))) + 1(\Delta H°_f(NH_3(g)))] - [1(\Delta H°_f(NH_4Cl(g)))]$$
$$= [1(-92.3 \text{ kJ}) + 1(-45.9 \text{ kJ})] - [1(-314.4 \text{ kJ})]$$
$$= [-138.2 \text{ kJ}] - [-314.4 \text{ kJ}]$$
$$= +176.2 \text{ kJ then}$$

Reactant/Product	$S°$(J/mol K from Appendix IIB)
$NH_4Cl(s)$	94.6
$HCl(g)$	186.9
$NH_3(g)$	192.8

Be sure to pull data for the correct formula and phase.

$$\Delta S°_{\text{rxn}} = \sum n_p S°(\text{products}) - \sum n_r S°(\text{reactants})$$
$$= [1(S°(HCl(g))) + 1(S°(NH_3(g)))] - [1(S°(NH_4Cl(g)))]$$
$$= [1(186.9 \text{ J/K}) + 1(192.8 \text{ J/K})] - [1(94.6 \text{ J/K})]$$
$$= [379.7 \text{ J/K}] - [94.6 \text{ J/K}]$$
$$= +285.1 \text{ J/K}$$

then $T = 273.15 + 25 °C = 298 \text{ K then } +285.1\frac{J}{K} \times \frac{1 \text{ kJ}}{1000 \text{ } J} = +0.2851 \text{ kJ/K then}$

$$\Delta G^{\circ} = \Delta H^{\circ}_{rxn} - T\Delta S^{\circ}_{rxn} = +176.2 \text{ kJ} - \left((298 \text{ K}) \left(+0.2851 \frac{\text{kJ}}{\text{K}} \right) \right) = +91.2 \text{ kJ} = +9.12 \times 10^4 \text{ J; so the}$$

reaction is nonspontaneous. It can be made spontaneous by raising the temperature.

Check: The units (kJ, J/K, and kJ) are correct. The reaction requires the breaking of a bond, so we expect that this will be an endothermic reaction. We expect a positive entropy change because we are increasing the number of moles of gas. Because the positive enthalpy term dominates at room temperature, the reaction is nonspontaneous. The second term can dominate if we raise the temperature high enough.

(c)   **Given:** $3 \text{ H}_2(g) + \text{Fe}_2\text{O}_3(s) \rightarrow 2 \text{ Fe}(s) + 3 \text{ H}_2\text{O}(g)$ at $25\,^{\circ}\text{C}$ **Find:** ΔH°_{rnx}, ΔS°_{rxn}, ΔG°_{rxn}, and spontaneity. Can temperature be changed to make it spontaneous?

Conceptual Plan: $\Delta H^{\circ}_{rxn} = \sum n_p H^{\circ}_f(\text{products}) - \sum n_r H^{\circ}_f(\text{reactants})$ then

$\Delta S^{\circ}_{rxn} = \sum n_p S^{\circ}(\text{products}) - \sum n_r S^{\circ}(\text{reactants})$ then $^{\circ}\text{C} \rightarrow \text{K}$ then $\text{J/K} \rightarrow \text{kJ/K}$ then

$$\text{K} = 273.15 + \,^{\circ}\text{C} \qquad \frac{1 \text{ kJ}}{1000 \text{ J}}$$

$\Delta H^{\circ}_{rxn}, \Delta S^{\circ}_{rxn}, T \rightarrow \Delta G^{\circ}$

$$\Delta G = \Delta H_{rxn} - T\Delta S_{rxn}$$

Solution:

Reactant/Product	ΔH°_f (kJ/mol from Appendix IIB)
$\text{H}_2(g)$	0.0
$\text{Fe}_2\text{O}_3(s)$	-824.2
$\text{Fe}(s)$	0.0
$\text{H}_2\text{O}(g)$	-241.8

Be sure to pull data for the correct formula and phase.

$$\begin{aligned}
\Delta H^{\circ}_{rxn} &= \sum n_p \Delta H^{\circ}_f(\text{products}) - \sum n_r \Delta H^{\circ}_f(\text{reactants}) \\
&= [2(\Delta H^{\circ}_f(\text{Fe}(s))) + 3(\Delta H^{\circ}_f(\text{H}_2\text{O}(g)))] - [3(\Delta H^{\circ}_f(\text{H}_2(g))) + 1(\Delta H^{\circ}_f(\text{Fe}_2\text{O}_3(s)))] \\
&= [2(0.0 \text{ kJ}) + 3(-241.8 \text{ kJ})] - [3(0.0 \text{ kJ}) + 1(-824.2 \text{ kJ})] \\
&= [-725.4 \text{ kJ}] - [-824.2 \text{ kJ}] \\
&= +98.8 \text{ kJ then}
\end{aligned}$$

Reactant/Product	S° (J/mol K from Appendix IIB)
$\text{H}_2(g)$	130.7
$\text{Fe}_2\text{O}_3(s)$	87.4
$\text{Fe}(s)$	27.3
$\text{H}_2\text{O}(g)$	188.8

Be sure to pull data for the correct formula and phase.

$$\begin{aligned}
\Delta S^{\circ}_{rxn} &= \sum n_p S^{\circ}(\text{products}) - \sum n_r S^{\circ}(\text{reactants}) \\
&= [2(S^{\circ}(\text{Fe}(s))) + 3(S^{\circ}(\text{H}_2\text{O}(g)))] - [3(S^{\circ}(\text{H}_2(g))) + 1(S^{\circ}(\text{Fe}_2\text{O}_3(s)))] \\
&= [2(27.3 \text{ J/K}) + 3(188.8 \text{ J/K})] - [3(130.7 \text{ J/K}) + 1(87.4 \text{ J/K})] \\
&= [621.0 \text{ J/K}] - [479.5 \text{ J/K}] \\
&= +141.5 \text{ J/K}
\end{aligned}$$

then $T = 273.15 + 25\,^{\circ}\text{C} = 298 \text{ K}$ then $+141.5 \frac{\text{J}}{\text{K}} \times \frac{1 \text{ kJ}}{1000 \text{ J}} = +0.1415 \text{ kJ/K}$ then

$$\Delta G^{\circ} = \Delta H^{\circ}_{rxn} - T\Delta S^{\circ}_{rxn} = +98.8 \text{ kJ} - \left((298 \text{ K}) \left(+0.1415 \frac{\text{kJ}}{\text{K}} \right) \right) = +56.6 \text{ kJ} = +5.66 \times 10^4 \text{ J; so the}$$

reaction is nonspontaneous. It can be made spontaneous by raising the temperature.

Check: The units (kJ, J/K, and kJ) are correct. The reaction requires the breaking of a bond, so we expect that this will be an endothermic reaction. We expect a positive entropy change because there is no change in the number of moles of gas, but the product gas is more complex. Because the positive enthalpy term dominates at room temperature, the reaction is nonspontaneous. The second term can dominate if we raise the temperature high enough. This process is the opposite of rusting, so we are not surprised that it is nonspontaneous.

(d)   **Given:** $N_2(g) + 3H_2(g) \rightarrow 2NH_3(g)$ at 25 °C
       Find: ΔH_{rxn}°, ΔS_{rxn}°, ΔG_{rxn}°, and spontaneity. Can temperature be changed to make it spontaneous?
       Conceptual Plan: $\Delta H_{rxn}^\circ = \sum n_p H_f^\circ(\text{products}) - \sum n_r H_f^\circ(\text{reactants})$ **then**

$\Delta S_{rxn}^\circ = \sum n_p S^\circ(\text{products}) - \sum n_r S^\circ(\text{reactants})$ **then °C → K then J/K → kJ/K then**

$$K = 273.15 + °C \qquad \frac{1\,kJ}{1000\,J}$$

$\Delta H_{rxn}^\circ, \Delta S_{rxn}^\circ, T \rightarrow \Delta G^\circ$

$$\Delta G = \Delta H_{rxn} - T\Delta S_{rxn}$$

Solution:

Reactant/Product	ΔH_f°(kJ/mol from Appendix IIB)
$N_2(g)$	0.0
$H_2(g)$	0.0
$NH_3(g)$	−45.9

Be sure to pull data for the correct formula and phase.

$$\begin{aligned}
\Delta H_{rxn}^\circ &= \sum n_p \Delta H_f^\circ(\text{products}) - \sum n_r \Delta H_f^\circ(\text{reactants}) \\
&= [2(\Delta H_f^\circ(NH_3(g)))] - [1(\Delta H_f^\circ(N_2(g))) + 3(\Delta H_f^\circ(H_2(g)))] \\
&= [2(-45.9\,kJ)] - [1(0.0\,kJ) + 3(0.0\,kJ)] \\
&= [-91.8\,kJ] - [0.0\,kJ] \\
&= -91.8\,kJ \text{ then}
\end{aligned}$$

Reactant/Product	S°(J/mol K from Appendix IIB)
$N_2(g)$	191.6
$H_2(g)$	130.7
$NH_3(g)$	192.8

Be sure to pull data for the correct formula and phase.

$$\begin{aligned}
\Delta S_{rxn}^\circ &= \sum n_p S^\circ(\text{products}) - \sum n_r S^\circ(\text{reactants}) \\
&= [2(S^\circ(NH_3(g)))] - [1(S^\circ(N_2(g))) + 3(S^\circ(H_2(g)))] \\
&= [2(192.8\,J/K)] - [1(191.6\,J/K) + 3(130.7\,J/K)] \\
&= [385.6\,J/K] - [583.7\,J/K] \\
&= -198.1\,J/K
\end{aligned}$$

then $T = 273.15 + 25\,°C = 298\,K$ then $-198.1 \dfrac{J}{K} \times \dfrac{1\,kJ}{1000\,J} = -0.1981\,kJ/K$ then

$$\Delta G^\circ = \Delta H_{rxn}^\circ - T\Delta S_{rxn}^\circ = -91.8\,kJ - \left((298\,K)\left(-0.1981\frac{kJ}{K}\right)\right) = -32.8\,kJ = -3.28 \times 10^4\,J;\text{ so the}$$

reaction is spontaneous.

Check: The units (kJ, J/k, and kJ) are correct. The reaction requires the breaking of a bond, so we expect that this will be an endothermic reaction. We expect a positive entropy change because we are increasing the number of moles of gas. Because the negative enthalpy term dominates at room temperature, the reaction is spontaneous. The second term can dominate if we raise the temperature high enough.

17.59 (a)   **Given:** $N_2O_4(g) \rightarrow 2NO_2(g)$ at 25 °C   **Find:** ΔG_{rxn}° and spontaneity and compare to Problem 17.57
       Determine which method would show how free energy changes with temperature.
       Conceptual Plan: $\Delta G_{rxn}^\circ = \sum n_p \Delta G_f^\circ(\text{products}) - \sum n_r \Delta G_f^\circ(\text{reactants})$ **then compare to Problem 17.57**
       Solution:

Reactant/Product	ΔG_f°(kJ/mol from Appendix IIB)
$N_2O_4(g)$	99.8
$NO_2(g)$	51.3

Be sure to pull data for the correct formula and phase.

$$\begin{aligned}
\Delta G_{rxn}^\circ &= \sum n_p \Delta G_f^\circ(\text{products}) - \sum n_r \Delta G_f^\circ(\text{reactants}) \\
&= [2(\Delta G_f^\circ(NO_2(g)))] - [1(\Delta G_f^\circ(N_2O_4(g)))] \\
&= [2(51.3\,kJ)] - [1(99.8\,kJ)] \\
&= [102.6\,kJ] - [99.8\,kJ] \\
&= +2.8\,kJ
\end{aligned}$$

So the reaction is nonspontaneous. The value is similar to that in Problem 17.57.

Check: The units (kJ) are correct. The free energy of the products is greater than that of the reactants; so the answer is positive, and the reaction is nonspontaneous. The answer is the same as that in Problem 17.57 within the error of the calculation.

(b)   **Given:** $NH_4Cl(s) \rightarrow HCl(g) + NH_3(g)$ at 25 °C
Find: ΔG_{rxn}° and spontaneity and compare to Problem 17.57
Conceptual Plan: $\Delta G_{rxn}^{\circ} = \sum n_p \Delta G_f^{\circ}(\text{products}) - \sum n_r \Delta G_f^{\circ}(\text{reactants})$ **then compare to Problem 17.57**
Solution:

Reactant/Product	$\Delta G_f^{\circ}(\text{kJ/mol from Appendix IIB})$
$NH_4Cl(s)$	-202.9
$HCl(g)$	-95.3
$NH_3(g)$	-16.4

Be sure to pull data for the correct formula and phase.

$$\begin{aligned}
\Delta G_{rxn}^{\circ} &= \sum n_p \Delta G_f^{\circ}(\text{products}) - \sum n_r \Delta G_f^{\circ}(\text{reactants}) \\
&= [1(\Delta G_f^{\circ}(HCl(g))) + 1(\Delta G_f^{\circ}(NH_3(g)))] - [1(\Delta G_f^{\circ}(NH_4Cl(g)))] \\
&= [1(-95.3\,kJ) + 1(-16.4\,kJ)] - [1(-202.9\,kJ)] \\
&= [-111.7\,kJ] - [-202.9\,kJ] \\
&= +91.2\,kJ
\end{aligned}$$

So the reaction is nonspontaneous. The result is the same as that in Problem 17.57.

Check: The units (kJ) are correct. The answer matches the one in Problem 17.57.

(c)   **Given:** $3\,H_2(g) + Fe_2O_3(s) \rightarrow 2\,Fe(s) + 3\,H_2O(g)$ at 25 °C
Find: ΔG_{rxn}° and spontaneity and compare to Problem 17.57
Conceptual Plan: $\Delta G_{rxn}^{\circ} = \sum n_p \Delta G_f^{\circ}(\text{products}) - \sum n_r \Delta G_f^{\circ}(\text{reactants})$ **then compare to Problem 17.57**
Solution:

Reactant/Product	$\Delta G_f^{\circ}(\text{kJ/mol from Appendix IIB})$
$H_2(g)$	0.0
$Fe_2O_3(s)$	-742.2
$Fe(s)$	0.0
$H_2O(g)$	-228.6

Be sure to pull data for the correct formula and phase.

$$\begin{aligned}
\Delta G_{rxn}^{\circ} &= \sum n_p \Delta G_f^{\circ}(\text{products}) - \sum n_r \Delta G_f^{\circ}(\text{reactants}) \\
&= [2(\Delta G_f^{\circ}(Fe(s))) + 3(\Delta G_f^{\circ}(H_2O(g)))] - [3(\Delta G_f^{\circ}(H_2(g))) + 1(\Delta G_f^{\circ}(Fe_2O_3(s)))] \\
&= [2(0.0\,kJ) + 3(-228.6\,kJ)] - [3(0.0\,kJ) + 1(-742.2\,kJ)] \\
&= [-685.8\,kJ] - [-742.2\,kJ] \\
&= +56.4\,kJ
\end{aligned}$$

So the reaction is nonspontaneous. The value is similar to that in Problem 17.57.

Check: The units (kJ) are correct. The answer is the same as that in Problem 17.57 within the error of the calculation.

(d)   **Given:** $N_2(g) + 3\,H_2(g) \rightarrow 2\,NH_3(g)$ at 25 °C
Find: ΔG_{rxn}° and spontaneity and compare to Problem 17.57
Conceptual Plan: $\Delta G_{rxn}^{\circ} = \sum n_p \Delta G_f^{\circ}(\text{products}) - \sum n_r \Delta G_f^{\circ}(\text{reactants})$ **then compare to Problem 17.57**
Solution:

Reactant/Product	$\Delta G_f^{\circ}(\text{kJ/mol from Appendix IIB})$
$N_2(g)$	0.0
$H_2(g)$	0.0
$NH_3(g)$	-16.4

Be sure to pull data for the correct formula and phase.

$$\Delta G_{rxn}^{\circ} = \sum n_p \Delta G_f^{\circ}(\text{products}) - \sum n_r \Delta G_f^{\circ}(\text{reactants})$$

$$= [2(\Delta G_f^\circ(NH_3(g)))] - [1(\Delta G_f^\circ(N_2(g))) + 3(\Delta G_f^\circ(H_2(g)))]$$
$$= [2(-16.4\,kJ)] - [1(0.0\,kJ) + 3(0.0\,kJ)]$$
$$= [-32.8\,kJ] - [0.0\,kJ]$$
$$= -32.8\,kJ$$

So the reaction is spontaneous. The result is the same as that in Problem 17.57.

Check: The units (kJ) are correct. The answer matches the one in Problem 17.57.
Values calculated by the two methods are comparable. The method using ΔH° and ΔS° is longer, but it can be used to determine how ΔG° changes with temperature.

17.61   **Given:** $2\,NO(g) + O_2(g) \rightarrow 2\,NO_2(g)$ **Find:** ΔG_{rxn}° and spontaneity at (a) 298 K, (b) 715 K, and (c) 855 K

Conceptual Plan: $\Delta H_{rxn}^\circ = \sum n_p H_f^\circ(\text{products}) - \sum n_r H_f^\circ(\text{reactants})$ then

$\Delta S_{rxn}^\circ = \sum n_p S^\circ(\text{products}) - \sum n_r S^\circ(\text{reactants})$ then $J/K \rightarrow kJ/K$ then $\Delta H_{rxn}^\circ, \Delta S_{rxn}^\circ, T \rightarrow \Delta G^\circ$

$$\frac{1\,kJ}{1000\,J} \qquad \Delta G = \Delta H_{rxn} - T\Delta S_{rxn}$$

Solution:

Reactant/Product	ΔH_f°(kJ/mol from Appendix IIB)
$NO(g)$	91.3
$O_2(g)$	0.0
$NO_2(g)$	33.2

Be sure to pull data for the correct formula and phase.

$$\Delta H_{rxn}^\circ = \sum n_p \Delta H_f^\circ(\text{products}) - \sum n_r \Delta H_f^\circ(\text{reactants})$$
$$= [2(\Delta H_f^\circ(NO_2(g)))] - [2(\Delta H_f^\circ(NO(g))) + 1(\Delta H_f^\circ(O_2(g)))]$$
$$= [2(33.2\,kJ)] - [2(91.3\,kJ) + 1(0.0\,kJ)]$$
$$= [66.4\,kJ] - [182.6\,kJ]$$
$$= -116.2\,kJ \text{ then}$$

Reactant/Product	S°(J/mol K from Appendix IIB)
$NO(g)$	210.8
$O_2(g)$	205.2
$NO_2(g)$	240.1

Be sure to pull data for the correct formula and phase.

$$\Delta S_{rxn}^\circ = \sum n_p S^\circ(\text{products}) - \sum n_r S^\circ(\text{reactants})$$
$$= [2(S^\circ(NO_2(g)))] - [2(S^\circ(NO(g))) + 1(S^\circ(O_2(g)))]$$
$$= [2(240.1\,J/K)] - [2(210.8\,J/K) + 1(205.2\,J/K)] \quad \text{then} -146.6\frac{J}{K} \times \frac{1\,kJ}{1000\,J} = -0.1466\,kJ/K$$
$$= [480.2\,J/K] - [626.8\,J/K]$$
$$= -146.6\,J/K$$

(a) $\Delta G^\circ = \Delta H_{rxn}^\circ - T\Delta S_{rxn}^\circ = -116.2\,kJ - (298\,K)\left(-0.1466\frac{kJ}{K}\right) = -72.5\,kJ = -7.25 \times 10^4\,J$; so the reaction is spontaneous.

(b) $\Delta G^\circ = \Delta H_{rxn}^\circ - T\Delta S_{rxn}^\circ = -116.2\,kJ - (715\,K)\left(-0.1466\frac{kJ}{K}\right) = -11.4\,kJ = -1.14 \times 10^4\,J$; so the reaction is spontaneous.

(c) $\Delta G^\circ = \Delta H_{rxn}^\circ - T\Delta S_{rxn}^\circ = -116.2\,kJ - (855\,K)\left(-0.1466\frac{kJ}{K}\right) = +9.1\,kJ = +9.1 \times 10^3\,J$; so the reaction is nonspontaneous.

Check: The units (kJ) are correct. The enthalpy term dominates at low temperatures, making the reaction spontaneous. As the temperature increases, the decrease in entropy starts to dominate and in the last case the reaction is nonspontaneous.

17.63   Because the first reaction has Fe_2O_3 as a product and the reaction of interest has it as a reactant, we need to reverse the first reaction. When the reaction direction is reversed, ΔG changes.

$Fe_2O_3(s) \rightarrow 2\,Fe(s) + 3/2\,O_2(g)$ $\hspace{3cm}$ $\Delta G° = +742.2\,kJ$

Because the second reaction has 1 mole of CO as a reactant and the reaction of interest has 3 moles of CO as a reactant, we need to multiply the second reaction and the ΔG by 3.

$3[\,CO(g) + 1/2\,O_2(g) \rightarrow CO_2(g)\,]$ $\hspace{2.5cm}$ $\Delta G° = 3(-257.2\,kJ) = -771.6\,kJ$

Hess's law states that the ΔG of the net reaction is the sum of the ΔG of the steps.

The rewritten reactions are:

$Fe_2O_3(s) \rightarrow 2\,Fe(s) + \cancel{3/2\,O_2(g)}$ $\hspace{3cm}$ $\Delta G° = +742.2\,kJ$

$3\,CO(g) + \cancel{3/2\,O_2(g)} \rightarrow 3\,CO_2(g)$ $\hspace{2.5cm}$ $\Delta G° = -771.6\,kJ$

$\overline{Fe_2O_3(s) \rightarrow 3\,CO(g) \rightarrow 2\,Fe(s) + 3\,CO_2(g)}$ $\hspace{1.5cm}$ $\Delta G°_{rxn} = -29.4\,kJ$

Free Energy Changes, Nonstandard Conditions, and the Equilibrium Constant

17.65   (a)   **Given:** $I_2(s) \rightarrow I_2(g)$ at 25.0 °C   **Find:** $\Delta G°_{rxn}$

Conceptual Plan: $\Delta G°_{rxn} = \sum n_p \Delta G°_f(\text{products}) - \sum n_r \Delta G°_f(\text{reactants})$

Solution:

Reactant/Product	$\Delta G°_f(kJ/mol$ from Appendix IIB$)$
$I_2(s)$	0.0
$I_2(g)$	19.3

Be sure to pull data for the correct formula and phase.

$$\Delta G°_{rxn} = \sum n_p \Delta G°_f(\text{products}) - \sum n_r \Delta G°_f(\text{products})$$
$$= [\,1(\Delta G°_f(I_2(g)))\,] - [\,1(\Delta G°_f(I_2(s)))\,]$$
$$= [\,1(19.3\,kJ)\,] - [\,1(0.0\,kJ)\,]$$
$$= +19.3\,kJ;\ \text{so the reaction is nonspontaneous.}$$

Check: The units (kJ) are correct. The answer is positive because gases have higher free energy than do solids and the free energy change of the reaction is the same as free energy of formation of gaseous iodine.

(b)   **Given:** $I_2(s) \rightarrow I_2(g)$ at 25.0 °C (i) $P_{I_2} = 1.00$ mmHg; (ii) $P_{I_2} = 0.100$ mmHg   **Find:** ΔG_{rxn}

Conceptual Plan: °C $\rightarrow$ K and mmHg $\rightarrow$ atm then $\Delta G°_{rxn}, P_{I_2}, T \rightarrow \Delta G_{rxn}$

$$K = 273.15 + °C \hspace{1cm} \frac{1\,atm}{760\,mmHg} \hspace{1cm} \Delta G_{rxn} = \Delta G°_{rxn} + RT\,\ln Q \hspace{0.3cm} \text{where } Q = P_{I_2}$$

Solution: $T = 273.15 + 25.0\,°C = 298.2\,K$ and (i) $1.00\ \cancel{mmHg} \times \dfrac{1\,atm}{760\ \cancel{mmHg}} = 0.001\underline{3}1579\,atm$

then $\Delta G_{rxn} = \Delta G°_{rxn} + RT\,\ln Q = \Delta G°_{rxn} + RT\,\ln P_{I_2} =$

$+19.3\,kJ + \left(\left(8.314\dfrac{J}{K\cdot mol} \right) \left(\dfrac{1\,kJ}{1000\,J} \right) (298.2\,K)\ \ln(0.001\underline{3}1579) \right) = +2.9\,kJ;$ so the reaction is nonspontaneous.

(ii) $0.100\ \cancel{mmHg} \times \dfrac{1\,atm}{760\ \cancel{mmHg}} = 0.000\underline{1}31579\,atm$ then

$\Delta G_{rxn} = \Delta G°_{rxn} + RT\,\ln Q = \Delta G°_{rxn} + RT\,\ln P_{I_2} =$

$+19.3\,kJ + \left(\left(8.314\dfrac{J}{K\cdot mol} \right) \left(\dfrac{1\,kJ}{1000\,J} \right) (298.2\,K)\ \ln(0.000\underline{1}31579) \right) = -2.9\,kJ;$ so the reaction is spontaneous.

Check: The units (kJ) are correct. The answer is positive at higher pressure because the pressure is higher than the vapor pressure of iodine. Once the desired pressure is below the vapor pressure (0.31 mmHg at 25.0 °C), the reaction becomes spontaneous.

(c)   Iodine sublimes at room temperature because there is an equilibrium between the solid and the gas phases. The vapor pressure is low (0.31 mmHg at 25.0 °C), so a small amount of iodine can remain in the gas phase, which is consistent with the free energy values.

17.67 **Given:** $CH_3OH(g) \rightleftharpoons CO(g) + 2 H_2(g)$ at 25 °C, $P_{CH_3OH} = 0.855$ atm, $P_{CO} = 0.125$ atm, $P_{H_2} = 0.183$ atm
Find: ΔG
Conceptual Plan:
$\Delta G_{rxn}^{\circ} = \sum n_p \Delta G_f^{\circ}(\text{products}) - \sum n_r \Delta G_f^{\circ}(\text{reactants})$ then °C → K then $\Delta G_{rxn}^{\circ}, P_{CH_3OH}, P_{CO}, P_{H_2}, T \to \Delta G$

$K = 273.15 + \text{°C} \quad \Delta G_{rxn} = \Delta G_{rxn}^{\circ} + RT \ln Q \quad \text{where } Q = \dfrac{P_{CO}P_{H_2}^2}{P_{CH_3OH}}$

Solution:

Reactant/Product	ΔG_f°(kJ/mol from Appendix IIB)
$CH_3OH(g)$	−162.3
$CO(g)$	−137.2
$H_2(g)$	0.0

Be sure to pull data for the correct formula and phase.

$$\begin{aligned}
\Delta G_{rxn}^{\circ} &= \sum n_p \Delta G_f^{\circ}(\text{products}) - \sum n_r \Delta G_f^{\circ}(\text{reactants}) \\
&= \left[1(\Delta G_f^{\circ}(CO(g))) + 2(\Delta G_f^{\circ}(H_2(g))) \right] - \left[1(\Delta G_f^{\circ}(CH_3OH(g))) \right] \\
&= \left[1(-137.2 \text{ kJ}) + 2(0.0 \text{ kJ}) \right] - \left[1(-162.3 \text{ kJ}) \right] \\
&= \left[-137.2 \text{ kJ} \right] - \left[-162.3 \text{ kJ} \right] \\
&= +25.1 \text{ kJ}
\end{aligned}$$

$T = 273.15 + 25 \text{ °C} = 298 \text{ K}$ then $Q = \dfrac{P_{CO}P_{H_2}^2}{P_{CH_3OH}} = \dfrac{(0.125)(0.183)^2}{0.855} = 0.00489605$ then

$\Delta G_{rxn} = \Delta G_{rxn}^{\circ} + RT \ln Q = +25.1 \text{ kJ} + \left(\left(8.314 \dfrac{J}{K \cdot mol} \right) \left(\dfrac{1 \text{ kJ}}{1000 \text{ J}} \right) (298 \text{ K}) \ln (0.00489605) \right) = +11.9 \text{ kJ};$

so the reaction is nonspontaneous.

Check: The units (kJ) are correct. The standard free energy for the reaction was positive, and the fact that Q was less than 1 made the free energy smaller. But the reaction at these conditions is still not spontaneous.

17.69 (a) **Given:** $2 CO(g) + O_2(g) \rightleftharpoons 2 CO_2(g)$ at 25 °C **Find:** K
Conceptual Plan:
$\Delta G_{rxn}^{\circ} = \sum n_p \Delta G_f^{\circ}(\text{products}) - \sum n_r \Delta G_f^{\circ}(\text{reactants})$ then °C → K then $\Delta G_{rxn}^{\circ}, T \to K$

$K = 273.15 + \text{°C} \qquad \Delta G_{rxn}^{\circ} = -RT \ln K$

Solution:

Reactant/Product	ΔG_f°(kJ/mol from Appendix IIB)
$CO(g)$	−137.2
$O_2(g)$	0.0
$CO_2(g)$	−394.4

Be sure to pull data for the correct formula and phase.

$$\begin{aligned}
\Delta G_{rxn}^{\circ} &= \sum n_p \Delta G_f^{\circ}(\text{products}) - \sum n_r \Delta G_f^{\circ}(\text{reactants}) \\
&= \left[2(\Delta G_f^{\circ}(CO_2(g))) \right] - \left[2(\Delta G_f^{\circ}(CO(g))) + 1(\Delta G_f^{\circ}(O_2(g))) \right] \\
&= \left[2(-394.4 \text{ kJ}) \right] - \left[2(-137.2 \text{ kJ}) + 1(0.0 \text{ kJ}) \right] \\
&= \left[-788.8 \text{ kJ} \right] - \left[-274.4 \text{ kJ} \right] \\
&= -514.4 \text{ kJ} \qquad\qquad\qquad T = 273.15 + 25 \text{ °C} = 298 \text{ K then}
\end{aligned}$$

$\Delta G_{rxn}^{\circ} = -RT \ln K$ Rearrange to solve for K.

$$K = e^{\frac{-\Delta G_{rxn}^{\circ}}{RT}} = e^{\dfrac{-(-514.4 \text{ kJ}) \times \frac{1000 \text{ J}}{1 \text{ kJ}}}{\left(8.314 \frac{J}{K \cdot mol} \right)(298 \text{ K})}} = e^{207.623} = 1.48 \times 10^{90}$$

Check: The units (none) are correct. The standard free energy for the reaction was very negative, so we expect a very large K. The reaction is spontaneous, so mostly products are present at equilibrium.

(b) **Given:** $2\,H_2S(g) \rightleftharpoons 2\,H_2(g) + S_2(g)$ at 25 °C **Find:** K

Conceptual Plan:

$\Delta G°_{rxn} = \sum n_p \Delta G°_f(\text{products}) - \sum n_r \Delta G°_f(\text{reactants})$ then °C → K then $\Delta G°_{rxn}, T \to K$

$K = 273.15 + °C \qquad \Delta G°_{rxn} = -RT \ln K$

Solution:

Reactant/Product	$\Delta G°_f$(kJ/mol from Appendix IIB)
$H_2S(g)$	−33.4
$H_2(g)$	0.0
$S_2(g)$	79.7

Be sure to pull data for the correct formula and phase.

$$\Delta G°_{rxn} = \sum n_p \Delta G°_f(\text{products}) - \sum n_r \Delta G°_f(\text{reactants})$$
$$= [2(\Delta G°_f(H_2(g))) + 1(\Delta G°_f(S_2(g)))] - [2(\Delta G°_f(H_2S(g)))]$$
$$= [2(0.0\,kJ)] + 1(79.7\,kJ)] - [2(-33.4\,kJ)]$$
$$= [79.7\,kJ] - [-66.8\,kJ]$$
$$= +146.5\,kJ \qquad\qquad T = 273.15 + 25\,°C = 298\,K \text{ then}$$

$\Delta G°_{rxn} = -RT \ln K$ Rearrange to solve for K.

$$K = e^{\frac{-\Delta G°_{rxn}}{RT}} = e^{\frac{-146.5\,kJ \times \frac{1000\,J}{1\,kJ}}{\left(8.314\frac{J}{K\cdot mol}\right)(298\,K)}} = e^{-59.1305} = 2.09 \times 10^{-26}$$

Check: The units (none) are correct. The standard free energy for the reaction was positive, so we expect a small K. The reaction is nonspontaneous, so mostly reactants are present at equilibrium.

17.71 **Given:** $CO(g) + 2\,H_2(g) \rightleftharpoons CH_3OH(g)$ $K_p = 2.26 \times 10^4$ at 25 °C

Find: $\Delta G°_{rxn}$ at (a) standard conditions; (b) at equilibrium; and (c) $P_{CH_3OH} = 1.0$ atm, $P_{CO} = P_{H_2} = 0.010$ atm

Conceptual Plan: °C → K then (a) $K, T \to \Delta G°_{rxn}$ then (b) at equilibrium, $\Delta G_{rxn} = 0$ then

$K = 273.15 + °C \qquad \Delta G°_{rxn} = -RT \ln K$

(c) $\Delta G°_{rxn}, P_{CH_3OH}, P_{CO}, P_{H_2}, T \to \Delta G$

$\Delta G_{rxn} = \Delta G°_{rxn} + RT \ln Q \qquad$ where $Q = \dfrac{P_{CH_3OH}}{P_{CO}P_{H_2}^2}$

Solution: $T = 273.15 + 25\,°C = 298\,K$ then

(a) $\Delta G°_{rxn} = -RT \ln K = -\left(8.314\dfrac{J}{K\cdot mol}\right)\left(\dfrac{1\,kJ}{1000\,J}\right)(298\,K)\ln(2.26\times10^4) = -24.8\,kJ$

(b) at equilibrium, $\Delta G_{rxn} = 0$

(c) $Q = \dfrac{P_{CH_3OH}}{P_{CO}P_{H_2}^2} = \dfrac{1.0}{(0.010)(0.010)^2} = 1.0\times10^6$ then

$\Delta G_{rxn} = \Delta G°_{rxn} + RT \ln Q = -24.8\,kJ + \left(8.314\dfrac{J}{K\cdot mol}\right)\left(\dfrac{1\,kJ}{1000\,J}\right)(298\,K)\ln(1.0\times10^6) = +9.4\,kJ$

Check: The units (kJ) are correct. The K was greater than 1, so we expect a negative standard free energy for the reaction. At equilibrium, by definition, the free energy change is zero. Because the conditions give a $Q > K$, the reaction needs to proceed in the reverse direction, which means that the reaction is spontaneous in the reverse direction.

17.73 (a) **Given:** $2\,CO(g) + O_2(g) \rightleftharpoons 2\,CO_2(g)$ at 25 °C **Find:** K at 525 K

Conceptual Plan: $\Delta H°_{rxn} = \sum n_p H°_f(\text{products}) - \sum n_r H°_f(\text{reactants})$ then

$\Delta S°_{rxn} = \sum n_p S°(\text{products}) - \sum n_r S°(\text{reactants})$ then J/K → kJ/K then $\Delta H°_{rxn}, \Delta S°_{rxn}, T \to \Delta G$

$\dfrac{1\,kJ}{1000\,J} \qquad \Delta G = \Delta H_{rxn} - T\Delta S_{rxn}$

then $\Delta G°_{rxn}, T \to K$

$\Delta G°_{rxn} = -RT \ln K$

Solution:

Reactant/Product	ΔH_f°(kJ/mol from Appendix IIB)
$CO(g)$	−110.5
$O_2(g)$	0.0
$CO_2(g)$	−393.5

Be sure to pull data for the correct formula and phase.

$$\begin{aligned}\Delta H_{rxn}^\circ &= \sum n_p \Delta H_f^\circ(\text{products}) - \sum n_r \Delta H_f^\circ(\text{reactants}) \\ &= [2(\Delta H_f^\circ(CO_2(g)))] - [2(\Delta H_f^\circ(CO(g))) + 1(\Delta H_f^\circ(O_2(g)))] \\ &= [2(-393.5\text{ kJ})] - [2(-110.5\text{ kJ}) + 1(0.0\text{ kJ})] \text{ then} \\ &= [-787.0\text{ kJ}] - [-221.0\text{ kJ}] \\ &= -566.0\text{ kJ}\end{aligned}$$

Reactant/Product	S°(J/mol K from Appendix IIB)
$CO(g)$	197.7
$O_2(g)$	205.2
$CO_2(g)$	213.8

Be sure to pull data for the correct formula and phase.

$$\begin{aligned}\Delta S_{rxn}^\circ &= \sum n_p S^\circ(\text{products}) - \sum n_r S^\circ(\text{reactants}) \\ &= [2(S^\circ(CO_2(g)))] - [2(S^\circ(CO(g))) + 1(S^\circ(O_2(g)))] \\ &= [2(213.8\text{ J/K})] - [2(197.7\text{ J/K}) + 1(205.2\text{ J/K})] \\ &= [427.6\text{ J/K}] - [600.6\text{ J/K}] \text{ then} \\ &= -173.0\text{ K/K}\end{aligned}$$

$$-173.0\frac{\cancel{\text{J}}}{\text{K}} \times \frac{1\text{ kJ}}{1000\,\cancel{\text{J}}} = -0.1730\text{ kJ/K then}$$

$$\Delta G^\circ = \Delta H_{rxn}^\circ - T\Delta S_{rxn}^\circ = -566.0\text{ kJ} - (525\,\cancel{\text{K}})\left(-0.1730\frac{\text{kJ}}{\cancel{\text{K}}}\right) = -475.2\text{ kJ} = -4.752 \times 10^5\text{ J then}$$

$\Delta G_{rxn}^\circ = -RT\ln K$ Rearrange to solve for K.

$$K = e^{\frac{-\Delta G_{rxn}^\circ}{RT}} = e^{\frac{-(-4.752\times10^5\,\cancel{\text{J}})}{\left(8.314\frac{\cancel{\text{J}}}{\cancel{\text{K}}\cdot\text{mol}}\right)(525\,\cancel{\text{K}})}} = e^{10\underline{8}.870} = 1.91 \times 10^{47}$$

Check: The units (none) are correct. The free energy change was very negative, indicating a spontaneous reaction. This results in a very large K.

(b) **Given:** $2\,H_2S(g) \rightleftharpoons 2\,H_2(g) + S_2(g)$ at 25 °C **Find:** K at 525 K

Conceptual Plan: $\boldsymbol{\Delta H_{rxn}^\circ = \sum n_p H_f^\circ(\text{products}) - \sum n_r H_f^\circ(\text{reactants})}$ **then**

$\boldsymbol{\Delta S_{rxn}^\circ = \sum n_p S^\circ(\text{products}) - \sum n_r S^\circ(\text{reactants})}$ **then J/K → kJ/K then** $\boldsymbol{\Delta H_{rxn}^\circ, \Delta S_{rxn}, T \rightarrow \Delta G}$

$\frac{1\text{ kJ}}{1000\text{ J}}$ $\qquad \Delta G = \Delta H_{rxn} - T\Delta S_{rxn}$

then $\boldsymbol{\Delta G_{rxn}^\circ, T \rightarrow K}$

$\Delta G_{rxn}^\circ = -RT\ln K$

Solution:

Reactant/Product	ΔH_f°(kJ/mol from Appendix IIB)
$H_2S(g)$	−20.6
$H_2(g)$	0.0
$S_2(g)$	128.6

Be sure to pull data for the correct formula and phase.

$$\begin{aligned}\Delta H_{rxn}^\circ &= \sum n_p \Delta H_f^\circ(\text{products}) - \sum n_r \Delta H_f^\circ(\text{reactants}) \\ &= [2(\Delta H_f^\circ(H_2(g))) + 1(\Delta H_f^\circ(S_2(g)))] - [2(\Delta H_f^\circ(H_2S(g)))] \\ &= [2(0.0\text{ kJ}) + 1(128.6\text{ kJ})] - [2(-20.6\text{ kJ})] \\ &= [128.6\text{ kJ}] - [-41.2\text{ kJ}] \\ &= +169.8\text{ kJ}\end{aligned}$$

Reactant/Product	$S°$ (J/mol K from Appendix IIB)
$H_2S(g)$	205.8
$H_2(g)$	130.7
$S_2(g)$	228.2

Be sure to pull data for the correct formula and phase.

$$
\begin{aligned}
\Delta S°_{rxn} &= \sum n_p S°(\text{products}) - \sum n_r S°(\text{reactants}) \\
&= [2(S°(H_2(g))) + 1(S°(S_2(g)))] - [2(S°(H_2S(g)))] \\
&= [2(130.7 \text{ J/K}) + 1(228.2 \text{ J/K})] - [2(205.8 \text{ J/K})] \\
&= [489.6 \text{ J/K}] - [411.6 \text{ J/K}] \\
&= +78.0 \text{ J/K}
\end{aligned}
$$

then $+78.0\dfrac{J}{K} \times \dfrac{1 \text{ kJ}}{1000 \text{ J}} = +0.0780 \text{ kJ/K}$ then

$$
\Delta G° = \Delta H°_{rxn} - T\Delta S°_{rxn} = +169.8 \text{ kJ} - (525 \text{ K})\left(+0.0780\dfrac{\text{kJ}}{\text{K}}\right) = +128.\underline{85} \text{ kJ} = +1.28\underline{85} \times 10^5 \text{ J} \text{ then}
$$

$$\Delta G°_{rxn} = -RT \ln K \quad \text{Rearrange to solve for } K.$$

$$
K = e^{\frac{-\Delta G°_{rxn}}{RT}} = e^{\dfrac{-1.28\underline{85} \times 10^5 \text{ J}}{\left(8.314\frac{J}{K \cdot mol}\right)(525 \text{ K})}} = e^{-29.\underline{5}199} = 1.51 \times 10^{-13}
$$

Check: The units (none) are correct. The free energy change is positive, indicating a nonspontaneous reaction. This results in a very small K.

17.75   **Given:** table of K_p versus temperature   **Find:** $\Delta H°_{rxn}$, $\Delta S°_{rxn}$
Conceptual Plan: Plot ln K verus $1/T$. The slope will be $-\Delta H°_{rxn}/R$, and the intercept will be $\Delta S°_{rxn}/R$.

Solution: Plot ln K verus $1/T$. Because $\ln K = -\dfrac{\Delta H°_{rxn}}{R}\dfrac{1}{T} + \dfrac{\Delta S°_{rxn}}{R}$, the negative of the slope will

be $-\dfrac{\Delta H°_{rxn}}{R}$ and the intercept will be $\dfrac{\Delta S°_{rxn}}{R}$. The slope can be determined by measuring $\Delta y/\Delta x$ on

the plot or by using functions such as "add trendline" in Excel. Because the slope is $-60\underline{9}2.2$ K,

$$
\Delta H°_{rxn} = -slope\ R = -(-60\underline{9}2.2 \text{ K})\left(8.314\dfrac{J}{K \cdot mol}\right)\left(\dfrac{1 \text{ kJ}}{1000 \text{ J}}\right) = 50.\underline{6}1 \text{ kJ/mol} = 50.6 \text{ kJ/mol. Because the}
$$

intercept is $27.\underline{1}36 \text{ K}^{-1}$, $\Delta S°_{rxn} = intercept\ R = \left(\dfrac{27.\underline{1}36}{\text{K}}\right)\left(8.314\dfrac{J}{K \cdot mol}\right) = 225.\underline{6}09 \text{ J/K} = 226 \text{ J/K.}$

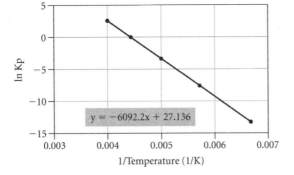

Check: The units are correct (kJ/mole and J/K). The plot is very linear. The numbers are typical for reactions. Because the slope is negative, the enthalpy change must be positive.

17.77   **Given:** $\Delta H°_{rxn} = -25.8 \text{ kJ/mol}$, $K = 1.4 \times 10^3$ at 298 K   **Find:** K at 655 K
Conceptual Plan: $\Delta H°_{rxn}, K_1, T_1, T_2 \rightarrow K_2$

$$\ln K = -\dfrac{\Delta H°_{rxn}}{R}\dfrac{1}{T} + \dfrac{\Delta S°_{rxn}}{R}$$

Solution: Because $\ln K = -\dfrac{\Delta H^\circ_{rxn}}{R}\dfrac{1}{T} + \dfrac{\Delta S^\circ_{rxn}}{R}$, $\ln K_1 + \dfrac{\Delta H^\circ_{rxn}}{R}\dfrac{1}{T_1} = \dfrac{\Delta S^\circ_{rxn}}{R} = \ln K_2 + \dfrac{\Delta H^\circ_{rxn}}{R}\dfrac{1}{T_2}$.

Rearrange to solve for K_2.

$$\ln K_2 = \ln K_1 + \dfrac{\Delta H^\circ_{rxn}}{R}\left(\dfrac{1}{T_1} - \dfrac{1}{T_2}\right) = \ln(1.4 \times 10^3) + \dfrac{\dfrac{-25.8\ \text{kJ}}{\text{mol}} \times \dfrac{1000\ \text{J}}{1\ \text{kJ}}}{8.314\dfrac{\text{J}}{\text{K}\cdot\text{mol}}}\left(\dfrac{1}{298\ \text{K}} - \dfrac{1}{655\ \text{K}}\right) = 1.\underline{5}6852\ \text{so}$$

$$K_2 = e^{1.\underline{5}6852} = 4.\underline{7}9954 = 4.8$$

Check: The units are correct (none). Because the reaction is exothermic, we expect the K to decrease with increasing temperature.

Cumulative Problems

17.79   (a)   + because vapors have higher entropy than do liquids

         (b)   − because solids have less entropy than do liquids

         (c)   − because there is only one microstate for the final macrostate and there are six microstates for the initial macrostate

17.81   (a)   **Given:** $N_2(g) + O_2(g) \rightarrow 2\,NO(g)$ **Find:** ΔG°_{rxn} and K_p at 25 °C

            Conceptual Plan: $\Delta H^\circ_{rxn} = \sum n_p H^\circ_f(\text{products}) - \sum n_r H^\circ_f(\text{reactants})$ then

            $\Delta S^\circ_{rxn} = \sum n_p S^\circ(\text{products}) - \sum n_r S^\circ(\text{reactants})$ then °C → K then J/K → kJ/K then

$$K = 273.15 + °C \qquad \dfrac{1\ \text{kJ}}{1000\ \text{J}}$$

            $\Delta H^\circ_{rxn}, \Delta S^\circ_{rxn}, T \rightarrow \Delta G$ then $\Delta G^\circ_{rxn}, T \rightarrow K$

$$\Delta G = \Delta H_{rxn} - T\Delta S_{rxn} \qquad \Delta G^\circ_{rxn} = -RT \ln K$$

            Solution:

Reactant/Product	ΔH°_f(kJ/mol from Appendix IIB)
$N_2(g)$	0.0
$O_2(g)$	0.0
$NO(g)$	91.3

            Be sure to pull data for the correct formula and phase.

$$\begin{aligned}
\Delta H^\circ_{rxn} &= \sum n_p \Delta H^\circ_f(\text{products}) - \sum n_r \Delta H^\circ_f(\text{reactants})\\
&= [2(\Delta H^\circ_f(NO(g)))] - [1(\Delta H^\circ_f(N_2(g))) + 1(\Delta H^\circ_f(O_2(g)))]\\
&= [2(91.3\ \text{kJ})] - [1(0.0\ \text{kJ}) + 1(0.0\ \text{kJ})]\\
&= [182.6\ \text{kJ}] - [0.0\ \text{kJ}]\\
&= +182.6\ \text{kJ then}
\end{aligned}$$

Reactant/Product	S°(J/mol K from Appendix IIB)
$N_2(g)$	191.6
$O_2(g)$	205.2
$NO(g)$	210.8

            Be sure to pull data for the correct formula and phase.

$$\begin{aligned}
\Delta S^\circ_{rxn} &= \sum n_p S^\circ(\text{products}) - \sum n_r S^\circ(\text{reactants})\\
&= [2(S^\circ(NO(g)))] - [1(S^\circ(N_2(g))) + 1(S^\circ(O_2(g)))]\\
&= [2(210.8\ \text{J/K})] - [1(191.6\ \text{J/K}) + 1(205.2\ \text{J/K})]\\
&= [421.6\ \text{J/K}] - [396.8\ \text{JK}]\\
&= +24.8\ \text{J/K}
\end{aligned}$$

            then $T = 273.15 + 25\ °C = 298\ \text{K}$ then $+24.8\dfrac{\text{J}}{\text{K}} \times \dfrac{1\ \text{kJ}}{1000\ \text{J}} = +0.0248\ \text{kJ/K}$ then

$$\Delta G^\circ = \Delta H^\circ_{rxn} - T\Delta S^\circ_{rxn} = +182.6\ \text{kJ} - (298\ \text{K})\left(0.0248\dfrac{\text{kJ}}{\text{K}}\right) = +175.2\ \text{kJ} = +1.752 \times 10^5\ \text{J}$$

            then $\Delta G^\circ_{rxn} = -RT \ln K$ Rearrange to solve for K.

$$K = e^{\frac{-\Delta G^\circ_{rxn}}{RT}} = e^{\frac{-1.752 \times 10^5 \, \cancel{J}}{\left(8.314 \frac{\cancel{J}}{K \cdot mol}\right)(298 \, K)}} = e^{-70.\underline{7}144} = 1.95 \times 10^{-31}; \text{ so the reaction is nonspontaneous, and at}$$

equilibrium, mostly reactants are present.

Check: The units (kJ and none) are correct. The enthalpy is twice the enthalpy of formation of NO. We expect a very small entropy change because the number of moles of gas is unchanged. Because the positive enthalpy term dominates at room temperature, the free energy change is very positive and the reaction in the forward direction is nonspontaneous. This results in a very small K.

(b)   **Given:** $N_2(g) + O_2(g) \rightarrow 2\,NO(g)$ **Find:** ΔG°_{rxn} at 2000 K
     Conceptual Plan: Use results from part (a) $\Delta H^\circ_{rxn}, \Delta S_{rxn}, T \rightarrow \Delta G$ then $\Delta G^\circ_{rxn}, T \rightarrow K$

$$\Delta G = \Delta H_{rxn} - T\Delta S_{rxn} \qquad \Delta G^\circ_{rxn} = -RT \ln K$$

Solution: $\Delta G = \Delta H_{rxn} - T\Delta S_{rxn} = +182.6 \, kJ - (2000 \, K)\left(0.0248 \frac{kJ}{K}\right) = +133.0 \, kJ = +1.330 \times 10^5 \, J$

then $\Delta G^\circ_{rxn} = -RT \ln K$ Rearrange to solve for K.

$$K = e^{\frac{-\Delta G^\circ_{rxn}}{RT}} = e^{\frac{-1.330 \times 10^5 \, \cancel{J}}{\left(8.314 \frac{\cancel{J}}{K \cdot mol}\right)(2000 \, K)}} = e^{-7.998557} = 3.36 \times 10^{-4}; \text{ so the forward reaction is becoming more}$$

spontaneous.

Check: The units (kJ and none) are correct. As the temperature rises, the entropy term becomes more significant. The free energy change is reduced, and the K increases. The reaction is still nonspontaneous.

17.83   **Given:** $C_2H_4(g) + X_2(g) \rightarrow C_2H_4X_2(g)$ where X = Cl, Br, and I
     Find: $\Delta H^\circ_{rxn}, \Delta S^\circ_{rxn}, \Delta G^\circ_{rxn}$, and K at 25 °C and spontaneity trends with X and temperature
     Conceptual Plan: $\Delta H^\circ_{rxn} = \sum n_p H^\circ_f(\text{products}) - \sum n_r H^\circ_f(\text{reactants})$ then
     $\Delta S^\circ_{rxn} = \sum n_p S^\circ(\text{products}) - \sum n_r S^\circ(\text{reactants})$ then °C → K then J/K → kJ/K then

$$K = 273.15 + \text{°C} \qquad \frac{1 \, kJ}{1000 \, J}$$

$\Delta H^\circ_{rxn}, \Delta S^\circ_{rxn}, T \rightarrow \Delta G$ then $\Delta G^\circ_{rxn}, T \rightarrow K$

$$\Delta G = \Delta H_{rxn} - T\Delta S_{rxn} \qquad \Delta G^\circ_{rxn} = -RT \ln K$$

Solution:

Reactant/Product	ΔH°_f(kJ/mol from Appendix IIB)
$C_2H_4(g)$	52.4
$Cl_2(g)$	0.0
$C_2H_4Cl_2(g)$	−129.7

Be sure to pull data for the correct formula and phase.

$\Delta H^\circ_{rxn} = \sum n_p \Delta H^\circ_f(\text{products}) - \sum n_r \Delta H^\circ_f(\text{reactants})$
$= [1\Delta H^\circ_f(C_2H_4Cl_2)(g)))] - [1(\Delta H^\circ_f(C_2H_4(g))) + 1(\Delta H^\circ_f(Cl_2(g)))]$
$= [1(-129.7 \, kJ)] - [1(52.4 \, kJ) + 1(0.0 \, kJ)]$
$= [-129.7 \, kJ] - [52.4 \, kJ]$
$= -182.1 \, kJ$ then

Reactant/Product	S°(J/mol K from Appendix IIB)
$C_2H_4(g)$	219.3
$Cl_2(g)$	223.1
$C_2H_4Cl_2(g)$	308.0

Be sure to pull data for the correct formula and phase.

$\Delta S^\circ_{rxn} = \sum n_p S^\circ(\text{products}) - \sum n_r S^\circ(\text{reactants})$
$= [1(S^\circ(C_2H_4Cl_2(g)))] - [1(S^\circ(C_2H_4(g))) + 1(S^\circ(Cl_2(g)))]$
$= [1(308.0 \, J/K)] - [1(219.3 \, J/K) + 1(223.1 \, J/K)]$
$= [308.0 \, J/K] - [442.4 \, J/K]$
$= -134.4 \, J/K$

then $T = 273.15 + 25\,°C = 298\,K$ then $-134.4\dfrac{J}{K} \times \dfrac{1\,kJ}{1000\,J} = -0.1344\,kJ/K$ then

$$\Delta G° = \Delta H°_{rxn} - T\Delta S°_{rxn} = -182.1\,kJ - (298\,K)\left(-0.1344\dfrac{kJ}{K}\right) = -142.0\,kJ = -1.420 \times 10^5\,J \text{ then}$$

$$\Delta G°_{rxn} = -RT \ln K \quad \text{Rearrange to solve for } K.$$

$$K = e^{\frac{-\Delta G°_{rxn}}{RT}} = e^{\dfrac{-(-1.420 \times 10^5\,J)}{\left(8.314\frac{J}{K \cdot mol}\right)(298\,K)}} = e^{57.\underline{3}14} = 7.78 \times 10^{24}; \text{ so the reaction is spontaneous.}$$

Reactant/Product	$\Delta H°_f$ (kJ/mol from Appendix IIB)
$C_2H_4(g)$	52.4
$Br_2(g)$	30.9
$C_2H_4Br_2(g)$	38.3

Be sure to pull data for the correct formula and phase.

$$\begin{aligned}
\Delta H°_{rxn} &= \sum n_p \Delta H°_f(\text{products}) - \sum n_r \Delta H°_f(\text{reactants}) \\
&= [1(\Delta H°_f(C_2H_4Br_2(g)))] - [1\Delta H°_f(C_2H_4(g))) + 1(\Delta H°_f(Br_2(g)))] \\
&= [1(38.3\,kJ)] - [1(52.4\,kJ) + 1(30.9\,kJ)] \\
&= [38.3\,kJ] - [83.3\,kJ] \\
&= -45.0\,kJ \text{ then}
\end{aligned}$$

Reactant/Product	$S°$ (J/mol K from Appendix IIB)
$C_2H_4(g)$	219.3
$Br_2(g)$	245.5
$C_2H_4Br_2(g)$	330.6

Be sure to pull data for the correct formula and phase.

$$\begin{aligned}
\Delta S°_{rxn} &= \sum n_p S°(\text{produts}) - \sum n_r S°(\text{reactants}) \\
&= [1(S°(C_2H_4Br_2(g)))] - [1(S°(C_2H_4(g))) + 1(S°(Br_2(g)))] \\
&= [1(330.6\,J/K)] - [1(219.3\,J/K)] + 1(245.5\,J/K)] \\
&= [330.6\,J/K] - [466.8\,J/K] \\
&= -134.2\,J/K
\end{aligned}$$

then $-134.2\dfrac{J}{K} \times \dfrac{1\,kJ}{1000\,J} = -0.1342\,kJ/K$

then $\Delta G° = \Delta H°_{rxn} - T\Delta S°_{rxn} = -45.0\,kJ - (298\,K)\left(-0.1342\dfrac{kJ}{K}\right) = -5.\underline{0}084\,kJ = -5.\underline{0}084 \times 10^3\,J$

then $\Delta G°_{rxn} = -RT \ln K$

Rearrange to solve for K. $K = e^{\frac{-\Delta G°_{rxn}}{RT}} = e^{\dfrac{-(-5.\underline{0}084 \times 10^3\,J)}{\left(8.314\frac{J}{K \cdot mol}\right)(298\,K)}} = e^{2.021495} = 7.5; \text{ so the reaction is spontaneous.}$

Reactant/Product	$\Delta H°_f$ (kJ/mol from Appendix IIB)
$C_2H_4(g)$	52.4
$I_2(g)$	62.42
$C_2H_4I_2(g)$	66.5

Be sure to pull data for the correct formula and phase.

$$\begin{aligned}
\Delta H°_{rxn} &= \sum n_p \Delta H°_f(\text{products}) - \sum n_r \Delta H°_f(\text{reactants}) \\
&= [1(\Delta H°_f(C_2H_4I_2(g))) - [1\Delta H°_f(C_2H_4(g))) + 1(\Delta H°_f(I_2(g)))] \\
&= [1(66.5\,kJ)] - [1(52.4\,kJ) + 1(62.42\,kJ)] \\
&= [66.5\,kJ] - [114.\underline{8}2\,kJ] \\
&= -48.\underline{3}2\,kJ \text{ then}
\end{aligned}$$

Reactant/Product	$S°$(J/mol K from Appendix IIB)
$C_2H_4(g)$	219.3
$I_2(g)$	260.69
$C_2H_4I_2(g)$	347.8

Be sure to pull data for the correct formula and phase.

$$\Delta S°_{rxn} = \sum n_p S°(\text{products}) - \sum n_r S°(\text{reactants})$$
$$= [1(S°(C_2H_4I_2(g)))] - [1(S°(C_2H_4(g))) + 1(S°(I_2(g)))]$$
$$= [1(347.8 \text{ J/K})] - [1(219.3 \text{ J/K}) + 1(260.69 \text{ J/K})]$$
$$= [347.8 \text{ J/K}] - [479.\underline{99} \text{ J/K}]$$
$$= -132.2 \text{ J/K}$$

then $-132.2 \dfrac{J}{K} \times \dfrac{1 \text{ kJ}}{1000 \, J} = -0.1322 \text{ kJ/K}$

then $\Delta G° = \Delta H°_{rxn} - T\Delta S°_{rxn} = -48.\underline{32} \text{ kJ} - (298 \text{ K})\left(-0.1322\dfrac{\text{kJ}}{\text{K}}\right) = -8.\underline{9}244 \text{ kJ} = -8.\underline{9}244 \times 10^3 \text{ J}$

then $\Delta G°_{rxn} = -RT \ln K$ Rearrange to solve for K.

$$K = e^{\frac{-\Delta G°_{rxn}}{RT}} = e^{\frac{-(-8.\underline{9}244 \times 10^3 \, J)}{\left(8.314\frac{J}{K \cdot mol}\right)(298 \, K)}} = e^{3.6021} = 37; \text{ so the reaction is spontaneous.}$$

Cl_2 is the most spontaneous in the forward direction; K for Cl_2 is 7.78×10^{24}, K for Br_2 is 7.5, K for I_2 is 37. This means Cl_2 is the most spontaneous and Br_2 is the least. The entropy change in the reactions is very constant. The spontaneity is determined by the standard enthalpy of formation of the dihalogenated ethane. Higher temperatures make the forward reactions less spontaneous.

Check: The units (kJ and none) are correct. The enthalpy change becomes less negative as we move to larger halogens. The enthalpy term dominates at room temperature, and the free energy change is the same sign as the enthalpy change. The more negative the free energy change, the larger the K.

17.85   (a)   **Given:** $N_2O(g) + NO_2(g) \rightleftharpoons 3 \, NO(g)$ at 298 K   **Find:** $\Delta G°_{rxn}$
Conceptual Plan: $\Delta G°_{rxn} = \sum n_p \Delta G°_f(\text{products}) - \sum n_r \Delta G°_f(\text{reactants})$
Solution:

Reactant/Product	$\Delta G°_f$(kJ/mol from Appendix IIB)
$N_2O(g)$	103.7
$NO_2(g)$	51.3
$NO(g)$	87.6

Be sure to pull data for the correct formula and phase.

$$\Delta G°_{rxn} = \sum n_p \Delta G°_f(\text{produts}) - \sum n_r \Delta G°_f(\text{reactants})$$
$$= [3(\Delta G°_f(NO(g)))] - [1(\Delta G°_f(N_2O(g))) + 1(\Delta G°_f(NO_2(g)))]$$
$$= [3(87.6 \text{ kJ})] - [1(103.7 \text{ kJ}) + 1(51.3 \text{ kJ})]$$
$$= [262.8 \text{ kJ}] - [155.0 \text{ kJ}]$$
$$= +107.8 \text{ kJ}$$

The reaction is nonspontaneous.

Check: The units (kJ) are correct. The standard free energy for the reaction was positive, so the reaction is nonspontaneous.

(b)   **Given:** $P_{N_2O} = P_{NO_2} = 1.0$ atm initially   **Find:** P_{N_2O} when reaction ceases to be spontaneous
Conceptual Plan: Reaction will no longer be spontaneous when $Q = K$, **so** $\Delta G°_{rxn}, T \rightarrow K$.
$$\Delta G°_{rxn} = -RT \ln K$$

Then solve the equilibrium problem to get gas pressures. Because $K \ll 1$, **the amount of NO generated will be very, very small compared to 1.0 atm; within experimental error,** $P_{N_2O} = P_{NO_2} = 1.0$ **atm. Simply solve for** P_{NO}.

$$K = \frac{P_{NO}^3}{P_{N_2O}P_{NO_2}}$$

Solution: $\Delta G^\circ_{rxn} = -RT \ln K$ Rearrange to solve for K.

$$K = e^{\frac{-\Delta G^\circ_{rxn}}{RT}} = e^{\frac{-(+107.8\,\cancel{kJ}) \times \frac{1000\,J}{1\,\cancel{kJ}}}{\left(8.314\frac{J}{K \cdot mol}\right)(298\,K)}} = e^{-43.\underline{5}103} = 1.27 \times 10^{-19} \quad \text{Because } K = \frac{P^3_{NO}}{P_{N_2O}P_{NO_2}}, \text{rearrange to solve}$$

for P_{NO}. $P^3_{NO} = \sqrt[3]{K\,P_{N_2O}P_{NO_2}} = \sqrt[3]{(1.27 \times 10^{-19})(1.0)(1.0)} = 5.0 \times 10^{-7}$ atm

Note that the assumption that P_{N_2O} was very, very small was valid.

Check: The units (atm) are correct. Because the free energy change was positive, the K was very small. This leads us to expect that very little NO will be formed.

(c)   **Given:** $N_2O(g) + NO_2(g) \rightleftharpoons 3\,NO(g)$ **Find:** temperature for spontaneity
Conceptual Plan: $\Delta H^\circ_{rxn} = \sum n_p H^\circ_f(\text{products}) - \sum n_r H^\circ_f(\text{reactants})$ then

$\Delta S^\circ_{rxn} = \sum n_p S^\circ(\text{products}) - \sum n_r S^\circ(\text{reactants})$ then J/K → kJ/K then $\Delta H^\circ_{rxn}, \Delta S_{rxn} \rightarrow T$

$$\frac{1\,kJ}{1000\,J} \qquad \qquad \Delta G = \Delta H_{rxn} - T\Delta S_{rxn}$$

Solution:

Reactant/Product	ΔH°_f(kJ/mol from Appendix IIB)
$N_2O(g)$	81.6
$NO_2(g)$	33.2
$NO(g)$	91.3

Be sure to pull data for the correct formula and phase.

$$\begin{aligned}
\Delta H^\circ_{rxn} &= \sum n_p \Delta H^\circ_f(\text{products}) - \sum n_r \Delta H^\circ_f(\text{reactants}) \\
&= [3(\Delta H^\circ_f(NO(g)))] - [1(\Delta H^\circ_f(N_2O(g))) + 1(\Delta H^\circ_f(NO_2(g)))] \\
&= [3(91.3\,kJ)] - [1(81.6\,kJ) + 1(33.2\,kJ)] \\
&= [273.9\,kJ] - [114.8\,kJ] \\
&= +159.1\,kJ \text{ then}
\end{aligned}$$

Reactant/Product	S°(J/mol K from Appendix IIB)
$N_2O(g)$	220.0
$NO_2(g)$	240.1
$NO(g)$	210.8

Be sure to pull data for the correct formula and phase.

$$\begin{aligned}
\Delta S^\circ_{rxn} &= \sum n_p S^\circ(\text{products}) - \sum n_r S^\circ(\text{reactants}) \\
&= [3(S^\circ(NO(g)))] - [1(S^\circ(N_2O(g))) + 1(S^\circ(NO_2(g)))] \\
&= [3(210.8\,J/K)] - [1(220.0\,J/K) + 1(240.1\,J/K)] \\
&= [632.4\,J/K] - [460.1\,J/K] \\
&= +172.3\,J/K
\end{aligned}$$

then $+172.3\dfrac{J}{K} \times \dfrac{1\,kJ}{1000\,J} = +0.1723$ kJ/K. Because $\Delta G = \Delta H_{rxn} - T\Delta S_{rxn}$, set $\Delta G = 0$ and rearrange to

solve for T. $T = \dfrac{\Delta H_{rxn}}{\Delta S_{rxn}} = \dfrac{+159.1\,\cancel{kJ}}{0.1723\dfrac{\cancel{kJ}}{K}} = +923.4$ K

Check: The units (K) are correct. The reaction can be made more spontaneous by raising the temperature because the entropy change is positive (increase in the number of moles of gas).

17.87   (a)   **Given:** $ATP(aq) + H_2O(l) \rightarrow ADP(aq) + P_i(aq)$ $\Delta G^\circ_{rxn} = -30.5$ kJ at 298 K   **Find:** K
Conceptual Plan: $\Delta G^\circ_{rxn}, T \rightarrow K$

$$\Delta G^\circ_{rxn} = -RT \ln K$$

Solution: $\Delta G^\circ_{rxn} = -RT \ln K$ Rearrange to solve for K.

$$K = e^{\frac{-\Delta G^\circ_{rxn}}{RT}} = e^{\frac{-(-30.5 \text{ kJ}) \times \frac{1000 \text{ J}}{1 \text{ kJ}}}{\left(8.314 \frac{\text{J}}{\text{K} \cdot \text{mol}}\right)(298 \text{ K})}} = e^{12.3104} = 2.22 \times 10^5$$

Check: The units (none) are correct. The free energy change is negative, and the reaction is spontaneous. This results in a large K.

(b)   **Given:** oxidation of glucose drives reforming of ATP
Find: ΔG°_{rxn} of oxidation of glucose and moles ATP formed per mole of glucose
Conceptual Plan: Write a balanced reaction for glucose oxidation then
$\Delta G^\circ_{rxn} = \sum n_p \Delta G^\circ_f (\text{products}) - \sum n_r \Delta G^\circ_f (\text{reactants})$ **then** $\Delta G^\circ_{rxn} s \rightarrow$ **moles ATP/mole glucose.**

$$\frac{\Delta G^\circ_{rxn} \text{ glucose oxidation}}{\Delta G^\circ_{rxn} \text{ ATP hydrolysis}}$$

Solution: $C_6H_{12}O_6(s) + 6 O_2(g) \rightarrow 6 CO_2(g) + 6 H_2O(l)$

Reactant/Product	ΔG°_f (kJ/mol from Appendix IIB)
$C_6H_{12}O_6(s)$	-910.4
$O_2(g)$	0.0
$CO_2(g)$	-394.4
$H_2O(l)$	-237.1

Be sure to pull data for the correct formula and phase.

$$\begin{aligned}
\Delta G^\circ_{rxn} &= \sum n_p \Delta G^\circ_f (\text{products}) - \sum n_r \Delta G^\circ_f (\text{reactants}) \\
&= [6(\Delta G^\circ_f(CO_2(g))) + 6(\Delta G^\circ_f(H_2O(l)))] - [1(\Delta G^\circ_f(C_6H_{12}O_6(s))) + 6(\Delta G^\circ_f(O_2(g)))] \\
&= [6(-394.4 \text{ kJ}) + 6(-237.1 \text{ kJ})] - [1(-910.4 \text{ kJ}) + 6(0.0 \text{ kJ})] \\
&= [-3789.0 \text{ kJ}] - [-910.4 \text{ kJ}] \\
&= -2878.6 \text{ kJ}
\end{aligned}$$

So the reaction is very spontaneous. $\dfrac{2878.6 \dfrac{\text{kJ generated}}{\text{mole glucose oxidized}}}{30.5 \dfrac{\text{kJ needed}}{\text{mole ATP reformed}}} = 94.4 \dfrac{\text{mole ATP reformed}}{\text{mole glucose oxidized}}$

Check: The units (mol) are correct. The free energy change for the glucose oxidation is large compared to the ATP hydrolysis, so we expect to reform many moles of ATP.

17.89   (a)   **Given:** $2 CO(g) + 2 NO(g) \rightarrow N_2(g) + 2 CO_2(g)$ **Find:** ΔG°_{rxn} and effect of increasing T on ΔG
Conceptual Plan: $\Delta G^\circ_{rxn} = \sum n_p \Delta G^\circ_f (\text{products}) - \sum n_r \Delta G^\circ_f (\text{reactants})$
Solution:

Reactant/Product	ΔG°_f (kJ/mol from Appendix IIB)
$CO(g)$	-137.2
$NO(g)$	87.6
$N_2(g)$	0.0
$CO_2(g)$	-394.4

Be sure to pull data for the correct formula and phase.

$$\begin{aligned}
\Delta G^\circ_{rxn} &= \sum n_p \Delta G^\circ_f (\text{products}) - \sum n_r \Delta G^\circ_f (\text{reactants}) \\
&= [1(\Delta G^\circ_f(N_2(g))) + 2(\Delta G^\circ_f(CO_2(g)))] - [2(\Delta G^\circ_f(CO(g))) + 2(\Delta G^\circ_f(NO(g)))] \\
&= [1(0.0 \text{ kJ}) + 2(-394.4 \text{ kJ})] - [2(-137.2 \text{ kJ}) + 2(87.6 \text{ kJ})] \\
&= [-788.8 \text{ kJ}] - [-99.2 \text{ kJ}] \\
&= -689.6 \text{ kJ}
\end{aligned}$$

Because the number of moles of gas is decreasing, the entropy change is negative; ΔG will become less negative with increasing temperature.

Check: The units (kJ) are correct. The free energy change is negative because the carbon dioxide has such a low free energy of formation.

(b)   **Given:** $5 H_2(g) + 2 NO(g) \rightarrow 2 NH_3(g) + 2 H_2O(g)$ **Find:** ΔG°_{rxn} and effect of increasing T on ΔG
Conceptual Plan: $\Delta G^\circ_{rxn} = \sum n_p \Delta G^\circ_f(\text{products}) - \sum n_r \Delta G^\circ_f(\text{reactants})$
Solution:

Reactant/Product	$\Delta G^\circ_f(\text{kJ/mol from Appendix IIB})$
$H_2(g)$	0.0
$NO(g)$	87.6
$NH_3(g)$	-16.4
$H_2O(g)$	-228.6

Be sure to pull data for the correct formula and phase.

$$\begin{aligned}
\Delta G^\circ_{rxn} &= \sum n_p \Delta G^\circ_f(\text{products}) - \sum n_r \Delta G^\circ_f(\text{reactants}) \\
&= [2(\Delta G^\circ_f(NH_3(g))) + 2(\Delta G^\circ_f(H_2O(g)))] - [5(\Delta G^\circ_f(H_2(g))) + 2(\Delta G^\circ_f(NO(g)))] \\
&= [2(-16.4\,\text{kJ}) + 2(-228.6\,\text{kJ})] - [5(0.0\,\text{kJ}) + 2(87.6\,\text{kJ})] \\
&= [-490.0\,\text{kJ}] - [175.2\,\text{kJ}] \\
&= -665.2\,\text{kJ}
\end{aligned}$$

Because the number of moles of gas is decreasing, the entropy change is negative; so ΔG will become less negative with increasing temperature.

Check: The units (kJ) are correct. The free energy change is negative because ammonia and water have such a low free energy of formation.

(c)   **Given:** $2 H_2(g) + 2 NO(g) \rightarrow N_2(g) + 2 H_2O(g)$ **Find:** ΔG°_{rxn} and effect of increasing T on ΔG
Conceptual Plan: $\Delta G^\circ_{rxn} = \sum n_p \Delta G^\circ_f(\text{products}) - \sum n_r \Delta G^\circ_f(\text{reactants})$
Solution:

Reactant/Product	$\Delta G^\circ_f(\text{kJ/mol from Appendix IIB})$
$H_2(g)$	0.0
$NO(g)$	87.6
$N_2(g)$	0.0
$H_2O(g)$	-228.6

Be sure to pull data for the correct formula and phase.

$$\begin{aligned}
\Delta G^\circ_{rxn} &= \sum n_p \Delta G^\circ_f(\text{products}) - \sum n_r \Delta G^\circ_f(\text{reactants}) \\
&= [1(\Delta G^\circ_f(N_2(g))) + 2(\Delta G^\circ_f(H_2O(g)))] - [2(\Delta G^\circ_f(H_2(g))) + 2(\Delta G^\circ_f(NO(g)))] \\
&= [1(0.0\,\text{kJ}) + 2(-228.6\,\text{kJ})] - [2(0.0\,\text{kJ}) + 2(87.6\,\text{kJ})] \\
&= [-457.2\,\text{kJ}] - [175.2\,\text{kJ}] \\
&= -632.4\,\text{kJ}
\end{aligned}$$

Because the number of moles of gas is decreasing, the entropy change is negative; so ΔG will become less negative with increasing temperature.

Check: The units (kJ) are correct. The free energy change is negative because water has such a low free energy of formation.

(d)   **Given:** $2 NH_3(g) + 2 O_2(g) \rightarrow N_2O(g) + 3 H_2O(g)$ **Find:** ΔG°_{rxn} and effect of increasing T on ΔG
Conceptual Plan: $\Delta G^\circ_{rxn} = \sum n_p \Delta G^\circ_f(\text{products}) - \sum n_r \Delta G^\circ_f(\text{reactants})$
Solution:

Reactant/Product	$\Delta G^\circ_f(\text{kJ/mol from Appendix IIB})$
$NH_3(g)$	-16.4
$O_2(g)$	0.0
$N_2O(g)$	103.7
$H_2O(g)$	-228.6

Be sure to pull data for the correct formula and phase.

$$\begin{aligned}
\Delta G^\circ_{rxn} &= \sum n_p \Delta G^\circ_f(\text{products}) - \sum n_r \Delta G^\circ_f(\text{reactants}) \\
&= [1(\Delta G^\circ_f(N_2O(g))) + 3(\Delta G^\circ_f(H_2O(g)))] - [2(\Delta G^\circ_f(NH_3(g))) + 2(\Delta G^\circ_f(O_2(g)))] \\
&= [1(103.7\,\text{kJ}) + 3(-228.6\,\text{kJ})] - [2(-16.4\,\text{kJ}) + 2(0.0\,\text{kJ})]
\end{aligned}$$

$$= [-582.1 \text{ kJ}] - [-32.8 \text{ kJ}]$$
$$= -549.3 \text{ kJ}$$

Because the number of moles of gas is constant, the entropy change will be small and slightly negative; so the magnitude of ΔG will decrease with increasing temperature.

Check: The units (kJ) are correct. The free energy change is negative because water has such a low free energy of formation. The entropy change is negative once the $S°$ values are reviewed ($\Delta S_{rxn} = -9.6 \text{ J/K}$).

17.91   With one exception, the formation of any oxide of nitrogen at 298 K requires more moles of gas as reactants than are formed as products. For example, 1 mole of N_2O requires 0.5 mole of O_2 and 1 mole of N_2. One mole of N_2O_3 requires 1 mole of N_2 and 1.5 moles of O_2, and so on. The exception is NO, where 1 mole of NO requires 0.5 mole of O_2 and 0.5 mole of N_2: $\frac{1}{2}N_2(g) + \frac{1}{2}O_2(g) \rightarrow NO(g)$. This reaction has a positive ΔS because what is essentially mixing of the N and O has taken place in the product.

17.93   **Given:** $X_2(g) \rightarrow 2X(g)$; $P_{initial\ X_2} = 755$ torr, $P_{final\ X} = 103$ torr at 298 K; $P_{initial\ X_2} = 748$ torr, $P_{final\ X} = 532$ torr at 755 K   **Find:** $\Delta H°_{rxn}$

Conceptual Plan: At each temperature $P_{initial\ X_2}, P_{final\ X} \rightarrow K$ then $K_1, K_2, T_1, T_2 \rightarrow \Delta H°_{rxn}$,

$$\text{Use stoichiometry to calculate } [X_2]_{final} \text{ and } K = \frac{P_X^2}{P_{X_2}} \qquad \ln K = -\frac{\Delta H°_{rxn}}{R}\frac{1}{T} + \frac{\Delta S°_{rxn}}{R}$$

Solution: Use stoichiometry of $X_2(g) \rightarrow 2X(g)$; $P_{final\ X_2} = P_{initial\ X_2} - \frac{1}{2}P_{final\ X}$ so that at $T_1 = 298$ K,

$P_{final\ X_2} = P_{initial\ X_2} - \frac{1}{2}P_{final\ X} = 755 \text{ torr} - \frac{1}{2}(103 \text{ torr}) = 703.5$ torr and at $T_{12} = 755$ K, $P_{final\ X_2} = P_{initial\ X_2} - \frac{1}{2}$

$P_{final\ X} = 748 \text{ torr} - \frac{1}{2}(532 \text{ torr}) = 482$ torr. Then $K = \frac{P_X^2}{P_{X_2}}$ so that at $T_1 = 298$ K,

$$K_1 = \frac{P_X^2}{P_{X_2}} = \frac{(103)^2}{703.5} = 15.08031 \text{ and at } T_2 = 755 \text{ K}, K_2 = \frac{P_X^2}{P_{X_2}} = \frac{(532)^2}{482} = 587.1867.$$

Because $\ln K = -\dfrac{\Delta H°_{rxn}}{R}\dfrac{1}{T} + \dfrac{\Delta S°_{rxn}}{R}$, $\ln K_1 + \dfrac{\Delta H°_{rxn}}{R}\dfrac{1}{T_1} = \dfrac{\Delta S°_{rxn}}{R} = \ln K_2 + \dfrac{\Delta H°_{rxn}}{R}\dfrac{1}{T_2}$.

Rearrange to solve for $\Delta H°_{rxn}$.

$$\Delta H°_{rxn} = \frac{\ln \dfrac{K_2}{K_1}}{\left(\dfrac{1}{T_1} - \dfrac{1}{T_2}\right)}R = \frac{\ln\left(\dfrac{587.1867}{15.08031}\right)}{\left(\dfrac{1}{298 \text{ K}} - \dfrac{1}{755 \text{ K}}\right)} \times 8.314\frac{J}{K \cdot mol} \times \frac{1 \text{ kJ}}{1000 \text{ J}} = 14.9889 \text{ kJ/mol} = 15.0 \text{ kJ/mol}$$

Note that an alternate way to generate the expressions for the two equilibrium constants is to use an ICE table.

Check: The units are correct (kJ/mol). Because K increases with increasing temperature, the reaction is expected to be endothermic.

17.95   (a) $\Delta S_{univ} > 0$. The process is spontaneous. It is slow unless a spark is applied.

(b) $\Delta S_{univ} > 0$. Although the change in the system is not spontaneous, the overall change, which includes such processes as combustion or water flow to generate electricity, is spontaneous.

(c) $\Delta S_{univ} > 0$. The acorn oak/tree system is becoming more ordered, so the processes associated with growth are not spontaneous. But they are driven by spontaneous processes such as the generation of heat by the sun and the reactions that produce energy in the cell.

17.97   At equilibrium ($P_{H_2O} = 18.3$ mmHg), $\Delta G_{rxn} = 0$.

$$Q = P_{H_2O}^6 = \left(\frac{18.3 \text{ mmHg}}{\dfrac{760 \text{ mmHg}}{1 \text{ atm}}}\right)^6 = (0.0240 7894 \text{ atm})^6 = 1.949059 \times 10^{-10} \text{ and}$$

$\Delta G_{rxn} = \Delta G°_{rxn} + RT \ln Q = 0$ Rearrange to solve for $\Delta G°_{rxn}$ (at $P_{H_2O} = 760$ mmHg = standard conditions).

$$\Delta G°_{rxn} = \Delta G_{rxn} - RT \ln Q = 0 - \left(8.314\frac{J}{K \cdot mol}\right)\left(\frac{1 \text{ kJ}}{1000 \text{ J}}\right)(298 \text{ K}) \ln(1.949059 \times 10^{-10}) = 55.3948 \frac{\text{kJ}}{\text{mol}}$$

$$= 55.4 \frac{\text{kJ}}{\text{mol}}$$

Challenge Problems

17.99   (a)   **Given:** glutamate (aq) + $NH_3(aq) \rightarrow$ glutamine(aq) + $H_2O(l)$ $\Delta G°_{rxn} = 14.2$ kJ at 298 K   **Find:** K

          Conceptual Plan: $\Delta G°_{rxn}, T \rightarrow K$

$$\Delta G°_{rxn} = -RT \ln K$$

          Solution: $\Delta G°_{rxn} = -RT \ln K$ Rearrange to solve for K.

$$K = e^{\frac{-\Delta G°_{rxn}}{RT}} = e^{\frac{-(+14.2 \,\cancel{kJ}) \times \frac{1000 \, J}{1 \, \cancel{kJ}}}{\left(8.314 \frac{\cancel{J}}{K \cdot mol}\right)(298 \, \cancel{K})}} = e^{-5.7\underline{3}142} = 3.24 \times 10^{-3}$$

          Check: The units (none) are correct. The free energy change is positive, and the reaction is nonspontaneous. This results in a small K.

   (b)   **Given:** pair ATP hydrolysis with glutamate/NH_3 reaction   **Find:** show coupled reactions, $\Delta G°_{rxn}$ and K

          Conceptual Plan: Use the reaction mechanism shown, where A = NH_3 and B = glutamate $(C_5H_8O_4N^-)$. Then calculate $\Delta G°_{rxn}$ by adding free energies of reactions then $\Delta G°_{rxn}, T \rightarrow K$.

$$\Delta G°_{rxn} = -RT \ln K$$

          Solution:

$NH_3(aq) + ATP(aq) + \cancel{H_2O(l)} \rightarrow \cancel{NH_3 - P_i(aq)} + ADP(aq)$	$\Delta G°_{rxn} = -30.5$ kJ
$\cancel{NH_3 - P_i(aq)} + C_5H_8O_4N^-(aq) \rightarrow C_5H_9O_3N_2^-(aq) + \cancel{H_2O(l)} + P_i(aq)$	$\Delta G°_{rxn} = +14.2$ kJ

$NH_3(aq) + C_5H_8O_4N^-(aq) + ATP(aq) \rightarrow C_5H_9O_3N_2^-(aq) + ADP(aq) + P_i(aq)$	$\Delta G°_{rxn} = -16.3$ kJ

          then $\Delta G°_{rxn} = -RT \ln K$ Rearrange to solve for K.

$$K = e^{\frac{-\Delta G°_{rxn}}{RT}} = e^{\frac{-(-16.3 \,\cancel{kJ}) \times \frac{1000 \, J}{1 \, \cancel{kJ}}}{\left(8.314 \frac{\cancel{J}}{K \cdot mol}\right)(298 \, \cancel{K})}} = e^{6.5\underline{7}902} = 7.20 \times 10^{2}$$

          Check: The units (none) are correct. The free energy change is negative, and the reaction is spontaneous. This results in a large K.

17.101   (a)   **Given:** $\frac{1}{2} H_2(g) + \frac{1}{2} Cl_2(g) \rightarrow HCl(g)$, define standard state as 2 atm   **Find:** $\Delta G°_f$

          Conceptual Plan: $\Delta G°_f, P_{H_2}, P_{Cl_2}, P_{HCl}, T \rightarrow$ new $\Delta G°_f$

$$\Delta G_{rxn} = \Delta G°_{rxn} + RT \ln Q \quad \text{where } Q = \frac{P_{HCl}}{P_{H_2}^{1/2} P_{Cl_2}^{1/2}}$$

          Solution: $\Delta G°_{rxn} = -95.3$ kJ/mol and $Q = \dfrac{P_{HCl}}{P_{H_2}^{1/2} P_{Cl_2}^{1/2}} = \dfrac{2}{2^{1/2} \, 2^{1/2}} = 1$ then

$$\Delta G_{rxn} = \Delta G°_{rxn} + RT \ln Q = -95.3 \frac{kJ}{mol} + \left(8.314 \frac{\cancel{J}}{K \cdot mol}\right)\left(\frac{1 \, kJ}{1000 \, \cancel{J}}\right)(298 \, \cancel{K}) \ln(1) =$$

          -95.3 kJ/mol $= -95,300$ J/mol

          Because the number of moles of reactants and products are the same, the decrease in volume affects the entropy of both equally; so there is no change in $\Delta G°_f$.

          Check: The units (kJ) are correct. The Q is 1, so $\Delta G°_f$ is unchanged under the new standard conditions.

   (b)   **Given:** $N_2(g) + \frac{1}{2} O_2(g) \rightarrow N_2O(g)$, define standard state as 2 atm   **Find:** $\Delta G°_f$

          Conceptual Plan: $\Delta G°_f, P_{N_2}, P_{O_2}, P_{N_2O}, T \rightarrow$ new $\Delta G°_f$

$$\Delta G_{rxn} = \Delta G°_{rxn} + RT \ln Q \quad \text{where } Q = \frac{P_{N_2O}}{P_{N_2} P_{O_2}^{1/2}}$$

          Solution: $\Delta G°_f = +103.7$ kJ/mol and $Q = \dfrac{P_{N_2O}}{P_{N_2} P_{O_2}^{1/2}} = \dfrac{2}{2 \, 2^{1/2}} = \dfrac{1}{\sqrt{2}}$ then

$$\Delta G_{rxn} = \Delta G°_{rxn} + RT \ln Q = 103.7 \frac{kJ}{mol} + \left(8.314 \frac{\cancel{J}}{K \cdot mol}\right)\left(\frac{1 \, kJ}{1000 \, \cancel{J}}\right)(298 \, \cancel{K}) \ln\left(\frac{1}{\sqrt{2}}\right) =$$

          $+102.8$ kJ/mol $= +102,800$ J/mol

The entropy of the reactants (1.5 mol) is decreased more than the entropy of the product (1 mol). Because the product is relatively more favored at lower volume, ΔG_f° is less positive.

Check: The units (kJ) are correct. The Q is less than 1, so ΔG_f° is reduced under the new standard conditions.

(c)   **Given:** $1/2\, H_2(g) \rightarrow H(g)$, define standard state as 2 atm   **Find:** ΔG_f°

Conceptual Plan: $\Delta G_f^\circ, P_{H_2}, P_H, T \rightarrow$ **new** ΔG_f°

$$\Delta G_{rxn} = \Delta G_{rxn}^\circ + RT \ln Q \quad \text{where } Q = \frac{P_H}{P_{H_2}^{1/2}}$$

Solution: $\Delta G_f^\circ = +203.3$ kJ/mol and $Q = \dfrac{P_H}{P_{H_2}^{1/2}} = \dfrac{2}{2^{1/2}} = \sqrt{2}$ then

$$\Delta G_{rxn} = \Delta G_{rxn}^\circ + RT \ln Q = +203.3\frac{kJ}{mol} + \left(8.314\frac{J}{K \cdot mol}\right)\left(\frac{1\ kJ}{1000\ J}\right)(298\ K)\ln(\sqrt{2}) =$$
$+204.2$ kJ/mol $= +204{,}200$ J/mol

The entropy of the product (1 mol) is decreased more than the entropy of the reactant (1/2 mol). Because the product is relatively less favored, ΔG_f° is more positive.

Check: The units (kJ) are correct. The Q is greater than 1, so ΔG_f° is increased under the new standard conditions.

17.103   **Given:** $K = 3.9 \times 10^5$ at 300 K and $K = 1.2 \times 10^{-1}$ at 500 K   **Find:** $\Delta H_{rxn}^\circ, \Delta S_{rxn}^\circ$
Conceptual Plan: Plot ln K verus $1/T$. The slope will be $-\Delta H_{rxn}^\circ/R$, and the intercept will be $\Delta S_{rxn}^\circ/R$.

Solution: Plot ln K versus $1/T$. Because $\ln K = -\dfrac{\Delta H_{rxn}^\circ}{R}\dfrac{1}{T} + \dfrac{\Delta S_{rxn}^\circ}{R}$, the negative of the slope will be $-\dfrac{\Delta H_{rxn}^\circ}{R}$ and the intercept will be $\dfrac{\Delta S_{rxn}^\circ}{R}$. The slope can be determined by measuring $\Delta y/\Delta x$ on the plot or by using functions such as "add trendline" in Excel. Because the slope is $+11246$ K,
$\Delta H_{rxn}^\circ = -slope\ R =$

$$-(+11246\ K)\left(8.314\frac{J}{K \cdot mol}\right)\left(\frac{1\ kJ}{1000\ J}\right) =$$

-93.499 kJ/mol $= -93$ kJ/mol

Because the intercept is $-24.612\ K^{-1}$,

$$\Delta S_{rxn}^\circ = intercept\ R = \left(\frac{-24.612}{K}\right)\left(8.314\frac{J}{K \cdot mol}\right)$$
$= -204.62$ J/K $= -2.0 \times 10^2$ J/K.

Note: This problem can also be solved using the mathematically equivalent equation in Problems 17.93 and 17.94.

Check: The units are correct (kJ/mole and J/K mol). The numbers are typical for reactions. Because the slope is positive, the enthalpy change must be negative. The entropy change is expected to be negative because the number of moles of gas is decreasing.

17.105   **Given:** ΔH°_{vap} table   **Find:** ΔS_{vap}; then compare values
Conceptual Plan: $^\circ C \rightarrow K$ then $\Delta H^\circ_{vap}, T \rightarrow \Delta S_{vap}$

$$K = 273.15 + ^\circ C \qquad \Delta S_{vap} = \frac{-\Delta H_{vap}}{T}$$

Solution:

Diethyl ether: $T = 273.15 + 34.6 = 307.8$ K then $\Delta S_{vap} = \dfrac{\Delta H_{vap}}{T} = \dfrac{26.5\ kJ}{307.8\ K} = 0.0861$ kJ/K $= 86.1$ J/K

Acetone: $T = 273.15 + 56.1 = 329.3$ K then $\Delta S_{vap} = \dfrac{\Delta H_{vap}}{T} = \dfrac{29.1\ kJ}{329.3\ K} = 0.0884$ kJ/K $= 88.4$ J/K

Benzene: $T = 273.15 + 79.8 = 353.0$ K then $\Delta S_{vap} = \dfrac{\Delta H_{vap}}{T} = \dfrac{30.8\ kJ}{353.0\ K} = 0.0873$ kJ/K $= 87.3$ J/K

Chloroform: $T = 273.15 + 60.8 = 334.0$ K then $\Delta S_{vap} = \dfrac{\Delta H_{vap}}{T} = \dfrac{29.4\ kJ}{334.0\ K} = 0.0880$ kJ/K $= 88.0$ J/K

Ethanol: $T = 273.15 + 77.8 = 351.0\text{ K}$ then $\Delta S_{vap} = \dfrac{\Delta H_{vap}}{T} = \dfrac{38.6\text{ kJ}}{351.0\text{ K}} = 0.110\text{ kJ/K} = 110.\text{ J/K}$

Water: $T = 273.15 + 100 = 373.15\text{ K}$ then $\Delta S_{vap} = \dfrac{\Delta H_{vap}}{T} = \dfrac{40.7\text{ kJ}}{373.15\text{ K}} = 0.109\text{ kJ/K} = 109\text{ J/K}$

The first four values are very similar because they have similar intermolecular forces between molecules (dispersion forces and/or dipole–dipole interactions). The values for ethanol and water are higher because the intermolecular forces between molecules are stronger due to hydrogen bonding. Because of this, more energy is dispersed when these interactions are broken.

Conceptual Problems

17.107   (c)   The spontaneity of a reaction says nothing about the speed of a reaction. It only states which direction the reaction will go as it approaches equilibrium.

17.109   (b)   has the largest decrease in the number of microstates from the initial to the final state. In (a), there are initially $\dfrac{9!}{4!\,4!\,1!} = 90$ microstates and $\dfrac{9!}{3!\,3!\,3!} = 1680$ microstates at the end; so $\Delta S > 0$. In (b), there are initially $\dfrac{9!}{4!\,2!\,3!} = 1260$ microstates and $\dfrac{9!}{6!\,3!\,0!} = 84$ microstates at the end; so $\Delta S < 0$. In (c), there are initially $\dfrac{9!}{3!\,4!\,2!} = 1260$ microstates and $\dfrac{9!}{3!\,4!\,2!} = 1260$ microstates at the end; so $\Delta S = 0$. Also, the final state in (b) has the least entropy.

17.111   (c)   Because the vapor pressure of water at 298 K is 23.78 mmHg or 0.03129 atm. As long as the desired pressure (0.010 atm) is less than the equilibrium vapor pressure of water, the reaction will be spontaneous.

17.113   The relationship between ΔG°_{rxn} and K is $\Delta G^{\circ}_{rxn} = -RT \ln K$. When $K > 1$, the natural log of K is positive and $\Delta G^{\circ}_{rxn} < 0$. When $Q = 336$, the second term in $\Delta G_{rxn} = \Delta G^{\circ}_{rxn} + RT \ln Q$ is positive; so at high temperature, this term can dominate and make $\Delta G_{rxn} > 0$.

18 Electrochemistry

Review Questions

18.1   Electrochemistry uses spontaneous redox reactions to generate electrical energy.

18.3   Oxidation is the loss of electrons that corresponds to an increase in oxidation state. Reduction is the gain of electrons that corresponds to a decrease in oxidation state. Balancing redox reactions can be more complicated than balancing other types of reactions because both the mass (or number of each type of atom) and the charge must be balanced. Redox reactions occurring in aqueous solutions can be balanced by using a special procedure called the half-reaction method of balancing. In this procedure, the overall equation is broken down into two half-reactions: one for oxidation and one for reduction. The half-reactions are balanced individually and then added together so that the number of electrons generated in the oxidation half-reaction is the same as the number of electrons consumed in the reduction half-reaction.

18.5   In all electrochemical cells, the electrode where oxidation occurs is called the anode. In a voltaic cell, the anode is labeled with a negative $(-)$ sign. The anode is negative because the oxidation reaction that occurs at the anode releases electrons. Electrons flow from the anode to the cathode (from negative to positive) through the wires connecting the electrodes.

18.7   A salt bridge is a pathway by which ions can flow between the half-cells and complete an electrical circuit and maintain electroneutrality without the solutions in the half-cells totally mixing.

18.9   The standard cell potential E°_{cell} or standard emf is the cell potential under standard conditions (1 M concentration for reactants in solution and 1 atm pressure for gaseous reactants). The cell potential is a measure of the overall tendency of the redox reaction to occur spontaneously. The more positive the cell potential, the more spontaneous the reaction. A negative cell potential indicates a nonspontaneous reaction.

18.11   Inert electrodes, such as platinum (Pt) or graphite, are used as the anode or cathode (or both) when the reactants and products of one or both of the half-reactions are in the same phase.

18.13   The overall cell potential for any electrochemical cell will always be the sum of the half-cell potential for the oxidation reaction and the half-cell potential for the reduction reaction or $E^\circ_{cell} = E^\circ_{cathode} - E^\circ_{anode}$.

18.15   In general, any reduction half-reaction will be spontaneous when paired with the reverse of a half-reaction below it in the table.

18.17   We have seen that a positive standard cell potential (E°_{cell}) corresponds to a spontaneous oxidation–reduction reaction. We also know (from Chapter 17) that the spontaneity of a reaction is determined by the sign of ΔG°. Therefore, E°_{cell} and ΔG° must be related. We also know from Section 17.9 that ΔG° for a reaction is related to the equilibrium constant (K) for the reaction. Because E°_{cell} and ΔG° are related, E°_{cell} and K must also be related. The equation that relates the three quantities is $\Delta G^\circ_{rxn} = -RT \ln K = -n F E^\circ_{cell}$.

18.19   The Nernst equation relates concentration and cell potential as follows: $E_{cell} = E^\circ_{cell} - \dfrac{0.0592 \text{ V}}{n} \log Q$, where E_{cell} is the cell potential in V, E°_{cell} is the standard cell potential in V, n is the number of moles of electrons transferred in the redox reaction, and Q is the reaction quotient. Increasing the concentration of the reactants decreases the value of Q, so it increases the cell potential. Increasing the concentration of the products increases the value of Q, so it decreases the cell potential.

18.21   A concentration cell is a voltaic cell in which both half-reactions are the same but in which a difference in concentration drives the current flow (because the cell potential depends not only on the half-reactions occurring in the cell, but also on the concentrations of the reactants and products in those half-reactions).

Therefore, $E_{cell} = -\dfrac{0.0592 \text{ V}}{n} \log Q$.

18.23   Lead–acid storage batteries consist of six electrochemical cells wired in series in which each cell produces 2 V for a total of 12 V. Each cell contains a porous lead anode where oxidation occurs according to $Pb(s) + HSO_4^-(aq) \rightarrow PbSO_4(s) + H^+(aq) + 2\,e^-$ and a lead(IV) oxide cathode where reduction occurs according to $PbO_2(s) + HSO_4^-(aq) + 3\,H^+(aq) + 2\,e^- \rightarrow PbSO_4(s) + 2\,H_2O(l)$. As electric current is drawn from the battery, both the anode and the cathode become coated with $PbSO_4(s)$ and the solution becomes depleted of $HSO_4^-(aq)$. If the battery is run for a long time without being recharged, too much $PbSO_4(s)$ develops on the surface of the electrodes and the battery goes dead. The lead–acid storage battery can be recharged, however, by running electric current through it in reverse. The electric current, which must come from an external source such as an alternator in a car, causes the preceding reaction to occur in reverse, converting the $PbSO_4(s)$ back to $Pb(s)$ and $PbO_2(s)$, recharging the battery.

18.25   Fuel cells are like batteries, but the reactants must be constantly replenished. Normal batteries lose their ability to generate voltage with use because the reactants become depleted as electric current is drawn from the battery. In a fuel cell, the reactants—the fuel—constantly flow through the battery, generating electric current as they undergo a redox reaction.

The most common fuel cell is the hydrogen–oxygen fuel cell. In the cell, hydrogen gas flows past the anode (a screen coated with a platinum catalyst) and undergoes oxidation: $2\,H_2(g) + 4\,OH^-(aq) \rightarrow 4\,H_2O(l) + 4\,e^-$. Oxygen gas flows past the cathode (a similar screen) and undergoes reduction: $O_2(g) + 2\,H_2O(l) + 4\,e^- \rightarrow 4\,OH^-(aq)$. The half-reactions sum to the following overall reaction: $2\,H_2(g) + O_2(g) \rightarrow 2\,H_2O(l)$. Notice that the only product is water.

18.27   Applications of electrolysis include breaking water into hydrogen and oxygen, converting metal oxides to pure metals, production of sodium from molten sodium chloride, and plating metals onto other metals (e.g., plating silver onto a less expensive metal).

18.29   The anion is oxidized. The cation is reduced to the metal.

18.31   In the electrolysis of aqueous NaCl solutions, two different reduction half-reactions are possible at the cathode, the reduction of Na^+ and the reduction of water.

$$2\,Na^+(l) + 2\,e^- \rightarrow 2\,Na(s) \qquad\qquad E^\circ_{red} = 2.71 \text{ V}$$
$$2\,H_2O(l) + 2\,e^- \rightarrow H_2(g) + 2\,OH^-(aq) \qquad E_{red} = -0.41 \text{ V } ([OH^-] = 10^{-7} \text{ M})$$

The half-reaction that occurs most easily (the one with the least negative, or most positive, half-cell potential) is the one that actually occurs.

18.33   We can determine the number of moles of electrons that have flowed in a given electrolysis cell by measuring the total charge that has flowed through the cell, which in turn depends on the magnitude of current and the time the current has run. Because 1 ampere = 1 C/s, if we multiply the amount of current (in A) flowing through the cell by the time (in s) the current flowed, we can find the total charge that passed through the cell in that time: current (C/s) × time (s) = charge (C). The relationship between charge and the number of moles of electrons is given by Faraday's constant, which corresponds to the charge in coulombs of 1 mol of electrons; so $F = 96{,}485$ C/mole e^-. These relationships can be used to solve problems involving the stoichiometry of electrolytic cells.

18.35   Moisture must be present for many corrosion reactions to occur. The presence of water is necessary because water is a reactant in many oxidation reactions [such as $4\,Fe^{2+}(aq) + O_2(g) + (4 + 2n)\,H_2O(l) \rightarrow 2\,Fe_2O_3 \cdot n\,H_2O(s) + 8\,H^+(aq)$], and because charge (either electrons or ions) must be free to flow between the anodic and cathodic regions.

Additional electrolytes promote more corrosion. The presence of an electrolyte (such as sodium chloride) on the surface of iron promotes rusting because it enhances current flow. This is why cars rust so quickly in cold climates where roads are salted and in areas directly adjacent to beaches where saltwater mist is present.

The presence of acids promotes corrosion. Because H^+ ions are involved as a reactant in the reduction of oxygen, lower pH enhances the cathodic reaction and leads to faster corrosion.

Problems by Topic

Balancing Redox Reactions

18.37 **Conceptual Plan: Separate the overall reaction into two half-reactions: one for oxidation and one for reduction.** → **Balance each half-reaction with respect to mass in the following order: (1) Balance all elements other than H and O, (2) balance O by adding H_2O, and (3) balance H by adding H^+.** → **Balance each half-reaction with respect to charge by adding electrons. (The sum of the charges on both sides of the equation should be made equal by adding electrons as necessary.)** → **Make the number of electrons in both half-reactions equal by multiplying one or both half-reactions by a small whole number.** → **Add the two half-reactions, canceling electrons and other species as necessary.** → **Verify that the reaction is balanced with respect to both mass and charge.**

Solution:

(a)

Separate:	$K(s) \rightarrow K^+(aq)$	and	$Cr^{3+}(aq) \rightarrow Cr(s)$
Balance elements:	$K(s) \rightarrow K^+(aq)$	and	$Cr^{3+}(aq) \rightarrow Cr(s)$
Add electrons:	$K(s) \rightarrow K^+(aq) + e^-$	and	$Cr^{3+}(aq) + 3e^- \rightarrow Cr(s)$
Equalize electrons:	$3 K(s) \rightarrow 3 K^+(aq) + 3 e^-$	and	$Cr^{3+}(aq) + 3 e^- \rightarrow Cr(s)$
Add half-reactions:	$3 K(s) + Cr^{3+}(aq) + \cancel{3e^-} \rightarrow 3 K^+(aq) + \cancel{3e^-} + Cr(s)$		
Cancel electrons:	$3 K(s) + Cr^{3+}(aq) \rightarrow 3 K^+(aq) + Cr(s)$		

Check:

Reactants	Products
3 K atoms	3 K atoms
1 Cr atom	1 Cr atom
+3 charge	+3 charge

(b)

Separate:	$Al(s) \rightarrow Al^{3+}(aq)$	and	$Fe^{2+}(aq) \rightarrow Fe(s)$
Balance elements:	$Al(s) \rightarrow Al^{3+}(aq)$	and	$Fe^{2+}(aq) \rightarrow Fe(s)$
Add electrons:	$Al(s) \rightarrow Al^{3+}(aq) + 3e^-$	and	$Fe^{2+}(aq) + 2 e^- \rightarrow Fe(s)$
Equalize electrons:	$2 Al(s) \rightarrow 2 Al^{3+}(aq) + 6 e^-$	and	$3 Fe^{2+}(aq) + 6 e^- \rightarrow 3 Fe(s)$
Add half-reactions:	$2 Al(s) + 3 Fe^{2+}(aq) + \cancel{6e^-} \rightarrow 2 Al^{3+}(aq) + \cancel{6e^-} + 3 Fe(s)$		
Cancel electrons:	$2 Al(s) + 3 Fe^{2+}(aq) \rightarrow 2 Al^{3+}(aq) + 3 Fe(s)$		

Check:

Reactants	Products
2 Al atoms	2 Al atoms
3 Fe atoms	3 Fe atoms
+6 charge	+6 charge

(c)

Separate:	$BrO_3^-(aq) \rightarrow Br^-(aq)$	and	$N_2H_4(g) \rightarrow N_2(g)$
Balance non H & O elements:	$BrO_3^-(aq) \rightarrow Br^-(aq)$	and	$N_2H_4(g) \rightarrow N_2(g)$
Balance O with H_2O:	$BrO_3^-(aq) \rightarrow Br^-(aq) + 3 H_2O(l)$	and	$N_2H_4(g) \rightarrow N_2(g)$
Balance H with H^+:	$BrO_3^-(aq) + 6 H^+(aq) \rightarrow Br^-(aq) + 3 H_2O(l)$	and	$N_2H_4(g) \rightarrow N_2(g) + 4 H^+(aq)$

Add electrons:

$$BrO_3^-(aq) + 6 H^+(aq) + 6 e^- \rightarrow Br^-(aq) + 3 H_2O(l) \text{ and } N_2H_4(g) \rightarrow N_2(g) + 4 H^+(aq) + 4 e^-$$

Equalize electrons:

$$2 BrO_3^-(aq) + 12 H^+(aq) + 12 e^- \rightarrow 2 Br^-(aq) + 6 H_2O(l) \text{ and } 3 N_2H_4(g) \rightarrow 3 N_2(g) + 12 H^+(aq) + 12 e^-$$

Add half-reactions: $2 BrO_3^-(aq) + \cancel{12 H^+(aq)} + 3 N_2H_4(g) + \cancel{12 e^-} \rightarrow 2 Br^-(aq) + 6 H_2O(l) + 3 N_2(g)$
$+ \cancel{12 H^+(aq)} + \cancel{12 e^-}$

Cancel electrons & others: $2 BrO_3^-(aq) + 3 N_2H_4(g) \rightarrow 2 Br^-(aq) + 6 H_2O(l) + 3 N_2(g)$

Check:

Reactants	Products
2 Br atoms	2 Br atoms
6 O atoms	6 O atoms
12 H atoms	12 H atoms
6 N atoms	6 N atoms
−2 charge	−2 charge

18.39 **Conceptual Plan: Separate the overall reaction into two half-reactions: one for oxidation and one for reduction.** → **Balance each half-reaction with respect to mass in the following order: (1) Balance all elements other than H and O, (2) balance O by adding H_2O, and (3) balance H by adding H^+.** → **Balance each half-reaction with respect to charge by adding electrons. (The sum of the charges on both sides of the equation should be made equal by adding electrons as necessary.)** → **Make the number of electrons in both half-reactions equal by multiplying one or both half-reactions by a small whole number.** → **Add the two half-reactions, canceling electrons and other species as necessary.** → **Verify that the reaction is balanced with respect to both mass and charge.**
Solution:

(a) Separate: $\qquad\qquad\qquad\qquad PbO_2(s) \rightarrow Pb^{2+}(aq)$ $\qquad$ and $\quad I^-(aq) \rightarrow I_2(s)$
Balance non H & O elements: $PbO_2(s) \rightarrow Pb^{2+}(aq)$ $\qquad$ and $\quad 2\,I^-(aq) \rightarrow I_2(s)$
Balance O with H_2O: $PbO_2(s) \rightarrow Pb^{2+}(aq) + 2\,H_2O(l)$ $\qquad$ and $\quad 2\,I^-(aq) \rightarrow I_2(s)$
Balance H with H^+: $PbO_2(s) + 4\,H^+(aq) \rightarrow Pb^{2+}(aq) + 2\,H_2O(l)$ $\qquad$ and $\quad 2\,I^-(aq) \rightarrow I_2(s)$
Add electrons: $PbO_2(s) + 4\,H^+(aq) + 2\,e^- \rightarrow Pb^{2+}(aq) + 2\,H_2O(l)$ $\quad$ and $\quad 2\,I^-(aq) \rightarrow I_2(s) + 2\,e^-$
Equalize electrons: $PbO_2(s) + 4\,H^+(aq) + 2\,e^- \rightarrow Pb^{2+}(aq) + 2\,H_2O(l)$ $\quad$ and $\quad 2\,I^-(aq) \rightarrow I_2(s) + 2\,e^-$
Add half-reactions: $PbO_2(s) + 4\,H^+(aq) + \cancel{2\,e^-} + 2\,I^-(aq) \rightarrow Pb^{2+}(aq) + 2\,H_2O(l) + I_2(s) + \cancel{2\,e^-}$
Cancel electrons & others: $PbO_2(s) + 4\,H^+(aq) + 2\,I^-(aq) \rightarrow Pb^{2+}(aq) + 2\,H_2O(l) + I_2(s)$

Check:

Reactants	Products
1 Pb atom	1 Pb atom
2 O atoms	2 O atoms
4 H atoms	4 H atoms
2 I atoms	2 I atoms
+2 charge	+2 charge

(b) Separate: $\qquad\qquad\qquad\quad MnO_4^-(aq) \rightarrow Mn^{2+}(aq)$ $\quad$ and $\quad SO_3^{2-}(aq) \rightarrow SO_4^{2-}(aq)$
Balance non H & O elements: $MnO_4^-(aq) \rightarrow Mn^{2+}(aq)$ $\quad$ and $\quad SO_3^{2-}(aq) \rightarrow SO_4^{2-}(aq)$
Balance O with H_2O: $MnO_4^-(aq) \rightarrow Mn^{2+}(aq) + 4\,H_2O(l)$ $\quad$ and $\quad SO_3^{2-}(aq) + H_2O(l) \rightarrow SO_4^{2-}(aq)$
Balance H with H^+:
$\quad MnO_4^-(aq) + 8\,H^+(aq) \rightarrow Mn^{2+}(aq) + 4\,H_2O(l)$ and $SO_3^{2-}(aq) + H_2O(l) \rightarrow SO_4^{2-}(aq) + 2\,H^+(aq)$
Add electrons: $MnO_4^-(aq) + 8\,H^+(aq) + 5\,e^- \rightarrow Mn^{2+}(aq) + 4\,H_2O(l)$ $\quad$ and
$\qquad\qquad\qquad\qquad\qquad SO_3^{2-}(aq) + H_2O(l) \rightarrow SO_4^{2-}(aq) + 2\,H^+(aq) + 2\,e^-$
Equalize electrons: $2\,MnO_4^-(aq) + 16\,H^+(aq) + 10\,e^- \rightarrow 2\,Mn^{2+}(aq) + 8\,H_2O(l)$ $\quad$ and
$\qquad\qquad\qquad\qquad 5\,SO_3^{2-}(aq) + 5\,H_2O(l) \rightarrow 5\,SO_4^{2-}(aq) + 10\,H^+(aq) + 10\,e^-$
Add half-reactions: $2\,MnO_4^-(aq) + 6\,\cancel{16}\,H^+(aq) + \cancel{10\,e^-} + 5\,SO_3^{2-}(aq) + \cancel{5}\,H_2O(l) \rightarrow$
$\qquad\qquad\qquad\qquad 2\,Mn^{2+}(aq) + 3\,\cancel{8}\,H_2O(l) + 5\,SO_4^{2-}(aq) + \cancel{10}\,H^+(aq) + \cancel{10\,e^-}$
Cancel electrons: $2\,MnO_4^-(aq) + 6\,H^+(aq) + 5\,SO_3^{2-}(aq) \rightarrow 2\,Mn^{2+}(aq) + 3\,H_2O(l) + 5\,SO_4^{2-}(aq)$

Check:

Reactants	Products
2 Mn atoms	2 Mn atoms
23 O atoms	23 O atoms
6 H atoms	6 H atoms
5 S atoms	5 S atoms
−6 charge	−6 charge

(c) Separate: $\qquad\qquad\qquad\qquad S_2O_3^{2-}(aq) \rightarrow SO_4^{2-}(aq)$ $\qquad$ and $\quad Cl_2(g) \rightarrow Cl^-(aq)$
Balance non H & O elements: $S_2O_3^{2-}(aq) \rightarrow 2\,SO_4^{2-}(aq)$ $\qquad$ and $\quad Cl_2(g) \rightarrow 2\,Cl^-(aq)$
Balance O with H_2O: $S_2O_3^{2-}(aq) + 5\,H_2O(l) \rightarrow 2\,SO_4^{2-}(aq)$ $\qquad$ and $\quad Cl_2(g) \rightarrow 2\,Cl^-(aq)$
Balance H with H^+: $S_2O_3^{2-}(aq) + 5\,H_2O(l) \rightarrow 2\,SO_4^{2-}(aq) + 10\,H^+(aq)$ and $Cl_2(g) \rightarrow 2\,Cl^-(aq)$
Add electrons:
$\quad S_2O_3^{2-}(aq) + 5\,H_2O(l) \rightarrow 2\,SO_4^{2-}(aq) + 10\,H^+(aq) + 8\,e^-$ $\quad$ and $\quad Cl_2(g) + 2\,e^- \rightarrow 2\,Cl^-(aq)$
Equalize electrons:
$\quad S_2O_3^{2-}(aq) + 5\,H_2O(l) \rightarrow 2\,SO_4^{2-}(aq) + 10\,H^+(aq) + 8\,e^-$ $\quad$ and $\quad 4\,Cl_2(g) + 8\,e^- \rightarrow 8\,Cl^-(aq)$
Add half-reactions:
$\quad S_2O_3^{2-}(aq) + 5\,H_2O(l) + 4\,Cl_2(g) + \cancel{8\,e^-} \rightarrow 2\,SO_4^{2-}(aq) + 10\,H^+(aq) + \cancel{8\,e^-} + 8\,Cl^-(aq)$
Cancel electrons: $S_2O_3^{2-}(aq) + 5\,H_2O(l) + 4\,Cl_2(g) \rightarrow 2\,SO_4^{2-}(aq) + 10\,H^+(aq) + 8\,Cl^-(aq)$

Check:

Reactants	Products
2 S atoms	2 S atoms
8 O atoms	8 O atoms
10 H atoms	10 H atoms
8 Cl atoms	8 Cl atoms
-2 charge	-2 charge

18.41   **Conceptual Plan: Separate the overall reaction into two half-reactions: one for oxidation and one for reduction.** → **Balance each half-reaction with respect to mass in the following order: (1) Balance all elements other than H and O, (2) balance O by adding H_2O, (3) balance H by adding H^+, and (4) neutralize H^+ by adding enough OH^- to neutralize each H^+. Add the same number of OH^- ions to each side of the equation.** → **Balance each half-reaction with respect to charge by adding electrons. (The sum of the charges on both sides of the equation should be made equal by adding electrons as necessary.)** → **Make the number of electrons in both half-reactions equal by multiplying one or both half-reactions by a small whole number.** → **Add the two half-reactions, canceling electrons and other species as necessary.** → **Verify that the reaction is balanced with respect to both mass and charge.**
Solution:

(a)   Separate: $ClO_2(aq) \rightarrow ClO_2^-(aq)$ and $H_2O_2(aq) \rightarrow O_2(g)$
Balance non H & O elements: $ClO_2(aq) \rightarrow ClO_2^-(aq)$ and $H_2O_2(aq) \rightarrow O_2(g)$
Balance O with H_2O: $ClO_2(aq) \rightarrow ClO_2^-(aq)$ and $H_2O_2(aq) \rightarrow O_2(g)$
Balance H with H^+: $ClO_2(aq) \rightarrow ClO_2^-(aq)$ and $H_2O_2(aq) \rightarrow O_2(g) + 2\,H^+(aq)$
Neutralize H^+ with OH^-:

$$ClO_2(aq) \rightarrow ClO_2^-(aq) \text{ and } H_2O_2(aq) + 2\,OH^-(aq) \rightarrow O_2(g) + \underbrace{2\,H^+(aq) + 2\,OH^-(aq)}_{2\,H_2O(l)}$$

Add electrons: $ClO_2(aq) + e^- \rightarrow ClO_2^-(aq)$ and $H_2O_2(aq) + 2\,OH^-(aq) \rightarrow O_2(g) + 2\,H_2O(l) + 2\,e^-$
Equalize electrons:

$$2\,ClO_2(aq) + 2\,e^- \rightarrow 2\,ClO_2^-(aq) \text{ and } H_2O_2(aq) + 2\,OH^-(aq) \rightarrow O_2(g) + 2\,H_2O(l) + 2\,e^-$$

Add half-reactions:

$$2\,ClO_2(aq) + 2\,\cancel{e^-} + H_2O_2(aq) + 2\,OH^-(aq) \rightarrow 2\,ClO_2^-(aq) + O_2(g) + 2\,H_2O(l) + 2\,\cancel{e^-}$$

Cancel electrons: $2\,ClO_2(aq) + H_2O_2(aq) + 2\,OH^-(aq) \rightarrow 2\,ClO_2^-(aq) + O_2(g) + 2\,H_2O(l)$

Check:

Reactants	Products
2 Cl atoms	2 Cl atoms
8 O atoms	8 O atoms
4 H atoms	4 H atoms
-2 charge	-2 charge

(b)   Separate: $MnO_4^-(aq) \rightarrow MnO_2(s)$ and $Al(s) \rightarrow Al(OH)_4^-(aq)$
Balance non H & O elements: $MnO_4^-(aq) \rightarrow MnO_2(s)$ and $Al(s) \rightarrow Al(OH)_4^-(aq)$
Balance O with H_2O: $MnO_4^-(aq) \rightarrow MnO_2(s) + 2\,H_2O(l)$ and $Al(s) + 4\,H_2O(l) \rightarrow Al(OH)_4^-(aq)$
Balance H with H^+:
$MnO_4^-(aq) + 4\,H^+(aq) \rightarrow MnO_2(s) + 2\,H_2O(l)$ and $Al(s) + 4\,H_2O(l) \rightarrow Al(OH)_4^-(aq) + 4\,H^+(aq)$
Neutralize H^+ with OH^-: $MnO_4^- + \underbrace{4\,H^+(aq) + 4\,OH^-(aq)}_{2\,4\,H_2O(l)} \rightarrow MnO_2(s) + 2\,\cancel{H_2O(l)} + 4\,OH^-(aq)$

and $Al(s) + 4\,\cancel{H_2O(l)} + 4\,OH^-(aq) \rightarrow Al(OH)_4^-(aq) + 4\,H^+(aq) + 4\,OH^-(aq)$
$$\overbrace{}^{4\,H_2O(l)}$$
Add electrons: $MnO_4^-(aq) + 2\,H_2O(l) + 3\,e^- \rightarrow MnO_2(s) + 4\,OH^-(aq)$ and
$$Al(s) + 4\,OH^-(aq) \rightarrow Al(OH)_4^-(aq) + 3\,e^-$$
Equalize electrons: $MnO_4^-(aq) + 2\,H_2O(l) + 3\,e^- \rightarrow MnO_2(s) + 4\,OH^-(aq)$ and
$$Al(s) + 4\,OH^-(aq) \rightarrow Al(OH)_4^-(aq) + 3\,e^-$$

Add half-reactions:
$MnO_4^-(aq) + 2\,H_2O(l) + 3\,\cancel{e^-} + Al(s) + 4\,\cancel{OH^-(aq)} \rightarrow MnO_2(s) + 4\,\cancel{OH^-(aq)} + Al(OH)_4^-(aq) + 3\,\cancel{e^-}$
Cancel electrons: $MnO_4^-(aq) + 2\,H_2O(l) + Al(s) \rightarrow MnO_2(s) + Al(OH)_4^-(aq)$

Check:	Reactants	Products
	1 Mn atom	1 Mn atom
	6 O atoms	6 O atoms
	4 H atoms	4 H atoms
	1 Al atom	1 Al atom
	-1 charge	-1 charge

(c)   Separate: $Cl_2(g) \rightarrow Cl^-(aq)$ and $Cl_2(g) \rightarrow ClO^-(aq)$
Balance non H & O elements: $Cl_2(g) \rightarrow 2\,Cl^-(aq)$ and $Cl_2(g) \rightarrow 2ClO^-(aq)$
Balance O with H_2O: $Cl_2(g) \rightarrow 2\,Cl^-(aq)$ and $Cl_2(g) + 2\,H_2O(l) \rightarrow 2\,ClO^-(aq)$
Balance H with H^+: $Cl_2(g) \rightarrow 2\,Cl^-(aq)$ and $Cl_2(g) + 2\,H_2O(l) \rightarrow 2\,ClO^-(aq) + 4\,H^+(aq)$
Neutralize H^+ with OH^-:
$Cl_2(g) \rightarrow 2\,Cl^-(aq)$ and $Cl_2(g) + 2\,\cancel{H_2O}(l) + 4\,OH^-(aq) \rightarrow 2\,ClO^-(aq) + \underline{4\,H^+(aq) + 4\,OH^-(aq)}$
$$2\,\cancel{4}\,H_2O(l)$$

Add electrons: $Cl_2(g) + 2\,e^- \rightarrow 2\,Cl^-(aq)$ and $Cl_2(g) + 4\,OH^-(aq) \rightarrow 2\,ClO^-(aq) + 2\,H_2O(l) + 2\,e^-$
Equalize electrons: $Cl_2(g) + 2\,e^- \rightarrow 2\,Cl^-(aq)$ and $Cl_2(g) + 4\,OH^-(aq) \rightarrow 2\,ClO^-(aq) + 2\,H_2O(l) + 2\,e^-$
Add half-reactions: $Cl_2(g) + \cancel{2\,e^-} + Cl_2(g) + 4\,OH^-(aq) \rightarrow 2\,Cl^-(aq) + 2\,ClO^-(aq) + 2\,H_2O(l) + \cancel{2\,e^-}$
Cancel electrons: $2\,Cl_2(g) + 4\,OH^-(aq) \rightarrow 2\,Cl^-(aq) + 2\,ClO^-(aq) + 2\,H_2O(l)$
Simplify: $Cl_2(g) + 2\,OH^-(aq) \rightarrow Cl^-(aq) + ClO^-(aq) + H_2O(l)$

Check:	Reactants	Products
	2 Cl atoms	2 Cl atoms
	2 O atoms	2 O atoms
	2 H atoms	2 H atoms
	-2 charge	-2 charge

Voltaic Cells, Standard Cell Potentials, and Direction of Spontaneity

18.43   **Given:** voltaic cell overall redox reaction
Find: Sketch voltaic cell, labeling anode, cathode, all species, and direction of electron flow
Conceptual Plan: Separate the overall reaction into two half-cell reactions and add electrons as needed to balance reactions. Put the anode reaction on the left (oxidation = electrons as product) and the cathode reaction on the right (reduction = electrons as reactant). Electrons flow from anode to cathode.
Solution:

(a) $2\,Ag^+(aq) + Pb(s) \rightarrow 2\,Ag(s) + Pb^{2+}(aq)$ separates to $2\,Ag^+(aq) \rightarrow$ $2\,Ag(s)$ and $Pb(s) \rightarrow Pb^{2+}(aq)$; then add electrons to balance to get the cathode reaction—$2\,Ag^+(aq) + 2\,e^- \rightarrow 2\,Ag(s)$— and the anode reaction—$Pb(s) \rightarrow Pb^{2+}(aq) + 2\,e^-$.

Because we have $Pb(s)$ as the reactant for the oxidation, it will be our anode. Because we have $Ag(s)$ as the product for the reduction, it will be our cathode. Simplify the cathode reaction, dividing all terms by 2.

(b) $2\,ClO_2(g) + 2\,I^-(aq) \rightarrow 2\,ClO_2^-(aq) + I_2(s)$ separates to $2\,ClO_2(g)$ $\rightarrow 2\,ClO_2^-(aq)$ and $2\,I^-(aq) \rightarrow I_2(s)$; then add electrons to balance to get the cathode reaction—$2\,ClO_2(g) + 2\,e^- \rightarrow 2\,ClO_2^-(aq)$—and the anode reaction—$2\,I^-(aq) \rightarrow I_2(s) + 2\,e^-$.
Because we have $I^-(aq)$ as the reactant for the oxidation, we will need to use Pt as our anode. Because we have $ClO_2^-(aq)$ as the product for the reduction, we will need to use Pt as our cathode. Because $ClO_2(g)$ is our reactant for the reduction, we need to use an electrode assembly like that used for a SHE. Simplify the cathode reaction, dividing all terms by 2.

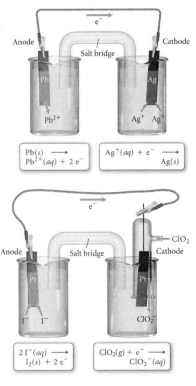

(c) $O_2(g) + 4\,H^+(aq) + 2\,Zn(s) \rightarrow 2\,H_2O(l) + 2\,Zn^{2+}(aq)$ separates to $O_2(g) + 4\,H^+(aq) \rightarrow 2\,H_2O(l)$ and $2\,Zn(s) \rightarrow 2\,Zn^{2+}(aq)$; then add electrons to balance to get the cathode reaction— $O_2(g) + 4\,H^+(aq) + 4\,e^- \rightarrow 2\,H_2O(l)$—and the anode reaction— $2\,Zn(s) \rightarrow 2\,Zn^{2+}(aq) + 4\,e^-$.

Because we have $Zn(s)$ as the reactant for the oxidation, it will be our anode. Because we have $H_2O(l)$ as the product for the reduction, we will need to use Pt as our cathode. Because $O_2(g)$ is our reactant for the reduction, we need to use an electrode assembly like that used for a SHE. Simplify the anode reaction, dividing all terms by 2.

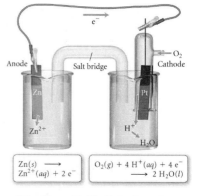

18.45   **Given:** overall reactions from Problem 18.43   **Find:** E°_{cell}

Conceptual Plan: Use Table 18.1 to look up half-reactions from the solution of Problem 18.43. Calculate the standard cell potential by subtracting the electrode potential of the anode from the electrode potential of the cathode: $E^\circ_{cell} = E^\circ_{cathode} - E^\circ_{anode}$.

Solution:

(a) $Ag^+(aq) + e^- \rightarrow Ag(s)$ $E^\circ_{red} = 0.80\,V = E^\circ_{cathode}$ and $Pb(s) \rightarrow Pb^{2+}(aq) + 2\,e^-$ $E^\circ_{red} = -0.13\,V = E^\circ_{anode}$. Then $E^\circ_{cell} = E^\circ_{cathode} - E^\circ_{anode} = 0.80\,V - (-0.13\,V) = 0.93\,V$.

(b) $ClO_2(g) + e^- \rightarrow ClO_2^-(aq)$ $E^\circ_{red} = 0.95\,V = E^\circ_{cathode}$ and $2\,I^-(aq) \rightarrow I_2(s) + 2\,e^-$ $E^\circ_{red} = +0.54\,V = E^\circ_{anode}$. Then $E^\circ_{cell} = E^\circ_{cathode} - E^\circ_{anode} = 0.95\,V - 0.54\,V = 0.41\,V$.

(c) $O_2(g) + 4\,H^+(aq) + 4\,e^- \rightarrow 2\,H_2O(l)$ $E^\circ_{red} = 1.23\,V$ and $Zn(s) \rightarrow Zn^{2+}(aq) + 2\,e^-$ $E^\circ_{red} = -0.76\,V = E^\circ_{anode}$. Then $E^\circ_{cell} = E^\circ_{cathode} - E^\circ_{anode} = 1.23\,V - (-0.76\,V) = 1.99\,V$.

Check: The units (V) are correct. All of the voltages are positive, which is consistent with a voltaic cell.

18.47   **Given:** voltaic cell drawing

Find: (a) Determine electron flow direction, anode, and cathode; (b) write balanced overall reaction and calculate E°_{cell}; (c) label electrodes as $+$ and $-$; and (d) find directions of anions and cations from salt bridge.

Conceptual Plan: Look at each half-cell and write a reduction reaction by using electrode and solution composition and adding electrons to balance. Look up half-reactions and standard reduction potentials in Table 18.1. Because this is a voltaic cell, the cell potentials must be assigned to give a positive E°_{cell}. Calculate the standard cell potential by subtracting the electrode potential of the anode from the electrode potential of the cathode, $E^\circ_{cell} = E^\circ_{cathode} - E^\circ_{anode}$, choosing the electrode assignments to give a positive E°_{cell}.

(a)   **Label the electrode where the oxidation occurs as the anode. Label the electrode where the reduction occurs as the cathode. Electrons flow from anode to cathode.**

(b)   **Take two half-cell reactions and multiply the reactions as necessary to equalize the number of electrons transferred. Add the two half-cell reactions and cancel electrons and any other species.**

(c)   **Label anode as $(-)$ and cathode as $(+)$.**

(d)   **Cations will flow from the salt bridge toward the cathode, and the anions will flow from the salt bridge toward the anode.**

Solution:

left side: $Fe^{3+}(aq) \rightarrow Fe(s)$; right side: $Cr^{3+}(aq) \rightarrow Cr(s)$. Add electrons to balance $Fe^{3+}(aq) + 3\,e^- \rightarrow Fe(s)$ and right side—$Cr^{3+}(aq) + 3\,e^- \rightarrow Cr(s)$. Look up cell standard reduction potentials: $Fe^{3+}(aq) + 3\,e^- \rightarrow Fe(s)$ $E^\circ_{red} = -0.036\,V$ and $Cr^{3+}(aq) + 3\,e^- \rightarrow Cr(s)$ $E^\circ_{red} = -0.73\,V$. To get a positive cell potential, the second reaction is the oxidation reaction (anode). $E^\circ_{cell} = E^\circ_{cathode} - E^\circ_{anode} = -0.036\,V - (-0.73\,V) = +0.69\,V$ ((a), (c), and (d)). (b) Add two half-reactions with the second reaction reversed. $Fe^{3+}(aq) + \cancel{3\,e^-} + Cr(s) \rightarrow Fe(s) + Cr^{3+}(aq) + \cancel{3\,e^-}$ Cancel electrons to get $Fe^{3+}(aq) + Cr(s) \rightarrow Fe(s) + Cr^{3+}(aq)$.

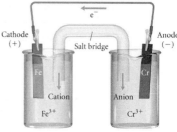

Check: All atoms and charge are balanced. The units (V) are correct. The cell potential is positive, which is consistent with a voltaic cell.

18.49   **Given:** overall reactions from Problem 18.43   **Find:** line notation
Conceptual Plan: Use the solution from Problem 18.43. Write the oxidation half-reaction components on the left and the reduction on the right. A double vertical line ($\mid\mid$), indicating the salt bridge, separates the two half-reactions. Substances in different phases are separated by a single vertical line ($\mid$), which represents the boundary between the phases. For some redox reactions, the reactants and products of one or both of the half-reactions may be in the same phase. In these cases, the reactants and products are separated from each other with a comma in the line diagram. Such cells use an inert electrode, such as platinum (Pt) or graphite, as the anode or cathode (or both).
Solution:

(a)   Reduction reaction: $Ag^+(aq) + e^- \rightarrow Ag(s)$; oxidation reaction: $Pb(s) \rightarrow Pb^2 + (aq) + 2\,e^-$
so $Pb(s) \mid Pb^{2+}(aq) \mid\mid Ag^+(aq) \mid Ag(s)$

(b)   Reduction reaction: $ClO_2(g) + e^- \rightarrow ClO_2^-(aq)$; oxidation reaction: $2\,I^-(aq) \rightarrow I_2(s) + 2\,e^-$
so $Pt(s) \mid I^-(aq) \mid I_2(s) \mid\mid ClO_2(g) \mid ClO_2^-(aq) \mid Pt(s)$

(c)   Reduction reaction: $O_2(g) + 4\,H^+(aq) + 4\,e^- \rightarrow 2\,H_2O(l)$; oxidation reaction: $Zn(s) \rightarrow Zn^{2+}(aq) + 2\,e^-$
so $Zn(s) \mid Zn^{2+}(aq) \mid\mid O_2(g) \mid H^+(aq), H_2O(l) \mid Pt(s)$

18.51   **Given:** $Sn(s) \mid Sn^{2+}(aq) \mid\mid NO_3^-(aq), H^+(aq), H_2O(l) \mid NO(g) \mid Pt(s)$
Find: Sketch voltaic cell, labeling anode, cathode, all species, direction of electron flow, and E°_{cell}
Conceptual Plan: Separate overall reaction into two half-cell reactions, knowing that the oxidation half-reaction components are on the left and the reduction half-reaction components are on the right. Add electrons as needed to balance reactions. Multiply the half-reactions by the appropriate factors so that an equal number of electrons are transferred. Add the half-cell reactions and cancel electrons. Put anode reaction on the left (oxidation = electrons as product) and cathode reaction on the right (reduction = electrons as reactant). Electrons flow from anode to cathode. Look up half-reactions in Table 18.1. Calculate the standard cell potential by subtracting the electrode potential of the anode from the electrode potential of the cathode:
$E^\circ_{cell} = E^\circ_{cathode} - E^\circ_{anode}.$
Solution: Oxidation reaction (anode): $Sn(s) \rightarrow Sn^{2+}(aq) + 2\,e^-$ $E^\circ_{red} = -0.14$ V; reduction reaction (cathode): $NO_3^-(aq) + 4\,H^+(aq) + 3\,e^- \rightarrow NO(g) + 2\,H_2O(l)$ $E^\circ_{red} = 0.96$ V. $E^\circ_{cell} = E^\circ_{cathode} - E^\circ_{anode} = 0.96$ V $- (-0.14$ V$) = 1.10$ V. Multiply

the first reaction by 3 and the second reaction by 2 so that 6 electrons are transferred. $3\,Sn(s) \rightarrow 3\,Sn^{2+}(aq) + 6\,e^-$ and $2\,NO_3^-(aq) + 8\,H^+(aq) + 6\,e^- \rightarrow 2\,NO(g) + 4\,H_2O(l)$. Add the two half-reactions and cancel electrons. $3\,Sn(s) + 2\,NO_3^-(aq) + 8\,H^+(aq) + \cancel{6\,e^-} \rightarrow 3\,Sn^{2+}(aq) + \cancel{6\,e^-} + 2\,NO(g) + 4\,H_2O(l)$. So balanced reaction is $3\,Sn(s) + 2\,NO_3^-(aq) + 8\,H^+(aq) \rightarrow 3\,Sn^{2+}(aq) + 2\,NO(g) + 4\,H_2O(l)$.

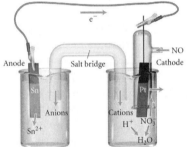

Check: All atoms and charge are balanced. The units (V) are correct. The cell potential is positive, which is consistent with a voltaic cell.

18.53   **Given:** overall reactions   **Find:** spontaneity in forward direction
Conceptual Plan: Separate the overall reaction into two half-cell reactions and add electrons as needed to balance reactions. Look up half-reactions in Table 18.1. Calculate the standard cell potential by subtracting the electrode potential of the anode from the electrode potential of the cathode: $E^\circ_{cell} = E^\circ_{cathode} - E^\circ_{anode}.$ **If** $E^\circ_{cell} = 0$, **the reaction is spontaneous in the forward direction.**
Solution:

(a) $Ni(s) + Zn^{2+}(aq) \rightarrow Ni^{2+}(aq) + Zn(s)$ separates to $Ni(s) \rightarrow Ni^{2+}(aq)$ and $Zn^{2+}(aq) \rightarrow Zn(s)$.
Add electrons. $Ni(s) \rightarrow Ni^{2+}(aq) + 2\,e^-$ and $Zn^{2+}(aq) + 2\,e^- \rightarrow Zn(s)$. Look up cell potentials. Ni is oxidized, so $E^\circ_{red} = -0.23$ V $= E^\circ_{anode}$. Zn^{2+} is reduced, so $E^\circ_{cathode} = -0.76$ V. Then $E^\circ_{cell} = E^\circ_{cathode} - E^\circ_{anode} = -0.76$ V $- (-0.23$ V$) = -0.53$ V, so the reaction is nonspontaneous.

(b) $Ni(s) + Pb^{2+}(aq) \rightarrow Ni^{2+}(aq) + Pb(s)$ separates to $Ni(s) \rightarrow Ni^{2+}(aq)$ and $Pb^{2+}(aq) \rightarrow Pb(s)$.
Add electrons. $Ni(s) \rightarrow Ni^{2+}(aq) + 2\,e^-$ and $Pb^{2+}(aq) + 2\,e^- \rightarrow Pb(s)$. Look up cell potentials. Ni is oxidized, so $E^\circ_{red} = -0.23$ V $= E^\circ_{anode}$. Pb^{2+} is reduced, so $E^\circ_{red} = -0.13$ V $= E^\circ_{cathode}$. Then $E^\circ_{cell} = E^\circ_{cathode} - E^\circ_{anode} = -0.13$ V $- (-0.23$ V$) = +0.10$ V, so the reaction is spontaneous.

(c) $Al(s) + 3\,Ag^+(aq) \rightarrow Al^{3+}(aq) + 3\,Ag(s)$ separates to $Al(s) \rightarrow Al^{3+}(aq)$ and $3\,Ag^+(aq) \rightarrow 3\,Ag(s)$.
Add electrons. $Al(s) \rightarrow Al^{3+}(aq) + 3\,e^-$ and $3\,Ag^+(aq) + 3\,e^- \rightarrow 3\,Ag(s)$. Simplify the Ag reaction to $Ag^+(aq) + e^- \rightarrow Ag(s)$. Look up cell potentials. Al is oxidized, so $E^\circ_{red} = -1.66$ V E°_{anode}. Ag^+ is reduced, so $E^\circ_{red} = 0.80$ V $= E^\circ_{cathode}$. Then $E^\circ_{cell} = E^\circ_{cathode} - E^\circ_{anode} = 0.80$ V $- (-1.66$ V$) = +2.46$ V, so the reaction is spontaneous.

(d) $Pb(s) + Mn^{2+}(aq) \rightarrow Pb^{2+}(aq) + Mn(s)$ separates to $Pb(s) \rightarrow Pb^{2+}(aq)$ and $Mn^{2+}(aq) \rightarrow Mn(s)$.
Add electrons. $Pb(s) \rightarrow Pb^{2+}(aq) + 2\,e^-$ and $Mn^{2+}(aq) + 2\,e^- \rightarrow Mn(s)$. Look up cell potentials.
Pb is oxidized, so $E^{\circ}_{red} = -0.13\ V = E^{\circ}_{anode}$. Mn^{2+} is reduced, so $E^{\circ}_{red} = -1.18\ V = E^{\circ}_{cathode}$. Then
$E^{\circ}_{cell} = E^{\circ}_{cathode} - E^{\circ}_{anode} = -1.18\ V - (-0.13\ V) = -1.05\ V$, so the reaction is nonspontaneous.

Check: The units (V) are correct. If the voltage is positive, the reaction is spontaneous. If the voltage is negative, the reaction is nonspontaneous.

18.55   For a metal to be able to reduce an ion, it must be below it in Table 18.1 (need positive $E^{\circ}_{cell} = E^{\circ}_{cathode} - E^{\circ}_{anode}$). So we need a metal that is below Mn^{2+} but above Mg^{2+}. Aluminum is the only one in the table that meets those criteria.

18.57   In general, metals whose reduction half-reactions lie below the reduction of H^+ to H_2 in Table 18.1 will dissolve in acids, while metals above it will not. (a) Al and (c) Pb meet that criterion. To write the balanced redox reactions, pair the oxidation of the metal with the reduction of H^+ to $H_2(2\,H^+(aq) + 2\,e^- \rightarrow H_2(g))$. For Al, $Al(s) \rightarrow Al^{3+}(aq) + 3\,e^-$. To balance the number of electrons transferred, we need to multiply the Al reaction by 2 and the H^+ reaction by 3. So $2\,Al(s) \rightarrow 2\,Al^{3+}(aq) + 6\,e^-$ and $6\,H^+(aq) + 6\,e^- \rightarrow 3\,H_2(g)$. Adding the two reactions: $2\,Al(s) + 6\,H^+(aq) + 6\,e^- \rightarrow 2\,Al^{3+}(aq) + 6\,e^- + 3\,H_2(g)$. Simplify to $2\,Al(s) + 6\,H^+(aq) \rightarrow 2\,Al^{3+}(aq) + 3\,H_2(g)$. For Pb, $Pb(s) \rightarrow Pb^{2+}(aq) + 2\,e^-$. Because each reaction involves two electrons, we can add the two reactions. $Pb(s) + 2\,H^+(aq) + 2\,e^- \rightarrow Pb^{2+}(aq) + 2\,e^- + H_2(g)$. Simplify to $Pb(s) + 2\,H^+(aq) \rightarrow Pb^{2+}(aq) + H_2(g)$.

18.59   Nitric acid (HNO_3) oxidizes metals through the following reduction half-reaction: $NO_3^-(aq) + 4\,H^+(aq) + 3\,e^- \rightarrow NO(g) + 2\,H_2O(l)\ E^{\circ}_{red} = 0.96\ V$. Because this half-reaction is above the reduction of H^+ in Table 18.1, HNO_3 can oxidize metals (copper, for example) that cannot be oxidized by HCl. (a) Cu, which is below nitric acid in the table, will be oxidized, but (b) Au, which is above nitric acid in the table (and has a reduction potential of 1.50 V), will not be oxidized. To write the balanced redox reactions, pair the oxidation of the metal with the reduction of nitric acid: $(NO_3^-(aq) + 4\,H^+(aq) + 3\,e^- \rightarrow NO(g) + 2\,H_2O(l))$. For Cu, $Cu(s) \rightarrow Cu^{2+}(aq) + 2\,e^-$. To balance the number of electrons transferred, we need to multiply the Cu reaction by 3 and the nitric acid reaction by 2. So $3\,Cu(s) \rightarrow 3\,Cu^{2+}(aq) + 6\,e^-$ and $2\,NO_3^-(aq) + 8\,H^+(aq) + 6\,e^- \rightarrow 2\,NO(g) + 4\,H_2O(l)$. Adding the two reactions: $3\,Cu(s) + 2\,NO_3^-(aq) + 8\,H^+(aq) + 6\,e^- \rightarrow 3\,Cu^{2+}(aq) + 6\,e^- + 2\,NO(g) + 4\,H_2O(l)$. Simplify to $3\,Cu(s) + 2\,NO_3^-(aq) + 8\,H^+(aq) \rightarrow 3\,Cu^{2+}(aq) + 2\,NO(g) + 4\,H_2O(l)$.

18.61   **Given:** overall reactions   **Find:** E°_{cell} and spontaneity in forward direction
Conceptual Plan: Separate the overall reaction into two half-cell reactions and add electrons as needed to balance reactions. Look up half-reactions in Table 18.1. Calculate the standard cell potential by subtracting the electrode potential of the anode from the electrode potential of the cathode: $E^{\circ}_{cell} = E^{\circ}_{cathode} - E^{\circ}_{anode}$. If $E^{\circ}_{cell} > 0$, the reaction is spontaneous in the forward direction.
Solution:

(a) $2\,Cu(s) + Mn^{2+}(aq) \rightarrow 2\,Cu^+(aq) + Mn(s)$ separates to $2\,Cu(s) \rightarrow 2\,Cu^+(aq) + Mn(s)$ and $Mn^{2+}(aq) \rightarrow Mn(s)$.
Add electrons. $2\,Cu(s) \rightarrow 2\,Cu^+(aq) + 2\,e^-$ and $Mn^{2+}(aq) + 2\,e^- \rightarrow Mn(s)$. Simplify the Cu reaction to $Cu(s) \rightarrow Cu^+(aq) + e^-$. Look up cell potentials. Cu is oxidized, so $E^{\circ}_{red} = +0.52\ V = E^{\circ}_{anode}$. Mn^{2+} is reduced, so $E^{\circ}_{red} = -1.18\ V = E^{\circ}_{cathode}$. Then $E^{\circ}_{cell} = E^{\circ}_{cathode} - E^{\circ}_{anode} = -1.18\ V - 0.52\ V = -1.70\ V$, so the reaction is nonspontaneous.

(b) $MnO_2(s) + 4\,H^+(aq) + Zn(s) \rightarrow Mn^{2+}(aq) + 2\,H_2O(l) + Zn^{2+}(aq)$ separates to $MnO_2(s) + 4\,H^+(aq) \rightarrow Mn^{2+}(aq) + 2\,H_2O(l)$ and $Zn(s) \rightarrow Zn^{2+}(aq)$. Add electrons. $MnO_2(s) + 4\,H^+(aq) + 2\,e^- \rightarrow Mn^{2+}(aq) + 2\,H_2O(l)$ and $Zn(s) \rightarrow Zn^{2+}(aq) + 2\,e^-$. Look up cell potentials. Zn is oxidized, so $E^{\circ}_{red} = -0.76\ V = E^{\circ}_{anode}$. Mn is reduced, so $E^{\circ}_{red} = 1.21\ V = E^{\circ}_{cathode}$. Then $E^{\circ}_{cell} = E^{\circ}_{cathode} - E^{\circ}_{anode} = 1.21\ V - (-0.76\ V) = +1.97\ V$, so the reaction is spontaneous.

(c) $Cl_2(g) + 2\,F^-(aq) \rightarrow 2\,Cl^-(aq) + F_2(g)$ separates to $Cl_2(g) \rightarrow 2\,Cl^-(aq)$ and $2\,F^-(aq) \rightarrow F_2(g)$.
Add electrons. $Cl_2(g) + 2\,e^- \rightarrow 2\,Cl^-(aq)$ and $2\,F^-(aq) \rightarrow F_2(g) + 2\,e^-$. Look up cell potentials. F^- is oxidized, so $E^{\circ}_{red} = 2.87\ V = E^{\circ}_{anode}$. Cl is reduced, so $E^{\circ}_{red} = 1.36\ V = E^{\circ}_{cathode}$. Then $E^{\circ}_{cell} = E^{\circ}_{cathode} - E^{\circ}_{anode} = 1.36\ V - 2.87\ V = -1.51\ V$, so the reaction is nonspontaneous.

Check: The units (V) are correct. If the voltage is positive, the reaction is spontaneous.

18.63   (a) Pb^{2+}. The strongest oxidizing agent is the one with the reduction reaction that is closest to the top of Table 18.1 (most positive, least negative reduction potential).

Cell Potential, Free Energy, and the Equilibrium Constant

18.65   **Given:** overall reactions   **Find:** ΔG°_{rxn} and spontaneity in forward direction

Conceptual Plan: Separate the overall reaction into two half-cell reactions and add electrons as needed to balance reactions. Look up half-reactions in Table 18.1. Calculate the standard cell potential by subtracting the electrode potential of the anode from the electrode potential of the cathode: $E^{\circ}_{cell} = E^{\circ}_{cathode} - E^{\circ}_{anode}$. Then calculate ΔG°_{rxn} using $\Delta G^{\circ}_{rxn} = -nF\,E^{\circ}_{cell}$.

Solution:

(a) $Pb^{2+}(aq) + Mg(s) \rightarrow Pb(s) + Mg^{2+}(aq)$ separates to $Pb^{2+}(aq) \rightarrow Pb(s)$ and $Mg(s) \rightarrow Mg^{2+}(aq)$.
Add electrons. $Pb^{2+}(aq) + 2\,e^{-} \rightarrow Pb(s)$ and $Mg(s) \rightarrow Mg^{2+}(aq) + 2\,e^{-}$. Look up cell potentials.
Mg is oxidized, so $E^{\circ}_{red} = -2.37\ V = E^{\circ}_{anode}$. Pb^{2+} is reduced, so $E^{\circ}_{red} = -0.13\ V = E^{\circ}_{cathode}$. Then
$E^{\circ}_{cell} = E^{\circ}_{cathode} - E^{\circ}_{anode} = -0.13\ V - (-2.37\ V) = +2.24\ V.\ n = 2$, so $\Delta G^{\circ}_{rxn} = -nF\,E^{\circ}_{cell} =$

$$-2\ \cancel{mol\ e^{-}} \times \frac{96{,}485\ C}{\cancel{mol\ e^{-}}} \times 2.24\ V = -2 \times 96{,}485\,\cancel{C} \times 2.24\frac{J}{\cancel{C}} = -4.32 \times 10^{5}\ J = -432\ kJ.$$

(b) $Br_{2}(l) + 2\,Cl^{-}(aq) \rightarrow 2\,Br^{-}(aq) + Cl_{2}(g)$ separates to $Br_{2}(g) \rightarrow 2\,Br^{-}(aq)$ and $2\,Cl^{-}(aq) \rightarrow Cl_{2}(g)$.
Add electrons. $Br_{2}(g) + 2\,e^{-} \rightarrow 2\,Br^{-}(aq)$ and $2\,Cl^{-}(aq) \rightarrow Cl_{2}(g) + 2\,e^{-}$. Look up cell potentials.
Cl is oxidized, so $E^{\circ}_{red} = 1.36\ V = E^{\circ}_{anode}$. Br is reduced, so $E^{\circ}_{red} = 1.09\ V = E^{\circ}_{cathode}$. Then
$E^{\circ}_{cell} = E^{\circ}_{cathode} - E^{\circ}_{anode} = 1.09\ V - 1.36\ V = -0.27\ V.\ n = 2$, so $\Delta G^{\circ}_{rxn} = -nF\,E^{\circ}_{cell} = -2\ \cancel{mol\ e^{-}}$

$$\times \frac{96{,}485\ C}{\cancel{mol\ e^{-}}} \times -0.27\ V = -2 \times 96{,}485\ \cancel{C} \times -0.27\frac{J}{\cancel{C}} = 5.2 \times 10^{4}\ J = 52\ kJ.$$

(c) $MnO_{2}(s) + 4\,H^{+}(aq) + Cu(s) \rightarrow Mn^{2+}(aq) + 2\,H_{2}O(l) + Cu^{2+}(aq)$ separates to
$MnO_{2}(s) + 4\,H^{+}(aq) \rightarrow Mn^{2+}(aq) + 2\,H_{2}O(l)$ and $Cu(s) \rightarrow Cu^{2+}(aq)$. Add electrons.
$MnO_{2}(s) + 4\,H^{+}(aq) + 2\,e^{-} \rightarrow Mn^{2+}(aq) + 2\,H_{2}O(l)$ and $Cu(s) \rightarrow Cu^{2+}(aq) + 2\,e^{-}$. Look up cell
potentials. Cu is oxidized, so $E^{\circ}_{red} = 0.34\ V = E^{\circ}_{anode}$. Mn is reduced, so $E^{\circ}_{red} = 1.21\ V = E^{\circ}_{cathode}$. Then
$E^{\circ}_{cell} = E^{\circ}_{cathode} - E^{\circ}_{anode} = 1.21\ V - 0.34\ V = +0.87\ V.\ n = 2$, so

$$\Delta G^{\circ}_{rxn} = -nF\,E^{\circ}_{cell} = -2\ \cancel{mol\ e^{-}} \times \frac{96{,}485\ C}{\cancel{mol\ e^{-}}} \times 0.87\ V = -2 \times 96{,}485\ \cancel{C} \times 0.87\frac{J}{\cancel{C}} = -1.7 \times 10^{5}\ J =$$

$$-1.7 \times 10^{2}\ kJ.$$

Check: The units (kJ) are correct. If the voltage is positive, the reaction is spontaneous and the free energy change is negative.

18.67   **Given:** overall reactions from Problem 18.65   **Find:** K

Conceptual Plan: $^{\circ}C \rightarrow K$ then $\Delta G^{\circ}_{rxn}, T = K$

$$K = 273.15 + {^{\circ}C} \qquad \Delta G^{\circ}_{rxn} = -RT \ln K$$

Solution: $T = 273.15 + 25\ {^{\circ}C} = 298\ K$ then

(a) $\Delta G^{\circ}_{rxn} = -RT \ln K$. Rearrange to solve for K.

$$K = e^{\frac{-\Delta G^{\circ}_{rxn}}{RT}} = e^{\dfrac{-(-432\ \cancel{kJ}) \times \frac{1000\ J}{1\ \cancel{kJ}}}{\left(8.314\frac{\cancel{J}}{K \cdot mol}\right)(298\ \cancel{K})}} = e^{174.364} = 5.31 \times 10^{75}$$

(b) $\Delta G^{\circ}_{rxn} = -RT \ln K$. Rearrange to solve for K.

$$K = e^{\frac{-\Delta G^{\circ}_{rxn}}{RT}} = e^{\dfrac{-(-52\ \cancel{kJ}) \times \frac{1000\ J}{1\ \cancel{kJ}}}{\left(8.314\frac{\cancel{J}}{K \cdot mol}\right)(298\ \cancel{K})}} = e^{-20.998} = 7.77 \times 10^{-10}$$

(c) $\Delta G^{\circ}_{rxn} = -RT \ln K$. Rearrange to solve for K.

$$K = e^{\frac{-\Delta G^{\circ}_{rxn}}{RT}} = e^{\dfrac{-(-170\ \cancel{kJ}) \times \frac{1000\ J}{1\ \cancel{kJ}}}{\left(8.314\frac{\cancel{J}}{K \cdot mol}\right)(298\ \cancel{K})}} = e^{68.616} = 6.3 \times 10^{29}$$

Check: The units (none) are correct. If the voltage is positive, the reaction is spontaneous and the free energy change is negative and the equilibrium constant is large.

18.69   **Given:** $Ni^{2+}(aq) + Cd(s) \rightarrow$ **Find:** K
Conceptual Plan: **Write two half-cell reactions and add electrons as needed to balance the reactions. Look up half-reactions in Table 18.1. Calculate the standard cell potential by subtracting the electrode potential of the anode from the electrode potential of the cathode:** $E°_{cell} = E°_{cathode} - E°_{anode}$, **then** $°C \rightarrow K$ **then** $E°_{cell}, n, T \rightarrow K$.

$$K = 273.15 + °C \qquad \Delta G°_{rxn} = -RT \ln K = -n F E°_{cell}$$

Solution: $Ni^{2+}(aq) + 2e^- \rightarrow Ni(s)$ and $Cd(s) \rightarrow Cd^{2+}(aq) + 2e^-$. Look up cell potentials. Cd is oxidized, so $E°_{red} = -0.40$ V $= E°_{anode}$. Ni^{2+} is reduced, so $E°_{red} = -0.23$ V $= E°_{cathode}$. Then $E°_{cell} = E°_{cathode} - E°_{anode} = -0.23$ V $- (-0.40$ V$) = +0.17$ V. The overall reaction is $Ni^{2+}(aq) + Cd(s) \rightarrow Ni(s) + Cd^{2+}(aq)$. $n = 2$ and $T = 273.15 + 25$ °C $= 298$ K then $\Delta G°_{rxn} = -RT \ln K = -n F E°_{cell}$. Rearrange to solve for K.

$$K = e^{\frac{n F E°_{cell}}{RT}} = e^{\dfrac{2\ \cancel{mol\ e^-} \times \dfrac{96{,}485\ \cancel{C}}{\cancel{mol\ e^-}} \times 0.17\ \dfrac{\cancel{J}}{\cancel{C}}}{\left(8.314\ \dfrac{\cancel{J}}{K\cdot mol}\right)(298\ \cancel{K})}} = e^{13.241} = 5.6 \times 10^5$$

Check: The units (none) are correct. If the voltage is positive, the reaction is spontaneous and the equilibrium constant is large.

18.71   **Given:** $n = 2$ and $K = 25$ **Find:** $\Delta G°_{rxn}$ and $E°_{cell}$
Conceptual Plan: $K, T \rightarrow \Delta G°_{rxn}$ **and** $\Delta G°_{rxn}, n \rightarrow E°_{cell}$

$$\Delta G°_{rxn} = -RT \ln K \qquad\qquad \Delta G°_{rxn} = -n F E°_{cell}$$

Solution: $\Delta G°_{rxn} = -RT \ln K = -\left(8.314\dfrac{J}{K\cdot mol}\right)(298\ K)\ln 25 = -7.\underline{9}7500 \times 10^3$ J $= -8.0$ kJ and

$\Delta G°_{rxn} = -n F E°_{cell}$. Rearrange to solve for $E°_{cell}$.

$$E°_{cell} = \frac{\Delta G°_{rxn}}{-n F} = \frac{-7.\underline{9}7500 \times 10^3\ J}{-2\ \cancel{mol\ e^-} \times \dfrac{96{,}485\ C}{\cancel{mol\ e^-}}} = 0.041\frac{V\cdot\cancel{C}}{\cancel{C}} = 0.041\ V$$

Check: The units (kJ and V) are correct. If $K > 1$, the voltage is positive and the free energy change is negative.

Nonstandard Conditions and the Nernst Equation

18.73   **Given:** $Sn^{2+}(aq) + Mn(s) \rightarrow Sn(s) + Mn^{2+}(aq)$
Find: (a) $E°_{cell}$; (b) E_{cell} when $[Sn^{2+}] = 0.0100$ M; $[Mn^{2+}] = 2.00$ M; and (c) E_{cell} when $[Sn^{2+}] = 2.00$ M; $[Mn^{2+}] = 0.0100$ M
Conceptual Plan: **(a) Separate the overall reaction into two half-cell reactions and add electrons as needed to balance the reactions. Look up half-reactions in Table 18.1. Calculate the standard cell potential by subtracting the electrode potential of the anode from the electrode potential of the cathode:** $E°_{cell} = E°_{cathode} - E°_{anode}$.
(b) and (c) $E°_{cell}, [Sn^{2+}], [Mn^{2+}], n \rightarrow E_{cell}$

$$E_{cell} = E°_{cell} - \frac{0.0592\ V}{n}\log Q \quad \text{where } Q = \frac{[Mn^{2+}]}{[Sn^{2+}]}$$

Solution:
(a)   Separate the overall reaction to $Sn^{2+}(aq) \rightarrow Sn(s)$ and $Mn(s) \rightarrow Mn^{2+}(aq)$. Add electrons. $Sn^{2+}(aq) + 2e^- \rightarrow Sn(s)$ and $Mn(s) \rightarrow Mn^{2+}(aq) + 2e^-$. Look up cell potentials. Mn is oxidized, so $E°_{red} = -1.18$ V $= E°_{anode}$. Sn^{2+} is reduced, so $E°_{red} = -0.14$ V $= E°_{cathode}$. Then $E°_{cell} = E°_{cathode} - E°_{anode} = -0.14$ V $- (-1.18$ V$) = +1.04$ V.

(b) $Q = \dfrac{[Mn^{2+}]}{[Sn^{2+}]} = \dfrac{2.00\ \cancel{M}}{0.0100\ \cancel{M}} = 200.$ and $n = 2$ then

$$E_{cell} = E°_{cell} - \frac{0.0592\ V}{n}\log Q = 1.04\ V - \frac{0.0592\ V}{2}\log 200. = +0.97\ V$$

(c)　　$Q = \dfrac{[\text{Mn}^{2+}]}{[\text{Sn}^{2+}]} = \dfrac{0.0100 \text{ M}}{2.00 \text{ M}} = 0.00500$ and $n = 2$ then

$$E_{\text{cell}} = E^{\circ}_{\text{cell}} - \dfrac{0.0592 \text{ V}}{n} \log Q = 1.04 \text{ V} - \dfrac{0.0592 \text{ V}}{2} \log 0.00500 = +1.11 \text{ V}$$

Check: The units (V, V, and V) are correct. The Sn^{2+} reduction reaction is above the Mn^{2+} reduction reaction, so the standard cell potential will be positive. Having more products than reactants reduces the cell potential. Having more reactants than products raises the cell potential.

18.75　　**Given:** $\text{Pb}(s) \rightarrow \text{Pb}^{2+}(aq, 0.10 \text{ M}) + 2 \text{ e}^-$ and $\text{MnO}_4^-(aq, 1.50 \text{ M}) + 4 \text{ H}^+(aq, 2.0 \text{ M}) + 3 \text{ e}^- \rightarrow \text{MnO}_2(s) +$
　　　　$2 \text{ H}_2\text{O}(l)$　　**Find:** E_{cell}
　　　　Conceptual Plan: Look up half-reactions in Table 18.1. Calculate the standard cell potential by subtracting the electrode potential of the anode from the electrode potential of the cathode: $E^{\circ}_{\text{cell}} = E^{\circ}_{\text{cathode}} - E^{\circ}_{\text{anode}}$**.**
　　　　Equalize the number of electrons transferred by multiplying the first reaction by 3 and the second reaction by 2. Add the two half-cell reactions and cancel the electrons.
　　　　Then $E^{\circ}_{\text{cell}}, [\text{Pb}^{2+}], [\text{MnO}_4^-], [\text{H}^+], n \rightarrow E_{\text{cell}}$**.**

$$E_{\text{cell}} = E^{\circ}_{\text{cell}} - \dfrac{0.0592 \text{ V}}{n} \log Q \quad \text{where } Q = \dfrac{[\text{Pb}^{2+}]^3}{[\text{MnO}_4^-]^2 [\text{H}^+]^8}$$

　　　　Solution: Pb is oxidized, so $E^{\circ}_{\text{red}} = -0.13 \text{ V} = E^{\circ}_{\text{anode}}$. Mn is reduced, so $E^{\circ}_{\text{red}} = 1.68 \text{ V} = E^{\circ}_{\text{cathode}}$.
　　　　Then $E^{\circ}_{\text{cell}} = E^{\circ}_{\text{cathode}} - E^{\circ}_{\text{anode}} = 1.68 \text{ V} - (-0.13 \text{ V}) = +1.81 \text{ V}$. Equalizing the electrons:
　　　　$3 \text{ Pb}(s) \rightarrow 3 \text{ Pb}^{2+}(aq) + 6 \text{ e}^-$ and $2 \text{ MnO}_4^-(aq) + 8 \text{ H}^+(aq) + 6 \text{ e}^- \rightarrow 2 \text{ MnO}_2(s) + 4 \text{ H}_2\text{O}(l)$. Adding the two
　　　　reactions: $3 \text{ Pb}(s) + 2 \text{ MnO}_4^-(aq) + 8 \text{ H}^+(aq) + \cancel{6 \text{ e}^-} \rightarrow 3 \text{ Pb}^{2+}(aq) + \cancel{6 \text{ e}^-} + 2 \text{ MnO}_2(s) + 4 \text{ H}_2\text{O}(l)$. Cancel
　　　　the electrons: $3 \text{ Pb}(s) + 2 \text{ MnO}_4^-(aq) + 8 \text{ H}^+(aq) \rightarrow 3 \text{ Pb}^{2+}(aq) + 2 \text{ MnO}_2(s) + 4 \text{ H}_2\text{O}(l)$. So $n = 6$ and

$$Q = \dfrac{[\text{Pb}^{2+}]^3}{[\text{MnO}_4^-]^2 [\text{H}^+]^8} = \dfrac{(0.10)^3}{(1.50)^2 (2.0)^8} = 1.\underline{7}361 \times 10^{-6}. \text{ Then}$$

$$E_{\text{cell}} = E^{\circ}_{\text{cell}} - \dfrac{0.0592 \text{ V}}{n} \log Q = 1.81 \text{ V} - \dfrac{0.0592 \text{ V}}{6} \log 1.\underline{7}361 \times 10^{-6} = +1.87 \text{ V}.$$

　　　　Check: The units (V) are correct. The MnO_4^- reduction reaction is above the Pb^{2+} reduction reaction, so the standard cell potential will be positive. Having more reactants than products raises the cell potential.

18.77　　**Given:** Zn/Zn^{2+} and Ni/Ni^{2+} half-cells in voltaic cell; initially, $[\text{Ni}^{2+}] = 1.50 \text{ M}$ and $[\text{Zn}^{2+}] = 0.100 \text{ M}$
　　　　Find: (a) initial E_{cell}, (b) E_{cell} when $[\text{Ni}^{2+}] = 0.500 \text{ M}$, and (c) $[\text{Ni}^{2+}]$ and $[\text{Zn}^{2+}]$ when $E_{\text{cell}} = 0.45 \text{ V}$
　　　　Conceptual Plan:
　　(a)　　**Write two half-cell reactions and add electrons as needed to balance reactions. Look up half-reactions in**
　　　　　Table 18.1. Calculate the standard cell potential by subtracting the electrode potential of the anode from
　　　　　the electrode potential of the cathode: $E^{\circ}_{\text{cell}} = E^{\circ}_{\text{cathode}} - E^{\circ}_{\text{anode}}$**. Choose the direction of the half-cell**
　　　　　reactions so that $E^{\circ}_{\text{cell}} > 0$**. Add two half-cell reactions and cancel electrons to generate overall reaction.**
　　　　　Define Q **based on overall reaction. Then** $E^{\circ}_{\text{cell}}, [\text{Ni}^{2+}], [\text{Zn}^{2+}], n \rightarrow E_{\text{cell}}$**.**

$$E_{\text{cell}} = E^{\circ}_{\text{cell}} - \dfrac{0.0592 \text{ V}}{n} \log Q$$

　　(b)　　**When** $[\text{Ni}^{2+}] = 0.500 \text{ M}, [\text{Zn}^{2+}] = 1.100 \text{ M}$**. (Because the stoichiometric coefficients for** $\text{Ni}^{2+} : \text{Zn}^{2+}$
　　　　　are 1:1 and the $[\text{Ni}^{2+}]$ **drops by 1.00 M, the other concentration must rise by 1.00 M.) Then**
　　　　　$E^{\circ}_{\text{cell}}, [\text{Ni}^{2+}], [\text{Zn}^{2+}], n \rightarrow E_{\text{cell}}$**.**

$$E_{\text{cell}} = E^{\circ}_{\text{cell}} - \dfrac{0.0592 \text{ V}}{n} \log Q$$

　　(c)　　$E^{\circ}_{\text{cell}}, E_{\text{cell}}, n \rightarrow [\text{Zn}^{2+}]/[\text{Ni}^{2+}] \rightarrow [\text{Ni}^{2+}], [\text{Zn}^{2+}]$

$$E_{\text{cell}} = E^{\circ}_{\text{cell}} - \dfrac{0.0592 \text{ V}}{n} \log Q \quad [\text{Ni}^{2+}] + [\text{Zn}^{2+}] = 1.50 \text{ M} + 0.100 \text{ M} = 1.60 \text{ M}$$

　　Solution:
　　(a)　　$\text{Zn}^{2+}(aq) + 2 \text{ e}^- \rightarrow \text{Zn}(s)$ and $\text{Ni}^{2+}(aq) + 2 \text{ e}^- \rightarrow \text{Ni}(s)$. Look up cell potentials. For Zn, $E^{\circ}_{\text{red}} = -0.76 \text{ V}$.
　　　　For Ni, $E^{\circ}_{\text{red}} = -0.23 \text{ V}$. To get a positive E°_{cell}, Zn is oxidized; so $E^{\circ}_{\text{red}} = -0.76 \text{ V} = E^{\circ}_{\text{anode}}$. Ni^{2+} is reduced,
　　　　so $E^{\circ}_{\text{red}} = -0.23 \text{ V} = E^{\circ}_{\text{cathode}}$. Then $E^{\circ}_{\text{cell}} = E^{\circ}_{\text{cathode}} - E^{\circ}_{\text{anode}} = -0.23 \text{ V} - (-0.76 \text{ V}) = +0.53 \text{ V}$. Adding

the two half-cell reactions: $Zn(s) + Ni^{2+}(aq) + 2e^- \rightarrow Zn^{2+}(aq) + 2e^- + Ni(s)$. The overall reaction is

$Zn(s) + Ni^{2+}(aq) \rightarrow Zn^{2+}(aq) + Ni(s)$. Then $Q = \dfrac{[Zn^{2+}]}{[Ni^{2+}]} = \dfrac{0.100}{1.50} = 0.0666667$ and $n = 2$. Then

$$E_{cell} = E^{\circ}_{cell} - \frac{0.0592 \text{ V}}{n}\log Q = 0.53 \text{ V} - \frac{0.0592 \text{ V}}{2}\log 0.0666667 = +0.56 \text{ V}.$$

(b) $\quad Q = \dfrac{[Zn^{2+}]}{[Ni^{2+}]} = \dfrac{1.100}{0.500} = 2.20$ then

$$E_{cell} = E^{\circ}_{cell} - \frac{0.0592 \text{ V}}{n}\log Q = 0.53 \text{ V} - \frac{0.0592 \text{ V}}{2}\log 2.20 = +0.52 \text{ V}$$

(c) $\quad E_{cell} = E^{\circ}_{cell} - \dfrac{0.0592 \text{ V}}{n}\log Q$ so $0.45 \text{ V} = 0.53 \text{ V} - \dfrac{0.0592 \text{ V}}{2}\log Q \rightarrow 0.08 \cancel{V} = \dfrac{0.0592 \cancel{V}}{2}\log Q \rightarrow$

$\log Q = 2.70270 \rightarrow Q = 10^{2.70270} = 504.31$. Then $Q = 504.31 = \dfrac{[Zn^{2+}]}{1.60 \text{ M} - [Zn^{2+}]}$. Solving for

$[Zn^{2+}]: (504.31)(1.60 \text{ M} - [Zn^{2+}]) = [Zn^{2+}] \rightarrow [Zn^{2+}] = \dfrac{806.896 \text{ M}}{505.31} = 1.59683 \text{ M} = 1.60 \text{ M}$ then

$[Ni^{2+}] = 1.60 \text{ M} - 1.59683 \text{ M} = 0.003 \text{ M}$

Check: The units (V, V, and M) are correct. The standard cell potential is positive, and because there are more reactants than products, this raises the cell potential. As the reaction proceeds, reactants are converted to products; so the cell potential drops for parts (b) and (c).

18.79 **Given:** Zn/Zn^{2+} concentration cell, with $[Zn^{2+}] = 2.0 \text{ M}$ in one half-cell and $[Zn^{2+}] = 1.0 \times 10^{-3} \text{ M}$ in other half-cell

Find: Sketch a voltaic cell, labeling the anode, the cathode, the reactions at electrodes, all species, and the direction of electron flow.

Conceptual Plan: In a concentration cell, the half-cell with the higher concentration is always the half-cell where the reduction takes place (contains the cathode). The two half-cell reactions are the same but reversed. Put anode reaction on the left (oxidation = electrons as product) and cathode reaction on the right (reduction = electrons as reactant). Electrons flow from anode to cathode.

Solution:

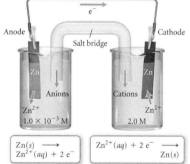

Check: The figure looks similar to the right side of Figure 18.12.

18.81 **Given:** Sn/Sn^{2+} concentration cell with $E_{cell} = 0.10 \text{ V}$

Find: ratio of $[Sn^{2+}]$ in two half-cells

Conceptual Plan: Determine n, then $E^{\circ}_{cell}, E^{\circ}_{cell}, n \rightarrow Q$ = ratio of $[Sn^{2+}]$ in two half-cells.

$$E_{cell} = E^{\circ}_{cell} - \frac{0.0592 \text{ V}}{n}\log Q$$

Solution: Because $Sn^{2+}(aq) + 2 e^- \rightarrow Sn(s), n = 2$. In a concentration cell, $E^{\circ}_{cell} = 0 \text{ V}$. So

$$E_{cell} = E^{\circ}_{cell} - \frac{0.0592 \text{ V}}{n}\log Q, \text{ so } 0.10 \text{ V} = 0.00 \text{V} - \frac{0.0592 \text{ V}}{2}\log Q \rightarrow 0.10 \cancel{V} = -\frac{0.0592 \cancel{V}}{2}\log Q \rightarrow$$

$$\log Q = -3.\underline{3}784 \rightarrow Q = 10^{-3.\underline{3}784} = 4.2 \times 10^{-4} = \frac{[Sn^{2+}](ox)}{[Sn^{2+}](red)}.$$

Check: The units (none) are correct. Because the concentration in the reduction reaction half-cell is always greater than the concentration in the oxidation half-cell in a voltaic concentration cell, the Q or ratio of two cells is less than 1.

Batteries, Fuel Cells, and Corrosion

18.83   **Given:** alkaline battery   **Find:** optimum mass ratio of Zn to MnO_2
        Conceptual Plan: Look up alkaline battery reactions. Use stoichiometry to get mole ratio. Then

$$Zn(s) + 2OH^-(aq) \rightarrow Zn(OH)_2(s) + 2e^-$$
$$\frac{1 \text{ mol Zn}}{2 \text{ mol } MnO_2}$$
$$2\,MnO_2(s) + 2\,H_2O(l) + 2\,e^- \rightarrow 2\,MnO(OH)(s) + 2\,OH^-(aq)$$

mol Zn $\rightarrow$ g Zn Zn then mol MnO_2 $\rightarrow$ g MnO_2.

$$\frac{65.38 \text{ g Zn}}{1 \text{ mol Zn}} \qquad\qquad \frac{1 \text{ mol } MnO_2}{86.94 \text{ g } MnO_2}$$

    Solution: $\dfrac{1 \text{ mol Zn}}{2 \text{ mol } MnO_2} \times \dfrac{65.38 \text{ g Zn}}{1 \text{ mol Zn}} \times \dfrac{1 \text{ mol } MnO_2}{86.94 \text{ g } MnO_2} = 0.3760 \,\dfrac{\text{g Zn}}{\text{g } MnO_2}$

Check: The units (mass ratio) are correct. Because more moles of MnO_2 are needed and the molar mass is larger, the ratio is less than 1.

18.85   **Given:** $CH_4(g) + 2\,O_2(g) \rightarrow CO_2(g) + 2\,H_2O(g)$ **Find:** E°_{cell}
        Conceptual Plan: $\Delta G^\circ_{rxn} = \sum n_p \Delta G^\circ_f(\text{products}) - \sum n_r \Delta G^\circ_f(\text{reactants})$ **and determine n then**
        $\Delta G^\circ_{rxn}, n \rightarrow E^\circ_{cell}$

$$\Delta G^\circ_{rxn} = -n\,F\,E^\circ_{cell}$$

Solution:

Reactant/Product	ΔG°_f(kJ/mol from Appendix IIB)
$CH_4(g)$	-50.5
$O_2(g)$	0.0
$CO_2(g)$	-394.4
$H_2O(g)$	-228.6

Be sure to pull data for the correct formula and phase.

$$\Delta G^\circ_{rxn} = \sum n_p \Delta G^\circ_f(\text{products}) - \sum n_r \Delta G^\circ_f(\text{reactants})$$
$$= [1(\Delta G^\circ_f(CO_2(g))) + 2(\Delta G^\circ_f(H_2O(g)))] - [1(\Delta G^\circ_f(CH_4(g))) + 2(\Delta G^\circ_f(O_2(g)))]$$
$$= [1(-394.4 \text{ kJ}) + 2(-228.6 \text{ kJ})] - [1(-50.5 \text{ kJ}) + 2(0.0 \text{ kJ})]$$
$$= [-851.6 \text{ kJ}] - [-50.5 \text{ kJ}]$$
$$= -801.1 \text{ kJ} = -8.011 \times 10^5 \text{ J}$$

Also, because one C atom goes from an oxidation state of -4 to $+4$ and four O atoms are going from 0 to -2, $n = 8$ and $\Delta G^\circ_{rxn} = -nF\,E^\circ_{cell}$. Rearrange to solve for E°_{cell}.

$$E^\circ_{cell} = \frac{\Delta G^\circ_{rxn}}{-n\,F} = \frac{-8.011 \times 10^5 \text{ J}}{-8 \text{ mole} \times \dfrac{96,485 \text{ C}}{\text{mole}}} = 1.038\,\frac{\text{V} \cdot \cancel{\text{C}}}{\cancel{\text{C}}} = 1.038 \text{ V}$$

Check: The units (V) are correct. The cell voltage is positive, which is consistent with a spontaneous reaction.

18.87   When iron corrodes or rusts, it oxidizes to Fe^{2+}. For a metal to be able to protect iron, it must be more easily oxidized than iron or be below it in Table 18.1. (a) Zn and (c) Mn meet that criterion.

Electrolytic Cells and Electrolysis

18.89   **Given:** electrolytic cell sketch
        Find: (a) Label the anode and cathode and indicate half-reactions, (b) indicate direction of electron flow, and (c) label battery terminals and calculate minimum voltage to drive reaction.

Conceptual Plan: (a) Write two half-cell reactions and add electrons as needed to balance reactions. Look up half-reactions in Table 18.1. Calculate the standard cell potential by subtracting the electrode potential of the anode from the electrode potential of the cathode: $E°_{cell} = E°_{cathode} - E°_{anode}$. **Choose the direction of the half-cell reactions so that** $E°_{cell} < 0$. **(b) Electrons flow from anode to cathode. (c) Each half-cell reaction moves forward, so direction of the concentration changes can be determined.**

Solution:

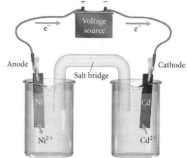

(a) $Ni^{2+}(aq) + 2e^- \rightarrow Ni(s)$ and $Cd^{2+}(aq) + 2e^- \rightarrow Cd(s)$. Look up cell potentials. For Ni, $E°_{red} = -0.23$ V. For Cd, $E°_{red} = -0.40$ V. To get a negative cell potential, Ni is oxidized; so $E°_{red} = -0.23$ V $= E°_{anode}$. Cd^{2+} is reduced, so $E°_{red} = -0.23$ V $= E°_{cathode}$. Then $E°_{cell} = E°_{cathode} - E°_{anode} = -0.40$ V $- (-0.23$ V$) = -0.17$ V. Because oxidation occurs at the anode, Ni is the anode and the reaction is $Ni(s) \rightarrow Ni^{2+}(aq) + 2e^-$. Because reduction takes place at the cathode, Cd is the cathode and the reaction is $Cd^{2+}(aq) + 2e^- \rightarrow Cd(s)$.

(c) Because reduction is occurring at the cathode, the battery terminal closest to the cathode is the negative terminal. Because the cell potential from part (a) is $= -0.17$ V, a minimum of 0.17 V must be applied by the battery.

Check: The reaction is nonspontaneous because the reduction of Ni^{2+} is above Cd^{2+}. Electrons still flow from the anode to the cathode. The reaction can be made spontaneous with the application of electrical energy.

18.91   **Given:** electrolysis of molten KBr   **Find:** write half-reactions
Conceptual Plan: Write two half-cell reactions, taking the cation and the anion to their elemental forms at high temperatures and adding electrons as needed to balance reactions.
Solution: KBr breaks apart to K^+ and Br^-. $K^+(l) + e^- \rightarrow K(l)$ and $2\,Br^-(l) \rightarrow Br_2(g) + 2\,e^-$.

Check: The cation is reduced, and the anion is oxidized. Mass and charge are balanced in the reactions.

18.93   **Given:** electrolysis of mixture of molten KBr and molten LiBr   **Find:** write half-reactions
Conceptual Plan: Write two half-cell reactions, taking the cation and the anion to their elemental forms at high temperatures and adding electrons as needed to balance reactions. Look up cation cell potentials to see which one generates the more spontaneous reaction.
Solution: KBr breaks apart to K^+ and Br^-. $K^+(l) + e^- \rightarrow K(l)$ and $2\,Br^-(l) \rightarrow Br_2(g) + 2\,e^-$. LiBr breaks apart to Li^+ and Br^-. $Li^+(l) + e^- \rightarrow Li(l)$ and $2\,Br^-(l) \rightarrow Br_2(g) + 2\,e^-$. The anions are the same, so the anode reaction is $2\,Br^-(l) \rightarrow Br_2(g) + 2\,e^-$. Looking up the cation reduction potentials, for K^+ $E°_{red} = -2.92$ V and for Li^+ $E°_{red} = -3.04$ V. Because the reduction potential is more positive for K^+, the reaction at the cathode is $K^+(l) + e^- \rightarrow K(l)$.

Check: The cation is reduced, and the anion is oxidized. Mass and charge are balanced in the reactions. The cation that is higher in Table 18.1 will be the reaction at the cathode.

18.95   **Given:** electrolysis of aqueous solutions   **Find:** write half-reactions
Conceptual Plan: Write two half-cell reactions, taking the cation and the anion to their elemental forms at standard conditions and adding electrons as needed to balance reactions. Look up cell potentials and compare to the cell potentials for the electrolysis of water to see which one generates the more spontaneous reaction.
Solution: The hydrolysis of water reactions are as follows: at a neutral pH, $2\,H_2O(l) \rightarrow O_2(g) + 4\,H^+(aq) + 4\,e^-$ where $E = -0.82$ V, and when $[OH^-] = 10^{-7}$ M. $2\,H_2O(l) + 2\,e^- \rightarrow H_2(g) + 2\,OH^-(aq)$ where $E = -0.41$ V.

(a) NaBr breaks apart to Na^+ and Br^-. $Na^+(aq) + e^- \rightarrow Na(s)$ and $2\,Br^-(aq) \rightarrow Br_2(l) + 2\,e^-$. Looking up the half-cell potentials, for Na^+ $E°_{red} = -2.71$ V and for Br^- $E°_{anode} = E°_{red} = +1.09$ V. Because -0.82 V is more positive than $E°_{anode} = -1.09$ V, the oxidation reaction will be $2\,H_2O(l) \rightarrow O_2(g) + 4\,H^+(aq) + 4\,e^-$ at the anode. Because -0.41 V is more positive than -2.71 V, the reduction reaction will be $2\,H_2O(l) + 2\,e^- \rightarrow H_2(g) + 2\,OH^-(aq)$ at the cathode.

(b) PbI_2 breaks apart to Pb^{2+} and I^-. $Pb^{2+}(aq) + 2\,e^- \rightarrow Pb(s)$ and $2\,I^-(aq) \rightarrow I_2(s) + 2\,e^-$. Looking up the half-cell potentials, for Pb^{2+} $E°_{red} = -0.13$ V and for I^- $E°_{anode} = E°_{red} = -0.54$ V. Because $E°_{anode} = -0.54$ V is more positive than -0.82 V, the oxidation reaction will be $2\,I^-(aq) \rightarrow I_2(s) + 2\,e^-$ at the anode. Because -0.13 V is more positive than -0.41 V, the reduction reaction will be $Pb^{2+}(aq) + 2\,e^- \rightarrow Pb(s)$ at the cathode.

(c) Na_2SO_4 breaks apart to Na^+ and SO_4^{2-}. $Na^+(aq) + e^- \rightarrow Na(s)$ and $SO_4^{2-}(aq) + 4\,H^+(aq) + 2\,e^- \rightarrow H_2SO_3(aq) + H_2O(l)$. Notice that both of these are reductions. Because S is in such a high oxidation state ($+6$), it cannot be oxidized. Looking up the half-cell potentials, for Na^+ $E^\circ_{red} = -2.71$ V and for SO_4^{2-} $E^\circ_{red} = 0.20$ V. Because sodium sulfate solutions are neutral and not acidic, even though 0.20 V is more positive than -0.41 V and -2.71 V, the reduction reaction will not be: $SO_4^{2-}(aq) + 4H^+(aq) + 2e^- \rightarrow H_2SO_3(aq) + H_2O(l)$ at the cathode, which only occurs in acidic solutions. Thus, the reduction reaction will be $Na^+(aq) + e^- \rightarrow Na(s)$. Because only one oxidation reaction is possible, the oxidation reaction will be $2H_2O(l) \rightarrow O_2(g) + 4H^+(aq) + 4e^-$ at the anode.

Check: The most positive reactions are the reactions that will occur.

18.97   **Given:** electrolysis cell to electroplate Cu onto a metal surface
          Find: Draw a cell and label the anode and cathode and write half-reactions.
          Conceptual Plan: Write two half-cell reactions and add electrons as needed to balance reactions. The cathode reaction will be the reduction of Cu^{2+} to the metal. The anode will be the reverse reaction.
          Solution:

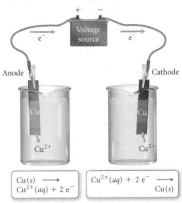

    Check: The metal to be plated is the cathode because metal ions are converted to $Cu(s)$ on the surface of the metal.

18.99   **Given:** Cu electroplating of 325 mg Cu at a current of 5.6 A; $Cu^{2+}(aq) + 2e^- \rightarrow Cu(s)$ **Find:** time
          Conceptual Plan: mg Cu $\rightarrow$ g Cu $\rightarrow$ mol Cu $\rightarrow$ mol e^- $\rightarrow$ C $\rightarrow$ s

$$\frac{1\,g}{1000\,mg} \quad \frac{1\,mol\,Cu}{63.55\,g\,Cu} \quad \frac{2\,mol\,e^-}{1\,mol\,Cu} \quad \frac{96,485\,C}{1\,mol\,e^-} \quad \frac{1\,s}{5.6\,C}$$

    Solution: $325 \; \cancel{mg\,Cu} \times \dfrac{1\,\cancel{g\,Cu}}{1000\,\cancel{mg\,Cu}} \times \dfrac{1\,\cancel{mol\,Cu}}{63.55\,\cancel{g\,Cu}} \times \dfrac{2\,\cancel{mol\,e^-}}{1\,\cancel{mol\,Cu}} \times \dfrac{96,485\,\cancel{C}}{1\,\cancel{mol\,e^-}} \times \dfrac{1\,s}{5.6\,\cancel{C}} = 180$ s.

    Check: The units (s) are correct. Because far less than a mole of Cu is electroplated, the time is short.

18.101   **Given:** Na electrolysis, 1.0 kg in one hour   **Find:** current
          Conceptual Plan: $Na^+(l) + e^- \rightarrow Na(l)$ $\dfrac{kg\,Na}{h} \rightarrow \dfrac{g\,Na}{h} \rightarrow \dfrac{mol\,Na}{h} \rightarrow \dfrac{mol\,e^-}{h} \rightarrow \dfrac{C}{h} \rightarrow \dfrac{C}{min} \rightarrow \dfrac{C}{s}$

$$\frac{1000\,g}{1\,kg} \quad \frac{1\,mol\,Na}{22.99\,g\,Na} \quad \frac{1\,mol\,e^-}{1\,mol\,Na} \quad \frac{96,485\,C}{1\,mol\,e^-} \quad \frac{1\,h}{60\,min} \quad \frac{1\,min}{60\,s}$$

    Solution:

$$\frac{1.0\,\cancel{kg\,Na}}{1\,\cancel{h}} \times \frac{1000\,\cancel{g\,Na}}{1\,\cancel{kg\,Na}} \times \frac{1\,\cancel{mol\,Na}}{22.99\,\cancel{g\,Na}} \times \frac{1\,\cancel{mol\,e^-}}{1\,\cancel{mol\,Na}} \times \frac{96,485\,C}{1\,\cancel{mol\,e^-}} \times \frac{1\,\cancel{h}}{60\,\cancel{min}} \times \frac{1\,\cancel{min}}{60\,s} = 1.2 \times 10^3 \frac{C}{s} = 1.2 \times 10^3 \text{ A}$$

    Check: The units (A) are correct. Because the amount per hour is so large, we expect a very large current.

Cumulative Problems

18.103   **Given:** $MnO_4^-(aq) + Zn(s) \rightarrow Mn^{2+}(aq) + Zn^{2+}(aq)$, 0.500 M $KMnO_4$, and 2.85 g Zn
          Find: balance equation and volume $KMnO_4$ solution
          Conceptual Plan: Separate the overall reaction into two half-reactions: one for oxidation and one for reduction. $\rightarrow$ Balance each half-reaction with respect to mass in the following order: (1) Balance all elements other than H and O, (2) balance O by adding H_2O, and (3) balance H by adding H^+. $\rightarrow$ Balance each half-reaction

with respect to charge by adding electrons. (The sum of the charges on both sides of the equation should be made equal by adding electrons as necessary.) → Make the number of electrons in both half-reactions equal by multiplying one or both half-reactions by a small whole number. → Add the two half-reactions, canceling electrons and other species as necessary. → Verify that the reaction is balanced with respect to both mass and charge.

Then g Zn → mol Zn → mol MnO_4^- → mol $KMnO_4$ → L $KMnO_4$ → mL $KMnO_4$.

$$\frac{1 \text{ mol Zn}}{65.38 \text{ g Zn}} \quad \frac{2 \text{ mol } MnO_4^-}{5 \text{ mol Zn}} \quad \frac{1 \text{ mol } KMnO_4}{1 \text{ mol } MnO_4^-} \quad \frac{1 \text{ L } KMnO_4}{0.500 \text{ mol } KMnO_4} \quad \frac{1000 \text{ mL } KMnO_4}{1 \text{ L } KMnO_4}$$

Solution:

Separate: $\qquad\qquad\qquad MnO_4^-(aq) \rightarrow Mn^{2+}(aq) \qquad$ and $\quad Zn(s) \rightarrow Zn^{2+}(aq)$

Balance non H & O elements: $MnO_4^-(aq) \rightarrow Mn^{2+}(aq) \qquad$ and $\quad Zn(s) \rightarrow Zn^{2+}(aq)$

Balance O with H_2O: $MnO_4^-(aq) \rightarrow Mn^{2+}(aq) + 4 H_2O(l) \qquad$ and $\quad Zn(s) \rightarrow Zn^{2+}(aq)$

Balance H with H^+: $MnO_4^-(aq) + 8 H^+(aq) \rightarrow Mn^{2+}(aq) + 4 H_2O(l) \quad$ and $\quad Zn(s) \rightarrow Zn^{2+}(aq)$

Add electrons: $MnO_4^-(aq) + 8 H^+(aq) + 5 e^- \rightarrow Mn^{2+}(aq) + 4 H_2O(l)$ and $\quad Zn(s) \rightarrow Zn^{2+}(aq) + 2 e^-$

Equalize electrons:

$2 MnO_4^-(aq) + 16 H^+(aq) + 10 e^- \rightarrow 2 Mn^{2+}(aq) + 8 H_2O(l)$ and $5 Zn(s) \rightarrow 5 Zn^{2+}(aq) + 10 e^-$

Add half-reactions:

$2 MnO_4^-(aq) + 16 H^+(aq) + \cancel{10 e^-} + 5 Zn(s) \rightarrow 2 Mn^{2+}(aq) + 8 H_2O(l) + 5 Zn^{2+}(aq) + \cancel{10 e^-}$

Cancel electrons: $2 MnO_4^-(aq) + 16 H^+(aq) + 5 Zn(s) \rightarrow 2 Mn^{2+}(aq) + 8 H_2O(l) + 5 Zn^{2+}(aq)$

$$2.85 \text{ g } \cancel{Zn} \times \frac{1 \text{ mol } \cancel{Zn}}{65.38 \text{ g } \cancel{Zn}} \times \frac{2 \text{ mol } \cancel{MnO_4^-}}{5 \text{ mol } \cancel{Zn}} \times \frac{1 \text{ mol } \cancel{KMnO_4}}{1 \text{ mol } \cancel{MnO_4^-}} \times \frac{1 \text{ L } \cancel{KMnO_4}}{0.500 \text{ mol } \cancel{KMnO_4}} \times \frac{1000 \text{ mL } KMnO_4}{1 \text{ L } \cancel{KMnO_4}}$$

$= 34.9$ mL $KMnO_4$

Check:

Reactants	Products
2 Mn atoms	2 Mn atoms
8 O atoms	8 O atoms
16 H atoms	16 H atoms
5 Zn atoms	5 Zn atoms
+14 charge	+14 charge

The units (mL) are correct. Because far less than a mole of zinc is used, less than a mole of permanganate is consumed; so the volume is less than a liter.

18.105 **Given:** beaker with Al strip and Cu^{2+} ions **Find:** Draw sketch after Al is submerged for a few minutes.
Conceptual Plan: Write two half-cell reactions and add electrons as needed to balance reactions. Look up half-reactions in Table 18.1. Calculate the standard cell potential by subtracting the electrode potential of the anode from the electrode potential of the cathode: $E°_{cell} = E°_{cathode} - E°_{anode}$. **If** $E°_{cell} > 0$, **the reaction is spontaneous in the forward direction and Al will dissolve and Cu will deposit.**
Solution: $Al(s) \rightarrow Al^{3+}(aq)$ and $Cu^{2+}(aq) \rightarrow Cu(s)$. Add electrons.
$Al(s) \rightarrow Al^{3+}(aq) + 3 e^-$ and $Cu^{2+}(aq) + 2 e^- \rightarrow Cu(s)$. Look up cell potentials.
Al is oxidized, so $E°_{anode} = E°_{red} = -1.66$ V. Cu^{2+} is reduced, so $E°_{cathode} = E°_{red} = 0.34$ V.
Then $E°_{cell} = E°_{cathode} - E°_{anode} = 0.34$ V $- (-1.66$ V) + 2.00 V, so the reaction is spontaneous. Al will dissolve to generate $Al^{3+}(aq)$, and $Cu(s)$ will deposit.

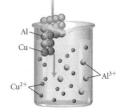

Check: The units (V) are correct. If the voltage is positive, the reaction is spontaneous; so Al will dissolve and Cu will deposit.

18.107 **Given:** (a) 2.15 g Al, (b) 4.85 g Cu, and (c) 2.42 g Ag in 3.5 M HI
Find: If metal dissolves, write a balanced reaction and the minimum amount of HI needed to dissolve the metal.
Conceptual Plan: In general, metals whose reduction half-reactions lie below the reduction of H^+ **to H_2 in Table 18.1 will dissolve in acids, while metals above it will not. Stop here if metal does not dissolve. To write the balanced redox reactions, pair the oxidation of the metal with the reduction of** H^+ **to**

H_2 $(2\,H^+(aq)\,+\,2\,e^-\rightarrow H_2(g))$. **Balance the number of electrons transferred. Add the two reactions. Cancel electrons. Then g metal $\rightarrow$ mol metal $\rightarrow$ mol $H^+\rightarrow$ L HI $\rightarrow$ mL HI.**

$$\mathcal{M} \qquad \frac{x\ \text{mol}\ H^+}{y\ \text{mol metal}}\quad \frac{1\ \text{L HI}}{3.5\ \text{mol HI}}\quad \frac{1000\ \text{mL HI}}{1\ \text{L HI}}$$

Solution:

(a)   Al meets this criterion. For Al, $Al(s)\rightarrow Al^{3+}(aq)\,+\,3\,e^-$. We need to multiply the Al reaction by 2 and the H^+ reaction by 3. So $2\,Al(s)\rightarrow 2\,Al^{3+}(aq)\,+\,6\,e^-$ and $6\,H^+(aq)\,+\,6\,e^-\rightarrow 3\,H_2(g)$. Adding the half-reactions: $2\,Al(s)\,+\,6\,H^+(aq)\,+\,\cancel{6\,e^-}\rightarrow 2\,Al^{3+}(aq)\,+\,\cancel{6\,e^-}\,+\,3\,H_2(g)$. Simplify to $2\,Al(s)\,+\,6\,H^+(aq)\rightarrow 2\,Al^{3+}(aq)\,+\,3\,H_2(g)$. Then

$$2.15\,\text{g}\ \cancel{Al}\times\frac{1\ \text{mol}\ \cancel{Al}}{26.98\ \text{g}\ \cancel{Al}}\times\frac{6\ \text{mol}\ \cancel{H^+}}{2\ \text{mol}\ \cancel{Al}}\times\frac{1\ \cancel{\text{L HI}}}{3.5\ \cancel{\text{mol HI}}}\times\frac{1000\ \text{mL HI}}{1\ \cancel{\text{L HI}}}=68.3\ \text{mL HI}.$$

(b)   Cu does not meet this criterion, so it will not dissolve in HI.

(c)   Ag does not meet this criterion, so it will not dissolve in HI.

Check: Only metals with negative reduction potentials will dissolve. The volume of acid needed is fairly small because the amount of metal is much less than 1 mole and the concentration of acid is high.

18.109   **Given:** $Pt(s)\mid H_2(g,\,1\,\text{atm})\mid H^+(aq,\,?\,M)\mid\mid Cu^{2+}(aq,\,1.0\,M)\mid Cu(s)$, $E_{cell}=355\,\text{mV}$ **Find:** pH
   Conceptual Plan: Write half-reactions from line notation. Look up half-reactions in Table 18.1. Calculate the standard cell potential by subtracting the electrode potential of the anode from the electrode potential of the cathode: $E^{\circ}_{cell}=E^{\circ}_{cathode}-E^{\circ}_{anode}$. Add the two half-cell reactions and cancel the electrons. Then mV $\rightarrow$ V then $E^{\circ}_{cell},\,E_{cell},\,P_{H_2},\,[Cu^{2+}],\,n\rightarrow[H^+]\rightarrow$ pH.

$$\frac{1\ \text{V}}{1000\ \text{mV}}\qquad\qquad E_{cell}=E^{\circ}_{cell}-\frac{0.0592\ \text{V}}{n}\log Q\qquad \text{pH}=-\log\,[H^+]$$

   Solution: The half-reactions are $H_2(g)\rightarrow 2\,H^+(aq)\,+\,2\,e^-$ and $Cu^{2+}(aq)\,+\,2\,e^-\rightarrow Cu(s)$. H is oxidized, so $E^{\circ}_{red}=0.00\,\text{V}=E^{\circ}_{anode}$. Cu is reduced, so $E^{\circ}_{red}=0.34\,\text{V}=E^{\circ}_{cathode}$. Then $E^{\circ}_{cell}=E^{\circ}_{cathode}-E^{\circ}_{anode}$

$=0.34\,\text{V}-0.00\,\text{V}=+0.34\,\text{V}$. Adding the two reactions: $H_2(g)\,+\,Cu^{2+}(aq)\,+\,\cancel{2\,e^-}\rightarrow 2\,H^+(aq)\,+\,\cancel{2\,e^-}\,+\,Cu(s)$.

Cancel the electrons: $H_2(g)\,+\,Cu^{2+}(aq)\rightarrow 2\,H^+(aq)\,+\,Cu(s)$. Then $355\ \cancel{\text{mV}}\times\dfrac{1\ \text{V}}{1000\ \cancel{\text{mV}}}=0.355\,\text{V}$. So $n=2$

and $Q=\dfrac{[H^+]^2}{P_{H_2}[Cu^{2+}]}=\dfrac{(x)^2}{(1)(1.0)}=x^2$. Then $E_{cell}=E^{\circ}_{cell}-\dfrac{0.0592\ \text{V}}{n}\log Q$. Substitute values and solve for

x. $0.355\,\text{V}=0.34\,\text{V}-\dfrac{0.0592\ \text{V}}{2}\log x^2\rightarrow 0.01\underline{5}\ \cancel{\text{V}}=-\dfrac{0.0592\ \cancel{\text{V}}}{2}\log x^2\rightarrow -0.5\underline{0}676=\log x^2\rightarrow$

$x^2=10^{-0.5\underline{0}676}=0.3\underline{1}134\rightarrow x=0.5\underline{5}798$ then pH $=-\log\,[H^+]=-\log\,[0.5\underline{5}798]=0.2\underline{5}338=0.3$

   Check: The units (none) are correct. The pH is acidic, which is consistent with dissolving a metal in acid.

18.111   You should be wary of the battery because the most a pairing of half-cell reactions can generate is 5 to 6 V, not 24 V.

18.113   **Given:** Mg oxidation and Cu^{2+} reduction; initially, $[Mg^{2+}]=1.0\times10^{-4}\,M$ and $[Cu^{2+}]=1.5\,M$ in 1.0 L half-cells
   Find: (a) initial E_{cell}, (b) E_{cell} after 5.0 A for 8.0 h, and (c) how long battery can deliver 5.0 A
   Conceptual Plan:

(a)   **Write the two half-cell reactions and add electrons as needed to balance reactions. Look up half-reactions in Table 18.1. Calculate the standard cell potential by subtracting the electrode potential of the anode from the electrode potential of the cathode: $E^{\circ}_{cell}=E^{\circ}_{cathode}-E^{\circ}_{anode}$. Add the two half-cell reactions, cancel electrons, and determine n. Then $E^{\circ}_{cell},\,[Mg^{2+}],\,[Cu^{2+}],\,n\rightarrow E_{cell}$.**

$$E_{cell}=E^{\circ}_{cell}-\frac{0.0592\ V}{n}\log Q$$

(b)   **hr $\rightarrow$ min $\rightarrow$ s $\rightarrow$ C $\rightarrow$ mol $e^-\rightarrow$ mol Cu reduced $\rightarrow[Cu^{2+}]$ and**

$$\frac{60\ \text{min}}{1\ \text{hr}}\quad\frac{60\ \text{s}}{1\ \text{min}}\quad\frac{5.0\ \text{C}}{1\ \text{s}}\quad\frac{1\ \text{mol}\ e^-}{96,485\ \text{C}}\quad\frac{1\ \text{mol}\ Cu^{2+}}{2\ \text{mol}\ e^-}\ \ \text{because}\ V=1.0\,\text{L}\quad[Cu^{2+}]_f=[Cu^{2+}]_i-\frac{\text{mol}\ Cu^{2+}\ \text{reduced}}{1.0\ \text{L}}$$

   mol Cu reduced $\rightarrow$ mol Mg oxidized $\rightarrow[Mg^{2+}]$

$$\frac{1\ \text{mol Mg oxidized}}{1\ \text{mol}\ Cu^{2+}\text{reduced}}\ \ \text{because}\ V=1.0\,\text{L}\quad[Mg^{2+}]_f=[Mg^{2+}]_i+\frac{\text{mol Mg oxidized}}{1.0\ \text{L}}$$

(c) $[Cu^{2+}]\rightarrow$ **mol $e^-\rightarrow$ C $\rightarrow$ s $\rightarrow$ min $\rightarrow$ hr**

$$\frac{2\ \text{mol}\ e^-}{1\ \text{mol}\ \cancel{Cu^{2+}}}\quad\frac{96,485\ \text{C}}{1\ \text{mol}\ e^-}\quad\frac{1\ \text{s}}{5.0\ \text{C}}\quad\frac{1\ \text{min}}{60\ \text{s}}\quad\frac{1\ \text{hr}}{60\ \text{min}}$$

Solution:

(a) Write half-reactions and add electrons. $Cu^{2+}(aq) + 2e^- \rightarrow Cu(s)$ and $Mg(s) \rightarrow Mg^{2+}(aq) + 2e^-$.

Look up cell potentials. Mg is oxidized, so $E_{red}^\circ = -2.37\ V = E_{anode}^\circ$. Cu^{2+} is reduced, so

$E_{red}^\circ = 0.34\ V = E_{cathode}^\circ$. Then $E_{cell}^\circ = E_{cathode}^\circ - E_{anode}^\circ = 0.34\ V - (-2.37\ V) = +2.71\ V$. Add

the two half-cell reactions: $Cu^{2+}(aq) + \cancel{2e^-} + Mg(s) \rightarrow Cu(s) + Mg^{2+}(aq) + \cancel{2e^-}$. Simplify to

$Cu^{2+}(aq) + Mg(s) \rightarrow Cu(s) + Mg^{2+}(aq)$. So $Q = \dfrac{[Mg^{2+}]}{[Cu^{2+}]} = \dfrac{1.0 \times 10^{-4}}{1.5} = 6.\underline{6}667 \times 10^{-5}$ and $n = 2$.

Then $E_{cell} = E_{cell}^\circ - \dfrac{0.0592\ V}{n}\log Q = 2.71\ V - \dfrac{0.0592\ V}{2}\log(6.\underline{6}667 \times 10^{-5}) = +2.8\underline{3}361\ V = +2.83\ V$.

(b) $8.0\ \cancel{hr} \times \dfrac{60\ \cancel{min}}{1\ \cancel{hr}} \times \dfrac{60\ \cancel{s}}{1\ \cancel{min}} \times \dfrac{5.0\ \cancel{C}}{1\ \cancel{s}} \times \dfrac{1\ \cancel{mol\,e^-}}{96,485\ \cancel{C}} \times \dfrac{1\ mol\ Cu^{2+}}{2\ \cancel{mol\,e^-}} = 0.7\underline{4}623\ mol\ Cu^{2+}$ and

$[Cu^{2+}]_f = [Cu^{2+}]_i - \dfrac{mol\ Cu^{2+}\ reduced}{1.0\ L} = 1.5\ M - \dfrac{0.7\underline{4}623\ mol\ Cu^{2+}}{1.0\ L} = 0.7\underline{5}377\ M\ Cu^{2+}$ and

$0.7\underline{4}623\ \cancel{mol\ Cu^{2+}} \times \dfrac{1\ mol\ Mg\ oxidized}{1\ \cancel{mol\ Cu^{2+}\ reduced}} = 0.7\underline{4}623\ mol\ Mg\ oxidized$ and

$[Mg^{2+}]_f = [Mg^{2+}]_i + \dfrac{mol\ Mg\ oxidized}{1.0\ L} = 1.0 \times 10^{-4}\ M + \dfrac{0.7\underline{4}623\ mol\ Mg\ oxidized}{1.0\ L}$

$= 0.7\underline{4}633\ M\ Mg^{2+}$

$Q = \dfrac{[Mg^{2+}]}{[Cu^{2+}]} = \dfrac{0.7\underline{4}633}{0.7\underline{5}377} = 0.9\underline{9}013$ and $n = 2$ then

$E_{cell} = E_{cell}^\circ - \dfrac{0.0592\ V}{n}\log Q = 2.71\ V - \dfrac{0.0592\ V}{2}\log 0.9\underline{9}013 = +2.7\underline{1}013\ V = +2.71\ V$

(c) In 1.0 L, there are initially 1.5 moles of Cu^{2+}. So

$1.5\ \cancel{mol\ Cu^{2+}} \times \dfrac{2\ \cancel{mol\,e^-}}{1\ \cancel{mol\ Cu^{2+}}} \times \dfrac{96,485\ \cancel{C}}{1\ \cancel{mol\,e^-}} \times \dfrac{1\ \cancel{s}}{5.0\ \cancel{C}} \times \dfrac{1\ \cancel{min}}{60\ \cancel{s}} \times \dfrac{1\ hr}{60\ \cancel{min}} = 16\ hr.$

Check: The units (V, V, and hr) are correct. The Cu^{2+} reduction reaction is above the Mg^{2+} reduction reaction, so the standard cell potential will be positive. Having more reactants than products increases the cell potential. As the reaction proceeds, the potential drops. The concentrations drop by $\frac{1}{2}$ in 8 hours [part (b)], so all of it is consumed in 16 hours.

18.115 **Given:** water electrolysis at 7.8 A; $H_2(g)$: $V = 25.0\ L$, $P = 25.0\ atm$; $T = 25\ ^\circ C$ **Find:** time

Conceptual Plan: $^\circ C \rightarrow K$ then $V, P, T \rightarrow n$ then write half-reactions

$\qquad\qquad\qquad K = ^\circ C + 273.15 \qquad\quad PV = nRT$

mol H_2 $\rightarrow$ mol e^- $\rightarrow$ C $\rightarrow$ s $\rightarrow$ min $\rightarrow$ hr

$\qquad\dfrac{2\ mol\ e^-}{1\ mol\ H_2}\quad \dfrac{96,485\ C}{1\ mol\ e^-}\quad \dfrac{1\ s}{7.8\ C}\quad \dfrac{1\ min}{60\ s}\quad \dfrac{1\ hr}{60\ min}$

Solution: $T = 25\ ^\circ C + 273.15 = 298\ K$, then $PV = nRT$ Rearrange to solve for n.

$n = \dfrac{PV}{RT} = \dfrac{25.0\ \cancel{atm} \times 25.0\ \cancel{L}}{0.08206\ \dfrac{\cancel{L}\cdot\cancel{atm}}{mol\cdot \cancel{K}} \times 298\ \cancel{K}} = 25.\underline{5}583\ mol\ H_2.$ The hydrolysis of water reactions are as follows:

$2\ H_2O(l) \rightarrow O_2(g) + 4\ H^+(aq) + 4\ e^-$ and $2\ H_2O(l) + 2\ e^- \rightarrow H_2(g) + 2\ OH^-(aq)$

$25.\underline{5}583\ \cancel{mol\ H_2} \times \dfrac{2\ \cancel{mol\,e^-}}{1\ \cancel{mol\ H_2}} \times \dfrac{96,485\ \cancel{C}}{1\ \cancel{mol\,e^-}} \times \dfrac{1\ \cancel{s}}{7.8\ \cancel{C}} \times \dfrac{1\ \cancel{min}}{60\ \cancel{s}} \times \dfrac{1\ hr}{60\ \cancel{min}} = 176\ hr$

Check: The units (hr) are correct. Because we have 25 L of gas at 25 atm and 25 °C, we expect ~25 moles of gas (remember that 1 mole of gas at STP = 22.4 L). A very long time is expected because we have so many moles of gas to generate.

18.117   **Given:** $Cu(s) \mid CuI(s) \mid I^-(aq, 1.0\ M) \mid\mid Cu^+(aq, 1.0\ M) \mid Cu(s)$, $K_{sp}(CuI) = 1.1 \times 10^{-12}$ **Find:** E_{cell}

Conceptual Plan: Write half-reactions from line notation. Because this is a concentration cell, $E_{cell}^\circ = 0.00$ V.
Then $K_{sp}, [I^-] \rightarrow [Cu^+](ox)$ then $E_{cell}^\circ, [Cu^+](ox), [Cu^+](red), n \rightarrow E_{cell}$.

$$K_{sp} = [Cu^+][I^-] \qquad\qquad\qquad E_{cell} = E_{cell}^\circ - \frac{0.0592\ V}{n}\log Q$$

Solution: The half-reactions are $Cu(s) \rightarrow Cu^+(aq) + e^-$ and $Cu^+(aq) + e^- \rightarrow Cu(s)$. Because this is a concentration cell, $E_{cell}^\circ = 0.00$ V and $n = 1$. Because $K_{sp} = [Cu^+][I^-]$, rearrange to solve for $[Cu^+](ox)$.

$$[Cu^+]_{ox} = \frac{K_{sp}}{[I^-]} = \frac{1.1 \times 10^{-12}}{1.0} = 1.1 \times 10^{-12}\ M\ \text{then}$$

$$Q = \frac{[Cu^+]_{ox}}{[Cu^+]_{red}} = \frac{1.1 \times 10^{-12}}{1.0} = 1.1 \times 10^{-12}\ \text{then}$$

$$E_{cell} = E_{cell}^\circ - \frac{0.0592\ V}{n}\log Q = 0.00\ V - \frac{0.0592\ V}{1}\log(1.1 \times 10^{-12}) = 0.71\ V$$

Check: The units (V) are correct. Because $[Cu^+]_{ox}$ is so low and $[Cu^+]_{red}$ is high, the Q is very small; so the voltage increase compared to the standard value is significant.

18.119   **Given:** (a) disproportionation of $Mn^{2+}(aq)$ to $Mn(s)$ and $MnO_2(s)$ and (b) disproportionation of $MnO_2(s)$ to $Mn^{2+}(aq)$ and $MnO_4^-(s)$ in acidic solution   **Find:** ΔG_{rxn}° and K

Conceptual Plan: Separate the overall reaction into two half-reactions: one for oxidation and one for reduction. →
Balance each half-reaction with respect to mass in the following order: (1) Balance all elements other than
H and O, (2) balance O by adding H_2O, and (3) balance H by adding H^+. → Balance each half-reaction with
respect to charge by adding electrons. (The sum of the charges on both sides of the equation should be made
equal by adding electrons as necessary.) → Make the number of electrons in both half-reactions equal by mul-
tiplying one or both half-reactions by a small whole number. → Add the two half-reactions, canceling electrons
and other species as necessary. → Verify that the reaction is balanced with respect to both mass and charge.
Look up half-reactions in Table 18.1. Calculate the standard cell potential by subtracting the electrode potential
of the anode from the electrode potential of the cathode: $E_{cell}^\circ = E_{cathode}^\circ - E_{anode}^\circ$.
Then calculate ΔG_{rxn}° using $\Delta G_{rxn}^\circ = -n F E_{cell}^\circ$. Finally, $^\circ C \rightarrow K$ then $\Delta G_{rxn}^\circ, T \rightarrow K$.

$$K = 273.15 + {}^\circ C \qquad\qquad \Delta G_{rxn}^\circ = -RT \ln K$$

Solution:

(a)   Separate: $Mn^{2+}(aq) \rightarrow MnO_2(s)$ and $Mn^{2+}(aq) \rightarrow Mn(s)$

Balance non H & O elements: $Mn^{2+}(aq) \rightarrow MnO_2(s)$ and $Mn^{2+}(aq) \rightarrow Mn(s)$

Balance O with H_2O: $Mn^{2+}(aq) + 2\,H_2O(l) \rightarrow MnO_2(s)$ and $Mn^{2+}(aq) \rightarrow Mn(s)$

Balance H with H^+: $Mn^{2+}(aq) + 2\,H_2O(l) \rightarrow MnO_2(s) + 4\,H^+(aq)$ and $Mn^{2+}(aq) \rightarrow Mn(s)$

Add electrons: $Mn^{2+}(aq) + 2\,H_2O(l) \rightarrow MnO_2(s) + 4\,H^+(aq) + 2\,e^-$ and $Mn^{2+}(aq) + 2\,e^- \rightarrow Mn(s)$

Equalize electrons: $Mn^{2+}(aq) + 2\,H_2O(l) \rightarrow MnO_2(s) + 4\,H^+(aq) + 2\,e^-$ and $Mn^{2+}(aq) + 2\,e^- \rightarrow Mn(s)$

Add half-reactions: $Mn^{2+}(aq) + 2\,H_2O(l) + Mn^{2+}(aq) + \cancel{2\,e^-} \rightarrow MnO_2(s) + 4\,H^+(aq) + \cancel{2\,e^-} + Mn(s)$

Cancel electrons: $2\,Mn^{2+}(aq) + 2\,H_2O(l) \rightarrow MnO_2(s) + 4\,H^+(aq) + Mn(s)$

Look up cell potentials. Mn is oxidized in the first half-cell reaction, so $E_{anode}^\circ = E_{red}^\circ = +1.21$ V. Mn is reduced in the second half-cell reaction, so $E_{cathode}^\circ = E_{red}^\circ = -1.18$ V. Then $E_{cell}^\circ = E_{cathode}^\circ - E_{anode}^\circ = -1.18\ V - 1.21\ V = -2.39$ V and $n = 2$ so

$$\Delta G_{rxn}^\circ = -n F E_{cell}^\circ = -2\ \cancel{mole^-} \times \frac{96{,}485\ C}{\cancel{mole^-}} \times -2.39\ V = -2 \times 96{,}485\ \cancel{C} \times -2.39\,\frac{J}{\cancel{C}} =$$

$4.61\underline{1}98 \times 10^5\ J = 461$ kJ and $T = 273.15 + 25\ ^\circ C = 298$ K then $\Delta G_{rxn}^\circ = -RT \ln K$. Rearrange to solve for K.

$$K = e^{\frac{-\Delta G_{rxn}^\circ}{RT}} = e^{\frac{-4.61\underline{1}98 \times 10^5\ J}{\left(8.314\,\frac{J}{K \cdot mol}\right)(298\ K)}} = e^{-1\underline{8}6.149} = 1.43 \times 10^{-81}$$

Check:

Reactants	Products
2 Mn atoms	2 Mn atoms
2 O atoms	2 O atoms
4 H atoms	4 H atoms
+4 charge	+4 charge

The units (kJ and none) are correct. If the voltage is negative, the reaction is nonspontaneous and the free energy change is very positive and the equilibrium constant is extremely small.

(b)   Separate: $MnO_2(s) \rightarrow Mn^{2+}(aq)$ and $MnO_2(s) \rightarrow MnO_4^-(aq)$
Balance non H & O elements: $MnO_2(s) \rightarrow Mn^{2+}(aq)$ and $MnO_2(s) \rightarrow MnO_4^-(aq)$
Balance O with H_2O: $MnO_2(s) \rightarrow Mn^{2+}(aq) + 2\,H_2O(l)$ and $MnO_2(s) + 2\,H_2O(l) \rightarrow MnO_4^-(aq)$
Balance H with H^+:
$MnO_2(s) + 4\,H^+(aq) \rightarrow Mn^{2+}(aq) + 2\,H_2O(l)$ and $MnO_2(s) + 2\,H_2O(l) \rightarrow MnO_4^-(aq) + 4\,H^+(aq)$
Add electrons: $MnO_2(s) + 4\,H^+(aq) + 2\,e^- \rightarrow Mn^{2+}(aq) + 2\,H_2O(l)$ and $MnO_2(s) + 2\,H_2O(l) \rightarrow$
$$MnO_4^-(aq) + 4\,H^+(aq) + 3\,e^-$$
Equalize electrons: $3\,MnO_2(s) + 12\,H^+(aq) + 6\,e^- \rightarrow 3\,Mn^{2+}(aq) + 6\,H_2O(l)$ and
$$2\,MnO_2(s) + 4\,H_2O(l) \rightarrow 2\,MnO_4^-(aq) + 8\,H^+(aq) + 6\,e^-$$
Add half-reactions: $3\,MnO_2(s) + 4\,\cancel{12}\,H^+(aq) + \cancel{6\,e^-} + 2\,MnO_2(s) + \cancel{4}\,H_2O(l) \rightarrow$
$$3\,Mn^{2+}(aq) + 2\,\cancel{6}\,H_2O(l) + 2\,MnO_4^-(aq) + 8\,\cancel{H^+(aq)} + \cancel{6\,e^-}$$
Cancel electrons & species: $5\,MnO_2(s) + 4\,H^+(aq) \rightarrow 3\,Mn^{2+}(aq) + 2\,H_2O(l) + 2\,MnO_4^-(aq)$ Look up cell potentials. Mn is reduced in the first half-cell reaction, so $E^\circ_{cathode} = E^\circ_{red} = 1.21$ V. Mn is oxidized in the second half-cell reaction, so $E^\circ_{anode} = E^\circ_{red} = +1.68$ V. Then $E^\circ_{cell} = E^\circ_{cathode} - E^\circ_{anode} = 1.21$ V $- 1.68$ V $= -0.47$ V and $n = 6$ so

$$\Delta G^\circ_{rxn} = -nF\,E^\circ_{cell} = -6\,\cancel{mol\,e^-} \times \frac{96{,}485\text{ C}}{\cancel{mol\,e^-}} \times -0.47\text{ V} = -6 \times 96{,}485\,\cancel{C} \times -0.47\frac{J}{\cancel{C}}$$

$$= 2.7\underline{2}09 \times 10^5\text{ J} = 270\text{ kJ} = 2.7 \times 10^2\text{ kJ and } T = 273.15 + 25\,^\circ\text{C} = 298\text{ K then } \Delta G^\circ_{rxn} = -RT\ln K$$

Rearrange to solve for K. $K = e^{\frac{-\Delta G^\circ_{rxn}}{RT}} = e^{\frac{-2.7209 \times 10^5\,\cancel{J}}{\left(8.314\frac{\cancel{J}}{\cancel{K}\cdot mol}\right)(298\,\cancel{K})}} = e^{-109.82} = 2.0 \times 10^{-48}$

Check:

Reactants	Products
5 Mn atoms	5 Mn atoms
10 O atoms	10 O atoms
4 H atoms	4 H atoms
+4 charge	+4 charge

The units (kJ and none) are correct. If the voltage is negative, the reaction is nonspontaneous, the free energy change is very positive, and the equilibrium constant is extremely small. The voltage is less than in part (a), so the free energy change is not as large and the equilibrium constant is not as small.

18.121   **Given:** metal, M, 50.9 g/mol, 1.20 g of metal reduced in 23.6 minutes at 6.42 A from molten chloride
Find: empirical formula of chloride
Conceptual Plan: min $\rightarrow$ s $\rightarrow$ C $\rightarrow$ mol e^- and g M $\rightarrow$ mol M then mol e^-, mol M $\rightarrow$ charge $\rightarrow$ MCl_x

$$\frac{60\text{ s}}{1\text{ min}} \quad \frac{6.42\text{ C}}{1\text{ s}} \quad \frac{1\text{ mol }e^-}{96{,}485\text{ C}} \qquad \frac{1\text{ mol M}}{50.9\text{ g M}} \qquad \frac{1\text{ mol }e^-}{1\text{ mol M}}$$

Solution: $23.6\,\cancel{min} \times \dfrac{60\,\cancel{s}}{1\,\cancel{min}} \times \dfrac{6.42\,\cancel{C}}{1\,\cancel{s}} \times \dfrac{1\text{ mol }e^-}{96{,}485\,\cancel{C}} = 0.094\underline{2}190$ mol e^- and

$1.20\,\cancel{g\,M} \times \dfrac{1\text{ mol M}}{50.9\,\cancel{g\,M}} = 0.023\underline{5}756$ mol M then $\dfrac{0.094\underline{2}190\text{ mol }e^-}{0.023\underline{5}756\text{ mol M}} = 3.99\underline{6}46\dfrac{e^-}{M}$

So the empirical formula is MCl_4.

Check: The units (none) are correct. The result was an integer within the error of the measurements. The formula is typical for a metal salt. It could be vanadium, which has a +4 oxidation state.

18.123   **Given:** 0.535g impure Sn; dissolve to form Sn^{2+} and titrate with 0.0344 L of 0.0448 M NO_3^- to generate NO
Find: percent by mass Sn
Conceptual Plan: Use balanced reaction from Problem 18.40(c) then
L $\rightarrow$ mol NO_3^- $\rightarrow$ mol Sn $\rightarrow$ g Sn $\rightarrow$ percent by mass Sn.

$$M = \frac{mol}{L} \qquad \frac{3\text{ mol Sn}}{2\text{ mol }NO_3^-} \qquad \frac{118.71\text{ g Sn}}{1\text{ mol Sn}} \qquad \text{percent by mass Sn} = \frac{\text{g Sn}}{\text{g sample}} \times 100\%$$

Solution: $2\,NO_3^-(aq) + 8\,H^+(aq) + 3\,Sn^{2+}(aq) \rightarrow 2\,NO(g) + 4\,H_2O(l) + 3\,Sn^{4+}(aq)$

$0.0344\,\cancel{L} \times \dfrac{0.0448\,\cancel{mol\,NO_3^-}}{1\,\cancel{L}} \times \dfrac{3\,\cancel{mol\,Sn}}{2\,\cancel{mol\,NO_3^-}} \times \dfrac{118.71\text{ g Sn}}{1\,\cancel{mol\,Sn}} = 0.27\underline{4}4195$ g Sn then

$$\text{percent by mass Sn} = \frac{\text{g Sn}}{\text{g sample}} \times 100\% = \frac{0.27\underline{4}4195 \text{ g Sn}}{0.535 \text{ g sample}} \times 100\% = 51.3\% \text{ by mass Sn}$$

Check: The units (% by mass) are correct. The result was a number less than 100%.

18.125 **Given:** 1.25 L of a 0.552 M HBr solution, convert H^+ to $H_2(g)$ for 73 minutes at 11.32 A **Find:** pH
Conceptual Plan: min $\rightarrow$ s $\rightarrow$ C $\rightarrow$ mol e^- $\rightarrow$ mol H^+ consumed and L, M $\rightarrow$ mol H^- initially then

$$\frac{60 \text{ s}}{1 \text{ min}} \quad \frac{11.32 \text{ C}}{1 \text{ s}} \quad \frac{1 \text{ mol } e^-}{96,485 \text{ C}} \quad \frac{2 \text{ mol } H^+}{2 \text{ mol } e^-} \qquad\qquad M = \frac{\text{mol}}{L}$$

mol H + initially, mol H^+ consumed $\rightarrow$ mol H^+ remaining then mol H^+ remaining, L $\rightarrow$ $[H^+] \rightarrow$ pH

$$\text{mol } H^+ \text{ initially} - \text{mol } H^+ \text{ consumed} = \text{mol } H^+ \text{ remaining} \qquad\qquad M = \frac{\text{mol}}{L} \quad pH = -\log [H^+]$$

Solution: $2 H^+(aq) + 2 e^- \rightarrow H_2(g)$

$$73 \text{ min} \times \frac{60 \text{ s}}{1 \text{ min}} \times \frac{11.32 \text{ C}}{1 \text{ s}} \times \frac{1 \text{ mol } e^-}{96,485 \text{ C}} \times \frac{2 \text{ mol } H^+}{2 \text{ mol } e^-} = 0.5\underline{1}3879 \text{ mol } H^+ \text{ consumed and}$$

$$1.25 \text{ L } H^+ \times \frac{0.552 \text{ mol } H^+}{1 \text{ L } H^+} = 0.690 \text{ mol } H^+ \text{ initially then mol } H^+ \text{ initially} - \text{mol } H^+ \text{ consumed} =$$

$$\text{mol } H^+ \text{ remaining} = 0.690 \text{ mol } H^+ \text{ initially} - 0.5\underline{1}3879 \text{ mol } H^+ \text{ consumed} = 0.1\underline{7}612 \text{ mol } H^+ \text{ remaining}$$

$$\text{then } [H^+] = \frac{0.1\underline{7}612 \text{ mol } H^+}{1.25 \text{ L}} = 0.1\underline{4}0897 \text{ M } H^+ \text{ then pH} = -\log [H^+] = = -\log 0.1\underline{4}0897 = 0.85$$

Check: The units (none) are correct. The result is a pH higher than the initial pH $[-\log (0.552) = 0.258]$, as is expected.

18.127 **Given:** MnO_2/Mn^{2+} electrode at pH 10.24 **Find:** $[Mn^{2+}]$ to get half-cell potential = 0.00 V
Conceptual Plan: pH $\rightarrow$ $[H^+]$ then

$$pH = -\log [H^+]$$

Write half-cell reactions and look up half-reactions in Table 18.1. Define Q based on half-cell reaction. Then $E^\circ_{\text{half-cell}}$, n, $[H^+] \rightarrow [Mn^{2+}]$.

$$E_{\text{half-cell}} = E^\circ_{\text{half-cell}} - \frac{0.0592 \text{ V}}{n} \log Q$$

Solution: Because $pH = -\log [H^+]$ so $[H^+] = 10^{-pH} = 10^{-10.24} = 5.\underline{7}5440 \times 10^{-11}$ M

$$MnO_2(s) + 4 H^+(aq) + 2 e^- \rightarrow Mn^{2+}(aq) + 2 H_2O(l) \; E^\circ_{\text{half-cell}} = 1.21 \text{ V}, n = 2, \text{ and } Q = \frac{[Mn^{2+}]}{[H^+]^4}$$

$$E_{\text{half-cell}} = E^\circ_{\text{half-cell}} - \frac{0.0592 \text{ V}}{n} \log Q \text{ so } 0.00 \text{ V} = 1.21 \text{ V} - \frac{0.0592 \text{ V}}{2} \log \frac{[Mn^{2+}]}{(5.\underline{7}5440 \times 10^{-11})^4} \rightarrow$$

$$1.21 \text{ V} = \frac{0.0592 \text{ V}}{2} \log \frac{[Mn^{2+}]}{(5.\underline{7}5440 \times 10^{-11})^4} \rightarrow 40.8\underline{7}84 = \log \frac{[Mn^{2+}]}{(5.\underline{7}5440 \times 10^{-11})^4} \rightarrow$$

$$\frac{[Mn^{2+}]}{(5.\underline{7}5440 \times 10^{-11})^4} = 10^{40.8\underline{7}84} = 7.\underline{5}5750 \times 10^{40} \rightarrow [Mn^{2+}] = 0.8\underline{2}8664 \text{ M} = 0.83 \text{ M } Mn^{2+}$$

Check: The units (M) are correct. The standard half-cell potential is very large and positive. Most of the shift toward 0.00 V is due to the fourth-order dependence in $[H^+]$ at a high pH, so the $[Mn^{2+}]$ is close to 1 M. So the concentration is reasonable.

Challenge Problems

18.129 **Given:** hydrogen–oxygen fuel cell; 1.2×10^3 kWh of electricity/month **Find:** V of $H_2(g)$ at STP/month
Conceptual Plan: Write half-reactions. Look up half-reactions at pH 7. The reaction on the left is the oxidation. Calculate the standard cell potential by subtracting the electrode potential of the anode from the electrode potential of the cathode: $E^\circ_{\text{cell}} = E^\circ_{\text{cathode}} - E^\circ_{\text{anode}}$. Add the two half-cell reactions and cancel the electrons. Then kWh $\rightarrow$ J $\rightarrow$ C $\rightarrow$ mol e^- $\rightarrow$ mol H_2 $\rightarrow$ V.

$$\frac{3.60 \times 10^6 \text{ J}}{1 \text{ kWh}} \quad \frac{1 \text{ C}}{1.23 \text{ J}} \quad \frac{1 \text{ mol } e^-}{96,485 \text{ C}} \quad \frac{2 \text{ mol } H_2}{4 \text{ mol } e^-} \quad \text{at STP} \quad \frac{22.414 \text{ L}}{1 \text{ mol } H_2}$$

Solution: $2 H_2(g) + 4 OH^-(aq) \rightarrow 4 H_2O(l) + 4 e^-$ where
$E^\circ_{red} = -0.83 \text{ V} = E^\circ_{anode}$ and $O_2(g) + 2 H_2O(l) + 4 e^- \rightarrow 4 OH^-(aq)$ where
$E^\circ_{red} = 0.40 \text{ V} = E^\circ_{cathode}$. $E^\circ_{cell} = E^\circ_{cathode} - E^\circ_{anode} = 0.40 \text{ V} - (-0.83 \text{ V}) = 1.23 \text{ V} = 1.23 \text{ J/C}$ and $n = 4$. Net reaction is $2 H_2(g) + O_2(g) \rightarrow 2 H_2O(l)$. Then

$$1.2 \times 10^3 \text{ kWh} \times \frac{3.60 \times 10^6 \text{ J}}{1 \text{ kWh}} \times \frac{1 \text{ C}}{1.23 \text{ J}} \times \frac{1 \text{ mole}^-}{96{,}485 \text{ C}} \times \frac{2 \text{ mol } H_2}{4 \text{ mole}^-} \times \frac{22.414 \text{ L}}{1 \text{ mol } H_2} = 4.1 \times 10^5 \text{ L}$$

Check: The units (L) are correct. A large volume is expected because we are trying to generate a large amount of electricity.

18.131 **Given:** Au^{3+}/Au electroplating; surface area $= 49.8 \text{ cm}^2$, Au thickness $= 1.00 \times 10^{-3}$ cm, density $= 19.3 \text{ g/cm}^3$; at 3.25 A **Find:** time

Conceptual Plan: Write the half-cell reaction and add electrons as needed to balance reactions. Then surface area, thickness $\rightarrow$ V $\rightarrow$ g Au $\rightarrow$ mol Au $\rightarrow$ mol e^- $\rightarrow$ C $\rightarrow$ s

$V = surface\ area \times thickness$ $\quad \frac{19.3 \text{ g Au}}{1 \text{ cm}^3 \text{ Au}} \quad \frac{1 \text{ mol Au}}{196.97 \text{ g Au}} \quad \frac{3 \text{ mol } e^-}{1 \text{ mol Au}} \quad \frac{96{,}485 \text{ C}}{1 \text{ mol } e^-} \quad \frac{1 \text{ s}}{3.25 \text{ C}}$

Solution: Write the half-reaction and add electrons. $Au^{3+}(aq) + 3 e^- \rightarrow Au(s)$.

$V = surface\ area \times thickness = (49.8 \text{ cm}^2)(1.00 \times 10^{-3} \text{ cm}) = 0.0498 \text{ cm}^3$ then

$$0.0498 \text{ cm}^3 \text{ Au} \times \frac{19.3 \text{ g Au}}{1 \text{ cm}^3 \text{ Au}} \times \frac{1 \text{ mol Au}}{196.97 \text{ g Au}} \times \frac{3 \text{ mole}^-}{1 \text{ mol Au}} \times \frac{96{,}485 \text{ C}}{1 \text{ mole}^-} \times \frac{1 \text{ s}}{3.25 \text{ C}} = 435 \text{ s}$$

Check: The units (s) are correct. Because the layer is so thin, there is far less than a mole of gold; so the time is not very long. To be an economical process, it must be fairly quick.

18.133 **Given:** $C_2O_4^{2-} \rightarrow CO_2$ and $MnO_4^-(aq) \rightarrow Mn^{2+}(aq)$, 50.1 mL of MnO_4^- to titrate 0.339 g $Na_2C_2O_4$; 4.62 g U sample titrated by 32.5 mL MnO_4^-; and $UO^{2+} \rightarrow UO_2^{2+}$ **Find:** percent U in sample

Conceptual Plan: Separate the overall reaction into two half-reactions: one for oxidation and one for reduction. $\rightarrow$ Balance each half-reaction with respect to mass in the following order: (1) Balance all elements other than H and O, (2) balance O by adding H_2O, and (3) balance H by adding H^+. $\rightarrow$ Balance each half-reaction with respect to charge by adding electrons. (The sum of the charges on both sides of the equation should be made equal by adding electrons as necessary.) $\rightarrow$ Make the number of electrons in both half-reactions equal by multiplying one or both half-reactions by a small whole number. $\rightarrow$ Add the two half-reactions, canceling electrons and other species as necessary. $\rightarrow$ Verify that the reaction is balanced with respect to both mass and charge. Then

mL $MnO_4^- \rightarrow$ L MnO_4^- and g $Na_2C_2O_4 \rightarrow$ mol $Na_2C_2O_4 \rightarrow$ mol MnO_4^- then

$\frac{1 \text{ L } MnO_4^-}{1000 \text{ mL } MnO_4^-}$ $\qquad \frac{1 \text{ mol } Na_2C_2O_4}{134.00 \text{ g } Na_2C_2O_4} \quad \frac{2 \text{ mol } MnO_4^-}{5 \text{ mol } Na_2C_2O_4}$

L MnO_4^-, mol $MnO_4^- \rightarrow$ M MnO_4^- then write U half-reactions and balance as above. $\rightarrow$

$M = \frac{\text{mol } MnO_4^-}{L}$

Make the number of electrons in both half-reactions equal by multiplying one or both half-reactions by a small whole number. $\rightarrow$ Add the two half-reactions, canceling electrons and other species as necessary. $\rightarrow$ Verify that the reaction is balanced with respect to both mass and charge.

Then mL MnO_4^-, M $MnO_4^- \rightarrow$ mol $MnO_4^- \rightarrow$ mol U $\rightarrow$ g U then g U, g sample $\rightarrow$ % U.

$M = \frac{\text{mol } MnO_4^-}{L}$ $\quad \frac{5 \text{ mol U}}{2 \text{ mol } MnO_4^-} \quad \frac{238.03 \text{ g U}}{1 \text{ mol U}}$ $\qquad \text{percent U} = \frac{\text{g U}}{\text{g sample}} \times 100\%$

Solution:

Separate: $\qquad\qquad\qquad\qquad MnO_4^-(aq) \rightarrow Mn^{2+}(aq) \qquad$ and $\quad C_2O_4^{2-}(aq) \rightarrow CO_2(g)$

Balance non H & O elements: $MnO_4^-(aq) \rightarrow Mn^{2+}(aq) \qquad$ and $\quad C_2O_4^{2-}(aq) \rightarrow 2 CO_2(g)$

Balance O with H_2O: $MnO_4^-(aq) \rightarrow Mn^{2+}(aq) + 4 H_2O(l) \quad$ and $\quad C_2O_4^{2-}(aq) \rightarrow 2 CO_2(g)$

Balance H with H^+: $MnO_4^-(aq) + 8 H^+(aq) \rightarrow Mn^{2+}(aq) + 4 H_2O(l)$ and $C_2O_4^{2-}(aq) \rightarrow 2 CO_2(g)$

Add electrons: $MnO_4^-(aq) + 8 H^+(aq) + 5 e^- \rightarrow Mn^{2+}(aq) + 4 H_2O(l)$ and $C_2O_4^{2-}(aq) \rightarrow 2 CO_2(g) + 2 e^-$

Equalize electrons:

$2 MnO_4^-(aq) + 16 H^+(aq) + 10 e^- \rightarrow 2 Mn^{2+}(aq) + 8 H_2O(l)$ and $5 C_2O_4^{2-}(aq) \rightarrow 10 CO_2(g) + 10 e^-$

Add half-reactions:

$2 MnO_4^-(aq) + 16 H^+(aq) + \cancel{10 e^-} + 5 C_2O_4^{2-}(aq) \rightarrow 2 Mn^{2+}(aq) + 8 H_2O(l) + 10 CO_2(g) + \cancel{10 e^-}$

Cancel electrons: $2 MnO_4^-(aq) + 16 H^+(aq) + 5 C_2O_4^{2-}(aq) \rightarrow 2 Mn^{2+}(aq) + 8 H_2O(l) + 10 CO_2(g)$

then $50.1 \text{ mL MnO}_4^- \times \dfrac{1 \text{ L MnO}_4^-}{1000 \text{ mL MnO}_4^-} = 0.0501 \text{ L MnO}_4^-$

$0.399 \text{ g Na}_2\text{C}_2\text{O}_4 \times \dfrac{1 \text{ mol Na}_2\text{C}_2\text{O}_4}{134.00 \text{ g Na}_2\text{C}_2\text{O}_4} \times \dfrac{2 \text{ mol MnO}_4^-}{5 \text{ mol Na}_2\text{C}_2\text{O}_4} = 0.0011\underline{9}10448 \text{ mol MnO}_4^-$

$M = \dfrac{0.0011\underline{9}10448 \text{ mol MnO}_4^-}{0.0501 \text{ L}} = 0.023\underline{7}73345 \text{ M MnO}_4^-$

Separate: $\qquad\qquad\qquad$ $MnO_4^-(aq) \to Mn^{2+}(aq)$ $\quad$ and $\quad$ $UO^{2+}(aq) \to UO_2^{2+}(aq)$

Balance non H & O elements: $\quad MnO_4^-(aq) \to Mn^{2+}(aq)$ $\quad$ and $\quad$ $UO^{2+}(aq) \to UO_2^{2+}(aq)$

Balance O with H_2O: $MnO_4^-(aq) \to Mn^{2+}(aq) + 4\,H_2O(l)$ $\quad$ and $\quad$ $UO^{2+}(aq) + H_2O(l) \to UO_2^{2+}(aq)$

Balance H with H^+:

$MnO_4^-(aq) + 8\,H^+(aq) \to Mn^{2+}(aq) + 4\,H_2O(l)$ and $UO^{2+}(aq) + H_2O(l) \to UO_2^{2+}(aq) + 2\,H^+(aq)$

Add electrons: $MnO_4^-(aq) + 8\,H^+(aq) + 5\,e^- \to Mn^{2+}(aq) + 4\,H_2O(l)$ and

$\qquad\qquad\qquad\qquad\qquad\qquad UO^{2+}(aq) + H_2O(l) \to UO_2^{2+}(aq) + 2\,H^+(aq) + 2\,e^-$

Equalize electrons: $2\,MnO_4^-(aq) + 16\,H^+(aq) + 10\,e^- \to 2\,Mn^{2+}(aq) + 8\,H_2O(l)$ and

$\qquad\qquad\qquad\qquad\qquad 5\,UO^{2+}(aq) + 5\,H_2O(l) \to 5\,UO_2^{2+}(aq) + 10\,H^+(aq) + 10\,e^-$

Add half-reactions: $2\,MnO_4^-(aq) + 6\;\cancel{16}\,H^+(aq) + \cancel{10\,e^-} + 5\,UO^{2+}(aq) + \cancel{5\,H_2O(l)} \to$

$\qquad\qquad\qquad\qquad 2\,Mn^{2+}(aq) + 3\;\cancel{8}\,H_2O(l) + 5\,UO_2^{2+}(aq) + \cancel{10\,H^+(aq)} + \cancel{10\,e^-}$

Cancel electrons & species: $2\,MnO_4^-(aq) + 6\,H^+(aq) + 5\,UO^{2+}(aq) \to 2\,Mn^{2+}(aq) + 3\,H_2O(l) + 5\,UO_2^{2+}(aq)$

$32.5 \text{ mL MnO}_4^- \times \dfrac{0.023\underline{7}73345 \text{ mol MnO}_4^-}{1000 \text{ mL MnO}_4^-} \times \dfrac{5 \text{ mol U}}{2 \text{ mol MnO}_4^-} \times \dfrac{238.03 \text{ g U}}{1 \text{ mol U}} = 0.45\underline{9}7750 \text{ g U then}$

$\text{percent U} = \dfrac{\text{g U}}{\text{g sample}} \times 100\% = \dfrac{0.45\underline{9}7750 \text{ g U}}{4.63 \text{ g sample}} \times 100\% = 9.93\%$

Check: first reaction

Reactants	Products
2 Mn atoms	2 Mn atoms
28 O atoms	28 O atoms
16 H atoms	16 H atoms
10 C atoms	10 C atoms
+4 charge	+4 charge

second reaction

Reactants	Products
2 Mn atoms	2 Mn atoms
13 O atoms	13 O atoms
6 H atoms	6 H atoms
5 U atoms	5 U atoms
+14 charge	+14 charge

The reactions are balanced. The units (%) are correct. The percentage is between 0 and 100%.

18.135 $\quad$ The overall cell reaction for the first cell is $2\,Cu^+(aq) \to Cu^{2+}(aq) + Cu(s)$. The overall cell reaction for the second cell is $Cu^+(aq) \to Cu^{2+}(aq)$. The biggest difference in $E°$ is because $n = 1$ for the first cell and $n = 2$ for the second cell. Because $\Delta G_{rxn}° = -n\,F\,E_{cell}°$ for the first cell,

$\Delta G_{rxn}° = -1 \text{ mol } e^- \times \dfrac{96{,}485 \text{ C}}{\text{mol } e^-} \times 0.364 \text{ V} = -1 \times 96{,}485 \text{ C} \times 0.364\,\dfrac{J}{C} = -35.1 \text{ kJ, and for the}$

second cell,

$\Delta G_{rxn}° = -2 \text{ mol } e^- \times \dfrac{96{,}485 \text{ C}}{\text{mol } e^-} \times 0.182 \text{ V} = -2 \times 96{,}485 \text{ C} \times 0.182\,\dfrac{J}{C} = -35.1 \text{ kJ}$. Thus, $\Delta G_{rxn}°$ is the same.

Conceptual Problems

18.137 $\quad$ (a) $\quad$ Looking for anion reductions that are between the reduction potentials of Cl_2 and Br_2. The only one that meets that criterion is the dichromate ion.

18.139 $\quad$ Because $K < 1$, the reaction must be nonspontaneous under standard conditions. Therefore, $E_{cell}°$ is negative and $\Delta G_{rxn}°$ is positive.

19 Radioactivity and Nuclear Chemistry

Review Questions

19.1 Radioactivity is the emission of subatomic particles or high-energy electromagnetic radiation by the nuclei of certain atoms. Radioactivity was discovered in 1896 by the French scientist Antoine-Henri Becquerel (1852–1908). Becquerel placed crystals—composed of potassium uranyl sulfate, a compound known to phosphoresce—on top of a photographic plate wrapped in black cloth. The photographic plate showed a bright exposure spot where the crystals had been. Becquerel, Marie Curie, and Pierre Curie received the Nobel Prize for the discovery of radioactivity.

19.3 A is the mass number (number of protons + neutrons), Z is the atomic number (number of protons), and X is the chemical symbol of the element.

19.5 An alpha particle has the same symbol as a helium nucleus, ^4_2He. When an element emits an alpha particle, the number of protons in its nucleus decreases by two and the mass number decreases by four, transforming it into a different element.

19.7 Gamma rays are high-energy (short-wavelength) photons that have a symbol of $^0_0\gamma$. A gamma ray has no charge and no mass. When a gamma-ray photon is emitted from a radioactive atom, it does not change the mass number or the atomic number of the element. Gamma rays, however, are usually emitted in conjunction with other types of radiation.

19.9 Electron capture occurs when a nucleus assimilates an electron from an inner orbital of its electron cloud. Like positron emission, the net effect of electron capture is the conversion of a proton into a neutron: $^1_1\text{p} + {}^{\ 0}_{-1}\text{e} \rightarrow {}^1_0\text{n}$. When an atom undergoes electron capture, its atomic number decreases by one because it has one less proton and its mass number is unchanged.

19.11 For the lighter elements, the N/Z ratio of stable isotopes is about one (equal numbers of neutrons and protons). However, beyond about $Z = 20$, the N/Z ratio of stable nuclei begins to get larger (reaching about 1.5). Above $Z = 83$, stable nuclei do not exist. If the N/Z ratio is too high, it indicates the presence of too many neutrons and so the neutrons will be converted to protons via beta decay. If the N/Z ratio is too low, it indicates the presence of too many protons and therefore position emission or electron capture will be used to convert protons to neutrons.

19.13   (a)   Film-badge dosimeters consist of photographic film held in a small case that is pinned to clothing and are standard for most people working with or near radioactive substances. These badges are collected and processed (or developed) regularly to monitor a person's exposure. The more exposed the film has become in a given period of time, the more radioactivity to which the person has been exposed.

   (b)   A Geiger-Müller counter (commonly referred to as a Geiger counter) is an instrument that can detect radioactivity instantaneously. Particles emitted by radioactive nuclei pass through an argon-filled chamber. The energetic particles create a trail of ionized argon atoms. An applied high voltage between a wire within the chamber and the chamber itself causes these newly formed ions to produce an electrical signal that can be detected on a meter or turned into an audible click. Each click corresponds to a radioactive particle passing through the argon gas chamber. This clicking is the stereotypical sound most people associate with a radiation detector.

(c)   A scintillation counter is another instrument that can detect radioactivity instantaneously. In this device, the radioactive emissions pass through a material (such as NaI or CsI) that emits ultraviolet or visible light in response to excitation by energetic particles. The radioactivity excites the atoms to a higher energy state. The atoms release this energy as light, which is then detected and turned into an electrical signal that can be read on a meter.

19.15   Radiocarbon dating, a technique devised in 1949 by Willard Libby at the University of Chicago, is used by archeologists, geologists, anthropologists, and other scientists to estimate the ages of fossils and artifacts. Carbon-14 is constantly formed in the upper atmosphere by the neutron bombardment of nitrogen and then decays back to nitrogen by beta emission with a half-life of 5,730 years. The continuous formation of carbon-14 in the atmosphere and its continuous decay back to nitrogen-14 produces a nearly constant equilibrium amount of atmospheric carbon-14, which is oxidized to carbon dioxide and incorporated into plants by photosynthesis. The C-14 then makes its way up the food chain and ultimately into all living organisms. As a result, all living plants, animals, and humans contain the same ratio of carbon-14 to carbon-12 (^{14}C:^{12}C) as is found in the atmosphere. When a living organism dies, however, it stops incorporating new carbon-14 into its tissues. The ^{14}C:^{12}C ratio then decreases with a half-life of 5,730 years. The accuracy of carbon-14 dating can be checked against objects whose ages are known from historical sources.

To make C-14 dating more accurate, scientists have studied the carbon-14 content of western bristlecone pine trees, which can live up to 5,000 years. The tree trunk contains growth rings corresponding to each year of the tree's life, and the wood laid down in each ring incorporates carbon derived from the carbon dioxide in the atmosphere at that time. The rings thus provide a record of the historical atmospheric carbon-14 content and allow for corrections to carbon-14 concentrations due to atmospheric changes.

The maximum age that can be estimated from carbon-14 dating is about 50,000 years—beyond that, the amount of carbon-14 becomes too low to measure accurately.

19.17   Nuclear fission is the splitting of an atom into smaller products. The process emits enormous amounts of energy. Three researchers in Germany—Lise Meitner (1878–1968), Fritz Strassmann (1902–1980), and Otto Hahn (1879–1968), repeating uranium bombardment experiments by Fermi, performed careful chemical analysis of the products. The nucleus of the neutron-bombarded uranium atom is split into barium, krypton, and other smaller products. A nuclear equation for a fission reaction, showing how uranium breaks apart into the daughter nuclides, is $^{235}_{92}U + ^{1}_{0}n \rightarrow ^{140}_{56}Ba + ^{93}_{36}Kr + 3^{1}_{0}n$ + energy. The process produces three neutrons, which have the potential to initiate fission in three other U-235 atoms. Scientists quickly realized that a sample rich in U-235 could undergo a chain reaction in which neutrons produced by the fission of one uranium nucleus would induce fission in other uranium nuclei. The result would be a self-amplifying reaction capable of producing an enormous amount of energy—an atomic bomb. However, to make a bomb, a critical mass of U-235—enough U-235 to produce a self-sustaining reaction—would be necessary. Because an enormous amount of energy is produced, it can be used to produce steam to drive turbines (as shown in Figure 19.11).

19.19   The advantages of using fission to generate electricity are that (1) a typical nuclear power plant generates enough electricity for a city of about 1 million people and uses about 50 kg of fuel per day (as opposed to a coal-burning power plant that uses about 2,000,000 kg of fuel to generate the same amount of electricity) and that (2) a nuclear power plant generates no air pollution and no greenhouse gases. (Coal-burning power plants also emit carbon dioxide, a greenhouse gas.) The disadvantages are as follows: (1) the danger of nuclear accidents, such as overheating and the release of radiation, and (2) waste disposal because the products of the reaction are radioactive and have long half-lives.

19.21   This difference in mass between the products and the reactants is known as the mass defect. The energy corresponding to the mass defect, obtained by substituting the mass defect into the equation $E = mc^2$, is known as the nuclear binding energy, the amount of energy that would be required to break apart the nucleus into its component nucleons. The nuclear binding energy per nucleon peaks at a mass number of 60. The significance of this is that the nuclides with mass numbers of about 60 are among the most stable.

19.23   An extremely high temperature is required for fusion to occur. To date, no material can withstand these temperatures. The U.S. Congress has reduced funding for these projects so that it is less likely that a viable process will be developed.

19.25   In a single-stage linear accelerator, a charged particle such as a proton is accelerated in an evacuated tube. The accelerating force is provided by a potential difference between the ends of the tube. In multi stage linear accelerators such as the Stanford Linear Accelerator (SLAC) at Stanford University (Figure 19.14), a series of tubes of increasing length are connected to a source of alternating voltage, as shown in Figure 19.15. The voltage alternates in such a way that as a positively charged particle leaves a particular tube, that tube becomes positively charged, repelling the particle to the next tube. At the same time, the tube the particle is now approaching becomes negatively charged, pulling the particle toward it. This continues throughout the linear accelerator, allowing the particle to be accelerated to velocities up to 90% of the speed of light. Linear accelerators can be used to conduct nuclear transmutations, making nuclides that don't normally exist in nature.

19.27   The energy associated with radioactivity can ionize molecules. When radiation ionizes important molecules in living cells, problems can develop. The ingestion of radioactive materials, especially alpha and beta emitters, is particularly dangerous because the radioactivity is then inside the body and can do even more damage. The effects of radiation is divided into three different types: acute radiation damage, increased cancer risk, and genetic effects.

19.29   The biological effectiveness factor, or RBE (relative biological effectiveness), corrects the dosage (in rads) for the type of radiation. It is a correction factor that is usually multiplied by the dose in rads to obtain the dose in a unit called the rem (roentgen equivalent man). So dose in rads $\times$ biological effectiveness factor = dose in rem. The biological effectiveness factor for alpha radiation, for example, is much higher than for gamma radiation.

Problems by Topic

Radioactive Decay and Nuclide Stability

19.31   **Conceptual Plan: Begin with the symbol for a parent nuclide on the left side of the equation and the symbol for a particle on the right side (except for electron capture). $\rightarrow$ Equalize the sum of the mass numbers and the sum of the atomic numbers on both sides of the equation by writing the appropriate mass number and atomic number for the unknown daughter nuclide. $\rightarrow$ Using the periodic table, deduce the identity of the unknown daughter nuclide from the atomic number and write its symbol.**
Solution:

(a)   U-234 (alpha decay) $^{234}_{92}U \rightarrow ^{?}_{?}? + ^{4}_{2}He$ then $^{234}_{92}U \rightarrow ^{230}_{90}? + ^{4}_{2}He$ then $^{234}_{92}U \rightarrow ^{230}_{90}Th + ^{4}_{2}He$

(b)   Th-230 (alpha decay) $^{230}_{90}Th \rightarrow ^{?}_{?}? + ^{4}_{2}He$ then $^{230}_{90}Th \rightarrow ^{226}_{88}? + ^{4}_{2}He$ then $^{230}_{90}Th \rightarrow ^{226}_{88}Ra + ^{4}_{2}He$

(c)   Pb-214 (beta decay) $^{214}_{82}Pb \rightarrow ^{?}_{?}? + ^{0}_{-1}e$ then $^{214}_{82}Pb \rightarrow ^{214}_{83}? + ^{0}_{-1}e$ then $^{214}_{82}Pb \rightarrow ^{214}_{83}Bi + ^{0}_{-1}e$

(d)   N-13 (positron emission) $^{13}_{7}N \rightarrow ^{?}_{?}? + ^{0}_{+1}e$ then $^{13}_{7}N \rightarrow ^{13}_{6}? + ^{0}_{+1}e$ then $^{13}_{7}N \rightarrow ^{13}_{6}C + ^{0}_{+1}e$

(e)   Cr-51 (electron capture) $^{51}_{24}Cr + ^{0}_{-1}e \rightarrow ^{?}_{?}?$ then $^{51}_{24}Cr + ^{0}_{-1}e \rightarrow ^{51}_{23}?$ then $^{51}_{24}Cr + ^{0}_{-1}e \rightarrow ^{51}_{23}V$

Check: (a) $234 = 230 + 4, 92 = 90 + 2$, and thorium is atomic number 90. (b) $230 = 226 + 4, 90 = 88 + 2$, and radium is atomic number 88. (c) $214 = 214 + 0, 82 = 83 - 1$, and bismuth is atomic number 83. (d) $13 = 13 + 0, 7 = 6 + 1$, and carbon is atomic number 6. (e) $51 + 0 = 51, 24 - 1 = 23$, and vanadium is atomic number 23.

19.33   **Given:** Th-232 decay series: $\alpha, \beta, \beta, \alpha$ **Find:** balanced decay reactions
Conceptual Plan: Begin with the symbol for a parent nuclide on the left side of the equation and the symbol for a particle on the right side (except for electron capture). $\rightarrow$ Equalize the sum of the mass numbers and the sum of the atomic numbers on both sides of the equation by writing the appropriate mass number and atomic number for the unknown daughter nuclide. $\rightarrow$ Using the periodic table, deduce the identity of the unknown daughter nuclide from the atomic number and write its symbol. $\rightarrow$ Use the product of this reaction to write the next reaction.
Solution:

Th-232 (alpha decay) $^{232}_{90}Th \rightarrow ^{?}_{?}? + ^{4}_{2}He$ then $^{232}_{90}Th \rightarrow ^{228}_{88}? + ^{4}_{2}He$ then $^{232}_{90}Th \rightarrow ^{228}_{88}Ra + ^{4}_{2}He$

Ra-228 (beta decay) $^{228}_{88}Ra \rightarrow ^{?}_{?}? + ^{0}_{-1}e$ then $^{228}_{88}Ra \rightarrow ^{228}_{89}? + ^{0}_{-1}e$ then $^{228}_{88}Ra \rightarrow ^{228}_{89}Ac + ^{0}_{-1}e$

Ac-228 (beta decay) $^{228}_{89}Ac \rightarrow ^{?}_{?}? + ^{0}_{-1}e$ then $^{228}_{89}Ac \rightarrow ^{228}_{90}? + ^{0}_{-1}e$ then $^{228}_{89}Ac \rightarrow ^{228}_{90}Th + ^{0}_{-1}e$

Th-228 (alpha decay) $^{228}_{90}Th \rightarrow ^{?}_{?}? + ^{4}_{2}He$ then $^{228}_{90}Th \rightarrow ^{224}_{88}? + ^{4}_{2}He$ then $^{228}_{90}Th \rightarrow ^{224}_{88}Ra + ^{4}_{2}He$

Thus, the decay series is $^{232}_{90}Th \rightarrow ^{228}_{88}Ra + ^{4}_{2}He, ^{228}_{88}Ra \rightarrow ^{228}_{89}Ac + ^{0}_{-1}e, ^{228}_{89}Ac \rightarrow ^{228}_{90}Th + ^{0}_{-1}e, ^{228}_{90}Th \rightarrow ^{224}_{88}Ra + ^{4}_{2}He$.

Check: $232 = 228 + 4, 90 = 88 + 2$, and radium is atomic number 88. $228 = 228 + 0, 88 = 89 - 1$, and actinium is atomic number 89. $228 = 228 + 0, 89 = 90 - 1$, and thorium is atomic number 90. $228 = 224 + 4, 90 = 88 + 2$, and radium is atomic number 88.

19.35   **Conceptual Plan: Equalize the sum of the mass numbers and the sum of the atomic numbers on both sides of the equation by writing the appropriate mass number and atomic number for the unknown species. → Using the periodic table and the list of particles, deduce the identity of the unknown species from the atomic number and write its symbol.**

    Solution:

    (a) $^{?}_{?}? \rightarrow {}^{217}_{85}\text{At} + {}^{4}_{2}\text{He}$ becomes $^{221}_{?}? \rightarrow {}^{217}_{85}\text{At} + {}^{4}_{2}\text{He}$ then $^{221}_{87}\text{Fr} \rightarrow {}^{217}_{85}\text{At} + {}^{4}_{2}\text{He}$

    (b) $^{241}_{94}\text{Pu} \rightarrow {}^{241}_{95}\text{Am} + {}^{?}_{?}?$ becomes $^{241}_{94}\text{Pu} \rightarrow {}^{241}_{95}\text{Am} + {}^{0}_{-1}?$ then $^{241}_{94}\text{Pu} \rightarrow {}^{241}_{95}\text{Am} + {}^{0}_{-1}\text{e}$

    (c) $^{19}_{11}\text{Na} \rightarrow {}^{19}_{10}\text{Ne} + {}^{?}_{?}?$ becomes $^{19}_{11}\text{Na} \rightarrow {}^{19}_{10}\text{Ne} + {}^{0}_{1}?$ then $^{19}_{11}\text{Na} \rightarrow {}^{19}_{10}\text{Ne} + {}^{0}_{+1}\text{e}$

    (d) $^{75}_{34}\text{Se} + {}^{?}_{?}? \rightarrow {}^{75}_{33}\text{As}$ becomes $^{75}_{34}\text{Se} + {}^{0}_{-1}? \rightarrow {}^{75}_{33}\text{As}$ then $^{75}_{34}\text{Se} + {}^{0}_{-1}\text{e} \rightarrow {}^{75}_{33}\text{As}$

    Check: (a) $221 = 217 + 4, 87 = 85 + 2$, and francium is atomic number 87. (b) $241 = 241 + 0, 94 = 95 - 1$, and the particle is a beta particle. (c) $19 = 19 + 0, 11 = 10 + 1$, and the particle is a positron. (d) $75 = 75 + 0, 34 - 1 = 33$, and the particle is an electron.

19.37   (a)   Mg-26: stable, N/Z ratio is close to 1, acceptable for low Z atoms

    (b)   Ne-25: not stable, N/Z ratio is much too high for low Z atom

    (c)   Co-51: not stable, N/Z ratio is less than 1, much too low

    (d)   Te-124: stable, N/Z ratio is acceptable for this Z

19.39   Sc, V, and Mn each have an odd number of protons. Atoms with an odd number of protons typically have fewer stable isotopes than those with an even number of protons.

19.41   (a)   Mo-109, $N = 67, Z = 42, N/Z = 1.6$, beta decay, because N/Z is too high

    (b)   Ru-90, $N = 46, Z = 44, N/Z = 1.0$, positron emission, because N/Z is too low

    (c)   P-27, $N = 15, Z = 12, N/Z = 0.8$, positron emission, because N/Z is too low

    (d)   Rn-196, $N = 110, Z = 86, N/Z = 1.3$, positron emission, because N/Z is too low

19.43   (a)   Cs-125, $N/Z = 70/55 = 1.3$; Cs-113, $N/Z = 58/55 = 1.1$; Cs-125 will have the longer half-life because it is closer to the proper N/Z

    (b)   Fe-62, $N/Z = 36/26 = 1.4$; Fe-70, $N/Z = 44/26 = 1.7$; Fe-62 will have the longer half-life because it is closer to the proper N/Z

The Kinetics of Radioactive Decay and Radiometric Dating

19.45   **Given:** U-235, $t_{1/2}$ for radioactive decay = 703 million years   **Find:** t to 10.0% of initial amount

    Conceptual Plan: Radioactive decay implies first-order kinetics, $t_{1/2} \rightarrow k$ then

$$t_{1/2} = \frac{0.693}{k}$$

    $m_{\text{U-235 0}}, m_{\text{U-235 }t}, k \rightarrow t$

        $\ln N_t = -kt + \ln N_0$

    Solution: $t_{1/2} = \dfrac{0.693}{k}$ Rearrange to solve for k. $k = \dfrac{0.693}{t_{1/2}} = \dfrac{0.693}{703 \times 10^6 \text{ yr}} = 9.8\underline{5}7752 \times 10^{-10} \text{ yr}^{-1}$

    Because $\ln m_{\text{U-235 }t} = -kt + \ln m_{\text{U-235 0}}$, rearrange to solve for t.

$$t = -\frac{1}{k}\ln\frac{m_{\text{U-235 }t}}{m_{\text{U-235 0}}} = -\frac{1}{9.8\underline{5}7752 \times 10^{-10} \text{ yr}^{-1}}\ln\frac{10.0\%}{100.0\%} = 2.34 \times 10^9 \text{ yr}$$

    Check: The units (yr) are correct. The time is just over three half-lives, when 1/8 of the original amount will be left.

19.47   **Given:** $t_{1/2}$ for isotope decay = 3.8 days; 1.55 g isotope initially   **Find:** mass of isotope after 5.5 days

    Conceptual Plan: Radioactive decay implies first-order kinetics, $t_{1/2} \rightarrow k$ then

$$t_{1/2} = \frac{0.693}{k}$$

    $m_{\text{isotope 0}}, t, k \rightarrow m_{\text{isotope }t}.$

        $\ln N_t = -kt + \ln N_0$

Solution: $t_{1/2} = \dfrac{0.693}{k}$ Rearrange to solve for k. $k = \dfrac{0.693}{t_{1/2}} = \dfrac{0.693}{3.8 \text{ days}} = 0.1\underline{8}237 \text{ day}^{-1}$ Because

$\ln N_t = -kt + \ln N_0 = -(0.1\underline{8}237 \text{ day}^{-1})(5.5 \text{ day}) + \ln(1.55 \text{ g}) = -0.5\underline{6}478 \rightarrow N_t = e^{-0.56478} = 0.57 \text{ g}$.

Check: The units (g) are correct. The amount is consistent with a time between one and two half-lives.

19.49   **Given:** F-18 initial decay rate $= 1.5 \times 10^5/s$, $t_{1/2}$ for F-18 $= 1.83$ h   **Find:** t to decay rate of $2.5 \times 10^3/s$
Conceptual Plan: Radioactive decay implies first-order kinetics, $t_{1/2} \rightarrow k$ then Rate$_0$, Rate$_t$, $k \rightarrow t$.

$$t_{1/2} = \dfrac{0.693}{k} \qquad\qquad \ln\dfrac{\text{Rate}_t}{\text{Rate}_0} = -kt$$

Solution: $t_{1/2} = \dfrac{0.693}{k}$ Rearrange to solve for k. $k = \dfrac{0.693}{t_{1/2}} = \dfrac{0.693}{1.83 \text{ h}} = 0.37\underline{8}689 \text{ h}^{-1}$ Because

$\ln\dfrac{\text{Rate}_t}{\text{Rate}_0} = -kt$ rearrange to solve for t.

$t = -\dfrac{1}{k}\ln\dfrac{\text{Rate}_t}{\text{Rate}_0} = -\dfrac{1}{0.37\underline{8}689 \text{ h}^{-1}}\ln\dfrac{2.5 \times 10^3/\text{s}}{1.5 \times 10^5/\text{s}} = 10.8 \text{ h}$

Check: The units (h) are correct. The time is between five and six half-lives, and the rate is just over $1/2^6$ of the original amount.

19.51   **Given:** boat analysis, C-14/C-12 $= 72.5\%$ of living organism   **Find:** t
Other: $t_{1/2}$ for decay of C-14 $= 5730$ years
Conceptual Plan: Radioactive decay implies first-order kinetics, $t_{1/2} \rightarrow k$ then 72.5% of $m_{\text{C-14 0}}$, $k \rightarrow t$.

$$t_{1/2} = \dfrac{0.693}{k} \qquad\qquad \ln N_t = -kt + \ln N_0$$

Solution: $t_{1/2} = \dfrac{0.693}{k}$ Rearrange to solve for k. $k = \dfrac{0.693}{t_{1/2}} = \dfrac{0.693}{5730 \text{ yr}} = 1.2\underline{0}942 \times 10^{-4} \text{ yr}^{-1}$ then

$[\text{C-14}]_t = 0.725[\text{C-14}]_0$ Because $\ln m_{\text{C-14 }t} = -kt + \ln m_{\text{C-14 0}}$, rearrange to solve for t.

$t = -\dfrac{1}{k}\ln\dfrac{m_{\text{C-14 }t}}{m_{\text{C-14 0}}} = -\dfrac{1}{1.2\underline{0}942 \times 10^{-4} \text{ yr}^{-1}}\ln\dfrac{0.725 \, m_{\text{C-14 0}}}{m_{\text{C-14 0}}} = 2.66 \times 10^3 \text{ yr}$

Check: The units (yr) are correct. The time to 72.5% decay is consistent with a time less than one half-life.

19.53   **Given:** skull analysis, C-14 decay rate $= 15.3$ dis/min $\cdot$ gC in living organisms and 0.85 dis/min $\cdot$ gC in skull
Find: t **Other:** $t_{1/2}$ for decay of C-14 $= 5730$ years
Conceptual Plan: Radioactive decay implies first-order kinetics, $t_{1/2} \rightarrow k$ then Rate$_0$, Rate$_t$, $k \rightarrow t$.

$$t_{1/2} = \dfrac{0.693}{k} \qquad\qquad \ln\dfrac{\text{Rate}_t}{\text{Rate}_0} = -kt$$

Solution: $t_{1/2} = \dfrac{0.693}{k}$ Rearrange to solve for k. $k = \dfrac{0.693}{t_{1/2}} = \dfrac{0.693}{5730 \text{ yr}} = 1.2\underline{0}942 \times 10^{-4} \text{ yr}^{-1}$

Because $\ln\dfrac{\text{Rate}_t}{\text{Rate}_0} = -kt$, rearrange to solve for t.

$t = -\dfrac{1}{k}\ln\dfrac{\text{Rate}_t}{\text{Rate}_0} = -\dfrac{1}{1.2\underline{0}942 \times 10^{-4} \text{ yr}^{-1}}\ln\dfrac{0.85 \text{ dis/min} \cdot \text{gC}}{15.3 \text{ dis/min} \cdot \text{gC}} = 2.39 \times 10^4 \text{ yr}$

Check: The units (yr) are correct. The rate is 6% of initial value, and the time is consistent with a time just more than four half-lives.

19.55   **Given:** rock analysis, 0.438 g Pb-206 to every 1.00 g U-238, no Pb-206 initially   **Find:** age of rock
Other: $t_{1/2}$ for decay of U-238 to Pb-206 $= 4.5 \times 10^9$ years
Conceptual Plan: Radioactive decay implies first-order kinetics, $t_{1/2} \rightarrow k$ then

$$t_{1/2} = \dfrac{0.693}{k}$$

g Pb-206 $\rightarrow$ mol Pb-206 $\rightarrow$ mol U-238 $\rightarrow$ g U-238 then $m_{\text{U-238 }t}$, $k \rightarrow t$.

$$\dfrac{1 \text{ mol Pb-206}}{206 \text{ g Pb-206}} \qquad \dfrac{1 \text{ mol U-238}}{1 \text{ mol Pb-206}} \qquad \dfrac{238 \text{ g U-238}}{1 \text{ mol U-238}} \qquad\qquad \ln N_t = -kt + \ln N_0$$

Solution: $t_{1/2} = \dfrac{0.693}{k}$ Rearrange to solve for k. $k = \dfrac{0.693}{t_{1/2}} = \dfrac{0.693}{4.5 \times 10^9 \text{ yr}} = 1.\underline{5}4 \times 10^{-10} \text{ yr}^{-1}$ then

$0.438 \text{ g Pb-206} \times \dfrac{1 \text{ mol Pb-206}}{206 \text{ g Pb-206}} \times \dfrac{1 \text{ mol U-238}}{1 \text{ mol Pb-206}} \times \dfrac{238 \text{ g U-238}}{1 \text{ mol U-238}} = 0.50\underline{6}039 \text{ g U-238}$ Because

$\ln \dfrac{m_{\text{U-238 }t}}{m_{\text{U-238 }0}} = -kt$, rearrange to solve for t.

$t = -\dfrac{1}{k} \ln \dfrac{m_{\text{U-238 }t}}{m_{\text{U-238 }0}} = -\dfrac{1}{1.\underline{5}4 \times 10^{-10} \text{ yr}^{-1}} \ln \dfrac{1.00 \text{ g U-238}}{(1.00 + 0.50\underline{6}039) \text{ g U-238}} = 2.7 \times 10^9 \text{ yr}$

Check: The units (yr) are correct. The amount of Pb-206 is less than half the initial U-238 amount, and time is less than one half-life.

Fission and Fusion

19.57   **Given:** U-235 fission induced by neutrons to Xe-144 and Sr-90   **Find:** number of neutrons produced
Conceptual Plan: Write the species given on the appropriate side of the equation. → Equalize the sum of the mass numbers and the sum of the atomic numbers on both sides of the equation by writing the stoichiometric coefficient in front of the desired species.
Solution: $^{235}_{92}\text{U} + ^{1}_{0}\text{n} \rightarrow ^{144}_{54}\text{Xe} + ^{90}_{38}\text{Sr} + ?^{1}_{0}\text{n}$ becomes $^{235}_{92}\text{U} + ^{1}_{0}\text{n} \rightarrow ^{144}_{54}\text{Xe} + ^{90}_{38}\text{Sr} + 2^{1}_{0}\text{n}$, so two neutrons are produced.

Check: $235 + 1 = 144 + 90 + 2, 92 + 0 = 54 + 38 + 0$, and no other particle is necessary to balance the equation.

19.59   **Given:** fusion of two H-2 atoms to form He-3 and one neutron   **Find:** balanced equation
Conceptual Plan: Write the species given on the appropriate side of the equation. → Equalize the sum of the mass numbers and the sum of the atomic numbers on both sides of the equation by writing the stoichiometric coefficient in front of the desired species.
Solution: $2\,^{2}_{1}\text{H} \rightarrow ^{3}_{2}\text{He} + ^{1}_{0}\text{n}$

Check: $2(2) = 3 + 1, 2(1) = 2 + 0$, and no other particle is necessary to balance the equation.

19.61   **Given:** U-238 bombarded by neutrons to form U-239, which undergoes two beta decays to form Pu-239
Find: balanced equations
Conceptual Plan: Write the species given on the appropriate side of the equation. → Equalize the sum of the mass numbers and the sum of the atomic numbers on both sides of the equation by writing the stoichiometric coefficient in front of the desired species. → Use the product of this reaction to write the next reaction until the process is complete.
Solution: $^{238}_{92}\text{U} + ?^{1}_{0}\text{n} \rightarrow ^{239}_{92}\text{U}$ becomes $^{238}_{92}\text{U} + ^{1}_{0}\text{n} \rightarrow ^{239}_{92}\text{U}$ then
beta decay $^{239}_{92}\text{U} \rightarrow ^{239}_{?}? + ^{0}_{-1}\text{e}$ becomes $^{239}_{92}\text{U} \rightarrow ^{239}_{93}? + ^{0}_{-1}\text{e}$ then $^{239}_{92}\text{U} \rightarrow ^{239}_{93}\text{Np} + ^{0}_{-1}\text{e}$ then
beta decay $^{239}_{93}\text{Np} \rightarrow ^{239}_{?}? + ^{0}_{-1}\text{e}$ becomes $^{239}_{93}\text{Np} \rightarrow ^{239}_{94}? + ^{0}_{-1}\text{e}$ then $^{239}_{93}\text{Np} \rightarrow ^{239}_{94}\text{Pu} + ^{0}_{-1}\text{e}$
The entire process is $^{238}_{92}\text{U} + ^{1}_{0}\text{n} \rightarrow ^{239}_{92}\text{U}, \,^{239}_{92}\text{U} \rightarrow ^{239}_{93}\text{Np} + ^{0}_{-1}\text{e}, \,^{239}_{93}\text{Np} \rightarrow ^{239}_{94}\text{Pu} + ^{0}_{-1}\text{e}$.

Check: $238 + 1 = 239, 92 + 0 = 92$, and no other particle is necessary to balance the equation.
$239 = 239 + 0, 92 = 93 - 1$, and neptunium is atomic number 93. $239 = 239 + 0, 93 = 94 - 1$, and plutonium is atomic number 94.

19.63   **Given:** Rf-257 synthesized by bombarding Cf-249 with C-12   **Find:** balanced equation
Conceptual Plan: Write the species given on the appropriate side of the equation. → Equalize the sum of the mass numbers and the sum of the atomic numbers on both sides of the equation by writing the stoichiometric coefficient in front of the desired species. → Use the product of this reaction to write the next reaction until the process is complete.
Solution: $^{249}_{98}\text{Cf} + ^{12}_{6}\text{C} \rightarrow ^{257}_{?}\text{Rf} + ?^{?}_{?}?$ becomes $^{249}_{98}\text{Cf} + ^{12}_{6}\text{C} \rightarrow ^{257}_{104}\text{Rf} + 4^{1}_{0}\text{n}$.

Check: $249 + 12 = 257 - 4, 98 + 6 = 104 + 0$, and rutherfordium is atomic number 104, and four neutrons are needed to balance the equation.

Energetics of Nuclear Reactions, Mass Defect, and Nuclear Binding Energy

19.65 **Given:** 1.0 g of matter converted to energy   **Find:** energy
Conceptual Plan: g $\rightarrow$ kg $\rightarrow$ E

$$\frac{1 \text{ kg}}{1000 \text{ g}} \quad E = mc^2$$

Solution: $1.0 \text{ g} \times \dfrac{1 \text{ kg}}{1000 \text{ g}} = 0.0010 \text{ kg}$ then $E = mc^2 = (0.0010 \text{ kg})\left(2.9979 \times 10^8 \dfrac{\text{m}}{\text{s}}\right)^2 = 9.0 \times 10^{13} \text{ J}$

Check: The units (J) are correct. The magnitude of the answer makes physical sense because we are converting a large quantity of amus to energy.

19.67 **Given:** (a) O-16 = 15.9949145 amu, (b) Ni-58 = 57.935346 amu, and (c) Xe-129 = 128.904780 amu
Find: mass defect and nuclear binding energy per nucleon
Conceptual Plan: $_{Z}^{A}X$, isotope mass $\rightarrow$ mass defect $\rightarrow$ nuclear binding energy per nucleon

$$\text{mass defect} = Z(\text{mass } _{1}^{1}\text{H}) + (A - Z)(\text{mass } _{0}^{1}\text{n}) - \text{mass of isotope} \qquad \frac{931.5 \text{ MeV}}{(1 \text{ amu})(A \text{ nucleons})}$$

Solution: mass defect $= Z(\text{mass } _{1}^{1}\text{H}) + (A - Z)(\text{mass } _{0}^{1}\text{n}) - \text{mass of isotope}$

(a)   O-16 mass defect $= 8(1.00783 \text{ amu}) + (16 - 8)(1.00866 \text{ amu}) - 15.9949145 \text{ amu} =$

$0.1370055 \text{ amu} = 0.13701 \text{ amu}$ and $0.1370055 \text{ amu} \times \dfrac{931.5 \text{ MeV}}{(1 \text{ amu})(16 \text{ nucleons})} = 7.976 \dfrac{\text{MeV}}{\text{nucleon}}$

(b)   Ni-58 mass defect $= 28(1.00783 \text{ amu}) + (58 - 28)(1.00866 \text{ amu}) - 57.935346 \text{ amu} =$

$0.543694 \text{ amu} = 0.54369 \text{ amu}$ and $0.543694 \text{ amu} \times \dfrac{931.5 \text{ MeV}}{(1 \text{ amu})(58 \text{ nucleons})} = 8.732 \dfrac{\text{MeV}}{\text{nucleon}}$

(c)   Xe-129 mass defect $= 54(1.00783 \text{ amu}) + (129 - 54)(1.00866 \text{ amu}) - 128.904780 \text{ amu} =$

1.16754 amu and $1.16754 \text{ amu} \times \dfrac{931.5 \text{ MeV}}{(1 \text{ amu})(129 \text{ nucleons})} = 8.431 \dfrac{\text{MeV}}{\text{nucleon}}$

Check: The units (amu and MeV/nucleon) are correct. The mass defect increases with an increasing number of nucleons, but the MeV/nucleon does not change by as much (on a relative basis).

19.69 **Given:** $_{92}^{235}\text{U} + _{0}^{1}\text{n} \rightarrow _{54}^{144}\text{Xe} + _{38}^{90}\text{Sr} + 2_{0}^{1}\text{n}$, U-235 = 235.043922 amu, Xe-144 = 143.9385 amu, and Sr-90 = 89.907738 amu   **Find:** energy per g of U-235
Conceptual Plan: mass of products and reactants $\rightarrow$ mass defect $\rightarrow$ mass defect/g of U-235 then

$$\text{mass defect} = \Sigma \text{mass of reactants} - \Sigma \text{mass of products} \qquad \frac{\text{mass defect}}{235.043922 \text{ g U-235}}$$

g $\rightarrow$ kg $\rightarrow$ E

$$\frac{1 \text{ kg}}{1000 \text{ g}} \quad E = mc^2$$

Solution: mass defect $= \Sigma \text{mass of reactants} - \Sigma \text{mass of products}$; notice that we can cancel a neutron from each side to get $_{92}^{235}\text{U} \rightarrow _{54}^{144}\text{Xe} + _{38}^{90}\text{Sr} + _{0}^{1}\text{n}$ and

mass defect $= 235.043922 \text{ g} - (143.9385 \text{ g} + 89.907738 \text{ g} + 1.00866 \text{ g}) = 0.189024 \text{ g}$

then $\dfrac{0.189024 \text{ g}}{235.043922 \text{ g U-235}} \times \dfrac{1 \text{ kg}}{1000 \text{ g}} = 8.04207 \times 10^{-7} \dfrac{\text{kg}}{\text{g U-235}}$ then

$E = mc^2 = \left(8.04207 \times 10^{-7} \dfrac{\text{kg}}{\text{g U-235}}\right)\left(2.9979 \times 10^8 \dfrac{\text{m}}{\text{s}}\right)^2 = 7.228 \times 10^{10} \dfrac{\text{J}}{\text{g U-235}}$

Check: The units (J) are correct. A large amount of energy is expected per gram of fuel in a nuclear reactor.

19.71 **Given:** $2_{1}^{2}\text{H} \rightarrow _{2}^{3}\text{He} + _{0}^{1}\text{n}$, H-2 = 2.014102 amu, and He-3 = 3.016029 amu   **Find:** energy per g reactant
Conceptual Plan: mass of products and reactants $\rightarrow$ mass defect $\rightarrow$ mass defect/g of H-2 then

$$\text{mass defect} = \Sigma \text{mass of reactants} - \Sigma \text{mass of products} \qquad \frac{\text{mass defect}}{2(2.014102 \text{ g H-2})}$$

$$g \rightarrow kg \rightarrow E$$

$$\frac{1\,kg}{1000\,g} \quad E = mc^2$$

Solution: mass defect $= \sum$mass of reactants $- \sum$mass of products and mass defect $= 2(2.014102\,g) - (3.016029\,g + 1.00866\,g) = 0.003515\,g$

then $\dfrac{0.003515\,g}{2(2.014102\,g\,\text{H-2})} \times \dfrac{1\,kg}{1000\,g} = 8.72597 \times 10^{-7}\dfrac{kg}{g\,\text{H-2}}$ then

$$E = mc^2 = \left(8.72597 \times 10^{-7}\dfrac{kg}{g\,\text{H-2}}\right)\left(2.9979 \times 10^{8}\dfrac{m}{s}\right)^2 = 7.84 \times 10^{10}\dfrac{J}{g\,\text{H-2}}$$

Check: The units (J) are correct. A large amount of energy is expected per gram of fuel in a fusion reaction.

Effects and Applications of Radioactivity

19.73 **Given:** 75 kg human exposed to 32.8 rad and falling from chair **Find:** energy absorbed in each case
Conceptual Plan: rad, kg $\rightarrow$ J and assume $d = 0.50$ m chair height then mass, $d \rightarrow$ J

$$1\,rad = \dfrac{0.01\,J}{1\,kg\,\text{body tissue}} \qquad\qquad E = F \cdot d = mgd$$

Solution: $32.8\,rad = 32.8\dfrac{0.01\,J}{1\,kg\,\text{body tissue}} \times 75\,kg = 25\,J$ and

$$E = F \cdot d = mgd = 75\,kg \times 9.8\dfrac{m}{s^2} \times 0.50\,m = 370\,kg\dfrac{m^2}{s^2} = 370\,J$$

Check: The units (J and J) are correct. Allowable radiation exposures are low because the radiation is very ionizing and thus damaging to tissue. Falling may have more energy, but it is not ionizing.

19.75 **Given:** $t_{1/2}$ for F-18 $= 1.83$ h, 65% of F-18 makes it to the hospital, traveling at 60.0 miles/hour
Find: distance between hospital and cyclotron
Conceptual Plan: $t_{1/2} \rightarrow k$ then $m_{\text{F-18}\,0}, m_{\text{F-18}\,t}, k \rightarrow t$ then h $\rightarrow$ mi

$$t_{1/2} = \dfrac{0.693}{k} \qquad\qquad \ln\dfrac{m_{\text{F-18}t}}{m_{\text{F-18}0}} = -kt \qquad \dfrac{60.0\,mi}{1\,h}$$

Solution: $t_{1/2} = \dfrac{0.693}{k}$ Rearrange to solve for k. $k = \dfrac{0.693}{t_{1/2}} = \dfrac{0.693}{1.83\,h} = 0.378689\,h^{-1}$ Because

$\ln\dfrac{m_{\text{F-18}t}}{m_{\text{F-18}0}} = -kt$, rearrange to solve for t.

$$t = -\dfrac{1}{k}\ln\dfrac{m_{\text{F-18}\,t}}{m_{\text{F-18}\,0}} = -\dfrac{1}{0.378689\,h^{-1}}\ln\dfrac{0.65\,m_{\text{F-18}\,0}}{m_{\text{F-18}\,0}} = 1.1376\,h \text{ then}$$

$$1.1376\,h \times \dfrac{60.0\,mi}{1\,h} = 68\,mi$$

Check: The units (mi) are correct. The time is less than one half-life, so the distance is less than 1.83 times the speed of travel.

Cumulative Problems

19.77 **Given:** incomplete reactions **Find:** balanced reaction and energy (in J/mol reactant)
Conceptual Plan: Equalize the sum of the mass numbers and the sum of the atomic numbers on both sides of the equation by writing the appropriate mass number and atomic number for the unknown species. $\rightarrow$ Using the periodic table and the list of particles, deduce the identity of the unknown species from the atomic number and write its symbol. Then mass of products & reactants $\rightarrow$ mass defect in g $\rightarrow$ mass defect in kg $\rightarrow$ E

$$\text{mass defect} = \sum\text{mass of reactants} - \sum\text{mass of products} \qquad \dfrac{1\,kg}{1000\,g} \qquad E = mc^2$$

Solution:

(a) $^{?}_{?}? + ^{9}_{4}\text{Be} \rightarrow ^{6}_{3}\text{Li} + ^{4}_{2}\text{He}$ becomes $^{1}_{?}? + ^{9}_{4}\text{Be} \rightarrow ^{6}_{3}\text{Li} + ^{4}_{2}\text{He}$ then $^{1}_{1}\text{H} + ^{9}_{4}\text{Be} \rightarrow ^{6}_{3}\text{Li} + ^{4}_{2}\text{He}$

mass defect $= \Sigma$mass of reactants $- \Sigma$mass of products and

mass defect $= (1.00783 \text{ g} + 9.012182 \text{ g}) - (6.015122 \text{ g} + 4.002603 \text{ g}) = 0.002287 \text{ g}$

then $\dfrac{0.002287 \text{ g}}{2 \text{ mol reactants}} \times \dfrac{1 \text{ kg}}{1000 \text{ g}} = 1.1435 \times 10^{-6} \dfrac{\text{kg}}{\text{mol reactants}}$ then

$E = mc^2 = \left(1.1435 \times 10^{-6} \dfrac{\text{kg}}{\text{mol reactants}}\right)\left(2.9979 \times 10^8 \dfrac{\text{m}}{\text{s}}\right)^2 = 1.03 \times 10^{11} \dfrac{\text{J}}{\text{mol reactants}}$

Check: $1 + 9 = 6 + 4, 1 + 4 = 3 + 2$. The units (J) are correct.

(b) $^{209}_{83}\text{Bi} + ^{64}_{28}\text{Ni} \rightarrow ^{272}_{111}\text{Rg} + ^{?}_{?}?$ becomes $^{209}_{83}\text{Bi} + ^{64}_{28}\text{Ni} \rightarrow ^{272}_{111}\text{Rg} + ^{1}_{0}?$ then $^{209}_{83}\text{Bi} + ^{64}_{28}\text{Ni} \rightarrow ^{272}_{111}\text{Rg} + ^{1}_{0}\text{n}$

mass defect $= (208.980384 \text{ g} + 63.927969 \text{ g}) - (272.1535 \text{ g} + 1.00866 \text{ g}) = -0.253807 \text{ g}$

(Note: Because this is negative, energy must be put in.)

then $\dfrac{0.253807 \text{ g}}{2 \text{ mol reactants}} \times \dfrac{1 \text{ kg}}{1000 \text{ g}} = 1.269035 \times 10^{-4} \dfrac{\text{kg}}{\text{mol reactants}}$ then

$E = mc^2 = \left(1.269035 \times 10^{-4} \dfrac{\text{kg}}{\text{mol reactants}}\right)\left(2.9979 \times 10^8 \dfrac{\text{m}}{\text{s}}\right)^2 = 1.141 \times 10^{13} \dfrac{\text{J}}{\text{mol reactants}}$

Check: $209 + 64 = 272 + 1, 83 + 28 = 111 + 0$. The units (J) are correct.

(c) $^{179}_{74}\text{W} + ^{?}_{?}? \rightarrow ^{179}_{73}\text{Ta}$ becomes $^{179}_{74}\text{W} + ^{0}_{-1}? \rightarrow ^{179}_{73}\text{Ta}$ then $^{179}_{74}\text{W} + ^{0}_{-1}\text{e} \rightarrow ^{179}_{73}\text{Ta}$

mass defect $= (178.94707 \text{ g} + 0.00055 \text{ g}) - 178.94593 \text{ g} = 0.00169 \text{ g}$ then

$\dfrac{0.00169 \text{ g}}{2 \text{ mol reactants}} \times \dfrac{1 \text{ kg}}{1000 \text{ g}} = 8.45 \times 10^{-7} \dfrac{\text{kg}}{\text{mol reactants}}$ then

$E = mc^2 = \left(8.45 \times 10^{-7} \dfrac{\text{kg}}{\text{mol reactants}}\right)\left(2.9979 \times 10^8 \dfrac{\text{m}}{\text{s}}\right)^2 = 7.59 \times 10^{10} \dfrac{\text{J}}{\text{mol reactants}}$

Check: $179 + 0 = 179, 74 - 1 = 73$. The units (J) are correct.

19.79 **Given:** (a) RU-114, (b) Ra-216, (c) Zn-58, and (d) Ne-31 **Find:** Write a nuclear equation for the most likely decay.
Conceptual Plan: Referring to the Valley of Stability graph in Figure 19.5, decide on the most likely decay mode depending on N/Z (too large = beta decay, too low = positron emission). → Write the symbol for the parent nuclide on the left side of the equation and the symbol for a particle on the right side. → Equalize the sum of the mass numbers and the sum of the atomic numbers on both sides of the equation by writing the appropriate mass number and atomic number for the unknown daughter nuclide. → Using the periodic table, deduce the identity of the unknown daughter nuclide from the atomic number and write its symbol.
Solution:

(a) RU-114 ($N/Z = 1.6$) will undergo beta decay $^{114}_{44}\text{Ru} \rightarrow ^{?}_{?}? + ^{0}_{-1}\text{e}$ then $^{114}_{44}\text{Ru} \rightarrow ^{114}_{45}? + ^{0}_{-1}\text{e}$ then $^{114}_{44}\text{Ru} \rightarrow ^{114}_{45}\text{Rh} + ^{0}_{-1}\text{e}$

(b) Ra-216 ($N/Z = 1.4$) will undergo positron emission $^{216}_{88}\text{Ra} \rightarrow ^{?}_{?}? + ^{0}_{+1}\text{e}$ then $^{216}_{88}\text{Ra} \rightarrow ^{216}_{87}? + ^{0}_{+1}\text{e}$ then $^{216}_{88}\text{Ra} \rightarrow ^{216}_{87}\text{Fr} + ^{0}_{+1}\text{e}$

(c) Zn-58 ($N/Z = 0.9$) will undergo positron emission $^{58}_{30}\text{Zn} \rightarrow ^{?}_{?}? + ^{0}_{+1}\text{e}$ then $^{58}_{30}\text{Zn} \rightarrow ^{58}_{29}? + ^{0}_{+1}\text{e}$ then $^{58}_{30}\text{Zn} \rightarrow ^{58}_{29}\text{Cu} + ^{0}_{+1}\text{e}$

(d) Ne-31 ($N/Z = 2$) will undergo beta decay $^{31}_{10}\text{Ne} \rightarrow ^{?}_{?}? + ^{0}_{-1}\text{e}$ then $^{31}_{10}\text{Ne} \rightarrow ^{31}_{11}? + ^{0}_{-1}\text{e}$ then $^{31}_{10}\text{Ne} \rightarrow ^{31}_{11}\text{Na} + ^{0}_{-1}\text{e}$

Check: (a) $114 = 114 + 0, 44 = 45 - 1$, and rhodium is atomic number 45. (b) $216 = 216 + 0, 88 = 87 + 1$, and francium is atomic number 87. (c) $58 = 58 + 0, 30 = 29 + 1$, and copper is atomic number 29.
(d) $31 = 31 + 0, 10 = 11 - 1$, and sodium is atomic number 11.

19.81 **Given:** Bi-210, $t_{1/2} = 5.0$ days, 1.2 g Bi-210, 209.984105 amu, 5.5% absorbed
Find: beta emissions in 13.5 days and dose (in Ci)
Conceptual Plan: $t_{1/2} \rightarrow k$ then $m_{\text{Bi-210 0}}, t, k \rightarrow m_{\text{Bi-210 }t}$ then

$$t_{1/2} = \dfrac{0.693}{k} \qquad \ln N_t = -kt + \ln N_0$$

$g_0, g_t \rightarrow$ **g decayed** $\rightarrow$ **mol decayed** $\rightarrow$ **beta decays then day** $\rightarrow$ **h** $\rightarrow$ **min** $\rightarrow$ **s then**

$$g_0 - g_t = \text{g decayed} \quad \frac{1 \text{ mol Bi-210}}{209.984105 \text{ g Bi-210}} \quad \frac{6.022 \times 10^{23} \text{ beta decays}}{1 \text{ mol Bi-210}} \qquad \frac{24 \text{ h}}{1 \text{ day}} \quad \frac{60 \text{ min}}{1 \text{ h}} \quad \frac{60 \text{ s}}{1 \text{ min}}$$

beta decays, s $\rightarrow$ **beta decays/s** $\rightarrow$ **Ci available** $\rightarrow$ **Ci absorbed**

$$\text{take ratio} \qquad \frac{1 \text{ Ci}}{\dfrac{3.7 \times 10^{10} \text{ decays}}{s}} \qquad \frac{5.5 \text{ Ci absorbed}}{100 \text{ Ci emitted}}$$

Solution: $t_{1/2} = \dfrac{0.693}{k}$ Rearrange to solve for k. $k = \dfrac{0.693}{t_{1/2}} = \dfrac{0.693}{5.0 \text{ days}} = 0.1\underline{3}86 \text{ day}^{-1}$ Because

$\ln m_{\text{Bi-210}\,t} = -kt + \ln m_{\text{Bi-210}\,0} = -(0.1\underline{3}86 \text{ day}^{-1})(13.5 \text{ day}) + \ln(1.2 \text{ g}) = -1.6\underline{8}88 \rightarrow$

$m_{\text{Bi-210}\,t} = e^{-1.6888} = 0.1\underline{8}474 \text{ g}$ then $g_0 - g_t = \text{g decayed} = 1.2 \text{ g} - 0.1\underline{8}474 \text{ g} = 1.0\underline{1}53 \text{ g Bi-210}$

then $1.0\underline{1}53 \text{ g Bi-210} \times \dfrac{1 \text{ mol Bi-210}}{209.984105 \text{ g Bi-210}} \times \dfrac{6.022 \times 10^{23} \text{ beta decays}}{1 \text{ mol Bi-210}} = 2.9\underline{1}17 \times 10^{21} \text{ beta decays} =$

2.9×10^{21} beta decays then $13.5 \text{ day} \times \dfrac{24 \text{ h}}{1 \text{ day}} \times \dfrac{60 \text{ min}}{1 \text{ h}} \times \dfrac{60 \text{ s}}{1 \text{ min}} = 1.1\underline{6}64 \times 10^6 \text{ s}$ then

$$\dfrac{2.9117 \times 10^{21} \text{ beta decays}}{1.1\underline{6}64 \times 10^6 \text{ s}} \times \dfrac{1 \text{ Ci}}{\dfrac{3.7 \times 10^{10} \text{ decays}}{s}} = 6.\underline{7}468 \times 10^4 \text{ Ci emitted} \times \dfrac{5.5 \text{ Ci absorbed}}{100 \text{ Ci emitted}} = 3700 \text{ Ci}$$

Check: The units (decays and Ci) are correct. The amount that decays is large because the time is over three half-lives and we have a relatively large amount of the isotope. Because the decay is large, the dosage is large.

19.83 **Given:** Ra-226 (226.05402 amu) decays to Rn-224, $t_{1/2} = 1.6 \times 10^3$ yr, 25.0 g Ra-226, $T = 25.0\,°C$, $P = 1.0$ atm
Find: V of Rn-224 gas produced in 5.0 day
Conceptual Plan: **day** $\rightarrow$ **yr then** $t_{1/2} \rightarrow k$ **then** $m_{\text{Ra-226}\,0}\,t, k \rightarrow m_{\text{Ra-226}\,t}$

$$\frac{1 \text{ yr}}{365.24 \text{ day}} \qquad t_{1/2} = \frac{0.693}{k} \qquad \ln N_t = -kt + \ln N_0$$

then $g_0, g_t \rightarrow$ **g decayed** $\rightarrow$ **mol decayed** $\rightarrow$ **mol Rn-224 formed then** $°C \rightarrow K$ **then** $P, n, T \rightarrow V$

$$g_0 - g_t = \text{g decayed} \quad \frac{1 \text{ mol Ra-226}}{226.05402 \text{ g Ra-226}} \frac{1 \text{ mol Rn-224}}{1 \text{ mol Ra-226}} \qquad K = °C + 273.15 \qquad PV = nRT$$

Solution: $5.0 \text{ day} \times \dfrac{1 \text{ yr}}{365.24 \text{ day}} = 0.01\underline{3}690 \text{ yr}$ then $t_{1/2} = \dfrac{0.693}{k}$ Rearrange to solve for k.

$k = \dfrac{0.693}{t_{1/2}} = \dfrac{0.693}{1.6 \times 10^3 \text{ yr}} = 4.\underline{3}3125 \times 10^{-4} \text{ yr}^{-1}$ Because

$\ln m_{\text{Ra-226}\,t} = -kt + \ln m_{\text{Ra-226}\,0} = -(4.\underline{3}3125 \times 10^{-4} \text{ yr}^{-1})(0.01\underline{3}690 \text{ yr}) + \ln(25.0 \text{ g}) = 3.2\underline{1}887 \rightarrow$

$m_{\text{Ra-226}\,t} = e^{3.21887} = 24.\underline{9}99854 \text{ g}$ then

$g_0 - g_t = \text{g decayed} = 25.0 \text{ g} - 24.\underline{9}99854 \text{ g} = 0.\underline{0}00146 \text{ g Ra-226}$ then

$0.\underline{0}00146 \text{ g Ra-226} \times \dfrac{1 \text{ mol Ra-226}}{226.05402 \text{ g Ra-226}} \times \dfrac{1 \text{ mol Rn-224}}{1 \text{ mol Ra-226}} = 6.\underline{4}586 \times 10^{-7} \text{ mol Rn-224}$ then

$T = 25.0\,°C + 273.15 = 298.2 \text{ K}$ then $PV = nRT$. Rearrange to solve for V.

$V = \dfrac{nRT}{P} = \dfrac{6.\underline{4}586 \times 10^{-7} \text{ mol} \times 0.08206 \dfrac{\text{L} \cdot \text{atm}}{\text{mol} \cdot \text{K}} \times 298.2 \text{ K}}{1.0 \text{ atm}} = 1.\underline{5}804 \times 10^{-5} \text{ L} = 1.6 \times 10^{-5} \text{ L}$

Two significant figures are reported as requested in the problem.

Check: The units (L) are correct. The amount of gas is small because the time is so small compared to the half-life.

19.85 **Given:** $^0_1e + ^0_{-1}e \rightarrow 2^0_0\gamma$ **Find:** energy (in kJ/mol)
Conceptual Plan:
mass of products and reactants $\rightarrow$ **mass defect (g)** $\rightarrow$ **kg** $\rightarrow$ **kg/mol** $\rightarrow$ E **(J/mol)** $\rightarrow$ E **(kJ/mol)**

$$\text{mass defect} = \Sigma \text{mass of reactants} - \Sigma \text{mass of products} \qquad \frac{1 \text{ kg}}{1000 \text{ g}} \, 2 \text{ mol} \qquad E = mc^2 \qquad \frac{1 \text{ kJ}}{1000 \text{ J}}$$

Solution: mass defect $= \Sigma$mass of reactants $- \Sigma$mass of products $= (0.00055 \text{ g} + 0.00055 \text{ g}) - 0 \text{ g} =$

0.00110 g then $\dfrac{0.00110 \text{ g}}{2 \text{ mol}} \times \dfrac{1 \text{ kg}}{1000 \text{ g}} = 5.50 \times 10^{-7} \dfrac{\text{kg}}{\text{mol}}$ then

$E = mc^2 = \left(5.50 \times 10^{-7} \dfrac{\text{kg}}{\text{mol}}\right)\left(2.9979 \times 10^8 \dfrac{\text{m}}{\text{s}}\right)^2 = 4.9\underline{4}307 \times 10^{10} \dfrac{\text{J}}{\text{mol}} \times \dfrac{1 \text{ kJ}}{1000 \text{ J}} = 4.94 \times 10^7 \dfrac{\text{kJ}}{\text{mol}}$

Check: The units (kJ/mol) are correct. A large amount of energy is expected per mole of mass lost. The photon is in the gamma ray region of the electromagnetic spectrum.

19.87 **Given:** ^{3}He $= 3.016030$ amu **Find:** nuclear binding energy per atom
Conceptual Plan: $^A_Z X$, **isotope mass** $\rightarrow$ **mass defect** $\rightarrow$ **nuclear binding energy per nucleon**

$\text{mass defect} = Z(\text{mass } ^1_1\text{H}) + (A - Z)(\text{mass } ^1_0\text{n}) - \text{mass of isotope} \quad \dfrac{931.5 \text{ MeV}}{1 \text{ amu}}$

Solution: mass defect $= Z(\text{mass } ^1_1\text{H}) + (A - Z)(\text{mass } ^1_0\text{n}) - \text{mass of isotope}$
He-3 mass defect $= 2(1.00783 \text{ amu}) + (3 - 2)(1.00866 \text{ amu}) - 3.016030 \text{ amu} = 0.0082\underline{9} \text{ amu}$

and $0.0082\underline{9}$ amu $\times \dfrac{931.5 \text{ MeV}}{1 \text{ amu}} = 7.72 \text{ MeV}$

Check: The units (MeV) are correct. The number of nucleons is small, so the MeV is not that large.

19.89 **Given:** ^{247}Es and five neutrons made by bombarding ^{238}U **Find:** identity of bombarding particle
Conceptual Plan: Begin with the symbols for the nuclides given. $\rightarrow$ **Equalize the sum of the mass numbers and the sum of the atomic numbers on both sides of the equation by writing the appropriate mass number and atomic number for the unknown daughter nuclide.** $\rightarrow$ **Using the periodic table, deduce the identity of the unknown nuclide from the atomic number and write its symbol.**
Solution:
$^{238}_{92}\text{U} + ^?_?? \rightarrow ^{247}_{99}\text{Es} + 5^1_0\text{n}$ then $^{238}_{92}\text{U} + ^{14}_7? \rightarrow ^{247}_{99}\text{Es} + 5^1_0\text{n}$ then $^{238}_{92}\text{U} + ^{14}_7\text{N} \rightarrow ^{247}_{99}\text{Es} + 5^1_0\text{n}$

Check: $238 + 14 = 247 + 5(1), 92 + 7 = 99 + 5(0)$, and nitrogen is atomic number 7.

19.91 **Given:** $t_{1/2}$ for decay of ^{238}U $= 4.5 \times 10^9$ years, 1.6 g rock, 29 dis/s all radioactivity from U-238
Find: percent by mass ^{238}U in rock
Conceptual Plan: $t_{1/2} \rightarrow k$ and $\text{s} \rightarrow \text{min} \rightarrow \text{h} \rightarrow \text{day} \rightarrow \text{yr}$ then Rate, $k \rightarrow N \rightarrow \text{mol } ^{238}\text{U} \rightarrow \text{g } ^{238}\text{U}$

$t_{1/2} = \dfrac{0.693}{k} \quad \dfrac{1 \text{ min}}{60 \text{ s}} \quad \dfrac{1 \text{ h}}{60 \text{ min}} \quad \dfrac{1 \text{ day}}{24 \text{ h}} \quad \dfrac{1 \text{ yr}}{365.24 \text{ day}} \qquad \text{Rate} = kN \quad \dfrac{1 \text{ mol dis}}{6.022 \times 10^{23} \text{ dis}} \quad \dfrac{238 \text{ g } ^{238}\text{U}}{1 \text{ mol } ^{238}\text{U}}$

then g ^{238}U, g rock $\rightarrow$ percent by mass ^{238}U

$\text{percent by mass } ^{238}\text{U} = \dfrac{\text{g } ^{238}\text{U}}{\text{g rock}} \times 100\%$

Solution: $t_{1/2} = \dfrac{0.693}{k}$ Rearrange to solve for k. $k = \dfrac{0.693}{t_{1/2}} = \dfrac{0.693}{4.5 \times 10^9 \text{ yr}} = 1.5\underline{4} \times 10^{-10} \text{ yr}^{-1}$ and

$1 \text{ s} \times \dfrac{1 \text{ min}}{60 \text{ s}} \times \dfrac{1 \text{ h}}{60 \text{ min}} \times \dfrac{1 \text{ day}}{24 \text{ h}} \times \dfrac{1 \text{ yr}}{365.24 \text{ day}} = 3.1\underline{6}889554 \times 10^{-8} \text{ yr}$. Rate $= kN$. Rearrange to solve for N.

$N = \dfrac{\text{Rate}}{k} = \dfrac{\dfrac{29 \text{ dis}}{3.1\underline{6}889554 \times 10^{-8} \text{ yr}}}{1.5\underline{4} \times 10^{-10} \text{ yr}^{-1}} = 5.9\underline{4}25 \times 10^{18} \text{ dis}$ then

$5.9\underline{4}25 \times 10^{18} \text{ dis} \times \dfrac{1 \text{ mol dis}}{6.022 \times 10^{23} \text{ dis}} \times \dfrac{238 \text{ g } ^{238}\text{U}}{1 \text{ mol } ^{238}\text{U}} = 2.3\underline{4}86 \times 10^{-3} \text{g } ^{238}\text{U}$ then

$\text{percent by mass } ^{238}\text{U} = \dfrac{\text{g } ^{238}\text{U}}{\text{g rock}} \times 100\% = \dfrac{2.3\underline{4}86 \times 10^{-3} \text{ g } ^{238}\text{U}}{1.6 \text{ g rock}} \times 100\% = 0.15\%$

Check: The units (%) are correct. The mass percent is low because the dis/s is low.

19.93 **Given:** $V = 1.50$ L, $P = 745$ mmHg, $T = 25.0\,°C$, 3.55% Ra-220 by volume, $t_{1/2} = 55.6$ s
Find: number of alpha particles emitted in 5.00 min
Conceptual Plan: mmg $\rightarrow$ atm and $°C \rightarrow$ K then $P, V, T \rightarrow n_{Total} \rightarrow n_{Ra\text{-}220}$ and min $\rightarrow$ s

$$\frac{1\ \text{atm}}{760\ \text{mmHg}} \qquad K = °C + 273.15 \qquad PV = nRT \qquad \frac{3.55\ \text{mol Ra-220 particles}}{100\ \text{mol gas particles}} \qquad \frac{60\ \text{s}}{1\ \text{min}}$$

then $t_{1/2} \rightarrow k$ then $n_{Ra\text{-}220\,0}, t, k \rightarrow n_{Ra\text{-}220\,t} \rightarrow$ **number of particles remaining** $\rightarrow$ **particles emitted**

$$t_{1/2} = \frac{0.693}{k} \qquad \ln N_t = -kt + \ln N_0 \qquad \frac{6.022 \times 10^{23}\ \text{particles}}{1\ \text{mol}}$$

Solution: $745\ \text{mmHg} \times \dfrac{1\ \text{atm}}{760\ \text{mmHg}} = 0.9802632$ atm and $T = 25.0\,°C + 273.15 = 298.2$ K then

$PV = nRT$. Rearrange to solve for n.

$$n = \frac{PV}{RT} = \frac{0.9802632\ \text{atm} \times 1.50\ \text{L}}{0.08206\dfrac{\text{L} \cdot \text{atm}}{\text{mol} \cdot \text{K}} \times 298.2\ \text{K}} = 0.06008898\ \text{mol gas particles}$$

then $0.06008898\ \text{mol gas particles} \times \dfrac{3.55\ \text{mol Ra-220 particles}}{100\ \text{mol gas particles}} = 0.002133159\ \text{mol Ra-220 particles}$

$5.00\ \text{min} \times \dfrac{60\ \text{s}}{1\ \text{min}} = 300.\ \text{s}$ then $t_{1/2} = \dfrac{0.693}{k}$. Rearrange to solve for k.

$$k = \frac{0.693}{t_{1/2}} = \frac{0.693}{55.6\ \text{s}} = 0.01246403\ \text{s}^{-1}\ \text{Because}$$

$\ln m_{Ra\text{-}220\,t} = -kt + \ln m_{Ra\text{-}220\,0} = -(0.01246403\ \text{s}^{-1})(300.\ \text{s}) + \ln (0.002133159\ \text{mol}) = -9.889360 \rightarrow$

$m_{Ra\text{-}220\,t} = e^{-9.889360} = 5.071139 \times 10^{-5}$ mol alpha particles remaining

The number of alpha particles emitted would be the difference between this and the initial number of moles.
$0.002133158\ \text{mol} - 0.00005071139\ \text{mol} = 0.002082447\ \text{mol}$

$$0.002082447\ \text{mol} \times \frac{6.022 \times 10^{23}\ \text{particles}}{1\ \text{mol}} = 1.254050 \times 10^{21}\ \text{particles} = 1.25 \times 10^{21}\ \text{particles}$$

Check: The units (particles) are correct. The number of particles is far less than a mole because we have far less than a mole of gas.

19.95 **Given:** $^{0}_{+1}e + ^{0}_{-1}e \rightarrow 2^{0}_{0}\gamma$ **Find:** wavelength of gamma ray photons
Conceptual Plan:
mass of products and reactants $\rightarrow$ mass defect (g) $\rightarrow$ kg $\rightarrow$ kg/mol $\rightarrow$ E (J/mol) $\rightarrow$ E (kJ/mol)

$$\text{mass defect} = \Sigma\text{mass of reactants} - \Sigma\text{mass of products} \qquad \frac{1\ \text{kg}}{1000\ \text{g}} \quad 2\ \text{mol} \qquad E = mc^2 \qquad \frac{1\ \text{kJ}}{1000\ \text{J}}$$

This energy is for 2 moles of γ, so $E(\text{J/2 mol }\gamma) \rightarrow E(\text{J}/\gamma \text{ photon}) \rightarrow \lambda$.

$$\frac{1\ \text{mol }\gamma}{6.022 \times 10^{23}\ \gamma\ \text{photons}} \qquad E = \frac{hc}{\lambda}$$

Solution: mass defect $= \Sigma\text{mass of reactants} - \Sigma\text{mass of products} = (0.00055\ \text{g} + 0.00055\ \text{g}) - 0\ \text{g} = 0.00110\ \text{g}$

then $\dfrac{0.00110\ \text{g}}{2\ \text{mol}} \times \dfrac{1\ \text{kg}}{1000\ \text{g}} = 5.50 \times 10^{-7}\dfrac{\text{kg}}{\text{mol}}$ then

$$E = mc^2 = \left(5.50 \times 10^{-7}\frac{\text{kg}}{\text{mol}}\right)\left(2.9979 \times 10^8\frac{\text{m}}{\text{s}}\right)^2 = 4.94307 \times 10^{10}\frac{\text{J}}{\text{mol}} \times \frac{1\ \text{kJ}}{1000\ \text{J}} = 4.94 \times 10^7\frac{\text{kJ}}{\text{mol}}$$

$$E = 4.94307 \times 10^{10}\frac{\text{J}}{\text{mol}\ \gamma} \times \frac{1\ \text{mol}\ \gamma}{6.022 \times 10^{23}\ \gamma\ \text{photons}} = 8.20835 \times 10^{-14}\frac{\text{J}}{\gamma\ \text{photons}}\ \text{then}\ E = \frac{hc}{\lambda}\ \text{Rearrange to solve}$$

for λ. $\lambda = \dfrac{hc}{E} = \dfrac{(6.626 \times 10^{-34} \, \text{J} \cdot \text{s})\left(2.9979 \times 10^8 \dfrac{\text{m}}{\text{s}}\right)}{8.20835 \times 10^{-14} \dfrac{\text{J}}{\gamma \text{ photons}}} = 2.42 \times 10^{-12} \, \text{m} = 2.42 \, \text{pm}$

Check: The units (m or pm) are correct. A large amount of energy is expected per mole of mass lost. The photon is in the gamma ray region of the electromagnetic spectrum.

19.97 **Given:** $^2_1\text{H} + {}^2_1\text{H} \rightarrow {}^3_2\text{He} + {}^1_0\text{n}$ releases 3.3 MeV; $^2_1\text{H} + {}^2_1\text{H} \rightarrow {}^3_1\text{H} + {}^1_1\text{p}$ releases 4.0 MeV

Find: the energy change for $^3_2\text{He} + {}^1_0\text{n} \rightarrow {}^3_1\text{H} + {}^1_1\text{p}$ and why this can happen at a much lower temperature

Conceptual Plan: Use Hess's law to calculate the energy change and give the two reactions.

Solution:

$^3_2\text{He} + {}^1_0\text{n} \rightarrow {}^2_1\text{H} + {}^2_1\text{H} \qquad \Delta E = 3.3 \text{ MeV}$

$\dfrac{{}^2_1\text{H} + {}^2_1\text{H} \rightarrow {}^3_1\text{H} + {}^1_1\text{p} \qquad \Delta E = -4.0 \text{ MeV}}{}$

$^3_2\text{He} + {}^1_0\text{n} \rightarrow {}^3_1\text{H} + {}^1_1\text{p} \qquad \Delta E = -0.7 \text{ MeV}$

The energy change is much less, and there is no coulombic barrier for collision with a neutron; so the process can occur at lower temperatures.

Check: The units (MeV) are correct. Because one reaction releases energy and one requires energy, the resulting energy change is much smaller in magnitude.

Challenge Problems

19.99 (a) **Given:** 72,500 kg Al(s) and $10 \, \text{Al}(s) + 6 \, \text{NH}_4\text{ClO}_4(s) \rightarrow 4 \, \text{Al}_2\text{O}_3(s) + 2 \, \text{AlCl}_3(s) + 12 \, \text{H}_2\text{O}(g) + 3 \, \text{N}_2(g)$ and 608,000 kg O$_2(g)$ that reacts with hydrogen to form gaseous water

Find: energy generated $(\Delta H^\circ_{\text{rxn}})$

Conceptual Plan: Write a balanced reaction for $\text{O}_2(g)$ then

$\Delta H^\circ_{\text{rxn}} = \sum n_\text{p} \Delta H^\circ_\text{f}(\text{products}) - \sum n_\text{r} \Delta H^\circ_\text{f}(\text{reactants})$ then

kg $\rightarrow$ g $\rightarrow$ mol $\rightarrow$ energy then add the results from the two reactions.

$\dfrac{1000 \text{ g}}{1 \text{ kg}} \qquad \mathcal{M} \qquad \Delta H^\circ_{\text{rxn}}$

Solution:

Reactant/Product	ΔH°_f (kJ/mol from Appendix IIB)
Al(s)	0.0
NH$_4$ClO$_4(s)$	-295
Al$_2$O$_3(s)$	-1675.7
AlCl$_3(s)$	-704.2
H$_2$O(g)	-241.8
N$_2(g)$	0.0

Be sure to pull data for the correct formula and phase.

$\Delta H^\circ_{\text{rxn}} = \sum n_\text{p} \Delta H^\circ_\text{f}(\text{products}) - \sum n_\text{r} \Delta H^\circ_\text{f}(\text{reactants})$

$= [4(\Delta H^\circ_\text{f}(\text{Al}_2\text{O}_3(s))) + 2(\Delta H^\circ_\text{f}(\text{AlCl}_3(s))) + 12(\Delta H^\circ_\text{f}(\text{H}_2\text{O}(g))) + 3(\Delta H^\circ_\text{f}(\text{N}_2(g)))] +$
$\quad -[10(\Delta H^\circ_\text{f}(\text{Al}(s))) + 6(\Delta H^\circ_\text{f}(\text{NH}_4\text{ClO}_4(s)))]$

$= [4(-1675.7 \text{ kJ}) + 2(-704.2 \text{ kJ}) + 12(-241.8 \text{ kJ}) + 3(0.0 \text{ kJ})] - [10(0.0 \text{ kJ}) + 6(-295 \text{ kJ})]$

$= [-11012.8 \text{ kJ}] - [-1770. \text{ kJ}]$

$= -9242.8 \text{ kJ}$

then $72,500 \text{ kg Al} \times \dfrac{1000 \text{ g Al}}{1 \text{ kg Al}} \times \dfrac{1 \text{ mol Al}}{26.98 \text{ g Al}} \times \dfrac{9242.8 \text{ kJ}}{10 \text{ mol Al}} = 2.483703 \times 10^9 \text{ kJ}$

balanced reaction: $\text{H}_2(g) + \frac{1}{2}\text{O}_2(g) \rightarrow \text{H}_2\text{O}(g) \; \Delta H^\circ_{\text{rxn}} = \Delta H^\circ_\text{f}(\text{H}_2\text{O}(g)) = -241.8 \text{ KJ/mol}$ then

$$608{,}000 \text{ kg } O_2 \times \frac{1000 \text{ g } O_2}{1 \text{ kg } O_2} \times \frac{1 \text{ mol } O_2}{32.00 \text{ g } O_2} \times \frac{241.8 \text{ kJ}}{0.5 \text{ mol } O_2} = 9.1\underline{8}84 \times 10^9 \text{ kJ So the total is}$$

$2.48\underline{3}703 \times 10^9 \text{ kJ} + 9.1\underline{8}84 \times 10^9 \text{ kJ} = 1.16\underline{7}2103 \times 10^{10} \text{ kJ} = 1.167 \times 10^{10} \text{ kJ}.$

Check: The units (kJ) are correct. The answer is very large because the reactions are very exothermic and the weight of reactants is so large.

(b)   **Given:** $^1_1\text{H} + ^{-1}_{-1}\text{p} + ^0_{+1}\text{e} \rightarrow ^0_0\gamma$ **Find:** mass of antimatter to give same energy as in part (a)

Conceptual Plan: Because the reaction is an annihilation reaction, no matter will be left; so the mass of antimatter is the same as the mass of the hydrogen. So kJ $\rightarrow$ J $\rightarrow$ kg $\rightarrow$ g

$$\frac{1000 \text{ J}}{1 \text{ kJ}} \qquad E = mc^2 \qquad \frac{1000 \text{ g}}{1 \text{ kg}}$$

Solution: $1.16\underline{7}2103 \times 10^{10} \text{ kJ} \times \dfrac{1000 \text{ J}}{1 \text{ kJ}} = 1.16\underline{7}2103 \times 10^{13} \text{ J}$. Because $E = mc^2$, rearrange to solve for m.

$$m = \frac{E}{c^2} = \frac{1.16\underline{7}2103 \times 10^{13} \text{ kg} \frac{m^2}{s^2}}{\left(2.9979 \times 10^8 \frac{m}{s}\right)^2} = 1.299 \times 10^{-4} \text{ kg} \times \frac{1000 \text{ g}}{1 \text{ kg}} = 0.1299 \text{ g total matter, } 0.0649 \text{ g each of}$$

matter and antimatter

Check: The units (g) are correct. A small mass is expected because nuclear reactions generate a large amount of energy.

19.101   **Given:** $^{235}_{92}\text{U} \rightarrow ^{206}_{82}\text{Pb}$ and $^{232}_{90}\text{Th} \rightarrow ^{206}_{82}\text{Pb}$ **Find:** decay series

Conceptual Plan: Write the species given on the appropriate side of the equation. $\rightarrow$ Equalize the sum of the mass numbers and the sum of the atomic numbers on both sides of the equation by writing the stoichiometric coefficient in front of the desired species.

Solution: $^{235}_{92}\text{U} \rightarrow ^?_{82}\text{Pb} + ^?_2\text{He} + ?^0_{-1}\text{e}$ becomes $^{235}_{92}\text{U} \rightarrow ^{207}_{82}\text{Pb} + 7^4_2\text{He} + 4^0_{-1}\text{e}.$

$^{232}_{90}\text{Th} \rightarrow ^?_{82}\text{Pb} + ^?_2\text{He} + ?^0_{-1}\text{e}$ becomes $^{232}_{90}\text{Th} \rightarrow ^{208}_{82}\text{Pb} + 6^4_2\text{He} + 4^0_{-1}\text{e}.$

U-235 forms Pb-207 in 7 α-decays and 4 β-decays, and Th-232 forms Pb-208 in 6 α-decays and 4 β-decays.

Check: $235 = 207 + 7(4) + 4(0)$, and $92 = 82 + 7(2) + 4(-1)$. $232 = 208 + 6(4) + 4(0)$, and $90 = 82 + 6(2) + 4(-1)$. The mass of the Pb can be determined because alpha particles are large and need to be included as integer values. To make the masses balance requires more alpha particles than can be supported by the number of protons in the total equation. To account for this, an appropriate number of beta decays are added.

19.103   **Given:** $\text{H}^{38}_{17}\text{Cl}(g) \rightarrow ^{38}_{18}\text{Ar}(g) + ^0_{-1}\text{e} + \frac{1}{2}\text{H}_2(g)$, 0.40 mol $\text{H}^{38}_{17}\text{Cl}(g)$; $V = 6.24$ L, $P = 1650$ mmHg, $t_{1/2} = 80.0$ min

Find: T

Conceptual Plan: Because $t = 2\,t_{1/2}$, then $N_{\text{Cl-38 }t} = \frac{1}{4}N_{\text{Cl-38 }0} \rightarrow n_{\text{gas }t}$ then mmHg $\rightarrow$ atm and

$$n_{\text{gas }t} = 3/2\, n_{\text{HCl }0} - 1/2\, n_{\text{HCl }t} \qquad \frac{1 \text{ atm}}{760 \text{ mmHg}}$$

$P, V, n_{\text{gas }t} \rightarrow T$,

$$PV = nRT$$

Solution: Because $t = 2\,t_{1/2}$, then $N_{\text{Cl-38 }t} = \frac{1}{4}N_{\text{Cl-38 }0} = \frac{1}{4}(0.40 \text{ mol}) = 0.10 \text{ mol}$. As the HCl disintegrates, it produces argon gas, beta particles and hydrogen gas, with a ratio of three particles produced (2 Ar, 1 H_2) for every two HCl molecules that decay. There were initially 0.40 mole of undisintegrated gas; now 0.10 mole HCl remains. So the total number of gas particles now in the container is

$n_{\text{gas }t} = 3/2\, n_{\text{HCl }0} - 1/2\, n_{\text{HCl }t} = 3/2(0.40 \text{ mol}) - 1/2\,(0.10 \text{ mol}) = 0.55 \text{ mol gas then } PV = nRT$. Rearrange to

solve for T. $T = \dfrac{PV}{nR} = \dfrac{1650 \text{ mmHg} \times \dfrac{1 \text{ atm}}{760 \text{ mmHg}} \times 6.24 \text{ L}}{0.55 \text{ mol} \times 0.08206 \dfrac{\text{L} \cdot \text{atm}}{\text{mol} \cdot \text{K}}} = 300.17 \text{ K} = 3.0 \times 10^2 \text{ K}$

Check: The units (K) are correct. The temperature is reasonable considering the volume of a gas at STP and the fact that most of the initial 0.40 mol has decomposed.

Conceptual Problems

19.105   **Given:** $^{21}_{9}\text{F} \rightarrow ^{?}_{?}? + ^{0}_{-1}\text{e}$ **Find:** missing nucleus

Conceptual Plan: Write the species given on the appropriate side of the equation. $\rightarrow$ Equalize the sum of the mass numbers and the sum of the atomic numbers on both sides of the equation by writing the stoichiometric coefficient in front of the desired species.

Solution: $^{21}_{9}\text{F} \rightarrow ^{?}_{?}? + ^{0}_{-1}\text{e}$ becomes $^{21}_{9}\text{F} \rightarrow ^{21}_{10}\text{Ne} + ^{0}_{-1}\text{e}$.

Check: $21 = 21 + 0$, and $9 = 10 - 1$. Neon is atomic number 10, and no other species are needed to balance the equation.

19.107   Nuclide A is more dangerous because the half-life is shorter (18.5 days); so it decays faster.

19.109   Iodine is used by the thyroid gland to make hormones. Normally, we ingest iodine in foods, especially iodized salt. The thyroid gland cannot tell the difference between stable and radioactive iodine and will absorb both. KI tablets work by blocking radioactive iodine from entering the thyroid. When a person takes KI, the stable iodine in the tablet gets absorbed by the thyroid. Because KI contains so much stable iodine, the thyroid gland becomes "full" and cannot absorb any more iodine—either stable or radioactive—for the next 24 hours.

20 Organic Chemistry

Review Questions

20.1 Most common smells are caused by organic molecules, molecules containing carbon combined with several other elements including hydrogen, nitrogen, oxygen, and sulfur.

20.3 Carbon is unique in the vast number of compounds it can form. Life needs diversity to exist, and carbon has the ability to form more compounds than any other element.

20.5 Silicon can form chains with itself. However, silicon's affinity for oxygen—the Si—O bond is 142 kJ/mol stronger than the Si—Si bond—coupled with the prevalence of oxygen in our atmosphere means that silicon–silicon chains are readily oxidized to form silicates (the silicon–oxygen compounds that compose a significant proportion of minerals). By contrast, the C—C bond (347 kJ/mol) and the C—O bond (359 kJ/mole) are nearly the same strength, allowing carbon chains to exist relatively peacefully in an oxygen rich environment. In other words, silicon's affinity for oxygen robs it of the rich diversity that catenation provides to carbon.

20.7 Hydrocarbons—compounds that contain only carbon and hydrogen—are the simplest organic compounds. However, because of the uniqueness of carbon, many different kinds of hydrocarbons exist. Hydrocarbons are commonly used as fuels. Candle wax, oil, gasoline, liquid propane (LP) gas, and natural gas are all composed of hydrocarbons. Hydrocarbons are also the starting materials in the synthesis of many different consumer products, including fabrics, soaps, dyes, cosmetics, drugs, plastic, and rubber.

20.9 A structural formula shows not only the numbers of each kind of atoms, but also how the atoms are bonded together. The condensed structural formula groups the hydrogen atoms together with the carbon atom to which they are bonded. Condensed structural formulas may show some of the bonds or none at all. The carbon skeleton formula shows the carbon–carbon bonds only as lines. Each end or bend of a line represents a carbon atom bonded to as many hydrogen atoms as necessary to form a total of four bonds. Space-filling or ball-and-stick models are three-dimensional representations that show how atoms are bonded together. The space-filling models show the relative size of the atoms that are bonded together.

20.11 Optical isomers are two molecules that are nonsuperimposable mirror images of each other. Optical isomers contain a carbon atom with four different substituent groups. Most properties of optical isomers are the same. The differences appear when they interact with polarized light and when they are placed in environments that can interact with one isomer and not the other.

20.13 Alkanes are often called saturated hydrocarbons because they are saturated (loaded to capacity) with hydrogen. Unsaturated hydrocarbons contain multiple bonds and therefore contain fewer hydrogen atoms.

20.15 Double bonds are composed of a sigma bond and a pi bond. Sigma bonds allow for free rotation about the bond. The pi bond restricts the rotational motion about the bond, resulting in isomers that are referred to as geometric isomers (same connectivity but different geometric positions). The two isomers are designated as *cis* (meaning "same side") and *trans* (meaning "opposite sides"). Cis–trans isomerism is common in alkenes and results in different boiling points and melting points and in different abilities to interact with rigid molecules.

20.17   The most common types of reactions of alkanes are as follows:

- Hydrocarbon combustion in the presence of oxygen to form carbon dioxide and water. An example is $C_8H_{18}(l) + 25/2\ O_2(g) \rightarrow 8\ CO_2(g) + 9\ H_2O(g)$.
- Halogen substitution in the presence of a halogen gas to form a halogenated alkane and a hydrogen halide. An example is $CH_4(g) + Cl_2(g) \rightarrow CH_3Cl(g) + HCl(g)$.

20.19   The most common types of reactions of alkenes are as follows:

- Hydrocarbon combustion in the presence of oxygen to form carbon dioxide and water. An example is $C_2H_4(g) + 3\ O_2(g) \rightarrow 2\ CO_2(g) + 2\ H_2O(g)$.
- Hydrogenation in the presence of hydrogen to form an alkane. An example is $C_2H_4(g) + H_2(g) \rightarrow C_2H_6(g)$.
- Halogen addition in the presence of a halogen gas or hydrogen halide to form a halogenated alkane. An example is $CH_2CHCH_2CH_3(g) + HCl(g) \rightarrow CH_3CClHCH_2CH_3(g)$.

20.21   The structure of benzene, C_6H_6, is a six-member ring where three pi bonds are delocalized on all six of the C—C bonds that form the ring. Benzene rings are represented as one or both of the Kekulé structures,

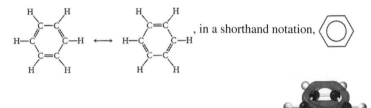

, in a shorthand notation,

or as a ball-and-stick diagram with molecular orbitals,

20.23   A functional group is a characteristic atom or group of atoms that is inserted into a hydrocarbon. Examples of functional groups are alcohols (—OH), halogens (—X, where X = F, Cl, Br, and I), and carboxylic acids (—COOH).

20.25   In organic chemistry, we think of oxidation and reduction from the point of view of the carbon atoms in the organic molecule. Thus, oxidation is the gaining of oxygen or the losing of hydrogen by a carbon atom. Reduction, then, is the loss of oxygen or the gaining of hydrogen by a carbon atom.

20.27   Aldehydes and ketones have the general structural formulas of RCHO and RCOR′, respectively, where R and R′ represent hydrocarbon groups. Both aldehydes and ketones contain a carbonyl group, an oxygen double-bonded to a carbon atom $\left(\begin{smallmatrix} O \\ \parallel \\ -C- \end{smallmatrix}\right)$. Ketones have an R group attached to both sides of the carbonyl, while aldehydes have one R group and a hydrogen atom. An exception is formaldehyde, which is an aldehyde with two H atoms attached to the carbonyl group. Acetaldehyde is $CH_3-\overset{\overset{\displaystyle O}{\parallel}}{C}-H$ and propanone is $CH_3-\overset{\overset{\displaystyle O}{\parallel}}{C}-CH_3$.

20.29   Carboxylic acids and esters have the following general structural formulas of RCOOH and RCOOR′, respectively, where R and R′ represent hydrocarbon groups. Both carboxylic acids and esters contain a carbonyl group, an oxygen double-bonded to a carbon atom $\left(\begin{smallmatrix} O \\ \parallel \\ -C- \end{smallmatrix}\right)$. In carboxylic acids, the molecule ends in $\overset{\overset{\displaystyle O}{\parallel}}{-C}-O-H$ and esters have $\overset{\overset{\displaystyle O}{\parallel}}{-C}-O-$ inserted into a hydrocarbon chain. Acetic acid is $CH_3-\overset{\overset{\displaystyle O}{\parallel}}{C}-O-H$ and ethyl ethanoate is

$CH_3-\overset{\overset{\displaystyle O}{\parallel}}{C}-O-CH_2-CH_3$.

20.31   Ethers have the general structural formula of ROR', where R and R' represent hydrocarbon groups. The two hydrocarbon groups are linked through an oxygen atom. Diethyl ether is $CH_3-CH_2-O-CH_2-CH_3$, and ethyl propyl ether is $CH_3-CH_2-O-CH_2-CH_2-CH_3$.

20.33   Polymers are long chainlike molecules composed of repeating units called monomers. Copolymers are long chainlike molecules composed of two different monomers.

Problems by Topic

Hydrocarbons

20.35   (a) C_5H_{12} is an alkane because it follows the general formula C_nH_{2n+2}, where $n = 5$.
         (b) C_3H_6 is an alkene because it follows the general formula C_nH_{2n}, where $n = 3$.
         (c) C_7H_{12} is an alkyne because it follows the general formula C_nH_{2n-2}, where $n = 7$.
         (d) $C_{11}H_{22}$ is an alkene because it follows the general formula C_nH_{2n}, where $n = 11$.

20.37 $CH_3-CH_2-CH_2-CH_2-CH_2-CH_2-CH_3$, $CH_3-\overset{\displaystyle |}{\underset{\displaystyle CH_3}{CH}}-CH_2-CH_2-CH_2-CH_3$,

$CH_3-CH_2-\overset{\displaystyle |}{\underset{\displaystyle CH_3}{CH}}-CH_2-CH_2-CH_3$, $CH_3-\overset{\displaystyle CH_3}{\overset{\displaystyle |}{\underset{\displaystyle |}{\underset{\displaystyle CH_3}{C}}}}-CH_2-CH_2-CH_3$,

$CH_3-CH_2-\overset{\displaystyle |}{\underset{\displaystyle CH_3}{CH}}-\overset{\displaystyle |}{\underset{\displaystyle CH_3}{CH}}-CH_3$, $CH_3-CH_2-\overset{\displaystyle CH_3}{\overset{\displaystyle |}{\underset{\displaystyle |}{\underset{\displaystyle CH_3}{C}}}}-CH_2-CH_3$,

$CH_3-\overset{\displaystyle |}{\underset{\displaystyle CH_3}{CH}}-CH_2-\overset{\displaystyle |}{\underset{\displaystyle CH_3}{CH}}-CH_3$, $CH_3-CH_2-\overset{\displaystyle |}{\underset{\displaystyle CH_2-CH_3}{CH}}-CH_2-CH_3$, and $CH_3-\overset{\displaystyle CH_3}{\overset{\displaystyle |}{\underset{\displaystyle |}{\underset{\displaystyle CH_3}{C}}}}-\overset{\displaystyle |}{\underset{\displaystyle CH_3}{CH}}-CH_3$

20.39   (a)   No, this molecule will not because all four of the substituents are Cl atoms.
         (b)   Yes, this molecule will because the third carbon has four different substituent groups.
         (c)   Yes, this molecule will because the second carbon has four different substituent groups.
         (d)   No, each carbon has, at most, three different substituent groups.

20.41   (a)   They are enantiomers because they are non-superimposable mirror images of each other.
         (b)   They are the same because you can get the second molecule by rotating the first molecule counterclockwise about the C—H bond.
         (c)   They are enantiomers because they are non-superimposable mirror images of each other.

Alkanes

20.43   **Given:** alkane structures   **Find:** name
         Conceptual Plan: Count the number of carbon atoms in the longest continuous carbon chain to determine the base name of the compound. Find the prefix corresponding to this number of atoms in Table 20.5 and add the ending *-ane* to form the base name. → Consider every branch from the base chain to be a substituent. Name each substituent according to Table 20.6. → Beginning with the end closest to the branching, number the base chain and assign a number to each substituent. (If two substituents occur at equal distances from each end, go to the next substituent to determine from which end to start numbering.) → Write the name of the compound in the following format: (subst. #)-(subst. name)(base name). → If there are two or more substituents, give each

one a number and list them alphabetically with hyphens between words and numbers. → **If a compound has two or more identical substituents, designate the number of identical substituents with the prefix *di*- (2), *tri*- (3), or *tetra*- (4) before the substituent's name. Separate the numbers with a comma, indicating the positions of the substituents relative to each other. The prefixes are not taken into account when alphabetizing.**
Solution:

(a) CH_3—CH_2—CH_2—CH_2—CH_3 has five carbons as the longest continuous chain. The prefix for 5 is penta-. There are no substituent groups on any of the carbons, so the name is pentane. Because this is a straight-chain molecule, it can be more specifically named *n*-pentane.

(b) $\boxed{CH_3\text{—}CH_2\text{—}CH\text{—}CH_3}$ has four carbons as the longest continuous chain. The prefix for 4 is but-, and
  $\quad\quad\quad\quad\quad\quad\quad\quad\overset{|}{CH_3}$

the base name is butane. The only substituent group is a methyl group. $\boxed{C^4H_3\text{—}C^3H_2\text{—}C^2H\text{—}C^1H_3}$ If
  $\quad\quad\quad\quad\quad\quad\quad\quad\quad\quad\quad\quad\quad\quad\quad\quad\quad\quad\quad\quad\quad\quad\quad\quad\overset{|}{\boxed{CH_3}}$

we start numbering the chain at the end closest to the methyl group, the methyl substituent is assigned the number 2. The name of the compound is 2-methylbutane.

(c)
  $\quad\quad\quad\quad\quad\quad\quad\quad\quad\quad CH_3$
  $\quad\quad\quad\quad\quad\quad\quad\quad\quad\quad |$
  $\quad\quad\quad\quad\quad CH_3\quad\quad CH\text{—}CH_3$
  $\quad\quad\quad\quad\quad |\quad\quad\quad\quad\quad |$
  $\boxed{CH_3\text{—}CH\text{—}CH_2\text{—}CH\text{—}CH_2\text{—}CH_2\text{—}CH_3}$ has seven carbons as the longest continuous chain. The prefix for 7 is hept-, and the base name is heptane. The substituent groups are methyl and isopropyl groups.

  $\quad\quad\quad\quad\quad\quad\quad\quad\quad\quad\boxed{CH_3}$
  $\quad\quad\quad\quad\quad\quad\quad\quad\quad\quad|$
  $\quad\quad\quad\quad\quad\boxed{CH_3}\quad\quad\boxed{CH\text{—}CH_3}$
  $\quad\quad\quad\quad\quad|\quad\quad\quad\quad\quad|$
  $\boxed{C^1H_3\text{—}C^2H\text{—}C^3H_2\text{—}C^4H\text{—}C^5H_2\text{—}C^6H_2\text{—}C^7H_3}$ If we start numbering the chain at the end closest to the methyl group, the methyl substituent is assigned the number 2 and the isopropyl group is assigned the number 4. Because *i* comes before *m*, the name of the compound is 4-isopropyl-2-methylheptane.

(d) $\boxed{CH_3\text{—}CH\text{—}CH_2\text{—}CH\text{—}CH_2\text{—}CH_3}$ has six carbons as the longest continuous chain. The prefix
  $\quad\quad\quad\quad\quad\quad\quad |\quad\quad\quad\quad\quad\quad |$
  $\quad\quad\quad\quad\quad\quad CH_3\quad\quad\quad CH_2\text{—}CH_3$

for 6 is hex-, and the base name is hexane. The only substituent groups are methyl and ethyl groups.

$\boxed{C^1H_3\text{—}C^2H\text{—}C^3H_2\text{—}C^4H\text{—}C^5H_2\text{—}C^6H_3}$ If we start numbering the chain at the end closest to the
  $\quad\quad\quad\quad\quad |\quad\quad\quad\quad\quad\quad\quad |$
  $\quad\quad\quad\quad\boxed{CH_3}\quad\quad\quad\boxed{CH_2\text{—}CH_3}$

methyl group, the methyl substituent is assigned the number 2 and the ethyl group is assigned the number 4. Because *e* comes before *m*, the name of the compound is 4-ethyl-2-methylhexane.

20.45 **Given:** alkane names **Find:** structure
  Conceptual Plan: Find the number of carbon atoms corresponding to the prefix of the base name in Table 20.5. →
  Draw the base chain and number the carbons from left to right. → **Using Table 20.6 and the prefix *di*- (2), *tri*- (3), or *tetra*- (4) before the substituent's name, determine each substituent.** → **Add the substituent to the proper carbon position in the chain.** → **Add hydrogen atoms to the base chain so that each carbon has four bonds.**
  Solution:

(a) 3-ethylhexane. The base name hexane indicates that there are six carbon atoms in the base chain.
  C^1—C^2—C^3—C^4—C^5—C^6. 3-ethyl designates that a —CH_2CH_3 group is in the third position.
  C^1—C^2—C^3—C^4—C^5—C^6. Add hydrogens to the base chain to get the final molecule
  $\quad\quad\quad\quad\overset{|}{\boxed{CH_2\text{—}CH_3}}$

  CH_3—CH_2—CH—CH_2—CH_2—CH_3.
  $\quad\quad\quad\quad\quad\quad |$
  $\quad\quad\quad\quad CH_2\text{—}CH_3$

(b) 3-ethyl-3-methylpentane. The base name pentane indicates that there are five carbon atoms in the base chain. C^1—C^2—C^3—C^4—C^5. 3-ethyl designates that a —CH_2CH_3 group is in the third position, and

3-methyl designates that a $—CH_3$ group is in the third position. $C^1—C^2—\overset{\overset{\boxed{CH_3}}{|}}{C^3}—C^4—C^5$. Add hydrogens to the base chain to get the final molecule $CH_3—CH_2—\overset{\overset{CH_3}{|}}{\underset{\underset{CH_2—CH_3}{|}}{C}}—CH_2—CH_3$.

(c)   2,3-dimethylbutane. The base name butane indicates that there are four carbon atoms in the base chain. $C^1—C^2—C^3—C^4$. 2,3-dimethyl designates $—CH_3$ groups in the second and third positions. $C^1—C^2—C^3—C^4$ with $\boxed{CH_3}$ $\boxed{CH_3}$. Add hydrogens to the base chain to get the final molecule $CH_3—\underset{\underset{CH_3}{|}}{CH}—\underset{\underset{CH_3}{|}}{CH}—CH_3$.

(d)   4,7-diethyl-2,2-dimethylnonane. The base name nonane indicates that there are nine carbon atoms in the base chain. $C^1—C^2—C^3—C^4—C^5—C^6—C^7—C^8—C^9$. 4,7-diethyl designates $—CH_2CH_3$ groups in the fourth and seventh positions, and 2,2-dimethyl designates two $—CH_3$ groups in the second position. $C^1—\overset{\overset{\boxed{CH_3}}{|}}{\underset{\underset{\boxed{CH_3}}{|}}{C^2}}—C^3—\underset{\underset{\boxed{CH_2—CH_3}}{|}}{C^4}—C^5—C^6—\underset{\underset{\boxed{CH_2—CH_3}}{|}}{C^7}—C^8—C^9$. Add hydrogens to the base chain to get the final molecule $CH_3—\overset{\overset{CH_3}{|}}{\underset{\underset{CH_3}{|}}{C}}—CH_2—\underset{\underset{CH_2—CH_3}{|}}{CH}—CH_2—CH_2—\underset{\underset{CH_2—CH_3}{|}}{CH}—CH_2—CH_3$.

20.47   Hydrocarbon combustion in the presence of oxygen forms carbon dioxide and water. Balance the reaction.

(a) $CH_3CH_2CH_3(g) + 5 O_2(g) \rightarrow 3 CO_2(g) + 4 H_2O(g)$

(b) $CH_3CH_2CH{=}CH_2(g) + 6 O_2(g) \rightarrow 4 CO_2(g) + 4 H_2O(g)$

(c) $2 CH{\equiv}CH(g) + 5 O_2(g) \rightarrow 4 CO_2(g) + 2 H_2O(g)$

20.49   Halogen substitution reactions remove a hydrogen atom from the alkane, replace it with a halogen atom, and generate a hydrogen halide. Assume one substitution on the hydrocarbon.

(a) $CH_3CH_3 + Br_2 \rightarrow CH_3CH_2Br + HBr$. Only one carbon-containing product is possible because the $C—C$ bond freely rotates.

(b) $CH_3CH_2CH_3 + Cl_2 \rightarrow [CH_3CH_2CH_2Cl$ and $CH_3CHClCH_3] + HCl$. Two carbon-containing products are possible (on either the end carbon or the middle carbon) because the $C—C$ bond freely rotates and the end carbons are equivalent before reaction.

(c) $CH_2Cl_2 + Br_2 \rightarrow CHBrCl_2 + HBr$. Only one carbon-containing product is possible because halogen substitution reactions only remove hydrogen atoms.

(d) $CH_3—\underset{\underset{CH_3}{|}}{CH}—CH_3 + Cl_2 \longrightarrow \left[CH_3—\overset{\overset{H}{|}}{\underset{\underset{CH_3}{|}}{C}}—CH_2Cl \text{ and } CH_3—\overset{\overset{Cl}{|}}{\underset{\underset{CH_3}{|}}{C}}—CH_3 \right] + HCl$

Two carbon-containing products are possible (on either an end carbon or the middle carbon) because the $C—C$ bond freely rotates and the end carbons are all equivalent before reaction.

Alkenes and Alkynes

20.51 $CH_2\!=\!CH\!-\!CH_2\!-\!CH_2\!-\!CH_2\!-\!CH_3$, $CH_3\!-\!CH\!=\!CH\!-\!CH_2\!-\!CH_2\!-\!CH_3$, and $CH_3\!-\!CH_2\!-\!CH\!=\!CH\!-\!CH_2\!-\!CH_3$ are the only structural isomers. Remember that cis–trans isomerism generates geometric isomers, not structural isomers.

20.53   **Given:** alkene structures   **Find:** name
     Conceptual Plan: Count the number of carbon atoms in the longest continuous carbon chain that contains the multiple bond to determine the base name of the compound. Find the prefix corresponding to this number of atoms in Table 20.5 and add the ending *-ene* to form the base name. → Consider every branch from the base chain to be a substituent. Name each substituent according to Table 20.6. → Beginning with the end closest to the multiple bond, number the base chain and assign a number to each substituent. → Write the name of the compound in the following format: (subst. #)-(subst. name)(base name). → If there are two or more substituents, give each one a number and list them alphabetically with hyphens between words and numbers. → If a compound has two or more identical substituents, designate the number of identical substituents with the prefix *di*- (2), *tri*- (3), or *tetra*- (4) before the substituent's name. Separate the numbers with a comma, indicating the positions of the substituents relative to each other. The prefixes are not taken into account when alphabetizing.
     Solution:

(a) $\boxed{CH_2\!=\!CH\!-\!CH_2\!-\!CH_3}$ has four carbons as the longest continuous chain. The prefix for 4 is but-, and the base name is butene. There are no substituent groups. $\boxed{C^1H_2\!=\!C^2H\!-\!C^3H_2\!-\!C^4H_3}$ Start numbering on the left because it is closer to the double bond. Because the double bond is between positions 1 and 2, the name of the compound is 1-butene.

(b) 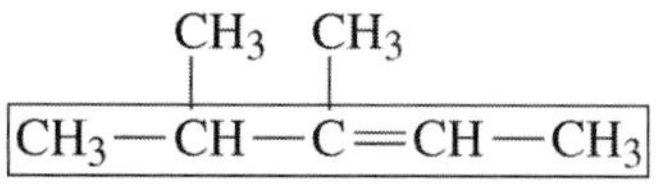
$$\begin{array}{cc} CH_3 & CH_3 \\ | & | \end{array}$$
$\boxed{CH_3\!-\!CH\!-\!C\!=\!CH\!-\!CH_3}$ has five carbons as the longest continuous chain. The prefix for 5 is pent-, and the base name is pentene. The substituent groups are two methyl groups. 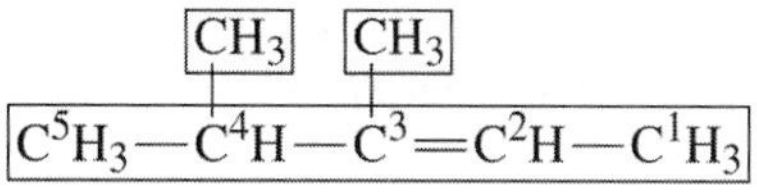
$$\boxed{CH_3} \quad \boxed{CH_3}$$
$$| \qquad\quad |$$
$\boxed{C^5H_3\!-\!C^4H\!-\!C^3\!=\!C^2H\!-\!C^1H_3}$ If we start numbering the chain at the end closest to the double bond, the double bond is between positions numbered 2 and 3 and the methyl substituents are assigned the numbers 3 and 4. The name of the compound is 3,4-dimethyl-2-pentene.

(c) $\boxed{CH_2\!=\!CH\!-\!CH\!-\!CH_2\!-\!CH_2\!-\!CH_3}$ has six carbons as the longest continuous chain. The
$$CH_3\!-\!CH$$
$$|$$
$$CH_3$$
prefix for 6 is hex-, and the base name is hexene. The only substituent group is an isopropyl group. 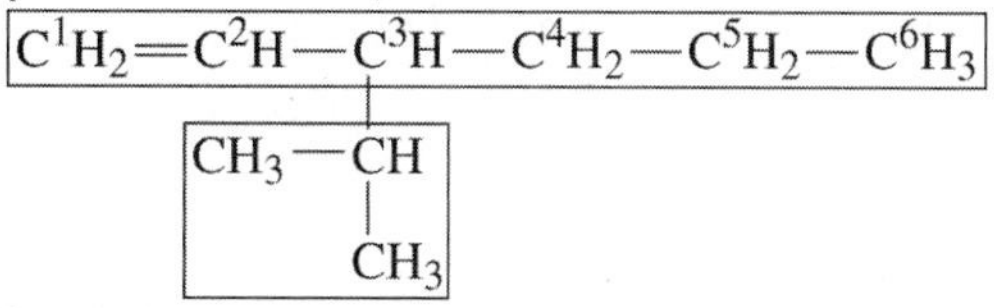
$\boxed{C^1H_2\!=\!C^2H\!-\!C^3H\!-\!C^4H_2\!-\!C^5H_2\!-\!C^6H_3}$ Start numbering on the left because it is closer to the double

bond. Because the double bond is between positions 1 and 2 and the isopropyl group is at position 3, the name of the compound is 3-isopropyl-1-hexene.

(d) 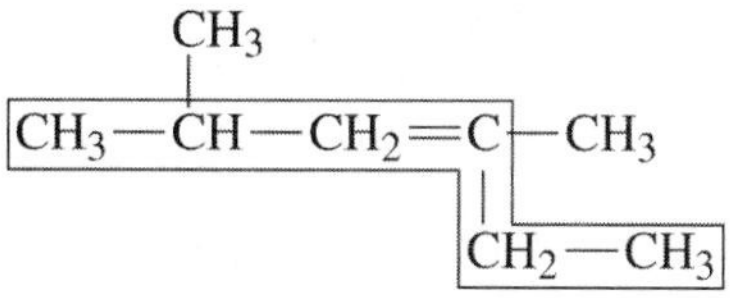
$$CH_3$$
$$|$$
$\boxed{CH_3\!-\!CH\!-\!CH_2\!=\!C}\!-\!CH_3$ has six carbons as the longest continuous chain. The prefix for 6 is
$$\boxed{|\!| \atop CH_2\!-\!CH_3}$$
hex-, and the base name is hexene. The only substituent groups are two methyl groups.
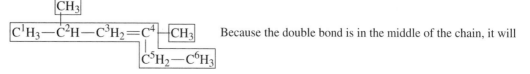
Because the double bond is in the middle of the chain, it will

be at position 3 in both numbering schemes. If we start numbering the chain at the end closest to the left methyl group (closest to an end), the methyl substituents are assigned the numbers 2 and 4. The name of the compound is 2,4-dimethyl-3-hexene.

20.55   **Given:** alkyne structures   **Find:** name

Conceptual Plan: Count the number of carbon atoms in the longest continuous carbon chain that contains the multiple bond to determine the base name of the compound. Find the prefix corresponding to this number of atoms in Table 20.5 and add the ending -*yne* to form the base name. → Consider every branch from the base chain to be a substituent. Name each substituent according to Table 20.6. → Beginning with the end closest to the multiple bond, number the base chain and assign a number to each substituent. → Write the name of the compound in the following format: (subst. #)-(subst. name)(base name). → If there are two or more substituents, give each one a number and list them alphabetically with hyphens between words and numbers. → If a compound has two or more identical substituents, designate the number of identical substituents with the prefix *di*- (2), *tri*- (3), or *tetra*- (4) before the substituent's name. Separate the numbers with a comma, indicating the positions of the substituents relative to each other. The prefixes are not taken into account when alphabetizing.

Solution:

(a) $\boxed{CH_3—C\equiv C—CH_3}$ has four carbons as the longest continuous chain. The prefix for 4 is but-, and the base name is butyne. There are no substituent groups. $\boxed{C^1H_3—C^2\equiv C^3—C^4H_3}$ Because the triple bond is in the middle of the molecule, it does not matter at which end the numbering is started. Because the triple bond is between positions 2 and 3, the name of the compound is 2-butyne.

(b) 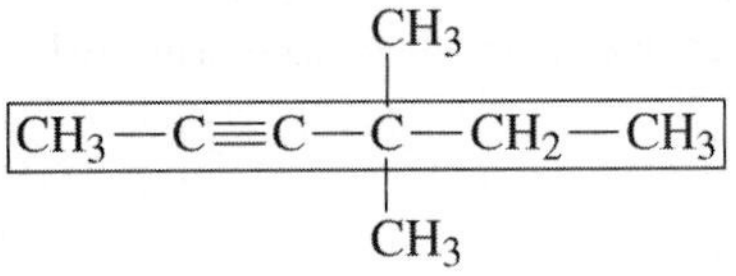
$$CH_3$$
$$|$$
$$\boxed{CH_3—C\equiv C—C—CH_2—CH_3}$$
$$|$$
$$CH_3$$
has six carbons as the longest continuous chain. The prefix for 6

is hex-, and the base name is hexyne. The only substituent groups are two methyl groups.

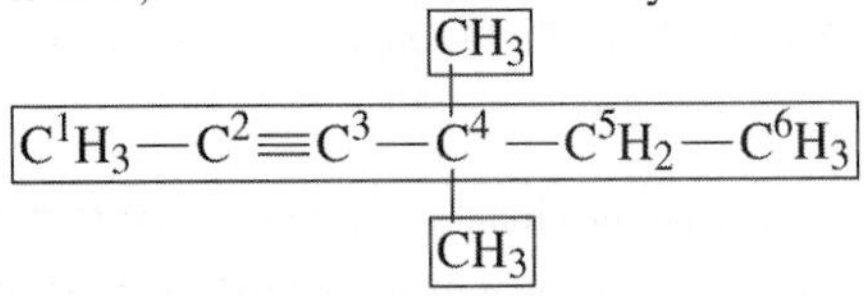

$$\boxed{CH_3}$$
$$|$$
$$\boxed{C^1H_3—C^2\equiv C^3—C^4—C^5H_2—C^6H_3}$$
$$|$$
$$\boxed{CH_3}$$
If we start numbering the chain at the end closest to the triple

bond, the triple bond is between positions 2 and 3 and the methyl substituents are assigned the numbers 4 and 4. The name of the compound is 4,4-dimethyl-2-hexyne.

(c) 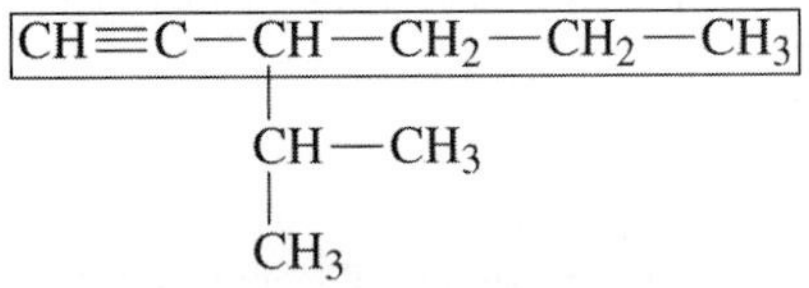 $\boxed{CH\equiv C—CH—CH_2—CH_2—CH_3}$ has six carbons as the longest continuous chain. The prefix for 6 is
$$|$$
$$CH—CH_3$$
$$|$$
$$CH_3$$

hex-, and the base name is hexyne. The only substituent group is an isopropyl group.

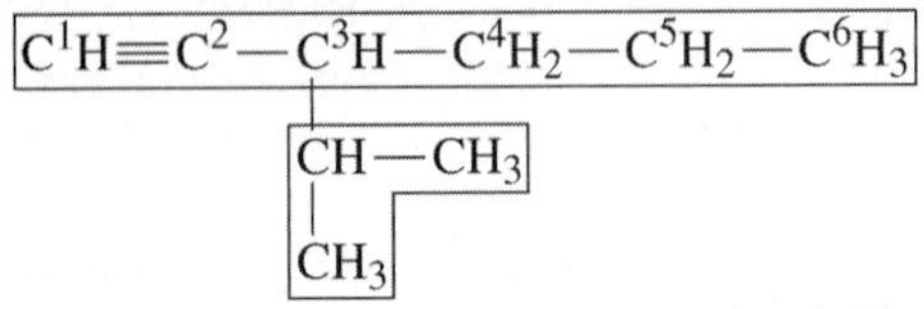

$$\boxed{C^1H\equiv C^2—C^3H—C^4H_2—C^5H_2—C^6H_3}$$ Start numbering at the end closest to the triple bond. The
$$\boxed{CH—CH_3}$$
$$|$$
$$\boxed{CH_3}$$

isopropyl substituent is assigned number 3. The name of the compound is 3-isopropyl-1-hexyne.

(d) 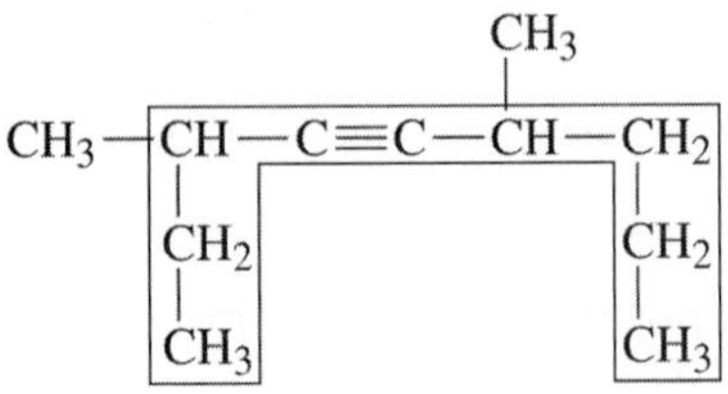
$$CH_3$$
$$|$$
$$CH_3—CH—C\equiv C—CH—CH_2$$
$$|\qquad\qquad\qquad|$$
$$CH_2\qquad\qquad CH_2$$
$$|\qquad\qquad\qquad|$$
$$CH_3\qquad\qquad CH_3$$
has nine carbons as the longest continuous chain. The prefix for 9 is non-,

and the base name is nonyne. The substituent groups are two methyl groups.

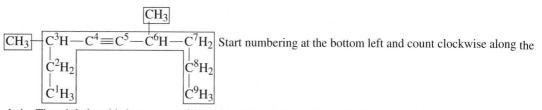

chain. The triple bond is between positions 4 and 5, and the methyl substituents are assigned the numbers 3 and 6. The name of the compound is 3,6-dimethyl-4-nonyne.

20.57   **Given:** hydrocarbon names   **Find:** structure
**Conceptual Plan: Find the number of carbon atoms corresponding to the prefix of the base name in Table 20.5. →
Draw the base chain and number the carbons from left to right. → Determine the multiple bond type from the
ending of the base name (*-ene* = double bond; *-yne* = triple bond) and place it in the appropriate position in
the chain. → Using Table 20.6 and the prefix *di-* (2), *tri-* (3), or *tetra-* (4) before the substituent's name, determine
each substituent. → Add the substituent to the proper carbon position in the chain. → Add hydrogen atoms to
the base chain so that each carbon has four bonds.**
Solution:

(a)   4-octyne. The base name octyne indicates that there are eight carbon atoms in the base chain.
$C^1 - C^2 - C^3 - C^4 - C^5 - C^6 - C^7 - C^8$. The *-yne* ending denotes a triple bond, and the 4 prefix places it
between the fourth and fifth positions. $C^1 - C^2 - C^3 - C^4 \equiv C^5 - C^6 - C^7 - C^8$. Add hydrogens to the base
chain to get the final molecule $CH_3 - CH_2 - CH_2 - C \equiv C - CH_2 - CH_2 - CH_3$.

(b)   3-nonene. The base name nonene indicates that there are nine carbon atoms in the base chain.
$C^1 - C^2 - C^3 - C^4 - C^5 - C^6 - C^7 - C^8 - C^9$. The -ene ending denotes a double bond, and the 3 prefix places
it between the third and fourth positions. $C^1 - C^2 - C^3 = C^4 - C^5 - C^6 - C^7 - C^8 - C^9$. Add hydrogens to the
base chain to get the final molecule $CH_3 - CH_2 - CH = CH - CH_2 - CH_2 - CH_2 - CH_2 - CH_3$.

(c)   3,3-dimethyl-1-pentyne. The base name pentyne indicates that there are five carbon atoms in the base chain.
$C^1 - C^2 - C^3 - C^4 - C^5$. The *-yne* ending denotes a triple bond, and the 1 prefix places it between the first
and second positions. $C^1 \equiv C^2 - C^3 - C^4 - C^5$. 3,3-dimethyl indicates that there are two $-CH_3$ groups

in the third position. $C^1 \equiv C^2 - \overset{\overset{\boxed{CH_3}}{|}}{C^3} - C^4 - C^5$. Add hydrogens to the base chain to get the final molecule

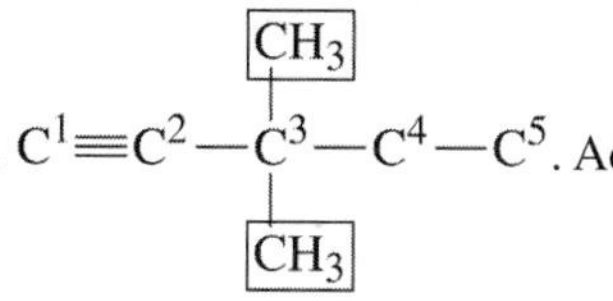

$CH \equiv C - \overset{\overset{CH_3}{|}}{\underset{\underset{CH_3}{|}}{C}} - CH_2 - CH_3$.

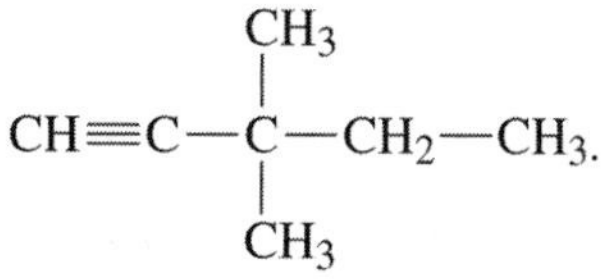

(d)   5-ethyl-3,6-dimethyl-2-heptene. The base name heptene indicates that there are seven carbon atoms in the base
chain. $C^1 - C^2 - C^3 - C^4 - C^5 - C^6 - C^7$. The -ene ending denotes a double bond, and the 2 prefix places it
between the second and third positions. $C^1 - C^2 = C^3 - C^4 - C^5 - C^6 - C^7$. 5-ethyl designates that there is a
$-CH_2CH_3$ group in the fifth position, and 3,6-dimethyl designates that there are $-CH_3$ groups in the third

and sixth positions. $C^1 - C^2 = \overset{\overset{}{}}{C^3} - C^4 - \overset{\overset{}{}}{C^5} - \overset{\overset{\boxed{CH_3}}{|}}{C^6} - C^7$. Add hydrogens to the base chain to get the final

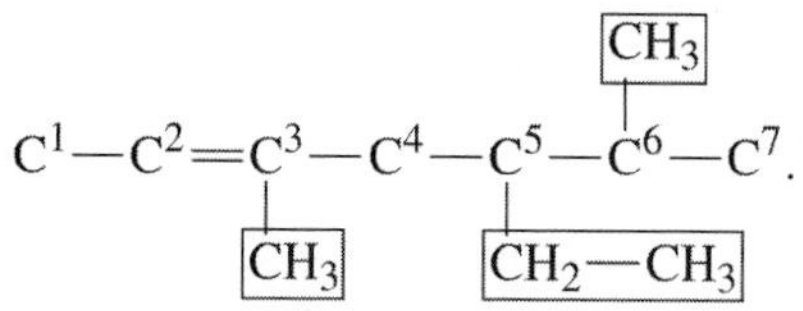

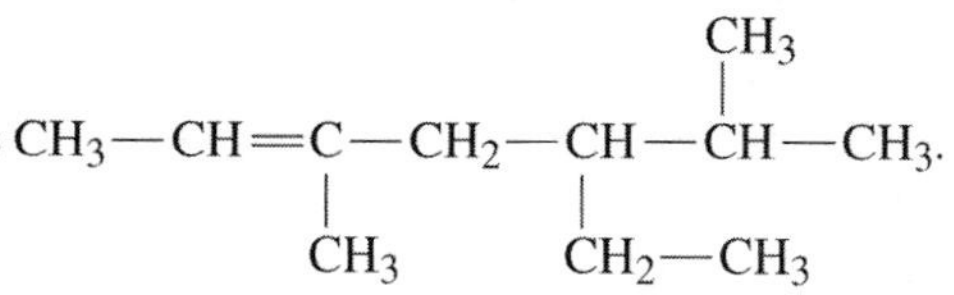

molecule $CH_3 - CH = \overset{\overset{}{}}{C} - CH_2 - \overset{\overset{}{}}{CH} - \overset{\overset{CH_3}{|}}{CH} - CH_3$.

20.59   Alkene addition reactions convert a double bond to a single bond and place the two halves of the other reactant on the two carbons that were in the double bond. Markovnikov's rule states that when a polar reagent is added to an unsymmetrical alkene, the positive end (the least electronegative part) of the reagent adds to the carbon atom that has the most hydrogen atoms. In most reactions of this type, the positive end of the reagent is hydrogen; therefore, the hydrogen atom goes to the carbon atom that already contains the most hydrogen atoms.

(a) $CH_3-CH=CH-CH_3 + Cl_2 \rightarrow$ $CH_3-\underset{\underset{Cl}{|}}{CH}-\underset{\underset{Cl}{|}}{CH}-CH_3$

(b) $CH_3-\underset{\underset{CH_3}{|}}{CH}-CH=CH-CH_3 + HBr \rightarrow$ $CH_3-\underset{\underset{CH_3}{|}}{CH}-CH_2-\underset{\underset{Br}{|}}{CH}-CH_3$ or $CH_3-\underset{\underset{CH_3}{|}}{CH}-\overset{\overset{Br}{|}}{CH}-CH_2-CH_3$

(c) $CH_3-CH_2-CH=CH-CH_3 + Br_2 \rightarrow$ $CH_3-CH_2-\underset{\underset{Br}{|}}{CH}-\underset{\underset{Br}{|}}{CH}-CH_3$

(d) $CH_3-\underset{\underset{CH_3}{|}}{CH}-CH=\overset{\overset{CH_3}{|}}{C}-CH_3 + HCl \rightarrow CH_3-\underset{\underset{CH_3}{|}}{CH}-CH_2-\underset{\underset{Cl}{|}}{\overset{\overset{CH_3}{|}}{C}}-CH_3$

Only one product is formed according to Markovnikov's rule.

20.61   Hydrogenation reactions convert a double bond to a single bond and place the hydrogen atoms on each of the two carbons that were in the double bond.

(a) $CH_2=CH-CH_3 + H_2 \rightarrow CH_3-CH_2-CH_3$

(b) $CH_3-\underset{\underset{CH_3}{|}}{CH}-CH=CH_2 + H_2 \longrightarrow CH_3-\underset{\underset{CH_3}{|}}{CH}-CH_2-CH_3$

(c) $CH_3-\underset{\underset{CH_3}{|}}{CH}-\underset{\underset{CH_3}{|}}{C}=CH_2 + H_2 \longrightarrow CH_3-\underset{\underset{CH_3}{|}}{CH}-\underset{\underset{CH_3}{|}}{CH}-CH_3$

Aromatic Hydrocarbons

20.63   **Given:** monosubstituted benzene structures   **Find:** name
Conceptual Plan: Consider the branch from the benzene ring to be a substituent. Name the substituent according to Table 20.6 or using the base name of a halogen with an *o* added at the end. → Write the name of the compound in the following format: (name of substituent)benzene.
Solution:

(a) $-CH_3$ or methyl is the substituent group, so the name is methylbenzene or toluene.

(b) $-Br$ or bromo is the substituent, so the name is bromobenzene.

(c) $-Cl$ or chloro is the substituent, so the name is chlorobenzene.

20.65   **Given:** hydrocarbon structures   **Find:** name
Conceptual Plan: Count the number of carbon atoms in the longest continuous carbon chain (if a multiple bond is present, make sure this chain contains the multiple bond) to determine the base name of the compound. Find the prefix corresponding to this number of atoms in Table 20.5 and add the ending *-ane* for an alkane, *-ene* for an alkene and *-yne* for an alkyne to form the base name. → Consider every branch from the base chain to be a substituent. Name each substituent according to Table 20.6. If a benzene ring is a substituent, use the phenyl group name. → Beginning with the end closest to the multiple bond, number the base chain and assign a number to each substituent. → Write the name of the compound in the following format: (subst. #)-(subst. name) (base name). → If there are two or more substituents, give each one a number and list them alphabetically with hyphens between words and numbers. → If a compound has two or more identical substituents, designate the number of identical substituents with the prefix *di-* (2), *tri-* (3), or *tetra-* (4) before the substituent's name. Sep-

arate the numbers with a comma, indicating the positions of the substituents relative to each other. The prefixes are not taken into account when alphabetizing.

Solution: Note: For simplicity, $\varnothing = \langle\hexagon\rangle$

(a)
$$CH_3{-}\overline{CH{-}CH_2{-}CH{-}CH_2{-}\overset{\varnothing}{CH}{-}CH_2{-}CH_3}$$
with $\overset{CH_3}{|}$ branch and $CH_2{-}CH_3$ branch, has nine carbons as the longest continuous chain.

The prefix for 9 is non- and the base name is nonane. The substituent groups are two methyl groups and a

phenyl group. $CH_3{-}\boxed{C^3H{-}C^4H_2{-}C^5H{-}C^6H_2{-}C^7H{-}C^8H_2{-}C^9H_3}$ with $\boxed{CH_3}$ and $\boxed{\varnothing}$ substituents and $C^2H_2{-}C^1H_3$ branch. Start numbering at the end

closest to the methyl group. The substituent groups are two methyl groups at positions 3 and 5 and a phenyl group at position 7. Because *m* comes before *p*, the name of the compound is 3,5-dimethyl-7-phenylnonane. If the numbering is in the opposite direction, the name will be 5,7-dimethyl-3-phenylnonane, but this is not the approved name because IUPAC rules indicate that the first group cited receives the lowest number.

(b) $\boxed{CH_3{-}\overset{\varnothing}{CH}{-}CH{=}CH{-}CH_2{-}CH_2{-}CH_2{-}CH_3}$ has eight carbons as the longest continuous chain. The prefix for 8 is oct-, and because there is a double bond, the base name is octene. The only substituent group

is a phenyl group. $\boxed{C^1H_3{-}\overset{\boxed{\varnothing}}{C^2H}{-}C^3H{=}C^4H{-}C^5H_2{-}C^6H_2{-}C^7H_2{-}C^8H_3}$ If we start numbering the chain at the end closest to the double bond, the double bond is between positions 3 and 4 and the phenyl substituent is assigned the number 2. The name of the compound is 2-phenyl-3-octene.

(c) $\boxed{CH_3{-}C{\equiv}C{-}\overset{\varnothing}{CH}{-}CH{-}CH{-}CH_2{-}CH_3}$ with CH_3 and CH_3 branches, has eight carbons as the longest continuous chain. The prefix for 8 is oct-, and because there is a triple bond, the base name is octyne. The substituent groups are two

methyl groups and a phenyl group. $\boxed{C^1H_3{-}C^2{\equiv}C^3{-}C^4H{-}C^5H{-}\overset{\boxed{\varnothing}}{C^6H}{-}C^7H_2{-}C^8H_3}$ with $\boxed{CH_3}$ and $\boxed{CH_3}$ substituents. Start

numbering at the end closest to the triple bond. The triple bond is between positions 2 and 3, the methyl groups are assigned numbers 4 and 5, and the phenyl group is assigned number 6. Because *m* is before *p*, the name of the compound is 4,5-dimethyl-6-phenyl-2-octyne.

20.67   **Given:** disubstituted benzene structures   **Find:** name
Conceptual Plan: Consider the branches from the benzene ring to be substituents. Name the substituent according to Table 20.6 or using the base name of a halogen with an *o* added at the end. → Number the benzene ring starting with the substituent that is first alphabetically and count in the direction that gets to the second substituent with the lower position number. → Write the name of the compound in the following format: (name of substituent)benzene. List them alphabetically with hyphens between words and numbers. → If a compound has two identical substituents, designate the number of identical substituents with the prefix *di-* (2) before the substituent's name. Separate the numbers with a comma, indicating the positions of the substituents relative to each other. → Alternative names are 1,2 = *ortho* or *o*; 1,3 = *meta* or *m*; and 1,4 = *para* or *p* replacing the numbers.
Solution:
(a)   Both of the substituent groups are —Br or bromo groups. Start by giving one bromo an assignment of 1 and count in either direction to give the second bromo group an assignment of 4. The name is 1, 4-dibromobenzene or *p*-dibromobenzene.

(b)   Both of the substituent groups are —CH_2CH_3 or ethyl groups. Start by giving the top ethyl an assignment of 1 and count in a clockwise direction to give the second ethyl group an assignment of 3. The name is 1,3-diethylbenzene or *m*-diethylbenzene.

(c)   One substituent group is —Cl or chloro, and the other is —F or fluoro. Start by giving the chloro group an assignment of 1 and count in a counterclockwise direction to give the fluoro group an assignment of 2. The name is 1-chloro-2-fluorobenzene or *o*-chlorofluorobenzene.

20.69   **Given:** substituted benzene names   **Find:** structures

           Conceptual Plan: Start with a benzene ring. Identify the structure of the substituent(s) according to Table 20.6 or using the base name of a halogen with an *o* added at the end. → If only one substituent is present, simply attach it to the benzene ring. → If a compound has two identical substituents, they are designated with the prefix *di*- (2) before the substituent's name. → If there are two substituents, place the first one on the benzene ring and count clockwise around the ring to determine where to attach the second substituent. Note: 1,2 = *ortho* or *o*; 1,3 = *meta* or *m*; and 1,4 = *para* or *p* replacing the numbers.

           Solution:

(a)   isopropylbenzene. The isopropyl group is —CH—CH_3. Attach this to a benzene ring to make
                                                        |
                                                       CH_3

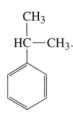

(b)   *meta*-dibromobenzene. There are two bromo or —Br groups. The positions are 1 and 3 because the designation is *meta*. Attach the —Br groups at the first and third positions to make

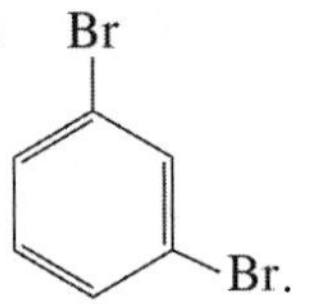

(c)   1-chloro-4-methylbenzene. One substituent group is —Cl or chloro, and the other is —CH_3 or methyl. Start by giving the chloro group an assignment of 1 and count in a clockwise direction to give the methyl group an assignment of 4. Attach the two groups to make Cl—〈 〉—CH_3.

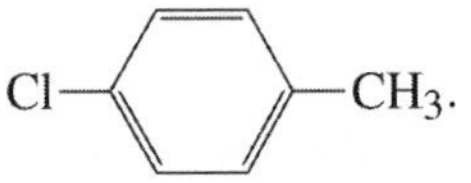

20.71   Aromatic substitution reactions substitute a halogen or alkyl group for a hydrogen (with a preference for the alkyl group) to generate a monosubstituted benzene and a hydrogen halide.

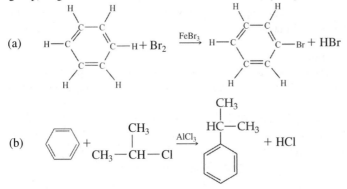

Alcohols

20.73   **Given:** alcohol structures   **Find:** name

Conceptual Plan: Alcohols are named like alkanes with the following differences: (1) The base chain is the longest continuous carbon chain that contains the —OH functional group. (2) The base name has the ending -*ol*. (3) The base chain is numbered to give the —OH group the lowest possible number. (4) A number indicating the position of the —OH group is inserted just before the base name.

Solution:

(a) $\boxed{CH_3-CH_2-CH_2}-OH$ The longest carbon chain has three carbons. The prefix for 3 is prop-, so the base name is propanol. The alcohol group is on the end, or first carbon, so the name is 1-propanol.

(b) 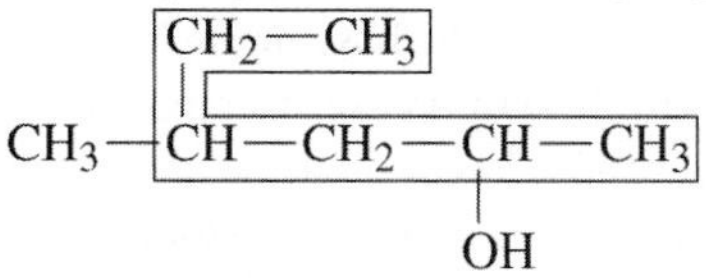 $CH_3-\boxed{CH-CH_2-CH-CH_3}$ with $\boxed{CH_2-CH_3}$ branch and OH. The longest carbon chain has six carbons. The prefix for 6 is hex-, so the base name is hexanol. There is a methyl substituent group.

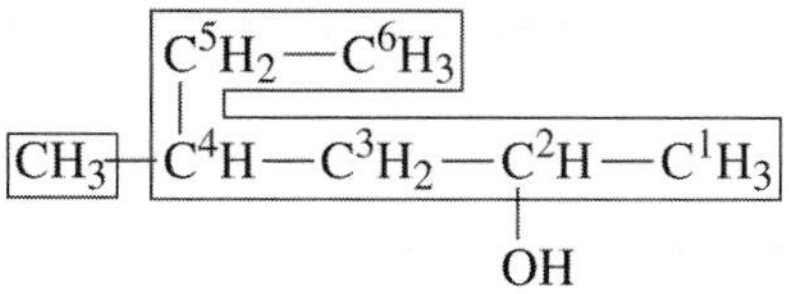

$\boxed{CH_3}-\boxed{C^4H-C^3H_2-C^2H-C^1H_3}$ with $\boxed{C^5H_2-C^6H_3}$ branch and OH. Start numbering at the end closest to the alcohol group. The alcohol group is assigned number 2, and the methyl group is assigned number 4. The name is 4-methyl-2-hexanol.

(c) 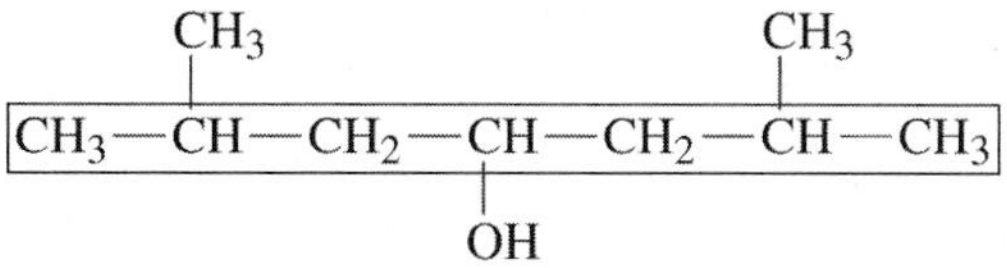 $\boxed{CH_3-CH-CH_2-CH-CH_2-CH-CH_3}$ with two CH_3 branches and OH. The longest carbon chain has seven carbons. The prefix for 7 is hept-, so the base name is heptanol. There are two methyl substituent groups.

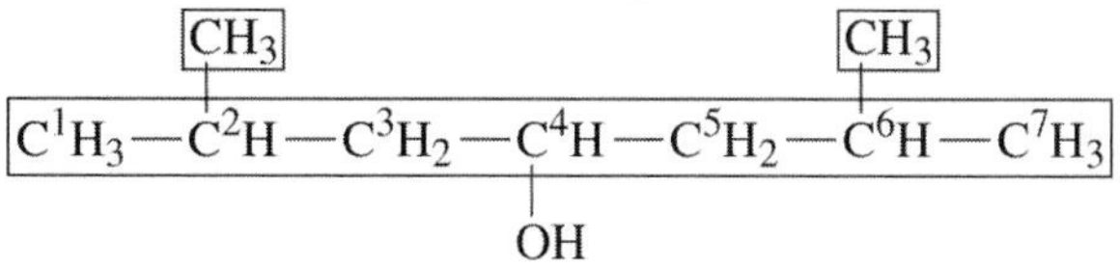

$\boxed{C^1H_3-C^2H-C^3H_2-C^4H-C^5H_2-C^6H-C^7H_3}$ with $\boxed{CH_3}$ branches and OH. It does not matter at which end the numbering is started because substitutions are symmetrically attached. The alcohol group is assigned number 4, and the methyl groups are assigned numbers 2 and 6. The name is 2,6-dimethyl-4-heptanol.

(d) 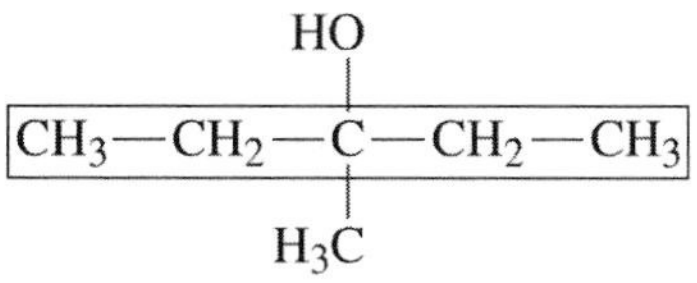 $\boxed{CH_3-CH_2-C-CH_2-CH_3}$ with HO and H_3C branches. The longest carbon chain has five carbons. The prefix for 5 is pent-, so the base name is pentanol. There is a methyl substituent group. $\boxed{C^1H_3-C^2H_2-C^3-C^4H_2-C^5H_3}$ with HO and $\boxed{H_3C}$ branches. It does not matter at which end the numbering is started because substitutions are symmetrically attached. The alcohol group is assigned number 3, and the methyl group is assigned number 3. The name is 3-methyl-3-pentanol.

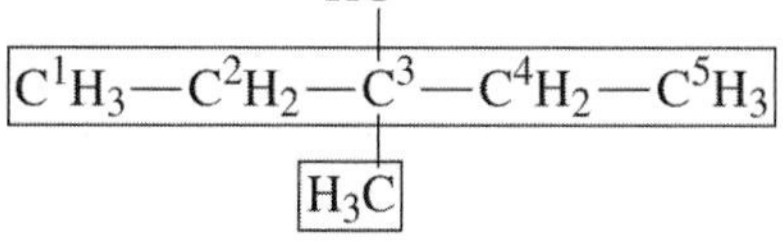

20.75   In a substitution reaction, an alcohol reacts with an acid, such as HBr, to form halogenated hydrocarbons and water. In an elimination (or dehydration) reaction, concentrated acids, such as H_2SO_4, react with alcohols to eliminate water, forming an alkene. In an oxidation reaction, carbon atoms gain oxygen atoms and/or lose hydrogen atoms. In these reactions, the alcohol becomes a carboxylic acid. Active metals, such as Na, react with alcohols to produce a strong base and eliminate hydrogen.

(a)   This is a substitution reaction, so $CH_3-CH_2-CH_2-OH + HBr \rightarrow CH_3-CH_2-CH_2-Br + H_2O$.

(b)   This is an elimination reaction, so $CH_3-CH-CH_2-OH$ (with CH_3) $\xrightarrow{H_2SO_4}$ $CH_3-C=CH_2$ (with CH_3) $+ H_2O$.

(c)   Sodium is an active metal, so $CH_3-CH_2-OH + Na \rightarrow CH_3-CH_2-O^-Na^+ + 1/2\ H_2$.

(d)   This is an oxidation reaction, so

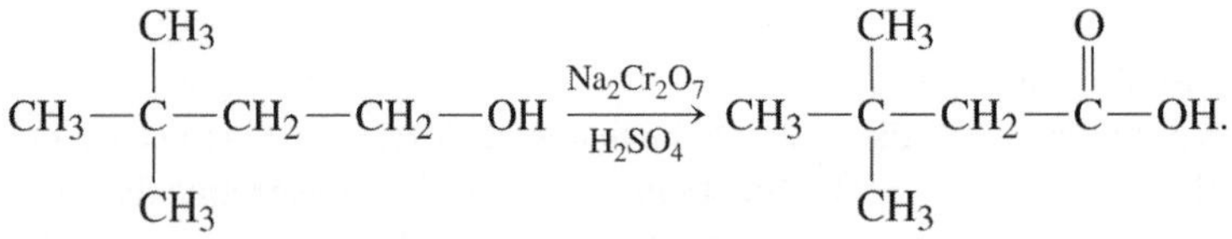

$$CH_3-\underset{\underset{CH_3}{|}}{\overset{\overset{CH_3}{|}}{C}}-CH_2-CH_2-OH \xrightarrow[\text{H}_2\text{SO}_4]{\text{Na}_2\text{Cr}_2\text{O}_7} CH_3-\underset{\underset{CH_3}{|}}{\overset{\overset{CH_3}{|}}{C}}-CH_2-\overset{\overset{O}{\|}}{C}-OH.$$

Aldehydes and Ketones

20.77   **Given:** aldehyde or ketone structures   **Find:** name
Conceptual Plan: Simple aldehydes are systematically named according to the number of carbon atoms in the longest continuous carbon chain that contains the carbonyl group. Form the base name from the name of the corresponding alkane by dropping the *-e* and add the ending *-al*. Simple ketones are systematically named according to the longest continuous carbon chain containing the carbonyl group. Form the base name from the name of the corresponding alkane by dropping the letter *-e* and adding the ending *-one*. For ketones, number the chain to give the carbonyl group the lowest possible number. Because an aldehyde always has a carbonyl terminal group, the aldehyde is always at the 1 position even though it is not specified with a number in the name.
Solution:

(a) 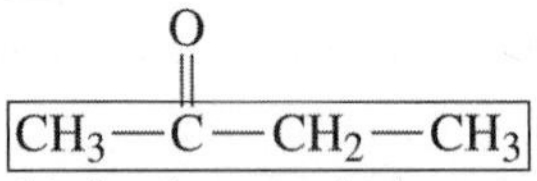 $\boxed{CH_3-\overset{\overset{O}{\|}}{C}-CH_2-CH_3}$ The longest carbon chain has four carbons. The prefix for 4 is but-, and because this is a ketone, the base name is butanone. The position of the carbonyl carbon does not need to be specified because there is only one place it can be in the molecule (if it were on the end carbon, it would be an aldehyde, not a ketone). Because there are no other substituent groups, the name is butanone.

(b) 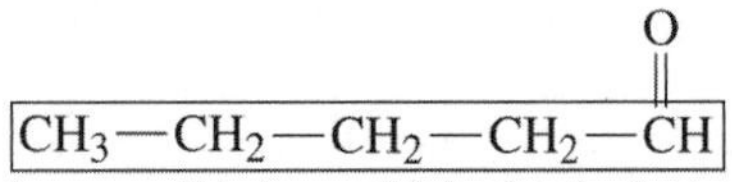 $\boxed{CH_3-CH_2-CH_2-CH_2-\overset{\overset{O}{\|}}{CH}}$ The longest carbon chain has five carbons. The prefix for 5 is pent-, and because this is an aldehyde, the base name is pentanal. Because there are no other substituent groups, the name is pentanal.

(c) 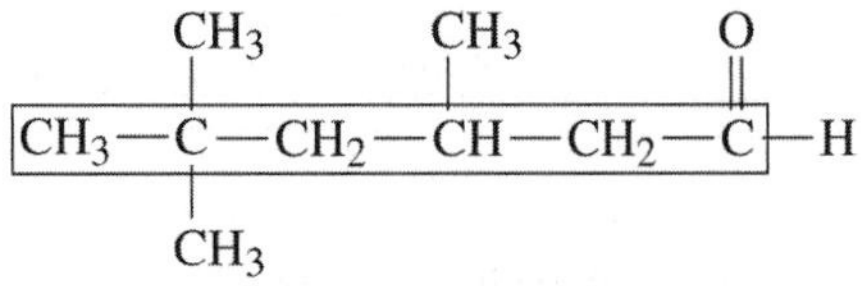 $\boxed{CH_3-\underset{\underset{CH_3}{|}}{\overset{\overset{CH_3}{|}}{C}}-CH_2-\overset{\overset{CH_3}{|}}{CH}-CH_2-\overset{\overset{O}{\|}}{C}}-H$ The longest carbon chain has six carbons. The prefix for 6 is hex-,

and because this is an aldehyde, the base name is hexanal. There are three methyl substituent groups.

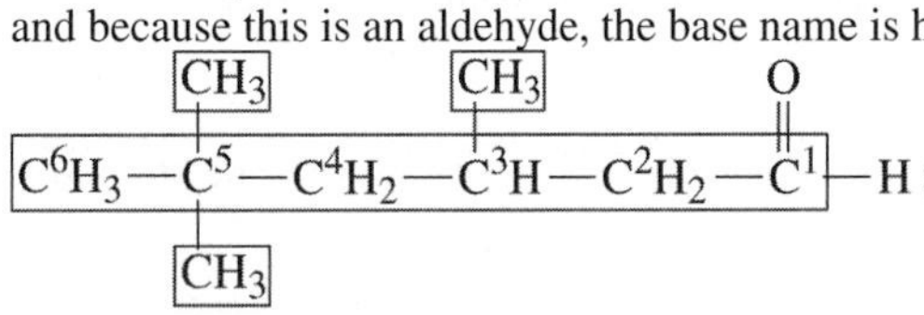

$\boxed{C^6H_3-\underset{\underset{\boxed{CH_3}}{|}}{\overset{\overset{\boxed{CH_3}}{|}}{C^5}}-C^4H_2-\overset{\overset{\boxed{CH_3}}{|}}{C^3H}-C^2H_2-\overset{\overset{O}{\|}}{C^1}}-H$ Start numbering at the carbonyl carbon. The methyl groups

are at positions 3, 5, and 5. The name of the molecule is 3,5,5-trimethylhexanal.

(d) 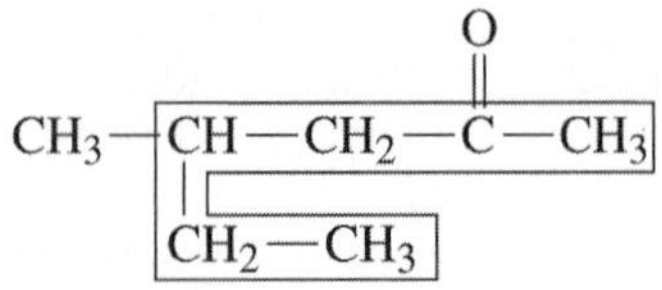 $CH_3-\boxed{\underset{\underset{\boxed{CH_2-CH_3}}{|}}{CH}-CH_2-\overset{\overset{O}{\|}}{C}-CH_3}$ The longest carbon chain has six carbons. The prefix for 6 is hex-, and

because this is a ketone, the base name is hexanone. There is one methyl substituent group.

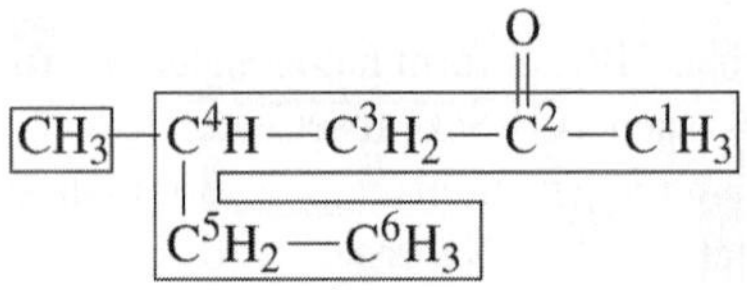

$\boxed{CH_3}-\boxed{\underset{\underset{\boxed{C^5H_2-C^6H_3}}{|}}{C^4H}-C^3H_2-\overset{\overset{O}{\|}}{C^2}-C^1H_3}$ Start numbering at the end closest to the carbonyl carbon. The

carbonyl carbon is in position 2, and the methyl group is at position 4. The name of the molecule is 4-methyl-2-hexanone.

20.79 This is an addition reaction,

$$CH_3-CH_2-CH_2-\overset{\displaystyle O}{\overset{\|}{C}}-H + H-C\equiv N \xrightarrow{\text{NaCN}} CH_3-CH_2-CH_2-\overset{\displaystyle OH}{\underset{\displaystyle H}{\overset{|}{\underset{|}{C}}}}-C\equiv N.$$

Carboxylic Acids and Esters

20.81 **Given:** carboxylic acid or ester structures **Find:** name
Conceptual Plan: Carboxylic acids are systematically named according to the number of carbon atoms in the longest chain containing the —COOH functional group. Form the base name by dropping the -e from the name of the corresponding alkane and adding the ending -oic acid. Esters are systematically named as if they were derived from a carboxylic acid by replacing the H on the OH with an alkyl group. The R group from the parent acid forms the base name of the compound. Change the -ic on the name of the corresponding carboxylic acid to -ate and drop acid. The R group that replaced the H on the carboxylic acid is named as an alkyl group with the ending -yl. The —COOH terminal group is always at the 1 position even though it is not specified with a number in the name.
Solution:

(a) $\boxed{CH_3-CH_2-CH_2-\overset{\displaystyle O}{\overset{\|}{C}}}-O-\boxed{CH_3}$ The carbon chain that has the carbonyl carbon has four carbons. The prefix for 4 is but-, so the base name is butanoate. The R group that replaces the H of the carboxylic acid is a methyl group. The name is methylbutanoate.

(b) $\boxed{CH_3-CH_2-\overset{\displaystyle O}{\overset{\|}{C}}}-OH$ The carbon chain has three carbons. The prefix for 3 is prop- and there are no other substituents. The name is propanoic acid.

(c) $\boxed{CH_3-\underset{\underset{\displaystyle CH_3}{|}}{CH}-CH_2-CH_2-CH_2-\overset{\displaystyle O}{\overset{\|}{C}}}-OH$ The carbon chain has six carbons. The prefix for 6 is hex- so the base name is hexanoic acid. There is one methyl substituent group.

$\boxed{C^6H_3-\underset{\underset{\boxed{CH_3}}{|}}{C^5H}-C^4H_2-C^3H_2-C^2H_2-\overset{\displaystyle O}{\overset{\|}{C^1}}}-OH$ Number the chain starting with the carbonyl carbon.

The methyl is in the fifth position. The name of the molecule is 5-methylhexanoic acid.

(d) $\boxed{CH_3-CH_2-CH_2-CH_2-\overset{\displaystyle O}{\overset{\|}{C}}}-O-\boxed{CH_2-CH_3}$ The carbon chain that has the carbonyl carbon has five carbons. The prefix for 5 is pent-, so the base name is pentanoate. The R group that replaces the H of the carboxylic acid is an ethyl group. The name is ethylpentanoate.

20.83 The reactions is a condensation reaction. The two noncarbonyl oxygen groups react linking the two molecules (or parts of a molecule) with the concomitant generation of a water molecule.

(a) $CH_3-CH_2-CH_2-\overset{\displaystyle O}{\overset{\|}{C}}-OH + CH_3-CH_2-OH \xrightarrow{\text{H}_2\text{SO}_4}$

$$CH_3-CH_2-CH_2-\overset{\displaystyle O}{\overset{\|}{C}}-O-CH_2-CH_3 + H_2O$$

(b)

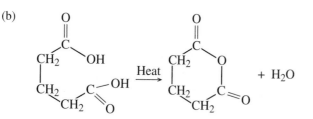

Ethers

20.85   **Given:** ether structures   **Find:** name
Conceptual Plan: Common names for ethers have the format (R group 1)(R group 2) ether, where the alkyl groups are those found in Table 20.5. List the names alphabetically. If the two R groups are different, use each of their names. If the two R groups are the same, use the prefix *di-*.
Solution:

(a) $\boxed{CH_3-CH_2-CH_2}-O-\boxed{CH_2-CH_3}$ The left carbon chain has three carbons, so it is a propyl group.

The right carbon chain has two carbons, so it is an ethyl group. Listing them alphabetically, the name is ethyl propyl ether.

(b) $\boxed{CH_3-CH_2-CH_2-CH_2-CH_2}-O-\boxed{CH_2-CH_3}$ The left carbon chain has five carbons, so it is a pentyl

group. The right carbon chain has two carbons, so it is an ethyl group. Listing them alphabetically, the name is ethyl pentyl ether.

(c) $\boxed{CH_3-CH_2-CH_2}-O-\boxed{CH_2-CH_2-CH_3}$ Both carbon chains have three carbons, so they are propyl

groups. The name is dipropyl ether.

(d) $\boxed{CH_3-CH_2}-O-\boxed{CH_2-CH_2-CH_2-CH_3}$ The left carbon chain has two carbons, so it is an ethyl

group. The right carbon chain has four carbons, so it is a butyl group. Listing them alphabetically, the name is butyl ethyl ether.

Amines

20.87   **Given:** amine structures   **Find:** name
Conceptual Plan: Amines are systematically named by alphabetically listing the alkyl groups that are attached to the nitrogen atom and adding an *-amine* ending. The alkyl groups are found in Table 20.5. Use the prefix *di-* or *tri-* if the alkyl groups are the same.
Solution:

(a) $\boxed{CH_3-CH_2}-N-\boxed{CH_2-CH_3}$ Both carbon chains have two carbons, so they are ethyl groups. The name
          $\quad\quad\quad\quad\quad | $
          $\quad\quad\quad\quad\quad H$
is diethylamine.

(b) $\boxed{CH_3-CH_2-CH_2}-N-\boxed{CH_3}$ The left carbon chain has three carbons, so it is a propyl group. The
          $\quad\quad\quad\quad\quad\quad\quad | $
          $\quad\quad\quad\quad\quad\quad\quad H$
right carbon chain has one carbon, so it is a methyl group. Listing them alphabetically, the name is methylpropylamine.

(c) $\boxed{CH_3-CH_2-CH_2}-N-\boxed{CH_2-CH_2-CH_2-CH_3}$ The left carbon chain has three carbons, so it is
          $\quad\quad\quad\quad\quad\quad\quad\overset{\boxed{CH_3}}{|}$
a propyl group. The top carbon chain has one carbon, so it is a methyl group. The right carbon chain has four carbons, so it is a butyl group. Listing them alphabetically, the name is butylmethylpropylamine.

20.89   (a)   This reaction is an acid–base reaction, $CH_3NHCH_3(aq) + HCl(aq) \rightarrow (CH_3)_2NH_2^+(aq) + Cl^-(aq)$.
          (b)   Because the acid is a carboxylic acid, this reaction is a condensation reaction, $CH_3CH_2NH_2(aq) + CH_3CH_2COOH(aq) \rightarrow CH_3CH_2CONHCH_2CH_3(aq) + H_2O(l)$.
          (c)   This reaction is an acid–base reaction, $CH_3NH_2(aq) + H_2SO_4(aq) \rightarrow CH_3NH_3^+(aq) + HSO_4^-(aq)$.

Polymers

20.91   Because the monomer is $F_2C=CF_2$, as the polymer forms, the double bond breaks to link with another monomer.

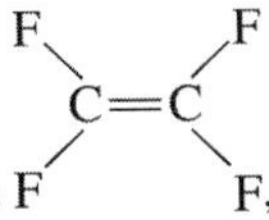

The structure is

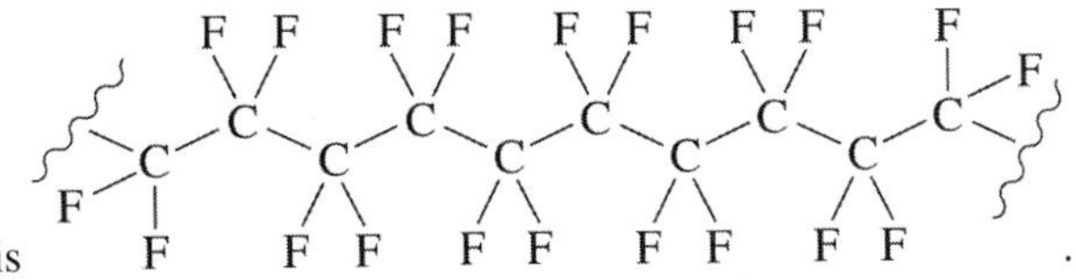

20.93   In a condensation polymer, atoms are eliminated. Here, water is eliminated. The structure is

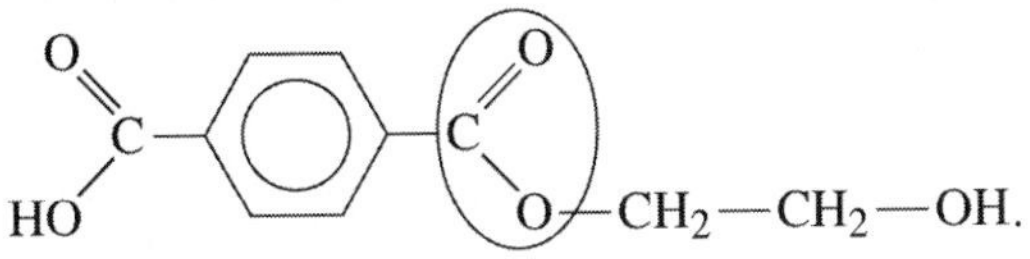

Cumulative Problems

20.95   **Given:** structures   **Find:** identify molecule type and name
Conceptual Plan: Look for functional groups to identify molecule type. Review the appropriate set of naming rules to name the molecule.
Solution:

(a) $CH_3-CH-CH_2-\overset{\displaystyle O}{\overset{\|}{C}}-O-CH_3$ The compound has the formula RCOOR′, so it is an ester.

CH_3

$\boxed{CH_3-CH-CH_2-\overset{\displaystyle O}{\overset{\|}{C}}}-O-\boxed{CH_3}$ The left carbon chain has four carbons, which translates to

CH_3

but- and a base name of butanoate. There is a methyl substituent on this chain. The right carbon chain

has one carbon, so it is a methyl group. $\boxed{C^4H_3-C^3H-C^2H_2-\overset{\displaystyle O}{\overset{\|}{C^1}}}-O-\boxed{CH_3}$ Number the left

CH_3

chain starting with the carbonyl carbon. The methyl group is at position 3. The name of the compound is
methyl-3-methylbutanoate.

CH_3

(b) $\boxed{CH_3-CH_2-CH-CH_2}-O-\boxed{CH_2-CH_3}$ The compound has the formula ROR′, so it is an ether.
The left carbon chain has four carbons, so it is a butyl group. There is a methyl substituent on this chain. The
right carbon chain has two carbons, so it is an ethyl group.

$\boxed{CH_3}$

$\boxed{C^4H_3-C^3H_2-C^2H-C^1H_2}-O-\boxed{CH_2-CH_3}$ Number the left chain starting with the carbon next to

the oxygen. The methyl substituent is in position 2, so the alkyl group is 2-methylbutyl. Listing the alkyl group
alphabetically, the name is ethyl 2-methylbutyl ether.

(c) 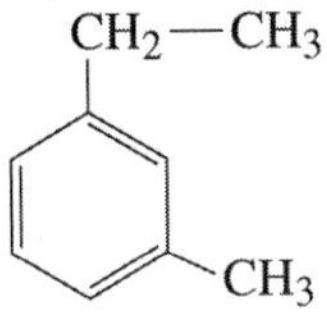   The molecule is a disubstituted benzene, so it is an aromatic hydrocarbon. The substituents
are an ethyl group and a methyl group. If the ethyl position is assigned the first position, the
methyl group is in the third position. Listing the functional groups alphabetically, the name is
1-ethyl-3-methylbenzene or *m*-ethylmethylbenzene.

(d) $CH_3-C\equiv C-CH-CH-CH_2-CH_3$ with CH_2-CH_3 and CH_3 substituents. The molecule contains a triple bond, so it is an alkyne. The longest carbon chain is 7. The prefix for 7 is hept-, and the base name is heptyne. The substituent groups are a methyl and ethyl group. $C^1H_3-C^2\equiv C^3-C^4H-C^5H-C^6H_2-C^7H_3$ with CH_2-CH_3 and CH_3. If we start numbering the chain at the end closest to the triple bond, the triple bond is between positions 2 and 3, the methyl substituent is assigned number 4, and the ethyl substituent is assigned number 5. Because e is before m, the name of the compound is 5-ethyl-4-methyl-2-heptyne.

(e) $CH_3-CH_2-CH_2-CH$ with O double bonded. The molecule has the formula RCHO, so it is an aldehyde. The longest carbon chain has four carbons. The prefix for 4 is but-, and because this is an aldehyde, the base name is butanal. Because there are no other substituent groups, the name is butanal.

(f) $CH_3-CH-CH_2$ with OH and H_3C substituents. The molecule has the formula ROH, so it is an alcohol. The longest carbon chain has three carbons. The prefix for 3 is prop-, so the base name is propanol. There is a methyl substituent group. $C^3H_3-C^2H-C^1H_2$ with OH and H_3C. Start numbering the carbon with the $-OH$ group. The alcohol group is assigned number 1, and the methyl group is assigned number 2. The name is 2-methyl-1-propanol.

20.97   **Given:** structures   **Find:** name
Conceptual Plan: Look for functional groups to identify molecule type. Review the appropriate set of naming rules to name the molecule.
Solution:

(a)   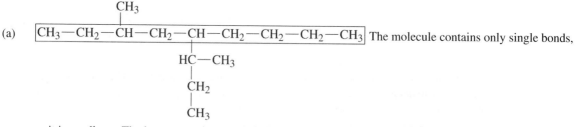 The molecule contains only single bonds, so it is an alkane. The longest continuous chain has nine carbons. The prefix for 9 is non-, and the base name is nonane. The substituent groups are a methyl group and an isobutyl group.

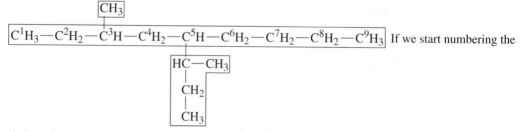 If we start numbering the chain at the end closest to the methyl group, the methyl substituent is assigned the number 3 and the isobutyl group is assigned the number 5. List the groups alphabetically to get the name of the compound, which is 5-isobutyl-3-methylnonane.

(b) 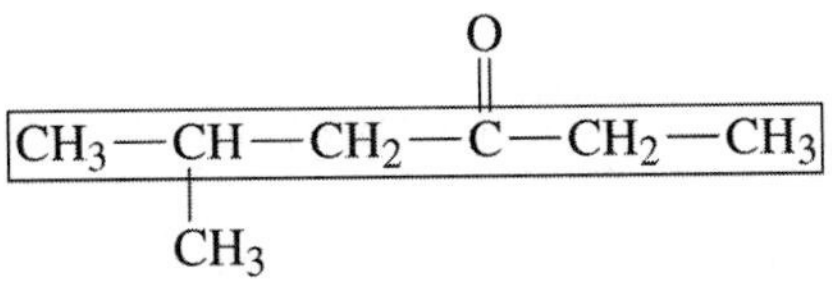 $\boxed{CH_3-CH-CH_2-\overset{\overset{O}{\parallel}}{C}-CH_2-CH_3}$ The molecule contains a carbonyl carbon in the interior of the

 CH$_3$

molecule, so it is a ketone. The longest carbon chain has six carbons. The prefix for 6 is hex-. Because this is a ketone, the base name is hexanone. There is one methyl substituent group.

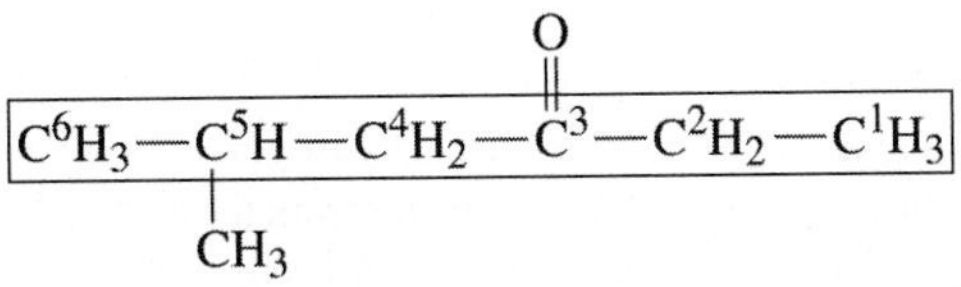

$\boxed{C^6H_3-C^5H-C^4H_2-\overset{\overset{O}{\parallel}}{C^3}-C^2H_2-C^1H_3}$ Start numbering at the right side of the molecule.

 CH$_3$

The carbonyl carbon is at position 3. The methyl group is at position 5. The name of the molecule is 5-methyl-3-hexanone.

 OH

(c) 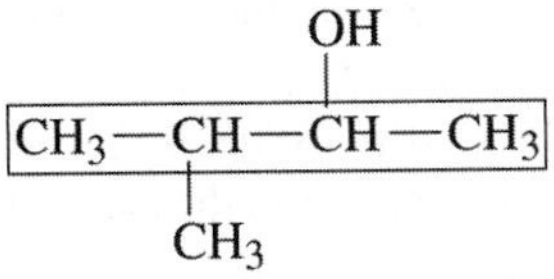 $\boxed{CH_3-CH-CH-CH_3}$ The molecule has the formula ROH, so it is an alcohol. The longest carbon

 CH$_3$

chain has four carbons. The prefix for 4 is but-, so the base name is butanol. The only substituent group is a

 OH

methyl group. $\boxed{C^4H_3-C^3H-C^2H-C^1H_3}$ Start numbering at the end with the —OH group. The alcohol

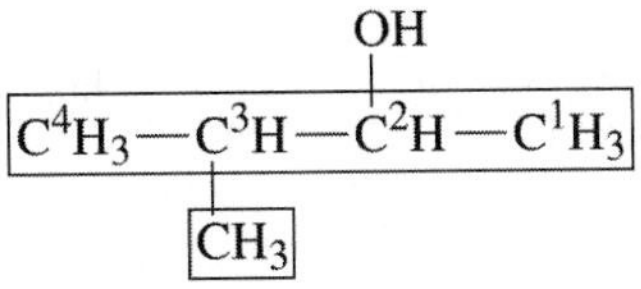

 $\boxed{CH_3}$

group is assigned number 2, and the methyl group is assigned number 3. The name is 3-methyl-2-butanol.

 CH$_3$

 CH$_2$ CH$_3$

(d) 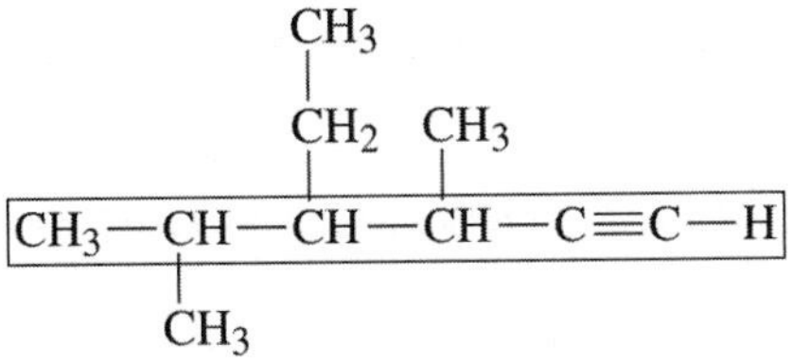 $\boxed{CH_3-CH-CH-CH-C\equiv C-H}$ The molecule contains a triple bond, so it is an alkyne. The longest

 CH$_3$

carbon chain is 6. The prefix for 6 is hex-, and the base name is hexyne. The substituent groups are two

 $\boxed{CH_3}$

 $\boxed{CH_2}$ $\boxed{CH_3}$

methyl groups and an ethyl group. $\boxed{C^6H_3-C^5H-C^4H-C^3H-C^2\equiv C^1-H}$ If we start numbering the

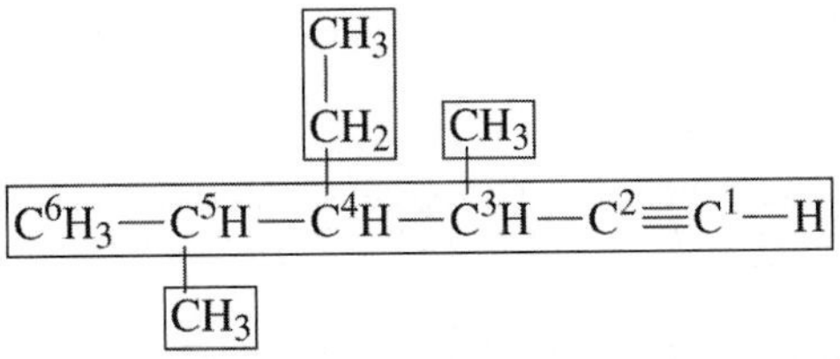

 $\boxed{CH_3}$

chain at the end closest to the triple bond, the triple bond is between position numbers 1 and 2, the methyl substituents are assigned numbers 3 and 5, and the ethyl substituent is assigned number 4. Because *e* comes before *m*, the name of the compound is 4-ethyl-3,5-dimethyl-1-hexyne.

20.99   (a)   These structures are structural isomers. The methyl and ethyl groups have been swapped.
 (b)   These structures are isomers. The second iodo group has been moved.
 (c)   These structures are the same molecule. The first methyl group is simply drawn in a different orientation.

20.101   **Given:** 15.5 kg 2-butene hydrogenation   **Find:** minimum g of H$_2$ gas
 **Conceptual Plan: Write a balanced equation kg C$_4$H$_8$ → g C$_4$H$_8$ → mol C$_4$H$_8$ → mol H$_2$ → g H$_2$.**

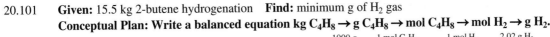

$$\frac{1000\ g}{1\ kg} \quad \frac{1\ mol\ C_4H_8}{56.10\ g\ C_4H_8} \quad \frac{1\ mol\ H_2}{1\ mol\ C_4H_8} \quad \frac{2.02\ g\ H_2}{1\ mol\ H_2}$$

Solution: $C_4H_8(g) + H_2(g) \rightarrow C_4H_{10}(g)$ then

$$15.5 \text{ kg } C_4H_8 \times \frac{1000 \text{ g } C_4H_8}{1 \text{ kg } C_4H_8} \times \frac{1 \text{ mol } C_4H_8}{56.10 \text{ g } C_4H_8} \times \frac{1 \text{ mol } H_2}{1 \text{ mol } C_4H_8} \times \frac{2.02 \text{ g } H_2}{1 \text{ mol } H_2} = 558 \text{ g } H_2$$

Check: The units (g) are correct. The magnitude of the answer (600) makes physical sense because we are reacting 15,500 g of butane. Hydrogen is lighter, and the molar ratio is 1:1; so the amount of hydrogen will be significant, but less than the mass of butene.

20.103   (a)   Combustion reaction—reaction with oxygen to form CO_2 and H_2O
         (b)   Alkane (halogen) substitution—reaction with halogen gas to substitute a hydrogen with a halogen
         (c)   Alcohol elimination—ROH reaction with concentrated acid to eliminate water from the alcohol to produce an alkene
         (d)   Aromatic (halogen) substitution—benzene reaction with halogen gas to substitute a hydrogen with a halogen

20.105   **Given:** alkene names   **Find:** structure
         Conceptual Plan: Find the number of carbon atoms corresponding to the prefix of the base name in Table 20.5. →
         Draw the base chain and number the carbons from left to right. → **Place the double bond in the appropriate**
         position in the chain. → **Using Table 20.6 and the prefix *di-* (2), *tri-* (3), or *tetra-* (4) before the substituent's name,**
         determine each substituent. → **Add the substituent to the proper carbon position in the chain.** → **Add hydrogen**
         atoms to the base chain so that each carbon has four bonds. → **Look for substitutions around double bonds to de-**
         termine whether cis–trans isomerism is applicable and look for carbons with four different alkyl groups attached.
         Solution:

         (a)   3-methyl-1-pentene. The base name pentene designates that there are five carbon atoms in the base chain.
               $C^1 - C^2 - C^3 - C^4 - C^5$. The -ene ending designates that there is a double bond, and the 1 prefix designates
               that it is between the first and second positions. $C^1 = C^2 - C^3 - C^4 - C^5$. 3-methyl designates that there is

               a $-CH_3$ group in the third position. $C^1 = C^2 - \overset{\overset{\boxed{CH_3}}{|}}{C^3} - C^4 - C^5$. Add hydrogens to the base chain to get the final

               molecule $CH_2 = CH - \overset{\overset{CH_3}{|}}{CH} - CH_2 - CH_3$. There is optical stereoisomerism because the third carbon has
               four different groups attached.

         (b)   3,5-dimethyl-2-hexene. The base name hexene designates that there are six carbon atoms in the base chain.
               $C^1 - C^2 - C^3 - C^4 - C^5 - C^6$. The -ene ending designates that there is a double bond, and the 2 prefix
               designates that it is between the second and third positions. $C^1 - C^2 = C^3 - C^4 - C^5 - C^6$. 3,5-dimethyl

               designates that there are $-CH_3$ groups in the third and fifth positions. $C^1 - C^2 = \overset{\overset{\boxed{CH_3}}{|}}{C^3} - C^4 - \overset{\overset{\boxed{CH_3}}{|}}{C^5} - C^6$.

               Add hydrogens to the base chain to get the final molecule $CH_3 - CH = \overset{\overset{CH_3}{|}}{C} - CH_2 - \overset{\overset{CH_3}{|}}{CH} - CH_3$. Cis–trans
               geometric isomerism is possible around the double bond because all of the groups surrounding the double
               bond are different.

         (c)   3-propyl-2-hexene. The base name hexene designates that there are six carbon atoms in the base chain.
               $C^1 - C^2 - C^3 - C^4 - C^5 - C^6$. The -ene ending designates that there is a double bond, and the 2 prefix
               designates that it is between the second and third positions. $C^1 - C^2 = C^3 - C^4 - C^5 - C^6$. 3-propyl

               designates that there is a $-CH_2CH_2CH_3$ group in the third position. $C^1 - C^2 = \overset{\overset{\boxed{CH_2 - CH_2 - CH_3}}{|}}{C^3} - C^4 - C^5 - C^6$.

               Add hydrogens to the base chain to get the final molecule $CH_3 - CH = \overset{\overset{CH_2 - CH_2 - CH_3}{|}}{C} - CH_2 - CH_2 - CH_3$. Cis–trans
               geometric isomerism is not possible around the double bond because the two groups on the third carbon are the same.

20.107 The 11 isomers are

$$\underset{HC-CH_2-CH_2-CH_3}{\overset{O}{\|}} = \text{aldehyde;} \qquad \underset{CH_3-C-CH_2-CH_3}{\overset{O}{\|}} = \text{ketone;}$$

$CH_2{=}C{-}CH_2{-}CH_2{-}OH$ = alcohol and alkene; $CH_3{-}CH{=}CH{-}CH_2{-}OH$ = alcohol and alkene;

$CH_3{-}CH_2{-}CH{=}CH{-}OH$ = alcohol and alkene; $CH_3{-}CH_2{-}O{-}CH{=}CH_2$ = ether and alkene;

$CH_3{-}O{-}CH_2{-}CH{=}CH_2$ = ether and alkene; $CH_3{-}O{-}CH{=}CH{-}CH_3$ = ether and alkene;

$$\underset{CH_3-CH_2-C=CH_2}{\overset{OH}{|}} = \text{alcohol and alkene;} \qquad \underset{CH_3-CH-CH=CH_2}{\overset{OH}{|}} = \text{alcohol and alkene; and}$$

$$\underset{CH_3-C=CH-CH_3}{\overset{OH}{|}} = \text{alcohol and alkene. There is also a twelfth isomer that is cyclic:}$$

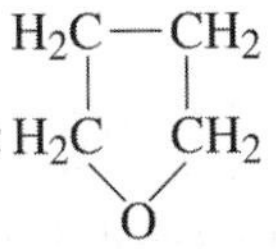

$\underset{O}{\overset{H_2C-CH_2}{H_2C\quad CH_2}}$ = ether.

20.109 In the acid form of the carboxylic acid, electron withdrawal by the C=O enhances acidity. The conjugate base, the carboxylate anion, is stabilized by resonance; so the two O atoms are equivalent and bear the negative charge equally. This resonance stability is not possible in an alcohol, where the carbonyl group is absent.

Challenge Problems

20.111 This is an internal condensation reaction. It requires only heat to cause it to happen.

(a)

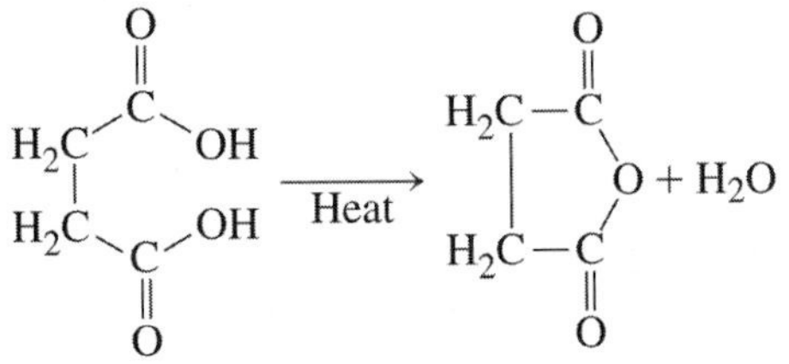

(b) This is a dehydration reaction, $CH_3{-}CH{-}CH_2{-}\underset{|}{\overset{CH_2-OH}{CH}}{-}CH_3 \xrightarrow{H_2SO_4}$

with $CH_2{-}CH_3$ below the first CH.

$$CH_3{-}CH{-}CH_2{-}\overset{CH_2}{\underset{CH_2-CH_3}{\overset{\|}{C}}}{-}CH_3 + H_2O.$$

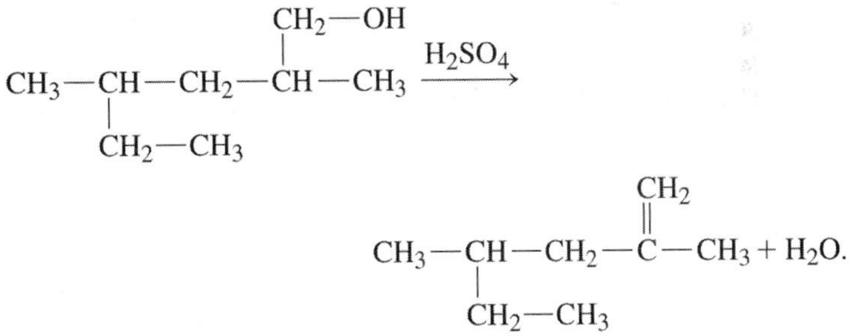

(c) This is an addition reaction that follows Markovnikov's rule,

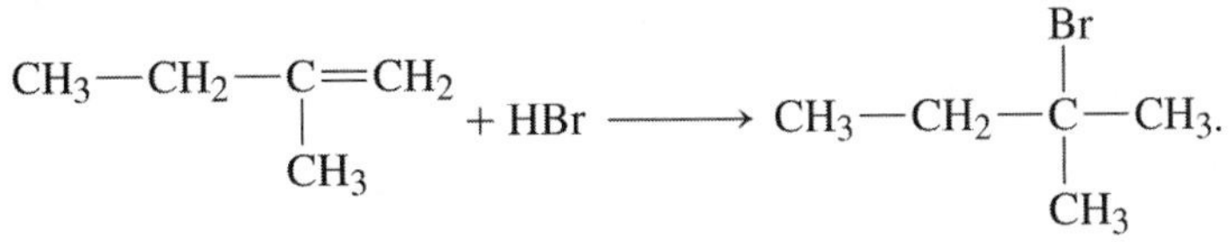

$$CH_3{-}CH_2{-}\underset{CH_3}{\overset{}{C}}{=}CH_2 + HBr \longrightarrow CH_3{-}CH_2{-}\underset{CH_3}{\overset{Br}{C}}{-}CH_3.$$

20.113 (a) Because there are six primary hydrogen atoms and two secondary hydrogen atoms, if they are equally reactive, we expect a ratio of 6:2 or 3:1.

(b) Assume that we generate 100 product molecules. The yield = (# hydrogen atoms)(reactivity). So $1° = 45 = (3)(\text{reactivity } 1°)$ and $2° = 55 = (1)(\text{reactivity } 2°)$. Taking the ratio, $2°:1° = 55:(45/3) = 55:15 = 11:3$. The 2° hydrogens are much more reactive.

20.115 The chiral products (that have four different groups on a carbon) are $CH_2Cl{-}CHCl{-}CH_2{-}CH_3$ (second carbon), $CH_2Cl{-}CH_2{-}CHCl{-}CH_3$ (third carbon), and $CH_3{-}CHCl{-}CHCl{-}CH_3$ (second and third carbons).

20.117   The first propagation step for F is very rapid and exothermic because of the strength of the H—F bond that forms. For I, the first propagation step is endothermic and slow because the H—I bond that forms is relatively weak. In addition, the C—F bond is much stronger than C—I (477 kJ versus 241 kJ).

20.119   The three structures that have no dipole moment are Cl—$\overset{\displaystyle H}{\underset{\displaystyle H}{C}}$—$\overset{\displaystyle H}{\underset{\displaystyle H}{C}}$—Cl, $\overset{H}{\underset{Cl}{}}$C=C$\overset{Cl}{\underset{H}{}}$, and Cl—C≡C—Cl.

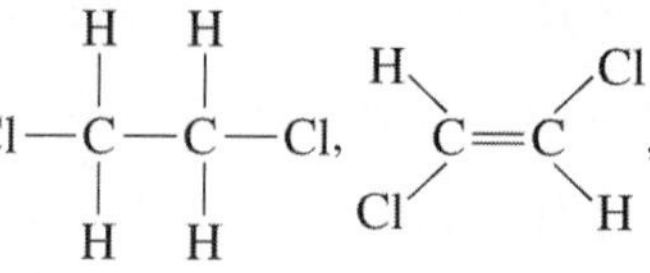

The structures with dipole moments are H—$\overset{\displaystyle H}{\underset{\displaystyle H}{C}}$—$\overset{\displaystyle Cl}{\underset{\displaystyle H}{C}}$—Cl, $\overset{Cl}{\underset{H}{}}$C=C$\overset{Cl}{\underset{H}{}}$, and $\overset{H}{\underset{H}{}}$C=C$\overset{Cl}{\underset{Cl}{}}$.

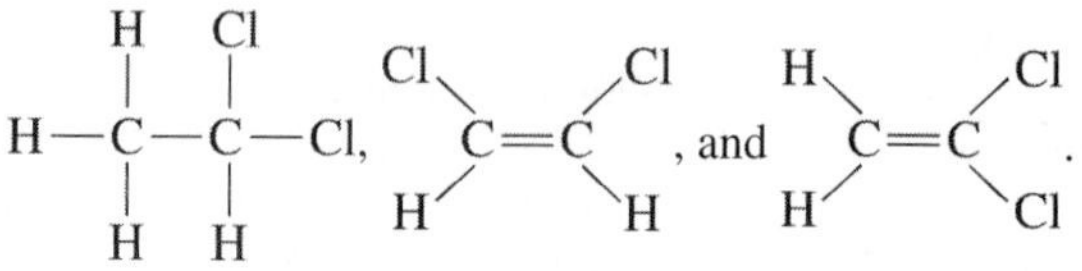

Conceptual Problems

20.121   For the structure to have only one product after a single bromination, all of the hydrogens must be equivalent.

The structure is CH$_3$—$\overset{\displaystyle CH_3}{\underset{\displaystyle CH_3}{C}}$—$\overset{\displaystyle CH_3}{\underset{\displaystyle CH_3}{C}}$—CH$_3$. To name it, find the longest carbon chain:

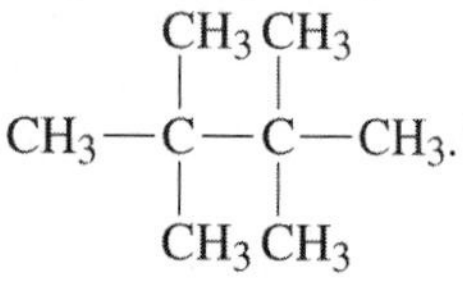

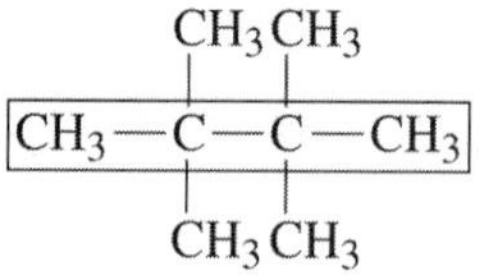

$\boxed{\text{CH}_3-\overset{\displaystyle CH_3}{\underset{\displaystyle CH_3}{C}}-\overset{\displaystyle CH_3}{\underset{\displaystyle CH_3}{C}}-\text{CH}_3}$. Because there are four carbons in this chain and it is an alkane, the base name is butane.

There are four methyl substituent groups. $\boxed{\text{C}^1\text{H}_3-\overset{\displaystyle \boxed{CH_3}}{\underset{\displaystyle \boxed{CH_3}}{\text{C}^2}}-\overset{\displaystyle \boxed{CH_3}}{\underset{\displaystyle \boxed{CH_3}}{\text{C}^3}}-\text{C}^4\text{H}_3}$ It does not matter how the chain is numbered because it is symmetrically substituted. The methyl groups are assigned the numbers 2, 2, 3, and 3. The name is 2,2,3,3-tetramethylbutane.

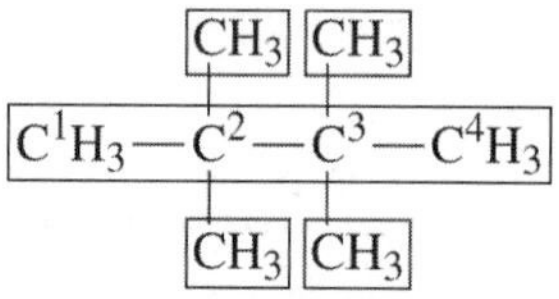

21 Biochemistry

Review Questions

21.1 Biochemistry is that area at the interface between chemistry and biology that strives to understand living organisms at the molecular level. It is the study of the chemistry occurring in living organisms. To help diabetics, Frederick Sanger discovered the detailed chemical structure of human insulin. This allowed the synthesis of insulin. This was further enhanced by biotechnology advances, particularly by Genentech, that allowed researchers to insert the human gene for insulin into the DNA of bacterial cells. This new way to manufacture human insulin revolutionized the treatment of diabetes because human insulin is now readily available.

21.3 A fatty acid is a carboxylic acid with a long hydrocarbon tail. The general structure for a fatty acid is as follows:

$$R - \overset{\displaystyle O}{\overset{\displaystyle \|}{C}} - OH$$

where R represents a hydrocarbon chain containing 3 to 19 carbon atoms.

21.5 Triglycerides are triesters composed of glycerol with three fatty acids attached. The general structure of a triglyceride is as follows:

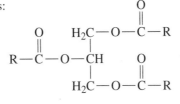

21.7 Phospholipids have the same basic structure as triglycerides except that one of the fatty acid groups is replaced with a phosphate group. Unlike a fatty acid, which is nonpolar, the phosphate group is polar. The phospholipid molecule has a polar region and a nonpolar region. The polar part of the molecule is hydrophilic, while the nonpolar part is hydrophobic.

Glycolipids have similar structures and properties. The nonpolar section of a glycolipid is composed of a fatty acid chain and a hydrocarbon chain. However, the polar section is a sugar molecule.

The structure of phospholipids and glycolipids makes them ideal components of cell membranes, where the polar parts can interact with the aqueous environments inside and outside the cell and the nonpolar parts interact with each other, forming a double-layered structure called a lipid bilayer. Lipid bilayers encapsulate cells and many cellular structures.

21.9 Carbohydrates have the general formula $(CH_2O)_n$. Structurally, carbohydrates are polyhydroxy aldehydes or ketones. Carbohydrates are responsible for short-term storage of energy in living organisms, and they make up the main structural components of plants.

21.11 Simple carbohydrates are the simple sugars, monosaccharides and disaccharides. Polysaccharides, which are long, chainlike molecules composed of many monosaccharide units bonded together, are complex carbohydrates.

21.13 Proteins are the workhorse molecules in living organisms and are involved in virtually all facets of cell structure and function. Most of the chemical reactions that occur in living organisms are enabled by enzymes, proteins that act as catalysts in biochemical reactions. (For e.g., amylase in saliva breaks down starches and chymotrypsin the stomach aids in digestion). Proteins are also the structural elements of muscle, skin, and cartilage (e.g., elastin in skin and keratin in hair). Proteins transport oxygen in the blood (e.g., hemoglobin), act as antibodies to fight disease, and function as hormones to regulate metabolic processes (e.g., insulin regulates glucose levels in the body).

21.15 The R groups, or side chains, differ chemically from one amino acid to another. When amino acids are strung together to make a protein, the chemical properties of the R group determine the structure and properties of the protein.

21.17

$$H_2N-\underset{\underset{CH_3}{|}}{\overset{\overset{H}{|}}{C}}-\overset{\overset{O}{\|}}{C}-OH \qquad \text{Alanine (Ala)} \qquad {}^+H_3N-\underset{\underset{CH_3}{|}}{\overset{\overset{H}{|}}{C}}-\overset{\overset{O}{\|}}{C}-O^-$$

21.19 Protein structure is critical to its function because cells contain receptors that fit a specific portion of the protein. If a protein was a different shape, it could not latch onto the receptors.

21.21 The primary structure of a protein is the amino acid sequence.

The secondary structure refers to the repeating patterns in the arrangement of protein chains such as alpha helices and beta sheets.

The tertiary structure refers to the large-scale twists and folds found in globular proteins but not fibrous proteins.

The quaternary structure refers to the arrangement of subunits in proteins that have more than one polypeptide chain.

21.23 In the α-helix structure, the amino acid chain is wrapped into a tight coil in which the side chains extend outward from the coil. The structure is maintained by hydrogen bonding interaction between NH and CO groups along the peptide backbone of the protein.

In the β-pleated sheet, the chain is extended and forms a zigzag pattern. The peptide backbones of neighboring chains interact with one another through hydrogen bonding to form zigzag-shaped sheets.

21.25 Nucleic acids are polymers. The individual units composing nucleic acids are called nucleotides. Each nucleotide has three parts: a sugar; a base; and a phosphate group, which serves as a link between sugars.

21.27 A codon is a sequence of three bases that codes for one amino acid.

A gene is a sequence of codons within a DNA molecule that codes for a single protein.

Genes are contained in structures within cells called chromosomes. There are 46 chromosomes in the nuclei of human cells.

21.29 When a cell is about to divide, the DNA unwinds and the hydrogen bonds joining the complementary bases break, forming two daughter strands. With the help of the enzyme DNA polymerase, a complement to each daughter strand with the correct complementary bases in the correct sequence is formed. The hydrogen bonds between the strands then re-form, resulting in two complete copies of the original DNA, one for each daughter cell.

Problems by Topic

Lipids

21.31 (a) No, it is an alkane chain.
(b) No, it is an amino acid.
(c) Yes, it is a saturated fatty acid.
(d) Yes, it is a steroid, cholesterol.

21.33　(a)　Yes, it is a saturated fatty acid.

(b)　No, it is an alkane.

(c)　No, it is an ether.

(d)　Yes, it is a monounsaturated fatty acid.

21.35

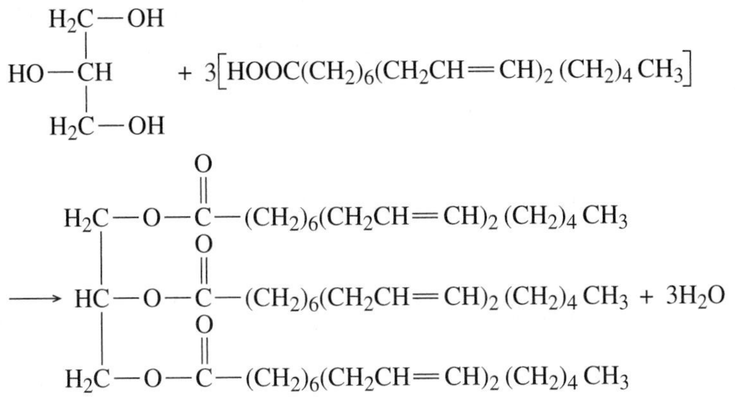

$$H_2C-OH$$
$$HO-CH \quad + 3\left[HOOC(CH_2)_6(CH_2CH=CH)_2(CH_2)_4CH_3\right]$$
$$H_2C-OH$$

$$\xrightarrow{\quad} \begin{array}{l} H_2C-O-\overset{\overset{\displaystyle O}{\|}}{C}-(CH_2)_6(CH_2CH=CH)_2(CH_2)_4CH_3 \\[4pt] HC-O-\overset{\overset{\displaystyle O}{\|}}{C}-(CH_2)_6(CH_2CH=CH)_2(CH_2)_4CH_3 \;+\; 3H_2O \\[4pt] H_2C-O-\overset{\overset{\displaystyle O}{\|}}{C}-(CH_2)_6(CH_2CH=CH)_2(CH_2)_4CH_3 \end{array}$$

The double bonds in the linoleic acid would most likely make this triglyceride liquid at room temperature.

Carbohydrates

21.37　(a)　Yes, it is a carbohydrate; it is a monosaccharide.

(b)　No, it is not a carbohydrate; it is a steroid: β-estradiol.

(c)　Yes, it is a carbohydrate; it is a disaccharide.

(d)　No, it is not a carbohydrate; it is a peptide.

21.39　(a)　The C—O double bond was on a terminal carbon prior to ring formation, so it is an aldose. There are six carbon atoms, so it is a hexose.

(b)　The C—O double bond is on a terminal carbon, so it is an aldose. There are five carbon atoms, so it is a pentose.

(c)　The C—O double bond is *not* on a terminal carbon, so it is a ketose. There are four carbon atoms, so it is a tetrose.

(d)　The C—O double bond was on a terminal carbon prior to ring formation, so it is an aldose. There are four carbon atoms, so it is a tetrose.

21.41　(a)　Five of the six carbon atoms have four different substituents attached; therefore, there are five chiral centers.

(b)　Three of the five carbon atoms have four different substituents attached; therefore, there are three chiral centers.

(c)　Only one of the four carbon atoms has four different substituents attached; therefore, there is only one chiral center.

(d)　Three of the four carbon atoms have four different substituents attached; therefore, there are four chiral centers.

21.43　Glucose

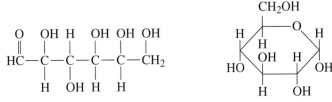

21.45   Hydrolysis is the addition of a water molecule across the glycosidic linkage to make monosaccharides.

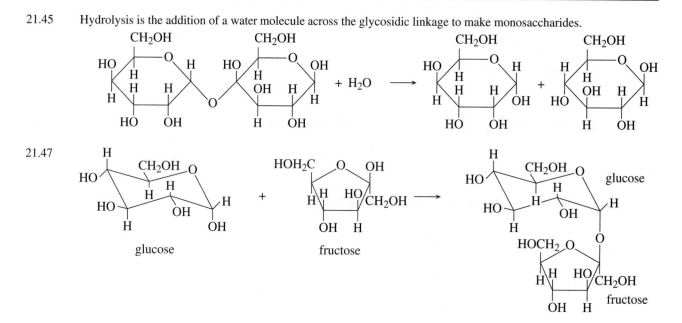

21.47

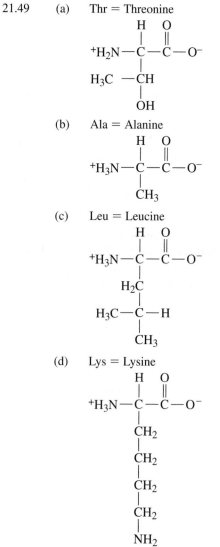

glucose   fructose

Amino Acids and Proteins

21.49   (a)   Thr = Threonine

(b)   Ala = Alanine

(c)   Leu = Leucine

(d)   Lys = Lysine

21.51   L – Alanine   D – Alanine

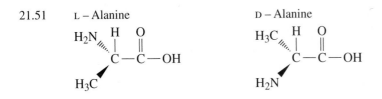

21.53   Because the N-terminal end and the C-terminal end are specific in a peptide, six tripeptides can be made from the three amino acids: serine, glycine, and cysteine.

Ser – Gly – Cys   Ser – Cys – Gly   Gly – Cys – Ser   Gly – Ser – Cys

Cys – Ser – Gly   Cys – Gly – Ser

21.55   Peptide bonds are formed when the carboxylic end of one amino acid reacts with the amine end of another amino acid.

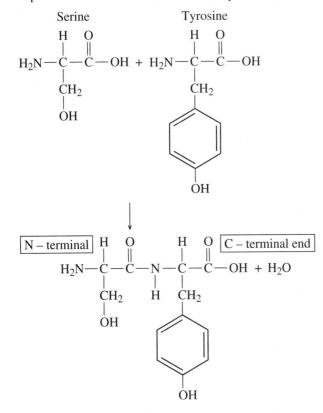

21.57   Peptide bonds are formed when the carboxylic end of one amino acid reacts with the amine end of another amino acid.

(a)   Gln – Met – Cys

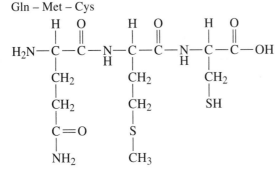

(b)   Ser – Leu – Cys

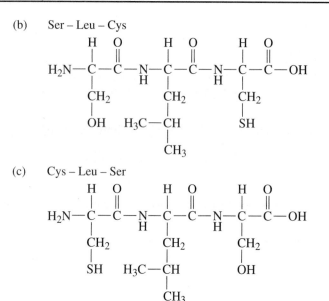

(c)   Cys – Leu – Ser

21.59   This would be a tertiary structure. Tertiary structure refers to the large-scale twists and folds of globular proteins. These are maintained by interactions between the R groups of amino acids that are separated by long distances in the chain sequence.

21.61   This would be a primary structure. Primary structure is simply the amino acid sequence. It is maintained by the peptide bonds that hold amino acids together.

Nucleic Acids

21.63   (a)   Yes, it is a nucleotide and the base is A, adenine.
         (b)   No, it is not a nucleotide; it is the base A, adenine.
         (c)   Yes, it is a nucleotide and the base is T, thymine.
         (d)   No, it is not a nucleotide; it is the amino acid methionine.

21.65   The two purine bases:

Adenine   Guanine

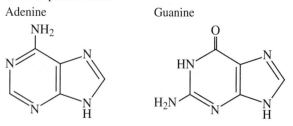

21.67   The bases pair in the specific arrangement: adenine always pairs with thymine, and guanine always pairs with cytosine.

The sequence

T G T A C G C

would have the complementary sequence

A C A T G C G

21.69   Every amino acid has a specific codon that is comprised of three nucleotides. So 154 amino acids would need 154 codons and 462 nucleotides.

Cumulative Problems

21.71 (a) Peptide bonds link amino acids to form proteins.
      (b) A glycosidic linkage is the connection between sugar molecules to form carbohydrates.
      (c) An ester linkage forms between the glycol molecule and a fatty acid to form lipids.

21.73 A codon is a sequence of three nucleotides with their associated bases. A nucleotide is a combination of a sugar group, a phosphate group, and a base. A codon codes for a specific amino acid. A gene is a sequence of codons that codes for a specific protein.

21.75 Write the Lewis structure of alanine:

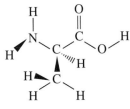

Determine the VSEPR geometry:
    N has three bonding pairs and one lone pair, which is trigonal pyramidal molecular geometry.
    C1 has four bonding pairs of electrons, which gives tetrahedral molecular geometry.
    C2 has three bonding pairs of electrons, which gives trigonal planar molecular geometry.
    C3 has four bonding pairs of electrons, which gives tetrahedral molecular geometry.
    O has two bonding pairs of electrons and two lone pairs, which gives bent molecular geometry.

21.77 Amino acids that are most likely to be involved in hydrophobic interactions have a nonpolar R group. These would be alanine, valine, leucine, isoleucine, and phenylalanine.

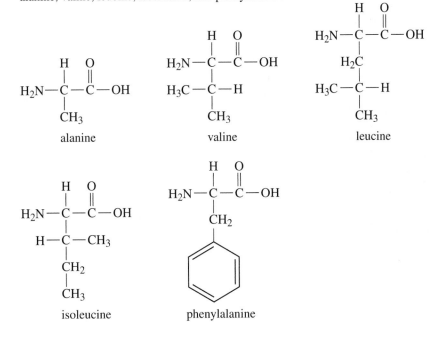

21.79   The reagents added break up the polypeptide but do not rearrange it. Therefore, by comparing the fragments from the two reagents, we can deduce the structure of the polypeptide.

Reagent 1:   Ala – Leu – Phe – Gly – Asn – Lys   Trp – Glu – Cys   Gly – Arg

Reagent 2:   Gly – Arg – Ala – Leu – Phe   Gly – Asn – Lys – Trp   Glu – Cys

So the sequence is   Gly – Arg – Ala – Leu – Phe – Gly – Asn – Lys – Trp – Glu – Cys

21.81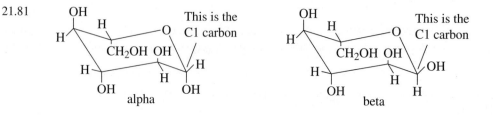

21.83   From the Gibbs–Helmholtz equation $\Delta G = \Delta H - T\Delta S$ when ΔG is negative, the process is spontaneous. At high temperature, the ΔS term (which is positive) will dominate, ΔG will be negative, and the DNA strand will be disrupted. At low temperature, the enthalpy of formation of the hydrogen bonds between the base pairs is a negative term and will dominate and cause ΔG to be negative; thus, the DNA strand will form.

Challenge Problems

21.85   The DNA chain forms with a phosphate linkage between the C5 carbon on one sugar and the C3 carbon on the next sugar. When the fake thymine is introduced into the cell, DNA replication cannot occur because the N chain is attached at the C_3 position instead of the $-$OH group and this will keep future phosphate linkages from forming between the sugars and stop the replication.

21.87   Rearrange the Michaelis–Menten equation to the form of a straight line: $y = mx + b$.

$$V_0 = \frac{V_{max}[\,glucose\,]}{K_t + [\,glucose\,]} \qquad \frac{1}{V_0} = \frac{K_t + [\,glucose\,]}{V_{max}[\,glucose\,]} \qquad \frac{1}{V_0} = \frac{K_t}{V_{max}[\,glucose\,]} + \frac{[\,\cancel{glucose}\,]}{V_{max}[\,\cancel{glucose}\,]}$$

$$\frac{1}{V_0} = \frac{K_t}{V_{max}[\,glucose\,]} + \frac{1}{V_{max}} \qquad \text{So } y = \frac{1}{V_0}; x = \frac{1}{[\,glucose\,]}; m = \frac{K_t}{V_{max}}; b = \frac{1}{V_{max}}$$

[glucose] (mM)	V_0 (μM/min)	1/[glucose] (L/mmol)	$1/V_0$ (min/μM)
0.5	12	2.0	0.083
1.0	19	1.0	0.053
2.0	27	0.50	0.037
3.0	32	0.33	0.031
4.0	35	0.25	0.029

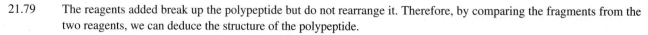

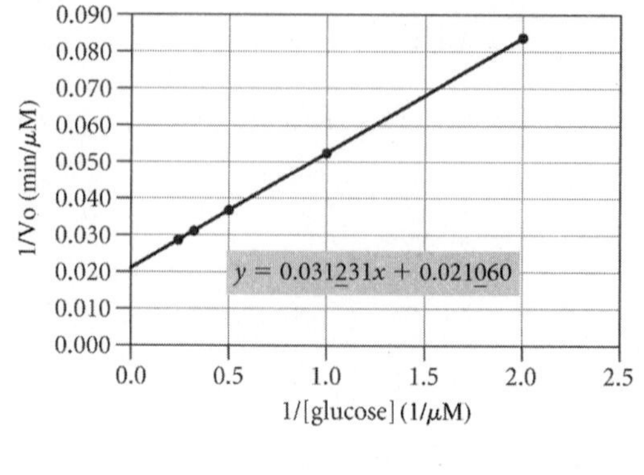

$y = 0.031231x + 0.021060$

$$y \text{ intercept} = b = 0.021\underline{0}60 \text{ min}/\mu M = \frac{1}{V_{max}}; \text{ so } V_{max} = \frac{1}{0.21\underline{0}60 \text{ min}/\mu M} = 47.5 \ \mu M/\text{min}$$

$$\text{slope} = m = 0.031\underline{2}31 \text{ min/mmol} = \frac{K_t}{V_{max}}; \text{ and}$$

$$K_t = (0.031\underline{2}361 \text{ min/mmol}) \times V_{max} = (0.031231 \ \text{min/mmol}) \times (47.5 \ \mu M/\text{min}) = 1.48 \text{ mmol}$$

21.89   At pH $= 2$: $H_3N^+CH_2COOH \rightleftharpoons H_3N^+CH_2COO^- + H^+ \quad pK_a = 2.3$

$$pH = pK_a + \log\frac{[A^-]}{[HA]} \qquad 2.0 = 2.3 + \log\frac{[H_3N^+CH_2COO^-]}{[H_3N^+CH_2COOH]} \qquad \frac{[A^-]}{[HA]} = 0.5$$

At pH $= 10$: $H_3N^+CH_2COO^- \rightleftharpoons H_2NCH_2COO^- + H^+ \quad pK_a = 9.6$

$$pH = pK_a + \log\frac{[A^-]}{[HA]} \qquad 10 = 9.6 + \log\frac{[H_2NCH_2COO^-]}{[H_3N^+CH_2COO^-]} \qquad \frac{[A^-]}{[HA]} = 2.5$$

The glycine is neutral when $[H_3N^+CH_2COOH] = [H_2NCH_2COO^-]$.

$H_3N^+CH_2COOH \rightleftharpoons \cancel{H_3N^+CH_2COO^-} + H^+ \quad pK_a = 2.3$
$\cancel{H_3N^+CH_2COO^-} \rightleftharpoons H_2NCH_2COO^- + H^+ \quad pK_a = 9.6$
$\overline{H_3N^+CH_2COOH \rightleftharpoons H_2NCH_2COO^- + 2\,H^+ \quad pK_a = 11.9}$

$$2\,pH = pK_a + \log\frac{[H_2NCH_2COO^-]}{[H_3N^+CH_2COOH]} \quad \text{but we have said that } [HNCH_2COO^-] = [H_3N^+CH_2COOH]$$

so $2\,pH = pK_a = 11.9$ and $pH = 5.95 = 6$

Conceptual Problems

21.91   A three-base codon codes for a single amino acid. If there were only three bases, there could be only six different three-base codon arrangements. Therefore, you could code only for 6 different amino acids, and you could not get the 20 different amino acids needed.

22 Chemistry of the Nonmetals

Review Questions

22.1   Boron nitride contains BN units that are isoelectronic with carbon in the sense that each BN unit contains eight valence electrons, or four per atom (just like carbon). Also, the size and electronegativity of the carbon atom is almost equal to the average of a boron atom and nitrogen atom. Therefore, BN forms a number of structures that are similar to those formed by carbon, including nanotubes.

22.3   The metallic character of the main-group elements increases going down each column due to the increasing size of the atoms and decreasing electronegativity.

22.5   Silica melts when heated above 1500 °C. After melting, if cooled quickly, silica does not crystallize back into the quartz structure. Instead, the Si atoms and O atoms form a randomly ordered or amorphous structure called a glass. A slow cooling will result in the crystalline quartz structure.

22.7   (a)   Orthosilicates (or nesosilicates) are minerals in which silicon bonds to oxygen in tetrahedral SiO_4^{4-} polyatomic anions (four electrons are gained, satisfying the octet rule for the oxygen atoms) that are isolated in the structure (not bonded to other tetrahedrons). The minerals require cations that have a total charge of 4+ to neutralize the negative charge.

   (b)   Amphiboles are minerals with double silicate chains. The repeating unit in the crystal is $Si_4O_{11}^{6-}$. Half of the tetrahedrons are bonded by two of the four corner O atoms, and half of the tetrahedrons are bonded by three of the four corners, bonding the two chains together. The bonding within the double chains is very strong, but the bonding between the double chains is not as strong, often resulting in fibrous-type minerals such as asbestos.

   (c)   Pyroxenes (or inosilicates) are minerals in which many of the silica tetrahedrons are bonded together forming chains. The formula unit for these chains is the SiO_3^{2-} unit, and the repeating unit in the structure is two formula units $(Si_2O_6^{4-})$. Two of the oxygen atoms are bonded to two silicon atoms (and thus to two other tetrahedrons) on two of the four corners of each tetrahedron. The silicate chains are held together by ionic bonding to metal cations that lie between the chains.

   (d)   Pyrosilicates (or sorosilicates) are minerals in which the silicate tetrahedrons can also form structures in which two tetrahedrons share one corner forming a disilicate ion, which has the formula $Si_2O_7^{6-}$. This group requires cations that balance the 6– charge on $Si_2O_7^{6-}$.

   (e)   Feldspars are a common aluminosilicate in which SiO_2 units become AlO_2^- upon substitution of aluminum. The negative charge is balanced by a positive counterion. This leads to a covalent network of $AlSi_3O_8^-$ or $Al_2Si_2O_8^{2-}$ units.

22.9   Boron tends to form electron-deficient compounds (compounds in which boron lacks an octet) because it has only three valence electrons. Examples of these compounds are boron halides such as BF_3 and BCl_3.

22.11   Solid carbon dioxide is referred to as dry ice because at 1 atmosphere, it goes directly from a solid to a gas, without going through a liquid or wet phase.

22.13   About 21% of Earth's atmosphere is composed of O_2.

22.15   Earth's early atmosphere was reducing (instead of oxidizing) and contained hydrogen, methane, ammonia, and carbon dioxide. About 2.7 billion years ago, cyanobacteria (blue-green algae) began to convert the carbon dioxide and water to oxygen by photosynthesis.

Problems by Topic

Silicates: The Most Abundant Matter in Earth's Crust

22.17   (a)   +4. In SiO_2, (Si ox state) + 2(O ox state) = 0. Because O ox state = −2, Si ox state = −2(−2) = +4.

      (b)   +4. In SiO_4^{4-}, (Si ox state) + 4(O ox state) = −4. Because O ox state = −2, Si ox state = −4 − 4(−2) = +4.

      (c)   +4. In $Si_2O_7^{6-}$, 2(Si ox state) + 7(O ox state) = −6. Because O ox state = −2, Si ox state $= \frac{1}{2}(-6 - 7(-2)) = +4$.

22.19   Three SiO_4^{4-} units have a total charge of −12. In trying to combine Ca^{2+} and Al^{3+} to get a total charge of +12, the only possible combination is to use 3 Ca^{2+} and 2 Al^{3+} ions. The formula unit is $Ca_3Al_2(SiO_4)_3$.

22.21   Because OH has a −1 charge, Al has a +3 charge, and Si_2O_5 has a −2 charge, 2(Al ox state) + 1(Si_2O_5 charge) + x (OH⁻ charge) = 0; so 2(+3) + 1(−2) + x(−1) = 0 and x = 4.

22.23   In Table 22.2, SiO_4 has a −4 charge and it belongs to the class of orthosilicates, where the silicate tetrahedra stand alone, not linked to each other. The −4 charge balances with a common oxidation state of Zr of +4.

22.25   Ca has a +2 charge, Mg has a +2 charge, Fe commonly has a charge of +2 or +3, and OH has a charge of −1. So far, the charge balance is 2(+2) + 4(+2) + 1(+2 or +3) + (Si_7AlO_{22} charge) + 2(−1) = 0 or Si_7AlO_{22} charge = −12 or −13. Looking at Table 22.2 and recalling that one AlO_2^- unit replaces one SiO_2 unit, $Si_4O_{11}^{6-}$ units are the best match. If we double the unit and replaced one-eighth of the Sis by Al, the new unit will be Si_7AlO_{22} and the charge will be 2(−6) − 1 = −13. The charge on Fe will be +3, and the class is amphibole, with a double chain structure.

Boron and Its Remarkable Structures

22.27   **Given:** $Na_2[B_4O_5(OH)_4] \cdot 3 H_2O$ = kernite, 1.0×10^3 kg of kernite-bearing ore, 0.98% by mass kernite, 65% B recovery   **Find:** g boron recovered

    Conceptual Plan: kg ore → g ore → g kernite → g B possible → g B recovered

$$\frac{1000\ \text{g}}{1\ \text{kg}} \quad \frac{0.98\ \text{g kernite}}{100\ \text{g ore}} \quad \frac{4(10.81)\ \text{g B}}{291.23\ \text{g kernite}} \quad \frac{65\ \text{g B recovered}}{100\ \text{g B}}$$

    Solution:

$$1.0 \times 10^3\ \text{kg ore} \times \frac{1000\ \text{g ore}}{1\ \text{kg ore}} \times \frac{0.98\ \text{g kernite}}{100\ \text{g ore}} \times \frac{4(10.81)\ \text{g B}}{291.23\ \text{g kernite}} \times \frac{65\ \text{g B recovered}}{100\ \text{g B}} = 950\ \text{g B recovered}$$

    Check: The units (g) are correct. The magnitude of the answer makes physical sense because there is very little mineral in the ore, a low recovery, and little B in the ore.

22.29   The bond angles are different in BCl_3 versus NCl_3 because B has three valence electrons and N has five valence electrons. The boron compound does not obey the octet rule: B has sp^2 hybridization and a trigonal planar geometry. The nitrogen compound does obey the octet rule and has a lone pair of electrons, sp^3 hybridization, and a trigonal pyramidal geometry (based on a tetrahedral geometry).

22.31   (a)   Looking at Figure 22.11, $B_6H_6^{2-}$ has six vertices and eight faces because it has an octahedral shape.

      (b) $B_{12}H_{12}^{2-}$ has the formula of a *closo*-borane $(B_nH_n^{2-})$ and has 12 vertices and 20 faces because it has an icosahedral shape.

22.33   *Closo*-boranes have the formula $B_nH_n^{2-}$ and form fully closed polyhedra with triangular sides. Each of the vertices in the polyhedra is a boron atom with an attached hydrogen atom. *Nido*-boranes, named from the Latin word for *net*, have the formula B_nH_{n+4}. They consist of a cage of boron atoms missing one corner. *Arachno*-boranes, named from the Greek word for *web*, have the formula B_nH_{n+6}. They consist of a cage of boron atoms that is missing two or three corners.

Carbon, Carbides, and Carbonates

22.35   Graphite is a good lubricant because the structure has strong covalent bonds within the sheets of the structure, but weak interactions between the sheets that allow the sheets to easily slide over one another. The diamond structure consists of carbon atoms connected to four other carbon atoms at the corners of a tetrahedron. This bonding extends throughout three dimensions, making giant molecules described as network covalent solids. There are no weak interactions in the diamond structure.

22.37   Regular charcoal is in large chunks that still resemble the wood from which it is made. Activated charcoal is treated with steam to break up the chunks into a finely divided powder with a high surface area.

22.39   Ionic carbides are composed of carbon and low-electronegativity metals such as the alkali metals and alkaline earth metals. Most of the ionic carbides contain the dicarbide ion, C_2^{2-}, commonly called the acetylide ion. Covalent carbides are composed of carbon and low-electronegativity nonmetals or metalloids. The most important covalent carbide is silicon carbide (SiC), a very hard material.

22.41   (a)   As the pressure is reduced, the solid will be converted directly to a gas.
              (b)   As the temperature is reduced, the gas will be converted to a liquid and then to a solid.
              (c)   As the temperature is increased, the solid will be converted directly to a gas.

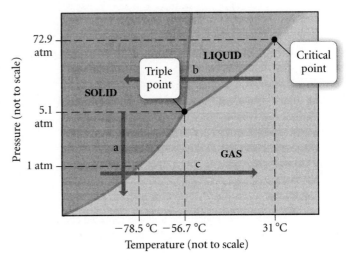

22.43   (a)   The reaction will be similar to that of iron oxide and carbon monoxide: $CO(g) + CuO(s) \xrightarrow{\text{Heat}} CO_2(g) + Cu(s)$. Note that the oxidation state of carbon increases by two and the oxidation state of Cu decreases by two.
              (b)   This reaction is discussed in Section 22.5, $SiO_2(s) + 3\,C(s) \xrightarrow{\text{Heat}} SiC(s) + 2\,CO(g)$.
              (c)   This reaction is discussed in Section 22.5, $CO(g) + S(s) \xrightarrow{\text{Heat}} COS(g)$.

22.45   (a) $+2$. In CO, (C ox state) + (O ox state) = 0. Because O ox state = -2, C ox state = $+2$.
              (b) $+4$. In CO_2, (C ox state) + 2(O ox state) = 0. Because O ox state = -2, C ox state = $-2(-2) = +4$.
              (c) $+4/3$. In C_3O_2, 3(C ox state) + 2(O ox state) = 0. Because O ox state = -2, C ox state = $1/3(-2(-2)) = +4/3$.

Nitrogen and Phosphorus: Essential Elements for Life

22.47   Even though there is a tremendous amount of N_2 gas in the atmosphere, the strong triple bond in elemental nitrogen renders it unusable by plants. To be used as fertilizer, elemental nitrogen has to be fixed, which means that it has to be converted into a nitrogen-containing compound such as NH_3.

22.49   White phosphorus consists of P_4 molecules in a tetrahedral shape, with the phosphorus atoms at the corners of the tetrahedron as shown in Figure 22.19. The bond angles between the three P atoms on any one face of the tetrahedron are small ($60°$) and strained, making the P_4 molecule unstable and reactive. When heated to about 300 °C in the absence of air, white phosphorus slightly changes its structure to a different allotrope called red phosphorus, which is amorphous. The general structure of red phosphorus is similar to that of white phosphorus, except that one of the bonds between two phosphorus atoms in the tetrahedron is broken, as shown in Figure 22.20. These two phosphorus atoms then link to other phosphorus atoms, making chains that can vary in structure. The red phosphorus structure is more stable because there is less strain when the linkage is changed.

22.51   **Given:** saltpeter (KNO_3) and Chile saltpeter ($NaNO_3$)   **Find:** mass percent N in each mineral
             Conceptual Plan: Mineral formula $\rightarrow$ mass percent nitrogen

$$\text{mass percent nitrogen} = \frac{14.01\ \text{g}}{\text{formula weight of mineral}} \times 100\%$$

             Solution: for KNO_3:

$$\text{mass percent nitrogen} = \frac{14.01\ \text{g}}{\text{formula weight of mineral}} \times 100\% = \frac{14.01\ \cancel{\text{g}}}{101.11\ \cancel{\text{g}}} \times 100\% = 13.86\%$$

and for $NaNO_3$:

$$\text{mass percent nitrogen} = \frac{14.01 \text{ g}}{\text{formula weight of mineral}} \times 100\% = \frac{14.01 \text{ g}}{85.00 \text{ g}} \times 100\% = 16.48\%.$$

Check: The units (%) are correct. The relative magnitude of the answers makes physical sense because K is heavier than Na.

22.53   **Given:** decomposition of hydrogen azide (HN_3) to its elements
Find: stability at room temperature, stability at any temperature
Conceptual Plan: Write the reaction. Evaluate ΔG°_{rxn} then determine if $\Delta G_{rxn} > 0$ at any T.

$$\Delta G = \Delta H_{rxn} - T\Delta S_{rxn}$$

Solution: $HN_3(g) \rightarrow 3/2\, N_2(g) + 1/2\, H_2(g)$, at 25 °C $\Delta G^\circ_{rxn} = -\Delta G^\circ_f(HN_3(g)) = -1422$ kJ/mol so the decomposition is spontaneous at room temperature. $\Delta H^\circ_{rxn} = -\Delta H^\circ_f(HN_3(g)) = -1447$ kJ/mol, and $\Delta S^\circ_{rxn} > 0$ because the decomposition generates more moles of gas. Because $\Delta G = \Delta H_{rxn} - T\Delta S_{rxn}$, we see that both the enthalpy and entropy term favor decomposition; so ΔG_{rxn} is always negative, and hydrogen azide is unstable at all temperatures.

Check: Azides are used as explosives, so it is not surprising that they are unstable.

22.55   (a)   This reaction is a decomposition reaction, $NH_4NO_3(aq) + \text{heat} \rightarrow N_2O(g) + 2\, H_2O(l)$.
(b)   This reaction is discussed in Section 22.6, $3\, NO_2(g) + H_2O(l) \rightarrow 2\, HNO_3(aq) + NO(g)$.
(c)   This reaction is discussed in Section 22.6, $2\, PCl_3(g) + O_2(g) \rightarrow 2\, POCl_3(l)$.

22.57   For N_3^-: $3(\text{N ox state}) = -1$, so N ox state $= -1/3$.
For $N_2H_5^+$: $2(\text{N ox state}) + 5(\text{H ox state}) = +1$. Because H ox state $= +1$, N ox state $= \frac{1}{2}(1 - 5(+1)) = -2$.
For NO_3^-: $1(\text{N ox state}) + 3(\text{O ox state}) = -1$. Because O ox state $= -2$, N ox state $= -1 - 3(-2) = +5$.
For NH_4^+: $1(\text{N ox state}) + 4(\text{H ox state}) = +1$. Because H ox state $= +1$, N ox state $= +1 - 4(+1) = -3$.
For NO_2^-: $1(\text{N ox state}) + 2(\text{O ox state}) = -1$. Because O ox state $= -2$, N ox state $= -1 - 2(-2) = +3$.
So $NO_3^- > NO_2^- > N_3^- > N_2H_5^+ > NH_4^+$.

22.59   P has five valence electrons, and Cl has seven valence electrons; so PCl_3 has $5 + 3(7) = 26$ valence electrons, and PCl_5 has $5 + 5(7) = 40$ valence electrons. Because the electronegativity of P is less than the electronegativity of Cl, the P is in the middle. So the Lewis dot structures are as follows:

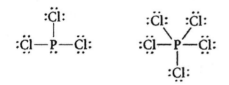

Notice that the PCl_5 violates the octet rule, which is allowed because P is in the third row of the periodic table and Cl is a fairly small atom. Because the PCl_3 has four electron groups surrounding the P atom, the orbital geometry is tetrahedral. Because there is one lone pair, the VSEPR geometry is trigonal pyramidal. Because the PCl_5 has five electron groups surrounding the P atom, the VSEPR geometry is trigonal bipyramidal.

22.61   **Given:** urea $(CO(NH_2)_2)$ reacts with water to produce 23 g ammonium carbonate   **Find:** g of urea needed
Conceptual Plan: Write a balanced equation.
$$\text{g } (NH_4)_2CO_3 \rightarrow \text{mol } (NH_4)_2CO_3 \rightarrow \text{mol } CO(NH_2)_2 \rightarrow \text{g } CO(NH_2)_2$$

$$\frac{1 \text{ mol } (NH_4)_2CO_3}{96.04 \text{ g } (NH_4)_2CO_3} \qquad \frac{1 \text{ mol } CO(NH_2)_2}{1 \text{ mol } (NH_4)_2CO_3} \qquad \frac{60.03 \text{ g } CO(NH_2)_2}{1 \text{ mol } CO(NH_2)_2}$$

Solution: $CO(NH_2)_2(aq) + 2\, H_2O(l) \rightarrow (NH_4)_2CO_3(aq)$ then

$$23 \text{ g } (NH_4)_2CO_3 \times \frac{1 \text{ mol } (NH_4)_2CO_3}{96.04 \text{ g } (NH_4)_2CO_3} \times \frac{1 \text{ mol } CO(NH_2)_2}{1 \text{ mol } (NH_4)_2CO_3} \times \frac{60.03 \text{ g } CO(NH_2)_2}{1 \text{ mol } CO(NH_2)_2} = 14 \text{ g } CO(NH_2)_2$$

Check: The units (g) are correct. The magnitude of the answer (14) makes physical sense because we are generating 23 g of $(NH_4)_2CO_3$. Because we are adding water to the urea, the amount of urea must be less than 23 g.

22.63   Tetraphosphorus hexaoxide, $P_4O_6(s)$, forms when the amount of oxygen is limited, and tetraphosphorus decaoxide, $P_4O_{10}(s)$, is formed when greater amounts of oxygen are available.

Oxygen

22.65   The major commercial production method of elemental oxygen is the fractionation of air. Air is cooled until its components liquefy. Then the air is warmed, and components such as N_2 and Ar are separated, leaving oxygen behind.

22.67   In oxides, the oxidation state of O is -2; in peroxides, it is -1; and in superoxides, it is $-\frac{1}{2}$. Alkali metals and alkaline earth metals have only two possible oxidation states—0 and $+1$ for the alkali metals and $+2$ for alkaline earth metals.

        (a) LiO_2 is a superoxide because 1(Li ox state) + 2(O ox state) = 0 and O ox state $= -\frac{1}{2}(1) = -1/2$.

        (b) CaO is an oxide because 1(Ca ox state) + 2(O ox state) = 0 and O ox state $= -1(+2) = -2$.

        (c) K_2O_2 is a peroxide because 2(K ox state) + 2(O ox state) = 0 and O ox state $= -\frac{1}{2}(2(+1)) = -1$.

Sulfur: A Dangerous but Useful Element

22.69   Initially, liquid sulfur becomes less viscous when heated because the S_8 rings have greater thermal energy, which overcomes intermolecular forces. Above 150 °C, the rings break and the broken rings entangle one another, causing greater viscosity. Above 180 °C, the intermolecular forces weaken and the solution becomes less viscous.

22.71   **Given:** 1.0 L of 5.00×10^{-5} M Na_2S **Find:** maximum g of metal sulfide that can dissolve
        Other: K_{sp} (PbS) $= 9.04 \times 10^{-29}$, K_{sp} (PbS) $= 2 \times 10^{-25}$
        Conceptual Plan: M $Na_2S \rightarrow$ M $S^{2-} \rightarrow$ M metal sulfide $\rightarrow$ mol metal sulfide $\rightarrow$ g metal sulfide

              $Na_2S(s) \rightarrow 2\,Na^+(aq) + S^{2-}(aq)$ ICE table $M = \dfrac{mol}{L}$ $\mathcal{M}$

        Solution: Because 1 S^{2-} ion is generated for each Na_2S, $[S^{2-}] = 5.00 \times 10^{-5}$ M Na_2S.

        (a) $PbS(s) \rightleftharpoons Pb^{2+}(aq) + S^{2-}(aq)$

Initial	0.00	5.00×10^{-5}
Change	$+S$	$+S$
Equil	S	$5.00 \times 10^{-5} + S$

                                        K_{sp} (PbS) $= [Pb^{2+}][S^{2-}] = 9.04 \times 10^{-29} = S(5.00 \times 10^{-5} + S)$

        Assuming $S << 5.00 \times 10^{-5}$, $9.04 \times 10^{-29} = S(5.00 \times 10^{-5})$ and $S = 1.81 \times 10^{-24}$ M. (Note that the assumption that S was very small was good.) Because there is 1.0 L of solution, we can dissolve at most

        1.8×10^{-24} moles of PbS then 1.8×10^{-24} ~~mol PbS~~ $\times \dfrac{239.3 \text{ g PbS}}{1 \text{ ~~mol PbS~~}} = 4.3 \times 10^{-22}$ g PbS.

        (b) $ZnS(s) \rightleftharpoons Zn^{2+}(aq) + S^{2-}(aq)$

Initial	0.00	5.00×10^{-5}
Change	$+S$	$+S$
Equil	S	$5.00 \times 10^{-5} + S$

                                          K_{sp} (ZnS) $= [Zn^{2+}][S^{2-}] = 2 \times 10^{-25} = S(5.00 \times 10^{-5} + S)$

        Assuming $S << 5.00 \times 10^{-5}$, $2 \times 10^{-25} = S(5.00 \times 10^{-5})$ and $S = 4 \times 10^{-21}$ M. (Note that the assumption that S was very small was good.) Because there is 1.0 L of solution, we can dissolve, at most, 4×10^{-21}

        moles of ZnS then 4×10^{-21} ~~mol ZnS~~ $\times \dfrac{97.48 \text{ g ZnS}}{1 \text{ ~~mol ZnS~~}} = 4 \times 10^{-19}$ g ZnS.

        Check: The units (g) are correct. The magnitude of the answer makes physical sense because the solubility constants are so small. The amount of PbS that can dissolve is less than the amount of ZnS because the solubility constant is much smaller.

22.73   **Given:** iron pyrite (FeS_2) roasted generating $S_2(g)$, 5.5 kg of iron pyrite   **Find:** balanced reaction and $V\,S_2$ at STP
        Conceptual Plan: Write the reaction then kg $FeS_2 \rightarrow$ g $FeS_2 \rightarrow$ mol $FeS_2 \rightarrow$ mol $S_2 \rightarrow V\,S_2$.

                               $\dfrac{1000 \text{ g}}{1 \text{ kg}}$ $\dfrac{1 \text{ mol } FeS_2}{119.99 \text{ g } FeS_2}$ $\dfrac{1 \text{ mol } S_2}{2 \text{ mol } FeS_2}$ $\dfrac{22.414 \text{ L } S_2}{1 \text{ mol } S_2}$

        Solution: $2\,FeS_2(s) \xrightarrow{\text{Heat}} 2\,FeS(s) + S_2(g)$ then

        5.5 ~~kg FeS_2~~ $\times \dfrac{1000 \text{ g ~~FeS_2~~}}{1 \text{ ~~kg FeS_2~~}} \times \dfrac{1 \text{ ~~mol FeS_2~~}}{119.99 \text{ g ~~FeS_2~~}} \times \dfrac{1 \text{ ~~mol S_2~~}}{2 \text{ ~~mol FeS_2~~}} \times \dfrac{22.414 \text{ L } S_2}{1 \text{ ~~mol S_2~~}} = 5.1 \times 10^2 \text{ L}\,S_2$

        Check: The units (L) are correct. The magnitude of the answer makes physical sense because there is much more than a mole of iron pyrite; so the volume of gas is many times 22 L.

Halogens

22.75 **(a)** +2. In XeF_2, 1(Xe ox state) + 2(F ox state) = 0. Because F ox state = −1, Xe ox state = −2(−1) = +2. Xe has eight valence electrons, and F has seven valence electrons; so XeF_2 has 8 + 2(7) = 22. Because the electronegativity of Xe is less than the electronegativity of F, the Xe is in the middle. So the Lewis dot

structure is $:\ddot{F}-\ddot{X}\ddot{e}-\ddot{F}:$

Notice that the XeF_2 violates the octet rule, which is allowed because Xe is in the fifth row of the periodic table and F is a fairly small atom. Because there are five electron groups surrounding the Xe atom, the orbital geometry is trigonal bipyramidal. Because there are three lone pairs, the VSEPR geometry is linear.

(b) +6. In XeF_6, 1(Xe ox state) + 6(F ox state) = 0. Because F ox state = −1, Xe ox state = −6(−1) = +6. Xe has eight valence electrons, and F has seven valence electrons; so XeF_6 has 8 + 6(7) = 50. Because the electronegativity of Xe is less than the electronegativity of F, the Xe is in the middle. So the Lewis dot structure is shown to the right. Notice that the XeF_6 violates the octet rule, which is allowed because Xe is in the fifth row of the periodic table and F is a fairly small atom. Because there are 7 electron groups surrounding the Xe atom, the VSEPR geometry is a distorted octahedral.

(c) +6. In $XeOF_4$, 1(Xe ox state) + 1(O ox state) + 4(F ox state) = 0. Because F ox state = −1 and O ox state = −2 Xe ox state = −(1(−2) + 4(−1)) = +6. Xe has eight valence electrons, O has six valence electrons, and F has seven valence electrons; so $XeOF_4$ has 8 + 1(6) + 4(7) = 42. Because the electronegativity of Xe is less than the electronegativity of F, the Xe is in the middle. Oxygen needs a double bond. So the Lewis dot structure is shown to the right. Notice that the $XeOF_4$ violates the octet rule, which is allowed because Xe is in the fifth row of the periodic table and O and F are fairly small atoms. Because there are six electron groups surrounding the Xe atom, the orbital geometry is octahedral. Because there is one lone pair, the VSEPR geometry is square pyramidal.

22.77 The red color is from molecular bromine (Br_2), so $Cl_2(g) + 2\,Br^-(aq) \rightarrow 2\,Cl^-(aq) + Br_2(l)$. Cl is being reduced, and Br is being oxidized; so Cl_2 is the oxidizing agent, and Br^- is the reducing agent.

22.79 **Given:** 55 g SiO_2 and 111 L of 0.032 M HF **Find:** limiting reagent and amount of excess reagent left
Conceptual Plan: Write the reaction then M HF, L HF → mol HF → mol SiO_2 → g SiO_2 then

$$M = \frac{mol}{L} \qquad \frac{1\ mol\ SiO_2}{4\ mol\ HF} \qquad \frac{60.09\ g\ SiO_2}{1\ mol\ SiO_2}$$

determine the limiting reagent then determine amount of excess reagent left.

$$\text{amount of excess reactant} = \text{initial amount} - \text{amount reacted}$$

Solution: $4\,HF(g) + SiO_2(s) \rightarrow SiF_4(g) + 2\,H_2O(l)$ then $\dfrac{0.032\ mol\ HF}{1\ \cancel{L}} \times 111\ \cancel{L\ HF} = 3.\underline{5}52\ mol\ HF$

then $3.\underline{5}52\ \cancel{mol\ HF} \times \dfrac{1\ mol\ SiO_2}{4\ \cancel{mol\ HF}} \times \dfrac{60.09\ g\ SiO_2}{1\ \cancel{mol\ SiO_2}} = 53.\underline{3}599\ g\ SiO_2$. Because 55 g of SiO_2 is available, HF is the

limiting reagent. The amount of excess SiO_2 is 55 g − 53.$\underline{3}$599 g = 2 g SiO_2.

Check: The units (g) are correct. The magnitude of the answer makes physical sense because the amount required to react all of the HF is just under the amount of glass available.

Cumulative Problems

22.81 **Given:** lignite coal 1 mol % S and bituminous coal 5 mol % S, 1.00×10^2 kg of coal
Find: g H_2SO_4 produced
Conceptual Plan: Table 22.3 Composition → $\mathcal{M}$ then

$$\mathcal{M} = \frac{(\text{mol }\%\ C)(12.01\ g/mol) + (\text{mol }\%\ H)(1.001\ g/mol) + (\text{mol }\%\ O)(16.00\ g/mol) + (\text{mol }\%\ S)(32.07\ g/mol)}{100\%}$$

kg coal → g coal → mol coal → mol S → mol H_2SO_4 → g H_2SO_4

$$\frac{1000\ g}{1\ kg} \qquad \mathcal{M} \qquad \frac{1\ or\ 5\ mol\ S}{100\ mol\ coal} \quad \frac{1\ mol\ H_2SO_4}{1\ mol\ S} \quad \frac{98.07\ g\ H_2SO_4}{1\ mol\ H_2SO_4}$$

Solution: lignite:

$$\mathcal{M} = \frac{(\text{mol}\%C)(12.01\ \text{g/mol})+(\text{mol}\%H)(1.001\ \text{g/mol})+(\text{mol}\%O)(16.00\ \text{g/mol})+(\text{mol}\%S)(32.07\ \text{g/mol})}{100\%} =$$

$$\frac{(71\ \text{mol}\%C)(12.01\ \text{g/mol})+(4\ \text{mol}\%H)(1.008\ \text{g/mol})+(23\ \text{mol}\%O)(16.00\ \text{g/mol})+(1\ \text{mol}\%S)(32.07\ \text{g/mol})}{100\%} =$$

$12.5678\ \text{g/mol}$

$$1.00 \times 10^2\ \text{kg-coal} \times \frac{1000\ \text{g-coal}}{1\ \text{kg-coal}} \times \frac{1\ \text{mol-coal}}{12.5678\ \text{g-coal}} \times \frac{1\ \text{mol-S}}{100\ \text{mol-coal}} \times \frac{1\ \text{mol}\ H_2SO_4}{1\ \text{mol-S}} \times \frac{98.07\ \text{g}\ H_2SO_4}{1\ \text{mol}\ H_2SO_4} =$$

$7.803 \times 10^3\ \text{g}\ H_2SO_4 = 8\ \text{kg}\ H_2SO_4$

bituminous:

$$\mathcal{M} = \frac{(\text{mol}\%C)(12.01\ \text{g/mol})+(\text{mol}\%H)(1.001\ \text{g/mol})+(\text{mol}\%O)(16.00\ \text{g/mol})+(\text{mol}\%S)(32.07\text{g/mol})}{100\%} =$$

$$\frac{(80\ \text{mol}\%C)(12.01\ \text{g/mol})+(6\ \text{mol}\%H)(1.008\ \text{g/mol})+(8\ \text{mol}\%O)(16.00\ \text{g/mol})+(5\ \text{mol}\%S)(32.07\ \text{g/mol})}{100\%} =$$

$12.55156\ \text{g/mol}$

$$1.00 \times 10^2\ \text{kg-coal} \times \frac{1000\ \text{g-coal}}{1\ \text{kg-coal}} \times \frac{1\ \text{mol-coal}}{12.55156\ \text{g-coal}} \times \frac{5\ \text{mol-S}}{100\ \text{mol-coal}} \times \frac{1\ \text{mol}\ H_2SO_4}{1\ \text{mol-S}} \times \frac{98.07\ \text{g}\ H_2SO_4}{1\ \text{mol}\ H_2SO_4} =$$

$3.9067 \times 10^4\ \text{g}\ H_2SO_4 = 40\ \text{kg}\ H_2SO_4$

Check: The units (kg and kg) are correct. The magnitude of the answer makes physical sense because sulfuric acid is heavier than sulfur and it more than balances out the sulfur content in the coal. Bituminous coal has a higher percentage of S than does lignite coal.

22.83   The strength of an acid is determined by the strength of the bond of the H to the rest of the molecule—the weaker the bond, the stronger the acid. Because Cl is more electronegative than I, it pulls density away from the O more than I does; so the O cannot interact with the H as strongly.

22.85   **Given:** HCl versus X = (a) Cl_2, (b) HF, and (c) HI   **Find:** ratio of effusion rates HCl/X
         Conceptual Plan: $\mathcal{M}(HCl), \mathcal{M}(X) \rightarrow$ **Rate (HCl)/Rate (X)**

$$\overset{..}{O} = \overset{+}{N} = \overset{..}{O} \text{ and } \quad \frac{\text{Rate (HCl)}}{\text{Rate }(X)} = \sqrt{\frac{\mathcal{M}(X)}{\mathcal{M}(HCl)}}$$

Solution: HCl: $\mathcal{M} = \dfrac{36.45\ \text{g}}{1\ \text{mol}} \times \dfrac{1\ \text{kg}}{1000\ \text{g}} = 0.03645\ \text{kg/mol}$

(a) Cl_2: $\mathcal{M} = \dfrac{70.90\ \text{g}}{1\ \text{mol}} \times \dfrac{1\ \text{kg}}{1000\ \text{g}} = 0.07090\ \text{kg/mol}$

$$\frac{\text{Rate (HCl)}}{\text{Rate }(Cl_2)} = \sqrt{\frac{\mathcal{M}(Cl_2)}{\mathcal{M}(HCl)}} = \sqrt{\frac{0.07090\ \text{kg/mol}}{0.03645\ \text{kg/mol}}} = 1.395$$

(b)   HF: $\mathcal{M} = \dfrac{20.00\ \text{g}}{1\ \text{mol}} \times \dfrac{1\ \text{kg}}{1000\ \text{g}} = 0.02000\ \text{kg/mol}$

$$\frac{\text{Rate (HCl)}}{\text{Rate (HF)}} = \sqrt{\frac{\mathcal{M}(HF)}{\mathcal{M}(HCl)}} = \sqrt{\frac{0.02000\ \text{kg/mol}}{0.03645\ \text{kg/mol}}} = 0.7407$$

(c)   HI: $\mathcal{M} = \dfrac{127.90\ \text{g}}{1\ \text{mol}} \times \dfrac{1\ \text{kg}}{1000\ \text{g}} = 0.12790\ \text{kg/mol}$

$$\frac{\text{Rate (HCl)}}{\text{Rate (HI)}} = \sqrt{\frac{\mathcal{M}(HI)}{\mathcal{M}(HCl)}} = \sqrt{\frac{0.12790\ \text{kg/mol}}{0.03645\ \text{kg/mol}}} = 1.873$$

Check: The units (none) are correct. The magnitude of the answers makes sense; the heavier molecules have the lower effusion rate (because they move slower).

22.87 Sodium peroxide is Na_2O_2, so the reaction is $4\,Na_2O_2 + 3\,Fe \rightarrow 4\,Na_2O + Fe_3O_4$. Note that the oxidation state of O is decreasing from -1 to -2, needing eight electrons, and the oxidation state of Fe is increasing from 0 to $+8/3$, generating eight moles of electrons for three moles of Fe.

22.89 **Given:** O_2^+, O_2, O_2^-, O_2^{2-}, and MO Theory **Find:** Explain bond lengths and state which are diamagnetic.
Conceptual Plan: Use the diagram from Chapter 10 and add the appropriate number of electrons to determine bond order and note whether they are diamagnetic.
Solution:
O_2^+ has 11 electrons in $n = 2$ shell; so its MO configuration is $\sigma_{2s}^2, \sigma_{2s}^{*2}, \sigma_{2p}^2, \pi_{2p}^4, \pi_{2p}^{*1}$, the bond order $=$

$$\frac{\text{bonding electrons} - \text{antibonding electrons}}{2} = \frac{8-3}{2} = 2.5, \text{ and it is paramagnetic (unpaired electron).}$$

O_2 has 12 electrons in $n = 2$ shell; so its MO configuration is $\sigma_{2s}^2, \sigma_{2s}^{*2}, \sigma_{2p}^2, \pi_{2p}^4, \pi_{2p}^{*2}$, and the bond order

$$= \frac{\text{bonding electrons} - \text{antibonding electrons}}{2} = \frac{8-4}{2} = 2.$$

O_2^- has 13 electrons in $n = 2$ shell; so its MO configuration is $\sigma_{2s}^2, \sigma_{2s}^{*2}, \sigma_{2p}^2, \pi_{2p}^4, \pi_{2p}^{*3}$, the bond order

$$= \frac{\text{bonding electrons} - \text{antibonding electrons}}{2} = \frac{8-5}{2} = 1.5, \text{ and it is paramagnetic (unpaired electron).}$$

O_2^{2-} has 14 electrons in $n = 2$ shell; so its MO configuration is $\sigma_{2s}^2, \sigma_{2s}^{*2}, \sigma_{2p}^2, \pi_{2p}^4, \pi_{2p}^{*4}$, the bond order

$$= \frac{\text{bonding electrons} - \text{antibonding electrons}}{2} = \frac{8-6}{2} = 1, \text{ and it is diamagnetic (all paired electrons).}$$

The bond order is decreasing because the electrons are being added to an antibonding orbital. As expected, the higher the bond order, the shorter the bond.

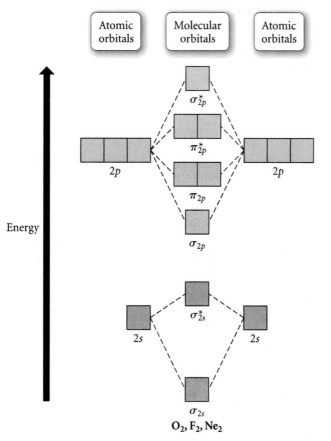

Check: The higher the bond order, the more electron orbital overlap and the shorter and stronger the bond. Molecular oxygen is paramagnetic, and because of the number of π_{2p} and π_{2p}^* orbitals, many configurations are paramagnetic.

22.91   **Given:** 1.0 mol C; C(diamond)$(s) \rightarrow$ C(g) **Find:** ΔH°_{rxn}
          Conceptual Plan: Consider the structure of diamond in Figure 22.14 to determine the number of bonds on each C atom. Write the reaction. Then $\Delta H^\circ_{rxn} = \sum n_p \Delta H^\circ_f(\text{products}) - \sum n_r \Delta H^\circ_f(\text{reactants})$ and using C—C bond energy in Table 9.3, calculate the energy of breaking the number of moles of bonds calculated earlier in this problem.
          Solution: Because each carbon in the diamond structure has four bonds and two carbons are involved in each bond, a total of 2.0 moles of C—C bonds are broken to generate 1.0 moles of C(g). C(diamond) $(s) \rightarrow$ C(g)

Reactant/Product	ΔH°_f(kJ/mol from Appendix IIB)
C(s)	1.88
C(g)	716.7

Be sure to pull data for the correct formula and phase.

$\Delta H^\circ_{rxn} = \sum n_p \Delta H^\circ_f(\text{products}) - \sum n_r \Delta H^\circ_f(\text{reactants})$
$= [1(\Delta H^\circ_f(C(g)))] - [1(\Delta H^\circ_f(C(\text{diamond})(s)))]$
$= [1(716.7 \text{ kJ})] - [1(1.88 \text{ kJ})]$
$= 714.8 \text{ kJ}$

Because the C—C bond energy in Table 9.3 is 347 kJ/mol, $2.0 \text{ mol C—C bonds} \times \dfrac{347 \text{ kJ}}{1 \text{ mol C—C bond}} = 690 \text{ kJ}.$

The C—C bond strength in Table 9.3 is an average bond strength for a wide variety of structures. This value, calculated from the bond energy, is too low because it doesn't include van der Waals attractions between C atoms not directly bonded to each other.

22.93   **Given:** SO$_2(g)$ + H$_2$O$(l) \rightarrow$ H$_2$SO$_3(aq)$; generation of 1 M SO$_2$ **Find:** ΔH°_{rxn}
          Conceptual Plan: Write the reaction. Then $\Delta H^\circ_{rxn} = \sum n_p \Delta H^\circ_f(\text{products}) - \sum n_r \Delta H^\circ_f(\text{reactants})$.
          Solution: SO$_2(g)$ + H$_2$O$(l) \rightarrow$ H$_2$SO$_3(aq)$ Assume that the reaction goes to completion.

Reactant/Product	ΔH°_f(kJ/mol from Appendix IIB)
H$_2$SO$_3(aq)$	−633
SO$_2(g)$	−296.8
H$_2$O(l)	−285.8

Be sure to pull data for the correct formula and phase.

$\Delta H^\circ_{rxn} = \sum n_p \Delta H^\circ_f(\text{products}) - \sum n_r \Delta H^\circ_f(\text{reactants})$
$= [1(\Delta H^\circ_f(H_2SO_3(aq)))] - [1(\Delta H^\circ_f(SO_2(g))) + 1(\Delta H^\circ_f(H_2O(l)))]$
$= [1(-633 \text{ kJ})] - [1(-296.8 \text{ kJ}) + 1(-285.8 \text{ kJ})]$
$= [-633 \text{ kJ}] - [-582.6 \text{ kJ}]$
$= -50. \text{ kJ}$

Check: The units (kJ) are correct. The magnitude of the answer (-50 kJ) makes sense because all of the heats of formation are negative; so most of the heats cancel each other out.

Challenge Problems

22.95   **Given:** iron oxides + CO$(g) \rightarrow$ Fe(s) + CO$_2(g)$, where (a) Fe$_3$O$_4$, (b) FeO, and (c) Fe$_2$O$_3$ **Find:** ΔH°_{rxn}
          Conceptual Plan: Write the reaction. Then $\Delta H^\circ_{rxn} = \sum n_p \Delta H^\circ_f(\text{products}) - \sum n_r \Delta H^\circ_f(\text{reactants})$.
          Solution:
          (a)   Fe$_3$O$_4(s)$ + 4 CO$(g) \rightarrow$ 3 Fe(s) + 4 CO$_2(g)$

Reactant/Product	ΔH°_f(kJ/mol from Appendix IIB)
Fe$_3$O$_4(s)$	−1118.4
CO(g)	−110.5
Fe(s)	0.0
CO$_2(g)$	−393.5

Be sure to pull data for the correct formula and phase.

$\Delta H^\circ_{rxn} = \sum n_p \Delta H^\circ_f(\text{products}) - \sum n_r \Delta H^\circ_f(\text{reactants})$
$= [3(\Delta H^\circ_f(Fe(s))) + 4(\Delta H^\circ_f(CO_2(g)))] - [1(\Delta H^\circ_f(Fe_3O_4(s))) + 4(\Delta H^\circ_f(CO(g)))]$
$= [3(0.0 \text{ kJ}) + 4(-393.5 \text{ kJ})] - [1(-1118.4 \text{ kJ}) + 4(-110.5 \text{ kJ})]$
$= [-1574.0 \text{ kJ}] - [-1560.4 \text{ kJ}]$
$= -13.6 \text{ kJ}$

(b) $FeO(s) + CO(g) \rightarrow Fe(s) + CO_2(g)$

Reactant/Product	ΔH_f°(kJ/mol from Appendix IIB)
$FeO(s)$	-272.0
$CO(g)$	-110.5
$Fe(s)$	0.0
$CO_2(g)$	-393.5

Be sure to pull data for the correct formula and phase.

$$\Delta H_{rxn}^\circ = \sum n_p \Delta H_f^\circ(\text{products}) - \sum n_r \Delta H_f^\circ(\text{reactants})$$
$$= [1(\Delta H_f^\circ(Fe(s))) + 1(\Delta H_f^\circ(CO_2(g)))] - [1(\Delta H_f^\circ(FeO(s))) + 1(\Delta H_f^\circ(CO(g)))]$$
$$= [1(0.0\,\text{kJ}) + 1(-393.5\,\text{kJ})] - [1(-272.0\,\text{kJ}) + 1(-110.5\,\text{kJ})]$$
$$= [-393.5\,\text{kJ}] - [-382.5\,\text{kJ}]$$
$$= -11.0\,\text{kJ}$$

(c) $Fe_2O_3(s) + 3\,CO(g) \rightarrow 2\,Fe(s) + 3\,CO_2(g)$

Reactant/Product	ΔH_f°(kJ/mol from Appendix IIB)
$Fe_2O_3(s)$	-824.2
$CO(g)$	-110.5
$Fe(s)$	0.0
$CO_2(g)$	-393.5

Be sure to pull data for the correct formula and phase.

$$\Delta H_{rxn}^\circ = \sum n_p \Delta H_f^\circ(\text{products}) - \sum n_r \Delta H_f^\circ(\text{reactants})$$
$$= [2(\Delta H_f^\circ(Fe(s))) + 3(\Delta H_f^\circ(CO_2(g)))] - [1(\Delta H_f^\circ(Fe_2O_3(s))) + 3(\Delta H_f^\circ(CO(g)))]$$
$$= [2(0.0\,\text{kJ}) + 3(-393.5\,\text{kJ})] - [1(-824.2\,\text{kJ}) + 3(-110.5\,\text{kJ})]$$
$$= [-1180.5\,\text{kJ}] - [-1155.7\,\text{kJ}]$$
$$= -24.8\,\text{kJ}$$

Fe_2O_3 is the most exothermic because it has the highest oxidation state and therefore oxidizes the most CO per mole of Fe.

Check: The units (kJ) are correct. The answers are negative, which means that the reactions are exothermic.

22.97   (a)   This is a linear molecule. Each C has four valence electrons, and each O has six valence electrons; so C_3O_2 has $3(4) + 2(6) = 24$ valence electrons. All of the bonds are double bonds, so the Lewis structure is
$\ddot{O}{=}C{=}C{=}C{=}\ddot{O}$

(b)   Because each carbon has two electron groups and needs to make 2 pi bonds; the hybridization is *sp*.

(c) $\ddot{O}{=}C{=}C{=}C{=}\ddot{O} + 2\,H{-}\ddot{O}{-}H \rightarrow$ H—O—C—C—C—O—H. The reaction involves breaking

2 C=C bonds and 2 O—H bonds, then making 2 C—H bonds, 2 C—C bonds, and 2 C—O bonds. The number of C=O bonds broken and created are equal, so they can be disregarded. Four O—H bonds are broken and two are formed, so two of the bonds may be cancelled out on the reactants side, and all may be canceled out on the products side. Thus, using values from Table 9.3,
$$\Delta H_{rxn}^\circ = \sum \Delta H (\text{bonds broken}) - \sum \Delta H (\text{bonds formed})$$
$$= [2(611\,\text{kJ}) + 2(464\,\text{kJ})] - [2(+414\,\text{kJ}) + 2(+347\,\text{kJ}) + 2(+360\,\text{kJ})] = [2150\,\text{kJ}] - [+2242\,\text{kJ}]$$
$$= -92\,\text{kJ}$$

22.99   (a)   Because the overall reaction is the sum of the two reactions, K_b for the overall reaction will be the product of the two individual K_bs; so $K_b = K_{b_1} \cdot K_{b_2} = (8.5 \times 10^{-7})(8.9 \times 10^{-16}) = 7.6 \times 10^{-22}$.

(b) $K_{a_1} = K_w / K_{b_1} = (1.0 \times 10^{-14})/(8.5 \times 10^{-7}) = 1.2 \times 10^{-8}$

(c)   **Given:** pH $= 8.5, 0.012$ mol N_2H_4 in 1.0 L   **Find:** $[N_2H_4], [N_2H_5^+],$ and $[N_2H_6^+]$
Other: $K_{b_1} = 8.5 \times 10^{-7}, K_{b_2} = 8.9 \times 10^{-16}$
Conceptual Plan: pH $\rightarrow$ pOH $\rightarrow$ [OH$^-$] and mol N_2H_4, L $\rightarrow$ M N_2H_4 then

$$\text{pH} + \text{pOH} = 14 \qquad \text{pOH} = -\log[OH^-] \qquad\qquad M = \frac{\text{mol}}{L}$$

$[OH^-], [N_2H_4]_0 \rightarrow [N_2H_4], [N_2H_5^+]$ and $[OH^-], [N_2H_4]_0 \rightarrow [N_2H_5^+], [N_2H_6^{2+}]$
ICE table   ICE table

Solution: pH + pOH = 14 so pOH = 14 $-$ pH = 14.0 $-$ 8.5 = 5.5 then $[OH^-] = 10^{-[OH^-]} = 10^{-5.5} =$
3.16×10^{-6}. There are 0.012 mol N_2H_4 in 1.0 L, so $[N_2H_4]_0 = 0.012$ M. Set up first ICE table.

$$N_2H_4(aq) + H_2O(l) \rightleftharpoons N_2H_5^+(aq) + OH^-(aq)$$

Initial	0.0012	0.00	3.16×10^{-6}
Change	$-x$	$+x$	
Equil	$0.0012 - x$	x	3.16×10^{-6}

Because the solution is buffered, $[OH^-]$ will be unchanged.

$K_{b_1} = \dfrac{[N_2H_5^+][OH^-]}{[N_2H_4]} = \dfrac{(x)(3.16 \times 10^{-6})}{0.012 - x} = 8.5 \times 10^{-7}$ Solve for x.

$(x)(3.16 \times 10^{-6}) = 8.5 \times 10^{-7}(0.012 - x) \rightarrow$
$x(3.16 \times 10^{-6} + 8.5 \times 10^{-7}) = 1.02 \times 10^{-8} \rightarrow x = 0.0025$ so $[N_2H_4] = 0.012$ M $-$ 0.0025 M $=$ 0.009 M
and $[N_2H_5^+] = 0.0025$ M. Then set up second ICE table.

$$N_2H_5^+(aq) + H_2O(l) \rightleftharpoons N_2H_6^{2+}(aq) + OH^-(aq)$$

Initial	0.0025	0.00	3.16×10^{-6}
Change	$-x$	$+x$	
Equil	$0.0025 - x$	x	3.16×10^{-6}

Because the solution is buffered, $[OH^-]$ will be unchanged.

$K_{b_2} = \dfrac{[N_2H_6^{2+}][OH^-]}{[N_2H_5^+]} = \dfrac{(x)(3.16 \times 10^{-6})}{0.0025 - x} = 8.9 \times 10^{-16}$ Assume $x \ll 0.0025$,

$\dfrac{(x)(3.16 \times 10^{-6})}{0.0025} = 8.9 \times 10^{-16}$ and $x = 7.0 \times 10^{-13}$ M $= [N_2H_6^{2+}]$. (Note that the assumption that x was
very small was good.)

Check: The units (M) are correct. The magnitude of the answer makes physical sense because only a portion
of the hydrazine dissociates and the product concentrations get smaller and smaller.

22.101   The acid is H—O—N = N—O—H, and the base is
$$H-\underset{\underset{H}{|}}{N}-\underset{\underset{O^-}{|}}{N^+} = O \leftrightarrow H-\underset{\underset{H}{|}}{N}-\underset{\overset{\|}{O}}{N^+}-O^-$$
. The acid is

weaker than nitrous acid because of electron donation by resonance in contributing structures such as
$$H-\underset{\underset{O^-}{|}}{N^+} = O \leftrightarrow H-\underset{\overset{\|}{O}}{N^+}-O^-$$
stabilize nitrous acid. $H_2N_2O_2$ is a weaker acid compared to nitrous acid because

it has fewer O atoms per N atom. The base is weaker than ammonia because of electron withdrawal by the electronegative nitro group.

Conceptual Problems

22.103   The triple bond in nitrogen is much stronger than the double bond in oxygen, so it is much harder to break. This makes
it less likely that the bond in nitrogen will be broken.

22.105   Sodium dihydrogen phosphate (NaH_2PO_4) can act as a weak base or a weak acid. A buffer can be made by mixing it
with Na_2HPO_4 or with Na_3PO_4, depending on the desired pH of the buffer solution.

22.107   F is extremely small and highly electronegative, so there is a huge driving force to fill the octet by adding an electron, giving a -1 oxidation state. Other halogens have access to the d orbital, which allows for more hybridization and oxidation state options.

22.109 SO_3 cannot be a reducing agent because the oxidation state of S is $+6$, the highest possible oxidation state for S. Reducing agents need to be able to be oxidized. SO_2 can be a reducing agent or an oxidizing agent because the oxidation state of S is $+4$.

Review Questions

23.1   The source of vanadium in some crude oil is the very animals from which the oil was formed. It appears that some extinct animals used vanadium for oxygen transport.

23.3   Even though over 2% of the Earth's mass is nickel, most of this is in the Earth's core, and because the core is so far from the surface, it is not accessible.

23.5   The mineral is the phase that contains the desired element(s), and the gangue is the undesired material. These are generally separated by physical methods.

23.7   Sodium cyanide has traditionally been used to leach gold by forming a soluble gold complex $(Au(CN)_2^-)$. The problem with this process is that cyanide is very toxic and the process has often resulted in the contamination of streams and rivers. New alternatives using thiosulfate ions $(S_2O_3^-)$ are being investigated to replace sodium cyanide.

23.9   The body-centered cubic unit cell consists of a cube with one atom at each corner and one atom in the center of the cube. In the body-centered unit cell, the atoms do not touch along each edge of the cube; rather, they touch along the diagonal line that runs from one corner through the middle of the cube to the opposite corner. The body-centered unit cell contains two atoms per unit cell because the center atom is not shared with any other neighboring cells. The coordination number of the body-centered cubic unit cell is 8, and the packing efficiency is 68%. The face-centered cubic unit cell is characterized by a cube with one atom at each corner and one atom in the center of each cube face. In the face-centered unit cell (like the body-centered unit cell), the atoms do not touch along each edge of the cube. Instead, the atoms touch along the face diagonal. The face-centered unit cell contains four atoms per unit cell because the center atoms on each of the six faces are shared between two unit cells. So there are $\frac{1}{2} \times 6 = 3$ face-centered atoms plus $\frac{1}{8} \times 8 = 1$ corner atoms, for a total of four atoms per unit cell. The coordination number of the face-centered cubic unit cell is 12, and its packing efficiency is 74%. In the face-centered structure, any one atom strongly interacts with more atoms than in either the simple cubic unit cell or the body-centered cubic unit cell.

23.11   Copper was one of the first metals used because it could be found in nature in its elemental form.

23.13   Bronze is an alloy of copper and tin. Brass is an alloy of copper and zinc.

Problems by Topic

The General Properties and Natural Distribution of Metals

23.15   Metals are opaque, are good conductors of heat and electricity, and are ductile and malleable, meaning that they can be drawn into wires and flattened into sheets.

23.17   Aluminum, iron, calcium, magnesium, sodium, and potassium are over 1% of the Earth's crust, as shown in Figure 22.2.

23.19   Hematite (Fe_2O_3) and magnetite (Fe_3O_4) are important mineral sources of iron. Cinnabar (HgS) is an important mineral source of mercury. Vanadinite $[Pb_5(VO_4)_3Cl]$ and carnotite $[K_2(UO_2)_2(VO_4)_2 \cdot 3H_2O]$ are important mineral sources of vanadium. Columbite $[Fe(NbO_3)_2]$ is an important mineral source of niobium.

Metallurgical Processes

23.21 $MgCO_3(s) \xrightarrow{\text{Heat}} MgO(s) + CO_2(g)$ and $Mg(OH)_2(s) \xrightarrow{\text{Heat}} MgO(s) + H_2O(g)$.

23.23 A flux is a material that will react with the gangue to form a substance with a low melting point. MgO is the flux in this reaction.

23.25 Hydrometallurgy is used to separate metals from ores by selectively dissolving the metal in a solution, filtering out impurities, and then reducing the metal to its elemental form.

23.27 The Bayer process is a hydrometallurgical process by which bauxite, a hydrated oxide is digested with hot concentrated NaOH under high pressure. The Al_2O_3 is selectively dissolved leaving the iron and silicon oxide as solids. The soluble form of aluminum is $Al(OH)_4^-$.

23.29 Sponge powdered iron contains many small holes in the iron particles due to escaping oxygen when the iron is reduced. Water atomized powdered iron has much smoother and denser particles as the powder is formed from molten iron.

Metal Structures and Alloys

23.31 (a) When one-half of the V atoms are replaced with Cr atoms, the composition will be 50% Cr by moles and 50% V by moles. To get percent by mass, assume 100 moles of alloy; so

$$\text{percent by mass Cr} = \frac{(\text{mol\% Cr})(52.00 \text{ g/mol})}{(\text{mol\% Cr})(52.00 \text{ g/mol}) + (\text{mol\% V})(50.94 \text{ g/mol})} \times 100\%$$

$$= \frac{(50 \text{ mol})(52.00 \text{ g/mol})}{(50 \text{ mol})(52.00 \text{ g/mol}) + (50 \text{ mol})(50.94 \text{ g/mol})} \times 100\% = 50.5\% \text{ Cr by mass}$$

and percent by mass V = 100.0% − 50.5% = 49.5% V.

(b) When one-fourth of the V atoms are replaced with Fe atoms, the composition will be 25% Fe by moles and 75% V by moles. To get percent by mass, assume 100 moles of alloy; so

$$\text{percent by mass Fe} = \frac{(\text{mol\% Fe})(55.85 \text{ g/mol})}{(\text{mol\% Fe})(55.85 \text{ g/mol}) + (\text{mol\% V})(50.94 \text{ g/mol})} \times 100\%$$

$$= \frac{(25 \text{ mol})(55.85 \text{ g/mol})}{(25 \text{ mol})(55.85 \text{ g/mol}) + (75 \text{ mol})(50.94 \text{ g/mol})} \times 100\% = 26.8\% \text{ Fe by mass}$$

and percent by mass V = 100.0% − 26.8% = 73.2% V.

(c) When one-fourth of the V atoms are replaced with Cr atoms and one-fourth of the V atoms are replaced with Fe atoms, the composition will be 25% Cr by moles, 25% Fe by moles, and 50% V by moles. To get percent by mass, assume 100 moles of alloy; so
percent by mass Cr

$$= \frac{(\text{mol\% Cr})(52.00 \text{ g/mol})}{(\text{mol\% Cr})(52.00 \text{ g/mol}) + (\text{mol\% Fe})(55.85 \text{ g/mol}) + (\text{mol\% V})(50.94 \text{ g/mol})} \times 100\%$$

$$= \frac{(25 \text{ mol})(52.00 \text{ g/mol})}{(25 \text{ mol})(52.00 \text{ g/mol}) + (25 \text{ mol})(55.85 \text{ g/mol}) + (50 \text{ mol})(50.94 \text{ g/mol})} \times 100\%$$

$$= 24.8\% \text{ Cr by mass}$$
percent by mass Fe

$$= \frac{(\text{mol\% Fe})(55.85 \text{ g/mol})}{(\text{mol\% Cr})(52.00 \text{ g/mol}) + (\text{mol\% Fe})(55.85 \text{ g/mol}) + (\text{mol\% V})(50.94 \text{ g/mol})} \times 100\%$$

$$= \frac{(25 \text{ mol})(55.85 \text{ g/mol})}{(25 \text{ mol})(52.00 \text{ g/mol}) + (25 \text{ mol})(55.85 \text{ g/mol}) + (50 \text{ mol})(50.94 \text{ g/mol})} \times 100\%$$

$= 26.6\%$ Fe by mass

and percent by mass V $= 100.0\% - (24.8\% + 26.6\%) = 48.6\%$ V.

23.33   Cr and Fe both form body-centered cubic structures. In addition, they are very close together in the periodic table (atomic numbers 24 and 26, respectively); so their respective atomic radii are probably close enough to form an alloy.

23.35   This phase diagram indicates that the solid and liquid phases are completely miscible. The composition is found by determining the x-axis value at the point.
A: solid with 20% Cr and 100% $-$ 20% = 80% Fe
B: liquid with 50% Cr and 100% $-$ 50% = 50% Fe

23.37   This phase diagram indicates that the solid and liquid phases are not completely miscible. Single phases only exist between the pure component and the red line. The composition will be a mixture of the two structures. The composition of the two phases is determined by moving to the left and right until reaching the red lines. The x-axis value is the composition of that structure. According to the lever rule, the phase that is closer to the point is the dominant phase.
A: solid with 20% Co and 100% $-$ 20% = 80% Cu overall composition. One phase will be the copper structure with ~4% Co, and the other phase will be the Co structure with ~7% Cu. According to the lever rule, there will be more of the Cu structure because point A is closer to the Cu structure phase boundary line.
B: solid single phase with Co structure and composition of 90% Co and 100% $-$ 90% = 10% Cu.

23.39   Because C (77 pm) is much smaller than Fe (126 pm), it will fill interstitial holes. Because Mn (130 pm) and Si (118 pm) are close to the same size as Fe, they will substitute for Fe in the lattice.

23.41   (a)   Because there are the same number of octahedral holes as there are metal atoms in a closest-packed structure and half are filled with N, the formula is Mo_2N.
        (b)   Because there are twice as many tetrahedral holes as there are metal atoms and all of the tetrahedral holes are occupied, the formula is CrH_2.

Sources, Properties, and Products of Some of the 3*d* Transition Metals

23.43   (a)   Zn because sphalerite is ZnS
        (b)   Cu because malachite is $Cu_2(OH)_2CO_3$
        (c)   Mn because hausmannite is Mn_3O_4

23.45   **Given:** Calcination of rhodochrosite ($MnCO_3$)   **Find:** heat of reaction
        Conceptual Plan: Write the reaction. Then $\Delta H^{\circ}_{rxn} = \sum n_p \Delta H^{\circ}_f(\text{products}) - \sum n_r \Delta H^{\circ}_f(\text{reactants}).$
        Solution: $MnCO_3(s) + \frac{1}{2}O_2(g) \xrightarrow{\text{Heat}} MnO_2(s) + CO_2(g)$

Reactant/Product	ΔH°_f(kJ/mol from Appendix IIB)
$MnCO_3(s)$	-894.1
$O_2(g)$	0.0
$MnO_2(s)$	-520.0
$CO_2(g)$	-393.5

Be sure to pull data for the correct formula and phase.

$\Delta H^{\circ}_{rxn} = \sum n_p \Delta H^{\circ}_f(\text{products}) - \sum n_r \Delta H^{\circ}_f(\text{reactants})$
$= [1\Delta H^{\circ}_f(MnO_2(s)) + 1\Delta H^{\circ}_f(CO_2(g))] - [1\Delta H^{\circ}_f(MnCO_3(s)) + 1/2\Delta H^{\circ}_f(O_2(g))]$
$= [1(-520.0 \text{ kJ}) + 1(-393.5 \text{ kJ})] - [1(-894.1 \text{ kJ}) + 1/2(0.0 \text{ kJ})]$
$= [-913.5 \text{ kJ}] - [-894.1 \text{ kJ}]$
$= -19.4 \text{ kJ}$

Check: The units (kJ) are correct. The answer is negative, which means that the reactions are exothermic.

23.47   When Cr is added to steel, it reacts with oxygen in steel to prevent it from rusting. A Cr steel alloy would be used in any situation where a steel might be easily oxidized, such as when it comes in contact with water.

23.49   rutile = TiO_2: The composition will be 33.3% Ti by moles and 66.7% O by moles. To get percent by mass, assume 100 moles of atoms; so

$$\text{percent by mass Ti} = \frac{(\text{mol\% Ti})(47.87 \text{ g/mol})}{(\text{mol\% Ti})(47.87 \text{ g/mol}) + (\text{mol\% O})(16.00 \text{ g/mol})} \times 100\%$$

$$= \frac{(33.3 \text{ mol})(47.87 \text{ g/mol})}{(33.3 \text{ mol})(47.87 \text{ g/mol}) + (66.7 \text{ mol})(16.00 \text{ g/mol})} \times 100\% = 59.9\% \text{ Ti by mass.}$$

ilmenite = $FeTiO_3$: The composition will be 20.0% Ti by moles, 20.0% Fe by moles, and 60.0% O by moles. To get percent by mass, assume 100 moles of atoms; so

percent by mass Ti =

$$\frac{(\text{mol\% Ti})(47.87 \text{ g/mol})}{(\text{mol\% Ti})(47.87 \text{ g/mol}) + (\text{mol\% Fe})(55.85 \text{ g/mol}) + (\text{mol\% O})(16.00 \text{ g/mol})} \times 100\%$$

$$= \frac{(20.0 \text{ mol})(47.87 \text{ g/mol})}{(20.0 \text{ mol})(47.87 \text{ g/mol}) + (20.0 \text{ mol})(55.85 \text{ g/mol}) + (60.0 \text{ mol})(16.00 \text{ g/mol})} \times 100\%$$

$$= 31.6\% \text{ Ti by mass.}$$

23.51   Titanium must be arc-melted in an inert atmosphere because the high temperature and flow of electrons would cause the metal to oxidize in a normal atmosphere.

23.53 TiO_2 is the most important industrial product of titanium and is often used as a pigment in white paint.

23.55   The Bayer process is a hydrometallurgical process used to separate the bauxite $(Al_2O_3 \cdot n\, H_2O)$ from the iron and silicon oxide with which it is usually found. In this process, the bauxite is digested with a hot, concentrated, aqueous NaOH solution under high pressure. The basic aluminum solution is separated from the oxide solids; then the aluminum oxide is precipitated out of the solution by neutralizing it. Calcination of the precipitate at temperatures greater than 1000 °C yields anhydrous alumina (Al_2O_3). Electrolysis is then used to reduce the Al to metal.

23.57   Carboloy steel contains cobalt and tungsten.

Cumulative Problems

23.59   **Given:** ilmenite, 2.0×10^4 kg of ore, 0.051% by mass ilmenite, 87% Fe recovery and 63% Ti recovery

Find: g Fe and Ti recovered

Conceptual Plan: ilmenite = $FeTiO_3$ kg ore → g ore → g ilmenite → g Ti possible → g Ti recovered

$$\frac{1000 \text{ g}}{1 \text{ kg}} \qquad \frac{0.051 \text{ g ilmenite}}{100 \text{ g ore}} \qquad \frac{1(47.87) \text{ g Ti}}{151.72 \text{ g ilmenite}} \qquad \frac{63 \text{ g Ti recovered}}{100 \text{ g Ti}}$$

and g ilmenite → g Fe possible → g Fe recovered

$$\frac{1(47.87) \text{ g Fe}}{151.72 \text{ g ilmenite}} \qquad \frac{87 \text{ g Fe recovered}}{100 \text{ g Fe}}$$

Solution: $2.0 \times 10^4 \text{ kg ore} \times \dfrac{1000 \text{ g ore}}{1 \text{ kg ore}} \times \dfrac{0.051 \text{ g ilmenite}}{100 \text{ g ore}} = 1.02 \times 10^4$ g ilmenite then

$$1.02 \times 10^4 \text{ g ilmenite} \times \frac{1(47.87) \text{ g Ti}}{151.72 \text{ g ilmenite}} \times \frac{63 \text{ g Ti recovered}}{100 \text{ g Ti}} = 2.0 \times 10^3 \text{ g Ti recovered} = 2.0 \text{ kg Ti recovered}$$

$$\text{and } 1.02 \times 10^4 \text{ g ilmenite} \times \frac{1(55.85) \text{ g Fe}}{151.72 \text{ g ilmenite}} \times \frac{87 \text{ g Fe recovered}}{100 \text{ g Ti}}$$

$$= 3.3 \times 10^3 \text{ g Fe recovered} = 3.3 \text{ kg Fe recovered}$$

Check: The units (g) are correct. The magnitude of the answers makes physical sense because there is very little mineral in the ore. The amount of Fe recovered is higher than Ti recovered because the atomic weight of Fe and the percent recovery of Fe are both higher than that of Ti.

23.61   Four atoms surround a tetrahedral hole, and six atoms surround an octahedral hole. The octahedral hole is larger because it is surrounded by a greater number of atoms.

23.63 Manganese has one more d electron orbital available for bonding than does chromium.

23.65 Ferromagnetic atoms, like paramagnetic atoms, have unpaired electrons. However, in ferromagnetic atoms, these electrons align with their spin oriented in the same direction, resulting in a permanent magnetic field.

23.67 The nuclear charge of the last three is relatively high because of the lanthanide series in which the $4f$ subshell falls between them and the other six metals of the group.

Challenge Problems

23.69   **Given:** cylinder, $h = 5.62$ cm after pressing, $r = 4.00$ cm; d (before pressing) $= 2.41$ g/mL, d (after pressing) $= 6.85$ g/mL, d (solid iron) $= 7.78$ g/mL
Find: (a) original height, (b) theoretical height if $d = d$(solid iron), and (c) % voids in component
Conceptual Plan:

(a)   **Because $d = \dfrac{m}{V}$, $V = \pi r^2 h$ and the mass of iron and the radius of iron are constant, $d \,\alpha\, \dfrac{1}{h}$.**

So d **(before pressing), d (after pressing), h (after pressing) $\rightarrow h$ (before pressing)**

$$\frac{d_1}{d_2} = \frac{h_2}{h_1}$$

(b) d **(solid iron), d (after pressing), h (after pressing) $\rightarrow h$ (solid iron)**

$$\frac{d_1}{d_2} = \frac{h_2}{h_1}$$

(c)   **Assume that d (air) $= 0$, so d (solid iron), d (after pressing) $\rightarrow$ % voids**

$$\% \text{ voids} = \frac{d_{\text{solid iron}} - d_{\text{after pressing}}}{d_{\text{after pressing}}} \times 100\%$$

Solution:

(a) $\dfrac{d_1}{d_2} = \dfrac{h_2}{h_1}$ Rearrange to solve for h_2. $h_2 = \dfrac{d_1 h_1}{d_2} = \dfrac{6.85 \,\frac{\text{g}}{\text{mL}} \times 5.62 \text{ cm}}{2.41 \,\frac{\text{g}}{\text{mL}}} = 16.0$ cm

(b) $\dfrac{d_1}{d_2} = \dfrac{h_2}{h_1}$ Rearrange to solve for h_2. $h_2 = \dfrac{d_1 h_1}{d_2} = \dfrac{6.85 \,\frac{\text{g}}{\text{mL}} \times 5.62 \text{ cm}}{7.78 \,\frac{\text{g}}{\text{mL}}} = 4.95$ cm

(c) $\% \text{ voids} = \dfrac{d_{\text{solid iron}} - d_{\text{after pressing}}}{d_{\text{after pressing}}} \times 100\% = \dfrac{7.78 \,\frac{\text{g}}{\text{mL}} - 6.85 \,\frac{\text{g}}{\text{mL}}}{6.85 \,\frac{\text{g}}{\text{mL}}} \times 100\% = 14\%$

Check: The units (cm, cm, and %) are correct. The higher the density, the smaller the height; so the answer for (a) is greater than 5.62 cm and the answer for (b) is smaller than 5.62 cm. Because the density of the pressed component is close to the density of solid iron, the volume of voids is low.

23.71 Because there are the same number of octahedral holes as there are metal atoms in a closest-packed structure and twice as many tetrahedral holes as there are metal atoms, there are a total of three holes for each metal atom and the formula would be LaH_3. Because the formula is $LaH_{2.76}$, the percentage of holes filled is $(2.76/3) \times 100\% = 92.0\%$ filled.

23.73 Using the data in Appendix II and $K_f([Cu(CN)_2]^-) = 1.0 \times 10^{24}$:

#1	$[Ag(CN)_2]^-(aq) \rightleftharpoons Ag^+(aq) + 2CN^-(aq)$	$K_1 = 1/K_f = 1/1 \times 10^{21} = 1 \times 10^{-21}$
#2	$Ag^+(aq) + e^- \rightleftharpoons Ag(s)$	$E° = 0.80 \text{ V} = E°_{\text{cathode}}$
#3	$Cu(s) \rightleftharpoons Cu^+(aq) + e^-$	$E° = 0.52 \text{ V} = E°_{\text{anode}}$
#4	$Cu^+(aq) + 2CN^-(aq) \rightleftharpoons [Cu(CN)_2]^-(aq)$	$K_f = 1.0 \times 10^{24}$

$$[Ag(CN)_2]^-(aq) + Cu(s) \rightleftharpoons [Cu(CN)_2]^-(aq) + Ag(s)$$

If we add reactions #2 and #3, we get $E^{\circ}_{cell} = E^{\circ}_{cathode} - E^{\circ}_{anode} = 0.80 \text{ V} - 0.52 \text{ V} = 0.28 \text{ V}$ and $n = 1$ so

$$\Delta G^{\circ}_{rxn} = -n F E^{\circ}_{cell} = -1 \text{ mole} \times \frac{96,485 \text{ C}}{\text{mole}} \times 0.28 \text{ V} = -96,485 \text{ C} \times 0.28 \frac{\text{J}}{\text{C}} = -2.\underline{7}0158 \times 10^4 \text{ J}$$

$$= -2\underline{7}.0158 \text{ kJ}$$

and $T = 298 \text{ K}$ then $\Delta G^{\circ}_{rxn} = -RT \ln K$. Rearrange to solve for K.

$$K_{2+3} = e^{\frac{-\Delta G^{\circ}_{rxn}}{RT}} = e^{\frac{-(-2.\underline{7}0158 \times 10^4 \text{ J})}{\left(8.314 \frac{\text{J}}{\text{K} \cdot \text{mol}}\right)(298 \text{ K})}} = e^{10.9742} = 5.43989 \times 10^4$$

The overall reaction is now the sum of the four reactions; the overall reaction $K = K_1 \times K_{2+3} \times K_f =$
$(1 \times 10^{-21}) \times (5.43989 \times 10^4) \times (1.0 \times 10^{24}) = \underline{5}.43989 \times 10^7 = 5.43989 \times 10^7$.

23.75   First, roast to form the oxide.
$4 \text{ CoAsS}(s) + 9 \text{ O}_2(g) \rightarrow 4 \text{ CoO}(s) + 4 \text{ SO}_2(g) + \text{As}_4\text{O}_6(s)$
Then reduce the oxide with coke.
$\text{CoO}(s) + \text{C}(s) \rightarrow \text{Co}(s) + \text{CO}(g)$
The oxides of arsenic are relatively volatile and can be separated, but they are poisonous. SO_2 and CO are also poisonous.

Conceptual Problems

23.77   Gold and silver are found in their elemental forms because of their low reactivity. Sodium and calcium are Group 1A and Group 2A metals, respectively, and are highly reactive as they readily lose their valence electrons to obtain octets.

24 Transition Metals and Coordination Compounds

Review Questions

24.1   A transition metal atom forms an ion by losing the ns^2 (valence shell) electrons first.

24.3   The +2 oxidation state is common because most of the transition metals have two electrons occupying the ns orbitals. These electrons are lost first by the metal.

24.5   The electronegativity of the transition elements generally increases across a row, following the main-group trend. However, in contrast to the main-group trend, electronegativity increases from the first transition row to the second. There is little electronegativity difference between the second and third transition row. There is a slight increase from silver to gold. Therefore, Au is the most electronegative of the transition metals.

24.7   A ligand can be considered a Lewis base because it donates a pair of electrons. The transition metal ion would be a Lewis acid because it accepts the pair of electrons.

24.9   Cis–trans isomerism occurs in square planar complexes of the general formula MA_2B_2 or octahedral complexes of the general formula MA_4B_2.

24.11   Because of the spatial arrangement of the ligands, the normally degenerate d orbitals are split in energy. The difference between these split d orbitals is the crystal field splitting energy. The magnitude of the splitting depends on the particular complex. In strong-field complexes, the splitting is large; in weak-field complexes, the splitting is small. The magnitude of the crystal field splitting depends in large part on the ligands attached to the central metal ion.

24.13 Zn^{2+} has a filled d subshell, and Cu^{2+} has nine d electrons. Because Zn^{2+} has this filled d subshell, the color of the compounds will be white. Cu^{2+} has an incomplete d subshell; so different ligands will cause a different crystal field splitting, and the compounds will have color.

24.15   Almost all tetrahedral complexes are high-spin because of reduced ligand–metal interactions. The d orbitals in a tetrahedral complex are interacting with only four ligands, as opposed to six in the octahedral complex; so the value of Δ is generally smaller.

Problems by Topic

Properties of Transition Metals

24.17   Identify the noble gas that precedes the element and put it in square brackets.
Determine the outer principal quantum level for the s orbital. Subtract 1 to obtain the quantum level for the d orbital. If the element is in the third or fourth transition series, include $(n - 2)f$ electrons in the configuration.
Count across the row to see how many electrons are in the neutral atom.
For an ion, remove the required number of electrons first from the s orbitals and then from the d orbitals.

(a) Ni; Ni²⁺

The noble gas that precedes Ni is Ar. Ni is in the fourth period; so the orbitals we use are $4s$ and $3d$, and Ni has 10 more electrons than Ar does.

Ni $[Ar]4s^23d^8$

Ni will lose electrons from the $4s$ and then from the $3d$.

Ni²⁺ $[Ar]4s^03d^8$

(b) Mn; Mn⁴⁺

The noble gas that precedes Mn is Ar. Mn is in the fourth period; so the orbitals we use are $4s$ and $3d$, and Mn has seven more electrons than Ar does.

Mn $[Ar]4s^23d^5$

Mn will lose electrons from the 4s and then from the $3d$.

Mn⁴⁺ $[Ar]4s^03d^3$

(c) Y; Y⁺

The noble gas that precedes Y is Kr. Y is in the fifth period; so the orbitals we use are $5s$ and $4d$, and Y has three more electrons than Kr does.

Y $[Kr]5s^24d^1$

Y will lose electrons from the $5s$ and then from the $4d$.

Y⁺ $[Kr]5s^14d^1$

(d) Ta; Ta²⁺

The noble gas that precedes Ta is Xe. Ta is in the sixth period; so the orbitals we use are $6s$, $5d$, and $4f$, and Ta has 19 more electrons than Xe does.

Ta $[Xe]6s^24f^{14}5d^3$

Ta will lose electrons from the $6s$ and then from the $5d$.

Ta²⁺ $[Xe]6s^04f^{14}5d^3$

24.19 (a) V   Highest oxidation state = +5. V = $[Ar]4s^23d^3$. Because V is to the left of Mn, it can lose all of the $4s$ and $3d$ electrons; so the highest oxidation state is +5

(b) Re   Highest oxidation state = +7. Re = $[Xe]6s^24f^{14}5d^5$. Re can lose all of the $6s$ and $5d$ electrons, so the highest oxidation state is +7.

(c) Pd   Highest oxidation state = +4. Pd = $[Kr]4d^{10}$. Metal ions with a d^8 electron configuration exhibit a square planar geometry. Pd can lose 4 electrons from the $5s$ and $4d$ orbitals, so the highest oxidation state is +4.

Coordination Compounds

24.21 (a) $[Cr(H_2O)_6]^{3+}$ H_2O is neutral, so Cr has an oxidation state of +3. Six H_2O molecules are attached to each Cr, so the coordination number is 6.

(b) $[Co(NH_3)_3Cl_3]^-$ NH_3 is neutral, and Cl has charge of 1−. The sum of the oxidation state of Co and the charge of chloride ion = −1. $x + (3(-1)) = 1-$, $x = +2$; therefore, the oxidation state of Co is +2. The three NH_3 molecules and the three Cl^- ions are bound directly to the Co atom; therefore, the coordination number is 6.

(c) $[Cu(CN)_4]^{2-}$ CN has a charge of 1−. The sum of the oxidation state of Cu and the charge of the cyanide ion = 2−. $x + (4(-1)) = 2-$, $x = +2$; therefore, the oxidation state of Cu is +2. The four cyanide ions are directly bound to the Cu atom; therefore, the coordination number is 4.

(d) $[Ag(NH_3)_2]^+$ NH_3 is neutral, so Ag has an oxidation number of +1. Two NH_3 molecules are attached to each Ag atom, so the coordination number is 2.

24.23 (a) $[Cr(H_2O)_6]^{3+}$ is hexaaquachromium(III) ion.

$[Cr(H_2O)_6]^{3+}$ is a complex cation.

Name the ligand: H_2O is aqua.

Name the metal ion: Cr^{3+} is chromium(III).

Name the complex ion by adding the prefixes to indicate the number of each ligand, followed by the name of each ligand and the name of the metal ion: hexaaquachromium(III) ion.

(b) $[Cu(CN)_4]^{2-}$ is tetracyanocuprate(II) ion.

$[Cu(CN)_4]^{2-}$ is a complex anion.

Name the ligand: CN^- is cyano.

Name the metal ion: Cu^{2+} is cuprate(II) because the complex is an anion.

Name the complex ion by adding the prefix to indicate the number of each ligand, followed by the name of each ligand and the name of the metal ion: tetracyanocuprate(II) ion.

(c) $[Fe(NH_3)_5Br]SO_4$ is pentaamminebromoiron(III) sulfate.

$[Fe(NH_3)_5Br]^{2+}$ is a complex cation, SO_4^{2-} is sulfate.

Name the ligands in alphabetical order: NH_3 is ammine; Br^- is bromo.

Name the metal cation: Fe^{3+} is iron(III).

Name the complex ion by adding prefixes to indicate the number of each ligand, followed by the name of the ligand and the name of the metal ion: pentaamminebromoiron(III).

Name the compound by writing the name of the cation before the anion. The only space is between the ion names: pentaamminebromoiron(III) sulfate.

(d) $[Co(H_2O)_4(NH_3)(OH)]Cl_2$ is amminetetraaquahydroxocobalt(III) chloride.

$[Co(H_2O)_4(NH_3)(OH)]^{2+}$ is a complex cation; Cl^- is chloride.

Name the ligands in alphabetical order: NH_3 is ammine; H_2O is aqua; OH^- is hydroxo.

Name the metal cation: Co^{3+} is cobalt(III).

Name the complex ion by adding prefixes to indicate the number of each ligand, followed by the name of the ligand and the metal ion: amminetetraaquahydroxocobalt(III).

Name the compound by writing the name of the cation before the anion. The only space is between the ion names: amminetetraaquahydroxocobalt(III) chloride.

24.25   (a)   Hexaamminechromium(III) is a complex ion with Cr^{3+} metal ion and six NH_3 ligands. $[Cr(NH_3)_6]^{3+}$

(b)   Potassium hexacyanoferrate(III) is a compound with 3 K^+ cations and a complex anion with Fe^{3+} metal ion and six CN^- ligands. $K_3[Fe(CN)_6]$

(c)   Ethylenediaminedithiocyanatocopper(II) is a compound with a Cu^{2+} metal ion, an ethylenediamine ligand, and two SCN^- ligands. $[Cu(en)(SCN)_2]$

(d)   Tetraaquaplatinum(II) hexachloroplatinate(IV) is a complex compound with a complex cation that contains a Pt^{2+} metal ion, four H_2O ligands, and a complex anion that contains a Pt^{4+} metal ion and six Cl^- ligands. $[Pt(H_2O)_4][PtCl_6]$

24.27   (a) $[Co(NH_3)_3(CN)_3]$ is triamminetricyanocobalt(III).

(b)   Because ethylenediamine is a bidentate ligand, you need three to have a coordination number of 6. $[Cr(en)_3]^{3+}$ is tris(ethylenediamine)chromium(III) ion.

Structure and Isomerism

24.29   In linkage isomers, the ligand coordinates to the metal in different ways.

pentaamminenitromanganese(III) ion   pentaamminenitritomanganese(III) ion

24.31   Coordination isomers occur when a coordinated ligand exchanges places with the uncoordinated counterion.

$[Fe(H_2O)_5Cl]Cl \cdot H_2O$ pentaaquachloroiron(II) chloride monohydrate

$[Fe(H_2O)_4Cl_2] \cdot 2H_2O$ tetraaquadichloroiron(II) dihydrate

24.33   Geometric isomers result when the ligands bonded to the metal have a different spatial arrangement.

(a)   No, an octahedral complex must have at least two different ligands to have geometric isomers.

(b)   Yes, there will be cis–trans isomers.

(c)   Yes, there will be fac–mer isomers.

(d)   No, a square planar complex must have at least two different ligands to have geometric isomers.

(e)   Yes, there will be cis–trans isomers.

24.35 (a) Square planar $[NiWXYZ]^{2+}$ would have three geometric isomers.

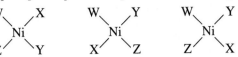

(b) Tetrahedral $[ZnWXYZ]^{2+}$ would have no geometric isomers, but would have stereoisomers.

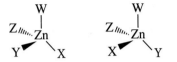

24.37 (a) $[Cr(CO)_3(NH_3)_3]^{3+}$ has a coordination number of 6 and is octahedral. There will be fac and mer isomers and no optical isomers because rotation of the mirror images is superimposable upon each other.

fac   mer

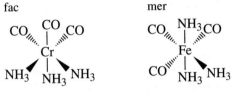

(b) $[Pd(CO)_2(H_2O)Cl]^+$ has a coordination number of 4 and is a d^8 complex, so it is square planar. There will be cis and trans isomers. There will be no optical isomers because rotation of the mirror images is superimposable upon each other.

cis   trans

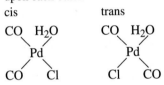

24.39 $[Cr(CO)_2(ox)_2]^-$ has a coordination number of 6 and is octahedral. There will be cis and trans isomers. The cis isomer has a mirror image that is nonsuperimposable.

trans   cis

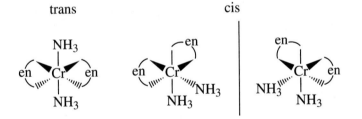

Bonding in Coordination Compounds

24.41 (a) Zn^{2+} d^{10} (b) Fe^{3+} d^5 high spin   low spin

(c) V^{3+} d^2 (d) Co^{2+} d^7 high spin

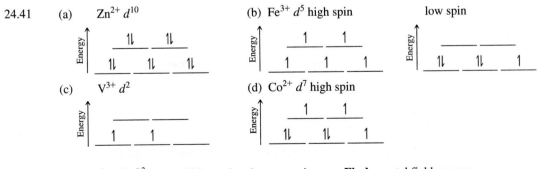

24.43 **Given:** $[CrCl_6]^{3-}$ $\lambda = 735$ nm absorbance maximum   **Find:** crystal field energy
Conceptual Plan: $\lambda \rightarrow \Delta$ (J) $\rightarrow$ kJ/ion $\rightarrow$ kJ/mol

$$\Delta = \frac{hc}{\lambda} \qquad \frac{1 \text{ kJ}}{1000 \text{ J}} \qquad \frac{6.02 \times 10^{23} \text{ ions}}{\text{mol}}$$

Solution: $\dfrac{(6.626 \times 10^{-34} \text{ J} \cdot \text{s})\left(3.00 \times 10^{8} \dfrac{\text{m}}{\text{s}}\right)}{(735 \text{ nm})\left(\dfrac{1 \times 10^{-9} \text{ m}}{\text{nm}}\right)} = 2.70\underline{4} \times 10^{-19} \text{ J/ion}$

$2.70\underline{4} \times 10^{-19} \dfrac{\text{J}}{\text{ion}} \times \dfrac{1 \text{ kJ}}{1000 \text{ J}} \times \dfrac{6.02 \times 10^{23} \text{ ions}}{\text{mol}} = 163 \text{ kJ/mol}$

Check: Cl is a weak-field ligand and would be expected to have a relatively small Δ, which is consistent with a value of 163 kJ.

24.45   The crystal field ligand strength would be $CN^{-} >$; $NH_{3} >$; F^{-}. The smaller the wavelength, the larger the energy and the greater the crystal field splitting observed. So $[Co(CN)_6]^{3-}$ would have a smaller wavelength than $[Co(NH_3)_6]^{3+}$, which would have a smaller wavelength than $[CoF_6]^{3-}$.
So $[Co(CN)_6]^{3-} = 290$ nm, which absorbs in the UV and would have a colorless solution.
$[Co(NH_3)_6]^{3+} = 440$ nm, which absorbs in the blue and would have an orange solution.
$[CoF_6]^{3-} = 770$ nm, which absorbs in the red and would have a green solution.

24.47 Mn^{2+} is d^5, and there are five unpaired electrons; so the crystal field splitting energy, Δ, is small compared to the energy to pair the electrons. Therefore, NH_3 induces a weak field with Mn^{2+}.

24.49   (a) $[RhCl_6]^{3-}$ Rh^{3+} d^6 Cl^- is a weak-field ligand, so the value of Δ will be small. This gives four unpaired electrons.

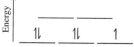

(b) $[Co(OH)_6]^{4-}$ Co^{2+} d^7 OH^- is a weak-field ligand, so the value of Δ will be small. This gives three unpaired electrons.

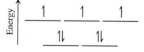

(c)   cis-$[Fe(en)(NO_2)_2]^+$ Fe^{3+} d^5 en and NO_2^- are strong-field ligands, so the value of Δ will be large. This gives one unpaired electron.

24.51 $[CoCl_4]^{2-}$ Co^{2+} is d^7, and Cl^- is a weak-field ligand; so the value of Δ will be small, and there will be three unpaired electrons. The crystal field splitting for a tetrahedral structure is as follows:

Applications of Coordination Compounds

24.53   Hemoglobin, cytochrome c, and chlorophyll all contain a porphyrin ligand.

24.55   Oxyhemoglobin is low-spin and a red color. Because it is low-spin, the crystal field splitting energy must be large. The red color means that the complex absorbs in the green region (~500 nm), which also indicates a large crystal field splitting energy; therefore, O_2 must be a strong-field ligand.
Deoxyhemoglobin is high-spin and a blue color. Because it is high-spin, the crystal field splitting energy must be small. The blue color means that the complex absorbs in the orange region (~600 nm), which also indicates a small crystal field splitting energy. Both of these are consistent with H_2O as a weak-field ligand.

Cumulative Problems

24.57   (a)   Cr $[\text{Ar}]4s^1 3d^5$ (b)   Cu $[\text{Ar}]4s^1 3d^{10}$

                       Cr^+ $[\text{Ar}]4s^0 3d^5$ Cu^+ $[\text{Ar}]4s^0 3d^{10}$

                       Cr^{2+} $[\text{Ar}]4s^0 3d^4$ Cu^{2+} $[\text{Ar}]4s^0 3d^9$

                       Cr^{3+} $[\text{Ar}]4s^0 3d^3$

24.59   (a)

$$\text{H}-\overset{\displaystyle ..}{\text{N}}-\text{H}$$
$$|$$
$$\text{H}$$

     (b) $\left[\ddot{\underset{..}{\text{S}}}=\text{C}=\ddot{\underset{..}{\text{N}}}\right]^-$ ligand can bond from either end.

     (c) $\text{H}\diagdown\overset{\ddot{\underset{..}{\text{O}}}}{}\diagup\text{H}$

24.61   An octahedral complex has six ligands.

     $\text{MA}_2\text{B}_2\text{C}_2$ will have cis–trans isomers: all cis; A trans, B,C cis; B trans, A,C cis; C trans, A,B cis; all trans.

     MAB_2C_3 will have fac–mer isomers.

     $\text{MA}_2\text{B}_3\text{C}$ will have fac–mer isomers.

     MAB_3C_2 will have fac–mer isomers.

     $\text{MA}_3\text{B}_2\text{C}$ will have fac–mer isomers.

     MA_2BC_3 will have fac–mer isomers.

     MA_3BC_2 will have fac–mer isomers.

     MABC_4 will have AB cis and trans.

     MAB_4C will have AC cis and trans.

     MA_4BC will have BC cis and trans.

24.63 $[\text{Fe(ox)}_3]^{3-}$ Fe^{3+} d^5 coordination number $= 6$, octahedral, structure has a nonsuperimposable mirror image.

24.65 $[\text{Mn(CN)}_6]^{3-}$ Mn^{3+} d^4 has a coordination number of 6 and is octahedral. CN^- is a strong-field ligand and will cause a large crystal field splitting energy, so the ion is low-spin.

 The ion will be paramagnetic with two unpaired electrons.

24.67   There are five geometric isomers, one of which is chiral.

1.   2.   3.

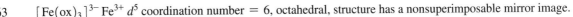

4.   5.

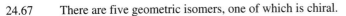

Challenge Problems

24.69   **Given:** 46.7% Pt, 17.0% Cl, 14.8% P, 17.2% C, 4.34% H   **Find:** formula, structures, and names for both compounds
     Conceptual Plan: % composition → pseudoformula → formula

$$n = \frac{g}{molar\ mass} \qquad \text{divide by smallest}$$

     Solution: $46.7\ \cancel{g\ Pt} \times \dfrac{1\ mol\ Pt}{195.1\ \cancel{g}} = 0.23936\ mol\ Pt$ $17.0\ \cancel{g\ Cl} \times \dfrac{1\ mol\ Cl}{35.45\ \cancel{g\ Cl}} = 0.47955\ mol\ Cl$

          $14.8\ \cancel{g\ P} \times \dfrac{1\ mol\ P}{30.97\ \cancel{g}} = 0.47788\ mol\ P$ $17.2\ \cancel{g\ C} \times \dfrac{1\ mol\ C}{12.01\ \cancel{g\ C}} = 1.4321\ mol\ C$

          $4.34\ \cancel{g\ H} \times \dfrac{1\ mol\ H}{1.008\ \cancel{g}} = 4.3056\ mol\ H$

          $Pt_{0.23936}Cl_{0.47955}P_{0.47788}C_{1.4321}H_{4.3056}$

          $Pt_{\frac{0.23936}{0.23936}}Cl_{\frac{0.47955}{0.23936}}P_{\frac{0.47788}{0.23936}}C_{\frac{1.4321}{0.23936}}H_{\frac{4.3056}{0.23936}}$

          $PtCl_2P_2C_6H_{18}$

          $[Pt(P(CH_3)_3)_2Cl_2]$

     cis–dichlorobis(trimethylphosphine)platinum(II)   *trans*–dichlorobis(trimethylphosphine)platinum(II)

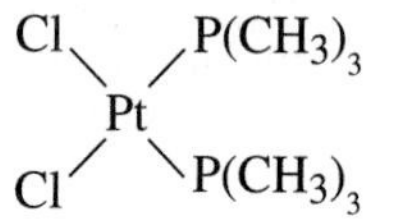

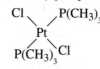

24.71

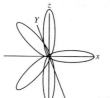

     The trigonal bipyramidal complex ion has lobes along the *x*-axis and between the *x*- and *y*-axes and along the *z*-axis. So the ligands will interact most strongly with z^2 orbital and then with the $x^2 - y^2$ and the *xy* orbitals and will not interact with the *xz* or the *yz* orbitals. So the crystal field splitting would look like the following:

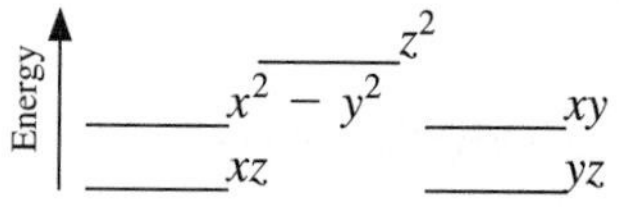

$$\underset{\text{Energy}}{\Bigg\uparrow} \qquad \begin{array}{ccc} & \overline{}\,z^2 & \\ \overline{}\,x^2 - y^2 & & \overline{}\,xy \\ \overline{}\,xz & & \overline{}\,yz \end{array}$$

24.73   (a)   **Given:** $K_{sp}(NiS) = 3 \times 10^{-16}$ **Find:** solubility in water
          Conceptual Plan: Write the reaction, prepare an ICE table, substitute into the equilibrium expression, and solve for *S*(molar solubility).
          Solution: $NiS(s) \rightleftharpoons Ni^{2+}(aq) + S^{2-}(aq)$

Initial		0.0	0.0
Change		S	S
Equil		S	S

          $K_{sp} = [Ni^{2+}][S^{2-}]$ $3 \times 10^{-16} = S^2$

          $S = 1.7 \times 10^{-8} = 2 \times 10^{-8}$

   (b)   **Given:** $K_{sp}(NiS) = 3 \times 10^{-16}$ $K_f[Ni(NH_3)_6]^{2+} = 2.0 \times 10^8$ **Find:** solubility in 3.0 M NH_3
          Conceptual Plan: Sum the reaction, prepare an ICE table, substitute into the equilibrium expression, and solve for *S*(molar solubility).

Solution:

Reaction 1: NiS(s) $\rightleftharpoons$ $\cancel{Ni}^{2+}(aq) + S^{2-}(aq)$ $\qquad$ $K_{sp}(NiS) = 3 \times 10^{-16}$

Reaction 2: $\cancel{Ni}^{2+}(aq) + 6\,NH_3(aq)$ $\rightleftharpoons$ $Ni(NH_3)_6^{2+}(aq)$ $\qquad$ $K_f = 2.0 \times 10^8$

Reaction 3: NiS(s) $+ 6\,NH_3(aq)$ $\rightleftharpoons$ $Ni(NH_3)_6^{2+}(aq) + S^{2-}(aq)$ $\quad$ $K = K_{sp}\,K_f = 6.0 \times 10^{-8}$

	3.0 M		0.0	0.0
Initial	3.0 M		0.0	0.0
Change	$-6S$		$+S$	$+S$
Equil	$3.0 - 6S$		S	S

$$K = \frac{[Ni(NH_3)_6]^{2+}[S^2]}{[NH_3]^6}\,6.0 \times 10^{-8} = \frac{(S)(S)}{(3.0 - S)^6}$$

Assume that $S << 3.0$.

$S^2 = 4.37 \times 10^{-5}$

$S = 6.6 \times 10^{-3}$

(c) $\quad$ NiS is more soluble in ammonia because the formation of the $[Ni(NH_3)_6]^{2+}$ complex ion is highly favorable. The formation removes Ni^{2+} ion from the solution, causing more of the NiS to dissolve.

24.75 $\quad$ If the complexes exhibit lability, there will be exchange of H_2O with the halide and some of the halide will be in solution. So prepare complexes with two different halides (e.g., Cl^- and Br^-) so that you have $[MCl_6]^{3-}$ and $[MBr_6]^{3-}$. Place both complexes in the same aqueous solution. If the complexes exhibit lability, a mixed complex containing both Cl^- and Br^- ligands, $[MCl_xBr_y]^{3-}$, will be formed because Cl^- and Br^- would be in solution when the ligands were exchanged with H_2O. If this mixed complex does not form, the $[MX_6]^{3-}$ complex does not exhibit lability.

24.77 $\quad$ **Given:** 0.10 M NaI, $Cd(OH)_2(s)$ $K_{sp} = 7.2 \times 10^{-15}$; $Cd(I)_4^{2-}(aq)$ $K_f = 2 \times 10^6$ $\quad$ **Find:** pH
$\quad$ **Conceptual Plan: Combine the solubility product reaction of $Cd(OH)_2(s)$ and the formation reaction of $Cd(I)_4^{2-}$ and determine the K for the new reaction. Prepare and ICE table and solve for S (the molar solubility). Then determine the $[OH^-]$, pOH, and pH.**
$\quad$ **Solution:** $Cd(OH)_2(s)$ $\rightleftharpoons$ $\cancel{Cd}^{2+}(aq) + 2\,OH^-(aq)$ $\qquad$ $K_{sp} = 7.2 \times 10^{-15}$

$\qquad\qquad$ $\cancel{Cd}^{2+}(aq) + 4\,I^-(aq)$ $\rightleftharpoons$ $Cd(I)_4^{2-}(aq)$ $\qquad$ $K_f = 2 \times 10^6$

$\qquad\qquad$ $Cd(OH)_2(s) + 4\,I^-(aq) \rightleftharpoons Cd(I)_4^{2-}(aq) + 2\,OH^-(aq)$ $\quad$ $K = 1.44 \times 10^{-8}$

	0.10 M	0	0
Initial	0.10 M	0	0
Change	$-4S$	$+S$	$+2S$
Equil	$0.10 - 4S$	S	$2S$

$$K = 1.44 \times 10^{-8} = \frac{[Cd(I)_4^{2-}][OH^-]^2}{[I^-]^4} = \frac{(S)(2S)^2}{(0.10 - 4S)^4} \quad \text{Assume that } S << 0.10.$$

$$1.44 \times 10^{-8} = \frac{4S^3}{(0.10)^4}$$

$S = \underline{7}.11 \times 10^{-5}$; $[OH^-] = 2S = 2(7.01 \times 10^{-5}) = 1.42 \times 10^{-4}$

pOH $= -\log(\underline{1}.42 \times 10^{-4}) = 3.85 = 3.9$ pH $= 14.0 - 3.9 = 10.1$

Conceptual Problems

24.79 $\quad$ Au would have the higher ionization energy. Ionization energy increases as you go down a group in the transition metals. Because there is a large increase in the number of protons and not a large increase in size, the ionization energy increases.